third edition

the Good Earth

Introduction to Earth Science

David McConnell
North Carolina State University

David Steer
University of Akron

Contributions by

Catharine Knight
University of Akron

Katharine Owens
University of Akron

THE GOOD EARTH: INTRODUCTION TO EARTH SCIENCE, THIRD EDITION

Published by McGraw-Hill Education, 2 Penn Plaza, New York, NY 10121.

This book is printed on acid-free paper.

1 2 3 4 5 6 7 8 9 0 RMN/RMN 1 0 9 8 7 6 5 4

ISBN 978-1-259-09499-6
MHID 1-259-09499-5

www.mhhe.com

Brief Contents

Contents

Preface

Teaching earth science can be viewed as content instruction, covering the principles of science and earth systems. But can it also be considered as an opportunity to *engage* students in the nature of scientific inquiry?

> *A traditional science instructor concentrates on teaching factual knowledge, with the implicit assumption that expert-like ways of thinking about the subject come along for free or are already present. But that is not what cognitive science tells us. It tells us instead that students need to develop these different ways of thinking by means of extended, focused, mental effort.*
>
> *Carl Wieman*
> *Nobel Prize winner*

For many, the wonder of Earth and its features is enough to drive learning. For these happy few, a readable book with lots of attractive photographs is almost all that is required. *But for many—in fact most—learning takes more than pretty words and pictures.* Providing high-quality teaching is the most cost-effective, tangible, and timely effort that geoscience instructors can make to improve student engagement, increase attendance, and add majors.

But how do we do that? There is extensive literature describing what effective teaching looks like, but most science instructors have not had access to these articles and books. Further, few of us were ever explicitly taught the components of good teaching. Instead, we were left to figure it out for ourselves on the basis of our classroom experiences as students.

The Good Earth was published to support both the traditional earth science class ***and*** to serve as an accessible resource for instructors seeking to apply effective teaching strategies to enhance learning.

The Good Earth *Difference*

We wrote *The Good Earth* to support an active learning approach to teaching and to provide the necessary resources for instructors moving through the transition from passive to active learning. Like you, we want our students to walk away from this course with an appreciation for science and the ability to make life decisions based on scientific reasoning.

Our goal was to write a book that was engaging for students but that also included resources that illustrated for instructors how to use teaching practices that have been shown to support student learning. The materials and methods discussed in the text and the accompanying *Instructor's Manual* have been tried and tested in our own classes. Our research shows that the integration of the materials and pedagogy provided in this book not only improved students' understanding of earth science as measured by standardized national tests, but it can also improve students' logical thinking skills by twice as much as a typical "traditional" lecture class. Such methods are overwhelmingly preferred by students and increase student attendance and satisfaction with the course. Finally, a significant point for us is that these methods make teaching class more fun for the instructor.

I love the voice the authors use. Reading the text is like listening to a very intelligent but down-to-earth friend explain a difficult topic. The authors are excellent at organizing and presenting the material. . . . The illustrations are superior to other texts in all ways.

Patricia Hartshorn
University of Michigan–Dearborn

Student-Centered Research

The Good Earth can be used as a text for a traditional, teacher-centered lecture-based course. In fact, we have taken great care to write a book that students would find more engaging than a typical text. But the greatest benefit will come when the book is used as part of an active-learning, student-centered course. For some instructors, it may simply be a matter of adding some of our exercises to an existing active-learning class environment. For others, the book and accompanying materials will give them an opportunity to add components as they gradually change their pedagogy. If you want a more interactive class, try one or all of the following three recommendations based on research findings:

1. Students learn key concepts better when they have opportunities to actively monitor their understanding during class. Rather than just standing up and talking, the instructor can break lectures into segments separated by brief exercises to make sure that students understand concepts before moving on. Students' understanding must be frequently challenged to provide an opportunity to identify misconceptions and replace them with improved, more realistic models.

   The Good Earth includes hundreds of Checkpoint exercises that can also be used as handout-ready PDF files (located on the text website along with answer keys). Practice makes perfect: the more opportunities students have to assess their learning and to practice the application of new skills, the better their performance. If you are concerned about reduced time for lecture, we have found that an emphasis on fostering deeper understanding and less content coverage in lecture, combined with greater student responsibility for reading, produced no decrease in content knowledge attainment and improved student comprehension of key concepts. Some exercises can be assigned as homework, and the answer key in the back of the book can help students to assess their self-directed-learning.

2. Students become better learners when we challenge them to answer questions that require the use of higher-order thinking skills (for example, analysis, synthesis, evaluation). Brain research shows that people become smarter when they experience cognitive challenges. However, it is important not to throw students into the deep end without any help. Instead, instructors need to step through a series of problems of increasing difficulty (scaffolding) so that they can train students to correctly apply their newly acquired thinking skills.

   Therefore, we have carefully created a series of color-coded **Checkpoint** exercises for each section of every chapter. The exercises are pitched at four skill levels: basic, intermediate, advanced, and superior, to give students and instructors an opportunity to scaffold student understanding of key concepts. The questions represent four levels of Bloom's taxonomy. Blue and green questions typically are comprehension and application-level questions. Yellow and red checkpoints typically require analysis, synthesis, or evaluation skills. It is not necessary to complete all the exercises; instructors can select the exercises that are most appropriate for their learning goals.

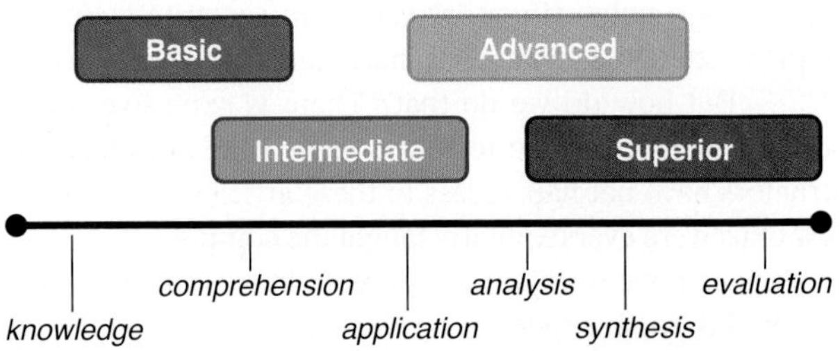

This was kind of a neat idea, and the questions [Checkpoints] do get quite challenging at higher orders. I feel these are good things for students to do while studying, with the idea that if they understand the higher order questions they will understand concepts better for exams. I thought these checkpoints have some very well-formulated questions in the chapters I reviewed.

Swarndeep Gill
California University of Pennsylvania

I like the fact that the authors are mindful and well versed in science education research and pedagogy. This aspect of the author's background is evident in the design of the Checkpoint questions.

The use of Concept Maps and Venn Diagrams is fairly cutting edge for introductory Earth Science textbooks that I am familiar with. This is probably the most innovative aspect of this book and distinguishes it from similar texts, even though the content is presented very similarly to other texts.

Jeffrey Templeton
Western Oregon University

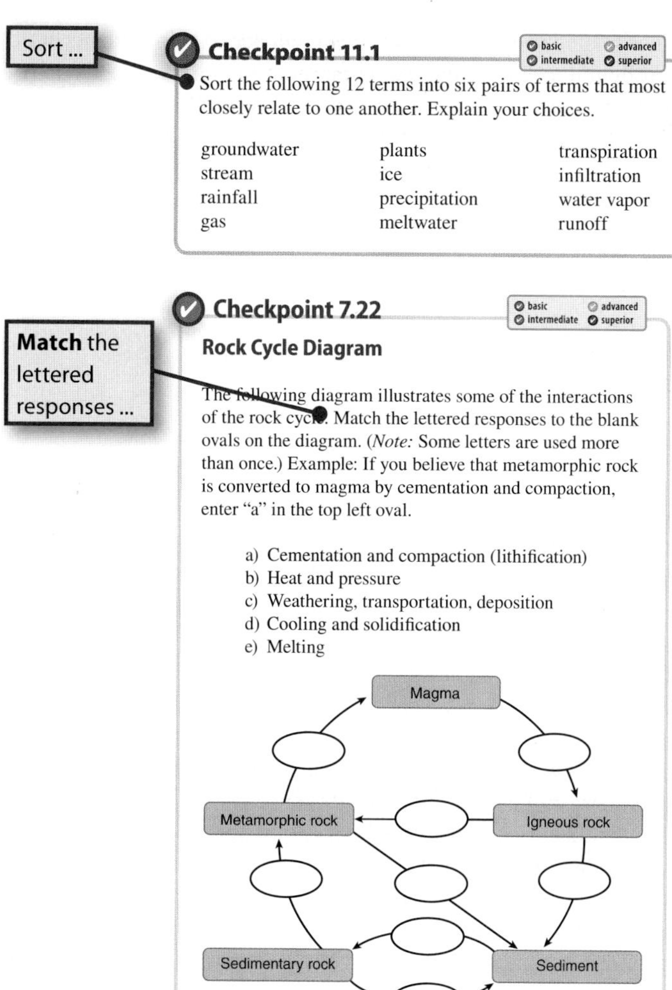

Sort ...

Checkpoint 11.1

basic intermediate advanced superior

Sort the following 12 terms into six pairs of terms that most closely relate to one another. Explain your choices.

groundwater	plants	transpiration
stream	ice	infiltration
rainfall	precipitation	water vapor
gas	meltwater	runoff

Match the lettered responses ...

Checkpoint 7.22

basic intermediate advanced superior

Rock Cycle Diagram

The following diagram illustrates some of the interactions of the rock cycle. Match the lettered responses to the blank ovals on the diagram. (*Note:* Some letters are used more than once.) Example: If you believe that metamorphic rock is converted to magma by cementation and compaction, enter "a" in the top left oval.

a) Cementation and compaction (lithification)
b) Heat and pressure
c) Weathering, transportation, deposition
d) Cooling and solidification
e) Melting

Evaluate the five most important factors ...

Checkpoint 12.12

basic intermediate advanced superior

Groundwater Evaluation Rubric

You are asked to help locate a new aquifer that will supply your town with water. In examining the potential sites, you recognize that several different factors will influence groundwater availability and at no single site are all of the factors optimal. You decide to create a scoring scheme to evaluate the five most important factors that will influence the availability of groundwater. The location that scores the highest according to the rubric will be selected for the well field. One factor is included as an example in the table below; identify five more.

Factors	Poor (1 point)	Moderate (2 points)	Good (3 points)
Depth to water table	Deep	Intermediate	Shallow

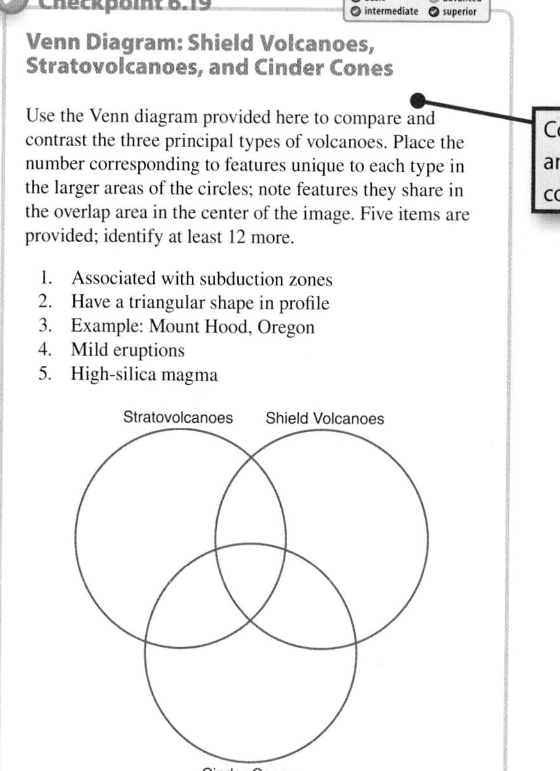

Checkpoint 6.19

basic intermediate advanced superior

Venn Diagram: Shield Volcanoes, Stratovolcanoes, and Cinder Cones

Use the Venn diagram provided here to compare and contrast the three principal types of volcanoes. Place the number corresponding to features unique to each type in the larger areas of the circles; note features they share in the overlap area in the center of the image. Five items are provided; identify at least 12 more.

1. Associated with subduction zones
2. Have a triangular shape in profile
3. Example: Mount Hood, Oregon
4. Mild eruptions
5. High-silica magma

Compare and contrast ...

> *I have to compliment you on putting together Checkpoint 3.3. This was probably the best evaluation tool I have seen for determining whether a student really understands the meaning of the words we use to describe the scientific methods (hypothesis, prediction, etc.).*
>
> *Neil Lundberg*
> *Florida State University*

3. Knowledge is socially constructed and people learn best in supportive social settings. Students do not enter our classrooms as empty vessels to be filled with knowledge. Instead, they actively construct mental models that assimilate new information with previous experiences. This construction of knowledge happens most readily when students work in small collaborative groups (three to four students), where they can talk and listen to peers as they build their understanding of new concepts. Students must be provided with opportunities to be self-reflective about their learning and to help them learn how to learn. Our research confirmed that students in classes where small groups worked to solve challenging problems outperformed students in classes where they worked on the same problems independently.

It is set up very user friendly and will make it easy for instructors to create an interactive learning environment. Also, the way the chapters and questions are laid out, students will know exactly what they should be getting from the chapter and how to test their knowledge and skills.

Jessica Kapp
University of Arizona

Whether you choose to use informal groups ("turn and talk to your neighbors") or formal groups determined by experiences (for example, number of science classes, scores on pretests, academic rank), collaborative learning is a powerful mechanism for maintaining attendance, increasing student-instructor dialogue, and enhancing learning. The Checkpoint exercises (especially advanced and superior level) and conceptests (conceptual multiple choice questions) provided with the book will give you many assignments that you can use as the basis for group work.

For detailed information regarding concept maps, Venn diagrams, Bloom's taxonomy, assessment, and so forth, please consult the *Instructor's Manual* on the text website: http://www.mhhe.com/thegoodearth3e

Tools for Teaching and Learning Science Literacy

Science can be thought of in three ways: as a body of knowledge, as the processes that people employ to explain the universe, and as a set of attitudes and values possessed by those who "do science." This latter aspect is often overlooked in college science textbooks. For each chapter of *The Good Earth*, the *Instructor's Manual* gives suggestions for incorporating into class discussion science attitudes and values such as open-mindedness, skepticism, persistence, and curiosity.

Additionally, the discussion of the **scientific method** is woven throughout the text. We emphasize three scientific themes throughout the text: 1) scientific literacy, 2) earth science and human experience, and 3) the science of global change. Numerous examples of human interaction with Earth serve as introductions to each chapter. Each chapter includes examples of the connection between science and technology, and builds on a context or event familiar to the student. We believe that links to students' past knowledge and experience are essential foundations upon which to build deeper understanding.

In addition to the theme of global change permeating the text, we devote a full chapter to the topic and do not duck the tough issues related to it. We use data and evidence to help students build their own understanding and assist them to realize that *"Much of what lies ahead for the good Earth is up to us. Know, care, act."*

I am pleased to see the final chapter on global change; most students assume that climate change is a political debate, so it is nice to see a textbook that discusses the science behind the news.

Bryan C. Wilbur
Pasadena City College

Ways to Direct Learning

Rather than put key vocabulary terms in bold, we put **key concepts** in bold font. Our rationale is that conceptual understanding is the goal; vocabulary terms alone may not lead to the understanding that we desire. Research suggests that listing key terms encourages the memorization of those terms, rather than the understanding of the associated concepts—rather like learning words in a foreign language but being unable to put together a sentence. To make students fluent in science, we chose to focus on a vocabulary that builds students' conceptual understanding of major ideas in earth science. These ideas were recommended by standards-setting groups, such as the American Association for the Advancement of Science (AAAS).

Students can use the Checkpoint surveys to self-evaluate their comprehension of the major concepts in the section. Self-evaluation is a life skill that persists far longer than the evaluation imposed by an outside party (that is, the instructor). We believe in ongoing assessment tied to each key concept while ideas are still fresh. In contrast, other texts may provide tools for assessment only at the end of the chapter, after all of the content has been covered.

National Committee on Science Education Standards and Assessment

National Research Council

LEARNING SCIENCE IS AN ACTIVE PROCESS. Learning science is something students do, not something that is done to them. In learning science, students describe objects and events, ask questions, acquire knowledge, construct explanations of natural phenomena, test those explanations in many different ways, and communicate their ideas to others. Science teaching must involve students in inquiry-oriented investigations in which they interact with their teachers and peers.

FOCUS AND SUPPORT INQUIRIES. Student inquiry in the science classroom encompasses a range of activities. Some activities provide a basis for observation, data collection, reflection, and analysis of firsthand events and phenomena. Other activities encourage the critical analysis of secondary sources—including media, books, and journals in a library.

ENCOURAGE AND MODEL THE SKILLS OF SCIENTIFIC INQUIRY, AS WELL AS THE CURIOSITY, OPENNESS TO NEW IDEAS, AND SKEPTICISM THAT CHARACTERIZE SCIENCE.

USE MULTIPLE METHODS AND SYSTEMATICALLY GATHER DATA ON STUDENT UNDERSTANDING AND ABILITY. Because assessment information is a powerful tool for monitoring the development of student understanding, modifying activities, and promoting student self-reflection, the effective teacher of science carefully selects and uses assessment tasks that are also good learning experiences.

Often students have some fundamental knowledge of earth science and, when reminded, are able to apply this information to the introduction of new concepts. Each chapter includes a **Self-Reflection Survey** to promote awareness of personal experiences.

Self-Reflection Survey: Section 1.1

Respond to the following questions as a means of uncovering what you already know about Earth and earth science.

1. Which of the following earth science phenom experienced? Which would you most like to ex Can you think of three more things to add to t
   - A volcanic eruption
   - A glacier
   - A river in flood
   - A cave system
   - An underground mine
   - A canyon
   - An earthquake
   - An erosional coastline (rocky cliffs)
   - A depositional coastline (beaches)
   - A hot desert
   - A continental divide
   - Rock layers with fossils
   - A big, assembled dinosaur skeleton
   - A meteor shower or comet
   - The aurora borealis (the northern lights)
   - A meteorite crater
   - A mountain range over 3,000 meters (over in elevation
   - The top of a cloud
2. What three questions about Earth would you able to answer by the end of this course?

Self-Reflection Survey: Section 17.1

Answer the following questions as a means of uncovering what you already know about global change.

1. Respond to the following questions taken from recent CNN and Gallup polls, and compare your answers to those of other respondents. (See footnote to compare responses.*)
   i) Which of the following statements comes closest to your view of global warming?
      a. Global warming is a proven fact and is mostly caused by emissions from cars and industrial facilities such as power plants and factories.
      b. Global warming is a proven fact and is mostly caused by natural changes that have nothing to do with emissions from cars and industrial facilities.
      c. Global warming is a theory that has not yet been proved.
      d. Unsure.
   ii) In thinking about the issue of global warming, sometimes called the *greenhouse effect*, how well do you feel you understand this issue?
      a. Very well.
      b. Fairly well.
      c. Not very well.
      d. Not at all.

Visuals are of great importance for understanding earth science concepts. *The Good Earth* features two-page **Snapshots** to emphasize an important concept in every chapter.

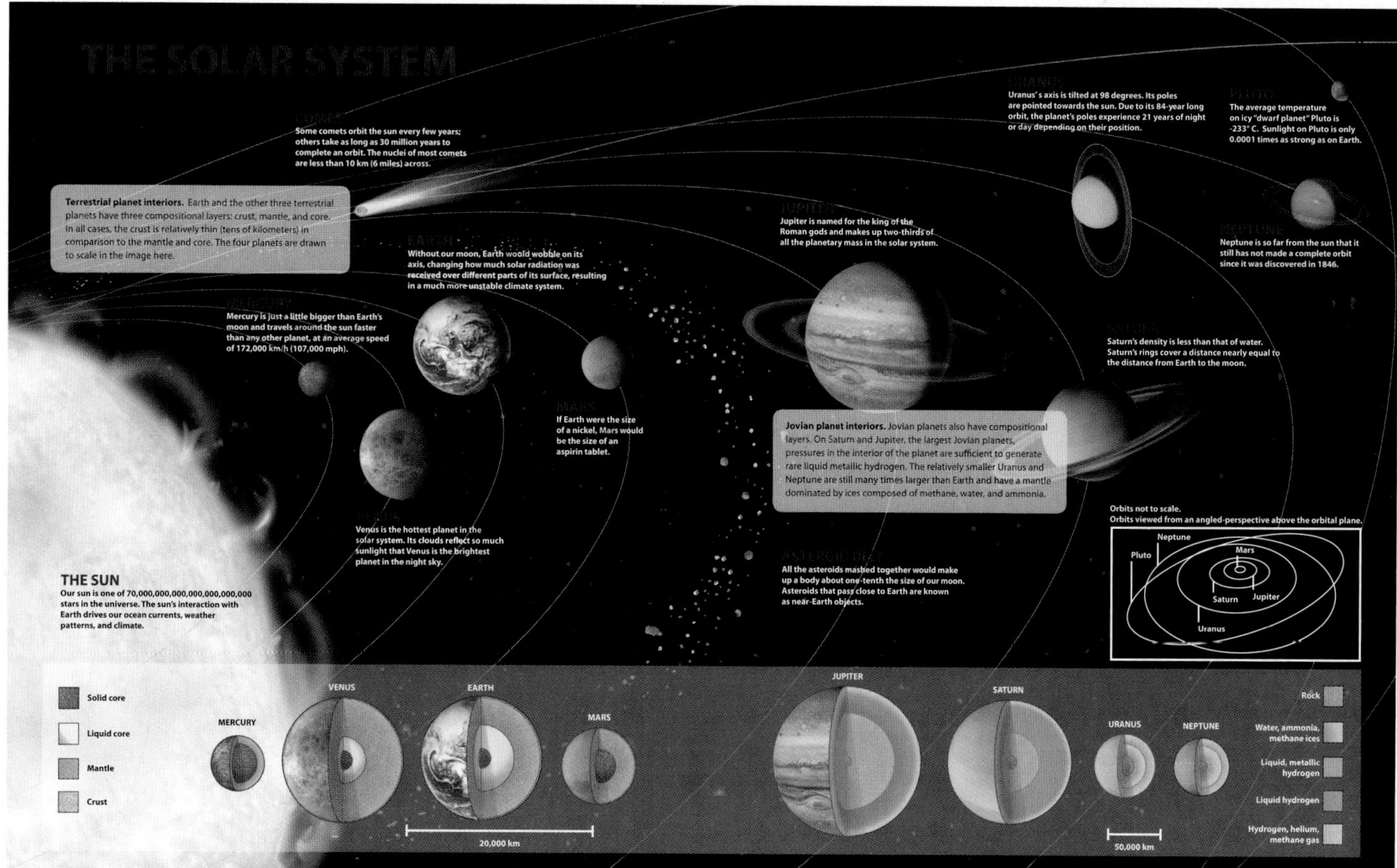

We frequently hear complaints that students don't get the **Big Picture** and become lost in the vocabulary or in trying to memorize facts. We responded to this concern by connecting a chapter-opening "Big Picture" question and photo to the end-of-chapter summary, titled **The Big Picture,** to help students link the key concepts before moving to a new chapter.

the big picture

When Mount St. Helens began rumbling in 1980, teams of scientists rushed to the mountain with truckloads of instruments to monitor the activity. Still, the May 18 eruption came as a surprise. Despite the experience of the scientists and the sophistication of the devices they deployed, little detailed information on the eruptive history of the volcano had been gathered beforehand and few monitoring instruments had been collecting data. That is no longer the case. In the past quarter-century, scientists have made a concerted effort to place a variety of instruments around the volcano, and even in space, to monitor every rumble and movement. Even with what they know today, it is unlikely that volcanologists would have predicted the precise time of the May 18 eruption. But they would have known enough to have more vigorously encouraged the authorities to move people farther from the volcano itself, dramatically reducing the loss of life.

Educating the public is an important factor in reducing the effects of hazards such as volcanoes. Education should provide a scientifically literate population with the necessary skills to critically respond to scientists' assertions. Deciding what evidence to dismiss and what to pay attention to might mean the difference between life and death for those who live in the shadow of an active volcano. The people living near Mount St. Helens in 1980 weighed the evidence and the accompanying call to action. Some heeded the call to evacuate, while others ignored the evidence provided by the volcanologists, chose to hold their ground, and paid for their decision with their lives.

Mount St. Helens is one of only a few US volcanoes with such a high degree of monitoring. However, the US Geological Survey plans to create a National Volcano Early Warning System that would identify the most threatening volcanic hazards, including the number of people and the extent of property endangered. A preliminary assessment of volcanic threat identified 55 volcanoes as high-threat or very-high-threat sites and recommended that each volcano have an extensive network of monitoring equipment to identify the first signs of unrest. Few such networks are currently deployed, and some of these volcanoes have no monitoring systems at all.

One of the volcanoes in the very-high-threat group is Mount Rainier, pictured looming over Tacoma, Washington, at the beginning of this chapter. At 4,392 meters (14,410 feet), Mount Rainier is the tallest and most imposing volcano in Washington. It is located about 70 kilometers (43 miles) southeast of Tacoma. What questions would you ask if you lived in Tacoma?

Historical records indicate that Mount Rainier does not erupt with the frequency of Mount St. Helens. The distance of the peak and the prevailing westerly winds make it unlikely that tephra would ever reach Tacoma. In addition, lava flows and pyroclastic debris would not extend beyond the foot of the mountain, staying tens of kilometers short of Tacoma. Still, large lahars have the potential to reach the northern suburbs of the city and enter neighboring Puget Sound. Even if Tacoma is safe, many smaller towns lie in stream valleys just a 10-minute trip from the volcano by lahar. It is the residents of towns such as Ashford, Packwood, and Orting (Figure 6.33) who need an early warning system for volcanoes.

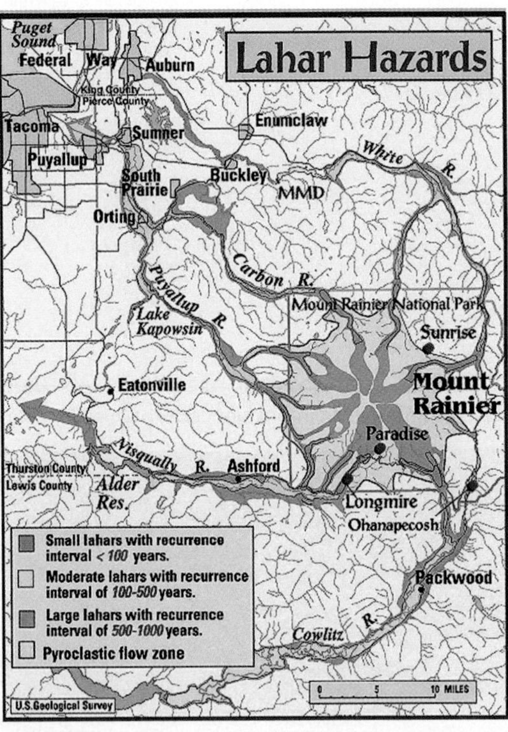

Figure 6.33 Lahar hazards associated with Mount Rainier, Washington.

ple was a magnitude-7.6 quake on October 8, 2005, in northern Pakistan, at the western end of the Himalayas. The earthquake demolished whole towns, killed 90,000 people, and left another 4 million homeless. The unrest continues; Earth at this very moment is shifting, rumbling, building, and decaying. We must carefully observe and prepare.

Volcanoes and Mountains: Concept Map

Complete the following concept map to evaluate your understanding of the interactions between the earth system and volcanoes and mountains. Match the following interactions with the lettered labels on the figure, using the information from this chapter.

Eruption melted ice on Nevado del Ruiz to cause fatal lahars.
Sulfur dioxide blocks incoming sunlight.
Added water causes partial melting of mantle.
Volcanoes add CO_2 and sulfur dioxide to atmosphere.
Commercial airlines are at risk from tephra clouds.
Solar radiation heats Tibetan plateau.
Rain strips CO_2 from atmosphere.

ive tsunami.
ities.
k zones for volcanoes; trees
rivers and lakes from
ng over Himalayas.
rocks in mountains.
Instrumentation of volcanoes.

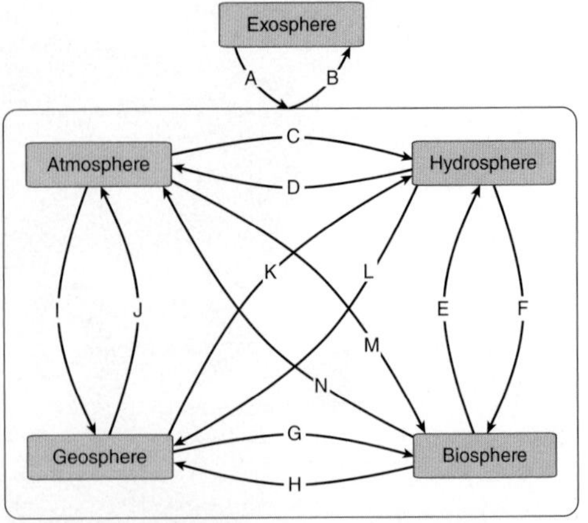

Numerous diagrams, photos, and tables support visual processes and concepts.

a.

b.

Figure 6.15 Hawaiian lava. **a.** A lava tube transports hot, fluid, low-viscosity basalt lava toward the front of a lava flow on Kilauea volcano, Hawaii. **b.** Walter's Kalapana Store and Drive-in was burned and buried within a few weeks in 1990 as lava from the Kilauea volcano invaded communities along the southern coast of Hawaii. Note the height of the original sign. How deep is the lava at this location?

a.

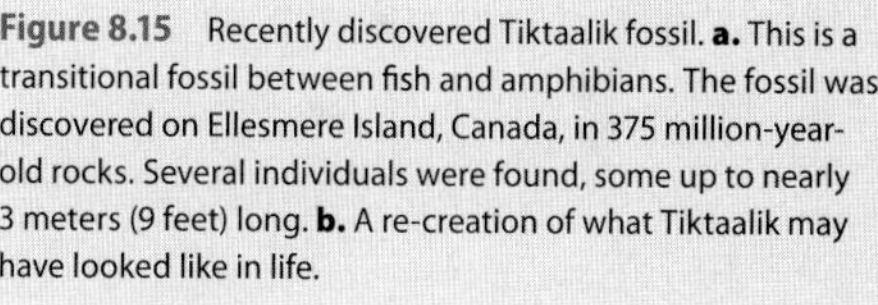

b.

Figure 8.15 Recently discovered Tiktaalik fossil. **a.** This is a transitional fossil between fish and amphibians. The fossil was discovered on Ellesmere Island, Canada, in 375 million-year-old rocks. Several individuals were found, some up to nearly 3 meters (9 feet) long. **b.** A re-creation of what Tiktaalik may have looked like in life.

To further aid in the understanding of earth processes, many figures include a simple drawing to portray a **Geologist's View.**

a.

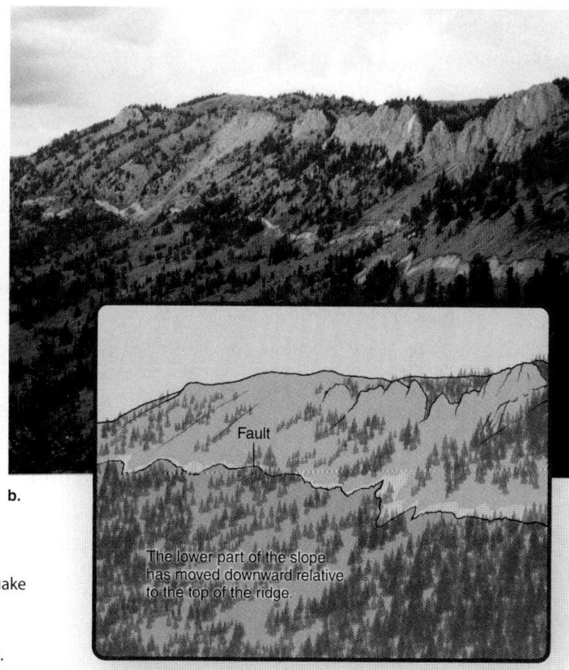
Fault
The lower part of the slope has moved downward relative to the top of the ridge.

b.

Geologist's View

Figure 5.5 Signs of movement on a fault. Movement on a 44-kilometer-long (27-mile-long) fault caused the Hebgen earthquake in Montana in 1959. **a.** The fault broke the surface near a ranch (background). **b.** The fault can be followed for several kilometers along the south flank of Kirkwood Ridge in the center of the image.

How Is This Text Organized?

The Good Earth covers the primary topics included in other earth science texts. However, there are a few notable differences in its content compared to other textbooks.

The Good Earth begins with an introduction (Chapter 1), then takes up the topic of astronomy (Chapters 2, 3), and moves on to solid earth (Chapters 4, 5, 6, 7, 8) and the surficial processes (Chapters 9, 10, 11, 12), which overlap with the hydrosphere (Chapters 11, 12, 13), before dealing with the atmosphere (Chapters 14, 15, 16) and finishing with a wrap-up chapter on global change (Chapter 17) that incorporates elements of all the previous chapters.

Astronomy is dealt with early in the text (Chapters 2 and 3) from the context of Earth's position in space. By beginning with Earth's place in the universe, we give students a "big picture," set the context for looking at the uniqueness of this planet in contrast to our neighbors in space, and hopefully, inspire a bit of wonder in the reader. In both chapters, we grab the reader's attention by emphasizing space from a human perspective. We believe this provides a more appealing beginning to an earth science class than the traditional several weeks spent discussing minerals, rocks, and weathering. Chapter 2, in particular, guides students to see methods that scientists employ as they build our knowledge of the planet and its place in the universe.

Plate tectonics appears early (Chapter 4). We introduce this important unifying concept at the beginning of the text and then use it as a foundation to introduce other solid earth topics (for example, earthquakes, volcanoes). Because an understanding of plate tectonics is pivotal to all the content that follows in subsequent chapters, we revisit this concept several times in subsequent chapters, thereby showing students the interrelationships among the other solid earth topics, such as rock formation, earthquakes, and volcanoes.

Driven by recent research findings, we have chosen to emphasize some topics that are discussed briefly or not at all in other earth science texts. We have included chapters on the threat of a collision with near-Earth objects (Chapter 3), Earth's climate system (Chapter 16), and global change (Chapter 17). In addition, the continuing debate about the teaching of creationism in the public schools has lead us to address this topic head-on in our treatment of geologic time (Chapter 8).

New in This Edition

One major change evident throughout the text is the addition of Chapter Learning Outcomes at the beginning of each chapter and the identification of key Learning Objectives at the start of each section in the chapter.

Additional updates to this edition include:

- Figures have been updated and/or replaced throughout the text to better illustrate key concepts and to provide updated data.
- References and discussions to recent significant events have been added:
  - the massive 2010 oil spill in the Gulf of Mexico after the blowout of the Deepwater Horizon drilling platform
  - the major earthquake and tsunami in Japan in 2011
  - east coast damage from superstorm (hurricane) Sandy in 2012
  - the destructive tornado that struck Moore, Oklahoma, in 2013
- Recent data on the human toll and economic costs of recent earthquakes
- Information about the recent sightings of Near Earth Objects
- New discussion on tools used by Earth Scientists
- Rewritten content on extra-solar planets and how planets formed
- A new more detailed account of the rejection of Wegener's drift hypothesis
- Addition of Harry Hess's contribution to the Seafloor Spreading Theory
- Expanded discussion on early earth evolution
- New statistics on weather hazards
- Updated information on recent changes in Arctic Ocean ice coverage
- Updated climate and emissions data
- Increased coverage on the factors affecting density of seawater
- An analogy of a water balloon is used to further explain the concept of Tidal Bulge

Digital Resources

McGraw-Hill offers various tools and technology products to support *The Good Earth,* 3rd Edition.

McGraw-Hill's Connect Plus (www.mcgrawhillconnect.com/Earth Science) is a web-based assignment and assessment platform that gives students the means to better connect with their coursework, with their instructors, and with the important concepts that they will need to know for success now and in the future. The following resources are available in Connect:

- Auto-graded assessments
- LearnSmart, an adaptive diagnostic tool
- Powerful reporting against learning outcomes and level of difficulty
- McGraw-Hill Tegrity Campus, which digitally records and distributes your lectures with a click of a button
- The full textbook as an integrated, dynamic eBook that you can also assign.
- Instructor Resources such as an Instructor's Manual, PowerPoints, and Test Banks.
- Image Bank that includes all images available for presentation tools.

With ConnectPlus, instructors can deliver assignments, quizzes, and tests online. Instructors can edit existing questions and author entirely new problems; track individual student performance—by question, assignment; or in relation to the class overall—with detailed grade reports; integrate grade reports easily with Learning Management Systems (LMS), such as WebCT and Blackboard; and much more.

By choosing Connect, instructors are providing their students with a powerful tool for improving academic performance and truly mastering course material. Connect allows students to practice important skills at their own pace and on their own schedule. Importantly, students' assessment results and instructors' feedback are all saved online, so students can continually review their progress and plot their course to success.

LEARNSMART® No two students are alike. Why should their learning paths be? LearnSmart uses revolutionary adaptive technology to build a learning experience unique to each student's individual needs. It starts by identifying the topics a student knows and does not know. As the student progresses, LearnSmart adapts and adjusts the content based on his or her individual strengths, weaknesses and confidence, ensuring that every minute spent studying with LearnSmart is the most efficient and productive study time possible.

LearnSmart also takes into account that everyone will forget a certain amount of material. LearnSmart pinpoints areas that a student is most likely to forget and encourages periodic review to ensure that the knowledge is truly learned and retained. In this way, LearnSmart goes beyond simply getting students to memorize material–it helps them truly retain the material in their long term memory. Want proof? Students who use LearnSmart are 35% more likely to complete their class; 13% more likely to pass their class; and have been proven to improve their performance by a full letter grade. To learn more log onto http://learnsmartadvantage.com

SMARTBOOK™ SmartBook is the first and only adaptive reading experience available for the higher education market. Powered by an intelligent diagnostic and adaptive engine, SmartBook facilitates the reading process by identifying what content a student knows and doesn't know through adaptive assessments. As the student reads, the reading material constantly adapts th ensure the student is focused on the content he or she needs the most to close any knowledge gaps. To see more about SmartBook, visit http://learnsmartadvantage.com

Tegrity Campus is a service that makes class time available all the time by automatically capturing every lecture in a searchable format for students to review when they study and complete assignments. With a simple one-click start and stop process, you capture all computer screens and corresponding audio. Students replay any part of any class with easy-to-use, browser-based viewing on a PC or Mac.

Educators know that the more students can see, hear, and experience class resources, the better they learn. With Tegrity Campus, students quickly recall key moments by using Tegrity Campus's unique search feature. This search helps students efficiently find what they need, when they need it, across an entire semester of class recordings. Help turn your students' study time into learning moments immediately supported by your lecture. To learn more about Tegrity, watch a 2-minute Flash demo at http://tegritycampus.mhhe.com.

create™ Create what you've only imagined. Introducing McGraw-Hill Create—a new, self-service website that allows you to create custom course materials—print and eBooks—by drawing upon McGraw-Hill's comprehensive, cross-disciplinary content. Add your own content quickly and easily. Tap into other rights-secured third party sources as well. Then, arrange the content in a way that makes the most sense for your course. Even personalize your book with your course name and information. Choose the best format for your course: color print, black and white print, or eBook. The eBook is now viewable on an iPad! And when you are finished customizing, you will receive a free PDF review copy in just minutes! Visit McGraw-Hill Create at www.mcgrawhillcreate.com today and begin building your perfect book.

CourseSmart Learn Smart. Choose Smart. CourseSmart is a new way for faculty to find and review eBooks. It's also a great option for students who are interested in accessing their course materials digitally and saving money. CourseSmart offers thousands of the most commonly adopted textbooks across hundreds of courses. It is the only place for faculty to review and compare the full text of a textbook online, providing immediate access without the environmental impact of requesting a print exam copy. At CourseSmart, students can save up to 50% off the cost of a print book, reduce their impact on the environment, and gain access to powerful Web tools for learning including full text search, notes and highlighting, and email tools for sharing notes between classmates.

To review comp copies or to purchase an eBook, go to www.coursesmart.com.

ACKNOWLEDGMENTS

The authors would like to express their appreciation for family, friends, colleagues, and students who provided encouragement throughout the writing process for the book. In particular, we would like to thank Tom Angelo who guided us through a detailed course on teaching and learning, and gave us many of the tools that helped us link together our teaching goals with appropriate learning exercises. We will always be grateful for what we learned under his thoughtful instruction. In addition, we are grateful for the enthusiasm and support of the McGraw-Hill development and production teams whose names appear on the copyright page and the support of all the reviewers who helped improve the quality of the text and illustrations.

We would like to thank the following individuals who wrote and/or reviewed learning goal-oriented content for **LearnSmart**.

Northern Arizona University, Sylvester Allred
Roane State Community College, Arthur C. Lee
State University of New York at Cortland, Noelle J. Relles
University of North Carolina at Chapel Hill, Trent McDowell
University of Wisconsin—Milwaukee, Tristan J. Kloss
University of Wisconsin—Milwaukee, Gina Seegers Szablewski
Elise Uphoff

Third Edition Reviewers

Special thanks and appreciation go out to all reviewers. This edition (through several stages of development) has enjoyed many constructive suggestions, new ideas, and invaluable advice provided by these reviewers:

Broward College, Neil M. Mulchan
Cal State Univ Northridge, Doug Fischer
Central Michigan University, Karen S. Tefend
Cerritos College, Tor Björn Lacy
East Los Angeles College, Randall J. Adsit
Emporia State University, Susan Aber
Florida State College, Jacksonville, Rob Martin
Florida State College, Jacksonville, Betty Gibson M.Ed.
Heartland Community College, Robert L. Dennison
Hillsborough Community College, Marianne O'Neal Caldwell
Indian River State College, Paul A. Horton
Johnson State College, Dr. Leslie Kanat
Laredo Community College, Glenn Blaylock
Laredo Community College, Sarah M. Fearnley
Lock Haven University, Thomas C. Wynn
Methodist University, Dr. John A. Dembosky
Middle Tennessee State University, Mark Abolins
Middle Tennessee State University, Dr. Clay Harris
Missouri State University, Jill Black
Montgomery County Community College, George Buchanan, P.G.
Murray State University, George W. Kipphut
Murray State University, Haluk Cetin
Northwest Missouri State University, Jeffrey D. Bradley
Purdue University, Lawrence W. Braile
Sierra College, Alejandro Amigo
Southwestern Illinois College, Stanley C. Hatfield
St. Petersburg College, Paul Cutlip
State College of Florida, Jay C. Odaffer
University of Arkansas, Fort Smith, Christopher Knubley
University of North Carolina—Greensboro, Michael Lewis
University of North Carolina—Greensboro, Jeffrey C. Patton
University of North Florida, Jane MacGibbon
Western Oregon University, Don Ellingson

Second Edition Reviewers

Bucks County Community College, Cristina Ramacciotti
California University of Pennsylvania, Swarndeep Gill
Cal State University–Northridge, Doug Fischer
Central Connecticut State University, Kristine Larsen
Charleston Southern University, Peter B. Jenkins
C. W. Post-Long Island University, Vic DiVenere
Eastern Michigan University, Maria-Serena Poli
Eastern Michigan University, Steven T. LoDuca
Florida State University, Neil Lundberg
Georgia Institute of Technology, L. Gregory Huey
Heartland Community College, Robert L. Dennison
Hillsborough Community College, Marianne O'Neal Caldwell
Ivy Tech Community College of Indiana, Donald L. Eggert
Kingsborough Community College, Cyrena Anne Goodrich
Methodist University, John A. Dembosky
Middle Tennessee State University, Mark Abolins
Middle Tennessee State University, Melissa Lobegeier
Missouri State University, Jill (Alice A.) Black
Murray State University, Haluk Cetin
North Carolina A & T State University, Godfrey A. Uzochukwu
Northern Oklahoma College, Eugene A. Young
Northwest Missouri State University, C. R. Rohs
Northwest Missouri State University, Jeffrey D. Bradley
The Ohio State University, Lindsay Schoenbohm
Pasadena City College, Bryan C. Wilbur
Pensacola Junior College, Kathleen Shelton
St. Petersburg College, Paul G. Cutlip
St. Petersburg College, William C. Culver
San Jose State University, Paula Messina
Santa Ana College, Claire M. Coyne
Santiago Canyon College, Debra Ann Brooks
State University of New York, College at Potsdam, Michael C. Rygel
Southwestern Illinois College, Stanley C. Hatfield
University of Dayton, Heidi S. McGrew
University of Dayton, Michael R. Sandy
University of Indianapolis, Thomas L. Chamberlin
University of Michigan–Dearborn, Patricia Hartshorn
Western Oregon University, Jeffrey Templeton

about the authors

The original version of The Good Earth was a product of a team of educators from the geosciences, science education, and cognitive psychology whose combined expertise created this text to teach essential earth science content in an engaging and cognitively supportive way. We wish to thank our colleagues Kathie Owens, Cathy Knight, and Lisa Park to their contributions to the textbook through the first two editions. The writing team has been reduced to the two principal authors for the third edition of the book.

David McConnell grew up in Londonderry, Northern Ireland, and was hooked on geology when he took his first course in high school with an inspirational teacher. His earliest geological exercises involved examining rocks along the rugged coastlines of Ireland. He graduated with a degree in geology from Queen's University, Belfast, before moving to the US to obtain graduate degrees from Oklahoma State and Texas A&M Universities. David spent much of his career at the University of Akron, Ohio, where he met David Steer, beginning a research partnership that eventually resulted in the book you are now holding. David relocated to North Carolina State University to build a geoscience education research group that continues to examine how to improve the student learning experience in large general education science classes.

David has taught a dozen different courses from introductory geoscience classes to advanced graduate courses. He has received several teaching awards, and he and his collaborators and graduate students have made many presentations and published articles on their educational research. When pressed for some personal information, David will tell you that he loves collecting vinyl records, is way too attached to Tottenham Hotspur football club, and enjoys spending weeks each summer hiking trails through a mountain range somewhere.

David Steer was fascinated with rocks as a child in Ohio. That interest was nurtured by his participation in a National Science Foundation–sponsored geology field camp for high school students that took him to the Black Hills of South Dakota. David's plan to become a geologist had to wait when he accepted an appointment to West Point and then served for a decade as an Army Corps of Engineers officer. While in the military, David attended Cornell University, earning a Master's of Engineering degree. He was then assigned to West Point Military Academy, where he taught physics. After leaving the service, David returned to Cornell University to pursue his early geological interests at the Ph.D. level, albeit in the field of geophysics. He began his appointment at the University of Akron in 1999.

Several years ago, David began employing student-centered learning techniques in his large introductory earth science classes. He has extensive experience in using conceptual questions, physical models, and other active learning techniques. His education research, allowing him to identify at-risk students very early in the course so that effective intervention can occur, has produced scholarly publications in the *Journal of Geoscience Education* and numerous national and regional conference presentations. David has been recognized for his extensive research and teaching scholarship at the institutional and national levels. He and David McConnell were recognized together as National Association of Geoscience Teachers Distinguished Speakers and travel the country making presentations about their educational research.

On a more personal note, David frequently experiments with using golf clubs as seismic energy sources and travels the country with his family with a goal of visiting every national park in the continental United States. David brings military discipline to the team and is one of the principal geoscience content writers. David made this comment about his participation: "Writing this text has been both rewarding and humbling. That endeavor constantly reminded me how much I still have to learn about our planet."

Contributing Authors

Catharine Knight originally hails from Minneapolis, Minnesota. Cathy began her career in teaching while a teenager, achieving national recognition in training her Shelties for obedience competition. Cathy has become an expert in effective teaching and learning, and in cognitive support of learning for humans, as well. With a master's degree and clinical certification in speech science and audiology from St. Cloud State University, Cathy brings a facility in the concepts of "hard science" to the science of learning and teaching: pedagogy. Her Ph.D. research in educational psychology and human development at Arizona State University and her research in cognitive development as a postdoctoral National Institutes of Health Research Fellow at the University of Denver began her dedication to making the science of cognitive development accessible, practical, and applicable to teachers and instructors in the real world.

She has devoted more than 25 years to the study of how students learn and develop, and how instructors can effectively teach, given the characteristics of both students and the concepts and content to be learned. This collaboration of earth science and pedagogical science results in a powerful tool to support teaching and learning in fundamentally new and excitingly effective ways.

When Cathy can grab some spare time from teaching, research, and writing, she kicks back with her Shelties or her cello, or best of all, on a Caribbean cruise ship where the only "requirement" is to do nothing!

Katharine Owens or Kathie, as she's called informally, is the other education member of the team. Kathie says that being a member of *The Good Earth* writing team is one of the highlights of her long career in education. Kathie started out teaching mathematics in junior high and, after getting her master's degree in science education at Texas A&M University where she learned a lot of geology, quickly found another love—teaching science in middle schools both in New York State and in Mississippi (Ed.D., University of Southern Mississippi). She reports that her interest in science began when she watched the *Apollo 8* astronauts circle the moon and greet everyone on "the good Earth" from their vantage point millions of miles away. When she was chosen as a Mississippi finalist in the Teacher in Space program and later as a Christa McAuliffe Fellow, she knew that for the rest of her teaching career, earth science would dominate. Currently, Kathie focuses on teacher education in science at the University of Akron, where she teaches methods courses to future teachers and develops science and technology lessons for the Akron Global Polymer Academy.

Kathie is convinced that how a subject is taught is equally as important as what is taught and that, if the instructor's methods make the content dull and boring or the students are not challenged to think through the content, much is lost. When she's away from her teaching job and education projects, she's traveling around the United States to add to her rock collection, tending her garden, playing with her grandchildren, or whipping up some goodies in the kitchen.

The Good Earth

Although we have long understood Earth's position in space, the unique nature of our planet was not fully appreciated until we were able to look at our home from some distance. The astronauts aboard the *Apollo 8* spacecraft were the first people to travel to the moon and were the first to glimpse our home planet from distant space. This view of Earth, commonly known as "Earthrise," was one of the most well-known images of the twentieth century. The photograph was taken by astronaut William Anders during *Apollo 8's* fourth orbit of the moon on Christmas Eve 1968. (The original image was actually rotated so that the moon's surface was near-vertical and to the right of Earth.) A few hours after snapping the photograph, the Apollo crew read the first 10 verses of the book of Genesis during a broadcast to Earth. At the end of the reading, Commander Frank Borman closed communications with ". . . Merry Christmas, and God bless all of you, all of you on the good Earth." For many at home, those early views of the planet from the inky darkness of space illustrated the unique wonders of the fragile environment we share on spaceship Earth.

"The materials of science are the materials of life itself. Science is part of the reality of living; it is the what, the how, and the why of everything in our experience. It is impossible to understand man without understanding his environment and the forces that have molded him physically and mentally. The aim of science is to discover and illuminate truth."

—Rachel Carson, marine biologist

Chapter 1

Introduction to Earth Science

Chapter Outline

the big picture

Tsunami coming onshore following the Tohoku earthquake, Japan, 2011.

See The Big Picture at the end of this chapter for the full story on this image.

1.1 Earth Science and the Earth System

Chapter Learning Outcomes

- Students will evaluate claims in a science-based argument.
- Students will describe the relationships among science, society, and government.
- Students will recognize that Earth is a complex system of interacting rock, water, air, and life.

The fourth-largest earthquake ever recorded buckled the seafloor east of Japan, at 2.46 P.M. on March 11, 2011. Within seconds the tremor was registered on the dense array of instruments, and alerts were issued to millions of citizens throughout the nation. Residents experienced severe shaking within a couple of minutes. Soon a tsunami warning was issued for coastal locations, giving people about 15 minutes to make it to safety. As earthquake-damaged structures blocked streets, people fled on foot, seeking higher ground. Katherine Heasley, an American teaching high school English in the harbor city of Kamaishi, didn't have classes that day. The long-lasting shaking alerted her to the need to move to a safer location. This is how she described what happened next:

> *An official-looking man was at the head of the column, directing us up the steep road to a recreation center. . . From the parking lot, I could see the estuary, the port and the long spit of land ending in a pier that separated the two. The first thing that concerned me was that the river was running very rapidly out to sea. Usually, even during low tide, the river was sluggish where it met the bay. The second thing that concerned me was that it wasn't the only thing running fast. Many of the larger fishing boats were absolutely speeding out of the harbor.*
>
> *Perhaps fifteen minutes after we got to the rec center, the tide turned. Kamaishi has a very sheltered bay, with the deepest breakwater in the world. The tsunami couldn't come in as the classic white-topped wave. But nothing could stop that much water. The water began flowing back, reversing the river's flow. It rose faster than I'd have thought possible. Soon, the pier was swamped. Those of us at the rec center watched helplessly as the water picked up an entire parking lot of cars on the pier and threw them into a building. And the water kept coming. It rose over the marina where the small fishing boats were kept. Then it swamped the area under the highway. Then it flowed into the parking lot of the keisatsu, or the main police station, just down the street from my apartment. And it kept. On. Coming. It threw the police cars into the keisatsu building. It began to hit houses, and the noise was indescribable as it lifted them off their foundations. It flowed across the highway and started rising up the hill we'd come up. There was a deafening crash as, across the bay, the derrick they used to load ships with Nippon Steel products was knocked down. Think about that: A derrick that regularly lifted tons of steel was no match for the tsunami.*
>
> *The water had turned into some kind of malevolent beast. Some days later, I thought about the Japanese Godzilla mythos, how a creature rises from the depths of the sea and destroys whole cities on land. I can't help but think I now know where that comes from in the collective unconscious of the Japanese people. They depend on the sea; they came across it to find their land, and many make their living off it. But once in a while, the sea comes to take back from the land. When that happens, there is nothing anyone can do except run. The tsunami broke down seawalls and ignored breakwaters. Nothing could stop it. Nothing.*

Much of what happened in Japan was eerily similar to what had occurred just seven years earlier in southesast Asia (Figure 1.1). The third-largest earthquake ever recorded buckled the seafloor

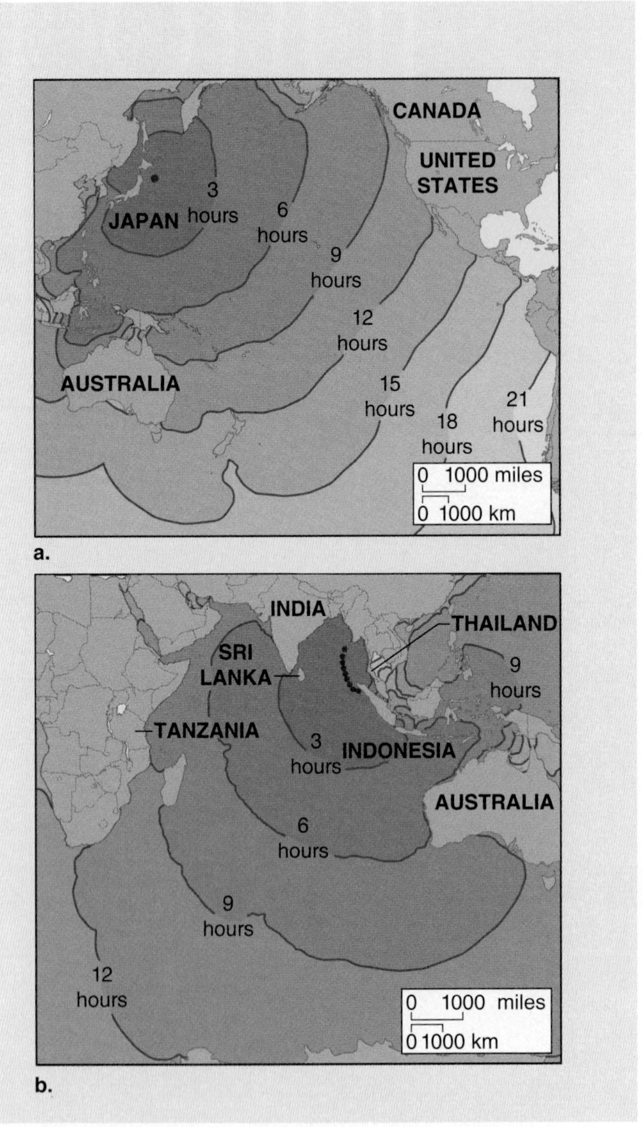

Figure 1.1 Travel times for recent tsunami. **a.** Time taken for 2011 tsunami to travel to locations along the margins of the Pacific Ocean. **b.** Time taken for 2004 tsunami to travel to locations along the margins of the Indian Ocean.

west of Sumatra, Indonesia, at 7:58 A.M. on the day after Christmas 2004. Hundreds of miles away, just before 10 A.M., Penny Smith saw her 10-year-old daughter, Tilly, staring at the sea off Mai Kao beach on the north end of Phuket Island, Thailand. The Smiths were enjoying a warm-weather vacation away from the cold, damp English winter. Tilly noticed that instead of the steady rhythm of breaking waves, the water was flowing down the beach—moving out to sea and not coming back in again. As the beach grew in size, everyone stood to watch, wondering just what strange natural phenomenon they were observing. Tilly knew. She remembered a recent school lesson about tsunami, a series of fast-moving, potentially dangerous ocean waves. Soon Tilly was yelling at her parents to get off the beach.

Meanwhile, about 30 kilometers (16 miles) south, Rick Von Feldt was passing the popular Patong beach in a taxi on his way to his hotel. He noticed people staring out to sea and wondered what they were looking at. A whale? Maybe some dolphins? He too noticed that the water had receded. Soon he was standing on an outdoor balcony of his hotel with a commanding view of the beach below and the bay beyond. He could see boats grounded on the beach and people milling about as if they were not quite sure what to do. This is how he described what happened next on his blog:

> *Suddenly—in front of our eyes—the bay begin(s) to fill. Rapidly. As if someone had turned on giant faucets—and it just seemed to rush in. In about 10–15 minutes this entire HUGE bay filled in. And then we saw the swell. It was perplexing—because the day was nice and sunny—and we could not figure out—or actually believe what was happening in front of our lives. But it did. The swell came—and we saw it rushing over the wall. And it kept risin(g)—and went higher than the palm trees along the edge. Local people started to cry—for they knew what it meant. We all stood there, stunned. People came running up the road—shrieking. "Water—the water" they were crying. The water receded slightly—and then, again with a vengeance. Rushed forward—rose again—and the 18 feet wall rolled over the front of the beach—the shops and everything in its path. We stood there in disbelief—not understanding WHY—but realizing that one of the most awful things that could happen—just had. But it wasn't over. It just keep coming and coming. It would recede—and then come again—rushing over the seawall.*

In both examples, images of the tsunami's destruction were soon being broadcast around the globe. The world looked on in horror as images of whole towns were wiped off the land (Figure 1.2). Piles of debris could be traced around the rim of the Indian Ocean or up and down the east coast of Japan. Before 2004, few would have anticipated the potential for tsunami linked to massive earthquakes as only three of these mega-earthquakes had ever been recorded worldwide. These types of destructive tsunami are so unusual in the modern era that earth science textbooks written 20 years ago would have briefly mentioned tsunami as part of the chapter on earthquakes, but here we are leading off the book comparing and contrasting two recent dramatic examples. As you can see from the first-hand descriptions above, there were similarities to the way the tsunami came on land, sweeping up all before them. However, there were also contrasts in the consequences of these events, mainly as a result of the awareness of the local populations and the warning systems that were in place.

Even though the earthquakes were of almost the same magnitude, the Japanese Tohoku earthquake resulted in about one-tenth of the casualties of the Indian Ocean earthquake. More than 230,000 people were missing or dead following the Indian Ocean tsunami; in contrast the death toll from the Japanese event

Figure 1.2 Destruction in Banda Aceh, Sumatra, Indonesia, from the Indian Ocean tsunami. The shoreline has been heavily eroded, and buildings have been almost completely destroyed.

Table 1.1 Human Toll and Economic Costs of Selected Recent Earthquakes

Date	Earthquake Location	Earthquake Magnitude	Deaths	Displaced	Economic Losses ($ Billions)
February 22, 2011	Christchurch, New Zealand	6.3	185	70,000	12
March 11, 2011	Tohoku, Japan	9.0	20,896	131,000	300
February 27, 2010	Maule, Chile	8.8	525	1,500,000	15–30
January 12, 2010	Haiti	7.0	316,000	1,300,000	7–14
September 30, 2009	Sumatra, Indonesia	7.5	1,117	451,000	2.3
May 12, 2008	Sichuan, China	7.9	87,587	>5,000,000	86
October 8, 2005	Kashmir, Pakistan	7.6	86,000	2,500,000	5
December 26, 2004	Sumatra, Indonesia	9.1	230,000	1,700,000	4.5
January 16, 1995	Kobe, Japan	6.9	5,502	31,000	200
January 17, 1994	Northridge, California	6.7	72	22,000	20

was closer to 21,000 (Table 1.1). Much of this can be ascribed to the extensive earthquake and tsunami warning systems that are in place in Japan. In contrast, in 2004 there was no tsunami early warning system in the Indian Ocean. The lack of effective communication systems in poor regions meant that thousands were killed along the coasts of India and Sri Lanka several hours after waves had already obliterated coastal towns in Indonesia. Economic losses were estimated to be more than 20 times higher in Japan than in Indonesia, largely as a consequence of the much more developed and industrialized Japanese coastline. The 2011 earthquake and tsunami produced the costliest natural disaster in history (~$300 billion). The resulting damage shut down industries for weeks, leading to a global slowdown in automobile manufacturing. The tsunami flooded the Fukushima nuclear power plant, resulting in both a meltdown in three of its six nuclear reactors and the release of dangerous radioactive gases. The disaster put a serious dent in the prospect of a nuclear-powered future as governments in many nations started backing away from plans for building new power plants. This destruction occurred in a country that is better prepared to deal with earthquakes than any nation on the planet. Much of the Japanese coastline is protected by 10-meter-high (33-foot-high) seawalls, specifically to protect against tsunami. Unfortunately, the seawalls were designed to protect against waves produced by earthquakes of smaller magnitude, the type and size that happen quite frequently along the coast of Japan.

Are you smarter than a fifth-grader such as Tilly Smith? What type of natural disaster are you most likely to experience in your lifetime? Are you ready? Will you, like Tilly, know what to do? How much responsibility do individuals have to protect themselves? How many resources should be devoted to trying to predict these events? What type of information should government agencies collect to protect their citizens from such hazards? How can this information be quickly communicated to those at risk? If we are to successfully answer these questions, we must understand the science of how Earth works and how to effectively integrate scientific findings with the needs of society. Scientists collect data on natural phenomena such as tsunami, but it is often politicians (and, indirectly, the people who elect them) who determine what actions should be taken to protect the public.

Your Introduction to Earth Science

In the remainder of this chapter, we will define *science* and describe how it is done by trained people doing basic research using an array of skills. These scientists ask questions and analyze data; some create maps or collect fossils, while others use sophisticated technologies to search for oil and gas, track ocean currents, or measure changes in the chemistry of Earth's atmosphere. We will explain the principles that scientists use to conduct investigations that weave together data collected from experiments and observations of the natural world. We will discuss the principal roles of the earth sciences in our lives, from finding ways to protect us from natural hazards to investigating the implications of climate changes for the future of humanity.

We finish the chapter by introducing the concept of global change and humans' impact on Earth. Global change is an idea that is currently generating research in a wide variety of disciplines relating to all components of the earth system, including geology, ecology, oceanography, and climatology. This work involves thousands of scientists across the globe and has implications for the

Checkpoint 1.1

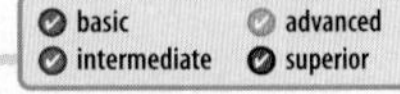

Good questions often produce answers that lead to yet more questions. Review the following statement and suggest some related questions that could clarify or expand the topic.

Students who work together in groups often learn more than students in the same class who work alone.

Self-Reflection Survey: Section 1.1

Respond to the following questions as a means of uncovering what you already know about Earth and earth science.

1. Which of the following earth science phenomena have you experienced? Which would you most like to experience? Can you think of three more things to add to the list?
   - A volcanic eruption
   - A glacier
   - A river in flood
   - A cave system
   - An underground mine
   - A canyon
   - An earthquake
   - An erosional coastline (rocky cliffs)
   - A depositional coastline (beaches)
   - A hot desert
   - A continental divide
   - Rock layers with fossils
   - A big, assembled dinosaur skeleton
   - A meteor shower or comet
   - The aurora borealis (the northern lights)
   - A meteorite crater
   - A mountain range over 3,000 meters (over 10,000 feet) in elevation
   - The top of a cloud
2. What three questions about Earth would you like to be able to answer by the end of this course?

long-term quality of life for you and your families and is likely to require challenging social decisions within your lifetime. Future economic, cultural, and political choices in all the world's nations will depend on the rate and degree of change. We will follow the theme of global change through many of the chapters of *The Good Earth* and use it to show the links among the components of the earth system. As you will see, there is little that happens on Earth that doesn't involve multiple earth system components.

1.2 The Scope of (Earth) Science

Learning Objectives

- Describe the principal earth system components.
- Write a one-sentence definition of the term *science.*
- Identify examples of the tools that scientists use to learn about Earth.

Earth System Basics

In *The Good Earth*, we introduce you to the study of earth science. Earth is a complex system of interacting rock, water, air, and life where the components and interactions cycle energy and mass throughout the system. *Earth science* can be broadly defined as the investigation of interactions among the four parts of the earth system— the atmosphere (air, weather), hydrosphere (water, ice), biosphere (plants, animals), and geosphere (land, rocks) (Figure 1.3).

Figure 1.3 The four components of the earth system: atmosphere, hydrosphere, biosphere, and geosphere. All components interact with the solar radiation and other elements from space. How many components are featured in each image?

Together, these components form an elegant support system for life. In addition, the sun and assorted features from space, collectively termed the *exosphere*, interact with the earth system and are sometimes considered a fifth earth system component. The historic 2011 tsunami involved three of the components— the hydrosphere (ocean), geosphere (seafloor earthquake), and biosphere (people, plants, animals; Figure 1.4). Throughout this book, we will examine the characteristics of each of the components through the lens of human experience. We will look at how the earth system affects us over a wide range of timescales and how we, in turn, affect different earth system components. We will also be interested in how these components interact with one another and how changes in one component influence processes in the others. Representatives of the earth science community spent some time considering the essential principles or big ideas that everyone should appreciate about the earth system. They ended up with nine "big ideas" that can be divided into a series of secondary concepts. The big ideas were as follows:

- Earth is a complex system of interacting rock, water, air, and life.
- Earth scientists use repeatable observations and testable ideas to understand and explain our planet.
- Earth is 4.6 billion years old.
- Earth is continuously changing.
- Earth is the water planet.
- Life evolves on a dynamic Earth and continuously modifies Earth.
- Humans depend on Earth for resources.

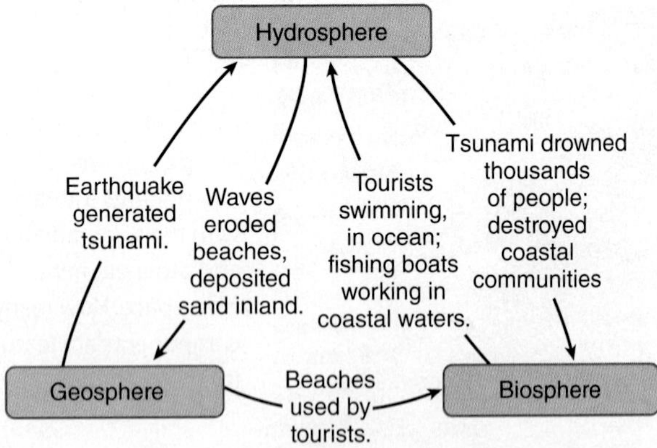

Figure 1.4 Interaction of earth system components involved in the tsunami described in this section. This diagram is a concept map. It includes key terms (in this case, three earth system components) and arrows that describe the interactions between the terms.

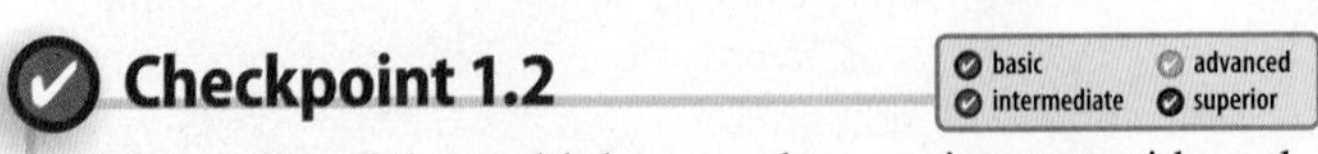

Make a list of the multiple ways that you interact with each of the four components of the earth system.

- Natural hazards pose risks to humans.
- Humans significantly alter Earth.

We already introduced you to the first idea at the start of this section, and we discuss the second item on the list in the rest of this chapter. The remainder of the points will be discussed in several chapters throughout the book. Take a few moments after you read each chapter to reflect on how many of these big ideas were represented in the material you just read.

Science and Discovery

The second word in the term *earth science* is just as important to us as the first. Much of what you learn in college about science will happen in this and perhaps one other course. Therefore, we want you to have a firm understanding of what science is—and what it is not. Science is not a list of facts to be memorized that have no relevance to your life. Just ask Tilly Smith's parents. The only way to understand how to think like a scientist is to learn to use the skills of scientific reasoning. So in this chapter, and throughout the book, we will give you lots of examples to show that science is a process, a way of thinking about the natural world.

Why do we care what you think about science? The United States is a world leader in scientific research and development. Government and corporate science programs flourish because of substantial investment in innovation and the discovery of new ideas. Even though survey results show that Americans are supportive of scientific research, most people have only a shaky grasp of underlying scientific principles. The National Science Foundation has conducted surveys that reveal that less than one-third of the adult population can define what it means to study something scientifically. Even if you do not make a career in science, it is important that you understand how to use scientific reasoning skills to make wise decisions as an informed citizen to help solve daily problems.

Science is a process of discovery that increases our body of knowledge. Earth science is like all sciences; some of it is known and can be learned, and much of it is still waiting to be discovered. But science has another less tangible but equally important element—the innate curiosity of the scientists as they search for answers. Increasingly, individual citizens have the opportunity to become involved in a variety of scientific research projects that are too large to complete without the participation of a large number of talented amateur scientists. These projects can involve thousands of observations made by citizens over a limited time interval. They may involve volunteers with little or no training in collecting data or citizen researchers working with scientists in an online community. Examples of some of these projects follow. Which would you most like to be involved with?

- *Stardust*: Volunteers help scientists find a handful of interstellar dust particles brought to Earth by the *Stardust* spacecraft and hiding among a million online images.
- IceWatchUSA: This project by the Nature Abounds organization seeks to compile local climate records by making seasonal observations of ice conditions on water bodies.

- World Water Monitoring Day: Tens of thousands of people in more than 40 nations use simple sample kits to collect information on the characteristics of local streams and lakes. Data are then entered into an online database to be compared with results from subsequent years.
- Project Budburst: This project seeks to track the timing of the leafing and flowering of native vegetation around North America.
- FeederWatch: This data set yields information on the changes in native and alien species of birds using backyard bird feeders and can serve as a proxy indicator of environmental change.
- Weather observers: Eleven thousand volunteers in the National Weather Service's cooperative observer program collect daily temperature and precipitation readings to aid in weather forecasting and to help measure long-term climate change.
- Did You Feel It?: This US Geological Survey (USGS) online program uses reports from people who feel earthquakes and graphs the data to create Community Internet Intensity Maps. For example, a recent 5.8 magnitude earthquake in Virginia generated more nearly 150,000 citizen reports from people in states up and down the east coast.

Earth scientists combine their basic knowledge of facts and concepts with technical skills to explore Earth and solve its mysteries. It is tempting to view science as a list of facts to be memorized and repeated. But the real essence of science is a detective story in which teams of investigators piece together evidence to generate well-founded explanations of the workings of our planet. Scientists constantly refine or challenge these explanations, causing some to be discarded while others gain wide acceptance. Our imaginations and the physical laws of nature present the only limits to science. Throughout this book, we will strive to give you an inside look at how science is done and to initiate you in the process of discovery. Whenever possible, we will feature real-life situations and pose questions that place you in the role of the scientist.

Tools Used by Earth Scientists

Earth scientists use direct measurements, indirect information, and models to better understand Earth. Direct measurements are collected at field locations by scientists or trained technicians

Figure 1.5 This instrument consists of numerous bottles that are submerged to collect water samples from the ocean.

(Figure 1.5). For example, they might determine the type of rocks present (see Chapter 7) to create a geologic map, collect water samples from drinking water wells (see Chapter 12), or gather samples of gases erupting from volcanoes (see Chapter 6). Samples are carefully analyzed and cataloged with information about their original location, the conditions under which they were obtained, and any other data that could affect understanding of the importance of that sample. In these cases, the scientist is directly measuring exactly what she is interested in measuring. The actual measurements may be obtained in the field or in a laboratory.

However, it is often not practical to measure some phenomena directly. In these cases, scientists use indirect measurements. Essentially, they measure something that they can then interpret to get a value for something else. For example, scientists cannot readily examine the features below the world's oceans, but they have been able to use a variety of methods to identify different properties of the rocks of the ocean floor. Measurements of the magnetic patterns of the ocean floor were used to determine that the age of the oceanic crust varied from place to place (see Chapter 4). Satellites measure variations in the height of the ocean surface which is related to the distribution of ridges and trenches on the ocean floor (see Chapter 13). Satellites can also make direct measurements of large regions that would be impossible to map on the surface. For example, scientists have used satellite measurements of Arctic sea ice coverage to show a steady decline over the last few decades (see Chapter 16).

Checkpoint 1.3

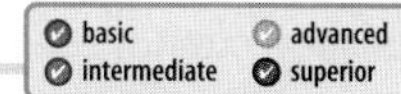

Three of the big ideas listed near the start of this section detail the interaction of humans and the earth system: (1) Humans depend on Earth for resources; (2) natural hazards pose risks to humans; and (3) humans significantly alter Earth. Take a few minutes and write what you can in support of each of these statements. Consider revising your responses as you progress through the semester to see if you can add more items and/or more information.

Figure 1.6 Wave tank used to understand processes occurring in coastal environments.

Checkpoint 1.4

basic | intermediate | advanced | superior

Scientific Analysis

Go to the US Geological Survey site (www.usgs.gov) and find an example of an earth science topic that USGS scientists have investigated.

1. Briefly describe the research, using no more than six sentences.
2. Identify:
   - The types of questions the scientists investigated.
   - The types of tools the scientists used.
   - An example of the data they collected.

In other cases, scientists use models to better understand Earth. Those models can be physical devices such as wave tanks (Figure 1.6), or they may be theoretical models. The latter are often computer models that simulate complex physical processes to allow investigators to examine the relationships between variables. For example, it has become commonplace for us to see the results of meteorological modeling on our evening news weather forecasts (see Chapter 15). Elsewhere atmospheric scientists use an array of models to predict the track of hurricanes, and ever more sophisticated models are being developed to predict future climate trends (see Chapter 17).

1.3 Doing Science

Learning Objectives

- Explain how scientists use observations and predictions to test hypotheses.
- Provide examples of inductive and deductive reasoning.
- Analyze the four basic principles of good science as applied to a real-world scientific investigation.
- Describe examples of poor scientific reasoning.

This information-rich world gives us ready access to all the facts we could ever want. In *The Good Earth*, we have cut down the volume of terms and the amount of new information in favor of addressing the skills needed to process that information—to separate the good from the bad, the significant from the trivial. Most people rate the ability to think critically—to analyze and evaluate information so as to make wise decisions—as more important than knowledge of facts or technical skills that can be easily memorized. Critical thinking is also something science does wonderfully well. The benefits of critical thinking extend beyond science to help us interpret information in a variety of forms, weigh the validity of competing claims, and make judgments on which course of action to pursue. Effective critical-thinking and decision-making skills can be applied to many aspects of life, including buying a house, planning a diet, or successfully managing your time in college.

Science advances by the application of the **scientific method, a systematic approach to answering questions** about the natural world. The scientific method implies that sufficient observation will reveal patterns that provide clues to the origin and history of Earth. We assume that the components of the universe interact in consistent, predictable ways. Scientists use their observations as an aid in predicting future events in the earth system and, in some cases, the universe.

From Observation to Hypothesis

All of us make observations that we use to mold our personal views of the cultural and physical worlds we inhabit. **Scientists also use observations to shape ideas.** Their ideas are known as hypotheses. **A hypothesis is a testable explanation of facts or observations.** For example, if we owned a Ford Mustang that broke down frequently, we might form a hypothesis that Mustangs or even all Fords were poorly built cars. Through experience, we test the limits of our personal world, allowing those limits to expand to accommodate a positive stimulus or shrink from a negative interaction. Suppose a friend is pleased with the performance of her Mustang. That might require us to modify our original interpretation. Personal observations may vary with the individual, but **valid scientific observations are empirical—that is, they can be measured and confirmed by others.** In that regard, we could collect data on a large number of Ford Mustangs and determine the average number of repairs per car over a specific period of time. We could then compare these data with repair rates for comparable vehicles to support or refute our original hypothesis. Further, others might be prompted to test a similarly large set of Mustangs to confirm our interpretation.

Inductive and Deductive Reasoning

The scientific method is not a single set of steps like a recipe. It can take a variety of forms but includes some or all of the following—making observations, forming and testing hypotheses, developing predictions, planning and conducting experiments, analyzing data, and evaluating results. **A scientific hypothesis is developed to provide a potential explanation of observations.** Hypotheses can be generated and tested using two basic procedures: inductive reasoning and deductive reasoning (Figure 1.7). **Inductive reasoning**

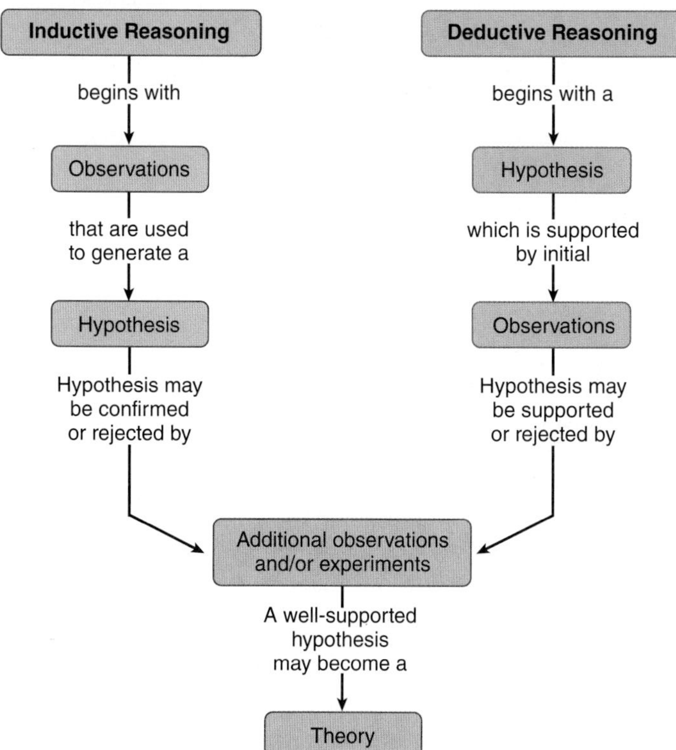

Figure 1.7 Concept map depicting two types of reasoning. Inductive reasoning results when scientists draw general conclusions from specific observations. Deductive reasoning occurs when scientists draw specific conclusions based on general principles.

occurs when scientists draw general conclusions from specific observations. The success of this method comes from recognizing patterns and identifying similarities between comparable systems.

Consider the following situation. Anne, an earth science professor, has students discuss the characteristics of different types of volcanoes with their neighbors during one of her classes. Later, she notices that the students do better on the exam questions about volcanoes than on other questions. She forms a hypothesis that students learn concepts better if they work together. She begins to assign daily exercises for the students to complete in small groups during class. Her average exam scores jump by 4.7 percent over previous classes. Anne is pleased that her students appear to be learning the material more thoroughly, and the students are pleased when she tells them that their grades are higher than expected. Anne has used **inductive reasoning, drawing a general conclusion from specific observations**.

Anne describes her teaching experiment to Don, an instructor in the physics department. He tells her that he too started using group assignments in class. Don tells Anne that he read an article in a physics journal by a colleague at another university that described how the use of groups resulted in improved student learning. Don had done some additional reading about the use of these methods in other science classes and determined that several instructors achieved similar improvements. Consequently, he concluded that if groups could improve learning in other science classes, they would probably also help his students. Don used **deductive reasoning** because he **used a general principle to reach a specific conclusion.**

Students frequently confuse inductive and deductive reasoning, so consider two other examples. What type of reasoning (inductive or deductive) is used in the following pair of statements and why?

1. All hurricanes form as low-atmospheric-pressure systems over oceans. Hurricane Harry is forming in the Atlantic. Hurricane Harry is a low-pressure system.
2. Three massive hurricanes caused large amounts of damage to the United States during the 2005 hurricane season. Hurricane Katrina had a pressure of 902 millibars (mbar); Hurricane Rita, 898 mbar; and Hurricane Wilma, 882 mbar. Therefore, massive hurricanes with air pressures of around 900 mbar or less will cause large amounts of damage if they make landfall.

The first scenario starts with a general statement about hurricanes and concludes with a specific statement about a single hurricane. Therefore, the reasoning is deductive—general to specific. The second scenario starts with specific data (air pressures) and ends with a general conclusion. This is inductive reasoning—specific to general. Most science involves components of both inductive and deductive reasoning.

From Hypothesis to Theory

The best hypotheses are logical and can be readily tested by experiment or by more observations. Continued observations over time will either confirm that a hypothesis is accurate or reveal that it is not quite right and needs to be either further refined or rejected. New information sometimes becomes available with the development of increasingly sophisticated technology and may lead to minor or major changes in existing hypotheses.

An initial hypothesis is a reasonable explanation on the basis of current science and needs further examination. After rigorous testing, bulked up with supporting facts and observations, a hypothesis may become a theory. The US National Academy of Sciences defines a **scientific theory as "a well-substantiated explanation of some aspect of the natural world that can incorporate facts, laws, inferences, and tested hypotheses."** Note that in science, a theory is not just an opinion or a guess; it is a well-supported explanation of a natural phenomenon. (For example, in Chapter 4 we will discuss the theory of plate tectonics.) An even higher standard of scientific scrutiny is reserved for laws. Scientific laws are statements that are so strongly supported by theory and observations that they are considered unchanging in nature. The law of gravity is an example.

Checkpoint 1.5

Scientists suggested the dinosaurs became extinct when an asteroid collided with Earth. They noted that ***the rare element iridium was present in 65-million-year-old rock layers around the world.*** The text in italic is an example of:

a) A hypothesis
b) A prediction
c) An observation
d) A theory

In our constantly changing world, hypotheses or even theories will be modified, and none can ever be completely proved. Widely accepted ideas will be confirmed and strengthened by the work of many scientists, but it is always possible that the next person to test the idea will discover a slightly different result and challenge part, or all, of the original hypothesis. **The willingness to continually question prevailing ideas and to modify or discard them as new information becomes available is the strength of science.** Science is an open book, a perpetual lie detector, limited only by the imagination and abilities of its practitioners. Given the complex nature of Earth, no scientist makes an observation, suggests a hypothesis, or develops a theory alone. The work of every scientist relies on the work of others who have gone before. Even Isaac Newton, whose law of gravity has withstood the test of time, noted, "If I have seen further, it is by standing on the shoulders of giants."

Checkpoint 1.6

basic / intermediate / advanced / superior

Observations, Hypotheses, and Mellinarks

Examine the images below. Based on your observations, form a hypothesis as to how many of the images in the bottom row represent Mellinarks.

What was the thought process you went through to arrive at an answer? Try to separate out the "thinking steps" that you took, identifying observations, predictions, and hypotheses. On a separate sheet of paper, briefly describe the steps.

All of these are Mellinarks.

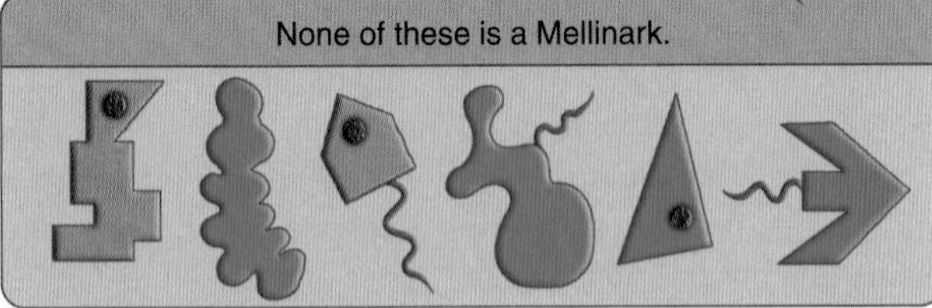
None of these is a Mellinark.

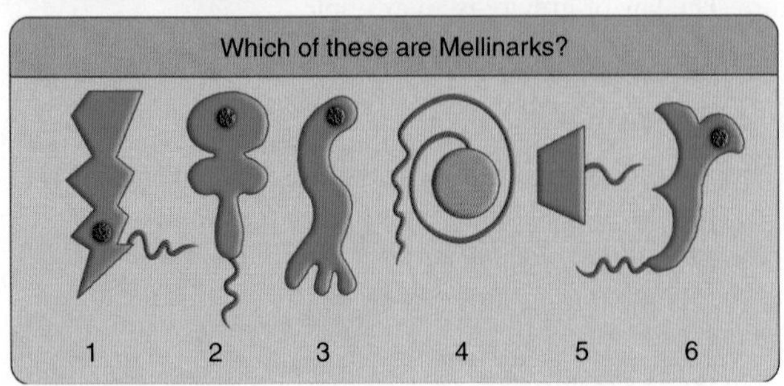
Which of these are Mellinarks?
1
2
3
4
5
6

Source: Question adapted from an article by Anton E. Lawson, *Journal of College Science Teaching*, May 1999, pp. 401–11.

The Characteristics of Good Science

Good scientific explanations follow four basic principles:

1. **Principle: Scientific explanations are provisional (tentative) and can and do change.** These changes are not always popular. A few years ago, new data on bodies orbiting at the fringe of the solar system resulted in the reclassification of Pluto to a dwarf planet. While science can change, we would emphasize that many of the key concepts discussed in this and other science books have been stable for a long time, some of them well established for centuries. Many of the changes occur in the details of the scientific explanations, rather than the major concepts themselves.
2. **Principle: Scientific explanations should be predictable and testable.** The daily weather forecast is a result of meteorologists using their knowledge of how air and moisture circulate through the atmosphere to predict short-term changes in weather patterns. The idea that hypotheses must be testable brings up a very important point about science. In science, one must be able to test a hypothesis or theory to determine if it could be false. That seems odd, does it not? Scientists construct a hypotheses or a theory to explain nature and then argue that the work is not science unless their idea has the potential to be found to be false. The central idea is this: science deals with the physical world. When we propose a hypothesis, we think it is true. Any additional experiments or observations should either support the hypothesis or show it to be false. Science progresses as hypotheses and theories are tested and shown to be supported (or not). If you cannot test a hypothesis, you are not doing science. Can you think of an idea that cannot be shown to be false (or true for that matter)?
3. **Principle: Scientific explanations are based on observations or experiments and are reproducible.** For example, scientists studying glaciers on different continents have noted that most are decreasing in size (see Chapter 16). These observations can be made routinely by photographing and measuring the dimensions of individual glaciers over several years or decades and are readily confirmed by others willing to visit the same locations. The important characteristic of science lies with the empirical, reproducible data used to support or refute a hypothesis. Scientific results are discussed openly at conferences and published in journals so that all ideas are exposed to review by other scientists (peers). This peer review process ensures that published research is original and adds to the body of valid scientific information. Rogue scientists who publish false data to advance their careers are often discredited by other scientists who are unable to reproduce the original results.
4. **Principle: A valid scientific hypothesis offers a well-defined natural cause or mechanism to explain a natural event.** Science looks for a cause for every effect. A tsunami is caused by an underwater earthquake; an earthquake is caused by movement of tectonic plates; tectonic plates move because Earth is releasing internal

Checkpoint 1.7

basic intermediate advanced superior

Identify the application of the four characteristics of good science in the passage that follows.

Science in Full View: The Hutchinson Gas Explosions

Jerry Clark heard the explosion and ran outside in time to see debris fall to the ground. Everyone in the city of Hutchinson, Kansas, heard or felt the blast at 10:45 A.M. on Wednesday, January 17, 2001. Glass covered downtown streets, and two stores were soon in flames. City emergency workers initially assumed that the blast was the result of a gas leak and so shut off gas supplies to the area. But the fire burned on. That evening, jets of gas erupted from the ground on the edge of the city east of the explosion site. A second fiery explosion on Thursday killed two people at a mobile home park near the gas jet site. By this time, no one knew what to expect next. The local police and National Guard quickly evacuated residents living near the second blast site, and the governor's office declared a state of emergency.

Both explosions and the gas jets were linked to abandoned wells, some drilled perhaps as much as a century earlier. It soon became clear that the gas was escaping through the wells from a source far below the city. Large underground caverns in the nearby Yaggy Corporation gas storage facility (see the figure) experienced a significant pressure drop on Wednesday morning as natural gas was being pumped into the storage caverns. Later investigations revealed that the source of the leak was a fist-sized hole in the casing of one of the wells used to pump gas into and out of the caverns. Kansas Gas Service (KGS) operated the Yaggy facility, one of a network of natural gas storage centers across the nation.

The Kansas Geological Survey immediately volunteered to help investigate the explosions. Survey geologists had previously produced a geologic map of the county around Hutchinson. This survey had provided sufficient information about local rocks to determine that it would be possible for gas to rapidly migrate the 10 kilometers (6 miles) from the Yaggy facility to Hutchinson. The scientists could do little more until they were able to analyze the situation more closely. At this early stage in the study, most investigators thought the gas had escaped from a leaking gas well, migrated upward, and then traveled toward Hutchinson in a rubblelike rock layer that capped the salt deposit containing the storage caverns (see the figure). KGS employees drilled numerous wells in an effort to allow any remaining gas to escape safely to the surface. The drilling revealed that the gas was not traveling along the rubble-like layer but was closer to the surface in a thin band of rocks about 100 meters (330 feet) below the city.

On January 30, the governor of Kansas sent a team of geologists, geophysicists, and engineers from the state Geological Survey to Hutchinson to find any remaining gas buildups. Survey representatives coordinated their work with local officials, KGS employees, and other state agencies and soon launched a website to keep residents and interested observers updated with the latest findings. The search for answers was made transparent as information was shared through the website or in public meetings where Geological Survey personnel answered questions and addressed concerns of local residents, officials, and reporters—even when sometimes the answer was "We don't know."

The scientists used ground imaging technology to examine the characteristics of the rock layers underlying Hutchinson and to search for possible locations where gas had collected. KGS drilled wells in two of those locations and vented large volumes of gas. Rock samples recovered from the gas-bearing zone contained a rock type known as dolomite. Geologists hypothesized that a series of connected fractures in the dolomite had served as a pathway for the gas to travel from the Yaggy facility to Hutchinson. Although they never recovered rock samples showing the presence of fractures, these scientists concluded that fractures in the dolomite layer represented the only reasonable passageway through which the gas could move quickly over that distance. Continued investigation revealed no remaining gas deposits, and on March 29, the Geological Survey's representatives told Hutchinson residents: "From a geological viewpoint, the city is safe."

The example of the Hutchinson gas explosions shows that scientific investigations are driven forward by the curiosity and persistence of scientists who systematically rule out potential solutions to arrive at an explanation. This example also illustrates that science doesn't have unlimited resources, personnel, or time. In some cases, it may be necessary to walk away and settle for the best answer available under the circumstances.

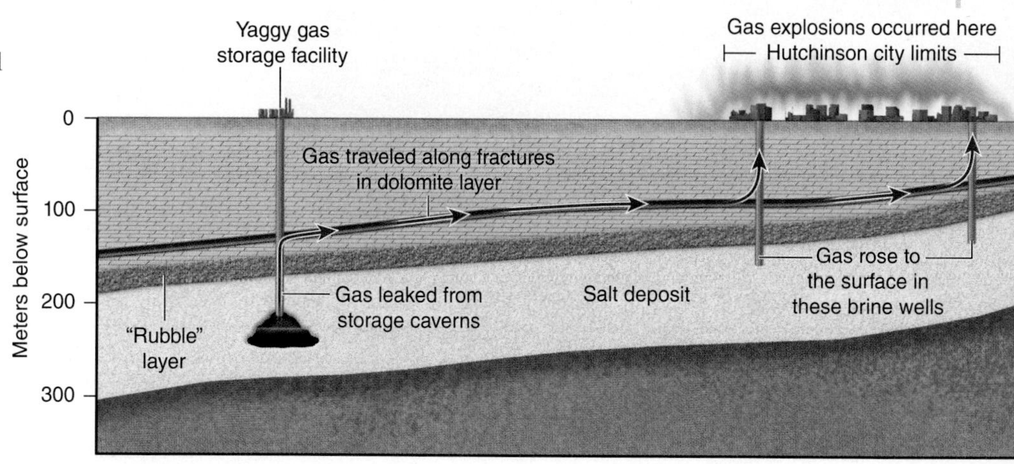

heat energy (more on all this in Chapter 4). Sometimes these cause-and-effect relationships are complicated, with multiple causes influencing the end result. For example, changes in Earth's climate are caused by effects from every component of the earth system including the composition of the atmosphere, oceanic circulation patterns, the combustion of fossil fuels, plant respiration, and our position relative to the sun. Ongoing research is attempting to determine just how much each of these factors influences climate change.

Let's see how these principles were applied in one of the most widely publicized scientific discoveries of recent years.

An Example of Good Science: The Alvarez Hypothesis

More than 30 years ago, a team of scientists led by the father-son pair Luis and Walter Alvarez suggested that the extinction of the dinosaurs was caused by a collision between Earth and an asteroid or comet. Walter Alvarez was a young paleontologist, a geologist specializing in using fossils to decipher the history of Earth. He was investigating the geologic history of the Mediterranean region, and his research took him to a large outcrop of rocks near the town of Gubbio in the Apennine Mountains of central Italy. There, he examined layers of rocks that represented the time in Earth's past that spanned the extinction of the dinosaurs. Dinosaurs (and many other species) died out 65 million years ago (over a period that may have lasted somewhere between 100,000 and 3 million years). Rocks formed before the extinction are classified as Cretaceous in age; those formed after are classified as Tertiary (more on these names in Chapter 8). Most species living on Earth during the Cretaceous became extinct prior to the start of the Tertiary time period. At Gubbio, a thin clay layer marked the Cretaceous-Tertiary boundary (abbreviated as the K-T boundary) between the different ages of rocks. The K-T boundary has subsequently been identified elsewhere (Figure 1.8). You may wonder why it wasn't known as the C-T boundary. Some other periods of geologic time also begin with a C, so, to avoid confusion, it was decided to use the letter K, taken from the German word for chalk 'kreide,' which is characteristic of rocks from that time.

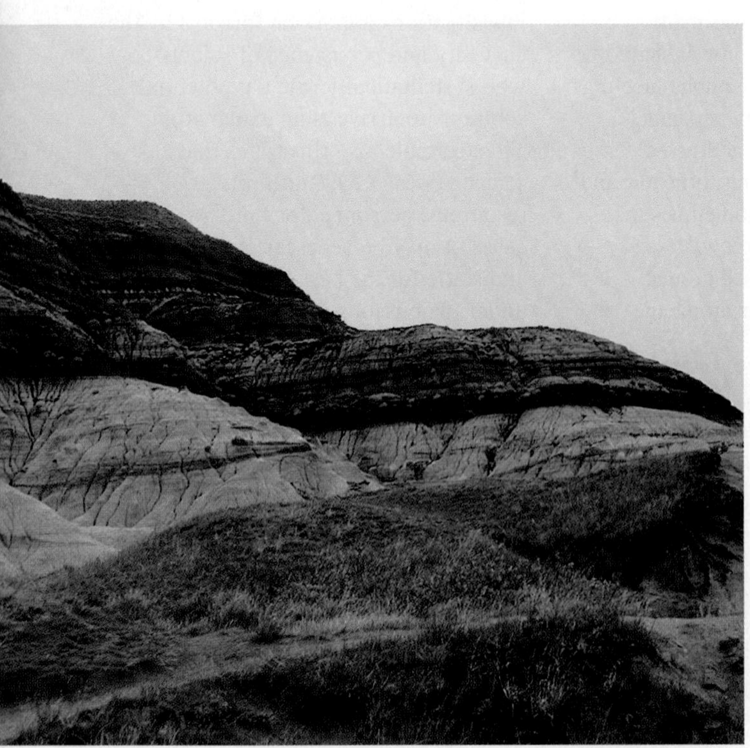

Figure 1.8 Badlands near Drumheller, Alberta, Canada, where erosion has exposed the K-T boundary. The boundary is located approximately where the light- and dark-colored rocks meet in the upper part of the outcrop.

At the time Alvarez was doing his field work, earth scientists had published a variety of different hypotheses seeking to explain the demise of the dinosaurs and other species. For example, some scientists believed the climate got too hot or cold for the dinosaurs, others thought that they were harmed by radiation from a supernova explosion, still others suggested that the evolution of flowering plants affected dinosaur eating habits, while yet others hypothesized that smaller organisms ate their eggs, causing a rapid population decline (provisional hypotheses; principle 1). However, some of these hypotheses could not be readily tested (violated principle 2), and none of the others had been widely accepted, so research continued. Walter Alvarez sought to estimate the rate at which species changed on either side of the K-T boundary by measuring the rate at which space dust had been deposited in the clay layer at the boundary (he used principle 3). Space dust falls to Earth daily and contains rare elements that can be readily measured in the lab, although in low concentrations of parts per billion.

Alvarez returned to the University of California, Berkeley, with samples of the clay layer. His physicist father, Luis Alvarez, suggested that his colleagues Helen Michel and Frank Asaro perform a chemical analysis of the clay material. The analysis revealed the rare metallic element iridium in the clay. Iridium is normally present in concentrations of 0.3 part per billion in rocks of Earth's crust, but Michel and Asaro found concentrations of 9 parts per billion, 30 times the expected amount. They found similar concentrations at other K-T boundary sites in Denmark and New Zealand (supporting data; principle 3). In seeking an explanation for the increase in iridium concentration over such a wide area, the Alvarez team recognized that objects such as asteroids and comets contained elevated levels of iridium and other rare metals. They hypothesized that a relatively large amount of iridium was deposited when an asteroid or comet collided with Earth (natural cause; principle 4; Figure 1.9). They published their hypothesis in 1980 in the journal *Science*, as a paper titled "Extraterrestrial cause for the Cretaceous-Tertiary extinction." The Alvarez hypothesis interpreted the data to suggest that a collision with an approximately 10-kilometer-wide (6 mile-wide) asteroid would have generated so much debris that it blocked incoming sunlight for several years. Vegetation would have died in the absence of light, leading to the deaths of plant-eating dinosaurs and the collapse of the global food chain. The decade following publication of the *Science* article saw a surge in research interest in the extinction event as scientists sought to find data that would support or refute the Alvarez hypothesis (continued study using principles 1–3 above). Within a dozen years, over 2,000 articles and books had been written on the topic.

Soon, several researchers had confirmed the presence of high concentrations of iridium in rocks at the Cretaceous-Tertiary boundary at multiple sites around the world (more data; principle 3). Scientists Alan Hildebrand and William Boynton discovered tsunami deposits in K-T boundary rocks in southern Texas, as well as thick layers of debris in deposits of the same age in Mexico and Haiti; and they predicted that an asteroid impact site should exist somewhere around the Gulf of Mexico (a prediction; principle 2). Soon it was discovered that two petroleum geologists had previously published the results of geophysical exploration in the Yucatan Peninsula, Mexico. They described a buried, near-circular feature (Chicxulub Crater) over 200 kilometers (125 miles) wide that Hildenbrand and Boynton now thought could be the possible impact site (another hypothesis; principle 4). The crater was linked to the K-T collision event as it lies in rocks that are older than the impact event and is covered by rocks that are less than 65 million years old. (For more on impacts, see Chapter 3.)

However, this story is not yet over. The provisional (tentative) nature of science makes it possible that other hypotheses may yet better explain some aspects of the extinction event. In the dinosaur extinction debate, some scientists suggest the source of the iridium was actually a massive series of volcanic eruptions that took place over an interval of half a million years around the same time as the impact event. More-recent research has discovered that iridium can be produced in relatively large quantities by some volcanic eruptions (another observation; principle 3). The span of the volcanic eruptions also better matches the more gradual die-off of some of the extinct species. Of course, it is possible that both events combined to kill the dinosaurs. Stay tuned: it is likely that the public's fascination with dinosaurs means that we haven't heard the last of this scientific debate.

Limitations of Science

We must keep in mind that scientific explanations may be limited by available technology or other factors. For example, prior to the invention of the telescope, knowledge about Earth's position in space was based on observations made with the naked eye. Astronomers such as Galileo used some of the first telescopes to make observations of planetary orbits that would support the hypothesis that the sun, not Earth, was the center of our solar system.

Although science is a powerful method for examining unresolved questions, *science cannot answer all questions.* Questions that center on ethics or theology often have more to do with cultural or social norms than with scientific concepts. For example, recent concerns about the potential for cloning humans can be separated into two distinct questions; one is scientific, the other ethical. Can we successfully clone humans? is a scientific question, and the current answer is no, although research suggests that a future response could be yes. Should we clone humans? is an ethical question. If the answer is no, we may never clone a person even if the scientific knowledge exists to do so.

The Characteristics of Bad Science

Poor scientific reasoning rarely reaches a public forum because of the checks and balances inherent in the scientific process. However, sometimes hypotheses are unveiled in the media before they can be rigorously tested. Unfortunately, on further analysis, some of these ideas may be proved wrong, prompting increased skepticism toward the scientific process and scientists in general. Some telltale

Figure 1.9 Asteroid impact with Earth. Luis and Walter Alvarez hypothesized that an event like this occurred 65 million years ago and contributed to the extinction of the dinosaurs.

signs indicate when an argument is not based on sound scientific thinking. Here are some things to look out for when people claim to disagree with a scientific hypothesis or offer unsupported explanations for natural phenomena:

- **An attack on the scientist, not the science.** Science does not advance based on the personalities of the scientists but on verification of facts and observations. (However, it can move backward when a scientist fabricates data to support a hypothesis. This happens only rarely.)
- **People who argue from authority.** Just because a person is important or powerful does not make him or her right. The extremely powerful Roman Catholic Church disagreed with Copernicus when he pointed out that Earth rotated around the sun, but it turned out that he was correct.
- **Confusion over cause and effect.** This type of thinking is often summarized by the Latin phrase *post hoc, ergo propter hoc*: "it happened after, so it was caused by." For example, a student may claim that he did well on an exam because he wore his "lucky" shirt.
- **The use of bad statistics.** Even the weakest scientific arguments may look appealing if supported by statistics that are based on a biased sample or on a sample size that is too small to be representative.

Scientists who don't engage in peer review to evaluate their research will not have their work published and may be discredited if their results cannot be reproduced by others. Alternatively, some hypotheses receive publicity before they have had an opportunity to be critically reviewed by experts. One example of a very public—but failed—hypothesis is described next.

An Example of Bad Science: Prediction of a Midcontinent Earthquake

Self-proclaimed climatologist and businessman Iben Browning proposed that an earthquake would occur on the New Madrid fault zone in southeastern Missouri on or around December 3, 1990. He based this prediction on the fact that New Madrid had been the site of an extraordinary series of major earthquakes (sometimes called the Mississippi Valley earthquakes) over a 3-month span from December 1811 to February 1812. Browning hypothesized that tidal forces due to the gravitational pull of the sun and moon could trigger another big earthquake on the New Madrid fault zone . . . or maybe one in Japan . . . or in California . . . well, somewhere in the Northern Hemisphere.

The hypothesis generated widespread media interest in the region and raised public anxiety sufficiently that many local schools closed in anticipation of an impending quake. Browning's claims were widely denounced by earthquake specialists, who were also frequently quoted in local newspapers. By the time the fateful day arrived, the hype had taken over, and the area was besieged with reporters who, as it turned out, were able to report that nothing happened.

This was bad science because Browning did not offer an accepted mechanism to explain the occurrence of the potential earthquake. Although tidal forces do exist, they had not been linked to earthquake activity. But this story also illustrates that even a clear, unambiguous message from experts (there's *not* going to be an earthquake!) can get lost in the shuffle. Scientists analyze situations; they do not write the newspaper stories or determine how schools and other public services should respond.

Checkpoint 1.8

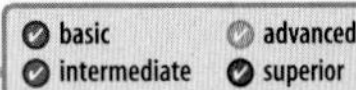

Employees at the Ripley's Believe It or Not! Museum in Myrtle Beach, South Carolina, declare that female visitors who come in contact with a pair of African fertility statues are more likely to become pregnant some time later. The statues, from the Boule tribe of the Ivory Coast, stand near the museum's entrance. Some visitors have volunteered the information that they gave birth 9 months after touching the statues and credit the statues. The museum notes that some couples travel from as far away as Texas to rub the statues.

a) What hypothesis is presented in this story?
b) Is the hypothesis supported by sufficient observations? Explain.
c) What prediction could be made to verify or falsify the hypothesis?

Source: Summarized from an article by Isaac J. Bailey, *Houston Chronicle,* October 15, 2000.

1.4 Science and Society

Learning Objectives

- Identify physical or chemical and social or cultural aspects of the earth system.
- Compare and contrast protection and adjustment procedures related to natural hazards.
- Explain the four principal roles that earth scientists play in society.
- Describe examples of how citizens interact with the natural environment at local, national, and global scales.

Why should you care about science—and earth science in particular? Most of us are removed from the process of science. (How many scientists do you know?) Past experiences have convinced some people that they will never understand science, whereas others may view the study of earth science as irrelevant to our comfortable twenty-first-century lifestyle. Many citizens are understandably bewildered by media reports that portray battling teams of scientists presenting opposing explanations for complex scientific problems. If the experts cannot agree, they reason, how can we be expected to make a decision? Besides, even if we understand environmental problems, we are often frustrated by the apparent inability of those responsible to do anything about them. This can range from simple individual actions (Why doesn't my neighbor recycle?) to issues of corporate responsibility (Why do companies produce air pollution?).

How can we become enlightened citizens capable of identifying problems that will affect us all and participate in their solutions? We have to combine the critical thinking we described in Section 1.3 with civic thinking that involves the analysis, planning, and evaluation of actions that may help society to arrive at solutions to these problems. In this context, *society* may refer to anything from a small town up to the global community. Here we suggest a simple three-step process: know, care, act.

- **Know.** We must take responsibility for our world by knowing how it works.
- **Care.** Our society works best when we care about how our actions affect others. But we should also be aware of how we are affected by the actions of others.
- **Act.** Do something. Make your opinion known. Go to a town meeting, blog about it, write a letter to your local paper, contact your congressperson or senator, vote. To quote anthropologist Margaret Mead: "Never doubt that a small group of thoughtful, committed citizens can change the world. Indeed, it is the only thing that ever has."

The Role of Earth Science

We have been interacting with Earth's environment since our human ancestors began to roam Earth's surface. Our demands on the planet have been magnified as technology developed and Earth's population increased. It has become increasingly necessary to monitor fundamental features of the environment around us. **The principal elements of the environment (air, water, soil) have specific chemical and physical characteristics that can be readily measured.** Scientists can measure the volume of dust in the air or the abundance of a chemical in a stream to determine if the air or water quality falls below community standards.

Social or cultural influences on decisions affecting the environment are more difficult to quantify than physical and chemical conditions. Consequently, environmental decisions are complex to evaluate and are often the subject of vigorous debate. Furthermore, the influence of these social and cultural factors changes with time as perceptions change. For example, our view of the role of wilderness has evolved in the 400 years since the earliest European settlers arrived on the North American continent. The early colonists considered the virgin forests home to unfriendly natives and mythical beasts, so they regarded wilderness with hostility. However, as the population expanded and the number of wilderness areas dwindled, the remaining natural lands began to be considered important cultural assets and were consequently protected by legislation such as the Wilderness Act (1964).

Given the complexities of people's relationships with our planet, earth scientists have several roles to play in modern society. These roles have become more crucial as our global population climbs past 7.1 billion people, with about 77 million more added each year. We are concerned about protecting life and property from the dangers of natural hazards, obtaining sufficient natural resources to maintain or improve our standard of living, and protecting the health of the natural environment. A final, more comprehensive goal, ensuring the future of our own species, has recently received increasing attention as we glimpse a future in which human actions modify the composition of the atmosphere and we recognize the global-scale devastation that may result from an asteroid impact event. All of these roles require a similar scientific approach. Acting as representatives for society at large, scientists seek to:

- Identify and measure natural processes and phenomena.
- Understand the natural processes involved.
- Monitor these variables over time and predict future trends or events.
- Engage citizens and their representatives in using this information to make effective decisions to meet society's needs.

Protecting Against Natural Hazards

Scientists play a vital role in understanding and determining the potential risks from natural phenomena that may harm people and damage property. Natural processes such as earthquakes, landslides, floods, volcanic eruptions, tornadoes, and hurricanes are considered hazards when they occur in populated areas (Figure 1.10). The detailed study of hazards in one area can help predict the potential risks elsewhere. For example, scientists used

Checkpoint 1.9

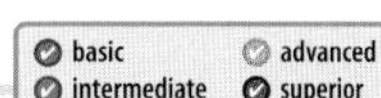

Read the following summary of the Comprehensive Environmental Response, Compensation, and Liability Act (CERCLA, 1980). Does this law involve the measurement of physical or chemical characteristics of the environment, or did it arise from social or cultural concerns, or were both important? Explain your answer.

CERCLA established prohibitions and requirements concerning closed and abandoned hazardous waste sites, provided for liability of persons responsible for releases of hazardous waste at these sites, and established a trust fund to provide for cleanup when no responsible party could be identified. The law authorized two kinds of response actions: short-term removals, where actions may be taken to address releases or threatened releases requiring prompt response; and long-term remedial response actions that permanently and significantly reduce the dangers associated with releases or threats of releases of hazardous substances that are serious, but not immediately life threatening.

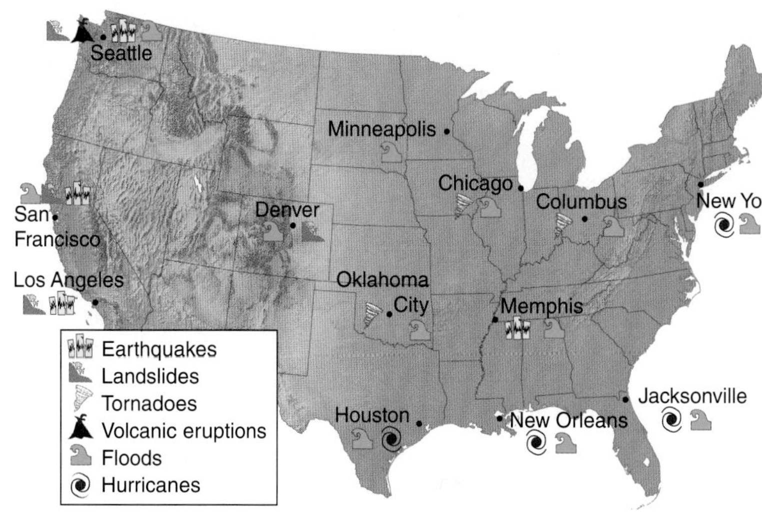

Figure 1.10 Principal natural hazards for some US cities. How does the type of hazard vary by region?

Figure 1.11 Hurricane Katrina approaching the Gulf Coast, August 2005. Katrina devastated coastal communities in Louisiana, Mississippi, and Alabama. The storm was responsible for more than 1,800 deaths and was the costliest natural disaster in US history.

Checkpoint 1.10

basic intermediate advanced superior

Is evacuation of a city before a hurricane an example of the prevention of a hazard or adjustment to a hazard?

a) Prevention b) Adjustment

What other examples of prevention or adjustment have been described in the chapter so far?

the information they learned from investigations of a 1980 volcanic eruption of Mount St. Helens in Washington state (see Chapter 6) to accurately predict the size and timing of the larger 1991 eruption of Mount Pinatubo in the Philippines. (Was this an example of inductive or deductive reasoning?)

Each decade the cost of property damage from natural hazards more than doubles (when adjusted for the rate of inflation). In some cases, the risks from natural hazards depend more on development decisions made before these events occur than on the natural phenomena themselves. Human beings are unlikely to be able to stop volcanoes from erupting or to banish earthquakes. However, the application of scientific knowledge and appropriate technology can help save lives and protect property. The principal advantage of technology is to provide accurate information to maximize the safety of people living in areas at risk for natural hazards. The effects of some potentially destructive phenomena can be partially offset by the following technological advances:

- Weather satellites are used to track hurricanes and predict landfall sites, allowing timely evacuation of residents (Figure 1.11).
- Doppler radar installations have more than doubled the amount of advance warning time for tornadoes.
- Networks of monitors linked by satellites measure stream flow and predict the magnitude and timing of floods, allowing emergency construction of levees and evacuation of residents.
- Engineering structures and strict building codes are deployed in areas at risk from large earthquakes to ensure that buildings, although damaged, will remain standing when the shaking subsides.
- Arrays of instruments at more than 150 volcanoes are monitored by the USGS National Volcano Early Warning system to predict the timing, location, and magnitude of future eruptions.

Mitigation represents actions that prevent or reduce the probability of a natural disaster or lessen the effects of the event. Although we cannot prevent most natural hazards from occurring, we can make adjustments that will minimize their impact through careful land use planning, the enforcement of building codes, and the purchase of insurance policies. These steps ensure that some areas at risk are not heavily populated, that key structures are built to withstand the hazard, or that funds are available to repair damages following a hazardous event. Floods and landslides are clearly linked to streams and slopes, allowing scientists to make local alterations to the environment in efforts to mitigate future hazards. For example, building levees to contain rising streams or reservoirs to store floodwaters can locally diminish or eliminate flooding. However, we should be aware that any alteration of a natural system has the potential to cause unanticipated changes. For example, building a levee may reduce flooding locally but actually increase the flood risk downstream where the stream is in its natural state. As the devastation of New Orleans following Hurricane Katrina amply illustrates, our confidence in engineered solutions such as levees and flood control structures often prompts increased development, ultimately resulting in greater loss of life and property when disaster inevitably strikes.

In assessing the risks associated with natural hazards, earth scientists must try to answer several questions: How often do such hazards occur? How large an area will be affected? How grave is the risk to people and property? What actions can be taken in both the short and long term to prevent some of these events or lessen their impact? Determining the correct answers to these questions requires knowledge of earth processes, the characteristics of the landscape, the type and distribution of rocks underlying a region, and the rocks' physical and chemical properties. These scientific questions are more than academic—they focus on the very safety and security of human lives. We will explore the science behind natural hazards in Chapters 5, 6, 10, 11, 13, and 15.

Finding and Sustaining Earth's Resources

Life on Earth requires the use of resources. The term *resources* covers everything we use, including such basic assets as air, soil, timber, and water; fuel resources such as coal, oil, and gas; and mineral resources (Figure 1.12), such as sand and gravel. These natural resources may be renewable or nonrenewable. Renewable resources are replenished constantly (wind, soil), on short-term timescales measured in months (crops) or over longer intervals of several years (timber). Nonrenewable resources either are lost following consumption (fossil fuels) or may be recycled to be used in other products (metals). In 1900, renewable resources (agriculture, food, forest materials) accounted for 41 percent of the consumption of US raw materials. Today, they represent less than 10 percent of total materials consumed (by weight).

The United States uses more mineral and energy resources than any other nation, and our economic growth is dependent on continued access to these resources. There are 1,300 billion barrels of oil available in global oil reserves. Each day, the world uses approximately 90 million barrels (33 billion barrels per year). How long before oil reserves are depleted, according to those numbers? Currently, the United States imports a substantial fraction (nearly 50 percent) of its oil, and we rely exclusively on other nations for more than a dozen critical mineral commodities. One-third of the world's people live in the rapidly expanding economies of China and India, placing even greater demands on global mineral and energy supplies. In an ideal world, the human race would develop into a sustainable society, a society that satisfies its desire for resources without completely consuming resources essential for future generations. However, given the pace of global economic development, it is unlikely that we will achieve sustainability in the near future.

Figure 1.12 Strip mining for the metal manganese in South Africa.

Earth scientists work to understand the distribution, abundance, and origin of resources. Will there be sufficient resources to support the growing global population 50 years from now? What steps can we take to preserve and protect the most heavily exploited resources? What alternative energy sources can be utilized to make the nation less dependent on foreign suppliers? Successfully answering these questions requires earth scientists to explore ever more remote parts of Earth's surface, including rain forests, rugged mountain ranges, and the deep ocean floor.

It is theoretically possible that Earth could support many times its current population, but such speculation takes no account of the quality of lives people would be required to lead to ensure sufficient food (and other resources) for all. Individual actions such as turning on a light switch or pouring a glass of water involve relatively modest resource use and require little thought except in the most extreme conditions. However, multiply those actions several billion-fold and divide some resources across international borders, and we can readily imagine situations where resource exploitation can have wide-ranging consequences (Figure 1.13). The distribution and exploitation of natural resources are examined in Chapters 2, 7, 9, 11, 12, 14, 16, and 17.

Protecting the Health of the Environment

The biosphere (plants, animals) has exhibited dramatic changes throughout Earth's history, but recent population growth has contributed to environmental change, albeit over a much shorter timescale. Global population more than quadrupled since 1900, and we will add several billion more people this century. As population has expanded, so has industrialization and consequently pollution of land, air, and water. Pollution is still readily visible in developing countries, but in the United States, its effects are muted and

Figure 1.13 Human-generated lights on Earth. These patterns indicate the distribution of population and serve as a proxy (substitute) indicator of energy consumption for different nations. Densely populated, developed regions (United States, Europe) show brighter lights than heavily populated, developing nations (India, China). Sparsely populated regions are dark (South American rain forest, central Australian desert).

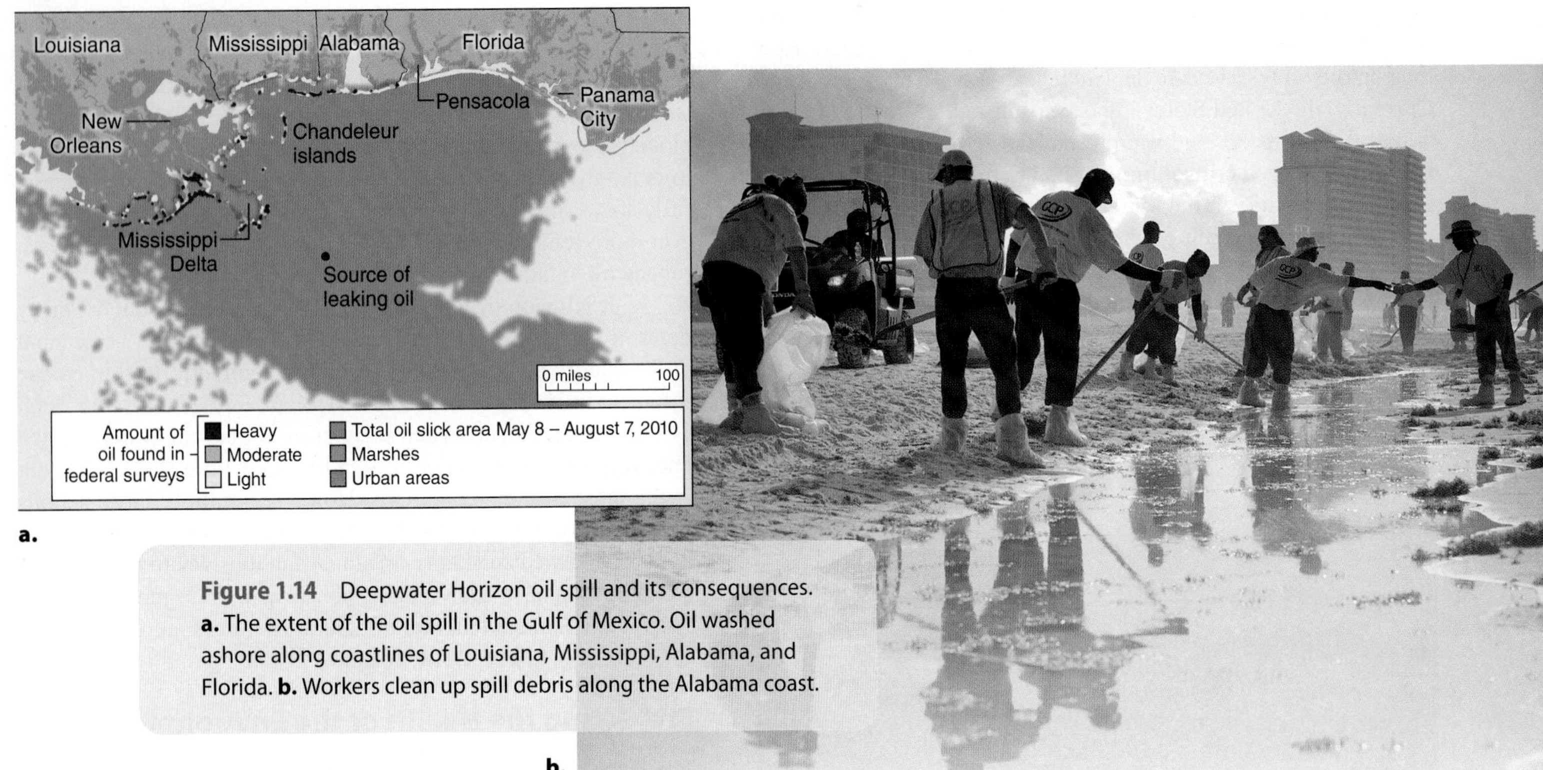

Figure 1.14 Deepwater Horizon oil spill and its consequences. **a.** The extent of the oil spill in the Gulf of Mexico. Oil washed ashore along coastlines of Louisiana, Mississippi, Alabama, and Florida. **b.** Workers clean up spill debris along the Alabama coast.

much more subtle, as indicated by reports of respiratory ailments and contaminated drinking water supplies. We can consider these threats as examples of slow-motion hazards. While they don't have the sudden impact of an earthquake or tornado, their long-term consequences have the potential to be much more devastating.

Human activities have the potential to endanger human life and natural ecosystems. For example, we have found a variety of ways to contaminate the hydrosphere. One of the most spectacular was the April 2010 explosion of the *Deepwater Horizon* oil rig, which resulted in the spill of 4.9 million barrels (206 million gallons) of oil into the Gulf of Mexico. The disaster continued for three months as crews struggled to cap the gushing well on the floor of the gulf, 1,600 meters (5,100 feet) below the sea surface. Eleven rig workers lost their lives in the initial explosion, and the spill eventually contaminated 1,064 kilometers (665 miles) of coastline in multiple states (Figure 1.14a). The oil spill itself, and the chemicals used to disperse the oil, combined to decimate marine species in much of the northeastern Gulf. British Petroleum, the company that leased the well, has spent approximately $40 billion cleaning up the region, and legal challenges remain. As we write this, three years after the disaster, it is too early to determine if the biological communities that were devastated by the spill had fully recovered. We will explore how the health of the environment is affected by both human and natural causes in Chapters 2, 6, 8, 9, 12, 13, 14, 15, 16, and 17.

Ensuring the Future of Human Life

The issues discussed so far occur at the local, regional, or national scale and involve events that are significant on timescales measured in hours to years. However, if we take a more global view, we can identify processes that have the potential to affect everyone, everywhere, for decades and perhaps centuries to come: the impact of a large asteroid and the effects of global climate change. Any program that attempts to address either asteroid impacts or global warming would be both complex and expensive, requiring cooperation among many nations and potentially taking decades to complete. Nevertheless, these threats cannot be ignored, and science has the potential to show the way to effective solutions. We examine the science behind these issues in Chapters 2, 3, 4, 6, 8, 13, 14, 16, and 17.

The **impact of a large asteroid with Earth represents a global-scale natural hazard** (Figure 1.9) that has the potential to end all life as we know it or to devastate a continent-sized area of the planet. Concerns about such an impact increased in recent years as we became aware that such events were more commonplace in the geological past than was previously thought. Although scientists have many ideas about how to stop an asteroid on a collision course with Earth, no mechanism yet exists for dealing with such an event. We will address issues related to asteroid impacts in Chapter 3.

An international panel of scientists has concluded that **global warming represents an alteration of global climate patterns resulting from human activity.** Most scientists think that carbon dioxide and other gases of human origin have altered global climates over the last century. Higher concentrations of carbon dioxide are associated with warmer temperatures. Warmer conditions have the potential to cause wholesale changes in natural systems around the world.

Scientific research on global change is an example of "big science" because it involves researchers around the world working on thousands of different projects, each contributing a small piece to a much larger puzzle. The 1990 Global Change Research Act required the federal government to implement a climate change research program. The US government budgets nearly $2 billion for climate change research each year through the Climate Change

Checkpoint 1.11

basic / intermediate / advanced / superior

Complete the following concept map to summarize the characteristics of the four principal roles that earth scientists play in society.

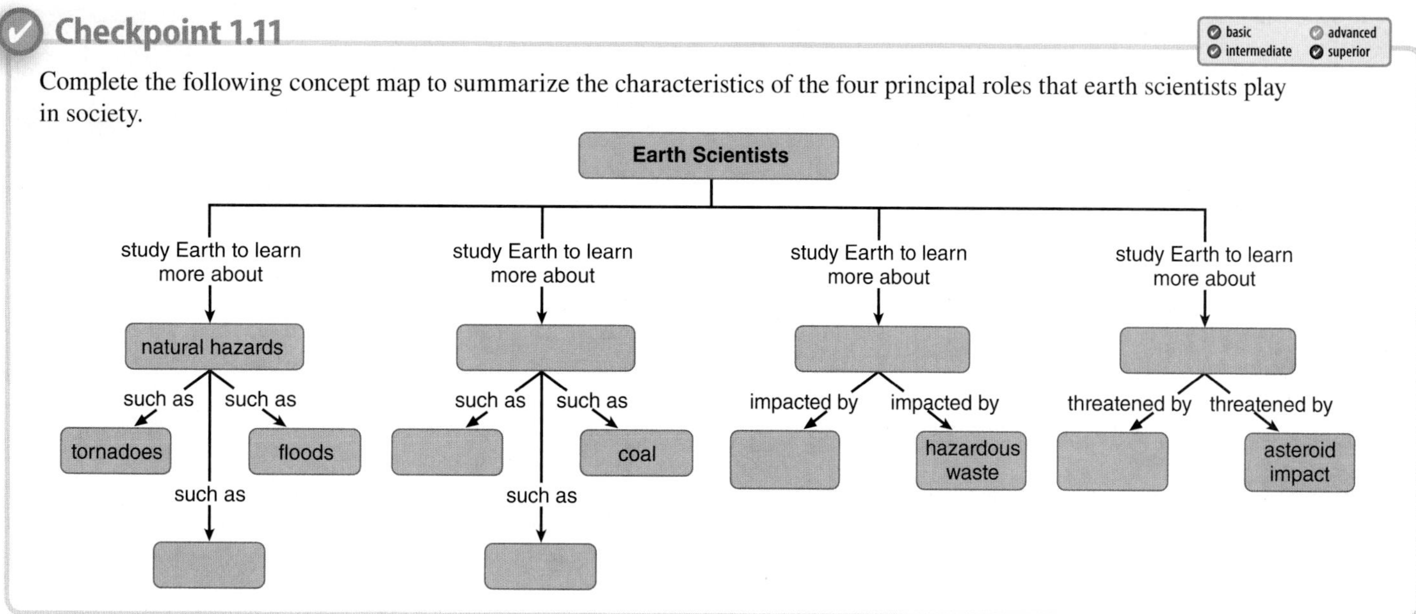

Science Program, which involves workers from more than a dozen government agencies. The investigations of the program involve hundreds of scientists working in research teams and examining many different topics that will contribute to our understanding of global change. Each team of scientists must make a research plan, collect data, make observations, draw conclusions, present their work at professional meetings, and write technical articles during the term of their research. Researchers seek to piece together a story about past and future global change by reading literally thousands of publications and synthesizing hundreds of ideas to build sophisticated computer models. This process represents a lot of hard work, and the process moves forward slowly in careful increments. This is the nature of science.

The Anthropocene: A New Time on Earth?

Nobel Prize–winning scientist Paul Crutzen suggested that human activity has produced such sweeping changes to the planet that we have entered a new period of Earth history, informally termed the Anthropocene. Crutzen and others point out that global temperature, sea level, and atmospheric chemistry were relatively stable for the last 8,000 years or more. This apparent stablity was disrupted in the last few centuries as the rapid increase in human population was accompanied by economic growth and industrialization, resulting in widespread resource exploitation and environmental change. Geologists differentiate distinct time intervals when they can recognize some characteristic features in the rock layers formed at that time. Some researchers suggest that thousands of years from now, future scientists will be able to recognize several key markers preserved in various parts of the earth system that will be readily interpreted as the result of changes during the Anthropocene time interval. Key changes include:

- Increasing erosion rates: More than one-third of Earth's land surface is exploited by humans. Layers of sediment on the seafloor will record activities such as agriculture and construction that remove more earth materials than natural processes.
- Changing atmospheric chemistry: Gas bubbles trapped in thick ice sheets will show higher concentrations of such gases as carbon dioxide, methane, and sulfur dioxide.
- Decreasing biotic diversity: Fewer species will be represented in fossil sites as many become extinct or less adaptable as single-variety forms used in agriculture replace the variability of natural species.
- Ocean chemistry: Surface ocean waters are becoming more acidic as they absorb higher concentrations of carbon dioxide. This hinders the growth of some species (e.g., corals) and dissolves the shells of others, removing them from the rock record.
- Sea level change: Short-term changes will be less than a meter over the next century but may be measured in tens of meters by the end of the millenium.

The concept of the Anthropocene represents significant global changes linked to all parts of the earth system. We will examine different aspects of global change research as we move through *The Good Earth*. By introducing the topic of global change here, we hope to give you an opportunity to look over the shoulders of the researchers to see how our understanding unfolds as scientists try to figure out how our home planet works.

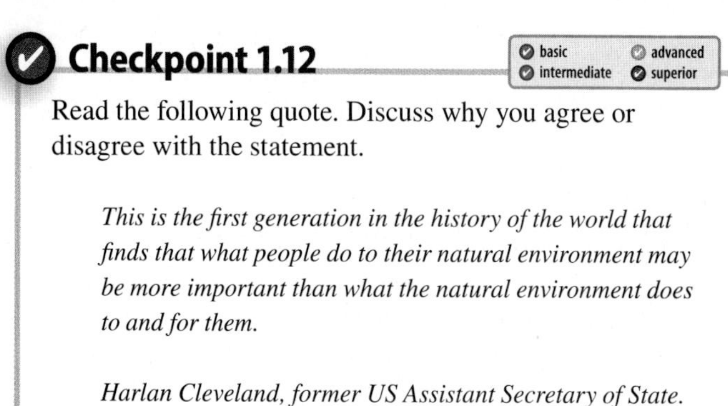

Checkpoint 1.12

basic / intermediate / advanced / superior

Read the following quote. Discuss why you agree or disagree with the statement.

This is the first generation in the history of the world that finds that what people do to their natural environment may be more important than what the natural environment does to and for them.

Harlan Cleveland, former US Assistant Secretary of State.

the big picture

Throughout this chapter, we have focused on the characteristics and practices of science. We started with examples of earth processes that challenge us to apply the methods of science to understand how our planet works so that we can avoid, or at least be prepared for, similar events in the future. Japan was reeling from the effects of the disastrous Tohoku earthquake and subsequent tsunami two days after the event. As time passed, fewer survivors were pulled from the rubble, and the vast scale of the destruction had become apparent. One fortunate survivor floated unexpectedly into this grim situation, his remarkable story providing a small moment of consolation for a country that had few glimmers of hope in the wake of a devastating disaster. Minutes after the tsunami warning was issued, Hiromitsu Shinkawa rushed home to collect his wife and gather a few belongings, but he was too late. His home in a coastal Japanese village was destroyed as the mass of water surged inland. He was separated from his wife and swept out to sea. He was fortunate to find part of the roof of his house floating by. After several attempts he was able to climb on board. He floated precariously on his makeshift raft for two days, lost among a dense field of debris in relatively calm seas. Despite the best efforts of rescuers, Hiromitsu wasn't spotted by a boat crew until he was 10 miles (16 kilometers) from the coast and had begun to despair of being found (Figure 1.15). Scientists will learn lessons from the deadly tsunami to be better prepared for the next one. In the wake of the Indian Ocean tsunami, governments of nations around the Indian Ocean worked to create a warning network to allow for timely evacuations in advance of the arrival of future tsunamis. This kind of intergovernment cooperation is necessary to address phenomena that have the potential to impact large regions of the globe. (We will describe the link between earthquakes and tsunami in greater detail in Chapter 5.)

The discussion of the Alvarez hypothesis for the extinction of the dinosaurs illustrated how scientists made predictions, tested (and rejected) hypotheses, and arrived at the conclusions that were most reasonable under the circumstances. We finished the chapter with an indication of how humans are affecting the earth, atmosphere, and oceans. More than 100 years ago, the English mathematician and philosopher William Whewell noted that:

> *The hypotheses we accept ought to explain phenomena which we have observed. But they ought to do more than this: our hypotheses ought to foretell phenomena which have not yet been observed.*

Figure 1.15 Hiromitsu Shinkawa is rescued two days after the March 2011 Tohoku tsunami. He was staying afloat on top of part of his roof 16 kilometers (10 miles) from land.

The Climate Change Science Program involves thousands of researchers seeking to explain the processes that control our current climate and to foretell Earth's future climate. Think of it as a global weather forecast for the next 100 years. We hope you will pardon us for revisiting this story throughout the book, but it will be one constant theme of science that you will hear, see, and read about in the years ahead so we figured we would get you ready for action.

Interaction	Letter
Plants absorb carbon dioxide.	
Earthquake destruction causes deaths.	
Wind blows sand.	
Spacecraft explore deep space.	
Continents deflect ocean currents.	
Plants release oxygen.	
Fish live in oceans.	
Asteroid impacts Earth.	
Volcano emits toxic gases.	
Animals drink water.	
Water evaporates from the oceans.	
Humans mine coal.	
Winds generate waves.	
A stream carves a canyon.	

Introduction to Earth Science: Concept Map

To evaluate your understanding of the interactions between the components of the earth system discussed in this chapter, complete the following concept map exercise.

Examine the following list of interactions between pairs of components in the earth system. Match each interaction with one of the lettered links in the concept map provided.

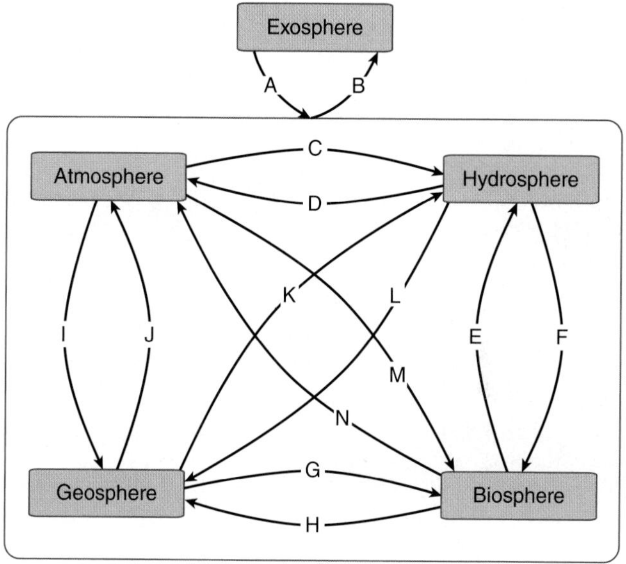

"We travel together, passengers on a little space ship, dependent on its vulnerable reserves of air and soil; all committed for our safety to its security and peace; preserved from annihilation only by the care, the work, and, I will say, the love we give our fragile craft."
—Adlai Stevenson, former governor of Illinois
"We have only one planet. If we screw it up, we have no place to go."
—J. Bennett Johnston, US Senator

Chapter 2

Earth in Space

Chapter Outline

the big picture

A planet you should know. How many more of these are there out there?

See The Big Picture box at the end of this chapter for the full story on this image.

2.1 Old Ideas, New Ideas

Chapter Learning Outcomes

- Students will explain the major processes and significant events that shaped formation of the solar system over billions of years.
- Students will describe the cause of the seasons and predict changes to future climate.
- Students will explain how and why there are various geologic boundaries within Earth.
- Students will describe the conditions necessary to support life on Earth.

Although Earth's position in the solar system seems obvious to us, it took us a few thousand years to arrive at this knowledge. For ancient civilizations without access to today's sophisticated lighting and heating technology, the activities of the sun were crucial to their daily routines. People observed the sun rising in the east and setting in the west (Figure 2.1) and inferred (wrongly!) that the sun revolved around Earth in a daily orbit. Further, the stars showed a similar pattern. **This early concept—that Earth lies at the center of our planetary system—is known as the geocentric orbit hypothesis.** Through research and observation, we now know that Earth rotates once each day so that we turn to face the sun in the morning and rotate away from its light as night descends.

Like almost everyone else living 2,300 years ago, the Greek philosopher Aristotle believed Earth was at the center of the universe and that the visible planets (Mercury, Venus, Mars, Jupiter, Saturn) and stars, including the sun, revolved around Earth (made a complete path around Earth) in a geocentric orbit (Figure 2.2). Aristotle's geocentric view of Earth's position in space dominated astronomy for almost two millennia but not without being challenged. Another Greek philosopher, Aristarchus, made some rudimentary calculations of the relative size of Earth, moon, and sun, and concluded that it was more probable that **Earth revolved around the larger sun in a heliocentric (sun-centered) orbit.** Furthermore, Aristarchus suggested that Earth rotated (turned around a central axis of rotation) and that the Earth-sun system was part of a much larger cosmos. Aristarchus was generally correct; however, as is often the case, his peers disregarded his novel ideas since they contradicted the widely held views of his time and there was no way to confirm his hypotheses. That would take another 1,800 years.

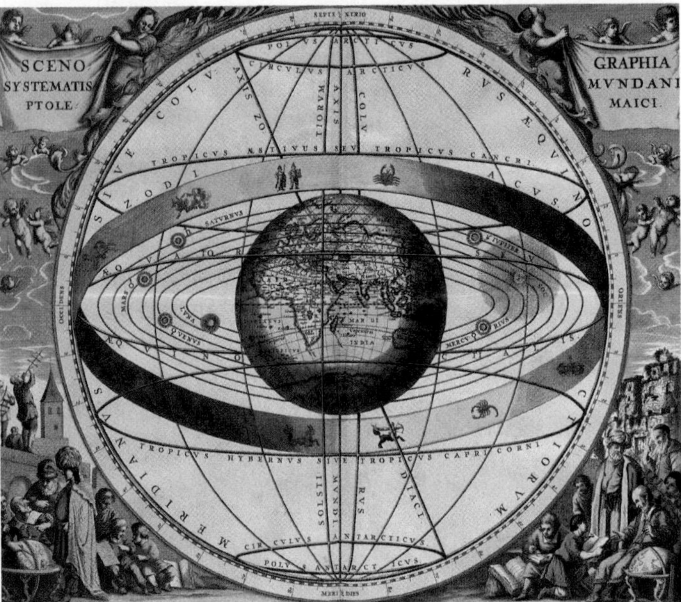

Figure 2.2 Ancient representation of the geocentric Earth. Note the position of the sun in orbit in the right center of the image.

The sixteenth-century scientist Nicolas Copernicus was the first person to expand on the heliocentric model sufficiently that it became a well-reasoned alternative to the geocentric view. But, as with many new scientific hypotheses, the technology did not yet exist to either confirm or reject Copernicus's ideas. It was not until 1609, when the Italian mathematician Galileo Galilei introduced the telescope into cosmic exploration, that observations could be made to test Copernicus's prediction and confirm the heliocentric hypothesis once and for all. As predicted by the heliocentric model,

Figure 2.1 Sunrise in the eastern sky and sunset in the west. The sun rises to the east (in this case, Myrtle Beach, South Carolina, left) as Earth rotates to face the sun, and sets to the west (Big Sur, California, right) as the planet slowly turns away from the sun at night. To people on Earth, it appears that we remain stationary and that the sun is revolving around us—this was the perception of the ancient Greeks and Romans. Today, we know that it is the sun that remains in the same location and Earth that turns on its axis.

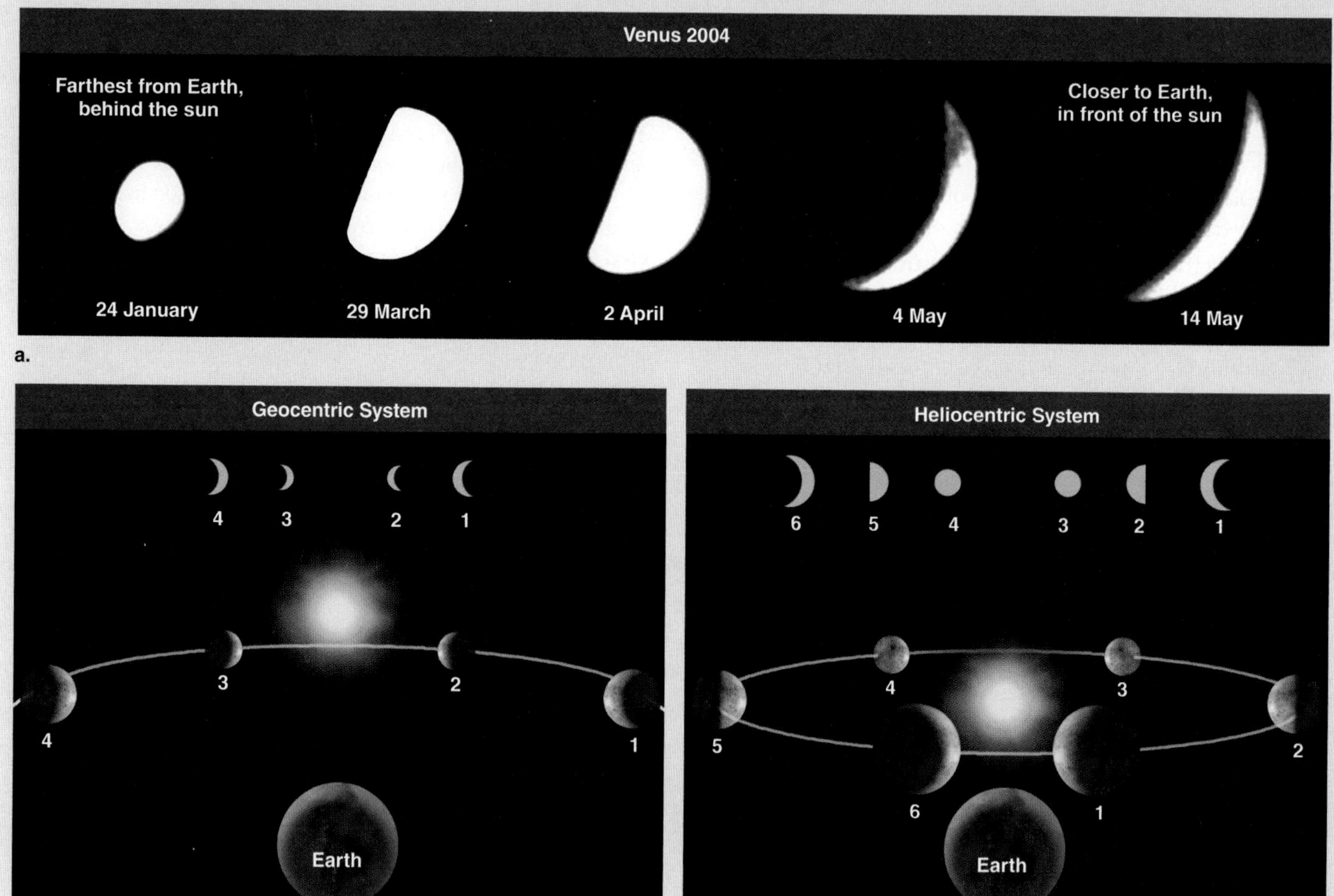

Figure 2.3 Phases of Venus in the heliocentric and geocentric systems. **a.** The phases of Venus from January 24 to May 14, 2004. Note how the size and appearance of the planet change. **b.** These changes cannot be explained by the geocentric model, where Venus lies between Earth and the sun. Consequently, the geocentric hypothesis is falsified by the phases of Venus. In the heliocentric model, Venus is either in front of the sun or behind the sun, relative to Earth. This is the pattern that Galileo observed and that we can see today.

Galileo observed that the appearance and relative size of Venus varied as its position changed relative to the sun and Earth (Figure 2.3). In the geocentric model, the sun was interpreted to revolve around Earth beyond Venus. Consequently, if Venus was located between Earth and the sun, an observer from Earth should only be able to see a small crescent of Venus lit by the more distant sun. In the heliocentric model, Venus constantly changes position relative to the sun, and an observer on Earth would see the full face of the planet lit when Venus was beyond the sun and progressively less of the planet as it moved between the sun and Earth (Figure 2.3).

Galileo's evidence was followed less than a century later by Isaac Newton's explanation of the force that held the planets in their orbits around the sun—gravity. As technology evolved, scientists discovered additional planets and moons, and they were able to make increasingly detailed observations about the characteristics of our neighborhood in space. Eventually, scientists were able to use their understanding of planetary motions to send spacecraft throughout our solar system to collect more data on these ancient worlds.

In this chapter, we will step back in time to describe the birth of the planet and travel even farther back to review the origin of the universe. We will describe our planet's safety features for this flight and explore the hostile space environment beyond our atmosphere. We will learn why energy from our sun makes life possible and that the inevitable demise of our nearest star will eventually result in the destruction of Earth. We will describe how internal and external energy sources may one day help reduce our dependence on fossil fuels and diminish the impact of global warming. Finally, we will consider why our home planet is the only one in our solar system capable of supporting life and why many people cannot correctly answer the question, Why is it warmer in summer and colder in winter? All of this should provide us with clues about where to look for life elsewhere in the universe.

Self-Reflection Survey: Earth in Space—Section 2.1

Answer the questions below as a means of uncovering what you already know about Earth's position in space.

1. Explain how we are influenced by Earth's position in space on a daily basis.
2. If you could make one trip into space, where would you most like to visit and why?
3. Think about some situation in your life where you changed how you thought about something. What circumstances were required for you to change your mind or point of view?

2.2 Origin of the Universe

Learning Objectives

- Explain the inflationary model of the origin of the universe known as the *Big Bang.*
- Predict the motion of distant stars, using the concept of Doppler shift.
- Compare and contrast examples of good and poor scientific process as it relates to understanding the origin of the Universe.

Earth is a small, rocky planet that circles the sun, one of the hundreds of billions of stars making up the Milky Way galaxy. A galaxy is a collection of stars, gases, and other matter bound together by the force of gravity. The Milky Way galaxy is one of billions of galaxies embedded in the much larger universe. Recent measurements by astronomers using the orbiting Hubble Space Telescope have cataloged about 3,000 other galaxies. Based on this small measurement, scientists project there are billions of galaxies in the universe, each galaxy having billions of stars. The universe itself comprises all of the energy and matter that physically exists.

Estimates of the age, scale, and origin of the universe are based on our understanding of the relative motions of distant galaxies. Current models suggest an age for the universe of nearly 14 billion years, but there is enough uncertainty that the age may lie anywhere between 11 and 20 billion years. Age estimates may change as scientists learn more about the characteristics of the most distant galaxies and stars.

Determining the Age and Size of the Universe

Brightness and Luminosity. For many decades, astronomers have used telescopes to study space. The first indication of the enormity of the universe came from measurements of the brightness of distant stars. The brightness of a star depends on the distance to the star and the amount of light energy it radiates (called luminosity).

From earlier discoveries, astronomers have identified a specific class of stars called *cepheid variables.* These stars pulsate (like the flashing of a road construction caution sign). Scientists can use modern telescopes to measure the time required for one of these pulsations, called the *period* of the pulsation. The pulsation period provides a good estimate of the cepheid luminosity. Scientists then use the brightness and luminosity to calculate our distance from the star. Edwin Hubble, for whom the Hubble Space Telescope is named, used data from these stars and distances to other galaxies to show that the universe extends far beyond the Milky Way. Hubble worked with Milton Humason to discover that galaxies are moving away from us; in other words, they discovered that **our universe is expanding.** (Humason had an interesting entry into science: he was a former janitor at California's Mount Wilson observatory with no formal education past the age of 14. He volunteered at the observatory, and his careful technique resulted in his being hired as a full-time staff member.)

The Doppler Effect. As technology improved, even more distant objects could be identified by using increasingly sophisticated telescopes. Unfortunately, technology was (and is) not advanced enough to measure pulsating stars in the most distant galaxies. Nevertheless, as is often the case in science, work in one area can unexpectedly contribute to the solution of some other problem. For example, Hubble's work with cepheid variables provided him with data needed to develop a new technique for measuring vast interstellar (between stars) distances on the basis of an everyday effect that we have all experienced.

You have noticed the changing frequency (pitch) of the horn on a passing emergency vehicle as you stand on the sidewalk or sit stationary in a car. You hear a higher pitch (higher frequency) when the vehicle approaches and a lower pitch (lower frequency) as the vehicle moves away. The siren on the vehicle always creates the same frequency sound—that is, if you were driving an ambulance or standing beside a stationary ambulance, the sound of the siren would always be the same (Figure 2.4a). The change in pitch experienced by an observer occurs when the source of sound is

Checkpoint 2.1

basic | intermediate | advanced | superior

Which of these lists of cosmic features is in the correct order of size, beginning with the largest?

a) Universe, galaxy, star, planet
b) Star, galaxy, universe, planet
c) Universe, planet, star, galaxy
d) Galaxy, universe, star, planet

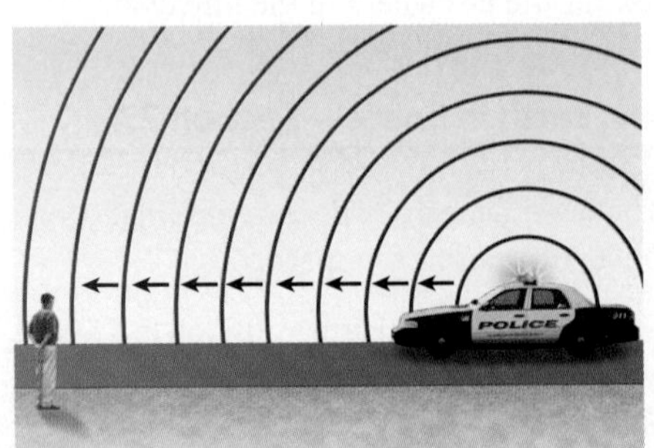

a. Observer hears normal frequency for the siren of stationary police car.

b. Observer hears higher frequency (shorter wavelength) siren for an approaching police car.

c. Lower frequency (longer wavelength) siren is heard for a police car moving away from observer.

Figure 2.4 The Doppler effect. **a.** When a police car is not moving, the frequency of its siren is normal. **b.** An approaching siren has a higher frequency. **c.** A receding siren has a lower frequency.

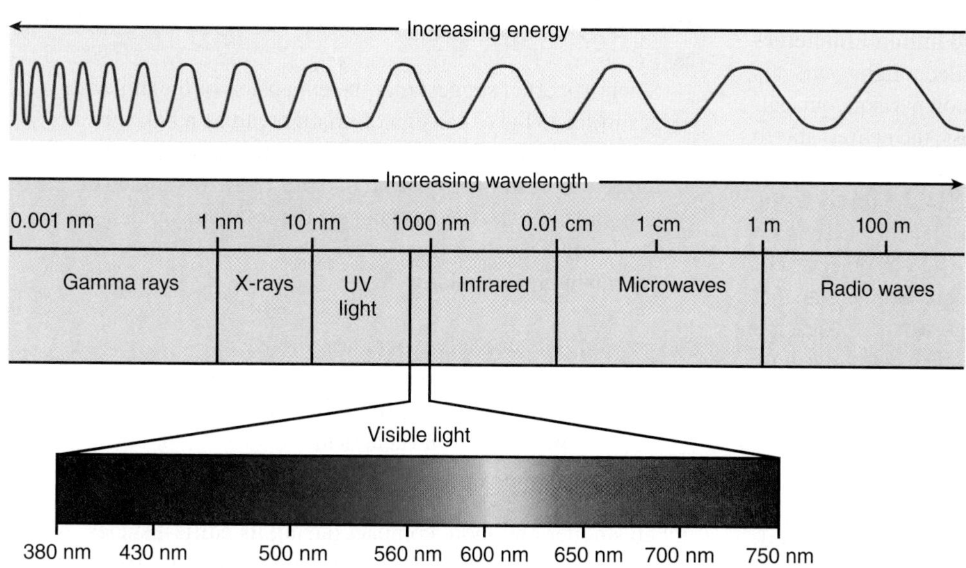

Figure 2.5 The electromagnetic spectrum. Radio waves can have wavelengths measured in hundreds of meters. In contrast, wavelengths for visible light are less than 1,000 nanometers (abbreviated as nm; 1,000 nm = 0.0001 centimeter (cm)) across but are 1 million times longer than the wavelength of gamma rays.

moving relative to the observer (Figure 2.4b and c). **This apparently changing frequency due to the relative motion of a sound source is called the Doppler effect** (after the mathematician who discovered it). If you knew the frequency of the vehicle siren and measured the frequency of an approaching siren, you could calculate the speed of the emergency vehicle. The same effect occurs with light, and Hubble used the Doppler effect on light to estimate Earth's distance from faraway stars. The same principle is applied every day by meteorologists to determine if storms and other weather phenomena are moving toward or away from their location.

In space, the velocity of light is always 3×10^8 meters per second (or 300,000 kilometers per second; 186,000 miles per second). In addition, the white light that we are so familiar with is actually a combination of the different colors of light that form a spectrum from violet to red. Each color has a different wavelength ranging between 380 and 750 nanometers (1 nanometer = 0.000000001 meter; Figure 2.5). Violet and blue have the shortest wavelengths, and red has the longest wavelength. Hubble analyzed the wavelengths of light from distant pulsating stars and noted that the wavelengths were typically longer—closer to the red end of the spectrum—when compared to light from closer stars.

This phenomenon, the shifting of the color of light from galaxies toward red, became known as *red shift*. Just as the frequency of the siren appears to change as an ambulance moves away from us, the wavelength of light appears to increase (undergo red shift) as stars in distant galaxies move away from us in the expanding universe. By calibrating the red shift data with information on the brightness of cepheid variables, scientists were able to use the size of the red shift to estimate the speed that individual galaxies were traveling away from us. Hubble also noted that **most (though not all) galaxies are moving away from us, and the farther away the galaxy, the greater the red shift (the faster they are moving away).** Astronomers used the amount of red shift to calculate the distance to the farthest galaxies.

Checkpoint 2.2

Suppose the light spectrum from a distant star shifted toward the blue end of the light spectrum. What would this imply?

a) The star is moving away from us.
b) The star is moving toward us.
c) The star is not moving relative to us.

Measuring Distances in Light-Years. The most distant objects so far observed in the universe are at least 13 billion light-years from Earth (Figure 2.6). **One light-year is the distance that light can**

Figure 2.6 Stars and galaxies in a small section of the universe. This deep-field view was taken with the Hubble Space Telescope and is a composite of hundreds of images collected over a 10-day period in 1995.

travel in one year and is equivalent to 9,500 billion kilometers (5,940 billion miles). So, even though it is called a light-*year,* it is actually a measure of distance, not time. In comparison, our galaxy is approximately 150,000 light-years across; the nearest star to our sun is 4.3 light-years away; and our modest little solar system is just a fraction of a light-year from one side to the other. Using light-years to measure distance has the added benefit of having a time component that allows us to identify the age of objects. For example, it takes 1 billion years to receive light from a star that is 1 billion light-years away. Think about it—when we observe light from the most distant stars, we are actually looking at light generated at least 13 billion years ago! So, the universe is really, really old.

Checkpoint 2.3

basic intermediate advanced superior

Scientists often suggest that the expansion of the universe is similar to the expansion of raisin bread as it bakes in an oven. As the loaf increases in size, individual raisins move farther apart in the expanding bread. During a homework assignment, two students suggest the following analogies for the universe, but these are not considered as good as the raisin bread analogy. Why?

a) The universe expands similarly to the concentric ripples formed when a rock is thrown into a pond.
b) The universe is similar to a Jell-O mold enclosing pieces of fruit (galaxies).

The Big Bang Theory

The discovery that distant galaxies are moving away from us yields clues about the origins of the universe. Because the universe is still expanding, the young universe clearly had to be much smaller than the one we see today. Initially, astronomers simply reversed the expansion of the universe to step back in time. By running the movie backward, it seemed clear that the universe must have been much smaller and more compact during its earliest stages. Astrophysicists are currently testing the hypothesis that **the universe began with a massive and rapid expansion called the *Big Bang*.** Prior to this expansion, there was no space or time.

The Big Bang expansion sent energy in all directions. Mathematical models indicate that the universe began as a

Figure 2.7 This false-color image from the National Aeronautics and Space Administration (NASA) Spitzer Space Telescope shows towering pillars of cool gas and dust that are incubators for the formation of new stars. Dozens of young stars can be seen inside the gas pillars.

rapid expansion (rather than an explosion). The models explain conditions back to a fraction of a second after initiation of the expansion. In 10^{-30} second (0.000000000000000000000000000001 second), the universe inflated from a point billions of times smaller than the tip of a pin to a huge volume about the size of our solar system. At this stage, the universe consisted of subatomic particles (protons, electrons) and free energy. Within a matter of minutes, these particles would have combined to form simple nuclei such as hydrogen and helium. More complex elements would not form for several hundreds of thousands of years and would require the high temperatures and pressures found in the cores of stars (see Section 2.3).

The Big Bang theory predicted that cosmic radiation would have been released in all directions everywhere around us. So where was it? The answer to that question came from two scientists in New Jersey who were working on an entirely different problem. In 1964, Arno Penzias and Robert Wilson were trying to reduce static noise in a radio experiment that involved using a big, dishlike radio antenna. Despite their best efforts, they could not eliminate a steady hissing background noise from their results. It was there at every hour of the day and appeared to come from everywhere. They determined that this static was made up of certain frequencies of microwave radiation, but they could not figure out its source. It even reached the point where they climbed into the dish and gave it a good cleaning. In desperation, they contacted Robert Dicke, a scientist just down the road at Princeton University who, coincidentally, was looking for the missing radiation predicted by the Big Bang theory. Dicke recognized that Penzias and Wilson's troublesome signal was the "lost" cosmic background radiation. The "lost" background radiation originated in the much smaller early universe. It has been spreading out ever since. As the universe expands, that radiation has become less energetic since the same amount of energy now occupies a much larger volume of space (see Figure 2.7). In 1978, Penzias and Wilson were awarded the Nobel Prize in physics for their unexpected discovery.

The Existence of Galaxies. The discovery of cosmic background radiation still did not answer questions related to the existence of "clumps" of matter (galaxies) in the universe. Again, the process of science kicked into gear. The original Big Bang model had to be refined to explain the existence of galaxies. Current versions of the theory suggest that at the time of the initial expansion, the mass-energy that now makes up the universe was not uniformly distributed. There were "gaps" and "bumps" that later formed regions where gravitational attraction pulled together clumps of gas and dust to form galaxies. Processes within these giant clouds of debris would result in material being pulled together to form massive stars and smaller planets (Figure 2.7).

The ultimate fate of the universe is up for grabs. Consult your favorite authoritative source (for example, www.nasa.gov) for an update on the current scientific debate about whether the universe will continue expanding indefinitely, stop expanding in a "big freeze," or eventually contract in a "big crunch."

Checkpoint 2.4

basic | intermediate | advanced | superior

Scientific Analysis

Explain how the development of concepts presented in this section exhibited the key characteristics of scientific explanations:

1. Provisional (tentative)
2. Based on observations
3. Predictable and testable
4. Offer natural causes for natural events

2.3 Stars and Planets

Learning Objectives

- Describe the sequence of events in the life cycle of a star such as the sun.
- Summarize the characteristics of the universe, stars, and planets and their principal relationships.
- Discuss how scientists search for extra-solar planets and how they might determine if these planets could support life.

Techno musician Moby had a hit record several years ago that was titled "We Are All Made of Stars." It turns out that the title is true. We *are* all made of stars; well, technically we are made of things that are made *in* stars. The cells in a human body are composed of a variety of different elements, but just eight of those elements account for more than 99 percent of each of us by weight. Living cells are composed mostly of water (hydrogen and oxygen) and basic organic compounds built around carbon. These three elements—hydrogen, oxygen, and carbon—account for more than 90 percent of your body by weight. Another five elements (nitrogen, calcium, phosphorus, potassium, and sulfur) get us over the 99 percent mark. Hydrogen was created during the original formation of the universe, and all of these other elements are produced during the life cycle of stars. Many more elements are present in our bodies in just trace amounts and are essential for good health. The life cycle of big stars represents a manufacturing process that churns out elements that combine to generate the complex compounds necessary for the formation of our planet and everything on it, including us. Instruments aboard spacecraft have detected over 70 different chemical compounds in clouds of cosmic debris, including molecules of common substances such as water, methane, and carbon dioxide. Scientists use observations from other stars and our own solar system coupled with simulations to deduce how the sun and planets formed.

How Stars Formed

If you look up at the night sky, you can see about 2,000 stars from any location on Earth. If you were to get yourself a nice telescope (about \$300), you would see several hundred thousand stars. If you invested in a really nice big telescope (about \$10,000), you could

Figure 2.8 The spiral galaxy NGC 4414. This is one of three main categories of galaxies (elliptical and irregular are the others). The galaxy's disk is about 56,000 light-years across. The system lies about 62 million light-years from Earth. As-yet undiscovered planets may orbit some sun-sized stars within the galaxy.

Checkpoint 2.5

Construct a timeline diagram that illustrates the life cycle of the sun.

see tens of thousands of galaxies, including hundreds of billions of stars (Figure 2.8). And those are just the stars that are close by. So how did all those stars form?

The Big Bang created clumps of cosmic debris that eventually formed galaxies throughout the universe. Within these galaxies, clouds of dust and gas coalesced, increasing the mass of the cloud and pulling in adjoining material. Eventually, the gravitational pull of these masses produced giant hot balls of glowing gas. After several several million years, the temperatures and pressures at the centers of these objects became so intense they fused together the nuclei of hydrogen atoms. This process, called *nuclear fusion,* occurs when hydrogen nuclei are mashed together under high temperatures and pressures to form helium. The result: stars were born.

Stars have several characteristics including luminosity, color, surface temperature, size, and mass. When plotted based on their temperature and luminosity (Figure 2.9), stars can be grouped into categories. Main sequence stars include the sun and other stars up to 3 times the size of our sun. Those stars all undergo fusion of hydrogen into helium in their cores. If the star is less than about 1.5 times the size of our sun (which is 1 solar mass), **heat energy comes from the nuclear fusion of four hydrogen atoms to form a single helium atom.** Larger main sequence stars also undergo hydrogen fusion but through a more complex sequence involving carbon, nitrogen, and oxygen.

In general, the more massive the star, the shorter its life span. Intermediate-sized stars, such as our sun, fall somewhere in the middle of the size range and last about 10 billion years. More massive stars, such as supergiants that are 70 to 100 times larger than our sun, last 10 to 100 million years.

The sun is composed exclusively of gases, with hydrogen and helium making up almost all of its mass. The fusion reactions are steadily consuming the sun's supply of hydrogen. According to our current understanding of the life cycle of stars, we are approximately halfway through the sun's life. As it burns through

Figure 2.9 Life cycle stages and types of stars (Hertsprung-Russell diagram). Stars have different degrees of luminosity (brightness) and temperature at various stages of their life cycle. $L_{\odot}$ is the luminosity of our sun.

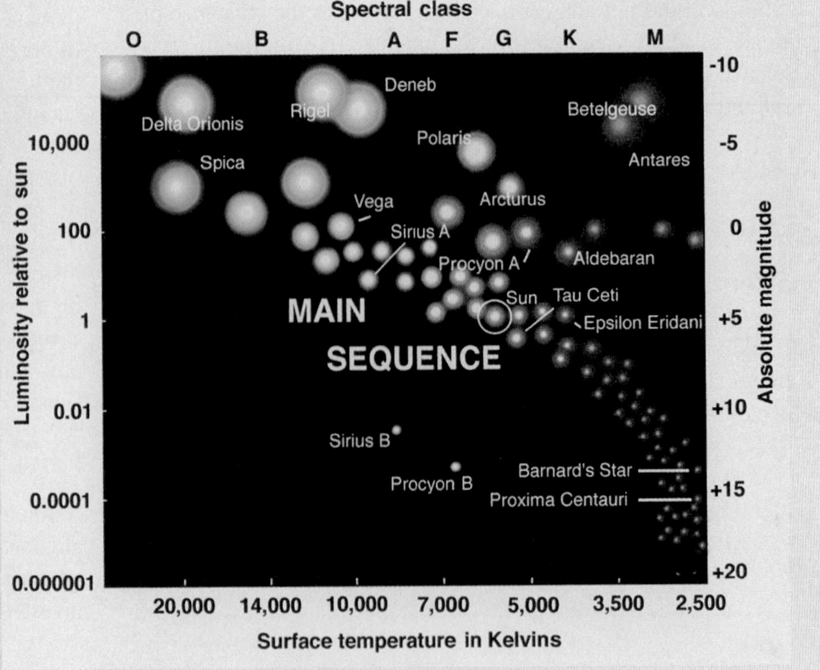

its hydrogen, the sun is slowly getting hotter. In the far distant future, about 2 billion years from now, this increasing heat will have caused Earth's oceans to evaporate and our planet will become more like Venus. In about 5 billion years, the outward pressure generated by fusion will no longer balance the effects of gravity, and the sun's core will begin to contract. Internal temperatures will temporarily rise and helium will fuse into carbon. Just as before, the energy from this fusion fuels an expansion of the star, this time producing what is called the *red giant* phase.

This expanding red giant sun will envelop Earth and will continue to expand until the helium is exhausted. This will not take long. The helium will be quickly converted to carbon, and that will be the end of the fusion. The outer layers of the sun will gradually disappear, leaving a hot core, termed a *white dwarf* star, that will be about the same size as Earth is today. Billions of years later, heat will be lost, and the white dwarf star will become a *black dwarf* star.

High-mass stars (greater than about 3 solar masses) burn brightly for less than 100 million years before they die a dramatic death. Their death throes also begin with a core contraction and the formation of a red supergiant. The greater mass of these supergiants means that their collapse will occur in several stages. Each stage forms more complex elements—helium fusing to carbon, carbon fusing to oxygen . . . eventually forming iron at the core of the dying star. The final stage of the star is marked by a *supernova* (Figure 2.10), a massive explosion that fuses together

Figure 2.10 Kepler's supernova. The astronomer Johannes Kepler and other observers noted the appearance of a new star in the sky on October 9, 1604. Modern images of this bubble-shaped cloud of gas and dust reveal that it is 14 light-years across. This is just one of six supernovae that have been observed in the Milky Way galaxy over the last millennium.

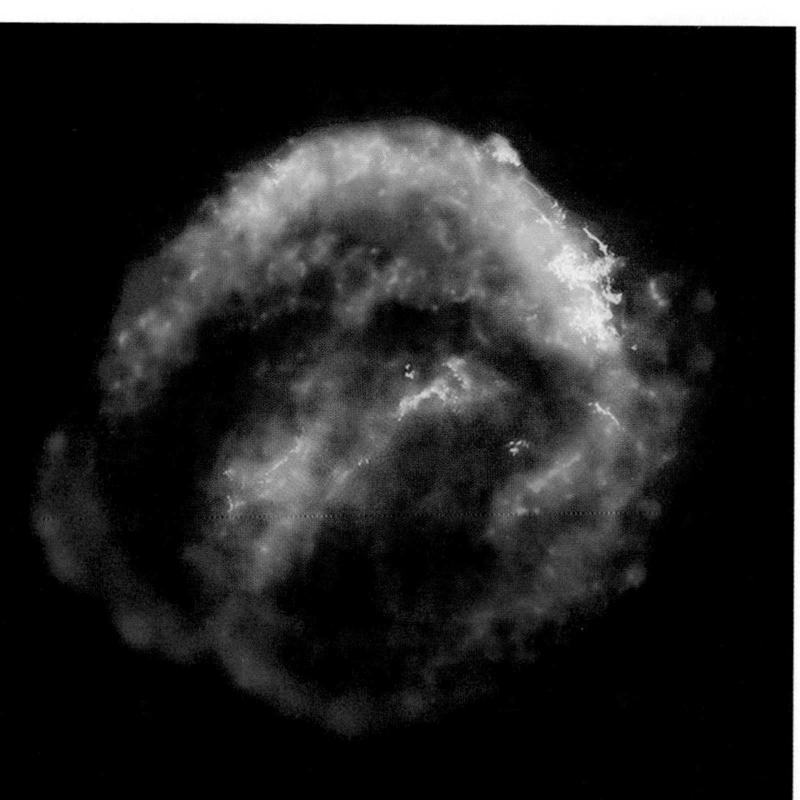

Checkpoint 2.6

basic · advanced · intermediate · superior

Which of the following statements is the most accurate? Explain the reason for your answer as well as why you did not choose either of the other two statements.

a) All stars and planets are about the same age.
b) Stars are approximately the same age as their orbiting planets.
c) The number of stars is declining as stars burn out.

Checkpoint 2.7

Characteristics of the Universe Exercise

Complete the following concept map by correctly adding the connecting phrases or terms provided to the appropriate locations. Some items may be used more than once; others may not be applicable to this diagram.

1. converts simple elements such as
2. present in
3. for example
4. extrasolar planets such as
5. orbits
6. formed early versions of
7. gas and dust formed
8. contains billions of
9. began to expand rapidly following the
10. destroyed in an explosion known as a
11. to more complex forms such as
12. are examples of red giants

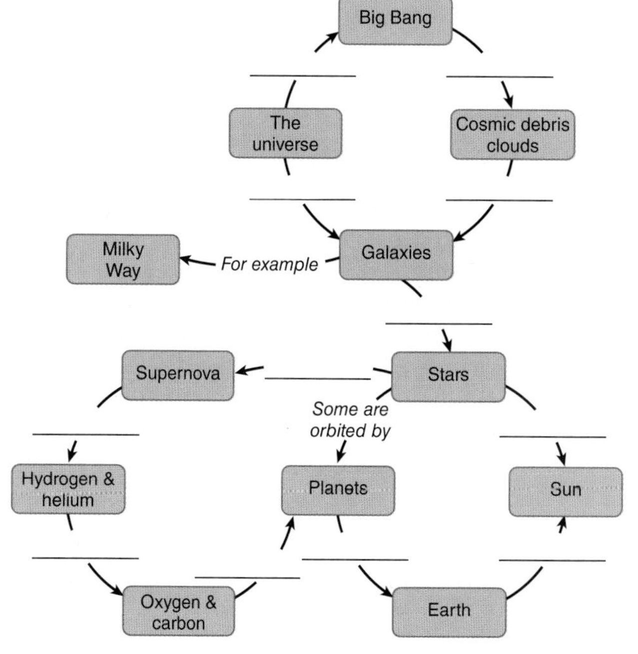

elements heavier than iron and blasts them throughout the universe. Eventually those heavy elements become part of a new solar system where they then are incorporated into planets or moons.

How Planets Formed

Our sun is one among billions of billions of stars. Just like every other star in our galaxy, when the sun formed, it was surrounded by a rotating disk of leftover debris. Gravity pulled together clumps of debris to form different-sized objects, the largest of which formed spherical bodies that we recognize as planets. Scientists know this because all planets orbit the sun in the same direction, and they all rotate the same direction as expected if they formed from a single, rotating disk of material. Lighter gases, such as hydrogen and helium, were blasted to the outer, colder parts of the young planetary system to form the icy, gas-rich planets. The heavier elements (such as carbon, oxygen, and iron) collected closer to the star to form the rocky inner planets. Smaller objects assumed a variety of shapes (personal favorite descriptor: potato-shaped) and mostly represent asteroids.

Did you know there are scientists who spend their day searching for planets outside of our solar system? Since they first reported the discovery of what are termed extra-solar planets in 1992, scientists have cataloged over 800 planets orbiting distant stars. Most of these planets are large, like our gas giant planets. There are five main methods that scientists use to find these planets; one uses direct observations, four use indirect measurements. The direct method involves blocking out the light of the star or using data from multiple telescopes to image the planet. For the most part, the planets are too far away to see directly using these methods. Therefore, scientists more frequently look for slight variations in either the color or the intensity of light coming from distant stars or for evidence of a gravitational field attributable to a planet. Some of their methods are limited because the information coming from these stars is faint, and it gets distorted in our atmosphere. That limitation is removed when satellites are used to monitor stars from space. One project, called the Kepler mission, involves simultaneous monitoring of over 100,000 stars. You can learn more about the science and other planet-finding projects at the NASA PlanetQuest site (http://planetquest.jpl.nasa.gov/). The vast size of some of the newly discovered planets, their proximity to their stars, and the fact that some have erratic orbits make most of them unlikely hosts for life. But it is a reasonable hypothesis that many of these planetary systems also contain smaller, Earthlike planets (one-half to twice the size of Earth and the right distance from their star to have liquid water). Improvements in technology over the next decade should enable scientists to identify the atmospheric chemistry of these planets to reveal whether they are capable of supporting life. Exciting discoveries may lie ahead!

Checkpoint 2.8

What planetary characteristics would you look for in an extrasolar planet that might have the potential to harbor life similar to that found on Earth?

2.4 Our Solar System

Learning Objectives

- Describe the processes that fuel stars.
- Predict how processes associated with the sun can affect the Earth system.
- Use historical data to make predictions about future sunspot activity.
- Summarize the principal characteristics of the solar system.
- Compare and contrast terrestrial and Jovian planets.
- Explain why the demotion of Pluto from planet to dwarf planet is consistent with the application of the scientific method.

Characteristics of the Sun

As we learned in Section 2.3, the sun is just one of billions of similar stars throughout our galaxy. The sun accounts for 99.8 percent of the mass of our solar system and dwarfs its orbiting planets. Earth's outer atmosphere receives an average of 1,368 watts of solar energy per square meter. This energy influences all aspects of the earth system. Variations in solar output make a modest contribution to global climate change. Approximately 10–20 percent of global warming is thought to be attributable to changes in solar energy output.

Just as Earth rotates every day, the sun rotates about once a month. However, this big ball of gas experiences differential rotation; that is, its equator rotates more rapidly than its polar regions (25 versus 36 days). Scientists are still trying to figure out why this happens, but they do know that the differential rotation and convection causes twisting of the sun's magnetic field lines.

These processes cause **disruptions in the sun's magnetic field that produce sunspots and solar flares**

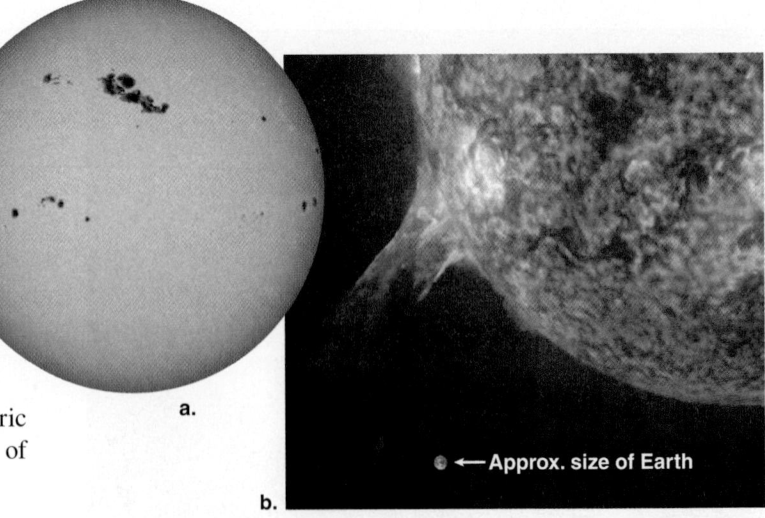

Figure 2.11 Sunspots and solar flares as imaged by the Solar and Heliospheric Observatory. **a.** Dark splotches on the photosphere, the sun's outermost layer, are sunspots. Sunspots have been recognized on the surface of the sun for several centuries. Individual sunspots may be as large as 50,000 kilometers (31,000 miles) in diameter, the approximate size of Neptune. **b.** Flares and other solar eruptions extend above the surface of the sun and are many times larger than Earth.

Checkpoint 2.9

basic intermediate advanced superior

What are the principal components of the sun? (*Hint:* See Section 2.3.)

a) Hydrogen and helium
b) Carbon and oxygen
c) Silicon and sodium
d) Nickel and iron

(Figure 2.11). Sunspots are dark blotches on the sun's outermost layer. The blotches represent slightly cooler areas of the sun's surface surrounded by brighter, hotter regions. The apparent movement of sunspots across the sun's face can be used to measure the periodicity of the sun's rotation. The number of sunspots can vary considerably but shows a long-term trend recognized as the sunspot cycle (Figure 2.12). The average number of sunspots varies over an 11-year cycle, from a handful of sunspots during a solar minimum to a few hundred at a time that are visible during the peak months known as the solar maximum. We have recently passed through the minimum of the cycle when very few sunspots were observed (Figure 2.12), and scientists have enough data to predict that this

Checkpoint 2.10

basic intermediate advanced superior

Sunspots, flares, and other emissions that generate charged particles from the sun's surface can have a negative impact on electrical systems on Earth. What would be the implications for this type of solar activity if the sun did *not* experience differential rotation?

a) There would be less sunspot activity.
b) There would be greater sunspot activity.
c) There would be no change in sunspot activity.

The sun is located approximately 150,000,000 kilometers (93,000,000 miles) from Earth. How long would it take for charged particles ejected from the sun to affect electrical systems on Earth, assuming the particles traveled at 1.6 million km/h?

a) A few minutes
b) A few hours
c) A few days
d) A few weeks

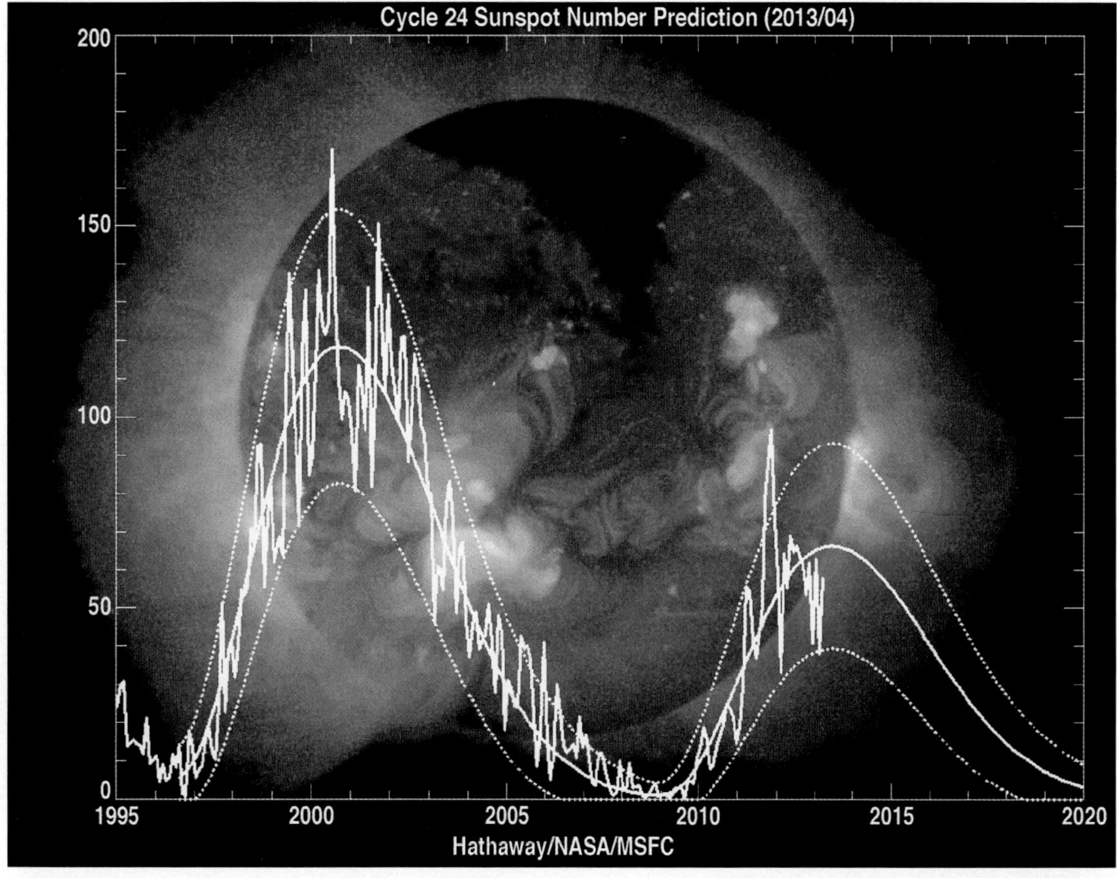

Figure 2.12 The current sunspot cycle. A record of sunspots has been kept since the middle of the eighteenth century, with each sunspot cycle numbered since then. The vertical axis on this figure shows the number of sunspots. Some sunspots are grouped so close together that the individual sunspots cannot be easily observed. The "Sunspot number" is the sum of the individual spots observed added to 10 times the number of groups since most groups average 10 spots. The smooth lines on the graph represent the predicted range of sunspot activity; the jagged lines show the actual activity.

cycle will hit a smaller maximum of about 70 sunspots in fall of 2013. Earth receives a slight increase (0.1 percent) in solar energy during a solar maximum and an equivalent decrease during a solar minimum. Solar flares, intense pulses of X-rays and ultraviolet radiation, are often associated with sunspots. These eruptions from the sun's surface can affect telecommunications and other technology systems on Earth.

Actions of Solar Wind. Space has its own weather system, but it doesn't have the moving air and clouds so familiar to us on Earth. Instead, **space weather is dominated by the solar wind, a constant stream of charged particles accelerated from the sun's outer layers by its magnetic field**. These charged particles travel at an average speed of 1.6 million kilometers per hour (1 million miles per hour). **The region of space affected by the solar wind is known as the heliosphere,** and it represents the volume of space in which our sun is the dominant influence. The heliosphere extends far beyond the planets of our solar system and may be likened to a giant bubble that shields us from harmful cosmic rays originating elsewhere in the universe. At several stages of your education, you were shown charts and diagrams of the sun and planets (like the Chapter Snapshot) that show Pluto as the farthest outpost of the solar system, like a lonely "Thanks for visiting" sign on the edge of a small town. In reality, the effect of the solar wind continues for some distance far beyond Pluto and the known planets, all the way to the edge of the heliosphere.

Earth has its own magnetic field that deflects the solar wind around our planet (Figure 2.13), protecting our atmosphere. Even though Earth's magnetic field keeps our protective atmosphere from eroding, we are still vulnerable to the harmful effects of occasional solar eruptions that hurl powerful pulses of dangerous X-rays, ultraviolet radiation, and charged particles toward Earth. **Intense streams of charged particles can disrupt Earth's magnetic field, generating electrical currents that result in power surges** and leading to blackout conditions as electrical systems shut down as well as threatening satellites. The economic costs of power outages are measured in billions of dollars. Disruptions in Earth's magnetic field caused by solar wind can also result in spectacular effects such as the dramatic light displays known as *aurora* in the upper atmosphere (Figure 2.14).

Every day, we depend on hundreds of satellites to provide information for a host of needs on Earth, including communications,

Checkpoint 2.11

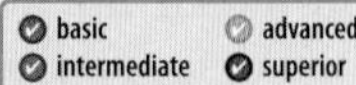

Use the graph of sunspot numbers (see Figure 2.12) to answer the following questions.

1. When was the most recent solar maximum (month/year)?
2. When was the most recent solar minimum (month/year)?
3. On a slow day in the lab, two graduate students made a bet about how many sunspots would occur two months later in spring semester, 2013. Chad estimated there would be 90 sunspots; Julie thought there would be 30. When the month rolled around, 63 sunspots were recorded. Does this mean that Chad and Julie do not understand the sunspot cycle? Explain your reasoning.

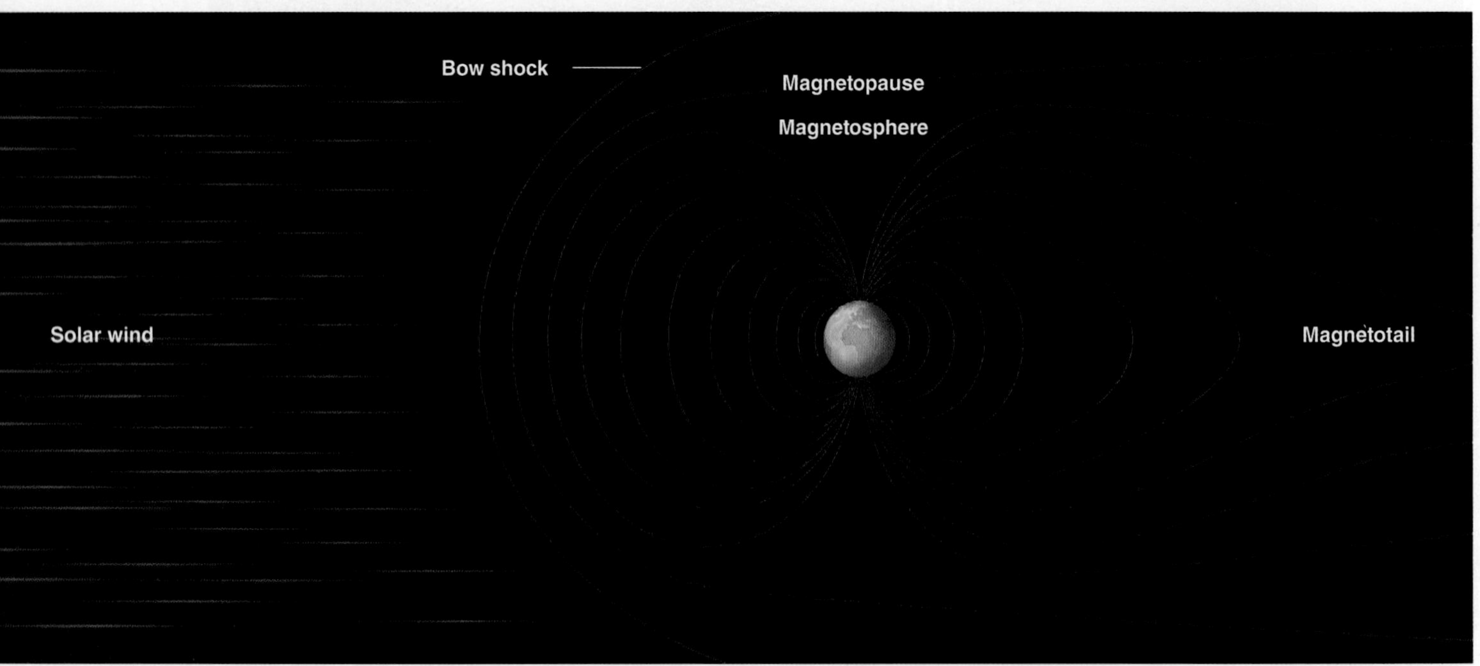

Figure 2.13 Deflection of Earth's magnetic field by the solar wind. Earth's magnetic field is compressed closer to the sun, meaning that the magnetic field lines are closer together on the side of Earth that is facing the sun.

Figure 2.14 Aurora seen from space, formed by the interaction of the solar wind with Earth's magnetic field. On Earth, these spectacular visual displays can be best observed at high latitudes, that is, closer to the poles.

defense radar, aircraft navigation, and weather forecasting. For example, Global Positioning Systems (GPS) could be as much as 50 meters (164 feet) off during solar storms. Satellites could be knocked out of action by concentrated streams of solar radiation. NASA has launched several satellites to monitor solar activity and provide notice of potentially damaging bursts of energy heading for Earth (Figure 2.15). Such warnings will be vital to future space exploration that could expose astronauts to the dangerous radiation generated by these solar emissions. People living outside the protection of Earth's magnetic field and atmosphere would be exposed to much higher levels of harmful radiation from the sun than the rest of us on Earth's surface.

Solar Power on Earth. Nonrenewable fossil fuels (coal, oil, natural gas) account for approximately 70 percent of US electric power. Concerns about the availability of imported oil and the potential for global warming have focused attention on alternative energy sources such as solar and geothermal energy (see Section 2.6) that are readily available in North America and don't produce greenhouse gases. Earth receives far more solar radiation than it needs to supply plentiful energy to the global population. Ideally, covering a sunny region about the size of New Mexico with high-efficiency solar cells should be enough to solve the world's energy problems. Unfortunately, current economic and technological limits prevent more extensive use of this promising resource. Solar energy accounts for just a fraction of 1 percent of US electricity generation.

Active (direct) solar energy can be used in two forms. Water or oil may be heated in solar collectors that use mirrors to focus the sunlight onto the liquid. The hot liquid is then used for heating. In contrast, photovoltaic cells convert sunlight directly to electricity. Current solar energy systems are too inefficient for widespread deployment. When coal is burned, approximately one-half the heat generated can be converted to electricity. Commercial solar (and geothermal) energy sources have typical efficiencies of 10–20 percent. Without an increase in efficiency, solar systems will continue to be several times as expensive as using coal or natural gas to produce electricity. However, continuing technological improvements and favorable government policies have resulted in a dramatic increase in the production of photovoltaic cells, and continued research holds promise for a future where solar energy systems will supply much of our daily electricity.

Eight, Nine, or Ten Planets?

In contrast to what you may have learned in elementary school, as of 2006, the sun is the centerpoint of a system of eight planets (see Chapter Snapshot). In order, with increasing distance from the sun, the planets are: **Mercury, Venus, Earth, Mars, Jupiter, Saturn, Uranus, and Neptune.** In recent years, there has been considerable debate in the space science community about exactly what constitutes a planet. In the summer of 2006, Pluto was shown the door of the planet club. The planets are now divided into two

Checkpoint 2.12

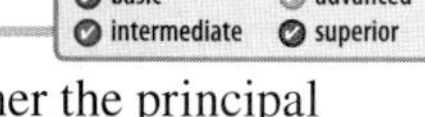

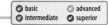

Create a concept map that links together the principal characteristics of the sun using the terms listed here, as well as any linking phrases you wish to create.

The sun	Helium
Solar wind	Nuclear fusion
Solar system	Sunspots
Magnetic field	X-rays
Star	Sunspot cycle
Hydrogen	Differential rotation

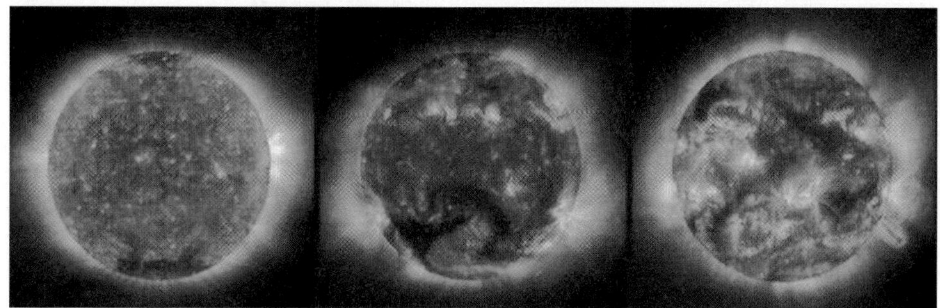

Figure 2.15 Satellite monitoring of solar activity. These three views of the sun were taken by the Solar and Heliospheric Observatory (SOHO) satellite in successive years using the Extreme Ultraviolet Imaging Telescope. The far-right image shows increased solar activity approaching a peak in the sunspot cycle, which could lead to damaging bursts of energy headed toward Earth.

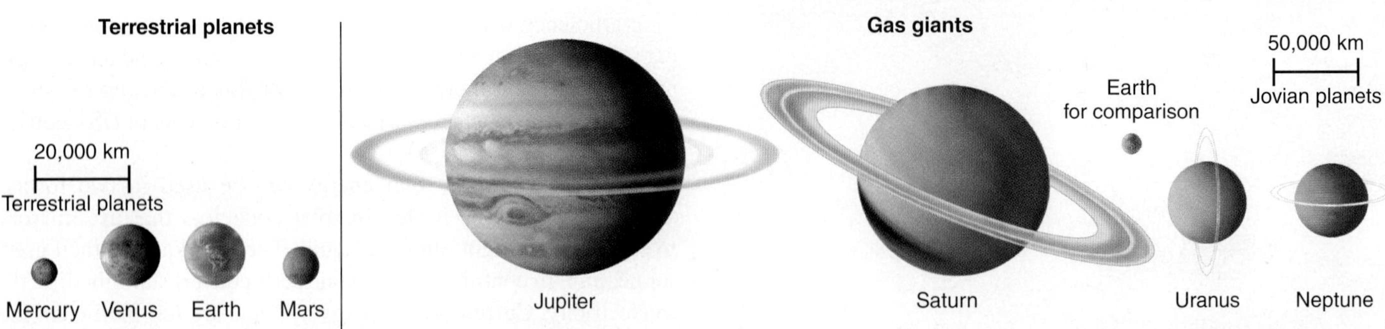

Figure 2.16 Relative sizes of the eight closest planets to the sun. Mercury, Venus, Earth, and Mars are called the terrestrial planets, while Jupiter, Saturn, Neptune, and Uranus are known as the Jovian planets. The terrestrial planets are closer to the sun—and to one another—than are the Jovian planets, which have much more space between them.

groups. The smaller planets—Mercury, Venus, Earth, and Mars—are nearest to the sun, share similar characteristics, and are termed the **terrestrial planets.** The much larger outlying planets—Jupiter, Saturn, Neptune, and Uranus—are called the **Jovian planets** (Jupiter-like; Chapter Snapshot, Figure 2.16). Smaller objects, similar in size to Pluto, are considered members of a newer, less prestigious group of **dwarf planets.**

The average distance from the sun to Earth represents one astronomical unit (AU; 1 AU = 150 million kilometers or 93 million miles). The planets' distances from the sun range from 0.4 AU for Mercury to 30 AU for Neptune (Table 2.1). We can read these distances on the page, but these units of distance will make little sense to most of us used to measuring travel in miles or hours. Even our most advanced spacecraft would take about a decade to travel beyond Neptune and 10,000 years to get to the edge of the heliosphere. In comparison, the terrestrial planets exist in a positively cozy little neighborhood. The four innermost planets all lie within 1.5 AU of the sun, essentially one planet per 0.3 AU. Mars, the farthest terrestrial planet from the sun, is separated from Jupiter, the nearest Jovian planet, by 3.7 AU. This gap contains the asteroid belt, which is composed of thousands of rocky or metallic bodies. The largest asteroids are almost the size of Pluto, and the smallest are little more than space pebbles.

Checkpoint 2.13

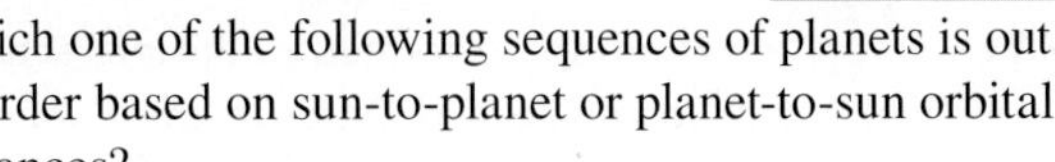

Which one of the following sequences of planets is out of order based on sun-to-planet or planet-to-sun orbital distances?

a) Mars, Jupiter, Saturn, Uranus, Neptune
b) Venus, Mars, Earth, Jupiter, Saturn
c) Uranus, Saturn, Jupiter, Mars, Earth
d) Jupiter, Mars, Earth, Venus, Mercury

The dwarf planets do not fit with either the terrestrial or Jovian planets. Pluto, smaller than Earth's moon, is composed of ice and rock, like the comets, and has an odd orbit that actually takes it closer to the sun than Neptune for part of its course. In 2005, Mike Brown, an astronomer of the California Institute of Technology, discovered an object orbiting the sun beyond Pluto. The International Astronomical Union (IAU), the body that coordinates the naming of celestial objects, would eventually name the object *Eris*, after the Greek god of conflict. The name was an ironic nod toward the debate that the discovery of Eris generated among scientists. On the basis of its brightness and orbital characteristics, many

Table 2.1 Characteristics of the Eight Planets

Planet	Size (radius), km	Orbital period	Distance from sun, million km [AU]	Principal atmospheric gases
Mercury	2,440	88 days	58 [0.4]	Helium, sodium
Venus	6,052	225 days	108 [0.7]	Carbon dioxide
Earth	6,378	365 days	150 [1]	Nitrogen, oxygen
Mars	3,397	687 days	228 [1.5]	Carbon dioxide
Jupiter	71,492	11.9 years	778 [5.2]	Hydrogen, helium
Saturn	60,268	29.5 years	1,427 [9.5]	Hydrogen, helium
Uranus	25,559	84 years	2,871 [19]	Hydrogen, helium
Neptune	24,746	165 years	4,497 [30]	Hydrogen, helium

basic intermediate advanced superior

Make four generalizations about the eight planets in our solar system, using the information in Table 2.1. For example, planets closer to the sun are smaller than those farther away.

scientists argue that Eris is larger and more massive than Pluto and that it is composed of ice and rock, just as Pluto is. Once it became apparent that Eris was bigger than Pluto, science began its slow, inevitable progress toward change. Either Eris had to be added to the club of nine planets, or the term *planet* had to be redefined in a way that would exclude Eris, and that would probably mean the end of the road for Pluto—everyone's favorite eccentric planet. Soon the world of astronomy was full of conflict over the fate of Pluto and all its icy companions.

Oddly, up to this time, there was no generally accepted definition for the term *planet.* The IAU had appointed a panel of astronomers that took 2 years to figure out that they could not agree on a definition. A smaller committee eventually proposed that a planet is *an object that orbits a star and is massive enough for gravity to pull its material into an approximately spherical shape.* By this definition, any roundish object that was about 800 kilometers (500 miles) or more in diameter and orbited the sun would qualify as a planet. This would have let in Eris and at least one asteroid (Ceres) and potentially dozens of other objects that advanced telescopes were beginning to identify beyond the orbit of Pluto.

To counter this generous new "big tent" definition, a more restrictive (little tent) option was proposed that would add a third criterion: that the object would be considered a planet *if it had cleared the neighborhood around its orbit.* By this definition, planets would be the only significant objects lying along the path of their orbits. Because Eris, Ceres, and Pluto all exist in neighborhoods with many other similarly sized objects, they would not be considered as planets but would be assigned to a new class of objects, the *dwarf planets.* While this definition is far from perfect (thousands of asteroids lie in the orbit of Jupiter, and Pluto passes within Neptune's orbit), it was adopted by a majority of scientists attending an IAU conference. So, for the time being, there are officially eight planets and a steadily increasing number of dwarf planets.

Types of Planets

Terrestrial Planets. The terrestrial planets are composed of rock and can be divided into different compositional layers. **For example, the interiors of Earth, its moon, and the other terrestrial planets can each be separated into three layers of contrasting composition and thickness: crust, mantle, and core** (Chapter Snapshot). The cores of terrestrial planets are composed of an iron and nickel mixture. Their composition is similar to that of metallic meteorites, which formed from debris that did not coalesce to form planets elsewhere in the solar system. As the young, hot Earth cooled, the heavier, metal-rich components sank to form the core, and the lighter, rocky materials made up the planet's outer layers represented by the crust and mantle.

Figure 2.17 Jupiter and four of its largest moons. From top to bottom: Io, Europa, Ganymede, and Callisto. Jupiter has 67 moons in all.

Jovian Planets. The Jovian planets are much larger than the terrestrial planets and are shrouded by dense gases (Chapter Snapshot and Figure 2.16) that create spectacular atmospheric effects. These characteristics led to their nickname, "the gas giants." We generally think of the surface of a terrestrial planet as the interface between the atmosphere and the rocky surface of the continents or the water in the oceans. Such a distinction is more difficult to draw for the **Jovian planets, which all have thick atmospheres over oceans cold enough to be formed of liquid hydrogen, helium, and/or methane.** Any rock present is buried under tremendous pressures at the core of the planet. Consequently, it is difficult to determine where the "surface" of a Jovian planet begins and the gassy atmosphere ends.

The average spacing between the orbits of Jovian planets is over 8 AU, about a 2-year trip if you could hitch a ride on a passing spacecraft. Not only are these planets much larger than terrestrial planets and much farther apart, but also they have much longer solar orbits (see the orbital periods in Table 2.1). The time it takes for a planet to complete a solar orbit increases with its distance from the sun. Mercury orbits the sun in a little less than 3 months, while it takes Neptune 165 years to finish a circuit.

Jovian planets have many moons (Jupiter, 67; Saturn, 62; Uranus, 27; Neptune, 13), and more are discovered almost every year. Figure 2.17 shows four of Jupiter's moons. In contrast, terrestrial planets are moon-poor; only Earth and Mars have moons (Earth, one; Mars, two, named Phobos and Deimos). Earth's moon plays an important role in generating tides (more about that in Chapter 13). **Most of the Jovian planets' moons revolve in consistent directions around their planets, evidence that the moons formed from the same mass of debris.** There are exceptions, such as Neptune's moon Triton, which orbits opposite the direction of the other moons. How could that happen? Scientists hypothesize that moons orbiting opposite to the norm either were

THE SOLAR SYSTEM

COMET

Some comets orbit the sun every few years; others take as long as 30 million years to complete an orbit. The nuclei of most comets are less than 10 km (6 miles) across.

Terrestrial planet interiors. Earth and the other three terrestrial planets have three compositional layers: crust, mantle, and core. In all cases, the crust is relatively thin (tens of kilometers) in comparison to the mantle and core. The four planets are drawn to scale in the image here.

Without our moon, Earth would wobble on its axis, changing how much solar radiation was received over different parts of its surface, resulting in a much more unstable climate system.

MERCURY

Mercury is just a little bigger than Earth's moon and travels around the sun faster than any other planet, at an average speed of 172,000 km/h (107,000 mph).

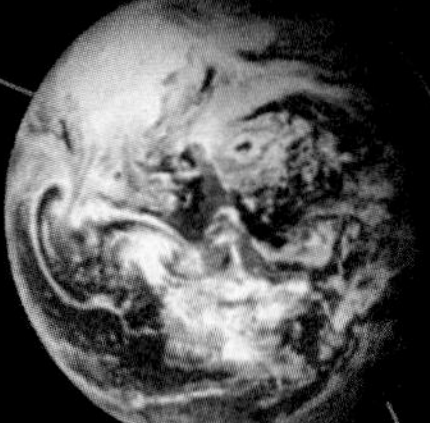

If Earth were the size of a nickel, Mars would be the size of an aspirin tablet.

Venus is the hottest planet in the solar system. Its clouds reflect so much sunlight that Venus is the brightest planet in the night sky.

THE SUN

Our sun is one of 70,000,000,000,000,000,000,000 stars in the universe. The sun's interaction with Earth drives our ocean currents, weather patterns, and climate.

Solid core

Liquid core

Mantle

Crust

MERCURY

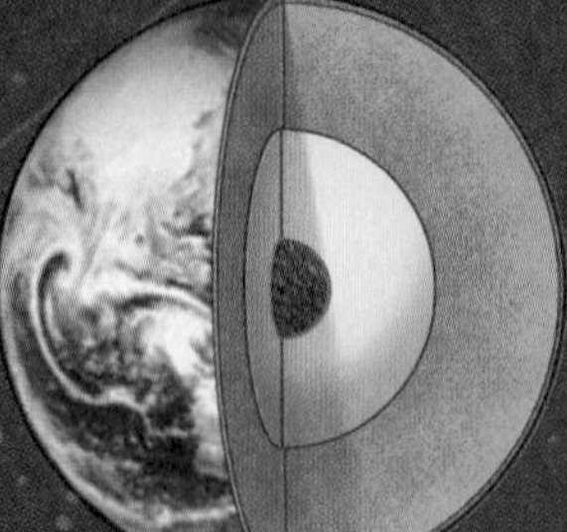

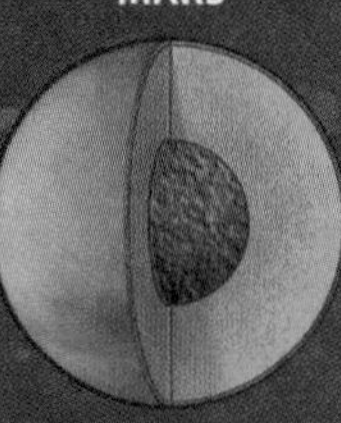

20,000 km

URANUS

Uranus' s axis is tilted at 98 degrees. Its poles are pointed towards the sun. Due to its 84-year long orbit, the planet's poles experience 21 years of night or day depending on their position.

PLUTO

The average temperature on icy "dwarf planet" Pluto is -233° C. Sunlight on Pluto is only 0.0001 times as strong as on Earth.

JUPITER

Jupiter is named for the king of the Roman gods and makes up two-thirds of all the planetary mass in the solar system.

NEPTUNE

Neptune is so far from the sun that it still has not made a complete orbit since it was discovered in 1846.

SATURN

Saturn's density is less than that of water. Saturn's rings cover a distance nearly equal to the distance from Earth to the moon.

Jovian planet interiors. Jovian planets also have compositional layers. On Saturn and Jupiter, the largest Jovian planets, pressures in the interior of the planet are sufficient to generate rare liquid metallic hydrogen. The relatively smaller Uranus and Neptune are still many times larger than Earth and have a mantle dominated by ices composed of methane, water, and ammonia.

ASTEROID BELT

All the asteroids mashed together would make up a body about one-tenth the size of our moon. Asteroids that pass close to Earth are known as near-Earth objects.

Orbits not to scale.
Orbits viewed from an angled-perspective above the orbital plane.

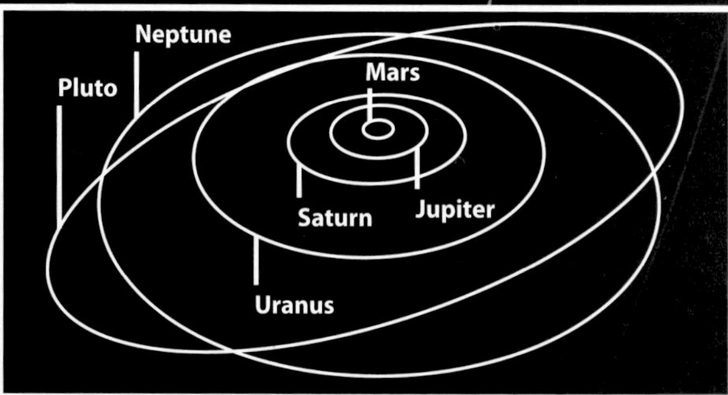

JUPITER

SATURN

URANUS

NEPTUNE

Rock

Water, ammonia, methane ices

Liquid, metallic hydrogen

Liquid hydrogen

Hydrogen, helium, methane gas

50,000 km

captured by the planet's gravitational fields or are evidence of early collisions between celestial bodies that could have reversed the direction of motion of some of the orbiting matter.

All of the Jovian planets have ring systems, of which Saturn's is the best example (Chapter Snapshot). Rings are composed of individual pieces of rocky debris or chunks of ice that range in size from micrometers to several meters in diameter. The rings are held in place by the contrasting gravitational pulls of the planets and their surrounding moons. Without the many moons that are characteristic of the Jovian planets, it is likely that the planets' gravity would have gradually pulled the material in the rings down to the planets' surfaces. Each piece is in its own orbit that is positioned such that the material does not fall into the planet or combine with other orbiting materials to form a larger body such as a moon.

Checkpoint 2.15

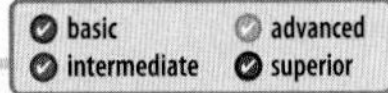

Venn Diagram: Terrestrial Versus Jovian Planets

Complete the following Venn diagram to compare and contrast the similarities and differences between the two major groups of planets in our solar system. Identify 12 characteristics that either are shared by both terrestrial and Jovian planets or are unique to one of the groups. Then place the number corresponding to each characteristic in the diagram. Characteristics 1 and 2 have been plotted on the diagram for you.

1. Orbits around sun
2. Gas giants
3.
4.
5.
6.
7.
8.
9.
10.
11.
12.

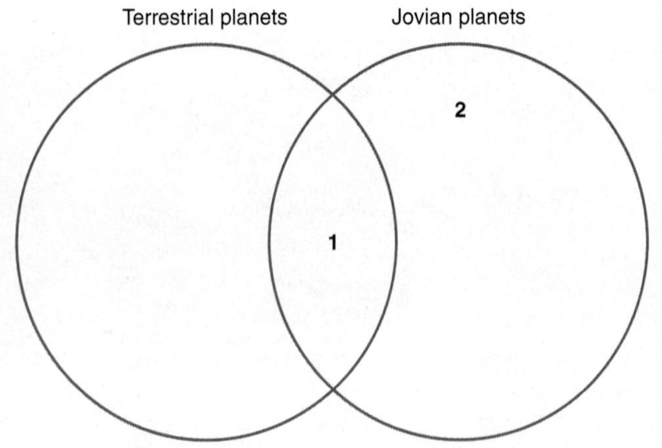

In 2006, the IAU defined a new class of celestial objects, dwarf planets. Explain why this is consistent with the nature of science described in Chaper 1.

2.5 Earth, the Sun, and the Seasons

Learning Objectives

- Describe the how the distribution of incoming solar radiation changes from the equator to the poles.
- Explain how Earth's orbit and rotational tilt angle affect the seasons.

Few scientific questions are more basic than "Why is it colder in winter than in summer?" (Think about it; what is your answer?)

When a small group of seniors graduating from a prestigious university were asked that question, nearly all of them answered incorrectly. The most common wrong explanation offered was that Earth is closer to the sun in summer and farther away in winter. But, in fact, the exact opposite is true for the Northern Hemisphere (where the students lived). **Earth's orbit is almost circular, but the planet comes a little closer to the sun during January and is slightly farther away in July** (Figure 2.18).

Heat from the sun is indeed the factor responsible for colder winters and warmer summers—but not because of Earth's orbital

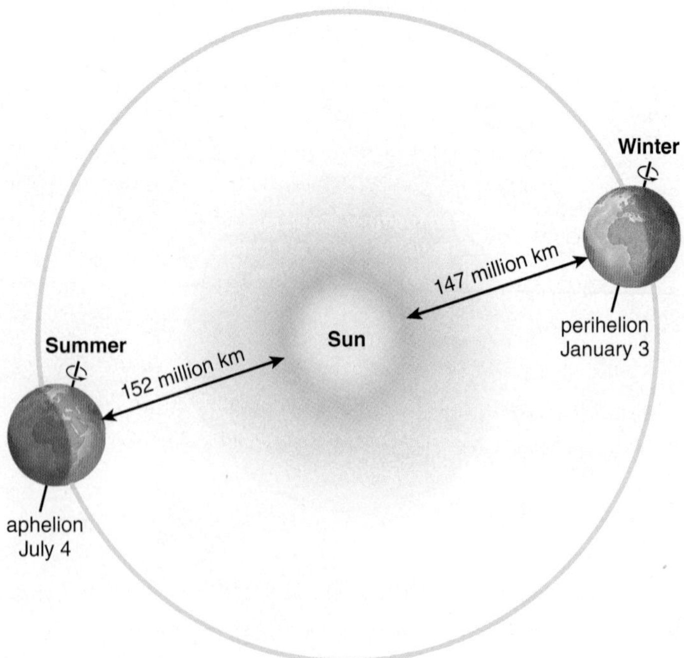

Figure 2.18 The path of Earth's orbit around the sun. In the Northern Hemisphere, Earth is farthest from the sun during early summer and closest during early winter.

pattern. It is not the amount but rather the distribution of solar radiation on Earth's surface that regulates the order of the seasons. The Northern Hemisphere receives more solar radiation in summer and less in winter. Why?

Distribution of Solar Radiation

To understand why different regions of Earth receive different amounts of solar radiation at different times of year, it is first important to understand that our planet rotates on its axis once every 23 hours, 56 minutes, 4.1 seconds. Further, Earth's rotation axis is tilted relative to its plane of orbit. The axis is tilted 23.5 degrees away from vertical (Figure 2.19). Because of this tilt, the amount of solar radiation hitting any location varies by season. Think of the axis as the shaft of an open umbrella; imagine the shaft of the umbrella is tilted and resting on your shoulder. You can twirl the umbrella by rotating the shaft. You can adjust how much rain reaches you by altering the angle of the shaft. Similarly, the amount of sunlight received by Earth varies with the tilt of the axis. Now imagine that you are carrying that umbrella as you are walking along a circle in a parking lot but always *facing* in the same direction (you would walk backward for one-half of the circle). The parking lot would represent the orbital plane of Earth, and the circle would be the orbit.

Insolation, the amount of solar radiation received by Earth, is greatest when the sun is directly overhead above a location and decreases as the sun's rays make a lower angle with respect to Earth's surface. When we are discussing insolation, three landmarks on Earth's surface are important to keep in mind: the equator, the Tropic of Cancer, and the Tropic of Capricorn (Figure 2.19a). The equator is an imaginary band encircling Earth at its widest circumference; it passes through northern South America, central Africa, and Indonesia. The Tropic of Cancer is located 23.5 degrees north of the equator, just north of the Hawaiian Islands and Cuba; and the Tropic of Capricorn is 23.5 degrees south of the equator, the same latitude as Rio de Janiero, Brazil, and Alice Springs in the center of Australia.

Approximately one-half of the solar radiation that illuminates Earth actually makes it to the planet's surface. The rest of the incoming sunlight is reflected back into space from clouds and particles in the atmosphere. There are two reasons why the sun provides less heat to the poles than to the equator. First, the sun's

Checkpoint 2.17

How do we define the length of a year on Earth?

a) A year is related to the revolution of Earth around the sun.
b) A year is related to the rotation of Earth on its axis.
c) A year is related to the rotation of the sun on its axis.
d) A year is related to the revolution of the sun around Earth.

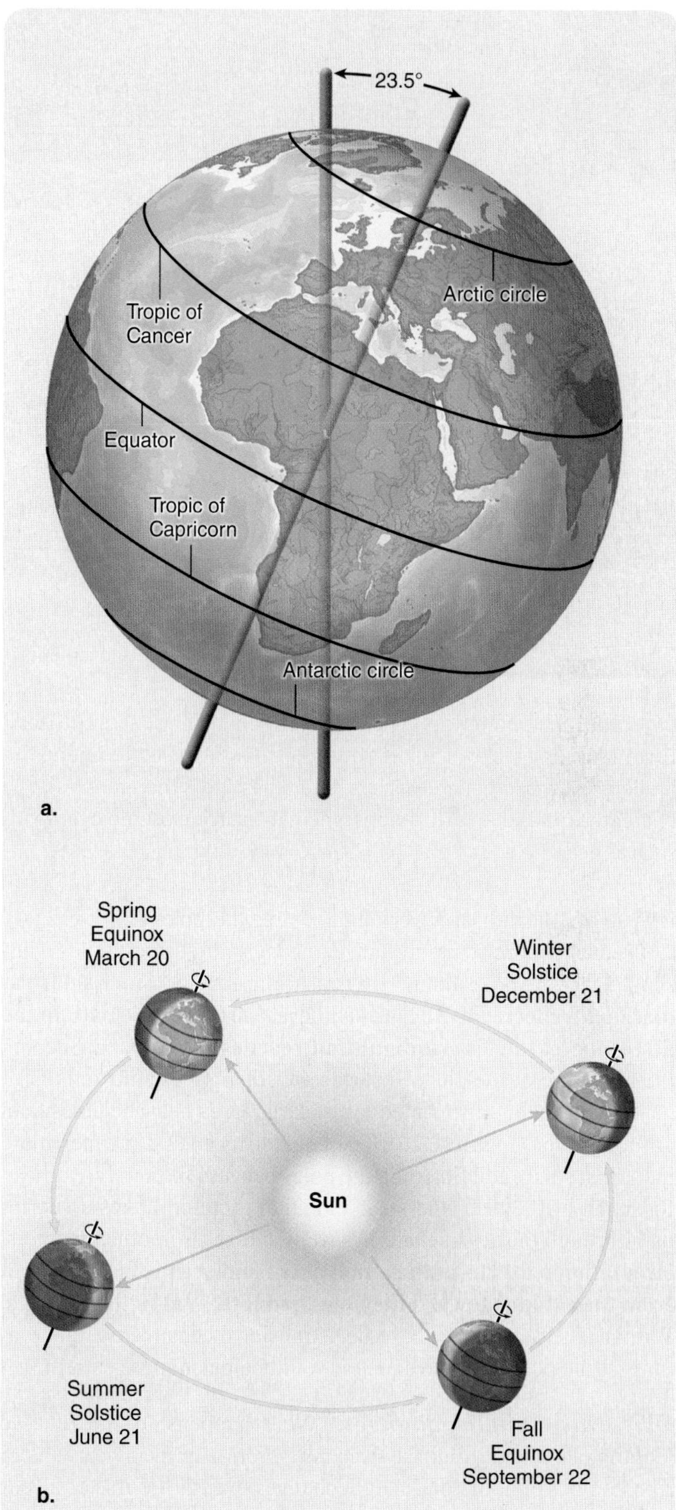

Figure 2.19 The sun's influence on Earth's seasons. **a.** Earth is tilted on an imaginary axis oriented at 23.5 degrees to vertical. The two tropics are 23.5 degrees of latitude north (Cancer) and south (Capricorn) of the equator. **b.** The sun is overhead at the Tropic of Cancer on June 21 and at the Tropic of Capricorn on December 21. It is overhead at the equator during the spring and fall equinoxes. Note that Earth's axis is always tilted in the same direction. Diagram is not to scale.

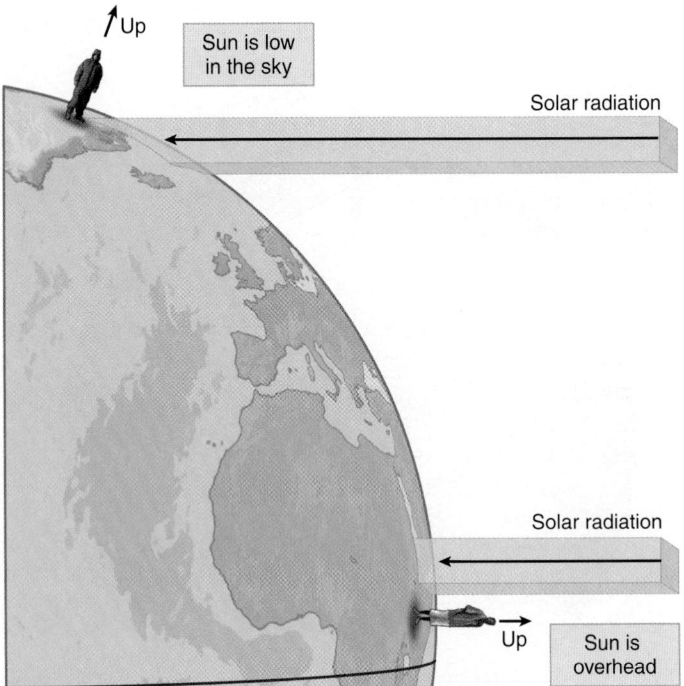

Figure 2.20 Differences in heat distribution. When sunlight strikes Earth's surface at a low angle, as occurs nearer the poles, its energy is diluted over a larger area. The resultant heating is therefore less nearer the poles than in locations where the sun is closer to overhead, such as near the equator.

rays strike Earth at a lower angle nearer the poles (Figure 2.20). Consequently, the sun's rays have a longer path through the atmosphere near the poles than do rays near the equator. **Since solar radiation is reflected by the atmosphere, the longer the path in the atmosphere, the less sunlight that reaches the surface.** Second, the curvature of the planet's surface results in the same amount of heat energy being distributed over a larger area near the poles than at the equator (Figure 2.20). The equator receives 2.5 times more incoming solar radiation than the poles because the sun is overhead for much of the year and solar radiation therefore takes a shorter path through the atmosphere. As we will discuss in other chapters, **circulation patterns in the atmosphere and oceans transfer heat from the equator toward the poles, moderating Earth's climate.** Without the actions of these heat transfer mechanisms, temperatures at the poles and the equator would be much more extreme.

Why Day Length Changes. What happens to the relative lengths of day and night as we move from summer into winter? The number of hours of daylight changes over the course of a year and by latitude. **The axial tilt places the sun directly overhead on the Tropic of Cancer at noon of the summer solstice** (June 21), our longest day of the year in the Northern Hemisphere. The greater amount of solar radiation reaching the Northern Hemisphere at this time accounts for the warm temperatures of summer. Day and night both last for 12 hours on the first days of spring and fall (equinoxes) when the sun is directly overhead at the equator (Figure 2.19). The hours of daylight increase northward during summer in the Northern Hemisphere and decrease southward in the Southern Hemisphere (where it is winter). At the North Pole, the sun rises above the horizon on the spring equinox and does not set until the fall equinox 6 months later (Figure 2.21). Imagine daylight 24 hours per day. This pattern is reversed during the winter, when the South Pole is illuminated for 24 hours and the North Pole is dark 24 hours per day (Figure 2.21).

When summer solstice is occurring in the Northern Hemisphere, what is happening in the Southern Hemisphere? The rays are directly overhead in the Northern Hemisphere while the sun's rays strike the Southern Hemisphere at a lower angle and transfer less energy, resulting in lower temperatures. Thus, in the Southern Hemisphere, winter begins in June, and summer (with the sun overhead at

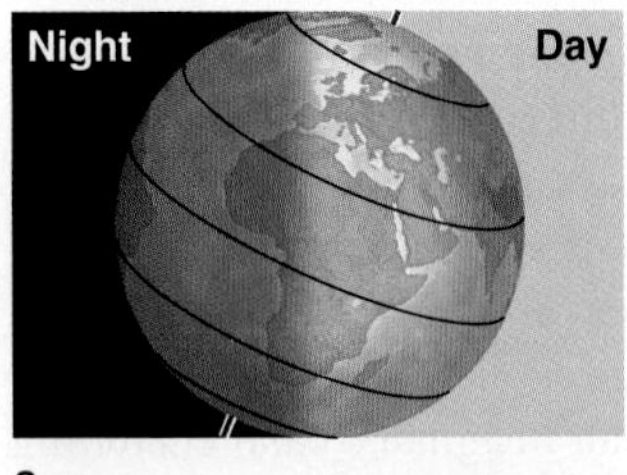

a.

Day
Night

b.

Figure 2.21 Differences in day length. In the Northern Hemisphere, the tilt of Earth's axis results in **a.** 24-hour daylight at the North Pole and almost complete daylight north of the Arctic Circle during summer in the Northern Hemisphere and **b.** perpetual darkness during winter. The situation is reversed south of the Antarctic Circle. Compare the length of time in daylight at the two tropics.

Checkpoint 2.18

basic · intermediate · advanced · superior

How would the amount of incoming solar radiation change at the equator if Earth's axis were vertical instead of tilted?

a) Incoming solar radiation would decrease.
b) Incoming solar radiation would be the same as at present.
c) Incoming solar radiation would increase.

Checkpoint 2.19

basic · intermediate · advanced · superior

Mars has a more asymmetric orbit of the sun than Earth. Mars is 20 percent closer to the sun during its winter than during its summer. How would Earth's climate be affected if Earth had a similarly eccentric orbit, being 20 percent closer to the sun during winter months in the Northern Hemisphere?

Checkpoint 2.20

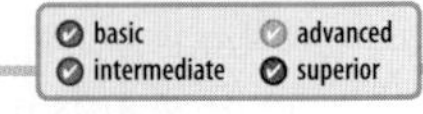

Imagine that it is your job to explain to a group of middle school students how the distribution of incoming solar radiation varies daily and seasonally on Earth's surface. Assuming you have a basketball and a flashlight to use as props, write a description of how you would have the students use the props in a demonstration.

the Tropic of Capricorn) begins on our winter solstice (December 21; Figure 2.19), the shortest day of the year in the Northern Hemisphere but the longest day along the Tropic of Capricorn.

2.6 The Unique Composition of Earth

Learning Objectives

- Describe the characteristics of Earth's internal layers.
- Explain the relationship between the planet's interior, the geothermal gradient, and transfer of heat that drives internal Earth processes.
- Discuss how various characteristics of Earth supported the evolution of life on our planet.

Earth's physical characteristics, size, and distance from the sun have contributed to its unique status as the only inhabited planet in our solar system. As noted in Section 2.4, Earth's interior can be separated into three layers of differing composition and thickness: the core, the mantle, and the crust. If we could slice Earth in half like an apple, we would see a bull's-eye pattern of concentric layers, with the core at the center and the crust around the outside (Chapter Snapshot and Figure 2.22). Each layer is made up of different materials and may be further subdivided according to these physical and compositional variations.

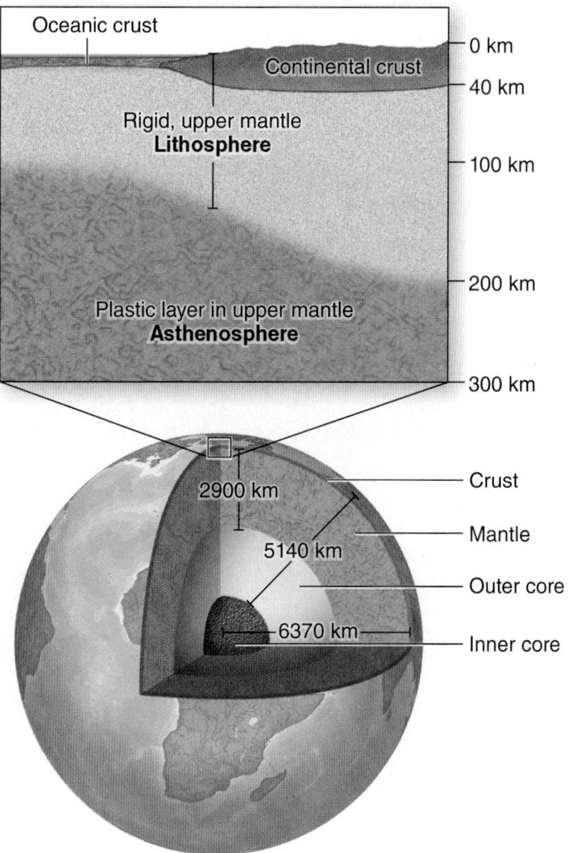

Figure 2.22 Compositional layers of Earth. Earth and the other terrestrial planets can be divided into three compositional layers: crust, mantle, and core. Earth's core can be divided further into a liquid outer core and a solid inner core.

Core, Mantle, and Crust

Earth's core is divided into two parts, a solid inner core and a partially melted outer core. The rocks of the core are largely composed of an iron and nickel mixture, metals that can be both molten and solid under the temperatures and pressures of the outer and inner cores, respectively. Scientists concluded that part of the core is liquid because some types of seismic waves (generated by earthquakes) do not travel through liquids and also cannot pass through the outer core. Earth's magnetic field (Figure 2.23) originates from slow-moving convection currents generated as the planet rotates on its axis and as molten material moves from hotter to cooler regions in the liquid outer core.

The mantle consists primarily of rocks composed of the elements oxygen (45 percent), silicon (22 percent), and magnesium (23 percent by weight). The mantle is a solid from the top of the outer core to the base of the crust. Even so, temperatures and pressures about 100 to 200 kilometers (62 to 124 miles) below the top of the mantle are close to the melting point of some minerals. The rocks in that region are hot enough that they can slowly flow (though they are not molten). These weaker rocks form a layer called the **asthenosphere** that is about 200 kilometers (124 miles) thick (Figure 2.22).

The crust is made of lower-density rocks that are composed chiefly of oxygen and silicon (75 percent by weight). **A rigid outer layer called the lithosphere, composed of the crust and uppermost mantle, overlays the asthenosphere** (Figure 2.22). The lithosphere is broken into large slabs termed *tectonic plates.* Although

Checkpoint 2.21

basic · intermediate · advanced · superior

What are the three compositional layers in Earth's interior?

a) Asthenosphere, lithosphere, core
b) Oceanic crust, continental crust, asthenosphere
c) Lithosphere, mantle, core
d) Crust, mantle, core

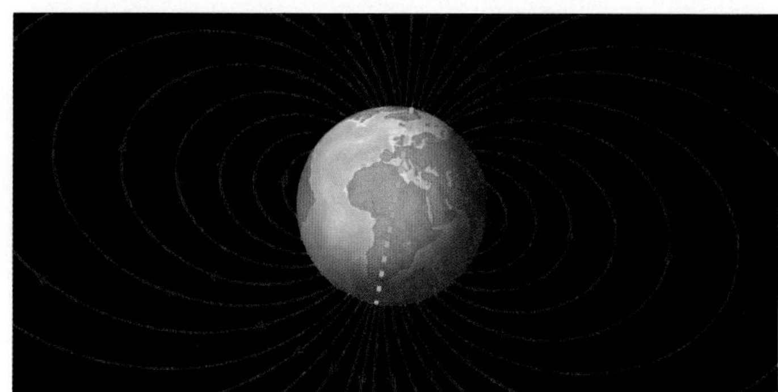

Figure 2.23 Earth's magnetic field. The inclination of Earth's magnetic field varies with latitude. The magnetic field is inclined downward in the Northern Hemisphere and upward (away from Earth's surface) in the Southern Hemisphere.

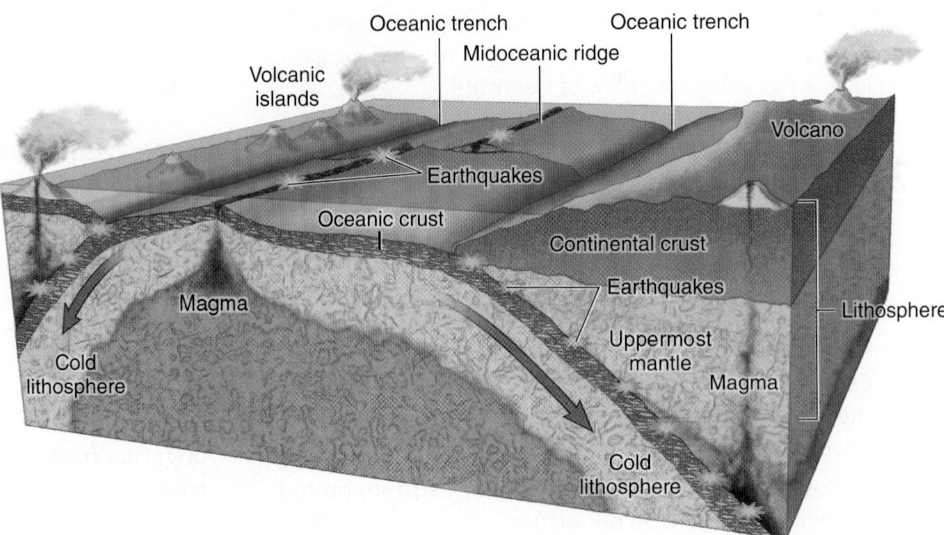

Figure 2.24 Comparison of physical and compositional layers near Earth's surface. Convection cells in the mantle are associated with oceanic ridges, regions of high heat flow on the ocean floor. Regions of cooler crust are often located along the margins of ocean basins, near oceanic trenches. Oceanic lithosphere forms at the oceanic ridge system and is consumed at trenches. (Features are not to scale.)

the crust and mantle parts of the lithosphere are composed of different rocks, the two of them together compose a tectonic plate (Figure 2.24; described in detail in Chapter 4).

Geothermal Gradient. Miners working deep underground have long recognized that Earth has a geothermal gradient—that is, its temperature increases with depth. The temperature in the crust increases by approximately 25°C per kilometer (72°F per mile) near the surface. The gradient varies with location and with depth and shows that the interior of the planet is much hotter than the exterior. In areas of volcanic activity where the gradient is higher, the surface is generally hotter at the surface than it is elsewhere. In South Africa, ice must be pumped down cooling systems in deep mines to reduce temperatures of over 50°C (122°F). (Imagine working in conditions as hot as the hottest day in Death Valley, California!)

If we can harness the heat from Earth's hot interior, we may be able to provide a source of plentiful, clean energy. Geothermal power uses hot water from underground sources to generate steam utilized to drive turbines and generate electricity. The water is then recycled back underground to recharge the geothermal reservoir and continue the process. Geothermal power is more commonly utilized in western states because of their underlying geology. States such as California, Utah, Nevada, and Hawaii have sites of recent volcanism fed by shallow magma chambers that provide ready heat sources for circulating groundwater. Recent studies have suggested that technologies used in oil exploration could be adapted for alternative "heat mining" operations that would provide wider access to geothermal energy resources in regions without volcanism. Heat mining requires water to be pumped down a deep well (5–10 kilometers, 3–6 miles), circulated through an underground reservoir where it would be heated, and then brought back to the surface through a second well (Figure 2.25). Earth's geothermal gradient would be sufficient to ensure that water would be heated to high temperatures over large regions of the continent, rather than just at locations associated with volcanism (Figure 2.26). Under this scenario, large western cities such as Tucson and Salt Lake City could be generating much of their electrical needs from geothermal power in the next few decades.

Checkpoint 2.22

basic intermediate advanced superior

Much of our understanding of the character of Earth's interior comes from analyzing seismic waves that travel through Earth. As these waves move through Earth's interior, they may pass through, bounce off (reflect), or bend (refract) at boundaries between different rock types. The time it takes a seismic wave to travel from a source in one location to a recording station at another location can be used to decipher the internal structure of Earth. Identify three similar methods that are commonly used to view the interior of objects in daily life without cutting or breaking them open.

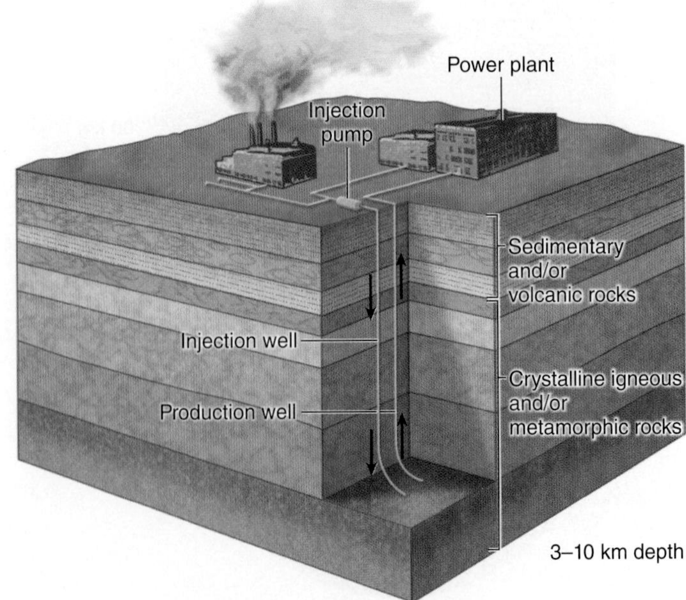

Figure 2.25 Paired deep wells would be drilled into artificially fractured rocks of a geothermal reservoir. Depending on the local geothermal gradient, the wells would be far enough below Earth's surface to result in heating of water to high temperatures.

Checkpoint 2.23

basic · intermediate · advanced · superior

The following graphs illustrate four idealized geothermal gradients for Earth.

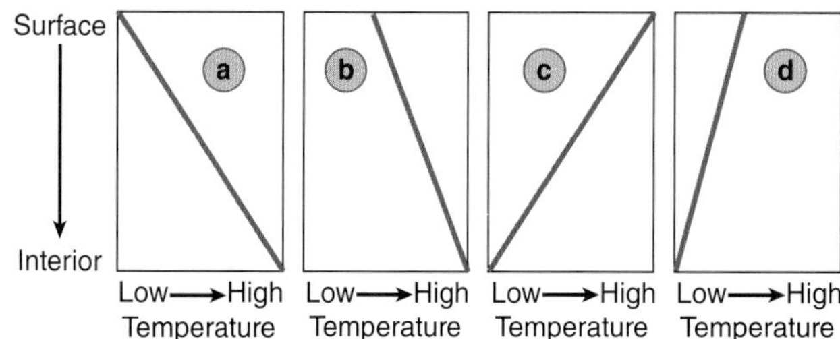

Bearing in mind that Earth is 4.6 billion years old, which plot is most likely to represent the present-day gradient?

Which plot is most likely to represent the gradient 2 billion years ago?

Checkpoint 2.24

basic · intermediate · advanced · superior

Earth's Layers Concept Map

Complete the following concept map by writing the correct terms in the appropriate blank locations as key terms or connecting phrases/terms. A partial list of the terms that you will need is provided here, but some of them may not be applicable to this diagram.

1. Compositional layers
2. Crust
3. One of three
4. Oceans
5. Is the source of Earth's
6. An upper rigid
7. Characteristic of terrestrial planet

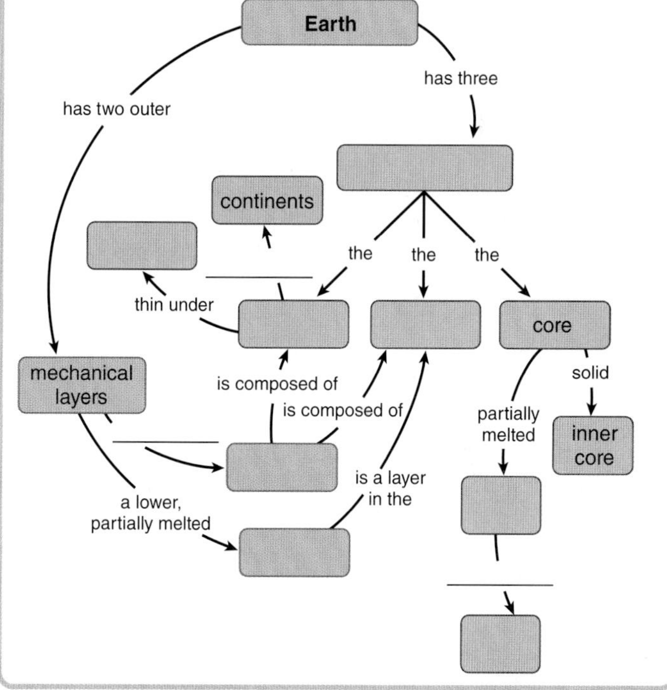

The processes that operate on the surface of Earth and within the planet's interior are driven by energy from different sources. External processes derive energy from solar radiation (see Section 2.5). Most of the heat-driving internal processes come from the radioactive decay of elements in Earth's interior, and a smaller fraction is heat that remains since the time Earth formed. Rocks are poor conductors of heat. Therefore, it takes billions of years to lose the planet's internal heat through the surface. **All terrestrial planets were much hotter when they formed and have cooled with time.** Mercury and Mars, relatively smaller planets, lost their heat hundreds of millions of years ago. In contrast, the greater mass of the larger terrestrial planets acts as insulation, serving to preserve their internal heat. Earth still has a hot interior, as evidenced by the more than 1,500 active volcanoes found around the world. Though no active volcanoes have been observed on Venus, it is thought to have a hot interior based on analyses of recent data. Those data are interpreted to suggest that volcanoes on Venus are 250,000 to 2.5 million years old. Even a 2.5 million year old Venusian volcano is considered active in a geologic sense, so scientists hypothesize Venus still has a hot interior.

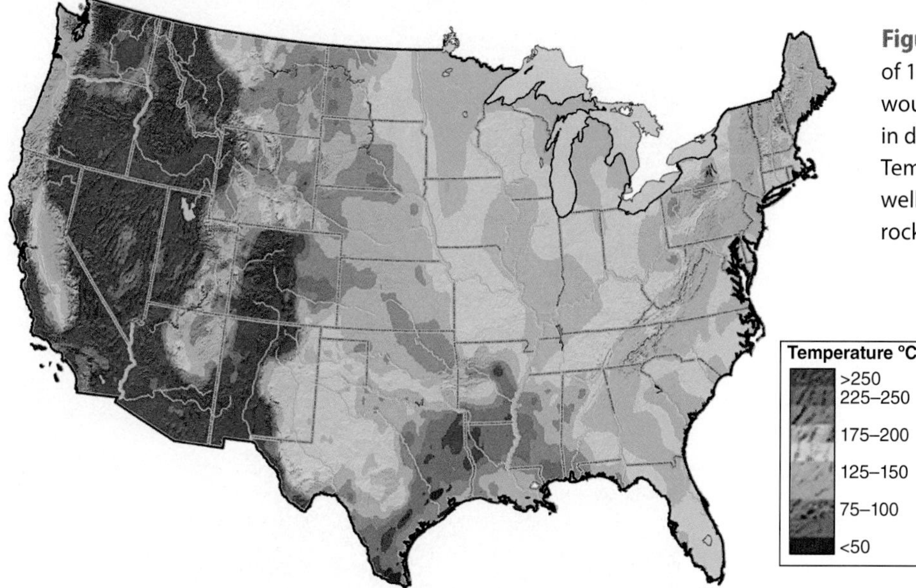

Figure 2.26 Estimated temperatures at the bottom of 10-kilometer-deep (6.2-mile-deep) wells. Waters would be superheated to more than 250°C (480°F) in deep wells throughout the western United States. Temperatures would be about 100°C (212°F) cooler in wells drilled in parts of many eastern states with older rocks exposed in the Appalachian Mountains.

Such geological processes as volcanic activity and the slow movement of tectonic plates are indications that heat is being transferred from the interior toward the surface. Heat transfer occurs in the mantle by conduction and convection. **Conduction is simply the transfer of energy from one location to another through vibrations of molecules.** For example, the uninsulated handle of a metal saucepan becomes hot when left on the stove because heat is transferred from the stove through the pan to the handle by conduction. **Convection is the transfer of heat through movement of hot material from one place to another.** For example, convection occurs in water as it boils on the stove. Convection is thought to occur in the mantle and plays a role in the movement of tectonic plates (see the arrows on Figure 2.24). Heat escapes from all parts of Earth's surface, though at most locations this occurs at rates so low as to be detectable by only the most sensitive instruments.

Why Is There Life on Earth?

Earth shares many features with other planets, so why does it support life while other planets do not? Before reading any farther, make a list of Earth's features that allow it to support life.

Does your list include any of the following: the existence of liquid water, gravity strong enough to allow for an atmosphere, a mixture of gases suitable for life, or a strong magnetic field to deflect harmful radiation?

Checkpoint 2.25

basic · intermediate · advanced · superior

Defining Earth's Characteristics

Read the sentences in the following table and circle the appropriate term in the right-hand columns that could be used to complete the sentence or fill in the blank. Explain the reason for your choices.

If Earth were farther from the sun, the planet would be ____________.	warmer	cooler
If Earth were a little closer to the sun, the oceans would be ____________.	smaller	larger
If Earth did not have a biosphere, the composition of Earth's atmosphere would be ____________.	the same	different
If Earth's biosphere had not evolved yet, we would have ____________ oxygen in the atmosphere.	less	more
If Earth did not have an atmosphere, its surface would have ____________ craters formed by meteor impacts.	fewer	more
If Earth did not have an atmosphere, life on the land surface would be ____________ developed than it is today.	less	more
If Earth were larger, its gravity would be ____________.	weaker	stronger
If Earth were smaller, its atmosphere would be ____________.	thicker	thinner
If Earth were smaller, it would have a ____________ interior.	hotter	colder

Liquid Water. Water is essential for life on Earth and presumably elsewhere in the universe. Even in the most extreme conditions on Earth, if water is present, life is too. Tube worms flourish in the dark, cold waters of the deep ocean floor alongside superheated volcanic vents; microbes have been recovered from ice cores near the South Pole; and bacteria and algae thrive in the scalding waters of hot springs in Yellowstone National Park. Water would evaporate on the blast furnace–hot surface of Venus, where temperatures average 464°C (867°F), and water would freeze on Mars, where temperatures resemble those of Earth's polar ice caps (−63°C, −81°F). But on Earth, the average temperature is a pleasant 15°C (59°F), making it possible for water to exist as a liquid.

The relatively mild temperatures on Earth are the result of the planet's distance from the sun combined with the presence of an atmosphere and a magnetic field. The atmosphere provides energy redistribution and insulation from the extreme cold of space. The magnetic field deflects solar radiation and keeps it from stripping away the atmosphere.

So why does Earth have liquid water while Venus and Mars do not? Scientists hypothesize that Venus, Earth, and Mars all had water when they originally formed. The closer proximity of Venus to the sun resulted in higher temperatures and, consequently, the conversion of more liquid water to water vapor by evaporation. The water molecules were split into their constituent oxygen and hydrogen atoms by intense ultraviolet radiation, with the lighter

Figure 2.27 Was there once water on Mars? This image shows what appear to be exposed layers of rocks that may have been deposited by water. Chapter 7 describes more recent findings about the geology of Mars.

Figure 2.28 Earth's fragile atmosphere. In this photo of a sunrise over the Caribbean Sea taken by the crew of space shuttle *Discovery,* the thin blue band is the stratosphere, and the red-orange color is the top of the troposphere, the lower part of the atmosphere where we live.

hydrogen lost to space. The original oceans on Venus likely evaporated over several million years, producing the dry, desertlike conditions now characteristic of the planet's surface.

Exploration on the Martian surface and interpretations of satellite data (Figure 2.27) support the hypothesis that Mars once had active rivers, lakes, and perhaps a small ocean. However, as usually occurs in science, there are lingering questions. The amount of water, how long it existed, and whether it still exists in large amounts are hotly debated topics (see www.nasa.gov for the latest). As its internal heat was lost, Mars was too far from the sun to sustain sufficient heat to maintain liquid water. Ongoing exploration missions have discovered substantial ice deposits just below the surface of Mars that have been preserved in the planet's present cold state.

Gravity and a Protective Atmosphere. Meteorites, moons, and planets are all held together by gravity. **Smaller planets have weaker gravity and thus thinner atmospheres; larger planets have stronger gravity fields and thus much thicker atmospheres.** Earth is large enough to have accumulated an atmosphere that protects the planet from all but the largest incoming space projectiles (comets, meteorites) and absorbs harmful radiation from the sun. Gravity holds Earth's layer of atmospheric gases close to the planet's surface (Figure 2.28). A fast-moving meteorite encounters gas molecules, mainly nitrogen and oxygen, as it plunges through Earth's atmosphere. Heat generated by the compression of the atmospheric gases is sufficient to melt the surface of the meteorite. This process continues until the rocky object is destroyed or until it plows into Earth's surface. The atmosphere thus reduces all but the largest space objects to relatively harmless debris.

Checkpoint 2.26

Venn Diagram Exercise: Earth and Mars

Complete the following Venn diagram to compare and contrast the similarities and differences between Earth and Mars. Identify 12 characteristics that either are shared by both planets or are different for each planet. Then place the numbers corresponding to the characteristics in the most suitable locations on the diagram.

1.
2.
3.
4.
5.
6.
7.
8.
9.
10.
11.
12.

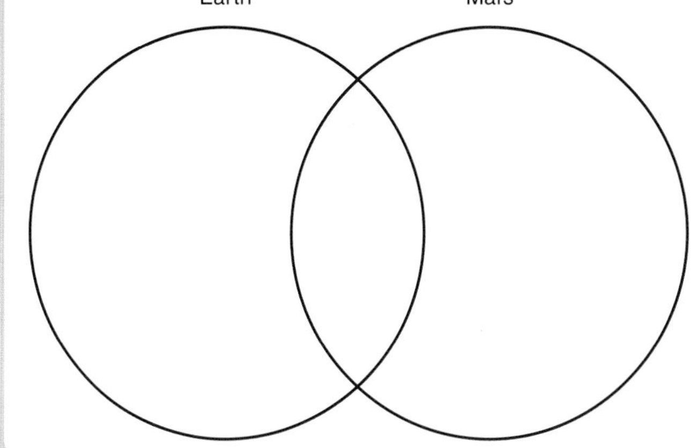

Short-wavelength solar radiation such as X-rays, gamma rays, and ultraviolet radiation would be extremely harmful to life on Earth, wiping out species and causing widespread mutations. Fortunately, our atmosphere intercepts and blocks these rays before they reach the planet's surface. Without a sufficiently thick atmosphere, life as we define it could not exist on land on Earth. Mars has a thin atmosphere that does little to protect the planet's surface. In contrast, Venus has a thick, dense atmosphere that serves as a protective blanket and would destroy most incoming meteorites and absorb harmful solar rays. However, that carbon dioxide–rich atmosphere causes extreme warming so that temperatures on Venus are too high to support life.

Life-Sustaining Gases. Earth's biosphere has moderated the composition of the atmosphere to make it more suitable for life. Vegetation absorbs carbon dioxide (a gas that is poisonous to humans in low concentrations) and produces oxygen, an essential gas for animal life. Earth's original atmosphere had higher concentrations of carbon dioxide, much like Venus, but much of that original carbon dioxide was removed by marine organisms and used to make rocks. Plant life built up the store of oxygen in Earth's atmosphere which fluctuated over time before settling to its current level (21 percent; see Chapter 14). The loss of oceans and the absence of a viable biosphere on Venus and Mars made it impossible to develop an oxygen-rich atmosphere.

The land and oceans absorb solar radiation to warm Earth's surface (Figure 2.29). Heat is radiated upward into the atmosphere where it is trapped by water vapor, carbon dioxide, and other gases to create a condition that has come to be known as the **greenhouse effect.** The greenhouse effect increases temperatures at Earth's surface by 33C° (a change of 59F°), ensuring that we have a livable planet. **Without the heat-absorbing property of the greenhouse gases, Earth would have an average surface temperature of −18°C (0°F) rather than the present average of +15°C (59°F).** Both Venus and Mars have atmospheres composed almost exclusively of carbon dioxide, a key greenhouse gas. The thin atmosphere on Mars absorbs relatively little heat, while the thick, dense atmosphere on Venus absorbs heat like a sponge absorbes water, resulting in a runaway greenhouse effect that has raised temperatures to their current extremes.

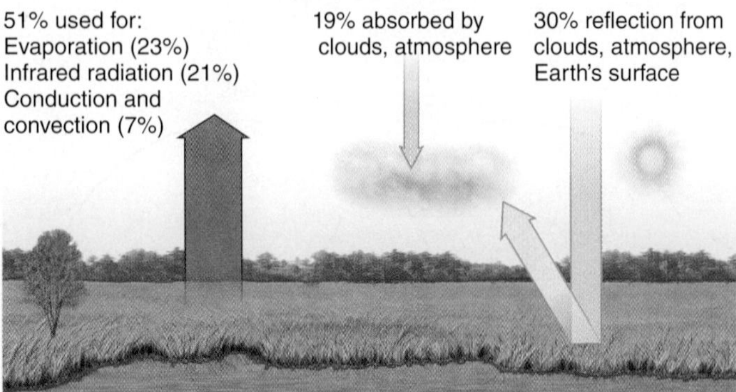

Figure 2.29 The solar energy budget. Approximately one-half of incoming solar radiation heats Earth's surface. Outgoing infrared radiation is absorbed by water vapor and carbon dioxide in the atmosphere.

A Strong Magnetic Field. **Convection in Earth's outer core generates a planetary magnetic field.** Were it not for the magnetic field, our outer atmosphere would have been steadily stripped away by the solar wind (Figure 2.13). A magnetic field is characteristic of planets that are large enough to still have hot interiors and that rotate relatively quickly. **Smaller planets with cool interiors (such as Mars) or larger planets that rotate slowly (e.g., Venus, one rotation every 243 days) may lack the currents in the core necessary to generate a strong magnetic field.** The envelope of protective gases that once surrounded Mars was swept away by the solar wind relatively early in that planet's history. Mars has a localized magnetic field in some regions that is associated with specific rock types having weak magnetic properties. In these locations, Mars retains thin, isolated patches of its original atmosphere. Venus has a weak "induced" magnetic field that is generated by the direct interaction of the solar wind with the planet's outer atmosphere.

So here we are, third rock from the sun, with plenty of water and big enough to have sufficient gravity to hold onto our atmosphere. The gases that make up the atmosphere support life and absorb ample heat to sustain livable conditions near Earth's surface. Finally, the planet's magnetic field protects us from the worst of space weather. So smile—you are living on the only habitable planet we know of, at least for now.

Checkpoint 2.27

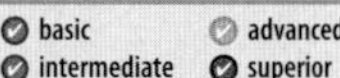

Which of the two following scenarios would be more likely to support life on Earth?

1. Earth is the same size as at present but has the orbit of Mars.
2. Earth has the same orbit as at present but is the size of Mars.

Explain the reasons you used to support this interpretation. Discuss how the four key characteristics of Earth (liquid water, gravity and atmosphere, life-sustaining gases, magnetic field) described in this section would vary in each scenario.

Checkpoint 2.28

Much space science research is concerned with what we would consider basic science: finding out about the origin of the universe, exploring other planets, studying how space phenomena affect Earth, and investigating the potential for life elsewhere in the universe. Often, space program research yields new discoveries with applications elsewhere. Each year, the federal government spends approximately $4 billion (0.15 percent of the total budget) on space science research. If you were in charge of the federal budget, would you increase or decrease funding for space science or continue to fund it at its current level? What are some aspects of this research on which you would place greater emphasis?

the big picture

This chapter shows how scientific ideas change with time. For example, the tools of technology have helped dismantle the geocentric view of Earth's position in the solar system and replaced that idea with a heliocentric view. Opposing ideas may create tension among scientists and confusion for the public, but new ideas are essential for the growth of science. Students of earth science should understand that this push and pull of conflicting ideas is an important strength of the scientific process. It is the scientific equivalent of free speech, allowing all ideas to be heard and, ultimately, enabling those best supported by data and analyses to prevail.

Our current understanding of the character of the universe illustrates how careful analysis of many observations led to hypotheses and eventually produced scientific knowledge. Often we represent nature by using a model that may fall short of the actual reality. However, models help us determine whether our current understanding of the physical laws of nature is sufficient to explain the major features of a system. Mathematical models of the origin of the universe or the gravitational attraction of the planets can replicate the major features identified by scientific observations. With time and improved technology, the finer details of these processes can be revealed.

Features of the universe such as stars and planets can be separated into systems that exhibit consistent characteristics. These systems and their components undergo changes (for example, rotation on axes, revolution around the sun) but also demonstrate constancy as they remain in the same orbits and relative positions. Earth science is full of examples of processes that cause changes on Earth, but those processes have been constantly operating for millions of years. Without this theme of constancy, it would be impossible to predict what Earth has in store for us in the future.

In the chapter opening image, we asked how many more planets like Earth are present in the universe. Recent evidence suggests that Earth is unique among the known planets because it is the only one to support life. This is a consequence of having plenty of water, gravity, an atmosphere rich in life-sustaining gases, and a magnetic field that shields us from damaging solar radiation. Earth's position in space provides us with just enough solar energy to sustain life. However, we are now beginning to identify planets beyond our solar system, and scientists have discovered hundreds of extrasolar planets orbiting stars. Some of these planets are about the same size as Earth, and some of the stars are not that different from our sun. It is perhaps inevitable that there are other Earth-like planets out there that are just the right distance from their star to harbor life.

Earth in Space: Concept Map

Complete the following concept map to identify various processes related to interactions of the earth system components discussed in this chapter. Examine the list of interactions between pairs of components in the following concept map. Match each interaction with one of the labeled links in the concept map. At least one interaction should be used twice.

Interaction	Letter
Revolve around	
Clumps of gas and dust formed	
Gravity controls orbit	
Produces heliosphere that contains	
Produce heavier elements	
For example	
Billions present in	
Only example with biosphere	
Age partially defined by light from	
Intermediate-sized example	
Supply light, heat for	
Magnetic field protects from solar wind	
Expanding outward from	
Forms solar system	

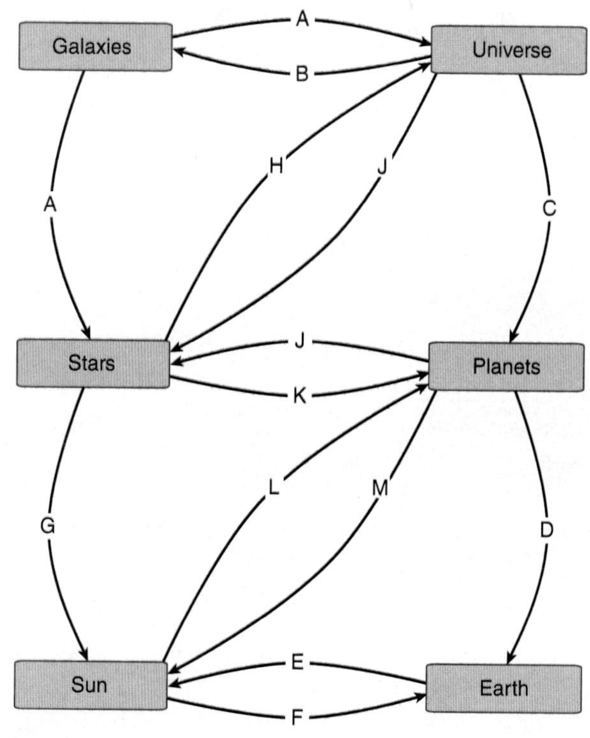

"So haste not, bright meteor; waste not strength, O fair planet, singing-sister."
—Hilda Doolittle
"Comets are the nearest thing to nothing that anything can be and still be something."
—National Geographic Society

Chapter 3

Near-Earth Objects

Chapter Outline

the big picture

What are the chances that an asteroid will land in your neighborhood?

See The Big Picture box at the end of this chapter for the full story on this image.

3.1 Chevy Asteroid

Chapter Learning Outcomes

- Students will be able to describe the characteristics of asteroids and comets, the major types of near-Earth objects (NEOs).
- Students will be able to explain the consequences of an NEO striking Earth's surface.
- Students will be able to discuss how scientists evaluate the risk from future collisions of NEOs with Earth.

A 1980 red Chevrolet Malibu Classic would seem like an odd exhibit for the American Museum of Natural History. You might consider the presence of the car even more surprising if you noticed the gaping hole passing from its trunk through the gas tank. Truth be told, it was the hole, not the car, that was really on exhibit at the museum. The hole formed on a 1992 fall evening when a 12-kilogram (30-pound) meteorite about the size of a football smashed through the car and embedded itself in Marie Knapp's driveway in Peekskill, New York. (Imagine that call to the insurance company!) The car belonged to Marie's daughter Michelle and quickly became an icon among the community of meteorite hunters willing to pay top dollar for these flying space rocks. A commercial meteorite sales company paid $10,000 for the car, which subsequently toured the world, including the natural history museum. The meteorite itself sold for nearly seven times more than the car.

In December 2004, NASA reported a 1-in-60 chance that the stadium-sized asteroid Apophis would collide with Earth in 2029. After a few more days of research that involved digging up older pictures of the asteroid, the astronomers said, "Never mind, it will not strike Earth, but it will be a near miss." When Apophis sails by Earth, it will be close enough to be visible to the naked eye. Apophis will be about 36,000 kilometers (22,500 miles) away, as close as satellites that orbit the planet. The asteroid's orbit will likely be altered by its close brush with Earth, making it interesting to see what happens when it returns again in April 2036 (which is well within the lifetime of most people reading this book).

So what, if anything, should be done about a possible change in the asteroid's orbit? Under the close gaze of the media, the scientific community must walk a fine line between appearing too secretive and scaring the pants off people. After the first Apophis report, NASA scientists were roundly criticized for not waiting until the additional research data were available. **Scientists use the phrase *near-Earth object* (NEO) to refer to objects such as asteroids and comets that approach Earth.** NEO impacts pose a real threat of global catastrophe. Consequently, it is important that a consistent protocol be followed for evaluating such threats so that we may protect people from both hysteria and flying rocks.

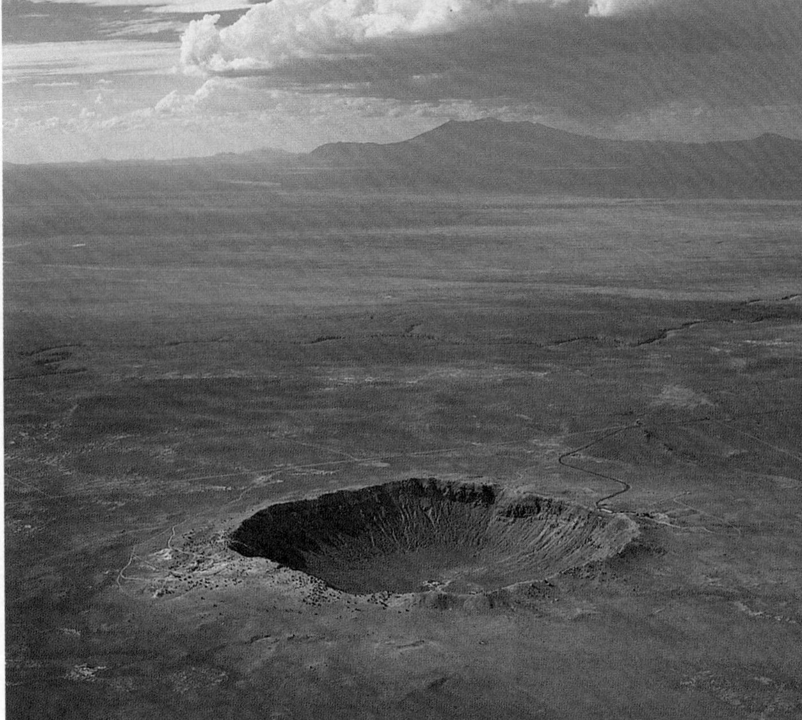

Figure 3.1 Barringer (aka "Meteor") Crater, near Winslow, Arizona. The crater is 1,200 meters (3,940 feet) wide and 175 meters (575 feet) deep. The crater rim rises up to 50 meters (164 feet) above the surrounding desert. The crater is thought to be 50,000 years old and to have formed when a meteorite crashed to Earth.

The Potential for NEO Impacts

If you think that a meteor could never affect you, think again. On the morning of February 15, 2013 a meteor exploded without warning near the city of Chelyabinsk, Russia. Chelyabinsk is a city in central Russia with a population of over 1 million. More than 1,000 people were injured, mostly from flying glass, when this bus-sized object exploded in midair with the force estimated equal to 33 Hiroshima nuclear bombs. Perhaps more interesting, this event is the most recorded in history because it was captured on numerous security cameras, dash-cams, and personal telephones. That same afternoon, a previously identified, 40-meter-diameter (130-foot-diameter) asteroid named DA14 dipped beneath geostationary satellites during its near miss with Earth. Search the web for "Chelyabinsk meteor 2013" or "DA14 asteroid," and you can watch some videos and find more information about that explosion in Russia and the near miss with Earth. But before you rush out to buy a hard hat, you should know that few of these cosmic visitors reach the surface of the planet intact, and most burn up harmlessly in the atmosphere. In Chapter 2, we explored why life as we know it exists on only Earth. In this chapter, we examine the threat to life on Earth from collisions with asteroids and comets. We also describe where asteroids and comets come from and explore their characteristics. Although there is some discussion among scientists, most agree that the impact of a large NEO with Earth caused a widespread extinction 65 million years ago. The demise of the dinosaurs at that time opened the door for mammals (like us) to become the planet's dominant organisms, so we can be thankful for NEO impact events! However, it would be foolish to ignore the evidence of past impacts (Figure 3.1) and the potential for similar future events. Approximately one impact every few hundred years is large enough to cause extensive destruction equivalent to a major natural disaster. We will discuss the telltale signatures of such impacts. Finally, we will review efforts to track larger NEOs and consider how we might cope with the discovery that a large rock has Earth in its crosshairs.

Self-Reflection Survey: Section 3.1

Consider the following questions as a means of uncovering what you already know about near-Earth objects.

1. Briefly describe three of your own personal perceptions about near-Earth objects based on media reports (newspapers, television), movies, or personal experiences.
2. What do you think a meteorite would look like and how might it differ from other rocks found on Earth's surface?
3. How much advance warning should a government agency provide if scientists discover that a large NEO is on a collision course for Earth?

3.2 Characteristics of Near-Earth Objects

Learning Objectives

- Compare and contrast the characteristics of asteroids and comets.
- Describe where NEOs come from and explain how this influences the frequency of their appearance near Earth.
- Identify similarities and differences between NEOs and planets.

Most asteroids and comets exist far from Earth and give us little trouble. But occasionally one is bumped out of its usual orbit and sent on a long, looping course around the sun. These small objects are difficult to detect in the vastness of space, and often we do not know they are there until they buzz by us, like a baseball whistling by your ear.

Several NEOs have approached Earth more closely than the distance to the moon in recent years (Figure 3.2, compare to the Apophis visit in 2029). Sometimes NEOs are not detected until *after* they have passed close to Earth. Scientists are currently focusing their detection efforts on the largest NEOs, those that could cause catastrophic global or continental-scale consequences, should they strike Earth.

Asteroids and Meteorites

Asteroids originate in the **asteroid belt,** a relatively dense jumble of cosmic debris that lies in orbit between Mars and Jupiter (see Chapter 2 Snapshot). Until about 200 years ago, nobody knew that these asteroids existed. Figure 3.3 shows the number of known asteroids in 1900 compared to 1999. About 90,000 asteroids have been cataloged, measuring from 940 kilometers (584 miles) in diameter to less than 50 meters (164 feet) across (Figure 3.4). Most of these objects will not approach Earth.

The gravitational attraction of nearby Jupiter jostles asteroids from their stable orbits, causing them to crash into one another. Such collisions can send small asteroids or crash debris

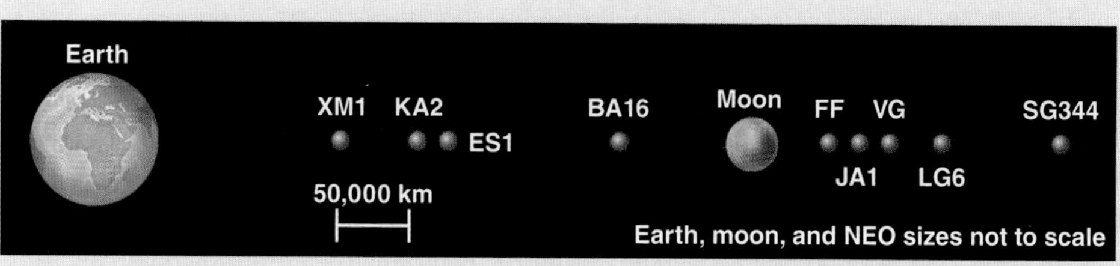

Figure 3.2 Proximity of NEOs to Earth and the moon. Several NEOs have approached Earth more closely than the distance to the moon in recent years. All of the NEOs shown here are several kilometers in diameter.

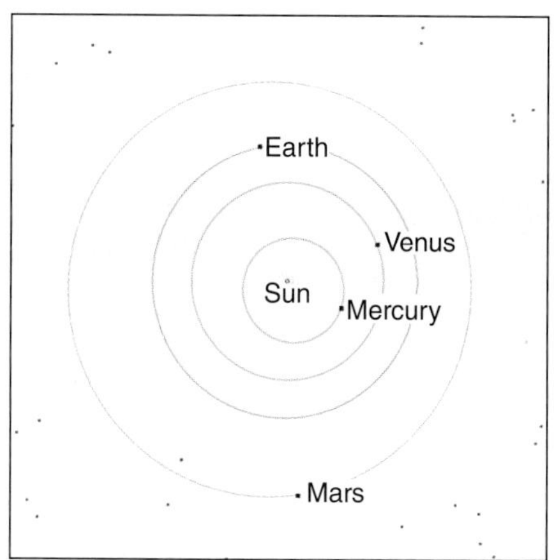

a. The inner solar system in 1900

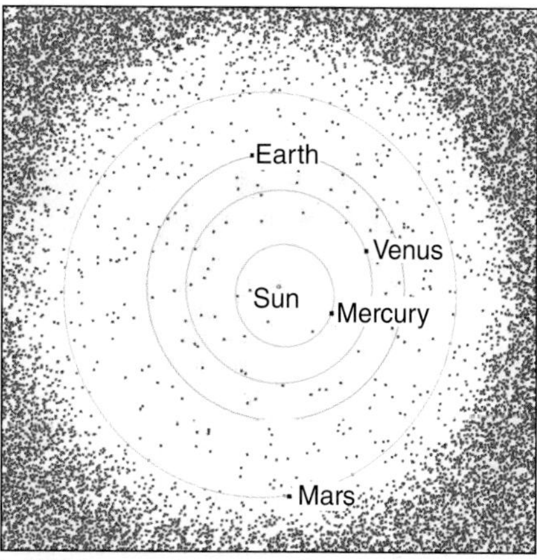

b. The inner solar system in 1999

Figure 3.3 Known asteroids in the inner solar system. Note the great increase in the number of identified asteroids from 1900 **a.** to 1999. **b.** Asteroids that do not approach Earth are indicated by green squares, and those that cross Earth's orbit are indicated by red squares.

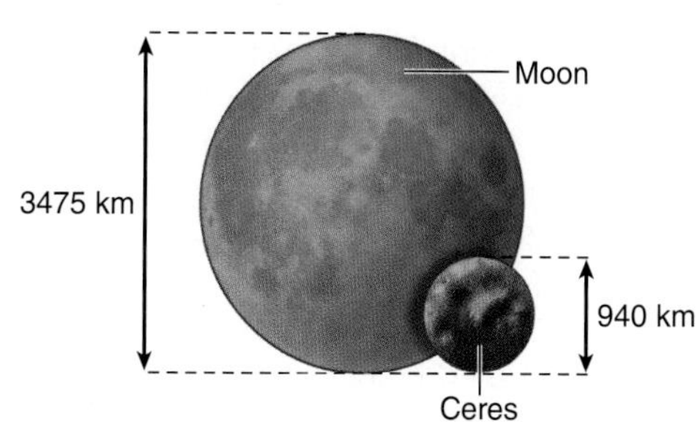

Figure 3.4 Asteroid sizes. Ceres, the largest asteroid, is about one-quarter of the diameter of the moon and is 18,000 times larger in diameter than the meteorite that formed Barringer (Meteor) Crater (see Figure 3.1). Ceres is in a stable orbit and poses no threat of a collision with Earth.

Checkpoint 3.1

basic | advanced | intermediate | superior

Which characteristics are true of *both* planets and asteroids?

a) They are approximately spherical.
b) There are thousands of examples.
c) They formed 1 to 2 billion years ago.
d) They are present in the solar system.

flying through space toward the inner planets. These materials follow eccentric orbits and may plunge into any of the terrestrial planets, leaving impact scars that remain visible for many thousands of years (Figure 3.1). It was only about 100 years ago that some asteroids were recognized as approaching Earth. When an asteroid comes toward Earth, it can miss completely, burn brightly in the atmosphere (a meteor), explode in the atmosphere, or strike Earth's surface to become a meteorite. Asteroids travel at such extreme speeds (average speed of 16 kilometers per second or 36,000 miles per hour) that even a relatively small object colliding with Earth can cause substantial damage.

Meteorites are composed of various materials. Stony meteorites, made of rocks similar to those found in Earth's crust or mantle, account for over 90 percent of known meteorites. The remainder are a combination of stony material and iron (known as stony-iron meteorites) or are composed of a mix of iron and nickel (known as iron meteorites). Interestingly, all of these materials are the same as those that make up the terrestrial planets. Therefore, scientists hypothesize that planets such as Earth and Mars formed from such material over 4 billion years ago. Rock and metals that were not incorporated into a planet were marooned in the asteroid belt. A very small number of meteorites found on Earth are composed of rocks similar to those on the moon or Mars. The impact of large asteroids with the moon or Mars caused these meteorites to be ejected into space. Later, some of those ejected fragments hit Earth.

Suppose you came across an odd-looking rock. How would you know if it was a meteorite? After all, some meteorites may be worth thousands of dollars per gram. Keep in mind that a meteorite will have undergone frictional heating on the way through our atmosphere, so it should have a dark surface that shows signs of melting. It should be solid and relatively smooth; holes are rare, as are jagged edges. Even a stony meteorite should have a higher concentration of metallic elements than most rocks, so it will feel heavy and will attract a magnet. Finally, a meteorite should be different from everything else around it. If you find several in the same place, they are probably not meteorites.

To learn more about the characteristics of NEOs, NASA sent an unmanned probe on a 5-year research mission to an asteroid known as 433 Eros. The probe landed on the surface of Eros in February 2001 (Figure 3.5). The goal of the Eros mission was to learn more about the geology, physical properties, and rotation of NEOs. Future efforts to destroy or deflect incoming asteroids will require an understanding of their composition and rotation sequence. Data from the probe showed that Eros closely matches meteorites of the stony type.

Checkpoint 3.2

Venn Diagram: Planets and Asteroids

Compare and contrast the characteristics of the planets (see Chapter 2) in our solar system with those of asteroids by placing the numbers from the list of characteristics in the most suitable locations on the diagram. Characteristics 4, 7, and 10 have been plotted for you.

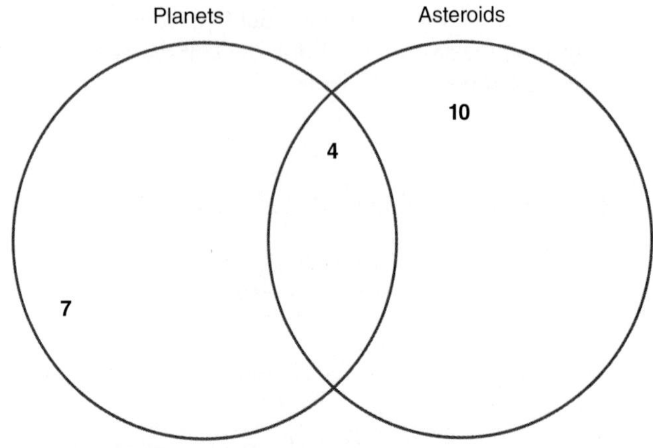

1. Radius greater than 500 kilometers
2. Essentially spherical
3. Orbit the sun
4. Have a gravitational field
5. Can rotate
6. May be made of materials similar to those found on Earth
7. Possess moons
8. Thousands of examples
9. Have atmospheres
10. Have less predictable orbits
11. Have a variety of shapes
12. Formed after the Big Bang over 4 billion years ago
13. Have craters
14. Some will collide with Earth
15. Example: Mars
16. Example: Eros

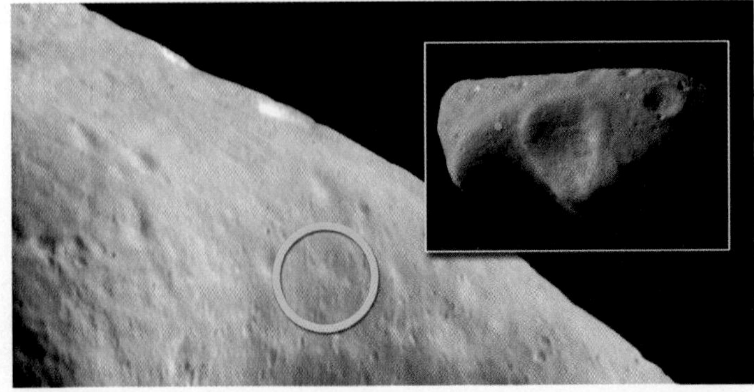

Figure 3.5 The asteroid 433 Eros. The inset (top right) shows the shape of the asteroid. The spacecraft's landing site is indicated by a circle.

Comets

The greatest fireworks show on July 4, 2005, occurred millions of miles from Earth and was viewed by a small but select audience at 80 observatories scattered around the world. The day before, the *Deep Impact* spacecraft had released a smaller "impactor" craft that steered itself into the path of an oncoming comet, Tempel 1. At 1:52 A.M. the following morning, the much larger comet smashed into the impactor like a tank running over a water-filled balloon. The collision produced a bright spray of dust and gases that was monitored at sites on Earth (Figure 3.6). It was a tremendous feat of planning to send a spacecraft millions of miles across space to launch a smaller probe to collide head-on into a target area 6 kilometers (3.7 miles) wide on the surface of a speeding comet. Some likened the task to hitting a bullet with a bullet. The *Deep Impact* mission provided information on the composition of the Tempel 1 comet (more about this later).

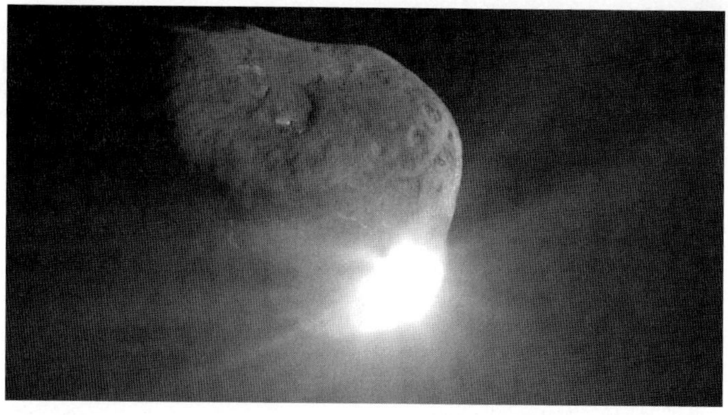

Figure 3.6 Comet collision. After comet Tempel 1 smashed into the "impactor" spacecraft, July 4, 2005, a flash occurred as debris and gas exploded into space.

Checkpoint 3.3

basic / intermediate / advanced / superior

Asteroid Crashes the Moon?

Read the following article and then answer the questions concerning the differences among observations, hypotheses, and predictions.

Asteroid Crashes the Moon?

Early in the morning of November 15, 1953, amateur astronomer Leon Stuart was looking at the surface of the moon through his homemade telescope. As he watched, he saw a bright flash that seemed to come from the lunar surface. He was able to snap a photograph of the event on a camera connected to the telescope. Stuart published the image in a journal for amateur astronomers and wrote a short article that interpreted the flash as the only known record of an asteroid crashing into the moon. Over the years since its publication, Stuart's interpretation has been challenged. Some scientists have suggested that what he actually observed was a "point meteor," an asteroid burning up in Earth's atmosphere as it moved directly toward the observer. In this interpretation, the moon would simply have been present in the background behind the burning meteor. Those who support this interpretation point out that the flash was observed for 8 seconds, much longer than would be expected for a meteorite impact.

Jump forward to 2001. Dr. Bonnie Buratti and her research assistant compared images of the moon's surface taken by the Clementine *spacecraft in 1994 with Stuart's original photograph. They estimated the size of a crater that would have been formed by the Stuart event on the basis of the size and brightness of the flash. They discovered a small, "fresh-looking" crater of the appropriate size in the center of their search area and inferred it to represent the crater formed by the meteorite impact. Their results were published in the astronomy journal* Icarus *in January 2003.*

Still, many astronomers were skeptical. Soon after publication of Buratti's article, John Westfall, an expert on the features of the lunar surface, examined archive photographs taken by the 100-inch Mount Wilson telescope in 1919 and found the same crater. Clearly it could not be the product of a 1953 meteorite impact if it was already present in 1919! The mystery continues.

Sources: *New York Times,* March 4, 2003, by Henry Fountain; *Toronto Star,* March 9, 2003; Jet Propulsion Lab, Cal Tech.

Questions

1. Which of the following represents a key *hypothesis* presented in this article?
   a) Meteorites don't collide with the moon very often.
   b) Craters on the moon are formed by meteorite impacts.
   c) A meteorite impact on the moon was photographed in 1953.
   d) Most young craters on the moon have a "fresh" appearance.
2. What was the principal *observation* used to support the hypothesis?
   a) An article appeared in an amateur astronomy journal.
   b) Leon Stuart photographed a bright flash on the moon's surface.
   c) Dr. Buratti found a recently formed crater on images of the moon taken by the *Clementine* spacecraft.
   d) Meteorites hit the moon once or twice a century.
3. What key *prediction* was tested in an attempt to prove the hypothesis?
   a) A small, fresh-looking crater would be present in a specific location.
   b) A bright flash would be visible on the moon's surface.
   c) Craters would be present on the moon.
   d) A meteorite could be observed at the location of Stuart's original photograph.
4. Which hypothesis was *falsified* in this article?
   a) The bright flash represented a point meteor.
   b) An asteroid collided with the moon on November 15, 1953.
   c) A crater photographed by the *Clementine* spacecraft was formed by the impact event.
   d) The Mount Wilson telescope observed the same crater in 1919.

basic · intermediate · advanced · superior

Checkpoint 3.4

Concept Maps of Asteroids

Review the following four concept maps, all of which describe some of the basic characteristics of asteroids. Your task is to identify which diagram best represents the most important features of asteroids. Rank the concept maps in order from best (1) to worst (4), and justify your rankings. Your justification should include a description of the criteria you used to decide what makes one map better than another.

Rank order: 1.___ 2. ___ 3.___ 4. ___

a.

Asteroids
are known as
meteorites
when found on
Earth's surface
impact
Asteroid belt
are composed of
rocks
up to
1000 km diameter
from the

b.

Asteroids
are examples of
near Earth objects
that come from the
Asteroid belt
located beyond the orbit of
Mars
are composed of
rocks and/or metals
with similar composition to the
interiors of planets
such as

c.

Asteroids
are up to
1000 km diameter
when found in the
Asteroid belt
on the margins of the
solar system
are composed of
rocks
from
planets
in the

d.

Asteroids
are known as
meteorites
which are a type of
near Earth object
that produce
impact craters
are composed of
rocks
similar to those found in
Earth
when they collide with

When Comets Collide with Planets. Astronomers estimate that there are many thousands of comets with diameters over 100 kilometers (62 miles), making them much more plentiful than large asteroids. Like asteroids, comets can and do occasionally collide with planets in our solar system. In 1994, a string of up to 20 separate parts of a comet known as Shoemaker-Levy 9 smashed into Jupiter over the span of a week. This was the first time scientists were able to observe a collision between two bodies in our solar system.

Although large comets do not collide with Earth as frequently as asteroids do, the consequences of an impact would be just as catastrophic. Comets travel at even faster speeds than asteroids and would have an impact velocity of about 50 kilometers per second (112,000 miles per hour) were they to strike Earth. A mysterious 1908 explosion in Tunguska, Siberia, is attributed to the air blast from a comet that disintegrated in the atmosphere a few kilometers above the land surface. The blast left no crater, but it flattened forests over an area of 2,100 square kilometers (810 square miles; Figure 3.7). The impact of such a comet today would have the potential to demolish a large urban area (for example, London or New York) and to generate other massive environmental disasters (as described in Section 3.4).

Figure 3.7 Map of location of 1908 Tunguska Event and image of trees felled by the Tunguska air blast, Siberia, 1908.

What Comets Are Made of. Although we can observe comets from Earth, they do not fall to the ground as meteorites do, so scientists have been able to make only educated guesses about their composition. Comets represent the remnants of cosmic debris left over from the formation of the solar system. Like distant Pluto, located far from the sun's heat, **much of a comet's mass is composed of ice, probably surrounding a rocky core.** Comets have long been compared to dirty snowballs—that is, mainly ice with some dust. However, analysis of materials produced by the collision with Tempel 1 revealed that the comet had more dust than ice, so comets may be better thought of as icy dirtballs. When a comet travels toward the center of our solar system, heat from the sun causes frozen ice to change from a solid to a gas. This then forms a trailing tail that points away from the sun in the direction of the solar wind (Figure 3.8).

Scientists used a telescope in Hawaii to analyze the light from the explosion on Tempel 1 (see Figure 3.6). The collision sprayed more than 1 million gallons of water and 20,000 tons of dust into space. Different molecules and atoms emit light at different frequencies. Scientists used an instrument called a *spectrometer* to identify the materials that made up the comet from the different frequencies of light produced by the collision. These data revealed that Tempel 1 is composed of common compounds such as water (H_2O), cyanide (HCN, a common interstellar gas), carbon dioxide (CO_2), and organic materials, many in amounts considerably greater than expected. Some scientists speculate that ancient comets may have brought these critical materials to an early, lifeless Earth.

The icy composition of comets prompted some scientists to suggest that the water in Earth's oceans may have been supplied by ancient comet impacts. However, recent analyses of two different comets have not been able to provide a definitive answer to this question. Scientists using infrared telescopes observed the comet C/1999 S4. They estimated that the chemistry of the water evaporating from that comet was similar to that of the water in Earth's oceans. In a separate study, chemists who used an array of radio telescopes to observe comet Hale-Bopp found evidence that the composition of Hale-Bopp's ice does *not* match that of the water in Earth's oceans. As is common in science, more studies are needed to resolve this issue.

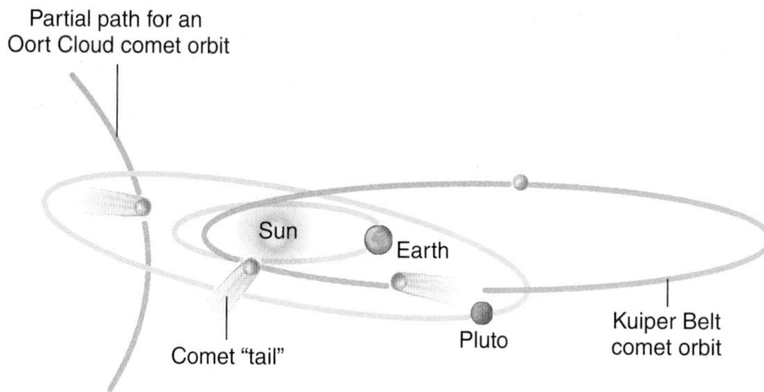

Figure 3.9 Comparison of comet orbit paths. Short-period comets from the Kuiper Belt orbit the sun in a plane similar to that of the planets. In contrast, comets from the Oort Cloud may intersect Earth's orbit at a high angle with an opposing sense of revolution around the sun.

The Two Types of Comets. As comets come from the outer reaches of our solar system, they approach the sun in wide elliptical orbits. Scientists recognize two types of comets, which are defined by how long it takes them to complete an orbit: short-period comets and long-period comets.

Short-period comets visit the inner solar system in time intervals of 200 years or less. They begin their journey in the Kuiper (pronounced "kie-per") Belt, a dense zone of comets that stretches from the neighborhood of Neptune to far beyond Pluto. These comets, collectively known as trans-Neptunian objects, circle the sun in the same direction and with a similar trajectory to the outer planets (Figure 3.9). Comet Borrelly, which circles Earth every 7 years, is an example of a short-period comet. NASA

Figure 3.8 A comet. As a comet nears the sun, it grows brighter and forms a trailing tail that always points away from the sun.

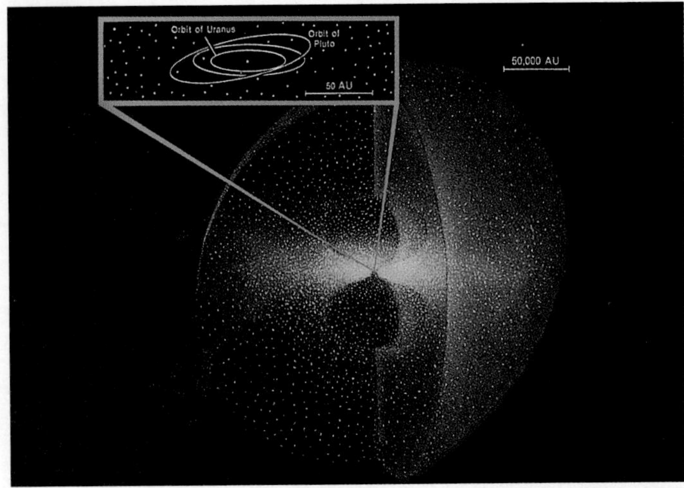

Figure 3.10 Comets that approach Earth originate in the Kuiper Belt or the Oort Cloud. The Kuiper Belt extends from the margins of our solar system outward and lies near the center of the much larger, spherical Oort Cloud.

scientists were able to coax the *Deep Space 1* spacecraft to fly within 2,200 kilometers (1,370 miles) of Borrelly in 2001, giving us our best up-close and personal view of a comet until the Tempel 1 collision in 2005. Halley's comet is another example of a famous short-period comet with an orbital period of 75 years.

Long-period comets originate at a great distance from Earth in a massive spherical cloud of cosmic debris known as the Oort Cloud, named for Dutch astronomer Jan Hendrick Oort (Figure 3.10). The margins of the Oort Cloud are located 50,000 astronomical units (AU) from the center of our solar system (for reference, Pluto is 39 AU from the sun), just at the edge of the heliosphere. The gravitational pull of passing stars or nebulae can disrupt the Oort Cloud, sending comets toward the center of our solar system. These comets have irregular elliptical orbits that may cross the orbits of the planets at a high angle (see Figure 3.9). Long-period comets may fly by Earth at time intervals of 200 years or more. The very bright comet Hale-Bopp passed by Earth in the mid-1990s but won't visit us again until around the year 4400. Figure 3.11 summarizes the characteristics of both types of comets in the form of a concept map.

Checkpoint 3.5

basic · intermediate · advanced · superior

Where is the Kuiper Belt relative to the asteroid belt?

a) The Kuiper Belt is closer to the sun than the asteroid belt.
b) The Kuiper Belt is farther from the sun than the asteroid belt.
c) The Kuiper Belt and the asteroid belt are located in the same region of the solar system.

Checkpoint 3.6

basic · intermediate · advanced · superior

Shoemaker Levy 9 broke into at least 21 fragments (up to 2 kilometers, or over 1 mile, in diameter) before hitting Jupiter in July 1994. Examine the image indicating surface changes associated with one of those impacts. Explain why those scars are so large (some are larger than the diameter of Earth) when the fragments were small compared to Jupiter.

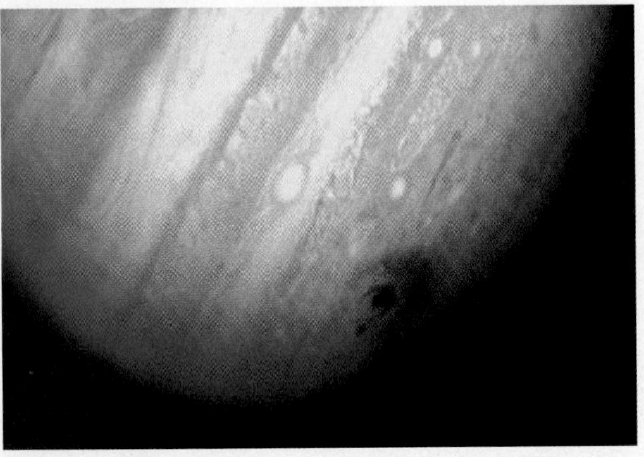

Checkpoint 3.7

basic · intermediate · advanced · superior

Venn Diagram: Planets and Comets

Complete the following Venn diagram to compare and contrast the similarities and differences between planets and comets (see Chapter 2). Add more characteristics to the following list, and place the numbers corresponding to the appropriate characteristics in the most suitable locations on the diagram. Two have been provided as examples.

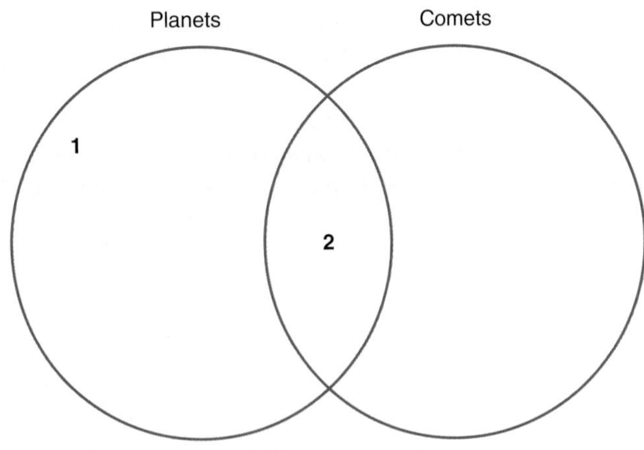

1. Diameter greater than 1,000 kilometers
2. Orbit the sun
3.
4.
5.
6.
7.
8.
9.
10.
11.
12.

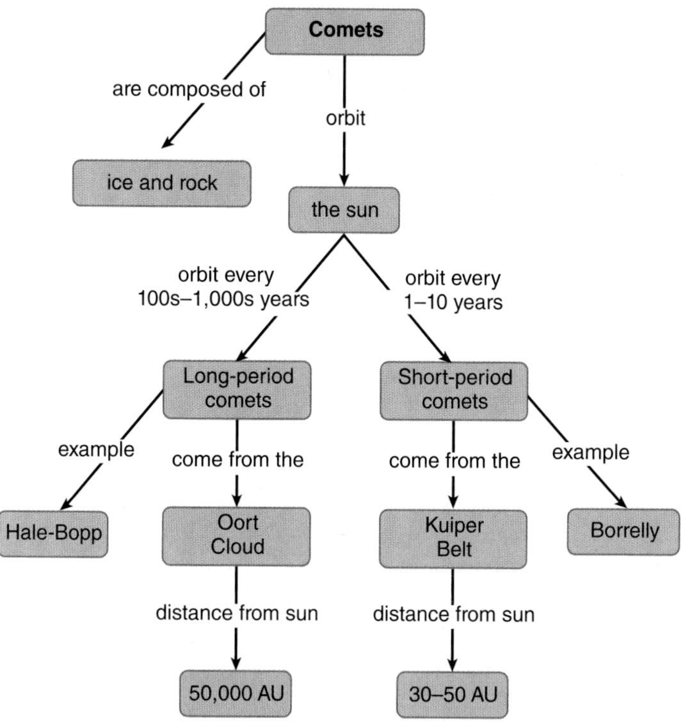

Figure 3.11 Concept map summarizing the characteristics of the two types of comets.

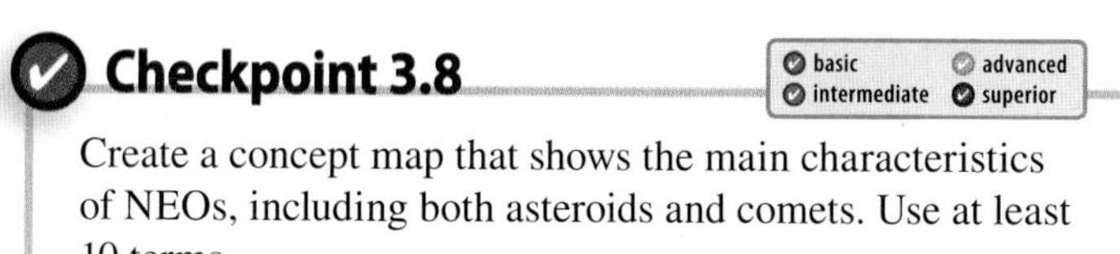

Checkpoint 3.8

Create a concept map that shows the main characteristics of NEOs, including both asteroids and comets. Use at least 10 terms.

3.3 Impact Features

Learning Objectives

- Explain how we can use the features of impact craters to determine the size of the impacting object.
- Describe how the characteristics of a typical impact crater would change moving a) horizontally and b) vertically away from the center of a crater.
- Discuss how scientists can use the 150 recognized impact craters to estimate how frequently Earth is struck by NEOs of different sizes.

An estimated 100 million kilograms (220 million pounds) of meteors strike Earth's atmosphere each year. The bulk of this material is in the form of small particles that burn up before reaching the surface. But when a comet or asteroid collides with a planet, an **impact crater** results. Impact craters are common on all the rocky terrestrial planets and their moons. A majority of these craters date from a period of intense bombardment soon after the formation of the solar system more than 3.9 billion years ago. On Earth, the oldest craters are no longer preserved because they have been worn away by millions of years of erosion and weathering (see Chapter 9) or have been covered up by later rock layers. More-recent impacts on Earth are preserved in relatively young rocks. On the moon, even the oldest craters are preserved in their original state because of the absence of wind and water.

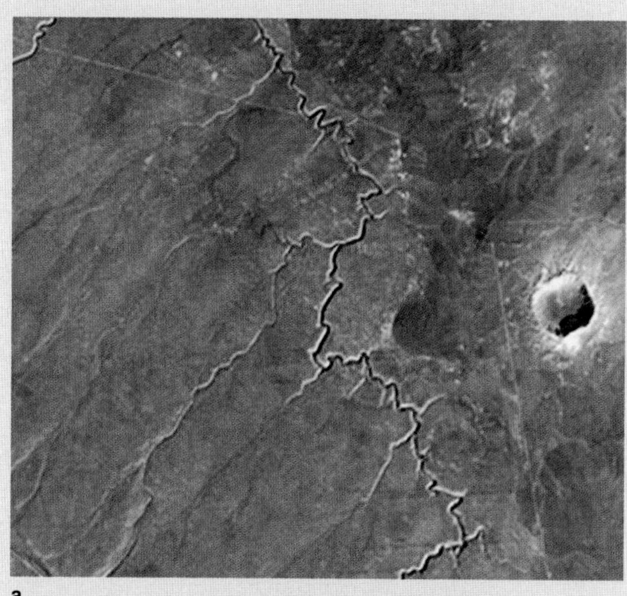

a.

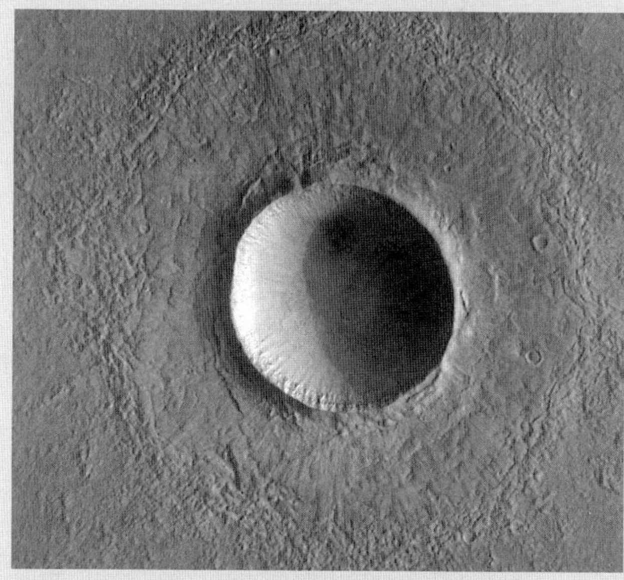

b.

Figure 3.12 Simple craters. **a.** Meteor Crater, Arizona, was the first impact crater recognized on Earth. The crater is 1,200 meters (3,940 feet) across. **b.** A Martian crater, 2,600 meters (8,530 feet) across, surrounded by a circular layer of debris known as an ejecta blanket.

Crater Characteristics

Impact craters come in two basic forms: simple and complex. Small, **simple craters** on Earth have a diameter of a few kilometers and are bowl-shaped. For example, Meteor Crater in Arizona is 1,200 meters (3,940 feet) across (Figures 3.1 and 3.12a). Large, **complex craters** with diameters of more than 4 kilometers (2.5 miles) have central peaks and ringlike structures along their

Checkpoint 3.9

basic · intermediate · advanced · superior

Meteor Crater (see Figure 3.1) was formed by an NEO approximately ___ meters in diameter.

a) 50 b) 600 c) 1,200

Checkpoint 3.10

basic · intermediate · advanced · superior

Imagine you are a scientist who recently discovered an impact crater beneath Chesapeake Bay along the Atlantic coast. The crater is 500 to 1,300 meters (1,640 to 4,265 feet) deep and approximately 90 kilometers (56 miles) across. What type of crater is it?

a) Simple b) Complex

margins where the crater rim collapsed inward (Figure 3.13). The central peak of a complex crater forms because the surface rebounds—that is, the center of the crater floor springs back after being violently compressed by the impacting NEO. (Think of dropping a stone in a pond.)

Crater size is largely a consequence of the size and velocity of the impacting meteorite or comet and the materials that make up the NEO and the impact site. As a rule of thumb, an **NEO impact gouges out a crater about 10 to 20 times larger than the colliding asteroid or comet.**

Both simple and complex craters often contain broken and fractured rocks, termed **breccia.** Material thrown out of the crater by the force of the impact is called **ejecta** and forms a layer surrounding the crater (Figures 3.12 and 3.14). Heat from the impact can cause melting of rocks on the crater floor (melt rocks). The force of the collision typically pulverizes the impacting body, although some small meteorite fragments may escape destruction. The extreme pressures generated by an impact can alter minerals in the rocks near the crater (Figure 3.15), creating features characteristic of an impact event.

Crater Distribution and Frequency. More than 150 impact sites have been identified on Earth's continents. Nearly one-half of these sites have craters over 10 kilometers (6 miles) across—the size widely accepted as posing a substantial threat to civilization (Figure 3.16 shows the locations of the impacts discussed in this section). By contrast, few impact sites have been identified in the oceans, which cover more than two-thirds of Earth's surface. This scarcity results either because impacts that occurred in the oceans were not large enough to form craters on the ocean floor or (more likely) because the impact sites have been destroyed or obscured by later geologic processes.

Geologists can determine when the craters formed by analyzing the age of the surrounding rocks and thus generate an

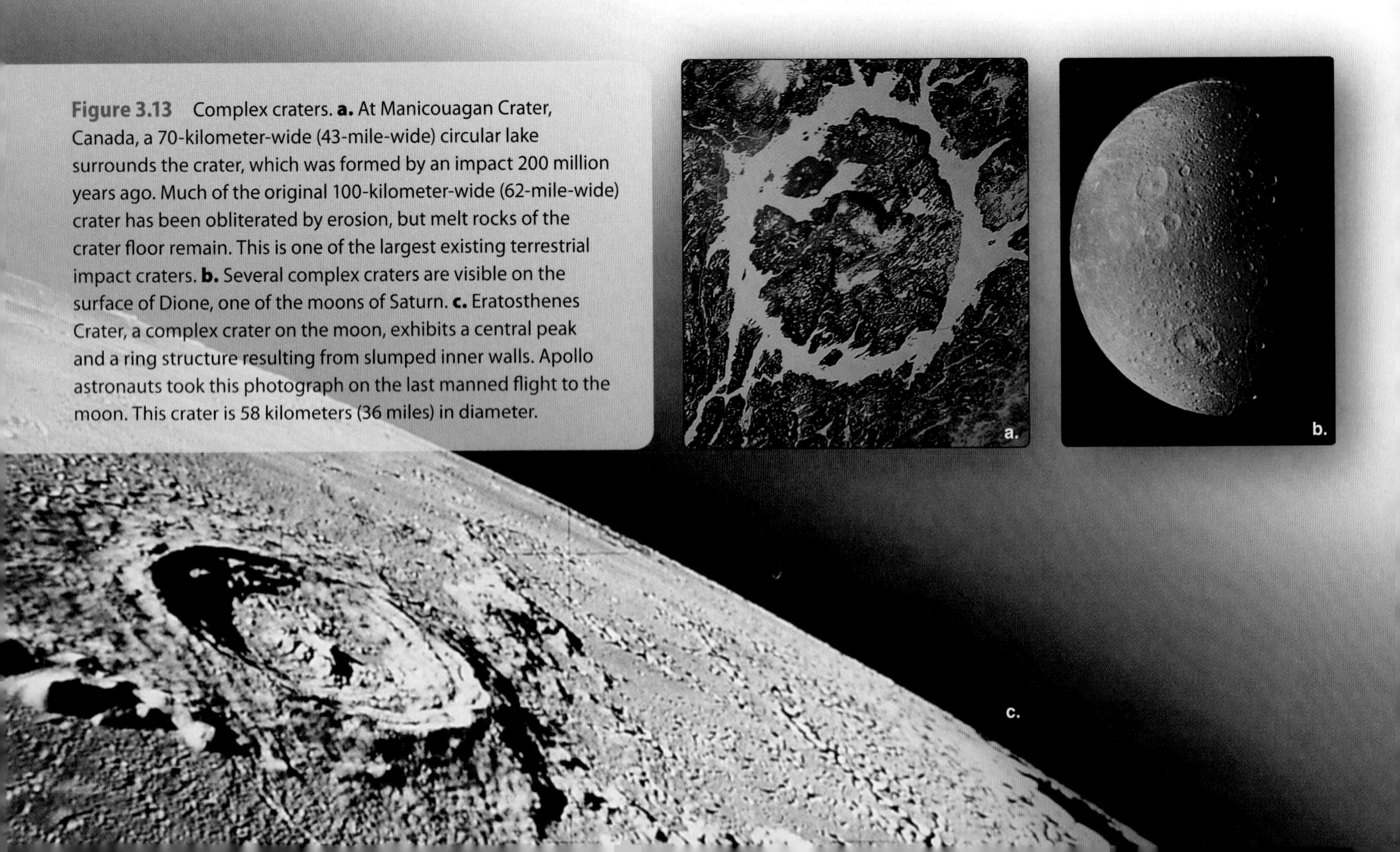

Figure 3.13 Complex craters. **a.** At Manicouagan Crater, Canada, a 70-kilometer-wide (43-mile-wide) circular lake surrounds the crater, which was formed by an impact 200 million years ago. Much of the original 100-kilometer-wide (62-mile-wide) crater has been obliterated by erosion, but melt rocks of the crater floor remain. This is one of the largest existing terrestrial impact craters. **b.** Several complex craters are visible on the surface of Dione, one of the moons of Saturn. **c.** Eratosthenes Crater, a complex crater on the moon, exhibits a central peak and a ring structure resulting from slumped inner walls. Apollo astronauts took this photograph on the last manned flight to the moon. This crater is 58 kilometers (36 miles) in diameter.

a. b. c.

estimate of how frequently Earth's surface is struck by NEOs. On the basis of the size and age distribution of craters, scientists infer that a 1-kilometer-wide (0.6 mile-wide) NEO collides with Earth every few 100,000 years. Collisions involving larger, more devastating NEOs 5 to 10 kilometers (3 to 6 miles) in diameter occur about once every 10–100 million years. These large-scale impacts leave a clear imprint on the geologic record. The most recent such event occurred 65 million years ago, forming the Chicxulub impact crater on the Yucatan Peninsula, Mexico (see Figure 3.16). The crater was discovered by geologists exploring for oil. The near-circular crater stood out on a gravity map the scientists created. This impact event is thought to be the primary cause of the worldwide extinction of the dinosaurs and 70 percent of all other species living at that time. (See discussion in Chapter 1.)

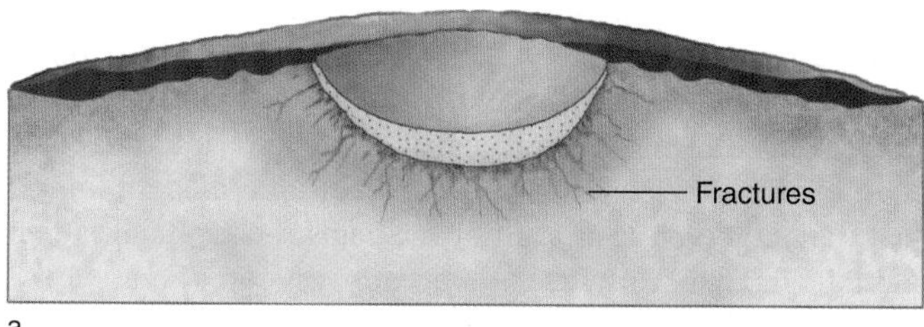

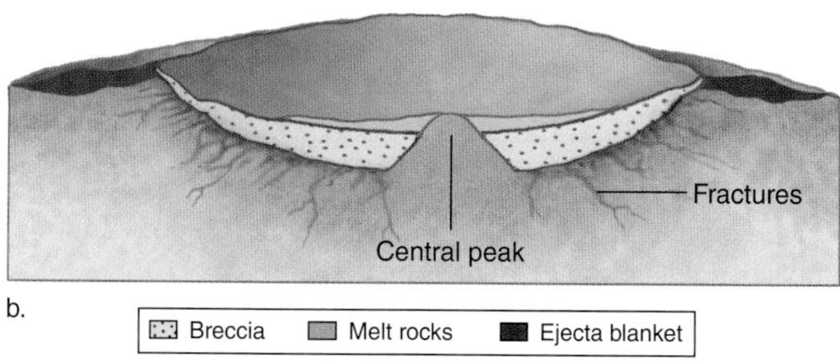

Figure 3.14 Crater features. **a.** Bowl-shaped simple craters exhibit fewer features and a smaller width-to-depth ratio than **b.** larger complex craters. Rocks smashed by the force of the collision are termed *breccia*. Material blasted out of the crater, known as ejecta, forms a layer surrounding the crater.

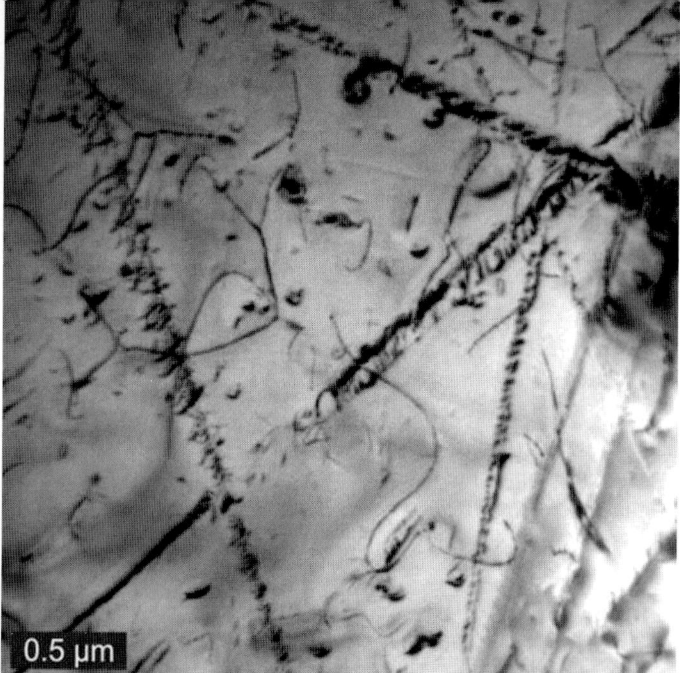

Figure 3.15 Shocked quartz. Evidence of a violent impact appears as damage to a mineral's crystalline structure. Note the linear features in this image of quartz in a rock sample from the Charlevoix impact crater in Quebec, Canada. The image was taken using a transmission electron microscope.

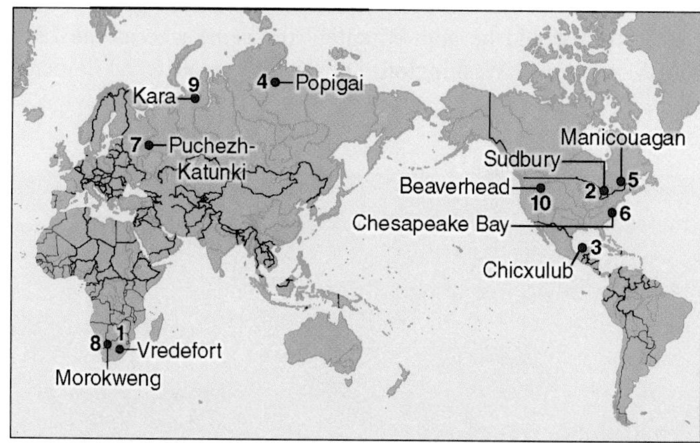

Figure 3.16 Sites of the 10 largest impact craters on Earth. Over 150 craters have been recognized. It is likely that exploration of less-accessible regions of Earth will yield many more examples.

Checkpoint 3.11

basic | advanced | intermediate | superior

Examine the accompanying image of the Uranius Tholus volcano from Mars (taken from NASA's *Viking 1* orbiter in 1977). The volcano is 60 kilometers (37 miles) across and rises about 3 kilometers (2 miles) from the surrounding plain. Describe the impact craters that you can identify from the image, and discuss the relative age of formation of the craters and the volcano. Non-impact-related craters often form during volcanic activity. How can you tell if a crater on Mars is formed by impacts or volcanism?

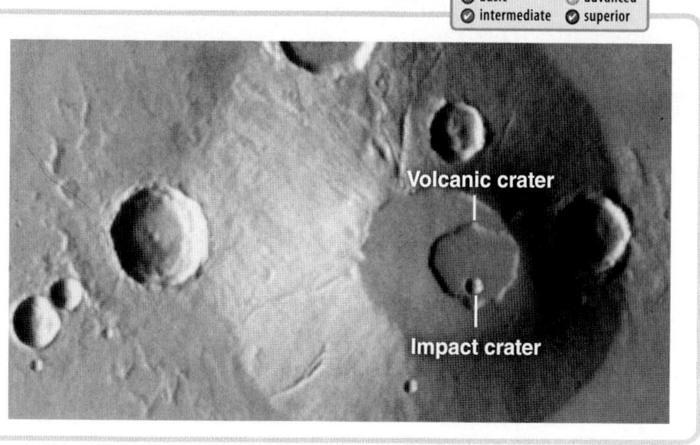

Checkpoint 3.12

basic advanced intermediate superior

Imagine that you are a member of a team of scientists who are drilling into a large buried crater approximately 200 kilometers (124 miles) wide. The crater was formed by an NEO impact that wiped out many species 251 million years ago. The drilling team will recover rock samples during the drilling process. You have some data to suggest that the crater is located between 500 and 1,500 meters (1,640 and 4,921 feet) below the surface. How will the drilling team know when the drill has cut through the overlying rock and reached the crater? What features will they observe?

3.4 Impact Hazards

Learning Objectives

- Describe the sequence of events, features, and consequences of the impact of a large NEO.
- Explain how an NEO collision in Chesapeake Bay would affect cities in the northeast.

When it comes to impact hazards, size matters. First, let us put the sizes of NEOs in perspective: A 50-meter-wide (164-feet-wide) NEO would be approximately the same size as the Lincoln Memorial in Washington, D.C., while an NEO 1 kilometer (0.6 mile) across could rest comfortably along the National Mall, between the Washington Monument and the Capitol building (Figure 3.17). For comparison, the whole District of Columbia could hide behind a 10-kilometer-wide (6-mile-wide) NEO.

As we learned in Section 3.3, an impact crater is 10 to 20 times larger than the size of the original NEO that produced it. Thus, NEOs just 50 to 100 meters (164 to 328 feet) across could level whole cities. A meteorite 1 kilometer (0.6 mile) in diameter is big enough to devastate most nations. And a large asteroid or comet with a diameter of 10 kilometers (6 miles) or more would be sufficient to produce global devastation.

Scientists believe that a 50-meter-wide (164-foot-wide) meteorite gouged out Meteor Crater in the Arizona desert (see Figure 3.1). The explosive force of even that relatively small meteorite was several times greater than the atomic bomb dropped on Hiroshima at the close of World War II. At the other end of the scale, the Chicxulub Crater in Mexico is approximately 200 kilometers (124 miles) in diameter and was formed by a meteorite approximately 10 to 15 kilometers (6 to 9 miles) wide. Almost every living thing in southern North America would have been killed by the collision. It is estimated that over 1 billion people would be killed worldwide as a direct result of a modern impact of that magnitude or as a result of the catastrophic changes that would follow. Thankfully, such events are very rare (Figure 3.18), but the devastating consequences for life on Earth make the risks much higher than the threat associated with smaller, more frequent collisions. (See the discussion in Chapter 1 on the development of the scientific hypothesis tied to the Chicxulub impact event.)

Figure 3.17 Relative sizes of NEOs. Compared to the National Mall and key buildings in Washington, D.C., a 50-meter-wide (164-foot-wide) NEO would be approximately the size of the Lincoln Memorial. A 1-kilometer-wide (0.6-mile-wide) NEO could rest on the National Mall between the Washington Monument and the Capitol building without touching either structure. A 10-kilometer-wide (6-mile-wide) NEO would cover the whole District of Columbia.

Checkpoint 3.13

basic advanced intermediate superior

Imagine that two identical asteroids are approaching Earth's orbit. One impacts Earth; the other, the moon. Both impacts produce craters in the same type of rocks. How will the craters compare?

a) The crater on the moon will be larger.
b) The crater on Earth will be larger.
c) The craters will have the same dimensions.

Checkpoint 3.14

Two 1-kilometer-wide (0.6-mile-wide) asteroids collide with Earth. The first asteroid strikes a desert area 1,000 kilometers (620 miles) from the city of Bang with a population of 1 million people. The other lands in the open ocean 1,000 kilometers from an identical city named Crash that also has 1 million people. Both cities are located along a low-lying coast similar to the Atlantic coast of the eastern United States. Predict which city will suffer the greater damage. Explain your choice.

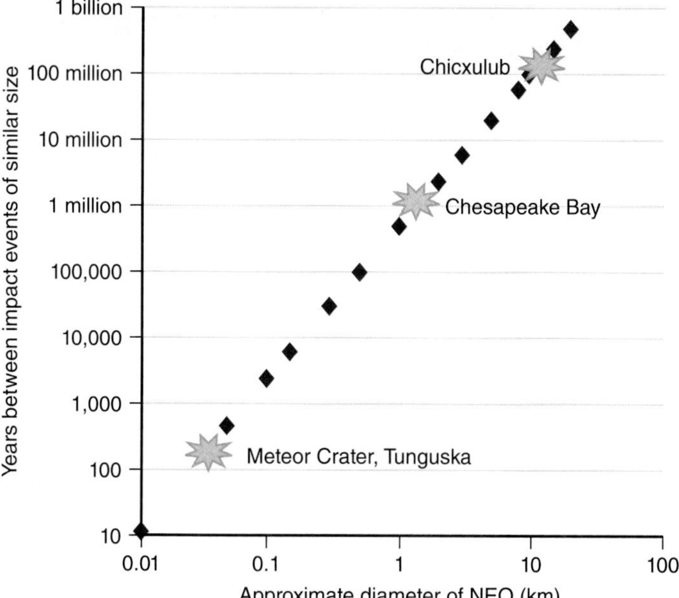

Figure 3.18 Frequency and sizes of impact events. The largest impacts occur at time intervals measured in hundreds of millions of years. Impacts large enough to destroy a large city or have substantial regional consequences occur every 100 to 1,000 years.

An Impact Event

It could be a day like any other. The collision of a large NEO with Earth may occur with no warning. Or it could be predicted decades in advance and watched anxiously by billions of people around the world. The entry of a large NEO into Earth's atmosphere would be accompanied by an atmospheric shock wave and heating of the speeding object as it compresses gas molecules in the atmosphere. About 100 kilometers (62 miles) above Earth's surface, heating would turn an approaching asteroid into a fireball and vaporize the outer surface to create a glowing dust trail (see Chapter Snapshot). For many people, this might be the first warning of their fate. An NEO would smash into Earth at such high speeds that those near the impact site would have just a few seconds between the first sighting of the object in the atmosphere and its explosion on impact. This is one sight you would not want to see.

For those within 1,200 kilometers (938 miles), the first effect of the collision would be the arrival of thermal radiation within a few seconds of the impact. A repeat of the impact at Chicxulub would produce sufficient heat to set clothing and grass on fire in Houston and Miami just 11 seconds after the impact. Any location near the impact site would be shaken by massive vibrations, greater than the largest earthquake ever recorded. Residents of cities throughout most of the United States would feel the effects of the shaking within a few minutes of the impact. Next, a rain of debris would begin to fall. The impact would pulverize rocks, ejecting a massive cloud of dust and melted rock fragments upward into the atmosphere (Chapter Snapshot). Within 11 minutes of the impact of a modern Chicxulub, 14 centimeters (5.6 inches) of material ejected from the crater would begin to fall on Memphis, Tennessee. Seven minutes later, Los Angeles would receive its own thin deposit of ejecta. Meanwhile, Houston would be buried under more than 35 centimeters (14 inches) of material, but the worst is yet to come.

Checkpoint 3.15

basic / intermediate / advanced / superior

Concept Map of NEO Impacts

Review the following concept map developed by a student, which describes the characteristics of NEO impacts. Score the concept map using the grading rubric, and redraw the diagram, making whatever changes you believe are appropriate to earn a 4 on the grading scale.

Grading Rubric

0 The concept map does not contain any information about NEO impact events.

1 The concept map contains some relevant terms, but several key terms are omitted and many linking phrases are either absent or inaccurate.

2 The concept map contains most of the relevant terms, but they are poorly organized and some linking phrases are absent or incorrect.

3 The concept map contains most of the relevant terms, but one or two key terms may be absent. The diagram is reasonably well organized, and almost all linking phrases are appropriate.

4 The concept map contains all of the relevant terms in a well-organized display that has appropriate linking phrases for each pair of terms.

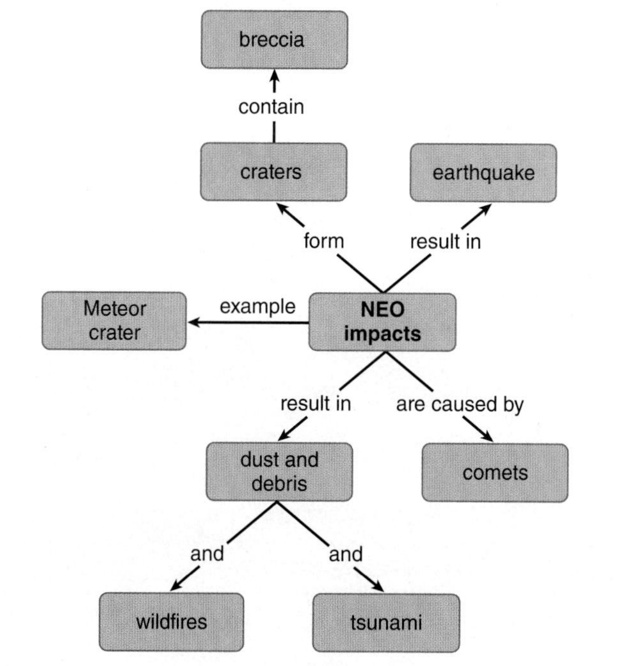

NEO IMPACT WITH EARTH

HOT FRAGMENTS

Pieces of molten rock blasted out of the crater would fall back to Earth to generate wildfires.

TSUNAMI

An impact event in the ocean would generate a tsunami, a giant wave that could drown coastal regions and travel far inland.

AIR BLAST

The collision would send out a powerful air blast that would flatten everything for hundreds of kilometers in every direction.

WILD FIRES

Would add smoke to the rapidly darkening skies.

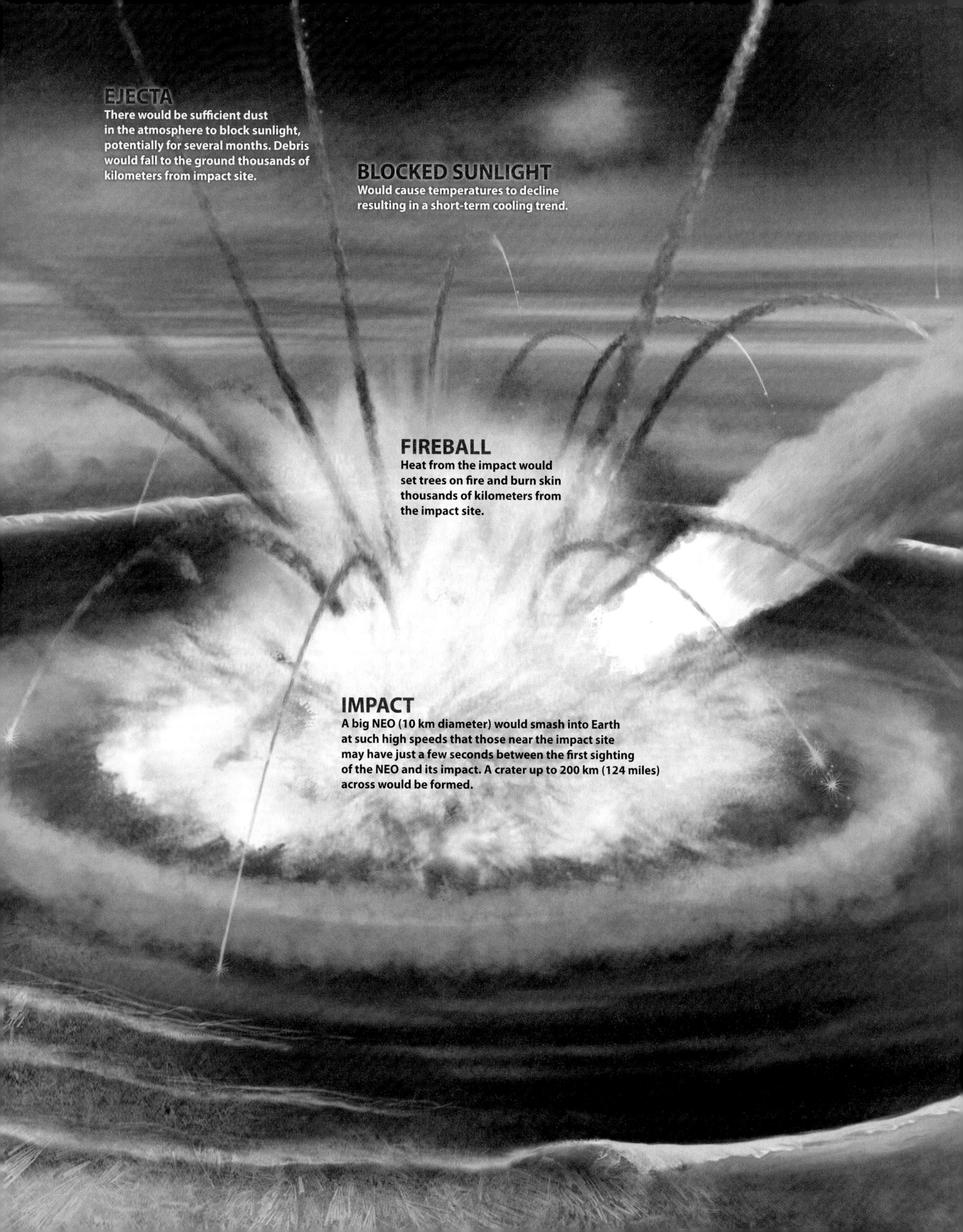

EJECTA

There would be sufficient dust in the atmosphere to block sunlight, potentially for several months. Debris would fall to the ground thousands of kilometers from impact site.

BLOCKED SUNLIGHT

Would cause temperatures to decline resulting in a short-term cooling trend.

FIREBALL

Heat from the impact would set trees on fire and burn skin thousands of kilometers from the impact site.

IMPACT

A big NEO (10 km diameter) would smash into Earth at such high speeds that those near the impact site may have just a few seconds between the first sighting of the NEO and its impact. A crater up to 200 km (124 miles) across would be formed.

Checkpoint 3.16

basic intermediate advanced superior

Scientists have recently discovered the existence of a 90-kilometer-wide (56-mile-wide) crater beneath Chesapeake Bay in the eastern United States. What would happen if the same impact event were to happen today? Examine the map that shows the location of Chesapeake Bay relative to major cities and physical landforms. Use the online program at www.lpl.arizona.edu/impacteffects/ to analyze how the impact of a 3-kilometer-wide (1.8-mile-wide) NEO would affect the cities pictured on the map. How would this event compare to the Chicxulub impact?

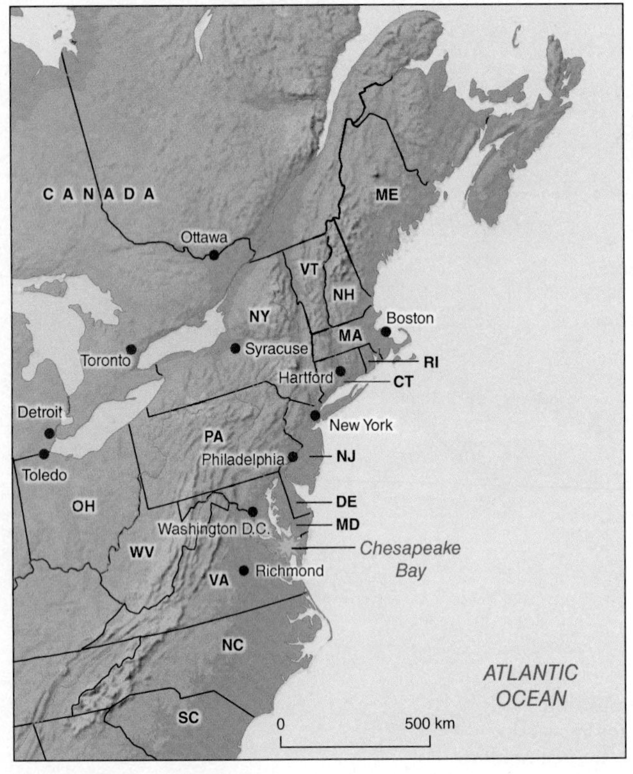

The collision would send out a powerful air blast that would flatten everything for hundreds of kilometers in every direction. Nearly an hour after the impact, the air blast would hit states around the Gulf Coast. Winds roaring at speeds of 870 km/h (543 mph), faster than the strongest tornadoes, would flatten major cities. Almost all trees would be ripped out of the ground, and homes and multistory buildings would collapse. The Chicxulub impact produced an air blast that felled forests 2,000 kilometers (1,243 miles) away in the interior of North America. Cities as far away as New York would experience winds equivalent to a category 4 hurricane, and the air blast would be strong enough to shatter glass windows in Anchorage, Alaska.

The impact would throw up sufficient dust to block sunlight, potentially for several months. Consequently, global temperatures would decline, resulting in a short-term cooling trend (see Chapter 1). Some scientists have estimated that Earth was in darkness for up to 6 months after the Chicxulub impact, which may have been sufficient to prevent photosynthesis for as long as the next year. Vegetation, lacking the ability to enter a dormant phase until conditions improved, would not have survived. Pieces of molten rock blasted out of the crater would fall back to Earth to generate colossal wildfires that would add smoke to the rapidly darkening skies. Tiny blobs of molten material would form glassy spheres, known as *spherules,* that are indicative of impact events. Some of these particles would travel fast enough to leave the atmosphere and orbit Earth before falling back to the surface.

An impact event in the open ocean would generate a tsunami, a series of giant waves that would drown coastal regions and travel far inland. Waves with heights measured in thousands of meters would be possible from a Chicxulub-sized event in the deep ocean. A 10-kilometer-wide (6-mile-wide) NEO would be over twice the average depth of the ocean floor. Scientists believe the tsunami associated with Chicxulub was less intense than it might have been because only a portion of the impact site was located in the shallow Gulf of Mexico along the margin of the Yucatan Peninsula. Even so, the impact is estimated to have generated a series of waves up to 300 meters (1,000 feet) high that pushed into the present Gulf Coast states and created sufficient backwash to carry forest debris from land as far as 500 meters (1,640 feet) offshore onto the open ocean.

Finally, gases derived from ocean waters or pulverized rocks would be added to the atmosphere. These gases, including sulfur dioxide, carbon dioxide, and water vapor, could linger in the atmosphere for years to decades, long after the dust had settled and the wildfires burned themselves out. In addition, injections of sufficient sulfur dioxide could alter the global hydrologic cycle by producing more-acidic precipitation, a condition known as *acid rain.* Whether leading to higher global temperatures or more acidic conditions, the effects on the biosphere could be significant—although not to the extent of the impact itself.

3.5 Beware of Flying Rocks

Learning Objectives

- Compare the size of impact craters to local or regional features of similar dimensions.
- Explain why the collision of an asteroid or comet with Earth is a natural hazard we can potentially prevent.
- Describe how scientists are trying to detect dangerous NEOs.
- Apply your understanding of the nature of science to the development of the Torino scale.
- Evaluate what characteristics of approaching asteroids would be the most important to consider in deciding which to destroy.

Even when scientists know the location, speed, and trajectory of Earth-orbiting objects, it is not easy to predict exactly where they will enter the atmosphere and crash to Earth. The 140-ton *Mir* space station, the largest constructed object to come back to Earth, crashed in the South Pacific Ocean in March 2001. Although most of its components burned up in the atmosphere, approximately 50 tons of debris splashed down in a zone covering thousands of square kilometers of ocean between New Zealand and Chile. There was sufficient concern about the reentry that the Russian space agency that operated *Mir* bought a $200 million insurance policy to guard against stray fragments causing harm. Undaunted by such uncertainty, a group of observers paid over $6,000 each for the privilege of flying over the splash-down zone in hopes of catching a view of blazing space debris zipping by their airplane window. Proving that money and good sense often do not go together, these spectators were fortunate that their unique form of Russian roulette did not end with a flaming space toilet knocking them into the ocean.

Predicting and Preventing Impact Events

Impact events are the only significant natural hazard that we have the potential to actually prevent. We do not have the technology to stop volcanic eruptions or earthquakes, but we are close to having the technical ability to prevent flying space rocks from smashing into our planet. With just a few days warning, we could readily anticipate the approximate impact site for an Earth-bound asteroid or comet. Given the object's size and speed, scientists could predict the potential consequences of the impact and make efforts to evacuate the region and prepare for the collision. Furthermore, if we know about potential impacts years or decades ahead of time, **it is possible that the NEO could be deflected away from Earth or destroyed** before it enters the atmosphere. Scientists use computer models to make sophisticated calculations of the future path of identified NEOs. With more observations, the hypothetical path of the NEO becomes more tightly constrained and less provisional. Efforts to prevent a collision could center on detonating an explosion on or near an asteroid or comet. At great distances, even a small nudge would be sufficient to avoid a collision, but a closer object might require the explosive force of a nuclear warhead to push it off track or break it into smaller, less threatening pieces. The key step is finding the object and correctly determining its path toward Earth.

There is just one caveat: **NEO hunters are focused mainly on big rocks over 1 kilometer (0.6 mile) in diameter** that would have the potential to create a continental- or global-scale catastrophe. Smaller NEOs with diameters greater than 50 meters (164 feet) could still survive the trip through Earth's atmosphere and pose a threat for substantial devastation, perhaps wiping out a major city. NEOs of about that size generated the Tunguska event and formed Meteor Crater. However, no active detection programs are presently in place for the millions of NEOs in this size range. Asteroids and comets of this size are routinely discovered during the search for larger NEOs. Essentially, we are accepting that a major city such as London or Los Angeles could be obliterated without warning, but we are doing our best to ensure that areas the size of Europe or the western United States would not be devastated. The reasons behind such a scale-dependent response are tied to the difficulty in finding NEOs, the available resources, current funding for NEO detection programs, and recognized levels of risk.

NEO Detection. Most NEOs approaching Earth are asteroids—small, dark, distant, fast-moving objects that reflect little sunlight and are therefore difficult to see. Imagine standing in a circular field surrounded by a high fence. A number of your friends are standing behind the fence, out of sight. They don't know where you are in the field; you don't know where they are behind the fence. Each person behind the fence has a never-ending supply of golf balls. All the golf balls are painted black. It is nighttime; there are no lights. You know that your friends will throw the golf balls as hard as they can over the fence. You also know that it is a pretty big field and the chance of any ball hitting you is slim, but you still need to keep your eyes open, just in case. The search for NEOs is even more challenging than trying to spot a black golf ball in the night.

Scientists are looking for the largest NEOs first, not only because they will be the easiest to find but also because they pose the greatest risk. There may be millions to hundreds of millions of years between strikes, but the consequences of such impacts would be so catastrophic that they could end human life. Smaller NEOs, although more likely to hit the planet, would have a more localized significance, causing severe regional devastation but having less consequence for the vast majority of life on Earth. Although NEO impacts are rare, they have the potential to kill very large numbers of people. As with many natural disasters, we often downgrade the potential threat until it happens. For example, the risk from tsunami is considered relatively small, but the Indian Ocean tsunami in 2004 killed more than 200,000 people. Current analysis suggests that a US resident is about as likely to be killed by a near-Earth object as to die as a result of a tsunami. That risk will decrease as more NEOs are discovered and their orbits mapped to make sure they will not collide with Earth.

Checkpoint 3.17

basic · advanced · intermediate · superior

A relatively small NEO with a diameter of 50 meters (164 feet) could generate a 500- to 1,000-meter-wide (164- to 328-foot-wide) crater; a 1-kilometer-diameter (0.62-mile-diameter) NEO would form a 10- to 20-kilometer-wide (6- to 12-mile-wide) crater; and the impact of a 10-kilometer-wide (6.2-mile-wide) NEO would result in a crater 100 to 200 kilometers (62 to 124 feet) across. Identify features around your campus or city that have similar dimensions to these three crater sizes.

Checkpoint 3.18

basic · advanced · intermediate · superior

Go to the NASA NEO site at http://neo.jpl.nasa.gov/. Select the "Close Approaches" icon. Review the "Upcoming Close Approaches to Earth" table. Find the three closest approaches and evaluate which one would be most threatening if it were to actually impact Earth. Support your choice.

Fewer than 100 people around the world are working at the few facilities equipped with the telescopes and automated cameras necessary to detect NEOs. These programs photograph the night sky at specific time intervals in search of objects that change location relative to the fixed background of stars. When NEOs are discovered, their paths are calculated and plotted. If a path approaches Earth and only a relatively small data set is available, scientists use archival (previously collected) data saved from previous NEO observations to expand the record and predict a more accurate orbit. They then calculate the distance of the object from Earth and the specific date of its approach. It would help if there were more people working to find NEOs, but only a limited number of telescopes are available, and most of those are in demand for other programs.

Evaluating Risks Using the Torino Scale. Scientists rank natural hazards (such as hurricanes, tornadoes, and earthquakes) using a variety of measurements or scales intended to reflect the potential dangers of a hazard. Astronomers have developed the **Torino scale to assess the potential risk from impact events** (Figure 3.19). A Torino scale value of 0 to 10 is assigned to an NEO, reflecting its potential to strike Earth and the consequences of that collision. A value of 0 (zero) represents an NEO that will either miss Earth or burn up in the atmosphere. Objects classified as yellow (2 to 4) or orange (5 to 7) are approaching close enough to Earth that it should be a priority to obtain more information about their orbits in order to exactly determine their potential for collision. NEOs classified at 8 to 10 on the Torino scale would be certain to strike Earth, and the actual number assigned to them would reflect the degree of destruction (8, local; 9, regional; 10, global). The level of destruction is mainly governed by the size and velocity of the impacting object.

The Torino Scale
Assessing Asteroid/Comet Impact Predictions

Category	Level	Description
No Hazard	0	The likelihood of collision is zero or is so low as to be effectively zero. Also applies to small objects such as meteors and bolides that burn up in the atmosphere as well as infrequent meteorite falls that rarely cause damage.
Normal	1	A routine discovery in which a pass near Earth is predicted that poses no unusual level of danger. Current calculations show the chance of collision is extremely unlikely with no cause for public attention or public concern. New telescope observations very likely will lead to reassignment to Level 0.
Meriting Attention by Astronomers	2	A discovery, which may become routine with expanded searches, of an object making a somewhat close but not highly unusual pass near Earth. While meriting attention by astronomers, no cause for public attention or public concern as an actual collision is very unlikely. New telescope observations very likely will lead to reassignment to Level 0.
	3	A close encounter, meriting attention by astronomers. Current calculations give a 1% or greater chance of collision capable of localized destruction. Most likely, new telescope observations will lead to reassignment to Level 0. Attention by the public and public officials is merited if the encounter is less than a decade away.
	4	A close encounter, meriting attention by astronomers. Current calculations give a 1% or greater chance of collision capable of regional devastation. Most likely, new telescope observations will lead to reassignment to Level 0. Attention by the public and public officials is merited if the encounter is less than a decade away.
Threatening	5	A close encounter posing a serious but still uncertain threat of regional devastation. Critical attention by astronomers is needed to detemine conclusively whether or not a collision will occur. If the encounter is less than a decade away, governmental contigency planning may be warranted.
	6	A close encounter by a large object posing a serious but still uncertain threat of a global catastrophe. Critical attention by astronomers is needed to detemine conclusively whether or not a collision will occur. If the encounter is less than three decades away, governmental contigency planning may be warranted.
	7	A very close encounter by a large object, which is occurring this century, poses an unprecedented but still uncertain threat of a global catastrophe. For such a threat in this century, international contigency planning is warranted, especially to determine urgently and conclusively whether or not a collision will occur.
Certain Collisions	8	A collision is certain, capable of causing localized destruction for an impact over land or possibly a tsunami if close offshore. Such events occur on average between once per 50 years and once per several thousand years.
	9	A collision is certain, capable of causing unprecedented regional devastation for a land impact or the threat of a major tsunami for an ocean impact. Such events occur on average between once per 10,000 years and once per 100,000 years.
	10	A collision is certain, capable of causing a global climatic catastrophe that may threaten the future of civilization as we know it, whether impacting land or ocean. Such events occur on average once per 100,000 years or less often.

Figure 3.19 The Torino scale.

Occasionally, astronomers identify an approaching asteroid that on initial examination has a slim chance of striking Earth (1 or 2 on the Torino scale). However, this tentative hypothesis has almost always been falsified as soon as additional data make it clear that these objects will miss us by a sizable distance. The chance of collision is then downgraded to 0. Notice how scientists use data, not opinions, to modify their initial interpretation. For example, astronomers spotted the 2-kilometer-wide (1.2-mile-wide) asteroid NT7 in July 2002. After 2 weeks of observation, it was ranked as a 1 on the Torino scale, predicting a slim chance that it would strike Earth in February 2019. Further analysis then provided enough data to calculate a more accurate trajectory, making it possible for scientists to downgrade the threat to 0.

As of February 2013, one NEO had a Torino rating of 1, and there were no cataloged objects with higher ratings. Asteroid 2007 VK184 with a diameter of 130 meters (425 feet) is estimated to have four trajectories that could potentially impact Earth between 2048 and 2057. The chance of a collision is about 1 in 1,820, or to put it another way, there is a 99.945 percent chance it will miss Earth. Until recently, Apophis had a Torino rating of 1 based on calculations that indicated a possible collision on April 13, 2036. As is common in science, however, additional observations and orbit modeling reduced that probability to less than 1 in 45,000. You can check the current Torino scale values for known NEOs at the Jet Propulsion Lab's Near-Earth Object Program site (http://neo.jpl.nasa.gov/risk/).

Scientists estimate that there are approximately 1,000 NEOs of 1-kilometer (0.6-mile) diameter or greater in our solar system. In 1998, NASA began the Spaceguard program with the goal of finding 90 percent of NEOs with a diameter of 1 kilometer (0.6 mile) or greater within a decade. We now know the orbits of more than 80 percent of the largest NEOs, and current search programs are looking for the remainder. Although none of the recognized objects is considered a real impact threat for Earth, until all of these NEOs are found, an unknown asteroid could smash into the planet tomorrow, and we would be none the wiser.

Discuss how the characteristics of the scientific method are apparent in the use of the Torino scale and the search for NEOs.

Checkpoint 3.20

basic · intermediate · advanced · superior

Asteroid Impact Risk Evaluation Rubric

It is 20 years in the future. Scientists have cataloged all the largest NEOs (diameter of 1 kilometer (0.6 mile) or greater) and have found that none of that size poses a threat to Earth. Ten years ago, they identified an 800-meter-wide (2,625-foot-wide) asteroid on a path for Earth. The asteroid was targeted with a space-based missile and was broken into several pieces. Unfortunately, three of the largest pieces are still heading for Earth.

Advances in tracking technologies now allow scientists to accurately pinpoint the location where such objects will strike the surface of the planet. Governments from around the world are supporting a mission to destroy the most dangerous of the three pieces of asteroid. You are on a team charged with choosing which piece to destroy first.

1. Your assignment is to create an evaluation rubric to assess the relative dangers from the broken asteroid pieces. You must find a method of ranking the risk of potential harm from each impact event. Consider what factors would contribute to the loss of lives and the high damage costs associated with an impact. One factor (size) has been done for you as an example (see accompanying table). After identifying at least four more factors, distinguish what characteristics related to impact hazards would make each of them a high-, moderate-, or low-risk phenomenon. For example, the larger the asteroid, the greater the risk. Consequently, large NEOs are given a high risk score (3 points), whereas smaller asteroids are viewed as low risk (1 point). Add your data to the table. You will then use the combination of factors with the highest cumulative score to identify which incoming object will be destroyed first.
2. After completing your rubric, your team is asked to double the score of the most important factor. Which factor would you choose? Why? Explain your choice.

Standards and Criteria

You will be assessed on your choice of:

- Relevant factors that would contribute to the potential for damaging impacts.
- Identification of what constitutes high-, moderate-, and low-risk situations for each factor.
- Your justification for choosing one factor in particular as the most significant.

Factors	Low risk (1 point)	Moderate risk (2 points)	High risk (3 points)
Size (diameter) of asteroid	Small (diameter approximately 25 meters (82 feet))	Intermediate (diameter 25–100 meters (82–328 feet))	Large (diameter more than 100 meters (328 feet))

the big picture

Scientists employ sophisticated data collection and analysis tools, including computer-enhanced telescopes, spacecraft, and mathematical models, to study the characteristics and origins of asteroids and comets. Beyond its scientific merits, this research has a significant social dimension. A major focus of this chapter is the role of science in informing citizens to guide decision making. Research findings raise vital questions that need to be addressed to help the public understand the scientific basis behind policy choices determining how federal funds should be spent. For example, should we devote millions of dollars to figuring out how to steer incoming NEOs away from Earth, or can we wait until we know that an asteroid or comet is coming our way? While large NEOs may come close to Earth, there are no known objects that are likely to collide with us in the near future.

A second emphasis of the chapter is the role of technology in identifying, detecting, and collecting data from NEOs. These data can provide an early warning of impending collisions, but they also allow scientists to understand the type of materials that make up the NEOs and consequently the potential character of an impact event. NASA's Spaceguard program is a prime example of this technology. The rationale is that sufficient warning time may enable us to deflect an NEO away from Earth or destroy it before it enters the atmosphere. Scientists use computer-linked telescopes to find these objects and computers running sophisticated models to map the future paths of identified NEOs. Technology has also been used to allow three different teams of scientists to hit asteroids and comets with tiny spacecraft millions of miles from home. This is good practice. If we do successfully identify an NEO on a collision path with us, we will have to launch some kind of device to try and deflect it away from Earth. The more target practice we get, the better.

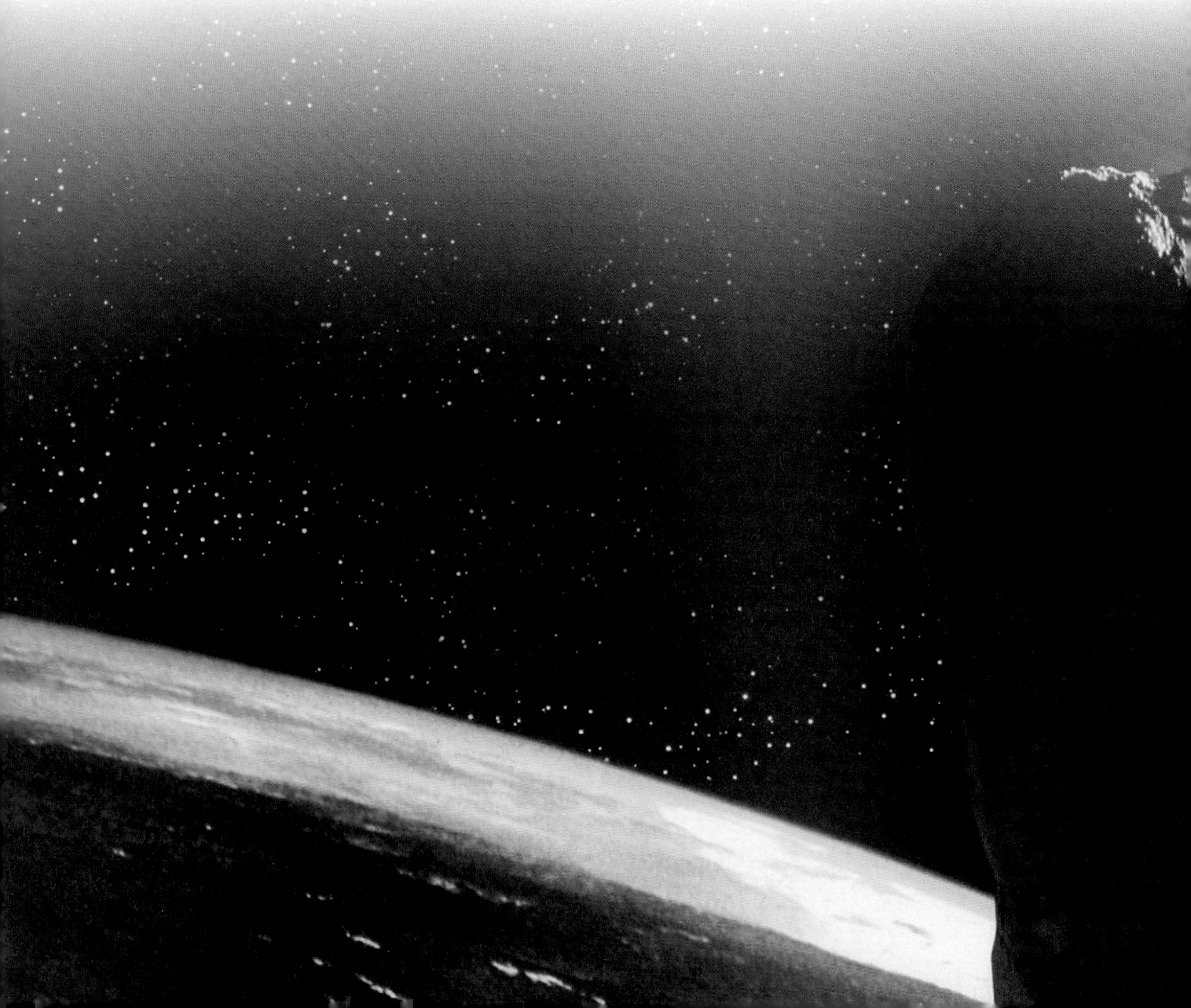

Finally, unlike most of the other hazards described in subsequent chapters, an NEO impact has the potential to produce global-scale devastation. Even those nations that would be fortunate enough to be located far beyond the immediate effects of an impact event are likely to experience physical and social consequences from altered climates, food and resources shortages, and economic hardships.

Near Earth Objects: Concept Map

Complete the following concept map exercise to evaluate your understanding of the interactions between NEOs and earth system components as discussed in this chapter. Examine the following figure and identify at least six potential interactions between earth system components and NEOs.

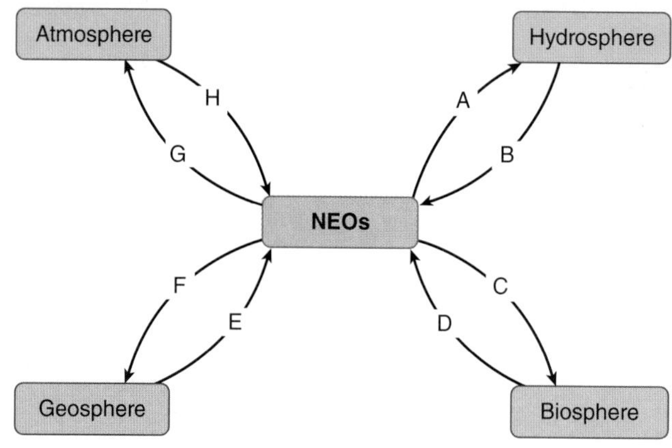

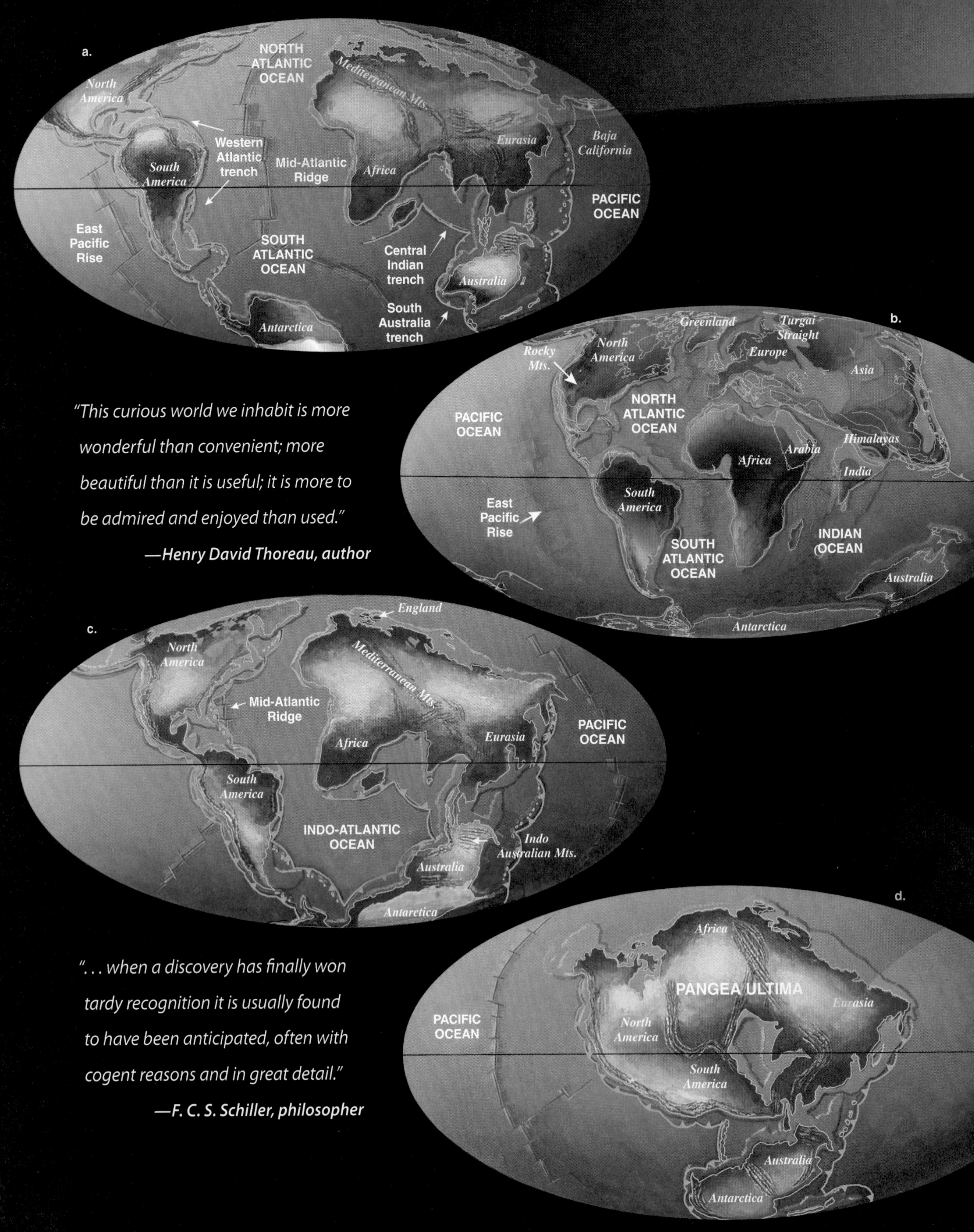

"This curious world we inhabit is more wonderful than convenient; more beautiful than it is useful; it is more to be admired and enjoyed than used."

—Henry David Thoreau, author

"... when a discovery has finally won tardy recognition it is usually found to have been anticipated, often with cogent reasons and in great detail."

—F. C. S. Schiller, philosopher

Chapter 4

Plate Tectonics

Chapter Outline

the big picture

Which of these images best represents what the geography of Earth will look like 50 million years in the future?

See The Big Picture box at the end of this chapter for the full story on this image.

4.1 Science and Santa Claus

Chapter Learning Outcomes

- Students will explain how and why Earth is constantly changing due to plate tectonic processes.
- Students will analyze maps to make predictions about earthquakes, volcanoes, mountains, seafloor topography, and heat flow.
- Students will sketch and label cross sections to illustrate the types of geologic processes and motions that occur in association with plate boundaries.

When I was 6 years old, I asked Santa Claus to bring me a volcano for Christmas. It seemed like an innocent enough request at the time. As I looked around, I couldn't help but notice that the relatively modest landscape around our house could benefit from an interesting landform or two. To my young imagination, no landform was more interesting than a volcano. Of course, I wanted a fully operational volcano, with magma included. Needless to say, I was underwhelmed by the toy trucks and socks that showed up under the tree that year. It was just a few years later that one of my schoolmates informed me that Santa Claus was, in fact, a mythical figure and that I should be taking any requests for volcanoes directly to my parents. As I adapted to this new version of Santa Claus reality, I underwent a personal paradigm shift.

A paradigm is a generally accepted view of how some aspect of the world works. **A paradigm shift occurs when we undergo a fundamental change in our view or rethink our understanding of a basic concept.** We discussed a famous example of a scientific paradigm shift in Chapter 2: the change from the geocentric to the heliocentric explanation of Earth's relationship to the sun. This chapter describes another key paradigm shift in our view of Earth.

Hey, Good Lookin'

What do you look like? Take a moment to think about it. Really! Stop reading and write down a short description of yourself.

Now do the same for Earth.

When you were describing yourself, did you mention your primary features—head, arms, legs—or did you jump ahead to secondary features such as hairstyle, eye color, or height? In terms of only our primary features, we all look pretty much alike. It's the same for the inner planets. They are all spherical, made of rocks, and orbiting the sun. Things get more interesting when we examine the secondary features. On Earth, the surface of the planet is clearly divided into two elevations (Figure 4.1). Most of the land surface lies within a few hundred meters of sea level, while much of the rest of the planet's surface lies at the bottom of the oceans at a depth of several thousand meters.

Looking more closely at the North American continent, we can observe that much of the eastern half of the continent is relatively flat and ringed with sandy beaches. In contrast, the western half is experiencing a geologic party time as characterized by numerous snow-capped mountains, chains of active volcanoes, and stretches of rugged, cliff-lined coast. What is it that makes California so different from the Carolinas? Why does one region experience frequent earthquakes while another does not? Why are the volcanoes of Hawaii relatively safe, while the volcanoes in the Pacific Northwest sometimes explode violently? Why does Earth have a bilevel surface?

Finding the reason why such phenomena occur required earth scientists to abandon accepted views and develop a new theory to explain such observations. That theory holds that Earth's surface can be divided into enormous mobile plates. The slow motion of those plates has opened and closed oceans and has caused continents to migrate across the globe. The most destructive earth hazards, such as volcanic eruptions and most earthquakes, are associated with processes occurring on the edges of these plates.

Dynamic processes that shape the surface topography of other planets and moons share some common elements with plate tectonic processes on Earth (for example, volcanism). However, they differ in many other respects. We need only look to our sister planet Venus to learn that the alternative to plate tectonics would be a cycle of cataclysmic events that would destroy everything on the surface of the planet.

In this chapter, we will discuss the development of the theory of plate tectonics, the most significant paradigm shift in earth science. Like all major paradigm shifts, it did not come easily. Plate tectonics grew out of a previous theory known as continental drift that had been proposed by scientists about 50 years earlier. It took decades of careful observations and data analyses to provide sufficient evidence for plate tectonics to evolve as the dominant paradigm in earth science. The story of the journey from continental drift to plate tectonics provides us with a great opportunity to examine how science is done. In the pages ahead, we will revisit themes from Chapters 1 and 2, and we will set the stage for Chapters 5 and 6, which discuss earthquakes, volcanoes, and mountains. Finally, we will consider how plate tectonics controls long-term global climate patterns.

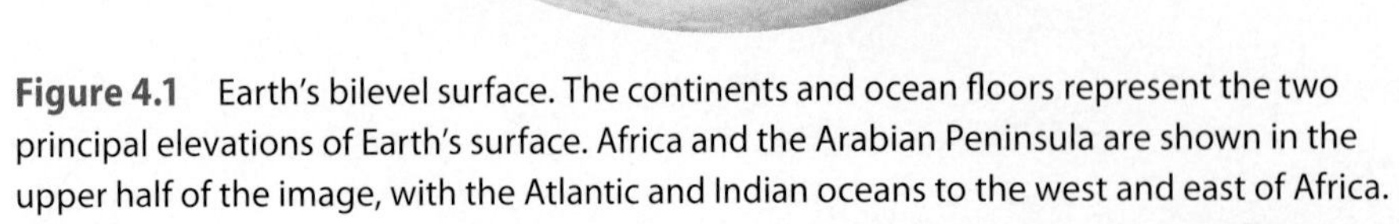

Figure 4.1 Earth's bilevel surface. The continents and ocean floors represent the two principal elevations of Earth's surface. Africa and the Arabian Peninsula are shown in the upper half of the image, with the Atlantic and Indian oceans to the west and east of Africa.

Self-Reflection Survey: Section 4.1

Answer the following questions as a means of uncovering what you already know about paradigm shifts and plate tectonics.

1. Name a state or province where you could
   a. find an active volcano.
   b. have a good chance to experience an earthquake.
   c. climb a mountain of over 3,300 meters (about 10,000 feet) in elevation.
2. Using examples that are not described in this chapter, identify and briefly describe
   a. a personal paradigm shift.
   b. a cultural or social paradigm shift.
   c. a scientific paradigm shift.

4.2 Continental Drift

Learning Objectives

- Discuss at least four observations that Wegener used to support his theory of continental drift.
- Describe some of the ideas that were used to make a case against continental drift.
- Explain how Wegener applied sound scientific principles to develop the theory of continental drift.

Centuries ago, as explorers drew the first maps of the Atlantic Ocean, the geographer Abraham Ortellus noted the matching shapes of the coastlines of Africa, Europe, and the Americas. In the third edition of his *Thesaurus Geographicus,* published in 1596, Ortellus pointed out that the pieces almost seemed capable of fitting together like a jigsaw puzzle. He suggested that America had been "torn away" from Europe and Africa and that the "projecting parts of Europe and Africa" would fit the "recesses" of America. In subsequent years, others commented on the complementary shapes of the coastlines, but these matching patterns were considered little more than a coincidence.

In the seventeenth and early eighteenth centuries, scientists recognized that Earth had a hot interior and that it was losing heat through its surface. Consequently, many believed that Earth was slowly cooling and contracting and that some features on Earth's surface such as mountain belts could be explained as a result of that contraction (similar to the wrinkles on a drying apple). They further proposed that the collapse of some sections of the surface explained the presence of ocean basins. This hypothesis, known as the "contracting Earth model," assumed that features on Earth's surface were approximately the same age. It also suggested that continents and oceans essentially would be fixed in place once they formed and that vertical crustal movements were responsible for differentiating oceans and continents.

Wegener's Theory

Alfred Wegener, a German meteorologist, was the first person to present a well-reasoned alternative explanation for the origin of continents and oceans. In a series of articles beginning in 1912 and later in a book, Wegener proposed his **continental drift theory.** The theory suggested that at some time in Earth's history, **the continents had come together to form a single supercontinent landmass he named Pangaea** (meaning "all lands"). Wegener suggested that Pangaea split apart into individual continents about 200 million years ago and the continents then gradually "drifted" to their current positions (Figure 4.2). The formation and breakup of Pangaea accounts only for the distribution of the continents over the last few hundred million years. Wegener did not assume that the continents had always been present as a single landmass following the formation of Earth.

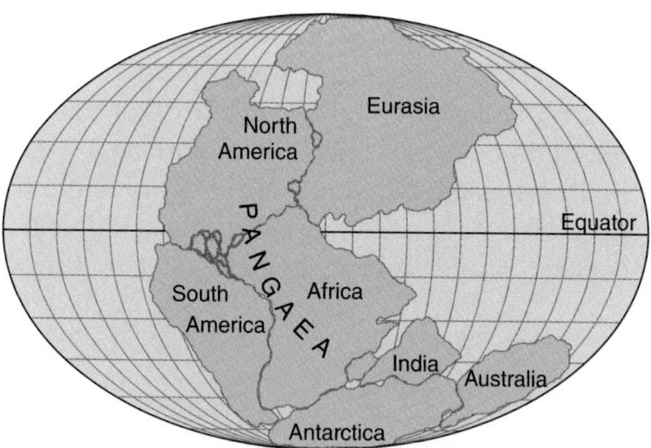

Pangaea, 250 MYA

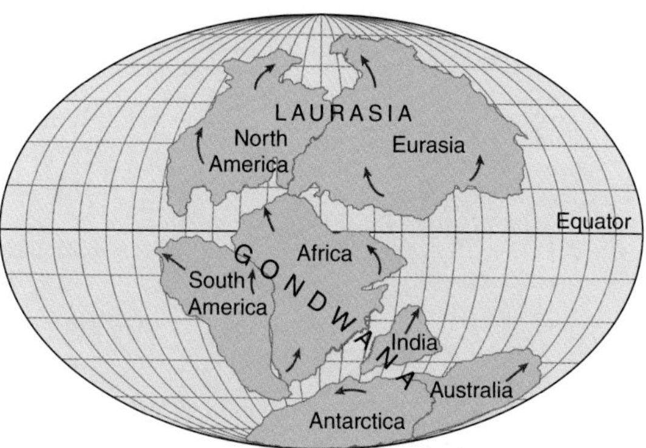

Laurasia and Gondwana, 210 MYA

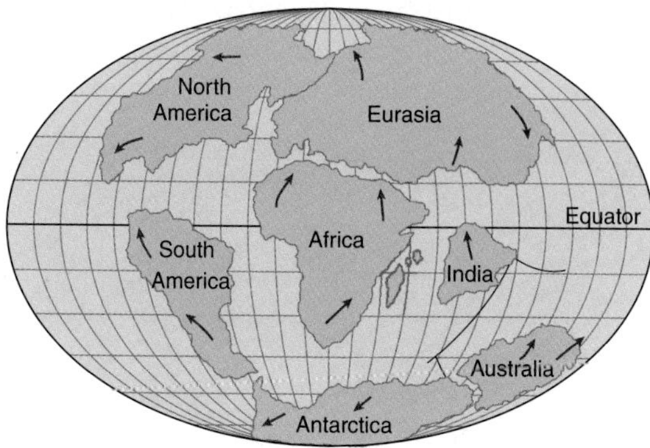

Most modern continents had formed by 65 MYA

Figure 4.2 Continental drift. Wegener hypothesized that the supercontinent Pangaea, made up of a northern group of continents known as Laurasia and a southern group of continents known as Gondwana, broke up about 200 to 250 million years ago (MYA) and that the individual continents eventually drifted to their present positions.

Wegener made the following principal observations:

Matching Features. Several geologic features on different continents would line up if the continents were reassembled as Pangaea. Wegener compared it to matching up the lines of print from a torn newspaper page. His observations included these:

- **Plant and animal fossils** found on separate continents form linked patterns if the continents are reassembled (Figure 4.3a). The animal fossils are those of freshwater amphibians and other similar species that would have died in salty ocean water. Therefore, they could not have crossed an open ocean between continents.
- Highlands in South Africa and Argentina (South America) and mountains in western Europe and eastern North America can be aligned to form **continuous mountain belts in a reassembled Pangaea** (Figure 4.3b).
- **Similar unusual sequences of rocks are found in Brazil, West Africa, and elsewhere** throughout the continents of the Southern Hemisphere. Later researchers were able to identify 2-billion-year-old rocks in Gabon, West Africa, that match up with rocks of the same age on the other side of the Atlantic Ocean in northeast Brazil (Figure 4.3c).

Fit of the Continents. The opposing coastlines of continents often fit together as do the pieces of a jigsaw puzzle. Later analyses would show that an even better fit occurs if we define the shape of a continent by the edge of the shallow water zone known as the *continental shelf* that surrounds each landmass. The classic fit is seen between the southwest coast of Africa and the east coast of South America (Figure 4.3a, c). Similar fits can be observed elsewhere, especially between the continents of the Southern Hemisphere.

Checkpoint 4.1

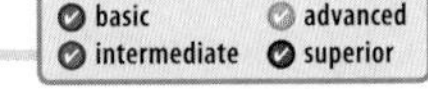

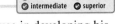

What type of reasoning did Wegener use in developing his continental drift theory? Explain your reasoning.

a) Inductive b) Deductive c) Both

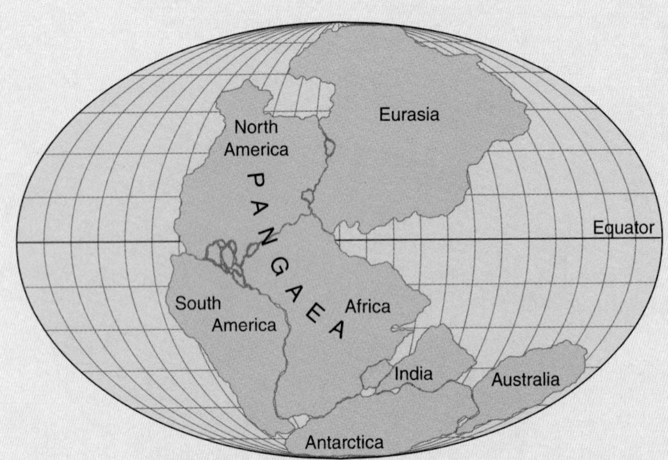

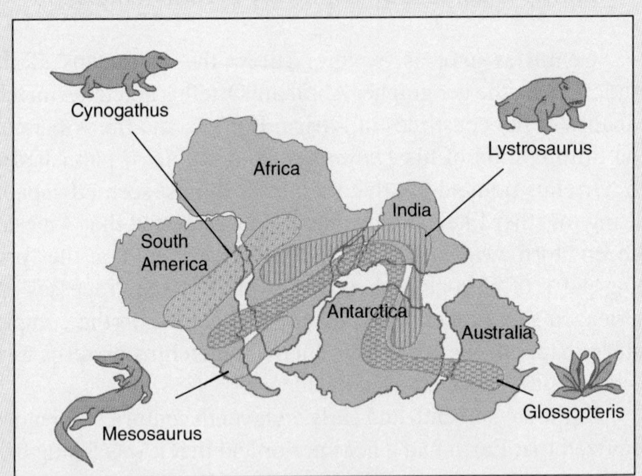

a. Fossil distribution

Figure 4.3 Wegener's evidence for Pangaea. Wegener used the similar shapes of some continental margins, matching rocks, mountains, and fossils, and evidence of glaciation on five continents to infer that the continents must have been joined together as a single supercontinent between 200 and 300 million years ago. It broke apart about 200 million years ago.

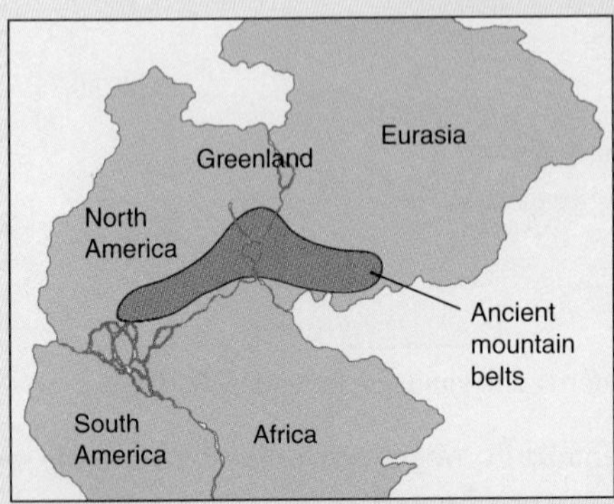

b. Match of mountain belts among North America, Europe, and Greenland

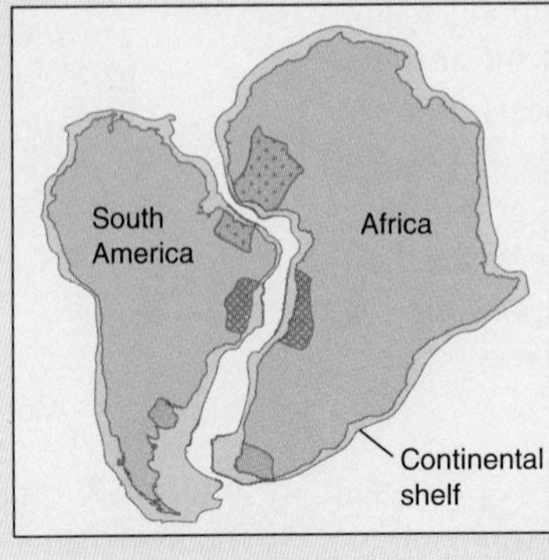

c. Fit of continents, matching rock units

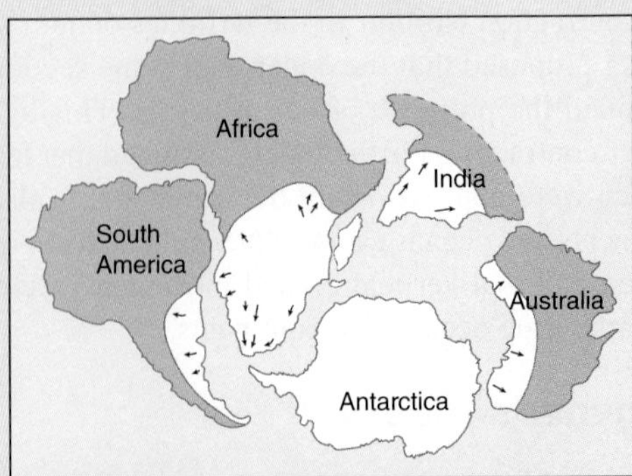

d. Glacial features. Arrows illustrate direction of ice movement determined by striations (grooves) cut by glaciers.

Paleoclimates (Ancient Climates). Scientists had long recognized that ice ages had occurred throughout Earth's history. Wegener documented ancient glacial deposits (formed about 300 million years ago) in India, Australia, South America, and southern Africa that all exhibited evidence of thick ice sheets similar to those found in Antarctica today (Figure 4.3d). This glaciation would have occurred as the continents were being assembled to form Pangaea. Such an event would have required a global ice age if the continents were in their present positions. However, the rock record reveals that coal was forming in tropical swamps in the eastern United States and Eurasia at the same time. Clearly, the rest of the world was not in a deep freeze. Such apparently **widespread glaciation could be best explained if certain continents were located close to the South Pole.** Likewise, when Pangaea was reassembled, it was found that areas in North America with coal deposits were located near the equator.

In hindsight, Wegener's theory looks like an open-and-shut case. It may be difficult to believe that his ideas were not championed by every reasonable earth scientist of his time. But, although some scientists embraced the evidence in favor of continental drift, others rejected it, holding fast to the contracting Earth model. Modern research on how our ideas change shows that old concepts (for example, the contracting Earth theory) are rejected when they cannot explain new observations. New theories (for example, continental drift) are accepted when they offer reasonable explanations for these new observations. For the old idea of the contracting Earth theory to be rejected, scientists first had to (1) recognize an inconsistency between the underlying theory and the new observations, (2) believe it necessary to reconcile the inconsistencies with existing concepts, and (3) be unable to make adjustments to incorporate the new observations into their existing theory.

Unfortunately for Wegener, many scientists of his day did not move past the second step in this process. For example, many fossils of the reptile *Mesosaurus* are found in both southern Africa and South America (Figure 4.3a). The fossil is preserved in rocks deposited in streams, so the reptile lived in freshwater environments. *Mesosaurus* could not have traveled across an ocean between the continents. Wegener proposed that the continents must have been located side by side or else *Mesosaurus* would have had to evolve separately on two continents at the same time, an unlikely explanation. Wegener's opponents were aware of this inconsistency with the contracting Earth model but reconciled it by suggesting that reptiles such as *Mesosaurus* migrated across narrow "land bridges" that originally connected the continents and later collapsed into the oceans. Therefore, they did not reach step 3, did not accept Wegener's observations as valid, and saw no significant challenge to their understanding of earth processes.

The opposition to Wegener's ideas was strongest in North America. Some rejected his ideas simply because, as a meteorologist, he was an outsider to the field of geology. Others were critical of the method he used. Wegener applied a deductive approach to formulate his hypothesis. That is, he came up with the idea that continents had drifted and then searched for confirming evidence based on the work of others. While this is a perfectly acceptable way to approach science, it was not the norm for the day. Most scientists studying the earth at that time used a more inductive, observation-based approach. A number of leading scientists contributed to a volume of articles published after a conference on continental drift in New York in 1928. Almost every article was hostile to the continental drift theory. The opposing scientists disputed the significance of many of the correlations that Wegener had relied on, suggesting that the matching features were not really that similar and that his interpretations of ancient climates were faulty. Another stumbling block was that Wegener was unable to propose an acceptable mechanism to explain *how the continents had moved*. The idea that huge masses of land could move such distances was an enormous intellectual leap. Skeptics could easily argue against continental drift when there was no viable explanation for how it had occurred. Wegener assumed that the continents had pushed through the rocks of the ocean floor, much as a plow cuts through the soil, but he recognized that there was no obvious

Checkpoint 4.2

Which of these lines of evidence were used to support Wegener's continental drift theory? (Select all that apply.)

a) The distribution of fossils
b) Fit of the continents
c) Match of mountain belts
d) Earthquake locations
e) Paleoclimate data

Checkpoint 4.3

basic advanced intermediate superior

Broken Plate Exercise

Examine the image of the broken (ceramic) plate and answer the questions that follow.

If you were asked to glue the pieces of the plate back together, which pieces would you match up with each other?

a) What features did you use to match the different pieces?
b) What equivalent features did Alfred Wegener use to reassemble the continents to form the supercontinent Pangaea?

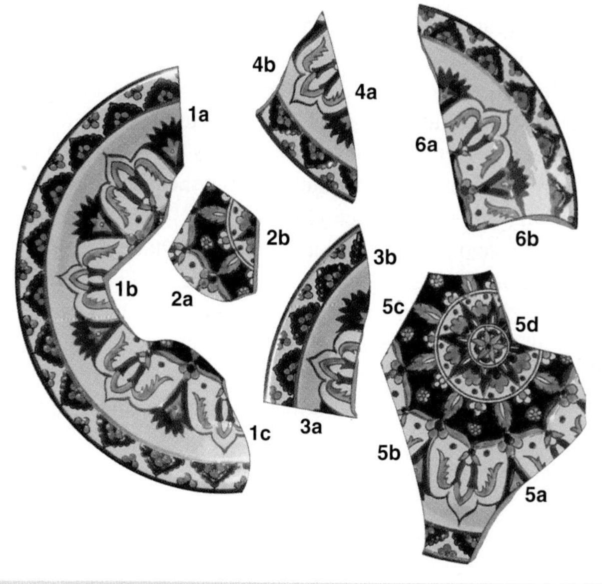

Checkpoint 4.4

basic · advanced · intermediate · superior

Write a paragraph that argues for or against the following statement:

The development of the theory of continental drift was consistent with the characteristics of good science.

explanation. To address this weakness in his theory, Wegener speculated that drift resulted from a combination of obscure forces. However, other scientists were able to calculate that these forces were too insignificant to move large landmasses. Consequently, the continental drift theory, although providing a compelling explanation for the distribution of common features on different continents, would have to wait another 50 years before some of its components would return in the theory of plate tectonics.

4.3 Evidence from the Seafloor

Learning Objectives

- Explain how the age of the seafloor varies relative to the locations of oceanic ridges and trenches.
- Compare and contrast global patterns of heat flow and the locations of volcanoes and earthquakes.
- Discuss how evidence from the exploration of oceans and continental margins supports the concept of seafloor spreading and contradicts the contracting Earth model.
- Predict how magnetic properties of rocks would vary in different scenarios.
- Describe how paleomagnetism can be used to provide support for the concept of seafloor spreading.

The amount of information about Earth's surface greatly increased after World War II, as scientists collected data by using new and improved technologies. Much of this technology was intended for national defense but had the added benefit of revealing new truths about Earth. For example, instruments installed to record vibrations that could pinpoint the locations of nuclear explosions were also used to identify the compositional layers in Earth's interior. The features of the seafloor were mapped very precisely to allow submarines to move around the oceans without being detected. Detailed surveys of Earth's magnetic and gravity fields were conducted to accurately predict the flight of long-range missiles carrying nuclear weapons.

Fifty years ago, it could be argued that scientists knew more about the surface of the moon than they did about the floors of Earth's oceans, but the new wave of research soon changed that perception. It became apparent that the old contracting Earth model could not explain the new observations, and a new group of earth scientists soon argued for a paradigm shift in our understanding of how Earth works. By the 1960s, new data about seafloor topography, the age and magnetic properties of the ocean floor rocks, and the global distribution of heat flow, volcanic activity, and earthquakes could no longer be reconciled with the contracting Earth model. A more dynamic explanation was needed.

Eventually, sufficient evidence had accumulated to provide a mechanism for Wegener's continental drift theory; this new mechanism became known as *seafloor spreading*. We divide the description of the evidence in support of seafloor spreading into two parts broadly recreating the chronological order in which the observations were made. Keep in mind that the original scientists did not have all the data and observations laid out before them that we enjoy today.

To understand how the theory of seafloor spreading developed, first we must describe some basic characteristics of the structure and processes of the ocean floor—namely, its topography, its age, and the distribution of heat flow, earthquakes, and volcanoes.

Seafloor Topography

The shapes of the bottoms of modern community swimming pools are in some ways similar to parts of the seafloor. Many community pools start with a "0 depth" area where toddlers can sit without danger of falling in over their heads. The depth then gradually increases up to the point of a flotation line stretched across the pool, where the water suddenly gets deeper. Likewise, the ocean floor gradually deepens from the coast (0 depth), moving seaward across the shallow continental shelf. The edge of the shelf is at approximately a 150-meter (490-foot) depth.

The seafloor gets much deeper (like the transition to the deep part of a swimming pool) beyond the edge of the continental shelf, gradually flattening out to form level areas known as *abyssal plains* at depths of approximately 4 kilometers (2.5 miles) below sea level (Figure 4.4). As we continue to move away from

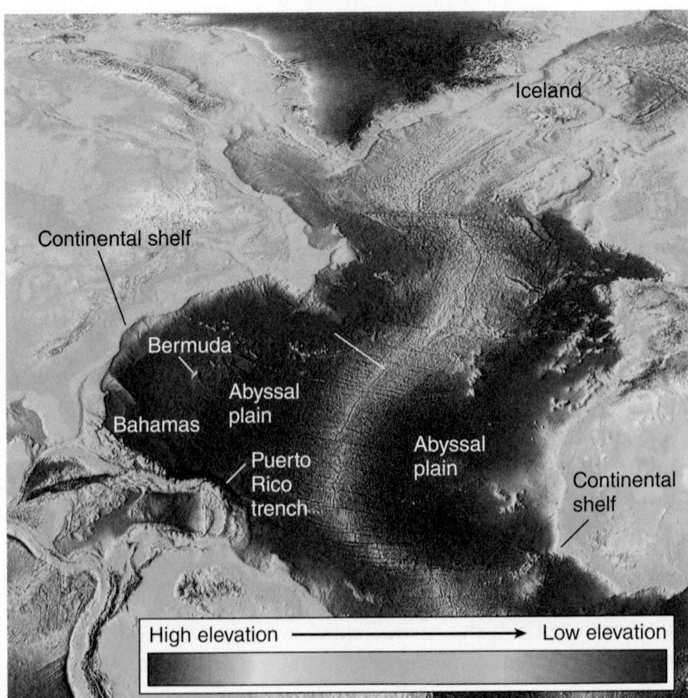

Figure 4.4 Features of the ocean floor. Traveling eastward from the Atlantic coast of North America, we cross a relatively narrow, shallow continental shelf before descending to the abyssal plain. Islands such as Bermuda that are located on the abyssal plain have a volcanic origin, while islands on the continental shelf, such as the Bahamas, are typically formed of sedimentary rocks.

the continents, the abyssal plains give way to an oceanic ridge system. **The oceanic ridge system consists of broad areas of shallower seafloor** that can be followed continuously through the world's major oceans, like a giant zipper that holds the outer layer of Earth in place (Figure 4.5). There are four principal regions: the Mid-Atlantic Ridge, the East Pacific Rise, the Antarctic Ridge, and the Indian Ridge, and many smaller connecting sections. In comparison to the depth at the abyssal plains, the ocean floor along the ridge system is shallower (less than 3 kilometers, or 1.9 miles, deep). Scientists surveying the ocean floor discovered that the **ridge system is a source of volcanic activity.** Volcanism can be observed firsthand where the oceanic ridge comes to the surface in Iceland. But volcanism does not occur exclusively at the oceanic ridges. In fact, the world's oceans are dotted with hundreds of small islands that represent the tops of volcanoes that grew upward from the ocean floor (for example, Bermuda; see Figure 4.4).

One last set of features found in most ocean basins is **narrow, deep oceanic trenches** (Figure 4.6) along the margins of

Checkpoint 4.5

basic intermediate advanced superior

After examining the diagram that follows, determine which pair of locations has oceanic trenches.

a) A, C
b) D, F
c) B, E
d) G, H

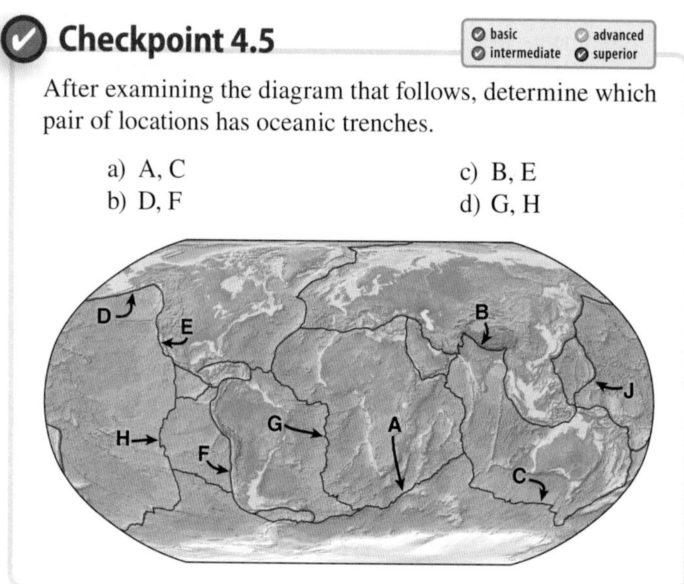

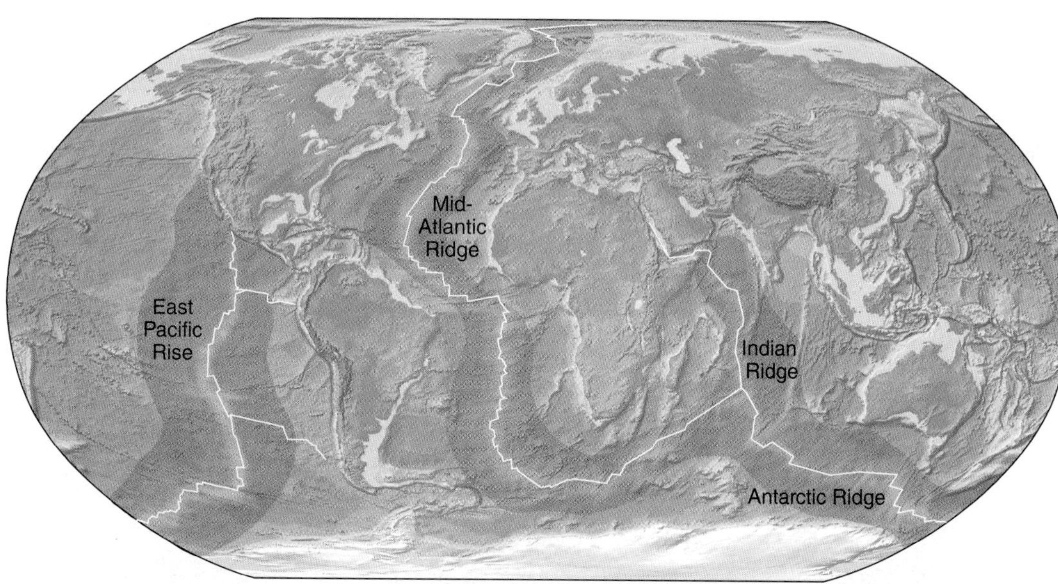

Figure 4.5 Oceanic ridge system. The purple areas indicate locations of the principal oceanic ridges, which cover approximately one-half of the ocean floor.

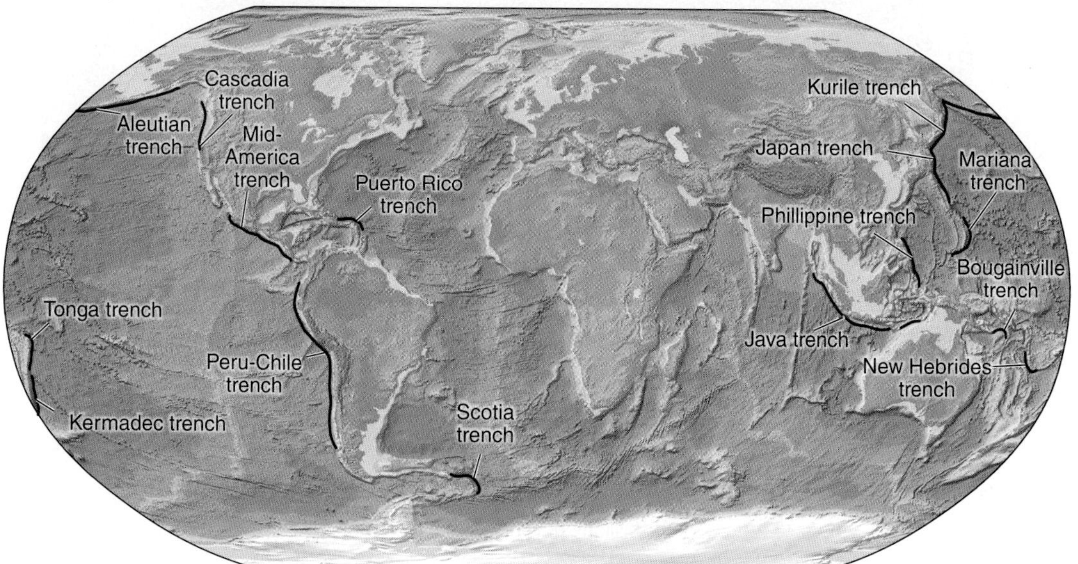

Figure 4.6 Distribution of trenches on the seafloor. Trenches are concentrated around the margins of the Pacific Ocean.

some continents (for example, western South America) or adjacent to volcanic island chains such as the Aleutian Islands, Alaska. These trenches are similar in shape to the valleys we see on land, but ocean trenches are much longer, some may be up to 11 kilometers (6.8 miles) deep, and they form by a completely different process than valleys do.

Notice that the ridges are located near the centers of most ocean basins, while trenches are located almost exclusively near ocean margins (Figures 4.5 and 4.6). An oceanic ridge appears to bisect the entire Atlantic Ocean. Likewise, Africa is surrounded on three sides by oceanic ridges. However, many trench systems are located in the western Pacific and northwest of Australia. In Section 4.4, we will learn that the ridges and trenches play key roles in the process of plate tectonics.

Checkpoint 4.6

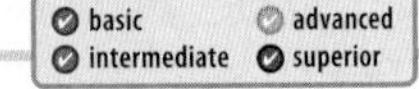

Regarding the relationship between age and the character of the ocean floor, which statement(s) is (are) *true?*

a) Deeper regions of the ocean floor tend to be younger.
b) The Pacific Ocean is larger than the Atlantic Ocean because it contains older oceanic floor.
c) The oldest oceanic crust is present only near trenches.
d) The youngest oceanic crust is near the ridges.

Age of the Ocean Floor

Analysis of rock samples from the ocean floor has revealed that the **oceanic crust is relatively young compared to the age of the continents.** The oldest oceanic rocks are less than 200 million years (Myr) old. (How these ages are actually determined is discussed in Chapter 8.) In contrast, the oldest known rocks on the continental crust are more than 4 billion years (4,000 million years) old. The **age of the ocean floor is consistently youngest at the oceanic ridges and oldest along the edges of oceans** (Figure 4.7). For example, oceanic rocks along the North American and African coastlines are approximately 180 million years old, whereas rocks adjacent to the ridge are less than 1 million years old. The age of the oceanic crust increases symmetrically moving away from the oceanic ridge system (Figure 4.7). From observation of the relatively young ocean floor, scientists suggested that Earth has a recycling system that constantly creates new oceanic rocks at oceanic ridges and destroys old oceanic floor elsewhere.

Heat Flow, Volcanoes, and Earthquakes

Scientists had long been aware that heat was escaping from Earth's interior; this was a key aspect of the contracting Earth model. Because they assumed a uniformly contracting Earth, early predictions were that heat flow would be relatively uniform around the world. However, scientists were surprised when surveys of the ocean floor revealed that heat flow was greatest along oceanic ridge systems. The ridge system was recognized as a place where **hot, melted rock, called magma,** was forcing its way into the oceanic crust with some erupting onto the seafloor (Figure 4.8a). With the

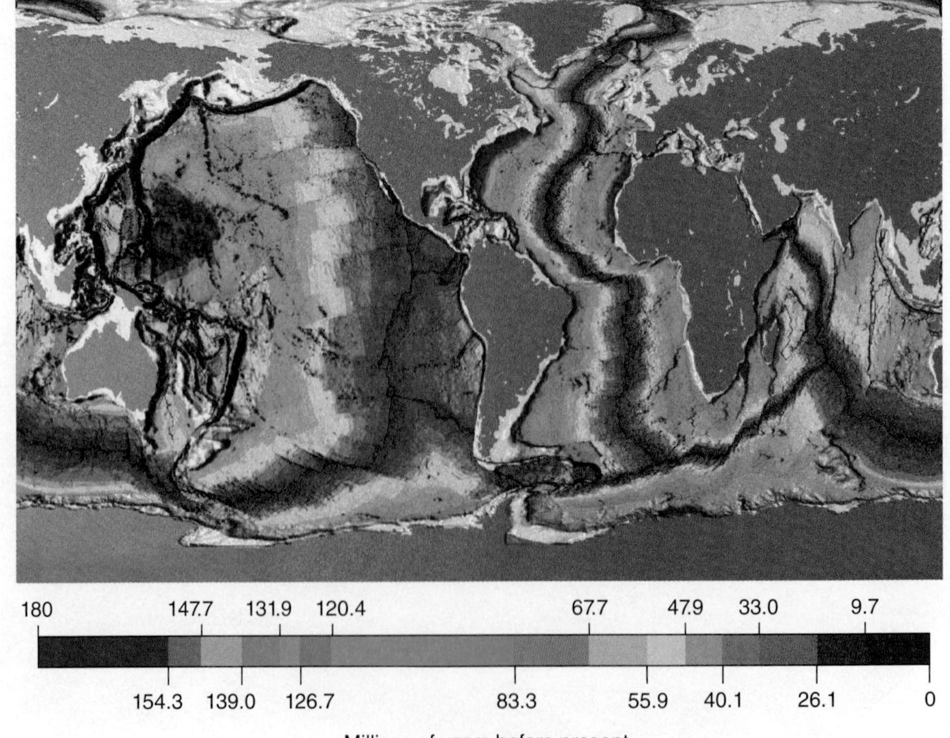

Figure 4.7 Age of oceanic crust. Ages range from young (less than 1 million years old) along the oceanic ridges (red) to old (180 million years old, blue) along the ocean margins (such as the northwest Pacific Ocean). The difference in oldest ages in the northern and southern Atlantic oceans has been interpreted to mean that the northern Atlantic Ocean began to form before the southern Atlantic Ocean.

exception of volcanic regions where magma comes to the surface, heat flow at any point on Earth's surface is too low to be noticeable. The region around the Pacific Ocean is called the Ring of Fire because of the many active volcanoes found there (Figure 4.8b). Most active volcanoes above sea level are located adjacent to oceanic trenches. Volcanoes are almost never present along continental margins that are not near a trench (compare Figures 4.6 and 4.8b).

When the locations of earthquakes were plotted on world maps, they were seen to occur more frequently in zones on or adjacent to oceanic ridges and trenches. The largest and deepest earthquakes were present in proximity to oceanic trenches in the Pacific and Indian oceans. Near ocean trenches, earthquakes occur from near the Earth's surface down to depths of 750 kilometers (465 miles; Figure 4.8c). The earthquakes are shallowest near the trench and get deeper on one side of the trench, defining an inclined zone called a *Wadati-Benioff zone,* after the researchers who initially recognized this pattern (Figure 4.9).

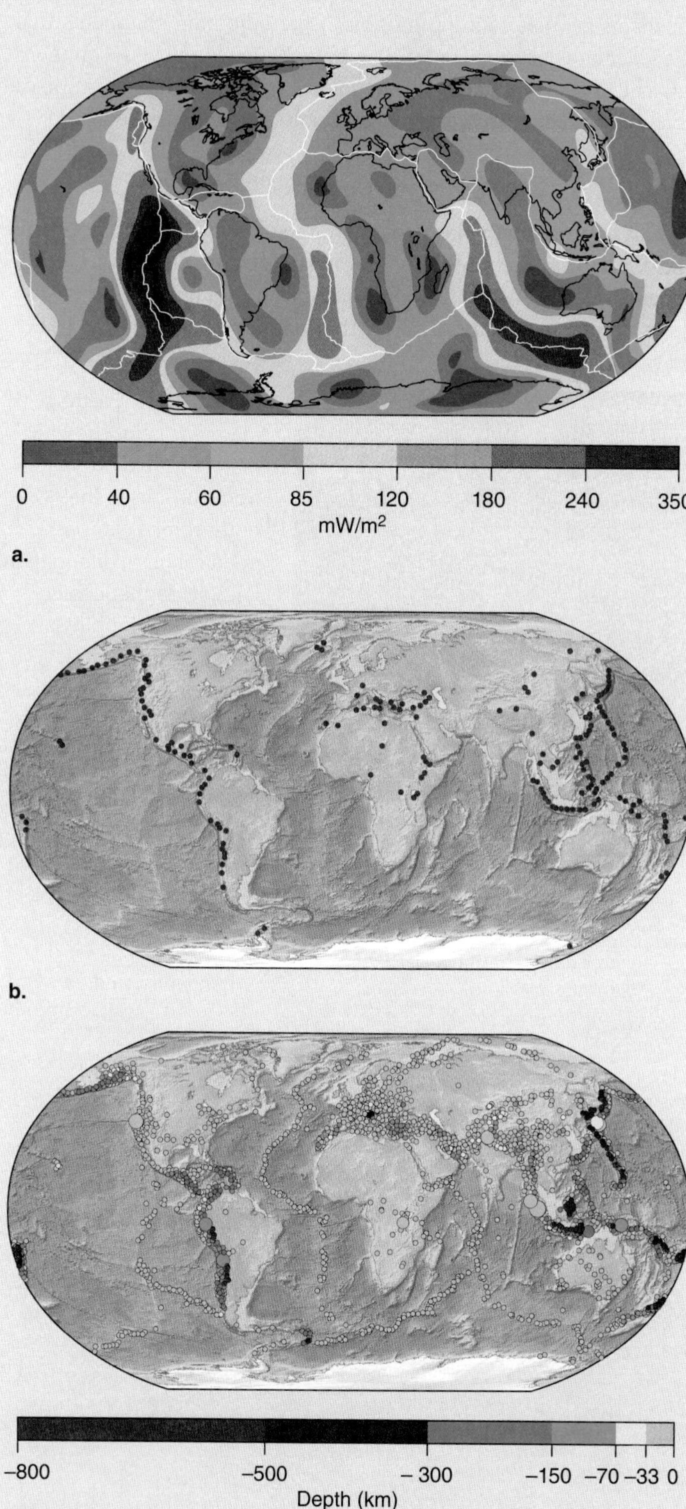

Figure 4.8 Distribution of global heat flow, volcanoes, and earthquakes. **a.** Global heat flow measured in milliwatts per square meter (mW/m^2). **b.** Global distribution of active volcanoes above sea level. **c.** Global distribution of earthquakes for 2005. Large circles indicate the locations of the largest earthquakes. Color represents depth.

Checkpoint 4.7

basic | intermediate | advanced | superior

Compare and contrast patterns of (1) topography of the ocean floor, (2) age of the ocean floor, (3) heat flow, (4) volcanic activity, and (5) earthquake activity for oceanic ridges and oceanic trenches, using Figures 4.5, 4.6, 4.7, and 4.8.

Checkpoint 4.8

basic | intermediate | advanced | superior

Explain how the following patterns can be interpreted to contradict the contracting Earth model: (1) topography of the ocean floor, (2) age of the ocean floor, (3) heat flow, (4) volcanic activity, and (5) earthquake activity in the ocean floor.

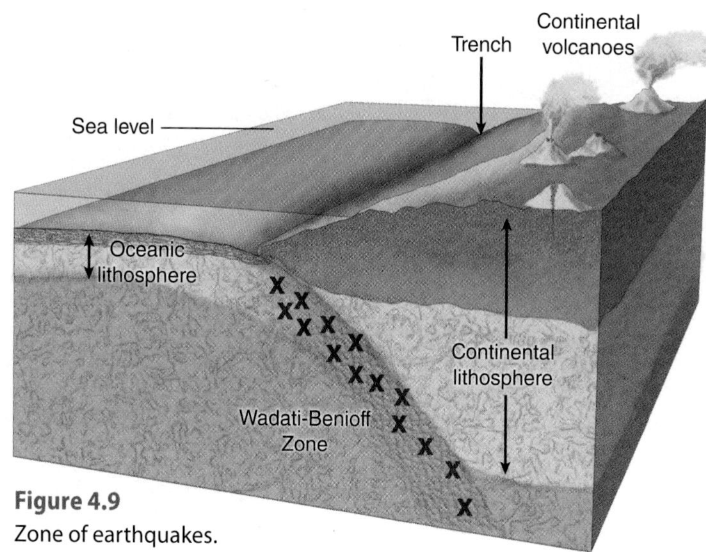

Figure 4.9
Zone of earthquakes.
Seismologists Kiyoo Wadati and Hugo Benioff discovered that earthquakes become progressively deeper as one moves away from oceanic trenches. The distribution of earthquakes defines the Wadati-Benioff zone. Red X's represent locations of earthquakes. Diagram is not to scale.

Seafloor Spreading Theory

Because the observations just described contradicted predictions of the contracting Earth model, scientists needed a better theory. Along came Harry Hess, a Captain of a US transport ship in World War II. Hess used the new sonar technology onboard his ship to collect seafloor elevation data as he traveled across the Pacific Ocean. He used those data, combined with more collected after the war, to suggest that the seafloor moved away from the oceanic ridges. Hess published his work in 1962, and it soon became one of the most referenced geology publications works ever written. The **seafloor spreading theory suggested that new oceanic lithosphere (crust and upper mantle) is being *continuously* formed along the ridge system by magma rising from below, and as this occurs, the existing rocks of the seafloor move away from the ridge** (Figure 4.10). The theory of seafloor spreading led scientists to conclude that the migration of hot magma from below the ridge heats the overlying seafloor, causing it to expand to produce the higher elevations of the oceanic ridge. The ocean floor was interpreted to act as a conveyor belt, gradually moving away from the ridge and creating a gap that was continuously filled with new magma from below.

Because new material is constantly being generated at ridges, old material must be destroyed somewhere else, or Earth would expand. Since Earth is not expanding, there had to be places on Earth's surface where older ocean floor was destroyed. The fact that the deepest earthquakes and some of the older ocean floor are adjacent to trenches led scientists to hypothesize that the **ocean floor descends into the mantle at ocean trenches.**

The concept of seafloor spreading provided a potential mechanism to explain the motion of the adjoining continents that Wegener had described decades earlier in connection with continental drift. A few years later, new evidence related to magnetization of ocean floor rocks strengthened the seafloor spreading theory.

Paleomagnetism

You have probably picked up a compass at some time in your life and know that the needle points north. A compass points north because the tips of the compass needles are magnetized, causing them to line up parallel with Earth's magnetic field (Figure 4.11).

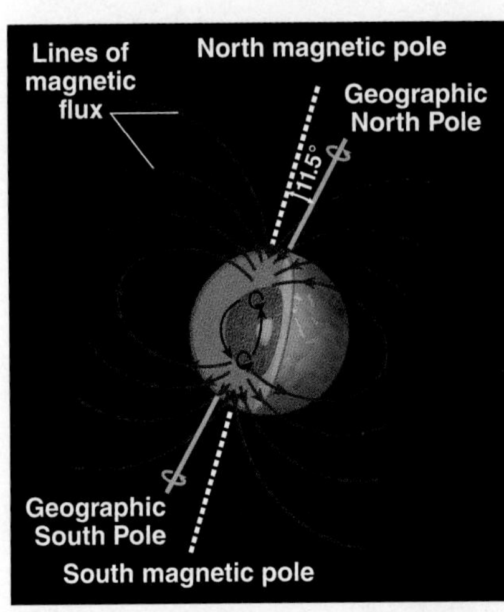

Figure 4.11 Earth's magnetic field. The magnetic field originates from currents in the outer core. A compass needle lines up along the lines of magnetic force (flux) so that it points toward the magnetic poles. The geographic and magnetic poles are not in the same location.

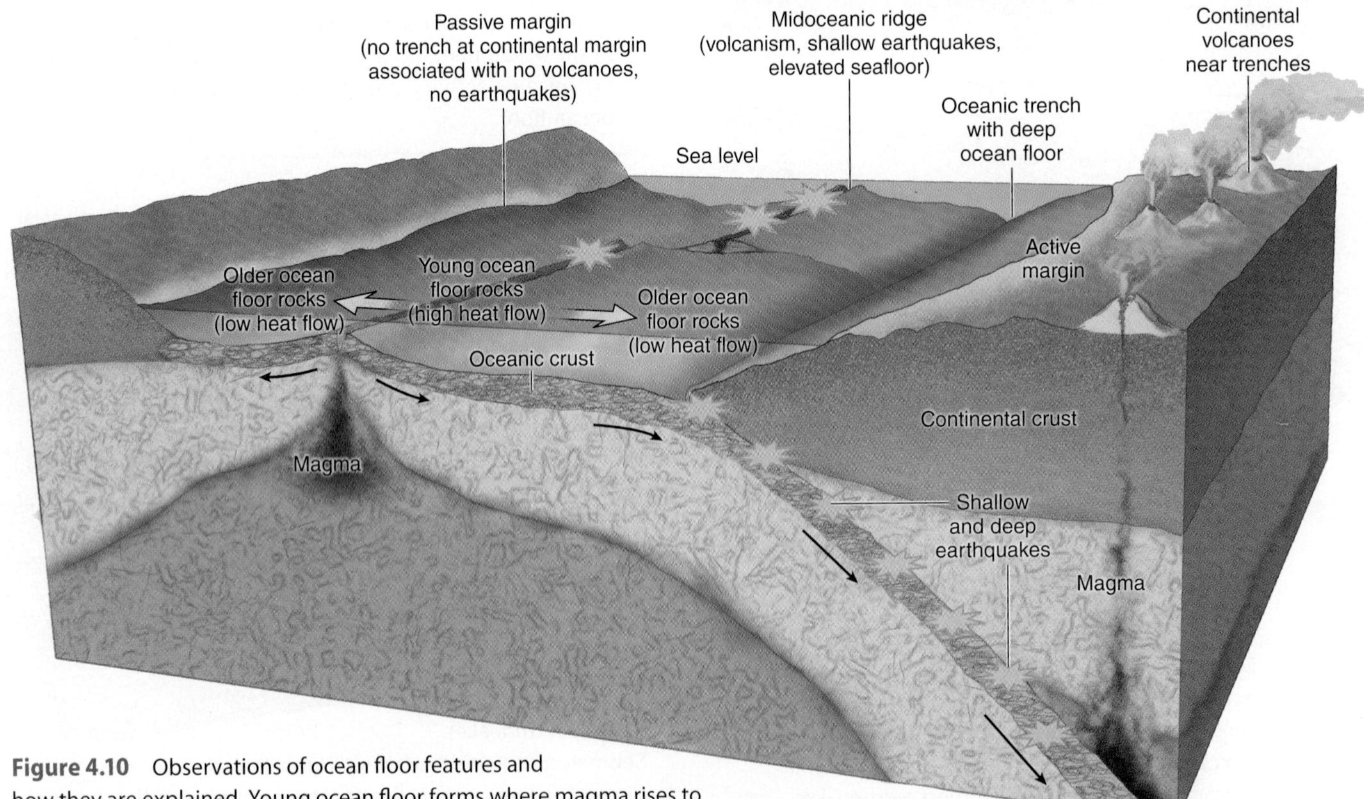

Figure 4.10 Observations of ocean floor features and how they are explained. Young ocean floor forms where magma rises to the surface at the ocean ridge, causing higher heat flow there. This process also pushes existing seafloor away from the ridge. Consequently, seafloor rocks become progressively older with increasing distance from the ridge. Volcanism and earthquakes are common along the margins of oceans where trenches are found. Seafloor rocks sink into the mantle adjacent to trenches, producing deep earthquakes and magma for volcanoes.

You will recall from Chapter 2 that Earth has three compositional layers—crust, mantle, and core—and that the molten outer core is the source for Earth's magnetic field. Planetary magnetic fields require three conditions:

1. A large reservoir of fluid that can conduct electricity—in this case, the iron-rich outer core.
2. Energy to move the fluid. Heat from radioactive decay and planetary formation generates vertical currents in the outer core.
3. Rotation to mix the fluids. The relatively rapid rotation of Earth creates spiraling currents to sustain flow patterns.

The magnetic field has negative (−) and positive (+) poles that are located near the geographic (North and South) poles (Figure 4.11).

Declination and Inclination of the Magnetic Field. The shape of **Earth's magnetic field varies with location and can be defined by its declination (or direction) and inclination** (Figure 4.11). *Declination* simply defines the angle between the direction the north arrow points on a compass at some location and the direction to the geographic North Pole from that location. If a compass needle could rotate in three dimensions, the needle's *inclination* would range from vertical at the magnetic poles to horizontal at the (magnetic) equator. The magnetic field is inclined downward in the Northern Hemisphere and upward in the Southern Hemisphere, and inclination gradually increases between the equator and the poles. In other words, the needle would point straight down at the magnetic north pole and straight up at the magnetic south pole.

Suppose a strong wind blew down some trees in a forest. Most of the trees would be aligned in the direction of the wind. Even without knowing anything about the original wind, you could figure out the direction the wind had blown by looking at the fallen trees. Geologists use a similar technique to determine where rocks formed relative to Earth's magnetic poles. The inclination of the magnetic field can be used as an estimate of a past (or "paleo") latitude. When volcanic rocks form on Earth's surface, their magnetic minerals act as miniature compasses. The **magnetic fields of atoms in magnetic minerals, especially those with high iron content, align parallel to Earth's magnetic field.** The inclinations of atoms in magnetized minerals in ancient lava flows can therefore be used to determine the original latitude of the cooling magma when the rock formed. **The study of the magnetic field preserved in these rocks is called *paleomagnetism*.**

Magnetic Field Reversals. Studies of continental rocks that form from magma have revealed that the direction of Earth's magnetic field sometimes points toward the geographic north and other times points toward the geographic south. Scientists interpret this contrast to mean that **the positive and negative magnetic poles have switched positions many times in the past** (Figure 4.12). The magnetic poles slowly dance around Earth's surface near the geographic poles and, from time to time, will completely reverse their positions. During these reversals, the poles flip from positive to negative and, thousands to millions of years later, change back again. Earth's magnetic field exhibits "normal" polarity when the negative magnetic pole and the geographic North Pole are adjacent, as they are today. The magnetic field would be considered to have "reverse" polarity if the magnetic polarities were opposite to the present-day pole positions.

Although scientists are not certain of the exact reason for these reversals of polarity, paleomagnetic data from different-aged rocks suggest that magnetic reversals have occurred frequently throughout the history of the planet. Each reversal lasts for an average of about 250,000 years, but some last much longer (37 million years) and others may last for just a few tens of thousands of years. Geophysicists who study the magnetic field have discovered that it can take as little as a few thousand years to change the field's polarity. The most recent reversal occurred 780,000 years ago, and the magnetic field has lost strength in the last few centuries, leading some to wonder if Earth is about to undergo another reversal of polarity. During a change in polarity, Earth may have a weaker than normal magnetic field and could even have multiple magnetic poles. Navigation that relied on compasses would be difficult, and

Checkpoint 4.9

basic · intermediate · advanced · superior

Which location has the greatest magnetic inclination value (that is, closest to vertical)?

a) Anchorage, Alaska
b) New York, New York
c) Miami, Florida

Checkpoint 4.10

basic · intermediate · advanced · superior

The magnetic north pole has migrated northward over the last century, so how did magnetic inclination readings at Chicago change between 1900 and 2000?

a) Inclination increased.
b) Inclination decreased.
c) Inclination remained constant.

a. Normal polarity

b. Reversed polarity

Figure 4.12 Polarity of Earth's magnetic field. **a.** The inclination of the magnetic field varies with latitude. The field is horizontal at the magnetic equator, steeper at high latitudes, and vertical at the magnetic poles. The magnetic field is inclined downward in the Northern Hemisphere and upward (away from Earth's surface) in the Southern Hemisphere. **b.** The magnetic field has reversed polarity in the past, meaning that the positive and negative polarities have switched positions.

radio communications would be adversely affected. However, such changes occur on geologic timescales and are at least several centuries or millennia in the future.

How Paleomagnetism Supports the Seafloor Spreading Theory. We can use paleomagnetism, this magnetic record in rocks, to deduce how the ocean floor has changed over time. Marine surveys of the ocean floor can measure the intensity (strength) of Earth's magnetic field, and the intensity values reveal the polarity of the rocks. The strength of Earth's magnetic field at any particular location will be slightly increased if magnetic minerals "frozen" in the rocks are aligned with Earth's present-day magnetic field or will be slightly weaker if they are aligned with a reverse-polarity field.

When scientists analyzed very precise magnetic field data recorded over the oceans, the data revealed "stripes" of high and low magnetic intensity corresponding to areas of normal and reverse polarity. These stripes are oriented parallel to adjacent oceanic ridges (Figure 4.13). **Ocean floor rocks reveal a symmetrical pattern of magnetic polarity reversals on either side of oceanic ridges.** For example, a sequence of magnetic reversals (a switch from normal to reverse polarity and back) in the western

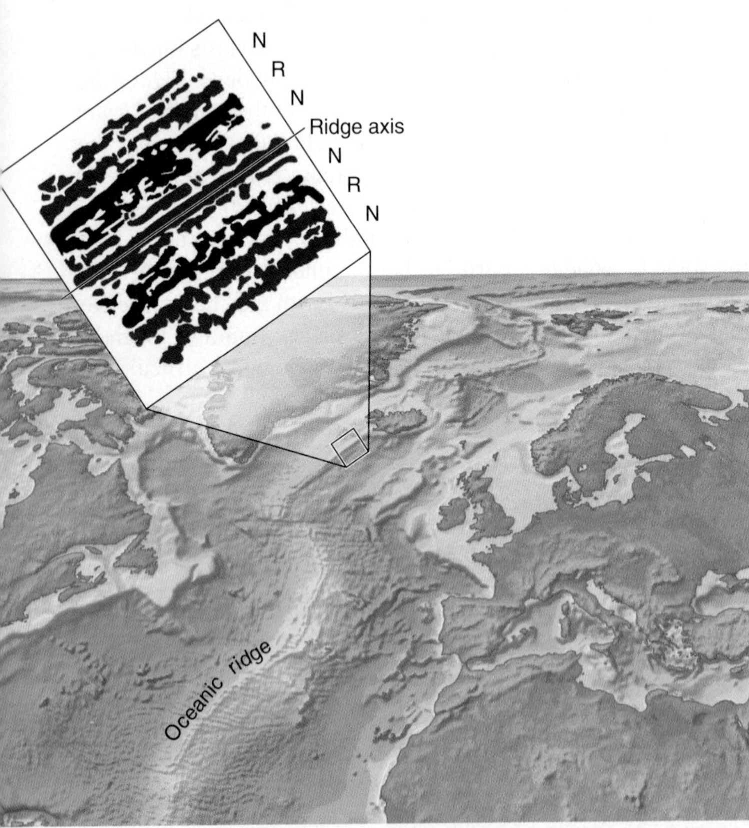

Figure 4.13 The polarity of Earth's magnetic field along the Mid-Atlantic Ridge, south of Iceland. Dark blue stripes indicate normal polarity (N) rocks; purple areas indicate reverse polarity (R). Note symmetrical patterns on either side of the dark band representing the center (axis) of the ridge.

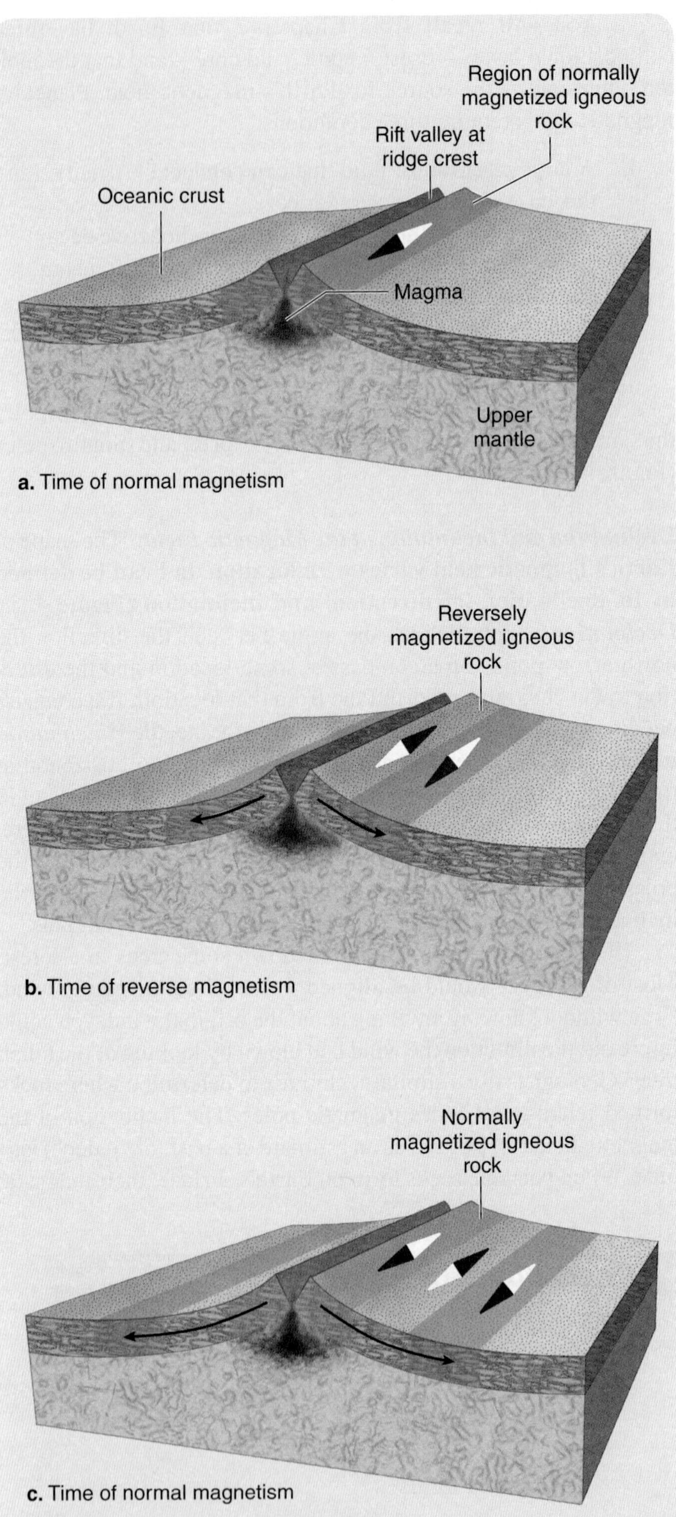

Figure 4.14 Evidence of seafloor spreading. **a.** Rocks with normal polarity (blue) form when the compass points to the magnetic north pole (as it does today). **b.** Conditions of reverse polarity (tan) represent periods when the compass arrow points to the magnetic south pole. **c.** Alternation of normal and reverse magnetic fields produces a striped pattern of magnetism in the ocean floor rocks (blue, tan, green). This evidence supports the seafloor spreading theory.

Checkpoint 4.11

basic · advanced · intermediate · superior

Inclination is determined for three lava flows preserved in a cliff as shown in the following image. What happened to the continent on which these rocks were formed? (Assume normal polarity throughout. Upper layers are youngest; lower layers are oldest.) The rocks in the cliff moved toward the ______ in the ______ Hemisphere.

a) Equator; Southern
b) Equator; Northern
c) Pole; Southern
d) Pole; Northern

Sedimentary rocks
Lava flows

Checkpoint 4.12

basic · advanced · intermediate · superior

Which magnetic property was more important in providing support for the seafloor spreading theory? Explain the reasoning behind your answer.

a) Magnetic inclination; the inclination of Earth's magnetic field varies with location.
b) Magnetic polarity; the north and south magnetic poles have switched positions throughout Earth's history.

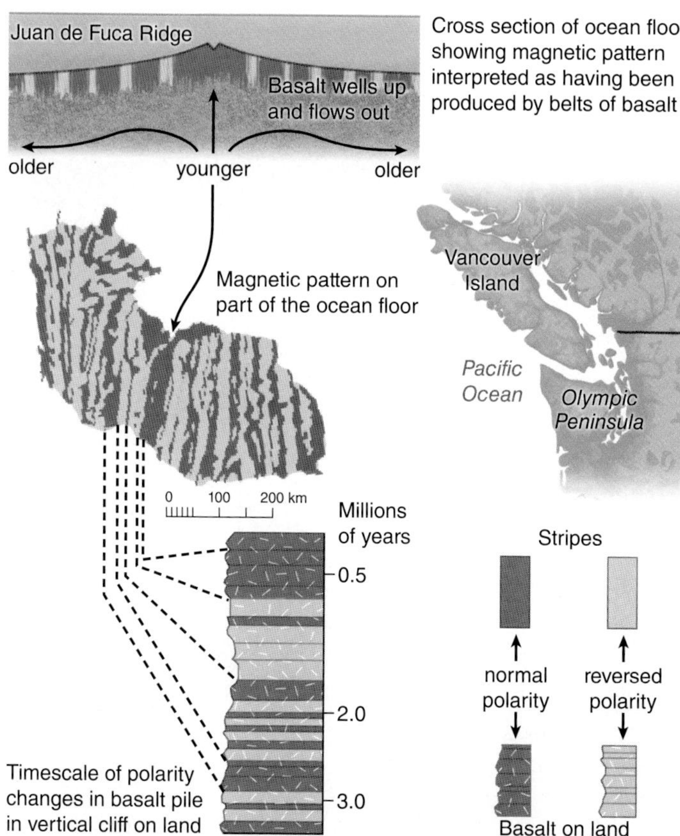

Figure 4.15 The polarity of the oceanic crust adjacent to the oceanic ridge separating the Juan de Fuca and Pacific plates, west of Washington. Dark gray pattern indicates normal polarity, and light gray represents reverse polarity. Note the symmetrical pattern of "stripes" on the ocean floor on either side of the oceanic ridge. These magnetic patterns can be matched to basalt on land to determine when the rocks in each stripe were formed. We can then use the distance of the stripes from the ridge to estimate the rate at which the plates are moving.

Atlantic, offshore from the Carolinas, can be matched with a similar sequence in the eastern Atlantic off the coast of West Africa. These patterns are interpreted to suggest that new crust was divided in half as it formed along the oceanic ridges, and each half moved in opposite directions away from the ridge (Figure 4.14). Scientists found the same sequences of reversals in each of the major oceans. The age and duration of magnetic reversals were initially determined by analyzing lava flows on the continents (Figure 4.15). The seafloor spreading theory predicted that ocean floor rocks would become progressively older with increasing distance from oceanic ridges. This was confirmed when the ages of some continental reversals were compared to those of oceanic reversals. Thus, the record of magnetic polarity in the rocks of the ocean floor provided unequivocal support for the concept of seafloor spreading.

4.4 Plate Tectonics

Learning Objectives

- Explain what is meant by the term *plate tectonics.*
- Identify three processes that cause rocks to melt.
- Use a blank map of the world to identify key features associated with plate tectonics.
- Explain how scientists can use data from ocean floor features to estimate the rate of plate motion.
- Draw and label a cross section through a tectonic plate to show two types of crust, the lithosphere and asthenosphere.
- Explain why Earth is the only terrestrial planet with active plate tectonics.

In Sections 4.2 and 4.3, we outlined the concepts of continental drift and seafloor spreading. The names of these concepts imply that both the continents and the ocean floors are in motion, but give little idea about how their motions are related. Our purpose in this section is to draw together all the threads of evidence, to show how the theory of plate tectonics explains many different land- and ocean-based observations. But first we must define the features and processes involved.

Key Layers and Processes

Lithosphere and Asthenosphere. Recall from Chapter 2 that **Earth's interior can be separated into the core (inner and outer),** the **mantle,** and the **crust.** The temperature and pressure for the first few hundred kilometers below Earth's surface are just right to cause most minerals to remain solid while allowing a small fraction to melt. As a consequence, two additional layers can be distinguished: the lithosphere and the asthenosphere (Figure 4.16).

The *lithosphere* is a rigid outer layer composed of the crust and the uppermost mantle. We should consider the lithosphere as the outermost layer of Earth, encompassing both the rocks of the continents (continental lithosphere) and the ocean floor (oceanic lithosphere). As we will see, it is events in the lithosphere that shape the planet's surface features and that generate earthquakes and volcanoes in a predictable pattern.

The *asthenosphere* is a layer beneath the lithosphere. It is about 100–200 kilometers (62–124 miles) thick. Although the asthenosphere is composed almost completely of solid rocks, it flows slowly because of the presence of just a small proportion (about 1 percent) of minerals that experience melting.

Melting of Rocks. Volcanic activity due to the melting of rocks in the lithosphere and asthenosphere is a critical process in plate tectonics. Thus, before going further, we must describe what causes rocks to melt. We know that ice melts to liquid water as its temperature rises above freezing. The higher temperature increases vibration between adjacent atoms or molecules and provides energy to break the bonds that hold the water molecules in the ice together. In the same way, certain processes break the bonds between the atoms and molecules that make up the minerals in rocks. Pressures within Earth's crust and mantle act to reduce vibrations in the atomic structure of minerals and thus to prevent melting. Three basic changes can produce melting of rocks: increasing temperature, decreasing pressure (on hot rocks), and chemical reactions following the addition of water (Figure 4.17).

Note that it is not necessary to completely melt a rock. The three processes break some of the bonds holding atoms or molecules together and lead to *partial melting* of some minerals and the production of magma. All three conditions that result in melting occur as a consequence of plate tectonics. (We will revisit these mechanisms in Chapter 6.)

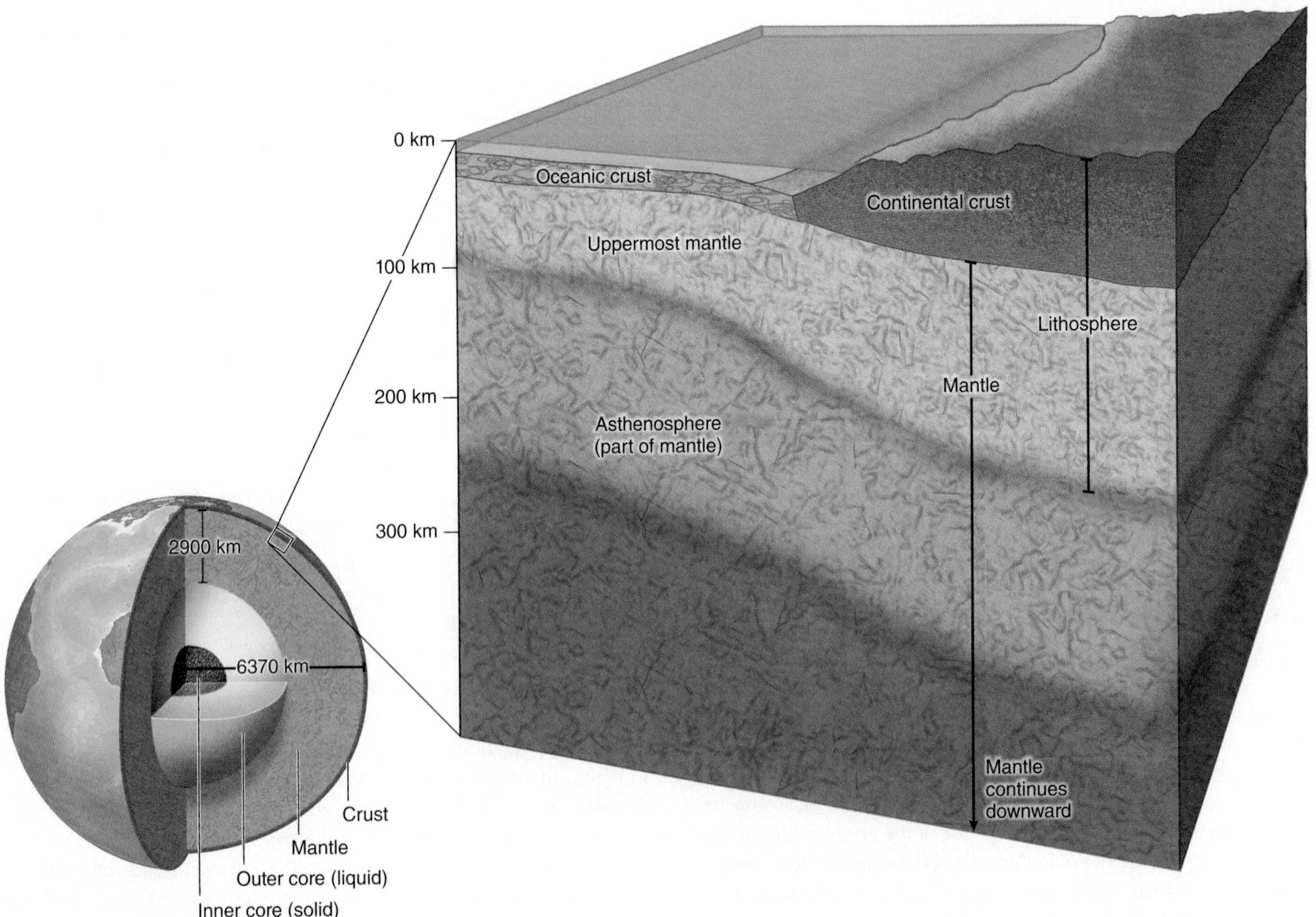

Figure 4.16 Lithosphere and asthenosphere. The outermost part of Earth is divided into two layers, the lithosphere (crust and rigid uppermost mantle) and the asthenosphere (plastic layer in the upper mantle).

The Process of Plate Tectonics

Earth's lithosphere is broken into a series of rigid, mobile **tectonic plates.** There are eight major plates (North American, South American, Pacific, Nazca, Eurasian, African, Antarctic, and Indian-Australian) and several smaller plates (for example, Caribbean, Arabian, Juan de Fuca; see Chapter Snapshot). Reinforcing the tentative nature of science, some scientists argue the Indian-Australian plate is actually two different plates with a boundary starting to form beneath the seafloor of the Indian Ocean. Discussions in this chapter use eight major plates since the location and characteristics of that boundary are not clearly defined. Plates are typically composed of both continental and oceanic lithosphere. For example, the African plate contains the continent of Africa as well as the southeastern Atlantic Ocean and the western Indian Ocean. Oceanic ridges and trenches represent the principal boundaries of the tectonic plates (compare Figures 4.5 and 4.6 and Chapter Snapshot). **The theory of plate tectonics states that interactions of the plates along these boundaries account for the formation of new lithosphere, mountains, earthquakes, and volcanoes and contribute to the gradual movement of continents and the opening of oceans.**

Formation of New Lithosphere. The oceanic lithosphere is thinnest below oceanic ridges, bringing the underlying asthenosphere closer to the surface. Melting occurs preferentially in the asthenosphere at this location as the thinner lithosphere results in a decrease in pressures. Consequently, partial melting below oceanic ridges is termed **decompression melting** (Figure 4.17b). Melting produces magma that is less dense than surrounding material and rises to the surface (Figure 4.18). The magma cools and solidifies

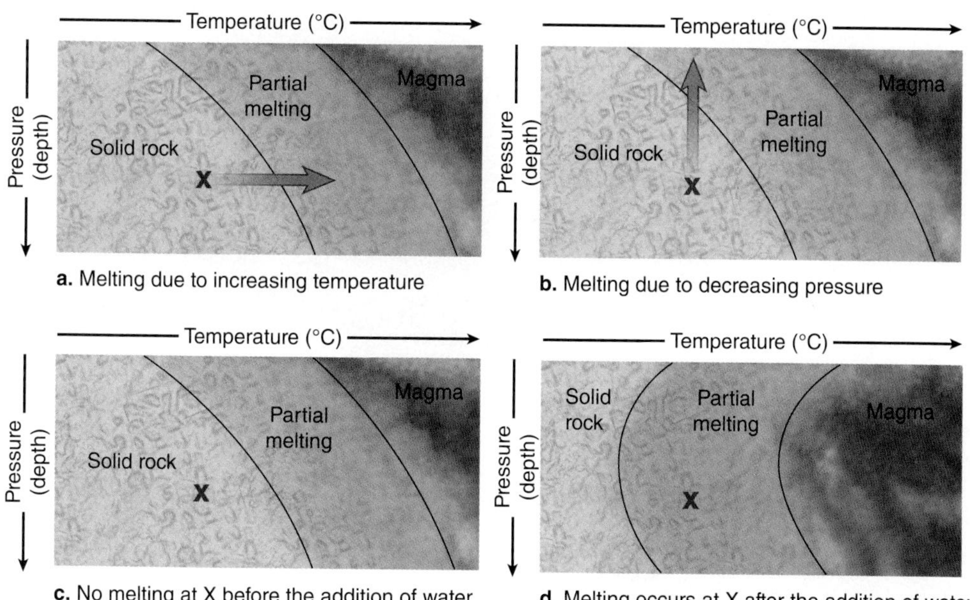

Figure 4.17 Why rocks melt. A rock (X) located in the crust or mantle may melt because of **a.** an increase in temperature, **b.** a decrease in pressure (decompression melting), or **c.** and **d.** the addition of water. The latter two processes are responsible for most melting. Increasing temperature may occur if a rock is buried more deeply or if it lies adjacent to a heat source such as a magma chamber. Pressure decreases when rocks rise through the mantle or crust. Water may be added to rocks by chemical reactions or seawater infiltration.

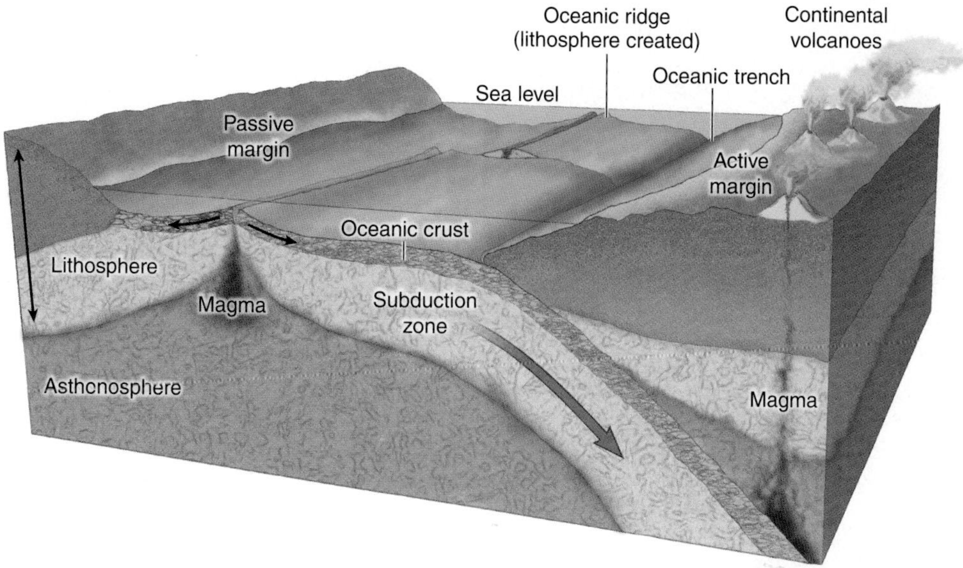

Figure 4.18 A cross section illustrating the plate tectonic cycle. Oceanic lithosphere is created by magma rising from the asthenosphere at the oceanic ridge. Plates move away from the ridge and may be consumed as they descend into the mantle at a subduction zone. Features not to scale.

PLATES OF THE WORLD

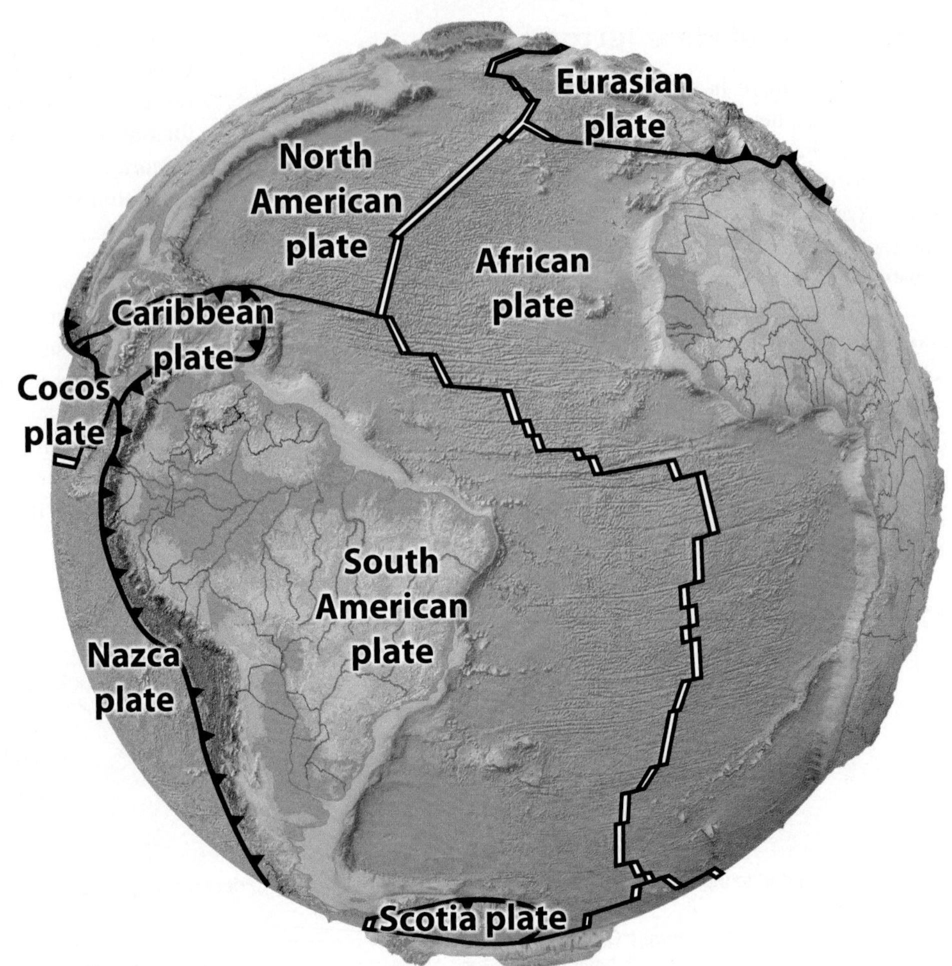
Eurasian plate
North American plate
African plate
Caribbean plate
Cocos plate
South American plate
Nazca plate
Scotia plate

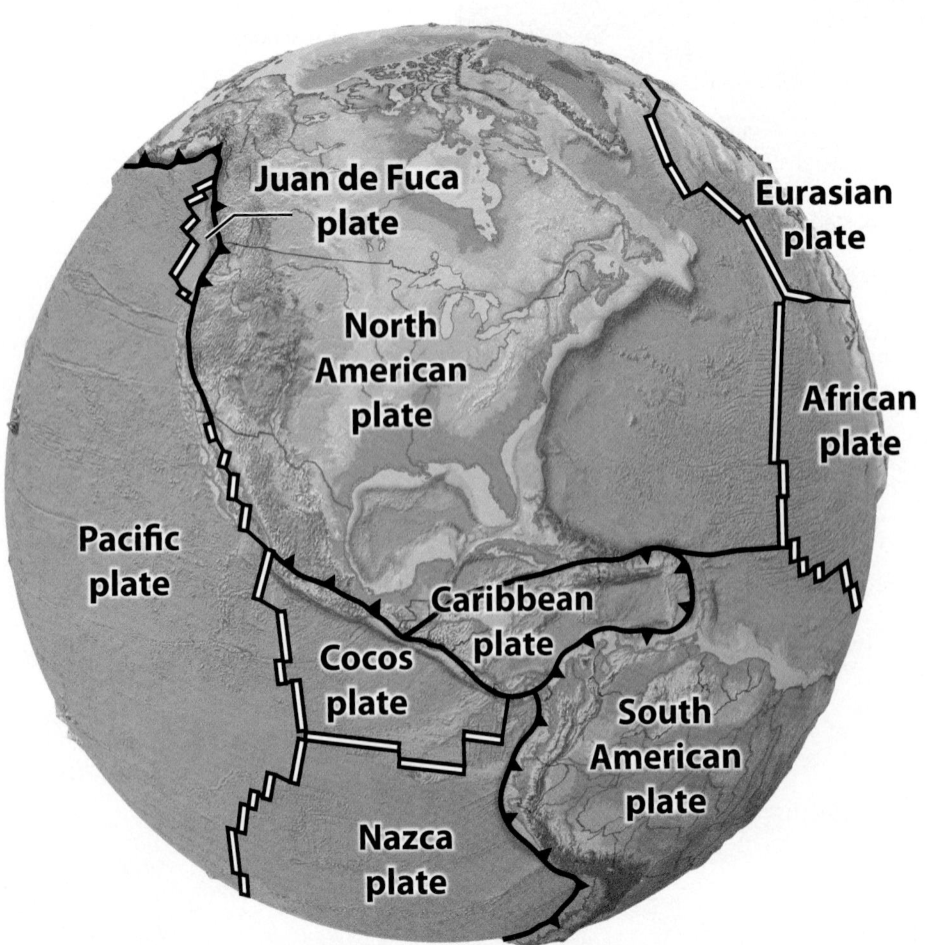
Juan de Fuca plate
Eurasian plate
North American plate
African plate
Pacific plate
Caribbean plate
Cocos plate
South American plate
Nazca plate

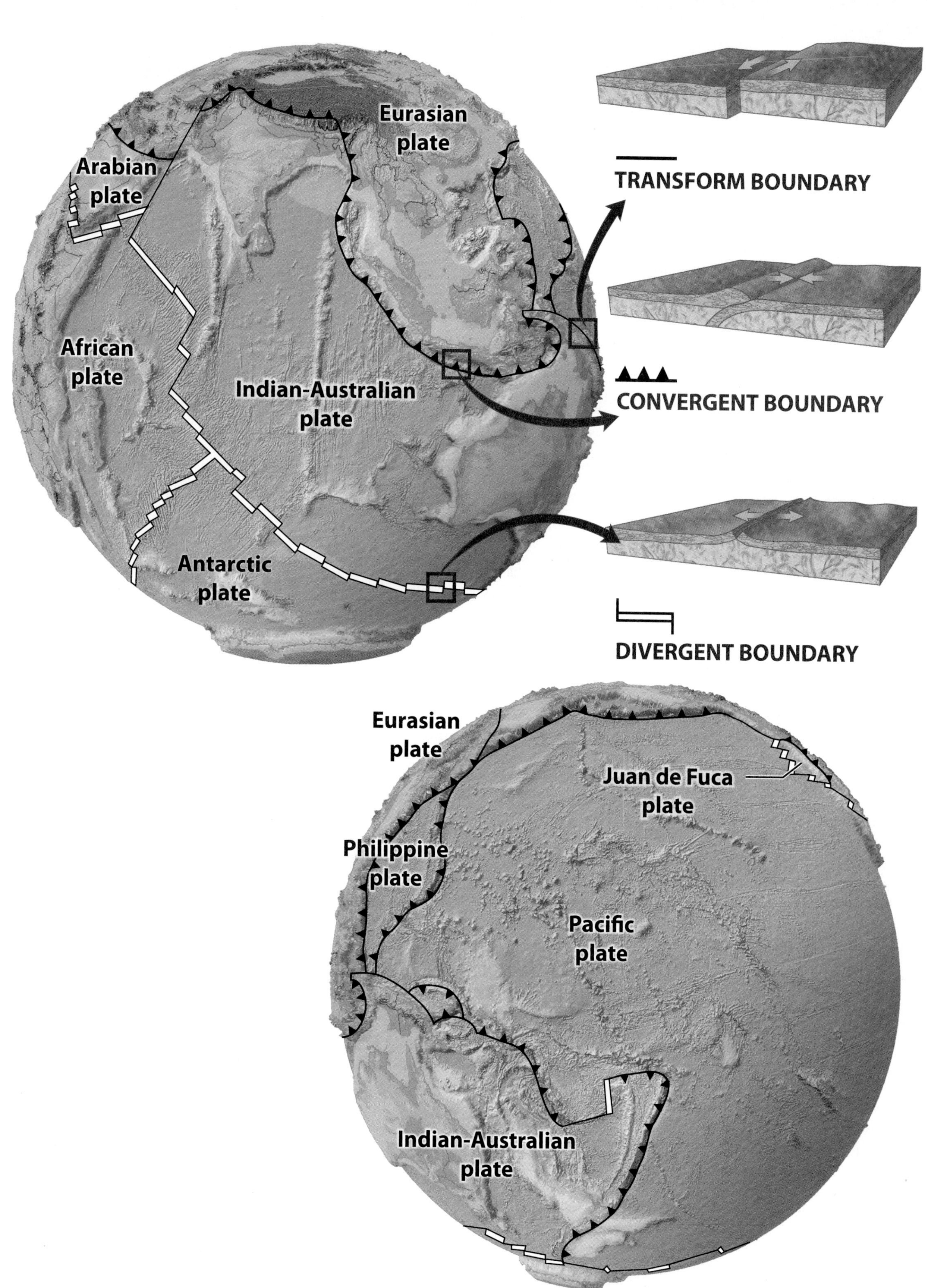

Eurasian plate
Arabian plate
African plate
Indian-Australian plate
Antarctic plate
TRANSFORM BOUNDARY
CONVERGENT BOUNDARY
DIVERGENT BOUNDARY
Eurasian plate
Juan de Fuca plate
Philippine plate
Pacific plate
Indian-Australian plate

Checkpoint 4.13

basic advanced
intermediate superior

Draw the approximate locations of the plate boundaries on the map to the right. Use different line symbols for the oceanic ridges and oceanic trenches. Name and label as many of the plates as you can.

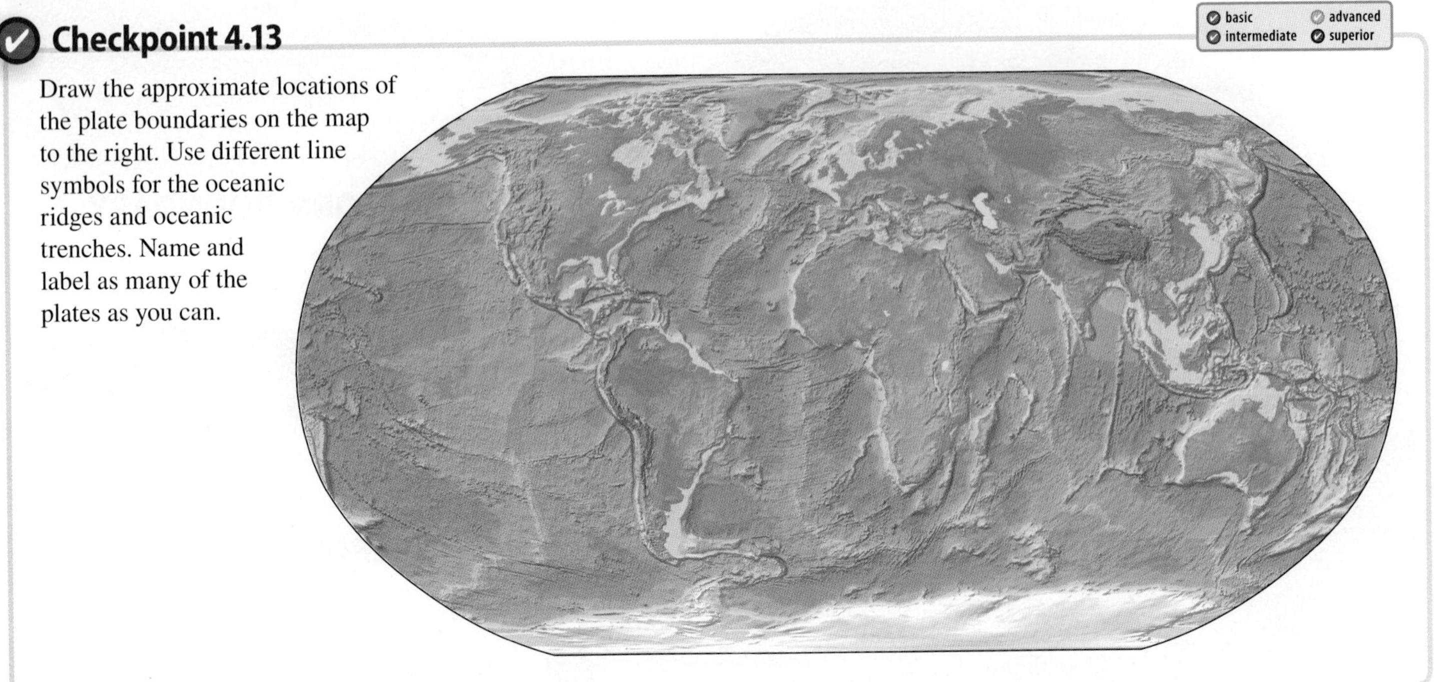

on and below the ocean floor to create new lithosphere that is added to the edge of each of the oceanic plates. This process is similar to the way your fingernail grows as new nail, added at the base, pushes the older part of the nail farther out. This forcing apart of the plates, called *ridge push*, is a process that contributes to the movement of tectonic plates.

Earthquakes, Volcanoes, and the Destruction of Lithosphere at Subduction Zones. To ensure that the size of Earth remains constant, the creation of new lithosphere at the oceanic ridges must be balanced by the destruction of lithosphere elsewhere. The theory of plate tectonics holds that **oceanic lithosphere is consumed as it descends into the mantle adjacent to trenches at regions termed subduction zones** (Figure 4.18). Since the plate is slightly more dense than surrounding material in the uppermost mantle, the plate may actually be pulled downward, a process called *slab pull.* Chemical reactions in the descending plate release water, which lowers the melting point of minerals in the overlying mantle to cause partial melting (Figure 4.17d). The magma generated either rises to the surface to create volcanoes or cools and solidifies within the crust. This addition of material thickens the continental crust adjacent to subduction zones and contributes to the formation of mountain belts.

The descending plate bends and fractures as it is pulled down into the mantle, causing shallow earthquakes to occur. Earthquakes happen only in the cold, brittle lithosphere. The lithosphere eventually loses its brittle nature as it warms up as it descends into the hotter mantle. The deepest earthquakes are thought to be the result of large-scale changes to mineral structures caused by increasing temperature and pressure deeper in the mantle.

Only oceanic lithosphere is consumed at subduction zones. Consequently, the ocean floor is continuously being created (oceanic ridges) and destroyed (subduction zones), while the continental lithosphere remains at Earth's surface. As we will see later, continents may be broken up into smaller pieces or may combine to form larger landmasses (such as Pangaea), but the mass of continental material does not change significantly.

Most continental margins are not plate boundaries. These continental margins are known as *passive margins* (Figure 4.18) and are generally free of volcanism and earthquake activity. Continental margins that represent plate boundaries are called *active margins* and are characterized by volcanoes and earthquakes. The Atlantic coastlines of North and South America are examples of passive margins, while the Pacific margins of each continent are active margins.

Rate of Plate Movements. Plate movements occur very slowly. Scientists originally determined the rates and directions of plate motions in the same way that you determine your average walking speed. First they determined the age of a seafloor rock, and then they divided that into the distance from an oceanic ridge system (where new material is created). This simple but effective calculation (distance divided by time) was compared to rates of plate motion determined by analyzing the contrasting ages of volcanic rocks on the Hawaiian Islands.

The Hawaiian Islands are not associated with a plate boundary but are formed in the interior of the Pacific plate above a plume of rising hot mantle rock (Figure 4.19). There are many of these rising plumes, and their locations remain essentially fixed relative to the overlying plates. The point where a plume reaches the surface is known as a *hot spot.* The Hawaiian Islands form as the Pacific plate moves over a hot spot, much as a conveyor belt moves above a heat source. The islands become progressively older with increasing distance from the hot spot. The relationship between the distance of the older islands from the hot spot (600 kilometers (km)) and the age of these islands (5 million years) yields the average rate of plate motion:

$$600 \text{ km}/5{,}000{,}000 \text{ yr} = 120 \text{ km/Myr} = 12 \text{ cm/yr (5 in/yr)}$$

Today, Global Positioning System satellites allow scientists to record tiny movements of known points on Earth's surface, thus determining the motions of plates (Figure 4.20). Different plates move at different rates, but each has a fairly constant rate of motion:

- Ultraslow spreading rates (less than 1 centimeter per year; 0.4 inch per year) are seen at the oceanic ridge in the Arctic Ocean.
- Slow rates (1 to 2 centimeters per year; 0.4 to 0.8 inch per year) are found for the North American and Eurasian plates along the ocean ridge in the northern Atlantic Ocean.
- Rapid spreading rates of more than 15 centimeters per year (6 inches per year) have been identified for the Pacific and Nazca plates along the East Pacific Rise.

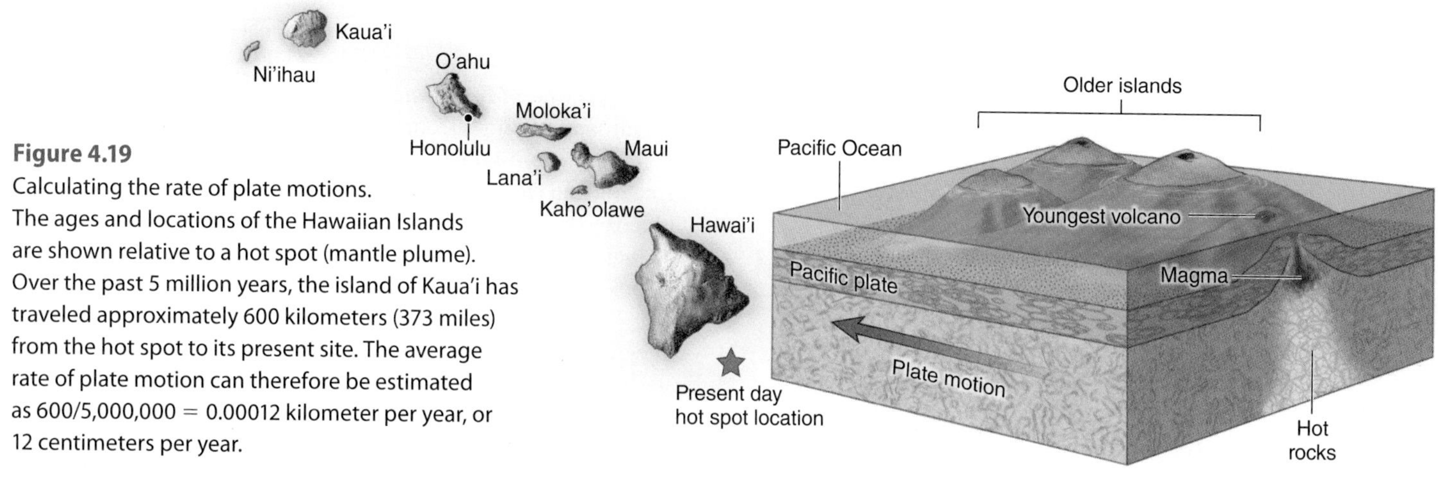

Figure 4.19 Calculating the rate of plate motions. The ages and locations of the Hawaiian Islands are shown relative to a hot spot (mantle plume). Over the past 5 million years, the island of Kaua'i has traveled approximately 600 kilometers (373 miles) from the hot spot to its present site. The average rate of plate motion can therefore be estimated as 600/5,000,000 = 0.00012 kilometer per year, or 12 centimeters per year.

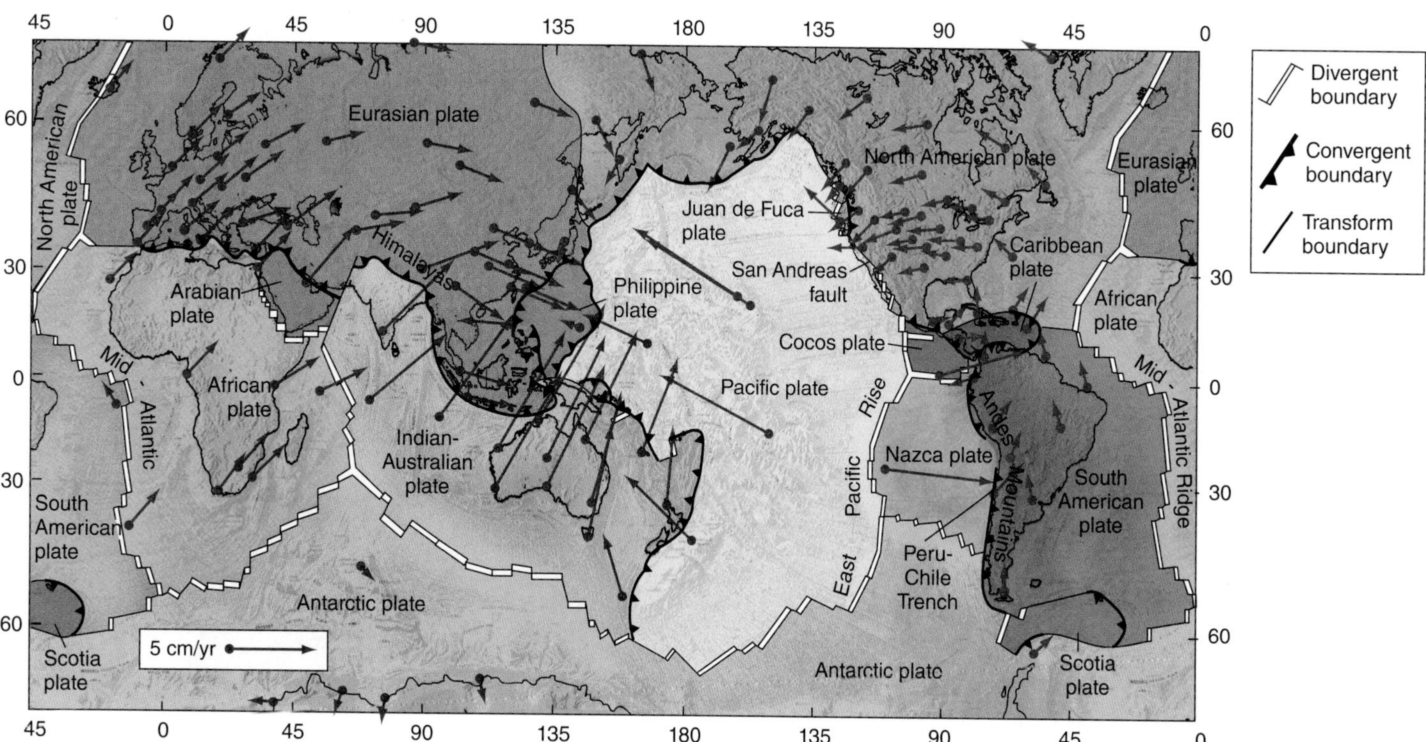

Figure 4.20 Motions of plates. Arrows show the directions and rates of plate motions determined by Global Positioning System satellites. Plate motions in the Pacific Ocean are nearly five times faster than in the Atlantic Ocean. The arrows point in the direction the plate is moving, and the lengths of the arrows represent rates. (Measurements are in centimeters per year.)

Checkpoint 4.14

basic · advanced · intermediate · superior

Using the map from Checkpoint 4.13, identify and label the map with as many of the features on the associated list as you can.

a) A subduction zone
b) A plate that is 99 percent oceanic lithosphere
c) A plate that is over 90 percent continental lithosphere
d) A pair of plates moving toward each other
e) A plate nearly surrounded by oceanic ridges
f) A plate moving approximately north
g) A plate moving approximately east
h) A fast-moving plate
i) A slow-moving plate
j) A passive continental margin
k) The site of the oldest oceanic crust
l) The location of rocks with normal polarity
m) A pair of plates moving away from each other
n) A hot spot
o) A location with very young rocks (0 to 5 million years)

Checkpoint 4.15

basic · advanced · intermediate · superior

Imagine that you were able to take a knife and slice through North America from the Atlantic Ocean to the Pacific Ocean. Draw and label a sketch illustrating the locations and positions of the lithosphere, asthenosphere, crust and mantle.

Current seafloor spreading rates are approximately five times higher for the East Pacific Rise than for the Mid-Atlantic Ridge. The oldest oceanic crust in both the Atlantic and Pacific oceans is the same age (about 180 million years; Figure 4.7), but the Pacific Ocean is much wider than the Atlantic Ocean because of more rapid plate motion. The slow-moving North American and African plates are driven almost exclusively by ridge push forces. In contrast, the more rapid motions for the plates below the Pacific Ocean can be attributed to a combination of ridge push and slab pull.

Although slow in comparison to human activities, plate movements are powerful. Over millions of years, the movements of plates have opened and closed oceans and have formed and broken apart continents. The slow movement of huge plates of lithosphere requires enormous amounts of energy. What supplies that energy?

Energy for Plate Movement. Most of the energy for plate movement ultimately comes from the heat flow from Earth's interior and the gradual cooling of Earth. Heat remaining from Earth's early formation and from decay of radioactive elements (see Section 8.4) within Earth slowly escapes from the interior through Earth's surface. Scientists infer that this heat is transferred by *convection* as hot rock rises below oceanic ridges and colder rock descends at subduction zones (see Figures 4.8a and 4.10). They point out that in the oceans and the atmosphere, warm water or air rises and cold water or air sinks to form convection cells that link vertical and horizontal motions of water and air. By analogy, scientists propose that warm material is rising below the oceanic ridges and cold, denser material is sinking along the subduction zones to form convection cells in the upper mantle (Figure 4.21). Since the rate of heat loss likely influences plate tectonic activity, greater heat loss during earlier times in Earth's history may have resulted in faster rates of plate motion and/or more plate boundaries.

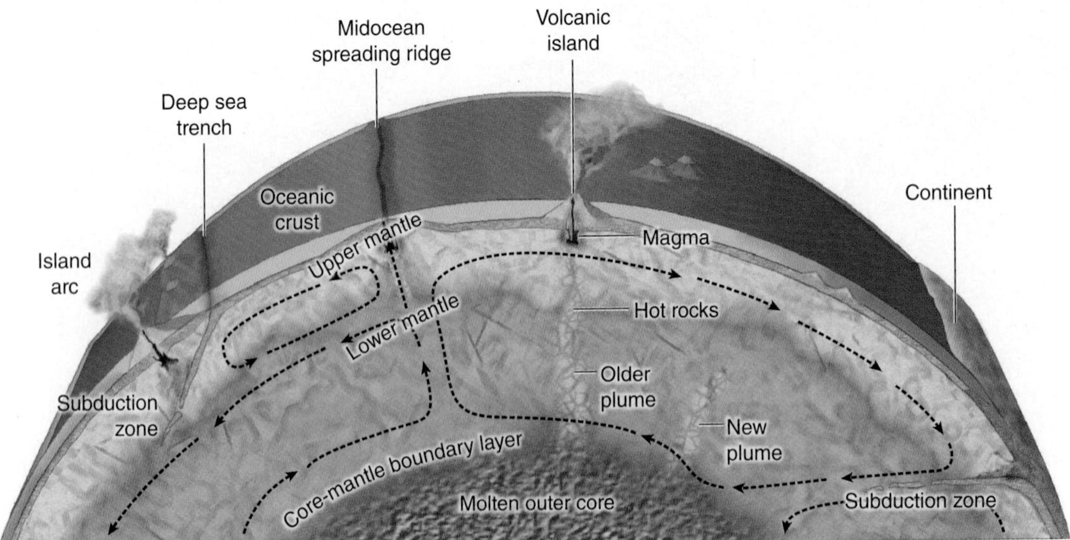

Figure 4.21 Mantle circulation system. The flow of mantle rock may occur as two-layer convection cells or as a result of whole-mantle convection. Both models have rock rising from the deep mantle, but subduction zones are interpreted to continue into the lower mantle or end in the upper mantle. Note that the crust, lithosphere, and mantle are not drawn to scale.

Plate movements are also aided by gravity. Recall that oceanic lithosphere is "pushed" away as new lithosphere forms at midocean ridges. At ocean trenches, plates are "pulled" into the subduction zones by the weight of the descending slab of lithosphere. Some plates appear to sink all the way to the core-mantle boundary. Scientists hypothesize that gravity pulling on the plate at subduction zones combines with warm rising rock near ridges to form circulation cells in the mantle, although they debate the details of this circulation (Figure 4.21).

An Ongoing Process. Plate tectonics supplies the big picture of how Earth works. It links the heat flow from Earth's interior and the locations of earthquakes and volcanoes to the distribution of continents and oceans.

Keep in mind that plate tectonics is an ongoing process. The map of plates (see Chapter Snapshot) is just a snapshot in time. Plates and plate boundaries change over time. Movement may stop at some locations, new boundaries will form, and plates will look different in the future. Two hundred million years ago, there was no Atlantic Ocean (see Figure 4.2). Two hundred million years in the future, the world will look much different than it does today.

Do Other Planets Have Plate Tectonics?

Plate tectonics requires not just a layered Earth with lithosphere and asthenosphere but also the excess internal heat necessary to drive the plates. Smaller terrestrial planets such as Mars cooled more rapidly than Earth and lost their internal heat much earlier in their history. There is no evidence that plate tectonics is occurring today on Mars, but surveys reveal that the oldest sections of the Martian crust have paleomagnetic patterns similar to those of Earth's ocean floors. Scientists interpret these observations to mean that Mars experienced plate tectonics very early in its history (4 billion years ago) when it was still hot. When the interior cooled, plate tectonics stopped.

But what about Venus? Venus is almost the same size as Earth and has a similar layered structure. However, Venus does not have plate tectonics. The reason lies in the distribution of lithosphere on Venus. Two-thirds of Earth's lithosphere is oceanic; the rest is continental. Much of Earth's heat is lost in creating new oceanic lithosphere along the oceanic ridge system. This could not occur if older oceanic lithosphere were not being destroyed along subduction zones. The shape and size of volcanoes on Venus suggest it has a thick lithosphere that hinders the motion of plates. Without the destruction of old lithosphere, there is no opportunity to create new lithosphere and thus to allow the planet's internal heat to escape along long-lived stable features such as oceanic ridges. Instead, some scientists hypothesize that heat builds up below the surface of Venus and is lost in periodic catastrophic volcanic-type events that destroy the existing lithosphere. The resulting magma slowly cools to form new lithosphere. Imagine the mantle boiling up to consume North America every 500 million years. In this scenario, it would be difficult for life to evolve as it has on Earth. This is yet another reason why Earth is unique among the planets of our solar system!

4.5 Plate Boundaries

Learning Objectives

- Explain how relative plate motions result in divergent, convergent, and transform plate boundaries.
- Compare and contrast the features of convergent and divergent plate boundaries.
- Explain the features found in association with a transform plate boundary.
- Interpret a plate configuration to predict the location of key geologic features (e.g., mountains, volcanoes, subduction zones), and sketch and label a representative cross section.
- Predict how the current distribution of plates, continents, and oceans will change millions of years in the future.
- Discuss how plate tectonic processes can alter global climate patterns.

Oceanic ridges and trenches, volcanoes, and earthquakes are just some of the phenomena present at the boundaries that separate the tectonic plates. The type and distribution of features are characteristic of the relative motions of the plates on either side of the boundary. We classify plate boundaries into three categories based on their relative plate motions (see Chapter Snapshot).

Divergent Plate Boundaries

Divergent plate boundaries occur where hot, rising mantle rock causes plates to move apart (Figures 4.18 and 4.22a and Chapter Snapshot). The evolution of a divergent plate boundary has three recognizable stages that we can loosely characterize as

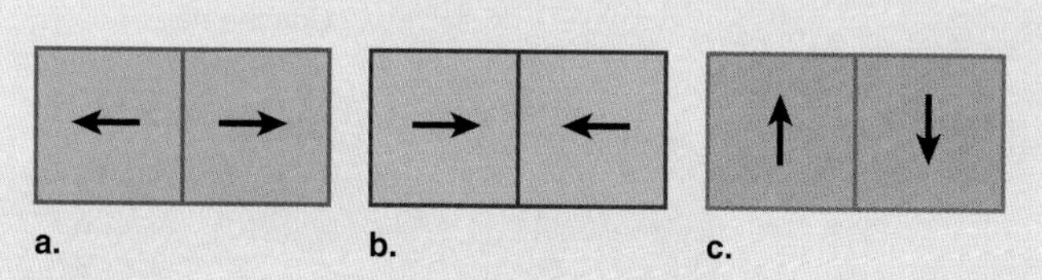

Figure 4.22 Three types of plate boundaries. **a.** At a divergent boundary, plates move apart. **b.** At a convergent boundary, plates move toward one another. **c.** At a transform boundary, plates move parallel to one another in opposite directions.

Checkpoint 4.16

Predict what the map of the world will look like (a) 5 million years in the future and (b) 50 million years in the future. Identify how the distribution of continents and oceans will change. Which plates will exhibit the greatest changes in comparison to today?

birth, youth, and maturity. The birth of a divergent boundary occurs when an existing piece of continental lithosphere begins to break apart. It may seem counterintuitive that the formation of an ocean begins with the breakup of a continent, but remember that the Atlantic Ocean owes its origin to the breakup of the supercontinent Pangaea. Such a location is characterized by thinning of the continental lithosphere, often accompanied by volcanic activity. This process is happening today on Africa in an area known as the East African Rift zone (Figures 4.23a and 4.24a). As the lithosphere breaks apart, it **forms a wide, steep-walled depression known as a rift valley**. The underlying asthenosphere is close to the surface below the rift valley, and decompression melting generates magma that forms volcanoes. Eventually, the continental crust in a rift valley separates to form a gap where rising magma creates new oceanic floor (Figure 4.24b). Inflow of seawater initially forms a narrow ocean (the youth stage), much like the Red Sea to the north of the East African Rift zone that separates the Arabian peninsula from Africa (Figure 4.24). It takes millions of years for narrow oceans to expand to form a mature ocean like the present-day Atlantic or Pacific Ocean because the rates of plate motions are so slow (Figure 4.23c).

What about a fourth stage—death? Eventually each oceanic ridge system will die; that is, it will stop producing new oceanic lithosphere. Ultimately, the ridge will be lost as the oceanic lithosphere descends into a subduction zone or is added to the margin of a continent by plate collision.

What do you think would happen if the East Pacific Rise, the oceanic ridge in the Pacific Ocean, stopped working? We would record a slow reduction in the size of the ocean basin. The Pacific and Nazca plates would be consumed by advancing subduction zones along the margins of the South American, North American, Eurasian, and Indian-Australian plates. Plate tectonics regulates Earth's heat loss, so we would expect to see a new divergent boundary start to form or other plate motions to accelerate, if a mature divergent boundary were to shut down.

Convergent Plate Boundaries

Oceanic lithosphere is consumed at subduction zones where it descends into the mantle beneath trenches. The plate junction where this occurs is a convergent boundary (Figure 4.22b). The descending plate has slowly traveled across the ocean floor to the trench for many millions of years. During this time, sediment collected on top of the oceanic lithosphere. Some of the sediment is scraped off the descending plate, and the rest is carried into the subduction zone. Water is abundant in the descending plate. It may be mixed with the sediment, present in fractures at the top of the oceanic crust, or combined in the atomic structure of minerals in the lithosphere. Increased pressures and temperatures in the descending plate compact the sediment, squeezing out water, and result in chemical reactions that flush water from minerals. Even though the addition of water reduces the melting temperatures of the rocks, the temperature of the descending lithosphere is too cold to melt because it has been near Earth's cold surface for millions of years.

In contrast, the temperatures are much higher in the region in the overriding plate immediately above the descending plate.

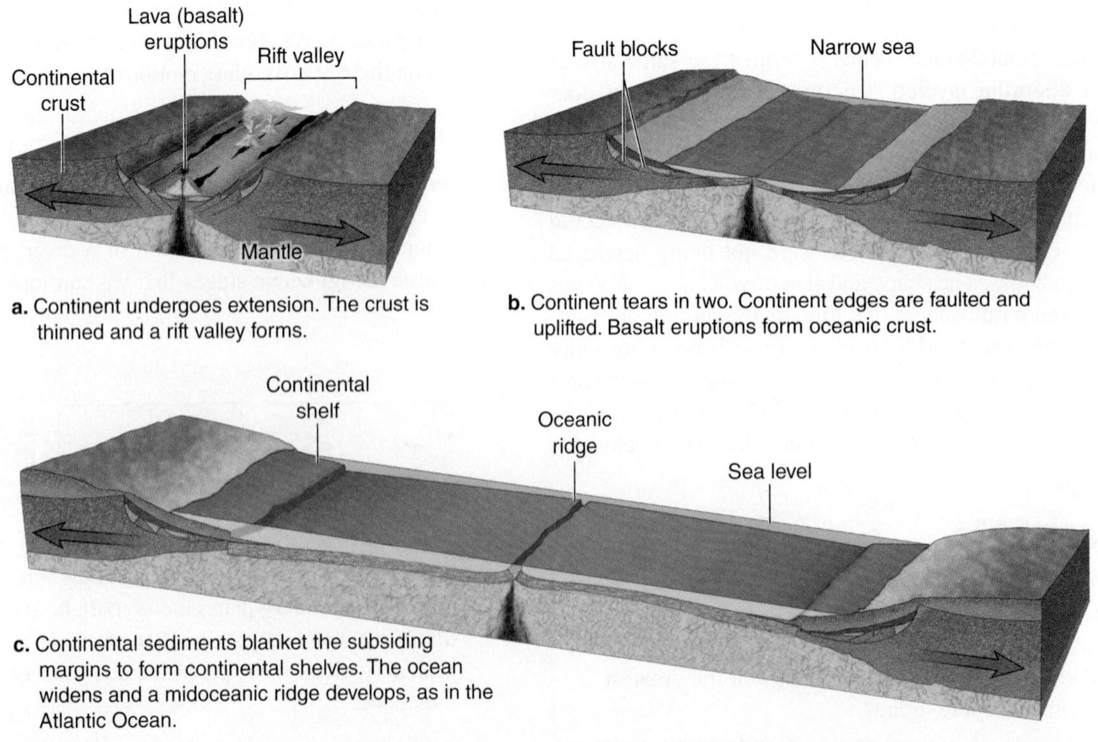

a. Continent undergoes extension. The crust is thinned and a rift valley forms.

b. Continent tears in two. Continent edges are faulted and uplifted. Basalt eruptions form oceanic crust.

c. Continental sediments blanket the subsiding margins to form continental shelves. The ocean widens and a midoceanic ridge develops, as in the Atlantic Ocean.

Figure 4.23 Cross sections illustrating the development of a divergent plate boundary. **a.** The continental lithosphere breaks up. **b.** A narrow ocean develops. **c.** A mature ocean basin forms.

This part of the lower lithosphere is termed the *mantle wedge* (Figure 4.25). The water that is expelled from the descending plate enters the hot mantle rocks of the mantle wedge. Recent research showing layered structures in and below the mantle wedge suggests that melting may even occur to depths as great as 450 kilometers (280 miles). Water that is forced out of the descending lithosphere reduces the melting temperature of the rocks in the wedge, resulting in partial melting to form magma. The magma rises through the overlying tectonic plate, and some of it reaches the surface to form volcanoes that may occur as a chain of islands or a range of mountains (Figure 4.25). (We will look at plate tectonics and magma generation in greater detail in

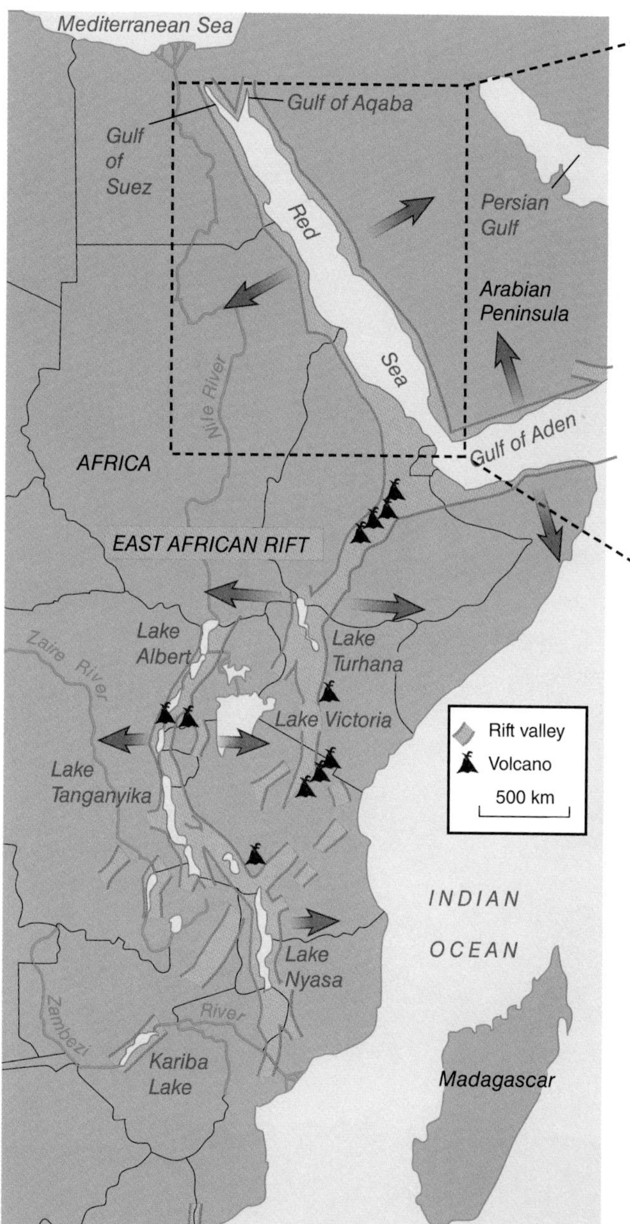

a.

b.

Figure 4.24
a. The East African Rift Valleys and the Red Sea. The red arrows indicate the relative motions of continental lithosphere on either side of this early-stage divergent boundary.
b. Satellite photo of the Red Sea.

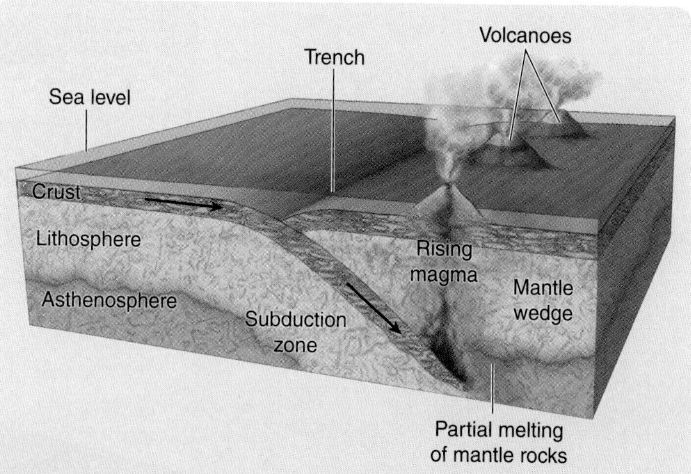

Figure 4.25 A cross section illustrating magma generation at a subduction zone. Magma is generated where water from the descending plate enters the overlying mantle wedge to cause partial melting of mantle rocks. The magma will rise through the crust, either solidifying or reaching the surface as a volcanic eruption. Reactions in the descending plate that generate the water typically begin to occur at depths of 100 to 150 kilometers (62 to 93 miles). Consequently, the distance of the volcanoes from the trench will depends on the slope of the subduction zone. The volcanic arc is closer to the trench for steeply inclined subduction zones and farther away where the descending plate slopes gently.

Chapter 6.) In addition to volcanoes, the largest and deepest earthquakes recorded are also concentrated along subduction zones (more on this in Chapter 5).

The density of lithospheric plates that meet across convergent boundaries plays a significant role in determining which plate descends into the mantle and which remains at the surface. When two oceanic plates collide, **the older of the two plates descends into the mantle because it is colder and denser** (heavier). Where oceanic and continental lithosphere collide, the oceanic plate descends because it is composed of denser minerals than the continental plate. **The less dense crust of the continental lithosphere does not descend into subduction zones** but is piled up to form high mountain belts at convergent boundaries. Remember, individual tectonic plates can be mostly oceanic lithosphere,

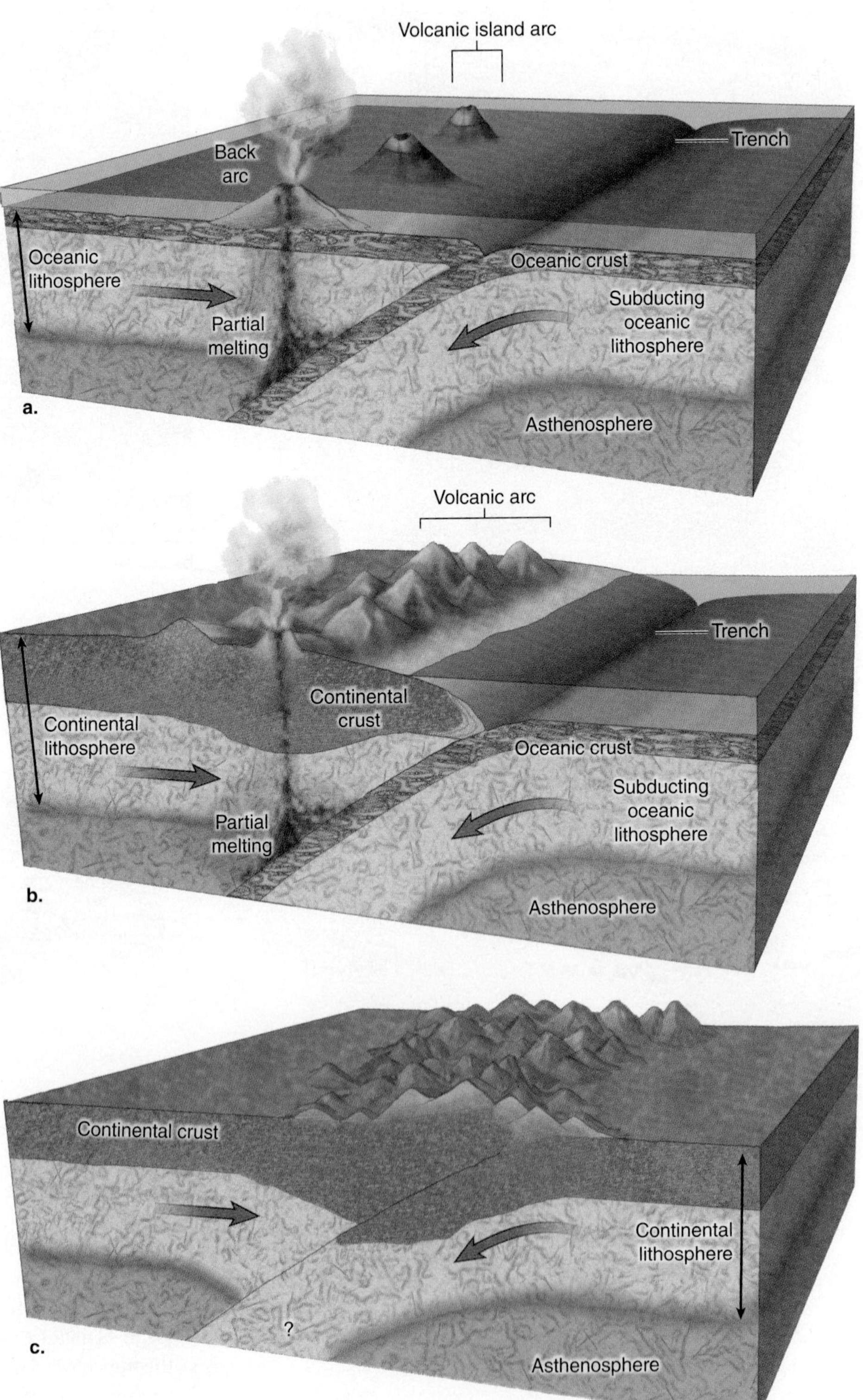

Figure 4.26 Three types of convergent plate boundaries (features are not to scale). **a.** When two oceanic plates collide, the plate with the older (cooler, denser) lithosphere descends into the subduction zone. A chain of volcanic islands (island arc) forms on the overriding plate. **b.** In an oceanic-continental collision, the oceanic plate descends, and a chain of volcanoes forms on the continental plate. **c.** A continental-continental collision is marked by a high mountain range but little volcanic activity. Crustal thickness is increased, but scientists still debate the fate of the mantle lithosphere in this type of collision. Features are not to scale.

mostly continental lithosphere, or both oceanic and continental lithosphere. When we discuss plate boundaries, we focus on the lithosphere actually involved in the collision. As such, convergent boundaries come in three varieties (Figure 4.26).

Oceanic Plate Versus Oceanic Plate Convergence. The magma produced by partial melting in the mantle wedge of the overriding oceanic plate adjacent to a subduction zone may reach the surface as volcanoes. As the volcanoes grow, they may rise above sea level to form a long chain of islands, known as a volcanic island arc (Figure 4.26a). The Aleutian Islands off the tip of Alaska were formed by magma generated when the Pacific plate descended below oceanic lithosphere on the edge of the North American plate (Chapter Snapshot). A 1995 volcanic eruption that devastated the Caribbean island of Montserrat was the result of subduction of the South American plate below an island arc along the edge of the Caribbean plate.

Oceanic Plate Versus Continental Plate Convergence. When oceanic lithosphere collides with continental lithosphere, the thinner, denser oceanic plate will descend into the subduction zone (Figure 4.26b). For example, on the western edge of South America, the oceanic lithosphere of the Nazca plate descends below the continental lithosphere of South America, forming an ocean-continent subduction zone. As these plates collide, the Andes Mountains, the second-highest mountain range in the world, are being formed (see Chapter Snapshot).

Continental Plate Versus Continental Plate Convergence. In these collision zones, the crustal part of the continental lithosphere is peeled away from the mantle part of the continental lithosphere, much as soil is scraped up by a bulldozer (Figure 4.26c). The resulting thickened pile of crust creates mountains with no volcanoes or deep earthquakes. The Himalayas form the tallest mountains in the world with several peaks over 8,000 meters

Checkpoint 4.17

basic intermediate advanced superior

On the following map, the green shaded areas labeled X, Y, and Z represent continents; assume the blue part of the map is ocean.

How many plates are present?

a) 3
b) 4
c) 5
d) 6

A B C D X Y Z

Volcano | Oceanic ridge | Oceanic trench

Checkpoint 4.18

basic intermediate advanced superior

Examine the map in Checkpoint 4.17. Draw arrows on the map showing the relative directions of plate motions. We could add a series of volcanoes associated with one of the plate boundaries. Which boundary would it be and why? Sketch two cross sections through the lithosphere along lines A–B and C–D, illustrating the characteristics of plate boundaries.

Checkpoint 4.19

basic intermediate advanced superior

Venn Diagram: Divergent and Convergent Plate Boundaries

Use the following Venn diagram to compare and contrast divergent and convergent plate boundaries. Three characteristics of plate boundaries are provided; identify at least 10 more. Then write the numbers of the features unique to either group in the larger areas of the left and right circles; note features that they share in the overlap area in the center of the image.

1. Rocks on either side of boundary are the same age.
2. Example: Nazca and South American plate boundary.
3. Associated with oceanic trenches.
4.
5.
6.
7.
8.
9.
10.
11.
12.
13.

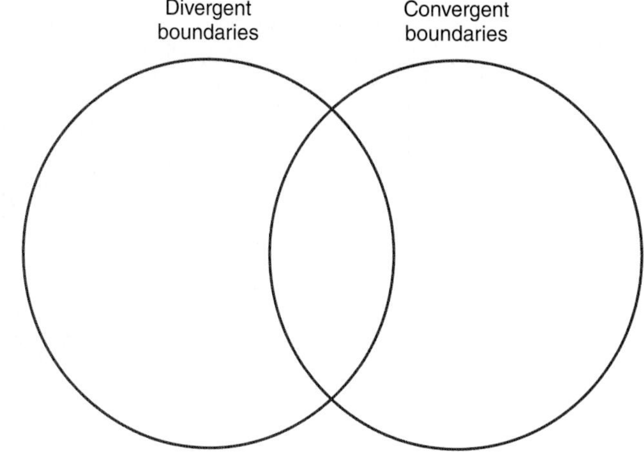

(26,247 feet). These mountains continue to grow today as a result of the ongoing continental collision between India and the Eurasian continent (Chapter Snapshot). The collision of the plates began over 40 million years ago, when India began its slow-motion crash into Asia as it finished its long journey northward after the breakup of Pangaea.

Mountain belts that formed as a result of collision between two plates of continental lithosphere represent the thickest sections of continental crust on the planet. For example, the crust under the Himalayas is approximately 70 kilometers (43 miles) thick, nearly twice the average crustal thickness under North America (Figure 4.27). Mountains formed along convergent boundaries will continue to rise as long as the plate tectonic forces responsible for the collision continue. The tallest mountains in the world are also among the youngest. The effects of weathering and erosion have worn down older mountain ranges where plate tectonic forces no longer dominate. For example, the 60-million-year-old Rocky Mountains reach altitudes of 4,300 meters (14,000 feet), while the older Appalachian Mountains (250 million years old) top out around 2,000 meters (6,600 feet). Both of these mountain ranges may once have been as tall as the Andes. Given enough time, mountains are worn down to the level of the surrounding land surface. (We will discuss the formation of mountains in Chapter 6.)

Figure 4.27 Thickness of Earth's crust. Most of the oceanic crust is less than 10 kilometers (6 miles) thick, while the continents are at least 30 kilometers (19 miles) thick. Less than 10 percent of the continental crust has a thickness of greater than 50 kilometers (31 miles) and is found exclusively below the Himalaya and Andes mountains.

Transform Plate Boundaries

Imagine trying to gift-wrap a basketball. It would be a difficult job, requiring many small scissor cuts to form the paper to the shape of the ball. Think of Earth as the basketball and the lithosphere as the wrapping paper. The rigid lithosphere does not readily conform to the curved surface. Many small adjustments in the shapes of the plates are needed to make them fit on the spherical Earth. Transform boundaries represent those geometric adjustments, much like the small scissor cuts needed to wrap the basketball.

If we are honest, we must admit that the transform boundary is a poor relative of the divergent and convergent boundaries. Nothing is created or destroyed. Plates just slide past their neighbors like traffic on a two-way street, moving in opposite but parallel directions (Figure 4.22c). Since the amount of lithosphere at the boundary does not change significantly over time, these boundaries are sometimes termed *conservative plate boundaries.* No volcanoes form at transform boundaries as there is no mechanism to melt rocks at these sites. Movement on the transform boundary can generate large earthquakes, as we will learn in Chapter 5.

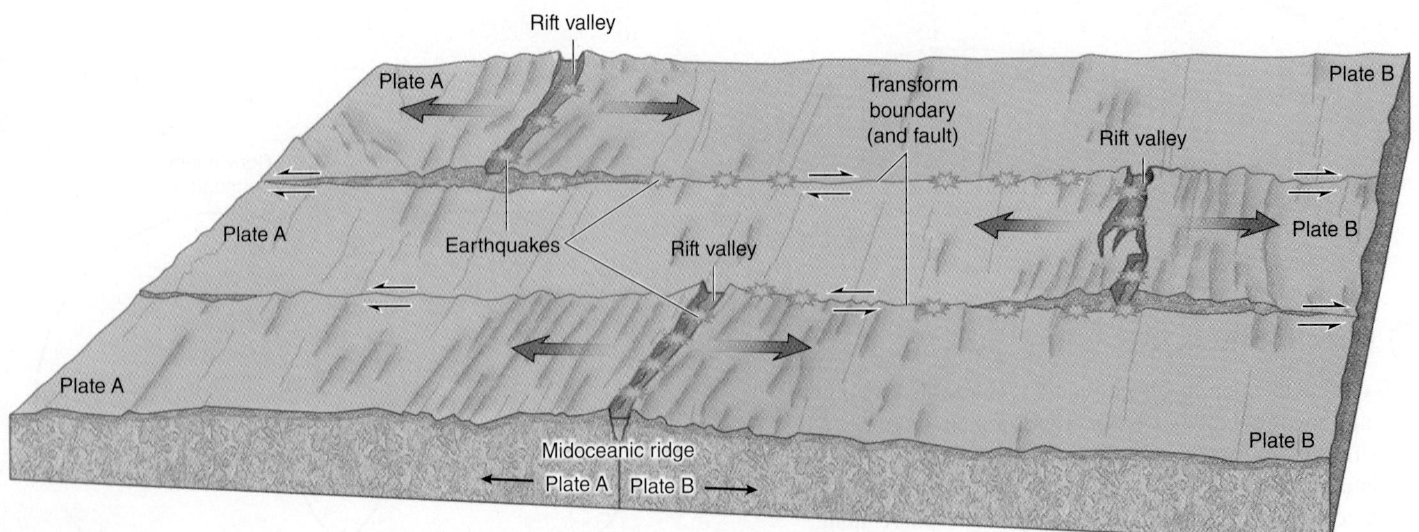

Figure 4.28 Features of transform boundaries that offset divergent boundaries. Plates A (tan) and B (green) move in opposite directions on either side of the transform boundary, causing earthquakes (yellow).

Transform boundaries can often be found between sections of divergent boundaries (Figure 4.28 and Chapter Snapshot). These features were discovered because of the earthquakes that occur as the plates slide past one another. The earthquakes only occur on sections of transform boundaries between segments of oceanic ridge (Figure 4.28). The topography of the seafloor can be determined by satellite measurements of the ocean surface. The topography clearly showed that oceanic ridges had offsets along transform boundaries.

Some transform boundaries, such as the San Andreas fault in California, occur on land. Like most transform boundaries, the San Andreas fault joins two other plate boundaries, a divergent boundary in the Gulf of California and the convergent Cascadia subduction zone (Figure 4.29). Land on the west side of the San Andreas fault, including Los Angeles and San Diego, is part of the Pacific plate. San Francisco lies east of the fault and is on the North American plate. Western California is moving slowly to the northwest relative to the rest of the state. Have you heard

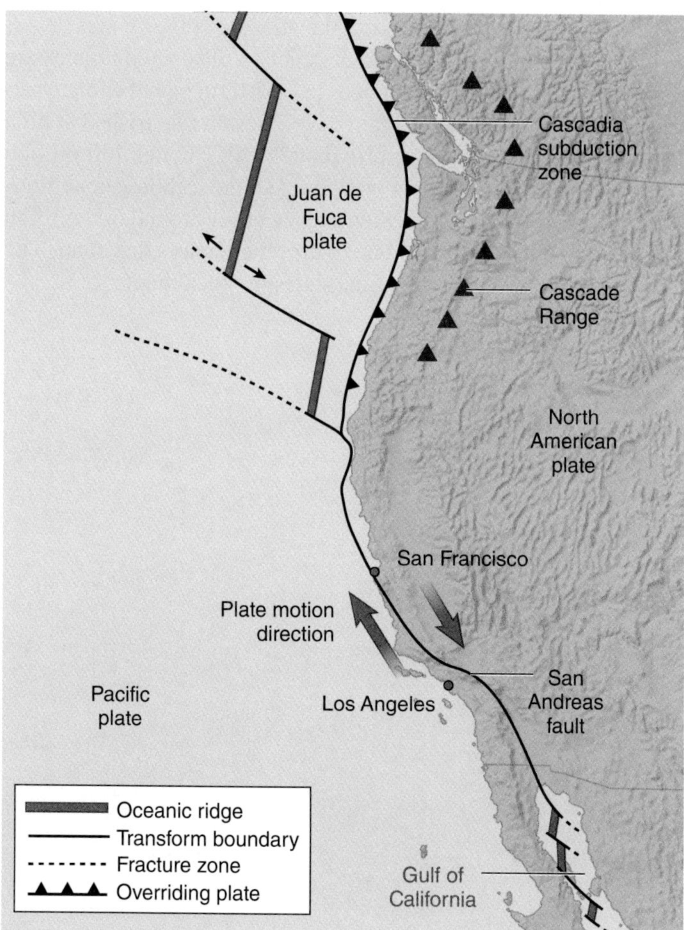

Figure 4.29 San Andreas fault. The San Andreas fault is a transform boundary that separates the North American and Pacific plates. The smaller Juan de Fuca plate lies between these two plates opposite Oregon, Washington, northern California, and part of British Columbia. The Pacific plate moves northwest relative to the North American plate. Over the next several million years, Los Angeles will migrate toward San Francisco.

Checkpoint 4.20

basic · intermediate · advanced · superior

Continuous Change of Earth's Surface Synthesis Exercise

The central image in the following figure shows a hypothetical plate configuration. Assume that research has revealed that the three pieces of continental crust are moving at different speeds, directions, or both. Plate A is traveling at 5 centimeters per year (50 kilometers per million years), B is also moving at 5 centimeters per year, and C is traveling west at 2 centimeters per year (20 kilometers per million years). The directions of A and B should be apparent from the plate configuration. Answer the questions that follow.

1. How many plates are present in the central figure?
2. Sketch and label a cross section through the lithosphere along the line X–Y in the central figure.
3. Fill in the upper and lower templates to show what the plate configuration looked like 4 million years earlier and what the plates will look like 15 million years in the future. Draw the configurations relative to the oceanic ridge (that is, assume it stays in the same place and everything else moves).

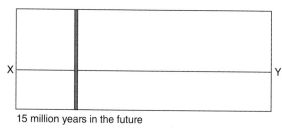

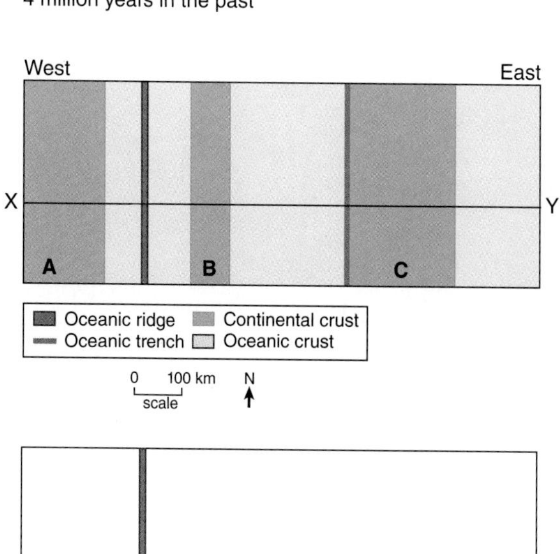

it said that California is going to fall off into the Pacific Ocean? Well, rest assured that it is not going to happen—but a portion of western California will gradually travel northward along the western edge of the North American plate, eventually colliding with Alaska millions of years from now, if plate motions remain stable that long.

Plate Tectonics and Climate

Plate tectonics plays both a direct and an indirect role in controlling global climate patterns. Today, the north-south alignment of most of the continents influences the paths of ocean currents that transfer warm waters to cooler, high latitudes and cold waters to warm regions. These patterns reduce the contrast in temperatures between the poles and the tropics. A world with east-west oriented clusters of continents that were located on either side of the equator would contribute to latitudinal oceanic currents that would result in more extreme climates.

Continents located over the poles provide a base for extensive continental ice sheets, increasing the reflectivity of the land surface that reduces the amount of solar radiation that would be absorbed at Earth's surface. The southern half of Pangaea, a landmass known as Gondwana, migrated back and forth over the South Pole at least twice during the last 500 million years, resulting in long ice ages each time. The rate at which plates move determines if these events occur at all and how long they last. Processes that occur at plate boundaries are also key contributors to global climate patterns.

Convergent plate boundaries are characterized by the formation of mountain ranges and volcanic activity. High mountains alter regional precipitation patterns and increase erosion rates. Geochemical reactions tied to weathering and erosion can capture atmospheric carbon dioxide. Faster weathering and erosion rates therefore lead to a reduction in greenhouse gas concentrations. In addition, mountains can host alpine glaciers, increasing the amount of sunlight reflected back into space. The haze of dust produced by volcanic eruptions at convergent boundaries can reduce the amount of incoming sunlight that reaches Earth's surface. Consequently, intervals of plate collisions, such as the formation of Pangaea, are often associated with cooling trends. In contrast, the breakup of a supercontinent such as Pangaea would have increased the length of the oceanic ridge, producing more carbon dioxide and reducing erosion, leading to a warming trend.

Faster spreading rates result in the more rapid formation of ocean floor rocks along oceanic ridges. These volcanic processes add carbon dioxide to the ocean and ultimately to the atmosphere, increasing the concentration of greenhouse gases and raising global temperatures. In addition, faster plate motions result in wider ridges and higher sea levels. Higher sea levels, in turn, contribute to lower continental erosion rates. A slowing of erosion can contribute to a climate warming trend by reducing the amount of carbon dioxide extracted from the atmosphere.

Cycles of formation and destruction of supercontinents are thought to have resulted in changes in long-term global climate patterns. Such changes help explain why dinosaurs flourished at high latitudes when Earth was approximately 10C° hotter 100 million years ago and how the planet was plunged into icehouse conditions of a "snowball Earth" with extensive ice cover around 600 million years ago. We will explore some of the other factors that contribute to climate change on shorter timescales in other chapters.

the big picture

Alfred Wegener shook up the science community when he proposed that continents drifted across Earth's surface. Wegener's work displayed a key characteristic of the nature of science—imagination. But science is a blend of creativity and logic, so what might start out as a product of imagination must eventually conform to principles of logical reasoning that support conclusions with solid evidence. Developing hypotheses to explain the workings of the world and putting them to the test of logic are creative but disciplined acts.

Many of Wegener's contemporaries met his ideas concerning continental drift with skepticism, which we recognize as a fundamental attitude of a scientific mind. Skepticism keeps scientists from rushing to instant judgment and embracing a new concept without first seeking the evidence behind it. Such verification can take years and is often supported by additional evidence obtained by new technologies. With additional observations and testing of predictions by many other scientists, Wegener's theory grew to become the theory of plate tectonics. Plate tectonics accounts for Earth's structure, numerous geologic phenomena, and processes of change over long intervals of time. Today, global positioning satellites allow scientists to make even more sophisticated observations of plate motion. These observations will inevitably improve as technology evolves and we learn even more about how Earth works. This is the way of science.

The processes and characteristics of plate boundaries are key to understanding the theory of plate tectonics. Remember, divergent boundaries are places where new plates form and move apart. Convergent boundaries are locations where plates collide, and transform boundaries are where plates slide past one another. The boundaries are locations where earthquakes, volcanoes, ocean trenches, and mountains may form, depending on the type of boundary. On the basis of what we know about the current directions and rates of plate motions, it is possible to make some predictions about how the plates will change over millions of years. The chapter opening image asked you to predict which configuration would most likely resemble the plate configuration on Earth 50 million years from now. Our best prediction is image (a). The Atlantic Ocean will continue to widen, but eventually a subduction zone is expected to form off the eastern coast of North and South America. Once the subduction zone becomes established, the Atlantic Ocean will inevitably begin to shrink as in image (c), and eventually another Pangaea-like landmass will be formed (image (d)). Notice, too, in image (c) that Africa collides with both Europe and southwest Asia in this scenario. For the sake of completeness, image (b) represents the distribution of continents 50 million years ago.

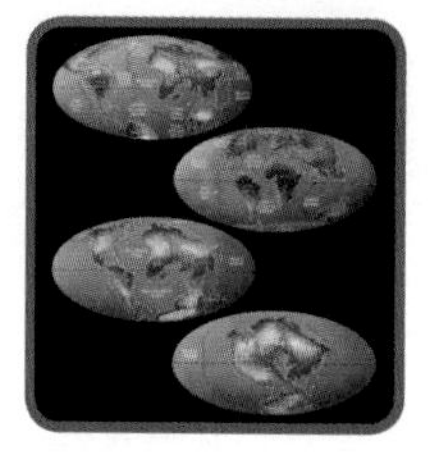

Plate Tectonics: Concept Map

Draw a concept map that illustrates the significant characteristics of the theory of plate tectonics. The majority of features related to plate tectonics are linked to the geosphere component of the earth system. Use the scoring rubric provided here to draw a concept map that can be scored as a 4.

Scoring Rubric	
0	The concept map does not contain any information about plate tectonics.
1	The concept map contains some terms that are significant to plate tectonics, but several key terms are omitted and many linking phrases are either absent or inaccurate.
2	The concept map contains most terms that are significant to plate tectonics, but they are poorly organized and some linking phrases are absent or incorrect.
3	The concept map contains most terms that are significant to plate tectonics, but one or two key term(s) may be absent. The diagram is reasonably well organized, and almost all linking phrases are appropriate.
4	The concept map contains all terms that are significant to plate tectonics in a well-organized display that has appropriate linking phrases for each pair of terms.

"We learn geology the morning after the earthquake, on ghastly diagrams of cloven mountains, upheaved plains, and the dry bed of the sea."

—Ralph Waldo Emerson, US author

Chapter 5

Earthquakes

Chapter Outline

the big picture

What is this man doing, and what does it have to do with earthquakes?

See The Big Picture box at the end of this chapter for the full story on this image.

5.1 Experiencing an Earthquake Firsthand

Chapter Learning Outcomes

- Students will describe the processes and consequences of earthquakes.
- Students will estimate the relative distance from different earthquakes, using indirect evidence.
- Students will compare and contrast different scales used to measure earthquakes.
- Students will evaluate the risk to society from earthquakes and their effects.

Waiting for an earthquake is a little like watching a well-crafted scary movie. We know what's coming, we just don't know when. As we focus on the obvious villain, we may be taken by surprise by an unexpected character. We can enjoy the thrill of suspense in movies, and even in the scariest moments, we can take comfort in the knowledge that it is an elaborate fiction. Unfortunately, with earthquakes the villain is real and reveals few clues about its identity. This makes preparing for an earthquake a difficult undertaking and predicting the timing of future quakes almost impossible.

During the twentieth century, over 2 million people were killed by earthquakes and associated phenomena. A handful of destructive earthquakes have already claimed nearly one-half that number in the twenty-first century (see Table 1.1). Many of the world's largest cities are located near tectonic plate boundaries that represent zones of frequent earthquake activity. Currently, over 30 million US citizens live in areas prone to destructive earthquake activity. Cities such as Los Angeles and San Francisco will inevitably suffer substantial damages and loss of life from a major earthquake in the future; we just don't know when.

Although scientists have identified the sources for many of the earthquakes that occurred in the past century, they can still be surprised by tremors that originate from unexpected locations that have no previous record of earthquake activity. That was the case on January 17, 1994, when the Northridge earthquake shook up much of southern California (Figure 5.1).

On the morning of the earthquake, I had just finished a conference in Los Angeles and was sound asleep in my hotel room. At 4:31 A.M.. my bed began to shake violently. As I struggled to wake up, my first reaction was to think that I had overslept and one of my colleagues was shaking me awake to make sure that I had time to get to the airport. Then my science-teacher brain kicked in, and I realized that I was experiencing an earthquake! By now, the bed was rocking back and forth and bouncing up and down. I got up and staggered to the window to see the water in the swimming pool sloshing about, creating surfable waves. I struggled to reach the doorway to the room, but the floor was vibrating so vigorously that I could barely stand up. I tried to stay upright in the doorway and ride out the earthquake. It only lasted for several seconds, but it seemed to go on for much longer. I had just experienced the Northridge earthquake.

There are many large earthquakes each year that result in less destruction, but the location of the Northridge earthquake made it one of the most expensive natural disasters in US history. The earthquake occurred below a densely populated city and caused more than $20 billion in damages. It was felt by 10 million people, of whom 57 were killed and another 9,000 injured. Coincidentally, exactly one year later, a slightly larger earthquake killed more than 5,000 people and destroyed 100,000 buildings in Kobe, Japan. Clearly, earthquakes present a serious threat to humans.

The Hollywood image of the ground fracturing open and people falling into a deep crevasse is not the real danger from an earthquake. Most lives are lost due to building collapse or tsunami (see Chapter 1). Consequently, the death and destruction that follow an earthquake often have as much to do with local building codes and public awareness of earthquake hazards as with the size of the earthquake itself. For example, nations around the rim of the Pacific Ocean have a sophisticated tsunami warning system to protect their citizens. This did not happen by chance; rather, a record of tsunami activity in the Pacific Ocean provided incentives for governments to collaborate to quickly identify threats from future events.

a.

b.

Figure 5.1 Damage from 1994 Northridge, California, earthquake. **a.** Sixteen people were killed when the upper two stories of this apartment building collapsed on top of the lower units. **b.** A parking garage built just 3 years earlier collapsed at California State University, Northridge.

Self-Reflection Survey: Section 5.1

Answer the following questions as a means of uncovering what you already know about earthquakes.

1. Have you ever experienced an earthquake? What was it like? Chances are, even if you have not experienced an earthquake, you have seen footage of earthquakes on television or the Internet. Describe what happened.
2. Where in North America are you most likely to experience an earthquake?
3. Imagine that scientists have predicted a 99 percent chance that a large earthquake will strike the region of the state where you live in the next 5 years. What action would you take? What if it were in the next 25 years?

In contrast, relatively few tsunami have been recorded for the Indian Ocean, and thus, surrounding nations had made little effort to share data on tsunami threats. This lack of a tsunami warning system resulted in massive loss of life from the 2004 Indian Ocean tsunami (see Chapter 1) that followed a giant earthquake off the coast of Sumatra. Though there was opportunity to warn the populace, individual governments did not have adequate disaster plans, failed to educate their people, and failed to communicate with other governments. This failure cost more than 200,000 lives.

Throughout this chapter, we will address several key questions related to earthquakes: Why do such catastrophic events occur? What factors cause earthquakes of similar size to produce very different results? Why do earthquakes occur in some areas but not in others?

We will also discuss the details of the Sumatra earthquake that caused the Indian Ocean tsunami and the more recent Tohoku earthquake in Japan. Apart from their significance as some of the largest earthquakes ever recorded, these events represent potential analogs for earthquakes that could occur along the coast of the Pacific Northwest. Until just a few years ago, no one knew this threat existed, but a combination of historical and geologic detective work brought to light this dangerous hazard and provided us with an excellent example of how science is done. We begin with that story.

5.2 The Science of Ghost Forests and Megathrust Earthquakes

Learning Objectives

- Describe the characteristics of a megathrust earthquake.
- Identify which of the world's largest cities are located relatively close to plate boundaries.
- Explain how researchers used the scientific method to discover that an ancient megathrust earthquake had occurred off the coast of Oregon and Washington.

Historical records indicate that on January 28, 1700, a ship carrying 30 tons of rice ran aground off the coast of Japan. The ship could not reach safe harbor because **a tsunami, a series of high ocean waves generated by vertical movement of the seafloor sometimes caused by an earthquake,** kept the ship at sea. The ship was subsequently caught in a storm and wrecked. The tsunami waves, some 3 meters (about 10 feet) high, damaged buildings in villages all along the Japanese coastline. Documentary evidence of damage reports was collected in a database of historical earthquake records in Japan. Japanese scientists analyzing these ancient records hypothesized that an earthquake had occurred a little before the arrival of the tsunami, but they could find no record of such an earthquake and were consequently puzzled by this apparent orphan tsunami. The consistent height of the tsunami waves at widespread sites was interpreted to indicate that the tsunami had originated from a distant location. Where was the earthquake that had caused it?

a.

b.

Figure 5.2 Ghost forests. **a.** These spruce trees in Resurrection Bay on the Kenai Peninsula, Alaska, are in an area that subsided 1 meter (about 3 feet) during the 1964 Alaskan earthquake. The subsidence dropped the shallow roots of these trees below high tide, where they were killed by repeated inundation in saltwater. **b.** A ghost forest along the Pacific Northwest coast.

Evidence from Trees

All of this might have remained a mystery if scientists working along the coast of the state of Washington had not made an intriguing discovery. They found groves of dead spruce and cedar trees at several locations. Although the trees had died, their stumps and dead trunks were preserved in present-day salt marshes. Scientists recognized that these ghost forests resembled stands of dead trees found on the Kenai Peninsula of Alaska (Figure 5.2). This section of the Alaskan coastline subsided (dropped) during a giant earthquake in 1964, and the trees drowned when their roots were submerged in saltwater. Could the Washington ghost forests be a clue to a past giant earthquake that caused a section of the Pacific Northwest coast to drop below sea level? There are no written records of such a large earthquake, but Native American tribes of the region tell stories of shaking and flooding of the coastal lands. Unfortunately, no specific dates can be linked to these accounts.

In an effort to determine when the trees died, scientists collected samples from the drowned spruce trees at two sites over 50 kilometers (31 miles) apart and determined their ages. These analyses returned a range of dates from 1680 to 1720 for both locations. The 40-year span was the greatest level of accuracy the radiocarbon dating technique they used could provide. (We will discuss how dating methods work in Chapter 8.) Evidently, whatever killed these trees had occurred about three centuries ago and had affected a broad section of the coastline. As evidence started to point toward a large historical earthquake, scientists began to look for a possible earthquake source.

Evidence from Plate Tectonics

Earthquakes occur when Earth's surface shakes because of the release of energy following the rapid movement of *pieces of the lithosphere* (crust and uppermost mantle) along zones of weakness called *faults*. The very largest earthquakes, informally called *megathrust earthquakes*, occur on faults that are hundreds or even a thousand kilometers in length. In contrast to the Northridge event described earlier, a megathrust earthquake causes shaking that lasts for several minutes, can lead to immense destruction, and is capable of generating tsunami that strike all around the surrounding ocean. In the Pacific Northwest, the only fault capable of producing a megathrust earthquake is a 1,100-kilometer-long (684-mile-long) *fault system* on the seafloor a few tens of kilometers off the coast of Washington and Oregon. (A fault system is a group of faults that move in a similar fashion.) These faults mark the Cascadia subduction zone, along the convergent boundary between the small Juan de Fuca plate and the much larger North American plate (Figure 5.3). Recall from Chapter 4 that earthquakes are larger and more frequent along convergent plate boundaries.

Because it is located on the seafloor, movement on the Cascadia subduction zone fault system would not only have caused extensive ground shaking but would also have displaced large volumes of ocean water to produce a tsunami. In about 15 minutes, waves 10 meters (33 feet) tall would have flooded the nearby coasts of Washington and Oregon, wiping out coastal settlements. Geologists found evidence for these local tsunami when they discovered Native American fire pits that had been covered by a layer of tsunami-transported sand (Figure 5.4a). Similar tsunami sand

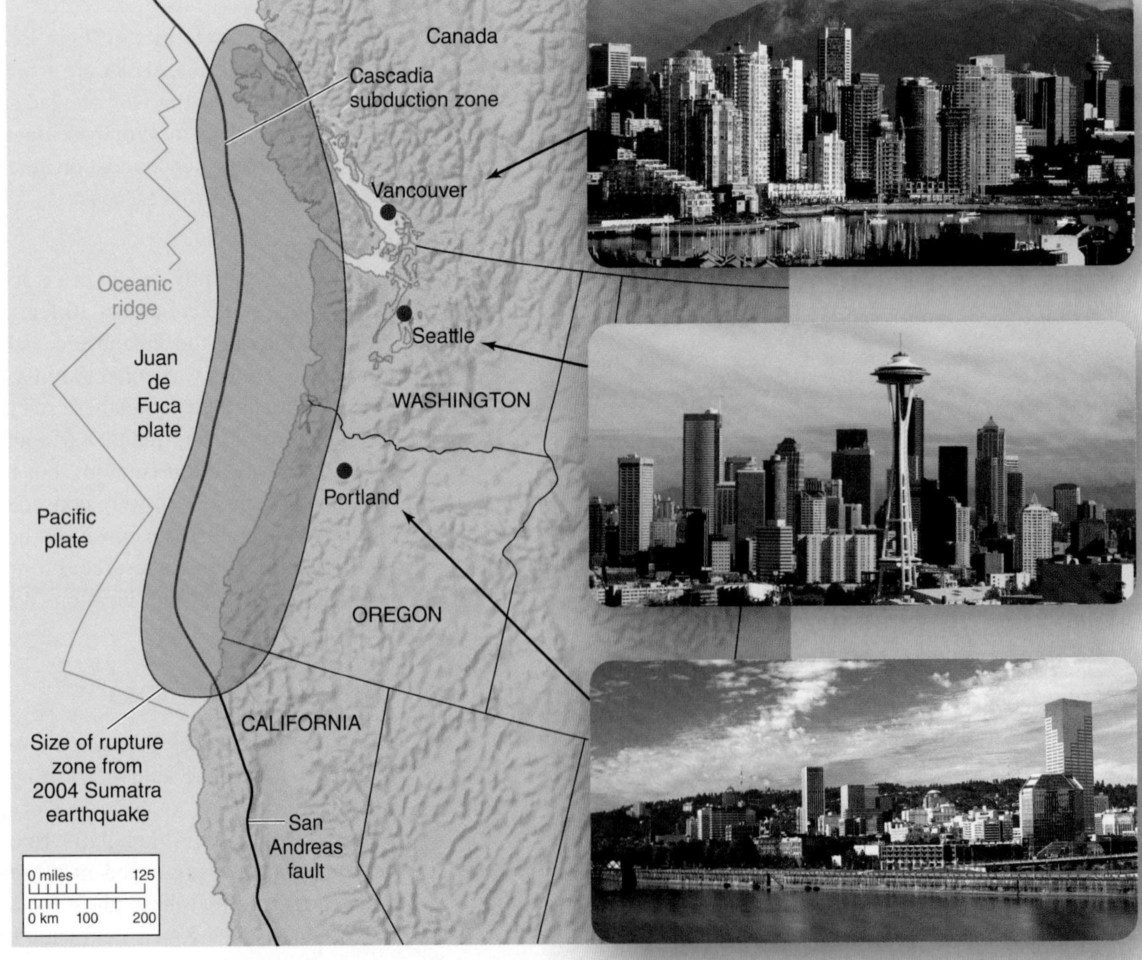

Figure 5.3 Location of the Cascadia subduction zone fault. The subduction zone represents a boundary between the Juan de Fuca plate and the North American plate. The Cascadia subduction zone is comparable in size to the fault system that produced the giant 2004 earthquake in Sumatra, which triggered the Indian Ocean tsunami disaster.

Checkpoint 5.1

basic · intermediate · advanced · superior

The map shows the locations of 15 of the world's largest urban areas. Compare it with a map of plate boundaries from the Chapter Snapshot in Chapter 4. On the basis of their locations relative to plate boundaries, which six cities are most likely to experience an earthquake?

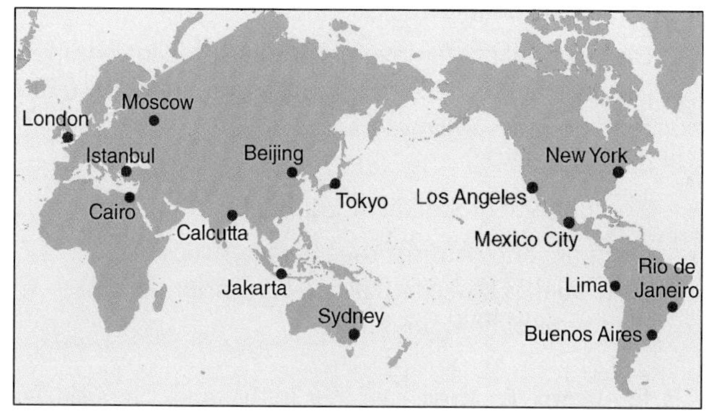

Checkpoint 5.2

basic · intermediate · advanced · superior

Using the map in Checkpoint 5.1, which of the following groups of cities are most likely to experience subduction zone earthquakes?

a) Istanbul, Los Angeles, Sydney
b) New York, Rio de Janeiro, Buenos Aires
c) Beijing, Tokyo, Calcutta
d) Jakarta, Mexico City, Lima

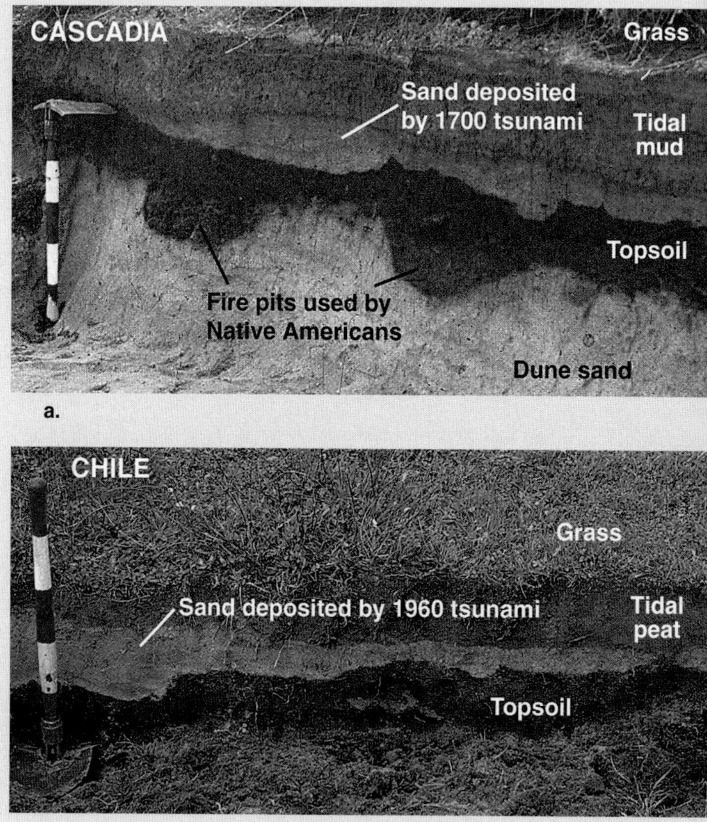

Figure 5.4 Evidence of tsunami. Tsunami sand deposits cover topsoil in **a.** Oregon (1700) and **b.** Chile (1960). Tsunami carried coastal sand inland to cover settlements and fields. The thickness of the sand layer decreases with distance from the coast.

deposits were soon identified at sites all along the coast of the Pacific Northwest, from northern California to British Columbia, Canada. These deposits were similar to sand layers formed along the coast of Chile in 1960 by a tsunami generated by a megathrust earthquake (Figure 5.4b). (Modern versions of these deposits were found along the coast of Sumatra and other nations surrounding the Indian Ocean after the 2004 tsunami.)

Linking the Evidence to the Orphan Tsunami

Observations and hypotheses regarding ghost forests and a possible earthquake in the Cascadia subduction zone were described in articles in scientific journals in the 1990s. Japanese researchers read these scientific articles and recognized that they had a detailed record of an orphan tsunami that fell in the time interval identified by radiometric dating. Consequently, the Japanese scientists offered a new, more detailed hypothesis: the tsunami originated on January 26, 1700, as a result of a great earthquake in the Cascadia subduction zone and later crashed into the east coast of Japan.

To test this hypothesis, dendrochronologists (scientists who study tree rings) tried to more tightly constrain when the trees died by taking samples from the roots of some red cedar stumps in the ghost forests. By comparing tree ring patterns in old living trees with those in the dead specimens, they were able to determine that the trees died between fall 1699 and spring 1700, when new tree rings would have begun to form. These dates were consistent with dates suggested from analysis of the Japanese tsunami records.

What These Findings Mean for the Future

Megathrust earthquakes on the Cascadia subduction zone occur relatively rarely—on average, every 500 years, although some events have been just a couple of centuries apart. However, as the citizens of southeast Asia and Japan can attest, a single event can have substantial economic and cultural implications for decades to come. A megathrust earthquake in the Pacific Northwest today would spell disaster for large cities such as Vancouver, Seattle, and Portland, which have populations in excess of 500,000 each (Figure 5.3). This threat is real enough to have resulted in stricter building code regulations for western Washington and Oregon. New structures in Seattle and Portland must now be much stronger than was required prior to the discovery of the potential for subduction zone earthquakes.

If these new building codes are not followed, many people could die in an earthquake. For example, following an earthquake in 1999 in Izmit, Turkey, engineers discovered that some contractors had failed to use proper materials to construct multistory apartment complexes. Those buildings collapsed, killing most of the occupants. In contrast, buildings in Seattle and the surrounding communities sustained relatively little structural damage following a strong 2001 earthquake because building codes had been followed. Nobody was killed, and only a handful of people received anything more than minor injuries.

Checkpoint 5.3

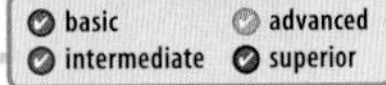
basic
advanced
intermediate
superior

The discovery of the potential for megathrust earthquakes on the Cascadia subduction zone is an example of how scientific explanations are developed. Briefly explain how the development of this hypothesis illustrated the following characteristics of scientific explanations:

1. It was provisional (tentative).
2. It was based on observations.
3. It was predictable and testable.
4. It offered a natural cause for natural events.

Checkpoint 5.4

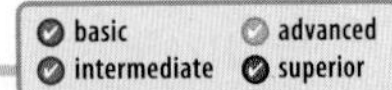
basic
advanced
intermediate
superior

Are earthquake insurance rates based on inductive or deductive reasoning? Explain your choice. (Review information on inductive and deductive reasoning from Chapter 1 if necessary.)

5.3 Faults, Earthquakes, and Plate Tectonics

Learning Objectives

- Define what is meant by the terms *focus*, *epicenter*, and *fault*.
- Explain how faults are classified.
- Discuss how plate movements measured in centimeters per year can result in earthquakes generated by faults that move several meters at a time.
- Describe the global distribution of earthquakes.
- Compare and contrast the characteristics of earthquakes that occur along convergent and divergent plate boundaries.

Faults are fractures, similar to the cracks in a sidewalk, **where two blocks of rock move past each other.** Faults are found anywhere in a tectonic plate where sections of lithosphere can move relative to one another (Figure 5.5). A majority of earthquakes occur on faults that are located along plate boundaries. Sometimes earthquakes occur where faults form for the first time, but most earthquakes occur at places where Earth's crust has ruptured often in the geologically recent past and where it may move again because of the forces associated with plate tectonics.

a.

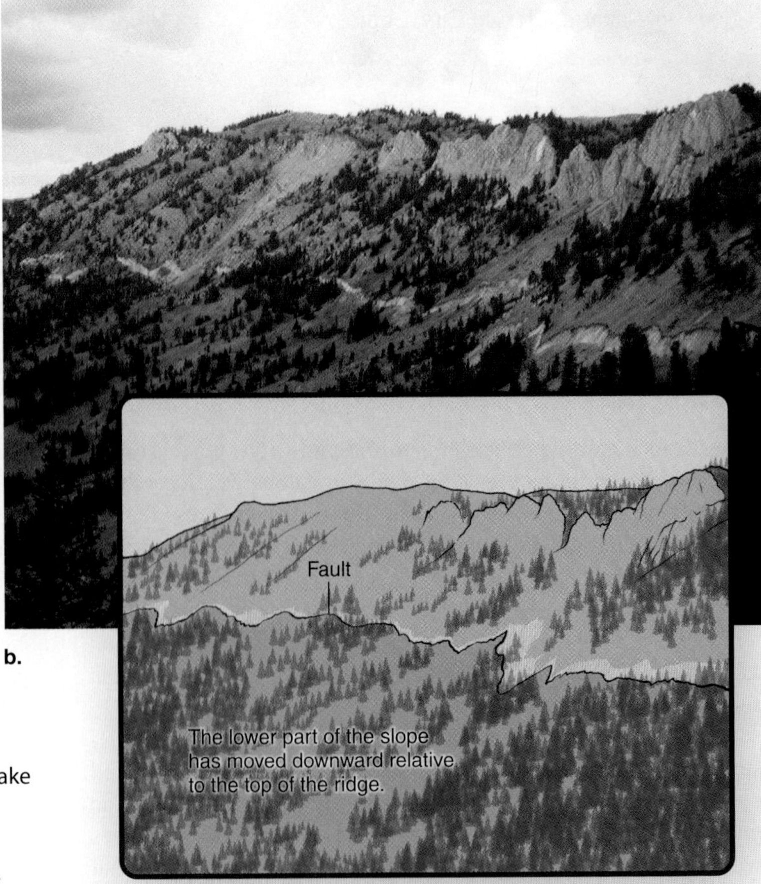
Fault
The lower part of the slope has moved downward relative to the top of the ridge.

b.

Geologist's View

Figure 5.5 Signs of movement on a fault. Movement on a 44-kilometer-long (27-mile-long) fault caused the Hebgen earthquake in Montana in 1959. **a.** The fault broke the surface near a ranch (background). **b.** The fault can be followed for several kilometers along the south flank of Kirkwood Ridge in the center of the image.

Common Features of Faults and Earthquakes

When describing earthquakes, geologists have identified several common features (Figure 5.6). **The location on the fault where movement begins (the earthquake source) is called the *focus*** (plural: foci). As an analogy, try tearing a piece of paper. The tear has to begin somewhere; that point is equivalent to the focus of the earthquake. Earthquake foci can occur at a range of depths to 750 kilometers (465 miles) below the surface. **The epicenter is the geographic location on Earth's surface directly above the earthquake focus.** The location of the epicenter provides the earthquake's name. For example, the 2001 earthquake near Seattle (see Section 5.2) was named the Nisqually earthquake because its focus was located 53 kilometers (33 miles) below the mouth of the Nisqually River in western Washington.

Geologists identify some faults by observing features on the land that have been displaced or rock layers that appear to have moved relative to one another (Figure 5.7). A change in

Checkpoint 5.5

basic · intermediate · advanced · superior

An earthquake occurred on the Erie fault 5 kilometers (3 miles) beneath San Gabriel. Damage from the earthquake was greatest in nearby Fremont. The farthest report of shaking was recorded in Stockton. Where was the earthquake's epicenter?

a) The Erie fault
b) San Gabriel
c) Fremont
d) Stockton

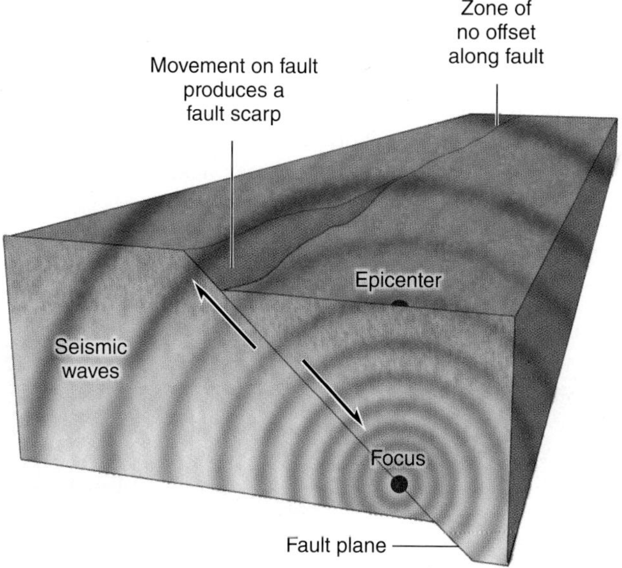

Figure 5.6 Earthquake features. The point on a fault surface where an earthquake begins is known as the focus. The location on Earth's surface directly above the focus is the earthquake's epicenter. It is common for only part of a fault to move during an earthquake. Movement begins at the focus and travels outward along the fault surface.

a.

b.

c.

Figure 5.7 Evidence of faults. **a.** A road displaced by a fault. **b.** A fence in California was offset 3 meters by the 1906 San Francisco earthquake. **c.** A geologist measures the offset of rock layers that formed a fault scarp in the 1964 Alaskan earthquake.

Checkpoint 5.6

basic · intermediate · advanced · superior

Examine the following images. Photo (a) was taken after the Izmit earthquake in Turkey in 1999. On the basis of your observation of the picture, locate the approximate position of the fault and classify the fault type. Cite the reasons, data, and inferences supporting your answer. Photo (b) was taken in the Grand Teton mountain range of northwestern Wyoming. Which location represents a fault scarp in the picture?

a.

b.

the elevation of the land surface across the fault creates a feature known as a *fault scarp* (Figure 5.7c; see Figures 5.5 and 5.6). The *fault plane* is the surface along which the slip or movement occurs during an earthquake. It is important to realize that many faults are deep enough in the lithosphere that there is no evidence on the surface that they even exist.

Directions of Fault Movement

Faults are classified according to the relative movements of rocks on either side of the fault plane. Some faults exhibit vertical movements, where the rocks on one side of an inclined fault move up or down relative to the rocks on the other side of the fault. If the block above the fault moves down, it is a *normal fault* (Figure 5.8a). If the block above the fault moves up, it is a *reverse fault* (Figure 5.8b). In many mountain belts, the reverse fault is inclined at a low angle (less than 30 degrees); in these cases, it is called a *thrust fault*. Repeated movements on groups of faults can build mountains. Alternatively, *strike-slip faults* exhibit horizontal movements, where one side of the fault moves left (or right) relative to the other (Figure 5.8c). The San Andreas fault system in California is a famous example of a series of faults that continues to experience mainly horizontal movements (Figure 5.9).

Each type of fault is most associated with a specific type of plate boundary. Normal faults are most common at divergent boundaries, where plates break apart, and reverse faults

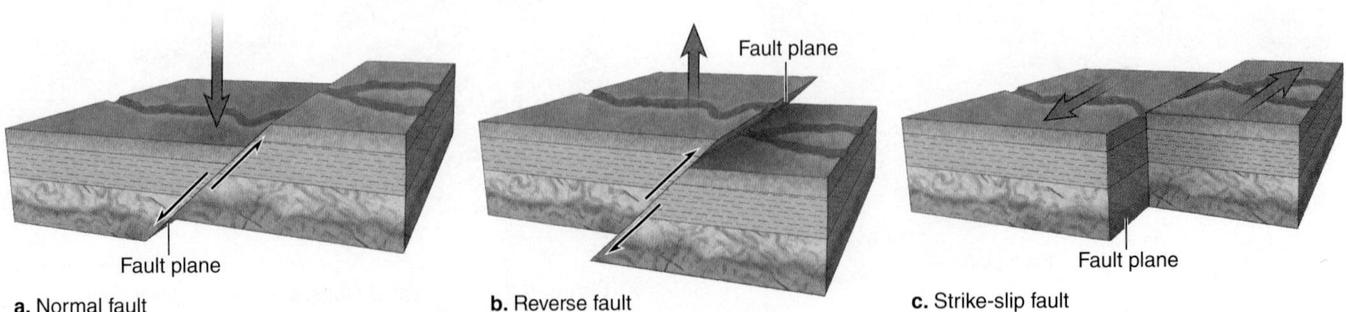

Figure 5.8 Three types of faults. **a.** In a normal fault, the block above the fault moves downward. **b.** In a reverse fault, the block above the fault moves up. The fault may offset the surface to form a steplike change in elevation known as a fault scarp. **c.** Vertical faults that show only horizontal movements are known as strike-slip faults.

characterize convergent boundaries, where plates collide. Strike-slip faults are common at transform boundaries, where two plates move sideways relative to each other.

Amounts of Fault Movement

A fault system may be hundreds of kilometers in length. However, only one of the individual faults may move at a time because the forces of plate tectonics are not equally spread across the entire fault system. Not even the whole fault needs to move to produce a damaging earthquake. In fact, only part of an individual fault in the fault system usually moves during a single earthquake. Individual faults tend to break along segments that are typically no more than a few tens of kilometers in length.

The blocks of rock that move during an earthquake are tens of kilometers thick and may be thousands of square kilometers in area. Consequently, even relatively small movements on faults can produce large earthquakes. For example, the huge 2004 Sumatra earthquake resulted in 20 meters (66 feet) of movement on a reverse fault. Most of this movement occurred on the southern end of the fault, but some movement was distributed over more than 1,000 kilometers (621 miles) of fault length. In other words, most of the eastern margin of the Indian Ocean moved upward. The size of the region that slipped to generate the Sumatra earthquake is about the same as the length of the Cascadia subduction zone (see Figure 5.3). The huge mass of rock that moved during the Sumatra earthquake was sufficient to cause a slight change in the shape of the planet and to speed up its rotation by 3 millionths of a second per day. In contrast, the 2011 Tohoku (Japan) earthquake was caused by movement of 30–40 meters (100–132 feet) over a fault segment that was about one-third the length of that broken in the Sumatra earthquake.

With the exception of megathrust earthquakes, movements on faults are almost always less than 5 meters (16 feet) per quake. For example, the Hebgen (Montana), Alaska, and San Francisco earthquakes were among the largest to strike North America in the last century, yet each managed only a few meters of fault movement. Keep in mind that it is not actual movement that causes most of the destruction from earthquakes but rather the violent shaking that occurs when such big blocks of Earth's lithosphere break.

Stress and Deformation

Movement on faults is related to the squeezing and deforming associated with plate tectonics. Think about bending a wooden pencil until it breaks. At first, the pencil bends (deforms). If you keep applying stress, you know that the pencil will suddenly snap, but you do not know exactly when or where. Stresses build up in rocks where plates interact, and the **rocks adjacent to the fault**

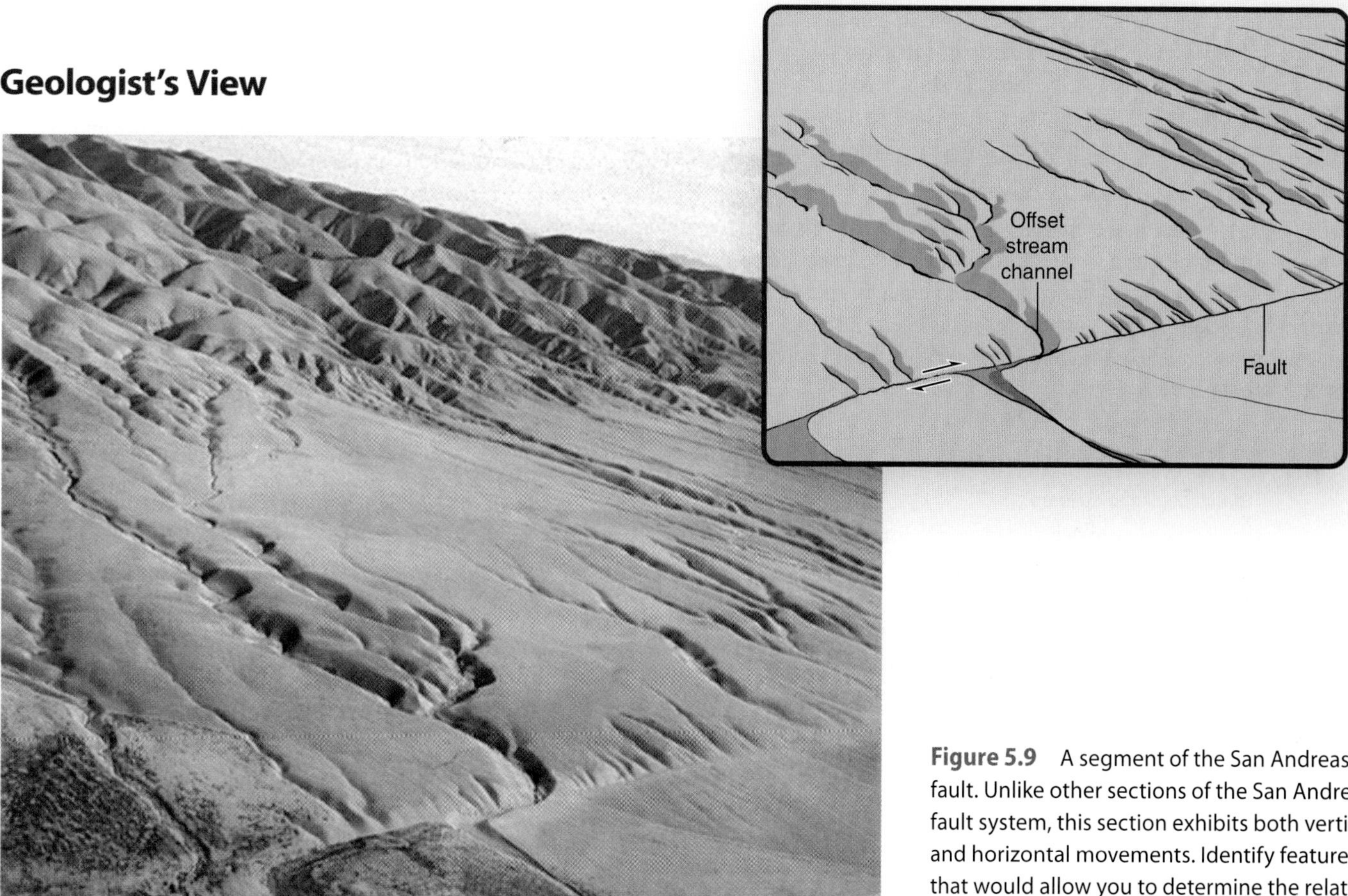

Figure 5.9 A segment of the San Andreas fault. Unlike other sections of the San Andreas fault system, this section exhibits both vertical and horizontal movements. Identify features that would allow you to determine the relative movements across the fault surface.

may be deformed prior to fault movement (Figure 5.10). Faults exhibit movement when stresses are high enough to overcome friction on a fault surface and to cause an existing fault to break again. (See In Further Depth box.)

After the break, there is a period of relative quiet along the fault while plate tectonic forces slowly build up to a level sufficient to break the fault again. **The length of time necessary to build up enough stress to cause a fault to break again is known as the *recurrence interval.*** The recurrence intervals for the mega-thrust earthquakes such as those on the Cascadia subduction zone are measured in centuries, but active fault systems may experience dangerous earthquakes at intervals measured in decades.

Scientists can analyze the deformation occurring in the vicinity of active faults by using various instruments that measure the distortion of rocks or the changes in distance between points on opposite sides of a fault.

- **Creep meters** survey changes in positions between two points on opposite sides of a fault, usually defined by laser sightings. The distance between the two points changes as deformation increases.
- **Strain meters** measure the distortion of an originally circular wall of a hole drilled in rock. Imagine holding a balloon between your two hands. As you push inward on the balloon, it bulges outward at the ends. Likewise, circular holes drilled in rock are distorted as deformation increases.
- **Satellites** of the Global Positioning System can be used to continuously monitor the location of receivers on the ground on either side of a fault. Distance changes between stations distributed over an area of hundreds of square kilometers can be determined to within a few centimeters.

Monitoring by instruments over months or years reveals changes in the surrounding rocks that are related to the buildup of stress along the fault. Changes in the deformation sequence of rocks over broad regions may help identify where stress is accumulating, but it does not tell us when an earthquake will occur or the location of its focus or epicenter.

Where to Expect Earthquakes

To preserve lives and property, scientists seek to determine the likelihood that an earthquake will occur at a given location. Researchers use statistical methods to predict the probability of future damaging earthquakes on particular faults that have sufficient record of prior earthquakes (Figure 5.11). Faults with a high deformation rate are usually more likely to move, but not always. For example, some faults that have been known to experience large earthquakes have not moved for decades or centuries. These segments of the fault zone are termed ***seismic gaps* and are considered potential sites for future earthquakes.**

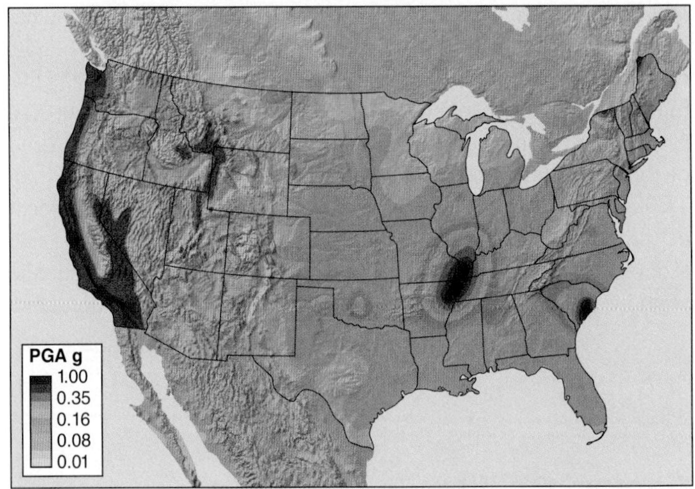
PGA g
1.00
0.35
0.16
0.08
0.01

Figure 5.11 Colors on this map show the levels of horizontal shaking that have a 2-in-100 chance of being exceeded in a 50-year period. Shaking is expressed as a percentage of *g* (*g* is the acceleration of a falling object due to gravity; see page 000 for full explanation). High values of probable ground motion (PGA *g*) correspond to areas with highest hazard (shown in brown). Earthquakes in these regions will typically be more frequent and generate higher *g* forces. Low values, shown in grays and light blues, correspond to areas of lowest hazard characterized by infrequent earthquakes.

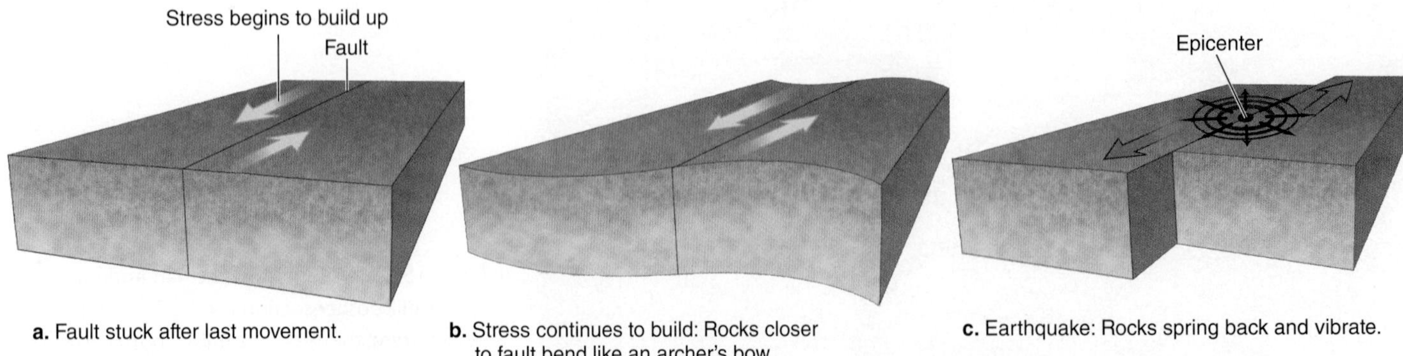
Stress begins to build up
Fault
Epicenter
a. Fault stuck after last movement.
b. Stress continues to build: Rocks closer to fault bend like an archer's bow.
c. Earthquake: Rocks spring back and vibrate.

Figure 5.10 Deformation of rocks before and after an earthquake. **a.** Plate tectonic processes apply stress to rocks around a fault. **b.** As stress builds up, the rocks on both sides of the fault may become distorted. **c.** When the amount of stress reaches the maximum the rock can bear, fault movement occurs, releasing the stress. Plate tectonics causes the cycle to repeat hundreds or thousands more times.

In northern Turkey, two fault segments of the North Anatolian fault system broke during separate large earthquakes in 1963 and 1967 (Figure 5.12). These earthquakes were separated by a 150-kilometer-long (93-mile-long) section of the fault that had experienced no recent movement. Eleven other destructive earthquakes had occurred along different segments of this fault system over the previous 60 years but never in the seismic gap. Then, in 1999, the Izmit earthquake occurred in this "gap," killing thousands. The North Anatolian fault, like the San Andreas fault system, is a strike-slip fault system that doubles as a plate boundary. The steady grind of plate movements makes earthquakes on these faults inevitable.

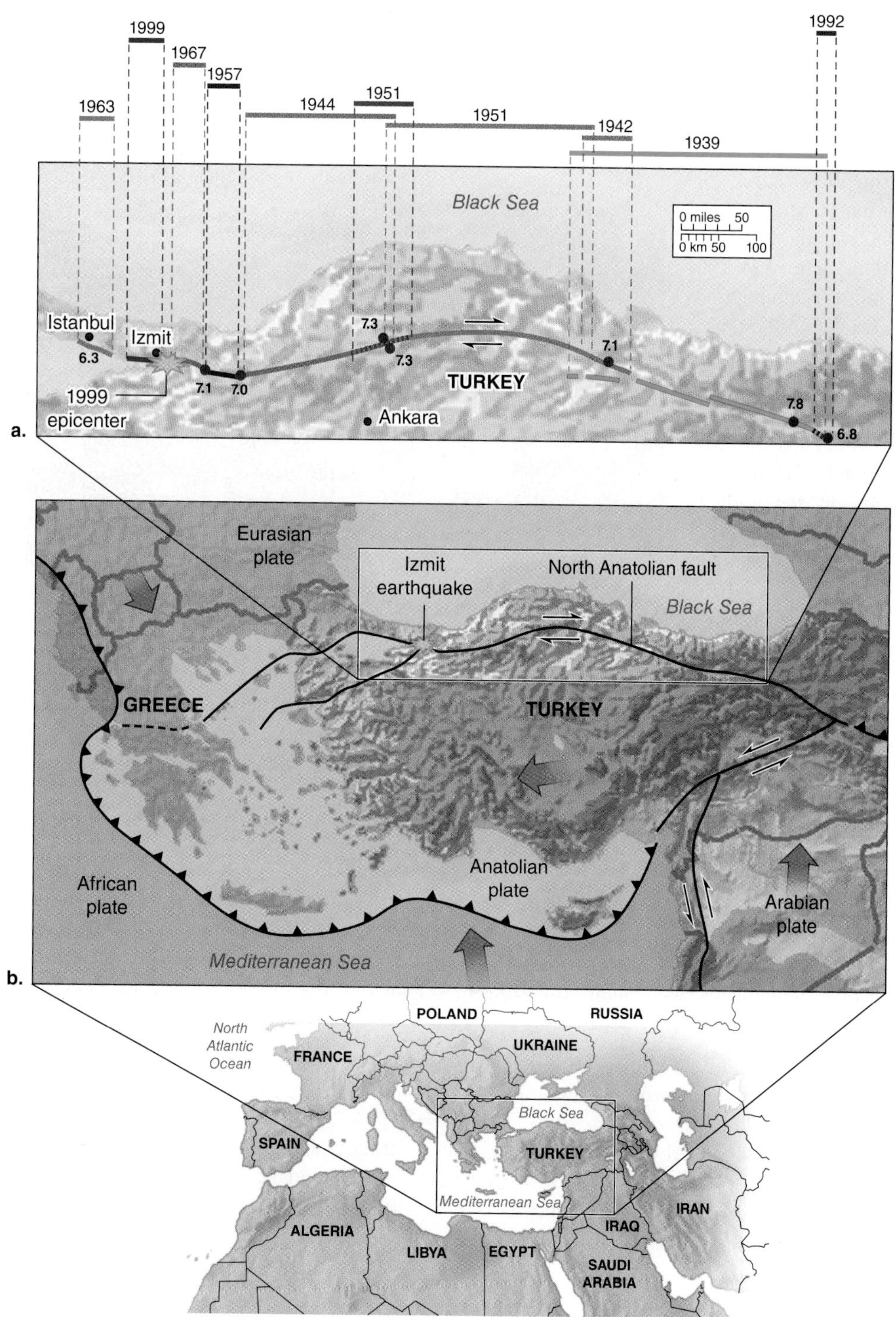

Figure 5.12 Plate tectonic setting for the Izmit earthquake. **a.** The earthquake sequence along segments of the North Anatolian fault, 1939–1999. A series of large earthquakes occurred on the fault system, each resulting from one section of the fault breaking. **b.** The small Anatolian plate is moving westward as it is wedged between the converging Arabian and African plates to the south and the Eurasian plate to the north. A subduction zone in the eastern Mediterranean Sea marks the boundary with the African plate. The other edges of the plate are made up of transform boundaries.

in Further Depth

Geophysicist Ross Stein works for the US Geological Survey Earthquake Hazards program. He has created a relatively simple machine to illustrate how stress builds up prior to movement on faults that produce earthquakes (Figure 1). Stein's device consists of elastic tubing attached to a brick that rests on some sandpaper on a table. Imagine that you are pulling on the elastic. The friction between the sandpaper and the brick initially causes the elastic to stretch without causing the brick to move. But at some point, you pull hard enough so that the brick moves forward and the tubing returns to near its original length. The more friction across the surface between the brick and the table, the more force must be applied to cause the brick to move. Replacing the sandpaper with talcum powder reduces the amount of friction, and thus, less force is needed to move the brick. So, depending on how the model is set up, movements may be slow and gradual, or they may be rapid and catastrophic.

This "brick and bungee model" is similar to what happens during an earthquake. Fault movements may be slow and gradual to generate only small earthquakes, or they may be rapid and catastrophic, causing widespread destruction. The surface of the brick in contact with the table is like the fault surface on which movement takes place. The forces that drive plate movements supply the energy (as when you pull on the tubing). Initially, crustal rocks accommodate some changes as stress builds up (tubing lengthens). However, eventually the stress is sufficient to move the brick (fault moves). Once the brick has slipped, you must build up additional force by pulling on the tubing again to cause another earthquake. Similarly, once a fault generates an earthquake, plate movements must build up stress again before another earthquake can occur.

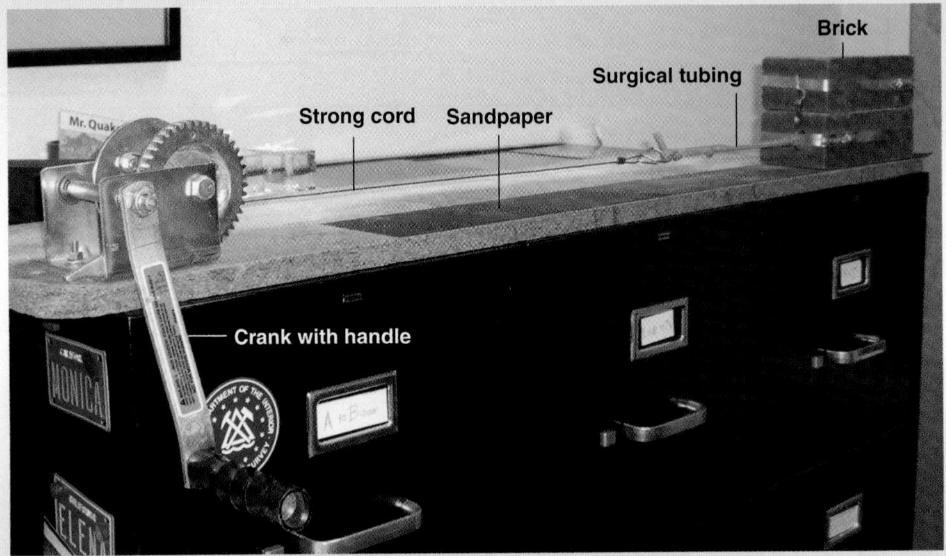

Figure 1 Ross Stein's earthquake machine. This simple device, nicknamed the "brick and bungee" model, effectively models the behavior of faults during earthquakes. A nonstretching cord is connected to the elastic surgical tubing. The tubing is connected to the brick that rests on sandpaper. An additional strip of sandpaper is on the base of the brick. As the crank is turned, the tubing stretches until sufficient stress is built up to move the brick. The brick lurches forward before coming to rest again.

World Distribution of Earthquakes. The distribution of earthquakes around the world has several characteristics that become apparent when a map of global earthquakes (Figure 5.13) is compared to a map of plate boundaries (see Chapter Snapshot in Chapter 4).

- Relatively few earthquakes occur in the interior of plates.
- Global oceanic ridge systems and transform boundaries are areas of only shallow earthquakes.
- The largest earthquakes are associated with convergent plate boundaries.
- Most earthquakes occur at shallow depths where the seismic energy is released closer to Earth's surface.
- Deep earthquakes (focal depth greater than 300 kilometers, or 186 miles) are found adjacent to subduction zones along convergent plate boundaries (for example, western South America).
- The only area where deep earthquakes are not present along the Pacific Rim is the San Andreas fault system, which marks a transform plate boundary in the western United States.

Earthquakes in the United States. There are several seismically active areas in the United States. Earthquakes occur most frequently along the western edge of the North American continent (see Figures 5.11 and 5.13). The largest earthquakes in US

Checkpoint 5.7

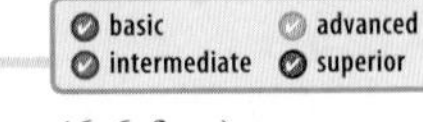

If the San Andreas fault moves 2 meters (6.6 feet) per big earthquake, and plate movement is 2.5 centimeters (0.025 meter per year, or 1 inch per year), how many years of plate motion must accumulate to produce one big earthquake? (Assume all plate motion is accommodated by movements on the San Andreas fault.)

a) 4 years
b) 20 years
c) 80 years
d) 200 years

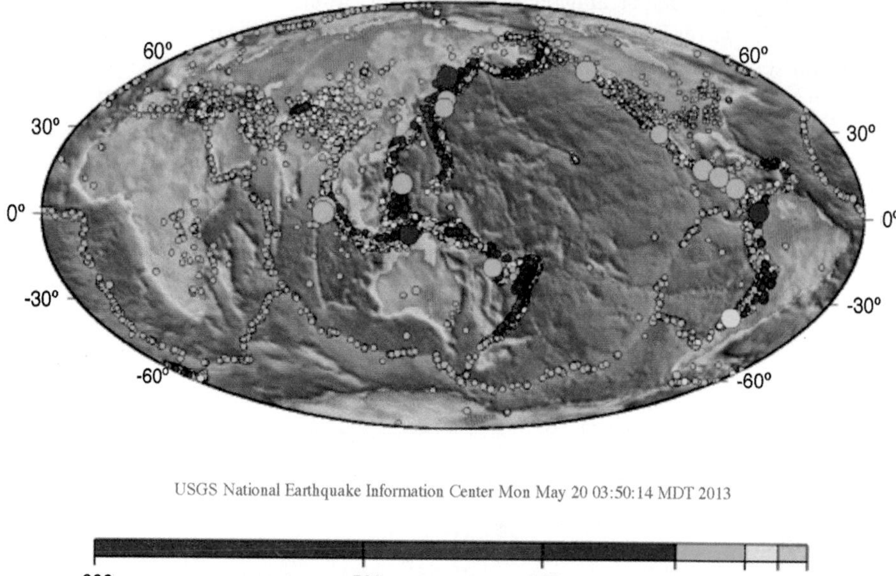

Figure 5.13 Global distribution of earthquakes for 2012. Large circles indicate locations of the largest earthquakes; colors indicate depth of earthquake foci. Compare and contrast with Figure 4.8c (Earthquakes in 2005) to identify consistent patterns of earthquake activity.

history have occurred on the southern coast of Alaska, along the convergent boundary between the Pacific and North American plates. However, California has suffered the greatest losses from earthquake activity because of the large amount of development that happens to be near the Pacific and North American transform plate boundary found there. The Juan de Fuca plate is subducting beneath the North American plate off the west coast of northern California, Oregon, and Washington, making that an area prone to earthquakes. Additionally, some very significant earthquakes have occurred on ancient faults located far from active plate boundaries in the central and southeastern United States in areas such as Memphis, Tennessee, and Charleston, South Carolina.

Checkpoint 5.8

basic | intermediate | advanced | superior

Venn Diagram: Plate Boundary Earthquakes

Use the Venn diagram provided to compare and contrast the characteristics of earthquakes that occur along convergent and divergent plate boundaries.

Convergent plate boundary

Divergent plate boundary

5.4 Seismic Waves and Earthquake Detection

Learning Objectives

- Define body waves and surface waves generated by earthquakes.
- Interpret seismograms to identify the arrivals of different types of seismic waves.
- Discuss how the location and size of an earthquake can be determined from data recorded by seismographs.
- Compare and contrast the characteristics of P, S, and surface waves.

Earthquakes release their energy as vibrations. These vibrations travel through rocks and are called seismic waves. They are recorded on instruments called **seismographs.** Think about the last time you were startled by a loud sound. Slight differences in the time and volume of the sound that reached each of your ears helped you orient on the direction, distance, and possible source of the sound. Just as your ear detects sound and your brain interprets those sounds, seismographs detect and record seismic waves. The printed record of the waves is a seismogram, and scientists interpret the data on seismograms to learn more about earthquakes.

Types of Seismic Waves

Like concentric ripples caused by dropping a pebble in a pond, seismic waves move outward from the earthquake's focus. But unlike ripples in the pond, seismic waves move away from the focus in essentially all directions, not just along Earth's surface. Observations at many stations all over Earth record data from earthquakes. These data show that there are two forms of seismic waves: **surface waves travel along Earth's surface, while body waves pass through Earth's interior** and move more quickly than surface waves (Figure 5.14).

Surface Waves. Surface waves can be further classified on the basis of their motion. Some of these waves, known as *Rayleigh waves,* cause the surface to move vertically in a wavelike motion and cause much of the destruction associated with earthquakes (Figure 5.15). Surface waves that move sideways but with

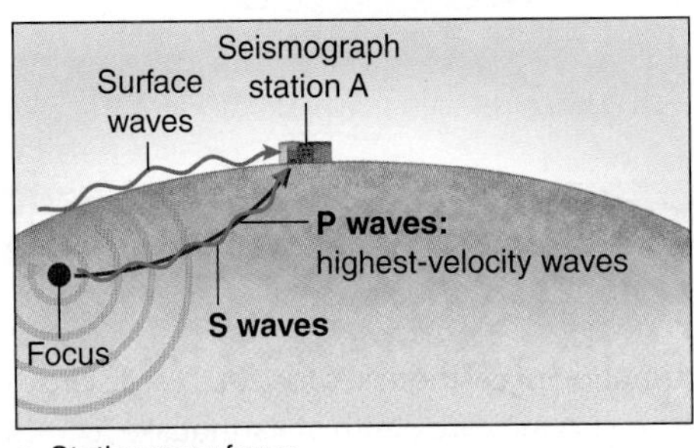

a. Station near focus

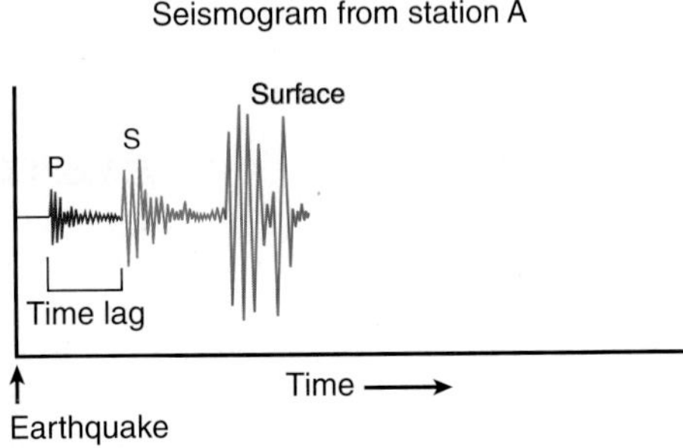

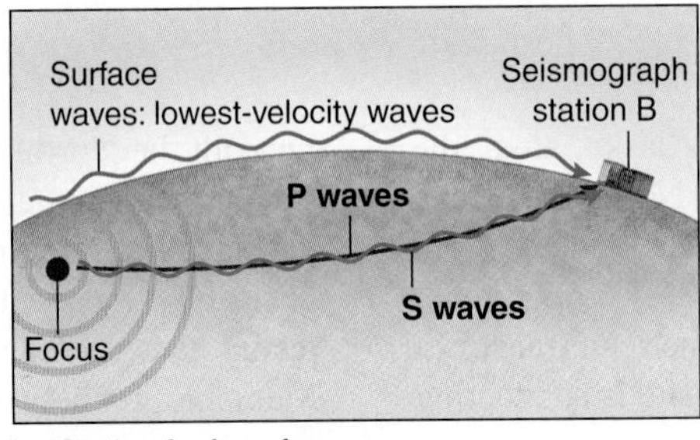

b. Station far from focus

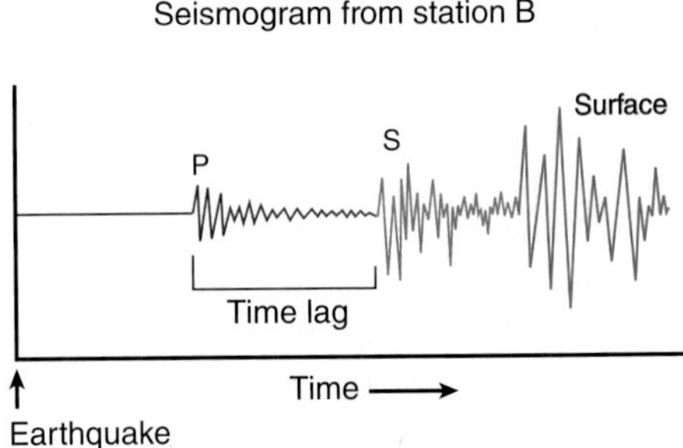

Figure 5.14 Travel paths for surface waves and body (P, S) waves. After an earthquake, body waves arrive at the seismograph station first because they take a more direct route and they travel at faster velocities than surface waves. (Note: Diagram is not to scale.) There is less difference in the arrival times for P and S waves for recording stations that are closer to the focus (a) than for more distant seismograph stations (b).

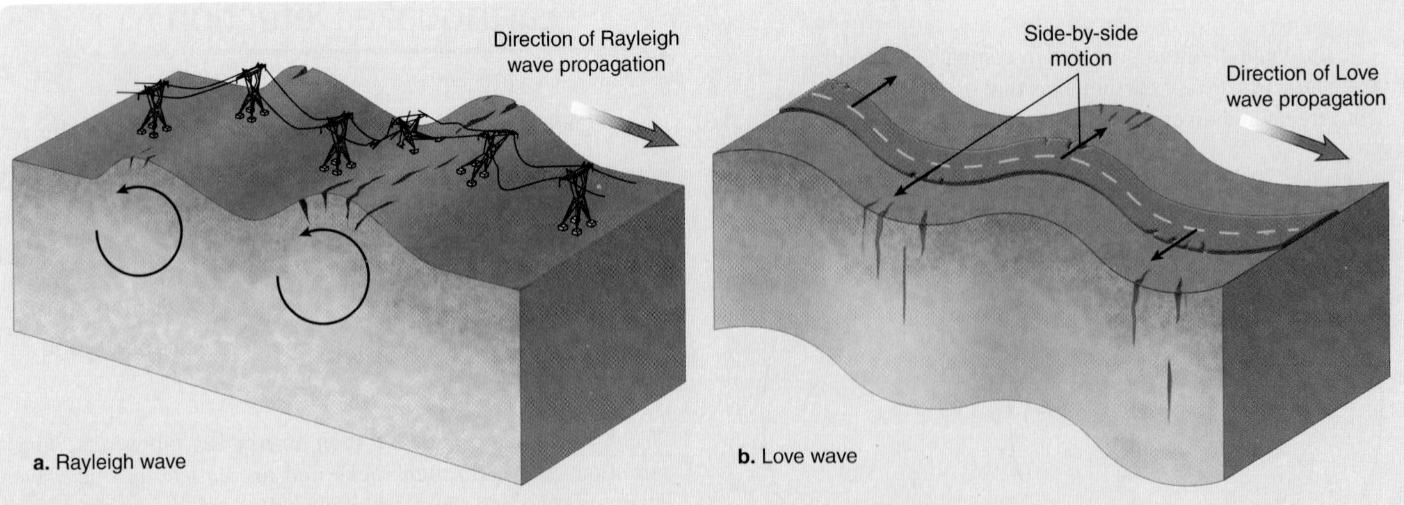

Figure 5.15 Two types of surface waves. **a.** Rayleigh waves produce vertical motions of the land surface. **b.** Love waves move sideways but not vertically.

no vertical movement are called *Love waves*. The combined effects of surface waves cause structures to be pushed upward, and then the surface moves horizontally, thus displacing a building from its foundation.

Body Waves. **Body waves can also be divided into two types: P and S waves.** The highest-velocity waves are known as *primary waves,* or *P waves*, because they are the first waves to arrive at a distant seismograph (Figure 5.14). P waves compress the material through which they pass in much the same way that a sound wave compresses air. Molecules vibrate parallel to the travel direction just as a vibration passes along a Slinky toy (Figure 5.16a). P waves travel at speeds of 4 to 6 kilometers per second (2.5 to 4 miles per second) in the crust. Because of different compositions and greater density of the materials, P waves travel at much faster speeds in the mantle and core (up to 12 kilometers per second, or 7.5 miles per second).

The second waves to arrive, the *shear waves,* or *S waves*, vibrate perpendicular to their travel direction, like the wave that passes along a rope when it is given a sharp snap downward (Figure 5.16b). S-wave velocities are 3 to 4 kilometers per second (2 to 2.5 miles per second) in the shallow crust. Since S waves cannot pass through liquids, they have been useful for inferring the physical state of rocks below Earth's surface (molten or solid). The velocities of both types of body waves are lower in loose materials (sand, gravel) and partially melted rock and higher in solid materials (rock). In general, the fastest seismic waves are P waves; S waves are intermediate in velocity; and surface waves are the slowest.

Checkpoint 5.9

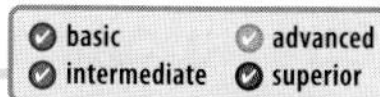

Examine the following cutaway view of a section of Earth's crust containing a fault. Answer the questions that follow, assuming that an earthquake originated at the focus shown on the fault surface.

1. Which locations would experience both body waves and surface waves?
   a) A and B only
   b) C and D only
   c) E only
   d) A, B, C, D, and E

2. Which locations would experience P and S waves?
   a) A and B only
   b) C and D only
   c) E only
   d) A, B, C, D, and E

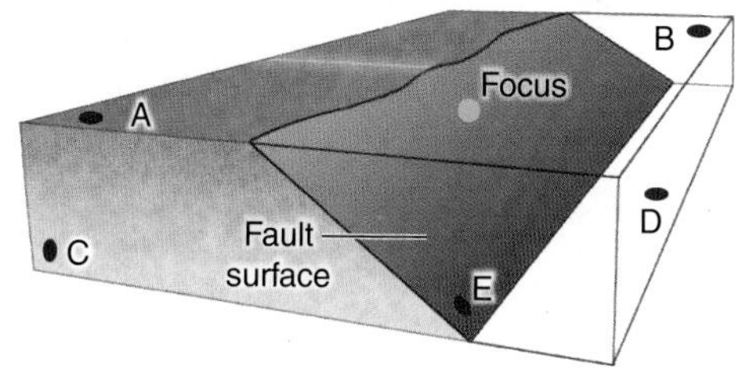

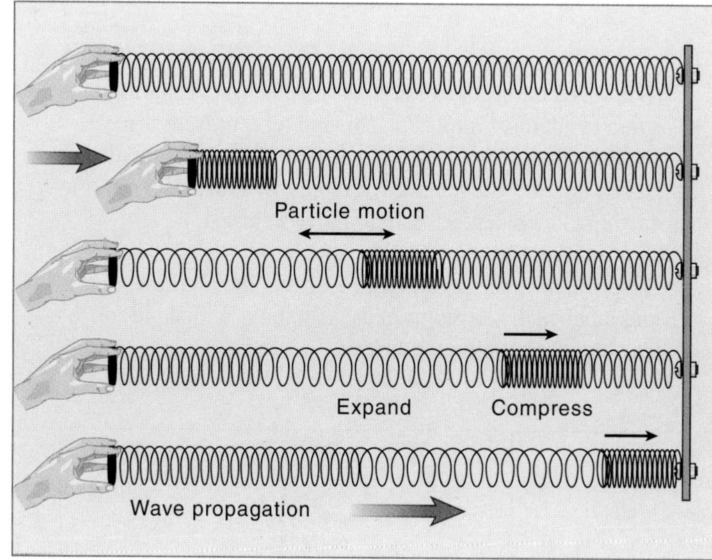

a. Primary wave

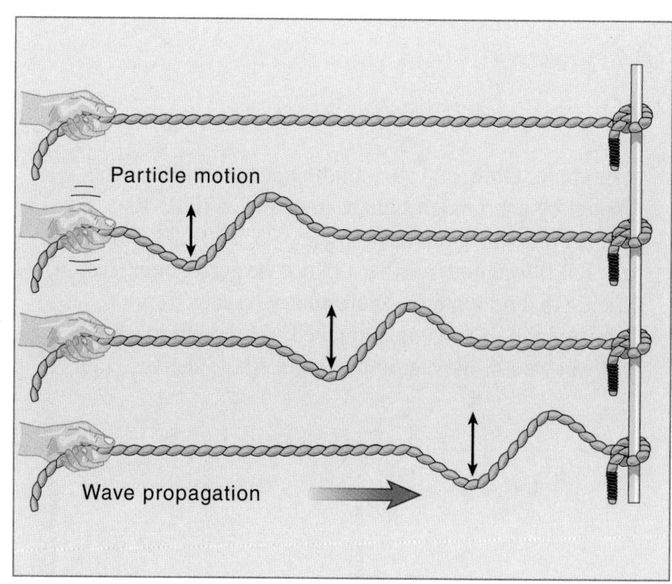

b. Secondary wave

Figure 5.16 P-wave and S-wave motions. **a.** P waves are similar to the passage of a vibration through a Slinky. The vibration occurs in the same direction that the wave travels. **b.** S-wave motion is analogous to a vibration moving along a rope. The vibration occurs perpendicular to the direction in which the wave travels.

Checkpoint 5.10

basic intermediate advanced superior

Examine the following seismogram and answer the questions.

1. Which letter represents the arrival of the first S waves?
2. Approximately how much time elapsed between the arrival of the first P and S waves?

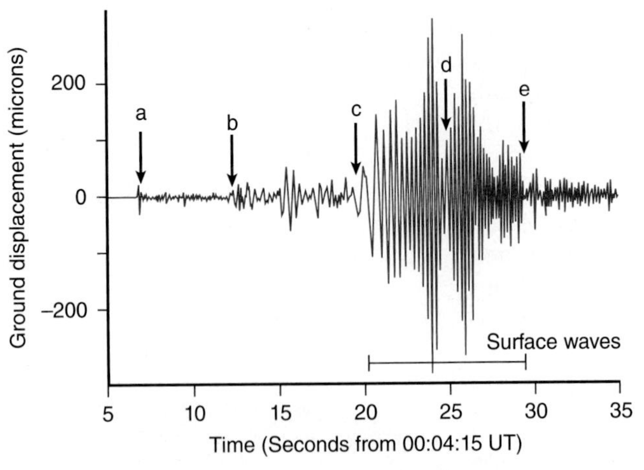

Determining Earthquake Location and Magnitude

The farther away a seismograph is from the earthquake focus, the longer it takes for seismic waves to reach the seismograph (Figure 5.14). Both P and S waves are generated at an earthquake focus as a result of movement on a fault. Which of these waves will arrive at the seismograph station first? P waves will arrive first because of their greater velocity. Surface waves are the last to arrive because P and S waves travel faster and take a more direct route through Earth.

To determine the location, time, and size of an earthquake, scientists use the times that the various waves arrive and their relative amplitudes (heights) as recorded on a seismogram. Because S waves are slower than P waves, the time interval between the arrival of P and S waves increases the farther the seismograph station is from the epicenter (Figure 5.14). Consequently, the difference in arrival time between P and S waves as shown on a seismogram can be used to determine the distance of the station from the earthquake source. Seismologists compare the time difference with established rate curves (Figure 5.17a) to determine distance from the earthquake.

Data from a single seismograph station are insufficient to determine the earthquake epicenter because a seismograph yields only the distance from each earthquake source. The epicenter could be located anywhere along a circular arc of the calculated distance from the seismograph station. Therefore, seismologists must use data from multiple recording stations to determine the epicenter. The common intersection point for several circles plotted relative to different stations represents the point on the surface

Checkpoint 5.11

Ashtabula Earthquake Exercise

Ashtabula, Ohio, and surrounding communities were shaken by a 4.5 magnitude earthquake at 10:03 P.M. eastern time on Thursday, January 25, 2001. The event was felt throughout northern Ohio, western Pennsylvania, Michigan, and Ontario. Preliminary damage reports from Ashtabula indicated cracked plaster and masonry, walls bowed or moved, items knocked off shelves, and a ruptured natural gas line that resulted in evacuation of some residents. Graphs (a), (b), and (c) represent three seismograms from the events as recorded at Ashtabula, Lakeland, and Cleveland. Cleveland was farthest from the earthquake epicenter; Ashtabula was closest.

Describe how the seismograms differ, and identify and label each seismogram as Ashtabula, Lakeland, or Cleveland. Explain your answer.

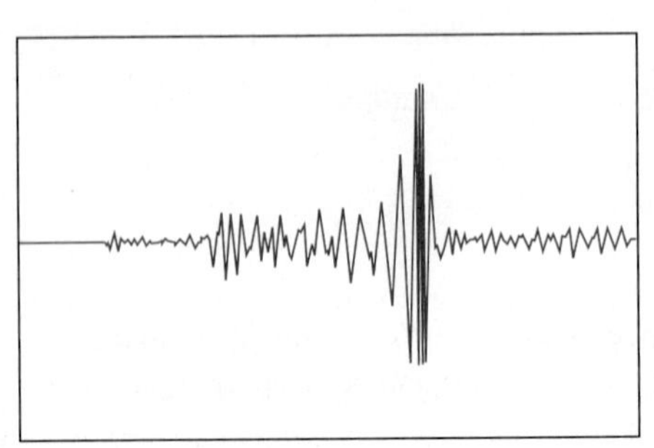

a.

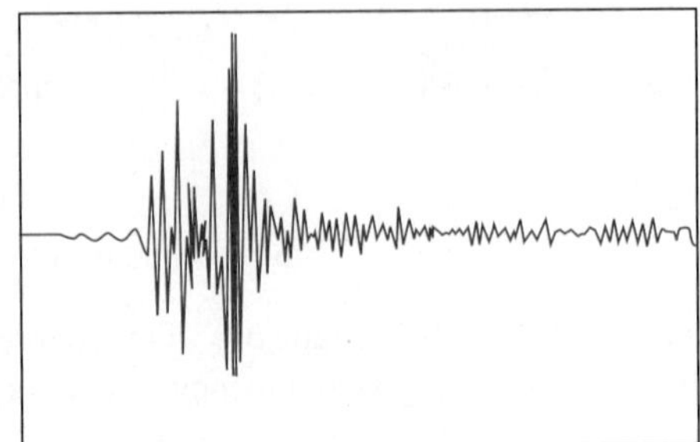

b.

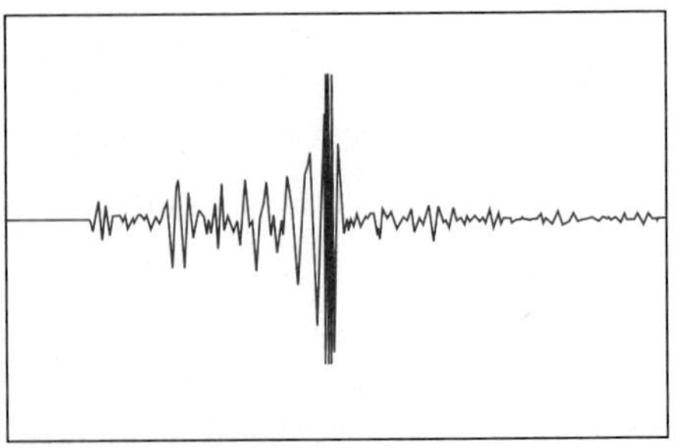

c.

above the earthquake source (Figure 5.17b). When three or more seismographs in different locations record the earthquake, the data can be used to determine the earthquake's epicenter.

The amplitude (height) of the seismic wave record is proportional to the magnitude of shaking associated with the earthquake—the greater the amplitude, the more violent the shaking. Depending upon how the magnitude of the earthquake is calculated, scientists will measure the amplitude of the P wave, S wave, or Rayleigh wave. The deeper an earthquake, the simpler the pattern of surface waves recorded on the seismogram. The most accurate method of determining the focal depth of an earthquake is to determine the difference in arrival time between the initial P waves and later P wave reflections that bounce off Earth's surface near the focus. The deeper the focus, the greater the time interval between the two seismic waves.

Checkpoint 5.12

basic · intermediate · advanced · superior

Venn Diagram: Seismic Waves

Place the numbers corresponding to the following phrases in the appropriate locations in the seismic wave Venn diagram provided.

1. Most damaging
2. First arrival
3. Last arrival
4. Body wave
5. Rayleigh wave
6. 4–6 kilometers per second in crust
7. Second arrival
8. Love wave
9. Particles move in direction of wave
10. Waves generated at time of earthquake
11. On Earth's surface
12. Determines magnitude

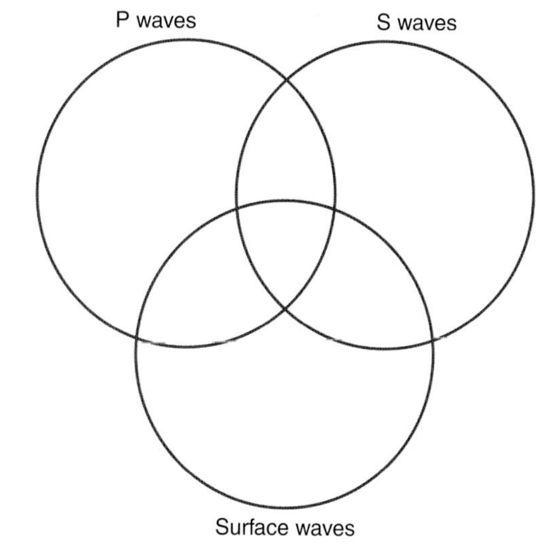

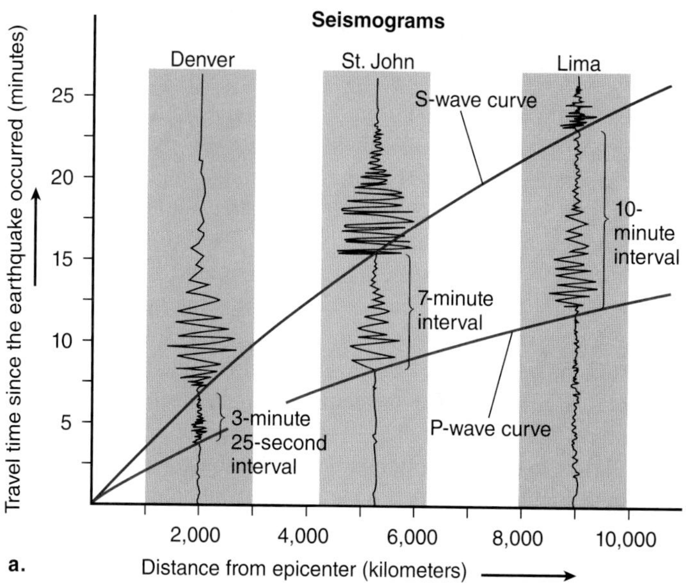

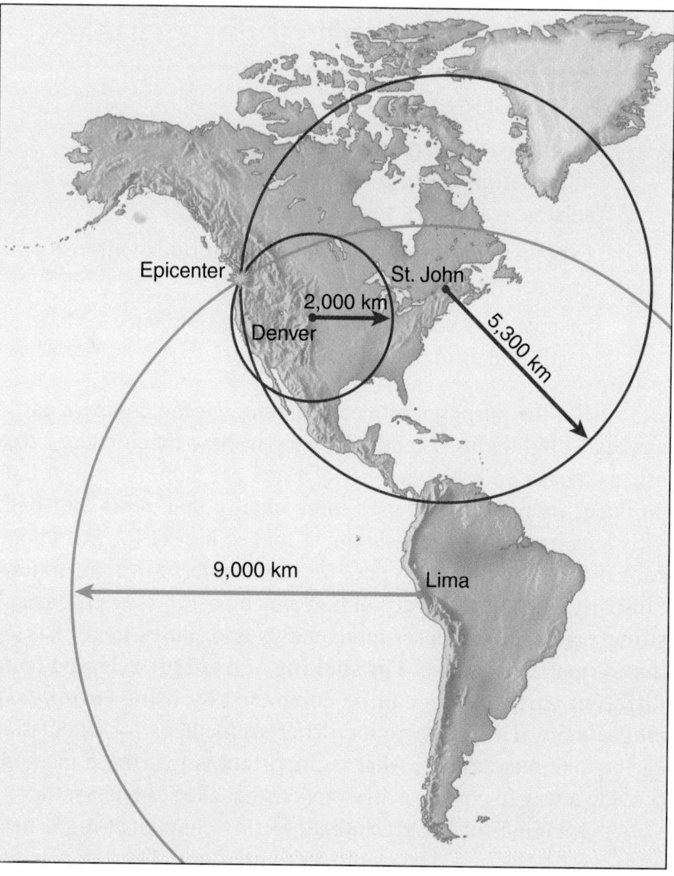

Figure 5.17 Determining earthquake location. **a.** The time interval between the arrival of the P and S waves increases with increasing distance from the earthquake focus. To determine the distance to the epicenter, the time interval is compared with the estimated rate curves (red and blue lines). **b.** Circles plotted at each station reflect the distance from the epicenter to that station. For example, an earthquake occurring 2,000 kilometers from Denver happens somewhere on the blue circle with a radius of 2,000 kilometers. The actual location can be found only when similar circles are plotted for at least two other stations. The epicenter is the intersection of the three circles.

Seismic Waves and Earthquake Warning Systems

No dead animals were found in the Yala wildlife reserves on the south coast of Sri Lanka following the 2004 Indian Ocean tsunami. Scientists believe that many animals (mammals, birds, insects, spiders) can detect Raleigh waves that would have reached the island several hours before the tsunami. The animals responded to their natural warning system by moving inland.

Scientists hope to have a human warning system in place that relies on technology gauging the amplitude of just the first 2-second-long record of body waves and sounds an alarm if it detects a large-magnitude earthquake. People in sites close to the epicenter would have less than 10 seconds to react to the event, just enough time to follow the standard drill to drop, cover, and hold on (drop and find cover under a table or desk and hold on during the subsequent shaking). Folks farther away might have several hours to reach higher ground to escape tsunami.

5.5 Measurement of Earthquakes

Learning Objectives

- Estimate the magnitude of an earthquake, using information about the amplitude and arrival times of seismic waves in a seismogram.
- Explain the difference between earthquake intensity and earthquake magnitude.
- Discuss why earthquake intensity would vary for two earthquakes in nearby locations.

For the purposes of measurement, we can compare earthquakes to lightbulbs. Lightbulbs vary in how much energy they use, as measured in the number of watts per bulb. A 75-watt bulb produces more light and uses more energy than a 40-watt bulb. This single, standard measurement makes it easy for us to compare different bulbs and select the right bulb for a light fixture. Similarly, earthquakes vary in how much energy they produce. A strong earthquake releases more energy and causes more shaking than a small earthquake. **The shaking and energy released from different earthquakes can be compared by using *earthquake magnitudes*.** Though there are different methods for calculating earthquake magnitudes, what is important is that these methods provide a way to compare different earthquakes to one another.

Imagine that you are reading this chapter at night near a 60-watt bulb in a desk lamp in your room. The lamp is the only light source in the room. The light coming from the lamp diminishes into shadows in the far corners of the room. If you used a light meter, you would be able to detect a range of light intensities at different locations throughout the room, with the brightest light near the lamp and lowest light intensity in the farthest corner of the room. In much the same way, we can observe that the intensity of damage associated with an earthquake is typically greatest near the epicenter and decreases with increasing distance from the source. ***Earthquake intensity* is a measure of the effects of earthquakes on people and buildings** and exhibits a range of values for each earthquake. Earthquake intensity can be used to compare and contrast damage resulting from earthquake activity.

Earthquake Magnitude

In the 1930s, Charles Richter developed the Richter scale to measure the magnitude of shallow earthquakes in California. His method was developed for a specific instrument but provided systematic methods to measure earthquake magnitude for different events. The scale was based on a reproducible method of interpreting seismogram records. These early measurements of magnitude simply relied on two key factors that could be readily determined from a seismogram: the difference in P- and S-wave arrival times and surface wave amplitude. These values were used in a standard formula that calculated the earthquake magnitude. This procedure was "standardized" because identical signals measured by different instruments could be compared to one another. As a result, all stations were able to determine the same earthquake magnitude for a given quake.

The Richter method provided consistent results if the earthquakes were no more than a few hundred kilometers from the seismograph stations and occurred at relatively shallow depths in the crust. However, as instruments became more sophisticated, it became clear that the original method could not accurately measure the magnitude of the largest earthquakes. Nor could it precisely account for deep earthquakes and those at large distances from the nearest seismograph stations. Thus, while the "Richter scale" has entered into everyday conversation as a measure of earthquake size, it has actually fallen out of use by the scientists who study earthquakes. Alternative methods that can take account of rock properties (such as the rock rigidity or stiffness), the area of the fault plane that actually moves, and the amount of movement on the fault provide more accurate measures of the energy released from an earthquake. Results from these more modern methods are similar to those indicated by the Richter scale at near distances. Today, scientists refer to measurement of earthquake magnitudes based on how they are calculated, rather than to Richter magnitudes. This is an example of technology changing how we interpret the world and its phenomena.

***Earthquake magnitude* is measured on a logarithmic scale in which each division represents a 10-fold increase in the ground motion** associated with the earthquake (Figure 5.18) and an approximate 32 times increase in energy released. For example, using a logarithmic scale, a magnitude 7 earthquake has 10 times as much ground motion (and releases over 32 times the energy) as a magnitude 6 earthquake. A magnitude 5 earthquake exhibits 100 times as much motion (about 1,000 times the energy) as a magnitude 3 earthquake (Table 5.1).

The descriptions of earthquakes range from minor to great, depending on their magnitude (Figure 5.18). Not surprisingly, minor earthquakes are more numerous than great ones. The description of an earthquake as "major" or "great" has specific connotations about its magnitude; it does not mean it was just a big earthquake.

The earthquake magnitude scale does not have a maximum value. Of the largest measured earthquakes discussed in this

Table 5.1 Comparison of Relative Amounts of Ground Motion and Energy Released from Earthquakes of Different Magnitudes

Magnitude	Ground motion	Energy
1	1	1
2	10	32
3	100	1,024
4	1,000	32,768
5	10,000	1,048,576
6	100,000	33,554,432
7	1,000,000	1,073,741,824
8	10,000,000	32,359,738,368
9	100,000,000	1,099,511,627,776

chapter, the 1960 Chile earthquake is the largest ever recorded, with a magnitude of 9.5, and the 1964 Alaskan quake is the second largest, rated at 9.2. The Tohoku earthquake was magnitude 9.0 and the 2004 Sumatra earthquake was magnitude 9.1. Keep in mind that earthquakes have been officially measured for less than a century. The largest earthquakes occur at intervals measured in hundreds of years, so there may well be larger magnitude earthquakes in the future.

Prior to the 2011 Tohoku earthquake, scientists were pretty certain that there was a correlation between fault length and earthquake magnitude; **the longer the fault rupture, the bigger the earthquake.** The 1906 San Francisco earthquake (magnitude 7.7) was caused by a rupture of 400 kilometers (250 miles) of the San Andreas fault, and shaking lasted for nearly 2 minutes. In contrast, the magnitude 6.7 Northridge earthquake was caused by displacement on a fault segment 14 kilometers (9 miles) long, and the duration of shaking was just 7 seconds. The relationship between fault length and magnitude had been interpreted to suggest that a magnitude-10 earthquake will probably never occur due to faulting. This is so because no fault on Earth is long enough to cause an earthquake of that size, and rocks would likely break before that much stress could accumulate. However, the Tohoku earthquake generated greater than expected fault slip (more than 30 meters or 100 feet) over a shorter than expected fault segment (hundreds of kilometers), causing scientists to reconsider their models for earthquake formation. In some cases, large earthquakes may occur or shorter faults but exhibit greater fault slip than normal.

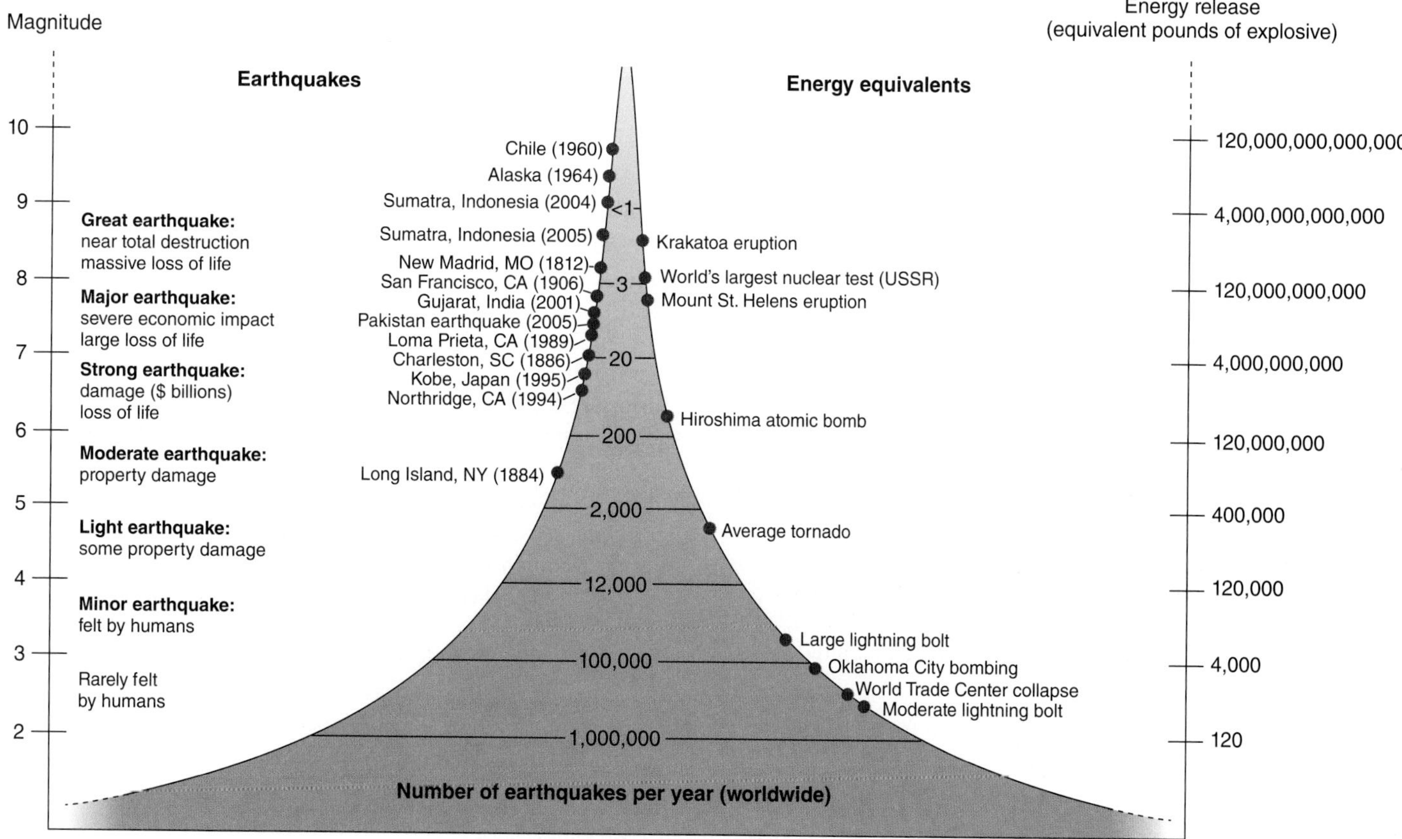

Figure 5.18 Earthquake magnitude scale. Earthquake magnitude is logarithmic; thus, a magnitude-5 earthquake has 10 times as much ground motion as a magnitude-4 earthquake. Earthquakes are classified from small to great, depending on their magnitude. Each year, there are fewer than 3 great earthquakes, 20 major earthquakes, 200 strong earthquakes, and approximately 800 moderate earthquakes.

Checkpoint 5.13

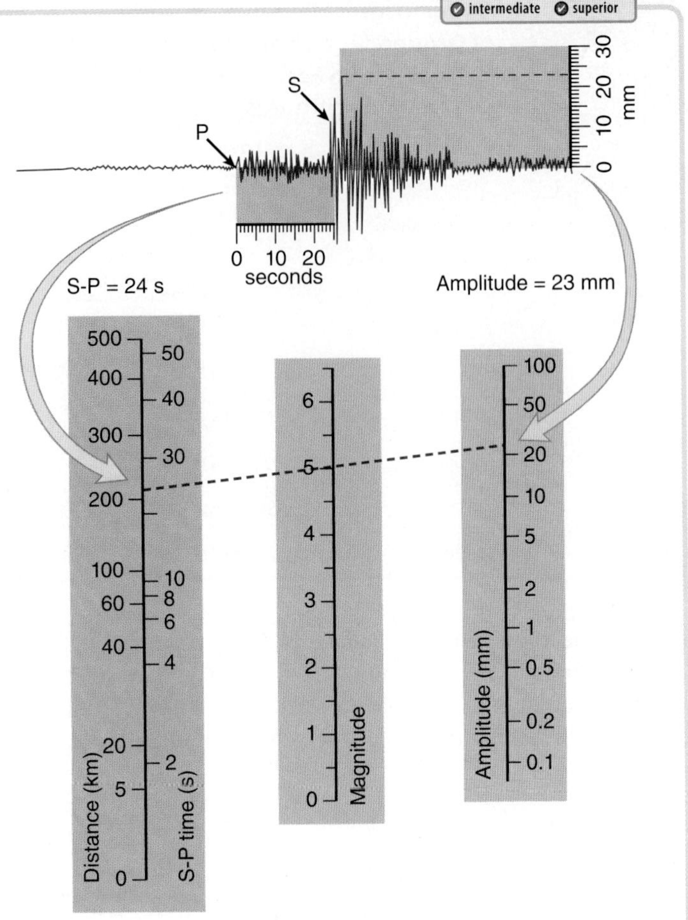

A graphical device, an earthquake nomogram, can be used to estimate the magnitude of an earthquake if we know our distance from the epicenter and can determine the amplitude of the seismic wave measured on a seismogram.

In the diagram, the plotted line represents a simulated earthquake that had a 24-second interval between the arrival of the P and S waves. This represents a distance of about 210 kilometers from the recording station. The earthquake produced a recorded seismic wave with 23 millimeters of amplitude. Drawing a line between the 24-second point on the S-P scale and 23 millimeters on the amplitude scale allows the magnitude to be determined (5). Answer the questions that follow.

1. Two earthquakes occurred 100 kilometers away with recorded amplitudes of a) 10 millimeters; b) 100 millimeters. What were the magnitudes of these earthquakes?
2. An earthquake of magnitude 6 has a recorded amplitude of 50 millimeters on the seismogram. How far away is the earthquake source, and what was the difference between the arrival times for S and P waves?

Checkpoint 5.14

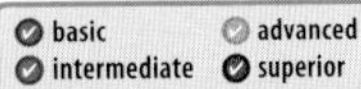

Three sites (L1, L2, L3) record earthquake magnitude and earthquake intensity for the same earthquake. L1 is located closest to the focus and L3 is farthest away. Where is the intensity greatest, and what happens to the earthquake magnitude calculated at the different sites?

a) Intensity is greatest at L1; calculated magnitude is the same at each site.
b) Intensity is greatest at L3; calculated magnitude is the same at each site.
c) Intensity is greatest at L1; calculated magnitude decreases with distance from the focus.
d) Intensity is greatest at L3; calculated magnitude decreases with distance from the focus.

Earthquake Intensity (Modified Mercalli Scale)

The ancient Chinese, Romans, and Egyptians wrote about the effects of earthquakes on society. Most of these accounts described the sensations and experiences people reported during the earthquake and the damage it caused. Eventually, a damage-based scale was developed to compare different earthquakes. One such scale, the *Modified Mercalli scale,* used both damage to structures and the experiences of people to define a measurement system (Table 5.2). The scale ranks earthquake intensity from I to XII (1 to 12), using Roman numerals.

The Modified Mercalli scale is useful in evaluating damage from regions having insufficient seismographs and in ranking historical earthquakes that occurred before the widespread use of seismographs (prior to World War II). There are several detailed accounts of historical events in places that we do not normally think of as being at risk from earthquakes. For example, three major earthquakes were centered in southeastern Missouri (New Madrid) over a 3-month period from December 1811 to February 1812. These data have been used to assemble seismic hazard maps (see Figure 5.11) that show significant hazards in parts of the eastern half of the United States. Have another look at Figure 5.11; are you living in or near an area at risk of significant shaking?

Fewer earthquakes have occurred in eastern North America than in the west, but their effects were felt farther from their sources. The energy from earthquakes in eastern North America is transferred through Earth more efficiently because the crust is less fractured than it is in the west. Earthquakes in the eastern United States typically occur on old fault systems representing plate boundaries that were active hundreds of millions of years ago. Movements on these faults are rare, but the size of these historical events, the potential for future earthquakes, and their greater reach can pose a real challenge to scientists trying to reduce earthquake hazards.

Although the intensity scale is relatively easy to use and is helpful in gauging the human impact of an earthquake, it is not widely applied in the scientific analysis of modern earthquakes for several reasons. First, the scale is based on damage, but the amount of damage depends on factors other than the earthquake itself. The greater the population, the greater the chance that some buildings will be damaged. This points out a potential flaw in using only the Modified Mercalli scale to determine earthquake magnitudes. For example, do you think the scale would accurately estimate the intensity in a sparsely populated area? Why or why not? With the Modified Mercalli system, earthquake intensity is underestimated in sparsely populated areas, where the damage is minimal because there are few buildings. A major earthquake in a remote region of Alaska may rank no more than a VI on the scale, the same as a much smaller earthquake with an epicenter of Savannah, Georgia, or Memphis, Tennessee. By contrast, earthquake intensity in highly populated areas might be overestimated because there are many buildings and people to interview about their experiences. Individual differences in interpretation can also be a factor. One person may define "considerable damage" or "total damage" differently from another person.

Buildings themselves are constructed with many different materials and contrasting building codes. Earthquake-prone areas such as California and Japan have strictly enforced regulations concerning the height, style, and strength of buildings. In comparison, other regions that have experienced large loss of life in earthquakes (India, Turkey, China) have vastly different building regulations and enforcement systems. For example, within a 4-day span in December 2003, two earthquakes of similar magnitude, one in California and another one in Iran, had dramatically different results. The San Simeon, California, earthquake resulted in the deaths of two people who were crushed by falling debris from an

Table 5.2 Modified Mercalli Scale for Earthquake Measurement

Index	Effects of earthquake on people and structures
I	Not felt by people.
II	Felt by people at rest on upper floors of buildings.
III	May be felt by people indoors. Vibrations similar to the passing of a truck. Hanging objects swing.
IV	Felt indoors by many, outdoors by few. Dishes, windows, doors rattle; walls make creaking sound. Sensation like heavy truck passing building.
V	Felt by nearly everyone; many awakened from sleep. Some dishes, windows broken; doors swing open or closed. Unstable objects overturned. Liquids slosh around in containers.
VI	Felt by all; many frightened. Windows, dishes, glassware broken. Books knocked off shelves. Some heavy furniture moved; a few instances of fallen plaster. Trees shaken. Damage slight.
VII	Difficult to stand. Drivers notice, large bells ring. Slight to moderate damage in ordinary structures; considerable damage in poorly built or badly designed structures. Some chimneys broken, falling plaster, bricks, tiles.
VIII	Difficult to steer vehicles. Branches broken from trees. Slight damage in buildings designed to withstand earthquakes; heavy damage in poorly constructed structures. Chimneys, columns, monuments, walls may fall.
IX	Considerable damage in specially designed structures. Damage great in substantial buildings; partial collapse. Buildings shifted off foundations, underground pipes broken, reservoirs damaged. General panic.
X	Some well-built wooden structures destroyed; most masonry and frame structures with foundations destroyed. Serious damage to dams and embankments; landslides.
XI	Few, if any (masonry) structures remain standing. Bridges destroyed. Rails bent greatly, underground pipelines out of service.
XII	Total damage, objects thrown into air, widespread rockslides and slope failure.

Checkpoint 5.15

basic · advanced · intermediate · superior

Compare Figure 5.18 and Table 5.2 to estimate the approximate magnitude associated with Modified Mercalli events of intensity I, IV, VII, and XI.

Checkpoint 5.16

basic · advanced · intermediate · superior

Examine the following Community Internet Intensity Map (CIIM) for the Whittier Narrows earthquake in southern California. This earthquake occurred 7 years before the Northridge earthquake (Figure 5.19). Compare and contrast the CIIMs for both events, and suggest possible explanations for the similarities and differences between the two maps.

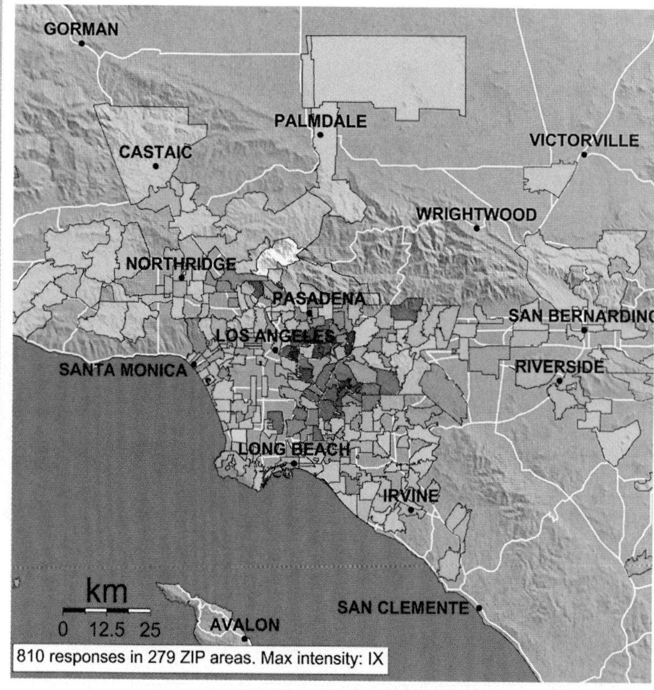

Intensity	I	II-III	IV	V	VI	VII	VIII	IX	X+
Shaking	Not felt	Weak	Light	Moderate	Strong	Very strong	Severe	Violent	Extreme
Damage	None	None	None	Very light	Light	Moderate	Moderate/Heavy	Heavy	Very heavy

old, unreinforced building. That region is relatively remote with a low population density. In stark contrast, the earthquake in central Iran destroyed almost every building in the city of Bam, killing 40,000 people. In the former case, there was slight-to-moderate damage to a few buildings. In the latter, almost every structure was flattened.

Finally, the intensity of an earthquake generates a range of values, depending upon the observer's location relative to the earthquake's epicenter. Damage generally decreases moving away from the epicenter, but that decrease is not the same in every direction. Can you name some of the factors that might affect this by analyzing Table 5.2? The range of values and lack of standardization associated with readings on the Modified Mercalli scale make it ineffective for the scientific comparison of earthquakes. However, the scale does provide helpful local data that can be used to identify which areas are most susceptible to shaking in regions of infrequent earthquake activity. In fact, the US Geological Survey generates Community Internet Intensity Maps, which show the distribution of intensities almost instantly following an earthquake through the use of an online questionnaire (http://earthquake.usgs.gov/earthquakes/dyfi). A computer program quickly generates an average intensity value for each ZIP code affected to produce a regional map (Figure 5.19).

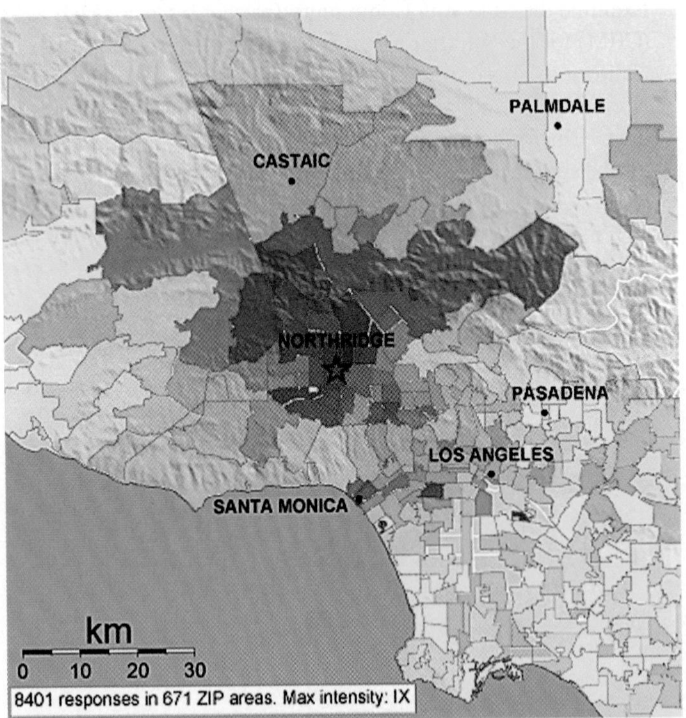

Intensity	I	II-III	IV	V	VI	VII	VIII	IX	X+
Shaking	Not felt	Weak	Light	Moderate	Strong	Very strong	Severe	Violent	Extreme
Damage	None	None	None	Very light	Light	Moderate	Moderate/Heavy	Heavy	Very heavy

Figure 5.19 Community Internet Intensity Map generated for the 1994 Northridge earthquake. A star marks the epicenter, and colored areas of the map indicate the earthquake intensity for each ZIP code area as reported by 8,401 responses to an online survey. The maximum measured intensity was IX, immediately adjacent to the epicenter.

5.6 Earthquake Hazards

Learning Objectives

- Interpret Community Internet Intensity Maps of earthquakes to predict the extent of strong ground shaking.
- Identify the geological and cultural characteristics that would contribute to extensive earthquake damage.
- Discuss how the use of an early warning system would affect citizens in a densely populated city.
- Evaluate the relative risk of earthquake damages for cities of different size and character in different geologic settings.

As the recent Tohoku earthquake and tsunami illustrated, we are often unprepared for natural hazards that inevitably occur. Protecting people and structures from an earthquake requires an understanding of the type of hazards associated with this natural phenomenon.

The United States had an example of what to expect with the magnitude-6.7 Northridge earthquake. About 100 to 200 earthquakes of this magnitude occur worldwide each year, but this time the strong earthquake affected a heavily developed, modern US city. At first, it might seem that a potential solution to reducing risk from earthquakes would be to simply identify the locations of faults and make sure nobody lives nearby. Unfortunately, that ship has sailed. California, our most heavily populated state, is laced with faults, not all of them equally dangerous. However, some of the most dangerous faults, including the San Andreas fault system, run right through California's most densely populated cities. In a state with thousands of kilometers of known faults (and some unknown), it only takes a small movement on a relatively short segment of one of those faults to cause trouble. Regardless of the potential risk, humans will continue to live in these areas, so it is important to scientists to educate residents concerning the nature of that risk.

Research predicts a 94 percent chance that one or more magnitude-7 or larger earthquakes will strike California sometime during the next few decades (and a 99.7 percent chance of a 6.7-magnitude or larger earthquake). Southern California, generally the area around Los Angeles, is forecast to be more likely to experience the magnitude-7 earthquake than northern parts of the state near San Francisco (82 percent versus 68 percent). If the weather forecast told you that there was an 82 percent chance of showers, wouldn't you carry a raincoat or an umbrella? Unfortunately, there is no "earthquake umbrella." The only ways to mitigate the effects of an earthquake are to build better structures and make the population aware of what to do when the inevitable occurs. That was the idea behind the Great Southern California Shakeout, an annual earthquake preparedness drill program involving millions of people in the effort to model what would happen in the event of a major earthquake. While many people limited their participation to crawling under a desk for a few minutes, emergency crews and other professionals participated in exercises over multiple days that simulated responses to collapsed buildings and other emergencies. Similar programs are now conducted throughout North America and around the world. More than 19 million people participated in drills in 2012.

We all know that shaking is associated with earthquakes, but these events have other characteristics that, though less obvious, have significant implications for any future actions we may take to mitigate potential damage and protect lives.

Ground Shaking

Consider the forces acting on your body when you slam on your car brakes to avoid a collision or when a roller coaster car goes into a steep descent. Scientists express these forces as a proportion of the *acceleration due to gravity* (*g*). For example, you would experience acceleration similar to gravity (1 *g*) if you were riding in a car that traveled 100 meters (330 feet) from rest in just 4 seconds. Astronauts in spaceships experience accelerations equivalent to three times gravity (3 *g*) during takeoff. A vertical acceleration of more than 1 *g* can overcome the downward pull of gravity, throwing objects into the air.

In the same way, we can measure the acceleration of objects caused by ground shaking during an earthquake. Rapid horizontal or vertical movements associated with earthquakes may shift homes off their foundations, destroy interchanges on elevated highways, and cause tall buildings to collapse or "pancake" as floors fall down onto one another.

Northridge experienced the strongest ground motions ever recorded in a US city. Vertical accelerations from the earthquake reached a maximum of 0.93 *g*—almost enough to lift objects off the ground! Houses built after California's strict building codes were enacted can withstand severe shaking from accelerations of 0.6 *g* with little more than chimney damage. However, many homes in California were built well before these building codes were adopted and could experience severe damage from strong earthquakes such as Northridge (Figures 5.1 and 5.20). Accelerations of just 0.1 *g* are sufficient to cause damage to homes in many areas of the United States that have no earthquake damage prevention building codes. The resulting motion is equivalent to a VII on the Modified Mercalli scale.

Regardless of how well homes are constructed, the degree of shaking will depend on the materials surrounding the homes' foundations. The shaking from an earthquake is greater for softer materials (sediment) than for bedrock, resulting in more destruction

Checkpoint 5.17

basic · intermediate · advanced · superior

Analyze the Community Internet Intensity Maps for the Northridge and Whittier Narrows earthquakes in Figure 5.19 and Checkpoint 5.16 and the Modified Mercalli scale in Table 5.2. Determine approximately how far from the epicenter ground shaking would have been strong enough to make it difficult for people to stand through both earthquakes.

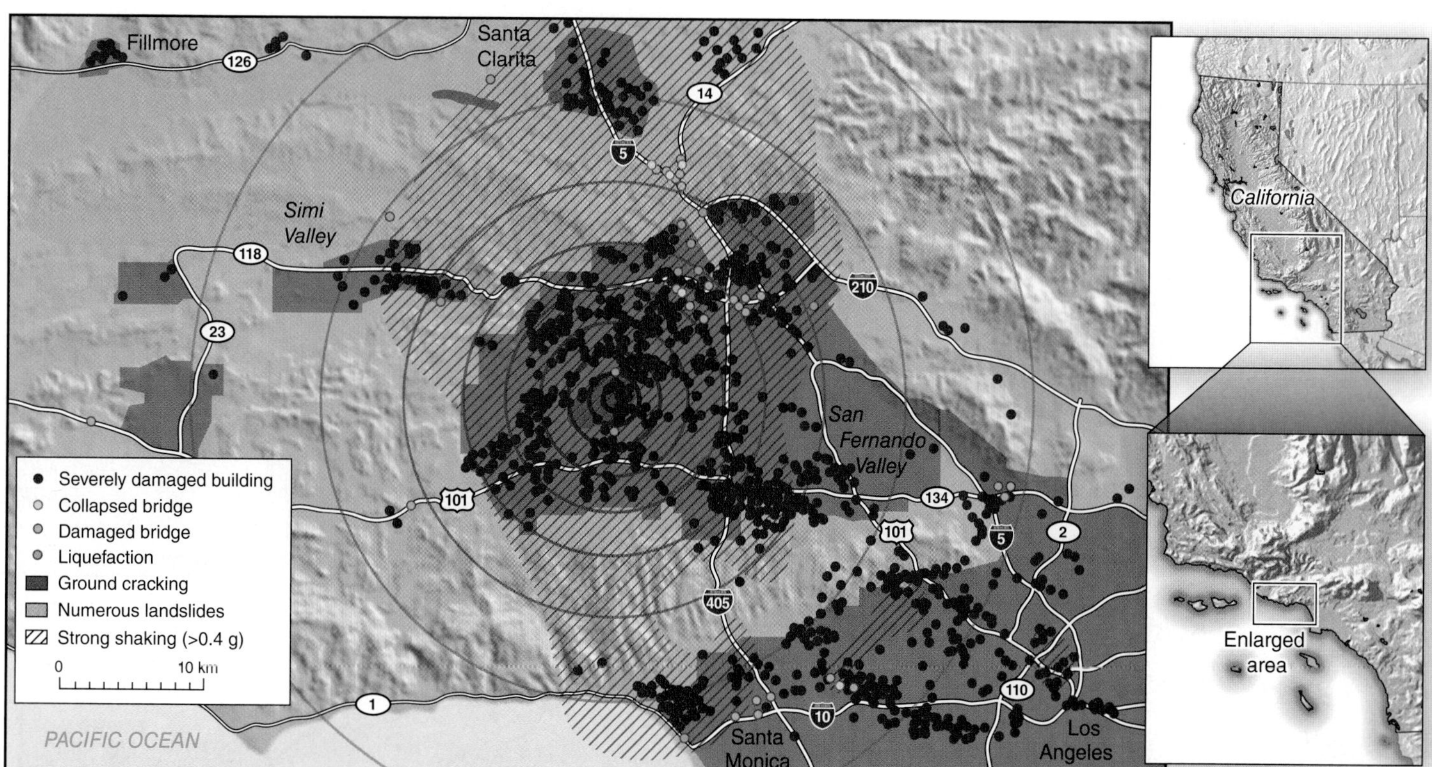

Figure 5.20 Effects of the Northridge earthquake, 1994. Ground shaking resulted in hundreds of severely damaged buildings (red circles), considerable areas of landslides, and more limited areas of liquefaction (blue circles). Before the earthquake, freeway overpasses were constructed to withstand ground acceleration of 0.4 *g*; however, much of the region experienced shaking in excess of 0.4 *g*.

in some areas and less in others that may be next to one another. For instance, the double-deck Cypress freeway was built in Oakland, California, in 1950. Some parts of the structure were built on sand and gravel, and other parts were located in soft mud. The section of the freeway constructed on the weaker mud experienced much more shaking and collapsed during the 6.9 magnitude Loma Prieta earthquake in 1989 (Figure 5.21). Other sections built on more stable material did not collapse at all.

Aftershocks

We tend to think of an earthquake as a single isolated event. But once an earthquake occurs (the main shock), it is followed by aftershocks, several of which may be felt by residents. The **aftershocks are earthquakes that occur on the same or nearby faults** for months or even years after the main shock. The larger the main earthquake, the more plentiful, and larger, the aftershocks are. Aftershocks decrease over time but can pose a significant hazard. The aftershock pattern from the 2004 Sumatra earthquake extended for more than 1,000 kilometers (621 miles) to the north of the epicenter (see Chapter Snapshot). These later earthquakes probably represent movements on several faults adjacent to the main fault plane.

Landslides

Earthquakes are often associated with mountains formed along convergent plate boundaries. **If slopes are steep and the rocks, soils, and materials forming the slope are not stable, landslides will occur** when the area is shaken. Landslides commonly follow earthquakes in California. For example, an estimated 11,000 landslides were generated as a result of the Northridge earthquake (Figure 5.22). Landslides in the mountains west of Los Angeles blocked many roads and aggravated transportation problems that had been caused by the earthquake. One unfortunate result of the landslides was the production of large volumes of airborne dust. Dust clouds blew into Simi Valley, where a cluster of more than 150 cases of valley fever (coccidioidomycosis) were diagnosed following the earthquake. Valley fever is a lung disorder that can

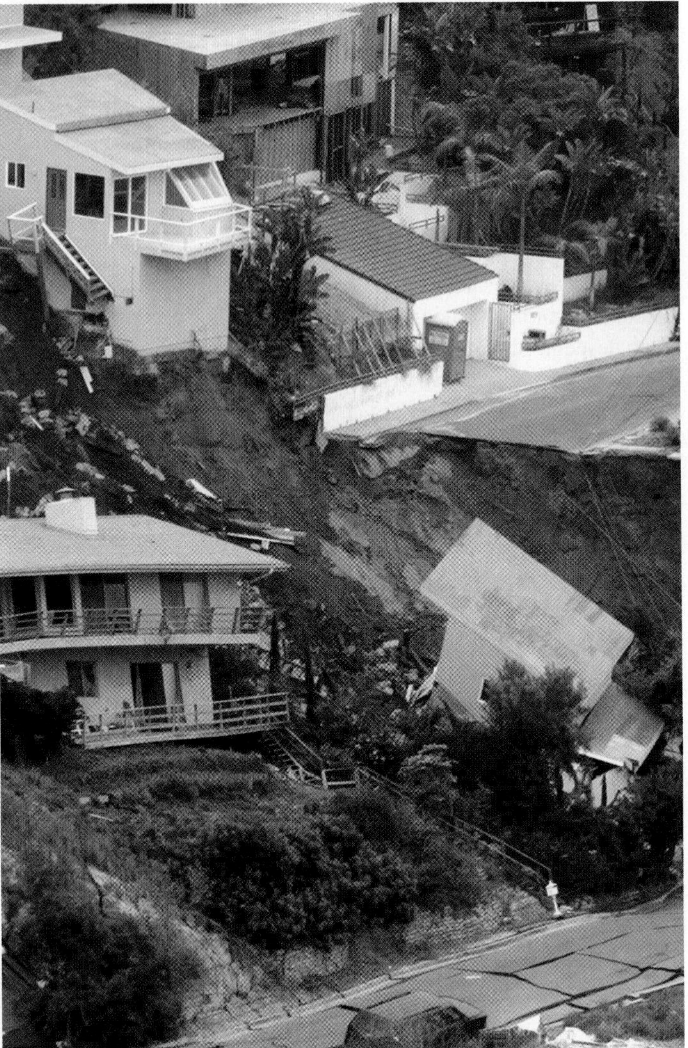

Figure 5.22 Landslide hazards from the 1994 Northridge earthquake. A landslide destroyed this home perched on high ground overlooking the Pacific Ocean and the Pacific Coast highway.

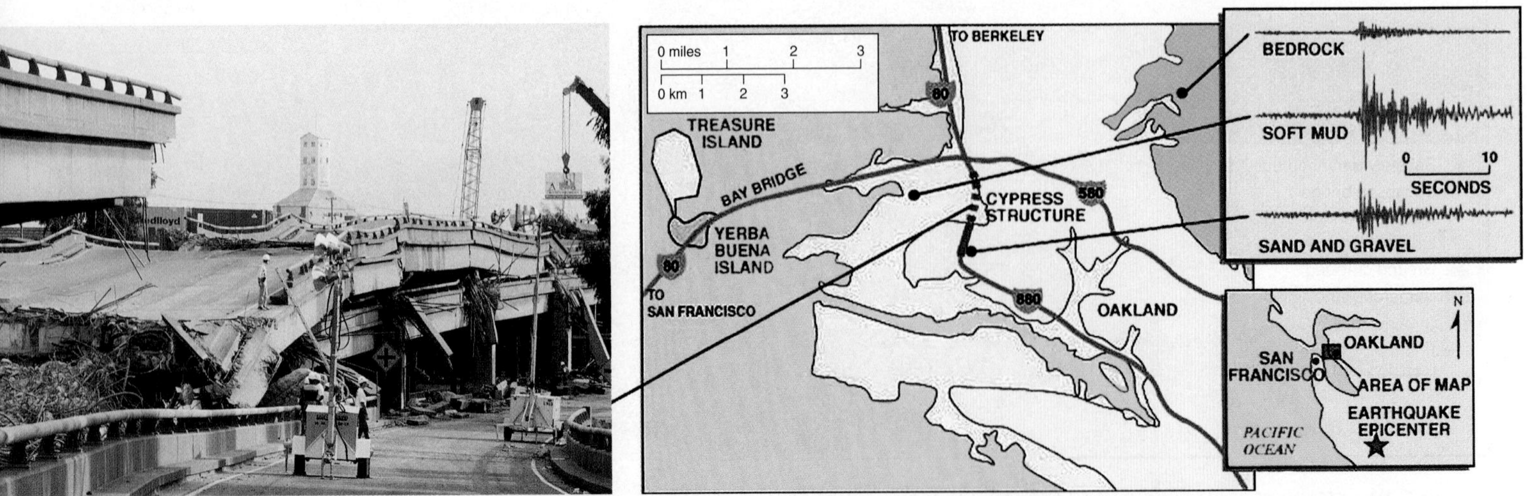

Figure 5.21 The double-deck Cypress freeway was constructed on two different types of underlying materials. During the 6.9 magnitude Loma Prieta earthquake, the section of the freeway constructed on the weaker mud experienced much more shaking and collapsed. Seismograms of the three different materials near the Cypress freeway illustrate that shaking was least for bedrock and greatest for mud.

only be contracted by humans who inhale dust containing fungal spores. Three valley fever fatalities occurred, accounting for about 5 percent of the deaths from the Northridge earthquake

Elevation Changes

When a fault undergoes vertical motion, large sections of Earth's surface can change elevation. For example, mountains east of Los Angeles were uplifted 0.3 meters (1 foot) by the Northridge earthquake. Such elevation changes are most dramatic when coastal lands become submerged below sea level or when previously inundated shorelines are uplifted. Small islands in the Indian Ocean located above the subduction zone responsible for the Sumatra earthquake were uplifted by 1 to 3 meters (3 to 10 feet), exposing surrounding coral reefs.

Liquefaction

Shaking resulting from earthquakes is exaggerated in areas where the underlying sediment is weak or saturated with water. **Liquefaction occurs when shaking of the ground causes compaction of the sediment. This increases water pressure, resulting in water-saturated materials being violently ejected at the surface.** The shaken, water-saturated sediment loses its strength as it is converted from a solid to a liquidlike mixture of water and sediment. The material may be expelled, causing the surface to collapse. Liquefaction features may be preserved where they cut across soil layers (sand dikes; Figure 5.23a) or where sediment was injected parallel to layers (sand sills). Scientists use these features to study the paleoseismology of regions with little current earthquake activity. The characteristics of ancient liquefaction features provide evidence for magnitude 6 and 7 earthquakes in many Midwestern states that today appear seismically quiet.

The ejection of material and water can lead to a decrease in elevation (subsidence) of several meters. A third of the city of Niigata, Japan, subsided by 2 meters (6.6 feet) due to compaction and the loss of water associated with liquefaction of sediments following a 1964 earthquake (Figure 5.23b). Part of San Francisco Bay was filled with debris (much of it from a large earthquake in 1906), creating over 80 hectares (200 acres) of new land. The Marina District was built on this fill. Liquefaction during the 1989 Loma Prieta earthquake destroyed much of the Marina District and was responsible for approximately $100 million of damage in the San Francisco area. Objects may collapse into the liquefied earth materials, only to be left partially buried (Figure 5.23c) when the expelled water flows away or infiltrates back into the ground.

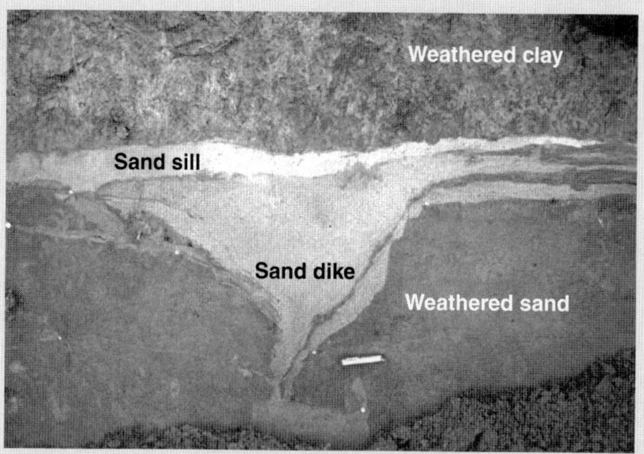

a.

b.

c.

Figure 5.23 Liquefaction effects. **a.** A sand dike and sill exposed in drainage ditch in southeastern Missouri. The sand dike intruded weathered sand; the sand sill emplaced below weathered clay. Layering within the dike and sill indicates that they formed during two or more events. For scale, the knife is 8 centimeters (3 inches) long. **b.** Leaning apartment buildings, Niigata, Japan, after a 7.4 magnitude earthquake in June 1964. One-third of the city subsided by up to 2 meters (6.6 feet) because land loses strength, compaction occurs, and buildings sink when the ground cannot hold their weight as liquefaction occurs. **c.** Car partially swallowed by liquefaction during the Christchurch, New Zealand earthquake (February, 2011).

Checkpoint 5.18

basic | advanced | intermediate | superior

Imagine that two inland cities of the same population experience two identical earthquakes (same magnitude, same focal depth, and same location relative to city). One city is devastated; the other suffers only light damage. Using only information from this section of the chapter, suggest two contrasting scenarios to explain why one city might be heavily damaged while the other escapes relatively unharmed.

Tsunami

Portions of the seafloor can move as a result of fault displacement, an ocean floor landslide, or a submarine volcanic eruption. This **movement can displace large volumes of ocean water to create a tsunami (see Chapter 1).** Megathrust earthquakes associated with subduction zones, such as the Tohoku or Sumatra earthquakes, occur when the overriding plate no longer slides over the descending plate as it moves downward into the subduction zone (see Chapter Snapshot). The leading edge of the overriding plate is carried a small distance downward into the subduction zone, causing the shape of the plate to become distorted as it buckles upward. Eventually, the "stuck" fault segment between the plates ruptures, generating a great earthquake and displacing the overlying water in the ocean to produce a tsunami. Tsunami wave forms can travel at the speed of a jet (up to 960 kilometers per hour; 600 miles per hour) and reach heights of tens of meters or more along coastlines to cause extensive damage (Figure 5.24).

Tsunamis caused by earthquakes occur along subduction zones around the perimeter of the Pacific Ocean and along the northwestern margin of the Indian Ocean. They can also occur in the Atlantic if earthquakes happen in the Caribbean subduction zone or from other seafloor displacements. The waves generated by such events can travel across thousands of kilometers of ocean to destroy coastal property on islands or on the other side of the ocean. The largest earthquake ever recorded, a magnitude-9.5 tremor, occurred along the coast of Chile on May 22, 1960. The earthquake and tsunami killed over 2,000 people. Sixty-one people died on the island of Hawaii 15 hours after the earthquake, and another 122 were killed in Japan 7 hours later. The height of the

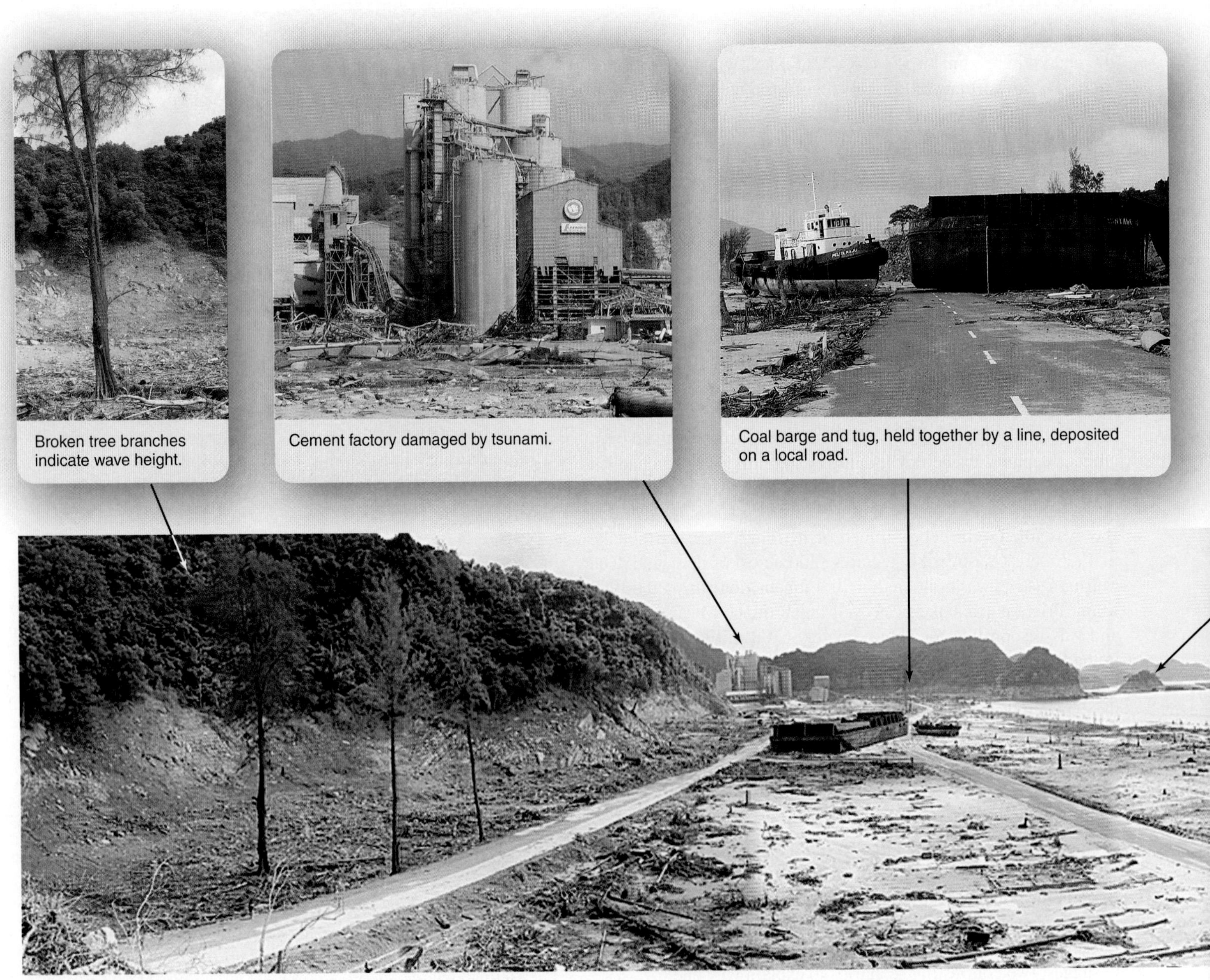

Broken tree branches indicate wave height.
Cement factory damaged by tsunami.
Coal barge and tug, held together by a line, deposited on a local road.

Figure 5.24 Tsunami damage, Lho Nga, northwestern Sumatra. The tsunami swept buildings from their foundations and knocked down trees. The base of the treeline on the slope to the left of the image is 31 meters (102 feet) above sea level and marks the maximum elevation that waves reached at this location.

Checkpoint 5.19

basic / intermediate / advanced / superior

Warning times associated with some natural hazards can be measured in months (volcanoes), days (hurricanes), or minutes (tornadoes). Scientists in Japan have designed an earthquake warning system using an extensive network of seismographs. They use the arrival of P waves to trigger an alarm that can stop high-speed trains before the arrival of the more damaging S waves and the later surface waves. How would such a system affect citizens in a densely populated city such as Los Angeles that is situated near numerous active faults?

Chile tsunami in Japan was similar to that recorded for the 1700 tsunami generated by fault movement on the Cascadia subduction zone (Figure 5.25). These wave heights were comparable to waves measured on coastlines throughout the Indian Ocean following the 2004 tsunami (see Chapter Snapshot).

Recall from the description of the tsunami in Chapter 1 that the first sign of the disaster may be water receding from the beach, only to be followed by a surge of water inland. The waves of a tsunami, like all waves, have high parts (crests) and low parts (troughs). Water levels may initially rise or fall, depending on which part of the wave arrives at the coast first. Sea level will fall between crests, with the second wave usually larger than the first. A tsunami consists of a series of waves that continue to batter coastlines for hours after the arrival of the first wave (Figure 5.26).

Ship overturned in factory harbor.

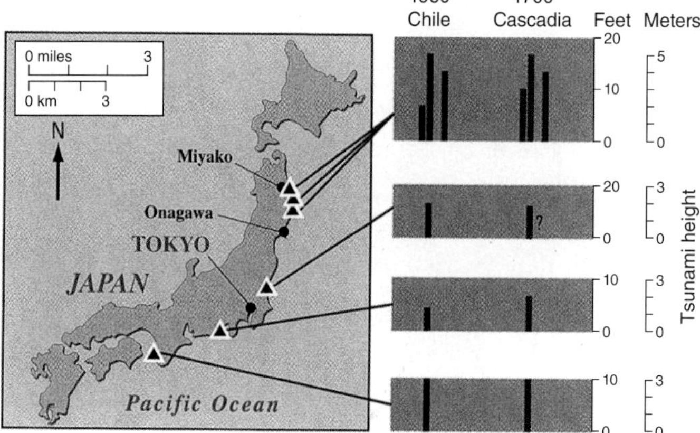

Figure 5.25 Chile and Cascadia tsunami heights recorded along the coast of Japan. Comparison of tsunami heights for four stations along the east coast of Japan following the Chile and Cascadia tsunami. Tsunami generated by subduction zone earthquakes in Chile and Cascadia were of similar size. Wave heights in Japan were determined from historical records.

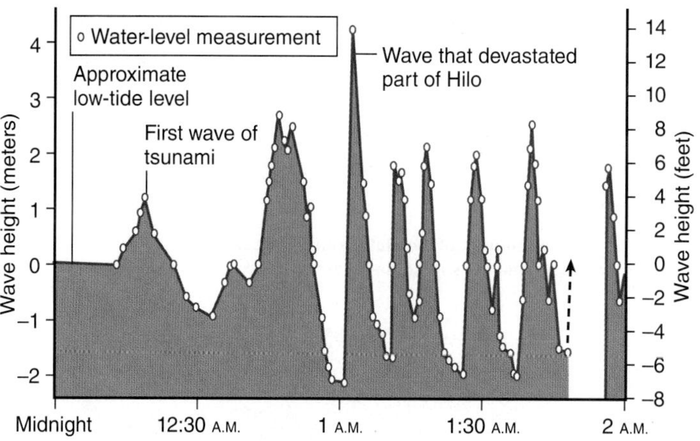

Figure 5.26 Multiple waves generated by the Chile tsunami of May 23, 1960, measured on the island of Hawaii. Water levels at Hilo were both above and below the low-tide level and correspond with crests and troughs of successive waves.

2004 TSUNAMI

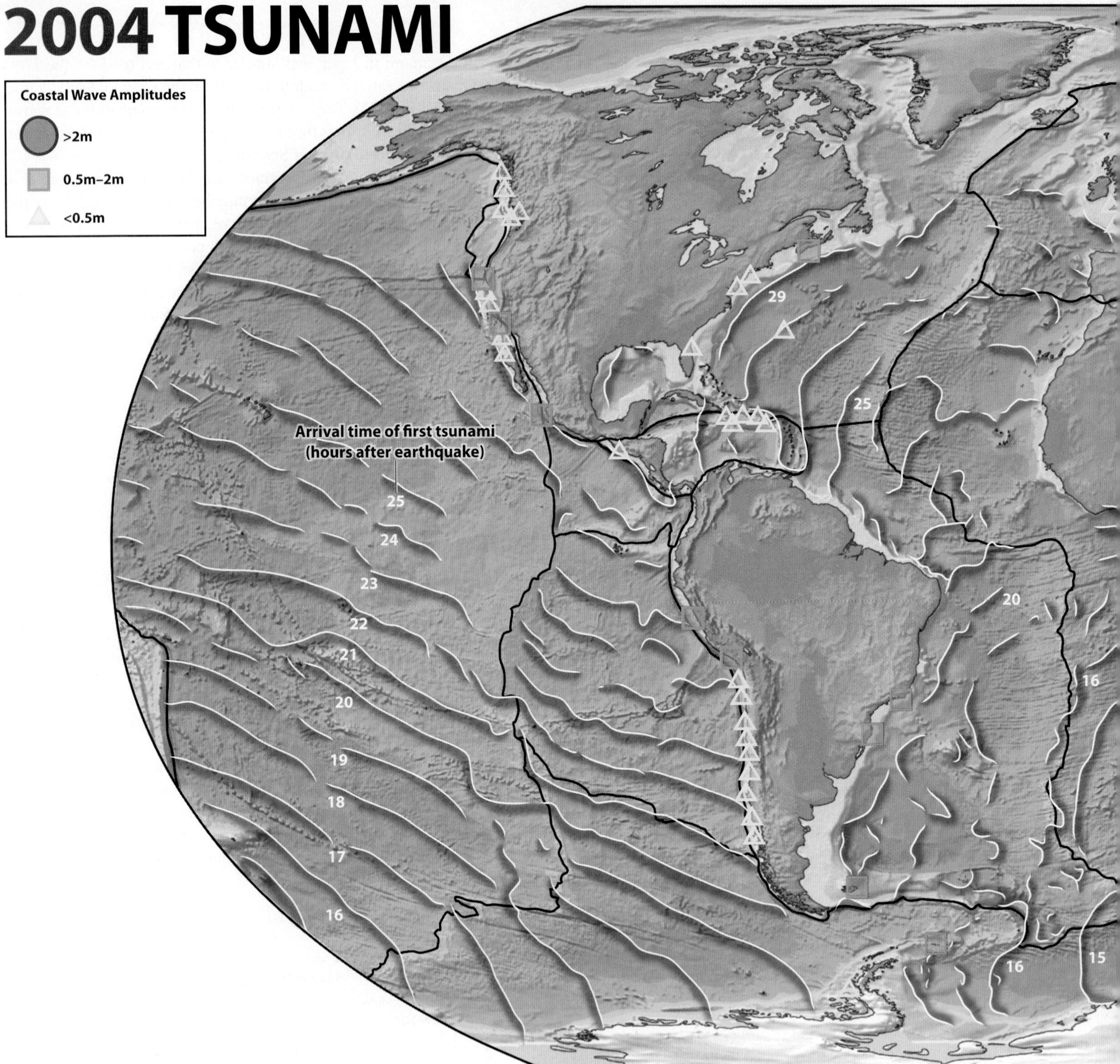

Coastal Wave Amplitudes
>2m
0.5m–2m
<0.5m
Arrival time of first tsunami
(hours after earthquake)
29
25
25
24
23
22
21
20
20
19
18
17
16
16
16
15

STEP 1

Friction along a segment of the plate boundary locks the overriding plate and the subducting plate.

Eurasian plate
Indian-Australian
plate

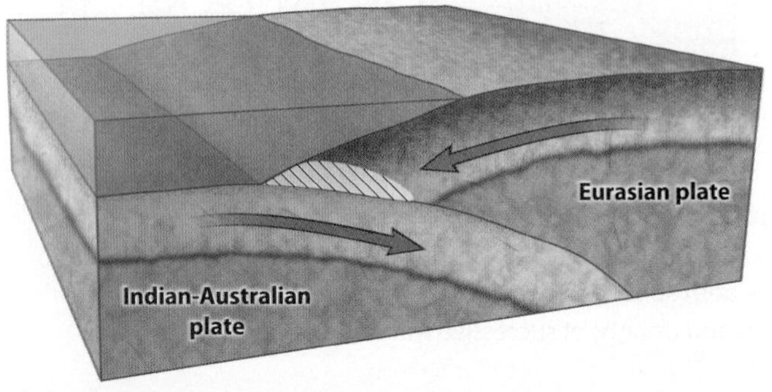

STEP 2

The shape of the overriding plate is distorted as it is pulled toward the subduction zone as the lower plate continues to descend. This can continue for hundreds of years.

Eurasian plate
Indian-Australian
plate

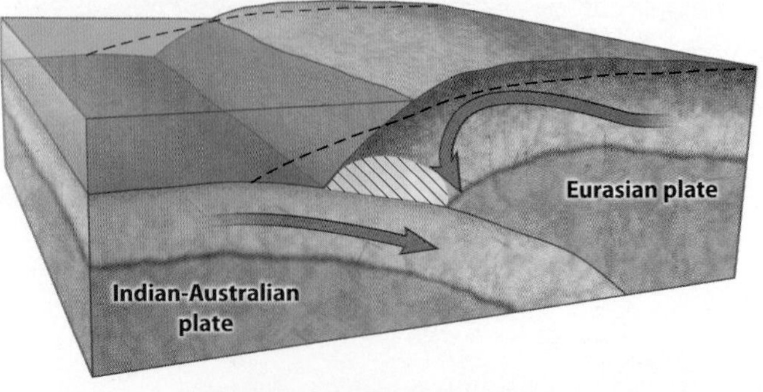

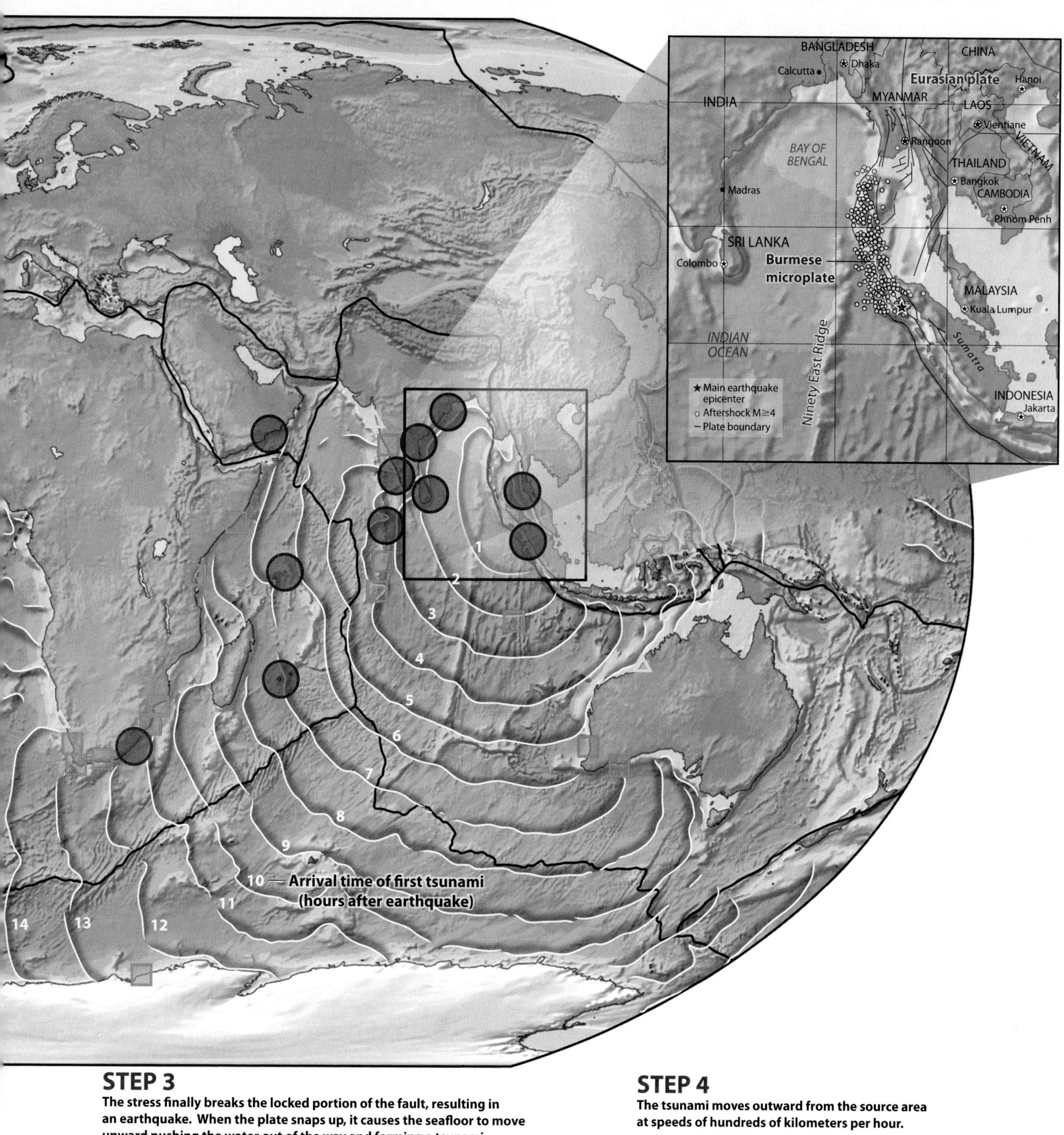

BANGLADESH
CHINA
Calcutta
Dhaka
Eurasian plate
Hanoi
INDIA
MYANMAR
LAOS
Vientiane
Rangoon
VIETNAM
BAY OF BENGAL
THAILAND
Bangkok
Madras
CAMBODIA
Phnom Penh
SRI LANKA
Colombo
Burmese microplate
MALAYSIA
Kuala Lumpur
INDIAN OCEAN
Ninety East Ridge
Sumatra
INDONESIA
Jakarta
★ Main earthquake epicenter
○ Aftershock M≥4
— Plate boundary
1
2
3
4
5
6
7
8
9
10 — Arrival time of first tsunami (hours after earthquake)
11
12
13
14

STEP 3

The stress finally breaks the locked portion of the fault, resulting in an earthquake. When the plate snaps up, it causes the seafloor to move upward pushing the water out of the way and forming a tsunami.

Eurasian plate
Indian-Australian plate

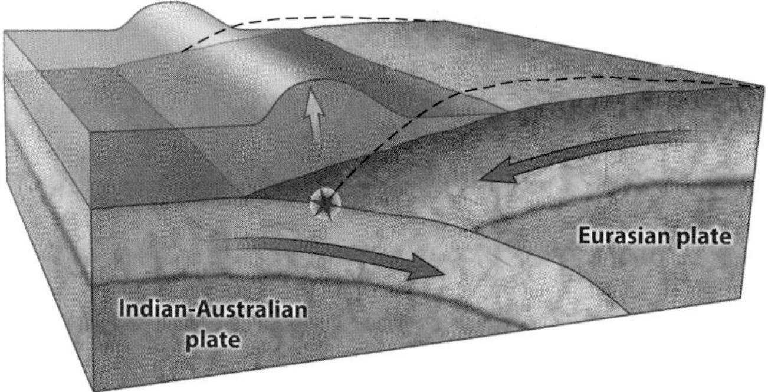

STEP 4

The tsunami moves outward from the source area at speeds of hundreds of kilometers per hour.

Eurasian plate
Indian-Australian plate

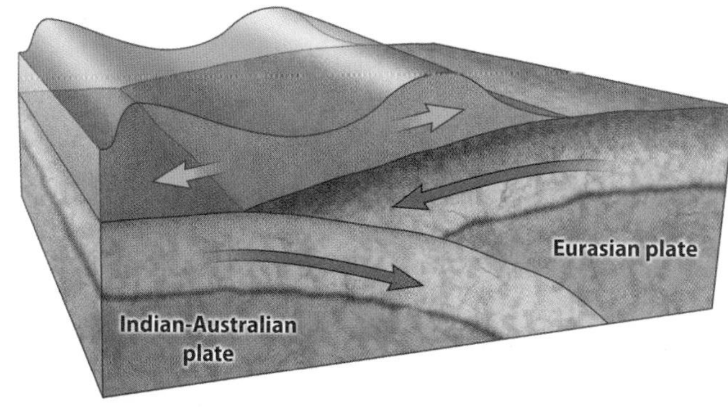

Checkpoint 5.20

Earthquake Risk Rubric Exercise

This exercise requires that you consider what combination of factors is most likely to contribute to the greatest risk of damage from a future earthquake.

After graduation, you get a job working for a county planning task force in California. The task force must examine the settings of several different cities and identify which is at **greatest risk for future earthquake damage** caused by movement on known faults. Your assignment is to fill out the **evaluation rubric** provided here by identifying factors that would influence the risk of damage from a future earthquake. The location that scores the highest on your scoring rubric will receive additional county funds to protect key structures from earthquake damage. The factors you identify may be either physical (such as nature of the local geology) or cultural (such as size of population centers).

Audience: You will create a scoring scheme that can be applied by most educated citizens. Specifically, you may consider that you are writing for city council members or a concerned citizens group.

Main point and purpose: To demonstrate your understanding of the principal factors that result in damage or loss of life from an earthquake.

Pattern and procedures: You must **identify four factors** and differentiate among characteristics that make them high-, moderate-, or low-risk phenomena. One factor, the proximity of the site to a fault, is included as an example (see accompanying table). If the fault were to move in the future, sites closest to the fault would suffer the most damage (high risk), whereas those farther away might experience little or no damage (low risk). Each high-risk factor receives 3 points; moderate-risk factors rate 2 points; and low-risk factors rate 1 point. The site with the highest cumulative score for all factors is at greatest risk from a future earthquake.

Standards and criteria: You will be assessed on your choice of:

1. Relevant factors that would contribute to the potential for earthquake-related losses.
2. Identification of high-, moderate-, or low-risk situations for each factor.
3. One factor as the most significant.
4. The applicability of your rubric to four sites in an idealized county in California.

What is the most important factor in your rubric? Distinguish which factor is the most significant under the circumstances of the exercise. The score for this factor will be doubled. Discuss your justification for choosing that factor.

Use your rubric in a hypothetical situation. Examine the information on the geology and characteristics of four cities shown on the following map, and rank them in order from greatest to least risk of damage in a future earthquake.

Rank:

1.
2.
3.
4.

We can continue to expect tsunami in coastal cities that border oceans, many of which have no protection from this type of event. Can technology help prevent future loss of life? Toward that end, the Pacific Tsunami Warning System (PTWS) network of stations was developed, and other systems are planned for the Atlantic and Indian oceans. The PTWS collects and analyzes information from ocean buoys and issues warnings or watches that predict tsunami arrival times for coastal areas in the Pacific. Such systems are designed to identify potentially damaging tsunami caused by earthquakes in or around the oceans. The first buoys for a comparable system have recently been placed in the Indian Ocean, but there are none in the Atlantic Ocean. Given that, which East Coast cities are at greatest risk of a tsunami from an earthquake in the Caribbean?

basic advanced
intermediate superior

Earthquake Risk Rubric			
Factors	**Low risk (1 point)**	**Moderate risk (2 points)**	**High risk (3 points)**
Proximity to fault	Far (>50 km)	Moderate (25–50 km)	Close (<25 km)

Pomeroy (pop. 50,000) Older industrial center of the county; founded in 1880 and built on floodplain sediments of the Mono River; transportation center for commercial barge traffic on river and truck traffic on highway.

Iron Hills (pop. 20,000) Old mining town founded in 1850 now caters to hikers and climbers because of surrounding mountains; underlain by granite bedrock.

Dunstable (pop. 60,000) County offices here, founded 1935, recent growth in last decade as several computer companies built new factories; city underlain by mix of bedrock and sediment.

Bell Harbor (pop. 100,000) Scenic coastal fishing community built on rocky cliffs overlooking ocean; population swelled in last few decades as it became a popular retirement destination; new hospitals and retirement homes built since 1980s.

the big picture

Earthquakes and their associated effects represent some of the most destructive natural hazards on Earth. In your lifetime, the west coast of the United States will certainly experience more strong earthquakes and, in all probability, at least one major earthquake. Furthermore, in the next few centuries, maybe even tomorrow, another great earthquake will devastate parts of California, Oregon, and/or Washington. Despite years of planning for such events, the human and economic costs will be tremendous. Even if it becomes possible to accurately predict earthquakes to within a specific year, it is unlikely that such an event could be pinpointed to within a few months, let alone weeks or days. Furthermore, we would probably be unable to collect sufficient data to predict earthquakes in areas of infrequent seismic activity. Imagine a devastating earthquake centered near a major city. Given the difficulty in predicting the timing of future earthquakes, we would be well advised to focus instead on technology-based engineering solutions that attempt to earthquake-proof buildings and important landmarks. Such buildings would be able to sustain damage from the largest earthquakes and remain standing, reducing the loss of life due to building collapse.

We began this chapter by comparing the recent large earthquakes and tsunami to a similar pairing of earthquake and tsunami that occurred along the Cascadia subduction zone in 1700. In the chapter-opening image, we asked, "What is this man doing, and what does it have to do with earthquakes?" The man, Guy Gelfenbaum, a USGS scientist, was a member of an international survey team that traveled to Sumatra just 1 month after the 2004 disaster to identify the characteristics of the tsunami. In the photo, he is measuring how much a section of the northwest coast of Sumatra had submerged as a result of the earthquake. The trees he is standing beside were on dry land before the coastline subsided and they were broken off by the tsunami.

The patterns scientists identified in Sumatra provided further data about a rare geologic event and match other patterns already in the geologic record. For example, by now you have probably realized that the broken-off trees in Sumatra are similar to the ghost forests we described along the coasts of Alaska and Oregon (see Figure 5.2). The sand deposits that were interpreted as tsunami deposits from previous earthquakes (see Figure 5.4) also have their parallels in the Indian Ocean tsunami (Figure 5.27). Scientists analyzing the details of the sand deposits can identify individual layers left behind by separate waves over an interval of a few hours. The survey team documented evidence for wave heights up to 30 meters (100 feet) along the 100-kilometer (60-mile) section of the coast of northwest Sumatra closest to the earthquake epicenter. This region sustained the greatest damage from the tsunami, accounting for about one-half the total casualties.

The comparison of features from several megathrust earthquakes or, more technically, great earthquakes, described in

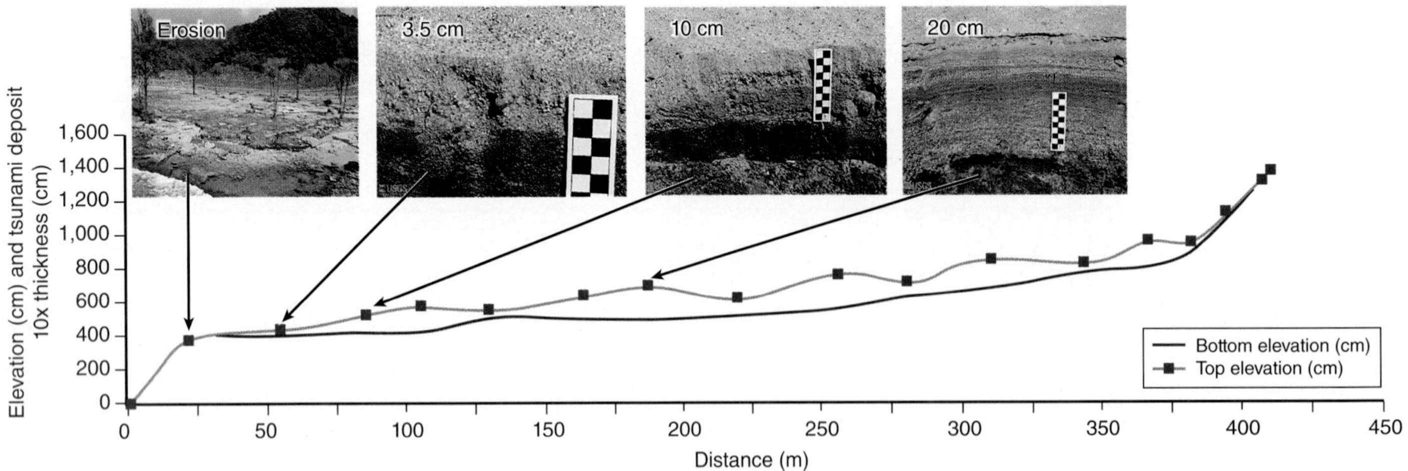

Figure 5.27 Sand deposits following the Indian Ocean tsunami. In this sediment profile across 410 meters (1,350 feet) of coastline in northwest Sumatra, the dark layers represent original topsoil. The left side (0) of the graph is sea level. The vertical distance between the red and blue lines indicates sand thickness. Sand deposited by the tsunami covers the soil to depths of up to 20 centimeters (8 inches) at this location and as much as 70 centimeters (28 inches) elsewhere. Sand nearest the coast was eroded and washed hundreds of meters inland by the waves.

this chapter illustrates that scientists build hypotheses by using observations. These hypotheses have predictive power but may not predict every aspect of future events. Incorporating observations of past earthquakes and the theory of plate tectonics assists geologists in predicting the location of future earthquakes but cannot predict the exact time, date, or strength of the next event. Scientists can help us understand the nature of earthquake hazards, especially of tsunami threats, but it is up to us as individuals, and to our social and political institutions, to use these data thoughtfully.

In the self-reflection survey, we asked what action you would take if you lived in a city with a 99 percent probability of a future large earthquake. By now, you have probably realized that the city in question is probably in California. Your hometown may have little risk of experiencing an earthquake but may be at risk for other hazards. What should be the responsible role of a resident of San Francisco (or Los Angeles, Seattle, etc.) in advance of a future earthquake? What should the city do? What about the state and federal governments? Without careful planning, a major or giant earthquake along the west coast has the potential to overwhelm authorities to an even greater degree than Hurricane Katrina affected cities along the Gulf of Mexico.

Earthquakes: Concept Map

Draw a concept map that illustrates the significant characteristics of earthquakes. Use the scoring rubric provided here to draw a concept map that can be scored as a "4."

Scoring Rubric	
0	The concept map does not contain any information about earthquakes.
1	The concept map contains some terms that are significant to the main features of earthquakes, but several key terms are omitted and many linking phrases are either absent or inaccurate.
2	The concept map contains most terms that are significant to the main features of earthquakes, but they are poorly organized and some linking phrases are absent or incorrect.
3	The concept map contains most terms that are significant to the main features of earthquakes, but one or two key terms may be absent. The diagram is reasonably well organized, and almost all linking phrases are appropriate.
4	The concept map contains all terms that are significant to explain the main features of earthquakes in a well-organized display that has appropriate linking phrases for each pair of terms.

"Civilization exists by geological consent, subject to change without notice."

—Will Durant, philosopher and historian

"One ought never to turn one's back on a threatened danger and try to run away from it. If you do that, you will double the danger. But if you meet it promptly and without flinching, you will reduce the danger by half."

—Winston Churchill, UK prime minister during World War II

Chapter 6

Volcanoes and Mountains

Chapter Outline

the big picture

What questions would you ask about that mountain if you lived here?

See The Big Picture box at the end of this chapter for the full story on this image.

6.1 The Volcano Commandos

Chapter Learning Outcomes

- Students will be able to describe the processes that form various types of volcanoes.
- Students will be able to describe the products of volcanic eruptions.
- Students will be able to evaluate the risks associated with a volcanic eruption.
- Students will be able to describe the processes that result in the formation and erosion of mountains.

Let's face it: The closest many scientists get to danger is heating up their lunch in the office microwave—but not so for the volcanologists of the rapid-response Volcano Disaster Assistance Program (VDAP). OK, so they are not exactly the X-Men—but they do use their unique superpowers (knowledge of volcanoes) to leap instantly into action (well, it might take a few days) from their secret base (actually, it's Vancouver, Washington) to help people in distress around the world.

The VDAP was formed in 1985 by the US Geological Survey (USGS) and the US Office of Foreign Disaster Assistance after a disastrous eruption of the Nevado del Ruiz volcano in Colombia, which killed more than 23,000 people (Figure 6.1). Like a team of volcano commandos, the VDAP scientists regularly find themselves heading off to some remote location threatened by a rumbling volcano. Their mission is to help save lives and reduce economic losses by finding answers to critical questions such as: When will the eruption occur? How big will it be? What will be the characteristics of the eruption?

Currently, more than 1,500 volcanoes around the world are classified as active, meaning that they might erupt. Eruptions can last anywhere from a matter of hours to decades and can produce a variety of hazards, from lava to mudflows to poisonous gases. We can get a sense of what these eruptions might look like by consulting historical records of earlier eruptions that have been

Figure 6.1 Aftermath of the 1985 Nevado del Ruiz eruption. The town of Armero was destroyed by mudflows following an eruption of the Nevado del Ruiz volcano in Colombia. More than a 2-hour interval ensued between the eruption and the arrival of the mudflows in Armero. Had the residents been warned, they could have readily evacuated to higher ground and the number of people killed (23,000) would have been much reduced.

9:08 P.M. Eruption from summit of Nevado del Ruiz. Snow and ice melted and mixed with debris from the eruption to form mudflows that moved downslope and filled major river channels.

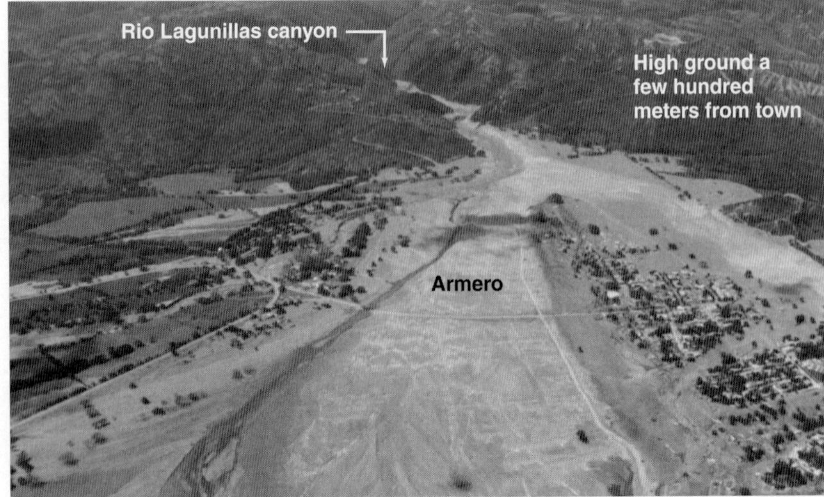

11:30 P.M. Mudflows reached Armero at the mouth of Rio Lagunillas canyon, 74 kilometers (46 miles) from the volcano.

Buildings were carried away or buried by mudflows up to 5 meters (16 feet) deep.

reported for approximately one-third of these volcanoes. On average, about one eruption per month causes significant damage and casualties somewhere in the world. Two or three eruptions each decade may be considered major disasters.

Protecting against volcanic hazards requires a variety of different monitoring instruments and involves teams of scientists with different specializations. VDAP personnel work with local scientists to rapidly deploy a suite of sophisticated devices, often linked to a mobile volcano observatory. As they collect critical data, VDAP team members also analyze the geology surrounding the volcano to unravel the history of eruptions in the region. Within a matter of weeks, sometimes less, the investigators can begin to form hypotheses about the potential character of an expected eruption—information that local government agencies can use to create plans to protect nearby residents. These volcanic emergencies provide USGS scientists with an opportunity to test their field equipment and the methods they will use to predict the character of future US volcanic eruptions.

The Speedy Lavas of Nyiragongo

A volcanic emergency occurred in January 2002 in the city of Goma, located along the eastern border of the Democratic Republic of Congo. People once considered Goma a base from which to visit the rare mountain gorillas in the nearby Virunga National Park. Travelers would relax in the luxurious gardens of the lakeside resorts and enjoy the spectacular views of twin volcanoes (Nyiragongo and Nyamuragira) 20 kilometers (12 miles) to the north. That was before civil wars in Congo and neighboring Rwanda sent waves of refugees surging back and forth across the border. Before long the economy collapsed, making the region one of the poorest in Africa. Sadly, just when it looked like things could not get much worse, the volcano nearer the city, Nyiragongo, erupted. Goma was cut in half by lava flows up to 2 kilometers (1.2 miles) wide that reached all the way to nearby Lake Kivu (Figure 6.2).

It was not a surprise that Nyiragongo erupted. The volcano had erupted about a dozen times in the preceding century, most recently in a nearly 2-year-long series that began in 1994. In the months before the 2002 eruption, the area's lone volcanologist, Dieudonne Wafula, had noted changes in the shape of the volcano's crater, increased volcanic gas emissions, and more frequent earthquakes. On January 17, lava began to flow from cracks in the ground high on the slopes of the volcano. Inhabitants of nearby villages knew to head for higher ground to avoid lava that flowed at speeds of 30 kilometers per hour (19 miles per hour) or faster. A network of fissures ultimately extended to within 2 kilometers (1.2 miles) of Goma, yielding slower-moving flows that eventually engulfed the city. Fortunately, Dr. Wafula had warned area residents of what to do if an eruption occurred (Run!). So, most of Goma's 450,000 citizens fled the city, while 147 were killed.

It is rare for lava to invade a city; most cities located near a volcano are more likely to be covered in ash than lava. But although the residents of Goma learned a harsh lesson about living near volcanoes, the consequences could have been far worse. For example, if fissures had opened up first in Goma—a real possibility for future events—residents could have been trapped in the narrow city streets and overwhelmed by the lava. In addition, the colder,

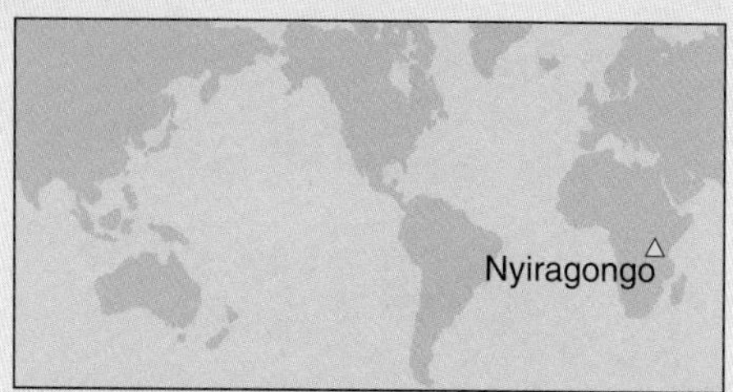

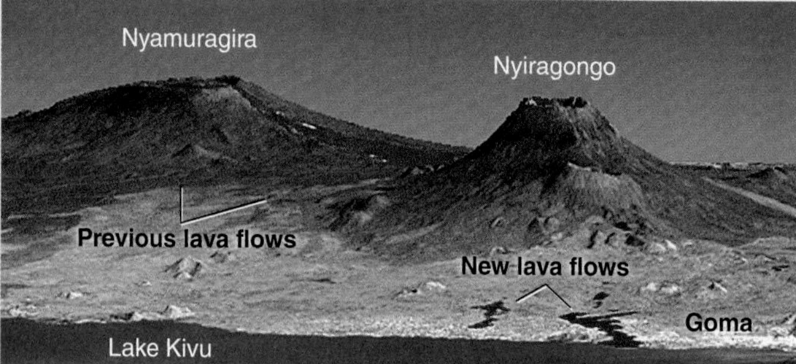

a.

b.

c.

Figure 6.2 Impact of the Nyiragongo eruption. **a.** A three-dimensional perspective from Lake Kivu shows the locations of lava flows in Goma. **b.** Burned buildings in Goma, with the profile of the Nyiragongo volcano in the background. **c.** Smoke and steam rising from lava flows in downtown Goma.

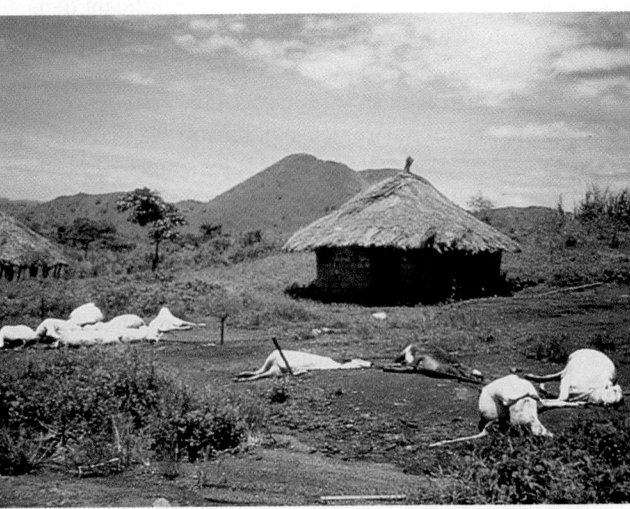

Figure 6.3 Effects of volcanic processes at Lake Nyos. **a.** A surge of carbon dioxide gas from the deep waters of this crater lake in Cameroon, West Africa, produced waves up to 25 meters (83 feet) high that stripped vegetation from around the lakeshore. **b.** The poisonous gas flowed downhill, overwhelming people and livestock in nearby villages.

isolated deep water of Lake Kivu stores high concentrations of carbon dioxide gas. Small carbon dioxide emissions from the lake routinely kill people and animals in low-lying areas surrounding Nyiragongo. Thus, if sufficient lava had flowed into the lake and caused the deep and shallow waters to mix, the carbon dioxide could have come to the surface, killing thousands of residents throughout the city who were beyond the reach of the lava flows. For instance, the 1986 escape of carbon dioxide from Lake Nyos in Cameroon, West Africa, resulted in the deaths of 1,700 people living on the surrounding slopes (Figure 6.3).

At the time of the Nyiragongo eruption, Goma's single operating seismograph station could do little more than record the event. It was not until after the 2002 eruption that USGS scientists installed the network of modern seismic instruments needed to estimate the depths and locations of earthquakes linked to volcanic activity. Like the residents of Goma, approximately 500 million people worldwide live within striking distance of active volcanoes. Despite this potential danger, **just 20 active volcanoes worldwide have a complete network of monitoring instruments,** and fewer than 200 have any instrumentation at all.

Although most of the volcanic activity that occurs on Earth takes place along the oceanic ridge system, this chapter will focus mainly on volcanoes that occur on land because they pose the greatest threat to humans. In particular, we will examine the 1980 eruption of Mount St. Helens, one of the most destructive volcanic eruptions in modern US history. We will also discuss the processes that build up mountains, whether of volcanic or tectonic origin, and learn that volcanic activity and mountain building can dramatically affect global climates in very different ways.

Self-Reflection Survey: Section 6.1

Answer the following questions as a means of uncovering what you already know about volcanoes and mountains.

1. What would be some of the pros and cons of working as a volcano expert for the rapid-response Volcano Disaster Assistance Program?
2. Imagine that you were to move to a city near a volcano. What three questions would you want to ask a volcano scientist?
3. Identify all the ways that people interact with mountains.

6.2 Magma Viscosity

Learning Objectives

- Predict how viscosity would vary for different materials or conditions.
- Explain what factors control the viscosity of magma or lava.
- Discuss why the violence of volcanic eruptions is largely controlled by magma viscosity.

Imagine that you were to pour water into one bowl and honey into another bowl. The water would flow quickly with little resistance to flow, while the honey would flow slowly. **The resistance to flow is called *viscosity*.** Low-viscosity fluids flow easily, while high-viscosity fluids resist flow (flow slowly). Does honey

Figure 6.4 Comparison of viscosities. The differing viscosities of the molten states of the three types of volcanic igneous rocks are shown here in relation to the viscosities of some common household materials. Units are in poise (0.1 kilogram per millisecond), a standard measure of viscosity (10 Poise = 1 Pascal-second). Values are approximate.

have a higher or lower viscosity than water? In this example, honey has the higher viscosity. We can reduce viscosity by heating a material and increase viscosity by cooling it. For example, warm honey flows more easily than honey that has been stored in a refrigerator.

Viscosity and Heat

Viscosity is an important property influencing volcanic activity, particularly as it relates to the flow of magma, molten rock below Earth's surface, and lava, molten rock on Earth's surface. Just as with honey, **the viscosity of magma or lava increases as its temperature decreases.** Magma works its way upward through pipes and fractures in Earth's crust because it is warmer and less dense than the surrounding material. As it comes closer to the surface, it gradually cools and its viscosity increases.

Low-viscosity magmas can flow quickly and may form lava flows that cover thousands of square kilometers. In contrast, high-viscosity lavas flow slowly and typically cover only small areas. On Earth's surface, lava solidifies (cools and hardens) to form volcanic igneous rocks. If magma solidifies within the crust, plutonic igneous rocks form (see Chapter 7). There are three common types of magma—basaltic, andesitic, and rhyolitic—classified on the basis of their compositions. Each has a different viscosity. The difference in viscosity between low-viscosity basaltic magma and high-viscosity rhyolitic magma is similar to the difference between water and thick peanut butter. Figure 6.4 compares the various viscosities of magma to those of some common household items.

Viscosity and Chemical Composition

Magma contains considerable volumes of dissolved gases, most commonly water vapor, carbon dioxide, and sulfur dioxide. These gases are dissolved in the magma under pressure deep in the crust. As the magma rises, the pressure decreases and the gas comes out of solution as bubbles that help push the magma to the surface. A similar phenomenon occurs when you take the cap off a bottle of any carbonated drink. Before you open the bottle, no gas bubbles are visible in the liquid. Once it is opened, the pressure inside the bottle decreases, and the gas bubbles are released. Many of us have been in the situation where an uncontrolled escape of gas bubbles causes a container of soda to overflow or spray its contents over those standing nearby.

While it is tempting to consider that lava is just magma at Earth's surface, there is a key difference between these hot, fluid materials. Magma approaching the surface releases gases as pressure decreases. The loss of gases changes the composition of the material, resulting in some differences in the rocks formed from solidified magma compared to hardened lava. However, for our purposes, we will assume that the three common types of lava correspond to three types of magma with similar names.

Checkpoint 6.1

basic / intermediate / advanced / superior

Place the following four materials—maple syrup, milk, peanut butter, frozen yogurt—in the correct positions (A, B, C, D) for their relative viscosity.

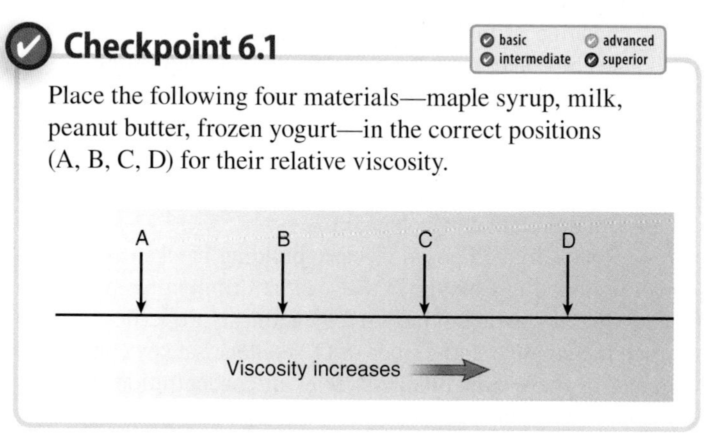

Checkpoint 6.2

How would the viscosity of motor oil in your car's engine change from the time you turn the key in the ignition to when you have driven 50 kilometers (30 miles)? Explain the reasoning behind your choice.

a) Viscosity would increase.
b) Viscosity would decrease.
c) Viscosity would stay the same.

basic advanced intermediate superior

How would you classify the viscosity of the magma that produced the eruption of Nyiragongo and the violence of the eruption itself?

a) Low-viscosity magma; violent eruption
b) High-viscosity magma; violent eruption
c) High-viscosity magma; mild eruption
d) Low-viscosity magma; mild eruption

The chemical content (composition) of magma is more important than temperature in determining viscosity. Especially significant is the amount of *silica* (SiO_2), a combination of the elements silicon and oxygen, present in the magma (Table 6.1). Silica combines with other elements in the magma (for example, iron, magnesium, potassium) to form long, branching chains of atoms. These chains get tangled together in the magma, making flow difficult and resulting in higher viscosity. In magmas with less silica, elements combine to form smaller, less complex structures, resulting in lower viscosity. The difference between the flow of low-silica and high-silica magmas is often compared to pouring peas or cooked spaghetti through a funnel. The peas will pass through the funnel easily, but the strands of spaghetti will become knotted and require greater effort to push through the funnel.

Viscosity and Volcanic Eruptions

The escape of gases from magma helps drive volcanic eruptions, but the relationship between escaping gas and the violence of volcanic eruptions is not as straightforward as it might at first appear. Would a violent eruption occur when gas can escape easily? No. In fact, it is exactly the opposite. **More-violent eruptions occur where gases cannot escape easily.**

To model this for yourself, take two liquids of contrasting viscosity—water and a milk shake work well—and blow through a straw into each. The water will bubble readily—that is, gas (the air you blow through the straw) escapes easily without much effort. In contrast, it will require a lot of lung power to get the milk shake to bubble. (Actually, the milk shake is more likely to spatter than bubble, so make sure nobody is standing too close if you try this.) The thicker the milk shake, the more violent the result.

The moral of this story is that gases can escape relatively easily from low-viscosity magmas, but high-viscosity magmas cannot lose their gases as readily.

To sum up, **the viscosity of magma increases with increasing silica content and decreasing temperature** (Table 6.1).

Table 6.1 Lava Type, Silica Content, and Viscosity

Lava type	Silica content (% weight)	Magma temperature	Viscosity rank
Basaltic	Less than 53	1,100–1,250°C	Low
Andesitic	53–68	800–1,100°C	Intermediate
Rhyolitic	More than 68	700–850°C	High

basic advanced intermediate superior

Construct a concept map that summarizes the information presented in this section of the chapter.

Gas pressures rise to higher levels in magmas with higher viscosity, resulting in more-violent eruptions. So viscosity—and therefore, the violence of eruptions—is determined by composition. But what controls composition? We will address that question in Section 6.3.

6.3 Magma Sources and Magma Composition

Learning Objectives

- Explain the concept of partial melting, using an analogy to common materials or objects.
- Compare and contrast the compositions and sources of the andesitic, basaltic and rhyolitic magma.
- Use your knowledge of plate tectonic processes and magma generation to predict the characteristics of magma produced at several active volcanoes around the world.

As we discussed in Chapter 4, **most of the world's active volcanoes above sea level are located along convergent plate boundaries.** Many of these are concentrated around the Pacific Ocean in a region known as the Ring of Fire (Figure 6.5). In the United States, more than 50 volcanoes have erupted in the past 200 years, most of them in the Aleutian Islands of Alaska, overlying the subduction zone where the Pacific plate descends below the North American plate (see Figure 4.26a). Although not as common, other active **volcanoes may be located in the interior of tectonic plates above hot spots** or may be associated with divergent plate boundaries where most magma forms. For example, the volcanoes that make up the Hawaiian Islands and the volcanic center of Yellowstone National Park in Wyoming are located above hot spots in plate interiors. Nyiragongo (described in Section 6.1) is one of several volcanoes located along the East African Rift system, an early-stage divergent plate boundary (see Figure 4.24a). **Different plate boundaries produce different magma compositions because each setting generates magma from melting a different source rock.**

Imagine a simple experiment involving a cookie containing chocolate chips and peanuts. You place the cookie in sunlight. As it warms, the chocolate chips melt, but the peanuts remain intact. The contrasting chemical and physical characteristics of the chocolate chips and peanuts result in their different behaviors. A similar process occurs when rocks are heated; some minerals melt, while others remain relatively unchanged.

Rocks are composed of basic building blocks called minerals (discussed in Chapter 7) that melt at different temperatures. Each mineral has a unique set of chemical properties. As we learned in Section 6.2, the silica (SiO_2) content is a key component in many of these minerals and varies in concentration, depending on which groups of minerals make up the rock (Table 6.1).

Checkpoint 6.5

basic · intermediate · advanced · superior

In addition to the chocolate chip cookie analog, describe another common material or object that models the partial melting of minerals with different silica contents.

The silica content of the minerals determines which ones melt first. **Silica-rich minerals have a lower melting temperature than silica-poor minerals.** Consequently, when physical conditions favor melting, the silica-rich minerals (chocolate chips) are the first to melt and the silica-poor minerals (peanuts) melt last. This **limited melting, involving only some minerals, is termed *partial melting*.** The resulting new magma that is produced may solidify into a rock that contains more minerals with low melting temperatures (silica-rich minerals) than the original source rock that experienced the partial melting.

The composition of the source rock and the plate tectonic setting under which melting occurs produce a range of magma types. These can be divided into three principal types of magma—basaltic, andesitic, and rhyolitic (Figure 6.6)—that solidify to form the volcanic rocks basalt, andesite, and rhyolite, respectively.

Basaltic magmas are characteristic of divergent plate boundaries (for example, East African Rift zone, oceanic ridges) **and oceanic hot spots** (for example, the Hawaiian Islands). They have low silica content (see Table 6.1) and, consequently, low viscosity. As we discussed in Chapter 4, partial melting occurs below oceanic ridges because of decompression melting of rocks

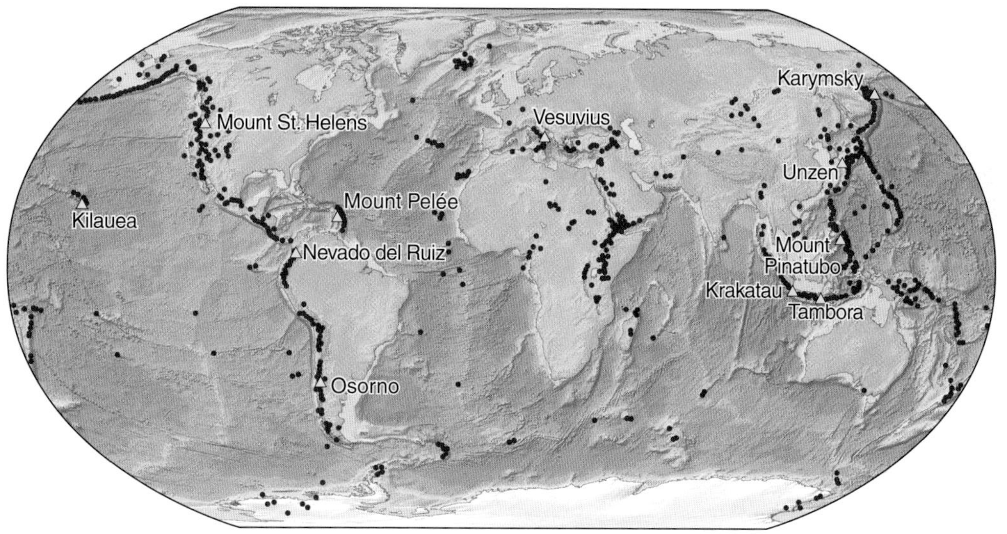

Figure 6.5 Distribution of the world's active volcanic zones. Most volcanoes are associated with subduction zones located around the rim of the Pacific Ocean or the northeast Indian Ocean. Some volcanoes are found along the oceanic ridge in the Atlantic Ocean, on Iceland or along the beginnings of a divergent boundary in East Africa. Isolated volcanoes in the interiors of plates are typically associated with hot spots above mantle magma sources. The yellow triangles correspond to volcanoes mentioned in this chapter. Red dots represent active volcanoes.

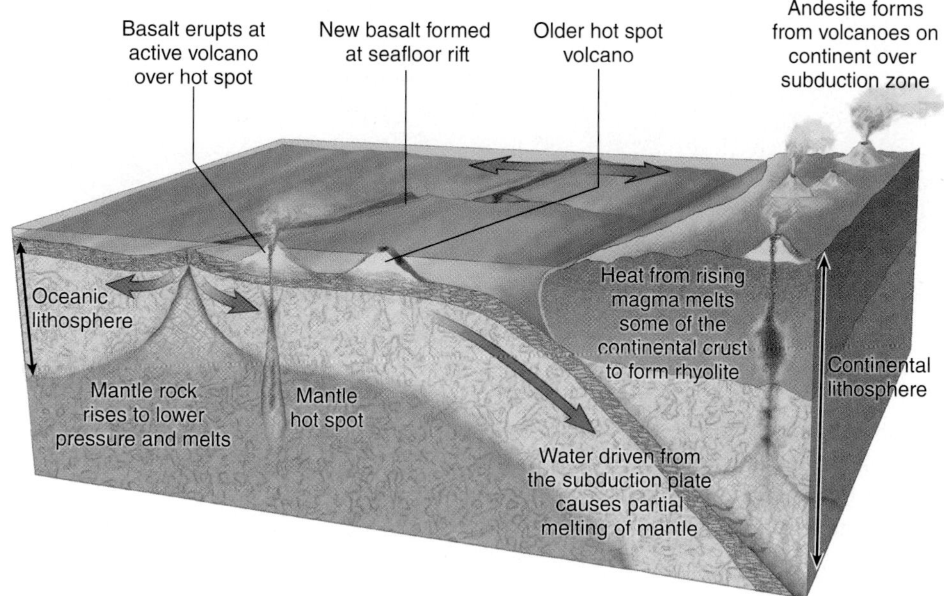

Figure 6.6 Formation of different magmas. Basalt is produced from partial melting of the mantle below volcanic islands, in the interiors of oceanic plates and oceanic ridges. It is also present in eruptions that occur in continental rift zones. Andesite is derived from the partial melting as a result of the introduction of water to the mantle wedge above descending plates of oceanic lithosphere. Rhyolite is formed by the partial melting of the continental crust due to the additional heat from rising andesitic magma.

from the asthenosphere. Melting under hot spots is due to elevated temperatures. These magmas are derived from a silica-poor mantle source. Consequently, the ocean floor and hot spot volcanoes are made up of low-silica basalt.

Andesitic magma is generated along convergent plate boundaries, where oceanic lithosphere descends into the mantle at subduction zones. Water that is expelled from the descending plate enters the hot rocks in the lower lithosphere of the mantle wedge. The presence of water reduces the temperature necessary for these rocks to melt and results in partial melting to form magma. As that magma rises through the overlying crust, slight mixing likely occurs as the crustal rocks are heated and may melt, adding to the volume of magma. This mixture generates the intermediate silica content characteristic of andesitic magma. Because of its increased silica content, andesitic magma has higher viscosity than basaltic magma. Consequently, volcanoes with

Checkpoint 6.6

basic · intermediate · advanced · superior

Venn Diagram: Magma Composition and Magma Sources

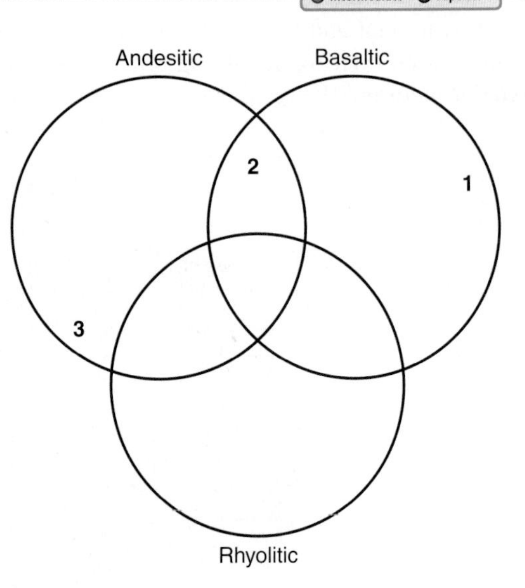

Use the Venn diagram provided to compare and contrast the compositions and sources of the three principal types of magma. Write the numbers corresponding to features unique to each magma type in the larger areas of each circle; note features that they share in the overlap areas. Identify at least seven more features.

1. Low silica content
2. From a mantle source
3. Example: Aleutian Island volcanoes
4.
5.
6.
7.
8.
9.
10.

Checkpoint 6.7

basic · intermediate · advanced · superior

Decade Volcanoes, Part I

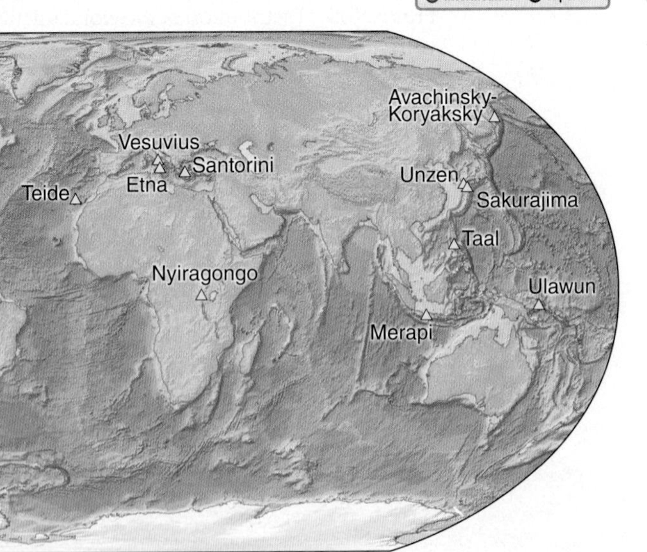

The Decade Volcano initiative is part of a United Nations program aimed at better utilizing science and emergency management to reduce the severity of natural disasters by directing attention to a small subset of active volcanoes around the world. By studying this set of volcanoes, science and land-use planning teams are attempting to learn the best ways to reduce the risks to life and property from volcano-related hazards everywhere. Answer the following questions, using the volcanoes illustrated on the map. It may be useful to review a map of plate boundaries (see Chapter 4 Snapshot) before attempting to answer the questions.

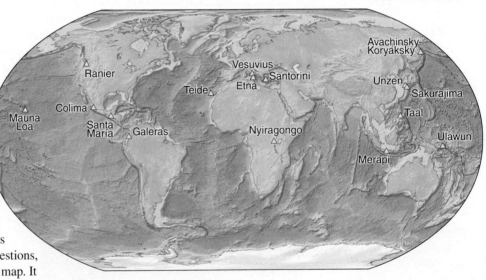

a) Underline five volcanoes that formed above subduction zones of five different plates.
b) Are any of these volcanoes associated with a present-day divergent plate boundary? Yes___ No___
c) Name a volcano that generates low-viscosity magma. ______________________
d) Other than Mauna Loa, which volcano is most likely to have formed above a hot spot? ______________________

Checkpoint 6.8

basic | intermediate | advanced | superior

Go to **http://volcano.si.edu/reports_weekly.cfm** to obtain the Smithsonian/USGS weekly volcanic activity report. Examine the world map and compare the location of each volcano with a map of plates (see Chapter 4 Snapshot) to predict the type of magma generated at each location. Write a one-sentence summary for each location to describe the volcanic activity. Is there any relationship between the type of activity at each volcano and the plate tectonic setting?

andesitic magma are more likely to erupt violently, just like the milk shake in the experiment described in Section 6.2. As a result, volcanoes on plates overlying subduction zones (for example, Mount St. Helens, Washington; Nevado del Ruiz, Colombia) yield some of the most spectacular and dangerous volcanic eruptions.

Rhyolitic magma, the most viscous magma, has a high silica content and is formed from partial melting of the continental crust. The average composition of continental crust is similar to the composition of andesite. But where does the heat that melts the continental crust come from? We cannot attribute the melting to decreasing pressure or to the addition of water, as we can for basaltic and andesitic magmas. Instead, the formation of rhyolitic magma results from melting of lower crust rocks due to the heat of magma that is rising from the mantle (Figure 6.6). This magma heat source may come from a plume of hot magma below a hot spot or from partial melting in the mantle wedge above a subduction zone. Either way, once produced, the rhyolitic magma can rise upward along fracture systems. The high-silica magma produced is so viscous that it is relatively immobile and is more likely to solidify within the crust without making it to the surface as lava. Where it does reach the surface, it can form thick flows, such as the rhyolite found in Yellowstone National Park, Wyoming, which formed above a hot spot.

6.4 The Mount St. Helens Eruption

Learning Objectives

- Predict the type of magma associated with the 1980 eruption of Mount St. Helens.
- Explain three types of observations scientists used to track events preceding the eruption of Mount St. Helens.
- Describe differences in the perceptions of scientists, government agencies, local businesses, and the general public prior to the eruption.
- Discuss how scientists use the volcano explosivity index (VEI) to classify eruptions and why casualties may not necessarily correlate with VEI rank.

As we discovered in Section 6.3, most active volcanoes are located along convergent plate boundaries. In this section, we will examine the characteristics of an eruption of one of these volcanoes. On May 18, 1980, Mount St. Helens in Washington generated one of the most destructive volcanic eruptions in modern US history. The eruption resulted in 57 deaths and approximately \$1 billion in damage. Mount St. Helens is one of several major volcanoes that are located within the Cascade Mountains (Figure 6.7). Its eruption provided scientists with a preview of what could happen if other volcanoes in the Cascade range erupt in the future. In this section, we use Mount St. Helens as a case study to illustrate the elements of volcanic eruptions and the work of geoscientists.

As we discussed in Chapters 4 and 5, the Cascade volcanoes formed as a result of the melting of the material in the mantle wedge in the Cascadia subduction zone just offshore from Washington and Oregon. Major cities such as Seattle in Washington, and Portland in Oregon, with combined populations of over 1 million people, are located less than 100 kilometers (62 miles) from these volcanoes. Many of the cities that are threatened by great earthquakes along subduction zones are also threatened by volcanic hazards. Smaller cities (Spokane, Washington; Boise, Idaho) in the region are also in danger as they are in the downwind path of Cascade volcanic eruptions.

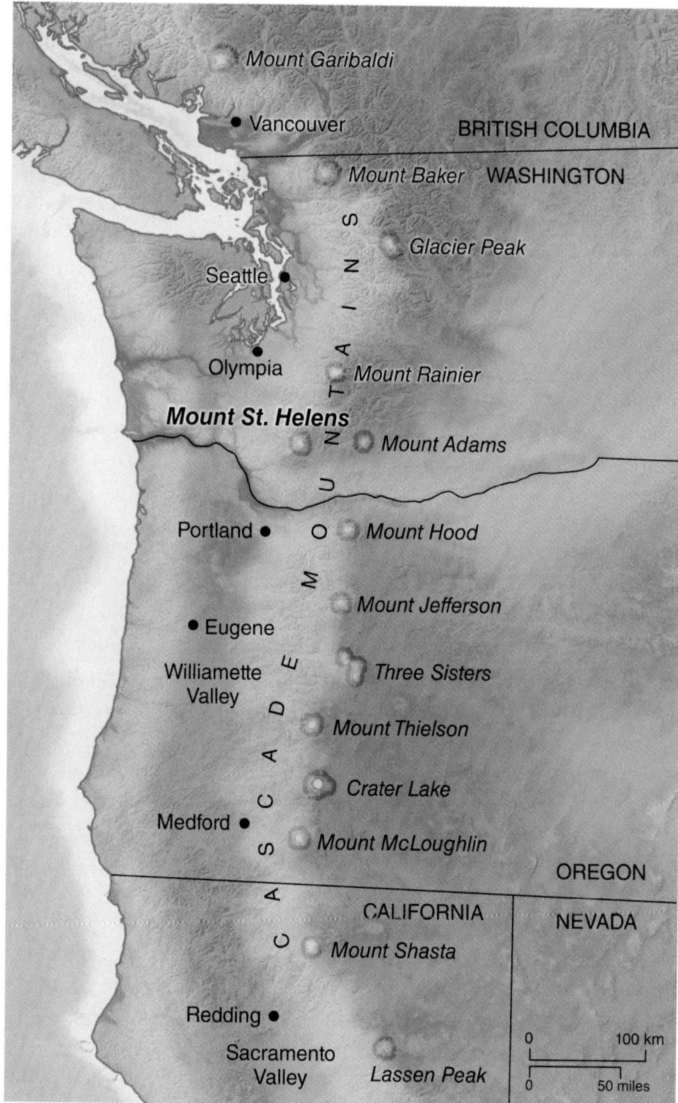

Figure 6.7 Mount St. Helens and other Cascade Range volcanoes.

Checkpoint 6.9

basic advanced intermediate superior

Predict the type of magma associated with Mount St. Helens.

a) Andesitic
b) Basaltic
c) Rhyolitic

Prior Activity

Authors of a 1978 report characterized Mount St. Helens as "more active and more explosive during the last 4,500 years than any other volcano in the conterminous United States." Within 2 years, the volcano would erupt again. Geologists have determined that Mount St. Helens has experienced eruptions at intervals of 100 to 200 years since the fifteenth century. In March 1980, after an inactive period of 123 years, the volcano again began to show signs of unrest. The first flurry of activity was characterized by small eruptions of ash and steam, as rising magma reacted with groundwater below the volcano to produce steam explosions. These events represented the first eruptions in the continental United States in 63 years. Other steam-driven eruptions followed, creating a new crater on the summit of the volcano, but no magma reached the surface. Knowing that about one-half of these periods of unrest lead to volcanic eruptions, teams of scientists began to study changes in the volcano in an effort to determine if an eruption was likely to occur. What type of data do you think they collected to identify signals of an eruption?

Prior to the beginning of unrest, only one instrument, a seismograph, was directly monitoring Mount St. Helens, but scientists soon installed a network of instruments around the volcano. **Three common indicators** are typically observed during the days or weeks prior to an eruption: (1) **increased frequency of earthquakes,** especially a specific type of low-frequency tremor associated with the migration of magma; (2) **rapid changes in the shape of the volcano;** and (3) **large-scale emissions of common volcanic gases,** especially sulfur dioxide.

Increased Frequency of Earthquakes. In the 5 years preceding the 1980 eruption, only 44 earthquakes were recorded near Mount St. Helens. In contrast, in 1 week in the middle of March 1980, over 100 earthquakes were recorded, and by April, reports of **50 to 100 earthquakes per day, several greater than magnitude 4, were common.** Local residents would have been unaware of most of these small earthquakes, but the largest shocks were felt by those living near the volcano. The tremors generated small avalanches on the snow-covered upper slopes of the volcano. Most earthquakes were identified as occurring a few kilometers below the volcano's north slope.

Change in Shape. By April, scientists observed a growing bulge on the north flank of the volcanic cone (Figure 6.8). The bulge was nearly 2 kilometers (1.2 miles) across and was growing at a rate of nearly 2 meters (6.6 feet) per day before the final eruption. **The rising bulge caused the north flank to grow by approximately 150 meters (490 feet) from its original elevation.** Scientists interpreted these observations as evidence that magma was building up beneath the summit of the volcano in advance of a major eruption.

a.

b.

c.

Figure 6.8 Prelude to eruption. These views of Mount St. Helens are from the north: the arrow shows the same reference point in all three images. **a.** North flank has no noticeable bulge on March 29. **b.** A bulge is noticeable on April 11. **c.** A large bulge is apparent on May 2.

The bulge pushed out the north flank of the volcano, causing its slopes to become steeper and generating concern that a major landslide could break free and race downslope in a matter of minutes.

Emission of Volcanic Gases. Scientists at observation stations on Mount St. Helens slopes or flying over the volcano in helicopters were able to collect samples of volcanic gases (steam, sulfur dioxide, carbon dioxide, and hydrogen sulfide). Sulfur dioxide levels jumped from barely measurable to approximately 10 to 20 tons per day during small eruptions and settled at levels of about 1 ton per day between explosions.

The May 18 Eruption

Even though scientists had plenty of data and observations at their disposal, the eruption of Mount St. Helens on May 18 came as a big surprise. (How is such a surprise consistent with the nature of science?) Despite the sophisticated array of instruments and the team of experts studying the mountain, there had been no clear signal that the eruption would occur that day—no rise in earthquake frequency, no change in gas emissions, and no sudden increase in the size of the bulge (Figure 6.9a).

At 8:32 A.M., a magnitude-5.1 earthquake occurred, centered about 1 kilometer (0.6 mile) below the north flank of the volcano. The earthquake was the apparent trigger for the eruption. A few seconds later, the north flank began to heave and break away in a series of enormous blocks that collapsed and slid downslope, merging to form a massive debris avalanche, the largest landslide in US recorded history (Figure 6.9b and c). The avalanche surged downward at high speed, and debris, tens of meters thick, filled the nearby valley of the north fork of the Toutle River. The debris avalanche removed several cubic kilometers of rock from the north flank of the volcano. This released pressure on the magma rising from below, much like opening a can of well-shaken soda. The result was a catastrophic lateral (sideways) eruption that blasted outward with such velocity that it soon overran the debris avalanche (Figure 6.9c). Though much of the area was evacuated, several geologists and area residents perished.

The blast was heard in surrounding states and hundreds of kilometers away in parts of northern California and Canada.

Checkpoint 6.10

basic · intermediate · advanced · superior

Earthquakes and Mount St. Helens

Volcanic eruptions are often preceded by earthquake activity as magma rises upward through the crust underlying the volcano. Examine the graphs of earthquake events averaged per week for 1980 (graph a) and for 1981–1992 (graph b). How did earthquake activity differ for the two intervals?

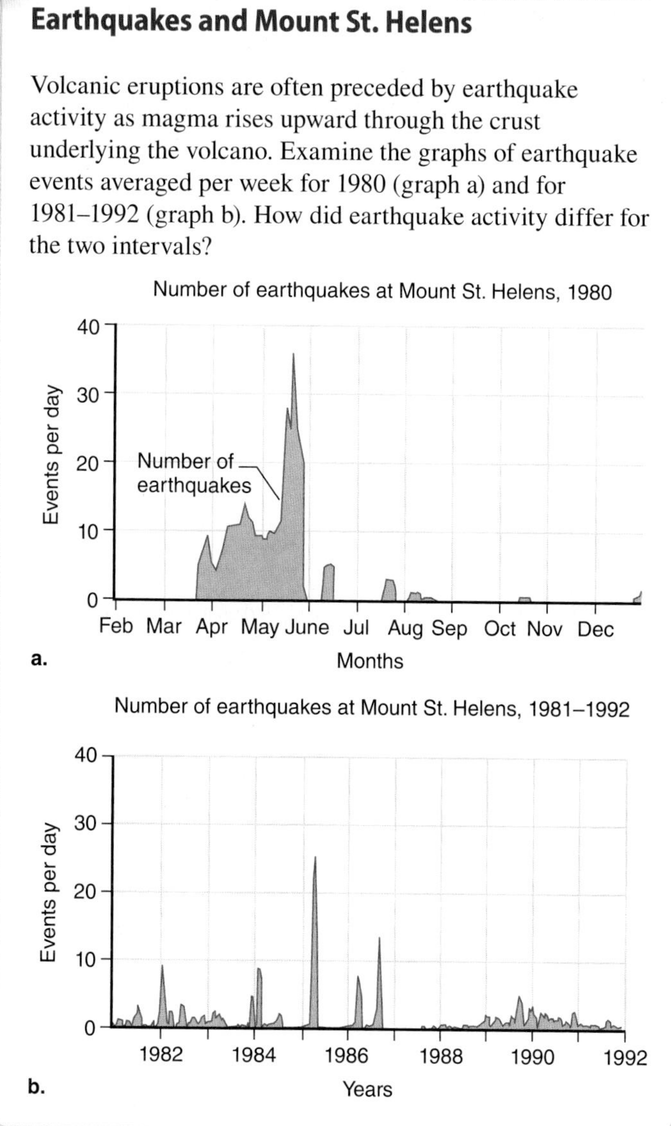

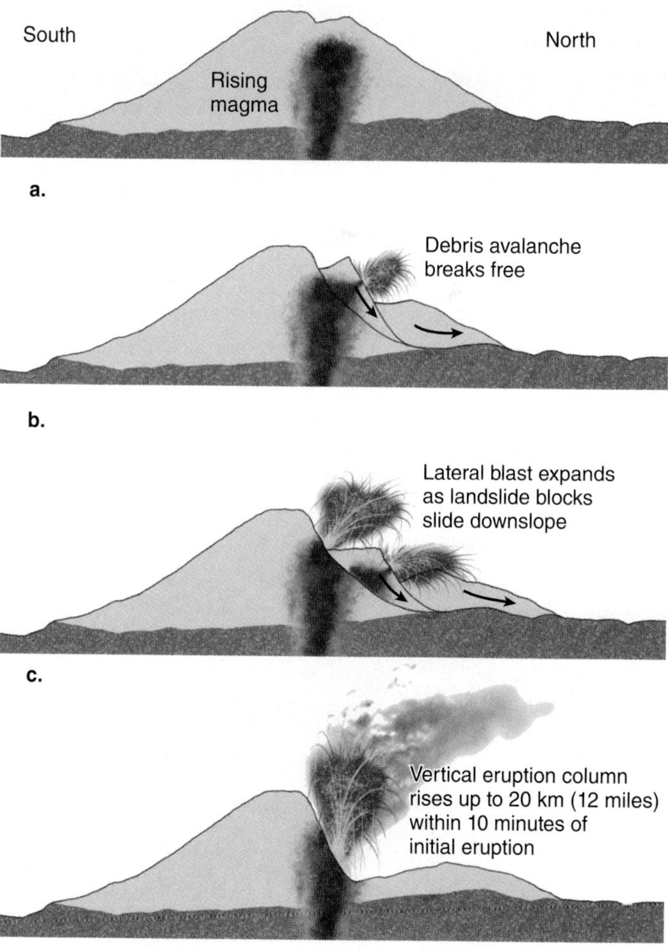

Figure 6.9 Eruption stages of Mount St. Helens. **a.** The volcano prior to eruption. **b.** Early moments of debris avalanche as landslide blocks begin to move downslope and eruption is first sighted. **c.** Lateral blast occurs as rocks are removed from above the magma source. **d.** Eruption takes classic vertical form following debris flow.

Figure 6.10 Vertical eruption column. This view is from the south side of Mount St. Helens.

It removed several hundred meters of the north side of the volcano before finally settling into a more familiar vertical eruption style (Figures 6.9d and 6.10). Within 10 minutes, the debris cloud from the eruption reached an altitude of 20 kilometers (12.4 miles), darkening the morning sky. The landscape to the north of the volcano was converted to a wasteland blanketed in gray volcanic debris. Ash from the eruption was deposited in a measurable layer as far away as Oklahoma.

Checkpoint 6.11 (basic, intermediate, advanced, superior)

Identify elements of the scientific method in the analysis of Mount St. Helens that preceded the eruption.

How Does Mount St. Helens Compare to Other Eruptions?

To compare the relative sizes of volcanic eruptions, scientists use the **volcanic explosivity index (VEI)** (Table 6.2). The volume of erupted material, duration of the event, and height ash reaches in the atmosphere contribute to determining the explosivity rank of an eruption. The most consistent results are generally obtained by estimating the total volume of erupted material. The VEI is split into eight divisions, with the first rank describing the smallest volume of debris and the eighth rank representing the greatest volume. Each interval of the VEI represents a 10-times increase in volume over the previous rank. The 1980 eruption of Mount St. Helens produced 1 cubic kilometer (0.24 cubic mile) of material and had a VEI of 5.

VEI scores do not refer to all eruptions of a volcano but to specific events, in much the same way that a fault may generate multiple earthquakes with different magnitudes. For example, Mount St. Helens produced more than 20 additional eruptions after May 18; almost all had a VEI of 2. In historical times, only 22 eruptions with a VEI of 5 or greater have been recorded, an average of about one large eruption every 20 years. Like all averages, there is no guarantee that these events will occur on a regular basis. Eleven years after Mount St. Helens erupted, Mount Pinatubo in the Philippines erupted and produced 10 times more volcanic material (VEI 6).

The eruption of the volcanic island of Krakatau, Indonesia, in 1883 also had a VEI of 6 and produced the loudest noise ever heard in historic times. The eruption was heard 3,000 kilometers (1,864 miles) away, comparable to a person in Indianapolis, Indiana hearing an explosion from Los Angeles.

Often the explosion itself is not the major cause of destruction or loss of life. For example, tsunami produced by the Krakatau eruption killed thousands of people on neighboring islands and contributed to most of the 36,000 deaths resulting from the eruption. Table 6.3 lists the five most deadly volcanic eruptions

Table 6.2 Volcanic Explosivity Index (VEI)

VEI rank	Volume of erupted material	Example
1	0.00001 km^3	Nyiragongo, 2002
2	0.001 km^3	Galeras, Colombia, 1992
3	0.01 km^3	Nevado del Ruiz, 1985
4	0.1 km^3	Mount Pelée, Martinique, 1902
5	1 km^3	Mount St. Helens, 1980
6	10 km^3	Vesuvius, A.D. 79; Krakatau, 1883; Mount Pinatubo, 1991
7	100 km^3	Tambora, 1815
8	1,000 km^3	Toba, Sumatra (74,000 years ago); Yellowstone (640,000 years ago)

Table 6.3 World's Deadliest Volcanic Eruptions

	Location	Year	Death toll
1	Tambora, Indonesia	1815	92,000 died, mostly from starvation due to crop failures resulting from loss of sunlight following the eruption.
2	Krakatau, Indonesia	1886	36,000 dead from tsunami.
3	Mount Pelée, Martinique	1902	30,000 killed by pyroclastic flows.
4	Nevado del Ruiz, Colombia	1985	25,000 buried when mudflows engulfed Armero and other towns at the base of the mountain.
5	Unzen, Japan	1792	15,000 died when the volcano collapsed and an associated tsunami inundated coastal settlements.

Checkpoint 6.12

basic · advanced · intermediate · superior

The People of Mount St. Helens

Examine the following descriptions to compare the actions of scientists, government agencies, businesses, and the general public in the weeks preceding the eruption of Mount St. Helens. Then answer these questions:

- How did the different constituencies perceive the threat of an eruption? Create a diagram that illustrates how the different constituencies perceived the threat from the volcano over the 2-month period from March 20 to May 18.
- What do you think would have happened if the main eruption had not occurred for another 2 months?

Keep in mind that the job description of USGS scientists prevents them from recommending specific actions to lessen risk but requires them to focus on assessing natural processes and forecasting geologic scenarios. Local, regional, and federal governments are responsible for instituting measures to protect populations and structures, while taking into account cultural and social factors.

Early Days (March 20–April 2)

Scientists: USGS adds more seismometers and warns of earthquakes related to volcanic activity; gas and ash samples are interpreted to indicate a magma source near the surface; more instruments are added.

Government agencies: The US Forest Service (USFS) warns the public to stay away and closes Spirit Lake; Federal Aviation Administration (FAA) imposes flight restrictions; evacuation within 24-kilometer (15-mile) radius; National Guard sets up 29 roadblocks; USFS moves possessions of staff from evacuated area.

Business: 300 loggers return to work on northwest flank of volcano.

Public: Sightseers crowd nearest viewing sites; citizens express frustration about being kept from property. Local cabin owner Harry Truman refuses to leave, saying, "I think the whole damn thing is overexaggerated. . . . You couldn't pull me out with a mule team."

Unrest Continues (April 3–April 18)

Scientists: Geologists identify the greatest hazards of potential eruption as flooding or mudflows; costs of monitoring volcano and maintaining safety forces continue to rise; USGS reduces number of scientists at volcano from 30 to 10.

Government agencies: Governor declares state of emergency; roadblocks moved farther from volcano; some roadblocks removed because of a reduction in unrest and public harassment; National Guard staffing reduced.

Business: Loggers request greater access to restricted areas; local businesses threaten to sue if roadblocks are not moved closer to mountain.

Public: On one day, 109 planes reported inside restricted air space on; tourists flock for close-up view of the mountain; climbers evade roadblocks and reach peak.

Danger Threatens ... Maybe (April 19–May 3)

Scientists: Bulging of north flank becomes more obvious; seismic activity increases; high concentrations of sulfur dioxide gas noted; geologist comments, "It's very dramatic to see this much ground motion. It can't be anything but some type of dramatic change going on inside the mountain."

Government agencies: Red Cross takes disaster workers off standby alert and asks county to return all extra shelter equipment. Governor closes red zone within 12 kilometers (7 miles) of mountain except for scientists and law enforcement; areas within 25 kilometers (16 miles) are accessible only to landowners and loggers.

Business: Timber company considers blocking active logging roads to restrict tourist access, citing concerns over potential problems with congestion, fires, and accidents with logging trucks.

Public: Emergency services personnel frustrated because public appears unaware of the danger: "The mountain looks so serene, so people can't fathom 4,000 vertical feet of earth, rock, and ice plunging into (Spirit Lake) in less than 2 minutes."

The End Is Nigh? (May 4–May 17)

Scientists: USGS geologist: "(The deformation) is continuing at a very high rate. Sometime it has to go. We just don't know how much longer it can last." USGS stops taking measurements high on mountain because of increased danger from sudden avalanches.

Government agencies: FAA officials issue a pilot's warning that an ash plume extended 32 kilometers (20 miles) to the north-northeast and was "of extreme hazard to aircraft"; governor grants request for 50 landowners to retrieve possessions.

Business: Logger quoted: "We're logging 10 miles away from the peak. . . . I don't see any hazard. I just came back from Hawaii, where they run tourist buses right up to the edge of a venting volcano."

Public: Tourists from around the world come to see the mountain; owners of property in evacuation zone near volcano demand access to homes.

May 18

At 8:32 A.M. an earthquake shakes Mount St. Helens volcano, resulting in a giant landslide from the area of the bulge on the north flank. Approximately 650 square kilometers (250 square miles) of surrounding lands are damaged by a lateral blast, and 150 million cubic meters (5,300 million cubic feet) of material are deposited directly by lahars (volcanic mudflows) into the river channels. The initial volcanic blast destroys forest lands and property within the cordon of roadblocks. Explosive eruptions continue until October, but none reaches the magnitude of the May 18 event.

in recorded history. Although some of the very largest eruptions have been responsible for huge death tolls, the greatest numbers of deaths from eruptions in the twentieth century were associated with two eruptions much smaller than Mount St. Helens. Compare the data in Tables 6.2 and 6.3 to figure out which two volcanoes we are referring to. In both cases, some of the products of the eruptions overran unsuspecting residents in nearby cities. We will examine these phenomena more closely in Section 6.5.

6.5 Products of Volcanic Eruptions

Learning Objectives

- Compare the scale and distance traveled of the Mount St. Helens eruption products to features in or around your location.
- Compare and contrast the different characteristics of the eruptions of Mount St. Helens and Nyiragongo.
- Discuss whether it would be preferable to organize the recovery after a major earthquake or a significant volcanic eruption.
- Analyze the potential for volcanic hazards in the area around Mount Shasta, California.

The effects of the Mount St. Helens eruption were immediately experienced close to the volcano and soon felt in states hundreds of kilometers away. The eruption blasted nearly 400 meters (1,300 feet) from the top of the volcano (Figure 6.11). Structures near the volcano were blown over by the force of the blast or carried away by mudflows. Some structures downwind simply collapsed under the weight of volcanic ash.

The products of the Mount St. Helens eruption are similar to volcanic materials that could be generated by most explosive volcanoes (see Chapter Snapshot). In addition to the debris avalanche

Trees were knocked down up to 27 kilometers (17 miles) from Mount St. Helens by the blast associated with the eruption. Where would a volcano in your region be located if it were 27 kilometers from your home or college?

described in Section 6.4, **these products can be divided into airborne elements (lateral blast, tephra, volcanic gases) and flows on the ground surface (lava, pyroclastic flows, lahars).**

Airborne Elements

Lateral Blast Effects. Volcanic blasts, driven by gas-rich magmas with intermediate viscosity, are common from volcanoes similar to Mount St. Helens. Most blasts are directed vertically, but **sideways, or lateral, blasts** have the potential to be much more destructive because they can devastate large areas immediately adjacent to volcanoes. In the Mount St. Helens eruption, the lateral blast was directed to the north, and almost all the local damage associated with the eruption (downed trees, destroyed buildings, landslides) was located on the north face of the mountain (Figure 6.12). The force of the blast completely demolished any objects within 12 kilometers (7 miles) of the crater and knocked down trees on the surrounding national forest lands as far as 27 kilometers (17 miles) away.

Tephra. The Mount St. Helens eruption blew apart the volcanic crater and blasted molten rock high into the atmosphere. **Particles of all sizes that are blasted into the air by a volcanic eruption are termed *tephra*** (Figure 6.13). The vertical column of tephra from a large eruption (see Figure 6.10) reaches an elevation of 20 kilometers (12 miles) within a few minutes and can continue for hours or even days. The principal airborne product is typically a fine volcanic ash that spreads out in the

a.

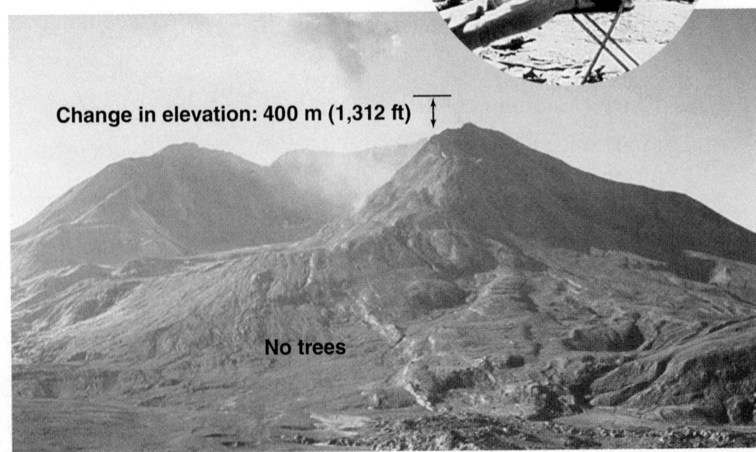

b.

Figure 6.11 Before and after the Mount St. Helens eruption. These images were taken from Johnston's Ridge, located 10 kilometers (6 miles) from the volcano. The site is named for geologist David Johnston (inset), who died at this location on the morning of the eruption. **a.** "Before" image taken on May 17. **b.** "After" image taken on October 9, 1980. Note that the forest and treeline in the center of image (a) are gone in image (b). All objects, including trees, within 12 kilometers (7 miles) of the crater were completely destroyed.

atmosphere and can be carried hundreds of kilometers downwind from the volcano. Volcanic ash from Mount St. Helens traveled eastward at a velocity of 100 kilometers per hour (62 miles per hour), reaching Spokane in eastern Washington by noon. By the next day, ash deposits were detected in the central United States. A few days later, wind wind was carrying ash across the Atlantic Ocean.

Tephra has the potential to produce significant economic losses and severely affect everyday operations. Volcanic ash covers everything, and a layer 0.5 meter (1.6 feet) thick can add sufficient weight, especially when wet, to cause roofs to collapse. Ash is composed of fine particles of rock and fragmented glass that can infiltrate most openings and damage or clog machinery (Figure 6.13b). This is especially a threat to commercial aircraft because ash abrades the windshield and causes jet engines to fail. Fine tephra deposits also become slippery when wet, producing an effect similar to ice on roads. In addition, this wet material is highly conductive and can cause blackouts by short-circuiting electrical systems. The eruption of Mount Tambora (VEI 7) in 1815 threw so much material into the atmosphere that it blocked sunlight and temperatures subsequently plunged across much of North America and Europe. Crops failed and bizarre weather was the norm in 1816, which became known as the "year without a summer."

There are hundreds of skeletons of fossil rhinos, horses, camels, and birds preserved in a 12-million-year-old volcanic ash layer near Creighton in northeast Nebraska. The ash was deposited by the eruption of a supervolcano 1,600 kilometers (1,000 miles) away in present-day Idaho. Supervolcanoes are hundreds to thousands of times larger than volcanoes like Mount St. Helens and produce truly massive eruptions (VEI 8; Table 6.2). The Bruneau-Jarbidge supervolcano was responsible for the ashfall deposits. The supervolcano formed as the North American continent tracked over a mantle hot spot that today is represented by the hot springs and geysers of Yellowstone National Park. Yellowstone lies at the end of a track of volcanism that began 16 million years ago. Since then, the continent has traveled 600 kilometers

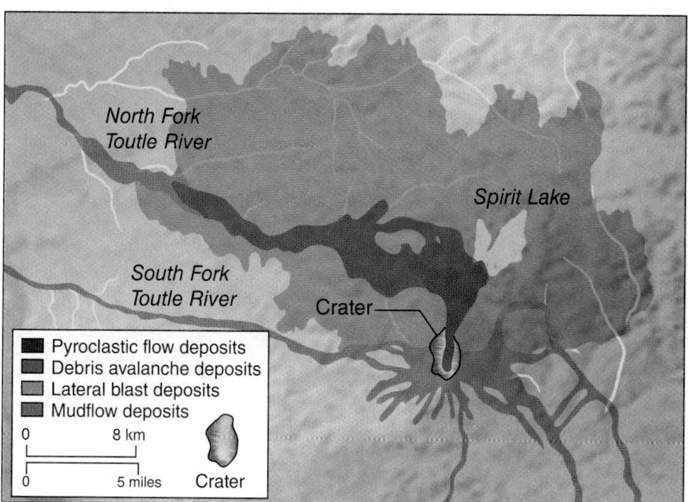

Figure 6.12 Areas affected by the Mount St. Helens eruption. The volcano's crater is illustrated in the bottom half of the map. With the exception of some lahars, all devastation occurred to the northern side of the volcano, consistent with the northward-directed lateral blast.

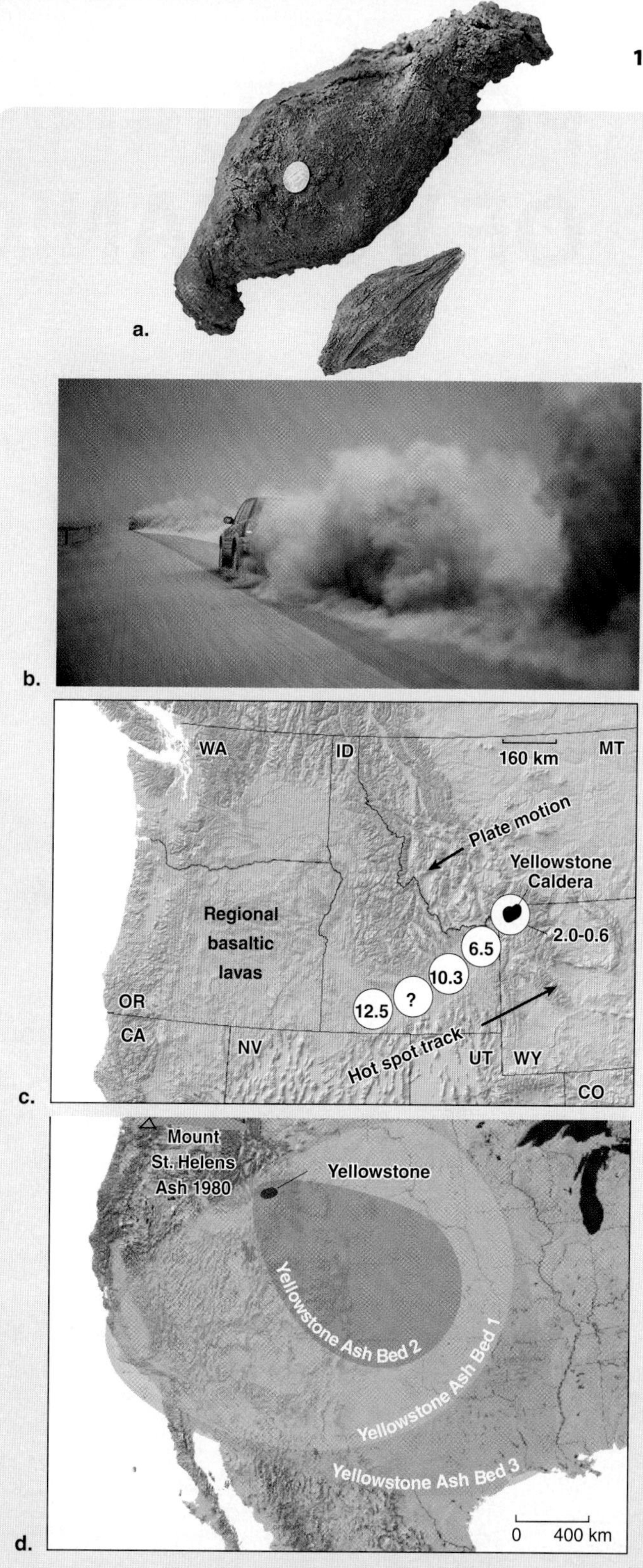

Figure 6.13 Tephra. **a.** Lava bombs form from blobs of molten material that cool and solidify while they are airborne. Volcanic ash is the name given to tephra particles that are less than 2 millimeters (0.08 inch) across. **b.** Vehicles disturb volcanic ash 2 weeks after eruption of Mount Pinatubo, Philippines, 1991. **c.** Yellowstone hot spot track. North America is moving to the southwest relative to the hot spot. Numbers correspond to dates of supervolcano eruptions in millions of years; 12.5 corresponds to the Bruneau-Jarbridge eruption. **d.** Distribution of ash deposits following Yellowstone and Mount St. Helens eruptions. The thickest deposits are immediately downwind from the volcano.

POTENTIAL FEATURES OF VOLCANIC ERUPTION

PREVAILING WIND

TEPHRA AND GASES
Larger volcanic rock fragments fall closer to the volcano vent; smaller particles are carried thousands of kilometers downwind in the eruption cloud. Gases such as water vapor, carbon dioxide, sulfur dioxide, and hydrogen chloride are released from rising magma.

PYROCLASTIC FLOW
A dense avalanche of a mixture of hot volcanic debris and gases that flows rapidly (100 km/h) down the slope of a volcano. The flow loses energy when it reaches the base of the volcano and soon comes to rest.

LANDSLIDE
Steep volcanic slopes may collapse to form fast-moving landslides known as debris avalanches. These materials may combine with rivers to form lahars.

LAVA FLOW
Molten rock that flows over the surface during a volcanic eruption. Depending on the viscosity of the lava, it may flow down the slope of a volcano (low viscosity) or form a volcanic dome (high viscosity).

MAGMA
A mixture of melted rock, crystallized minerals, rock fragments, and dissolved gases.

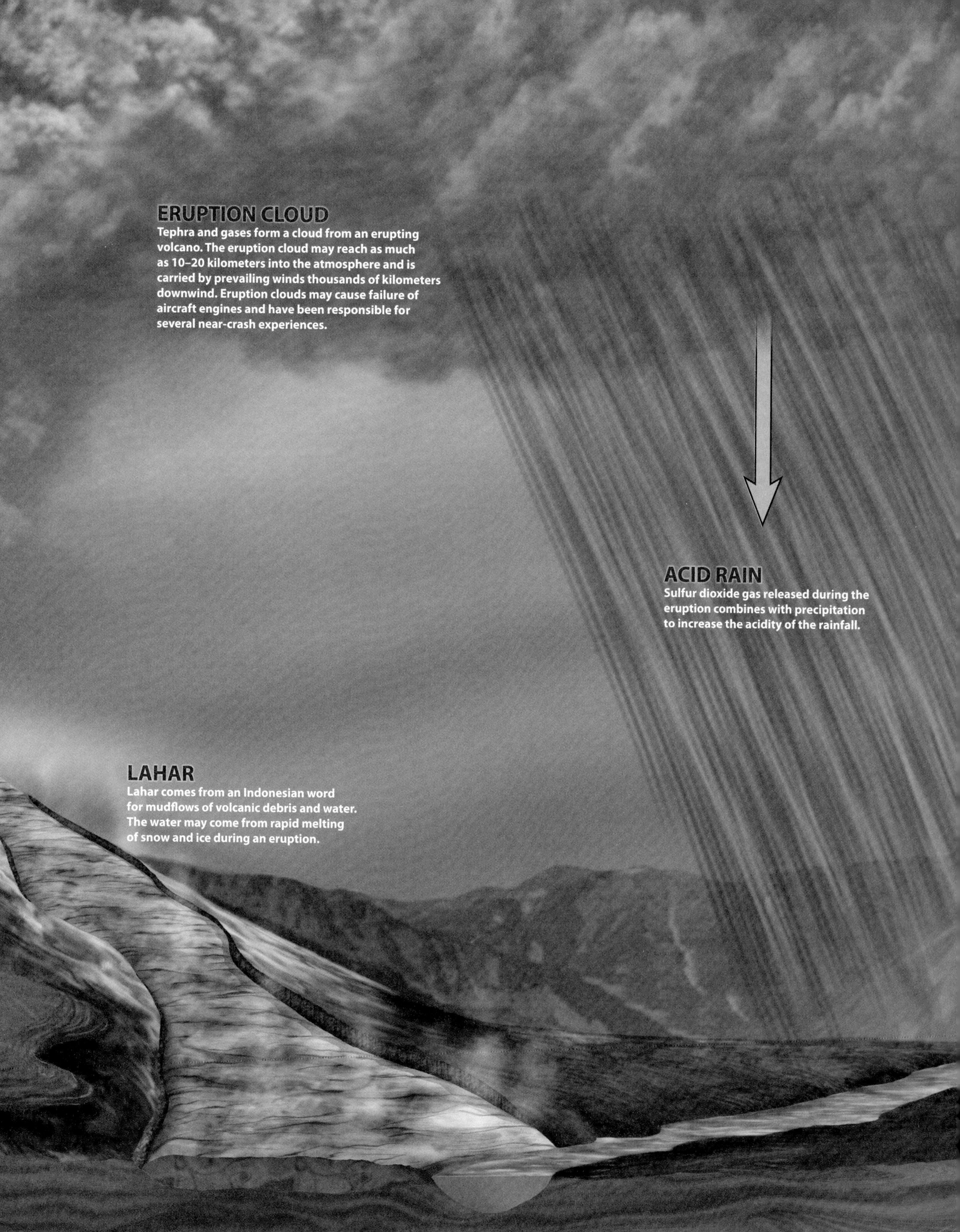
ERUPTION CLOUD
Tephra and gases form a cloud from an erupting volcano. The eruption cloud may reach as much as 10–20 kilometers into the atmosphere and is carried by prevailing winds thousands of kilometers downwind. Eruption clouds may cause failure of aircraft engines and have been responsible for several near-crash experiences.
ACID RAIN
Sulfur dioxide gas released during the eruption combines with precipitation to increase the acidity of the rainfall.
LAHAR
Lahar comes from an Indonesian word for mudflows of volcanic debris and water. The water may come from rapid melting of snow and ice during an eruption.

Figure 6.14 Trees killed by carbon dioxide. High concentrations of carbon dioxide were measured in the soil below these trees on the slopes of Mammoth Mountain in California. The gas is believed to come from a magma source below the volcano.

(373 miles) to the southwest, leaving a wide trail of volcanic rocks across southern Idaho and into northwest Wyoming (Figure 6.13c). Every few million years, the hot spot produced a supervolcano eruption that devastated much of western North America. The most recent series of eruptions occurred 2.1, 1.3, and 0.64 million years ago. Ash from each of these supereruptions has been identified over much of the western United States (Figure 6.13d). These massive eruptions ejected debris high into the atmosphere, punching holes through the ozone layer in the stratosphere. Volcanic debris produced by the eruptions would have blocked sunlight and reduced global temperatures by 5 to 15°C (9 to 27°F) for several years following each event.

Volcanic Gases. The escape of dissolved gases drives the volcanic eruptions that blast tephra high into the atmosphere. Remember that the three principal volcanic gases are water vapor, carbon dioxide, and sulfur dioxide. Carbon dioxide and sulfur dioxide can affect global climate patterns.

Sulfur dioxide has a short-term effect on global climate similar to that of volcanic ash. The 1991 eruption of Mount Pinatubo (VEI 6) released millions of tonnes of sulfur dioxide into the stratosphere, helping to block incoming sunlight for several months and cooling large regions of Earth by 0.5°C (0.9°F) for the next year.

The world's volcanoes release more than 130 million tonnes (143 million tons) of carbon dioxide into the atmosphere each year. This seems like a lot, but it is less than 1 percent of the annual volume of the gas generated by human activities. In almost all cases, the volcanic gas is so diluted in air that it presents no real threat. Only under exceptional circumstances in which carbon dioxide makes up at least 7 percent of air (about a 200-times increase in its normal concentration) does it cause death to humans and other organisms (Figure 6.14).

In rare cases, a series of massive eruptions may release huge volumes of carbon dioxide. This is interpreted to have happened during a period from 120 to 80 million years ago, when the breakup of Pangaea shifted into high gear with seafloor spreading between Europe and North America. Since increases in carbon dioxide are linked to rising temperatures, this addition of carbon dioxide contributed to global hothouse conditions that lasted for tens of millions of years. Elsewhere, large lava plateaus were built up on the seafloor at this time (see Section 6.6). Both of these processes required the production of large volumes of magma and thus the release of more carbon dioxide. Much of the additional carbon dioxide found its way to the atmosphere, resulting in concentrations that were about three times the present level. The rise in carbon dioxide contributed to an exaggerated long-term greenhouse effect that resulted in a 10°C (18°F) increase in global temperatures. During this time, there were no polar ice caps and palm trees were growing north of the Arctic Circle—all because of volcanic gases.

Surface Effects

Lava. The low-viscosity basalt lava flows that are common from Hawaiian volcanoes and the eruption of Nyiragongo can travel distances of up to 50 kilometers (31 miles) from their source. Hawaiian lavas can travel considerable distances as they flow much of that length in well-defined lava tubes at speeds of 30 kilometers per hour (19 miles per hour) or greater (Figure 6.15a). When lavas flow over open ground, the speed of even low-viscosity lavas is a relatively slow 1 kilometer per hour (0.6 miles per hour), much slower than walking speed. On Hawaii, recent flows have built piles of hardened lava up to 10 meters (33 feet) thick (Figure 6.15b).

In marked contrast with the Nyiragongo eruption, lava posed little hazard at Mount St. Helens because it was produced in only modest quantities and did not travel far due to its viscosity (Figure 6.16). (Given the characteristics of the eruption, do you think magma at Mount St. Helens had a higher or lower viscosity than at Nyiragongo?) The slow-moving lava cooled and solidified in the volcano's crater, gradually building up a lava dome. The lava

Geologist's View

a.

b.

Figure 6.15 Hawaiian lava. **a.** A lava tube transports hot, fluid, low-viscosity basalt lava toward the front of a lava flow on Kilauea volcano, Hawaii. **b.** Walter's Kalapana Store and Drive-in was burned and buried within a few weeks in 1990 as lava from the Kilauea volcano invaded communities along the southern coast of Hawaii. Note the height of the original sign. How deep is the lava at this location?

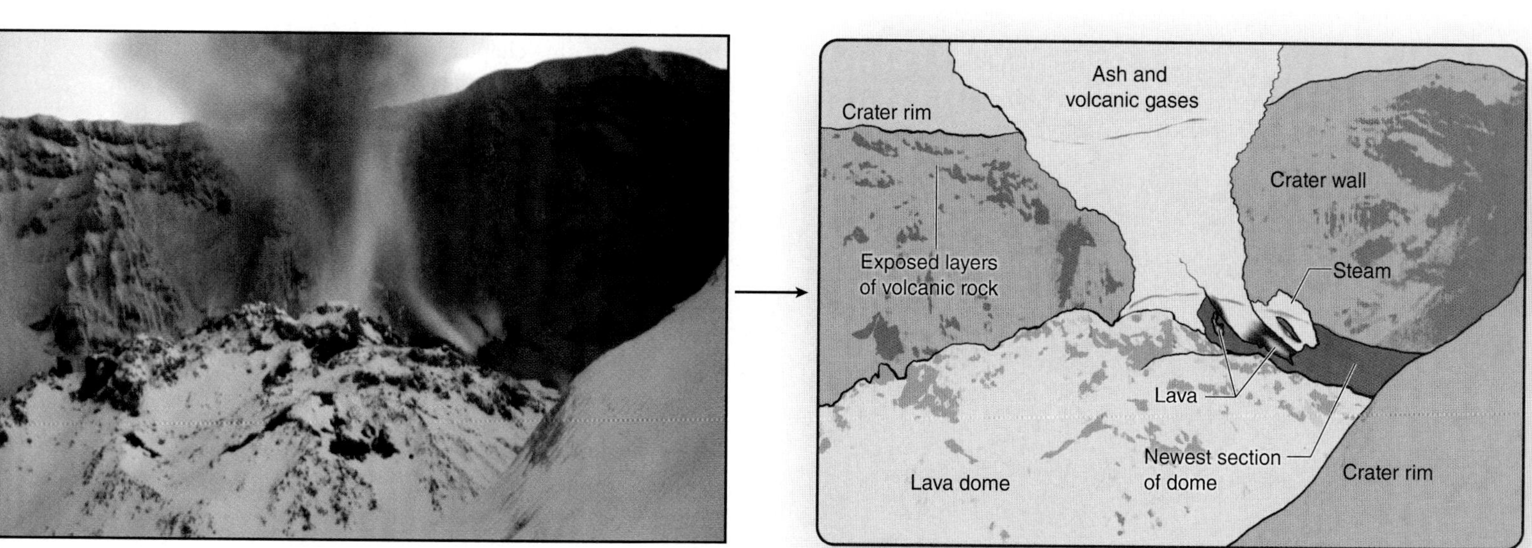

Figure 6.16 Lava dome in the Mount St. Helens crater. As more lava comes to the surface, the dome cracks and expands, releasing small volumes of gas and tephra (ash).

dome increased in size during the following decade, reaching an elevation of more than 300 meters (985 feet) and a basal width of over 1 kilometer (0.6 mile).

Checkpoint 6.14

basic · intermediate · advanced · superior

Venn Diagram: Mount St. Helens and Nyiragongo Compared

Use the Venn diagram provided here to compare and contrast the eruptions of Nyiragongo and Mount St. Helens. Write the numbers corresponding to features unique to either group in the larger areas of the left and right circles; note features they share in the overlap area in the center of the image.

1. Located near a convergent plate boundary.
2. Located near an early-stage divergent plate boundary.
3. Produced significant lava flows.
4. Eruption followed a century of inactivity.
5. Several eruptions in the last century.
6. Few monitoring instruments prior to unrest.
7. Volcanic gases released prior to main eruption.
8. Frequent earthquakes associated with unrest.
9. Unrest lasted for approximately 2 months before eruption.
10. Eruption occurred in daylight.
11. Volcano located within 20 kilometers (12 miles) of large city.
12. Volcanic activity subsided after about 1 week.
13. Low-viscosity magma.
14. USGS geologists aided in interpretation of volcanic activity.
15. Death toll less than 100.
16. Death toll more than 100.
17. Shape of volcano changed prior to eruption.
18. Eruption characterized by a massive lateral blast.

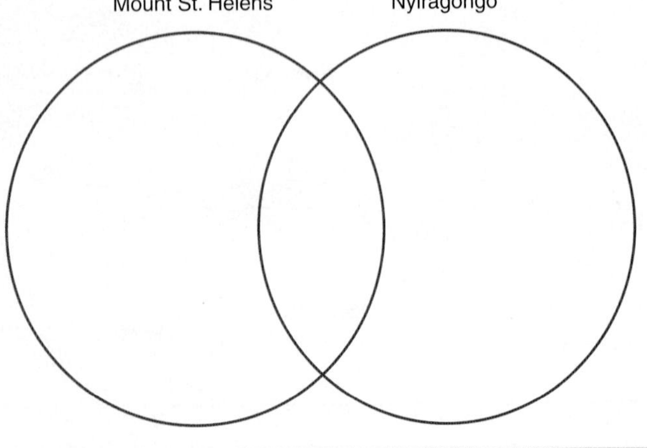

Although lava flows in modern cities are rare, they are not without precedent. The island of Hawaii is divided into nine hazard zones based on the potential for dangerous lava flows. Fortunately, while lava flows can cause considerable destruction, they rarely result in loss of life—Nyiragongo being an obvious exception. However, they can destroy communities (see Figure 6.2). Rare successful efforts to stop lava have focused on such strategies as blowing up lava tubes to spread out the flow and cool it faster, and pumping large volumes of water on advancing flows.

Pyroclastic Flows. Pyroclastic flows are mixtures of hot gases and ash, cinders, and other volcanic debris (Figure 6.17). These materials combine to **form a dense, hot cloud that races down the volcano's slope,** incinerating all in its path. Pyroclastic flows can reach speeds of 100 kilometers per hour (62 miles per hour) and temperatures of 200 to 700°C (392 to 1,292°F). They may be powerful enough to transport volcanic rocks the size of car tires. Because of their speed and high temperatures, pyroclastic flows are among the most dangerous hazards of a volcanic eruption. In 1902, a pyroclastic flow from the Mount Pelée volcano swept down on the city of St. Pierre on the Caribbean island of Martinique and killed almost every one of its 30,000 residents (Figure 6.17b, see Table 6.3). Even volcanologists are not immune to pyroclastic flows; several were among the 41 people killed when a pyroclastic flow began suddenly as a result of the collapse of a lava dome during a 1991 investigation of the Mount Unzen volcano in Japan.

Lahars. Lahars are **mudflows formed when fine-grained tephra mixes with water from snowmelt or streams.** Lahars flow mainly down stream channels at a velocity typically ranging from 16 to 40 kilometers per hour (10 to 25 miles per hour). Lahars carry thousands of tons of debris and are capable of destroying nearly any building or structure in their path. They can also occur years after an eruption.

Lahars associated with the Mount St. Helens eruption destroyed hundreds of homes and other structures in stream valleys around the volcano (Figure 6.18). Debris was swept downstream into the Columbia River, reducing channel depth by two-thirds and disrupting river traffic. The eruption of Nevado del Ruiz, described at the beginning of this chapter, produced about 1/100th of the volcanic debris of the 1980 Mount St. Helens eruption, but the heat of the eruption melted a large volume of ice, creating lahars that overwhelmed whole settlements and claimed thousands of lives (Figure 6.1, see Table 6.3).

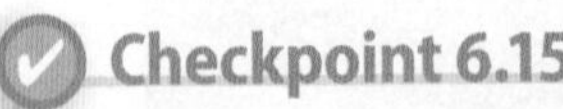

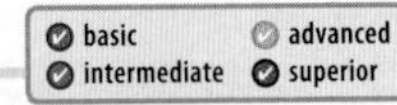

Compare and contrast the consequences of a major earthquake and a volcanic eruption with a VEI of 5. Imagine that you were mayor of a city located within 30 kilometers (19 miles) of the epicenter of the earthquake or the vent of the volcano and that you had evacuated the residents prior to the earthquake or eruption. Which aftermath would you prefer to deal with and why?

Geologist's View

Figure 6.17 Pyroclastic flows. **a.** A pyroclastic flow descends the flank of Mount St. Helens. **b.** Remnants of St. Pierre, Martinique, following a pyroclastic flow from Mount Pelée in 1902 that killed 30,000 people.

a.

b.

Figure 6.18 Lahar deposits along Muddy River, Mount St. Helens. Arrows indicate the level of mudflow on trees adjacent to the stream. Circled person is shown for scale. Flows reached a depth of 20 meters (66 feet) in places.

Checkpoint 6.16

basic · intermediate · advanced · superior

Volcanic Hazards Associated with Mount Shasta, California

Read the following description of the eruption history of Mount Shasta and answer the questions that follow.

Mount Shasta Volcanic History

Mount Shasta in northern California is a stratovolcano composed of multiple overlapping cones. Eruptions produced andesite lava, pyroclastic flows, and lahars. During the last millennium, Mount Shasta has erupted on the average at least once every 250 years. The most recent eruption occurred in A.D. 1786.

Lava flows issued from the summit and from vents on the slopes of the volcano. Individual lava flows are up to 13 kilometers (8 miles) long. No lava flows extended for more than 20 kilometers (12 miles) from the central summit.

Pyroclastic flows from the summit and at the Shastina vent (see map) have traveled distances of more than 20 kilometers (12 miles). Other vents produced flows that extended up to 10 kilometers (6 miles). Eruptions from the summit crater produced lahars that reached more than 20 kilometers and spread around the base of the volcano. The largest lahars entered the McCloud and Sacramento Rivers.

Tephra deposits cover the ground within 25 kilometers (16 miles) of the summit. A massive debris avalanche occurred around 300,000 years ago. The debris avalanche flowed more than 64 kilometers (40 miles) through the Shasta valley and covered more than 675 square kilometers (260 square miles).

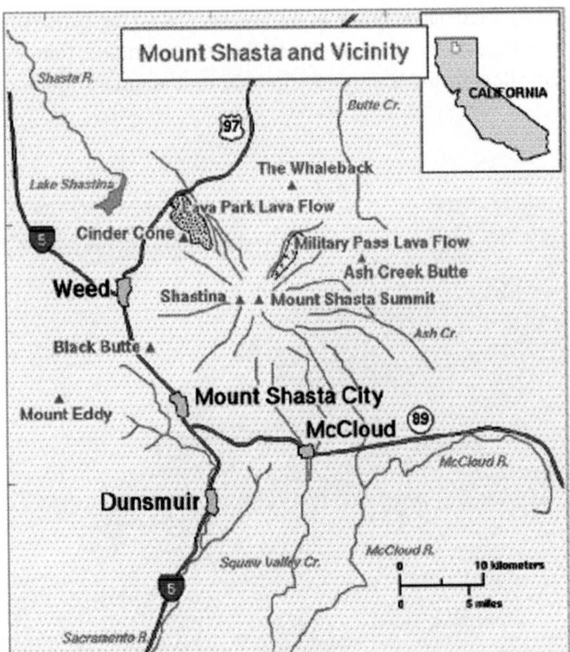

After reviewing the volcanic history of Mount Shasta and examining the map of the vicinity, identify potential volcanic hazards for the area surrounding Mount Shasta. Remember that the eruption characteristics and products of future events will be similar to those of historical eruptions.

1. Show the possible extent of selected hazards on the map of Mount Shasta and vicinity.
2. Evaluate whether the cities of Weed, Mount Shasta City, McCloud, and Dunsmuir will face the same types of hazards from a future eruption of Mount Shasta. Which city is at greatest risk? Explain your choice.

6.6 Volcanoes and Volcanic Landforms

Learning Objectives

- Predict the types of volcanoes found at a variety of plate tectonic settings.
- Compare and contrast the characteristics of the three major types of volcanoes.
- Identify and describe at least three other types of volcanic landforms that are not volcanoes.
- Create a summary diagram that synthesizes information about magma composition, the characteristics of volcanic landforms, and plate tectonic setting.

Three Classes of Volcanic Cones

Volcanoes are separated into three basic types: shield volcanoes, stratovolcanoes, and cinder cone volcanoes.

Shield Volcanoes. The Hawaiian Islands are composed of volcanoes that rise from the floor of the Pacific Ocean with **broad, gently sloping sides built up from thousands of fluid, low-viscosity lava flows** (Figure 6.19). The highest mountain outside the Himalayas would be the Mauna Loa shield volcano that makes up much of the island of Hawaii (8,700 meters; 28,000 feet)—if elevation were measured from the seafloor, but it is not. In profile view, these broad, gently sloping volcanoes resemble the shield of a Roman soldier, hence their name—shield volcanoes. The low-silica mobile lavas described in Section 6.5 create shield volcanoes. They are characteristically found above oceanic hot spots (see Figure 4.18) or in association with divergent plate boundaries (see Figure 4.22).

Stratovolcanoes. A majority of the world's active volcanoes are steep-sided, cone-shaped stratovolcanoes (also known as composite cones; Figure 6.20) that form along convergent plate boundaries. **These volcanoes have a violent style of eruption that blasts debris several kilometers into the atmosphere.** They are called composite cones because they are composed of many layers of volcanic debris, pyroclastic flow deposits, and lavas. Most of our familiar images of imposing volcanoes represent stratovolcanoes, but this type is smaller than the Hawaiian shield volcanoes. Stratovolcanoes develop from more-viscous magmas than shield volcanoes and hence have a narrower base and more steeply sloping sides (see Figure 6.20b).

Cinder Cone Volcanoes. The smallest volcanoes, **cinder cones are composed of coarse tephra produced when gas escapes rapidly from molten lava.** The expanding gases blast blobs of lava into the

a.

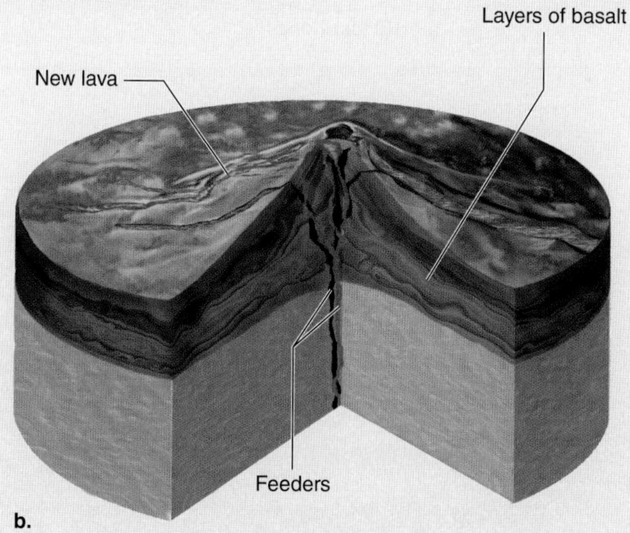

b.

Figure 6.19 Shield volcanoes. **a.** Mauna Loa in Hawaii exhibits the gentle slopes of a shield volcano. It rises approximately 8,700 meters (28,000 feet) from the ocean floor and is the world's largest active volcano. The peak of a second shield volcano, Mauna Kea, is present in the distance. **b.** Cutaway view of a shield volcano illustrating the layers of lava that build up the gently sloping cone.

Checkpoint 6.17

What type of volcano is Mount St. Helens?

a) Shield volcano
b) Stratovolcano
c) Cinder cone

a.

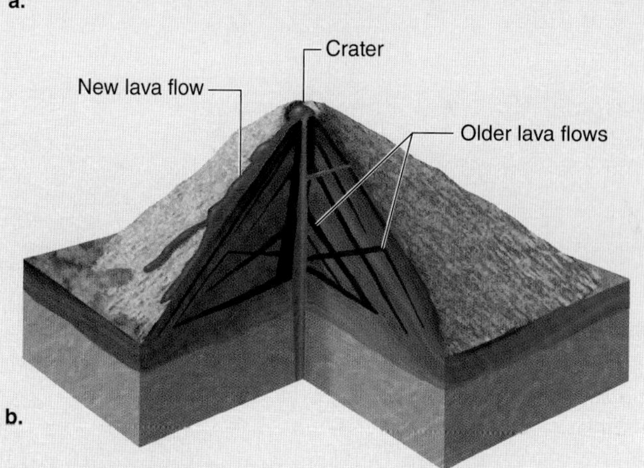

b.

Figure 6.20 Stratovolcanoes. **a.** Osorno volcano is a stratovolcano in the Andes Range in Chile. **b.** Cutaway view of a stratovolcano composed of alternating layers of lava (dark gray layers) and tephra (light colors).

air. As the blobs cool and solidify, they fall back to Earth around the volcanic vent to build up the cone. Cinder cones are the simplest types of volcano, rarely exceeding 400 meters (1,310 feet) in height (Figure 6.21). They often occur on the slopes of larger shield volcanoes or stratovolcanoes and are easily eroded because they are composed of loose, unconsolidated debris. Lava flows may form from the base of the cone after the eruption's gases have been expended.

Other Volcanic Landforms

Some volcanic landforms are generated when volcanoes collapse or when magma reaches the surface through a linear fissure (a long vertical crack in the crust), rather than through a central vent. Other volcanic features are characteristic of regions where groundwater and magma interact close to Earth's surface.

Checkpoint 6.18

basic advanced intermediate superior

Decade Volcanoes, Part II

Answer the following questions using the volcanoes illustrated on the map in Checkpoint 6.7. It may be useful to review the map of plate boundaries in Chapter 4 (see Chapter 4 Snapshot) before attempting to answer the questions.

1. Most of the volcanoes on this map are
   a) shield volcanoes
   b) stratovolcanoes
   c) cinder cone volcanoes
2. Name a stratovolcano not associated with the subduction of the Pacific plate.
3. Name a shield volcano.

Checkpoint 6.19

Venn Diagram: Shield Volcanoes, Stratovolcanoes, and Cinder Cones

Use the Venn diagram provided here to compare and contrast the three principal types of volcanoes. Place the number corresponding to features unique to each type in the larger areas of the circles; note features they share in the overlap area in the center of the image. Five items are provided; identify at least 12 more.

1. Associated with subduction zones
2. Have a triangular shape in profile
3. Example: Mount Hood, Oregon
4. Mild eruptions
5. Intermediate-silica magma

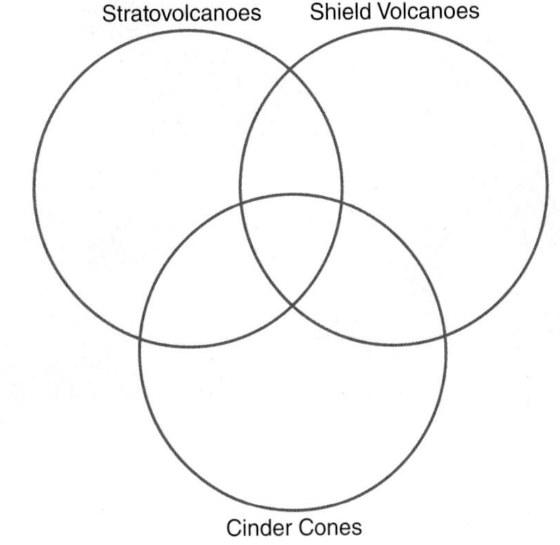

Figure 6.21 Cinder cone volcano. The 340-meter-tall (1,115-foot-tall) Sunset Crater in northern Arizona was formed less than 1,000 years ago.

Figure 6.22 A volcanic cone, crater, and caldera. The crater of the Karymsky volcano (Kamchatka, Russia) in the foreground is much smaller than the lake-filled caldera in the background.

Calderas. **A caldera is a crater that forms when a stratovolcano or shield volcano collapses into a shallow, empty magma chamber below the volcano** (Figure 6.22). Calderas vary in size but can be tens of kilometers across. An oval caldera approximately 80 kilometers (50 miles) long and 50 kilometers (31 miles) wide underlies Yellowstone National Park, Wyoming, and was formed by a supervolcano eruption 640,000 years ago (Figure 6.13). For comparison, the width of the crater in Mount St. Helens is less than 2 kilometers (1.2 miles). Keep in mind that the caldera represents the collapse of the ground above the magma chamber, not just the vent where the eruption occurred.

a.

b.

Figure 6.23 Lava plateau. **a.** Ancient lava flows are preserved as basalt layers in the Columbia River plateau, Washington. Individual layers are 10 to 20 meters (33 to 66 feet) thick. **b.** Lava associated with the Columbia River plateau covers 164,000 square kilometers (63,300 square miles).

Lava Plateaus. Low-viscosity lavas that reach the surface through fissures may accumulate to form **lava plateaus,** which are made up of many **layers of solidified basalt lava** (Figure. 6.23). These features represent the greatest volcanic eruptions on Earth. The largest examples, the Siberian Traps in northern Russia and the Ontong-Java plateau in the western Pacific Ocean, each cover an area larger than Alaska with piles of lava layers several thousand meters thick (Figure 6.24). The largest North American example of a lava plateau is the Columbia River plateau, which covers an area approximately equivalent to the size of the state of Washington. On the basis of their size and age, the lava plateaus may have produced lava at a rate equivalent to the total of the global oceanic ridge system. These massive eruptions released huge volumes of volcanic gases, contributing to elevated atmospheric carbon dioxide levels, and may have increased global temperatures, as described in Section 6.5.

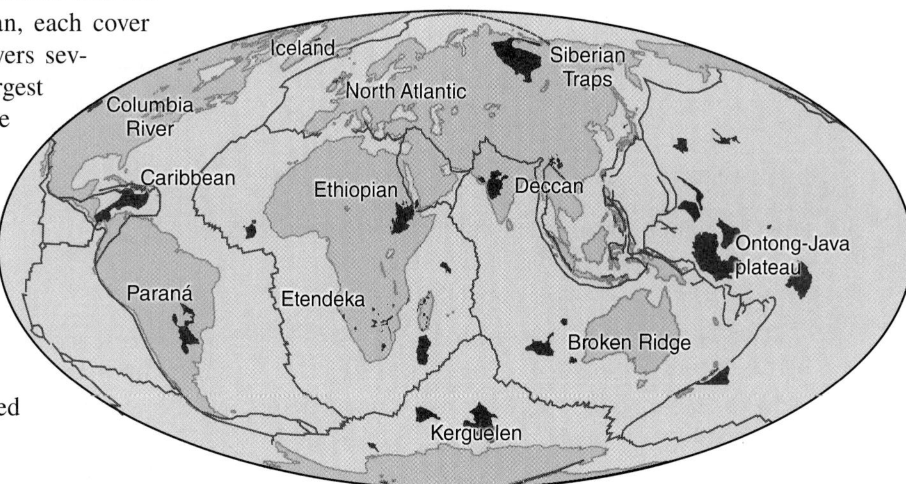

Figure 6.24 Lava plateaus of the world. Lava plateaus are equally prevalent on the ocean floor and on the continents. The largest examples are the Ontong-Java plateau and the Siberian Traps.

Figure 6.25 Volcanic features involving groundwater. Yellowstone National Park in Wyoming contains many examples of volcanic features, including **a.** geysers (Castle geyser shown here), **b.** hot springs, and **c.** mud volcanoes. **d.** Geologists sample gases from a fumerole at Mount Baker, Washington.

Checkpoint 6.20

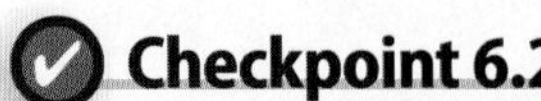

Create a concept map that relates the types of volcanic landforms with their magma composition and plate tectonic setting.

Checkpoint 6.21

Which mountain range contains the higher peaks?

a) Appalachian Mountains, North Carolina
b) Rocky Mountains, Colorado

Geysers, Hot Springs, Fumeroles, and Mud Volcanoes. Groundwater circulating near hot magma can result in the formation of several different features (Figure 6.25). Where this water is heated under pressure, it may erupt violently to form a *geyser* of steam and hot water. *Hot springs* form when heated waters rise to the surface. *Mud volcanoes* are produced when chemical reactions with heated, acidic waters convert rocks to soupy clays. These hot clays bubble and boil like oatmeal on a kitchen stove, hence the name *mud volcano. Fumeroles* form when volcanic gases escape in the absence of water. It is important to note that none of these four features produces lava.

6.7 Mountains: Why Are They There?

Learning Objectives

- Provide multiple explanations for the difference in the heights of mountain ranges on different continents.
- Discuss how you could use wood blocks floating in water as analogs to explain differences in the elevation blocks of continental crust.
- Draw and label a cross section across a continent to illustrate the concept of isostasy and its relationship to topography.

When asked why he wanted to climb Mount Everest (8,850 meters; 29,035 feet), British climber Sir George Mallory famously remarked, "Because it's there." But why is it there? Mount Everest is one of 14 peaks in the world that are over 8,000 meters (26,247 feet) in elevation. All are in the Himalayan Mountains in southern Asia. There are another 39 peaks over 7,000 meters (22,966 feet) high—all in the Himalayas, too. Aconcagua, a mountain in the Andes of western Argentina, is the tallest non-Himalayan peak in the world at 6,960 meters (22,835 feet). The tallest North American peak is Denali (formerly Mount McKinley) in Alaska at 6,194 meters (20,320 feet).

Why is there such a difference in the elevations of mountains across the world's continents? What is it about the Himalayas that makes them the dominant mountain range in the world? It all has to do with plate tectonics (big surprise!). But it also has to do with the same factors that control how high a marshmallow floats in a mug of hot chocolate or how much of an iceberg sticks out of the water. We will start our discussion by detailing the plate tectonic processes that transfer mass and heat to the crust. Then we will examine two more fundamental properties of materials—density and isostasy—that explain just how high a mountain will rise.

Mountains and Plate Tectonics

As we discussed in Chapter 4, the thickness of the continental crust varies across Earth's surface (see Figure 4.27). **The continental crust is generally thickest below mountain belts and thinnest at the margins of the continents** adjacent to the ocean basins. Mountains form where plates collide at convergent plate boundaries. The highest peaks in Asia (Everest), South America (Aconcagua), and North America (Denali) are all present in young mountain belts that were formed adjacent to convergent plate boundaries (Figure 6.26).

Convergent plate boundaries are characterized by thrust faults that stack up slices of continental crust. This increases crustal thickness where the plates collide (Figure 6.27). Imagine a big table scattered with hundreds of sheets of paper. What would happen if you and a friend began at opposite sides and pushed the loose sheets toward the middle of the table? Obviously, you would end up with a big stack of paper in the middle of the table. That stack of paper is analogous to mountains formed at the edge of continental plates adjacent to convergent boundaries. Each sheet represents a slice of crust carried along by a reverse fault.

The big difference in elevation between the Andes and the Himalayas can be explained as the difference between the styles of plate collisions. The Himalayas are the result of collision between two slabs of continental lithosphere, while the Andes formed as a result of convergence between oceanic lithosphere and a plate with continental lithosphere. In addition, the leading edge of the

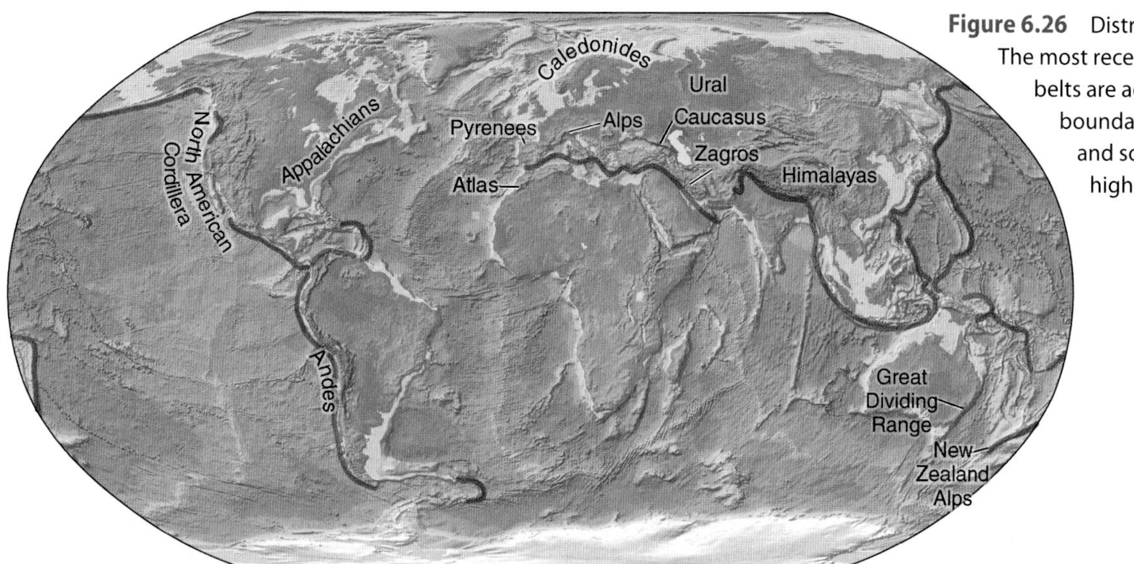

Figure 6.26 Distribution of mountain belts. The most recently formed mountain belts are adjacent to convergent plate boundaries (red lines) in the Americas and southern Asia (brown indicates higher elevation).

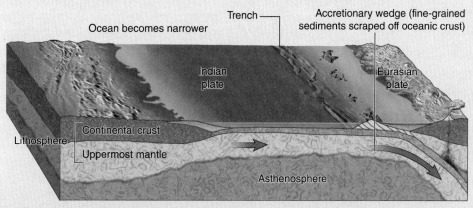

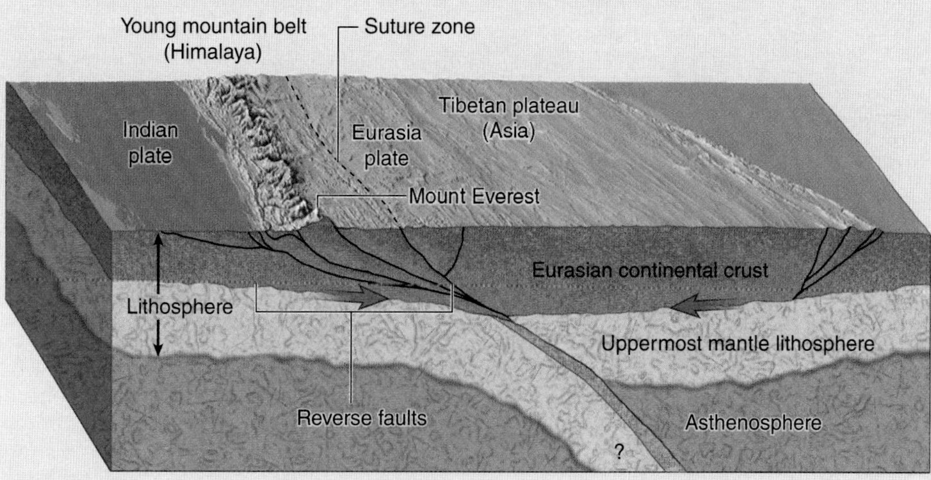

Figure 6.27 Features of a continent-continent collision zone. **a.** India moving north toward Asia with ongoing ocean-continent convergence. **b.** The edges of each continent combine to create a thick wedge of crust that produces high mountains adjacent to the suture zone where the two continental masses collided. The black lines represent faults that stack up and thicken the continental crust.

Checkpoint 6.22

basic · intermediate · advanced · superior

Suggest an explanation for why the tallest mountain in Australia (Kosciusko; 2,228 meters, 7,310 feet) is not even one-half as high as the tallest peaks on the other continents.

Indian plate has been wedged below the southern edge of the Eurasian plate. Consequently, the thickness of the continental crust has nearly doubled under the Himalayas.

Not all mountains are formed by adding mass to build up a pile of rocks. In some cases, they are formed by the addition of heat. Oceanic ridges form a submarine mountain range thousands of kilometers long in regions where high heat flow elevates the seafloor. Similar patterns are revealed in continental crust in Africa, where the highest peaks, such as Mount Kilimanjaro in Tanzania, are present in the East African Rift system, an early-stage divergent plate boundary (see Figure 4.24). These are typically areas of anomalously thin crust that are elevated because the rocks beneath expand as they are heated.

Earth's crust is about 40 kilometers (25 miles) thick under much of the continents and is thickest (70 kilometers; 43 miles) under the Himalaya Mountains. But if the crust is 30 kilometers (19 miles) thicker under the Himalayas, *why are the Himalayas not 30 kilometers higher than the average level of the continents?* To answer that question, we need to understand a little more about two fundamental properties of rocks—density and isostasy.

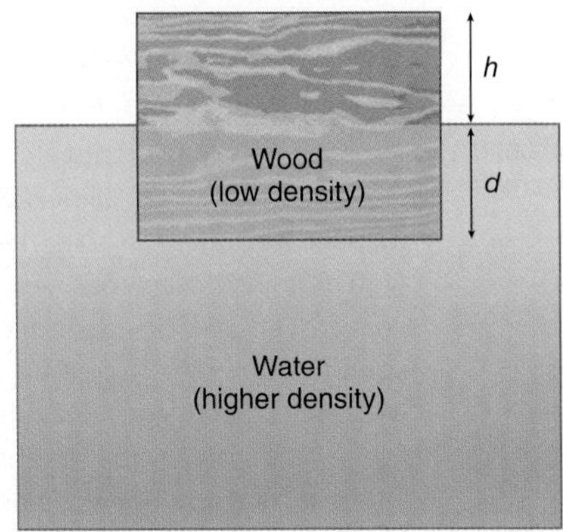

Figure 6.28 Demonstration of density. Wood floats because it has lower density than water. The relative heights of the block above (*h*) and below (*d*) the water level are determined by the relative densities of the wood and the water.

Density. Imagine that a small block of pine wood is floating in a water-filled container (Figure 6.28). Why does the wood float? The wood floats because it has lower density than the water. *Density* is measured in units of mass per volume. The density of water is 1 g/cm^3, and pine has a density of about 0.5 g/cm^3. That is, a volume of pine would weigh about one-half as much as an equal volume of water. Consequently, about one-half of the floating pine block would be above the water's surface, and the rest would be below.

Wood blocks float on more-dense water in much the same way that Earth's crust "floats" on a more-dense mantle. Earth's three compositional layers—crust, mantle, and core—are characterized by different densities. When Earth began to cool, the elements that made up the planet became differentiated, with the more-dense materials sinking to the core and the less-dense materials rising to form the crust. The composition of the continental and oceanic crusts differ from one another. The continental crust is less dense (2.7 g/cm^3), and the oceanic crust is more dense (3.0 g/cm^3). However, both are less dense than the underlying mantle part of the lithosphere (3.4 g/cm^3). Note that the density of the continental crust is only 80 percent as dense as the rocks in the upper mantle.

Isostasy. Let us take another look at the floating wood blocks. This time we will use a heavier wood with a density that is 0.8 times the density of water. The wood blocks will represent the crust, and the water will be a stand-in for the mantle. Some varieties of oak have a density of 0.8 g/cm^3, which means that a block of oak would float, but most of the block (80 percent) would be submerged. A smaller wood block would not float as high, but neither would it extend as far underwater. The difference in the elevation of the top of each block is much less than the difference in thickness of the blocks (Figure 6.29a). In much the same way, high mountains have a deep crustal "root" that extends much farther down into the mantle. Note the difference in the thickness and depth of each

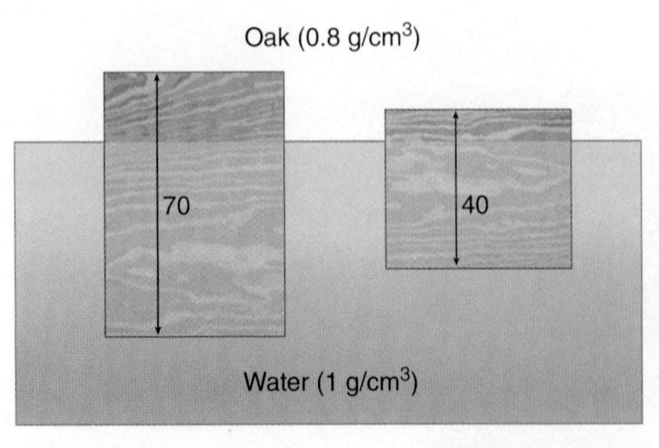

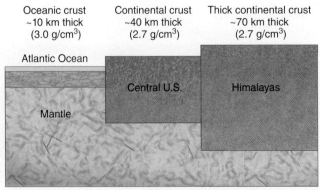

Figure 6.29 Demonstration of isostasy. **a.** Larger blocks of the same wood rise higher above the water but also extend farther below the surface. **b.** In the same way, thicker sections of crust below mountains have a deep crustal root that extends down farther into the mantle than do "normal" sections of continental crust. (Numbers in parentheses are densities.)

block of continental crust shown in Figure 6.29b; the difference in the depth of each block is greater than the difference in the surface elevation between these two blocks. Compare the crustal root to the foundation under a building. The taller the building, the deeper the foundation needs to be to ensure the stability of the structure. So now let us answer that question: *If the crust is 30 kilometers thicker under the Himalayas, why are the Himalayas not 30 kilometers higher than the average level of the continents?* Much of the 30-kilometer difference (80 percent) is actually in the crustal root. The remainder (6 kilometers) would be added to the average elevation of the continental surface, suggesting that the average height of the Himalayas and the adjacent Tibetan plateau should be around 6 to 7 kilometers (3.7 to 4.3 miles) (which is a close approximation of the actual height).

The elevation of the continental crust is dependent on both its thickness and its density when compared with the underlying mantle. The concept that there is **a balance between the topography of Earth's surface and the thickness and density of the underlying rocks is termed *isostasy*.** Higher elevations result from either thicker or less-dense crustal rocks; lower elevations result from either thinner or more-dense crust. The removal or addition of material at Earth's surface can result in isostatic changes in the whole column of underlying crust.

Checkpoint 6.23

basic / intermediate / advanced / superior

Different types of wood have different densities. For example, the density of pine (0.5 g/cm^3) is less than that of ebony (0.9 g/cm^3) but more than the density of balsa wood (0.14 g/cm^3). All would float in water but with different proportions of each block lying above and below the surface.

1. Draw a diagram to show what would happen if equal-sized blocks of each type of wood were added to a container of water.
2. What would happen to the blocks if we replaced the water with a liquid with higher density, such as corn syrup?

Checkpoint 6.24

basic / intermediate / advanced / superior

The following profile shows the topography of a continent. Draw the relative position of the base of the crust, taking into account the principles of isostasy. Label the continental crust, oceanic crust, and mantle. Explain your drawing.

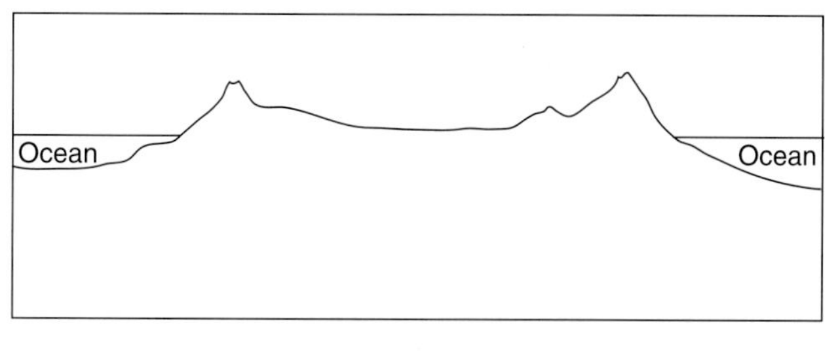

6.8 The Rise and Fall of Mountains and Temperatures

Learning Objectives

- Explain how density of the crust and age of mountain uplift contribute to the elevation of a mountain range.
- Describe a simple physical model that could serve as an analog for the relationship between erosion and change in elevation of mountains.
- Discuss how the erosion of the Himalayas contributed to a significant change in global temperatures.

Isostasy means that the elevation of a continent will increase as mass is added to the crust or will decrease as mass is removed. The amount of change depends on the relative densities of the crust and the mantle. Let us consider a simple example. Imagine that the wood blocks in Figure 6.30 are composed of several thin slices, each 20 units thick. The central block is made up of five slices, a total of 100 units in thickness. If we were to remove a slice to create an 80-unit pile, the top of the pile would decrease in elevation by only 4 units, while the base of the pile would rise by 16 units (80 percent). In contrast, if we added another 20-unit slice to the central block, it would increase in thickness to 120 units, but the top surface would rise by only 4 units and the base would drop another 16 units. **Isostasy compensates for added material by building a bigger submerged root and for lost material by raising the pile.** In each case, the proportion of the pile that is submerged is always 80 percent because the oak blocks have a density that is 80 percent the density of water.

During collision, the crustal thickening adds mass to the crust. Most of this increased thickness is taken up in an increase in the size of the crustal root. For every kilometer (1,000 meters; 3,280 feet) of material added to the crust by reverse faults, the elevation of the land surface rises by just 200 meters (656 feet). Erosion removes mass and the thickness of the crust decreases, but most of that decrease is accommodated by shrinking of the

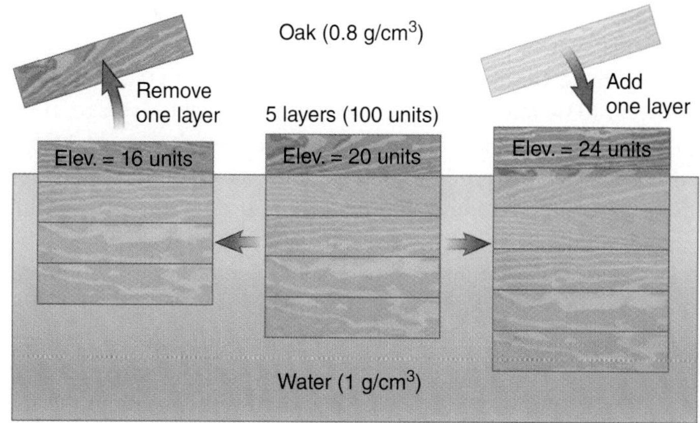

Figure 6.30 Elevation changes and isostasy. The central wood block is composed of five wood layers, each 20 units thick. If one wood layer is added or removed, the top of the wood block increases or decreases in elevation by just 4 units (20 percent of the thickness of the wood layer).

Checkpoint 6.25

basic · intermediate · advanced · superior

How would the elevation of mountains differ if Earth's crust were composed of denser rocks? Mountains would be

a) higher.
b) lower.
c) unchanged in elevation.

Checkpoint 6.26

basic · intermediate · advanced · superior

Use the information from this section to answer these questions:

1. Why are the 10 tallest US peaks all in Alaska?
2. Why are the Rocky Mountains taller than the Appalachian Mountains?
3. Why can we drive across the site of former mountains in Canada without rising in elevation?

Checkpoint 6.27

basic · intermediate · advanced · superior

Explain how you could create a wood block-and-water model to demonstrate that erosion can result in an increase in the elevation of the highest mountain peaks.

crustal root. For every kilometer of rock removed by erosion, the underlying crust rises so that the elevation of the land surface lowers by only 200 meters. This process is not instantaneous; rather, it happens over time.

In these simple models, we are assuming that material is added and removed equally across the mountain range. In these cases, erosion always results in a slow decrease in the elevation of the mountains. However, actual conditions may be more complex. For example, erosion in mountainous regions is often concentrated in stream valleys. This results in preferential erosion and deepening of valleys at the foot of the mountains, while erosion is almost nonexistent at the mountain peaks. Isostasy raises the whole mountain range to balance the material lost by erosion. If the high mountain peaks are not being eroded, they can end up actually higher than they were before (Figure 6.31). Consequently, erosion can cause an increase in the elevation of a mountain range! This is especially true in regions characterized by humid climates where stream erosion is concentrated in valleys.

The tallest mountain ranges are still being formed. They are supported partially by isostasy and partly by the forces associated with plate tectonics. India has been pushing into Asia for nearly 50 million years, so it is clear that **mountains must be long-lived features on Earth's surface.** Older mountains, such as the Appalachians in North America, were also formed by continent-continent collision. They were at much higher elevations when they formed but gradually eroded over time. The tallest peak in the Appalachians today is Mount Mitchell, North Carolina, at 2,037 meters (6,684 feet), more than 6 kilometers (3.7 miles) below the maximum elevation of the Himalayas.

Just how long does it take to wear away a mountain range? Scientists who study erosion have learned that erosion rates show considerable variation, but an erosion rate of 0.1 to 0.2 millimeter per year (0.04 to 0.08 inch per year) is a reasonable average. If the rocks of the Appalachians eroded at those rates, it would take 5 to 10 million years to erode 1 kilometer (3,280 feet) of rock. However, isostatic compensation would replace most of this rock, resulting in an overall elevation change of just 200 meters (656 feet). So the elevation of the mountain range would decrease by 1 kilometer every 25 to 50 million years. At that rate, it would take 150 to 300 million years to erode the 6-kilometer difference in elevation between the Himalayas and the Appalachians. This is merely a ballpark figure, but it fits well with what we know about the formation of the Appalachians; the plate collisions that formed the mountains ended around 280 million years ago. The combination of erosion and isostatic compensation can therefore account for the rate of erosion of the Appalachian Mountains. Without the influence of isostasy, the Appalachians would have been as flat as a pancake by now.

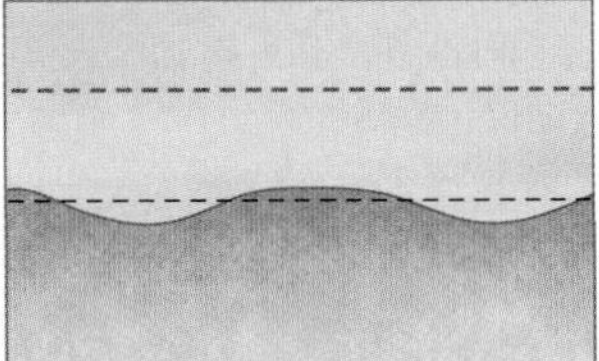

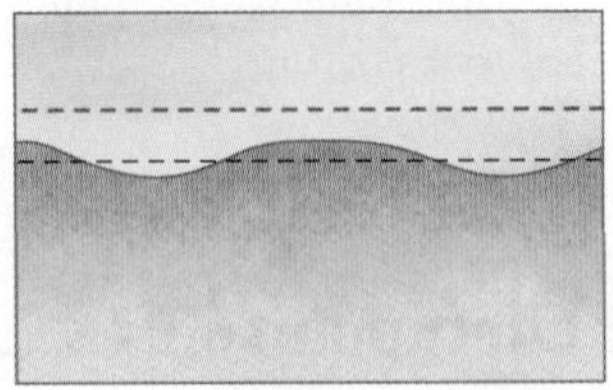

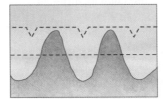

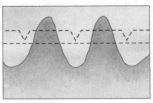

Figure 6.31 How erosion can cause uplift. Right: The loss of material uniformly across a region can lead to an overall decrease in elevation as in a plateau. Left: However, if erosion occurs exclusively in valleys, the original peaks will actually be lifted by isostatic compensation. **a.** Original elevations, prior to erosion. **b.** Erosion removes material from both settings. **c.** Isostatic compensation reduces the change in elevation for the plateau but results in a net loss of elevation. Compensation in the jagged peaks raises the whole range, resulting in higher elevations for the peaks (left). In the real world, erosion and isostatic compensation occur at the same time.

Mountains and Climate

Data from the Himalayas show that it takes tens of millions of years to form mountains, and the erosion of the Appalachians illustrates that it takes even longer to wear them away. Mountains therefore represent one of the largest and longest-lived Earth structures other than the continents and oceans. Mountains stretch for thousands of kilometers and reach several kilometers in elevation, forming a significant barrier for atmospheric circulation and influencing regional and perhaps global climate patterns. Nowhere is this more evident than in Asia, where the growth of the Himalayas has been responsible for generating the Asian monsoons, one of the most significant annual weather phenomena on the planet.

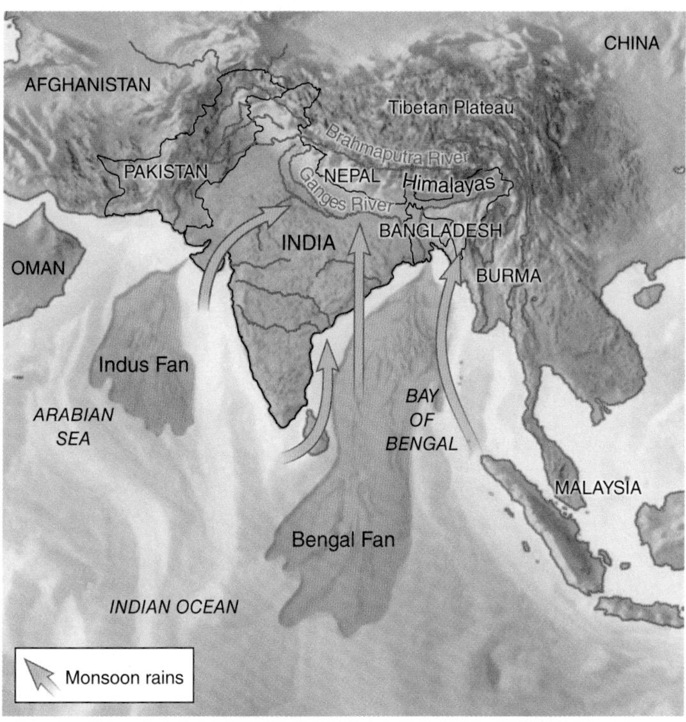

Figure 6.32 Monsoon weather patterns. Warm, wet air is drawn in from the ocean and rises along the front of the Himalayas to produce heavy monsoon rains that feed the Ganges and Brahmaputra rivers. The rivers both flow into the Bay of Bengal, producing the largest delta complex on Earth in Bangladesh.

Each year, the broad, flat Tibetan plateau, just north of the Himalaya Mountains, is heated by the summer sun. The overlying air warms and rises, pulling in cooler, moist air from the Indian Ocean (Figure 6.32). The moist air is forced to rise as it encounters the Himalayas, and as it rises, the moisture condenses to produce heavy monsoon rains. These rains flow into rivers that erode the base of the mountains, and that eroded sediment is carried back to the ocean by some of the largest rivers in the world (for example, Ganges, Brahmaputra). Every year, these rivers transport hundreds of millions of tons of sediment to the Bay of Bengal and the Arabian Sea, where it forms a pair of deep sea features known as the Bengal and Indus fans. These fans represent two of the largest deposits of sediment in the world (Figure 6.32).

This process of erosion and deposition has been operating for about 20 million years, since the mountains became tall enough to cause the monsoon air circulation. One consequence of this process is that rain has been stripping carbon dioxide from the atmosphere, and various chemical reactions have been combining it with minerals that were deposited in the deep sea fans. (We will discuss these reactions in greater detail in Chapter 9.) The steady removal of carbon dioxide contributed to reduction in the global greenhouse effect, cooling the planet by an estimated 5 C° (change of 9 F°) or more. If not for this event, Earth would be much warmer today because of the extra carbon dioxide that was added to the atmosphere by volcanic eruptions around 100 million years ago (see carbon dioxide discussion in Section 6.5). The uplift of the Himalayas helped reduce the atmospheric concentration of the greenhouse gas, returning global temperatures to a more moderate level. This process will continue as long as India continues to drive northward into Asia and as long as isostatic compensation ensures that erosion will bring deeply buried rocks to the surface. Other factors are also important in regulating the global supply of carbon dioxide, but volcanoes and mountains exert a long-term influence on the concentration of this greenhouse gas, which can explain some of the most dramatic shifts in temperature in the global climate record.

Checkpoint 6.28

Draw a concept map that illustrates how the erosion of the Himalayas resulted in a decrease in global temperatures.

the big picture

When Mount St. Helens began rumbling in 1980, teams of scientists rushed to the mountain with truckloads of instruments to monitor the activity. Still, the May 18 eruption came as a surprise. Despite the experience of the scientists and the sophistication of the devices they deployed, little detailed information on the eruptive history of the volcano had been gathered beforehand and few monitoring instruments had been collecting data. That is no longer the case. In the past quarter-century, scientists have made a concerted effort to place a variety of instruments around the volcano, and even in space, to monitor every rumble and movement. Even with what they know today, it is unlikely that volcanologists would have predicted the precise time of the May 18 eruption. But they would have known enough to have more vigorously encouraged the authorities to move people farther from the volcano itself, dramatically reducing the loss of life.

Educating the public is an important factor in reducing the effects of hazards such as volcanoes. Education should provide a scientifically literate population with the necessary skills to critically respond to scientists' assertions. Deciding what evidence to dismiss and what to pay attention to might mean the difference between life and death for those who live in the shadow of an active volcano. The people living near Mount St. Helens in 1980 weighed the evidence and the accompanying call to action. Some heeded the call to evacuate, while others ignored the evidence provided by the volcanologists, chose to hold their ground, and paid for their decision with their lives.

Mount St. Helens is one of only a few US volcanoes with such a high degree of monitoring. However, the US Geological Survey plans to create a National Volcano Early Warning System that would identify the most threatening volcanic hazards, including the number of people and the extent of property endangered. A preliminary assessment of volcanic threat identified 55 volcanoes as high-threat or very-high-threat sites and recommended that each volcano have an extensive network of monitoring equipment to identify the first signs of unrest. Few such networks are currently deployed, and some of these volcanoes have no monitoring systems at all.

One of the volcanoes in the very-high-threat group is Mount Rainier, pictured looming over Tacoma, Washington, at the beginning of this chapter. At 4,392 meters (14,410 feet), Mount Rainier is the tallest and most imposing volcano in Washington. It is located about 70 kilometers (43 miles) southeast of Tacoma. What questions would you ask if you lived in Tacoma?

Historical records indicate that Mount Rainier does not erupt with the frequency of Mount St. Helens. The distance of the peak and the prevailing westerly winds make it unlikely that tephra would ever reach Tacoma. In addition, lava flows and pyroclastic debris would not extend beyond the foot of the mountain, staying tens of kilometers short of Tacoma. Still, large lahars have the potential to reach the northern suburbs of the city and enter neighboring Puget Sound. Even if Tacoma is safe, many smaller towns lie in stream valleys just a 10-minute trip from the volcano by lahar. It is the residents of towns such as Ashford, Packwood, and Orting (Figure 6.33) who need an early warning system for volcanoes.

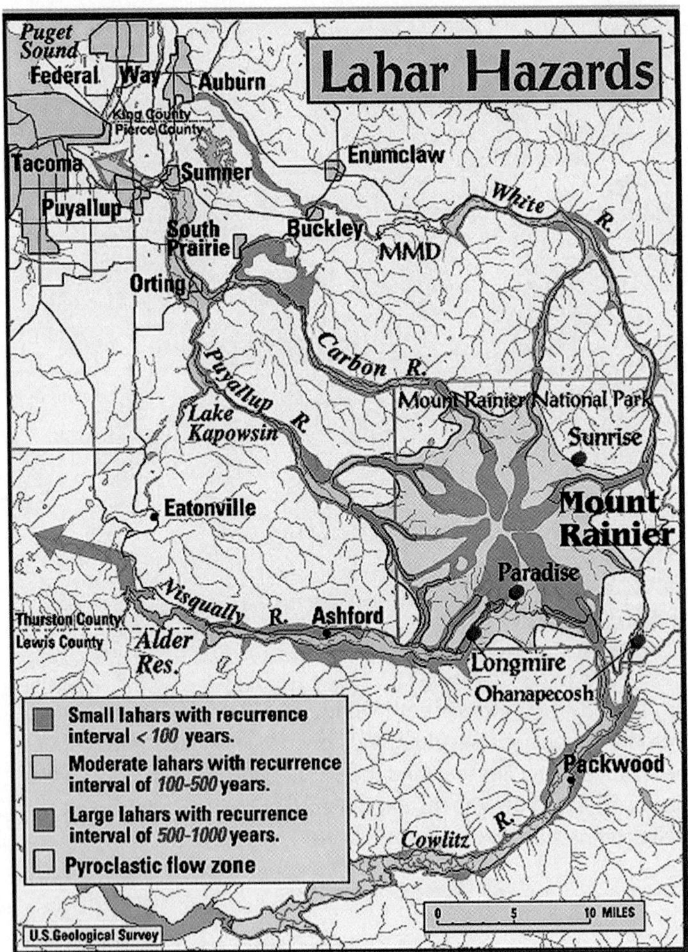

Figure 6.33 Lahar hazards associated with Mount Rainier, Washington.

Volcanoes and mountains are both part of the geosphere component of the earth system. However, both have the potential to influence, or be influenced by, all other components of the system. While scenic mountains may seem relatively benign, they are formed by movements on faults, movements that generate damaging earthquakes. Building a mountain range like the Himalayas involves thousands of faults that generate millions of earthquakes. Unfortunately, major earthquakes are still common in the Himalaya Mountains and other young mountain belts. A recent example was a magnitude-7.6 quake on October 8, 2005, in northern Pakistan, at the western end of the Himalayas. The earthquake demolished whole towns, killed 90,000 people, and left another 4 million homeless. The unrest continues; Earth at this very moment is shifting, rumbling, building, and decaying. We must carefully observe and prepare.

Volcanoes and Mountains: Concept Map

Complete the following concept map to evaluate your understanding of the interactions between the earth system and volcanoes and mountains. Match the following interactions with the lettered labels on the figure, using the information from this chapter.

Eruption melted ice on Nevado del Ruiz to cause fatal lahars.

Sulfur dioxide blocks incoming sunlight.

Added water causes partial melting of mantle.

Volcanoes add CO_2 and sulfur dioxide to atmosphere.

Commercial airlines are at risk from tephra clouds.

Solar radiation heats Tibetan plateau.

Rain strips CO_2 from atmosphere.

Krakatau eruption generated massive tsunami.

Tephra is carried downwind over cities.

Some 500 million people are in risk zones for volcanoes; trees are knocked down.

Industrial materials are swept into rivers and lakes from mudflows.

Monsoon rains result from air rising over Himalayas.

Weathering processes break down rocks in mountains.

Instrumentation of volcanoes.

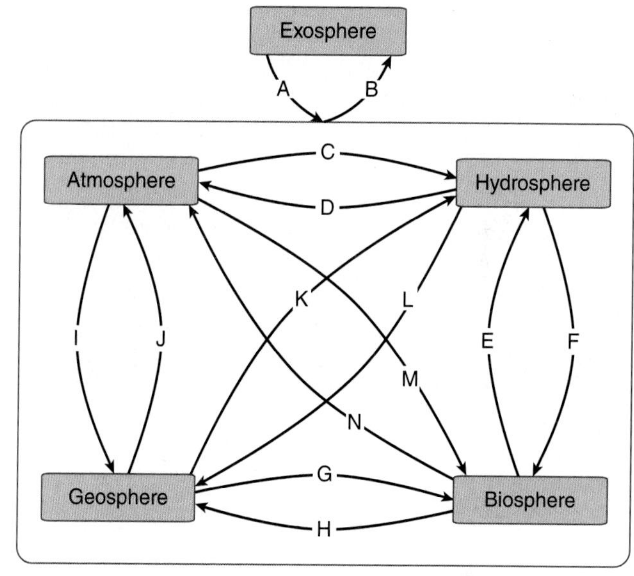

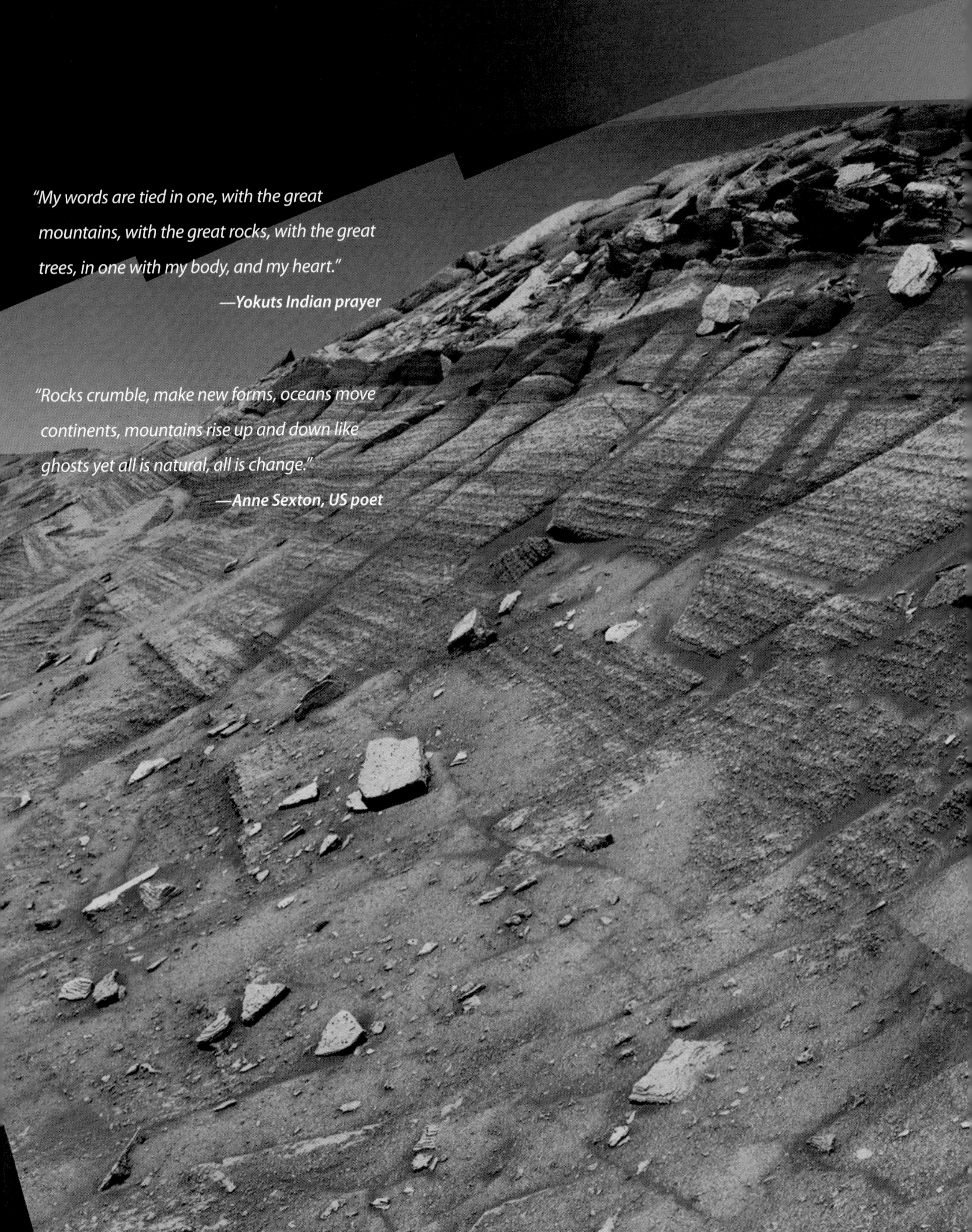

"My words are tied in one, with the great mountains, with the great rocks, with the great trees, in one with my body, and my heart."

—**Yokuts Indian prayer**

"Rocks crumble, make new forms, oceans move continents, mountains rise up and down like ghosts yet all is natural, all is change."

—**Anne Sexton, US poet**

Chapter 7

Rocks and Minerals

Chapter Outline

the big picture

Are the rocks on Mars the same as rocks on Earth?

See The Big Picture box at the end of this chapter for the full story on this image.

7.1 Earth Scientists: Nature Detectives

Chapter Learning Outcomes

- Students will characterize the principal components of rocks from the atomic to the outcrop scale.
- Students will classify rocks as igneous, sedimentary, or metamorphic on the basis of their chemical and physical properties.
- Students will identify the connections between humans and natural resources.

Like detectives, geologists use rocks to piece together the clues about a region's history. It is one thing to see a volcano erupt and determine that you are in an area of volcanic activity. It is a much greater challenge to examine the bedrock of an area to determine that a certain type of volcanic activity occurred there millions of years ago. Why do geologists need to uncover this kind of information?

Knowledge about the composition and distribution of rocks serves many purposes in today's society. For example, it enables us to locate important mineral and energy resources. Economic minerals and fossil fuels both form through a series of chemical reactions that occur gradually over millions of years under specific physical conditions in a select group of rocks. These conditions make it possible to predict where these resources may be found but also highlight the fact that the resources are nonrenewable and will not be replaced once used. Furthermore, as we learned in previous chapters, the analysis of basic rock types and their distribution has been used to match up continents during the assembly of Pangaea, to identify the potential for different volcanic hazards, and to recognize deposits indicative of a tsunami.

The science of geology sprang to life about 200 years ago amid debates about how rocks formed. At the beginning of the 1800s, the two major schools of thought about this issue were known as *Neptunism* and *Plutonism* (after the Roman gods of the sea and the underworld, respectively). According to the concept of *Neptunism*, rocks were formed in a worldwide ocean that subsequently receded. Neptunism presumed that rocks were originally deposited in a series of layers as a result of either chemical precipitation or deposition of material carried in suspension in the ocean. Neptunists had to be creative when it came to volcanoes. They proposed that volcanoes occurred when layers of coal burned, melting the overlying rocks to produce lava. *Plutonism*, on the other hand, held that some common rocks were originally melted by heat from Earth's interior. Plutonism presumed that some rocks were formed directly by molten material that cooled and solidified on or below Earth's surface. Rocks that were not originally melted were fused together by heat from the interior.

Both Neptunism and Plutonism had their good (and bad) points, but neither could account for all the features observed in rocks. However, as we will see, with a little modification, these concepts can explain some of the characteristics of the processes that form the most common classes of rocks.

a.

b.

Figure 7.1 Raw materials for brick manufacturing. **a.** Note the horizontal layers of the bedrock. Mining of this rock (shale) requires little more than heavy machinery such as power shovels, front-end loaders, and pit trucks. **b.** The different colors of these discarded bricks reflect differences in the composition of the raw materials used to manufacture them. Iron content is higher in dark red bricks and lower in light-colored bricks.

Figure 7.2 The brickmaking process. **a.** Shale or clay rock pieces are crushed. **b.** Large fragments are reduced to fine particles by grinding. **c.** Clay mixture is forced through a form to produce brick shape. **d.** Bricks are fired in a kiln at temperatures up to 1,100°C (2,010°F).

a. Fist-sized rock fragments enter plant after crushing

Where Do Bricks Come From?

The natural processes that make up rocks include similar steps to those involved in making bricks, except on a much longer timescale. Therefore, to aid in understanding the natural processes, we will first examine the manufacture of bricks, from raw materials to the finished product. Eight billion bricks are sold in the United States each year; most of these are used in home building. A typical new brick home is made of 20,000 bricks. Most bricks are made from rocks called *fireclay* or *shale* that can be readily excavated from rock layers at or near Earth's surface (Figure 7.1a). The raw materials may range from iron-rich rocks that produce the standard "brick" red to iron-poor fireclays that result in tan or cream-colored bricks (Figure 7.1b).

The raw materials first pass through a crusher that smashes the rock into fist-sized pieces (Figure 7.2a) before they are pulverized by a massive grinding wheel. The ground-up raw material is transported through a series of vibrating screens to separate out small particles about the size of grains of sugar and flour (Figure 7.2b). Next water is added, and sometimes ingredients such as manganese dioxide may be blended in to produce bricks with specific colors. The mixture has now been transformed from a dry powder to a more plastic form, much as flour and water are converted to dough prior to making bread. The mixture is then thoroughly blended in a pug mill—an industrial-sized kitchen mixer—by a series of rotating, bladed metal shafts. Next the materials are pushed through a metal form to generate the shape of the brick. The form contains a series of metal rods that create holes in the brick. As the mixture is extruded from the machine, it looks like a long block of chocolate (Figure 7.2c). The column of wet brick material is sliced into manageable lengths and carried on a conveyor belt to the cutting machine. The shorter columns of wet clay mixture are then cut into the correct sizes by a series of wire cutters, like a giant cheese slicer or bread cutter.

The wet clay is called "green" at this stage because it is immature and not yet "finished." Excess water must be removed in drying ovens before the bricks go to the kiln for final firing. The bricks would crack if they were dried too fast or might explode in the kiln later if they were not dried enough. The bricks are loaded onto kiln cars that slowly move through large dryers that look like long garages. The kiln cars take 2 or 3 days to pass through the dryers, where temperatures are around 200°C (390°F). The brick hardens and shrinks in size as the water content of the raw brick is reduced. The final stage in the production process occurs when the bricks are "fired" (or baked) as they move slowly through tunnel-like kiln ovens (Figure 7.2d). Temperatures in the ovens vary along their length but can reach as much as 1,100°C (2,010°F), depending on the composition of the brick. The key is to supply enough heat to cause the ingredients in the clay mixture to fuse together into a hard, nonabsorbent brick but not to supply so much heat that the brick begins to melt. Just as in rocks, the composition of the materials in the brick determines its properties. In addition, brick making involves physical disintegration, chemical changes, and thermal effects—all processes that occur during rock formation. As we shall see, some of these processes are more important in some groups of rocks than in others.

In this chapter, we will first examine the basic building blocks, or ingredients, that define the composition of all rocks and minerals. We will learn that rocks can be separated into different classes—igneous rocks, sedimentary rocks, and metamorphic rocks—on the basis of the processes that formed them. Some of these processes have been described in previous chapters and will serve as a foundation for interpreting the distribution and character of rocks at the planet's surface. We will also consider the processes contributing to the generation of energy and mineral resources in association with rock formation. Understanding how these finite resources form allows us to predict where to find them and to estimate how much we have left.

Self-Reflection Survey: Section 7.1

Answer the following questions as a means of uncovering what you already know about rocks and minerals.

1. Where can you find the nearest example of rock that is not part of the local bedrock? Where is the nearest location where you could find bedrock?
2. Identify three everyday objects that are made from rocks and minerals.
3. Can you think of another manufacturing process that creates materials that might be considered analogs for rocks?

b. Fine particles on conveyor after grinding

c. Long "chocolate bar" of wet clay mixture

d. Finished bricks exit kiln

7.2 Elements and Atoms: The Basic Building Blocks

Learning Objectives

- Classify an element on the basis of its atomic structure.
- Describe the features of a silicate mineral and explain how they are composed of silica tetrahedra.
- Explain how elements are joined together by bonds to form minerals.
- Summarize the relationships between atoms, elements, minerals and rocks.

Think back on the earth system discussions from Chapter 1. What makes up the geosphere? You might recall that it is mostly composed of rocks. To understand the geosphere and the various interactions that affect it, you need to understand something about rocks—what they are made of and how they form. In this and later sections, we are going to use some foods as analogs for the components of rocks, so you may want to go get a snack before you continue reading. Don't worry, we will wait. . . .

If you could walk outside and pick up a loose rock, what would you notice? (This might be harder to do in some places than others. Look for rocks used in landscaping, building stone, pavements, or gravel driveways.) In general, you are likely to observe that your rock is made of smaller pieces. **Rocks are made up of minerals.** Minerals are composed of elements arranged in characteristic atomic structures.

Geologists have recognized thousands of different minerals (more than 4,000), but only about 20 are commonly found in rocks and we will mention less than a dozen of those here. Individual rocks are typically composed of just a handful of minerals (Figure 7.3). For example, the rock granite is often found in mountain belts and is composed of the minerals quartz, feldspar, mica (such as biotite), and amphibole (also known as hornblende), with small amounts of a few other minerals. If we think of rocks as ice cream, the minerals would be the different ingredients (chocolate, marshmallows, nuts). Different combinations of minerals result in different types of rocks. For instance, the minerals quartz and feldspar show up in a lot of different rocks, and each rock type is characterized by different amounts of quartz and feldspar and different combinations of additional minerals.

Elements

So rocks are made of one or more minerals. What are minerals made of? **Minerals can be categorized by the chemical elements they contain.** Just as we are limiting this discussion of rocks to only a few minerals, we also need to focus on only a select group of elements. There are 92 naturally occurring elements on Earth, but 98 percent of the rocks in the crust below our feet are mainly composed of just eight elements. These eight elements are common ingredients in everyday objects or in the foods we eat (Table 7.1). For example, a bowl of breakfast cereal contains many of the common elements that are naturally extracted from rocks as minerals break down at Earth's surface to form soil (Figure 7.4). These elements are absorbed by crops growing in the soils and often find their way to our bodies via the foods we eat. Many of these elements are essential for good health.

The mineral quartz is composed of just two elements, silicon and oxygen. Other minerals may contain many elements.

Table 7.1 Most Common Elements in Continental Crust

Element	Ion	Percent by weight	Also found in
Oxygen (O)	O^{2-}	46.6	Air
Silicon (Si)	Si^{4+}	27.7	Window glass, computer chips
Aluminum (Al)	Al^{3+}	8.1	Cans, aircraft
Iron (Fe)	Fe^{2+}, Fe^{3+}	5.0	Meat, cornflakes, your car
Calcium (Ca)	Ca^{2+}	3.6	Milk, cheese, cement, antacids
Sodium (Na)	Na^{+}	2.8	Salt, bacon, cheese
Potassium (K)	K^{+}	2.6	Fish, fruit, nuts, fertilizer
Magnesium (Mg)	Mg^{2+}	2.1	Bread, nuts, salt
Others	–	1.5	–

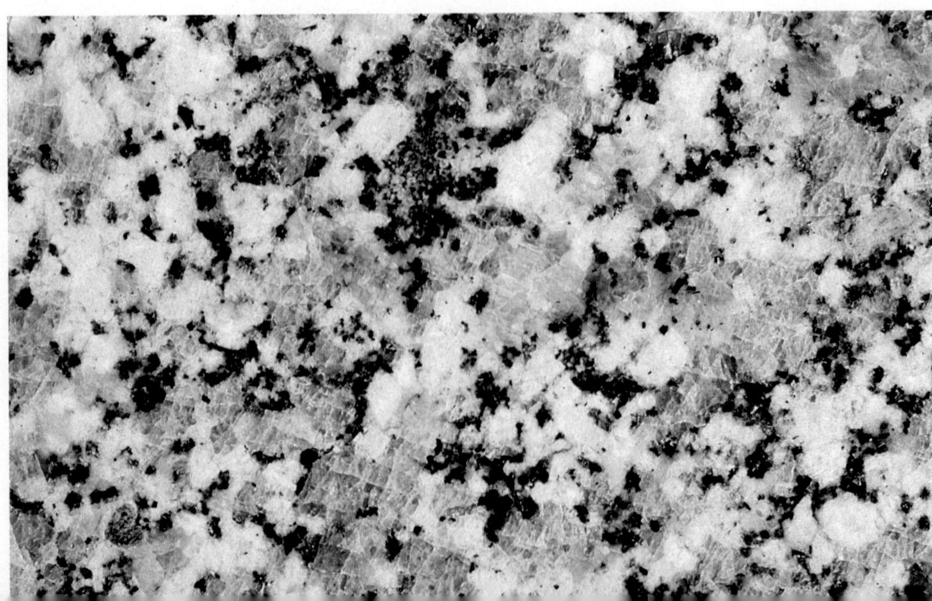

Figure 7.3 A rock composed of different minerals. In this piece of granite, four different minerals are apparent as grains of different sizes and colors. There are two dark minerals, hornblende (amphibole) and biotite; the tan mineral is feldspar, and the white mineral is quartz.

Figure 7.4 Elements in breakfast cereal. Many of the same elements that are common in Earth's crust are present in our daily breakfast cereal.

For example, the mineral amphibole is made up of calcium, magnesium, iron, aluminum, silicon, and oxygen. If we examine each of the common minerals in granite (Table 7.2), we notice that some elements are more common than others and are also among the most common in Earth's crust (Table 7.1).

Atoms

Elements are the last stop; they cannot be further divided into other materials, but they can be separated into individual atoms. **An atom is the smallest particle that retains the characteristics of an element.** Atoms are composed of one or more of the same three basic components—**neutrons, protons, and electrons.** Just as two bricks from different buildings may look alike, individual protons look the same, whether they come from oxygen or silicon or potassium. The same is true for neutrons and electrons. The protons and neutrons are present in the atom's nucleus, and the nucleus is surrounded by a swirling cloud of electrons (Figure 7.5).

In much the same way as a building would look different if different numbers of bricks were used, elements look different because the number of protons in each differs. **The number of protons in an atom is unique for each element and is called the element's *atomic number*.** For example, oxygen has 8 protons, and silicon has 14; consequently, their atomic numbers are 8 and 14, respectively. If we were to subtract a proton from or add a proton to oxygen, we would create a different element (nitrogen or fluorine, respectively).

Table 7.2	Most Common Elements in Granite Minerals
	Elements
Quartz	Oxygen, silicon
Feldspar	Oxygen, silicon, aluminum, calcium, sodium, potassium
Mica	Oxygen, silicon, aluminum, iron, potassium, magnesium, hydrogen
Amphibole	Oxygen, silicon, aluminum, iron, calcium, magnesium, hydrogen

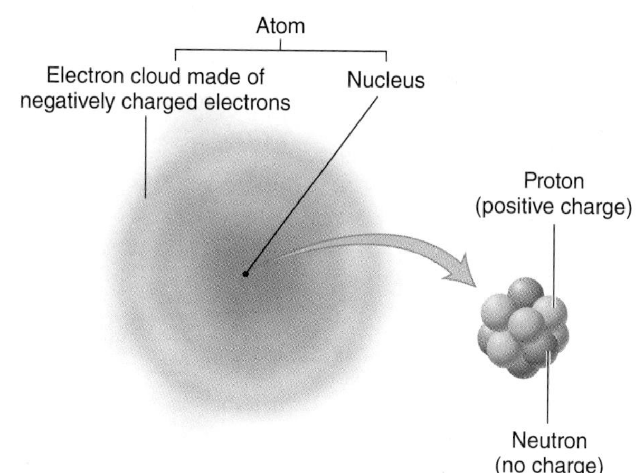

Figure 7.5 Model of an atom. The tiny, dense nucleus consists of positively charged protons and uncharged neutrons. Most of the volume of an atom is occupied by rapidly moving, negatively charged electrons, which can be represented as an electron cloud.

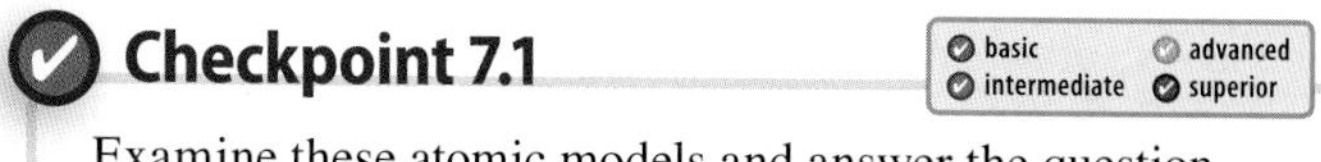

Checkpoint 7.1

Examine these atomic models and answer the question that follows. The filled black circles represent electrons. (Note: Electrons and nucleus are not drawn to scale.)

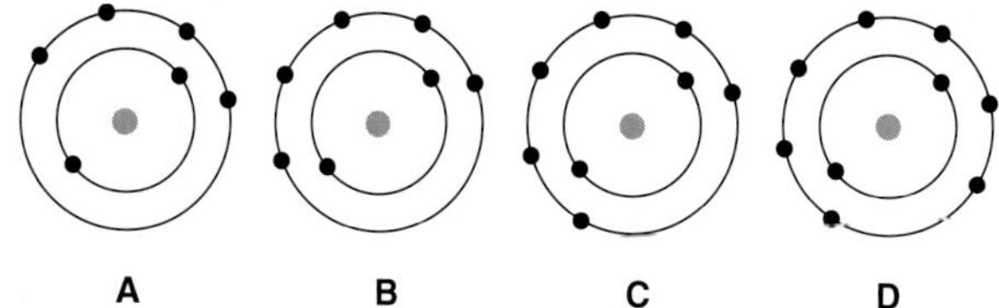

Assuming that the number of electrons is the same as the number of protons (which is not always the case), which picture represents the atom of carbon that has an atomic number of 6?

Protons have a positive electrical charge, while electrons are negatively charged. Neutrons are neutral; they have no charge. Electrons are in constant motion around the central nucleus. Usually, electrons are located in specific concentric zones, or energy levels. If the number of electrons (negative) equals the number of protons (positive), the atom has no net charge. For example, the elements helium and neon have full energy levels and equal numbers of protons and electrons (Figure 7.6). However, the outer energy levels of many elements are incomplete and may gain electrons from other elements to fill gaps or may lose electrons from a sparsely populated energy level (Figure 7.7).

Bonds. Atoms that lose or gain electrons are known as *ions* and have a positive or negative electrical charge. This is so because the number of electrons does not necessarily equal the number of protons. Ions of individual elements can become bonded together to form minerals. In the most common minerals, bonds form in two principal ways—**by balancing the electrical charges of different ions or by sharing electrons.** For instance, the salt that people shake onto their dinner is made up of the mineral halite. Halite is composed of the ions of two elements, sodium and chlorine. A sodium ion is missing an electron, so it has a positive charge (Na^+), making it a *cation*, while chlorine has an extra electron and has a negative charge (Cl^-), making it an *anion* (Figure 7.8a, b). Essentially, we can consider that a spare electron was transferred from sodium to chlorine. **The positive and negative charges balance**

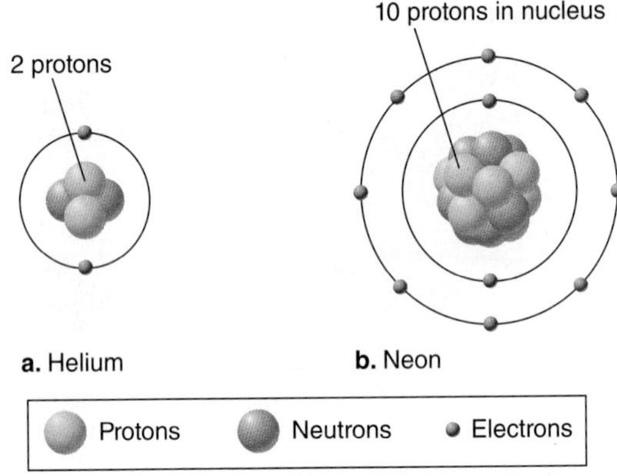

Figure 7.6 Atomic models of the distribution of protons, neutrons, and electrons for helium and neon. In all atoms, electrons lie in energy levels around the nucleus. There are a maximum of 2 electrons in the first energy level and a maximum of 8 electrons in the second energy level. In helium and neon, the number of protons equals the number of electrons, so these elements do not form ions and consequently do not bond with other elements.

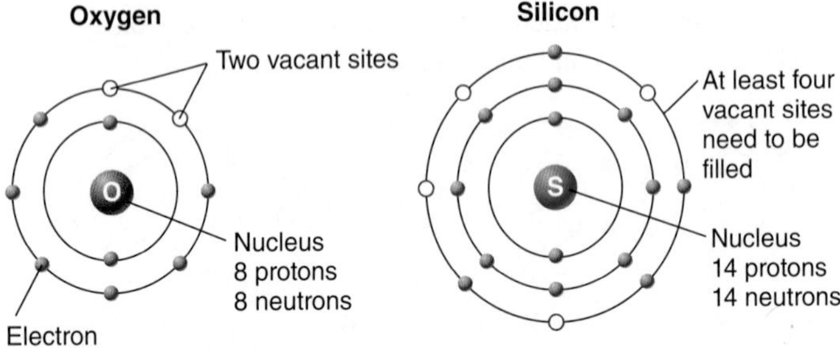

Figure 7.7 Oxygen and silicon atoms have gaps in their outermost energy levels. If electrons are gained to fill these gaps, the atoms will be transformed to a negatively charged ion. Alternatively, electrons may be lost. For example, if silicon lost its outer four electrons, it would become a positively charged ion (Si^{4+}).

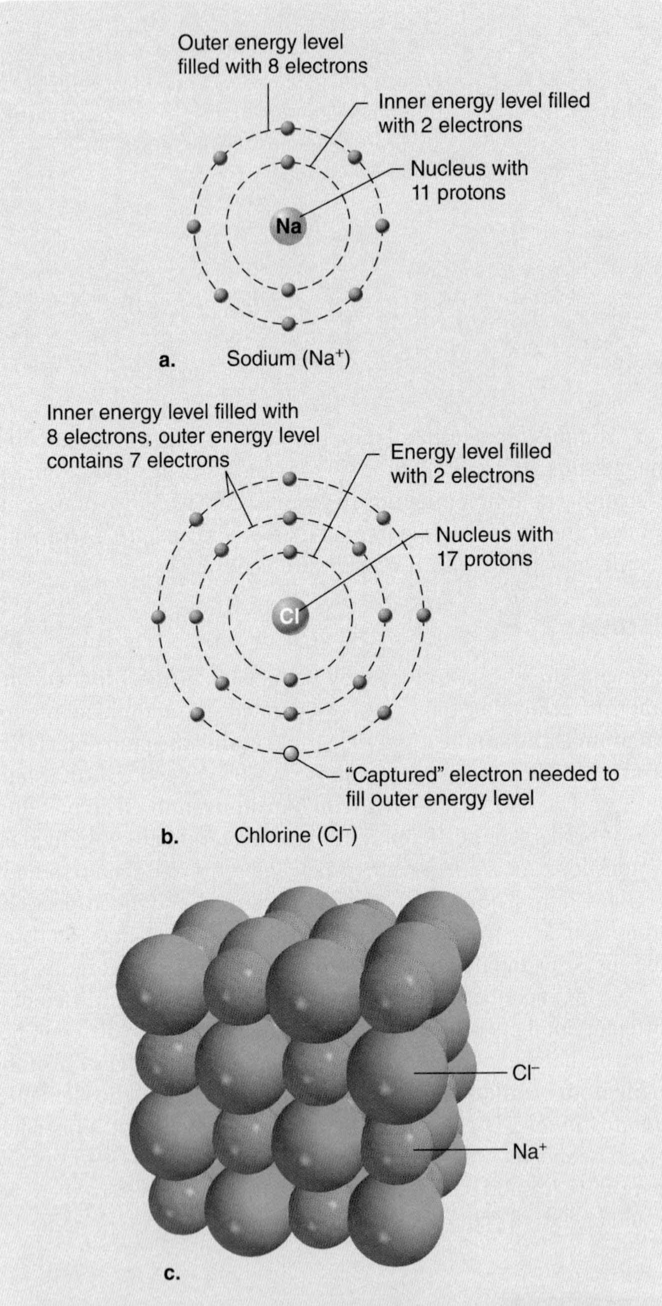

Figure 7.8 Atomic models of sodium and chlorine. **a.** Eleven electrons fill the energy levels of sodium, but the eleventh electron is often absent, giving the ion a positive charge (Na^+). **b.** Seventeen electrons fill the energy levels of chlorine. This results in a configuration that readily accepts an electron in the outer energy level, imparting a net negative charge to the ion (Cl^-). **c.** The two ions will bond together to form the atomic structure of the mineral halite (formula NaCl). Ionic bonds hold oppositely charged sodium ions to adjacent chlorine ions in a grid or cubic pattern.

to create an ionic bond due to electrical attraction between the oppositely charged ions (proving again that opposites attract). These ions pack themselves together in a gridlike pattern of alternating positively and negatively charged ions (Figure 7.8c).

Atoms and ions tend to gain, lose, or share electrons to get eight electrons in their outer shell. (In some cases, no electrons transfer from one atom to another.) For example, water (H_2O), one of the most common substances on the planet, is composed of two atoms of hydrogen (H) and one of oxygen (O). Each hydrogen atom has an electron that is shared with oxygen to give it a full energy level (Figure 7.9). Likewise, oxygen shares an electron with each of the hydrogen atoms. This type of sharing is called **covalent bonding** and typically occurs where atoms share electrons to achieve a stable structure.

A Special Bond: The Silicates. The two most common crustal elements, oxygen and silicon, combine to form a relatively unusual type of bond that is one-half ionic and one-half covalent. Four oxygen atoms bond covalently to a single silicon ion. **The bonding of silicon and oxygen atoms forms a pyramidlike shape known as the silica tetrahedron (or more accurately, the silica-oxygen tetrahedron),** in which the smaller silicon atom is surrounded by the four oxygen atoms (Figure 7.10). The resulting multiatom anion has a negative charge (SiO_4^{4-}) and will form ionic bonds with positively charged ions from most of the other common elements (see Table 7.1). Silicon and oxygen make up over 70 percent of the continental crust by weight, so we deduce that these elements would be present in many minerals. Minerals that contain both silicon and oxygen are known as **silicates** and feature atomic structures that are characterized by different arrangements of the silica tetrahedron. These basic building blocks, the tetrahedra, can be joined together (Figure 7.11) in combination with other

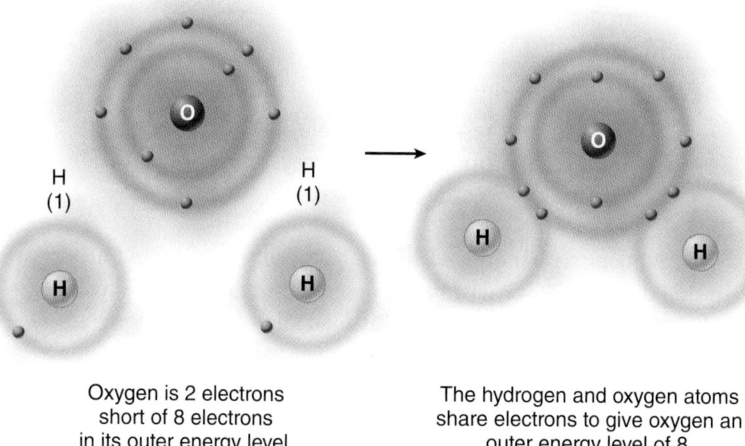

Figure 7.9 Covalent bonding of hydrogen and oxygen in water. Two hydrogen atoms each share an outer electron with an oxygen atom to fill each atom's outer energy level by covalent bonding.

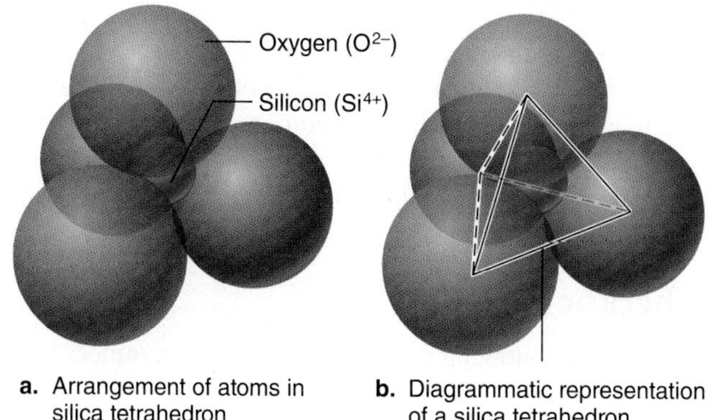

Figure 7.10 Silica tetrahedron. Silicon and oxygen join to form a pyramid (tetrahedron) by covalent bonding. **a.** Four oxygen atoms surround a smaller silicon atom. **b.** The combination of atoms is illustrated on diagrams by a pyramid (tetrahedron) shape (see Figure 7.11). The corners of the pyramid coincide with the centers of the oxygen atoms.

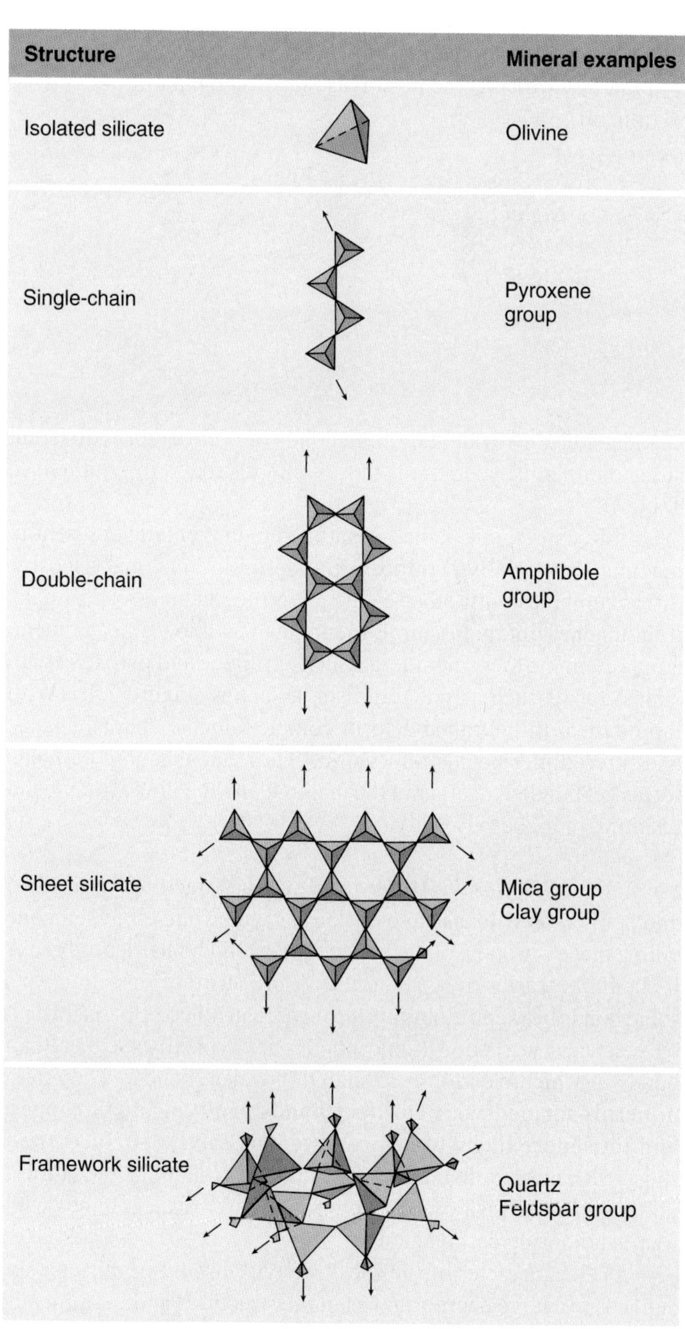

Structure
Mineral examples
Isolated silicate
Olivine
Single-chain
Pyroxene group
Double-chain
Amphibole group
Sheet silicate
Mica group
Clay group
Framework silicate
Quartz
Feldspar group

Figure 7.11 Common silicate structures. These are formed by combinations of silica tetrahedra and other atoms. Arrows indicate directions in which the structures repeat indefinitely.

Checkpoint 7.2

basic / intermediate / advanced / superior

Which of the following mineral formulas represents a silicate?

a) FeS_2
b) $KAlSi_3O_8$
c) Fe_2O_3
d) $CaSO_4 \bullet 2H_2O$

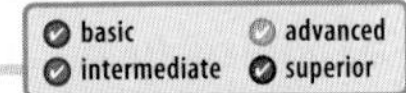

Checkpoint 7.3

The total electrical charges of the ions of the elements in the mineral olivine must balance. From the data in Table 7.1, which is the most reasonable formula for the mineral?

a) $MgSiO_2$
b) $MgSiO_4$
c) Mg_2SiO_4
d) Mg_4SiO_2

elements (using covalent or ionic bonds or both) to form different silicate minerals, such as all those minerals found in granite (see Table 7.2).

Recall from Chapter 6 that the amount of silica present in magma is especially significant in controlling magma viscosity. Silica combines with other elements in the magma (for example, iron, magnesium, potassium). In magmas with low silica content, atoms combine to form minerals such as olivine and pyroxene with simple forms such as pairs and single chains (Figure 7.11). With more silica, the tetrahedra form complex double chains, sheets, and three-dimensional frameworks. These larger, more complex forms get tangled together in the magma, making flow difficult and resulting in higher viscosity.

Strengths of Bonds in Minerals. Imagine attaching yourself to another person with a strip of Velcro or tying yourself to someone with a piece of rope. Which connection would be harder to break? It should be easier to separate the Velcro strip. In the same way, some bonds between atoms are stronger than others. Bonds formed by electrical attraction (ionic bonds) are typically weaker than bonds in which electrons are shared (covalent bonds). Therefore, **minerals formed with covalent bonds are typically stronger, and thus more likely to be preserved on Earth's surface, than those with only ionic bonds.** For example, a diamond, the hardest mineral, is formed of carbon atoms that share covalent bonds with four neighboring carbon atoms.

The character of the rocks we will discuss in this chapter depends to a large degree on the atomic structures of their minerals and the way the atoms are bonded together. Some minerals such as quartz are strong and show up in a variety of roles in different rock

Checkpoint 7.4

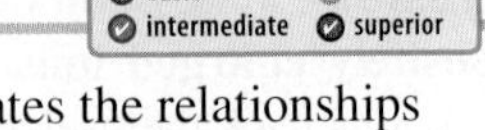

Construct a concept map that illustrates the relationships among atoms, elements, minerals, and rocks. Use the following six terms and add at least three more of your own choosing.

Atom	Bond	Electron
Mineral	Quartz	Ion

types. Other minerals such as halite will dissolve readily in water and consequently are often indicative of specific marine conditions. Finally, the combination of elements used to form minerals will determine the melting temperature of the rocks and the behavior of the magma produced.

7.3 Minerals

Learning Objectives

- Describe the principal features of a mineral.
- Explain how minerals can be identified on the basis of their physical properties such as crystal form, cleavage, hardness, color, luster, and streak.

Minerals are inorganic solids that occur naturally in Earth. (In this discussion, we will neglect some minerals that are made by humans in a lab.) Each mineral is made up of one or more elements that combine their atoms in a unique structure characteristic of that mineral. The consistent arrangement of atoms means that the mineral has a uniform chemical composition no matter where it is found. Stringing all that together, **minerals are naturally occurring inorganic solids of one or more elements that have a definite chemical composition with an orderly internal arrangement of atoms.** The most common minerals in Earth's crust are composed of the most common elements.

Mineral Characteristics

Suppose you encounter a food that you have never seen before and have no idea how it is made. How would you describe it? Chances are you might describe what it looked like, its texture, and something about its taste. You would remember those characteristics if you were served the same food again in the future. In a similar

Checkpoint 7.5

Which of the following cannot be classified as a mineral?

a) Salt
b) Ice
c) Diamond
d) Glass

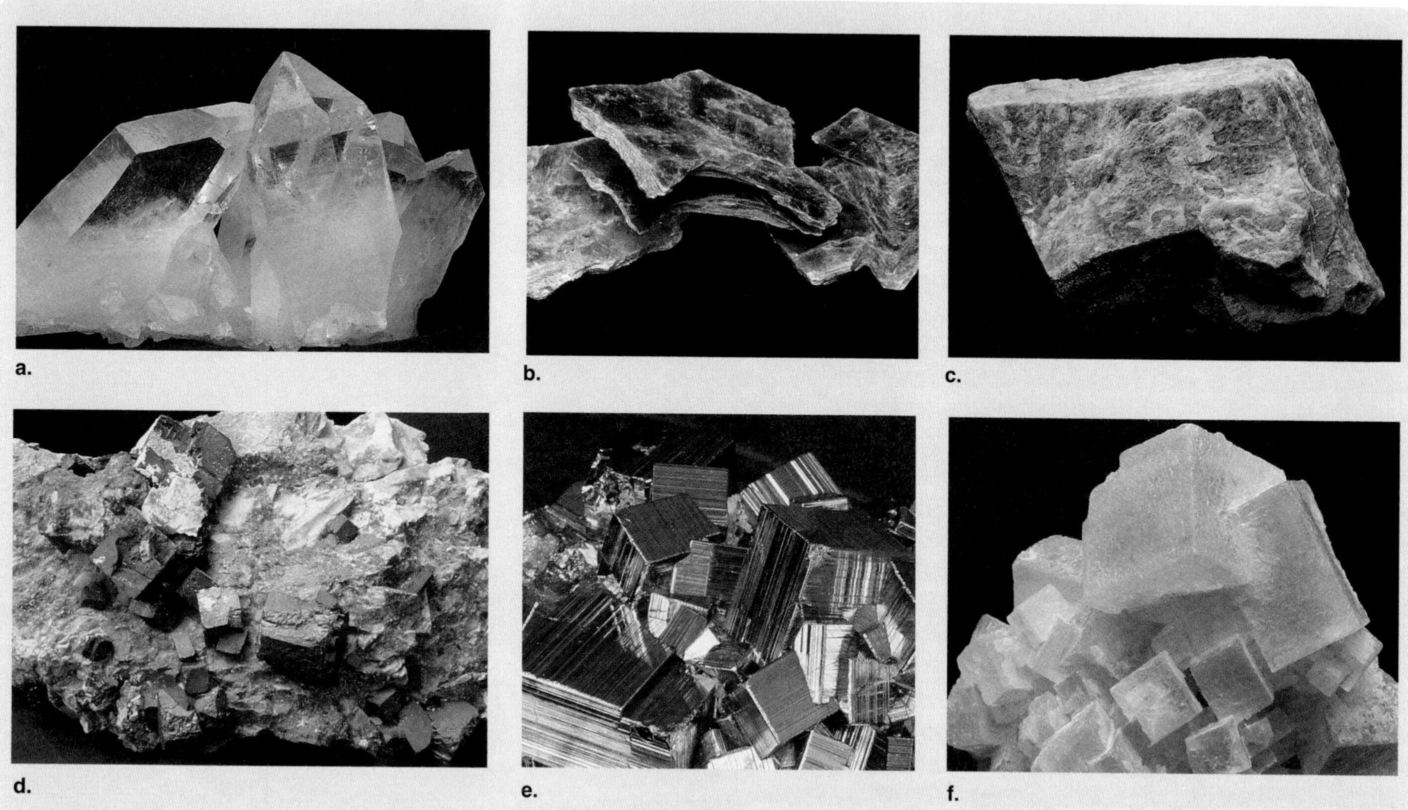

Figure 7.12 Diversity of minerals. **a.** Quartz; **b.** muscovite mica; **c.** orthoclase feldspar; **d.** galena; **e.** pyrite; **f.** halite. Galena, pyrite, and halite all exhibit a cubic crystal form. Galena and pyrite have a metallic luster.

way, geologists use various physical features to identify minerals. Although thousands of different minerals exist on Earth, not that many of them are frequently found in rocks. The same mineral found in different parts of the world has characteristic features and a consistent chemical composition. The most common features used to identify minerals are crystal form, cleavage, hardness, and color, as well as other secondary properties, such as luster and streak.

Crystal Form. If you take a magnifying glass and compare common table salt (a mineral) to pepper (a ground seed), you will notice some differences. (Try it. What do you see?) The salt will have a consistent cubic shape, while the pepper does not. When mineral crystals have lots of room to grow, they form specific shapes (Figure 7.12). For example, we often find well-formed crystals in caves where the mineral is precipitated out of solution and can grow outward from the cave wall into the open interior of the cave. **Crystal form is the shape of the crystal that a mineral forms when it is free to grow unimpeded.** The shape of a mineral crystal is related to the way the chemical bonds form. Atoms or molecules of the mineral align in a way to minimize the energy needed to form the crystal. Common shapes are prisms, pyramids, needles, cubes, and sheets (Figure 7.13). We saw in Section 7.2 that chlorine and sodium ions in halite are organized in a gridlike pattern; this is revealed by the cubic form of halite crystals (see Figure 7.12f).

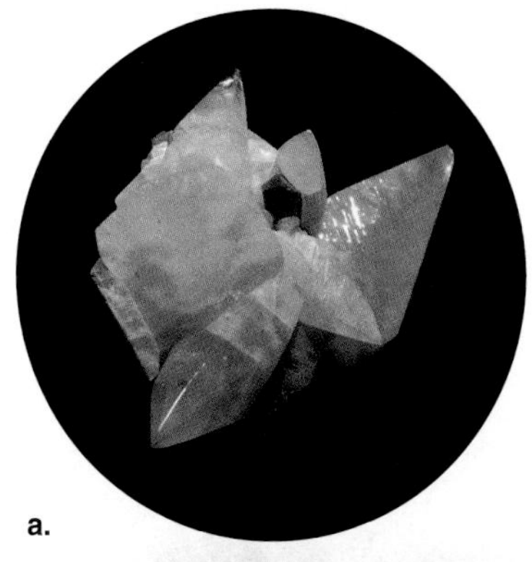

a.

b.

Figure 7.13 Crystal shapes include **a.** pyramid (calcite) and **b.** needlelike (tourmaline).

Cleavage. **The propensity of a mineral to break along one or more planes is called *cleavage*.** Depending on how their constituent atoms are arranged, minerals may break along specific surfaces of weakness called cleavage planes. Cleavage planes mark those parts of the mineral where ions are connected by relatively weak ionic bonds. In contrast, covalent bonds are strong and are less likely to form cleavage planes. Muscovite and biotite, two forms of the mineral mica, have a single series of parallel cleavage planes that causes them to separate into sheets, much like a stack of paper (see Figures 7.12b and 7.14a). In contrast, other minerals may have multiple cleavage planes that define sheets, cubes, or other shapes (Figure 7.14). Some minerals, such as quartz, have no cleavage planes and consequently break with an irregular fracture pattern.

Hardness. Which is harder, glass or a copper penny? Suppose that you do not already know and have a piece of each. What experiment could you do to determine which is harder? Minerals can be ranked by their relative hardness based on their ability to scratch one another. As it turns out, glass can scratch a penny, but a penny cannot scratch glass. **The resistance of a mineral to scratching is measured by its hardness.** Ten minerals make up the Mohs hardness scale (Table 7.3). Minerals not in the table are ranked relative to these. For example, a mineral that could scratch feldspar but not quartz would have a hardness of approximately 6.5. Quartz can scratch feldspar or any mineral with a lower Mohs scale number but cannot scratch topaz, corundum, or any mineral with a higher Mohs scale number. Softer minerals with low rankings on the scale are more likely to break down on Earth's surface, while harder minerals are more likely to survive. That is one reason why the sand on a beach is almost always quartz; most of the other minerals with lower rankings on the Mohs scale are broken down so that only quartz is left. For comparison, a copper penny would

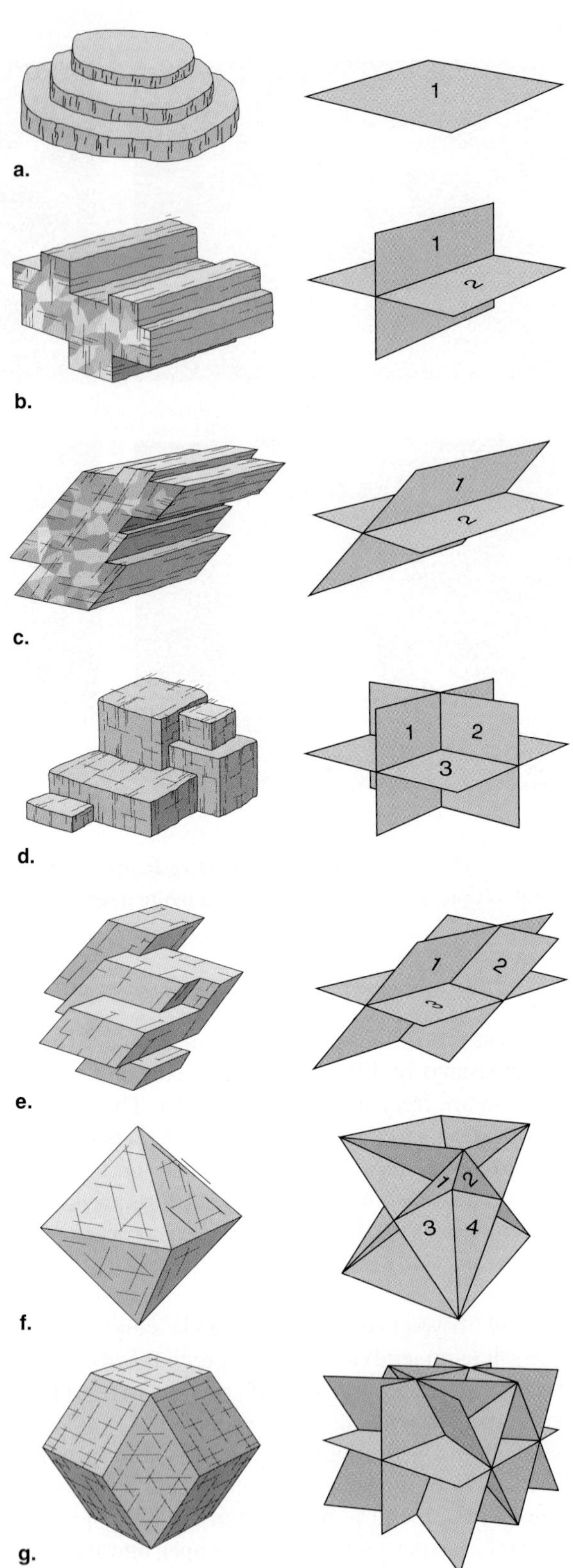

Figure 7.14 Types of cleavage. **a.** One direction; **b.** two directions that intersect at 90° angles; **c.** two directions that do not intersect at 90° angles; **d.** three directions, intersecting at 90° angles; **e.** three directions, not intersecting at 90° angles; **f.** four directions (for example, diamond); **g.** six directions. Some minerals have no cleavage planes (for example, quartz), while others may have several.

Table 7.3 Hardness Scales

	Mohs hardness scale	Mineral	Absolute hardness scale
Softest	1	Talc	1
	2	Gypsum	2
	3	Calcite	9
	4	Fluorite	21
	5	Apatite	48
	6	Feldspar	72
	7	Quartz	100
	8	Topaz	200
	9	Corundum	400
Hardest	10	Diamond	1,500

Checkpoint 7.6

basic · intermediate · advanced · superior

Finish the partially completed concept map for minerals provided here. How could you add additional levels to the concept map?

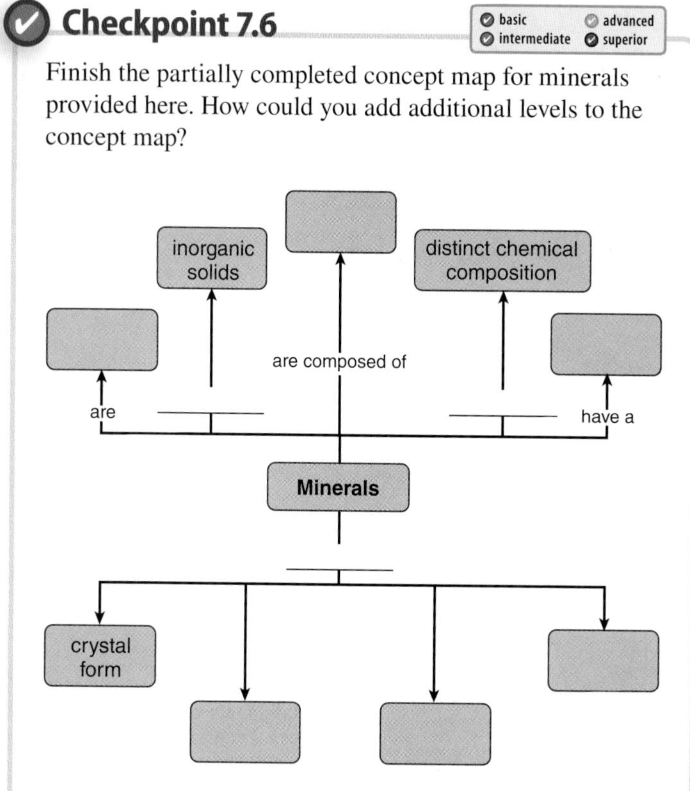

Checkpoint 7.7

basic · intermediate · advanced · superior

Examine the following images of the mineral halite, and identify how many cleavage planes are present.

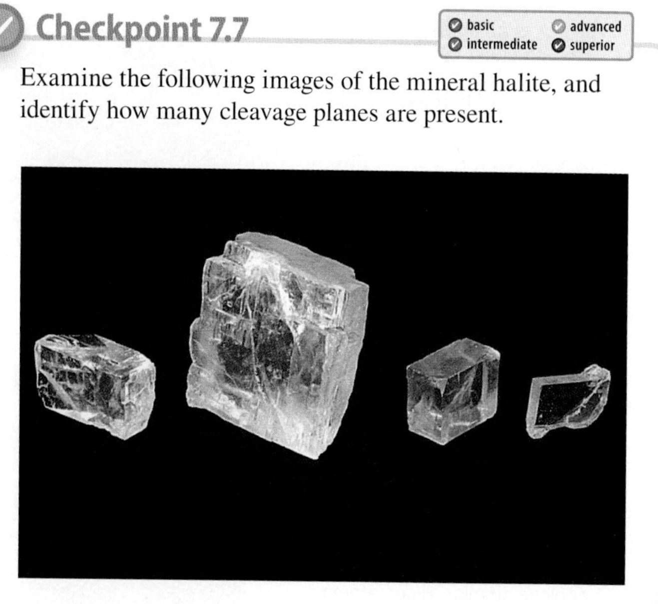

rank around 3 and glass would be a little harder than 6 on the Mohs scale. The scale is relative, meaning that it is based on one mineral's ability to scratch another, not on their intrinsic hardness. Therefore, another system exists, called the *absolute hardness scale* (Table 7.3). That scale denotes hardness based on how much pressure must be applied to a diamond point moved across a mineral before a scratch appears.

Color. Just as color might be used to describe a food (for example, a green apple), it is used to describe minerals, which come in a variety of colors (see Figures 7.3 and 7.12). Examples of common dark-colored minerals (black, dark brown, dark green) are amphibole, olivine, pyroxene, and biotite mica. The light-colored minerals (white, pink, gray, translucent) found in rocks that you might pick up around your house are typically quartz, feldspar, muscovite mica, gypsum, halite, and calcite. However, we must be careful when using color to identify minerals because some minerals exist in a wide range of colors. For instance, quartz occurs in over a dozen different colors. Minerals can also change color when they are exposed to changing natural conditions (rain, heat) on or near the surface of Earth.

Other Properties. Other mineral characteristics that geologists may use to distinguish between minerals include **luster (how light is reflected from a mineral**; see Figure 7.12) and **streak (the mark formed when a mineral is scratched across an unglazed piece of porcelain**; Figure 7.15a). For example, gold creates

a.

b.

Figure 7.15 Other properties of minerals that are useful in distinguishing one from another. **a.** A brown streak on white porcelain identifies this mineral as the metallic iron hematite. **b.** When weak hydrochloric acid is added to the mineral calcite, a characteristic reaction becomes visible.

A *rock* is an *aggregate* of *minerals*. The structure, texture, and types of minerals depend upon the conditions of formation of the rock.

Granite

Quartz (clear)

A mineral is a physically and chemically distinct part of a rock. It has properties you can see with an unaided eye, such as color and cleavage.

Biotite (black)

Feldspar (pink)

Each type of mineral also has its own orderly internal (crystalline) structure at the atomic level. This explains the physical properties seen at a larger scale.

In this example, showing quartz, large oxygen atoms enclose smaller silicon atoms.

We construct models of some crystal structures linking the centers of oxygen atoms together to form silica tetrahedra. This helps us envision the geometric beauty of a mineral's atomic arrangement.

Nucleus

An atom consists of a cloud of electrons at different energy levels enclosing a tiny nucleus.

An atomic nucleus is a swarm of protons and neutrons at the core of an atom.

Figure 7.16 Relationships among rocks, minerals, elements, and atoms.

a yellow streak, while iron sulfide (also known as pyrite or fool's gold) makes a black streak. Some minerals have specific traits. For example, the common mineral calcite creates bubbles of carbon dioxide gas when exposed to acids (Figure 7.15b); magnetite is magnetic; sulfur has a bad smell; and halite (salt) has a salty taste (if you are willing to lick a rock).

Figure 7.16 summarizes the relationships among rocks, minerals, elements, and atoms described in this section.

Checkpoint 7.8

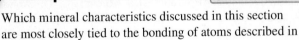

basic · intermediate · advanced · superior

Which mineral characteristics discussed in this section are most closely tied to the bonding of atoms described in Section 7.2?

7.4 Igneous Rocks

Learning Objectives

- Explain the relationship between the cooling rate of magma and the texture of igneous rocks.
- Classify and name images of common igneous rocks.
- Compare and contrast the characteristics of volcanic and plutonic igneous rocks.
- Discuss which igneous rocks would be found in different plate tectonic settings.

Igneous rocks form from the cooling of melted rock (magma), a concept that follows from the old idea of Plutonism. Igneous rocks can be classified as volcanic or plutonic according

Figure 7.17 Lava at the surface of a Hawaiian volcano. Lava is issuing from the vent in the right background and flowing toward the viewer. The black surface is where the rock is cooling and solidifying to form a crust. The crust breaks when moving over obstructions to show the extremely hot orange lava below.

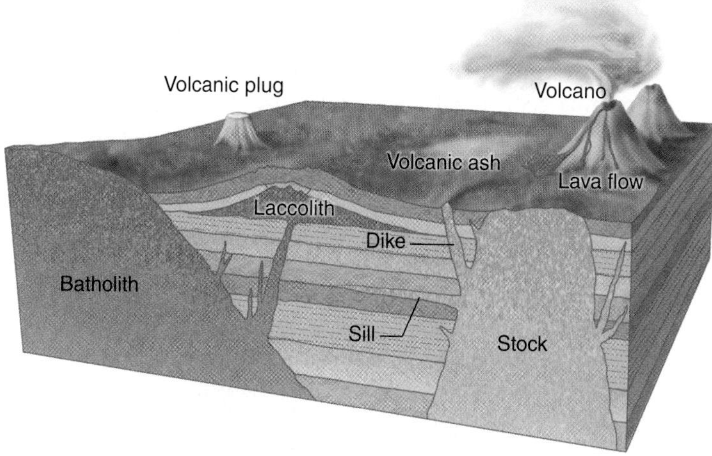

Figure 7.18 Igneous landforms. Volcanic processes form volcanoes, lava flows, and ash falls, while plutonic processes lead to batholiths, laccoliths, dikes, sills, and crystallized magma chambers (called stock).

to whether they form on Earth's surface or within Earth's interior. **Volcanic igneous rocks form when magma in Earth's interior rises to the surface through pipes or fractures in the crust.** Many of the gases trapped in magma (for example, CO_2, SO_2, and water vapor) escape as it nears the surface. At the surface, this slightly altered magma is termed *lava* (Figure 7.17). Landforms such as volcanoes and lava flows, discussed in Chapter 6, are the most easily recognized formations of igneous rocks.

Igneous rocks that cool below Earth's surface are termed *plutonic igneous rocks.* The features they form, termed plutons, remain hidden from view until erosion or tectonic uplift exposes them at the surface (Figure 7.18). Plutons come in a variety of shapes and sizes. They may represent a large reservoir of magma that cools to form a large mass (a batholith) or a thin vein of magma that solidifies as it flows through a near-vertical fracture (a dike) or along a mostly horizontal contact between adjacent layers of rock (a sill).

The Classification of Igneous Rocks

Examine the objects in Figure 7.19. Suppose you were asked to organize these objects in some logical way. How would you classify them? We will pause here for a snack; look at the figure. . . .

Most people organize the objects into classes labeled fasteners or paper clips and divide them by size and color. In a similar way, igneous rocks are classified on the basis of the size and arrangement of their mineral grains (known as texture) and their color, which is an indicator of chemical composition.

Texture. Recall that igneous activity occurs when magma rises toward the surface through pipes and fissures (fractures) in Earth's crust. **The size of the mineral crystals that form depends on how fast the magma cools and solidifies.** Plutonic rocks cool slowly far below the surface where the geothermal gradient ensures

Figure 7.19 Classification exercise. On the basis of your observations only, what features could you use to classify these objects into different groups? Be careful not to use your existing knowledge about properties that you cannot observe (for example, do not base your classification on the strength of material used to make them).

that it is relatively hot. Therefore, minerals that form in plutonic rocks have plenty of time to grow into large crystals as the magma cools slowly. In contrast, hot magmas undergo a rapid temperature drop of at least 700°C (1,292°F) when they reach the surface. Consequently, **volcanic rocks cool rapidly at the surface.** Individual mineral grains (crystals) do not have much time to grow when magma cools rapidly. Consequently, **the grain sizes in all volcanic rocks are small,** too small to be seen without a magnifying glass. The larger grains of plutonic igneous rocks make them easier to distinguish from volcanic rocks (Figure 7.20). Individual grains can be readily seen in specimens of plutonic rocks. Keep in mind that, despite the difference in mineral grain sizes, the same minerals are present in both volcanic and plutonic igneous rocks that can form from the same magma.

Color. The color of an igneous rock is representative of its composition—in particular, of its silica content (Table 7.4). Light-colored minerals such as quartz and feldspar contain more of the element silicon. Consequently, their abundant presence signals a silica-rich magma source. **Light-colored igneous rocks such as**

Figure 7.20 Comparison of volcanic and plutonic rocks. Plutonic rocks have larger grain sizes than volcanic rocks. Rocks with high silica content contain lighter-colored minerals than low-silica rocks. Ferromagnesian minerals are rich in iron and magnesium.

rhyolite and granite are formed from silica-rich magmas and contain abundant white, pink, or translucent silica-rich minerals. In contrast, **dark-colored silica-poor igneous rocks** such as basalt and gabbro are dominated by olivine and pyroxene, dark-colored minerals (black, brown, dark green) having a smaller proportion of silica. Rocks of an intermediate silica composition such as andesite and diorite lie somewhere between the light and dark rocks, depending on the minerals present. (Compare Table 7.4 with Table 6.1.)

Table 7.4 Silica Content of Igneous Rocks

Silica content	Volcanic rocks	Plutonic rocks	Common minerals
High	Rhyolite	Granite	Quartz, feldspar
Intermediate	Andesite	Diorite	Feldspar, amphibole, pyroxene
Low	Basalt	Gabbro	Pyroxene, feldspar, olivine

Correlation Between Rock Types and Magma Types. In Chapter 6, we discussed the variations in magma compositions and plate tectonic settings, but only in terms of the type of magma, not the specific igneous rocks. We labeled the magmas as basaltic, andesitic, and rhyolitic. The lavas they produced are named for the volcanic rocks—basalt, andesite, rhyolite. Where would we find their plutonic equivalents? The answer is simple—under the volcanic rocks (see Chapter Snapshot). If we were to drill through the basalt rocks that make up the new seafloor along oceanic ridges, we would find the plutonic rock gabbro and eventually peridotite, the mantle rock that undergoes partial melting to form basaltic magma. On rare occasions, we can observe the same association on the continents in places where thin slices of oceanic lithosphere have sometimes been trapped and preserved along convergent plate boundaries (Figure 7.21).

Checkpoint 7.9

basic · intermediate · advanced · superior

Geologists sometimes find a type of igneous rock known as porphyry, which contains both large and small crystals. Which is the best explanation for the formation of this rock?

a) The rock experienced a two-stage cooling process, with initial slow cooling at depth followed by rapid cooling at the surface.
b) The rock experienced a two-stage cooling process, with initial rapid cooling at depth followed by slow cooling at the surface.
c) The rock experienced a two-stage cooling process, with initial rapid cooling near the surface followed by slow cooling at depth.
d) The rock experienced a two-stage cooling process, with initial slow cooling near the surface followed by rapid cooling at depth.

Figure 7.21 Oceanic lithosphere on land. The uppermost part of mantle (peridotite) is visible at Table Mountain, western Newfoundland.

Checkpoint 7.10

basic · intermediate · advanced · superior

Name each of the four igneous rocks pictured here. Describe how each of these rocks formed.

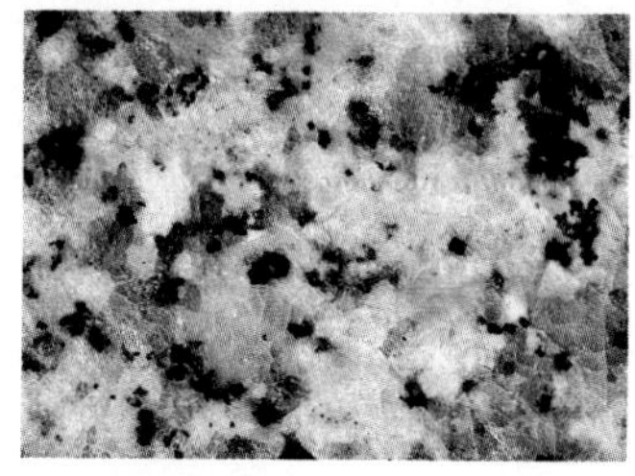

ORIGIN OF ROCKS

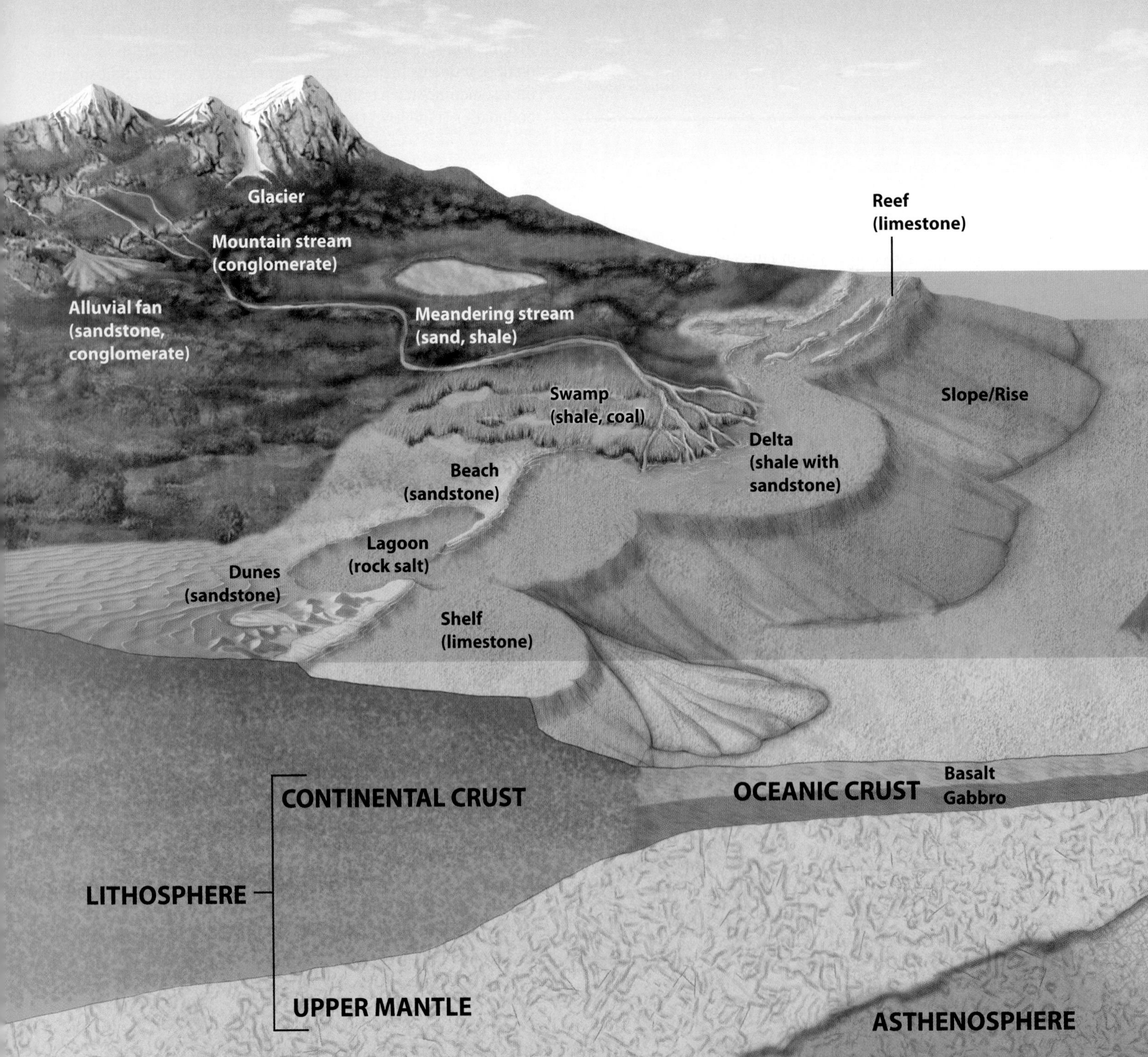
Glacier
Mountain stream
(conglomerate)
Alluvial fan
(sandstone,
conglomerate)
Meandering stream
(sand, shale)
Reef
(limestone)
Slope/Rise
Swamp
(shale, coal)
Delta
(shale with
sandstone)
Beach
(sandstone)
Lagoon
(rock salt)
Dunes
(sandstone)
Shelf
(limestone)
Basalt
Gabbro
OCEANIC CRUST
CONTINENTAL CRUST
LITHOSPHERE
UPPER MANTLE
ASTHENOSPHERE

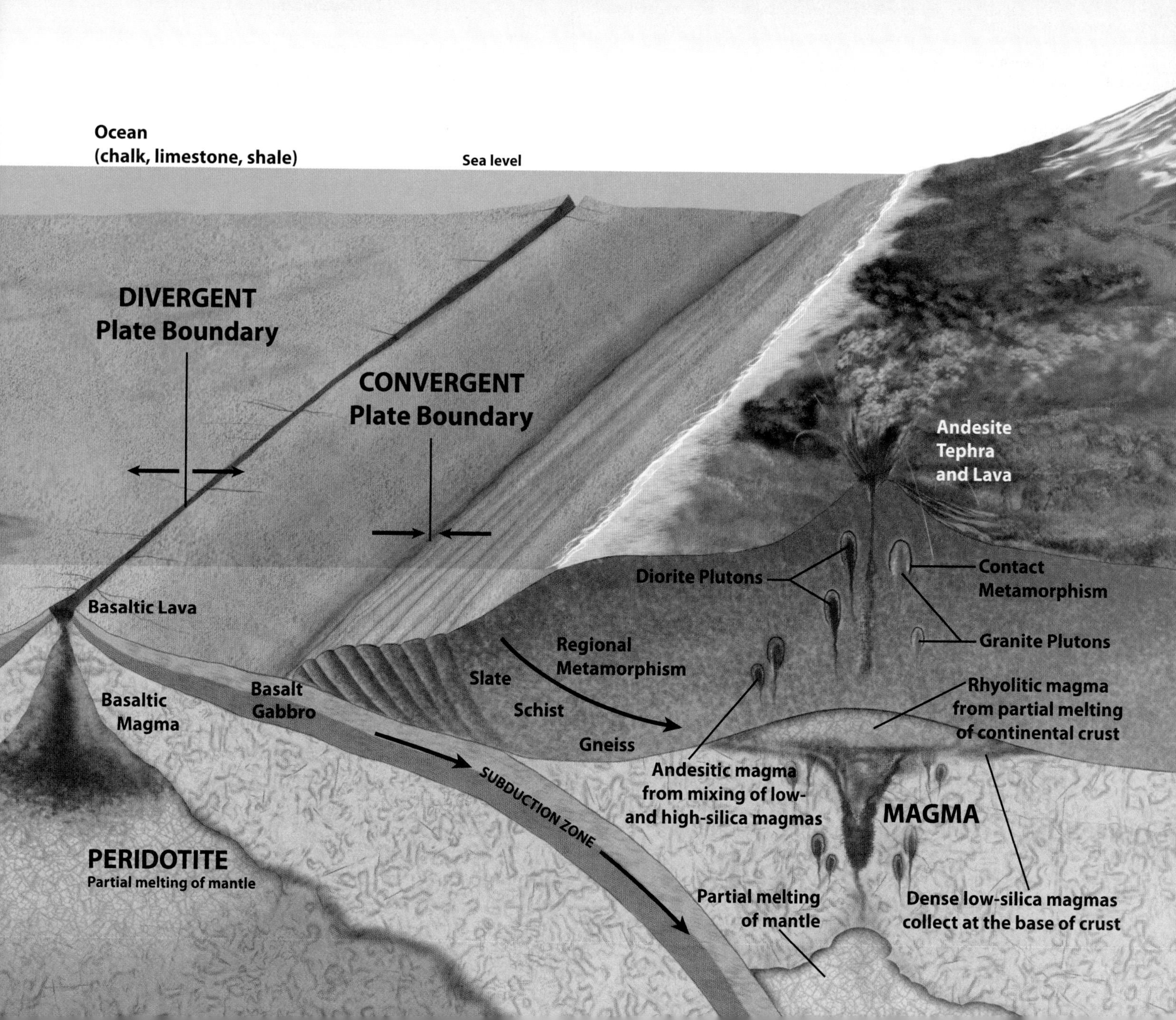
Ocean
(chalk, limestone, shale)
Sea level
DIVERGENT
Plate Boundary
CONVERGENT
Plate Boundary
Andesite
Tephra
and Lava
Diorite Plutons
Contact
Metamorphism
Basaltic Lava
Regional
Metamorphism
Granite Plutons
Slate
Basalt
Gabbro
Basaltic
Magma
Schist
Rhyolitic magma
from partial melting
of continental crust
Gneiss
Andesitic magma
from mixing of low-
and high-silica magmas
SUBDUCTION ZONE
MAGMA
PERIDOTITE
Partial melting of mantle
Partial melting
of mantle
Dense low-silica magmas
collect at the base of crust

Checkpoint 7.11

basic · intermediate · advanced · superior

Complete the following table by identifying which of the characteristics in the left-hand column are present in volcanic or plutonic igneous rocks or both and placing a check mark in the appropriate column(s). Do not place a check mark in either column if the characteristic is not present. One characteristic has been completed as an example.

Characteristic	Volcanic igneous rocks	Plutonic igneous rocks
May form from basaltic magma		
Form at Earth's surface		
Have texture	X	X
Made of small grains		
Granite is an example		
Form as a result of melting		
May form from rhyolitic magma		
May form in the presence of water		
Exposed at Earth's surface only after erosion		
Contain minerals		
Andesite is an example		
Classified based on density of minerals		
Dark-colored examples have low silica content		
Contain visible grains		

Checkpoint 7.12

basic · intermediate · advanced · superior

Finish the partially completed concept map for igneous rocks provided here by filling in the blanks with appropriate terms. Three of the appropriate terms are *magma*, *basalt*, and *plutonic* rocks.

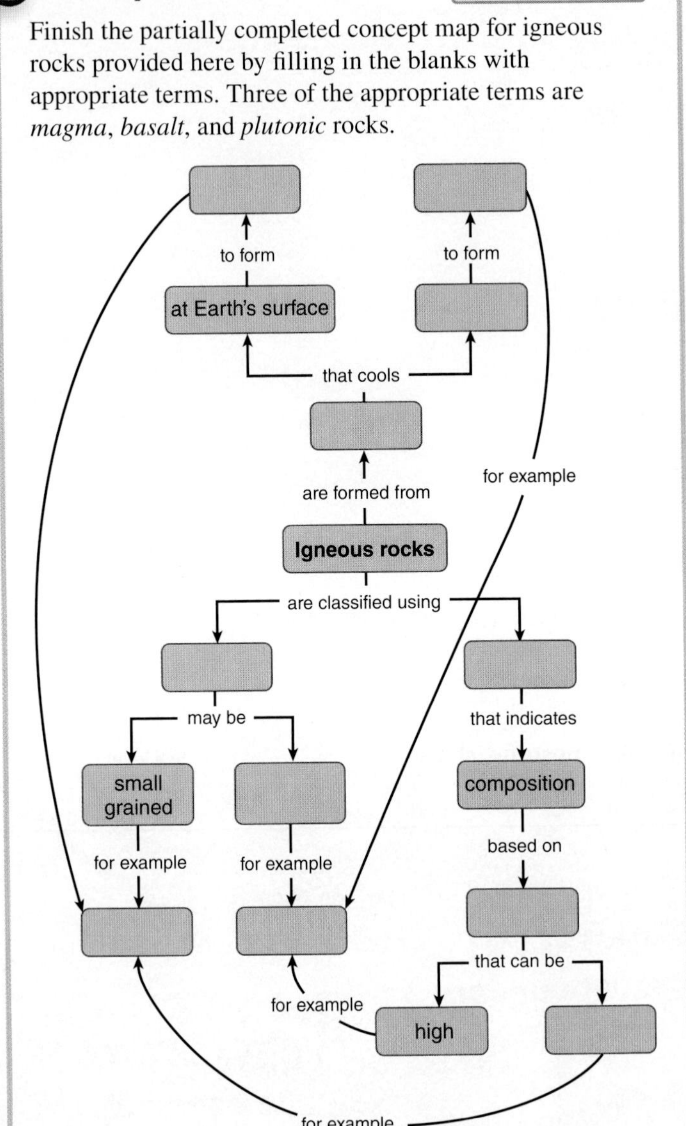

Andesite is a product of volcanic eruptions at convergent boundaries; the plutonic equivalent, diorite, would be found where ancient magma cooled and solidified below mountain ranges such as the Cascades and the Andes. Granite is a more common plutonic rock than diorite, and the (rhyolitic) magma that forms granite is more viscous and less likely to reach the surface. Rhyolitic magma forms from the partial melting of the continental crust in locations above subduction zones and over hot spots (see Chapter Snapshot). Granite plutons known as batholiths (see Figure 7.18) can account for huge volumes of continental crust and are often exposed by erosion, especially where the crust is thickened below mountains such as the Sierra Nevada range in California (Figure 7.22).

Figure 7.22 Batholiths. The peaks of much of the Sierra Nevada range in California are formed from granite batholiths that have been exposed by erosion.

in Further Depth

In Chapter 6, we told you that when a rock is heated, some minerals will melt while others remain relatively unchanged. Silica-rich minerals such as quartz melt first as they have a lower melting temperature than silica-poor minerals. But what happens in the reverse process—when magma cools?

As temperatures begin to drop below 1,200°C (2,190°F) in the magma chamber, some minerals with a *higher* melting temperature are the first to form crystals. What happens next was figured out by N. L. Bowen in a series of experiments about 80 years ago and is illustrated in Figure 1. Bowen observed a sequence of crystallization, Bowen's reaction series, which has two branches.

- The discontinuous branch represents the ferromagnesian minerals, so called because they are rich in iron (*ferro-*) and magnesium (*-magnesian*).
- The continuous branch represents a single type of feldspar that has sodium- and calcium-rich varieties.

Minerals crystallize from both branches at the same time. For example, the same magma will contain crystals of both olivine and plagioclase feldspar.

In an idealized magma, minerals with low silica content such as olivine and the calcium-rich variety of feldspar would be the first to form as the magma cooled (Figure 1). If temperatures remained high, these minerals would continue to form until they used up all the iron, magnesium, and calcium in the magma. However, it is more likely that the temperature of the magma would continue to decrease. In that case, two events are possible:

- The early-formed crystals could settle out and sink to the bottom of the magma chamber (crystal settling). The next mineral in Bowen's reaction series would then form (pyroxene and more calcium-rich feldspar; Figure 1).
- The early-formed ferromagnesian minerals would react with the remaining magma. These reactions would convert the early-formed minerals to the next mineral in the discontinuous branch of the series. For example, olivine would recrystallize to pyroxene. The early calcium-rich plagioclase feldspars would get steadily bigger but would begin to add more sodium into the crystal structure.

The reactions continue until the magma is used up. Early-formed magmas are rich in pyroxene, olivine, and plagioclase feldspar and have relatively low silica content. These form rocks such as basalt and gabbro (see Table 7.4). Any leftover magma forms silica-rich minerals such as quartz and potassium feldspar, which crystallize to form granite and rhyolite.

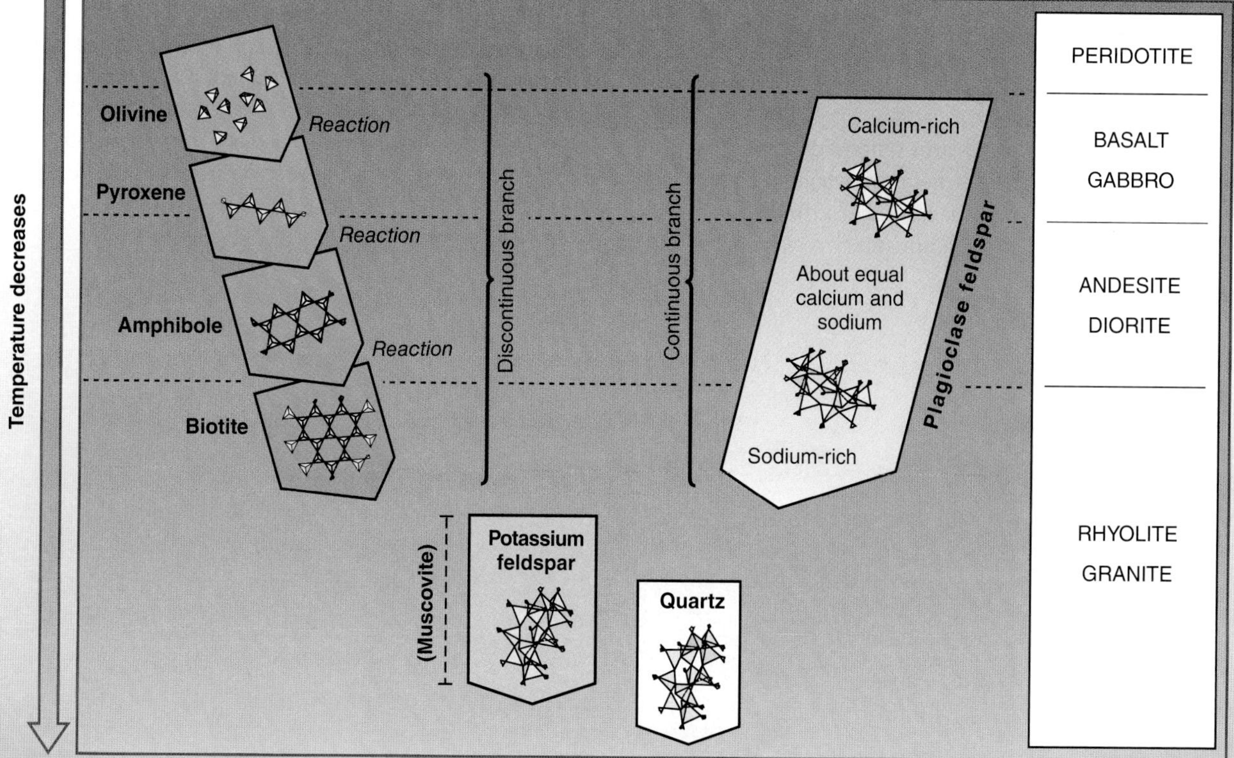

Figure 1 Bowen's reaction series. Silica-poor rocks such as basalt and gabbro form early in the series and are rich in olivine and calcium-rich plagioclase feldspar. Intermediate-silica rocks such as andesite and diorite form if magma solidifies in the middle of the series. Silica-rich magmas remain at the end of the series and are characterized by rocks such as rhyolite and granite, which feature minerals such as potassium feldspar and quartz. This is the unifying concept of how igneous rocks correlate with silica-based igneous minerals and cooling temperature. Compare this diagram with Figure 7.20.

7.5 Sedimentary Rocks

Learning Objectives

- Explain how clastic sedimentary rocks are formed.
- Describe how you could use the properties of a clastic sedimentary rock to make predictions about the conditions under which it formed.
- Compare and contrast the characteristics of chemical and biochemical sedimentary rocks.
- Discuss the geological conditions necessary for the formation of oil and gas deposits.

Take a minute and examine the rock formations in Figure 7.23. What do you notice?

Sedimentary rocks often form in a series of layers called **beds or strata.** These layers can be readily identified in nature on the basis of different colors, thicknesses, and resistance to erosion, properties that are linked to the composition and origin of the rocks. This layering is apparent in each of the images in Figure 7.23. Note that some layers stand out; this is so because they are more resistant to erosion than the others.

Sedimentary rocks form in a range of environments, from stream channels to the floor of the deep ocean (see Chapter Snapshot). Geologists study modern environments to understand how rocks form today and use these data to interpret the conditions under which rocks formed millions of years ago. Scientists can use this information to investigate past climates, mountain-building events, and tectonic plate movement.

Sedimentary rocks can be divided into three fundamental types based on the materials involved and the process by which each type forms:

- **Clastic sedimentary rocks** are composed of sediments—rock and mineral fragments that form when rocks break apart at or near Earth's surface. Clastic sedimentary rocks make up the majority of all sedimentary rocks.
- **Chemical sedimentary rocks** are crystallized from a solution (for example, seawater) as a result of changing conditions.
- **Biochemical sedimentary rocks** form by the actions of living organisms or are composed of the remains of dead organisms.

The processes involved in the formation of each of these rock types reveal echoes of the concept of Neptunism described at the beginning of the chapter.

Clastic Sedimentary Rocks

Clastic sedimentary rocks are formed from rock and mineral fragments (clasts). This process occurs in a series of steps: (1) generation of clasts due to the breakdown of an original rock by weathering; (2) transportation of the eroded material from the source area; and (3) lithification, the deposition and subsequent conversion of the material to rock.

Generation. Rocks on or near Earth's surface are physically broken into smaller pieces, and minerals undergo chemical changes to form weaker minerals. This process of disintegration and decomposition is termed *weathering* (see Chapter 9 for more details) and is influenced by the original rock type and by climatic conditions. Weathering is more rapid for rocks that contain minerals with well-defined cleavage planes or minerals that can dissolve in water. Climatic conditions characterized by high temperatures and lots of water promote rapid weathering.

Weathered material is known as *sediment* and is classified on the basis of its grain size. Small sediment particles are termed *clay* and *silt*, while larger grains form sand and gravel (Table 7.5). For comparison, sand-sized particles are typically about the same size as grains of sugar, and gravel-sized particles are about the same size as pieces of hard candy. Gravel and most sand-sized particles can be seen with the naked eye, but you would need a magnifying glass or microscope to identify individual silt- or clay-sized particles.

Figure 7.23 Sedimentary rocks from western parks. **a.** Kaibab limestone, Little Colorado River gorge, Grand Canyon National Park, Arizona; **b.** Entrada sandstone, Arches National Park, Utah; **c.** Cliff House sandstone, Cliff Canyon, Mesa Verde National Park, Colorado; **d.** White Rim sandstone, Canyonlands National Park, Utah.

a.

b.

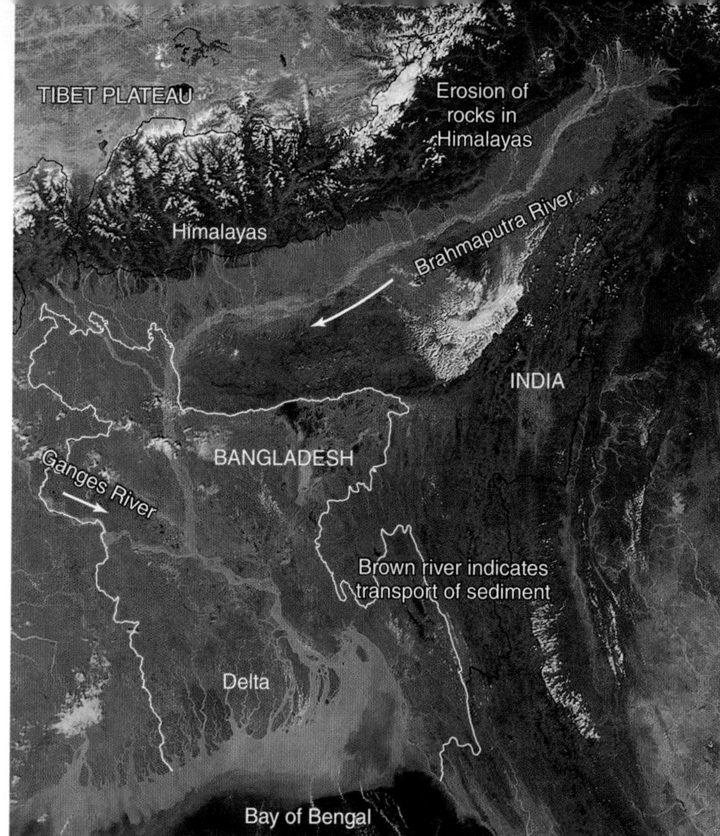

a.

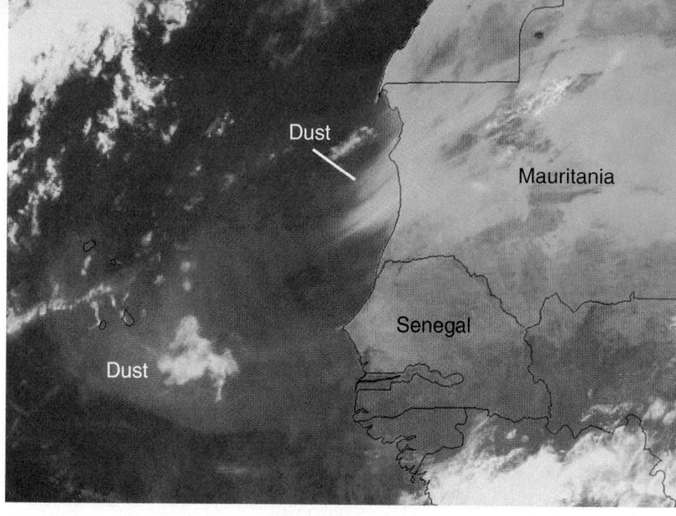

b.

Figure 7.24 Sediment transported by water and wind. **a.** The Brahmaputra and Ganges rivers are fed by sediment-filled streams from the slopes of the Himalayas. Much of the nation of Bangladesh is made up of a delta formed from sediment deposited by these two mighty rivers. Note the brown plume of sediment entering the Bay of Bengal. **b.** African dust carried from the Sahara desert over the Atlantic Ocean. Hundreds of millions of tons of dust are annually deposited on the other side of the ocean in the Caribbean Sea, Central America, and South America.

Table 7.5 Clastic Sediments and Sedimentary Rocks

Sediment	Grain size (diameter)	Rock	Grain size comparisons
Clay	Less than 0.0039 mm (less than 0.00015 in)	Shale, mudstone	Smaller than granulated sugar
Silt	0.0039 to 0.0625 mm (0.00015 to 0.0025 in)	Siltstone	Smaller than granulated sugar
Sand	0.0625 to 2 mm (0.0025 to 0.079 in)	Sandstone	Ranges from sugar to coarse salt
Gravel	More than 2 mm (more than 0.079 in)	Conglomerate	Ranges from rice grains to oranges

Transportation. The process by which **sediment is removed from its place of origin by running water, winds, or glaciers is known as *erosion.*** Erosion rates vary over several orders of magnitude depending on the material. For instance, granite erodes at rates as little as 0.002 millimeter per year (0.00008 inch per year) while unconsolidated sediment can be eroded from a stream bank at rates of meters per hour! A muddy river is an indication that sediment is being transported by water, while a cloud of dust is a sign that sediment is being carried by wind (Figure 7.24). The size of the sediment that can be transported depends on how fast the water flows or how hard the wind blows. If the velocity is high, a stream can move a lot of material; for example, a fast-moving stream in a flood can sometimes transport a boulder as large as an elephant.

c.

d.

The process of erosion shapes the landscape and contributes to the formation of many distinctive landforms, such as valleys, canyons, and mountains. As mentioned in Chapter 6, some of the most rapid erosion today is taking place in the Himalayan Mountains, filling the adjacent rivers with sediment. That sediment must eventually be deposited somewhere, which takes us to the next stage in the formation of clastic rocks—deposition and eventual lithification.

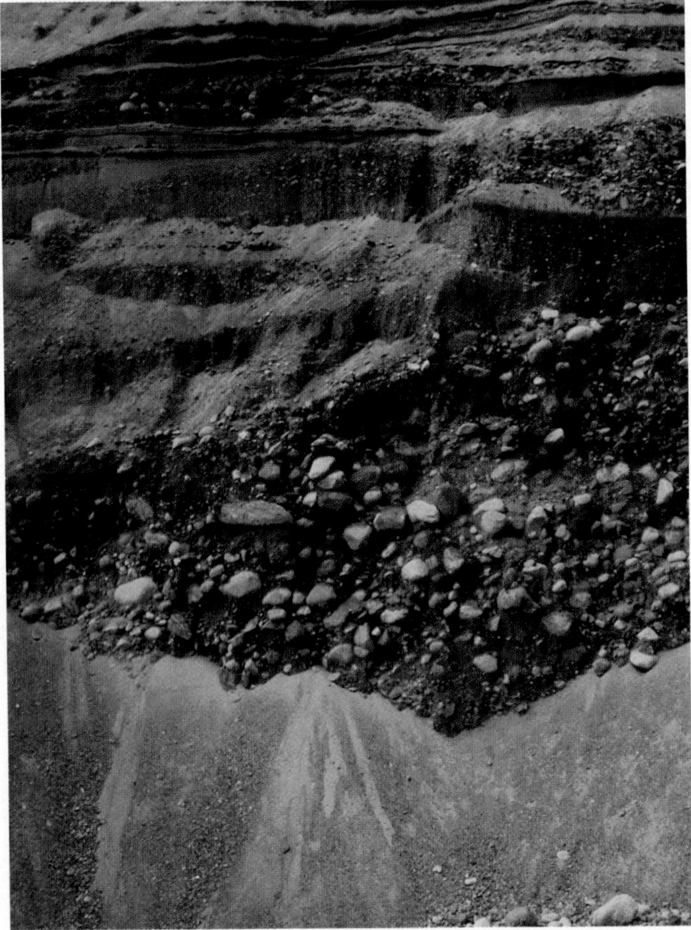

Figure 7.25 Clastic sediment deposit. A lens of gravel separates layers of sand. Pebbles can be seen in thin layers toward the top of the image. This is a typical sand and gravel deposit indicative of a sedimentary environment with alternating intervals of fast- and slow-flowing streams.

Lithification. Imagine leaves blowing in the wind on a late fall day. As the wind slows, the leaves fall. In a similar fashion, when the speed of a stream slows, the sediment it carries drops out. **Clastic sediments are deposited when the velocity of the transporting medium drops** (Figure 7.25). Large material is deposited first, followed by smaller and smaller particles as the velocity steadily decreases. Rivers dump much of their sediment where they enter the relatively quiet waters of an ocean or lake, creating the landform known as a delta (see Figure 7.24a). This material is often transported and redistributed along the coastline by shoreline currents to form beaches. The finest particles remain in suspension in the ocean before eventually being deposited offshore. The largest sediment deposits in the world are the deep sea fans in the Bay of Bengal and the Arabian Sea that formed over millions of years from sediment eroded from the Himalayas (see Figure 6.32). The Bengal fan is more than 2,500 kilometers (1,550 miles) long, hundreds of kilometers across, and several kilometers thick.

The sand, clay, and silt particles at the bottom of a pile of sediment become compacted, squeezing out water and forcing the grains closer together. Slowly, over hundreds of thousands of years, fluids circulating through the pile of sediment precipitate minerals to cement the grains together, converting the sediment into a rock. **The processes of compaction and cementation that convert sediment into a sedimentary rock are together termed *lithification*.** Clastic sedimentary rocks such as sandstone and siltstone have names that reflect the size of their grains (see Table 7.5; Figure 7.26). Different speeds of stream flow or wind serve to sort and concentrate sediment by weight (grain size). Consequently, we often find that the sand grains on a beach are the same size and that clastic rocks have a dominant grain size instead of being a random mix of sand, clay, and gravel-sized particles. The rate of clastic sediment formation is more rapid adjacent to active plate margins characterized by high elevations and faster weathering and erosion rates.

Checkpoint 7.13

Suppose you were given a plastic jar that is about one-third full of a mixture of dry gravel, sand, and silt. The jar is filled with water, sealed, and violently shaken. Draw a picture of what it would look like after the shaking stops and the jar is placed on the counter for a few minutes. How might the picture change after 10 hours?

a.

b.

Figure 7.26 Examples of clastic sedimentary rocks defined by grain size. **a.** Fine-grained sandstone has small sand grains cemented together that are visible to the naked eye. **b.** Conglomerate composed of gravel- and pebble-sized rounded grains.

Checkpoint 7.14

basic / intermediate / advanced / superior

The following photo shows a typical section of clastic sediments that geologists might study. What observations can you make about the grain size and arrangement of these sediments that would help determine their origin?

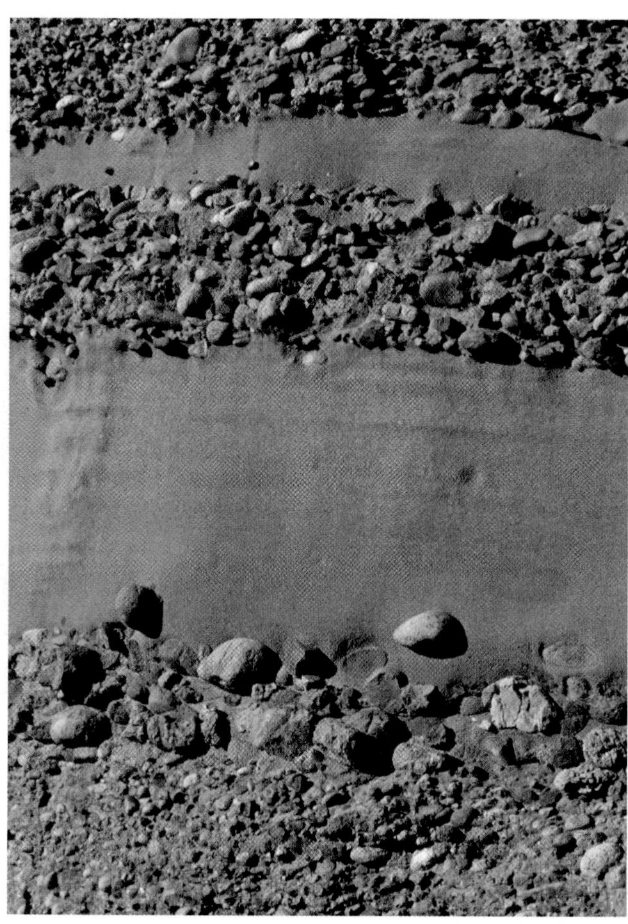

Note that the formation of clastic sedimentary rocks partially mirrors the brick manufacturing process. First, rocks are broken down to smaller sizes; these grains are sorted during transport to ensure that a rock is made up of consistent grain sizes. The mixture of water and sediment is deposited and compacted to expel excess water and air. With time and slightly increased temperature, the mixture hardens to form a rock (brick). The process of lithification applies to all types of sedimentary rocks formed by the processes we discuss in the next sections.

Chemical Sedimentary Rocks

Have you ever noticed the salty taste of your perspiration after completing an outdoor activity on a hot day? That taste is still there even after the perspiration is gone because the salt crystallizes as the water evaporates. In much the same way, **chemical sedimentary rocks form when minerals precipitate (crystallize out) from a solution as a result of changing physical conditions.** The most common solution is seawater, but minerals are also dissolved in lesser concentrations in groundwater and in freshwater lakes. Some minerals are dissolved from rocks on the continents by weathering and are carried in solution by steams to lakes and oceans. Millions of years ago, vast shallow tropical oceans covered much of what is now the North American continent. Those seas evaporated and left behind deposits that are the source for the bulk of the chemical sedimentary rocks mined in North America today.

The common table salt you sprinkle on your food is actually a mineral, halite. Salt forms when seawater evaporates. As more water evaporates, the salt concentration in the remaining water gradually rises to the point where the solution can no longer hold the salt. The salt begins to precipitate from the water and sink to the floor of the lake or ocean, forming a layer of halite (Figure 7.27). Evaporation typically occurs in restricted basins in hot, dry climates. The salt is later buried under other sedimentary rocks (perhaps clastic rocks). Scientists interpret thick salt deposits as an indication that there was a constant supply of seawater and

Figure 7.27 A chemical sedimentary deposit. A large lake, Lake Bonneville, covered much of Utah less than 10,000 years ago during the last ice age. Most of the lake dried up, and the water evaporated, leaving behind a flat lake bed covered in salt. Today, the Great Salt Lake near Salt Lake City is the last remnant of Lake Bonneville.

that hot climate conditions were stable at that location for millions of years. Different chemical compounds may form depending on the concentrations of elements present in solution. For example, gypsum (used in drywall) is another mineral dissolved in seawater that is precipitated when the water evaporates. These types of chemical sedimentary rocks are sometimes termed *evaporites* in recognition of the process that formed them.

Biochemical Sedimentary Rocks

Biochemical sedimentary rocks result from the actions of living organisms that cause minerals to be extracted from a solution or are composed of the remains of dead organisms. As such, these rocks represent a link between the biosphere and the geosphere. The actions of many organisms in seawater change the composition of the water, causing the mineral calcite (calcium carbonate, the principal ingredient in limestone) to be precipitated from solution. Calcite is made up of one atom of calcium, one atom of carbon, and three atoms of oxygen. Thus, the atomic structure of calcite traps carbon atoms, preventing the carbon from contributing to global warming.

Biochemical sedimentary rocks come from a wide variety of sources. For example, massive limestone coral reefs in the shallow tropical oceans around the world have formed because of the actions of the coral organisms (Figure 7.28). Other biochemical rocks are composed of the remains of dead organisms. Chalk, a type of limestone, forms from the skeletons of marine microorganisms known as coccolithophores. These tiny creatures are about the size of grains of clay or silt and live in the shallow, sunlit waters of the ocean (Figure 7.29). Unlike coral, coccolithophores are concentrated in colder, subpolar waters, sometimes in such great numbers as to change the color of the ocean and reflect back more incoming sunlight. The coccolithophore is covered by a mineral armor made up of rounded plates of calcite known as coccoliths. Scientists estimate that 1.4 billion kilograms (1.5 million tons) of coccoliths collect on the seafloor each year to form chalk. The shells of larger organisms, the types you might pick up on a beach, may be broken down, sorted by wave action, and cemented together to form a shell-bearing limestone known as coquina (Figure 7.30a).

The most recognizable example of a biochemical rock is coal (Figure 7.30b). **Coal forms from the compacted remains of dead plants,** such as those that grew rapidly in a tropical

Figure 7.28 Sources of biochemical sedimentary rocks. A coral reef in clear, shallow marine waters. Limestones and coquinas are rocks that commonly form in this type of environment.

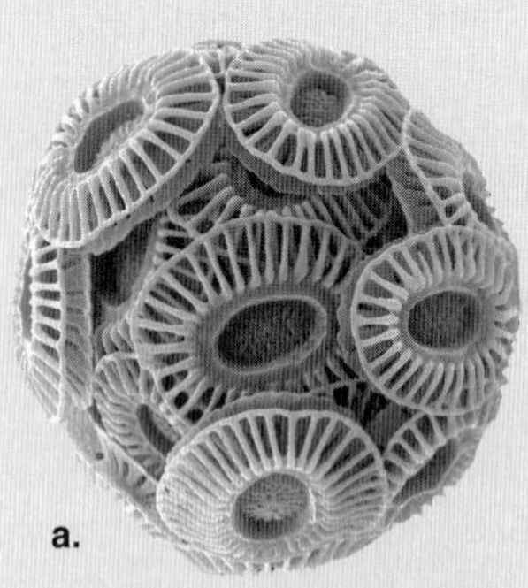

a.

b.

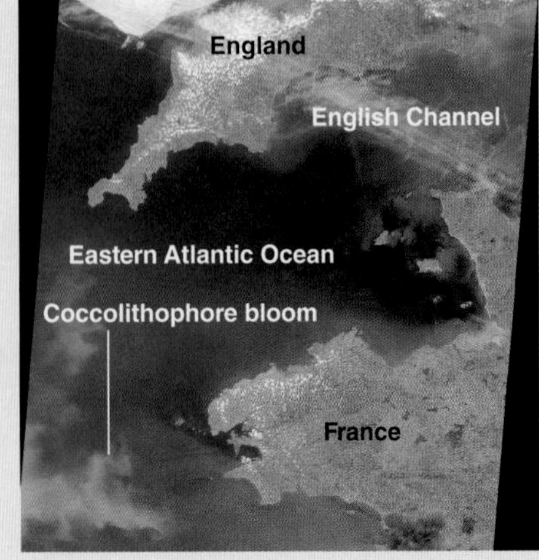

c.

Figure 7.29 Coccoliths and coccolithophores. **a.** Coccoliths are the tiny plates that surround a type of plankton known as a coccolithophore that lives in the shallow ocean. **b.** Billions of coccoliths combine to form layers of chalk. **c.** The presence of coccolithophores in the oceans can reflect more sunlight and change the color of the ocean water.

swamp environment. Such rocks are sometimes termed *organic sedimentary rocks*. What does the fact that large deposits of coal are found in West Virginia, Pennsylvania, and Kentucky tell you about the past environment there?

Figure 7.30 Two biochemical sedimentary rocks. **a.** Coquina is a form of limestone composed almost entirely of shell fragments. **b.** A massive coal seam in the Powder River basin, northern Wyoming. The seam is up to 60 meters (200 feet) thick in places. Note the large bulldozers for scale.

Sedimentary Rocks and Fossil Fuels

Current US energy use is weighted heavily toward fossil fuels (oil, natural gas, and coal), which account for approximately 90 percent of all energy used in the nation. Fossil fuels form from decayed organic material. Oil, coal, and natural gas are the most common products of this process. Oil and gas form from organic material in microscopic marine organisms, whereas coal forms from the decayed remains of land plants. The formation of these fossil fuels extracts carbon from the environment, but their consumption rapidly releases carbon dioxide, elevating global greenhouse gas concentrations (see Chapter 17).

Oil and Gas. The principal requirements in the development of oil and gas reserves are time, temperature, and rock type. The steps in the process are:

1. Organic-rich sediments are deposited and gradually buried to greater depths and converted to sedimentary rock (for example, shale).
2. Chemical reactions occur during burial under conditions of increasing temperature and pressure. The reactions occur at temperatures of 50 to 100°C (122 to 212°F).
3. The reactions change organic molecules to hydrocarbon molecules (chemical compounds of carbon and hydrogen). Higher temperatures "boil off" the hydrocarbons; lower temperatures are not sufficient to drive the chemical reactions. With increasing time (millions of years), the hydrocarbons become more mature, changing from heavy oils to lighter oils to natural gas.
4. Oil and gas migrate upward through fractures and spaces in sedimentary rocks. Some hydrocarbons escape at Earth's surface, but others are trapped and collect in sedimentary rocks below the surface to form commercial oil and gas deposits (Figure 7.31).

Global oil reserves will last for several decades at current consumption rates. The United States uses approximately 21 percent

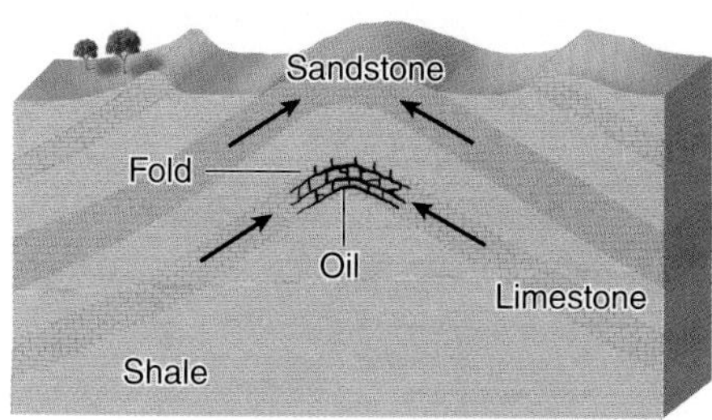

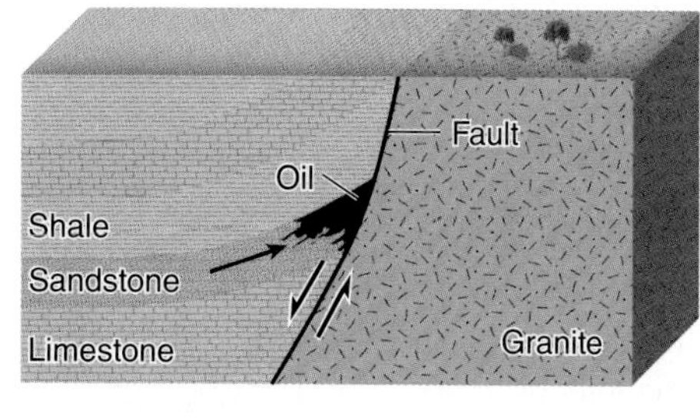

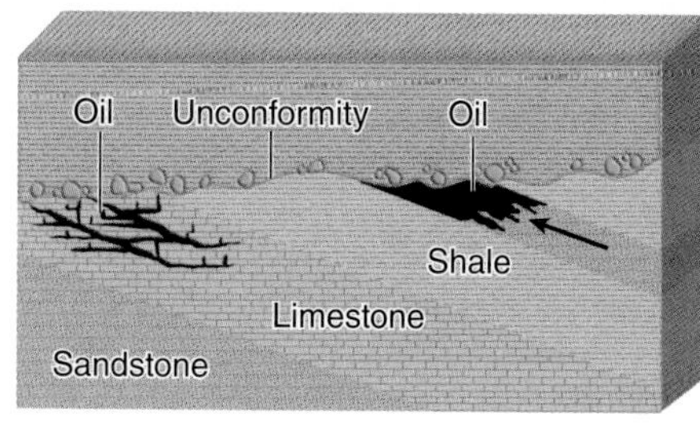

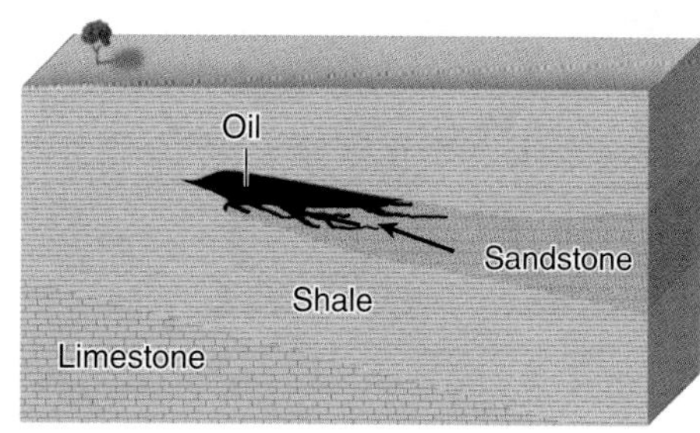

Figure 7.31 Oil deposits. Oil and gas are trapped in rocks below Earth's surface when they are juxtaposed with overlying or adjacent rocks that they cannot pass through. The hydrocarbons migrate (arrows) to the highest part of the reservoir rock until they meet an obstruction that prevents further movement.

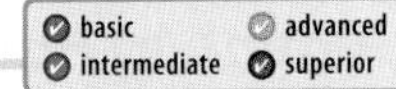

Checkpoint 7.15

basic · intermediate · advanced · superior

Venn Diagram: Chemical and Biochemical Sedimentary Rocks

Use the Venn diagram provided here to compare and contrast chemical and biochemical sedimentary rocks. Identify at least seven characteristics.

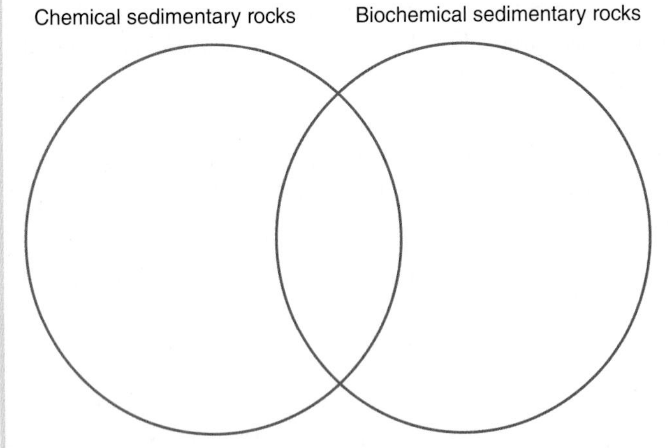

Checkpoint 7.16

basic · intermediate · advanced · superior

Similar organic-rich source rocks are present in rocks below two locations, Oiltown and Dryville. Oil and gas deposits formed in the overlying sedimentary rocks below Oiltown but did not form in rocks at Dryville. Suggest at least four potential explanations for this difference.

of the world's oil, more than any other nation (for comparison, China uses 11 percent). Recent oil discoveries have made the United States the world's third largest producer of oil and reduced imports to less than one-half of the oil we consume. The imports mostly come from other North American nations (Canada, Mexico). However, as more than one-half of the world's oil reserves are located in the Middle East, countries in that region (for example, Saudi Arabia, Kuwait, Iran) may play an increasingly important role in global oil supplies in the decades ahead unless we learn to drive vehicles using energy sources such as ethanol or electricity.

Coal. Coal, the carbon-rich residue of plants, can be classified by rank (carbon content). Coal matures by increasing rank with increasing heat and compaction with progressive burial that drives off water and other chemical constituents. Carbon content increases from low-grade peat (less than 25 percent carbon) to lignite (25 to 35 percent) to subbituminous (35 to 45 percent) and bituminous coal (45 to 86 percent) to anthracite (more than 86 percent). Peat is the least-mature form of coal, containing a large volume of fibrous plant matter. The higher the carbon content, the more heat that is released when the coal is burned. Small amounts of high-carbon coals produce the same heat as larger volumes of low-carbon coal. The volume of ash that remains after burning decreases with increasing rank.

There are three principal coal-producing regions in the United States (Figure 7.32). The first two, Appalachian basin states (Ohio, eastern Kentucky, West Virginia, Pennsylvania) and interior states (Illinois, Indiana, western Kentucky), produce high-rank bituminous coals and anthracite. These coals are produced from both surface and underground mines. Unfortunately, some of the bituminous coals have a high sulfur content and therefore contribute to air pollution. Given the stringent regulations on pollutants, some companies prefer to use lower-grade subbituminous coals from surface mines in Great Plains and Rocky Mountain states (Montana, Wyoming, North Dakota, South Dakota, Colorado).

Over 80 percent of the world's recoverable coal is found in just seven nations. The United States has the greatest reserves, accounting for 25 percent of the world's total, enough to last for more than 200 years at current consumption rates. This suggests that we will have a plentiful supply of electricity into the distant future, but it is of little help as a replacement fuel for refined oil products (gasoline) unless we can assume that automobiles of the future will run on electricity. (Bring on the electric car!) One drawback to this scenario is that burning coal is a major source of carbon dioxide, contributing to the warming of global climates. The continued exploitation of coal may require the widespread adoption of a means of trapping carbon dioxide emissions during power generation (for more on these options, see Chapter 17).

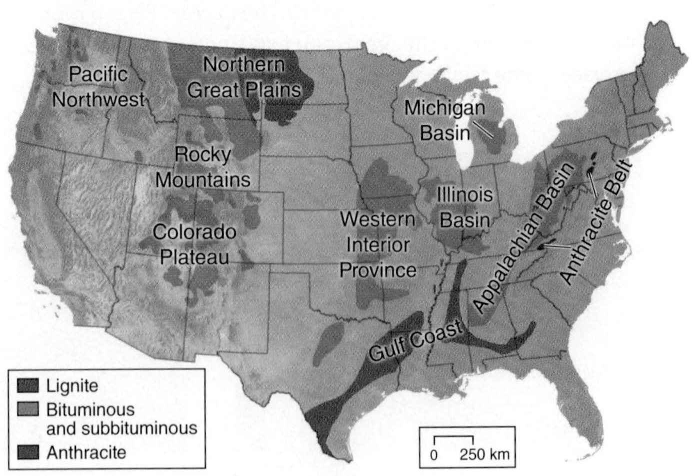

Figure 7.32 Coal-bearing areas of the United States. Eastern coal fields are mainly underground and typically produce bituminous coals. Western coal fields are mostly at the surface and produce subbituminous coal. The eastern coals have a higher carbon content but produce more pollution as a result of their higher sulfur content.

7.6 Metamorphic Rocks

Learning Objectives

- Define the terms *metamorphism* and *foliation.*
- Identify some everyday analogies for the formation of metamorphic rocks.
- Compare and contrast the features of rocks formed by contact metamorphism and regional metamorphism.

The word *metamorphosis* means "change" and was originally used to describe how vegetation changes from a seed to a full-grown plant. In earth science, **metamorphism relates to changes in mineral composition or texture that occur in solid rocks as a result of increasing pressure or temperature.** For example, limestone can become marble when heated near an igneous pluton. Notice that we referred to heating, not melting. If melting occurs, the end result is an igneous rock. Metamorphic rocks require a temperature range that is high enough to trigger the chemical reactions necessary to change minerals but not so high that the minerals will melt. Depending on the combination of pressure, temperature, and the chemical composition of circulating fluids, one rock has the potential to be converted into several different metamorphic rock types. Temperatures of at least 200°C (390°F) are generally needed before some of the minerals in a rock will undergo chemical reactions to cause compositional changes. Depending on their composition, most minerals in rocks near Earth's surface melt at temperatures ranging from 600 to 1,100°C (1,110 to 2,010°F). Consequently, the temperature "window" for metamorphism is from 200 to 1,100°C (390 to 2,010°F), depending on the minerals present in the original rock. Such temperatures are common near magma chambers or at deeper levels in the crust.

There are two main types of metamorphism: contact metamorphism and regional metamorphism. These types of metamorphism occur by different processes and create rocks with different characteristics.

Contact Metamorphism

Contact metamorphism occurs when rocks come in contact with a heat source (usually a magma body). Essentially, the rocks are cooked. A comparable change occurs in roasting meat or baking bread—that is, the composition does not change but the texture of the material does. Rocks do not conduct heat well (they are good insulators), so the zone of contact metamorphism is usually relatively narrow and occurs in the rock immediately surrounding the heat source. Marble is an example of a rock that may be formed by contact metamorphism (Figure 7.33a). Marble forms when limestone is heated to high temperatures. Both marble and

Figure 7.33 Rocks that may be formed by contact metamorphism. **a.** Marble. Inset shows large crystals of calcite viewed through a microscope. Crystals are 2 millimeters (0.08 inch) across. **b.** Quartzite. Inset shows quartz crystals approximately 0.5 millimeter (0.02 inch) in width. These are examples of rocks that may form as a result of contact metamorphism of limestone and sandstone, respectively.

Table 7.6 Metamorphic Rocks Based on Foliation and Texture (Grain Size)

	Grain size		
Foliation	Fine (<0.1 mm)	Medium (~0.1–4 mm)*	Coarse (>2 mm)*
No	Hornfels	Marble, quartzite	Marble, quartzite
Yes	Slate, phyllite	Schist	Gneiss

*Approximate sizes for comparison purposes only.

limestone may have the same composition, but marble typically has larger grains (Table 7.6).

Back to the bricks: The final stage in the brickmaking process requires the addition of heat. The green brick is heated to high temperatures comparable to those of metamorphism as it makes its way through the kiln. The heat is not sufficient to melt the materials in the brick, but it does cause them to fuse to produce a stronger product than the raw brick that went into the kiln. Similarly, metamorphic rocks are often stronger and more resistant to erosion and weathering than their sedimentary cousins.

Regional Metamorphism

Regional metamorphism occurs when rocks undergo increased temperatures and pressures typically associated with the plate tectonic processes that form mountains (see Chapter Snapshot). In these areas, sedimentary rocks may be buried to great depths (10 to 20 kilometers; 6 to 12 miles), and igneous rocks in the crust may end up 50 kilometers (31 miles) below the surface as a result of crustal thickening. Recall from Chapter 2 that the temperature in Earth's crust increases by approximately 25°C per kilometer (72°F per mile) near the surface. Assuming a surface temperature of 15°C (59°F), metamorphism would begin to occur in rocks at depths of approximately 7 to 8 kilometers (4 to 5 miles). The additional pressure causes sheetlike minerals (for example, mica) in the rock to physically rotate or to grow with a preferred alignment. The combination of both higher temperatures and pressures and the presence of fluids can cause some specific types of minerals to grow. Those minerals are aligned parallel to one another (like a stack of papers on a table) and perpendicular to the direction of pressure (like pressing down on the stack of papers). **This geometric orientation generates an alignment of minerals into sheets termed *foliations*** (Figure 7.34).

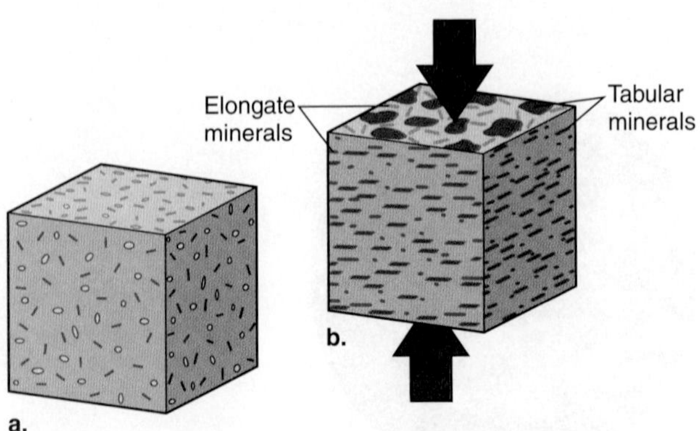

Figure 7.34 Foliation. Tabular or sheetlike minerals grow perpendicular to the direction of pressure to form a foliation. **a.** Randomly oriented minerals prior to metamorphism; **b.** directed pressure causes minerals to form a foliation.

Checkpoint 7.17

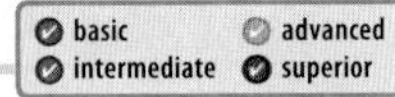

The conversion of bread to toast can be seen as an analog for the formation of a metamorphic rock by

a) Contact metamorphism.
b) Regional metamorphism.

Checkpoint 7.18

Metamorphic Rocks Defining Features Matrix

Identify which of the characteristics in the left-hand column of the following table are present in rocks formed by contact metamorphism, regional metamorphism, or both. Place an "X" in the two right-hand columns of the table where appropriate. Do not place a mark in either column if the characteristic is not present. One characteristic has been completed as an example.

	Rocks formed by	
Characteristic	Contact metamorphism	Regional metamorphism
Form at temperatures above 200°C		
May originally have been an igneous rock	X	X
Form as a result of increasing pressures		
May surround plutonic igneous rocks		
Slate is an example		
Form as a result of melting		
May underlie several adjacent states		
Found in mountain belts		
May originally have been a sedimentary rock		
May contain a foliation		
Marble is a possible example		
Form on Earth's surface		
Limestone is an example		
May have originally been a metamorphic rock		

Checkpoint 7.19

Venn Diagram: Contact and Regional Metamorphic Rocks

Use the Venn diagram provided here to compare and contrast metamorphic rocks formed by contact and regional metamorphism. Add at least eight items to the diagram.

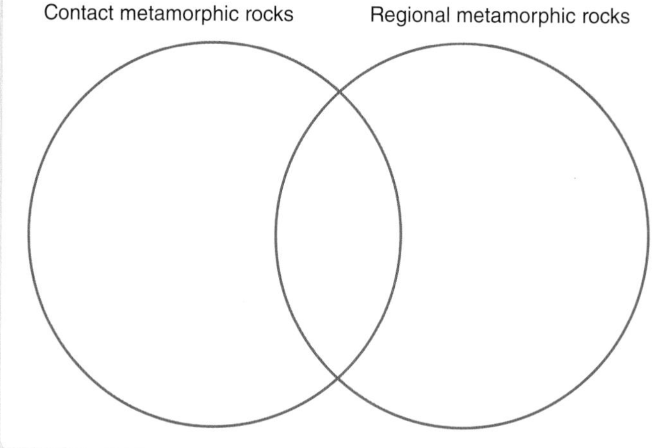

The higher the temperatures and pressures, the more intense the metamorphism. Rocks formed by regional metamorphism are classified by the size of minerals that form (a temperature- and pressure-dependent process) when they change

Checkpoint 7.20

Make a concept map that summarizes the characteristics of metamorphic rocks.

from the original rock to the metamorphic rock (see Table 7.6). Foliated metamorphic rocks, in order of increasing metamorphic grade (low to high temperature) and grain size (small to large), are slate, phyllite, schist, and gneiss. The grain size in slate is much too small to be seen with the naked eye, and the presence of millions of tiny crystals of sheetlike minerals means that slate splits easily into thin sheets (Figure 7.35a). If you had taken this class a century ago, your teacher would have been using a slate chalkboard. Mica, a common mineral in metamorphic rocks, can give some rocks (for example, phyllite, schist; Figure 7.35b) a shiny, silky appearance when present in larger grains. Gneiss (pronounced "nice") has the largest grain size, with quartz and feldspar often forming much of the rock. Light- and dark-colored minerals typically separate into darker and lighter bands in the rock (Figure 7.35c).

Not all rocks that undergo regional metamorphism form a foliation. For example, rocks that lack sheetlike minerals do not generate the parallel alignment of minerals necessary to create a foliation. Sandstone, composed mainly of quartz grains, is converted to quartzite by regional metamorphism; limestone forms marble. Neither metamorphic rock contains foliations (Figure 7.33).

Figure 7.35 Foliated metamorphic rocks. **a.** Foliations in an outcrop of slate in Antarctica. The foliations are oriented from top right to bottom left. **b.** Phyllite (top) and schist (bottom) both derive their shiny appearance from abundant mica. Schist has a larger grain size. **c.** Gneiss (bottom) formed from the metamorphism of granite with a similar composition (top).

7.7 The Rock Cycle and Mineral Resources

Learning Objectives

- Identify some common analogies for the formation of igneous, sedimentary, and metamorphic rocks.
- Describe how an element might move through the rock cycle, explaining the changes in conditions that it would experience.
- Explain how the rock cycle can be interpreted as an example of Earth operating as a constantly changing system.
- Identify at least four different processes that form mineral deposits, and give an example of one metal or commondity formed by each process.

As a society, we have recognized the need to protect our natural resources and have created extensive recycling programs. We collect papers, cans, and bottles, and we convert them to other products used in our daily lives. The materials of Earth are also "recycled" through a process known as the *rock cycle* (Figure 7.36). The rock cycle links together different materials and processes on Earth to form the components of the three principal groups of rocks.

Mineral resources become concentrated in Earth's crust as a result of specific geologic processes associated with the formation of rocks. Many of these resources are associated with a group of minerals called nonsilicates that form another important class of earth materials. Nonsilicate minerals do not have the silica-oxygen tetrahedron in their chemical structure. One subcategory of nonsilicates, called oxides, includes minerals with oxygen in

Checkpoint 7.21

Cooking an egg could be seen as an analog for the formation of

a) Igneous rock.
b) Metamorphic rock.
c) Sedimentary rock.

Concrete is formed by adding cement and water to a mixture of sand and gravel. This could be seen as an analog for the formation of what type of sedimentary rock?

a) Clastic
b) Chemical
c) Biochemical

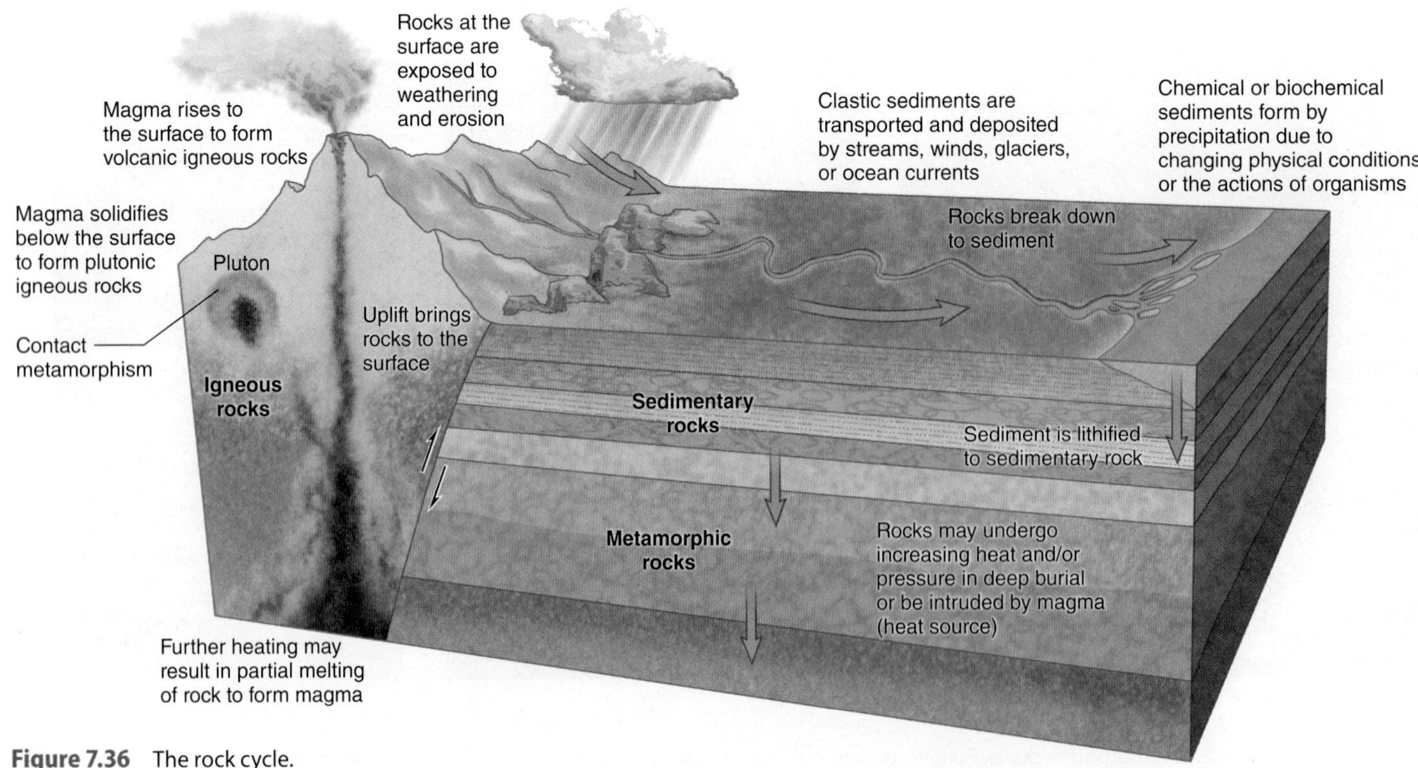

Figure 7.36 The rock cycle.

their chemical makeup. Hematite (Fe_2O_3) is an important example that is mined to produce iron. Sulfides include materials bound with sulfur such as pyrite, or fool's gold. Carbonates include CO_3 in their chemical structure and form materials such as limestone used in the construction industry. Gypsum (see earlier mention in Chemical Sedimentary Rocks) is a sulfate that has SO_4 in its chemical structure. An important halide, or material with chlorine or fluorine, is halite, or common salt. Last, phosphate-based nonsilicate minerals are commonly used in agricultural products such as fertilizer. Exploration for minerals requires that geologists recognize the telltale evidence that signals the presence of useful mineral deposits. Geologic processes that result in the concentration of mineral resources can be divided into those associated with chemical reactions resulting from changing temperatures and movement of fluids through rocks and those formed by the physical rearrangement of earth materials during erosion, transport, and deposition. The raw materials may be formed during one part of the rock cycle but concentrated during a later stage. Consequently, it becomes important to know not only how rocks form but also what might happen to them in the millions of years that follow.

Checkpoint 7.22

Rock Cycle Diagram

The following diagram illustrates some of the interactions of the rock cycle. Match the lettered responses to the blank ovals on the diagram. (*Note:* Some letters are used more than once.) Example: If you believe that metamorphic rock is converted to magma by cementation and compaction, enter "a" in the top left oval.

a) Cementation and compaction (lithification)
b) Heat and pressure
c) Weathering, transportation, deposition
d) Cooling and solidification
e) Melting

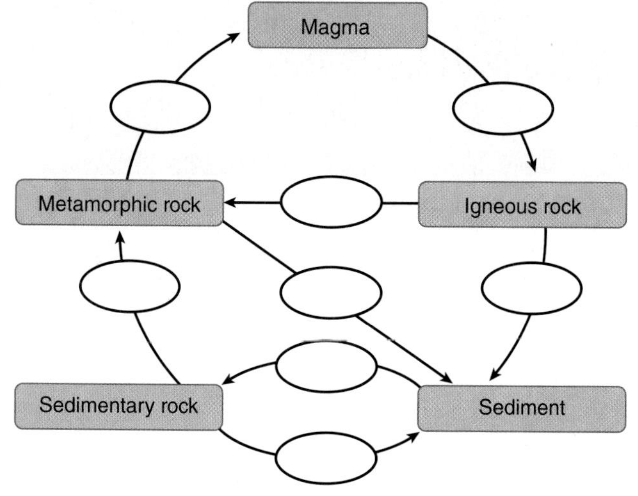

The Rock Cycle

Under the right conditions, any type of rock can become any other type of rock. Such changes involve a variety of interactions among the different components of the earth system, and they show that Earth is constantly changing. They also explain why geologists have found only a few rocks on Earth's surface that were formed close to the time the planet was created.

During their lifetime, minerals and elements in Earth's crust can be recycled through several different rocks. For example, follow the possible fate of a grain of quartz sand on a beach: As the sea level rises, that beach may become flooded, and the sand may be incorporated into a sedimentary rock such as sandstone. Over time, the forces of plate tectonics may cause the sandstone to be buried deep in the crust during a period of mountain building. The higher temperatures and pressures common in regional metamorphism may convert the sandstone to quartzite. Magma rising from the mantle may melt the adjacent quartzite, assimilating the quartz grain into the magma. The elements of the quartz grain (silicon, oxygen) may recombine with other elements to form new minerals as the magma solidifies to form granite within the crust. Erosion eventually exposes the rock, breaking it down to its constituent minerals and rock fragments (sediment), which are carried by streams to the coast, where the grains are deposited on a beach. And the cycle begins again.

Likewise, an igneous rock can remelt to form a new igneous rock, a sedimentary rock can re-erode and become another sedimentary rock, and a metamorphic rock can be heated and transformed over and over. Each cycle may last millions or even billions of years. Keep in mind that we are not adding new material to Earth, so any rocks forming today must be recycling elements from rocks formed at some time earlier in the planet's history.

Mineral Resources

The term *mineral resource* refers to nonfood, nonfuel resources such as metals (for example, aluminum, palladium) and industrial minerals (for example, gypsum, phosphate). The development of mineral resources depends on more than just the presence of a mineral deposit. The average concentration of minerals in the crust is insufficient to form an economically valuable mineral deposit. The concentration factor (CF) is the increase in the concentration of a mineral required to generate an economic deposit. For example,

Checkpoint 7.23

Use information at the Minerals Education Coalition (www.mineralseducationcoalition.org/minerals) or the USGS Minerals Yearbook (http://minerals.usgs.gov/minerals/pubs/commodity/myb/) to make a list of 10 different minerals that are used to manufacture objects that you would use everyday. Try to find at least three minerals that you have not heard of before.

copper makes up 55 parts per million (ppm, 0.0055 percent) of Earth's crust. Copper ores from the Bingham Canyon mine, Utah, are composed of 0.6 percent (6,000 ppm) copper, equivalent to a concentration factor of 109 (0.6/0.0055). Relatively rare minerals have large CF values (for example, gold, CF > 2,000), whereas more common elements have low CF values (for example, silicon, CF = 2).

High-Temperature Reactions. The high temperatures and abundant fluids associated with the formation of igneous and metamorphic processes can result in the concentration of a variety of metallic mineral resources. **A rock containing economic concentrations of metallic minerals is known as *ore*.** Many ores are formed as a result of igneous or metamorphic processes associated with plate tectonics. Consequently, metallic deposits are often concentrated near current or former plate boundaries. Hydrothermal and magmatic processes are some of the more common ways to form economic deposits of minerals.

Magma and water are the key ingredients in the generation of *hydrothermal mineral deposits* (Figure 7.37). One location where water is readily available is the ocean floor. Magma rises along the oceanic ridges. Cold bottom waters from

Checkpoint 7.24

basic · intermediate · advanced · superior

Element X and the Rock Cycle

The following graph illustrates the partial life cycle of a sample of an element (X) over several million years. At some points during its life, the sample experienced temperatures similar to those found at Earth's surface (A, G), while at other times, it experienced high temperatures of approximately 1,000 and 800°C, respectively (C, E).

Using what you know about rocks and how they form, write the life history of element X with special reference to what happened at and between times A–G on the graph. Assume that the element passed through several rock types during its life. To get you started, we provide the first few steps:

A	Element X is weathered from a rock at Earth's surface and is carried in solution by a stream.
A–B	Element X combines with another element to form a mineral in a pile of sediment and becomes part of a chemical sedimentary rock.
B	Temperature is sufficient to convert the sedimentary rock to a metamorphic rock.

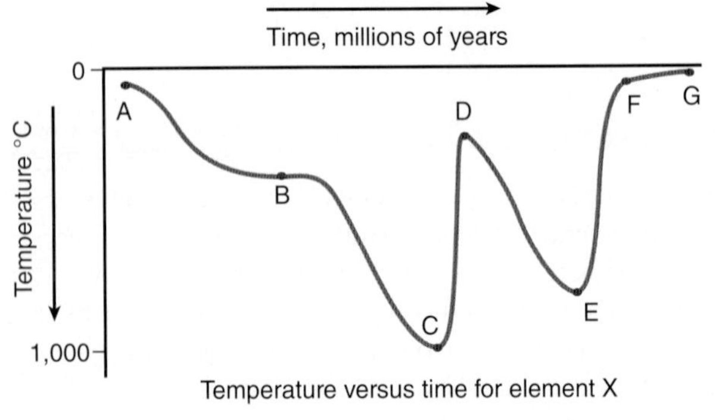

Temperature versus time for element X

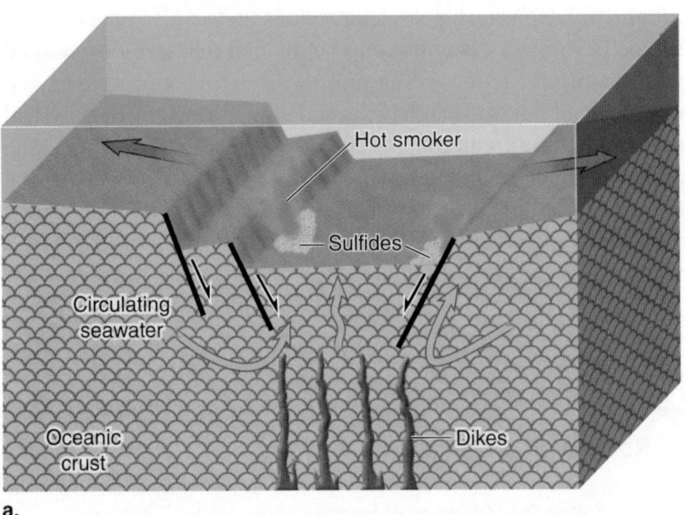

a.

b.

Figure 7.37 Geologic setting for the formation of hydrothermal mineral deposits along oceanic ridges. Cold seawater enters fractures formed where plates move apart in the rift at the crest of the oceanic ridge. **a.** Seawater reacts with hot rocks below the surface and leaches out metals. The metals are precipitated as sulfide deposits when the circulating seawater encounters the cold water. **b.** The plumes of precipitating metals are known as "hot smokers" as they resemble smoke billowing out of a chimney.

the ocean floor can infiltrate the oceanic crust through networks of narrow fractures. The waters are heated to high temperatures (300 to 400°C; 570 to 750°F) by the rising magma, and chemical reactions occur that leach out metals from the surrounding rocks. Fluids transport these mixtures of dissolved metals (manganese, iron, copper, zinc) to the ocean floor where they meet much colder waters. The sudden decrease in temperature causes the minerals to precipitate from solution to form deposits known as massive sulfides that are incorporated into sediments deposited along the ocean ridge system.

Hydrothermal deposits also form on land when metal-rich fluids are expelled from magma chambers to form veins that fill fractures in the surrounding rock. Quartz-rich fluids discharged from magma chambers often form associations with gold deposits. Alternatively, economically valuable mineral deposits can become concentrated when heated groundwater reacts with rocks below volcanoes or when metamorphic processes result in reactions that cause minerals to recrystallize into other varieties.

The cooling processes inside magma chambers may cause minerals to be concentrated in narrow zones within igneous rocks. These *magmatic mineral deposits* (for example, platinum-group minerals, chromium) may form at high temperatures before the magma solidifies. Once those minerals crystallize, they are heavier than the surrounding magma, so they sink to the bottom of the magma chamber by a process called *crystal settling*, forming an enriched layer (Figure 7.38). During the final phase of cooling and solidification of the magma, the residual melt mixture may contain relatively rare elements, such as lithium and beryllium. These enriched magmatic fluids may cool very slowly as they form veins in narrow fracture zones around the magma chamber. The extra-slow cooling may yield a rock known as a pegmatite that contains giant crystals of individual minerals that can be up to several meters across.

Low-Temperature Reactions. The largest US iron deposits represent ancient (more than 2 billion years old) chemical sedimentary rocks that were precipitated from seawater when the chemistry of Earth's oceans was considerably different from today. The greatest concentration of these deposits is in the Upper Peninsula of Michigan and northern Minnesota. The iron must be concentrated by additional geologic processes before these deposits can be considered an ore. Two potential concentration methods are:

- The leaching of silica from the rocks (leaving iron behind in greater relative concentrations) by weathering processes that occur on Earth's surface.
- Metamorphism that causes an increase in grain size and the formation of new iron minerals that are easier to separate from the rock.

Water flowing through rocks on or near the land surface may remove soluble minerals to leave behind sufficient concentrations of economic minerals to form an ore. These ***residual mineral deposits* are formed most rapidly in areas with rapid weathering rates** such as the tropics. Iron- and aluminum-rich laterite forms as a result of leaching of minerals from thick soils in tropical regions. The world's principal source of aluminum ore is from a form of laterite known as bauxite. Most of the world's bauxite is mined in Australia, Guinea, and Jamaica.

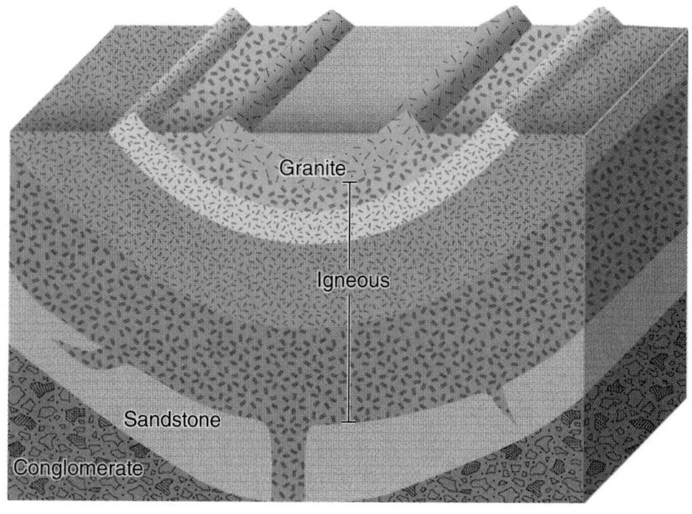

a.

b.

Figure 7.38 Layering of different minerals in an igneous pluton. **a.** This image is based on the character of the Bushveld complex in South Africa. Heavy, early-formed minerals sank to the bottom of the magma chamber (crystal settling), resulting in a layer containing crystals rich in platinum and chromium. **b.** The metal-rich layer is less than 1 meter (3 feet) thick but holds the world's greatest reserves of platinum-group metals.

Many sediments and rocks are mined for nonmetallic industrial minerals for a variety of uses. Although they are relatively inexpensive in comparison to metals, industrial minerals are commonly mined everywhere people live, and the sheer volume of materials produced makes them the most valuable mineral resources in the United States. Some common industrial minerals are chemical and biochemical sedimentary rocks formed by low-temperature reactions close to Earth's surface. For example:

- Limestone used for lime (steel production), cement, and crushed stone (construction). Limestone makes up two-thirds of all crushed stone, the most valuable nonfuel mineral resource in the United States, worth $14 billion annually.
- Gypsum is used in making wallboard, plaster of Paris, and cement.
- Salt deposits are mined for table salt and for use in water softeners, animal feed, and controlling ice on roadways.
- Phosphate rock is used in the manufacture of fertilizers.

Physical Rearrangement. **Placer deposits are found in present-day streams or in ancient stream deposits** preserved in rocks. Placer deposits represent a natural recycling of older mineral deposits. Minerals that are weathered out of veins may be carried downslope by streams. Stream flow serves to sort and concentrate the minerals. Metal-rich minerals are heavier than the rest of the material carried by the stream. Consequently, when flow velocity decreases, the heavy minerals are among the first materials to be deposited. Suitable sites for deposition are the insides of stream bends or at the stream mouth. Placer deposits can be mined relatively easily and have often held the imagination of would-be millionaires seeking to strike it rich quickly. The gold rushes of the West almost always began with thousands of hopeful miners descending upon streams with gold pans in search of placer deposits. Some common industrial minerals are clastic sediments or sedimentary rocks that are especially prized when they have been well sorted to create a deposit with consistent physical properties. Key deposits that are mined in almost every state include:

- Sand and gravel used for construction and produced by more than 6,000 operations nationwide.
- Sandstone for building stone (construction), as a source for silicon (computer chips), and for glassmaking.
- Clay used to make bricks and ceramics and in the manufacture of glossy paper, toothpaste, and antacid medications.

the big picture

Science presumes that the world is understandable and that scientific findings are consistent. That is, although science operates within some degree of uncertainty, we know that big ideas, such as the rock cycle, can explain the universal and consistent processes that reshape and recycle Earth's materials. The rock cycle applies not only on Earth but also throughout the solar system. In 2004, NASA landed two mobile, roving spacecraft on opposite sides of Mars to examine the planet's geology. Why travel 161,000,000 kilometers (100,000,000 miles) to look at rocks? Besides the adventure and the thrill of discovery, scientists knew that the best way to learn about Mars was to understand its most basic components—the rocks that make up the planet (see chapter-opening photograph). The rocks, and the minerals they contain, provided clues to the planet's origin and the processes that shaped its surface.

The space rovers *Spirit* and *Opportunity*, each about the size of a golf cart, were loaded with scientific instruments and traveled many kilometers from their landing sites. The landing sites were carefully chosen for having the best chance of showing evidence of former water-rich environments. *Opportunity* landed just a hop, skip, and jump from a light-colored outcrop in the wall of a crater that appeared to show layers of bedrock. On closer examination of the rocks, scientists directing the rover documented four observations in support of the hypothesis that these were sedimentary rocks that had been exposed to water:

1. *Opportunity*'s pictures of the surrounding terrain showed that the ground was covered with tiny spherical pebbles, dubbed "blueberries." Back on Earth, geologists in Utah couldn't believe their eyes. They had collected similar specimens from rocks at sites across the state. The "blueberries" are a type of rock formation known as a *concretion* that forms when minerals precipitate from groundwater as a result of reactions with the surrounding bedrock.

2. Close-up pictures of the outcrop revealed small, dime-sized cavities, similar to those found on Earth where salt crystals are dissolved out of the rock. Salt crystals form by precipitation as seawater evaporates. Later exposure to freshwater can cause the salt crystals to dissolve away, leaving only the cavities behind.

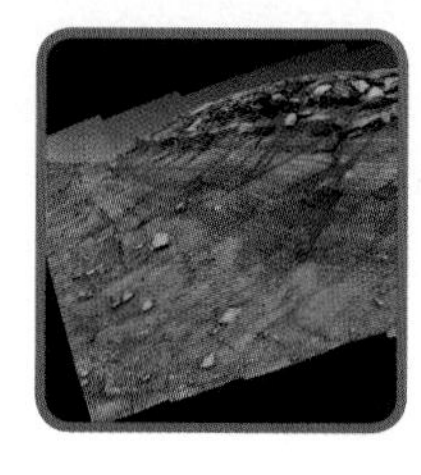

3. *Opportunity*'s instruments analyzed the chemical signatures of the rocks and found concentrations of chemicals (for example, sulfates) that characteristically form in the presence of water.

4. Finally, the layered bedrock contained finely rippled surfaces that are characteristic of wind action or running water on Earth.

We have a very modest collection of data on Martian geology in comparison with what we know about rocks on Earth. With such a small number of samples, scientists must be careful not to jump to conclusions. One outcrop may represent a small local geologic environment, or it may be representative of a much larger region. Additional observations are being collected by the more sophisticated *Curiosity* rover that landed on Mars in 2012. Scientists will hope to use the drills and probes of *Curiosity* to reveal more clues about the geologic history of Mars on the basis of our understanding of the rock cycle on Earth.

Rocks and Minerals: Concept Map

Complete the following concept map to evaluate your understanding of the interactions between the earth system and rocks and minerals. Label as many interactions as you can, using information from this chapter.

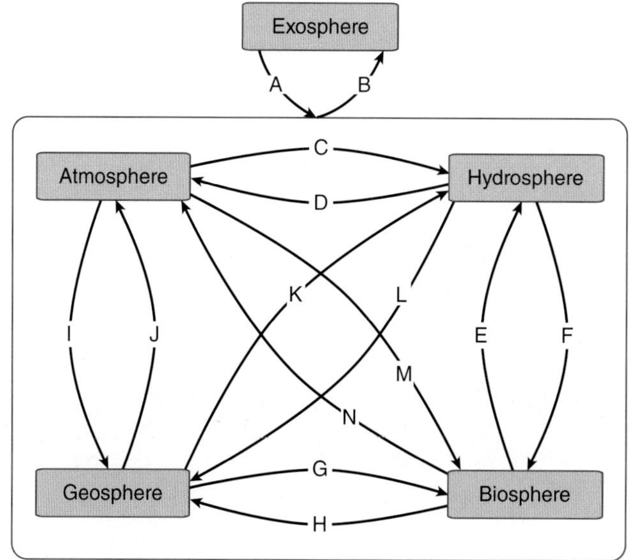

"All things in nature work silently. They come into being and possess nothing. They fulfill their function and make no claim. All things alike do their work, and then we see them subside. When they have reached their bloom, each returns to its origin. . . . This reversion is an eternal law. To know that law is wisdom."
—Lao-tsu
"Time, which measures everything in our idea, and is often deficient to our schemes, is to nature endless and as nothing."
—James Hutton

Chapter 8

Geologic Time

Chapter Outline

the big picture

Part of the Atacama Desert, Chile. How old is this landscape?

See The Big Picture box at the end of this chapter for the full story on this image.

8.1 Thinking About Time

Chapter Learning Outcomes

- Students will explain how and why Earth's materials provide a record of its history.
- Students will make connections between ancient environments and fossils.
- Students will describe the age of Earth and significant geologic events that shaped Earth in the past.

What does time look like? British artist and photographer Andy Goldsworthy represents time in ephemeral sculptures composed of natural materials such as snow, mud, leaves, and branches. Goldsworthy often uses his bare hands to shape these materials into short-lived sculptures that stand out against their surroundings. Giant snowballs melting in midsummer on a London street or a long line of green leaves stitched together and floating slowly downstream are transitory pieces of art destined to disappear back into nature (Figure 8.1a–c). Goldsworthy's longest-lasting sculptures are composed of rocks. Rocks also form the building blocks of the oldest preserved human structures, including the 4,800-year-old Egyptian pyramids and some sections of the Great Wall of China that were constructed 2,700 years ago (Figure 8.1b, c).

The human perspective of time ranges from the minutes-hours-days represented in much of Goldsworthy's work to the centuries-millennia of our most ancient structures. But to truly understand geologic time, we need to jump from the human time frame to what is known as "deep time," as represented by the geologic record. Deep time encompasses the history of Earth and is measured in intervals of millions to billions of years. Modern scientific methods have revealed that Earth is 4.6 billion (4,600 million) years old. In a geologic record that spans billions of years, our commonly used human dimensions of time are essentially indistinguishable. It is ironic that we have to reprogram ourselves to think in the timescales of Earth history, because our conception of time is shaped by Earth's own rhythms—our days determined by Earth's rotation on its axis, our years by its orbit around the sun. But what about longer time intervals? Think about what you have read in previous chapters; what other rhythms of the planet extend beyond human experience?

In this chapter, we will investigate the vast intervals of time in which geologists work. The geologic record visible in rocks is a portrait of a region that can be interpreted to tell much about the events that occurred there in the past. It can even tell us what organisms inhabited the region. It's just a matter of knowing how to interpret the clues. First, in a section on relative time, we will explain how the earliest geologists were able to use the scientific method to unravel the sequence of events that gave us our present Earth. A section describing the geologic timescale then reviews the history of Earth, especially the major changes in the biosphere over the last half-billion years. Next, we will explore the concept of numerical time as we examine how scientists discovered ways to date events and determine the age of Earth. The final section of the chapter considers how our knowledge of geologic rates of change has implications for life on Earth today.

Self-Reflection Survey: Section 8.1

Answer the following questions as a means of uncovering what you already know about geologic time.

1. How would you create a piece of art to illustrate the passage of time?
2. List the five most significant things that happened in the world in the last 50 years. Place them in chronological order, but don't worry about giving the exact dates when they occurred. Compare your list with those of some other students in class and build a single, cumulative list that includes all the events.
3. Using what you have learned so far, make a list of up to 10 geologic events or phenomena and place these items in the correct chronological order, from earliest to most recent. Include the dates of these events where possible. Here are two to get you started: a) Formation of Pangaea; b) Start of the uplift of the Himalaya Mountains.

Figure 8.1 Ephemeral and long-lived structures. **a.** A giant snowball by sculptor Andy Goldsworthy that was part of an installation of multiple snowballs that were placed around London in midsummer, 2000. Such short-lived creations quickly return to nature and disappear. Long-lasting rocks, on the other hand, are used in some of the oldest structures made by humans such as **b.** the oldest Egyptian pyramids (2650–2150 B.C.), and, **c.** the Great Wall of China (688 B.C.)

a.

b.

8.2 The History of (Relative) Time

Learning Objectives

- Explain how scientists apply specific principles to place the different rock units in the correct chronological order.
- Predict the relative ages of rock units and identify unconformities in photographs or diagrams of geologic features.
- Create and label a diagram that accurately illustrates the relative positions of a series of rock units formed by a reasonable sequence of geologic events.
- Describe how scientists can use fossils to interpret geologic history and match rock units found in different regions.

In seventeenth-century Europe, there were few widely read books, and none had more authority than the Bible. Science as we know it today was in its infancy, and the discipline of geology did not exist. In a time of plagues, famine, and natural disasters, the Bible was a source of comfort, and religious men and women were among the most learned people in any community. It is little wonder that the Bible served as the equivalent of a modern reference book and that everyone accepted its creation story.

In 1650, James Ussher completed a feat of considerable scholarship when he cross-referenced the ages of key historical figures with events in Middle Eastern cultures to estimate a date for the creation of Earth. Ussher, a gifted scholar and linguist who entered university at the age of 13, was an archbishop in the Church of Ireland (equivalent to the Episcopal Church in the United States). He determined that "the beginning of time, according to our Chronologie, fell upon the entrance of the night preceding the twenty third day of Octob[er] in the year . . . before Christ 4004." Or to put it in modern English, Ussher suggested that Earth was created on the evening of October 22, a little over 6,000 years from the end of the twentieth century. Similar chronologies had been produced by others, but Ussher's became the standard when his dates were included in the widely circulated King James version of the Bible. One consequence of this chronology was the establishment of dates for key biblical events, such as Noah's flood, which was determined to have occurred around 2,500 B.C.

Although Ussher's chronology seemed appropriate in the 1600s, it is no longer valid in the scientific framework in which we place Earth's history today. Our modern interpretation—that Earth is very old—represents another example of a significant paradigm shift in earth science. Almost all of the world's major religious faiths see no contradiction between their beliefs and the application of scientific principles to the history of Earth. However, this is one topic in earth science that is still greeted with skepticism, if not outright hostility, by a significant percentage of the adult population of the United States. Some faiths interpret the Bible literally and argue that our planet is a few thousand years old. Such literal explanations require that geologic data be interpreted to fit a biblical chronology. This reasoning is contrary to the hypothesis-testing method by which science progresses. Most scientists have no interest in attempting to disprove the existence of a Creator based on the age of Earth. Science seeks only to understand the natural world. One way geologists do this is by studying natural events in terms of relative time.

Checkpoint 8.1

basic | intermediate | advanced | superior

Place the following events that were described in the earlier chapters of the book in the correct relative chronological order, from earliest to most recent.

a) A tsunami struck Japan (1700).
b) An ice sheet was present in India.
c) An asteroid collided with the Yucatan Peninsula.
d) Mount Pinatubo erupted in the Philippines.
e) Wegener developed his continental drift theory.

c.

Relative Time

The study of **relative time** involves placing events in the order in which they occurred. In the context of geology, it means describing the order or sequence of geologic events. To describe the methods and philosophies geologists have used, we will use as our model the Grand Canyon in Arizona.

The Grand Canyon: Investigating Its Origin. The Grand Canyon is 15 to 30 kilometers (9 to 19 miles) across, and stair-steps downward over 1,500 meters (4,920 feet) from the canyon rim to the Colorado River below (Figure 8.2). Native Americans had been exploring the canyons of the West for many centuries, but the first person to make a detailed record of an expedition down the Colorado River and through the Grand Canyon was John Wesley Powell (Figure 8.3a). Powell was a truly exceptional man. He lost an arm in the battle of Shiloh during the Civil War, but thought nothing of leading a small team of men in three wooden boats down an unknown series of canyons in the most remote section of the West (Figure 8.3b). He actually did it twice, in 1869 and 1871, and made several stops along the way to document the landscape. Despite numerous hazards, the destruction of two of the boats, three of his men walking away from the expedition, and widespread reports of his death, Powell survived and was hailed as a national hero.

From observations made on his journeys, Powell produced publications that interpreted the geology of the Grand Canyon and the surrounding region. Powell's story, *Exploration of the Colorado River of the West,* was part adventure yarn, part scientific report, and a national best seller in its day. In his book, Powell describes one of the Native American beliefs that attributed the origin of the canyon to a native god, Tarvwoats, who carved the canyon to serve as a trail between the land of the living and the land of the dead. Tarvwoats later filled the canyon with a river to cover his trail. In contrast, Powell interpreted the rocks in the canyon walls to be the result of long-lived geological processes that he believed had been operating "through ages too long for man to compute." This is the accepted scientific view of the history of the Grand Canyon and the one that we will present data to support in this chapter.

In a recent book called *Grand Canyon: A Different View,* author Tom Vail offers a theology-based origin for the canyon. Vail's book suggests that the canyon's rock layers were deposited following Noah's flood, as recounted in the Bible, and are consequently only a few thousand years old. He went on to suggest that the canyon was formed in a few days by a rapidly eroding river that quickly carved through the relatively soft deposits left behind as floodwaters receded. Vail believes his interpretations are as valid as those of Powell and the many geologists who have subsequently studied the canyon. Vail has stated, "What they call science is theory just as what is in my book is theory. All the scientists here have as much a right to their opinion as anyone else."

A reader might consider these different views and then wonder: *Who is right?* It is important to recognize that these contrasting ideas about the canyon's origins depend on the conceptual

Figure 8.2 The Grand Canyon. The rock layers exposed in the Grand Canyon represent intervals of time stretching back thousands of millions of years. The canyon is up to 30 kilometers (19 miles) across in the well-layered sedimentary rocks but is much narrower in the metamorphic and igneous rocks of the inner gorge (center of image).

Figure 8.3 John Wesley Powell and his Colorado River expedition. **a.** A 35-year-old Powell just before his first trip down the Colorado River. **b.** An illustration of the second expedition running a rapid in the Grand Canyon.

framework people bring to the debate. Perhaps someone you know would agree with Tom Vail's interpretation of the Grand Canyon. Perhaps there are Native Americans or others who believe in the Tarvwoats version. However, whatever one believes to be true does not make that belief *scientifically* valid. Vail's perspective differs from that of scientists. Following Vail's logic, all ideas about the formation of the canyon would be equally valid, including the Tarvwoats interpretation. In contrast, the **scientific method argues that some ideas are more-valid explanations for geologic phenomena than others and, more important, that those ideas can be tested over and over with similar results.** Science can also be used to predict future events or processes with some accuracy. The claim that the Grand Canyon originated from a flood cannot be considered valid within a scientific framework because it cannot withstand the scrutiny of a disciplined scientific process.

What can we learn from present-day natural processes that will help us unravel the Grand Canyon's history? How does the theology-based story for the age of the rocks compare to the scientific explanation? How would you begin to figure out the age of rocks in a feature such as the Grand Canyon? Before we examine the geology of the canyon, we must take a step back to learn more about the clues that help us determine the relative ages of rocks.

Steno's Geologic Principles. Nicholas Steno, a Danish physician, began the process of determining the age of Earth from a scientific standpoint in the late seventeenth century, just a few years after Archbishop Ussher published his chronology. Steno was studying anatomy in the Italian province of Tuscany when, during the dissection of the head of a shark, he noted that the shark's teeth resembled objects found in local rocks. The common view of that time was that such objects grew naturally in the rocks and that it was just a coincidence that they resembled shark teeth. But Steno believed that the objects actually *were* preserved shark teeth. Although he was not the first person to make this interpretation, he was the first person to take it further.

Steno began thinking about how shark teeth, or any other object, could end up in rocks on a Tuscan hillside. He hypothesized that such objects were preserved when they sank to the bottom of a body of water and were covered up by sediment. He reasoned that if rock layers were deposited in water, they must have originally been horizontal. This observation—that **sedimentary rocks are deposited in nearly horizontal layers**—became known as the **principle of original horizontality** (Figure 8.4a). A consequence of this principle is that sedimentary rocks that are no longer horizontal must have undergone an episode of deformation after their formation (Figure 8.4b). For example, sedimentary layers are often tilted during the uplift associated with mountain formation (see Chapter 6). The principle of original horizontality is the geologists' first rule for reconstructing Earth's timescale.

Sometimes papers pile up on your desk for several weeks before you have time to put them away. The oldest papers are at the bottom of the pile, while the most recent additions are near the top. The same principle holds true for sedimentary rocks. The lowermost objects (layers) placed down first are on the bottom of the pile. This simple idea became Steno's second contribution to determining the sequence of geologic events on Earth. His **principle of superposition** states that, in a series of sedimentary rock layers that are not deformed, **the rocks at the bottom of the stack are the oldest and the rocks at the top are the youngest.** The same principle can be extended to many forms of volcanic igneous rocks. Volcanic eruptions can produce layers from lava flows or when ash and other debris fall back to Earth.

Steno applied his rules to some of the rocks in Tuscany. Prior to this time, most authorities saw Earth as a finished product that had once been covered in water by the biblical flood. Steno's work changed that perspective by using observations to create hypotheses that explained the landscape. His interpretations linked the origin of the layers to events that occurred before, during, and after the biblical flood. The fact that he based his rules on observations was critical to advancing science. In doing so, Steno's work over 300 years ago attempted to fit the components of our modern-day scientific method into a sequence of events constrained by the theological time frame of creation. Steno saw no contradiction between the formation of these rocks and the biblical timescale. But a century later, James Hutton introduced an alternative view.

Hutton's Geologic Principles. In the eighteenth century, the Scottish scientist James Hutton essentially invented our modern concept of geologic time. As was common for wealthy men in those days, Hutton was a gentleman farmer and dabbled in many branches of science. He noted that the landscape of his farmlands remained unchanged with the passage of time, even as rock was converted to sediment and sediment was washed from his fields into nearby streams. From this modest observation, he deduced two significant hypotheses. First, he suggested that the same slow-acting geologic processes that operate today must have operated in the past; therefore, **it must take a very long time for those processes (for example, weathering, erosion) to produce any significant change in the shape of Earth's surface.** This view of earth processes is now known as the *uniformitarian model,* and we will discuss it in greater depth in Section 8.5.

Second, Hutton noted that **all land should be worn flat unless some process acts to renew the landscape by forming new mountains,** which then renews the slow cycle of destruction. Hutton set out to find some examples of rocks that illustrated this idea of cyclical change. He found his evidence at an outcrop of rocks along the east coast of Scotland at a location known as Siccar Point. Here, he discovered nearly flat rocks overlying nearly vertical

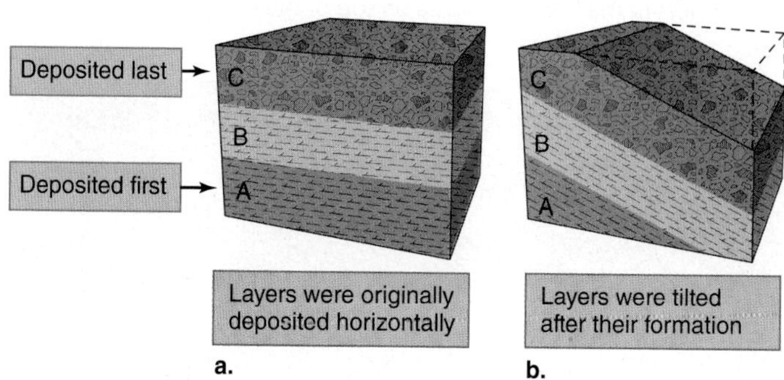

Figure 8.4 Steno's principles. Interpretations based on the principle of original horizontality are in blue; those based on the principle of superposition are in green. **a.** Layers A, B, and C must have been initially deposited as horizontal layers. A, as the lowest layer, must have been deposited first, and C was deposited last. **b.** Layers A, B, and C were tilted (deformed). Because they are parallel, it is likely that they were all deformed at the same time.

layers (Figure 8.5). He hypothesized that the lower layers of rock formed first (superposition) as flat layers (original horizontality) and were later pushed up to the surface and tilted. These layers were then worn down by erosion and buried beneath new horizontal deposits of sediment (Figure 8.6). When such events occur in the geologic record, the features they form are called *unconformities*. **Unconformities are evidence of breaks in the geologic record representing a gap in time.** Hutton thus had found an outcrop of rocks that illustrated his idea of cyclical processes.

Furthermore, in Hutton's example, the upper set of layers cuts across the lower layers and illustrates a third rule: the **principle of cross-cutting relationships.** This principle **states that older rocks may be cut by younger rocks or other geologic features** (Figure 8.6b). For example, when magma intrudes into a rock, it must be younger than the rock it cuts across (Figure 8.7), or when a fault cuts across rock layers, it must be younger than the layers themselves (Figure 8.7).

A variation on this theme is that younger rock units sometimes incorporate pieces of older rocks. For example, chunks of surrounding rock may collapse into a mass of magma as the molten rock forces its way upward through the crust. Often these chunks melt and become part of the magma, but sometimes the magma cools and solidifies to form an igneous rock before the other rock melts. These preserved chunks of rock are known as inclusions. **The principle of inclusions can be applied to identify older pieces of rock surrounded by younger igneous rocks** (Figure 8.8).

Hutton himself did not suggest an age for Earth, simply stating elegantly that the landforms he examined "had no vestige of a beginning—no prospect of an end." However, his message was clear: **Earth must be much older than the commonly accepted age of 6,000 years.** This was a hugely controversial idea at the time and was described in great detail in Hutton's thousand-page 1795 book *The Theory of the Earth*. Hutton's concept of an ancient Earth would not receive wide attention until after his death, when his ideas were summarized by a friend, John Playfair, and published in 1802 in a shorter form as *Illustrations of the Huttonian Theory of the Earth*. We will continue to trace the historical development of the concept of geologic time in the section Fossils and Chronology, but let us first pause to apply the rules we have just learned to the example of the Grand Canyon.

Figure 8.5 Outcrop of rocks at Siccar Point, Scotland. The gently tilted layers in the background lie on top of the vertical layers in the foreground.

The Grand Canyon: Applying the Rules. Identifying examples of the principles of original horizontality, superposition, cross-cutting relationships, and inclusions in nature allows geologists to reconstruct the geologic history of a rock sequence. Geologists do not stop at simply placing events in order; they also want to learn more about the conditions under which these rocks formed (see Chapter 7). Consequently, they must determine what types of rocks are present and what their characteristics tell us about their origins. For example, the rocks exposed in the Grand Canyon can be interpreted to reveal a complex geologic history.

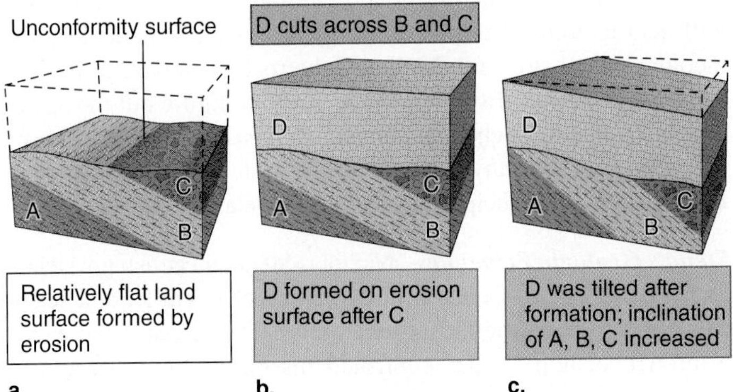

Figure 8.6 Hutton's principles. According to the principle of cross-cutting relationships, **a.** A, B, and C were subjected to weathering and erosion at Earth's surface. A significant time interval may have passed while erosion occurred. This eroded surface is known as an unconformity surface and is the physical expression of a gap in time. **b.** D lies above A, B, and C and cuts across these layers so it must have formed after C. **c.** D is no longer horizontal; therefore, all layers were tilted. Interpretations based on the principle of original horizontality are in blue; those based on the principle of superposition are in green; observations in red are made using the principle of cross-cutting relationships.

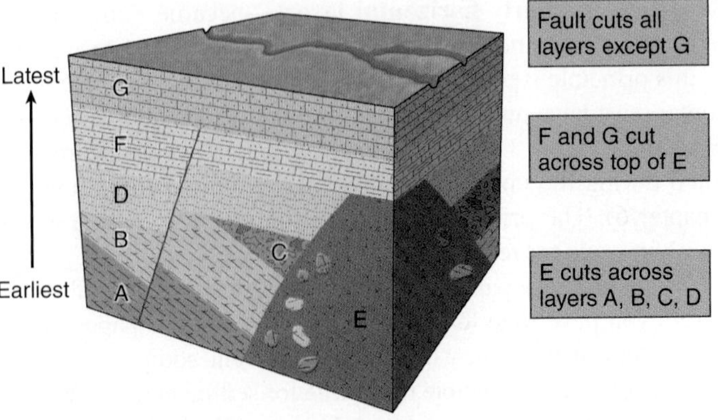

Figure 8.7 Cross-cutting relationships. The igneous pluton E formed after layers A–D and includes pieces of the layers (inclusions). F and G formed later because they lie atop all other rock units. The fault formed after F but before G. Finally, the river cut a valley through the upper part of G.

As we descend through the Grand Canyon, the rocks are defined by an obvious layering until we reach the inner gorge, close to the river. The walls of the inner gorge are composed of dark Vishnu Schist, which typically shows no layering—but is foliated—and lighter-colored Zoroaster Granite (Figures 8.9 and 8.10). The granite cuts across the foliations of the schist, so we can apply the principle of cross-cutting relationships to determine that

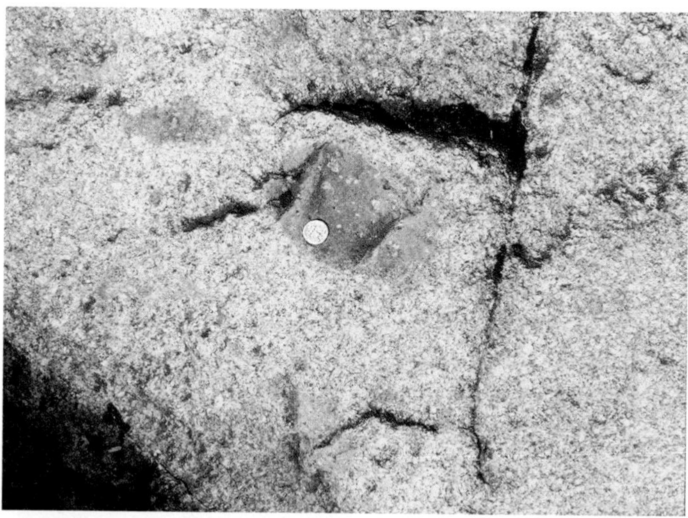

Figure 8.8 An inclusion. A dark gray rock is enclosed in a larger outcrop of the lighter-colored igneous plutonic rock in a streambed in Vermont. Note quarter for scale.

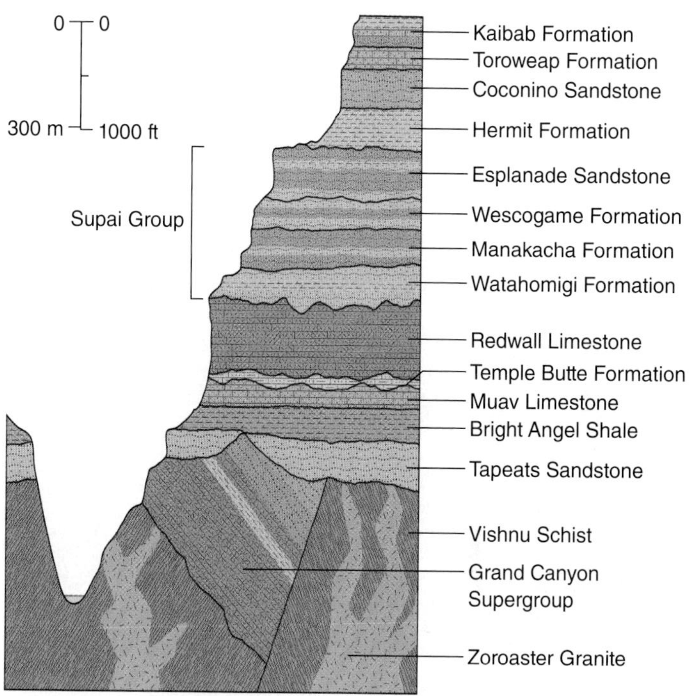

Figure 8.9 Sequence of rocks in the walls of the Grand Canyon.

a.

b.

Figure 8.10 Igneous and metamorphic rocks in the inner gorge of the Grand Canyon. **a.** Metamorphic and igneous rocks of the steep-walled inner canyon lie below the horizontally layered sedimentary rocks. **b.** The pink Zoroaster Granite cuts across steeply inclined foliations in the dark Vishnu Schist in the inner gorge.

Checkpoint 8.2

basic | advanced | intermediate | superior

Examine the following image of rock layers and answer questions 1 and 2 about relative time.

1. Which statement is most accurate?
   a) D is older than B.
   b) E is older than A.
   c) F is older than C.

2. An unconformity is most obviously present
   a) below the light-colored layer A.
   b) in the dark layer below E.
   c) between layers represented by D and F.
3. When did the tilting of the layers occur?
   a) After A was deposited
   b) Between deposition of layers E and A
   c) Before B was deposited
   d) Between deposition of layers C and E

Figure 8.11 Grand Canyon Supergroup. Tilted purple and gray rocks lie below the horizontal white Tapeats sandstone near the bottom of the canyon. Where is the unconformity surface? What principle would best apply to determine the order in which the tilted layers formed?

the granite must be younger than the schist. Therefore, the schist is the oldest rock in the canyon. Like all metamorphic rocks, the schist began as something else. Its texture and composition indicate that it was originally a sedimentary rock formed in a marine environment. Sometime after its formation and metamorphism, the schist was cut by the intrusion of the granite. Similar suites of rocks can be found today in the interiors of the Appalachian and Rocky Mountains, so from Hutton's principle of uniformitarianism, we conclude that **these rocks formed the roots of ancient mountain belts or volcanic arcs** (see Chapter Snapshot).

Elsewhere in the inner gorge, we can find the schist and granite lying below a tilted series of sedimentary rocks known as the Grand Canyon Supergroup (Figure 8.11). The rocks of the Supergroup formed under very different conditions from the underlying igneous and metamorphic rocks. A substantial period of erosion and uplift must have cut through and stripped away overlying rocks and brought the schist and granite to the land surface. This erosion surface would then have been submerged before the formation of the sedimentary rocks of the Supergroup. The contact between the lower schist and granite and the overlying Supergroup rocks is therefore an unconformity. Superposition and cross-cutting relations show that the unconformity is younger than the Zoroaster Granite and Vishnu Schist. Original horizontality allows one to interpret that the Supergroup rocks were tilted after their formation.

The horizontal layers of sedimentary rocks that characterize the upper section of the Grand Canyon (see Figure 8.2) lie directly above the schist and granite and the tilted Supergroup rocks of the inner gorge to form a second spectacular unconformity (Figure 8.11, Chapter Snapshot).

The lowermost section of horizontal layers is composed of Tapeats Sandstone. We can observe that the elevation of the base of the sandstone does not change much within the canyon, which suggests that this layer was deposited as sea level rose over a near-horizontal surface, with some isolated hills transformed into small islands. The sandstone is overlain by shale and limestone units (Bright Angel Shale, Muav Limestone; Figure 8.9). **The progression from sandstone to shale and limestone is typical of the sequence of rocks found along passive continental margins** (see Chapter 4) as sea level rises relative to land. We can compare modern environments and the types of fossils they contain with overlying formations of limestone (Redwall Limestone) and clastic rocks (Supai Group). These rocks are interpreted as indicative of shallow marine and delta conditions, respectively, formed as the sea level rose and fell. One exception to these near-shore marine settings is the 100-meter-thick (330-foot) cliff of Coconino Sandstone near the rim of the canyon (Figure 8.9). This layer contains patterns characteristic of **windblown sand deposits and desiccation (drying) cracks found in hot, dry deserts.**

The principle of superposition tells us that the upper layers (for example, Kaibab Formation; Figure 8.9) are the youngest in the canyon. But our story does not end with the limestone of the Kaibab Formation. These rocks were formed under shallow seas, and the presence of chemical sedimentary rocks such as gypsum suggests that from time to time, seawater was cut off from the sea and evaporated. Following deposition of these units, much of what is now northern Arizona must have undergone additional periods

Checkpoint 8.3

basic · intermediate · advanced · superior

Use the principles of original horizontality, superposition, cross-cutting relationships, and inclusions to determine the order of events for the idealized location shown in the following diagram.

a) Place the rock units in their order of formation, from oldest to youngest.

Youngest
1. _____
2. _____
3. _____
4. _____
5. _____
6. _____
7. _____
8. _____
9. _____
10. _____

Oldest

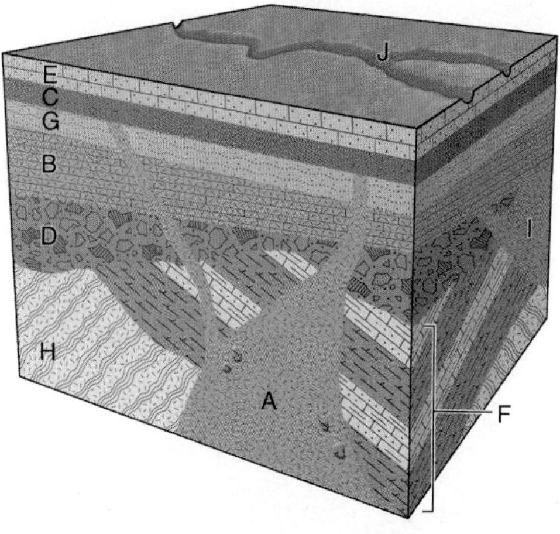

b) Examine the rock types identified by the symbols in the diagram, and determine which rock units best match the following descriptions.

Letter	Characteristic
_____	Interbedded layers of rocks that indicate alternating shallow marine environments and coastal swamps in tropical conditions
_____	Coarse-grained clastic sedimentary rocks overlying an erosional surface (unconformity surface)
_____	Granite
_____	A rock containing a foliation
_____	The most recently deposited sedimentary rock
_____	Sedimentary bed that has undergone contact metamorphism on its uppermost surface
_____	Basalt

Checkpoint 8.4

basic · intermediate · advanced · superior

Construct a diagram that illustrates a cross section of six rock units that would account for the features listed below (not in order). Clearly label your units. Remember, these events are not in order—you must determine the order of events based on the descriptions.

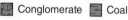

a) Rhyolite cross-cuts and covers all units except sandstone.
b) Dark, fine-grained igneous rock cross-cuts and covers conglomerate and older units.
c) Oldest rocks are made of black, biochemical layers that were later tilted.
d) Coarse-grained clastic rock is deposited immediately over coal.
e) Opaque chemical sedimentary rock is deposited directly over basalt.
f) River cuts partially into limestone.
g) Medium-grained clastic rock is deposited over small-grained, high-silica extrusive rock.

of uplift to push rocks that began below sea level to elevations in excess of 2,000 meters (6,500 feet; Chapter Snapshot). The Colorado River cut downward to form the canyon, perhaps keeping pace with the latest episode of uplift.

The **final event in the geologic history of the canyon was the formation of a series of cinder cone volcanoes** on its north rim. Lava flows from these volcanoes cascaded down the canyon's walls (Figure 8.12) and formed what are now rapids and waterfalls in the river below. (Which principle [or principles] tells us the lava flows were deposited after formation of the canyon?)

Figure 8.12 Cinder cone volcanoes. The Grand Canyon must have been present before the volcanoes formed because the lava from the volcanic eruptions covers the rocks of the canyon walls.

GEOLOGICAL HISTORY OF THE GRAND CANYON

1. SEDIMENTATION AND BURIAL

Layers of sediment were deposited in a marine environment and lithified to form sedimentary rocks.

2. UPLIFT AND MOUNTAIN BUILDING

Buried sedimentary rocks were converted to metamorphic rocks and magma was added to crust. Erosion began...

3. CONTINUAL EROSION

Erosion continued and removed overlying rocks to expose granite and schist.

4. RENEWED DEPOSITION

Igneous and metamorphic rocks of inner gorge were submerged and a series of overlying sedimentary rocks (Grand Canyon Supergroup) was deposited.

5. UPLIFT AND EROSION

A second episode of uplift tilted the older rocks and erosion created an unconformity surface.

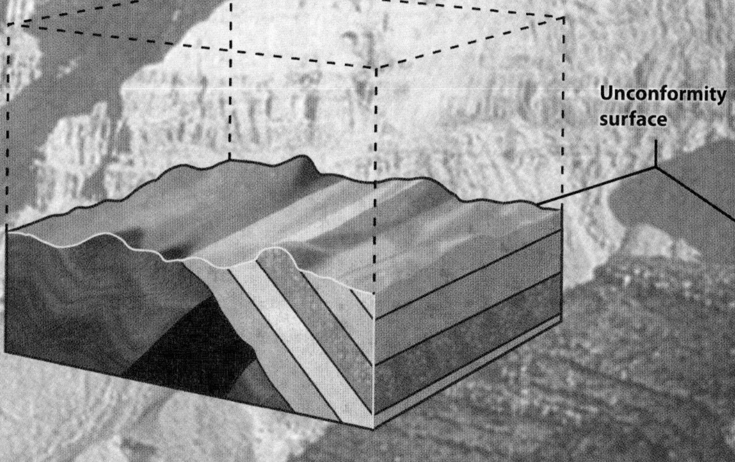

6. RENEWED DEPOSITION

Sea level rose to cover older rocks and younger sedimentary rocks were deposited. Several cycles of erosion and deposition occurred over many millions of years.

7. UPLIFT AND EROSION CUTS CANYON

A few million years ago the rocks were uplifted and the Colorado River cut the canyon.

8. VOLCANISM

Volcanism began more than one million years ago. Lava cascaded down the canyon walls to form dams along the river.

Fossils and Chronology

When we try to match rocks from different areas of any continent, we can rarely follow even the best exposures of layers of sedimentary rock for more than a few tens of kilometers before they disappear underground or are removed by erosion. Because of this, deciphering a region's geologic history becomes difficult. Consequently, if we are to determine the geologic history of a continent, we must use other methods to connect (or correlate) equivalent rock units separated by large distances. For example, how could we determine if the layers of Redwall Limestone forming cliffs in the Grand Canyon, Arizona, are equivalent in age to the limestone layers that were dissolved away to form Mammoth Cave in Kentucky? **The most straightforward method of correlating sedimentary rocks is to compare their fossils,** the remains or traces of organisms preserved in rocks.

The process of making something into a fossil is not an easy one. It takes fairly rapid burial in fine-grained sediment to cut off the three major "agents of destruction"—decay, dissolution, and disarticulation. This is why hard skeletons in the fossil record are more likely to have been preserved from marine environments than from on land. The technique of correlating rocks on the basis of their fossil content was first described about 200 years ago in England by William Smith, who was employed to survey the coal deposits in some mines of southern England. After repeated trips down several mine shafts, Smith realized that the alternating rock layers of coal, sandstone, limestone, and shale were present in a predictable order. Smith also noticed that these layers contained characteristic fossils. He observed more rock layers as part of his next job, which required him to survey possible routes for canals to transport the mined coal to market. He realized that the rock layers he observed were always present in a consistent order (superposition) and they always contained the same fossils. Smith understood that **fossils held the key to predicting the sequence of layers in a given location and to matching outcrops of similar rocks between different locations.** He later expanded his observations, eventually publishing the first accurate national geologic map of England in 1815.

Some fossils extend over long intervals of geologic time and are therefore found in many rock layers. Matching these fossils allows broad correlations to be made, but the wide range of the fossils makes it difficult to match individual layers between different regions. Geologists can make more precise correlations using **index fossils, species that existed for relatively short periods of geologic time and are found over large geographic areas.** Index fossils are useful because their appearance in the rock record represents a specific time interval that scientists can use to readily identify and correlate rocks between different regions. Fossils are especially important to understanding time sequences of rocks because the order in which the fossils evolved is unique and can be followed from species to species over time.

Paleontologists, who are geologists who study fossils, have used evidence from index fossils not only to correlate the rocks of individual continents but also to match rocks from around the

basic advanced
intermediate superior

Geologists look for similar rock types or fossils to tell them that geologic environments were similar between two widely spaced locations. What are some examples of modern environments that have characteristic types of plants and animals?

Checkpoint 8.6

basic advanced
intermediate superior

Outcrops of rock are examined in four different locations in a state. The rock types and the fossils they contain are illustrated in the following diagram. Which fossil would be the best choice to use as an index fossil for these rocks? Which fossil is least characteristic of a specific set of geologic conditions?

a) Fossil 1 b) Fossil 2 c) Fossil 3

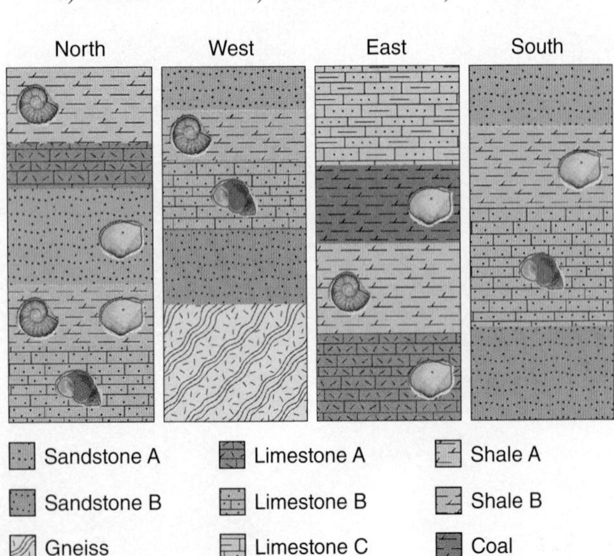

Examine the following illustration and predict which rock unit in the Grand Canyon is most likely to have formed in a depositional environment like the one pictured.

world. Consequently, they have been able to make interpretations about Earth history based on the characteristics of the rocks themselves and the fossils they contain. By the 1830s, fossils found in rocks throughout Europe had been used, along with other principles of relative dating, to group rocks together (see Section 8.3). Rock units from one part of a continent could then be matched with those in any other location because **similar rocks of the same age typically contain similar fossils.**

The Grand Canyon: Fossils. Fossils are preserved in many of the rock layers in the Grand Canyon (Figure 8.13), and the **character and distribution of these fossils support interpretations for the origins of Grand Canyon geology based on the rock types.** Some of the oldest rocks in the Grand Canyon Supergroup contain fossils of primitive algae known as stromatolites that grow in shallow marine waters. Similar stromatolites still grow on Earth today—modern-day equivalents of some of the earliest life to evolve. A little higher up in the section, the Muav Limestone contains fossils of trilobites, an extinct group of arthropods. Arthropods, such as modern-day crabs, have a hard outer skeleton divided into segments. Trilobites are found in rocks formed from sediment deposited on the seafloor.

The Muav Limestone also contains brachiopods, shelled animals that lived on the seafloor. We find different brachiopods higher in the section in the cliffs of the Redwall Limestone, in an assemblage of fossils that includes crinoids, corals, and cephalopods. Today, corals and crinoids grow on the seafloor. Corals form in relatively shallow, clear, warm waters and thus give us some clues about the depositional environment of the Redwall Limestone. Modern cephalopods have fleshy tentacles and include squid and the nautilus. A combination of such fossils found in the canyon can be used to match the Redwall Limestone with thick limestone deposits found in Indiana quarries and surrounding Kentucky's Mammoth Cave. The same fossils are found in the Rundle Limestone that forms steep cliffs in the Rocky Mountains near Banff, Canada. While these rocks are given different names in different locations, their fossils indicate that they were formed at the same time, when much of North America was covered by a shallow sea.

Higher in the canyon, the Supai Group and Hermit Formation (see Figure 8.9) include alternating layers containing either brachiopods or plant fossils, indicative of low-lying stream systems that were periodically flooded by shallow seas. The Coconino Sandstone contains no body fossils (that is, fossils representing the body of an organism), but **the tracks of ancient organisms** are preserved in many locations. The tracks are known as *trace fossils* and are similar to those produced by modern land-dwelling lizards, scorpions, and millipedes. Brachiopods and corals show up again in the Kaibab Limestone, along with occasional trilobites and cephalopods. Keep in mind that while some of the fossil types (for example, brachiopods) are found in multiple rock layers, the fossils in each layer are not exactly the same. That is, brachiopods make up a phylum, a large collection of organisms with similar traits. The phylum contains thousands of different species of brachiopods, and the ones found in the Kaibab Limestone are not the same species as those found in the Muav Limestone. (For comparison, humans are part of the phylum Chordata, which also includes fish, amphibians, snakes, and birds [all organisms with a backbone]—quite a diverse collection.)

Checkpoint 8.8

Limestone is present in multiple rock units at different elevations in the Grand Canyon. If you were handed six large samples of limestone, each containing fossils, which group of fossils would be most helpful in identifying the different units? Which would be least helpful? Explain your choices.

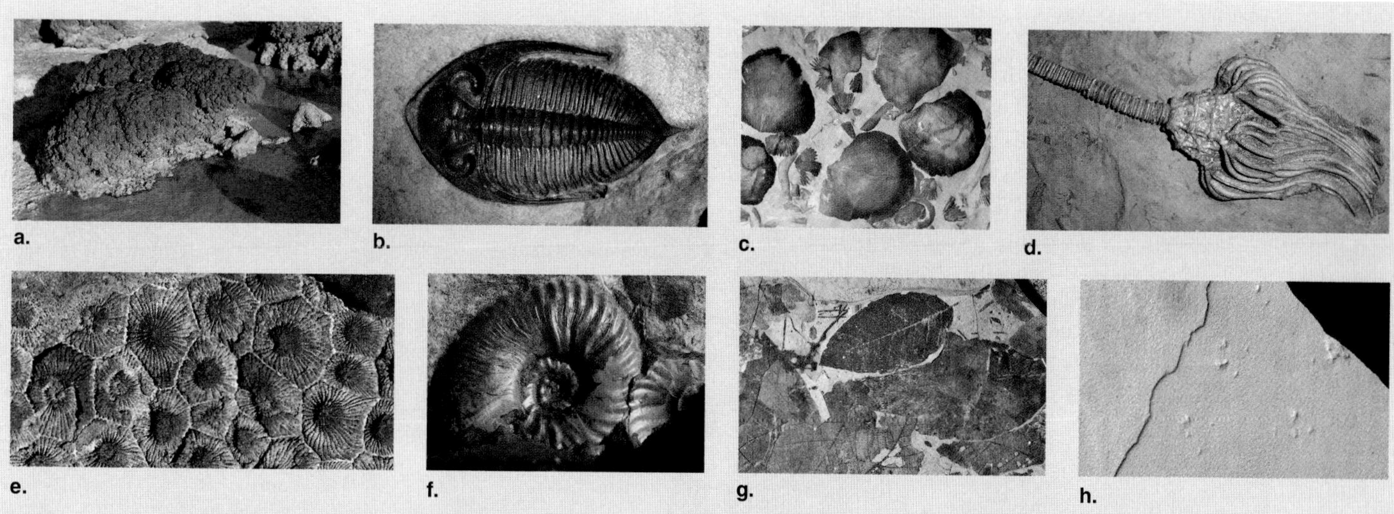

Figure 8.13 Examples of the types of fossils of the Grand Canyon: **a.** stromatolite, **b.** trilobite, **c.** brachiopod, **d.** crinoid, **e.** coral, **f.** cephalopod, **g.** plant, and **h.** tracks. Note that these images show the diversity of fossils in the canyon, but these specific fossil species may not be present. For example, trilobites are found in some of the rocks, but the particular species of trilobite shown here may not be present in Grand Canyon rocks.

8.3 Geologic Time

Learning Objectives

- Describe how and why Earth's history is divided into three long spans of time known as eons.
- Describe how the most recent eon differs significantly from previous eons.
- Explain the role of mass extinctions in defining transitions between eras in the most recent eon.
- Relate the fossil record to changes in the biosphere over much of Earth's history.

As discussed in Section 8.2, geologists use relative time to arrange geologic events in the correct sequence. But time can also be considered in purely numerical terms. Early in the twentieth century, scientists realized they could use the radioactive decay of isotopes to determine the actual age of some kinds of rocks and to put dates on the sequence of events. For convenience, we will incorporate the numerical times into this discussion of the geologic history of Earth, but we will hold the explanation of how these ages are determined until Section 8.4.

Evolution of Early Earth

Conditions on Earth more than 4 billion years ago would have been much different from today. Remember, Earth formed from accretion of materials orbiting the proto-solar system. This resulted in an Earth that was initially almost entirely molten. It was probably several hundred million years before Earth had a permanently solid crust. Scientists have determined that the oldest known rocks on Earth are nearly 4.3 billion years old and are found along the eastern shore of Hudson Bay, Canada. Even after the crust solidified, conditions on Earth would have been much different from today as heat left over from the planet's formation would have made conditions on Earth much warmer. There was little oxygen in the atmosphere, evidenced by the existence of minerals that would have broken down if oxygen had been present. Some scientists have interpreted the presence of extensive ancient volcanic rocks throughout the world to suggest that there was more volcanic activity than we observe today. To make matters worse, Earth was being regularly bombarded by asteroids and comets. In those early times, the environment in which life evolved was probably similar to the hot, acrid conditions we currently find in places such as the hot springs of Yellowstone National Park. Those scorching, acidic waters harbor heat-loving bacteria. Because bacteria are some of the most primitive life-forms we observe on Earth today, scientists argue that early life was probably similar to them.

The Geologic Timescale

Earth's past is divided into three big chunks of time known as eons. Eons are the largest intervals of the timescale, each over 500 million years long, and they mark major changes in the earth system (Figure 8.14). The oldest known rocks were formed during the early Archean eon. The first part of the Archean used to be called the Hadean eon, and some people still use that term to represent the earliest part of Earth's history from its initial molten state 4.6 billion years ago until the first rocks formed. However, as we discussed earlier, science is provisional and subject to change. The expansion of the range of the Archean to include the former Hadean is an example of such a change that occurred as earth scientists learned more about Earth's earliest history and have been able to refine the timescale with radiometric dates. The origin of life is still enigmatic to scientists; remember, early Earth was not like it is today. Over long periods of time, hundreds of million of years, things began to change. Our first evidence of possible life comes in the form of enhanced carbon concentrations in crustal rocks from Greenland that are 3.85 billion years old. While these enhanced carbon signatures are not visible cells, they do indicate that organic compounds were present that long ago. Making the jump from organic molecules to replicating life-forms is where the edge of our current scientific understanding lies. A recent research breakthrough in the laboratory has been able to determine a process by which basic small molecules such as amino acids could have polymerized to form self-replicating RNA molecules. This process involved basic steps of evaporation, heating, and condensation that could readily have occurred in small warm ponds on an early Earth. Next scientists will attempt to apply this newfound knowledge to move toward a comprehensive working model of how life originated.

About 3.5 billion years ago the first continents would have been differentiated, and some of the rocks preserved from this time contain single-celled bacteria-like fossils that are composed of small rods and spheres. The first primitive life was dominated by blue-green algae, a group of bacteria that formed mound-like structures similar to modern organisms known as stromatolites (Figure 8.13a).

About 2 billion years ago, during the Proterozoic eon ("earlier life"), oxygen began to accumulate in the atmosphere in greater quantities than before (as evidenced by oxidized minerals in ancient rocks), and life evolved beyond primitive bacteria (see Chapter 14 for more on the evolution of the atmosphere). Multicellular life did not show up until about 1.7 billion years ago. The earliest things that we might call animals appeared in the fossil record about 580 million years ago in the Ediacaran time. This was a good 40 million years before the Cambrian period, which is famous for its so-called "explosion" of life. The Ediacara animals were weird-looking marine creatures that were mostly soft-bodied.

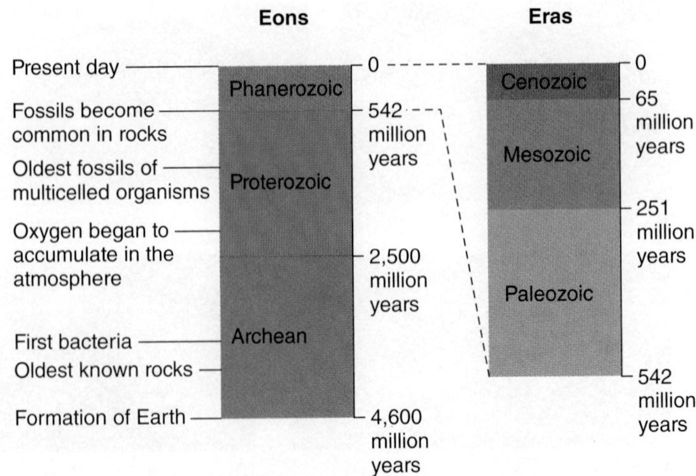

Figure 8.14 Eons and eras. Geologic time is divided into three long eons, the Archean, Proterozoic, and Phanerozoic. The Phanerozoic, the most recent eon, is characterized by abundant fossils, allowing it to be further subdivided into three eras. The eras are in turn broken down into shorter time intervals known as periods (see Table 8.1).

They have been thought to be related to jellyfish, mollusks, and sponges. However, their forms are so strange-looking that it is difficult for paleontologists to reconstruct just what they were.

The beginning of the last eon, the Phanerozoic ("life revealed"), marks the break between older rocks in which fossils are rare and younger rocks that contain abundant fossils formed from organisms with shells and hard skeletons. Things got more recognizable, but not by much. Many of the major animal groups can be found in the Cambrian period, even though they did not look anything like their descendants do today. The Phanerozoic eon represents just the last 12 percent of geologic time. Geologists often use the informal term *Precambrian* to refer to all geologic time prior to the Phanerozoic.

Rock units of the Phanerozoic eon have been assigned to three shorter eras (Paleozoic, Mesozoic, and Cenozoic) of varying lengths based on the characteristics of their fossils (Figure 8.14; Table 8.1). Paleozoic is derived from the Greek for "ancient life," Mesozoic means "middle life," and Cenozoic represents "new life." Each of the eras is further subdivided into shorter time intervals known as periods (Table 8.1). The beginning of the oldest period, the Cambrian, signaled an explosion of diversity within the biosphere. On average, **the number of species represented by fossils increases from the Cambrian period onward.** Although all major phyla were derived by the Cambrian, the diversity trends seen throughout the Phanerozoic eon suggest that diversity of organisms has increased through time. In fact, the majority of all species that have existed on Earth are now extinct. How could that be? Think about the length of time Earth has been around, and it becomes easier to imagine.

Life certainly was not the same in the past as it is today. The species that we see all around us evolved from common ancestors that originated in the late Proterozoic eon, when the earliest multicellular organisms first appeared. Paleozoic life was dominated by marine creatures such as trilobites (Figure 8.13b), brachiopods (Figure 8.13c), horn corals, and crinoids (sea lilies, Figure 8.13d) that lived on the bottom the shallow seas 542 to 251 million years ago. Most of these types of organisms are now extinct.

Primitive life on land first appeared in the Cambrian period, but land plants did not invade until 60 million years later, in the Silurian period. Vertebrate animals soon followed, represented by the now famous "walking fish" called Tiktaliik, which shares both fish and amphibian characteristics (Figure 8.15). The Tiktaliik fossil was found in the northern territories of Canada in 2006 and has an

Table 8.1 The Geologic Timescale

Era	Period		Remarks	Time, million years ago
CENOZOIC		Quaternary	Large ice sheets centered on Earth's poles. Modern humans appeared. Some large mammals (e.g., mastodons, mammoths) became extinct.	2.5–0
	Tertiary*	Neogene	Large mammals abundant (e.g., mastodons, mammoths); earliest human ancestors appeared. North and South America were connected.	23–2.5
		Paleogene	Age of mammals began. First large mammals on land and in the ocean (e.g., whales, elephants, horses, bears); widespread expansion of echinoids.	65.5–23
MESOZOIC		Cretaceous	Dinosaurs dominated Earth. First grasses and flowering plants appeared; dinosaurs wiped out by an extinction event at the end of the period; ammonites abundant.	145–65.5
		Jurassic	Atlantic Ocean began to form. First birds appeared; dinosaurs and flying reptiles (pterosaurs) common.	200–145
		Triassic	Pangaea began to split apart. Reptiles dominated Earth; earliest mammals and dinosaurs appeared; first conifers; modern corals developed.	251–200
PALEOZOIC		Permian	Much marine life was wiped out in the most massive extinction known, at the end of the Permian.	299–251
		Pennsylvanian† (Late Carboniferous)	Appalachian Mountains formed when North America collided with Africa. Insects and early reptiles on land; first evergreen trees appeared, earliest forests.	318–299
		Mississippian† (Early Carboniferous)	Shallow tropical oceans covered much of the interior of North America. Marine fossils in limestone common.	360–318
		Devonian	Age of fishes. First land vertebrate animals (tetrapods: amphibians); first seed plants, trees, and forests. Insects on land as well.	416–360
		Silurian	First fish with jaws appeared; much of North America was under a shallow tropical sea with abundant reefs. First primitive plants occur on land (ferns, mosses).	444–416
		Ordovician	North America was near the equator. Abundant marine life.	488–444
		Cambrian	Explosion of organisms with hard skeletons (which can be easily preserved as fossils) occurred at beginning of Cambrian; trilobites flourished.	542–488
PRECAMBRIAN			Fossils rare in Precambrian rocks. Soft-bodied organisms present in the youngest Precambrian rocks; dominated by single-celled life.	4,600–542

* Authorities have recently recommended that the name Tertiary be replaced by two new periods, the Neogene and Paleogene. It will take several years before this change is widely accepted.

† These periods are used widely in North America but are combined to form the Carboniferous period internationally.

Checkpoint 8.9

basic · intermediate · advanced · superior

Scientists Find Fossil Below K-T Boundary

The debate whether dinosaurs went extinct due to a large space rock that struck the Earth 65.5 million years ago (MYA) may have been answered with the discovery of a distinctive brow horn from a Ceratopsian dinosaur just 13 centimeters (5.1 inches) below the K-T boundary—the distinct layer of geological sediments separating the Cretaceous and Tertiary periods.

Rocks laid down 65.5 MYA show a thin layer abundant in rare elements like Iridium, spherules and shocked Quartz that could only have come from a meteorite impact. Since no fossils have ever been found in sediments above the K-T boundary, conventional wisdom has it that the end of dinosaurs came with an asteroid impact that caused firestorms, acid rain and a nuclear winter that blotted out the Sun.

But that theory had a hole in it. The fossil record showed an apparent lack of dinosaur fossils in the last few million years leading up to the impact, suggesting that the "three meter gap" proves that dinosaurs went extinct long before the catastrophic impact.

Scientists working in the Hell's Creek formation in the Montana badlands say they have resolved that dispute.

Writing in the Royal Society journal *Biology Letters,* the paleontologists report on the new discovery of the closest dinosaur fossil ever found to the K-T boundary.

The ancient remains uncovered in Montana belong to the last known dinosaur to ever walk the planet and gives weight to the theory that dinosaurs were in fact wiped out by an asteroid impact. All other dinosaur fossils found are either much older, or were unearthed after being washed from their original graves into much younger sediments, long after they died.

The fossil is most likely of an adult triceratops, a dinosaur growing up to 30 feet long and weighing up to 13 tons. The nearly 18-inch fossilized brow horn was found just 5 inches below the K-T boundary.

"This is the youngest dinosaur that has been discovered in situ. Others can be found in younger deposits, but those have been put there by geological processes and are actually much older," said Tyler Lyson, a paleontologist at Yale University.

The discovery undermines the theory that gained ground in the 1980s, that dinosaurs died out due to climate change or rising sea levels long before the planet was struck by a space rock. The theory carried some weight due to a lack of fossils found within the "three meter gap" of the K-T boundary.

The finding "demonstrates that dinosaurs did not go extinct prior to the impact and that at least some dinosaurs were doing very well right up until we had the impact," Lyson told the Guardian.

Gaps in the fossil record—which is patchy at best—are nothing new. Another, covering tens of millions of years, is clearly evident in the Hell's Creek formations some 60 meters (200 feet) below the K-T boundary.

But because similar dinosaur fossils are found both above and below this gap it is assumed the absence of fossils has more to do with geological processes, or simply blind prospecting luck, than any extinction event and subsequent miraculous reintroduction.

The "three meter gap" prior to the K-T boundary is unique because dinosaur fossils never reappear in the geological record.

Dr. Paul Barrett of the Natural History Museum said the discovery was strong evidence that dinosaurs were killed off in North America by a catastrophic event, but the evidence is not conclusive globally.

"It shows that in this part of the world dinosaurs were still viable and still roaming around at the time the meteorite hit. But what it doesn't tell us is what was going on in the rest of the world, and it could be that in other parts of the world dinosaurs were dying out at different rates and for different reasons because of other things going on at the time," he told BBC News.

He argues that just one brow horn discovery doesn't resolve the dispute over dinosaur extinction.

Source: Scientists Find Fossil Below K-T Boundary" by Lawrence LeBlond. RedOrbit.com, July 13, 2011. Reprinted by permission. http://www.redorbit.com/news/science/2078377/scientists_find_fossil_below_kt_boundary/

1. What are the key observations mentioned in this article (select all that apply)?
   a) Dinosaurs were wiped out by a meteor impact
   b) A dinosaur fossil was found within a few centimeters of the KT boundary
   c) No in-situ, non-bird dinosaur fossils are found above the KT boundary
   d) Climate change played a major role in the extinction of dinosaurs
2. What hypothesis is best supported from the observations?
   a) Dinosaurs across the planet were killed off by a meteor impact.
   b) Dinosaurs in North America were killed off by a meteor impact.
   c) Dinosaurs across the planet were not killed off by a meteor impact.
   d) Dinosaurs in North America were not killed off by a meteor impact.

Figure 8.15 Recently discovered Tiktaalik fossil. **a.** This is a transitional fossil between fish and amphibians. The fossil was discovered on Ellesmere Island, Canada, in 375 million-year-old rocks. Several individuals were found, some up to nearly 3 meters (9 feet) long. **b.** A re-creation of what Tiktaalik may have looked like in life.

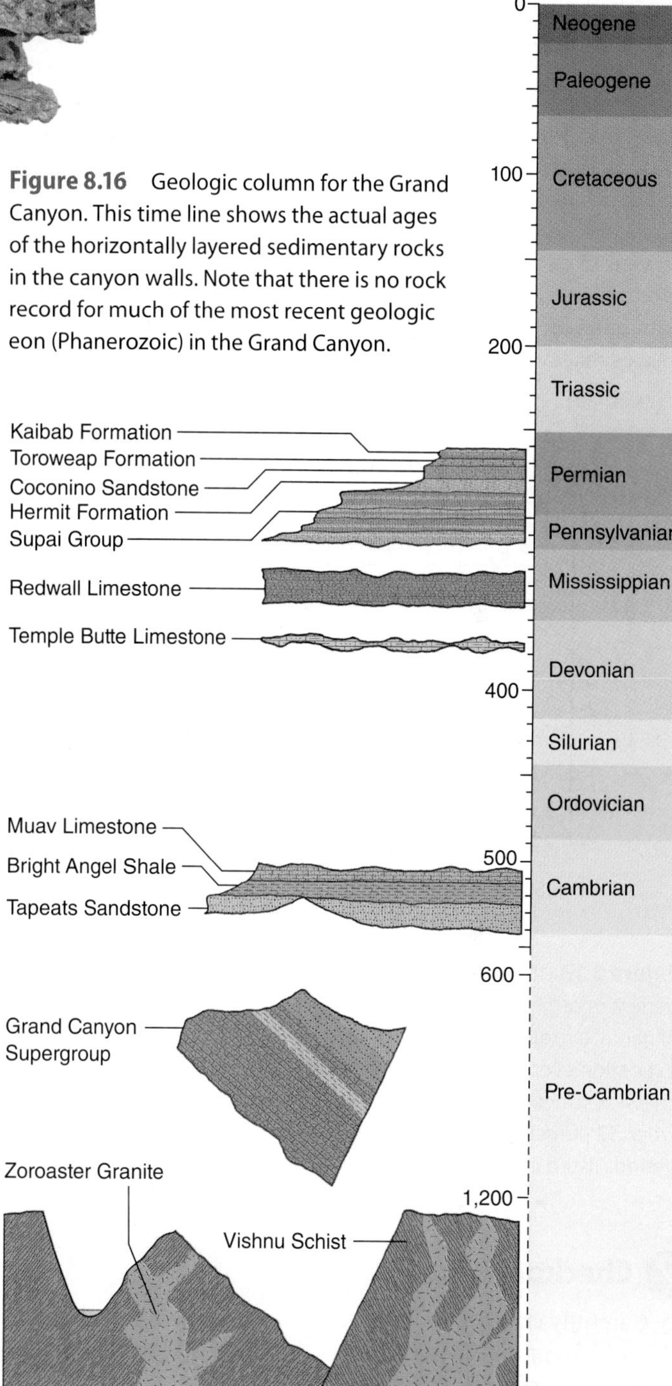

Figure 8.16 Geologic column for the Grand Canyon. This time line shows the actual ages of the horizontally layered sedimentary rocks in the canyon walls. Note that there is no rock record for much of the most recent geologic eon (Phanerozoic) in the Grand Canyon.

amphibian-like neck and head, including ears that could hear in air, with fishlike scales and fins. Paleontologists have found a series of transitional fossils that document the evolutionary steps from fish to amphibians. By mid-Paleozoic, the continental invasion was well underway, and many organisms found their way through estuaries into rivers and lakes and finally onto land. Those that had adaptations that allowed them to solve the problem of breathing air, moving out of water, and reproducing on land survived though the Paleozoic era.

Life changed—big time—during the Mesozoic. Dinosaurs evolved, as did organisms such as the giant squidlike creatures called ammonites. Some of their shells that have been found in the fossil record are the size of a kitchen table. During this time, flowering plants evolved, as did the birds and many of the insects that pollinated them (see Table 8.1). At the end of the Mesozoic era, the dinosaurs were replaced by the mammals that diversified and evolved into what we know as our modern fauna. Many of these mammal species, such as the woolly mammoth or the giant ground sloth, went extinct when climate conditions changed about 12,000 years ago, at the end of the last ice age.

The Grand Canyon: Geologic Record. Our first look at the thick section of sedimentary rocks preserved in the Grand Canyon (see Figure 8.2) suggests that each of the 12 periods of the Phanerozoic eon must be present in the layers of the canyon walls. However, closer examination of the fossil record reveals that the canyon's sedimentary rocks represent just five periods, all in the Paleozoic era, and none of those completely (Figure 8.16). It is a little like listening to a CD that skips—you know some songs should be there, but they are missing. In the case of the Grand Canyon, large parts of the geologic record were subsequently removed by erosion or were never preserved in the first place. Consequently, no matter how thorough our analysis of the canyon's rocks, it can never give us more than a partial record of Earth's history.

Mass Extinctions

While William Smith was at work in England, across the English Channel in France, Georges Cuvier was about to stir things up in the world of paleontology. Up until this time, most scientists, including Hutton, had believed that fossils were the remains of organisms that could still be found living somewhere on Earth. Many

Americans, notably Thomas Jefferson, were impressed when the fossil bones of a large elephant-like creature were discovered in what would become Kentucky. Jefferson wondered if Lewis and Clark would discover herds of these massive animals grazing on the plains during their exploration of the West. Cuvier, a professor at the National Museum of Natural History in Paris, was the first person to compare these fossil bones with those of modern elephants, and he concluded that they represented a different species, the mastodon, which no longer lived on Earth. Thus, Cuvier introduced the concept of extinction and was able to identify the fossils of many species that had disappeared from the living record. Paleontologists examining the fossil record have determined that 99.9 percent of all species that ever existed on Earth have become extinct.

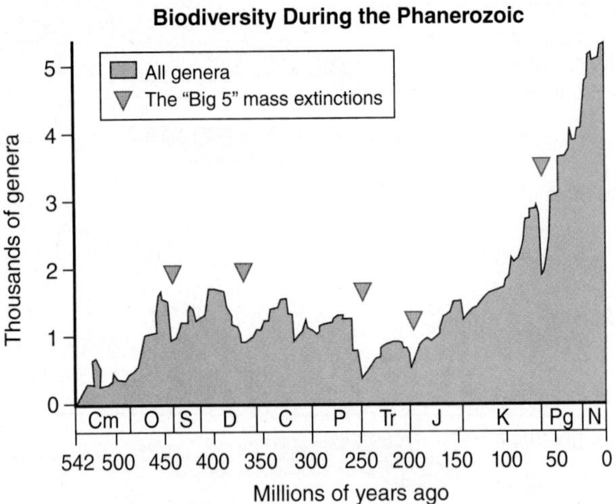

Figure 8.17 Biodiversity is measured here by the number of marine genera of organisms over the last 542 million years. (*Genera* is plural of *genus*, a group of similar species.) Decreasing biodiversity signals extinction events. Five events are highlighted, with the Permian-Triassic extinction representing the greatest drop in biodiversity, when 52 percent of marine genera died out. Letters correspond to periods listed in Table 8.1.

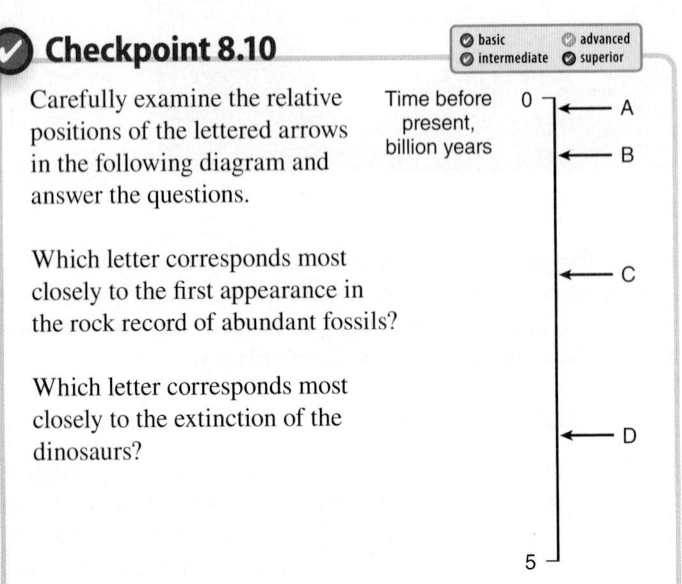

Checkpoint 8.10

Carefully examine the relative positions of the lettered arrows in the following diagram and answer the questions.

Which letter corresponds most closely to the first appearance in the rock record of abundant fossils?

Which letter corresponds most closely to the extinction of the dinosaurs?

Checkpoint 8.11

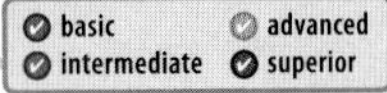

Ancient Leaves and Insect Extinctions

Read the following abbreviated version of a newspaper article and answer these questions.

a) What was the question being investigated by the scientists?
b) What observations did the scientists make during their investigations?
c) What was the principal conclusion of their research?

When a 6-mile-wide asteroid slammed into Earth 65 million years ago, it wiped out the dinosaurs, about 80% of the world's plant species, and all animals bigger than a cat. But what happened to the bugs?

It's been tough for scientists to determine how the insects fared because bugs rarely leave behind fossils. But a Denver paleontologist and his Smithsonian Institution colleagues found a way around the problem: By studying insect damage etched into thousands of fossil leaves, they determined that many plant-eating bugs perished in the big impact.

"These little insects are leaving their calling cards on the fossil leaves, and we have an excellent fossil record of leaves," said Kirk Johnson, curator of paleontology at the Denver Museum of Nature & Science. "So by looking at the insect damage on the leaves before and after the dinosaur extinctions, we can make a pretty good educated guess of what happened to the insects."

Johnson and his collaborators estimate that 55% to 60% of plant-eating insects were exterminated. Over the past 20 years, Johnson has collected 13,441 plant fossils from quarries in southwestern North Dakota. When the asteroid hit Mexico's Yucatan Peninsula, it threw up clouds of dust that traveled around the globe. Johnson pulled the fossils from rock layers directly above and below those sediments. At the time, southwestern North Dakota was a warm, forested plain with lots of broad-leafed trees.

Some leaves, now stored at the Denver museum and at Yale University, are up to a foot long. Individual leaf veins are visible, as are the diagnostic chomp marks, tunnels, and holes left by prehistoric beetles, grasshoppers, butterflies, caterpillars and moths. Some insects are specialists, relying on a single species of plant for sustenance; others are generalists that feed on several plant types. By analyzing insect-damaged leaves before and after the impact, the researchers determined that the generalists survived, while 70% of the specialists did not.

Source: Jim Erickson, *Rocky Mountain News* (Denver), February 22, 2002, page 7A.

Life on Earth has been repeatedly affected by **mass extinction events that resulted in the loss of large numbers of species** (Figure 8.17). So dramatic were these events that the types of fossils formed prior to the extinction events are substantially different from those found in rocks deposited following extinction. These contrasts in the fossil record were used to divide the Phanerozoic eon into three eras. The mass extinction that ended the reign of the dinosaurs occurred 65 million years ago, at the end of the Mesozoic era. This event is known as the **Cretaceous-Tertiary extinction** (termed the K-T event after the two periods that lie on either side of the time boundary). No dinosaurs, except perhaps birds, survived the event. Mammals, which had been a relatively minor group until that point, expanded after the extinction and are the ancestors of many of the organisms we know today. The K-T extinction is most often attributed to a large asteroid or comet colliding with Earth (see Chapter 3).

As spectacular as the K-T event was, it pales in comparison to an even more widespread disruption of the biosphere that occurred at the end of the Paleozoic era, 251 million years ago (see Figure 8.17). This event, known as the **Permian-Triassic extinction** (or P-T event)**,** killed off an estimated 96 percent of marine species and 70 percent of land species. That extinction event, often called "the great dying," may have taken a few million years. By studying such events, scientists hope to answer two basic questions: (1) What caused the mass extinction? and (2) Why did some species survive the event while others perished? In the case of the P-T extinction, scientists are closing in on a potential explanation that may help answer these questions. The P-T event was global in scale and affected both land and marine species. Marine organisms showed the greatest losses in response to regional environmental changes. Consider how each of the following events contributed to global changes to natural environments during the P-T event.

- **The supercontinent Pangaea was assembled during the Permian period, creating a single worldwide ocean.** This new configuration of continents would have reduced the area of the continental shelf, the shallow ocean floor around the continents that was home to a majority of marine species. Recent analyses have proposed that changes to oceanic circulation patterns, increased global temperatures, and a decrease in dissolved oxygen in the worldwide ocean may explain the extinction of marine species.

Checkpoint 8.12

basic · intermediate · advanced · superior

Geologic Time Metaphor

Suppose that all of geologic time were proportional to the length of a football field (100 yards). Earth would have formed at the opposing team's goal line (100 yards), and the present day would represent the home team's goal line (0 yards).

Metaphor equation

Metaphor value = (years before present / age of Earth) × metaphor maximum

Example

Oldest fossil bacteria = 3,500 million years old
Age of Earth = 4,600 million years
Metaphor maximum = 100 yards
Metaphor value = (3,500,000,000/4,600,000,000) × 100 = 76 yards

Key metaphor dimensions

100 yards = 4,600 million years
10 yards = 460 million years
1 yard = 46 million years
1 foot = 15.3 million years
1 inch = 1.3 million years

Calculate the yardage of the extinction at the end of the Paleozoic era. Then fill in the blank cell in the table and label the following figure.

Develop your own metaphor for geologic time and describe it. Choose some of the most significant geologic events from the geologic timescale and convert them into your own metaphor equation.

Don't try to be too detailed in your analysis. The intention here is to recognize the length of the geologic timescale and the relative positions of key events. Approximate lengths, distances, heights, widths, depths, sizes, time periods, and so on are okay as long as you recognize the relative proportions of the time intervals.

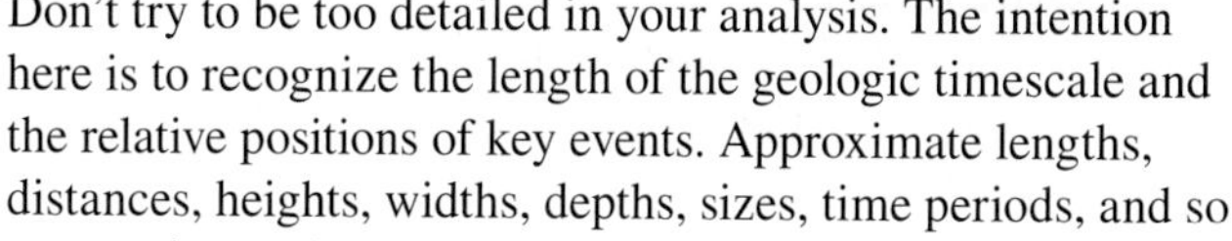

Distance from home goal line	Time, million years	Event
76 yards	3,500	Oldest fossil bacteria
26 yards	1,200	Oldest known animal fossil (jellyfish)
12 yards	542	Hard skeletons become common (fossils)
10 yards	458	First land plants (mosses)
	251	Widespread extinction ends Paleozoic era
1.4 yards	66	Dinosaurs become extinct
0.00036 inch	0.00051	Columbus landed, 1492

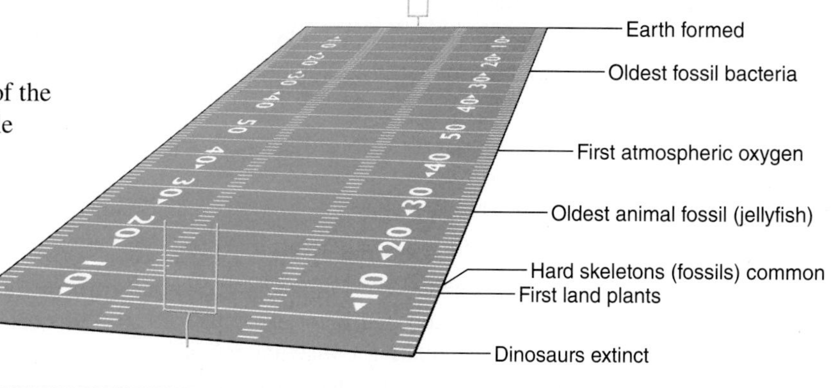

- About 251 million years ago, **thousands of eruptions took place over a period of 1 million years to form the Siberian Traps in northeastern Russia.** Over 1.6 million square kilometers (1 million square miles) were covered with lava, an area equivalent to the western half of the continental United States. How would that affect global ecosystems? Mapped deposits indicate that the Siberian eruptions had significantly more ash than modern-day basalt flows (see Chapter 6). The ash would have blocked incoming solar radiation and could have resulted in short-term cooling of the planet. Possibly more deadly were the sulfur dioxide and other gases emitted during these eruptions that would have contributed to widespread acid rain. Carbon dioxide emissions from the eruptions also may have exaggerated a global greenhouse effect, increasing long-term global temperatures.
- Scientists have discovered **chemical substances in P-T rocks that seem to point to an NEO impact event.** Some researchers use that evidence to argue that the extinction event might have accelerated when an asteroid collided with Earth approximately 251 million years ago. A potential impact site has not been unequivocally linked to this event. Keep in mind that a crater formed on the ocean floor would have been destroyed by plate tectonic processes. Imagine if a large asteroid or comet hit Earth (see Chapter 3). What would happen? If the object were large enough, fine debris could block incoming sunlight, causing rapid cooling of Earth.

Although none of these events alone would have caused a mass extinction on the scale of the P-T event, together their combined effects resulted in a massive change in Earth's physical and chemical environments. Species that were able to adapt to the changes survived to thrive in the following Mesozoic era, whereas those that failed to adapt breathed for the last time, and their remains became part of the fossil record. As more data related to these events are collected and analyzed, scientists will have a better understanding of when and why these extinction events occurred.

8.4 Numerical Time

Learning Objectives

- Explain why early attempts to estimate the age of Earth underestimated its true age.
- Describe how the radioactive decay of unstable isotopes can be used to determine the age of igneous and metamorphic rocks.
- Given isotopic parent and/or daughter information, determine the number of half-lives that have occurred and predict the age of a rock.
- Discuss how scientists can use the ages of igneous rocks to determine when layers of sedimentary rocks were formed.

The geologic timescale was established before scientists discovered a method for accurately calculating the actual age of rocks. Early attempts to determine the age of Earth estimated that our planet was several million years old. However, these efforts were based on an incomplete understanding of the rates of geologic processes and yielded ages that were much more recent than the actual time of Earth's formation. For example, over a century ago, scientists recognized that most of the salt in the oceans is delivered by streams that carry dissolved salts from the continents. Scientists were able to calculate the mass of salt in the oceans (16,000 trillion tons) and estimate the amount of salt contributed each year by the world's rivers (160 million tons). Dividing the latter number into the former yielded an age of 100 million years—old, but not old enough. What they missed was something we discussed in Chapter 7. The formation of chemical sedimentary rocks removes salt that would otherwise be dissolved in the ocean, thus reducing the salinity and resulting in an underestimation of the age of Earth.

Lord Kelvin, a prominent scientist in the 1800s, used an alternative, but no more accurate, method for calculating Earth's age. Knowing the volume of Earth, he used his knowledge of the physical properties of rocks to estimate how long it would take to cool the planet from an initial molten state. Kelvin repeated his calculations numerous times, making modest changes each time, to yield estimates of 30 to 100 million years. It turns out that Lord Kelvin's estimates were off because it had not yet been discovered that heat is produced from radioactive decay. In all, between 1890 and 1910, various scientists generated over 30 different estimates for Earth's age. While there was little agreement among these early methods, the results consistently predicted that the age of Earth was tens of millions of years. It was not until the discovery of the radioactive decay of uranium that science was able to provide a potential method for accurately determining the ages of certain rocks.

Radioactive Decay

Recall from Chapter 7 that the nucleus of each atom of each chemical element has a specific number of protons, known as the *atomic number;* that is how the element is defined. For example, the element potassium (K) always has 19 protons. Neutrons are also present in the nucleus. The combined number of protons and neutrons in the nucleus is the *mass number.* The number of neutrons varies in some elements. Atoms of the same element with different numbers of neutrons are called *isotopes.* For example, potassium has three isotopes with 20, 21, or 22 neutrons (Figure 8.18). While the atomic number of potassium is always 19, its mass number may be 39, 40, or 41, depending on the number of neutrons. These isotopes are identified as ^{39}K, ^{40}K, and ^{41}K.

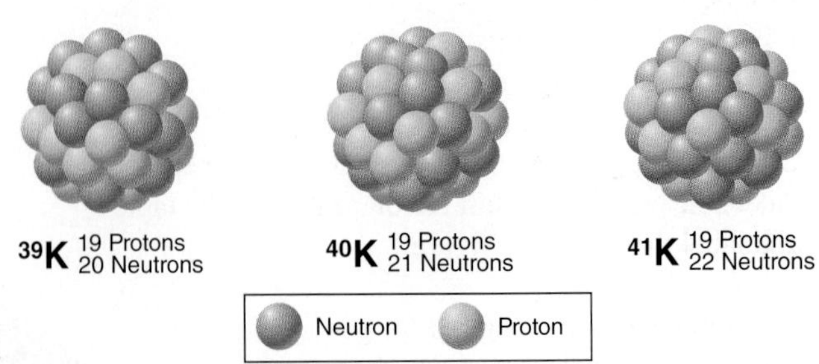

Figure 8.18 Isotopes of potassium. The three isotopes of potassium (K) are defined by contrasting numbers of neutrons in their atomic nucleus. The number of protons plus the number of neutrons equals the mass number of the isotope.

The positively charged protons in a nucleus repel one another, tending to make the nucleus unstable. When enough neutrons are present, these forces are minimized. Think of it as having buffers between the protons to keep them far enough apart to ensure that the nucleus is in stable state (it does not change). When a nucleus is unstable, it may spontaneously change to a more stable form through the process of radioactive decay. **Radioactive decay occurs when an unstable isotope changes to a new element** as a result of the emission or capture of subatomic particles (protons, neutrons, electrons; Figure 8.19). Decay is accompanied by the release of energy, some of which is converted to heat. Radioactive isotopes break down in Earth's interior, and this internal heat eventually reaches the surface, where it is released through the mechanism of plate tectonics and volcanism. Kelvin's estimate of Earth's age was so far off because he assumed that all of Earth's internal heat was left over from the planet's formation. The fact that radioactivity has been generating heat for billions of years meant that he seriously underestimated the age of Earth.

In analogy to parents and children, **the unstable original isotope is termed thea *parent isotope*, while the product of the decay is the *daughter*.** That daughter may be another radioactive isotope or an atom that is stable. For example, parent ^{40}K decays to daughter ^{40}Ar (argon) or ^{40}Ca (calcium), depending on whether protons are lost or gained, respectively (Figure 8.20). Radioactive decay starts and daughter atoms begin to accumulate in rocks as soon as a mineral forms. The age of the rock is taken as equivalent to the age of the mineral it contains. New minerals form when magma solidifies to form igneous rocks or when changing temperature results in the formation of metamorphic rocks.

Some neutrons or protons (or both) in the nucleus change during radioactive decay as the forces in the nucleus become more balanced. Protons may be transformed into neutrons when their positive charge is neutralized by the absorption of an electron. Neutrons may gain a positive charge and become protons by spitting out an electron (Figure 8.20). For example, the mass number of a parent isotope of potassium (40) stays the same, but the atomic number decreases by 1 when ^{40}K (19 protons, 21 neutrons) decays to ^{40}Ar (18 protons, 22 neutrons). In contrast, when ^{40}K decays to ^{40}Ca (20 protons, 20 neutrons), the atomic number increases by 1. Many decades of observations have shown that radioactive decay of specific isotopes occurs at a constant rate regardless of the physical or chemical conditions. By keeping track of the relative amounts of isotopes in individual minerals, we can calculate an age for the rock containing that mineral. Let's see how that is done.

Checkpoint 8.13

basic / intermediate / advanced / superior

Between 1860 and 1920, geologists attempted to estimate Earth's age by how long it would take for the thickest sequences of sedimentary rocks to form. Geologists examined sequences of rocks for each geologic period. From the estimated rates for the formation of these units, different scientists estimated ages for Earth ranging from 3 million years to 15 billion years. Explain why these estimates varied over such a wide range, based on what you know about how sedimentary rocks form.

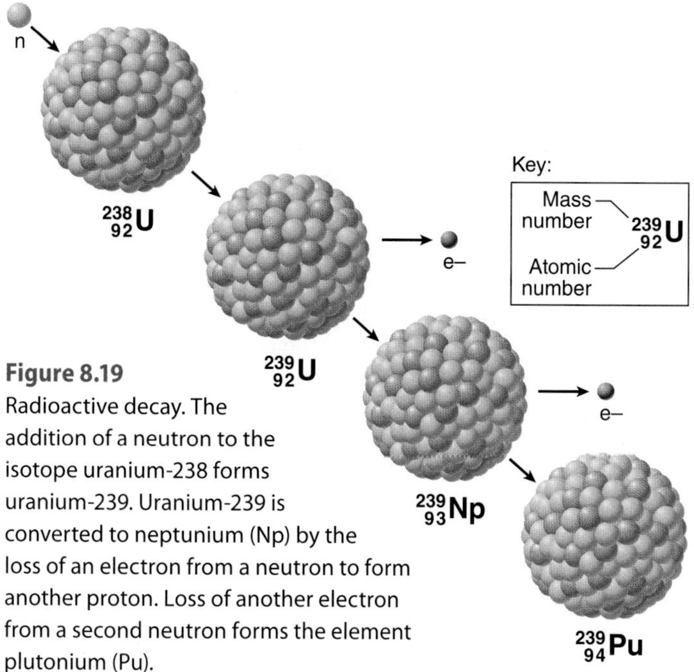

Figure 8.19 Radioactive decay. The addition of a neutron to the isotope uranium-238 forms uranium-239. Uranium-239 is converted to neptunium (Np) by the loss of an electron from a neutron to form another proton. Loss of another electron from a second neutron forms the element plutonium (Pu).

Half-Lives

Take a piece of paper, tear it in half, and set one half aside. Keep doing this every 5 seconds until you cannot tear it in half any more. How long did it take, and how many pieces did you end up with? This little exercise is analogous to radioactive decay. The half-life of the paper is 5 seconds. The half-life of an isotope is

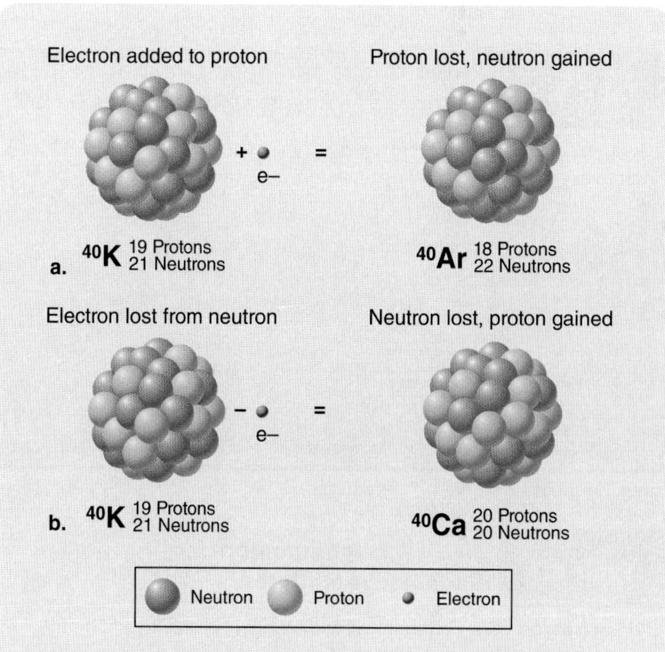

Figure 8.20 Electron absorption and emission changes the relative proportions of protons and neutrons in parent and daughter isotopes. An isotope of potassium (K) decays to form either **a.** argon or **b.** calcium as a result of the loss or gain of a proton, respectively.

the time taken for one-half of the parent isotopes to convert to daughter atoms. When a mineral originally forms, prior to decay, 100 percent of the parent isotope exists because radioactive decay has yet to occur (assuming no daughter element was present at the time of formation). As time passes, the amount of the parent isotope decreases, and the quantity of the daughter atoms increases (Figure 8.21). After one half-life, 50 percent of the parent remains, and the other 50 percent of the original atoms are converted to daughter atoms (Table 8.2). After two half-lives, the number of parent isotopes is again halved (25 percent), and the number of daughter atoms increases by an equivalent amount (to 75 percent).

The relative proportion (ratio) of parent isotopes and daughter atoms can be used to determine how many half-lives have passed since the formation of the mineral that contained the parent radioactive isotopes (Table 8.2; Figure 8.21). The half-lives for several of the most common radioactive isotopes are measured in millions or billions of years (Table 8.3). Isotopes with half-lives of billions or hundreds of millions of years can be used to determine the ages of the oldest rocks on Earth.

Let us look at an example of how dating can be done using half-lives: The calculated age of some rocks in northwestern Canada is approximately 4 billion years. The half-life for ^{238}U isotopes is 4.5 billion years; therefore, these isotopes would have experienced a little less than one half-life in a 4 billion-year-old rock. Such rocks should have a few more ^{238}U parent isotopes (55 percent) than ^{206}Pb daughter atoms (45 percent; see Tables 8.2 and 8.3). In comparison, the half-life for ^{235}U to ^{207}Pb decay is 0.7 billion years; ^{235}U isotopes in a 4 billion-year-old rock would have experienced nearly six half-lives and would have approximately 2 percent of the ^{235}U parent isotope remaining and 98 percent of ^{207}Pb. Not all radioactive isotopes decay over such long periods. The carbon isotope, ^{14}C, experiences a half-life every 5,730 years; ^{14}C is used to date archeological artifacts such as cloth fragments from Egyptian pyramids and some very young sedimentary rocks (younger than 50,000 years of age) that have previously living material trapped within them.

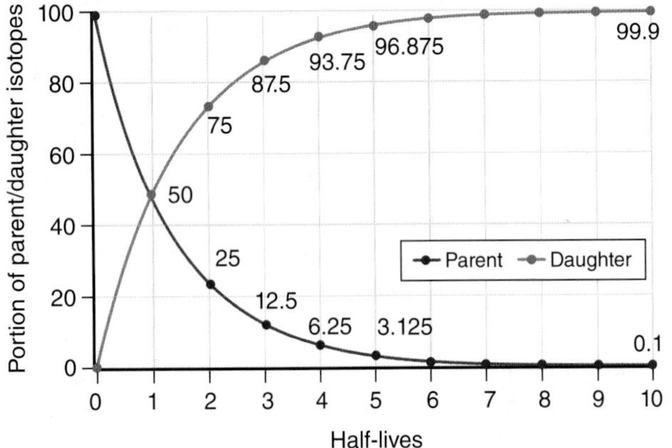

Figure 8.21 Change in percentage of parent and daughter isotopes undergoing radioactive decay. Notice that after about 10 half-lives, very little parent material remains.

Table 8.2 Proportion of Parent and Daughter Isotopes as a Result of Radioactive Decay

Number of half-lives	Parent isotopes (% total isotopes)	Daughter isotopes (% total isotopes)
0	100	0
1	50	50
2	25	75
3	12.5	87.5
4	6.25	93.75
5	3.125	96.875

Table 8.3 Half-Lives for Common Radioactive Isotopes

Parent isotope	Daughter product	Length of half-life
Uranium 238 (^{238}U)	Lead 206 (^{206}Pb)	4.5 billion years
Uranium 235 (^{235}U)	Lead 207 (^{207}Pb)	704 million years
Rubidium 87 (^{87}Rb)	Strontium 87 (^{87}Sr)	48.8 billion years
Potassium 40 (^{40}K)	Argon 40 (^{40}Ar)	1.25 billion years
Carbon 14 (^{14}C)	Nitrogen 14 (^{14}N)	5,730 years

Applying Both Relative and Numerical Time

Because sedimentary rocks are composed of recycled minerals from other rocks, radioactive isotopes found in rocks like sandstone cannot be used to date the formation of the sedimentary rock itself. The minerals in those rocks would simply yield the dates

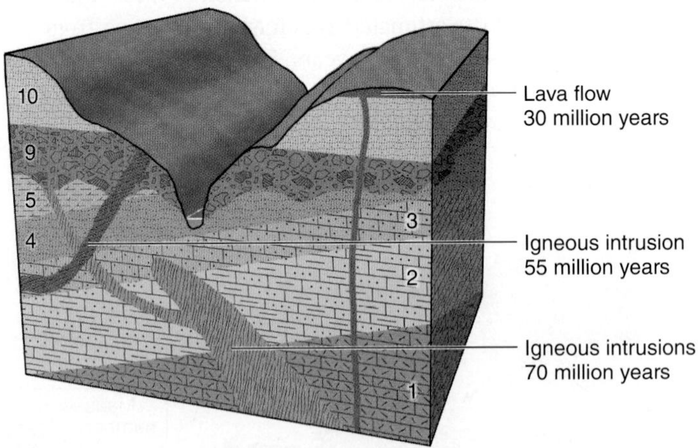

Figure 8.22 Relative time and numerical time elements can be used together to determine the ages of sedimentary rocks. From the information in this figure, we can estimate that beds 1 through 5 are greater than 70 million years old because they are cut by all the intrusions and must be older than the oldest intrusion. The bottom of bed 9 lies on an unconformity that eroded the oldest intrusion (70 million years old), but bed 9 is cut by a younger intrusion (55 million). Therefore, bed 9 must be 55 to 70 million years old. Bed 10 is younger than bed 9 (superposition), but the relationship between the 55 million years intrusion and bed 10 is unresolved because we can't tell from the information available if the bed truncates the intrusion or the intrusion cuts the bed. However, the bed is overlaid by a lava flow dated at 30 million years old, so we can estimate the age range for bed 10 as between 30 and 70 million years old.

for their igneous or metamorphic source rocks. Sedimentary rocks are arranged in relative chronological order by using the relative dating methods discussed in Section 8.2. Therefore, **the numerical age of sedimentary rocks is most accurately estimated by using the cross-cutting relationships for igneous intrusions and the principle of superposition for layers of volcanic ash or lava flows** (Figure 8.22). Dates can be obtained for the igneous rocks that will provide minimum or maximum ages for the formation of the adjoining sedimentary layers. Although highly effective, such techniques are limited to regions where there are enough igneous rocks to provide a sufficient range of ages to accurately estimate the ages of the associated sedimentary rocks.

One of the greatest challenges for earth scientists who worked to establish accurate dates for geologic events was to determine when organisms with shells or skeletons first became abundant on Earth. These were the first animal remains to be readily preserved as fossils in rocks across the world. These rocks represent the beginning of the Cambrian period at the start of the Phanerozoic eon. Geologists have long sought to limit the age of these rocks by dating the youngest igneous rocks immediately below this boundary or the oldest igneous rocks cutting across units found above the contact between the oldest Cambrian and the youngest Proterozoic rocks. There are several locations around the world where fossil-bearing Cambrian rocks are interlayered with volcanic rocks. Individual minerals within these volcanic rocks were dated at 521 to 539 million years old. Ash deposits, lavas, and plutons in the youngest Proterozoic rocks are more than 543 million years old (Figure 8.23). Recent analyses of volcanic rocks immediately adjacent to the oldest Cambrian rocks have yielded an estimated date of 542 million years ago for the beginning of the Cambrian period.

Checkpoint 8.14

basic / intermediate / advanced / superior

Radioactive isotopes in clastic sedimentary rocks always predict an age that is

a) older than the sedimentary rock.
b) younger than the sedimentary rock.
c) correct for the sedimentary rock.

The isotope of element X has 15 protons, 17 neutrons, and 15 electrons. The element has an atomic number of _____ and a mass number of _____.

a) 15; 32 c) 17; 47
b) 17; 15 d) 15; 30

If radioactive decay began with 400,000 parent isotopes, how many would be left after three half-lives?

a) 200,000 c) 50,000
b) 100,000 d) 25,000

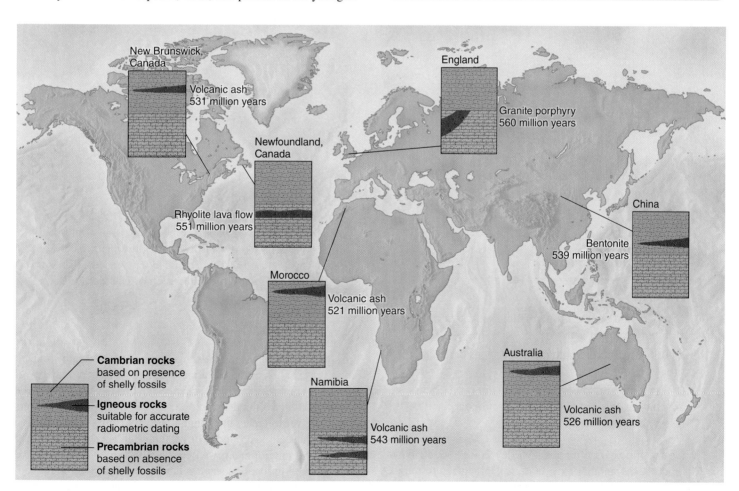

Figure 8.23 Deciphering sequences. Relative time and numerical time elements can be used together to determine the ages of sedimentary rocks. From this information, we can establish the age of the start of the Cambrian period as 542 million years ago.

Checkpoint 8.15

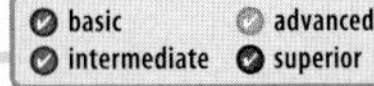

The half-life of a radioactive isotope is 500 million years. Scientists testing a rock sample discover that the sample contains three times as many daughter atoms as parent isotopes. What is the age of the rock?

a) 500 million years
b) 1,500 million years
c) 1,000 million years
d) 2,500 million years

Checkpoint 8.16

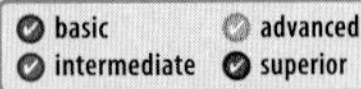

The following diagram represents three rock exposures containing fossils. Each exposure contains a layer of volcanic ash (in red) that has been dated by the analysis of $^{238}U/^{206}Pb$ isotopes.

a) Place the fossils in the correct order according to their relative ages, from oldest to youngest.
b) Explain how you would estimate the potential age ranges of the C, G, and K fossils on the basis of the ages determined for the three volcanic ash layers.

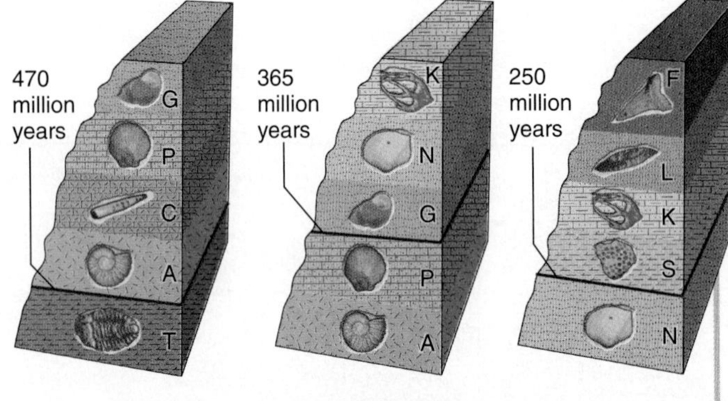

(Question adapted from J. Dodick and N. Orion, "Measuring student understanding of geological time," *Science Education,* 2003, vol. 87, pp. 708–31.)

a.

b.

Figure 8.24 Slow pace of geologic change. These two views of the Green River were photographed from the same location in Canyonlands National Park nearly a century apart. Image **a.** was taken during Powell's expedition in 1871, and image **b.** was taken in 1968. Note that the landforms have changed little over the time interval.

8.5 Rates of Change

Learning Objectives

- Compare and contrast the principles of catastrophism and uniformitarianism.
- Identify the approximate rates of a series of natural processes or events.
- Define high- and low-magnitude events and discuss which type of event had the greater impact on the history of life on Earth.

Nearly 100 years after John Wesley Powell set off down the Colorado River, another expedition was organized to retrace Powell's route and to take photographs from the same locations to identify any geologic changes in the river and its canyons. Although the later expedition did note some changes in vegetation, they saw no significant changes in the arrangement of the rocks, the route of the river channel, or the elevation of the canyon at the majority of locations (Figure 8.24). Small changes that had occurred could be attributed to local floods from side canyons that dumped boulders into the river to create new rapids or to the collapse of a section of cliff that had been undercut by the river. As we look around our natural environment, we occasionally see evidence of rapid change due to processes such as landslides. But more often, geologic changes, such as the movements of the plates, occur at rates that are too slow and gradual to be readily apparent on a human scale. These changes would only be evident over historical time.

Checkpoint 8.17

basic · intermediate · advanced · superior

We daily encounter evidence of things that have changed over time. For example, an instructor finds a stick of chalk that has become too small to use, or a student finds that jeans have become so worn that a hole has formed in the fabric. Identify three examples of everyday objects that change over time but at different rates. For example, something that is used up or worn out in a matter of days (for example, chalk), months (for example, jeans), or years.

Catastrophism

Prior to the acceptance of the scientific method, people recognized the grandeur of landforms, but because of the gradual nature of geologic processes, they could not identify how the landscape had developed. Since no plausible explanation for features such as mountains and oceans could be found, they were thought to have formed by catastrophic events at scales beyond human comprehension, a perspective known as *catastrophism*. **Catastrophism suggests that Earth has been (and can be) affected by short-duration, sometimes violent events that may be global in nature.** What is an example of a catastrophic event—one that is relatively rare but could have global consequences? Think back to Chapter 3, Near-Earth Objects. There is ample evidence that asteroid or comet impacts occurred during the geologic past, though infrequently; recall, for example, the K-T and P-T extinction events. Thus, it is necessary to recognize that some events are catastrophic in nature and that they are relatively rare. However, explanations for **catastrophic events that have no precedents and that cannot be explained by physical or chemical processes are not science.** For instance, the deposition of the rock layers in the Grand Canyon during a single flood and the carving of the canyon during a few days soon afterward would be catastrophic. However, we know that not all the rock units formed under water and that there is no evidence anywhere else on Earth of a similar event. Thus, a catastrophic origin is not a valid scientific explanation for the evolution of the Grand Canyon. Likewise, the concept of catastrophism cannot account for the vast majority of features on Earth.

Uniformitarianism

As mentioned in Section 8.2, the writings of James Hutton suggested that the **features on Earth's surface were formed by the same processes we see operating today.** This concept became known as **uniformitarianism,** often summarized by the phrase *the present is the key to the past* (Figure 8.25). This idea has a lot of currency in science. Scientists acknowledge that much of their work has been successful because they assume that nature's laws do not change over time. For example, to understand stars, we must assume that gravity, the speed of light, and the properties of atoms are invariant (unchanging) and follow nature's laws, not just on Earth but throughout the universe.

Some early geologists, most notably Charles Lyell, the first person to write the equivalent of a geology textbook, took the idea of uniformitarianism quite literally. Lyell assumed that not only were the slow, gradual processes operating on Earth today the same as in the past, but also the rates at which they operated were similar. As time passed, geologists adopted a less restrictive view of uniformitarianism that incorporated geologic evidence of catastrophic events tied to natural causes. In doing so, they recognized the importance of both relatively rare events that affected large areas (*high-magnitude events;* for example, tsunami, asteroid impact) and more-frequent processes that caused more-localized changes (*low-magnitude events;* for example, floods, soil formation).

a.

b.

Figure 8.25 Uniformitarianism. **a.** These mud cracks formed recently, while **b.** the mud cracks preserved in rocks are millions of years old. The concept of uniformitarianism would suggest that the ancient mud cracks formed under the same conditions necessary for the formation of the modern mud cracks.

The Grand Canyon: Applying Uniformitarianism. We can apply the concept of uniformitarianism to both the rocks of the Grand Canyon and the canyon itself. The 150-meter-thick (490-foot) Redwall Limestone forms steep cliffs throughout the canyon. Similar modern limestone environments have depositional rates of 1 to 3 centimeters (0.4 to 1.2 inches) every 1,000 years. Extrapolating this rate to the Redwall Limestone indicates that it would have taken approximately 5 to 15 million years to form, well within the 40-million-year interval of the Mississippian period (see Figure 8.16). The numerical ages of the lavas that flowed over the walls of the canyon, and therefore over the canyon itself, have been analyzed, and the oldest flows were found to be more than 1 million years old. The highest flows indicate that the largest lava dams must have formed giant plugs, some up to 60 stories (200 meters; 660 feet) high and over 30 kilometers (19 miles) long, that blocked the canyon on at least a dozen occasions. None of these dams remains. The river would have initially formed a reservoir before creating a waterfall as the reservoir level rose and eventually flowed over the lava dam. Water cascading down a waterfall would undercut the cliff that supports the waterfall itself, slowly backing its way upstream, a process known as headward erosion.

Headward erosion of 1 meter (3.3 feet) per year is observed at Niagara Falls, a modern analogy with softer underlying rocks. At these rates, it is unlikely any of the lava dams lasted more than 30,000 years. Scientists estimate that cumulatively, it would have taken about 250,000 years for the river to cut through all of the dams. The 1,500-meter-deep (4,900-foot) canyon would have taken longer to form through a combination of headward erosion and channel erosion of the Colorado River. Estimates place the time of canyon formation between 1 and 5 million years ago, lasting for about 1 to 4 million years, depending on a combination of factors, such as the volume and velocity of the ancient Colorado River, regional climate, and the rate of uplift of the land surface.

While this geologic evidence is compelling for those who accept the scientific process, it is not embraced by those who come to the Grand Canyon believing in the literal accuracy of the biblical account of creation. Such individuals believe the rocks in the canyon walls formed in a matter of weeks or months and the canyon was cut in a matter of days. This view is not supported by scientific evaluation, since such rates of deposition and erosion cannot be replicated in nature, nor can evidence of such rates be discerned in the geological past. Therefore, creationists' ideas fail to meet the standards of evidence demanded by the scientific method.

Nevertheless, despite the evidence of science, a large segment of the US population implicitly supports a brief time line for the evolution of the Grand Canyon. Five times since 1982, the Gallup polling organization has asked Americans a consistent series of questions about the origin and development of the human species. About 45 percent of respondents have consistently selected the statement closest to biblical creationism: "God created human beings pretty much in their present form at one time within the last 10,000 years or so." The remainder of those questioned chose statements involving much longer geologic timescales: 37 percent selected "Human beings have developed over millions of years from less advanced forms of life, but God guided this process," and 12 percent chose "Human beings have developed over millions of years from less advanced forms of life, but God had no part in this process." The other 6 percent of respondents reported that they had no opinion or did not know. Despite what some people believe, there are many lines of evidence that support the observation that Earth is very old. For instance, scientists can count annual markers such as layers of ice from cores drilled in Antarctica and find correlations with growth rings from tropical corals that go back several hundred thousand years. Many rock ages are supported by more than one radioisotope dating method. While there are some conflicting age dates in the literature, most of these examples have been found to be the result of contaminated samples or analyses that contained experimental errors. The science of geologic time is overwhelmingly strong, and a variety of data, evidence, and analyses are consistent the concept of deep time stretching back billions of years to Earth's origin.

What many scientists try to explain, but often gets overlooked, is that science is a process, while religion is a belief system. The media often represent science and religion as if they are at odds with each other, but they remain two different things. Consequently, you should no more expect a lecture on Newton's laws of motion and gravity in a church, synagogue, or mosque than you should expect to learn about the laws of Moses in your earth science class.

Checkpoint 8.18

Rates Time Line

Events happen on Earth over periods of time that vary from seconds to millions of years. Place each of the following events in the appropriate location on the time line provided here according to either its frequency (how often?) or the length of time over which it occurs (how long?).

1. The time between large eruptions of the same volcano
2. A season (for example, spring)
3. Time between great earthquakes on the San Andreas fault
4. Period required to form the Atlantic Ocean
5. Formation and decay of a tornado
6. Earth's orbit around the sun
7. Length of orbit for a long-period comet
8. Time between mass extinctions
9. Time required to carve the Grand Canyon
10. Growth of major US cities
11. Formation and decay of a hurricane

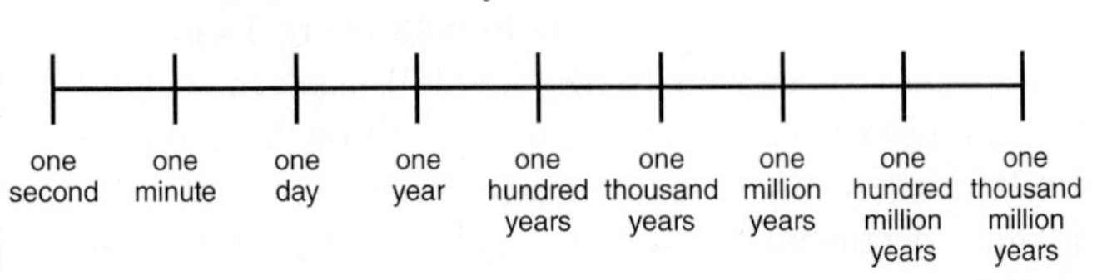

Checkpoint 8.19

We have just provided some examples of rare, high-magnitude events and common, low-magnitude events. Review the previous chapters and identify other examples of high- and low-magnitude events.

Checkpoint 8.20

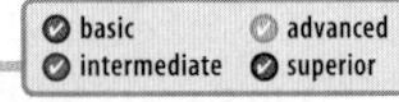

We have presented an Earth history stretching back 4.6 billion years. Has the history of life on Earth been more affected by rare, high-magnitude events or frequent, low-magnitude processes? Justify your choice.

the big picture

Our understanding of Earth's past rests on the nature of scientific inquiry. The scientific method argues that some ideas offer better explanations for natural phenomena than others and that those ideas can be tested over and over with similar results. Competing claims about the origin of the Grand Canyon have become a focal point of the latest battle over interpretations of Earth history. This chapter lays out evidence from expeditions into the Grand Canyon to help you formulate your own hypotheses as to its origin and age. Our understanding of Earth's origin, age, and rates of change rests on empirical evidence that has withstood the scrutiny of the scientific process.

In previous chapters, we have described the application of the scientific method in which predictions were made that could later be tested. When dealing with the past, our predictions deal with the rates at which events may occur or with what types of fossils should be present in particular assemblages of rocks. For example, we would predict that the fossils found in the rocks of the Grand Canyon should show a trend from more primitive forms in older rocks deep in the canyon to less primitive forms preserved higher in the canyon. Likewise, if these rocks were really formed at the same time as similar layers found in Canada, we would predict that we would find similar fossils in both locations, and in fact, we do.

The scientific method can also be used to falsify the claims of competing hypotheses. For example, if the Grand Canyon were carved in a matter of days, we would expect to see similarly rapid erosion rates today, but we do not. Indeed, the fact that modern rates of deposition are so incredibly slow leads to another prediction: It must have taken millions of years to form individual rock layers in the Grand Canyon. Relative time principles, coupled with our understanding of rock formation, can help us establish the sequence of events at a location. Furthermore, because the rocks of the Grand Canyon represent just a small slice of geologic time, Earth itself must be much older, just as radioactive dating shows us.

Much of this chapter has dealt with change, but there are some places on Earth with very slow erosion and weathering rates where change is barely perceptible. Cosmic radiation produces radioactive isotopes in rocks at Earth's surface. The longer the rocks remain at the surface, the more these isotopes are formed. Unfortunately, the isotopes form at an incredibly slow rate and so could not be accurately measured until the appropriate technology (accelerator mass spectrometry) came along at the end of the twentieth century. Scientists studying quartz pebbles from the very dry Atacama Desert in Chile (see chapter opening image) have discovered that the pebbles have remained at the land surface for more than 20 million years, making this the oldest landscape in the world. In most environments, rocks would be buried in soils or carried away in streams. But the Atacama Desert is the driest place on Earth (less than 3 millimeters [0.1 inch] of precipitation per year), and without water, there has been no erosion to remove the pebbles or weathering to break them down into finer particles to be removed by winds.

In this chapter, we read about John Wesley Powell's expedition down the Colorado River and the one that followed his route nearly 100 years later. Consider the change in the tools (that is, the technology) the two teams used to collect observations, make measurements, and map the location. Advances in technology—such as accelerator mass spectrometers—inspire new investigations and questions, which in turn motivate us to do further research. Imagine how the technologies of the twenty-second century will enable us to learn about changes in Earth.

Geologic Time: Concept Map

Complete the following concept map to evaluate your understanding of the interactions between the earth system and geologic time. Label as many interactions as you can, using information from this chapter.

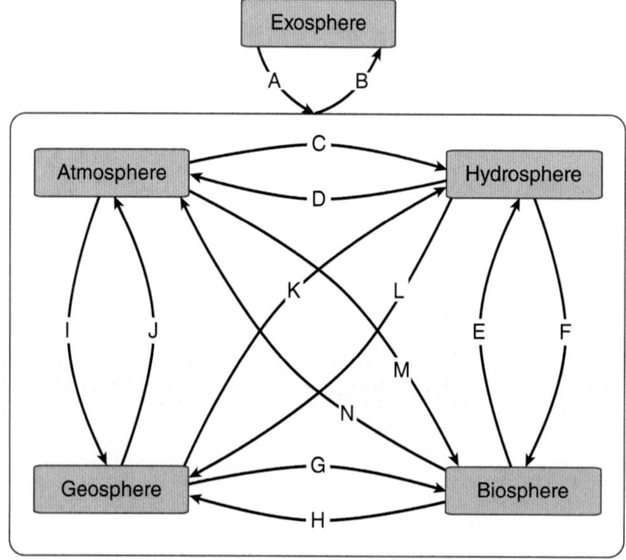

"For water continually dropping will wear hard rocks hollow."

—Plutarch, Greek historian

"When the soil is gone, men must go; and the process does not take long."

—Theodore Roosevelt, US president

Chapter 9

Weathering and Soils

Chapter Outline

the big picture

How did these volcanic rocks develop the appearance of Swiss cheese?

See the Big Picture box at the end of this chapter for the full story on this image.

9.1 The Dirt on Weathering

Chapter Learning Outcomes

- Students will relate weathering processes to landscape evolution.
- Students will explain the connection between weathering rates and geologic time.
- Students will describe connections between weathering and soil formation.
- Students will discuss how weathering processes are related to the degradation of cultural structures.

The morning routine is the same in countless kitchens around the world. Coffee beans are sliced up into tiny pieces in a coffee grinder. Water drips over the dry coffee grounds and is transformed into a cup of dark brown coffee. A spoonful of white sugar is added, quickly disappearing as it dissolves in the hot liquid. Each step in this process produces a chemical or physical change.

These commonplace actions are similar to the simple geologic processes that slowly wear down rocks at or near Earth's surface and build the soils below our feet. **Weathering is the physical, chemical, and biological breakdown of rocks and minerals;** we discussed these changes briefly in relation to the generation of clastic sedimentary rocks in Chapter 7. Rocks that are now exposed to weathering at Earth's surface were originally formed under much different temperature and pressure conditions, often in the absence of air and water. Weathering breaks down rocks and converts many of the minerals into new minerals that are stable under the low-temperature and low-pressure conditions at the surface. Weathering modifies the landscape around us and generates soil, one of our essential resources.

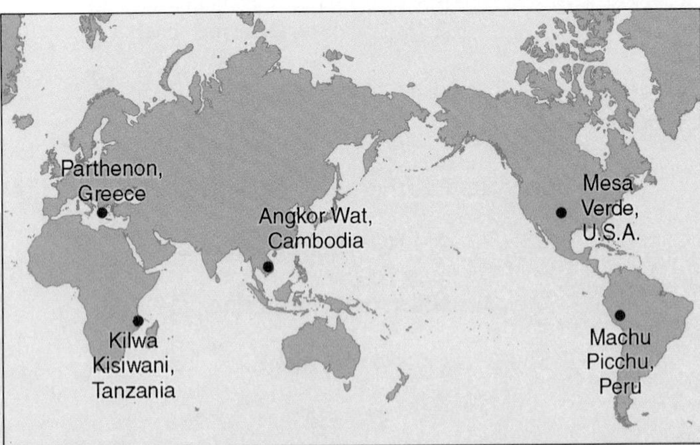

Figure 9.1 Five World Heritage Sites discussed in this chapter. Note that all sites are located in relatively warm climates. The sites include: **a.** the eleventh-century temple complex at Angkor Wat, Cambodia; **b.** the thirteenth-century cliff dwellings at Mesa Verde, Colorado; **c.** the 2,400-year-old Parthenon in Athens, Greece; **d.** Machu Picchu, Peru, constructed by the Incan culture high in the Andes Mountains in the fifteenth century; and **e.** a twelfth-century mosque in Kilwa Kisiwani, Tanzania.

Weathering of Cultural Sites

The first half of the chapter considers weathering processes in nature but also has implications for our shared cultural heritage. In November 1972, the United Nations Educational, Scientific and Cultural Organization (UNESCO) adopted an agreement called the Convention Concerning the Protection of the World Cultural and Natural Heritage. This international accord seeks to protect the cultural heritage of all nations as represented by monuments, archeological sites, and buildings of characteristic architecture. The convention established a World Heritage List that includes hundreds of cultural sites "of outstanding value to humanity." These sites are threatened by deterioration by a variety of factors, including weathering. Eight of these cultural sites are in the United States, including 800-year-old cliff dwellings at Mesa Verde, Colorado, and the Statue of Liberty, New York. We will consider natural weathering processes in the context of the deterioration and protection of five key structures on the World Heritage List (Figure 9.1).

The convention requires each country to develop scientific and technical studies and to conduct research to counteract the dangers that threaten its cultural heritage. It also provides economic assistance from the World Heritage Fund to help protect key sites "in accordance with modern scientific methods." Before any World Heritage Site receives financial assistance, it must first undergo a detailed scientific, economic, and technical evaluation. For example, a team of Japanese scientists has identified the causes of decay in building stone used in the construction of the eleventh-century temple complex at Angkor Wat, Cambodia (Figure 9.1a). Likewise, research shows that the aging of other historical buildings such as Greece's Parthenon (Figure 9.1c) is accelerated by the air pollution that is a consequence of life in modern cities.

The questions scientists seek to answer include: Why do some rocks (and building stones) show more deterioration than others? Why is weathering faster in some places than in others? What natural processes are most responsible for the deterioration of rocks and minerals (and World Heritage Sites)? What can be done to stop, or at least slow down, these natural processes? Which sites are in greatest need of this assistance from the World Heritage

a.

b.

Fund? Which is more threatened by weathering—a 900-year-old mosque along the coast of East Africa or a fifteenth-century archeological site in the mountains of South America? Which can be more readily protected?

Where Does Dirt Come From?

The second half of the chapter asks you to think about exactly what is under your feet? What would you find if you took a shovel and began digging a hole in the middle of a grassy lawn on your campus? If examined closely, the material that we commonly call "dirt" (and that is more properly termed *soil*) includes rock and mineral fragments, known as regolith. Regolith is produced by weathering of the solid bedrock underlying the regolith. **Soil is a mixture of regolith, water, air, and sufficient organic material to support plant life** (Figure 9.2). As such, soil contains components of all four parts of the earth system: geosphere (rock and mineral fragments), atmosphere (air), hydrosphere (water), and biosphere (organic material). The quantity and quality of soil resources depend on natural factors such as original rock material, weathering rates, and organic activity as well as human activity.

While soil forms slowly by weathering, soil erosion often occurs at rates that outpace soil formation. The loss of soil through erosion has been a persistent threat to US agriculture for centuries. By the end of the 1700s, the negative consequences of early farming techniques were becoming apparent in the abandoned fields of the eastern United States. Thomas Jefferson lamented poor farmers who ". . . run away to Alabama, as so many of our countrymen are doing, who find it easier to resolve on quitting their country, than to change the practices in husbandry (farming) to which they have been brought up." Today, the strategies that can be applied to reduce soil erosion on agricultural lands are not that different from those employed by Jefferson and his contemporaries. In the future, changing climate conditions may make it even more difficult for farmers to manage soil resources. It is for these situations that it becomes important for us to identify ways to protect soil from erosion.

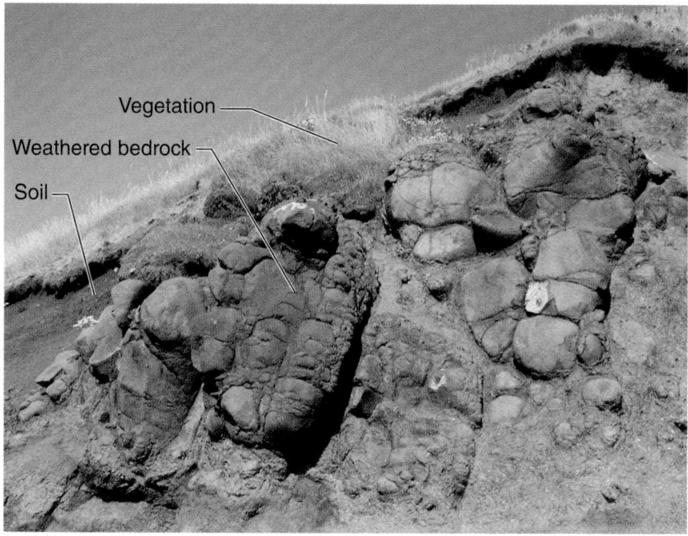

Figure 9.2 Soil and soil components. A thin soil overlying bedrock, Giant's Causeway, Northern Ireland. Soil is mostly composed of rock and mineral fragments (geosphere) with approximately equal parts water (hydrosphere) and air (atmosphere) and lesser amounts of organic material from the decay of plants or animals (biosphere). On average, soil is a mixture of 45 percent mineral fragments, 25 percent water, 25 percent air, and 5 percent organic material.

Self-Reflection Survey: Section 9.1

Answer the following questions as a means of uncovering what you already know about weathering and soils.

1. What factors should be considered in choosing a building or monument to add to the World Heritage Site list?
2. Which building would you select to preserve in the city or region where you live? Why?
3. Have you ever dug a hole in the ground that exposed some soil or observed excavation during building construction? Describe what the soil looked like. Did you reach solid rock?

c.

d.

e.

9.2 Physical Weathering

Learning Objectives

- Describe the processes that cause earth materials to physically break down.
- Compare and contrast mechanisms of physical weathering.

Physical weathering represents the disintegration of rocks and minerals into smaller pieces. People in northern states experience physical weathering firsthand by observing the potholes that form on paved streets during winter. Physical weathering can be further subdivided into unloading and wedging.

Unloading

Think about walking onto a diving board. Your weight causes the board to bend downward. The board "springs" back when you jump off because you remove the load on the board. Rocks below Earth's surface support the weight of the overlying mass of rock, much as a diving board supports you. Erosion strips away this overlying rock and decreases the load (or overlying pressure) on the buried rocks. It's hard to believe, but rocks are slightly elastic and can be compressed by the weight of a thick pile of overlying rocks (just as a diving board bends when you stand on it). **As overlying rocks erode, pressure is removed, a process known as *unloading*, and buried rocks expand upward.** This expansion creates pressure release cracks (joints) in the rocks that are parallel to the surface (Figure 9.3).

So what do you think happens as a result of these cracks? The cracks are zones of weakness that can cause slabs of rock to break off along the fractures. Sometimes, weathering exposes outcrops of previously buried rocks that are more resistant to weathering than those nearby. The resulting landform is a rounded hill, called an **exfoliation dome,** that stands out against the surrounding landscape (Figure 9.4; Chapter Snapshot). Exfoliation domes are most common in mountains, where erosion is more rapid. For example, exfoliation domes are readily observable in the Sierra Nevada, California (see Figure 7.22). Pressure release cracks are most likely to be observed in plutonic igneous rocks such as granite that lack any preexisting layering. Rocks with bedding planes (sedimentary rocks) or foliations (metamorphic rocks) are more likely to expand along these existing planes than to form new cracks as the overlying load is reduced.

Figure 9.4 An exfoliation dome. **a.** Bare rock and rounded surfaces at Independence Rock, Wyoming; **b.** inscriptions on the rock's surface were left by travelers on the Oregon Trail over 150 years ago.

Figure 9.3 Unloading and pressure release cracks. **a.** The weight of overlying material puts buried rocks under pressure. **b.** As the overlying rock is removed by erosion, the buried rocks expand upward to form pressure release cracks. The bare, rounded rock surfaces form an exfoliation dome. **c.** Pressure release fractures at Half Dome, Yosemite National Park, California.

Several kilometers

Batholith

a.

Sheet joints

Expansion

Uplift and erosion of region

b.

c.

Emigrants who traveled westward on the Oregon Trail in the 1840s were able to mark their journey using landmarks that had formed after thousands to millions of years of weathering. The rounded granite dome of Independence Rock, Wyoming, provided a site for so many inscriptions that it was dubbed the Register in the Desert (Figure. 9.4). The slow rate of geologic weathering processes in the dry western climate ensures that these natural landforms look the same today as they did when the first emigrants passed them over 150 years ago; however, the inscriptions these early travelers left behind are slowly fading because of continual weathering processes.

Checkpoint 9.1

basic · intermediate · advanced · superior

Examine the following photos. Both images show granite outcrops in the Sierra Nevada, California. Which outcrop contains pressure release cracks? Explain your choice.

a) Outcrop A
b) Outcrop B

a.

b.

Wedging

In areas where there are asphalt roads, traffic driving over the road surface causes the asphalt layer to bend and eventually to form tiny cracks. Water freezes in the cracks if the temperature drops below 0°C (32°F). Water increases in volume by 9 percent when it freezes. This greater volume of ice forces the cracks to expand, making them deeper and longer. The ice thaws when the temperature rises above freezing, and more water enters the cracks, only to freeze again and make the cracks even bigger. This freeze-thaw cycle continues, eventually letting water infiltrate below the asphalt and freeze, buckling the road surface upward. The daily wear and tear of traffic breaks up the asphalt, leaving behind a pothole (Figure 9.5).

Some of the same processes that form potholes on road surfaces also help break down rocks. All rocks have tiny cracks. When water gets into the cracks and freezes, it expands and causes a rock to break apart (Figure 9.6a). **This process is termed *wedging*,**

Figure 9.5 A pothole formed by wedging as water froze and expanded.

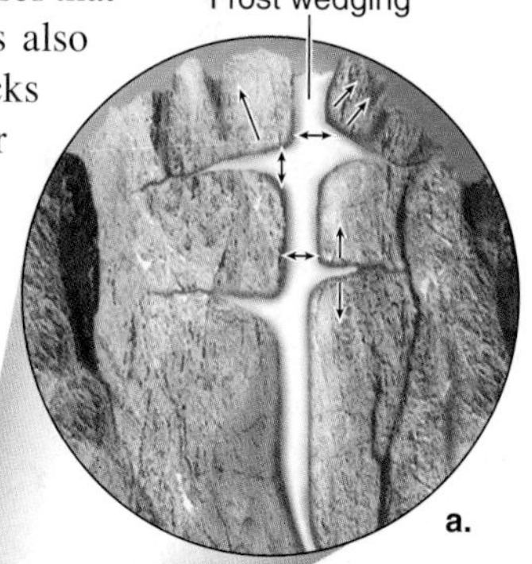

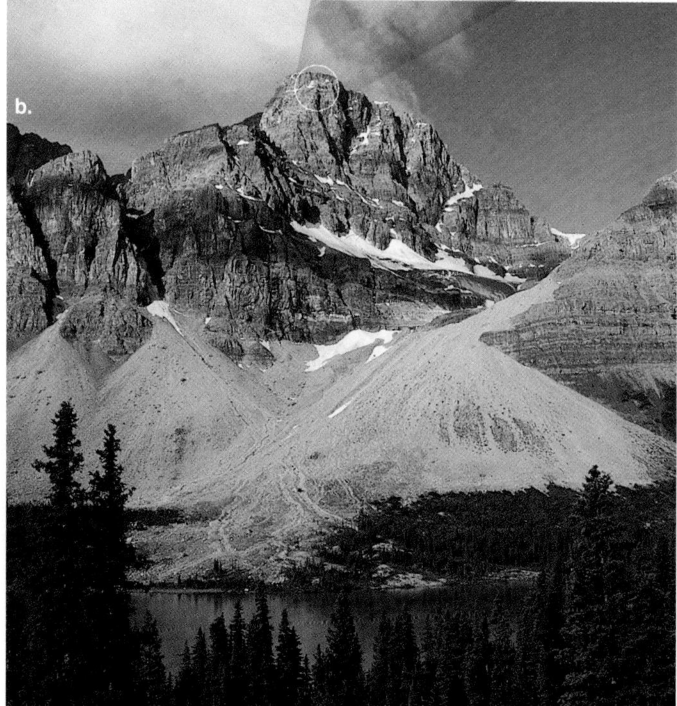

Figure 9.6 Frost wedging and talus. **a.** Frost wedging occurs as water goes through freeze-thaw cycles in cracks near the surface. **b.** Wedging breaks off rock fragments that collect at the base of the cliff to form a talus slope.

or more specifically, *frost wedging*. For frost wedging to occur, the temperature must alternate regularly above and below freezing (0°C, 32°F). Frost wedging breaks off rock fragments or blocks on exposed surfaces such as cliffs or mountain peaks (Chapter Snapshot). These angular blocks subsequently tumble down slopes to form a mound of broken rocks known as a *talus slope* (Figure 9.6b). Where would you predict that frost wedging would occur in North America? Areas that are almost continually covered in snow are too cold for most of the year to have ice melt, and many southern states do not get temperatures below freezing for long enough to have an active freeze-thaw cycle. However, states in the Midwest and Northeast have longer winters and springs, when temperatures may cycle back and forth from freeze to thaw frequently enough to make frost wedging a significant weathering process. Likewise, rocks at high elevations where temperatures are often below freezing at night may also be prone to wedging.

Wedging is caused by other substances besides water. A type of expansion process similar to frost wedging can occur as a result of the **growth of salt crystals in small openings in rocks**. This typically occurs when minerals such as gypsum or halite are precipitated as water containing salt moves through a rock. (In this context, precipitation represents the formation of solid material

Checkpoint 9.2

Outcrops of granite are examined in Alaska. They are all at similar elevations. Some are located in the dry interior of the state, others are located along the Pacific coast. The granites have identical compositions and textures. On the basis of the following information, which granite outcrop would weather most rapidly? Explain your choice.

a) Outcrop A; located at coast, contains fractures spaced 1 meter (3 feet) apart
b) Outcrop B; located at coast, does not have fractures
c) Outcrop C; located in interior, contains fractures spaced 1 meter (3 feet) apart
d) Outcrop D; located in interior, does not have fractures

Checkpoint 9.4

Imagine that you have been appointed to a team of researchers charged with determining which of the five World Heritage Sites in Figure 9.1 is at greatest risk from *physical* weathering. Identify at least three general questions you will ask as you begin to gather data for your study. Describe how you will use the information to plan your next steps.

Checkpoint 9.3

basic | advanced | intermediate | superior

Physical Weathering Concept Map

Review the following concept map describing the basic characteristics of physical weathering. Grade the concept map, using the rubric provided, and make whatever changes you believe are appropriate to earn a 4 on the grading scale.

Scoring Rubric	
0	Does not contain any information about physical weathering.
1	Contains some relevant terms, but several key terms are omitted and those that are included are poorly organized, with many linking phrases either absent or inaccurate.
2	Contains most of the relevant terms, but they could be better organized and some linking phrases are incorrect.
3	Contains almost all of the relevant terms, is reasonably well organized, and almost all linking phrases are appropriate.
4	Contains all relevant terms in a well-organized display that has appropriate linking phrases for each pair of terms.

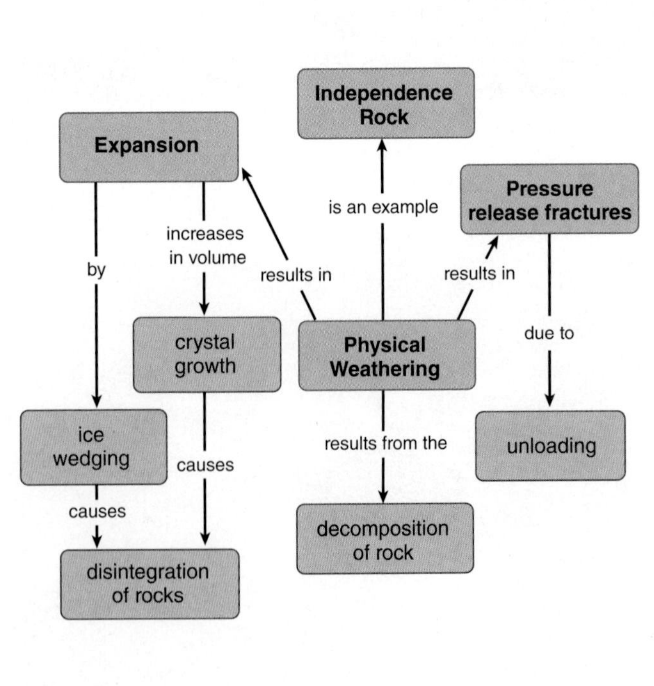

from a solution by a chemical reaction.) Salt crystals precipitate in regions with high evaporation rates or in coastal areas that are exposed to seawater (Chapter Snapshot). As the crystals grow on rocky surfaces, they push outward, breaking off individual mineral grains and causing the weathered surface of the rock to slowly disintegrate. If this weathering forms a hollow, it concentrates future weathering as water collects in the hollow. Weathering eventually produces small holes to give the rock a honeycomb appearance (Figure 9.7) or to form larger cavities known as *tafoni* that may be big enough to resemble small caves.

All of these physical weathering processes break up rocks and minerals into smaller pieces, creating greater surface area (Figure 9.8) for chemical weathering. This increased surface area then sets the stage for the topic of Section 9.3.

Figure 9.7 Honeycomb weathering. These holes formed when salt crystals grew on the surface of rock, gradually expanding the size of the hollows.

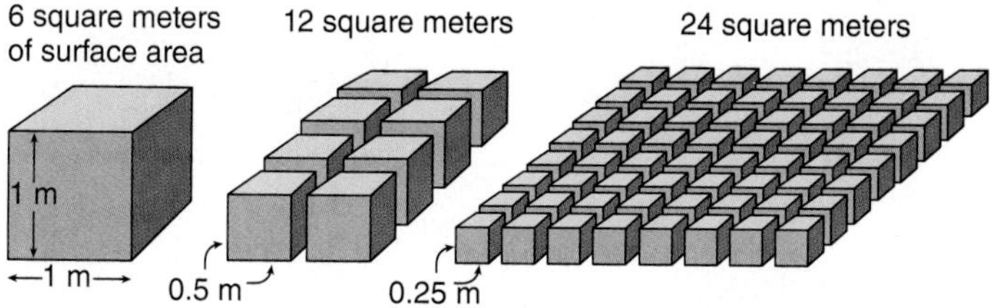

Figure 9.8 Increase in surface area. Physical weathering breaks rocks into smaller pieces, thus increasing the surface area over which chemical weathering can occur. In this example, the total surface area of the original block is doubled, then doubled again, by physical weathering, but the volume of rock present has not changed.

9.3 Chemical Weathering

Learning Objectives

- Describe the conditions that would cause earth materials to decompose by chemical weathering processes.
- Compare and contrast processes of chemical weathering.

Once physical weathering has increased the available surface area of rocks (and cars and buildings), chemical weathering can attack these surfaces. The processes at work during chemical weathering have a direct impact on our lives. The rust we see on our cars, the caves we visit on vacation, and even the clays that go into the manufacture of the glossy paper in our magazines and textbooks are all connected by chemical weathering. **Chemical weathering is the decomposition of rock as a result of the chemical breakdown of minerals.** A variety of chemical reactions result in changes in rock composition, but the three most common reactions are dissolution, hydrolysis, and oxidation. These reactions either remove minerals from a rock or change the mineral compositions. In most cases, strong minerals are replaced by weaker minerals, and this hastens the physical disintegration of the rock.

Dissolution

If you drop a sugar cube or a teaspoon of salt into a cup of hot water, either substance quickly disappears. The sugar and salt dissolve into the liquid; this process is dissolution. If the water were to evaporate, those materials would precipitate back out of the solution. Recall that some chemical sedimentary rocks, such as rock salt, and some biochemical rocks, such as limestone, form when minerals precipitate out of solution (Chapter 7). Dissolution is the reverse of this process, returning the minerals to solution. **Dissolution occurs when minerals in a rock are dissolved by water.** Although the effects of dissolution can be most readily observed in chemical sedimentary rocks, dissolution can also be effective in weathering clastic sedimentary rocks such as sandstone and siltstone if the grains are cemented by soluble minerals. The grains themselves remain relatively unaltered, but the mineral cement holding them in place may be dissolved, causing the rock to be converted back into loose sediment.

The amount of material that will be dissolved increases with the solubility of the rock type, the exposed surface area of the rock, and the volume of rainfall. Carbon dioxide (CO_2) makes up just a tiny fraction of the air we breathe, but it combines with rainwater to create a weak acid, carbonic acid (H_2CO_3; see Equation A). The bacterial decomposition of leaf litter is another source of carbon dioxide that can increase the acidity of groundwater.

Equation A

H_2O	+	CO_2	--->	H_2CO_3
rain		carbon dioxide (from air or bacterial decomposition)		carbonic acid (reacts with rocks)

Carbonic acid can dissolve certain rocks and minerals (for example, limestone or marble). The more acidic the carbonic acid, the more dissolution will occur. Air pollution from vehicles

WEATHERING

FROST WEDGING
Rocks break apart when water freezes and expands in rocks.

LICHENS
Release acids that break down bare rock surfaces in mountains.

ACID MINE DRAINAGE
Oxidation of metallic minerals produces acidic waters.

DECAYING PLANT MATERIAL
Microscopic organisms break down organic material to release gases.

SINKHOLE

TREE ROOTS
Wedge apart fractures in the bedrock.

CAVES
Minerals are dissolved to form caves. Some of the dissolved minerals are precipitated to form stalagmites and stalactites.

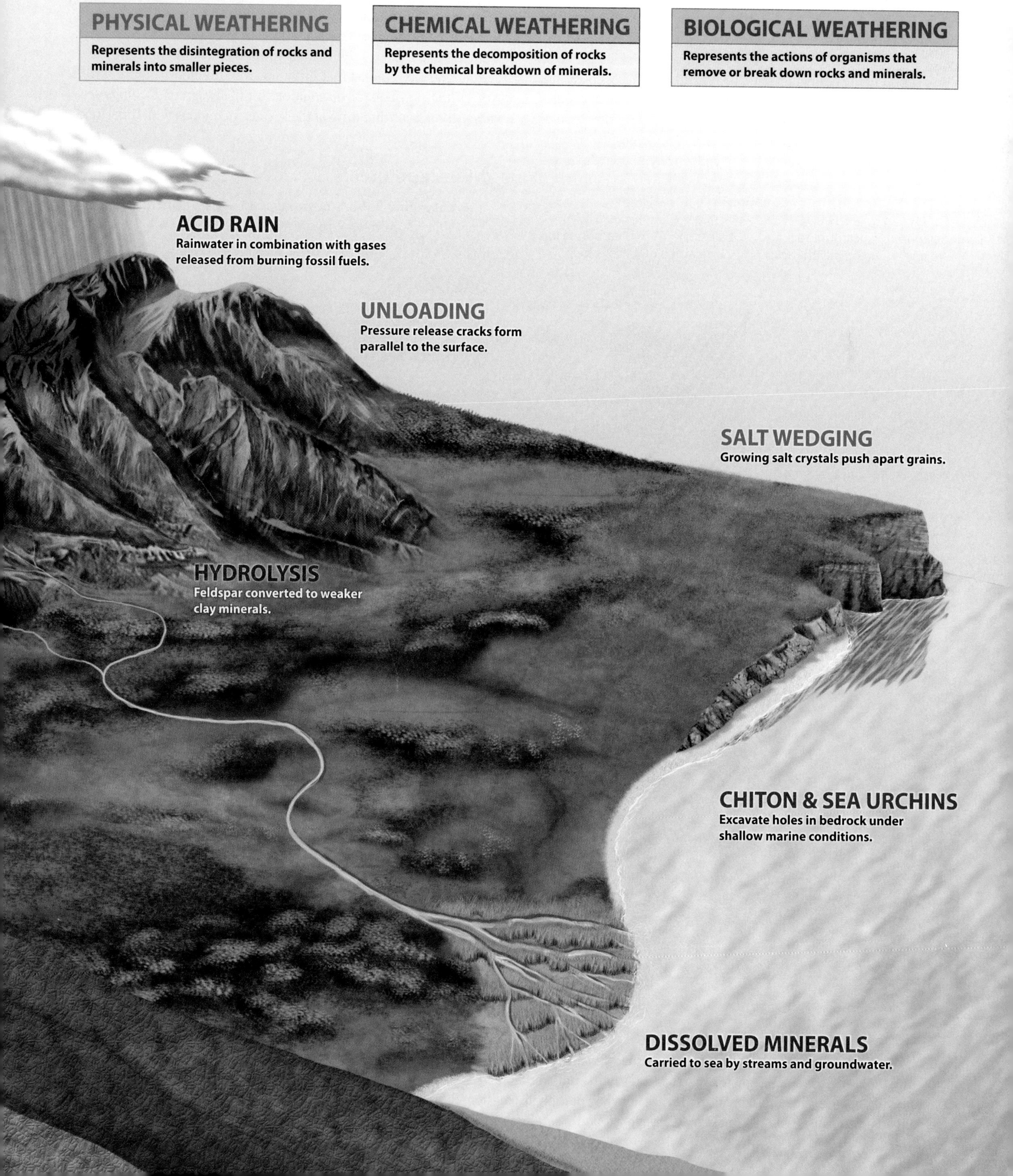
PHYSICAL WEATHERING
Represents the disintegration of rocks and minerals into smaller pieces.
CHEMICAL WEATHERING
Represents the decomposition of rocks by the chemical breakdown of minerals.
BIOLOGICAL WEATHERING
Represents the actions of organisms that remove or break down rocks and minerals.
ACID RAIN
Rainwater in combination with gases released from burning fossil fuels.
UNLOADING
Pressure release cracks form parallel to the surface.
SALT WEDGING
Growing salt crystals push apart grains.
HYDROLYSIS
Feldspar converted to weaker clay minerals.
CHITON & SEA URCHINS
Excavate holes in bedrock under shallow marine conditions.
DISSOLVED MINERALS
Carried to sea by streams and groundwater.

and industrial sources increases the acidity of the rainwater, accelerating chemical weathering in urban settings in comparison to neighboring rural areas (Chapter Snapshot). Dissolution can be seen in cities in the loss of inscriptions on marble gravestones and sculptures (Figure 9.9).

While other rocks may be more soluble, limestone is one of the most common sedimentary rocks at Earth's surface, so we are more likely to see the impact of dissolution in limestone than in other rock types (see Equation B). The landforms created by the dissolution of limestone include sinkholes, caves, and underground streams (Chapter Snapshot). The landforms are collectively known as *karst.*

Equation B

$CaCO_3$	+	H_2CO_3	—>	Ca^{++}	+	$2(HCO_3^-)$
limestone (rock)		carbonic acid (reacts with rocks)		calcium ion		bicarbonite ion
				(carried off in solution)		

Caves form when dissolution occurs along a series of fractures in limestone to create a larger opening. All rocks contain some cracks that can be attributed to processes acting during the formation or subsequent deformation of the rocks. The dissolved limestone is transported through the cave and may be precipitated to form new features, such as *stalactites* that grow downward from the cave ceiling and *stalagmites* that grow up from the floor (Figure 9.10). Not all the products of dissolution are below ground. *Sinkholes,* large depressions that form at the surface in regions with limestone

Checkpoint 9.5

basic advanced intermediate superior

Look at Figure 9.10b. According to this map, is there a cave near where you live? Why or why not?

Figure 9.9 Weathered gravestones and sculpture. **a.** Chemical weathering has erased inscriptions on the marble gravestone (right) but has had little effect on the slate gravestone (left). Each gravestone is more than 200 years old. **b.** Fine details have been dissolved away on this marble statue.

a.

b.

a.

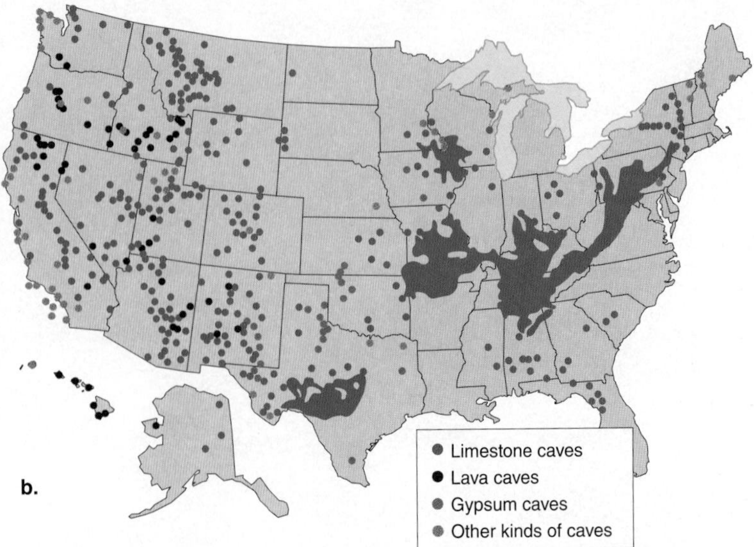

b.

Figure 9.10 Cave features and locations. **a.** At Carlsbad Caverns, New Mexico, the features hanging down from the ceiling are stalactites. Pillarlike features on the cave floor are stalagmites. The stalactites form from precipitation of minerals as limestone-rich water drips from the ceiling; the stalagmites form as the same water droplets fall to the cave floor and precipitate limestone. If stalactites and stalagmites meet, they form a column. **b.** Map of rocks containing caves in the United States. Most caves are found in limestone, many of them in the Appalachian Mountains of the eastern United States.

bedrock, often result from the collapse of the roof of an underlying cavern or by dissolution of rock along a series of intersecting fracture surfaces (Figure 9.11). Unfortunately, potential sinkholes can lurk undetected just below the surface and are often not evident until someone's garage is sitting at the bottom of one.

Hydrolysis

When a basketball team must replace a star player with someone from the bench, the replacement typically doesn't play as well, and the team's performance suffers. A comparable substitution occurs during hydrolysis when **hydrogen ions (H^+) in water replace other ions in silicate minerals,** leaving them weaker and more likely to break down by physical weathering processes. The hydrogen ion comes from the breakdown of carbonic acid into its two constituent ions, hydrogen and bicarbonate, as shown in the following equation:

Figure 9.11 A sinkhole. This sinkhole became known as the "December giant" after it formed in Alabama in 1972. It measures more than 50 meters (165 feet) deep and 100 meters (330 feet) across.

Equation C

$$H_2CO_3 \longrightarrow H^+ + HCO_3^-$$

H_2CO_3	⟶	H^+	+	HCO_3^-
carbonic acid (breaks down)		hydrogen ion		bicarbonate ion
		(two ions in solution)		

Recall from Chapter 7 that silicates are the largest family of rock-forming minerals and are found in all igneous rocks, including granite and basalt. The silicate mineral feldspar is the most common mineral in rocks on Earth's surface and ranks relatively high on the Mohs hardness scale (Chapter 7). Replacing these strong, common minerals with weaker minerals makes many rocks more susceptible to additional physical weathering. Feldspar reacts with water and the free hydrogen ions to form one or more types of weak clay minerals and additional ions that are dissolved in water, as shown in the following reaction:

Equation D

$$2KAlSi_3O_8 + 2H^+ + H_2O \longrightarrow Al_2Si_2O_5(OH)_4 + 2K^+ + 4SiO_2$$

$2KAlSi_3O_8$	$2H^+$	H_2O	$Al_2Si_2O_5(OH)_4$	$2K^+ + 4SiO_2$
feldspar (strong mineral)	hydrogen ions (in solution)	water	clay (weak mineral)	dissolved ions (in solution)

Hydrolysis is more rapid in silicate minerals with weak ionic bonds (characteristic of minerals such as feldspar and pyroxene) and is less effective in minerals with stronger covalent bonds (such as quartz). The rate of chemical weathering is consequently influenced by the mineral composition of the rocks and the relative abundance of hydrogen ions.

Generally, **the more hydrogen ions in the water, the more rapidly chemical weathering occurs due to hydrolysis *and* dissolution.** The pH of a solution is a measure of how many hydrogen ions are present. Pure water is considered neutral, with pH 7; acidic solutions have a pH of less than 7; alkaline solutions have higher pH values (pH > 7; Figure 9.12). The more

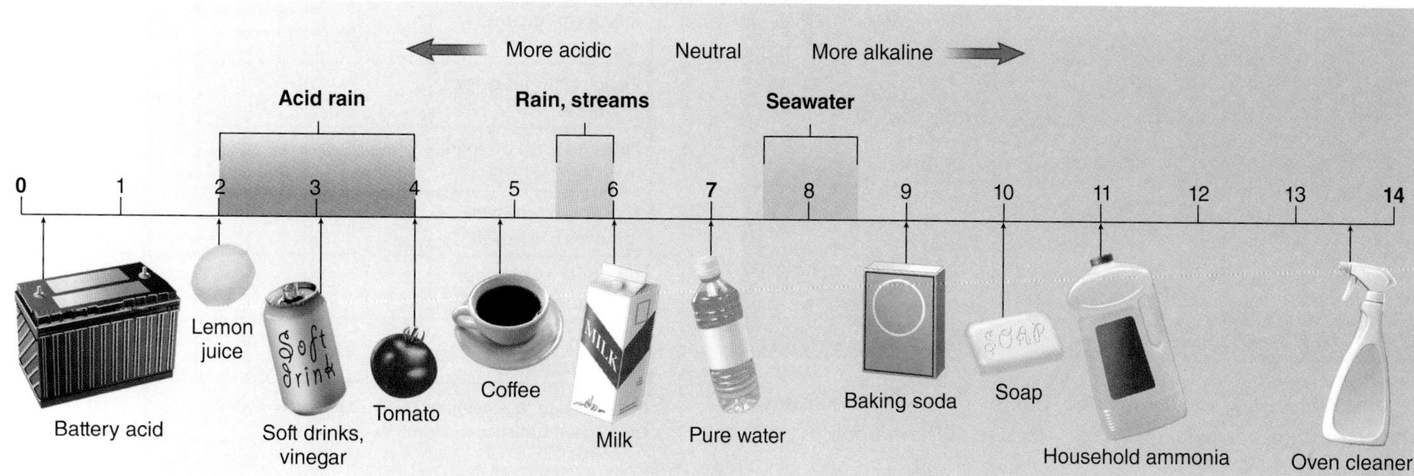

Figure 9.12 The pH scale. Values for some everyday materials are shown.

hydrogen ions, the lower the pH value. Like the magnitude scale for earthquakes, pH is measured on a logarithmic scale, so for each increment on the scale there is a 10-times increase or decrease in the concentration of hydrogen ions per volume of solution.

Because carbon dioxide and other gases dissolve in rainwater (see Equation A), all rain is mildly acidic (average pH about 5.6). The pH of rain water decreases significantly (that is, the solution becomes more acidic) with the addition of pollutants such as sulfur dioxide and nitrogen oxides that are generated from the burning of fossil fuels. **These pollutants combine with rain to yield sulfuric and nitric acid, producing an even more acidic solution termed *acid rain*.** This phenomenon is more likely to occur downwind from large industrial cities or coal-burning power plants. Acid rain can kill fish in lakes and destroy large areas of forest. Blocks of limestone are often added to lakes to combat the effects of the acid rain. The limestone blocks act as giant Rolaids™ antacid tablets to neutralize the acid, thus increasing the pH and returning it to near-normal levels capable of supporting life. However, unless the conditions that created the acid rain are altered, the lake will eventually return to acid conditions and the process will begin again.

Oxidation

The second most common element in the air we breathe, **oxygen, frequently reacts with iron or other metals in rocks during oxidation to form new mineral compounds.** A similar process results in the formation of rust on cars. Iron present in minerals, just as in steel, combines with oxygen in the atmosphere to form iron oxide (for example, the mineral hematite, or rust). Iron oxide is responsible for the red or yellow coloration commonly observed in some rocks and soils (Figure 9.13). These altered minerals are typically weaker than the original mineral.

Figure 9.13 Evidence of iron oxide. The color of the rock indicates the presence of iron oxide. This rock is located at Thermopolis, Wyoming.

The oxygen responsible for the reaction is typically dissolved in water, and thus oxidation reactions can occur well below Earth's surface. When water reacts with iron-rich minerals in mines, an environmental problem called *acid mine drainage* can result. Chemical reactions combined with microbial activity (see Section 9.4) may produce acidic water that exits the mine and

Checkpoint 9.6

basic | intermediate | advanced | superior

Weathering Analogies Matrix

Many simple occurrences in our daily lives are similar to geologic processes. The following table contains some everyday events that you may have experienced. Match these actions to specific weathering processes. Complete the table by placing an "x" in the columns on the right-hand side where appropriate. Identify which characteristic in the left-hand column is analogous (similar) to physical, chemical, or biological weathering. Two characteristics have been completed as examples. At this time, leave the Biological Weathering columns blank. We will return to this part of the exercise in the next section.

Characteristic	*Physical weathering*		*Chemical weathering*			*Biological weathering*	
	U	W	D	H	O	Ma	Mi
Paint on house gradually disappears.							
Spilled drink stains carpet.							
Groundhog digs a hole under your garage.							
Sugar disappears in hot coffee.							
Fungus forms on a fallen tree in the woods.							
Tree root pushes up paving slab.							
Rust forms on an old car.					X		
Bleach changes color of clothes.							
Potholes form on road in winter.		X					
Nail polish remover removes nail polish.							
Compost rots in your garbage can.							
Paperweight holds down stack of bills.							

U = unloading; **W** = wedging; **D** = dissolution; **H** = hydrolysis; **O** = oxidation; **Ma** = macro level; **Mi** = micro level.

Checkpoint 9.7

basic intermediate advanced superior

If you were to analyze the sand on a typical beach along the Atlantic coast, you would find that most of the sand grains are composed of the mineral quartz. In contrast, if you analyzed sand on some beaches in Hawaii, you might find that the dominant grains contain minerals such as pyroxene or olivine. Quartz, pyroxene, and olivine are all silicate minerals. Use weathering processes to explain why the compositions of these beaches differ.

a. White sand beach, Florida

b. Black sand beach, Hawaii

Checkpoint 9.8

basic intermediate advanced superior

Imagine that you have been appointed to a team of researchers charged with determining which of the five World Heritage Sites in Figure 9.1 is at greatest risk from *chemical* weathering. Identify at least three general questions you will ask as you begin to gather data for your study. Describe how you will use the information in combination with data on physical weathering to plan your next steps.

Figure 9.14 Acid mine drainage. When water reacts with iron-rich minerals and sulfur in mines, a distinctive orange or yellow color is evident in nearby streams.

produces a distinctive orange or yellow color in nearby streams (Figure 9.14). Scientists studying this problem have found that the action of microbes actually produces far more acidic drainage than the chemical process involving oxygen and water alone.

Linking Chemical and Physical Weathering Processes

Chemical and physical weathering can enhance each other. A chemically weathered rock is composed of weaker minerals and is therefore more easily broken into smaller pieces by physical processes. Physical weathering in turn helps break rock into smaller pieces that have more surface area on which chemical reactions can occur. Working in tandem over hundreds to thousands of years, physical and chemical weathering can dissolve a rock such as limestone or change a rock such as granite into clay and sand. This raises the questions: Can we stop weathering processes such as those seen at World Heritage Sites? If we cannot stop the processes, are there ways to slow them down or to minimize the human impact on them?

9.4 Biological Weathering and Decay

Learning Objectives

- Describe the biologic processes that cause earth materials to break down.
- Compare and contrast biological weathering processes.

Weathering produces small changes that can add up to big changes over long periods of time. It is a little like watching grass grow; it takes a while to see the result. Grass is an appropriate analogy here because biological processes are significant factors in weathering and soil development. Without the action of tiny microbes such as bacteria and fungi, we would not have the soil that supports life on Earth. We can divide biological weathering into weathering by organisms we can see (macroscopic processes) and weathering by organisms we cannot see with the naked eye (microscopic processes).

Figure 9.15 Macroscopic biological weathering. **a.** Tree roots force apart the rock along the original fracture. **b.** Chitons and **c.** sea urchins are responsible for excavating holes in bedrock under shallow marine conditions. **d.** Limestone rock along the coastline of the Bahamas undergoes biological weathering due to the activities of small organisms such as sea urchins and chitons.

Macroscopic Processes

Macroscopic biological weathering includes the **action of plant roots, animal burrows, termites, and other boring organisms that remove or break down rocks and minerals** (Figure 9.15 and Chapter Snapshot). Perhaps the most amazing example of this phenomenon is the wearing away of limestone rock by chitons (a type of mollusk) and sea urchins in tropical islands such as the Bahamas (Figure 9.15b and c). These little critters remove tens to thousands of tons of limestone bedrock every year by scraping rocks in the tidal margin. Holes and crevices formed by these determined organisms are evident at every level on the cliffs around the islands (Figure 9.15d).

Microscopic Processes

On a microscopic scale, microbial decay is the number one contributor to soil formation. There are so many bacteria in soil that it is difficult for scientists to count them accurately. Soil bacteria are also extremely diverse; up to 20,000 different species may be present in a single gram of soil. Typically, the biomass of bacteria in soil is between 400 and 5,000 kilograms per hectare (160 to 4,400 pounds per acre). Or, to put it another way, if you collected all the bacteria in 1 hectare (2.47 acres) of productive farm soil, it would weigh about as much as an elephant. A quick sniff of your kitchen garbage or garden compost pile can demonstrate the power of these tiny organisms in decaying solid matter. How do they do it? Microscopic biological weathering is primarily chemical decomposition of material that converts solid materials to gases whether or not water is present, as shown in the following equations:

Equation E1 (oxygen present)

CH_2O	+	O_2	—>	CO_2	+	H_2O
sugars in organic materials		oxygen in the air		carbon dioxide		water

Equation E2 (no free oxygen)

$2CH_2O$	—>	CO_2	+	CH_4
sugars in organic materials		carbon dioxide	and	methane gases

Reactions in the presence of oxygen (aerobic microbial decay) are much more rapid than the anaerobic decay that occurs in the absence of oxygen. The release of carbon dioxide from the decay of plant material is the reverse of the process of photosynthesis, during which plants convert the gas into plant material. While most macroscopic biological weathering works on rocks and minerals, microscopic biological weathering works mostly on organic material such as dead plant or animal matter.

Microscopic processes can combine with oxidation to break down bedrock. Lichen is a plantlike organism formed from a combination of fungi and algae that forms a crustlike growth on bare rocks. Lichens can break down bedrock because they release acids to get nutrients from the minerals within the rock. Therefore, physical, chemical, and biological weathering all work together as agents of destruction!

Checkpoint 9.9

basic / intermediate / advanced / superior

Describe an everyday example of microscopic or macroscopic biological weathering.

Checkpoint 9.10

basic / intermediate / advanced / superior

Finish the Weathering Analogies Matrix you started in Checkpoint 9.6. Identify analogies that correspond to biological weathering processes.

c.

d.

Checkpoint 9.11

basic intermediate advanced superior

Write a paragraph that argues for or against the following statement: *Biological weathering processes could be considered examples of physical or chemical weathering.*

Checkpoint 9.12

Imagine that you have been appointed to a team of researchers charged with determining which of the five World Heritage Sites in Figure 9.1 is at greatest risk from *biological* weathering. Identify at least three general questions you will ask as you begin to gather data for your study. Describe how you will use the information in combination with data on physical and chemical weathering to plan your next steps.

9.5 Weathering Rates

Learning Objectives

- Describe the primary factors controlling the weathering rates of rocks.
- Make connections between weathering rates and the mineral composition of the parent rock.
- Relate temperature, precipitation, and vegetation to soil formation processes.
- Apply concepts of biological, chemical, and physical weathering to a scenario.
- Taking into account factors associated with weathering rates, create an evaluation plan to protect culturally important sites.

As is apparent from our discussions in the previous sections, weathering of rocks on or near Earth's surface depends on the following factors:

1. **Rock composition.** Different minerals react to different chemical weathering processes. Because rocks are defined on the basis of their constituent minerals, we can make some generalizations about which rock types are most likely to resist or be vulnerable to weathering.
2. **Rock properties.** The presence of water in a rock promotes weathering. Any properties that make it easier for water to infiltrate a rock make it more susceptible to weathering.
3. **Climate.** Climate varies systematically with latitude and elevation. Some weathering processes cannot occur in some climate zones, while others may be accelerated by the presence of optimal climate conditions.

Rock Composition

Remember that rocks are composed of a variety of minerals. Weathering rates are greatest for rocks composed of minerals that simply dissolve in the presence of sufficient water, such as rock salt and gypsum. Weathering is slower where intermediate steps such as hydrolysis or oxidation change one mineral into another weaker form. Weathering rates are slowest in rocks composed of the most resistant minerals. Because it is very stable at low temperatures (such as Earth's surface conditions), **quartz is rarely affected by dissolution, hydrolysis, or oxidation.** In contrast, feldspar is readily altered to form clay in the presence of water. When a rock composed of quartz and feldspar weathers, the quartz will be the least affected mineral.

Imagine weathering in an area with granite bedrock in the interior of a continent. With the exception of the quartz, all the silicate minerals in the granite will be converted to soluble ions and tiny particles of clay minerals that may eventually be transported out to sea by rivers draining the region. The quartz grains will form sand particles that will also be carried downstream toward the coast. Stream velocity decreases where the stream reaches the ocean. At that point, the sand is deposited and may be transported along the coast by offshore currents to replenish local beaches. Much of the sand we observe on a beach is made up of the quartz grains that remain after chemical weathering converted other minerals to clay and the clay was washed away.

Rock Properties

Weathering rates are greater where the physical properties of the rock allow air and water to enter. **Porosity, the amount of space between grains in a rock, influences weathering rates** (see Chapter 12 for more details on porosity). Water and air accelerate the decomposition of high-porosity rocks in areas of moderate to high rainfall. Highly porous rocks may also be susceptible to frost wedging in cold climates because they hold water that freezes. In arid climates, porous rocks may experience cracking from minerals that grow as salt crystals in the rock pores. Fine-grained clastic sedimentary rocks, such as shale, are composed of compacted clay minerals. Water often penetrates these rocks along closely spaced surfaces between beds (layers). The combination of weak materials and abundant bedding surfaces makes shale a particularly "soft" rock that rarely stands out in the landscape. When dissolution occurs in limestone, it may create openings along fracture planes or bedding surfaces. After thousands of years of chemical weathering, those small cracks and openings become the passageways we walk through when we visit a limestone cave (Figure 9.10).

Igneous and metamorphic rocks have low porosity; consequently, they typically become weathered only on the surface or along fractures (Figure 9.16). Almost all rocks have some naturally occurring fractures or systems of cracks where physical and chemical weathering processes can occur. These surfaces act as passageways for water that promotes chemical weathering or serve as places where water may freeze to cause frost wedging. Consequently, any rocks that contain abundant fractures are typically weathered more rapidly than equivalent unfractured rocks. Intersecting systems of fractures may divide the rock into blocks. As weathering occurs, the edges and corners of the blocks disintegrate first, producing a rounded appearance (Figure 9.17). Pressure release cracks (see Figure 9.3), sometimes found down to depths of over 150 meters (500 feet), can serve as passageways for water into plutonic igneous rocks. **Weathering rates of igneous and metamorphic rocks are influenced by the relative abundance of the resistant mineral quartz and the presence or absence of rock structures such as joints or foliations**. For example, igneous rocks with little quartz (for example, basalt) and well-developed sets of fractures weather more rapidly than rocks that contain a higher proportion of quartz (for example, rhyolite) and few fractures.

Figure 9.16 Weathering in low-porosity rocks. Weathering rinds have not penetrated the interior of these dark gray volcanic igneous rocks from Mount Rainier National Park, Washington. (Note dime for scale.)

Climate

Climate is a third important factor in weathering rates. The climate of a region is defined by its average annual temperature and precipitation, and is reflected in its vegetation (more about this

a.

b.

Figure 9.17 Weathering along fractures in igneous rocks. **a.** Granite blocks at Pikes Peak, Colorado. **b.** Basalt blocks at the Giant's Causeway, Northern Ireland. (Notice how the degree of weathering decreases downward at each location.)

in Chapter 16). First, **warmer temperatures increase the rate of chemical reactions.** Second, chemical weathering reactions involved in dissolution and hydrolysis require water. The more quickly dissolution, hydrolysis, and oxidation occur, the more rapidly rocks weather. **The more water there is, the more these reactions are able to occur.** For example, dissolution of limestone around the Mediterranean Sea between Europe and Africa is more than twice as fast (8 centimeters, or 3 inches, per 1,000 years) in regions with 200 centimeters (80 inches) of yearly precipitation than in areas with 100 centimeters (40 inches) of annual rainfall. While limestone may dissolve readily in humid areas, it weathers slowly in arid areas.

Precipitation alone is not the only control on weathering potential. If the precipitation evaporates quickly, water does not enter the rock and therefore causes little or no chemical weathering. This is where vegetation, the third factor, plays a key role as a function of climate in controlling weathering rates. As we saw in Section 9.4, biological processes can promote weathering. In this case, vegetation can provide shade and thus prevent water from evaporating quickly, slowing the flow of water off the land surface and making it more likely that the water will infiltrate rocks to promote chemical weathering processes. Further, plant roots and decaying plant matter release carbon dioxide that can combine with water in the soil to form carbonic acid (see Section 9.3). This more acidic water infiltrates through the soil to the bedrock below, where chemical processes begin altering the rocks.

We can make some predictions about weathering in typical climate regions, assuming a consistent range of rock types and properties. For example, we would expect chemical weathering rates to be most rapid in tropical rain forests in equatorial regions of South America, Africa, and southeast Asia because these regions have both high temperatures and plenty of rainfall. Hot deserts such as the Sahara in North Africa and those found in areas of the southwestern United States have plenty of heat but insufficient water to cause significant physical or chemical weathering. Temperate mountains, such as the Rocky Mountains and the Sierra Nevada in the United States, have insufficient temperatures for rapid chemical weathering, but their high elevations allow freeze-thaw cycles necessary for physical weathering due to frost wedging. Also, biological activity by lichens, mosses, plants, and animals is prevalent in these regions. Finally, the polar and subpolar regions of Antarctica, Alaska, and Siberia are too cold to permit thawing for much of the year. Water in solid form (ice) is unable to react with rock, so neither chemical weathering from liquid water nor physical weathering from frost wedging dominates in those regions. However, erosional processes such as abrasion from glaciers and ice sheet movement are significant factors in removing rocks (see Chapter 16).

Checkpoint 9.13

basic / intermediate / advanced / superior

Assuming that the rock type is similar in each of the locations on the following map, predict whether physical or chemical weathering (or neither) dominates in each location. Explain your answers.

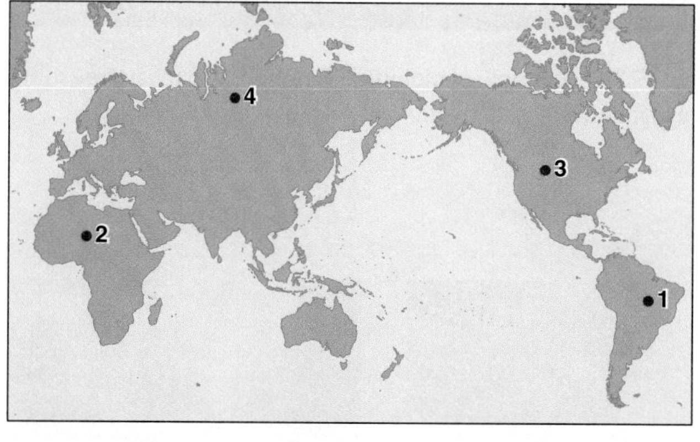

Weathering at World Heritage Sites

What about those five World Heritage Sites we mentioned in the introduction to this chapter (Figure 9.1)? Which weathering processes pose the greatest threat at each site? Which structures are most likely to suffer the greatest threat from weathering? Their climates vary from a humid tropical environment in Cambodia and Tanzania to a drier climate in southwestern Colorado and Greece (Table 9.1). Two of the structures are composed mostly

Checkpoint 9.14

basic / intermediate / advanced / superior

During a study of weathering, a scientist examines two tombstones marking graves in separate cemeteries. The inscription on one tombstone is almost unreadable, whereas the inscription on the other is sharp and clear. Provide three potential explanations for the differences in the present state of the inscriptions.

Table 9.1 Characteristics of the Featured World Heritage Sites

Location	Rainfall, cm/yr (in/yr)	Average temperature range	Rock type	Elevation	Age of structure (approximate)	Vegetation
Angkor Wat, Cambodia	140 (55)	21–35°C (70–95°F)	Sandstone	20 m (66 ft)	900–1,000 yr	Abundant
Kilwa Kisiwani, Tanzania	120 (47)	18–35°C (65–95°F)	Limestone	10 m (33 ft)	600–750 yr	Sparse
Machu Picchu, Peru	80 (31)	−1–24°C (30–75°F)	Granite	2,400 m (7,870 ft)	550 yr	Moderate
Mesa Verde, US	45 (18)	−7–29°C (20–85°F)	Sandstone	2,120 m (6,996 ft)	800–900 yr	Sparse
Parthenon, Greece	40 (16)	4–32°C (40–90°F)	Marble	250 m (820 ft)	2,440 yr	Sparse

of sandstone, one is made of granite blocks, and the others are limestone (blocks of coral) or marble (metamorphosed limestone). The oldest buildings are almost five times the age of the youngest structures. Construction styles range from blocks tightly placed together without mortar (Machu Picchu) to hand-carved blocks held together with a mortar made of mud (Mesa Verde) or lime (Kilwa Kisiwani). Two of the locations (Machu Picchu and Mesa Verde) are so remote that they remained undiscovered to the outside world until the last century, while the Parthenon lies in the heart of the city of Athens and is surrounded by 30 percent of the population of Greece. Chemical weathering processes will dominate at the sites in Cambodia and Tanzania that are characterized by plentiful rainfall and warm temperatures year-round. In addition, the relative age of the Angkor Wat structures and the presence of dense vegetation make that site especially susceptible to weathering.

A Case History: Angkor Wat, Cambodia. The city of Angkor Thom in central Cambodia was the national capital of the Khmer culture from the ninth to the twelfth centuries. It represents the largest settlement ever discovered in the preindustrial world. Many religious sites are preserved in the ancient city, which is today surrounded by tropical rainforest. The trees of the forest have invaded many of the sites, with tree roots forcing their way into the gaps between stone blocks (Figure 9.18). The most impressive buildings are in the temple complex at Angkor Wat, the largest religious structure in the world (see Figure 9.1a). The temple was largely constructed in the eleventh century using three types of sandstone. Weathering is most apparent in the gray sandstone blocks, a rock type that contains a variety of silicate minerals (quartz, feldspar, mica) and has a relatively high porosity. Blocks of the other two rock types—a red sandstone composed almost completely of quartz and a green sandstone with very low porosity—exhibit less evidence of weathering. (Why do you think that is?)

Checkpoint 9.15

basic · intermediate · advanced · superior

Analyze four of the World Heritage Sites described in Table 9.1. (Omit Angkor Wat, which is analyzed in the case history in this section.) Rank the locations from 1 through 4 on their potential for weathering. Justify your ranking scheme.

Figure 9.18 Example of biological weathering. Banyan tree roots have invaded a temple building at Angkor Wat, Cambodia.

a.

b.

Figure 9.19 Restoration at Angkor site, Cambodia. **a.** Weathering caused this pillar and wall to tilt, but **b.** restoration efforts have succeeded in restoring tilted walls to their original position.

Inspection of the gray sandstone blocks reveals that weathering rates are also dependent on the relative orientation of bed surfaces in the blocks. Some of the blocks were laid down during construction with their bedding surfaces vertical, while the remainder exhibit horizontal bedding surfaces. Water has infiltrated more readily into the blocks with vertical bedding, causing them to break apart and the surfaces of the blocks to flake off. Water is a constant presence at the Angkor Wat site during the wet season (May–October) when rainfall is abundant and the soil underlying the site is often saturated. Water seeps into the bases of walls and columns, increasing deterioration close to ground level. Hydrolysis has converted feldspar in the gray sandstone to clay minerals and can reduce the strength of the building foundations, causing the walls to tilt (Figure 9.19).

The final indignity to this ancient site is that much of its deterioration is attributable to bat droppings (guano). Thousands of bats inhabit the dark spaces of the ancient temple site. Rainfall leaches out certain chemicals in guano that precipitate in the sandstone as the mineral gypsum. The gypsum crystals expand within the sandstone blocks, exerting a crystallization pressure on the surrounding sand grains and causing the surface layers of the blocks to peel off in thin sheets.

Checkpoint 9.16

basic | intermediate | advanced | superior

Weathering Evaluation Rubric

This assignment requires you to identify the combination of factors most likely to contribute to weathering at a World Heritage Site. First, review the information on physical, chemical, and biological weathering and the Angkor Wat case history. Then, imagine you are a member of a US Geological Survey team that has been invited to assist the government of the small, impoverished nation of Sirinia. Your team will be working with the Department of Antiquities, a government agency responsible for managing over 100 historical sites representing over 1,500 years of history, back to the time of the ancient Cetacean Empire. The empire collapsed during a prolonged war from A.D. 1273 to 1314 and was subsequently divided into three nations, one of which is present-day Sirinia. Sirinia has an area of 200,000 square kilometers (77,220 square miles; smaller than Texas, bigger than California), and its elevation ranges from sea level along the coast to 5,265 meters (17,275 feet) inland at Mount Orcinus. Historical sites are distributed throughout the nation and are constructed from local building stone, which varies with location and includes varieties of sedimentary, metamorphic, and igneous rocks.

Audience: You will create a scoring scheme that can be applied by local volunteers for the Department of Antiquities. Specifically, you are writing for citizens who will volunteer to participate in a 1-week workshop during the summer before returning to their local regions to evaluate weathering at nearby monuments. The workshop will train them to recognize current weathering and the potential for future weathering. On the basis of the reports of these volunteers, the staff of the Department of Antiquities will send out a team of specialists to inspect the 10 sites determined to be deteriorating most rapidly as a result of weathering.

Main point and purpose: To demonstrate that you understand the principal factors that contribute to weathering of rocks.

Pattern and procedures: Identify five factors that influence weathering, and rank each factor's potential for weathering as high, moderate, or low. Record your data on the table provided; one factor, rock type, has been provided as an example. Each factor with a high weathering potential is scored as 3 points; low-potential factors rate 1 point. The site with the highest cumulative score for all factors would be at greatest risk for future weathering.

Standards and criteria: You will be assessed on your choice of:

1. Relevant factors that would contribute to weathering.
2. What constitutes high, moderate, or low weathering potential for each factor.
3. One factor in particular as the most significant. The score for this factor will be doubled. Discuss your justification for choosing the particular factor.

Factors	Low Weathering Potential (1 point)	Moderate Weathering Potential (2 points)	High Weathering Potential (3 points)
Rock type	Resistant rock (e.g., granite)	Less resistant rock (e.g., sandstone)	Weak rock (e.g., gypsum, limestone)

9.6 Soils: An Introduction

Learning Objectives

- Identify the four principal components in soils.
- Describe the factors determining the composition and color of horizons in a soil profile.
- Explain how soils are classified on the basis of their physical characteristics.

As we learned in Section 9.1, regolith is the collective name for rock and mineral fragments produced by weathering. Regolith includes materials from the ground surface down to the bedrock. Under the right conditions, the upper part of the regolith may form soil. **Soil is a stratified mixture of regolith that includes sufficient organic material, water, and air to support plant life** (see Figure 9.2).

Compare the pictures of the four soils in Figure 9.20. What are the similarities and differences among the images? From what you have learned so far about weathering, what factors might account for the differences? Take a minute to think about each question. (Research shows that we learn better when we base our first attempt to explain new information on what we already know.)

The accumulation of dead and decaying plants and animals provides organic material necessary for soils to form. That organic material is mixed with regolith in the uppermost soil layer by burrowing animals, worms, and insects such as termites and ants. At the same time, **water moving down through the soil dissolves ions, a process known as *leaching*.** These dissolved ions are transported downward to be deposited in lower layers by precipitation. Water can also pick up fine clay particles, transport them downward, and deposit them in lower layers. **These three processes—(1) organic activity, (2) leaching and precipitation, and (3) transport of clays—separate the soil into distinct layers with characteristic properties.** These layers are termed *horizons*.

Weathering processes cause soil to be divided into a series of distinct horizons collectively termed a *soil profile* (Figures 9.20 and 9.21). These distinctive horizons make it possible to readily differentiate soil from sediment.

Each horizon in a soil profile is designated by a letter. Beginning at the top, the horizons and the processes that are active in them are as follows:

- **O horizon.** Organic debris, dead leaves, and other plant and animal remains generally make up at least 30 percent of this layer. This layer, if present, is usually on the top (why?) but can be buried.
- **A horizon.** This layer consists of topsoil, dark organic material mixed with mineral grains by organic activity. Soluble ions and fine particles are carried downward, away from the A horizon, by leaching and clay transport.
- **E horizon.** This horizon consists of subsurface layers that have lost most of their minerals. It may be embedded in or replace the A horizon.
- **B horizon.** Ions that were dissolved from the A horizon are precipitated, and clays that are carried down from the A horizon are deposited here. Little organic material is present at this depth. The accumulation of iron oxide gives the soil a red color in areas where there is plenty of rain. In dry climates, calcium carbonate may accumulate to form a white layer (more on that in the discussion of soil types).
- **C horizon.** The lowest layer consists of soil parent material, either weathered bedrock (regolith) or unconsolidated sediments.

Soil-Forming Factors

Can you think of other factors that might affect the type of soil that forms and how rapidly it develops? Suppose that you live in an area where the main rock type is light-colored sandstone. How

Figure 9.20 Soils from different areas of the United States. What do you notice about these soils? Why do you think they look different?

would soil in that location differ from soil forming in an area dominated by red sandstone? At the very least, you might predict that the color of the soil would be different. The rock in an area likely plays some role in the type of soil that forms. Different rocks have different chemical properties, and so do the soils that form from those parent rocks. Sometimes the soil is blown in from another area or is composed of materials transported by streams or glaciers (as is common in the Midwest). In those cases, the parent material is not the underlying rock, but the rocks from where the transported material originated.

What other factors make soils different from one another? Consider a desert in the Southwest compared to an area in the northeastern part of the United States. Certainly **the temperature and the amount of rain (key components of climate) in a region affect soil formation.** As we learned in Section 9.5, high temperatures and plentiful rainfall increase the weathering rates and affect the type and thickness of the soils. Closely related to the amount of moisture present is the topography. **Steeper slopes cause water to flow more rapidly, transporting material away from the site and resulting in slower soil formation.** To earn the title *soil*, an accumulation of regolith must contain organic material. Therefore, the biologic activity occurring in an area also contributes to the soil characteristics. The more biologic activity, the more organic material that can be mixed with the regolith.

It takes time for weathering and the other factors mentioned here to form a soil. Bare rock surfaces exposed in cold climates (for example, Alaska) take thousands of years to develop even thin soils because chemical weathering rates are very slow in cold regions. The same rocks exposed to warm, wet climates develop soils in a few hundred years. Soil is absent on bare rock surfaces in northern Canada that were scraped clean by glaciers (massive sheets of moving ice) thousands of years ago. In contrast, soils are up to 1 meter (3 feet) thick in the temperate Midwest, where soil formation occurs at rates of approximately 1 millimeter (0.04 inch) every 10 years. The thickest soils may be over 30 meters (100 feet) thick in tropical regions that have year-round warm temperatures and plentiful rainfall.

Thick soils are not necessarily fertile soils. The fertility of the topsoil may change over time, depending on the relative rates of leaching and replacement of nutrients from weathering and decay of organic material. The heavy rainfall that contributes to the production of thick soils also carries away soil nutrients. Consequently, despite the dense vegetation that characterizes rain forests, only the upper few centimeters (inch) of soil are rich in nutrients. Slash-and-burn agricultural techniques are sometimes used in the rain forest to create farmlands by cutting and burning sections of rain forest to replace native vegetation with crops. These farming operations are short-lived as the soil nutrients are soon depleted. Once abandoned, the exposed soil is easily eroded by heavy rains. In temperate climates, weathering generally produces increasingly fertile topsoils as the A horizon becomes differentiated in a young soil. However, the long-term fertility of soils is controlled by the interaction of natural processes and human actions.

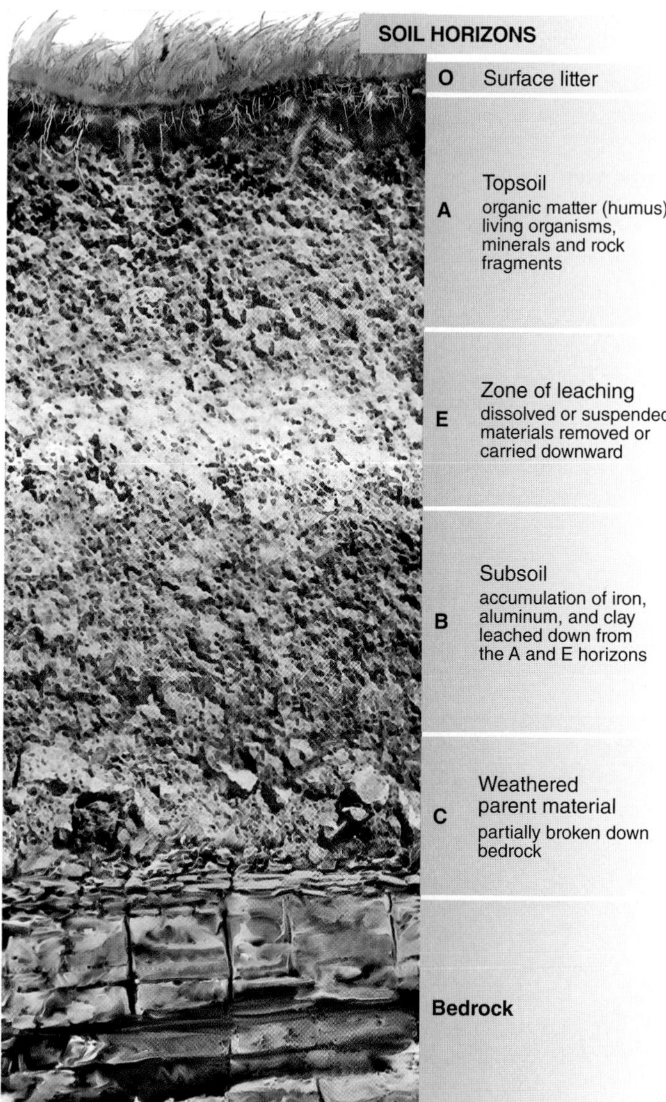

Figure 9.21 Soil profile. Each of the key horizons has distinct characteristics. The E horizon may be absent in many soil profiles.

Checkpoint 9.17

basic / intermediate / advanced / superior

How is the thickness of soil in a region related to weathering?

a) Weathering breaks down materials near Earth's surface and therefore reduces the thickness of soil.

b) Weathering increases the thickness of soil because it provides more materials to be incorporated into the soil.

c) Soil thickness is dependent on the character of the regolith and therefore is not related to weathering.

Checkpoint 9.18

basic / intermediate / advanced / superior

From what you learned about geologic time in Chapter 8, predict approximately when the first regolith formed on Earth. When do you think the first soils formed?

Soil Types

Soil scientists have identified thousands of soil types and developed a variety of classification systems that consider the texture or composition of the soil (Table 9.2; Figure 9.22). In the United States, the US Department of Agriculture (USDA) divides soils into 12 broad orders, differentiated by the soil's physical characteristics. These 12 soil order names are just the beginning; soils are further divided into suborders, groups, and subgroups to arrive at catchy names such as alfic fragiorthods or mollic eutroboralfs, which look like the result of someone dropping a bag of Scrabble® tiles. These names have meaning for soil scientists, but for our

Figure 9.22 **a.** Global soil orders map. **b.** Climate regions of the world. **c.** World vegetation map.

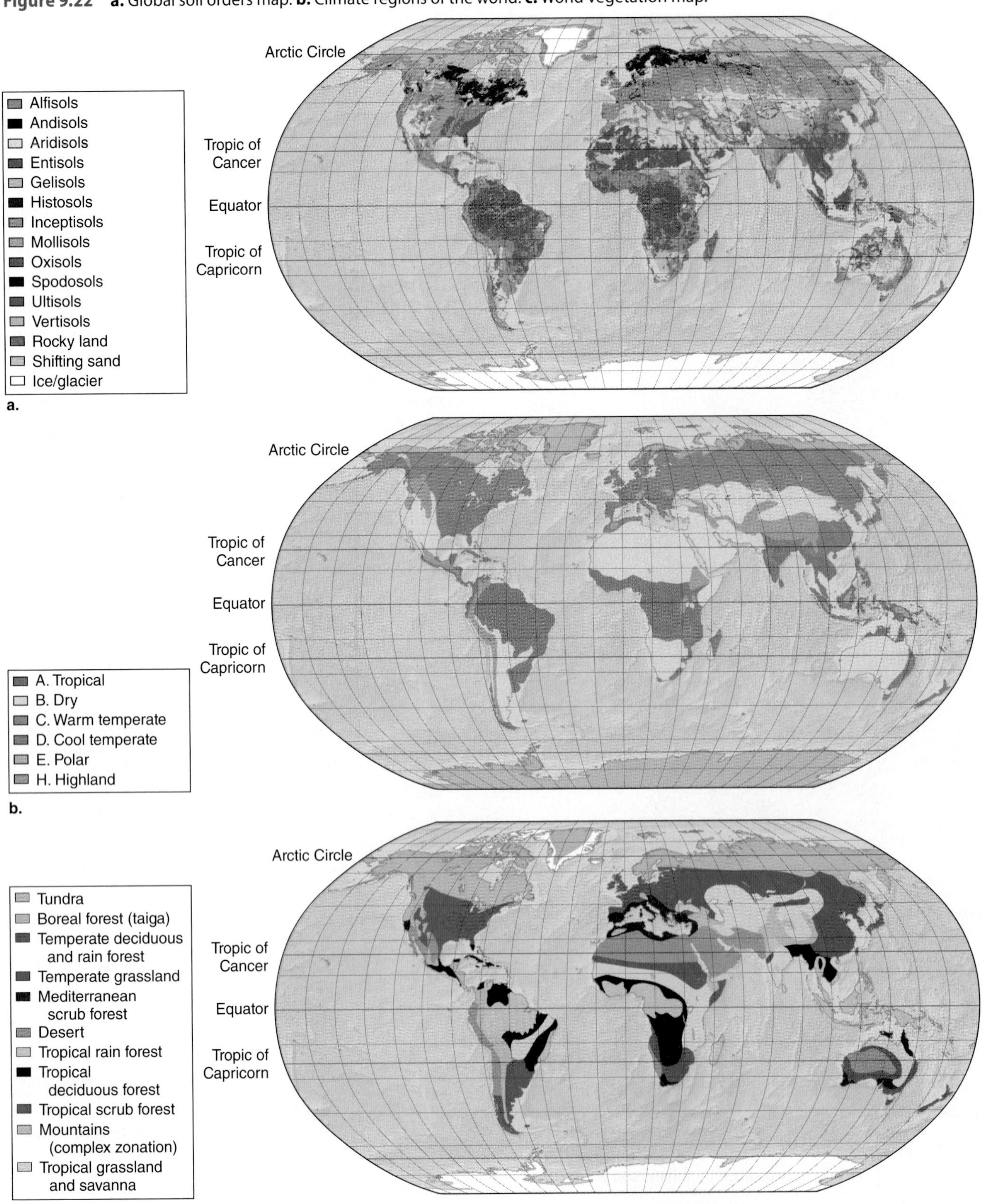

purposes, we will divide soils more simply into arid and temperate soils characteristic of relatively dry and wet climates, respectively (Figure 9.23). More generally, one can see connections within the earth system by analyzing the global soil, climate, and vegetation maps shown in Figure 9.22a–c. What do you notice? Can you see any patterns in common between the three figures? Hopefully you noticed similar patterns in the three maps. For example, notice the cool temperate climate across northern Canada and Russia. Those areas correspond to tundra and coniferous forests on the vegetation map. On the soils map, these areas appear to have gelisols and inceptisols for soils. Looking at the definitions for those soils in Table 9.2, we see such soils form in a wide range of temperature and moisture environments including tundra. That should make sense because those areas are at high latitude. So as shown in these maps, soil types are largely dependent on temperature, precipitation, and the type of vegetation in a region supplying organic matter for the soil.

Back to the United States, the relatively abundant precipitation in the temperate climate zones east of the Mississippi River results in the leaching of soluble ions from the A and B horizons. Less-soluble ions (iron, aluminum) remain behind in the B horizon (Figure 9.23). In tropical climates characterized by heavy rainfall, nearly all soluble ions are washed out of the soil and transported away in groundwater. This is similar to what happens when you make coffee in the morning. The water passes through the ground coffee beans and extracts the coffee and caffeine. If you were to use the same grounds again, the next pot of coffee would be weaker than the first. Eventually, if you repeated this process enough times, you would end up with a pot of hot water with no coffee flavor.

Checkpoint 9.19

basic · intermediate · advanced · superior

A *system* is defined as a group of independent but interrelated components comprising a unified whole. Support the following statement: *Soil is an example of a system, and soil type is controlled by the balance of inputs to and outputs from the system.*

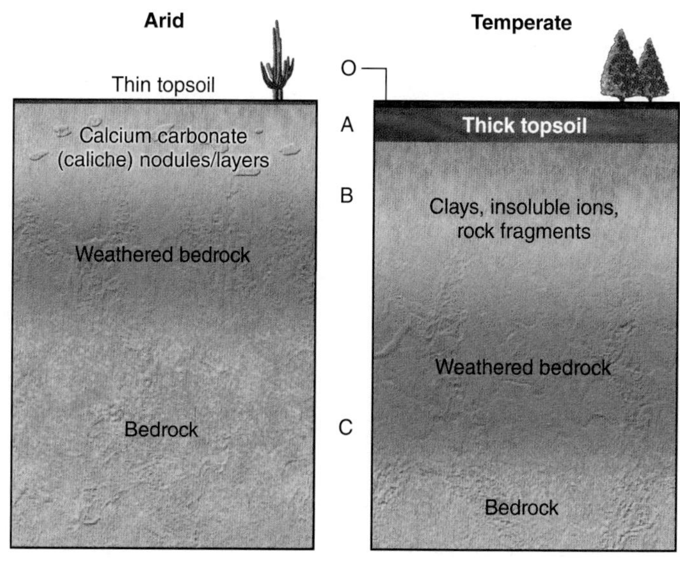

Figure 9.23 Idealized soil profiles for arid and temperate climates. The O horizon is absent in arid climates. While shown as equal in thickness here, the soil profile is thicker in environments with a plentiful supply of water.

Table 9.2 USDA Soil Classification System

Soil order	Description
Alfisols	Soils in semiarid to humid areas that have a clay and nutrient-enriched subsoil. They commonly have a mixed vegetative cover and are fertile and productive for most crops. Common soil type below farmlands in the Midwest.
Andisols	Soils that formed in volcanic parent material such as volcanic ash. These relatively young soils are common in the volcanic areas of Alaska, Hawaii, and the Pacific Northwest.
Aridisols	Soils found in dry regions with little rainfall. They may have a clay-enriched subsoil and often have deposits of salts or carbonates. These soils are common in the deserts of western states. They support vegetation such as cactus.
Entisols	Soils that have little or slight development and properties that reflect their parent material. They include soils on steep slopes, floodplains, and sand dunes. They occur in many environments.
Histosols	Dark soils that have slightly decomposed to well-decomposed organic materials derived from grasses, leaves, water-loving plants, and woody materials. These soils dominantly are very poorly drained and occur in low-lying areas.
Gelisols	Soils that commonly have a dark organic surface layer and mineral layers underlain by permafrost. These soils are common in the tundra regions of Alaska.
Inceptisols	Soils that have altered horizons but still retain some weatherable minerals. These soils occur in a wide range of temperature and moisture environments.
Mollisols	Soils that have a dark surface horizon. These soils formed from nutrient-rich parent material and are commonly found in grasslands.
Oxisols	Soils in humid, tropic, or subtropic areas that have clay and few weatherable minerals. They commonly are reddish or yellowish soils that do not have distinct horizons.
Spodosols	Soils in humid areas that have a light gray horizon over a reddish, aluminum- or iron-enriched horizon. They commonly have coniferous tree cover.
Ultisols	Soils that are in humid areas and have a clay-enriched subsoil that is low in nutrients. With soil amendments, they are productive for row crops.
Vertisols	Clayey soils that shrink and develop cracks as they dry and swell when they become moist. The shrinking and swelling can damage buildings and roads.

Soluble minerals can also be completely flushed from soils by plentiful rainfall, forming a soil called *laterite* (Figure 9.24). The leaching of all ions except aluminum in the B horizon of laterite can be so effective that it produces the aluminum-bearing ore *bauxite,* which can be economically mined. Even so, we must not forget that the bedrock that forms the regolith for the soil is more important to these soil-forming processes than climate. For example, there would be little iron or aluminum in areas where limestone creates the regolith because those elements are rarely found in limestone.

Figure 9.24 Laterite soil. Laterite is classified as a type of oxisol in the USDA soil orders. It develops in climates with high rainfall when intense leaching carries away everything except iron and aluminum oxides. The relatively uniform red coloration is often a feature of laterite soils.

Checkpoint 9.20

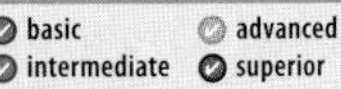

Soils developed on four lava flows on one of the Hawaiian Islands. The lavas were of four different ages, having formed over a span of several hundred years, the most recent flow being 10 years ago. USDA scientists define these four soils (in no particular order) as alfisols, entisols, inceptisols, and ultisols. Using the descriptions in the USDA soil classification scheme (Table 9.2) and your understanding of soil formation processes, describe how these four soil types might represent a chronological sequence from a young soil to a mature soil, all derived from the same parent material. Draw a soil profile for each soil type. Note that the profiles should change over time.

In dry regions (such as much of the United States west of the Mississippi River), calcium ions are leached as water passes downward through the A horizon. When evaporation exceeds precipitation, the water is actually drawn back to the surface. The calcium ions dissolved in the water are then precipitated in the B horizon, where they combine with carbonate ions to form nodules, coatings, or layers of calcium carbonate (limestone) termed *caliche* (see Figure 9.20c). When caliche impregnates an entire horizon, it is termed a *K horizon* in the soil profile. This material is sometimes so hard that a jackhammer is required to break it when construction requires an excavation.

9.7 Soil Erosion and Conservation

Learning Objectives

- Describe the conditions necessary for soil erosion.
- Compare soil forming rates to soil depletion rates through erosion.
- Discuss connections between human farming processes and soil erosion rates.
- Evaluate a soil conservation program, using concepts and processes of soil formation and erosion.

Soil is a semirenewable resource because it forms very slowly and, under ideal conditions, can be continually replenished,

Figure 9.25 Soil erosion features. **a.** The impact of a raindrop dislodges soil particles that can be subsequently transported away by flowing water. **b.** Water carved a series of small shallow channels known as rills in a plowed field in Iowa. Visible rills typically indicate an erosion rate of at least 30 tons per hectare (12 tons per acre). **c.** Wind erosion occurs as dust rises above an uncovered cultivated field. **d.** A massive dust storm approaches a town in the Great Plains during the Dust Bowl era.

a.

b.

given enough time. But when soil erosion outpaces soil formation, as often occurs, soil becomes a nonrenewable resource. Soil erosion rates are affected by climatic factors (water and wind) and by land use practices.

Erosion by Water and Wind

Erosion occurs when soil, sediment, or rock particles are detached from one location and transported away to be deposited elsewhere. These particles can be dislodged by the simple impact of raindrops (Figure 9.25a) or by the force of water or wind flowing over the surface. The loosened particles are then carried away by the flowing water or are bounced along the surface by the wind. Evidence of erosion can be seen as a network of fine channels cut into the surface or as clouds of dust (Figure 9.25b, c, d). In addition, gravity can cause material to move down slopes, and living organisms can detach and transport earth materials.

The amount of soil removed as a result of erosion due to flowing water or wind will vary, depending on:

- **Amount and frequency of rainfall.** Infrequent, intense rainfall removes more soil than steady, moderate precipitation. The lack of rainfall causes soils to dry out, making them lighter and more vulnerable to wind erosion.
- **Wind velocity.** Faster-moving winds generate more erosion than gentle breezes. Trees can shelter the soil from winds by acting as obstacles that reduce wind velocity and thus minimize wind erosion. Some of the highest average US wind speeds occur over the Great Plains states between North Dakota and Texas.
- **Character of the soil.** Soils with open texture due to the presence of organic materials allow water to infiltrate the soil rather than running over the surface, where it has the potential to cause erosion.
- **Vegetation cover.** Grass, crop debris, or other materials can act as a cover for soils, protecting them from the direct impact of raindrops, and their roots help hold the soil together. Loss of vegetation cover as a result of overgrazing by domestic animals or the removal of tree cover by excessive logging accelerates soil erosion.
- **Slope of the land surface.** Long, steep slopes lose more material to soil erosion from water than do shorter, gentler slopes.

Soil erosion by water or wind or both (Figure 9.26) reduces the area of land available for planting crops and lessens soil fertility. In regions where rainfall is abundant, most of the soil erosion is due to water running off fields. In areas where rainfall is scarce, wind is the main agent of erosion that carries away soil from farm fields.

Areas of the United States that are dominated by water and wind erosion can be roughly identified by drawing an imaginary line that runs from central Texas north to the border between North Dakota and Minnesota (Figure 9.26). East of that line, most regions receive more than 100 centimeters (40 inches) of rainfall per year, and water-related erosion dominates. In the states west of the line, wind erosion is more likely to occur because of consistently strong winds and a relatively dry climate. Wind erosion in parts of west Texas may account for the loss of approximately 34 tonnes of soil per hectare (15 tons per acre) per year.

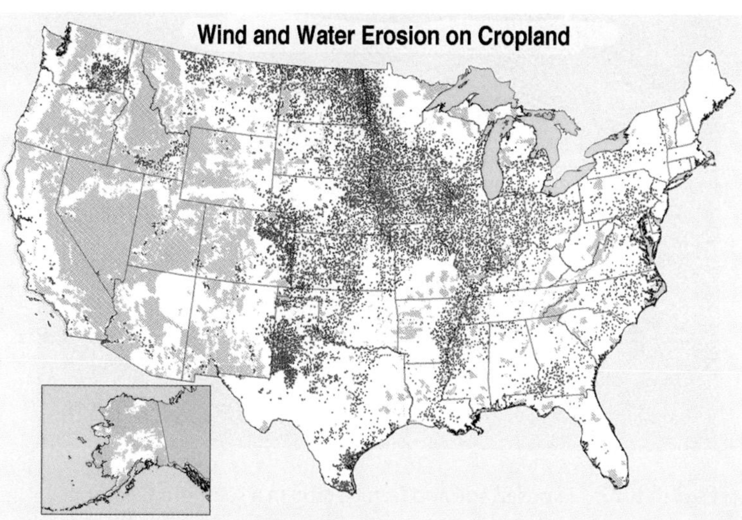

Figure 9.26 US patterns of significant soil erosion on croplands. Each dot represents erosion of 90,500 tonnes (100,000 tons) of soil due to water (blue dots) and wind (red dots).

c.

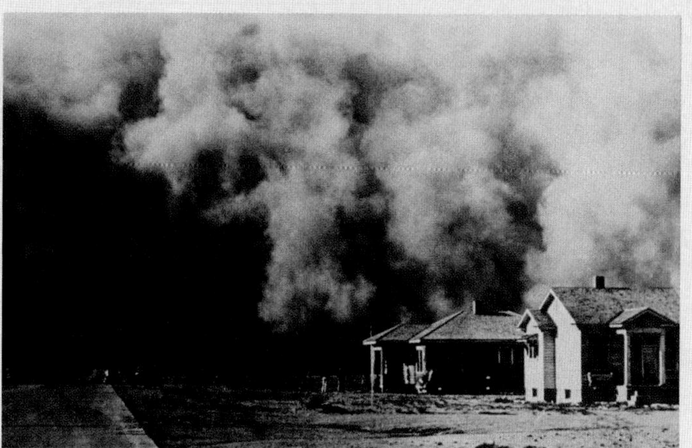

d.

Effects of Land Use Practices on Erosion

Recent studies indicate that humans are now more important as agents in moving soil and sediment than all other natural processes operating on Earth combined. Without appropriate safeguards, agriculture can accelerate soil loss. Poor soil management historically caused widespread soil erosion in North America and continues to cause soil erosion around the world. The highest erosion rates are in those regions of Africa, South America, and Asia where farming practices are largely unregulated and rainfall is abundant. The lowest erosion rates are generally found in Europe and North America, where governments encourage agricultural practices that preserve soils.

However, even in places where efforts are made to protect the soil, erosion can easily outpace soil formation. The average soil formation rate is about 0.9 tonne per hectare (0.4 ton per acre) per year, less than one-fifth of the soil lost each year from croplands. Each year, nearly 1.8 billion tonnes (2 billion tons) of soil is eroded from US lands that produce crops ranging from alfalfa to watermelon (Figure 9.27). The economic cost of soil erosion in the United States alone has been estimated at tens of billions of dollars a year. Soil erosion is generally higher where soil is exposed during the cultivation of crops. Less soil erosion occurs on grasslands used for grazing, and soil formation is more likely to dominate erosion in woodlands.

The Dust Bowl. The Dust Bowl represents the most dramatic example of soil erosion in American history. Reporter Robert Geiger introduced the term into the national lexicon on April 15, 1935; "Three little words, achingly familiar on a western farmer's tongue, rule life in the dust bowl of the continent—if it rains." Geiger's article followed Black Sunday, when huge dust clouds darkened the skies across the southern Great Plains (western Kansas, Oklahoma, north Texas, eastern Colorado; Figures 9.25d and 9.28).

Figure 9.27 Exposed soil and lettuce crop in a southern California field.

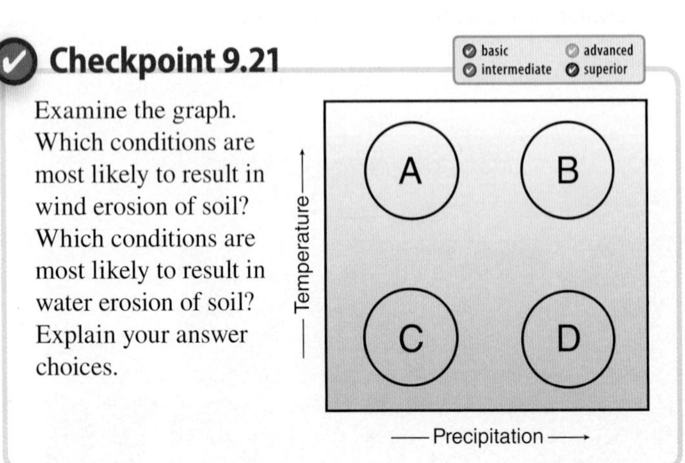

Checkpoint 9.21

Examine the graph. Which conditions are most likely to result in wind erosion of soil? Which conditions are most likely to result in water erosion of soil? Explain your answer choices.

Checkpoint 9.22

Venn Diagram: Water and Wind Soil Erosion

Complete the following Venn diagram to compare and contrast the factors that affect soil erosion due to water and wind. Identify characteristics that are shared by both systems (for example, characteristic 1) or are different for each group (for example, characteristic 2). Place the numbers in the most suitable locations on the diagram. Two have been inserted for you as examples.

1. Rate dependent on vegetation cover
2. Occurs in regions of warm, dry climate
3.
4.
5.
6.
7.
8.
9.
10.
11.
12.

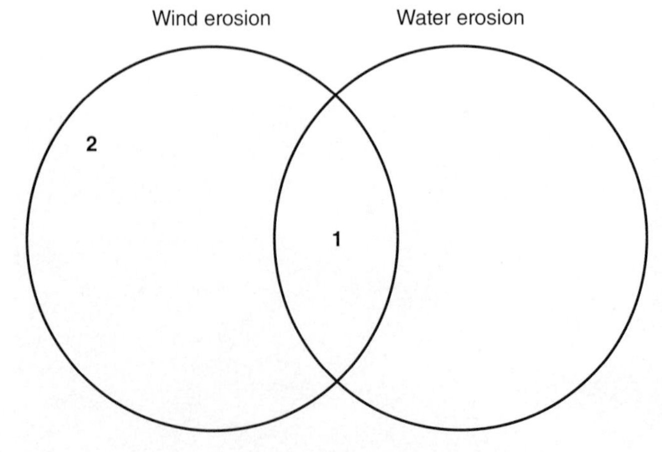

The Dust Bowl was created after farmers throughout the Great Plains had plowed native grasslands to convert them to wheat fields in the late 1920s. This followed a series of wetter-than-average years that produced bountiful harvests and the promise of an economic windfall to anyone who could plant a few acres. In subsequent years, when needed rains did not fall and crops failed to grow, wind erosion stripped an estimated 12 centimeters (5 inches) of topsoil from much of the region. This human-induced natural disaster is forever associated with the economic turmoil of the Great Depression that occurred at the same time.

The Dust Bowl provides a vivid but false perception of the nature of US soil erosion. Scientific measurements of soil losses indicate that the majority of current soil erosion occurs because of the effects of running water.

Soil Conservation

Because artificial soil formation is impractical and we wish to ensure a continued supply of agricultural products, we have little choice but to conserve the soil we have. Over 200 years ago, George Washington experimented with soil conservation strategies, and our modern methods do not differ much from his. These techniques, collectively known as *conservation tillage*, are designed to reduce water and wind erosion:

1. Keep soil covered to prevent erosion by leaving crop debris on fields (Figure 9.29a).
2. Reduce soil lost to surface runoff by limiting the effects of slopes through contour plowing (plowing across the slope; Figure 9.29b), terracing (Figure 9.29c), or both.
3. Provide shelter for fields exposed to prevailing winds by planting belts of trees (shelter belts) to break winds (Figure 9.29d).
4. Ensure a steady supply of nutrients by returning organic matter to the soil, adding fertilizers, and rotating crops (Figure 9.29e) so that essential nutrients are not depleted.

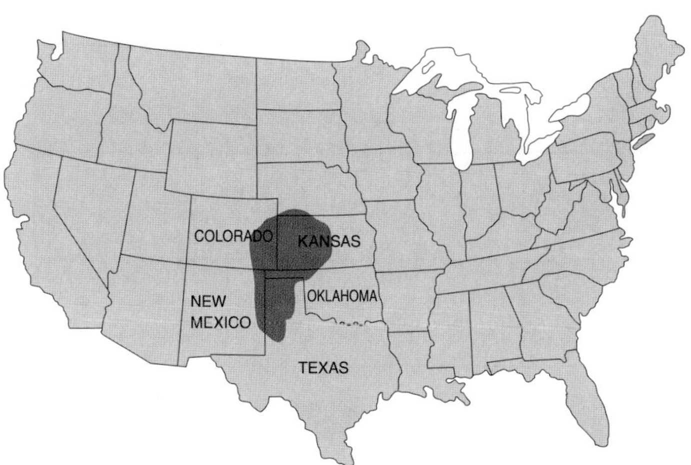

Figure 9.28 The Dust Bowl. Highlighting marks the areas of Great Plains states that lost the most soil during the Dust Bowl.

a.

b.

c.

d.

e.

Figure 9.29 Conservation tillage methods. **a.** Young soybean plants in Maryland are grown in the residue of a previous wheat crop that protects the soil from erosion and helps retain moisture for the new crop. **b.** Contour farming of cropland in southwest Iowa. **c.** Terraces in a California vineyard reduce the gradient of the slope and minimize soil lost to water erosion. **d.** A line of trees forms a windbreak to shelter cropland in Iowa. **e.** Strip cropping of alternating bands of corn, oats, and alfalfa reduces erosion in a field in northeast Iowa.

The federal government created several programs to encourage changes in farming practices on lands with potential for soil erosion. The most recent example, the *Conservation Reserve Program,* was begun as part of the 1985 farm bill to pay farmers not to cultivate potentially erodible farmland. Most of the contracts were issued to farmers in the Great Plains states (Figure 9.30). Government-sponsored conservation programs have helped reduce soil erosion rates by one-third over the past few decades.

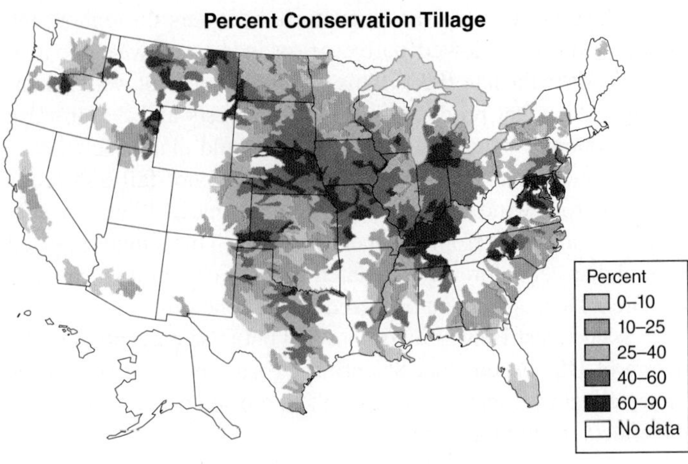

Figure 9.30 US farmland under conservation tillage.

Checkpoint 9.23

An average-sized US farm has an area of 176 hectares (440 acres). Design an experiment to get an accurate measurement of soil erosion for a farm of this size.

Checkpoint 9.24

basic advanced intermediate superior

Soil Conservation Program Evaluation Rubric

This assignment requires you to consider what combination of factors would contribute to the greatest risk of soil erosion on US farmland. Imagine that you work for the National Resource Conservation Service (NRCS). You are in charge of the program that pays farmers ***not*** to farm land that has the potential for soil erosion. Because of recent budget cuts, the NRCS must cut the amount of farmland in the program by one-third. Your job is to complete the evaluation rubric provided here, which will allow inspectors to rate the potential for erosion on all the farms currently supported by your program. The farms that score lowest on the rubric will have the least potential for soil erosion and will be cut from the program. Note that all of these farms will experience some soil erosion; your job is to determine which will have the least erosion. You must make a decision in the next 6 months, and you do not have the time, resources, or personnel to actually measure erosion rates directly in the field.

Audience: You will create a scoring scheme that can be applied by a team of federal soil erosion inspectors who have some basic science training and several years of experience in soil conservation. It is important that all of these inspectors use a standard system to assess the potential for soil erosion. Based on the reports of these inspectors, the staff of the NRCS will make recommendations about which farms to cut from the program.

Main point and purpose: To demonstrate that you understand the principal factors that contribute to soil erosion and conservation.

Pattern and procedures: Identify six factors and differentiate among characteristics that indicate their potential for erosion (high, moderate, or low risk). One factor, vegetation, is included as an example. Each high-risk factor receives 3 points; low-risk factors rate 1 point. The site with the highest cumulative score for all factors will be at greatest risk from future soil erosion.

Standards and criteria: You will be assessed on your choice of:

1. Relevant factors that would contribute to the potential for soil erosion
2. Identification of what constitutes high-, moderate-, or low-risk situations for each factor
3. One factor in particular as the most significant. The score for this factor will be doubled. Discuss your justification for choosing this factor.

Soil Conservation Program Evaluation Rubric

Factors	Low risk (1 point)	Moderate risk (2 points)	High risk (3 points)
Vegetation	Crop residue left on land, shelter belts present	Shelter belts present	No effort to keep soil covered or provide shelter belts

the big picture

In this chapter, we found that weathering breaks down and alters rocks and minerals at or near Earth's surface. Physical, chemical, and biological weathering breaks down rocks and minerals into smaller pieces and causes these fragments to decompose into other minerals. These processes reinforce one another. The more a rock is broken down by physical and biological weathering, the more susceptible it becomes to chemical weathering. Likewise, the more chemical weathering occurs, the weaker the rock becomes, making it easier to be broken down physically. The chapter opening image shows cliffs of jointed rhyolite at Mojave National Preserve, California, that are pockmarked with large, rounded cavities. This weathering pattern is known as *tafoni* (sometimes spelled taphoni) and forms from the growth of salt crystals (see Section 9.2).

Unlike volcanic eruptions or earthquakes, weathering does not produce much of a recognizable change over short periods of time and therefore is easy to overlook. But if we are to successfully preserve some of our most treasured cultural structures, we must consider the impact of the different types of weathering over timescales measured in centuries or millennia.

Weathering produces soils, but soil erosion due to human activities and natural processes greatly outpaces the rate of soil formation in many environments. For example, materials in soils that once covered Iowa farm fields are now on the bottom of the Gulf of Mexico, transported there by the streams of the Mississippi River drainage basin. Despite advances in technology that let us map soil erosion in great detail, modern technology can do little to prevent the loss of soils. The techniques that work today are pretty much the same ones that George Washington applied more than two centuries ago. The principal soil conservation techniques include keeping soil covered, limiting the effects of slopes through contour plowing and terracing, providing shelter for fields exposed to winds, and ensuring a steady supply of nutrients. Unfortunately, as basic as these recommendations are, many farmers on marginal lands around the world lack access to the resources necessary to do even these things. As the human population continues to expand, future wars may be fought not over oil or mineral resources but over access to soils to feed starving citizens.

Weathering and Soils: Concept Map

Complete the concept map to evaluate your understanding of the interactions between the earth system and weathering and soils. Label as many of the interactions as you can, using the information from this chapter.

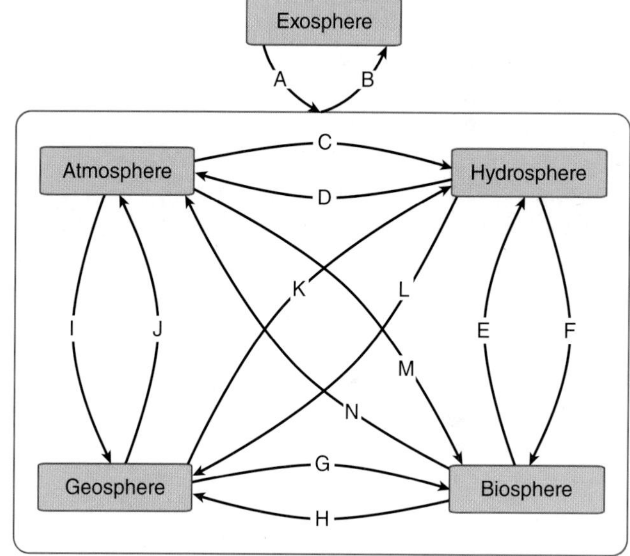

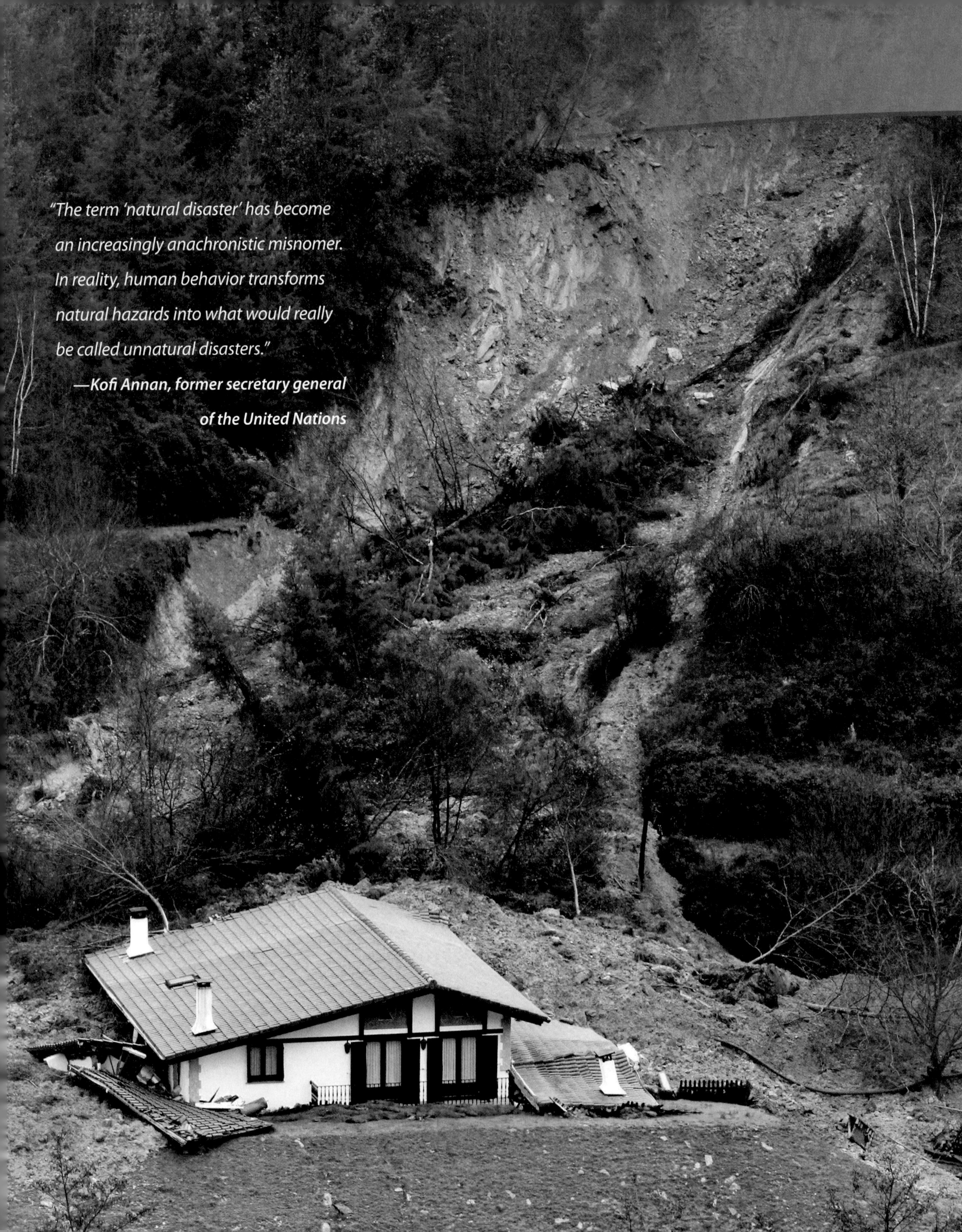

"The term 'natural disaster' has become an increasingly anachronistic misnomer. In reality, human behavior transforms natural hazards into what would really be called unnatural disasters."

—Kofi Annan, former secretary general of the United Nations

Chapter 10

Landslides and Slope Failure

Chapter Outline

the big picture

Would you live here?

See The Big Picture box at the end of this chapter for the full story on this image.

10.1 Mass Wasting: The Human Impact

Chapter Learning Outcomes

- Students will be able to describe the characteristics of common slope failure processes.
- Students will be able to evaluate the risk to society of a mass wasting event.

Resident Bob Hart describes the coastal community of La Conchita, California, as "a wonderful place. It's a gorgeous beach. The climate is one of the best in the world." Neighbor Jonathan Harrel agrees: "It is the greatest place in the world to live. And who has the money to go buy another home with a 180-degree view of the ocean?" Well, as it turns out, looks can be deceiving, and a home with a pleasant view of the ocean can become a death trap when it is located at the base of an unstable slope.

La Conchita, home to 700 people, covers just nine blocks along a narrow coastal strip fronted by the Pacific Ocean and backs up to a 180-meter-high (600-foot) bluff (Figure 10.1). In 1995, part of the slope collapsed, creating a landslide that destroyed nine homes and resulted in a partial evacuation of the town. Though no residents were killed, the economic impact on homeowners was significant as real estate prices plummeted. But just 5 years later, many people had forgotten what had happened. It seemed that the small community was on the rebound until the events of January 10, 2005.

The year began with heavy rains in southern California. The area around La Conchita received 40 centimeters (16 inches) of rain in 2 weeks, almost equivalent to the region's annual rainfall. On the morning of January 10, residents awoke to find that they could not leave town because of mudslides and flooding up and down the coastal highway. People were enjoying the unexpected holiday until just after noon. Resident Kathleen Wood described what happened next: "You heard what sounded like a thump. I looked up and saw dust and said, 'Here we go again.' It came down like lava down the mountains. It was explosive, like there was a stick of dynamite in there." In a matter of seconds, a 100-meter-wide (330-foot) portion of the 1995 landslide material mobilized and flowed down the slope, burying homes and blocking streets. Ten residents died as more than a dozen homes were destroyed and many others damaged.

What makes one landslide a natural phenomenon and another a natural disaster, or are they one and the same? Natural phenomena (related to the earth system) are constantly occurring. Phenomena such as landslides, volcanic eruptions, and earthquakes can cause rapid, unexpected changes to the earth system, while other processes such as weathering and rock formation produce slow, imperceptible changes. **Natural disasters occur when natural phenomena cause extensive damage to property or harm people or do both.** If there is no human impact, there is no disaster, just a natural phenomenon. Most case studies used in this chapter will involve disasters, but you should keep in mind that the processes involved can and do occur almost everywhere. By observing natural phenomena, we can prepare for and avoid natural disasters.

The Phenomenon of Mass Wasting

As we learned in Chapter 9, thick deposits of regolith can form in areas with rapid weathering processes. Regolith on slopes has the potential to fail by collapsing and moving downhill. Such failure is more likely when the material is not anchored to the underlying slope and is only loosely bound together. All other factors being

Figure 10.1 La Conchita. **a.** The community of La Conchita is situated along the coast of California between Santa Barbara and Ventura. Its 700 residents live in the narrow strip of land between the Pacific Ocean and a steep slope. **b.** Separate landslides destroyed homes on Vista Del Rincon Drive twice within a decade.

equal, the thicker the regolith, the more likely it will fail. **The downslope movement of material under the influence of gravity is termed *mass wasting*.** The terms *slope failure* and *mass wasting* are often used to describe the same processes. Mass wasting represents one of the most active processes in modifying the landscape in areas that have significant changes in elevation (Figure 10.2). The general term ***landslide* is used to describe many rapid forms of mass wasting.**

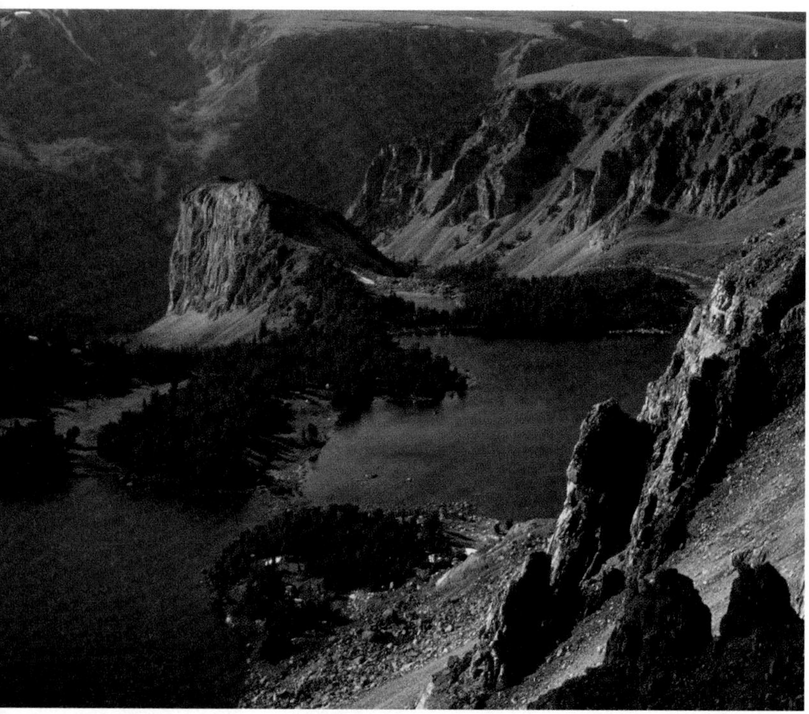

Figure 10.2 Mass wasting. Steep terrain with significant relief (change in elevation) is suitable for slope failure, as at this site in the Beartooth Mountains of southern Montana.

Examine the national map of landslide areas in Figure 10.3. This map shows areas with the greatest potential for landslides. Where do you see the areas of highest risk for failure? What do those areas have in common? **The greatest risks for slope failure are in mountainous regions, such as the Appalachians and Rockies, and in the states bordering the Pacific Ocean.** All of these areas have relatively steep slopes where materials may move rapidly downhill.

Are landslides and other slope failures common? Where and why do they occur? Can they be prevented? These are a few of the questions we will explore in this chapter. We will also discuss what can be done to protect people who live on or near steep slopes. Finally, we will describe how most mass wasting phenomena can be classified by examining the manner by which the material moves and the type of material involved.

Self-Reflection Survey: Section 10.1

Answer the following questions as a means of uncovering what you already know about landslides and slope failure.

1. Are there places in your community that are characterized by steep slopes? Have landslides occurred there?
2. Could Kofi Annan's quote (see chapter opening pages) be applied to the events at La Conchita on January 10, 2005? Explain your reasoning.
3. Why do you think people choose to live in places like La Conchita, even after landslides? Ventura County has invested time and resources to clean up after both landslides at La Conchita. Is it ever fair for the taxpayers of the county to say, "Enough! Don't use our money to bail out people who choose to live where there is a high risk of landslides"?

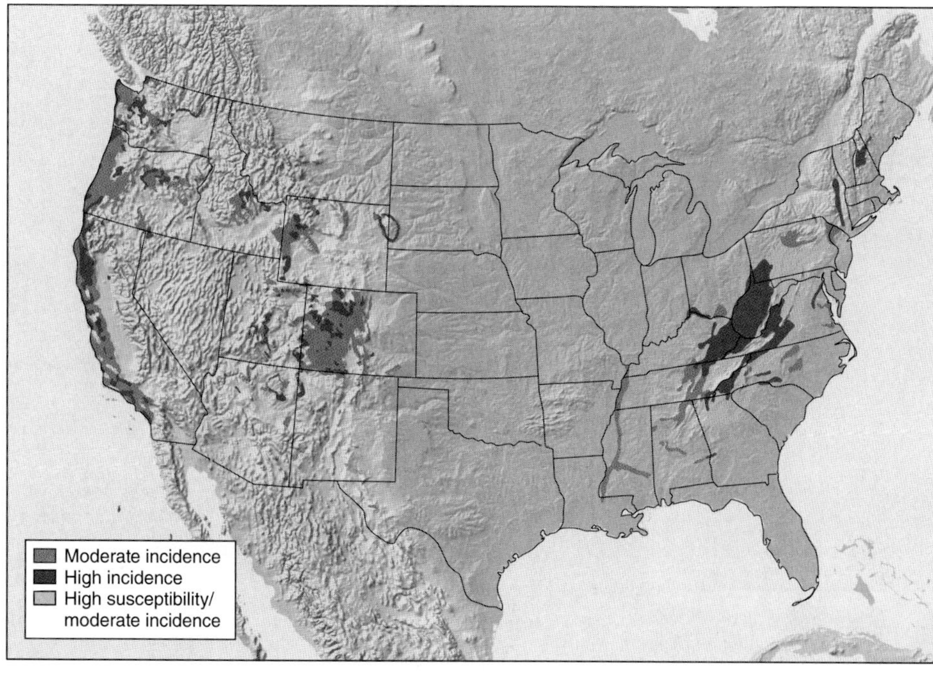

Figure 10.3 Landslide areas in the conterminous United States. In areas of high to moderate incidence (orange and brown), landslides occur more frequently. Areas of moderate incidence but high susceptibility (gold) have the potential for landslides, especially after heavy rains or earthquakes.

10.2 Factors Influencing Slope Failure

Learning Objectives

- Discuss the principal factors that result in mass wasting.
- Describe how human activities can contribute to greater damage from mass wasting events.
- Explain some strategies that could be applied to reduce the risk of future mass wasting.

The Sierra de Avila is a narrow mountain range that sweeps down to the Caribbean coastline of Venezuela (Figure 10.4). Nearly half a million people are crowded into a series of cities wedged into a thin coastal strip between the mountains and the sea. Historical records and geologic evidence reveal that this region has experienced catastrophic landslides and floods once or twice each century. The level ground between the mountains and the sea is typically the product of ancient landslides. This area serves as a case study for our discussion of the various factors influencing slope failure.

Slope Angle

Conditions along the north flank of the Sierra de Avila are almost ideal for producing devastating landslides. Elevation drops from more than 2,500 meters (8,200 feet) to sea level in less than 10 kilometers (6 miles), producing steep slopes (Figure 10.5) with angles of 30 to 60 degrees in places. Slope angle is a critical factor in determining the potential for landslides. Almost everyone remembers going down a slide on a playground as a child. You sat at the top and inched slowly forward until you started to slide, and down you went. The steeper and smoother the slide, the more quickly you reached the bottom. The force of gravity pulled you down the slope while the force of friction, the interaction between your clothes and the slide, slowed your movement. The same principles apply to materials moving down slopes during mass wasting events. **Gravity causes materials to move down slopes while friction acts to prevent or slow that movement.**

The Influence of Gravity

Gravity acts on all objects in the universe. On Earth, the force of gravity is always directed toward the center of the planet. When an object rests on a slope, the force of gravity can be resolved (divided) into two directions, called components. One component is parallel to the slope (g_s), and the other is perpendicular to the

Figure 10.5 Slope angle. The steep slopes of Sierra de Avila allow material to wash into narrow canyons that funnel debris toward coastal cities. The bare patches indicate where landslides occurred.

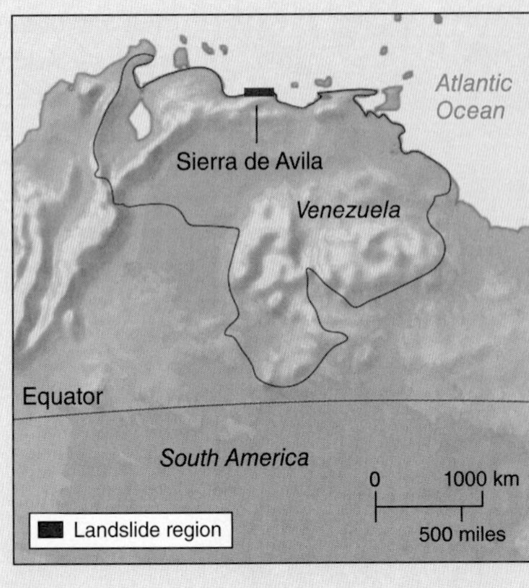

Figure 10.4 Site of Venezuelan landslides. **a.** The red rectangle on the map shows the location of most of the landslides that occurred in December 1999 in Venezuela. **b.** The view from the Caribbean Sea shows the mountains of the Sierra de Avila with the city of Caraballeda in the foreground.

slope (g_p; Figure 10.6). **The steeper the slope, the larger the component of gravity parallel to the slope, making it more likely the objects will move downhill,** setting the stage for slope failure. However, slope angle and gravity alone do not determine whether an object will move down the slope. Numerous steep slopes are not currently undergoing mass wasting. Can you think of some possible reasons why material would start moving (some triggers)?

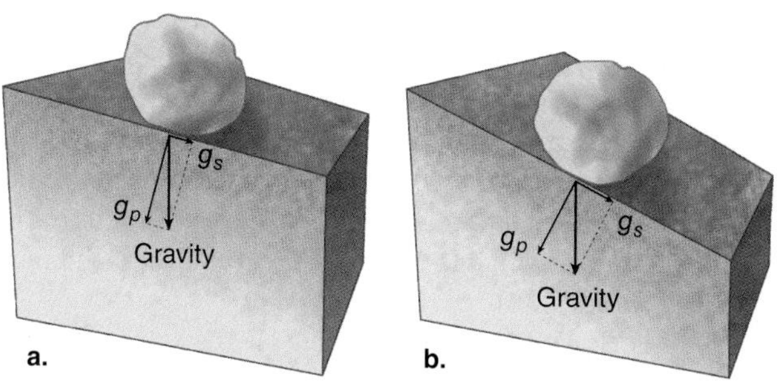

Figure 10.6 Gravity acting on a slope. **a.** Two components are in effect, one oriented parallel (g_s) and the other perpendicular (g_p) to the slope. **b.** The steeper the slope, the larger the component parallel to the slope and the greater the prospect for slope failure. Any process that acts to increase the force parallel to the slope (g_s) or decrease the force perpendicular to the slope (g_p) will increase the likelihood of slope failure.

The Effects of Water

Let's do another childhood flashback. Think about building a sand castle. What happens if you use dry sand? The sand castle falls apart immediately. Dry sand will not form a slope with an angle of more than about 35 degrees (Figure 10.7a). Children know that by adding a little water to the sand, they can sculpt structures with vertical walls. The surface tension created by a small amount of water holds the sand grains together, a phenomenon called *cohesion* (Figure 10.7b). **Slope failure is less likely if the unconsolidated materials on the slope are cohesive.**

What happens when too much water is added to the sand? Very wet sand is not suitable for sculpting sand castles because it flows like a liquid (Figure 10.7c). There are several reasons that excess water can increase the likelihood of a slope failure. Excess water that saturates the slope material will reduce cohesion between grains and allow them to move more freely. Further, the addition of water may promote instability by adding weight to a slope.

Finally, the water in the slope supports some of the weight of the overlying material, reducing the force (g_p) acting perpendicular to the slope and further reducing the friction that helps hold things in place. Imagine filling your sink halfway with dry sand and pressing a plate into the top of the sand. Now take a finger and try to slide the plate to one side. The sand will resist movement of the plate. Next, add water to the sand until the water comes just to the surface. Now the water is bearing much of the weight of the

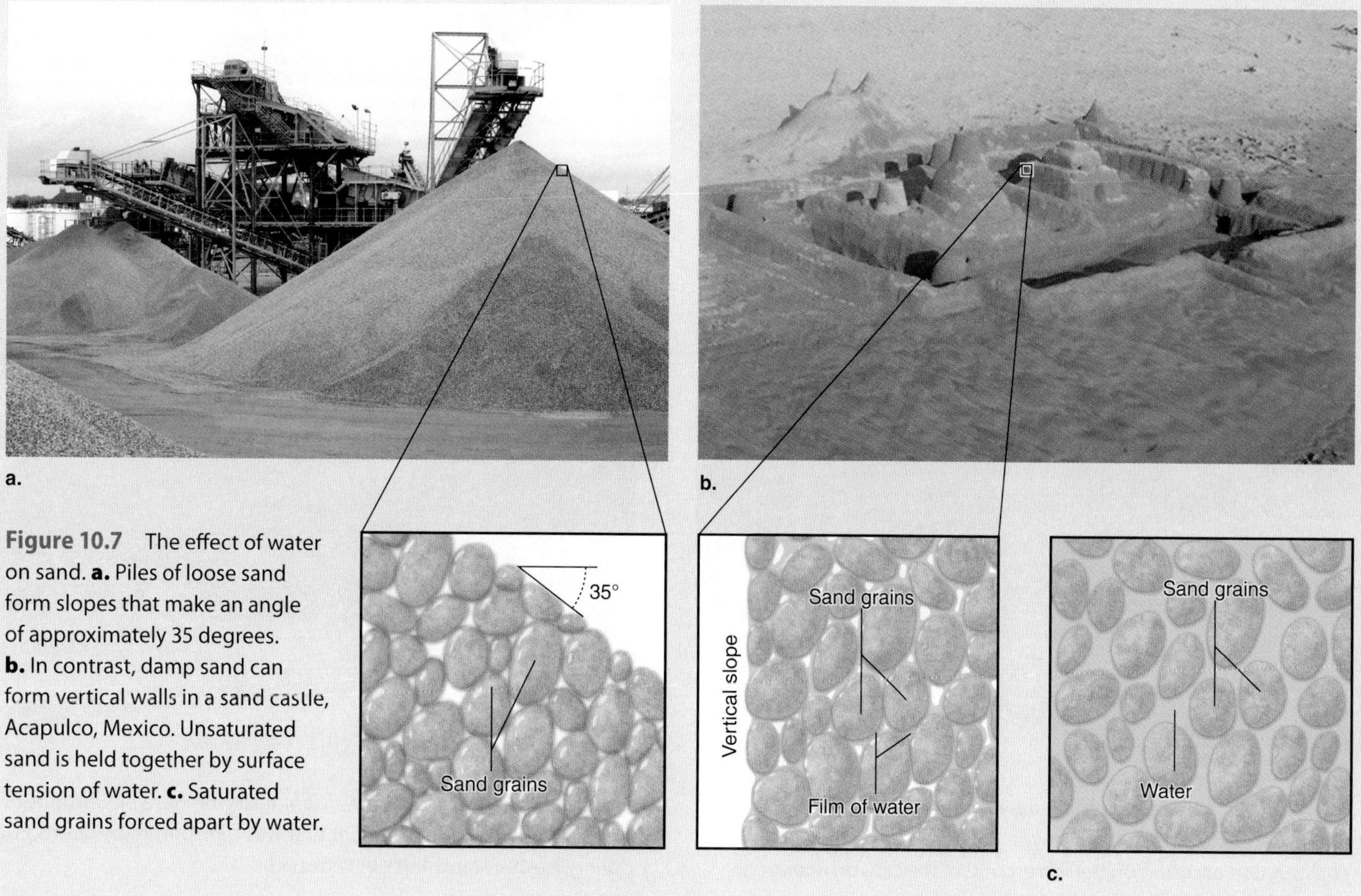

Figure 10.7 The effect of water on sand. **a.** Piles of loose sand form slopes that make an angle of approximately 35 degrees. **b.** In contrast, damp sand can form vertical walls in a sand castle, Acapulco, Mexico. Unsaturated sand is held together by surface tension of water. **c.** Saturated sand grains forced apart by water.

plate. Try pushing the plate again with your finger. This time it will move more easily. The water is supporting the plate, reducing the force acting perpendicular to the surface and making it easier to move the mass.

The effects of water on slope stability can be illustrated by comparing the mountains of the western United States with the Appalachian Mountains in the East. On the basis of elevation alone, we might expect the greatest landslide hazards to be in the western states, where peaks are more than twice as tall as the highest Appalachian Mountains. However, the greater landslide risk is in the eastern Appalachian states (see Figure 10.3). How would you account for this phenomenon? As we learned in Chapter 9, a major difference between the western and eastern United States is the amount of precipitation each region receives. Water is much more plentiful in the East compared to the West (with the exception of the Pacific Northwest). Remember, the **addition of water to slopes with loose materials almost always increases the likelihood of failure.**

Figure 10.8 Venezuelan canyons. View up a canyon from Caraballeda showing narrow canyon floor and steep sides. Most construction occurred on the flat ground of the canyon floor or on ancient deposits at the mouth of the canyon.

Case Study: Slope Failure in Venezuela

Between December 14 and 16, 1999, unseasonal storms dumped 91 centimeters (36 inches) of rain on the steep mountain slopes of northern Venezuela. This rainfall produced a series of floods and landslides that devastated communities along the coast and resulted in the deaths of at least 19,000 people. Such heavy precipitation, up to 7 centimeters (2.7 inches) per hour at times, is estimated to occur once every 1,000 years. For comparison, the 3-day rainfall total was equivalent to the average annual rainfall in midwestern US states such as Indiana and Ohio. Soil and regolith up to 3 meters (10 feet) thick received little protection from the shallow roots of vegetation and were washed off the slopes into nearby canyons (Figure 10.5). The narrow canyons funneled the mixtures of water and regolith downslope toward cities often located on the only available patch of flat ground—at the mouths of canyons (Figure 10.8).

Debris flows began to descend on the cities during the night of December 15, 1999. A slurrylike mixture of soil, rocks, and water flowed at velocities of 10 to 50 kilometers per hour (6 to 30 miles per hour). The flows were fast and powerful enough to carry bus-sized boulders (Figure 10.9).

Checkpoint 10.1

basic · intermediate · advanced · superior

The landslides in Venezuela occurred as a result of increasing urbanization along the narrow coastal strip adjacent to Sierra de Avila.

a) True   b) False

Figure 10.9 Landslide debris. Geologists measure dimensions of an 11-meter-long (36-foot) boulder that was transported by debris flows during the 1999 landslides in Venezuela.

These materials were eventually deposited in layers up to 5 meters (16 feet) thick on the canyon floor, at the mouth of the canyon, or just off the coast in the Caribbean Sea. Destruction was particularly significant in the city of Caraballeda, where single-story buildings were buried, apartment buildings suffered partial collapse, and the local marina was filled with a mixture of mud and boulders (Figure 10.10). Over 140,000 people were left homeless, and roads, communications, and utilities networks were badly damaged. Many of the people who were killed had lived in makeshift shantytowns built in dry river valleys and on steep slopes.

a.

b.

Figure 10.10 Destruction in Carabellada. **a.** Deposits several meters thick buried single-story buildings nearly to their roof lines. **b.** This apartment building partially collapsed after debris flowed through its second story.

Cities of the region had little in the way of urban planning, construction standards, or building inspections. Further, people were unable to evacuate the region by normal routes since the main road ran along the coast and was buried under numerous landslides. More than 70,000 survivors were rescued by sea or air. The cumulative cost of the damage was $2 billion, approximately the annual average cost of all the damage from US landslides combined.

The La Conchita and Venezuela examples illustrate that slope-destabilizing water can often contribute to natural processes; but excess water may also be a consequence of human activity such as leaking septic systems, overzealous irrigation, and poorly engineered dams. For example, many residents blamed the earlier 1995 slide at La Conchita on overwatering of crops on the land above the bluff. Human activity such as logging or natural factors such as forest fires may also remove vegetation that both shelters the slope and provides a network of roots to hold slope material in place. Devastating mudslides in Italy in 1998 followed heavy rains and were blamed on poor land management practices, including deforestation (widespread removal of trees) on surrounding slopes. Those mudslides killed more than 100 people in two villages. Earthquakes, vibrations caused by traffic, excessive loads from buildings and other structures, and poorly planned excavations can also trigger mass wasting.

Checkpoint 10.2

basic | advanced | intermediate | superior

List as many factors as you can that contributed to the debris flows in Venezuela in December 1999.

Checkpoint 10.3

basic | advanced | intermediate | superior

Venn Diagram: Comparison of La Conchita and Caraballeda Landslides

Use the Venn diagram to compare and contrast mass wasting events at La Conchita and Caraballeda. Write the features unique to either group in the larger areas of the left and right circles; note features that they share in the overlap area in the center of the image. Identify at least eight characteristics.

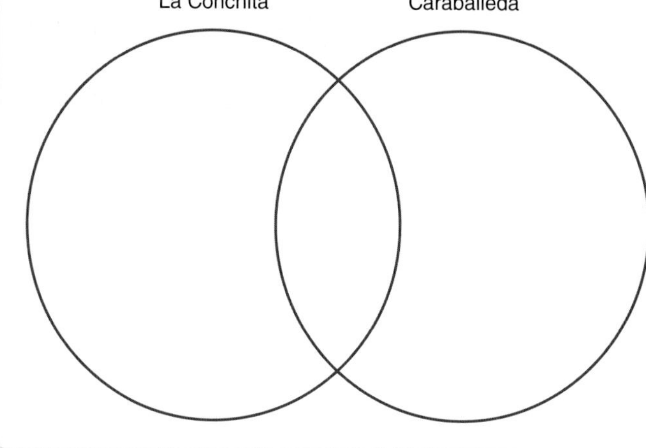

Figure 10.11 Road built on slope. Beartooth Highway traverses the east side of the Beartooth Mountains near Red Lodge, Montana.

Methods of Stabilizing Slopes

So, what can be done to minimize slope failures and their devastating consequences? One obvious solution is to avoid building structures on or near steep slopes. However, such realistic advice may not stop a homeowner in search of a spectacular view and is of little use to engineers who must build roads on steep slopes in mountainous terrain (Figure 10.11). Instead, efforts focus on improving slope drainage (to reduce the role of water) or attaching the slope material to bedrock with physical restraints (anchoring it). Such restraints may be as simple as covering slope material with wire mesh or cement, or bolting panels to the underlying bedrock (Figure 10.12). Examples of these restraints are visible along the Beartooth Highway, which climbs up the side of the Beartooth Mountains, Montana (see Figure 10.2), connecting the northeast entrance to Yellowstone National Park (Wyoming) with the city of Red Lodge, Montana. Persistent landslides on highways and connecting roads require changes to underlying slopes and periodic repairs to fix the road surface (Figure 10.13).

a.

b.

Figure 10.12 Slope stabilization methods. **a.** Wire mesh and **b.** concrete cover exposed slopes, and **c.** a retaining wall with drainage pipes protects the road. All images are from the Rocky Mountains.

c.

Checkpoint 10.4

basic / intermediate / advanced / superior

Landslide Risk Rubric

Create an evaluation rubric to use to review the potential landslide risk in San Bernardino County, California. Follow the guidelines below.

Your group is asked to create an evaluation rubric to identify factors that will influence the potential for damage from future landslides. This community is in a region where urban development has expanded right up to the foothills of the mountains. Consider the physical factors that contributed to the significant debris flows in Venezuela as you fill in the table to identify high-risk versus moderate-risk versus low-risk conditions. One factor, thickness of regolith, is completed in the table as an example. List as many additional factors as you can.

Audience: You will create a scoring scheme that can be applied by most educated citizens, specifically city council members or a concerned citizens group.

Main point and purpose: To demonstrate your understanding of the principal factors that result in potential damages or loss of life from a landslide event.

Pattern and procedures: Identify key factors and differentiate among characteristics that make them high-, moderate-, or low-risk phenomena. A landslide is most likely to occur, and therefore cause damage, at sites with the thickest regolith (high risk). Areas with no regolith may have less potential to cause damage (low risk). Each high-risk factor receives 3 points; low-risk factors rate 1 point. The site with the highest cumulative score for all factors would be at greatest risk of sustaining damage from a future landslide.

When you have completed the rubric, review your factors and identify the one you regard as most significant. Support your choice.

Standards and criteria: You will be assessed on your choice of:

1. Relevant factors that would contribute to landslide-related damages.
2. What constitutes high, moderate, or low risk for each factor.
3. One factor in particular as the most significant. The score for this factor will be doubled.

Factors	Low risk (1 point)	Moderate risk (2 points)	High risk (3 points)
Thickness of regolith cover on bedrock	No regolith	Thin regolith	Thick regolith

a.

b.

Figure 10.13 A landslide area in southwest Montana. **a.** Before precautions were taken, landslides were frequent. **b.** Repairs were made using supports added to the road above the slide area. The slope below the supports was also graded to reduce the risk of future mass wasting.

10.3 Slope Failure Processes

Learning Objectives

- Describe the similarities and differences between rockfalls and rockslides and between slumps and debris flows.
- Explain how mass wasting processes are classified on the basis of the type of material and type of movement.
- Evaluate the type of slope failure that would occur in a given scenario.
- Discuss the relationship between the type of mass wasting and weathering processes.

Mass wasting phenomena are directly linked to the weathering processes we described in Chapter 9. What does physical weathering do to a rock that might be related to mass wasting? Physical weathering processes break rock into fragments. Such fragments can be relatively small and may serve to loosen boulders that subsequently fall; or breaks may occur along preexisting joints or bedding surfaces, reducing cohesion and releasing large amounts of material that moves downslope.

What role do you think chemical weathering plays here? Chemical weathering weakens the rocks and leads to formation of regolith. If weathering is rapid, bare rock surfaces are less likely to develop, since they will be buried beneath a blanket of regolith. Thick layers of regolith in regions of rapid chemical weathering may fail due to the added weight of water delivered by precipitation. When physical weathering dominates (as in arid regions), rock is likely to be the material involved in mass wasting events. Regolith is more likely to be involved where chemical weathering processes dominate (as in humid areas). **Mass wasting can be classified according to the type of material moving down the slope (rock versus regolith) and the manner by which the material moves** (Table 10.1). Material is dislodged and moves nearly vertically during a fall, travels along an observable detachment surface during a slip, and moves as a fluid during a flow.

Rockfalls

Have you ever seen a sign announcing "Falling Rock Zone" along a highway? Did you look up to see if any rocks were headed your way? **A rockfall is simply the dislodging of a rock from a steep slope or cliff** (Chapter Snapshot). Rockfalls commonly occur when physical weathering (for example, frost wedging) loosens boulders from rocky cliffs in mountainous terrain. The boulders break off and fall, producing an apron of coarse debris (*talus*) at the base of the slope (Figure 10.14). Unless rockfalls trigger a rock avalanche (a condition in which more and more rocks are dislodged), they are rarely hazardous. This is so because they most frequently occur in isolated locations, and individual falls involve relatively small volumes of material. Activities that place people on or near rock slopes in mountainous areas can occasionally prove dangerous. Buildings and roads located below high slopes run the risk of being struck by boulders that break off cliffs and tumble hundreds of meters (Figure 10.15).

Even small volumes of debris can be deadly. On a warm afternoon in May 1999, dozens of tourists were relaxing around the plunge pool below the Sacred Falls in a state park on Oahu, Hawaii (Figure 10.16). Suddenly boulders were crashing down

Table 10.1 Types of Mass Wastings

	TYPE OF MOVEMENT		
Material	**Fall**	**Slip**	**Flow**
Rock	Rockfall Rock avalanche	Rockslide	
Regolith		Slump	Mudflow, creep, debris flow, debris avalanche

Figure 10.14 Talus. Rockfall debris collects at the base of a cliff in the Beartooth Mountains, Montana.

a.

b.

Figure 10.15 Rockfall hazards. **a.** Rockfall boulder on I-70, Glenwood Canyon, Colorado. **b.** Boulder in a house, Zion National Park, Utah.

around them, ricocheting off the steep, narrow canyon walls. Eight people were killed and 50 more injured when a group of rocks with an estimated volume of just 25 cubic meters (880 cubic feet) broke off the cliff 150 meters (490 feet) above. The steep, narrow-walled canyon provided little protection for the helpless victims.

Subsequent examination of the area by landslide experts from the US Geological Survey revealed numerous scars marking areas on the canyon walls where rocks had previously broken free and exposed bedrock below a tangle of vegetation. The scientists could find no evidence of a trigger for the event and concluded that long-term weakening of slope materials had been responsible for the rockfall. They noted that the trail to the falls was built on boulders that had fallen during previous landslides. They also determined that the only way to ensure no further injuries or deaths was to close public access to the canyon. Injured victims and relatives of those who died sued the state of Hawaii. The state maintained that the rockfall was an "act of God" and that signs in the park warned visitors that they entered at their own risk. The sign went on to state that the area was prone to falling rocks that could cause death or injury. However, an expert testified that the signs were too long-winded and that visitors could not be expected to read a 250-word description of potential hazards. The jury found Hawaii liable for not posting or maintaining adequate warning signs and awarded \$8.5 million to those affected by the rockfall.

Rockslides

How is a rockfall different from a rockslide? Consider this example: In 1925, at Gros Ventre Valley in northwest Wyoming, a sandstone layer with about twice the volume of the Caraballeda debris flows broke loose. The material raced down a slope, crossed a river, and continued 130 meters (425 feet) up the opposite slope. The movement was triggered by an earthquake that followed weeks of heavy rains that had saturated a slope underlain by weak shale

Figure 10.16 Sacred Falls, Hawaii, rockfall. **a.** The region is characterized by steep-walled, narrow canyons with slope angles of 70 to 80 degrees. Sacred Falls flows into a plunge pool surrounded by boulders (foreground) that have fallen off nearby cliffs. **b.** In May 1999, rocks broke away from a steep slope 150 meters (490 feet) above the canyon floor. Tourists were relaxing on boulders around the plunge pool and had no protection from the falling rocks.

a.

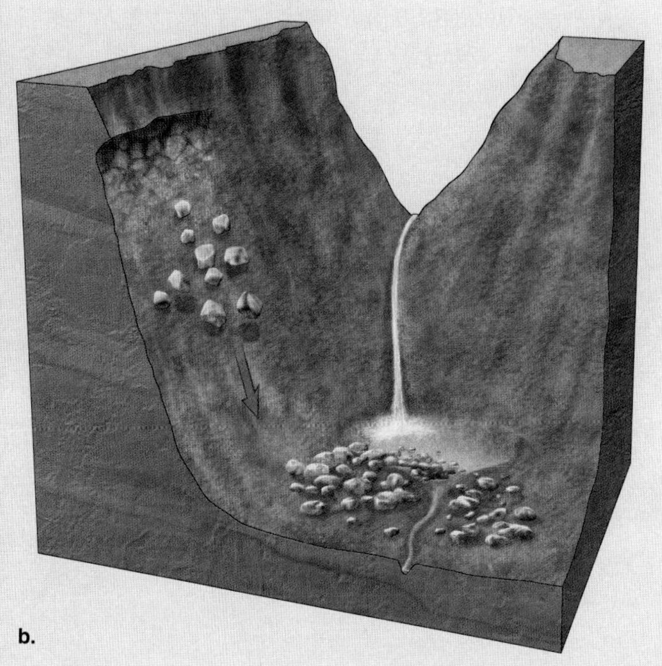
b.

LANDSLIDES

What type of landslide is this?
TOE
ROCKFALL
Rocks falling when dislodged from a cliff or steep slope as a result of physical weathering.
ROCKSLIDE
The rapid movement of a mass of rock downslope along a plane of weakness.

SCARP

Is this home in danger of a future landslide?

SLUMP

Downward rotational movement of blocks of unconsolidated material along curved surfaces.

CREEP

The very slow movement of unconsolidated material downslope under the influence of gravity.

Checkpoint 10.5

basic · intermediate · advanced · superior

Rock is to rockslide as ______ is to slump.

a) talus
b) regolith
c) lahar
d) debris flow

The mixture of materials pictured below is most likely to have been deposited by which mass wasting process?

a) Rockfall
b) Rockslide
c) Slump
d) Debris flow

The landslide pictured in the chapter opening illustration is a ______.

a) rockfall
b) rockslide
c) slump
d) debris flow

(Figure 10.17a). The whole event took just 3 minutes and formed a natural dam across the valley. The rocks exposed by the slide are still visible today (Figure 10.17b). **Rockslides are large-scale movements of rock traveling rapidly down a slope along a surface** (Chapter Snapshot). That sliding surface is typically along the original surface between sedimentary beds where the rock layers are tilted in the same direction as the slope (Figure 10.17). Alternatively, sliding may occur along a previously fractured surface such as pressure release fractures common on exfoliation domes (see Chapter 9).

Rockslides generate sheets of rock that are broken into smaller pieces as they move downslope. The leading edge of the rockslide is characterized by a jumbled collection of blocks (a rock avalanche), some up to hundreds of meters across, that collect at the bottom of the slide. A rock avalanche can sometimes trap a cushion of air under it as it moves, which allows it to travel long distances at high speeds.

Slumps

A slump is the movement of material down a slope on a curved (concave-upward) slip surface (Chapter Snapshot). Slumping typically occurs when unconsolidated regolith becomes saturated because of

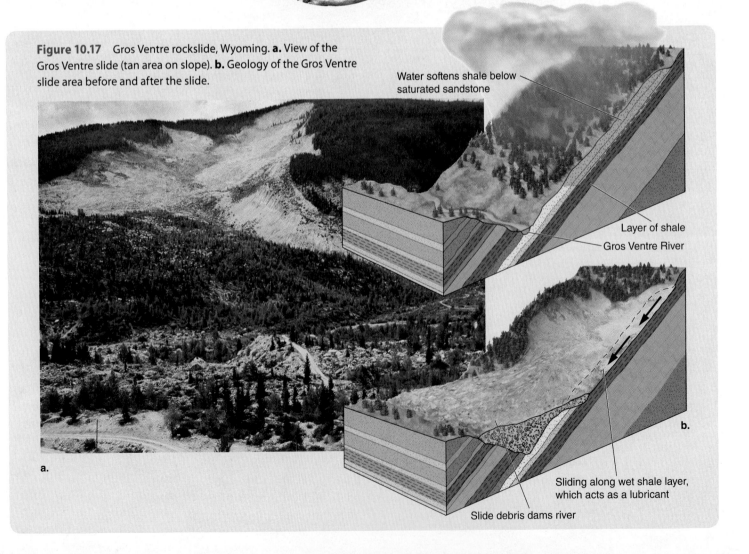

Figure 10.17 Gros Ventre rockslide, Wyoming. **a.** View of the Gros Ventre slide (tan area on slope). **b.** Geology of the Gros Ventre slide area before and after the slide.

heavy precipitation or when a slope is too steep for the material. The 1995 landslide at La Conchita, California, is classified as a slump. A clifflike scarp is usually found at the top of the slump where the regolith breaks away from the slope, and the toe is the terminus (end) of the slump (Figure 10.18).

Debris Flows and Mudflows

The landslides in Venezuela described in Section 10.2 are examples of flows. **When material flows downhill as a chaotic mixture of soil, rock, and water, it is called a *flow*.** The materials do not retain their original shape or coherence but move as a thick liquid and follow preexisting channels. Such flows occur when a relatively large volume of water is present in a mixture of coarse or fine-grained earth materials.

Flows are differentiated on the basis of the type of material involved. Debris flows, such as those in Venezuela, may travel at high speeds and involve unconsolidated regolith (most of which is coarser than sand) and particles up to the size of boulders. The 2005 landslide at La Conchita was a debris flow that formed when some of the material that was part of the 1995 slump became unstable and flowed downhill (Figure 10.18b).

Mudflows incorporate fine-grained sediment and typically follow stream channels. These fast-flowing, high-density flows are common following volcanic eruptions. When the mudflows involve volcanic debris, they are termed *lahars* (Figure 10.19). Lahars form from substantial volumes of fine-grained volcanic ash that mixes with water in streams (as described for volcanoes at Nevado del Ruiz, Colombia, and Mount St. Helens, Washington, in Chapter 6; see Figure 6.1).

Creep

One of the most destructive (in monetary terms) mass wasting processes is one of the slowest. Have you ever noticed a crack in a building foundation or have you ever visited an old house with slanted floors? Many times, those problems are caused when the

Figure 10.19 A lahar, a volcanic mudflow, generated when a tephra dam burst, releasing water from a lake near the crater of Mount Ruapehu, the largest volcano in New Zealand, 2007.

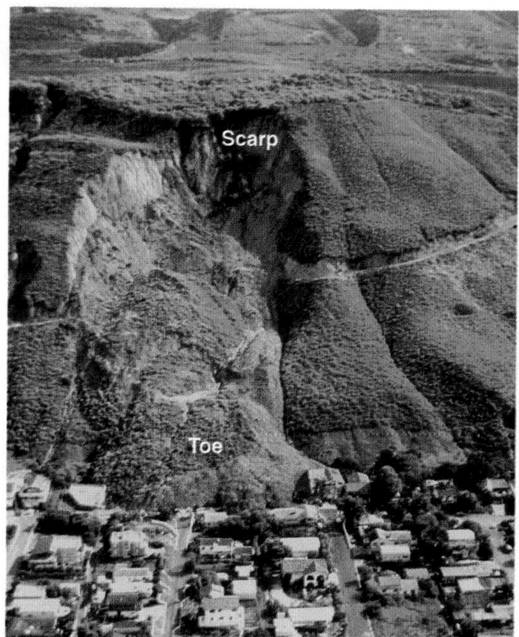

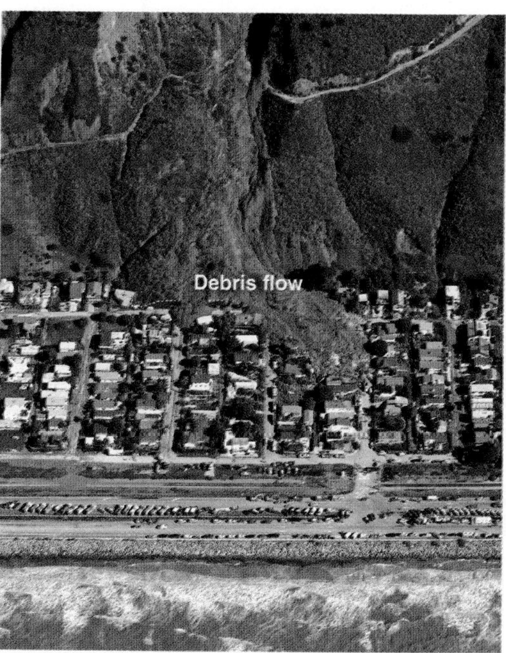

Figure 10.18 Elements of a slump. **a.** Unconsolidated material breaks away from the top of the slope, leaving a scarp surface. A slump may break into separate blocks, each with its own scarp surface. The toe of the slump overrides structures at the base of the slope. **b.** Material on the right side of the mass remobilized as a debris flow in 2005 at La Conchita, California.

foundation of the building moves. But why does it move? Frequently, the material around the foundation can slowly flow because of the added stress caused by the weight of the building. Building inspectors are supposed to make certain that developers compact loose soils before pouring a foundation. If this is not done properly, movement may occur. The same type of movement happens on gentle slopes when the force of gravity acts over long periods of time (Chapter Snapshot). **This slow progression of a material down a slope is called creep.** Many times the process is caused when clay gets wet and expands, lifting material slightly. When the clay dries, it drops the material slightly farther down the slope.

Checkpoint 10.6

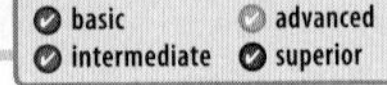

Landslide Hypothetical Example

Examine the following diagram and answer the accompanying questions. Assume that this diagram is constructed for an area in a midwestern state such as Indiana or Illinois.

The road cut in the diagram is likely to experience mass wasting by which process?

a) Rockfall
b) Rockslide
c) Slump
d) Debris flow

If the swimming pool on the right side of the diagram leaks, the underlying dirt fill is likely to experience mass wasting by which process?

a) Rockfall
b) Rockslide
c) Slump
d) Debris flow

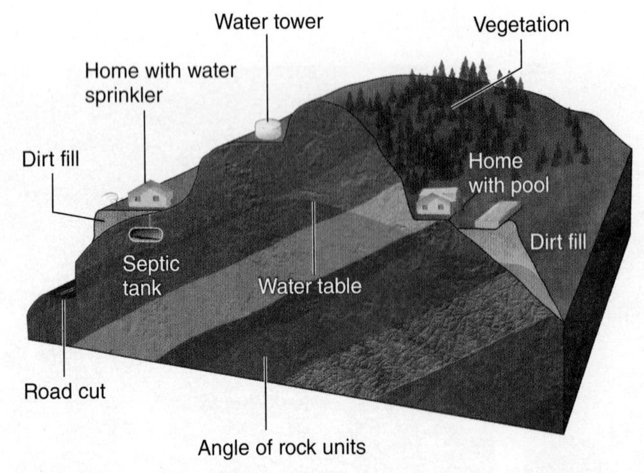

Checkpoint 10.7

Create a concept map that summarizes the characteristics of slope failure processes. Use no more than 12 terms and as many linking phrases as necessary.

Checkpoint 10.8

I-70 Highway Exercise

During the evening of November 24, 2004, an 18-wheeler truck flipped over, forcing the closure of the westbound lanes of I-70 in Glenwood Canyon, Colorado. I-70 is the principal east-west route across the mountains west of Denver, and a high volume of traffic was expected the next day, Thanksgiving. That accident probably saved lives because at about 7:30 the next morning, just up the road from the overturned truck, part of a cliff about 390 meters (1,300 feet) above the highway collapsed. A slab of rock 30 meters (100 feet) high by 10 meters (33 feet) across by 6 meters (20 feet) thick crumbled and crashed onto the roadway right between the accident scene and waiting vehicles. The massive landslide smashed holes in the highway and shut down the route for more than 24 hours. The rockfall damaged two bridges, mangled guardrails, and battered retaining walls.

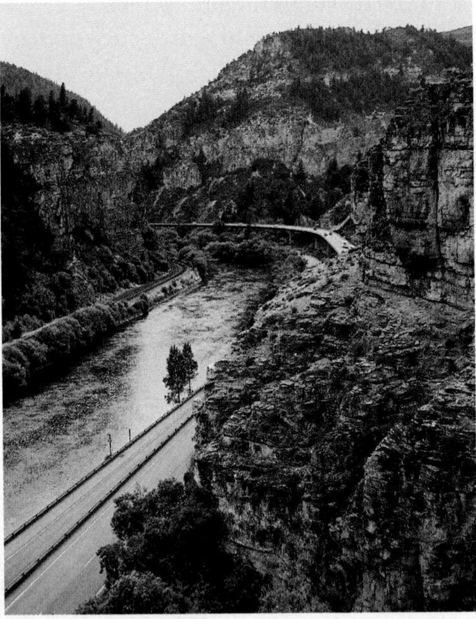

The construction of I-70 through Glenwood Canyon was a challenge because engineers had to replace a two-lane road with a modern four-lane highway in a narrow, deep canyon that also housed the Colorado River. Each day, 17,000 motorists travel the highway, risking injury or death from falling rocks. Two drivers died in separate incidents in 2003 when boulders crashed into their vehicles, and the Thanksgiving example illustrates the potential for significant loss of life from future incidents.

Imagine that you work for a consulting company hired by the Colorado Department of Transportation to create a plan for protecting motorists on I-70 in Glenwood Canyon from future landslide hazards. Resources are finite and the budget for addressing the problem is limited. What would you suggest the Department of Transportation do to minimize damage and injury?

the big picture

Mass wasting is one of the most significant processes in sculpting the landscape and is most active in areas of steep slopes that have a plentiful water supply. The theme of this chapter is the same as that of the previous one—change. But most of the changes discussed here are more catastrophic than the slow, gradual changes associated with weathering. In prescience, prehistory days, such events would have caused terror and might have been interpreted as "the gods are angry" with humankind. Science banishes this fear and ignorance. Science tells us that the world *is* understandable, that things and events occur in consistent patterns, and that people can discover the patterns in nature. For example, close examination of the slopes above La Conchita reveals several telltale signs that mass wasting has been active there for some time (Figure 10.20). A steep main scarp at the top of the slope indicates that the whole slope has moved in the past, a clear signal that mass movements could occur again.

Science exists for more than the purpose of theoretical discovery. It has a social value as well. Students of science should become thoughtful supporters of science, mathematics, and technology, understanding that scientific findings can enrich the quality of our lives and even prevent the loss of life. However, scientists can only alert the public and their elected officials of their findings. Individuals and local, state, or federal governments make the final decisions on how to address any perceived risk. For example, when landslides threaten, slopes can be made more stable by improving drainage or attaching the slope material to bedrock with physical restraints. Alternatively, supporting the base of the slope with retaining walls or planting vegetation on the slope can help inhibit failure.

Nevertheless, even well-intended efforts to protect residents may not be sufficient. After the 1995 La Conchita landslide, a retaining wall was built across the base of the slope to protect against additional small-volume landslides. The wall was easily overtopped by the 2005 debris flow, an event it was really not intended to stop (Figure 10.21). Ultimately, the cost to protect the small community was considered prohibitive by those who could have taken action to protect against future events. At that point, it was up to the individual residents to decide whether to stay or leave. Unfortunately, most citizens do not have sufficient experience with natural processes to make an informed decision, or they simply make a poor choice. For example, the chapter opening image shows a house built beside a stream in a Colorado valley. Forget for a moment the imminent threat of flooding in this narrow valley. The house is directly downslope from a landslide; in fact, it is built on the toe of an older slump. There is no retaining wall holding the slope material in place. It is just a matter of time before the slope fails again and dumps debris on the house or pushes it into the river.

Stories of natural disasters are made dramatic by their details. Everything in the reports, from the amount of rainfall, to the angle of the hillsides, to the photographs and maps showing the extent of damage, to the estimated cost of the destruction, relies on products of technology. Without sophisticated meteorological instruments, surveying tools, satellite mapping devices, and computers, these reports would lack the impact to allow people to understand and appreciate slope failure as a force of nature. With these data in hand, scientists can better recognize sites of potential future slope failure.

Figure 10.20 Evidence of mass wasting. Even before the slump and debris flow events at La Conchita, California, scars of several small slumps were apparent on the slopes above the town (yellow arrows). In addition, a large scarp surface indicated that the western two-thirds of the slope had moved as a single large mass sometime in the past. Debris flows were produced in the canyon to the west of town in 1995 and 2005.

Figure 10.21 Retaining wall failure. After the 1995 slump at La Conchita, a retaining wall was built to protect against small-volume landslides. That wall failed during the 2005 debris flow.

Landslides and Slope Failure: Concept Map

Complete the following concept map to evaluate your understanding of the interactions between the earth system and slope failure processes. Label as many interactions as you can, using information from this chapter.

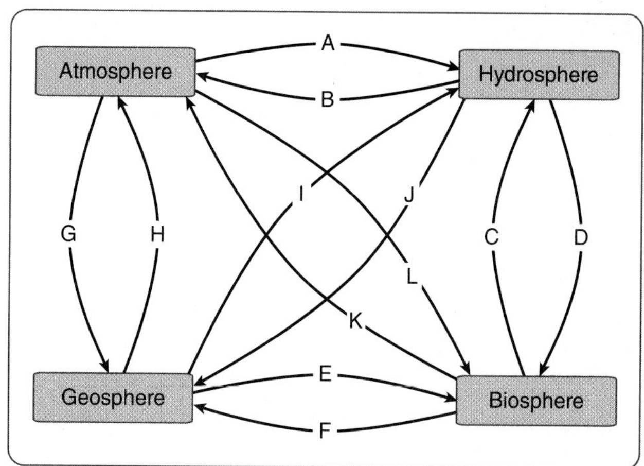

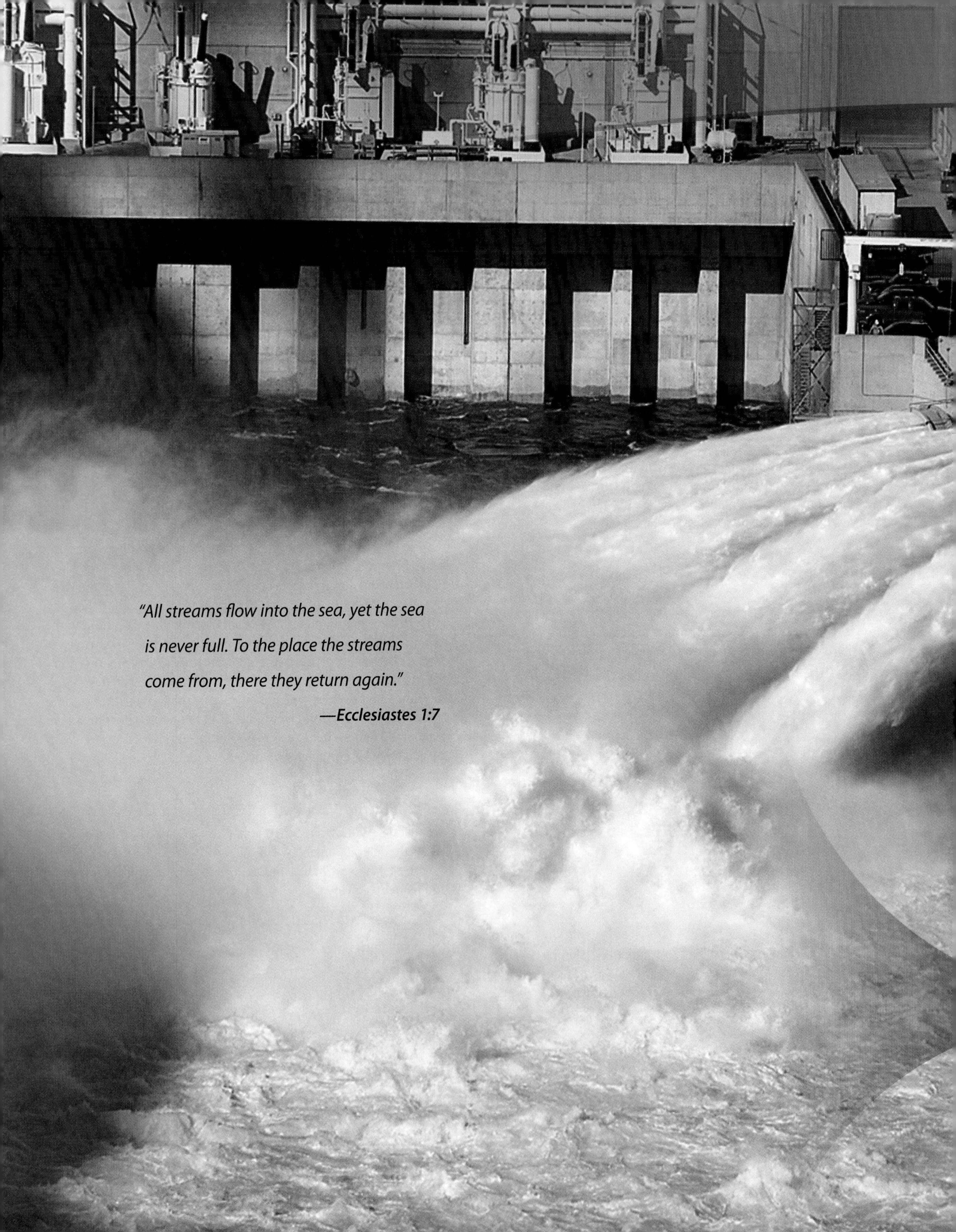
"All streams flow into the sea, yet the sea
is never full. To the place the streams
come from, there they return again."
—Ecclesiastes 1:7

Chapter 11

Streams and Floods

Chapter Outline

the big picture

What is going on here, and what does it have to do with streams and floods?

See The Big Picture box at the end of this chapter for the full story on this image.

11.1 Humans and Rivers

Chapter Learning Outcomes

- Students will recognize that water continuously cycles through the earth system.
- Students will connect the physical characteristics of a stream to the dynamic nature of stream flow and the underlying geology.
- Students will evaluate the risk to society from floods.

Sophia Pedro became the most famous new mother on Earth for a few days in March 2000. Like several of her neighbors, she had climbed a tree to avoid fast-rising floodwaters that had spread throughout much of the east African nation of Mozambique. Unlike her neighbors, Sophia was 9 months pregnant. Three days later, as a rescue helicopter was lifting her fellow tree dwellers to safety, Sophia, still in a tree, gave birth to daughter Rositha. Video footage of the dramatic mother and child rescue was broadcast around the world and served to inspire a tremendous outpouring of donations to charities helping the flood victims.

Sophia and Rositha were lucky. Seven hundred people died during the flood, and 500,000 were left homeless. Overall, this flood affected 1 million people in Mozambique and neighboring South Africa and Zimbabwe.

Geologists use the term *stream* to account for **any flow of water through a channel** (defined by its bed and banks), from the smallest creek to the largest river. As Sophia's story illustrates, rivers have the potential to cause death and destruction, but they also perform many positive functions. Streams bring water to irrigate crops, supply drinking water, provide coolant for power plants, transport barges, create ecosystems for wildlife, and serve as sites for recreation.

The Nile River: An Example of Stream Impact

The longest river on Earth, the Nile River, will serve as a model to illustrate the impact a stream can have on civilization. Stretching nearly 7,000 kilometers (4,350 miles) from the highlands of central Africa to the southern coast of the Mediterranean Sea, the Nile reaches across eight nations and 35 degrees of latitude (Figure 11.1). For a single North American river to have similar length, it would begin near the Arctic Circle in Alaska and flow to the Gulf of Mexico. (It is a common misconception that the Nile is the only river that flows north. Rivers flow downhill, not from north to south or any other particular direction.)

Egypt is a desert state with just a few centimeters of rainfall each year. However, with ready access to waters from the Nile River (Figure 11.2), Egypt became one of the earliest sites of civilization, characterized by organized agriculture, extensive cities, and a system of state governments. Even today, almost all of Egypt's agriculture and modern cities are located within the Nile valley. Most of the water that flows through Egypt originates as rainfall at elevations of around 1,800 meters (6,000 feet) much farther south, in other nations.

Civilization in Africa would have had a much rougher start without the Nile River. Humans began diverting water from the river to irrigate the desert more than 5,000 years ago. Farmers depended on seasonal variations that divided the year into times of flooding and times of low flow. Empires expanded and the pyramids were built during times of ample water; during prolonged droughts, empires collapsed and the region experienced famine and sandstorms. (How would your life change if you did not have access to an adequate water supply?) The history of the Nile illustrates that societies are stressed when water becomes scarce. Climatologists note that increasing global temperatures will result

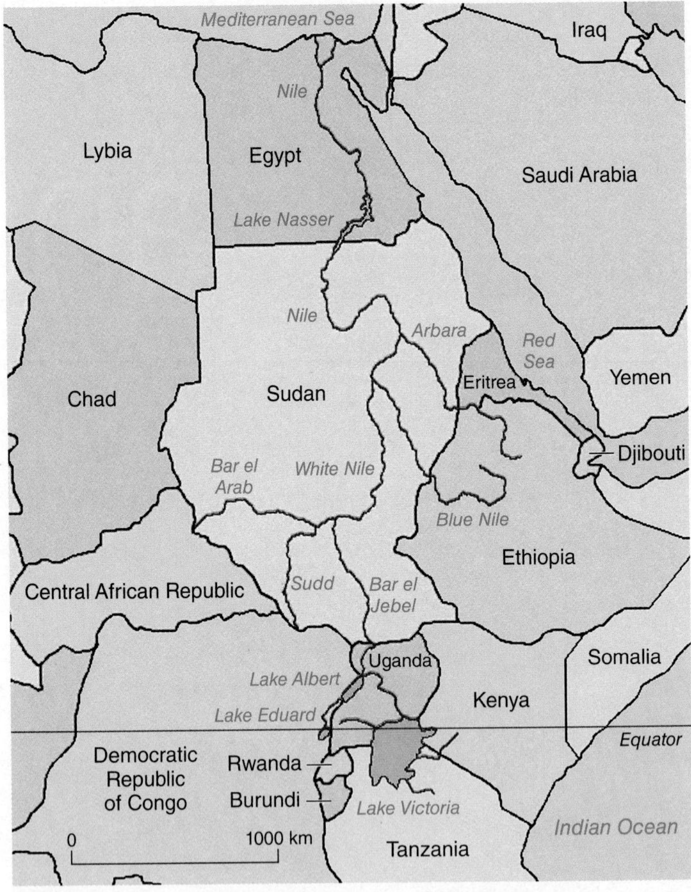

Figure 11.1 The Nile River basin. The most distant tributary of the Nile River is the Kagera River, just south of the equator in the nation of Burundi. The Kagera flows north through a series of lakes to become the White Nile and joins the Blue Nile and Arbara rivers in Sudan to form the Nile River. Lake Nasser formed behind a dam that was built at Aswan in southern Egypt to regulate the flow of the Nile and ensure a steady supply of water for agriculture and domestic uses. The Nile is 6,825 kilometers (4,241 miles) long and drops 1,800 meters (5,905 feet) in elevation from its source to its mouth.

Figure 11.2 Egypt and the Nile River. The river is the thin, sinuous ribbon that divides the eastern and western parts of Egypt. The border with Sudan to the south is at the bottom of the picture. The delta along the southern coast of the Mediterranean Sea is the triangular dark-green region near the top of the image.

in more evaporation, causing more precipitation at high latitudes and dryer conditions in some tropical areas. Some of the most significant future social changes will have more to do with declines in water supplies than with increasing temperatures.

Stream Management

While we are relatively powerless to prevent periods of drought or heavy rains, thoughtful management of our stream systems can help minimize the negative consequences of these events. The history of water in the twentieth century includes a record of extensive dam building where giant walls of concrete tamed big, wild rivers to provide steady water supplies, to control floods, and to generate electricity. Water power is by far the most widely used "green" energy source in the world. In the United States, it generates about eight times as much power as solar and wind energy combined. Just as in Egypt, the control of western rivers has allowed agriculture and cities to flourish in states characterized by desert climates. This clean energy source and steady water supply is threatened by climate change that is expected to lower water levels in reservoirs throughout the southwestern United States.

In this chapter, we will discuss where the water in rivers comes from when it is not raining, why our eyes may deceive us when we interpret how fast streams flow, and what we need to know to avoid having to give birth in a tree. By the time we are finished, you will have a new way to classify rivers that transport one of our most basic resources, freshwater. As we will discover, large stream systems are influenced by plate tectonic processes. We will begin by examining how stream systems are connected to the rest of the earth system and gradually focus on processes within a single stream before putting it all together to review the impact of floods on society.

Self-Reflection Survey: Section 11.1

Answer the following questions as a means of uncovering what you already know about streams and floods.

1. What river is nearest to where you live? Where does the river begin and where does it end?
2. How do people interact with streams? How do streams fit within the context of biology, economics, culture, politics, history, recreation, and aesthetics?
3. Use maps and other information sources to identify rivers that act as boundaries for US states or Canadian provinces. Find at least three examples of major rivers around the world that start in one country and end in another.

11.2 The Hydrologic Cycle

Learning Objectives

- Draw a labeled sketch of the water (hydrologic) cycle.
- Explain how water is distributed among different components of the earth system.
- Describe the origins of streams.

Where does the water in streams come from? It is tempting to think that stream channels just sit there waiting for rain to fall and fill them up. And yet, if you think about it, most of the time it doesn't rain. So how is it that there always seems to be water in large streams?

In reality, the distribution of water on Earth (and in streams) depends on the interactions among the four components of the Earth system: the geosphere, atmosphere, hydrosphere, and biosphere. **Water moves in and around the earth system, changing from one physical state (liquid, solid, water vapor) to another,**

in a sequence known as the hydrologic cycle (Figure 11.3). Any form of water (rain, snow, ice) that falls from the sky is precipitation. Most precipitation falls into the ocean, but some falls on the land surface. Precipitation may run off the land (in streams) or move down through the surface (infiltrate) to become groundwater. A majority of precipitation reenters the atmosphere as water vapor through the process of evaporation. Evaporation occurs directly from any place where water is exposed to the air. Water vapor also enters the air indirectly when it is released from plants, a process called *transpiration*. Figure 11.3 summarizes these processes, and in Chapter 14, we will discuss how much energy must be absorbed or released for water to change from one state to another (such as when ice melts or water freezes).

Oceans are the ultimate source and final destination for all water on or below the land surface (Figure 11.4a). Most water on Earth (about 97 percent) is in the oceans, a smaller amount (about 3 percent) resides in and on land (Figure 11.4b), mostly as freshwater (water having less than 0.05 percent salts). A very small amount of water (water vapor or ice) is in the atmosphere (less than 0.01 percent). Since most water is exposed to the atmosphere

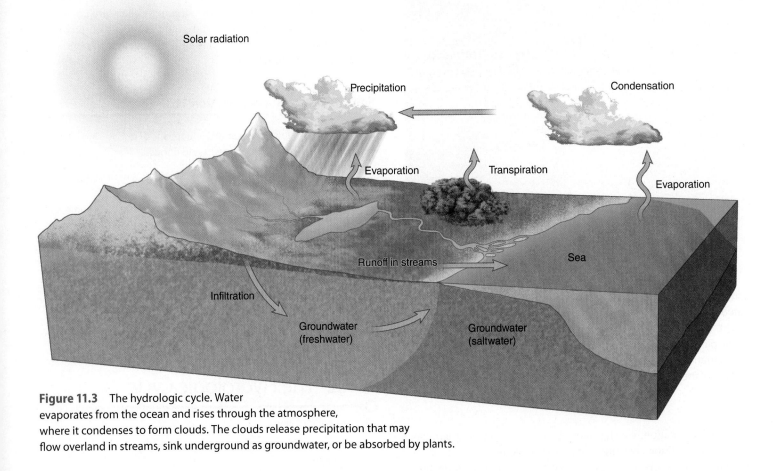

Figure 11.3 The hydrologic cycle. Water evaporates from the ocean and rises through the atmosphere, where it condenses to form clouds. The clouds release precipitation that may flow overland in streams, sink underground as groundwater, or be absorbed by plants.

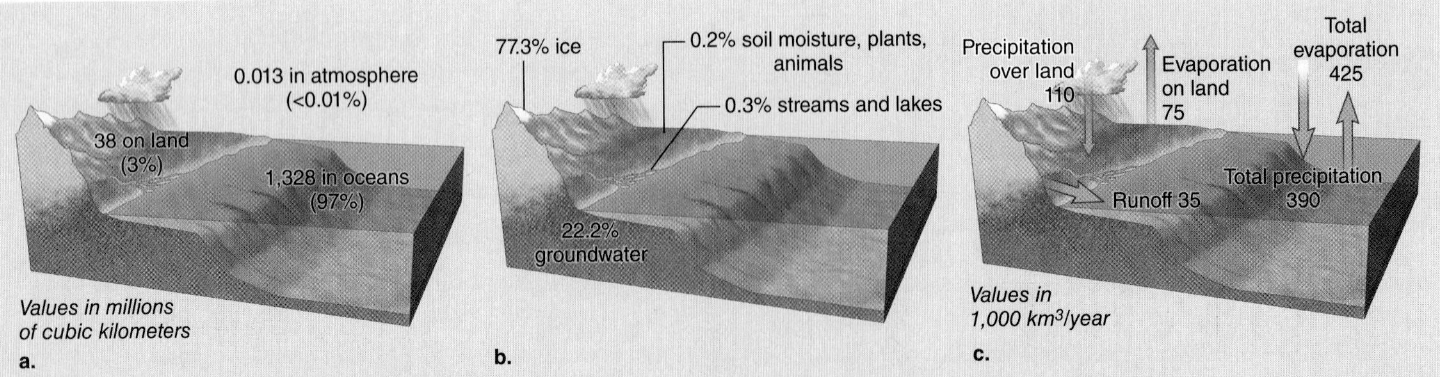

Figure 11.4 Water balance on Earth. **a.** The vast majority of Earth's water is in the oceans, with about 3 percent of Earth's water existing on land. **b.** Most of the water on land is stored as ice or groundwater. Streams and lakes are often supplied by water from melting ice in glaciers or from groundwater. **c.** Land areas receive more moisture by precipitation than they supply by evaporation. The difference is made up by evaporation from oceans. This excess water is returned to oceans by surface streams.

in oceans, the bulk of evaporation (85 percent) occurs there (Figure 11.4c). The greatest evaporation rates occur over areas of warm temperatures at low latitudes. The circular path of the hydrologic cycle links evaporation, transpiration, precipitation, runoff, and infiltration.

The Origin of Streams

Melting snow (known as meltwater) and rain run off the slopes of Long's Peak, Colorado, becoming a source of freshwater. That water flows into almost invisible depressions and carves out small channels that join to form a permanent stream, the source of the Colorado River (Figure 11.5a). More than 30 million people depend on water from the Colorado River, much of which originates as snowmelt. Earth's lakes store twice the freshwater of streams and serve as the source (or destination) for many rivers. For instance, the source of the Mississippi River is a relatively small lake (about 5 square kilometers; 2 square miles) in Minnesota.

Nearly one-third of the water falling as precipitation over land completes the circuit to the oceans by surface runoff in streams. (How long does it take for water that falls as precipitation to reach the ocean?) On average, **the residence time,** the **length of time that a given volume of water remains in streams,** is about 14 days. Because rates of water infiltration are slow compared to rates of runoff, only a small amount of the precipitation

Checkpoint 11.1

basic · intermediate · advanced · superior

Sort the following 12 terms into six pairs of terms that most closely relate to one another. Explain your choices.

groundwater	plants	transpiration
stream	ice	infiltration
rainfall	precipitation	water vapor
gas	meltwater	runoff

Checkpoint 11.2

basic · intermediate · advanced · superior

Imagine that it rained continuously all over the world for a month. If we were to measure the depth of the oceans over a 5-day period near the end of the month, what would we observe? Explain your choice.

a) Ocean depths rise steadily.
b) Ocean depths fall steadily.
c) Ocean depths stay the same.

a.

b.

Figure 11.5 Where streams come from. **a.** The Colorado River begins as a small stream on the slopes of Rocky Mountain National Park, Colorado. **b.** The Allegheny (left) and Monongahela (right) rivers join to form the Ohio River in Pittsburgh.

Checkpoint 11.3

basic · intermediate · advanced · superior

Use the hydrologic cycle to suggest a hypothesis about why rivers in South America carry approximately twice as much freshwater as do rivers in North America.

Checkpoint 11.4

basic · intermediate · advanced · superior

Draw a concept map that identifies the links among the components of the earth system and the hydrologic cycle.

soaks into the surface to join water stored in rocks and sediment. However, the volume of water stored in groundwater systems is about 70 times greater than the volume of water in streams and lakes (see Figure 11.4b). Likewise, ice is a huge freshwater storage system comprising more than three times the volume of water found in groundwater. Consequently, it is no surprise that many streams have their origins in glacial meltwaters (for example, Brahmaputra and Ganges rivers, India; see Figure 7.24a) or in springs that bring groundwater to the surface (for example, the Nile River). Finally, we should note that many large rivers form when two smaller rivers join together, such as when the Allegheny and Monongahela rivers join in Pittsburgh to form the Ohio River (Figure 11.5b).

Regardless of where or how a stream begins, water flows downslope and joins, or is joined by, other streams to form a network. For instance, the Mississippi River's source (called the headwaters) is a small stream about 1 meter (3.3 feet) deep and 10 meters (33 feet) wide when it leaves Lake Itasca in Minnesota. Many more streams (some of them large rivers in their own right, such as the Missouri and Ohio rivers) join the Mississippi as it makes its sinuous trek southward. The Mississippi River ends as a large river 5 meters (16 feet) deep and over 500 meters (1,640 feet) wide as it empties into the Gulf of Mexico at New Orleans. Section 11.3 explores the interconnected nature of streams in drainage basins and how flow in one part of a stream system may affect streams in the rest of the basin.

11.3 Drainage Networks and Patterns

Learning Objectives

- Define what is meant by the terms *base level, drainage basin,* and *drainage divide.*
- Estimate the approximate locations of drainage basins and drainage divides on a regional map.
- Discuss the origin and persistence of drainage basins in the context of other geological processes.
- Explain how drainage patterns are governed by the underlying geology of a region.

Think about a stream or river near you that flows all year long. Where does the water flowing in that stream end its journey? Approximately two-thirds (68 percent) of Earth's land surface contains rivers that eventually drain to the ocean. Geologists call **the lowest point to which a stream can flow the *base level.*** Base level for a stream can be a lake, reservoir, wetland, or a larger stream. Those places temporarily store surface water, thus representing a local base level. Eventually, much of the water in those locations will continue its journey to the ocean. The ocean is considered the ultimate base level.

The Drainage Basin or Watershed

So what determines how much water flows in a stream? Imagine placing two flat-bottomed, rectangular pans in your yard to collect rainwater. The pans have the same length and depth, but one is twice as wide as the other. How will the volume of water in each pan compare after the rain? Which one (if either) will have more water? The pan with the larger surface area will have a larger volume of water. The larger pan "drains" a larger area and so collects more precipitation. Similarly, **the amount of water in a stream channel is related to the size of the area it drains and the average precipitation over that area.**

The area drained by a stream and its smaller tributary streams is called a *drainage basin* or *watershed.* Drainage basins for the Mississippi and Nile rivers, two of the largest rivers in the world, occupy approximately 3 million square kilometers (1.16 million square miles), equivalent to one-half the land area of the United States. The two principal tributaries of the Mississippi are the Missouri and Ohio rivers. The drainage basins for each of these (and other) rivers are included within the larger Mississippi drainage basin (Figure 11.6). The Mississippi River is separated from streams that drain to the west by a drainage divide (a regional topographic high) located in the Rocky Mountains. Streams west of the Rockies flow to the Pacific Ocean or the Gulf of California. Streams in the eastern United States flow to the Gulf of Mexico or the Atlantic Ocean.

Drainage divides are found along the high ground separating drainage basins. For example, a drainage divide in Ohio separates streams that drain northward to the Great Lakes from streams that flow south to the Ohio River (Figure 11.7). The city of Akron in northeast Ohio straddles the divide between the Cuyahoga and Tuscarawas rivers. Most runoff directly north of Akron enters the Cuyahoga River, which drains to the north because it is part of the Great Lakes basin. Runoff south of Akron

Checkpoint 11.5

basic · intermediate · advanced · superior

Why is the volume of water in the Mississippi River about 10 times greater than the volume of water in the Nile River?

a) The Mississippi River drainage basin is 10 times bigger than the Nile basin.
b) The Mississippi River drainage basin receives more precipitation.
c) The Mississippi River is a longer stream.
d) There is less vegetation to absorb precipitation in the Mississippi River drainage basin.

flows into streams that empty into the Tuscarawas River. The Tuscarawas River drains south as part of the Muskingum drainage basin, and its water flows into the Ohio River, a major tributary of the Mississippi River. Clearly, precipitation falling on different sides of a drainage divide can take significantly different routes to its ultimate destination.

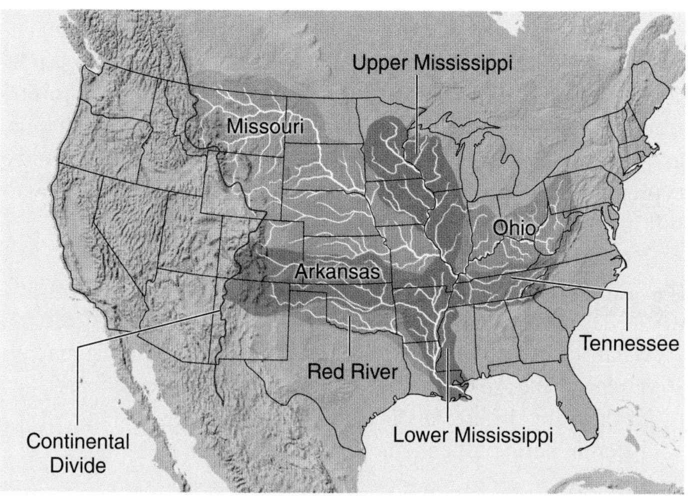

Figure 11.6 Mississippi River drainage basin. The different colors represent smaller drainage basins that make up the Mississippi River drainage basin.

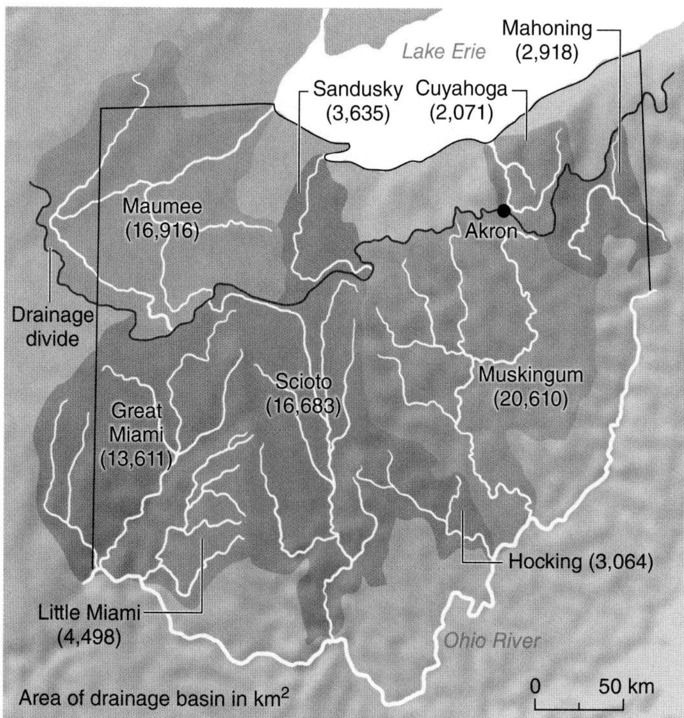

Figure 11.7 A drainage divide and selected drainage basins, Ohio. Streams north of the drainage divide flow to the Great Lakes and ultimately through the St. Lawrence Seaway to the Atlantic Ocean. Streams south of the divide flow to the Ohio River, which joins the Mississippi River and flows to the Gulf of Mexico. Note that the drainage divide is also visible on the map of the Mississippi River basin in Figure 11.6.

Checkpoint 11.6

basic · intermediate · advanced · superior

Rivers in Iowa flow to either the Mississippi River, which makes up the eastern state border, or the Missouri River, on the western state border. Draw the approximate position of the drainage divide for the Missouri and Mississippi basins in Iowa on the map provided here.

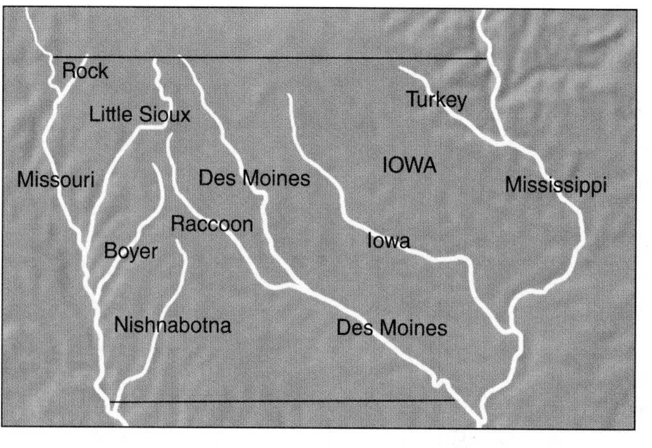

Checkpoint 11.7

basic · intermediate · advanced · superior

Examine the following map of rivers in southern Georgia. Draw the approximate boundaries of the drainage basins for the named streams. Note that the Savannah and St. Marys rivers mark parts of the boundary of the state. Do not include the Savannah or St. Marys rivers' drainage basins.

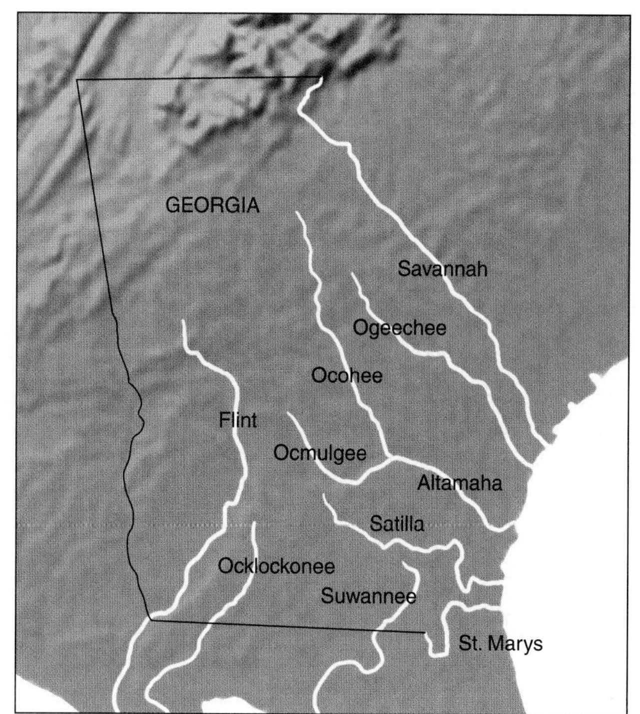

Evolution of Stream Systems

Large drainage basins are long-lived features on Earth. We have previously discussed the great length of time needed for the Colorado River to carve the Grand Canyon (millions of years). The largest rivers often drain relatively stable highlands formed by uplift linked to slow, gradual plate tectonic processes. For example, the Amazon River originated in its present form approximately 10 million years ago as the Andes Mountains grew tall along the northern and western margins of the South American continent. The northern half of the continent was originally drained by a north-flowing stream system that carried water to the Caribbean Sea (Figure 11.8a). These streams deposited light-colored sediment eroded from granite rocks in the continental interior.

As the mountains rose, shutting off drainage to the north, the Amazon developed as one of a series of rivers eroding the Andes Mountains and feeding a vast expanse of interior lakes and wetlands (Figure 11.8b). These rivers transported material from metamorphic rocks exposed in the mountains. The tropical swamp conditions were ideal for the development of layers of immature coal deposits (lignite) that alternated with clastic sedimentary rocks. As the western side of the continent continued to rise, the wetlands were eventually drained by the east-flowing Amazon River and its tributaries. The Andes sediments began showing up in layers deposited along the Atlantic continental shelf about 10 million years ago (Figure 11.8c). Scientists have interpreted this to mean that it took several million years for the modern Amazon River drainage basin to become established.

Drainage Patterns

Streams typically follow the path of least resistance, forming valleys where rock is most readily eroded or following the steepest slope gradient. Consequently, geologists have identified four basic stream patterns that relate to the underlying geology: dendritic, rectangular, radial, and trellis (Figure 11.9). **Dendritic (treelike) drainage patterns are characteristic of areas where the geology is relatively uniform,** as when rock layers are horizontal or where the underlying rocks have similar properties. When viewed on a map, dendritic streams intersect with a characteristic V pattern with the tip of the V pointing downstream. Dendritic patterns are typical of streams in the Mississippi River drainage basin and in Ohio streams (see Figure 11.7) and the Amazon River (Figure 11.8).

On a regional scale, streams sometimes intersect at right angles, forming trellis or rectangular drainage patterns. **Trellis drainage patterns develop in areas of folded layers of weak and resistant rocks.** Main streams occupy valleys, flowing parallel to the ridges, but occasionally cut across the geologic grain at water gaps. Smaller streams flow down the valley slopes perpendicular to the course of the main stream channel. **Rectangular drainage patterns** occur in regions where the streams are **controlled by joints,** fractures in the underlying bedrock. Streams exploit the fractures as planes of weakness in the rock. Finally, **radial patterns are typically found on volcanoes,** where streams flow directly downhill in all directions, away from the summit (Figure 11.9d).

Checkpoint 11.8

What type of drainage patterns can you observe in this map of part of the Appalachian Mountains and adjacent areas? Explain any differences in patterns that you observe in different parts of the map.

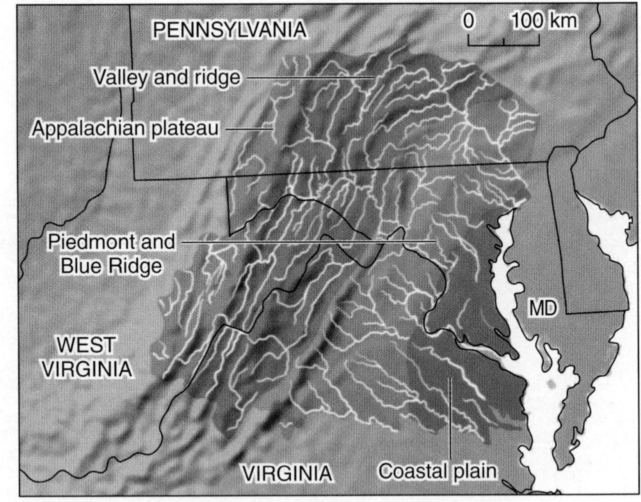

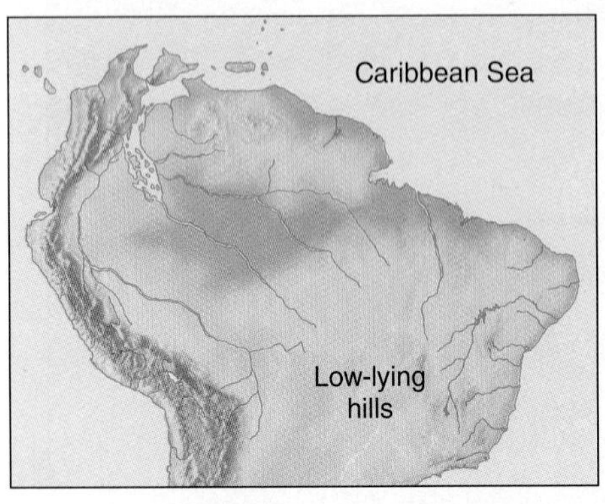

a.

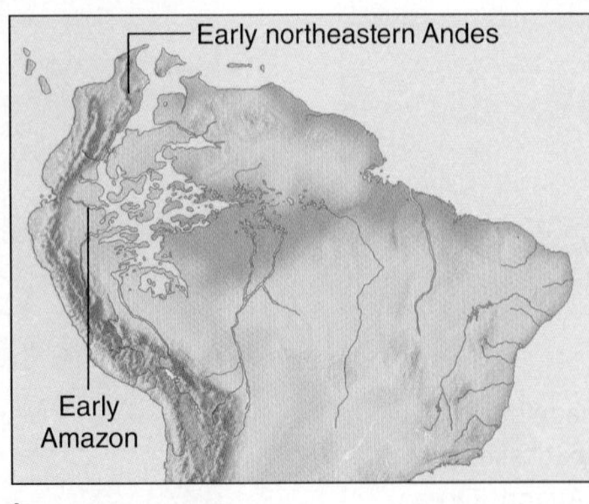

b.

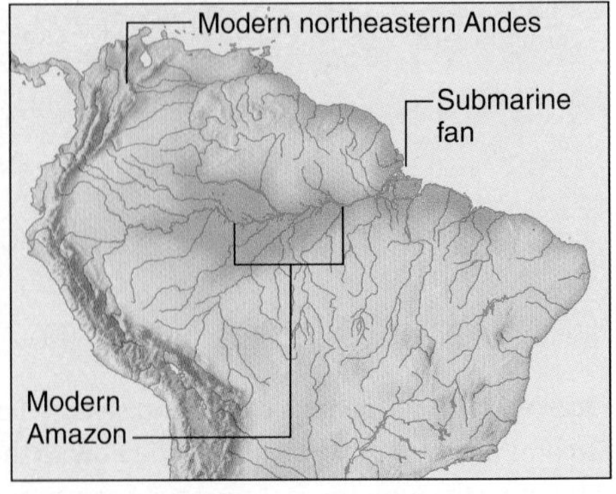

c.

Figure 11.8 The evolution of the Amazon River. **a.** Twenty-five million years ago, much of the northern half of South America was drained by a north-flowing stream system. **b.** About 15 million years ago, rivers fed an extensive interior basin filled with lakes and wetlands. **c.** The patterns of the modern river became established about 10 million years ago.

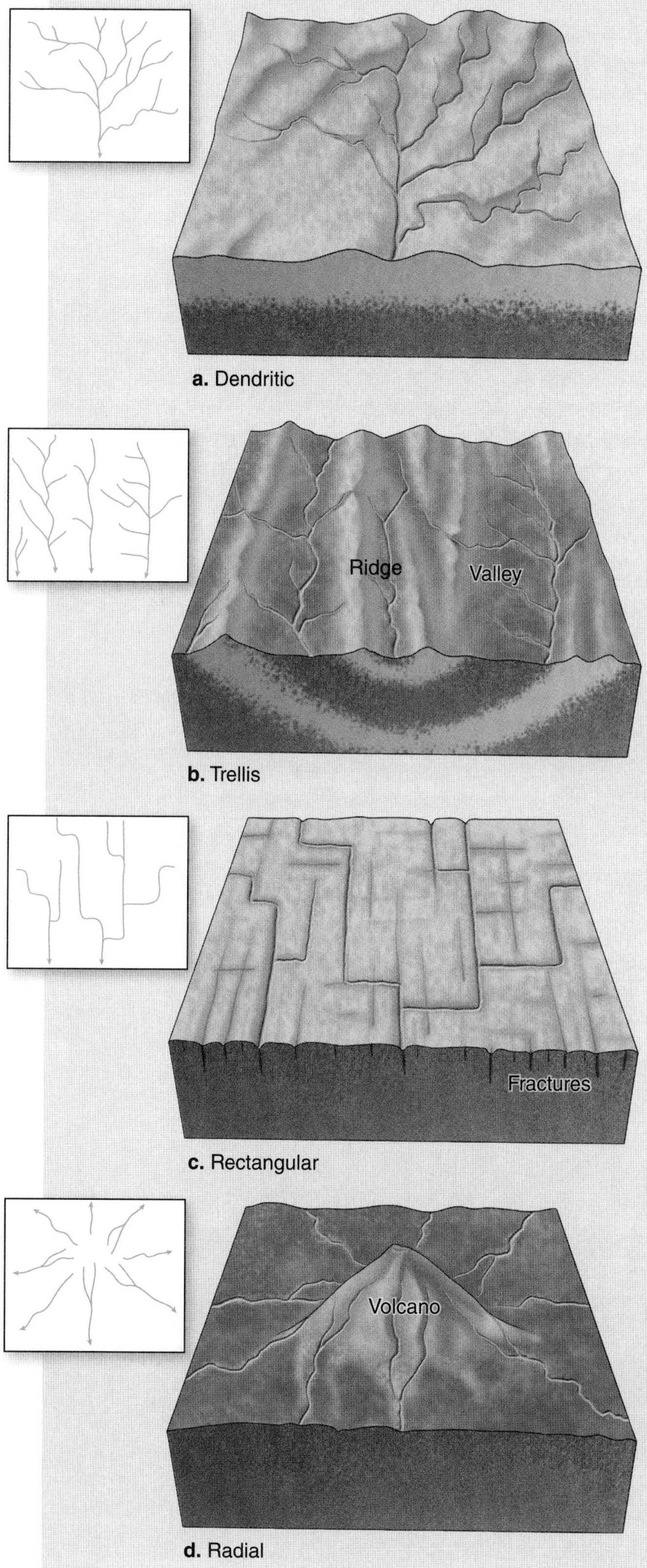

a. Dendritic

b. Trellis

Ridge

Valley

c. Rectangular

Fractures

d. Radial

Volcano

Figure 11.9 Stream patterns. The four principal types of drainage patterns are related to the underlying regional geology.

11.4 Factors Affecting Stream Flow

Learning Objectives

- Discuss how the characteristics of a stream would change moving from its headwaters to its mouth.
- Explain how the stream gradient, channel roughness, and wetted perimeter influence stream velocity.
- Define the term *stream discharge* and explain how it is calculated.

Think for a moment about the last time you saw a stream or river. How would you describe that river to a friend? Chances are, you would mention the speed and color of the water, the size of the stream, and perhaps the landforms around the stream. Hydrologists, scientists who study water, do much the same thing when they study streams and rivers. They examine a variety of characteristics, including how fast the water flows and the size, slope, and roughness of the stream channel. All of these factors affect how the stream flows and consequently influence the evolution of surrounding landforms.

Stream Gradient

Why do streams flow, anyway? The answer to that question seems rather obvious—water flows downhill. Therefore, streams simply carry water from higher elevations to lower elevations. **The slope of a stream (known as the stream gradient) is the change in elevation of the stream over a horizontal distance.** As you might expect, the highest gradients are found in steep-sided mountain stream valleys that may drop 40 to 60 meters for every kilometer (210 to 320 feet per mile) of stream length. The gradient would be 40 to 60 meters per 1,000 meters or, expressed as a percentage, 4 to 6 percent.

Where would you find the highest gradients for most streams—at the beginning (headwaters) or end (mouth) of the stream? The stream gradient gradually decreases along the length of a stream, from the headwaters (where it is steepest) to the mouth (Figure 11.10). Streams approaching the river mouth may decrease in elevation by as little as a few centimeters over 1 kilometer

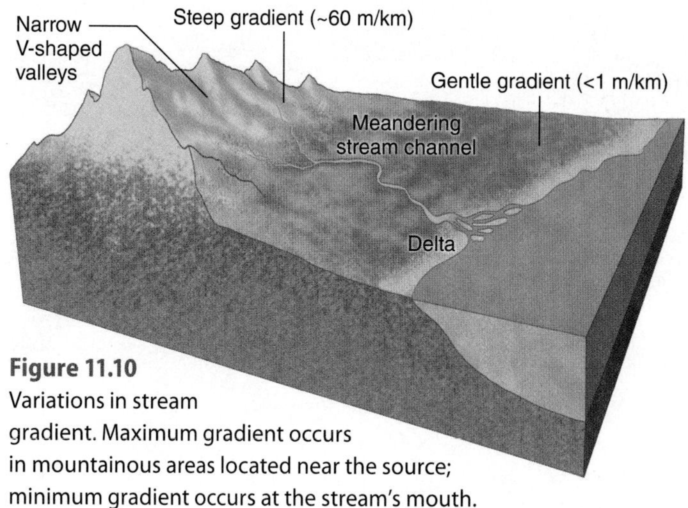

Figure 11.10 Variations in stream gradient. Maximum gradient occurs in mountainous areas located near the source; minimum gradient occurs at the stream's mouth.

(gradients of about 0.001 percent). The Nile River in Egypt has a gradient of about 8 centimeters per kilometer (5 inches per mile), and the last 1,000 kilometers (620 miles) of the Amazon River in Brazil drops just 1 centimeter per kilometer (less than half an inch per mile). So, how does the gradient relate to the stream velocity?

Stream Velocity

Anybody who has ever gone white-water rafting will tell you that much of the fun comes from being hurled from side to side as the raft bounces its way through the rapids of a mountain stream (Figure 11.11). That experience gives the impression that the raft, and hence all of the water in the stream, is moving fast. In contrast, if we walk or drive alongside rivers with low gradients, they appear to be moving relatively slowly—hence the phrase *lazy river.*

When we analyze the average velocity of water moving through a stream (the flow), it turns out that our senses are somewhat deceiving us. The average flow of water in the high-gradient part of a stream is typically slower than that through the low-gradient part of the stream. How can that be so? What makes a high-gradient stream different from a low-gradient stream? Look closely at Figure 11.11. How might the features you see there compare to what you would observe in a low-gradient river? The banks along the steep mountain streams typically have rocky channels. As one continues downstream, the streambed composition changes from gravel, to sand, and eventually to fine-grained silt and mud. **This characteristic of the stream channel is called the *channel roughness.***

How do you think the presence of boulders would affect the flow of the stream? Think of the water as going through a maze. The boulders cause the water to become turbulent (chaotic). It goes forward, then sideways, and sometimes even backward before it moves forward again. Compare that to the flow through the downstream portion of the river where the channel is smooth. Flow there is streamlined (laminar) because there are few obstacles in the channel to disrupt the flow. The channel roughness found in a high-gradient stream actually reduces the average stream velocity when compared a low-gradient stream.

But wait, there's more—a third characteristic, cross-sectional area, is involved in determining stream velocity. Cross-sectional area is determined by multiplying the stream's width by its depth, assuming a rectangular channel shape. (We are simplifying here. Most channels have more complex shapes that are carefully measured to determine cross-sectional area.) Consider cutting a slice through the cross section of a stream as represented in Figure 11.12. The bed and banks of the stream channel exert frictional drag on the passing water. The actual amount of drag is proportional to the **length of the surface of the channel banks and bed in contact with the water (that distance is called**

Figure 11.11 White-water rafting. A raft passes through Dark Canyon Rapids in Cataract Canyon of the Colorado River.

Checkpoint 11.9

Explain why stream velocity would change along the same section of a stream at different times of the year.

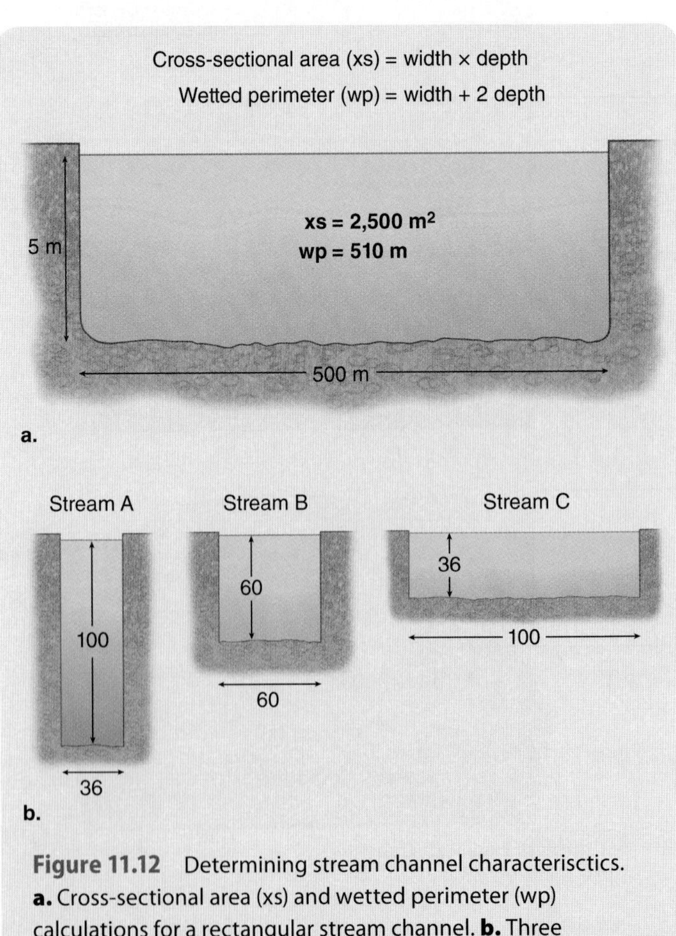

Figure 11.12 Determining stream channel characterisctics. **a.** Cross-sectional area (xs) and wetted perimeter (wp) calculations for a rectangular stream channel. **b.** Three channels with the same cross-sectional area (360 square meters; 3,230 square feet) but different lengths of wetted perimeters. If all other factors were equal, velocity would be highest for stream C.

the *wetted perimeter*). In streams with the same volume of water and same channel roughness, the channel with the larger wetted perimeter exerts more frictional drag on the water and so has a lower stream velocity (Figure 11.12b). Generally, the larger the stream discharge, the less significant the effect of frictional drag from the wetted perimeter.

The combined effects of these factors (gradient, channel roughness, and wetted perimeter) generally determine the average velocity of a stream. In simple cases, the steeper the gradient, the smoother the channel, and the shorter the wetted perimeter, the faster the velocity. However, these parameters often offset one another. For example, high-gradient streams often have greater channel roughness, while the smooth banks and bed of low-gradient streams exert less friction on the flowing water.

At its source, the Mississippi River has a cross-sectional area of 10 square meters (99 square feet). In contrast, closer to its mouth, it has a cross-sectional area of 2,500 square meters (26,240 square feet). The velocity of the Mississippi River at its source is 0.5 meter per second (1 mile per hour), while it cranks along at a merry 1.3 meters per second (3 miles per hour) as it passes New Orleans on its way into the Gulf of Mexico. Typically, cross-sectional areas of streams increase downstream as more and more water is added to the stream through tributaries. At the same time, channel roughness tends to decrease because particles on the streambed are smaller. Thus, **water in a stream moves faster as it moves downstream.**

Stream Discharge

In addition to cross-sectional area, just described, there are other ways of assessing a stream's size. In terms of length, the Nile River (6,825 kilometers; 4,240 miles) beats out the Amazon (6,450 kilometers; 4,008 miles) and Mississippi rivers (6,260 kilometers; 3,890 miles). However, we can also measure the volume of water that flows out through the stream. ***Stream discharge* is the volume (cubic meters or cubic feet) of water that passes a given point in one second.** The discharge is calculated by multiplying the cross-sectional area by the velocity of the stream flow (Figure 11.13).

$$\text{Discharge} = \text{width} \times \text{depth} \times \text{velocity}$$

$$\text{cubic meters per second } (m^3/s) = [\text{meters (m)}] \times [\text{meters (m)}] \times [\text{meters per second (m/s)}]$$

Imagine placing an empty 1-gallon milk jug in your kitchen sink. Adjust the flow of the faucet so that it takes 1 minute to fill the jug with water. The volume of water that flows out of the faucet in 1 minute is the discharge. If the gallon container fills in 60 seconds, the discharge from your faucet is 1 gallon per minute. Now repeat the experiment, but reduce the discharge from the faucet so that it takes 22 minutes to fill the gallon jug. You have just modeled the difference between stream discharge in the Amazon and Nile rivers. (For comparison, it would take about 10 minutes to fill the jug using the Mississippi River equivalent.)

Discharge from the Amazon River is about 200,000 cubic meters per second (7 million cubic feet per second) and accounts for 20 percent of all freshwater discharge from streams on Earth. To put it another way, 50 million gallons of water discharge into the Atlantic Ocean from the Amazon River every *second!* We will discuss how measurements of stream discharge become critical in tracking the potential for flooding in Section 11.6. But first, we will examine the impact of changes in stream velocity and discharge on a stream system.

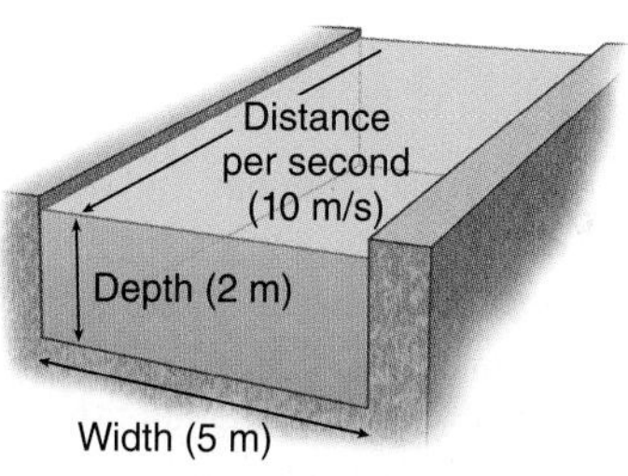

$$\text{Discharge} = 5 \times 2 \times 10 = 100\ m^3/s$$

Figure 11.13 Stream discharge. Discharge is a measure of the volume of water that passes a point in a given time and can be calculated by multiplying the area of the stream channel (width × depth) by the distance traveled in a given time (velocity).

Checkpoint 11.11

basic / advanced / intermediate / superior

Some scientists predict that global warming will result in a corresponding increase in evaporation from the oceans. How would this affect the discharge of the Amazon River?

a) Discharge would increase.
b) Discharge would decrease.
c) Discharge would stay the same.

Checkpoint 11.10

A stream channel narrows between support columns under a bridge. If discharge does not change, predict how stream velocity would be altered as water flowed under the bridge.

a) Velocity would increase.
b) Velocity would decrease.
c) Velocity would not change.

Checkpoint 11.12

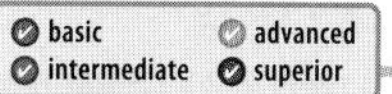

Create a concept map that illustrates the connections among the factors that influence stream flow. Include the following eight terms and up to four more of your own choosing.

discharge	velocity	wetted perimeter
depth	gradient	channel roughness
cross-sectional area	width	

11.5 The Work of Streams

Learning Objectives

- Make a labeled sketch of the materials that make up a stream's load and explain how they are transported.
- Explain why some streams transport more sediment than others.
- Describe where deposition is most likely to occur in association with streams.
- Interpret an image of a stream channel to identify locations of erosion and deposition, and predict the future path of the stream.

The action of streams is continually altering the land surface. Changes occur every day and may be almost imperceptible unless viewed on a timescale measured in decades or centuries.

Erosion

Streams wear away—or erode—the bed and banks of stream channels. Scientists observed an example of stream erosion following the eruption of Mount St. Helens in 1980 (see Chapter 6). That instance also served to answer the question: *Which comes first, the valley or the stream?* The eruption changed the landscape around the volcano, leaving behind thick deposits of ash and mud. Almost immediately, new streams began to form on the surface of these deposits. Initially, the streams cut relatively deep, narrow, channels due to stream erosion (Figure 11.14a). Subsequently, the stream erosion, in combination with mass wasting of the channel banks, gradually widened the stream channels (Figure 11.14b). While most stream channels are composed of much stronger material than the relatively soft unconsolidated volcanic debris, the nature of stream erosion and associated mass wasting is similar, serving to deepen and widen stream channels. Over geologic time, these actions sculpt valleys, gorges, and many other landforms we observe around us today. As we discussed in Chapter 8, the steady cutting of the Colorado River through the rocks of northern Arizona formed the Grand Canyon over millions of years.

All streams erode particles from channel beds and banks and carry them downstream. Larger sand and gravel particles that bounce or roll along the streambed are termed *bed load* (Figure 11.15). Finer particles that are light enough to be carried in suspension higher in the water column are termed *suspended load*. A form of erosion known as *abrasion* results from the impacts of these large and small particles against the bed and banks of the stream channel. You can think of abrasion as waterborne sandblasting. The force of flowing water in the channel can also push through cracks in rocks to break off chunks of bedrock. At the other end of the spectrum, the water can simply dissolve soluble bedrock such as limestone or rock salt through chemical

a.

b.

Figure 11.14 Effects of erosion on a stream channel. **a.** Approximately 5 months after the eruption of Mount St. Helens, this channel of the upper Muddy River had formed on the soft, unconsolidated volcanic materials. **b.** Note the increase in width and depth of the channel 12 months later.

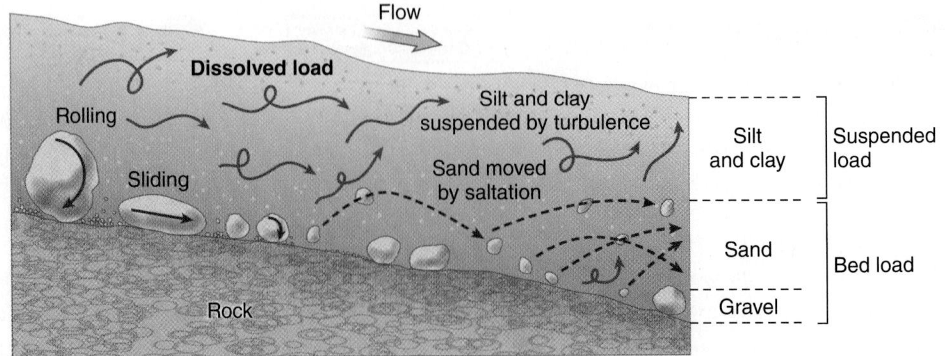

Figure 11.15 Load carried by a stream. Bed load rolls gravel-sized particles along the bed of the stream channel and bounces sand-sized particles downstream by saltation. Suspended and dissolved loads are carried higher in the stream channel by the flowing water.

weathering (see Chapter 9). This material is usually invisible in the stream and is known as the *dissolved load.* **Erosion produces the combination of bed load, suspended load, and dissolved load that can be collectively referred to as the *stream load.***

Transport

Erosion contributes to the stream load, and so does material delivered by discharge from tributary streams or from surface runoff from the land. How can you tell when a stream has a high sediment load? Think about what the water in a stream looks like after a heavy rain—it appears brown. That color is a result of fine material being transported as suspended load in the stream (Figure 11.16). The suspended load increases when discharge increases (Figure 11.17), because greater discharge is usually associated with higher average velocity that provides more energy to move particles. However, not all streams that carry a significant load are muddy. Streams in which the load is mainly composed of dissolved minerals may appear clear, just as a cup of hot water becomes transparent after a teaspoon of sugar dissolves.

Other factors that influence stream load are the geology and land use of the drainage basin. Some of the 300 million tons of sediment that the Mississippi River deposits in the Gulf of Mexico each year is stripped off fields as far away as Montana and Pennsylvania. You will recall that the Amazon River has 10 times the discharge of the Mississippi River, but it carries only three times the load. What factors might account for this difference? First, the soils of the Amazon basin are much different from those of the Mississippi River basin because of different parent materials and vastly different weathering histories (see Section 9.6, Soils: An Introduction). Also, much of the drainage basin of the Amazon is covered by natural vegetation, rain forest that protects the underlying soil from erosion. In contrast, much of the Mississippi River drainage basin has been cultivated for more than a century, reducing

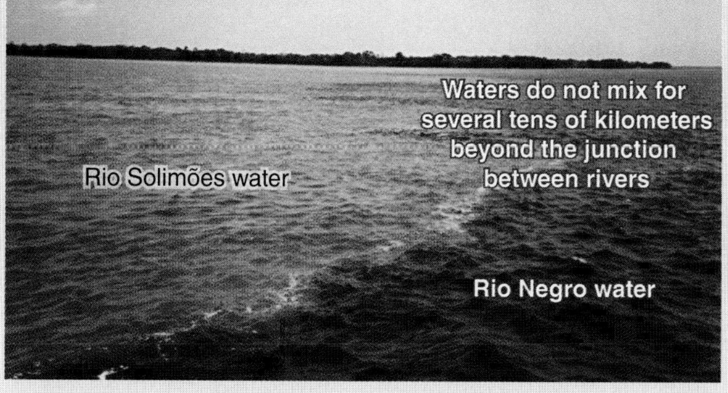

Figure 11.16 Evidence of stream load. View of the confluence of the Rio Solimões and Rio Negro to form the Amazon River near Manaus, Brazil (bottom photo). Note the contrast between the brown, sediment-laden Rio Solimões and its darker, sediment-light tributary. Flow is from the bottom of the image toward the top. Manaus is nearly 1,300 kilometers (800 miles) from the coast but is accessible by oceangoing ships because of the depth of the Amazon River. (Red colors in main image indicate vegetation imaged using the Advanced Spaceborne Thermal Emission and Reflection Radiometer.)

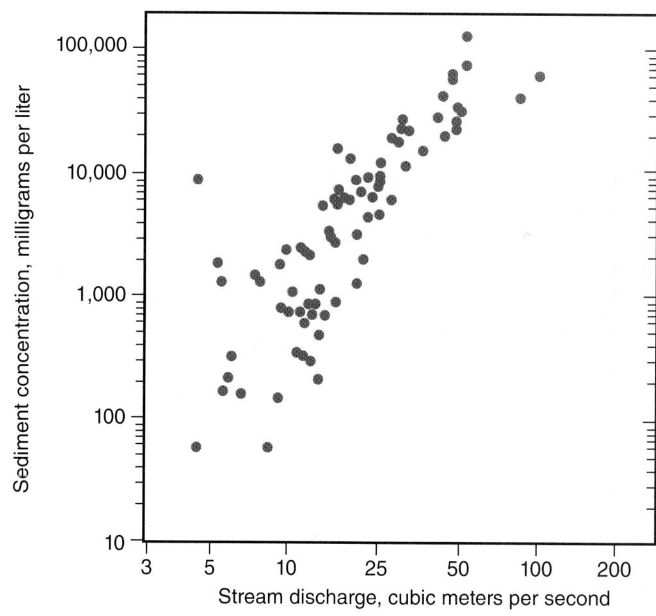

Figure 11.17 Sediment load versus stream discharge for the north fork of the Toutle River near Mount St. Helens, Washington. What is the relationship of sediment load to discharge?

the protection for the soil. Still, the Mississippi cannot match the sediment transport of China's Yellow River (Figure 11.18). That stream gets its name from its heavy load of fine, wind-deposited silt that is easily eroded and carried downstream. The Yellow and Mississippi rivers have similar lengths and discharges, but the Yellow River carries more than five times as much load as the Mississippi River.

Deposition

Now, imagine a stream that is flowing fast enough to transport all sizes of sediment, up to gravel-sized particles. What happens as the stream velocity decreases? Materials such as clay, silt, sand, and gravel that make up the stream load may be deposited in the streambed (just as on a windy fall day when blowing leaves settle from the sky as the wind dies down). Heavier particles drop out first, followed by lighter and lighter materials as the stream velocity decreases. When streams have heavy loads, sediment is often deposited in the stream channel itself to form sandbars or islands. These deposits cause the main channel to split into several smaller channels, forming a pattern known as a *braided stream* (Figure 11.19).

Other depositional features common near rivers are **the broad, flat lands alongside the stream channel, termed *floodplains*** (see Chapter Snapshot). Floodplains are areas along streams that are periodically covered with water and are actually part of the river system. Floodwaters carry sediments with them as they inundate the floodplain. When the water slows and finally recedes, this sediment is deposited in layers that, through time, leave broad, flat plains. Despite this threat of periodic flooding, floodplains have often been considered prime sites for urban and agricultural development because of their flatlands, fertile soils, and ready access to water. Though city zoning regulations often restrict development in floodplains, they are now home to more people and development than ever before. From a uniformitarian viewpoint, these are areas that will flood again, since past flooding formed them in the first place. As such, the areas are at high risk for damage or destruction when the stream inevitably exceeds its banks again.

Figure 11.18 The Yellow River, China. This river carries more sediment relative to its discharge than any other major river on Earth.

As we discussed in Section 11.4, the velocity of a stream tends to increase downstream with increasing discharge. However, velocity can also vary within any section of the stream channel. Within a cross section of a stream channel, **velocity is lowest**

Checkpoint 11.13

What statements are most likely true about a pebble found in a stream?

a) It formed from erosion of sedimentary rock in the streambed or bank.
b) It formed when sand and clay clumped together in the stream.
c) It is younger in age than the stream channel.
d) It may be composed of any type of rock.

Checkpoint 11.14

Consider the consequences of constructing a dam on a river that has a large stream load, such as the Yellow River. Assume the dam and its reservoir are located about two-thirds of the way down the river. How would stream flow conditions be altered above and below the dam and its reservoir? What would be the implications for erosion, transport, and deposition?

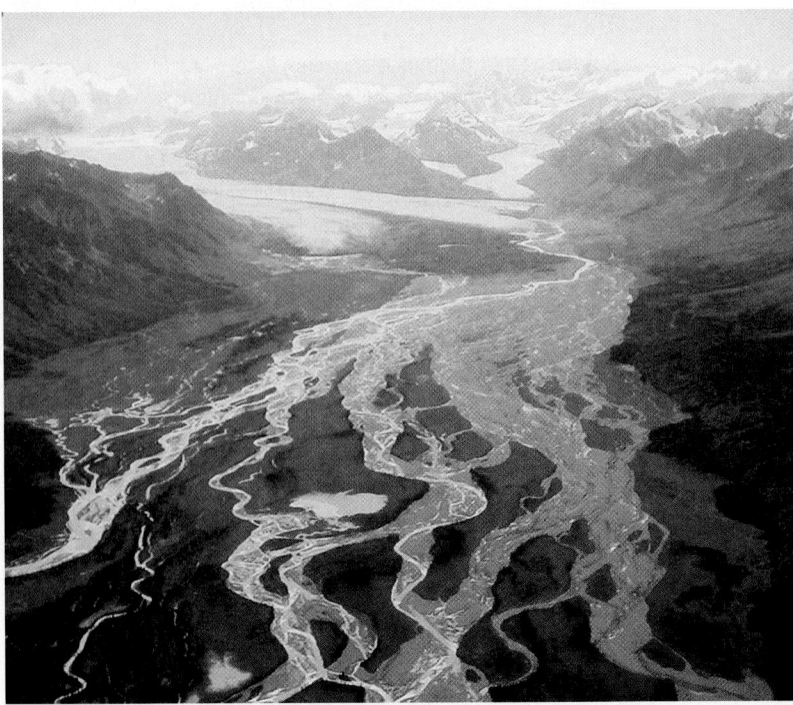

Figure 11.19 Wrangell-St. Elias National Park, Alaska. A sediment-rich, braided stream channel. These streams begin as melted ice from glaciers.

along the banks and bed of the stream, where friction is greatest; velocity is highest in the center of the stream, where friction is least (Figure 11.20).

The lower elevations of most streams have numerous **broad curves, known as meanders** (Figure 11.21a and Chapter Snapshot). As water moves around these long bends, the velocity within the stream channel varies. Think about what happens when you try to drive or run around a corner too fast. Typically, your momentum carries you toward the outside of the corner. If you are going too fast, you will drive off the road or run into a wall. The same thing happens to water traveling in a stream. It is hurled against the outside of a meander, and the stream load bumps into the channel and erodes the outer side of the channel at a *cut bank* (Figure 11.21). In contrast, stream velocity is least on the inside of bends. This decrease in velocity may allow coarse sediment such as sand and gravel to be added to the inner bank of the meander, building a feature known as a *point bar* (Figure 11.21).

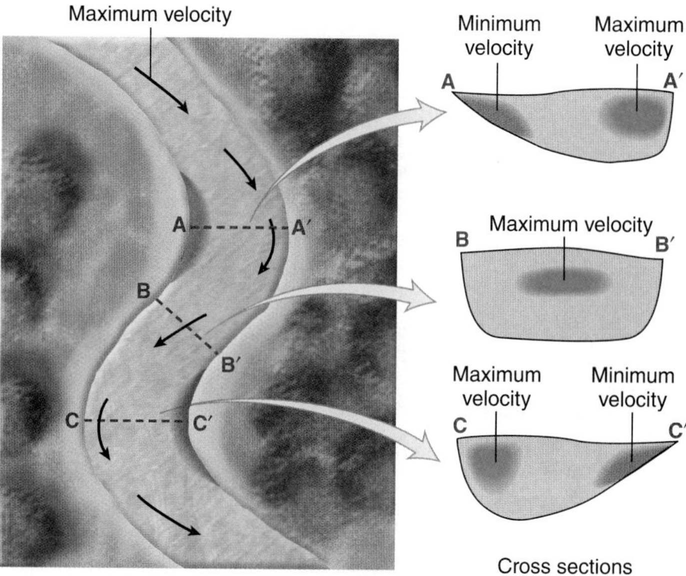

Figure 11.20 Variations in stream velocity. Stream velocity is greatest in the deeper parts of the stream and toward the outside of curves in the channel. Velocity is least on the inside of the curves.

As a result of erosion and deposition on opposite sides of its channel, **a stream can migrate across a valley floor in the direction of the erosion, a process called *lateral migration.*** Or during flooding, the river may rapidly cut through the narrow neck of a fully developed meander, causing dramatic changes in the course of the stream. The meander may be preserved as an oxbow lake, isolated from the stream channel (see Chapter Snapshot).

Suppose you owned some land along a meander and the river served as your property boundary. What would happen if the migrating stream slowly cut away the banks of your property and carried it away? Would that eroded sediment still belong to you? Could you follow it downstream and claim ownership wherever it settled? Darrell Turner knows the answer to that question. In the last 40 years, he has lost half his farm, at the rate of an average of 0.4 hectare (1 acre) a year, to the migration of the Sauk River in Washington. Farther north, in Newtok, Alaska, the entire community of 340 residents will have to move to avoid being swallowed up by the Ninglick River, which is chewing up the river bank at a rate of 30 meters (100 feet) a year.

Despite the loss of property, real estate laws generally do not protect owners from the slow loss of land by erosion on the outer banks of meanders. What was previously considered real estate is considered a loss when it is transported downstream and added to someone else's property along a point bar. It would be impossible to trace the constituent parts and to lay claim to them once

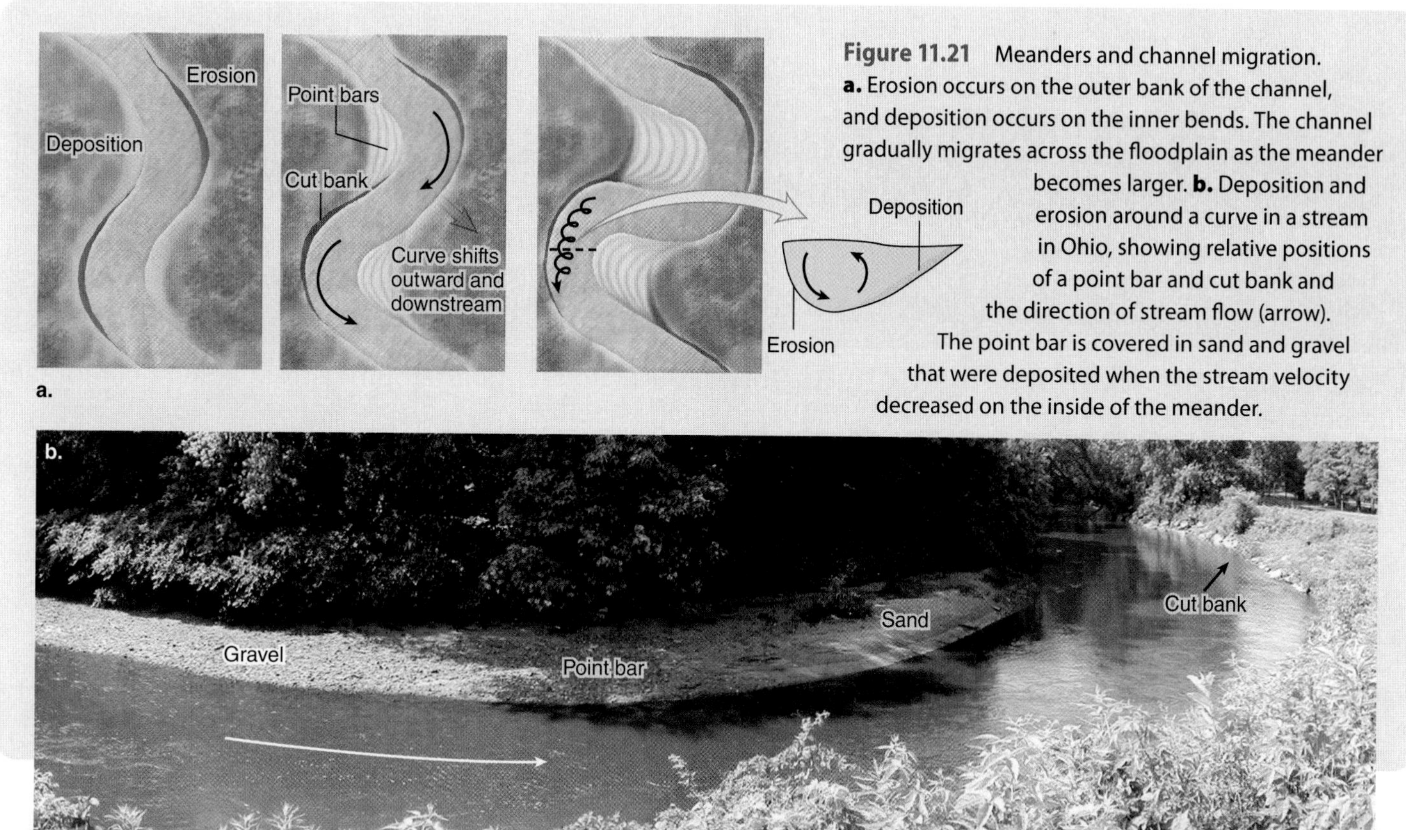

Figure 11.21 Meanders and channel migration. **a.** Erosion occurs on the outer bank of the channel, and deposition occurs on the inner bends. The channel gradually migrates across the floodplain as the meander becomes larger. **b.** Deposition and erosion around a curve in a stream in Ohio, showing relative positions of a point bar and cut bank and the direction of stream flow (arrow). The point bar is covered in sand and gravel that were deposited when the stream velocity decreased on the inside of the meander.

CHANNEL MIGRATION IN THE MAMORÉ RIVER

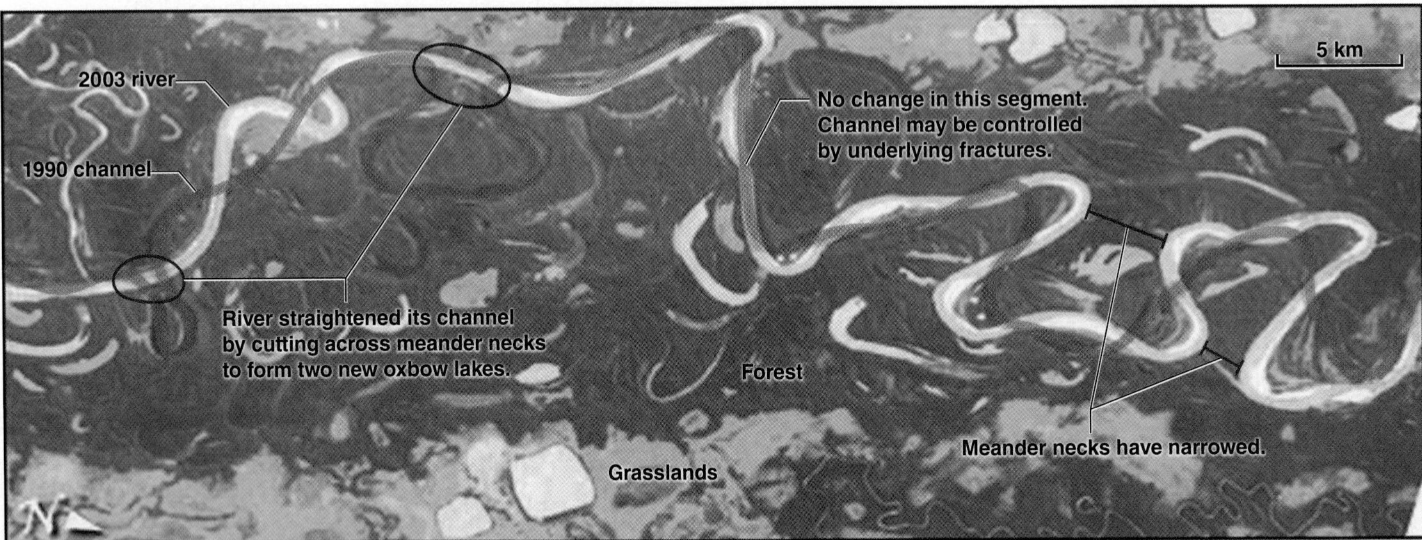

View of a section of the Mamoré River, Bolivia, taken from the International Space station. An image of the 1990 channel has been superimposed over the 2003 river to illustrate channel migration.

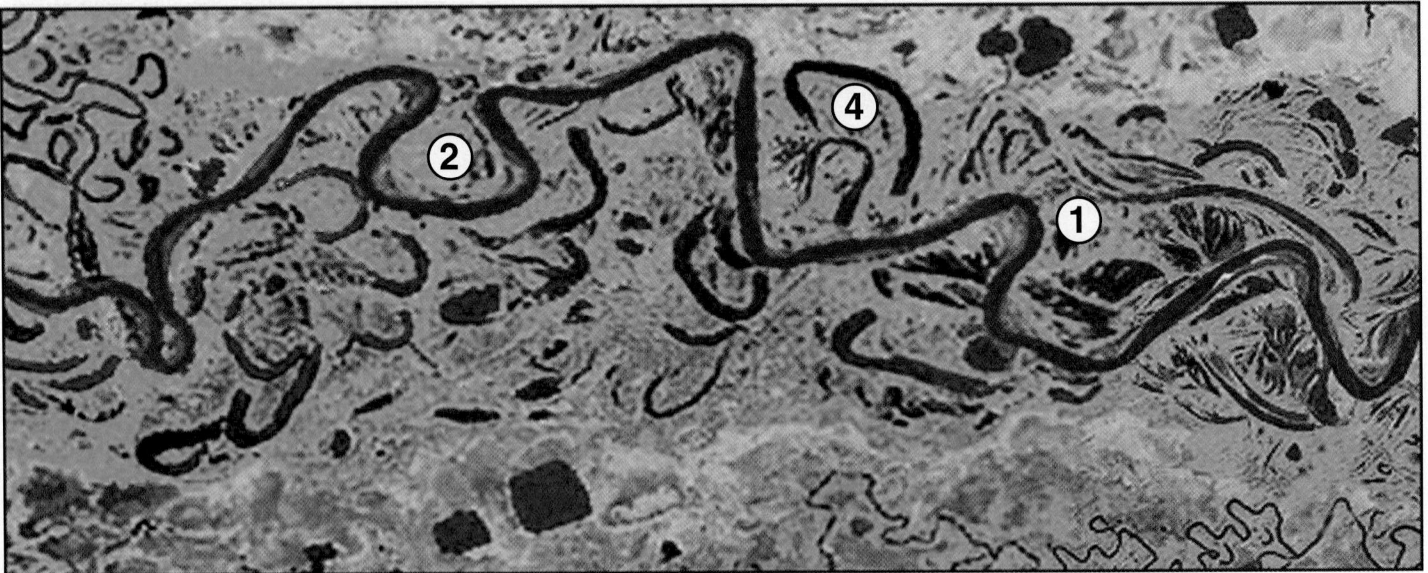

A section of the Mamoré River channel in 1990. Different parts of the stream channel represent different stages in channel migration (see numbered images below).

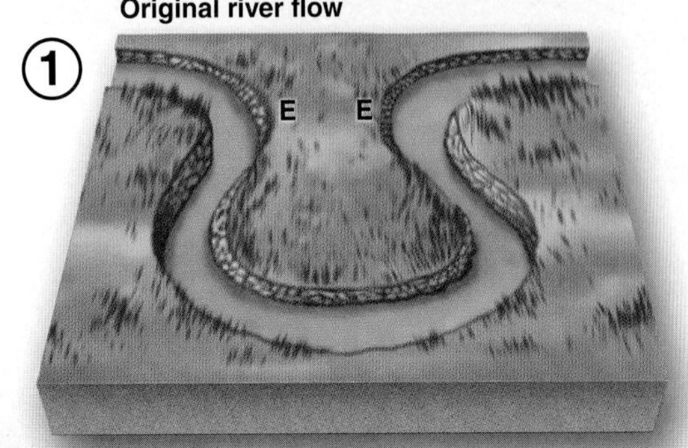

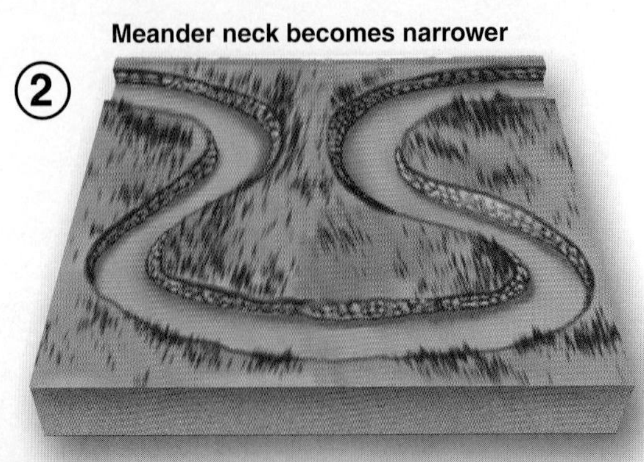

Creation of an Oxbow Lake

Erosion on the inside of meander bends (1, E) causes the meander neck to narrow (2) before the stream cuts across the neck (3). Eventually, the meander bend is disconnected from the straight channel segment to form an oxbow lake (4).

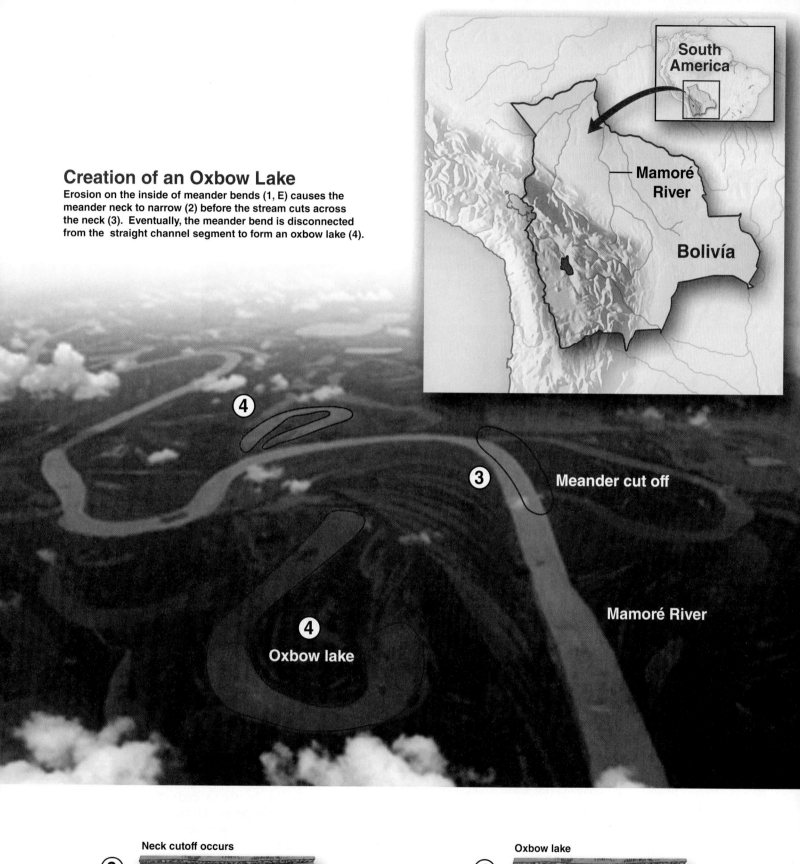

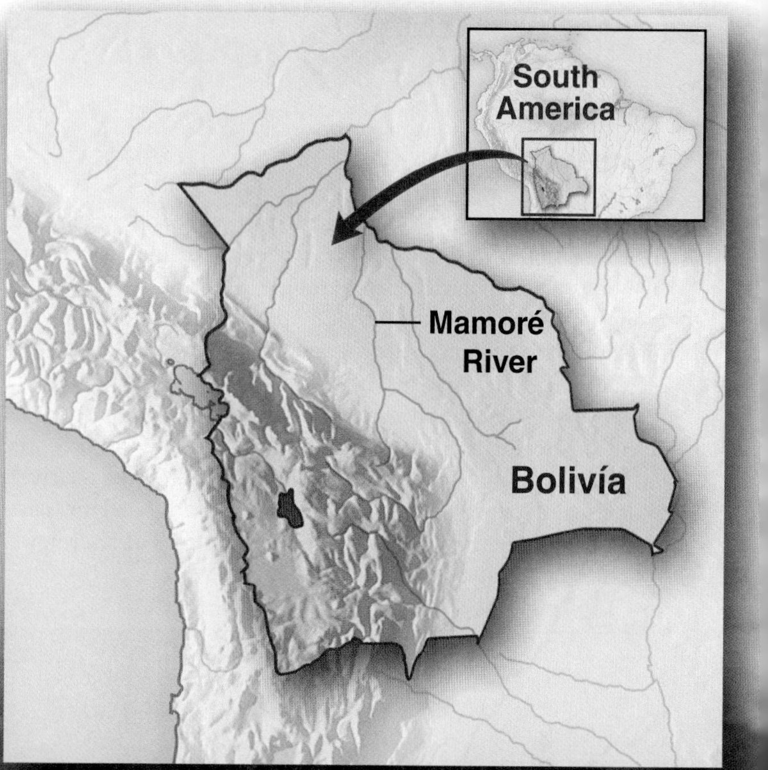

South America
Mamoré River
Bolivía

4
3
Meander cut off
Mamoré River
4
Oxbow lake

Neck cutoff occurs
3

Oxbow lake
4

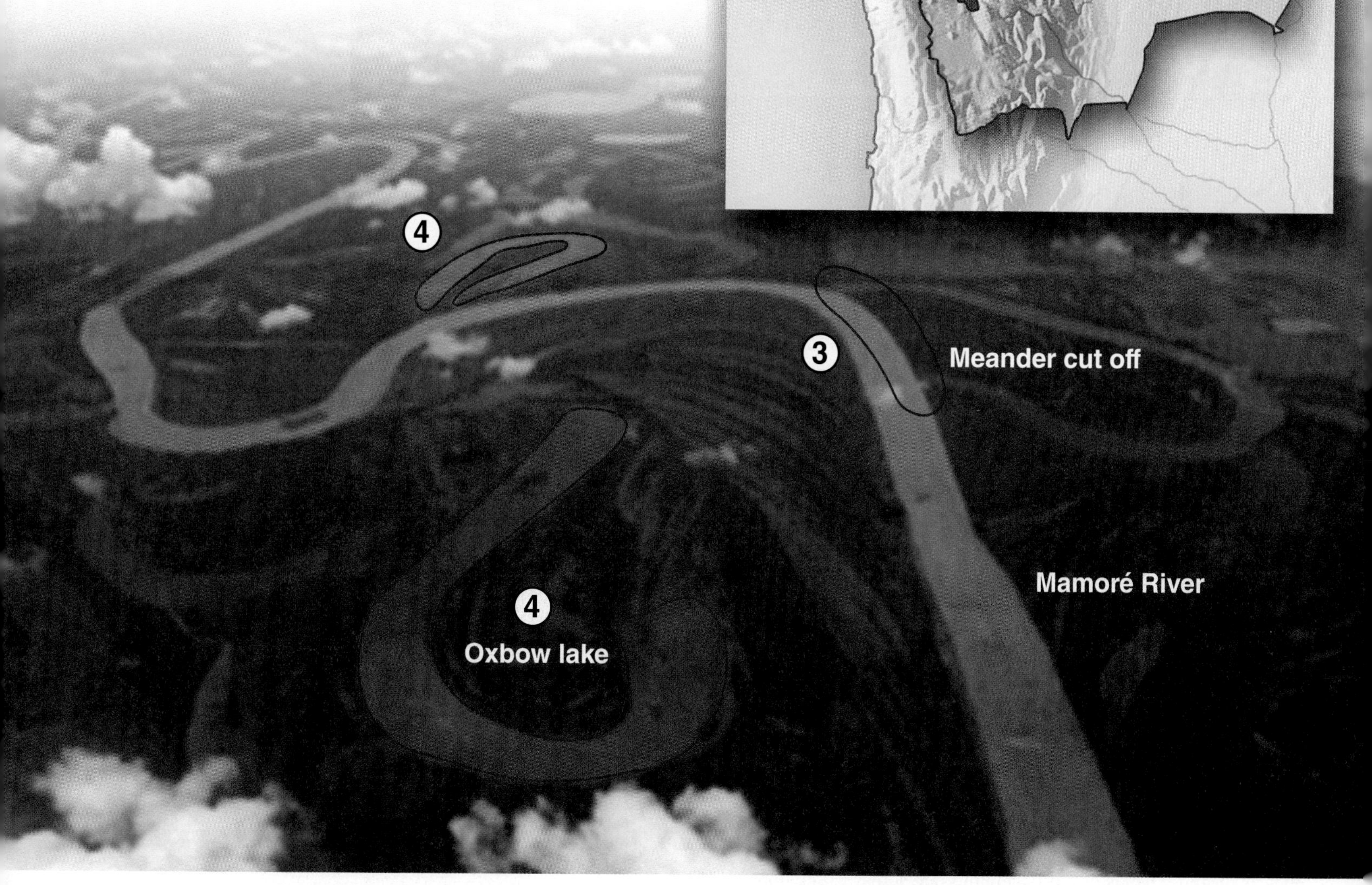

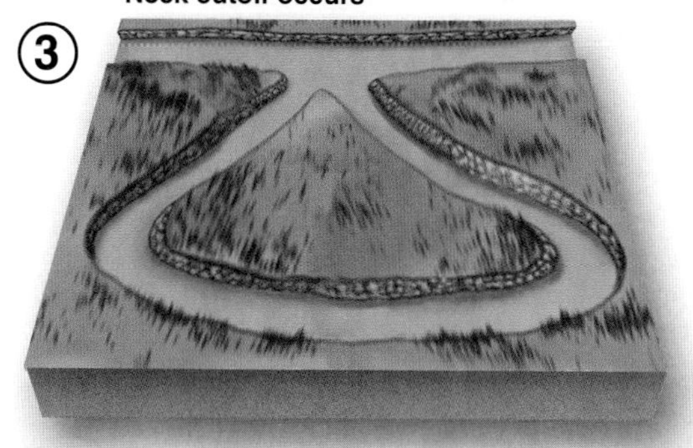

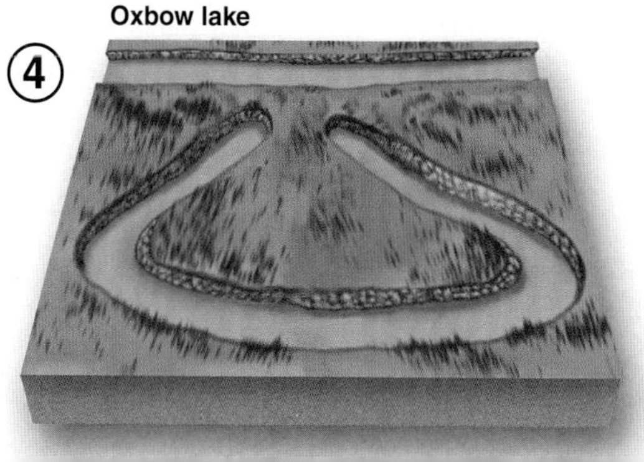

Checkpoint 11.15

Image Analysis: Mississippi River Radar Image

Examine the accompanying image of part of the lower course of the Mississippi River. This picture was taken by the Spaceborne Imaging Radar aboard the space shuttle. The long axis of the image is approximately 40 kilometers (25 miles) in length. North is to the top right. This section of the river represents part of the state borders of Arkansas, Louisiana, and Mississippi. Louisiana and Arkansas lie above (west of) the river, and Mississippi is below (east of) the river. This region is characterized by rich farmland (purple), where a variety of crops are grown. The gray regions bordering the river are undeveloped forested areas. The river is the black band that curves across the image from the top right-hand corner. The river flows from north to south.

1. Interpret the image and discuss the geologic history of this section of the river.
2. Identify where erosion and deposition are occurring along the stream channel, and label those locations E (erosion) and D (deposition) on the blank map to the left of the image.
3. Use the blank map to draw an earlier course of the channel.
4. Describe how velocity and depth change between points X and Y on the image, and plot how you expect velocity and depth to vary from X to Y on the graphs provided.

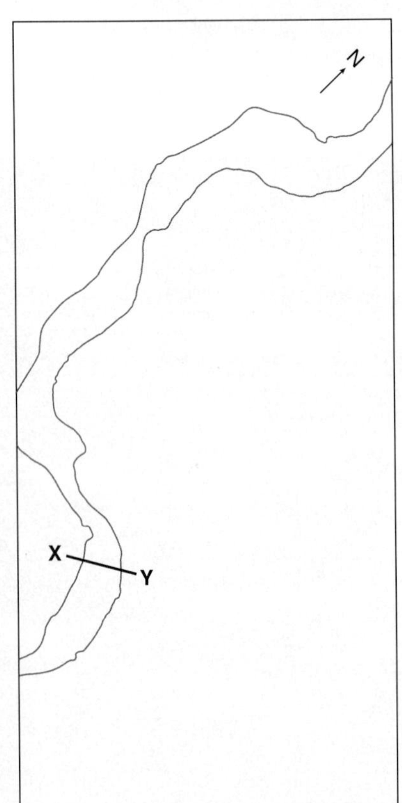

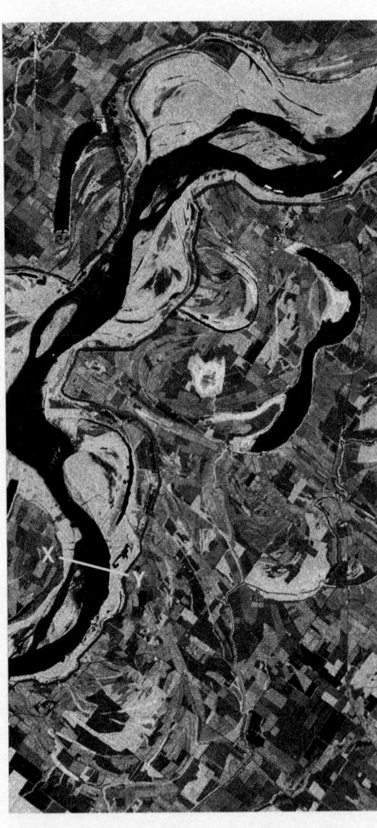

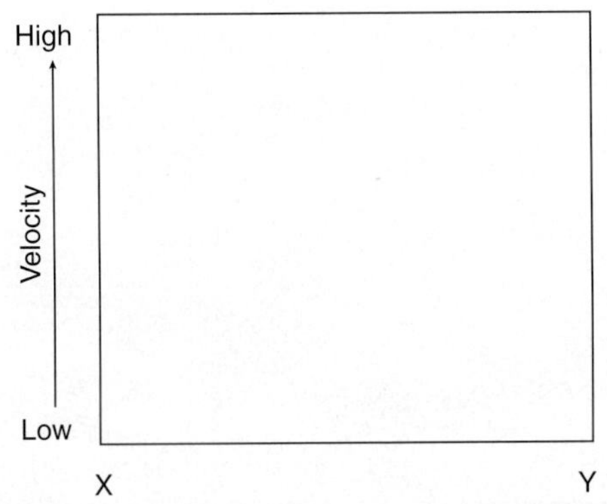

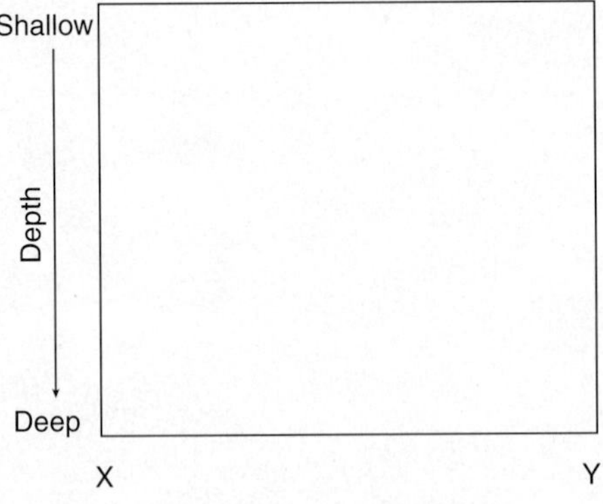

they settled on a channel bank farther downstream. This rule applies as meander bends progressively migrate across valley floors. In contrast, land originally encircled by a meander does not change ownership if a sudden shift in the stream channel isolates the property on the opposite bank of the stream. These sudden channel changes are common in floodplains (see Chapter Snapshot).

Since oceans are the ultimate base level, they are the eventual destination for much of the sediment transported by streams. What happens to the transported sediment when the stream velocity slows where the river meets the ocean? **Rivers dump much of their sediment to form a delta where they enter the relatively quiet waters of an ocean or lake** (Figure 11.22). The shape of the delta largely depends on the volume of transported sediment and the actions of coastal currents that may redistribute the sediment. The Nile River delta is eroding at its outer edges (see Figure 11.2), giving it a classic triangular shape that contrasts with the more irregular shape of the Mississippi River delta, where coastal erosion is lower and the course of the river has been controlled to ensure access for ship traffic. Continual dredging of the lower Mississippi River and the construction of levees and

Checkpoint 11.16

Write a story that describes the life cycle of a grain of sand that was eroded from the Andes Mountains and eventually deposited in the Atlantic Ocean. Include descriptions of erosion, transportation, and deposition processes.

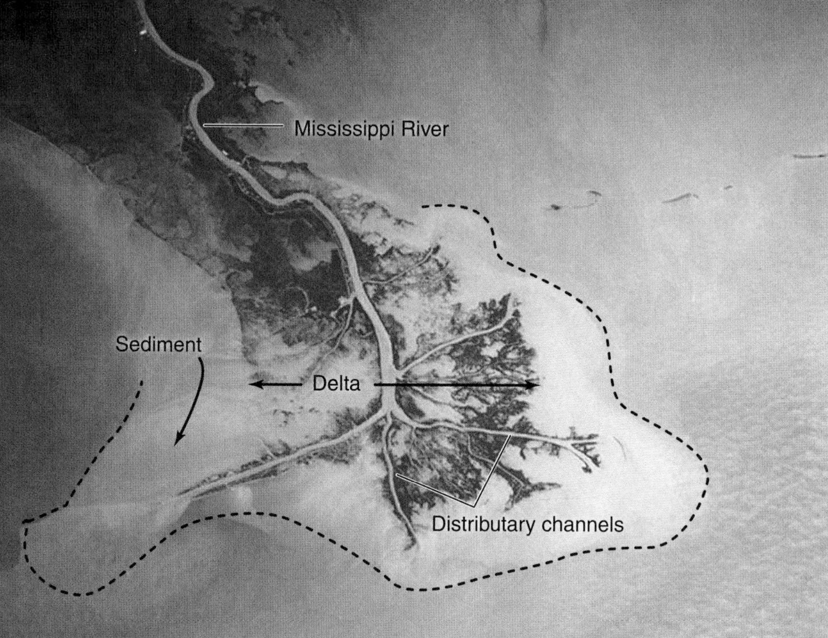

Figure 11.22 Mississippi River delta, Louisiana. Note the plume of sediment (dashed line) dumped by the distributary channels that form a "bird's-foot" delta. Constant dredging is required in order for oceangoing ships to navigate the main channel of the river.

other engineered structures have prevented the river from shifting its course near the coast. This has resulted in the delta now extending to near the edge of the continental shelf. Consequently, much of the sediment is lost to the coast as it is deposited in deeper waters of the Gulf of Mexico. Elsewhere, sediment in the Mississippi River basin is trapped in reservoirs behind dams. These factors have resulted in a reduction in the sediment delivered to the Louisiana coast and have contributed to increased coastal subsidence and higher erosion rates. The net result is that the coastline is more susceptible to damage from coastal storms and hurricanes such as Hurricane Katrina (see Chapter 15).

11.6 Floods

Learning Objectives

- Describe the principal natural causes of flood events.
- Explain how human activities can contribute to more frequent and more severe flooding.
- Discuss how scientists collect and analyze data to make accurate predictions of stream flow.
- Predict how the recurrence interval of floods is related to changes in stream discharge.

Many lives are lost and millions of dollars in damage occur every year when streams and rivers flow out of their banks, resulting in flooding. A **flood is the temporary overflow of a river onto adjacent lands not normally covered by water** (Figure 11.23). (This definition does not take into account coastal flooding resulting from hurricanes or tropical storms.) Floods occur when the amount of water on the land surface exceeds the volume of water that can be transported in stream channels and absorbed into the surrounding soil. Floods can occur for several reasons and can happen almost anywhere at any time.

Causes of Floods

The degree of flooding in a location is primarily related to the magnitude, timing, and type of precipitation, and human modifications of the physical landscape. Secondary factors that influence flood magnitude are the capability of the ground to absorb water, evaporation rates, and the physical characteristics of the stream system.

Precipitation. The most common cause of flooding is excess precipitation—rain, snow, or a combination of the two. Long-term precipitation that is greater than normal may gradually increase

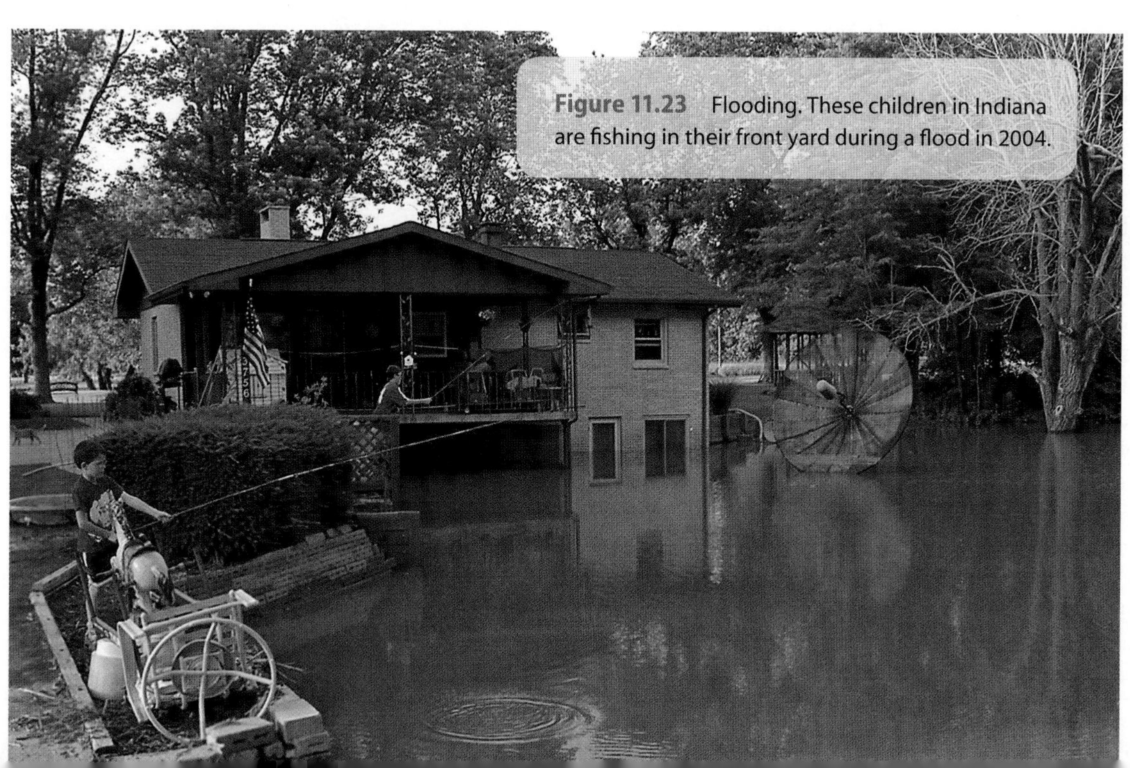

Figure 11.23 Flooding. These children in Indiana are fishing in their front yard during a flood in 2004.

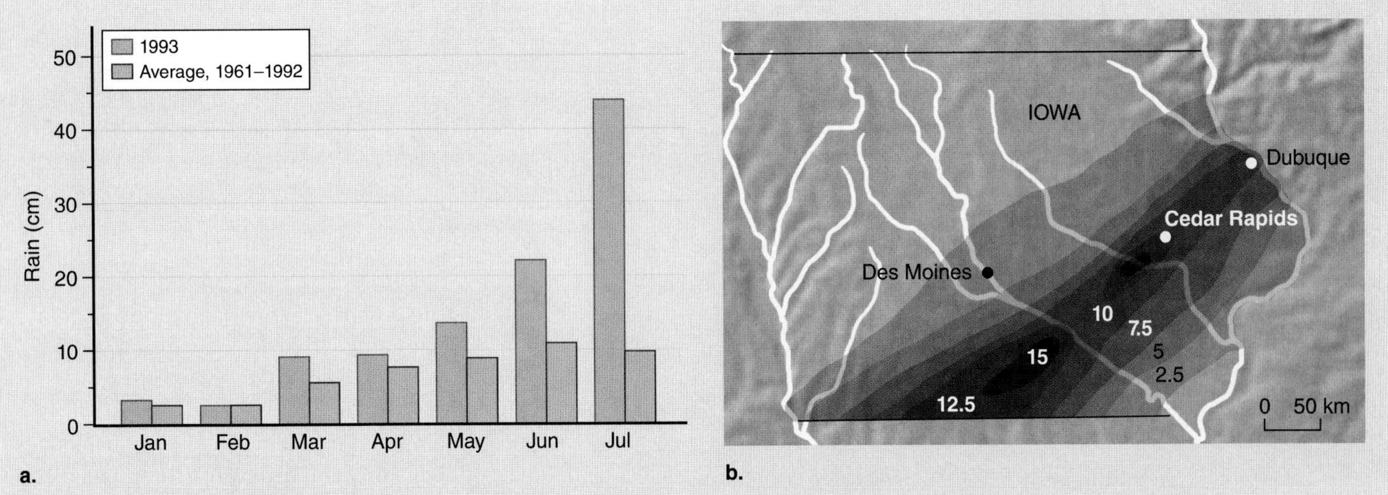

Figure 11.24 Precipitation leading to the 1993 Mississippi River flood. **a.** A graph of monthly rainfall for Cedar Rapids, Iowa, for January–July 1993 compared to average rainfall for those months. **b.** Distribution and magnitude of precipitation (in centimeters) associated with a single storm over southern Iowa, July 4 and 5, 1993.

the stream flow until a flood occurs. Alternatively, a flood can occur suddenly because of the addition of a large volume of water from a single heavy storm. Under normal conditions, some water infiltrates the soil where it may be used by plants or become part of the groundwater. However, once the ground has become saturated (holding all the water possible), excess water runs off to streams, rapidly increasing stream discharge and setting the stage for a flood.

In the case of the historic 1993 Mississippi River flood, both long-term and short-term precipitation came into play. First, from January through July of that year, several midwestern states received over 150 percent of their normal precipitation, and parts of North Dakota, Kansas, and Iowa received more than double their typical amounts (Figure 11.24a). Second, heavy storms added large amounts of rainfall all at once (Figure 11.24b). This excess precipitation filled storage reservoirs and saturated the ground, leading to the spatially largest flood in US history, one that covered 44,000 square kilometers (17,000 square miles) in nine states (Figure 11.25).

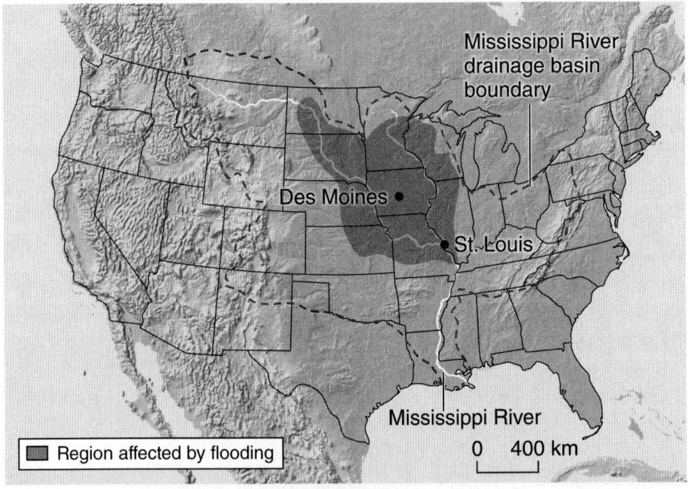

Figure 11.25 Area of the Mississippi River drainage basin affected by 1993 floods. Floodplains adjacent to the region's rivers were submerged by flooding.

However, flooding doesn't occur only in states that have a lot of precipitation. Some of the most frequent flooding occurs in arid southwestern states. The stream channels in these areas frequently have no surface water flowing in them during much of the year. In these dry regions, floods result from brief, intense storms rather than from prolonged rainfall.

When heavy precipitation is in the form of snow and ice, flooding may be simply delayed until temperatures rise sufficiently to cause rapid snowmelt. An abrupt, prolonged rise in temperatures may be all that is necessary to cause widespread flooding due to melting snow. Furthermore, the effects of flooding are exaggerated when ice accumulates to form an ice jam that blocks flow. This is a particular problem in areas where streams flow from warmer to colder regions (for example, from low to higher latitude). For example, north-flowing streams in mid-high latitudes may thaw in the south, near their source, but remain frozen closer to their northern mouth. As the water flows northward, it may be slowed or blocked by ice jams, causing a flood. This often happens during spring when the Red River is blocked by ice jams in northern Minnesota and North Dakota. The flow of ice and other debris often gets stuck under low bridges to form a temporary dam (Figure 11.26). The stream floods when unseasonably high temperatures cause rapid melting that submerges sections of towns along the river before the water is able to flow north into Canada.

Human Activities. Human activities can affect the amount of water that rapidly runs off a surface—and therefore can be a cause of flooding. For example, asphalt-paved parking lots, roads, and housing developments cover or alter natural surfaces that otherwise might absorb or slow the movement of water. These artificial surfaces are designed to rapidly transport rainfall through storm sewers and away from the developed area. These storm sewers can contribute to floods, as they empty water into natural streams and rivers that originally formed under lower flow conditions. As a consequence, many urban planners now require such water to be diverted to ponds designed to slow the release of water to the streams.

Checkpoint 11.17

basic · intermediate · advanced · superior

List five factors that influence flooding. Use one sentence to briefly describe the role of each factor in flooding.

Figure 11.26 Ice jam flooding along Red River, North Dakota. Ice and other debris blocked the flow of the Red River below this bridge in Fargo. The river has backed up, forming a temporary lake.

In addition, housing projects or agricultural fields sometimes replace natural wetlands (see Chapter 12). Natural wetlands are areas along rivers that normally act as water storage reservoirs; they absorb floodwaters and release them slowly after the flood crest has passed, thus reducing and slowing runoff. When wetlands are destroyed or altered, the surface no longer has a buffer, and runoff enters the stream more quickly. This in turn reduces the time between rainfall and flooding. To offset the loss of natural lands, developers are often required to replace lost wetlands. Much of the Mississippi River system was altered over the previous century by the draining of wetlands. States bordering the Mississippi have lost an average of two-thirds of their original wetlands; Iowa has lost 89 percent of its natural wetlands, mostly to agriculture.

Finally, the collapse of constructed dams has resulted in disastrous floods around the world. Flooding following the collapse of a dam in Johnstown, Pennsylvania, on May 31, 1889, killed more than 2,200 people. In Henan Province, China, in August 1975, several days of heavy rains caused a series of dams to burst, resulting in widespread flooding. Estimates are that as many as one-quarter of a million people drowned or died from disease or starvation following the Henan floods. Clearly, the human dimension associated with flooding should not be underestimated. We will further examine the interaction between people and floods in Section 11.7.

Estimating Floods: Measuring Stream Discharge and Stream Stage

Various measurement and data collection techniques enable scientists to assess the potential for flooding. Stream discharge varies seasonally and provides a key measure of the potential for flooding. For instance, when is discharge greatest in the streams near where you live? Many streams in North America have high discharge in spring when snow melts and rain is relatively frequent.

A record of typical stream discharge is required to understand the potential risk from flooding. Stream gauges, such as the one shown in Figure 11.27, are instruments that measure **the**

a.

b.

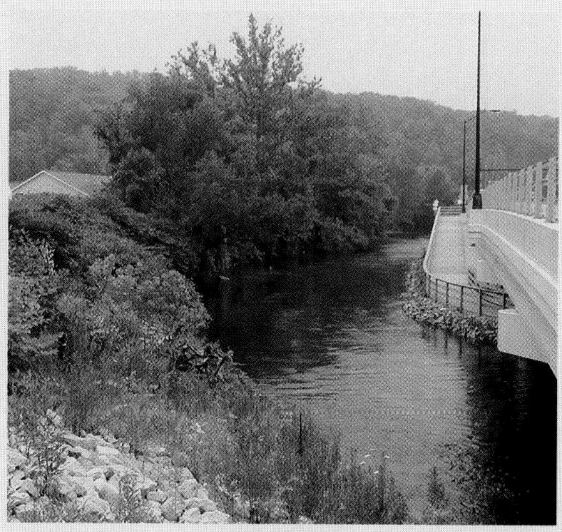

c.

Figure 11.27 A stream gauge installation on the Cuyahoga River, Akron, Ohio. **a.** Stream gauge instrument updates stream flow data several times each day via satellite. **b.** View upstream during high flow at stream gauge site on July 10, 2006. **c.** Same view on July 20.

stream stage, the depth of water in a stream channel relative to an easily identified starting point for measurement (which is not necessarily the bottom of the channel). The US Geological Survey (USGS) records stream discharge conditions from over 7,000 gauging stations across the country. (In comparison, Mozambique had just nine functioning stream gauges prior to the devastating floods of 2000.) The data from most of the gauging stations are relayed back to the USGS by satellites and are made available online, an example of how technology helps scientists inform the public.

Data on stream stage are relatively straightforward to collect because they are gathered simply by measuring the height (depth) of water in the stream channel. Discharge is typically estimated by using stage/discharge relations developed by on-site measurement of the stream's width, depth, and velocity during different flow conditions (Figure 11.28). For example, repeated observations may reveal that a stream stage of 2 meters (6.6 feet) is associated with discharge of 60 cubic meters per second (2,120 cubic feet per second). Gauging station locations are checked periodically to ensure that the shape of the stream channel has remained uniform. A graph that plots discharge over time, termed a *hydrograph* (Figure 11.29), illustrates changes in stream flow over intervals of days to years.

Stream discharge conditions along streams all across the United States are freely available at http://water.usgs.gov. Each stream gauge site contains graphs that display stream discharge and stage for the past week and provide access to data on historical stream flow values.

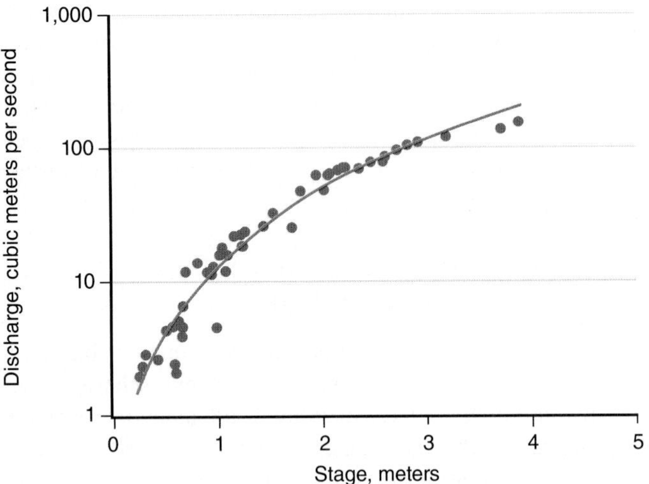

Figure 11.28 Stage and discharge relationships for the Cuyahoga River, Akron, Ohio. These data are for the stream gauge in Figure 11.27. Discharge increases with increasing stream stage.

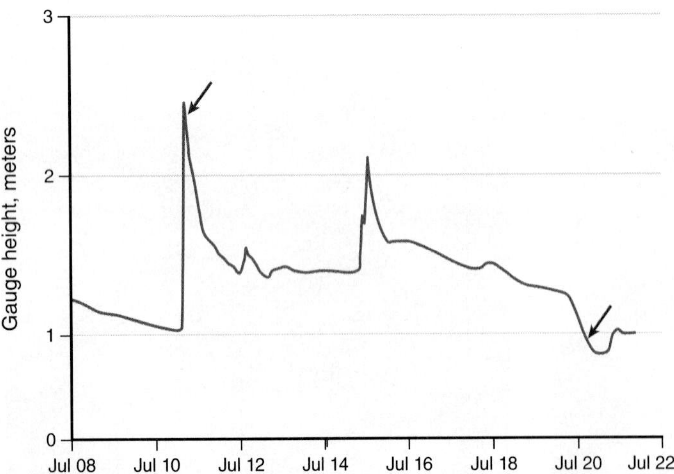

Figure 11.29 Hydrograph for the Cuyahoga River, Akron, Ohio. Graph shows stage measurements for the gauging station in Figure 11.27 for July 8–22, 2006. Figures 11.27b and c are represented by the red arrows. Flood stage at this location is nearly 3 meters (9 feet).

Determining Recurrence Interval

If city planners had some idea of how frequently large floods occur, they could take action to build structures to prevent future flood damage. In fact, information about past flooding stretching back several decades is available for many regions of the United States through the USGS. That historical information can be

Checkpoint 11.18

Four stream gauging station locations (A, B, C, D) are shown on the accompanying map. If the bedrock and topography are similar for each stream system, predict which station will record the greatest discharge. Explain your choice.

a) Station A
b) Station B
c) Station C
d) Station D

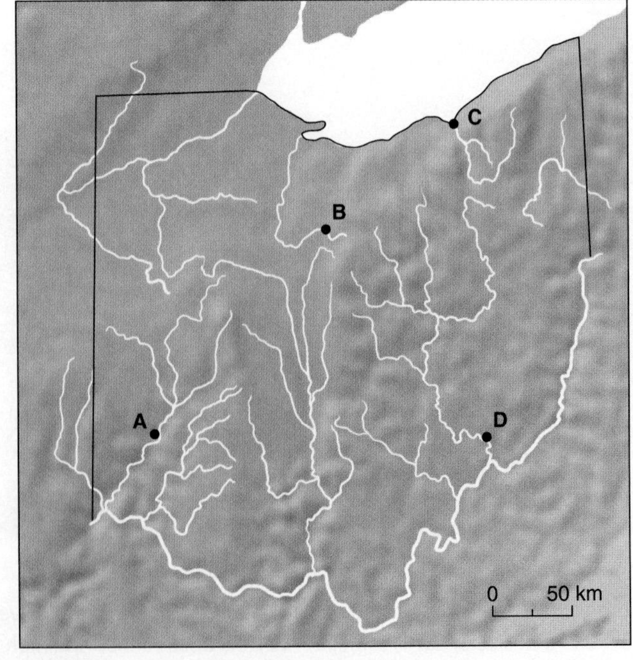

used to determine how frequently flooding has occurred in the past and to estimate the likelihood of a flood of a specific size in the future (recall the uniformitarianism approach discussed in Chapter 8).

So how do hydrologists estimate the size and frequency of future floods? First, they need historical discharge data for the river. Some of those data are derived from historical records and written accounts for floods that occurred before there were stream gauge stations. Better historical data are available for larger rivers where discharge has been accurately measured since the early 1900s. These discharge data are arranged in order from the largest flood discharge to the smallest, assigned a rank (in order by size of discharge), and graphed (Figure 11.30). **The average time in years between floods of the same size is called the *recurrence interval*.** Realize that the recurrence intervals are only averages. A 100-year flood happens, on average, once every 100 years. Over the time period from 1800 to 2000, two "100-year" floods could have occurred in 1899 and 1901 or in 1801 and 1999. The flood recurrence interval is simply the probability that an event of a

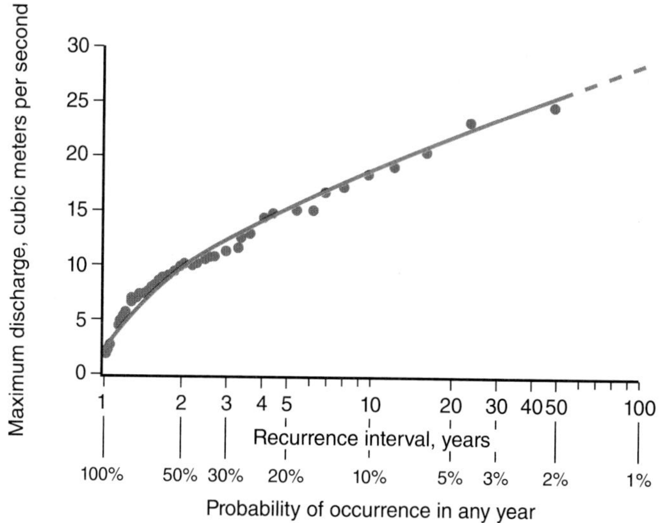

Figure 11.30 Recurrence intervals. Scientists can use graphs of maximum stream discharge versus recurrence interval to estimate the discharge of 100-year floods.

Checkpoint 11.19

basic · intermediate · advanced · superior

Flood Recurrence Interval: Mississippi River, Keokuk

The following table presents peak discharge data for the Mississippi River at Keokuk, Iowa, from 1943 to 1992. This data set ends the year prior to the 1993 Mississippi River flood. We will use these data to estimate the recurrence interval for the 1993 flood event.

1. Complete the table to the right by calculating the remaining recurrence intervals (RIs) for the flood events of 1965, 1969, 1976, and 1972, using the formula RI = $(N + 1)$/rank where N = number of readings (49) and rank = order of readings (largest = 1; smallest = 49). For example, the RI for 1960 is ([49+1]/3) = 16.7.
2. Plot the discharge versus recurrence interval on the grid provided. Draw a straight line through the plotted points with RI values of 2 or more, and estimate the size of 100-year and 500-year floods for this gauging station.

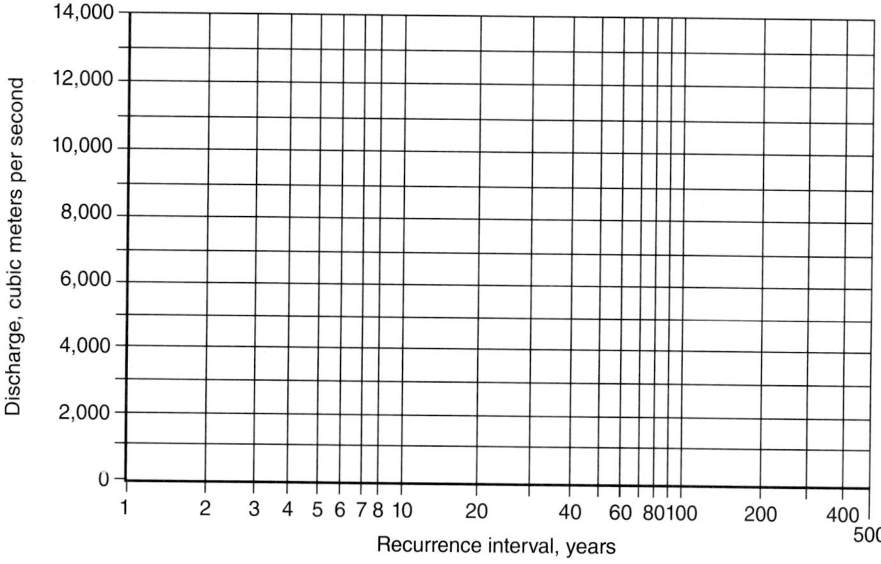

3. Compare the predicted flood discharge against the discharge of the 1993 Mississippi River flood, 12,630 m³/s (446,000 ft³/s). What is the recurrence interval for the 1993 flood?

Rank	Date	Discharge (m³/s)	RI
1	1973	9,741	50
2	1965	9,230	
3	1960	8,198	16.7
4	1986	7,589	12.5
5	1951	7,507	10
6	1974	7,362	8.3
7	1979	7,277	7.1
8	1944	7,249	6.2
9	1952	7,187	5.5
10	1969	7,164	
11	1975	7,136	4.5
12	1947	6,957	4.2
13	1986	6,824	3.8
14	1948	6,615	3.6
15	1982	6,371	3.3
16	1962	6,346	3.1
17	1983	6,343	2.9
18	1946	6,323	2.8
19	1967	6,264	2.6
20	1976	6,060	
25	1972	5,437	
30	1954	5,137	1.67
40	1970	3,964	1.25
49	1977	2,260	1.02

Checkpoint 11.20

basic intermediate advanced superior

An analysis of flood data in metropolitan areas over the last century suggests that floods caused by similar volumes of precipitation are actually larger and more devastating today than in the past, despite advances in flood monitoring. Provide some potential explanations for this apparent paradox.

Checkpoint 11.21

basic intermediate advanced superior

List and briefly explain four economic consequences of flooding.

particular size *may* happen, not a prediction that it *will* happen. The recurrence interval plots are extrapolated to estimate the probability of larger floods for which there are no data (see dashed line in Figure 11.30).

11.7 Flood Control

Learning Objective

- Give examples of adjustment and prevention actions associated with floods.

Economic losses following the 1993 flood in the Mississippi River drainage basin totaled billions of dollars, and indirect losses in the form of lost wages and production cannot be accurately calculated. The greatest economic losses occurred in cities located directly on the floodplain. For example, Des Moines, Iowa (Figure 11.31), located in the center of the flood region, became the largest US city to lose its water supply when its water treatment plant flooded during some of the hottest days of the summer. Water pipes, contaminated by floodwaters carrying sewage and agricultural chemicals, had to be flushed out before the municipal water supply could be reconnected. Hundreds of miles of roads built on the flat, wide floodplain were also closed. Flooding was estimated to have cost $500 million in road damage.

The flooding submerged 8 million acres of farmland. Production of corn and soybeans fell, and prices subsequently rose. Floods deposited thick layers of sand and mud in some fields (Figure 11.31b). Normally, barge traffic between Minneapolis and St. Louis carries 20 percent of the nation's coal, one-third of its petroleum, and one-half of its exported grain. Flooding halted barge traffic for 2 months, and carriers lost an estimated $1 million per day. (Why do you think a flood would stop barge traffic?)

Nearly 50 people died and 26,000 were evacuated as a result of the 1993 flooding in the midwestern United States. In contrast, more than 700 people were killed by the floods in 2000 in the Limpopo River drainage basin in Mozambique (Figure 11.32), described in the chapter introduction. Both floods were clearly massive. Was the difference in impact more attributable to the physical nature of the floods or to the social structures that existed to manage floods? Let us now consider the typical procedures that can be put in place to minimize damage and loss of life from flooding.

a.

b.

Figure 11.31 Effects of the Mississippi River flood. **a.** Flooding of the Raccoon River at Des Moines, Iowa, inundated the city. **b.** Flooded cornfields, Missouri.

Approaches to Flood Control

When dealing with floods, **we can either attempt to keep them from occurring (prevention) or recognize that they will occur and modify our lifestyle to deal with them effectively (adjustment).** To explain these two approaches, we will use as our model the Mississippi River, one of the most heavily engineered natural features in the world. Over the years, the character of the Mississippi's floodplain has changed to accommodate agriculture and urbanization.

Prevention. Scientists and engineers have attempted to control the flow of rivers in the Mississippi River basin. Can you predict what measures they take to control flooding? One method is the

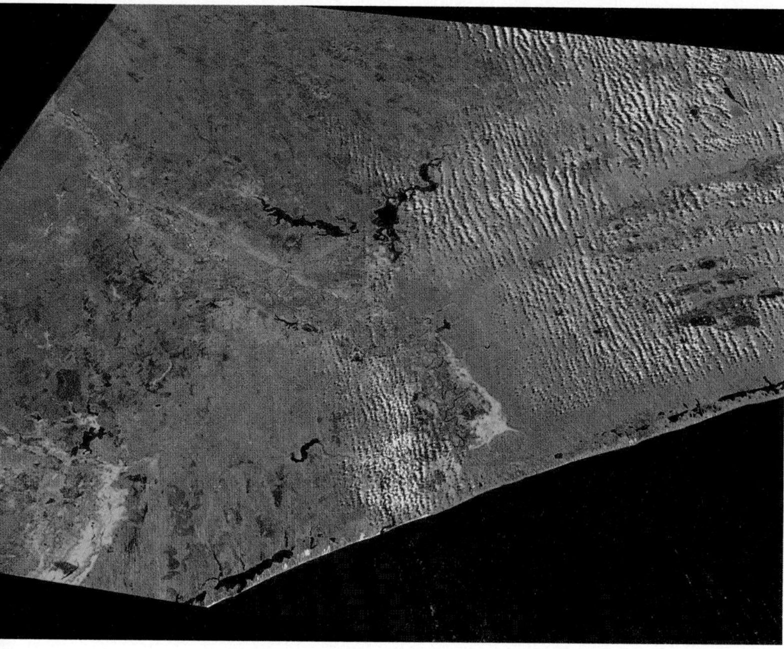
a.

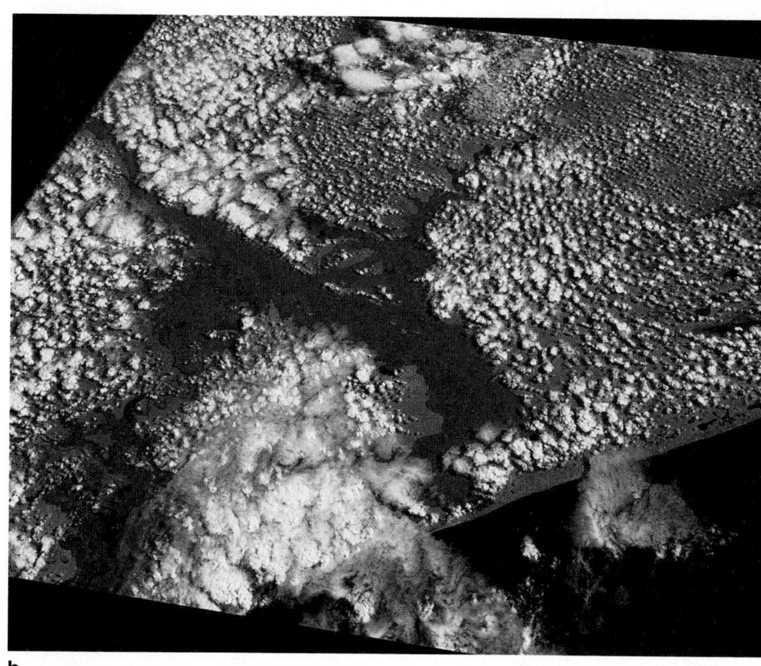
b.

Figure 11.32 Extent of the Mozambique flood. **a.** The size of the Limpopo River, Mozambique, in a normal year and **b.** during the floods of 2000. The floods swelled the width of the river to more than 130 kilometers (80 miles) in some locations.

construction of raised, earth-filled embankments along the stream bank (called *levees* or *floodwalls*) to hold back the river during a flood (Figure 11.33). The US Army Corps of Engineers was directed to construct flood control structures (dams, reservoirs, levees) on the Mississippi River after disastrous floods in the 1930s. However, levees can fail if the floodwater rises over the top of the structure or if the levee collapses under the weight of the water. While offering protection, these structures also serve to increase the volume of water held in the stream channel and thus may increase the impact of flooding. **Levees and floodwalls protect cities and fields on the floodplain from most floods,** but they may not protect against the largest floods. Floodplain residents may experience a false sense of security that can lead to more extensive development of flood-prone lands (the "levee effect").

Of those locations that measured river discharge during the 1993 Mississippi River flood, over one-third set discharge records and nearly one-half set stage records. Scientists who studied that flood point out that the river channel along the Mississippi and its tributaries had been altered to protect surrounding areas from flooding. Many of the alterations failed because the flood exceeded design expectations and because the river was constrained as never before. Over 9,300 kilometers (5,780 miles) of levees failed; of these, only 17 percent were federal levees, while up to 77 percent were locally constructed. Most levee breaks occurred south of St. Louis. St. Louis was protected by a tall floodwall, which developed a leak but held up over the length of the flood. Many smaller levees in rural areas failed.

Sometimes a community decides to construct a **floodway, or diversion channel, that will transport floodwaters away from inhabited areas.** Winnipeg, Canada, completed a floodway around

Figure 11.33 Levees. **a.** Artificial levees are raised embankments along a stream channel constructed to protect neighboring lands from rising floodwaters. **b.** A levee in Moorefield, West Virginia, protects homes from potential flooding of the South Branch of the Potomac River.

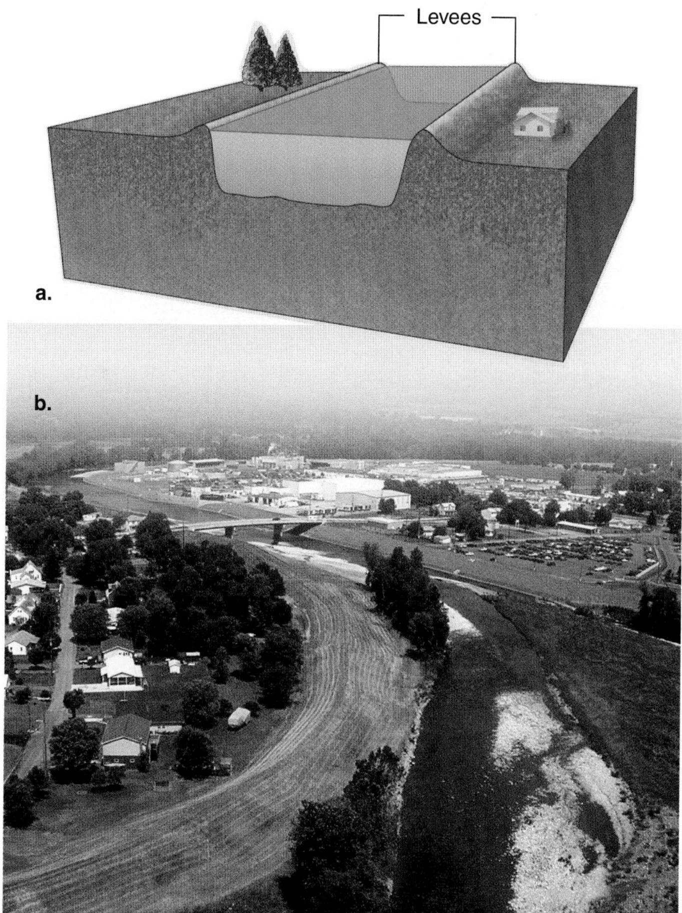

the city in 1968 (Figure 11.34). The floodway takes water from the Red River south of the city and loops around the developed region before dumping the water back into the river farther downstream. Winnipeg has been largely unaffected by many recent floods of the Red River, while cities upstream in North Dakota (Fargo, Grand Forks) have sustained significant damage.

The construction of dams for flood control follows the premise that floodwaters can be stored in reservoirs to be released slowly when the threat of flooding has receded. This plan works as long as storage capacity is sufficient to accommodate excess runoff. Unfortunately, in 1993, reservoirs in much of the Mississippi River basin were already near capacity following higher-than-normal precipitation and were largely ineffective in preventing flooding. Under normal conditions, the placement of large dams and reservoirs on the Mississippi River would hinder barge transportation. Therefore, the majority of structures designed to prevent flooding in the upper Mississippi River basin are actually located in the Missouri River basin. Many of the dams in the Missouri basin are located in a southern subbasin, the Kansas River basin, where 85 percent of stream flow is controlled (that is, it flows through dams; Figure 11.35). US reservoir capacity grew significantly from 1930 to 1970 but has remained almost uniform for the past few decades as new dam construction has slowed. Few rivers remain where new dams would have the potential to significantly diminish flood capacity.

Adjustment. If we know flooding can occur, what types of adjustments can we make to our lives? Two recent examples illustrate that the trend in flood control appears to be turning from prevention toward adjustment. Following the 1993 Mississippi River flood, the residents of Valmeyer, Illinois, decided to relocate their town to a bluff overlooking its original location on the floodplain 6 kilometers (4 miles) from the river. In Napa, California, voters approved a sales tax increase to fund a project that would modify

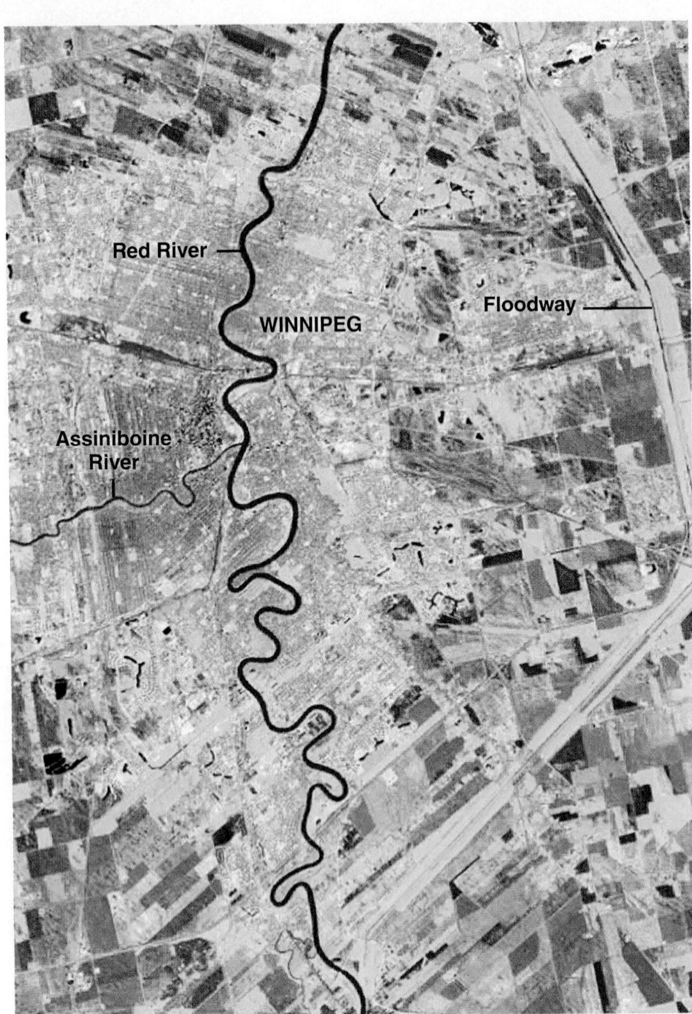

Figure 11.34 A floodway. Approximately 500,000 people live at the confluence of the Assiniboine and Red rivers in Winnipeg, Manitoba, Canada. A 47-kilometer (29-mile) floodway was constructed to counteract the threat of spring flooding from snowmelt in the Red River valley (infrared satellite image).

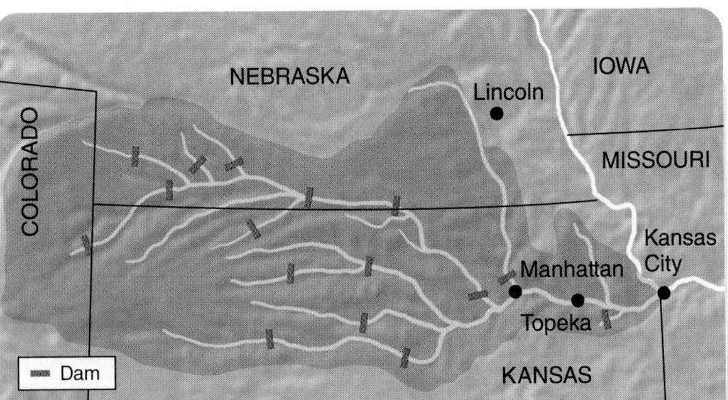

a.

b.

Figure 11.35 Dams in the Kansas River basin. **a.** These structures are intended to prevent flooding along the Mississippi River. **b.** Wilson Lake Dam, central Kansas.

land use patterns in the floodplain of the Napa River. The plan will remove some levees and restore wetlands along the bank where the river loops through the city's downtown. In addition, a floodway will reroute discharge during big floods. Community planners now more frequently turn to flood simulation models to support detailed zoning and urban planning, and avoid development in areas prone to flooding.

Although flood insurance does not protect people from harm or property from damage, it can reduce monetary loss for those who pay premiums. Typical residential home insurance does not cover losses associated with flooding. In 1968, the federal government established the National Flood Insurance Program to provide some financial protection for floodplain residents. However, most people do not take advantage of the program. Only 10 percent of midwestern residents living in flood-prone areas had flood insurance prior to the 1993 floods. Over 11 million buildings are located in flood-prone areas in the United States, but only about 18 percent of owners have purchased flood insurance.

The US government created the Federal Emergency Management Agency (FEMA) in 1979 to provide assistance to people and areas affected by natural disasters, including floods. Not all areas affected by flooding are declared disaster areas. Residents who don't buy flood insurance must gamble that they will receive disaster aid from FEMA to help cover some cleanup costs.

Checkpoint 11.22

basic | intermediate | advanced | superior

Flood Control Defining Features Matrix

Complete the following table by placing X's in the appropriate columns to identify whether the characteristics listed at the left represent prevention or adjustment measures for flooding. One characteristic has been completed as an example.

	Flood control by	
Characteristic	**Prevention**	**Adjustment**
Levee is constructed.	X	
Newspaper publishes flood evacuation route.		
New housing developments are elevated on pilings aboveground.		
Dredging removes sediment from streams.		
Flood zone maps are made available in local library.		
Buildings are relocated outside of flood zone.		
Dam is constructed upstream from community.		
Zoning regulations are enacted to prevent new construction in floodplain.		

Checkpoint 11.23

basic | intermediate | advanced | superior

Community Flood Rating System Activities

The National Flood Insurance Program (NFIP) developed a community rating system (CRS) to encourage additional community activities that would reduce flood losses. One goal of the program is to reduce the money spent by the NFIP to help cities recover from floods. The CRS evaluates 18 potential community activities that may contribute to better management of the floodplain. For each activity, a maximum number of points can be awarded, depending on its potential for reducing flood costs. Points can range from a low of 66 to a high of 3,200. We have listed six of the activities, along with six scores. Predict which activity matches which score by drawing lines from the activity column to the points column. Write justifications for your choices.

Activity	Points
Outreach methods—advising people of their flood hazard and what they can do about it (e.g., insurance availability)	66
Open-space preservation—maintaining flood-prone areas free of development	3,200
Acquisition and relocation—acquiring, relocating, or clearing buildings from the flood hazard area	315
Flood warning program—identifying flood threat and disseminating warnings to floodplain residents	2,800
Flood protection information—providing the community library or website with information on flood hazards and protection	900
Flood protection—protecting buildings from flooding through retrofitting (e.g., elevating structures) or small flood control projects (e.g., building small levees)	225

basic advanced intermediate superior

Checkpoint 11.24

Flood Risk Rubric

Several states in the Ohio River valley have a history of extensive flooding. Following graduation, you find yourself working for a state legislator who is part of a multistate task force seeking federal funding for flood protection for communities along a designated river corridor. The corridor stretches along the Ohio River from its source in Pittsburgh to its mouth in southern Illinois. The legislator asks you to be on a team to identify the key factors that contribute to flooding along the Ohio River. She requests that you make a list of factors and develop a scoring scheme that can be used to rank locations in terms of their future flood risk.

1. Your team is charged with creating an evaluation rubric to assess the relative risk of danger from future flooding. You must find a way to rank the risk of potential harm from each flood event. Consider what factors would contribute to the loss of lives and the high damage costs associated with flooding.
2. Complete your evaluation rubric by filling in the blanks on the accompanying table. Your rubric will be applied to several cities along the river, and the three that score the highest will receive funds to develop a flood mitigation plan. One factor is provided as an example. Identify at least five more, and determine the low-, intermediate-, and high-risk values for each one.
3. After completing the rubric, your team is asked to double the score of the most important factor. Which do they choose? Why? Explain their choice. Keep in mind that some factors, while important, may not show much change within a region.

Factors	Low risk (1 point)	Moderate risk (2 points)	High risk (3 points)
Average annual rainfall	Low (less than 75 cm/yr)	Intermediate (75–125 cm/yr)	High (more than 125 cm/yr)

the big picture

A stream is a system—a complex system with multiple parts, interconnections, and dimensions. A stream is actually a subsystem of an even more complex system, the hydrologic cycle, through which water moves among the four components of the larger earth system. In this chapter, we looked at the hydrologic cycle as it relates to the creation and maintenance of streams.

We can also think of a stream in its biological, economic, cultural, political, historical, and aesthetic dimensions. Perhaps no stream system in North America exemplifies these multiple perspectives more than the Colorado River drainage basin (Figure 11.36). The Colorado River drains parts of seven states and, as the only major stream flowing through much of the Southwest, represents a rare water source in an otherwise dry land. Water diversion structures siphon off the river's water to ensure a steady supply of domestic water for cities as far away as San Diego and to provide water to irrigate the winter vegetable crops of the Imperial Valley in southern California.

Much of the lower course of the Colorado River north and south of the Grand Canyon has been corralled behind massive dams. One consequence of dam construction was the alteration of the physical environment downstream from the largest dams. Regulated stream flows necessary for power generation are much lower than the peak stream flows prior to dam construction. Under predam conditions, spring floods eroded sandbars from the channel and formed beaches along the canyon walls. The rapid spring flows cleared the river of obstructions created by boulders deposited at the mouths of tributary streams. These processes, and the ecological environments they created, disappeared after dam construction. In a spectacular pair of experiments, scientists with the US Geological Survey have applied their knowledge of stream processes to produce controlled floods downstream from Glen Canyon Dam in efforts to restore some of the predam channel characteristics. The chapter opening image shows the rapid release of floodwaters from the base of the dam during one of these controlled flood events.

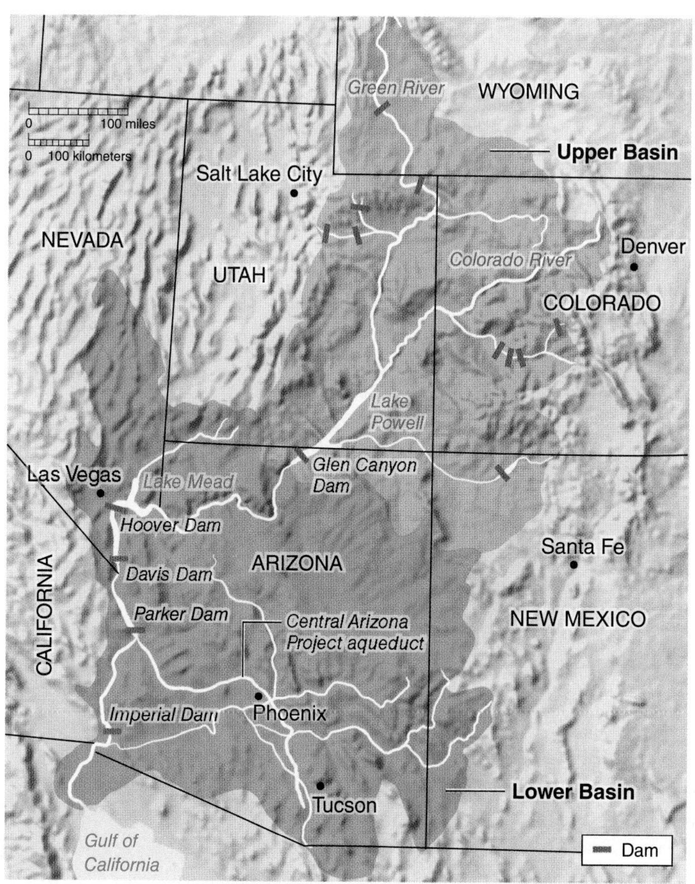

Figure 11.36 The principal hydrologic features of the Colorado River basin.

Variations in stream flow in the lower Colorado River were minimized after construction of the big dams. However, even the most regulated systems are subject to surprises. Heavier-than-normal snowfall in the winter of 1982 and rapid snowmelt the following spring quickly filled the reservoirs. Discharge from Glen Canyon Dam had to be increased to maximum capacity to avoid overtopping the dam and causing it to fail. As it was, the increased flow ripped out chunks of bedrock from spillway tunnels, threatening the survival of the dam. This little lesson reminds us that earth systems are complex. While we can understand stream processes, our records are still woefully short and probably miss some high-magnitude events that occur infrequently.

As citizens and voters, we need to appreciate not only that Earth can be understood but also that it can be altered in major ways by influencing any part of the interacting systems. In the pre–Hurricane Katrina world, the word *levee* was probably no more than a term in a textbook for most students. A levee is a marvelous piece of technology. Without it, the lands along rivers would undergo periodic, serious inundations by devastating floods. The failure of levees in New Orleans during Hurricane Katrina made our nation reevaluate its reliance on technology and what happens when we do not heed the warning signs of a flooding disaster. We must learn not to be too complacent about our technologies (all of them) or too ignorant of the science that led us to depend on them in the first place.

Streams and Floods: Concept Map

Complete the following concept map to evaluate your understanding of the interactions between the earth system and streams and floods. Label as many interactions as you can, using the information from this chapter.

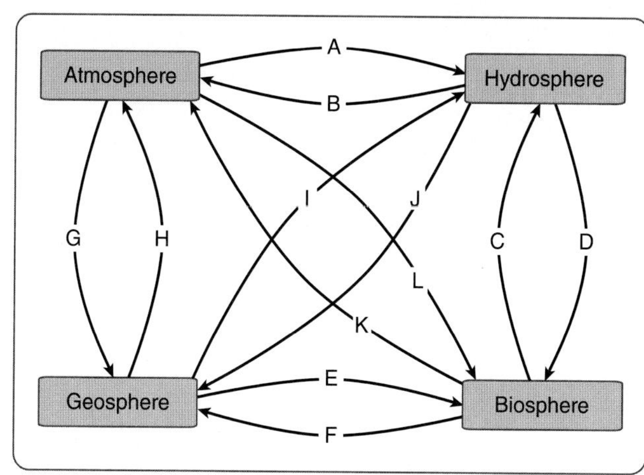

"We'll never know the worth of water till the well goes dry."

—Scottish proverb

"I derive more of my subsistence from the swamps which surround my native town than from the cultivated gardens in the village. . . . I enter a swamp as a sacred place, a sanctum sanctorum. *There is the strength, the marrow, of Nature."*

—Henry David Thoreau, US author/naturalist

Chapter 12

Groundwater and Wetlands

Chapter Outline

the big picture

Which is better, bottled water or tap water?

See The Big Picture box at the end of this chapter for the full story on these images.

12.1 Meet Your Drinking Water

Chapter Learning Outcomes

- Students will describe geologic conditions necessary for ready access to groundwater.
- Students will discuss how water enters and leaves the groundwater system.
- Students will explain the connections among groundwater, wetlands, and human activity.

More than a dozen brands of water are on display in my local grocery store. Their labels proclaim artesian water, natural spring water, sparkling mineral water, or, more basically, indicate that the water is from a community water system. On the basis of the bottles on the shelves, there is apparently a great demand for the waters of Europe. Pricey bottles from such places as Norway, Germany, Italy, England, and France are available. An attractive glass bottle of Swedish water can be purchased for $4. That works out to about $20 for 1 gallon (3.8 liters). Alternatively, you could spend 99 cents to get a plastic gallon jug of American water from a nearby bottling plant.

Consumption of bottled water in the United States has grown faster than for any other beverage and has more than doubled in the last decade. Each day, we drink more bottled water than coffee or milk. This rapid growth is occurring despite the fact that most people have easy access to an alternative source of clean, fresh drinking water that costs about 1,000 times less. It is called *tap water.* You can just turn on the faucet in your kitchen and get 3 gallons (11 liters) of water for about 1 cent. If you like that tap water taste but prefer to carry a water bottle, you are in luck. About one-quarter of all water sold commercially in the United States is little more than bottled tap water.

Where Drinking Water Comes From

Our drinking water, whether we buy it in bottles or get it out of the tap, originally comes from streams and lakes on Earth's surface or from groundwater. **Groundwater is water that is present in rocks or unconsolidated materials below Earth's surface.** Groundwater forges a link between surface water systems and the materials in Earth's crust.

In some cases, water simply flows through gaps in the rock, and there is little or no interaction between the water and the minerals in the surrounding rock. In other cases, such as in areas underlain by limestone, the groundwater can dissolve away the rock. This groundwater is what we typically consider *mineral water* because it has a relatively high proportion of dissolved minerals. The total dissolved solids (TDS) represent a measure of how many minerals are present in the water. To be labeled mineral water, bottled water must contain at least 250 parts per million (ppm) TDS. In contrast, the most popular brands of US bottled water are labeled *purified water* and contain less than 10 ppm TDS, essentially indicating that the water is free of all minerals. If your bottled water says purified on the label, this is probably an indication that it is simply municipal tap water that has undergone a second round of water treatment prior to bottling. While some would argue that it is the presence of minerals in bottled water that provides health benefits, there is little evidence to support this claim. Generally, it is not the natural minerals dissolved in the groundwater that we should be concerned about. Our water supplies are more likely to become contaminated as a result of human actions that allow manufactured chemicals to leak from storage tanks or to be washed off agricultural fields.

In the United States, there are 53,000 community water systems that provide clean water for domestic use from surface or groundwater sources. Despite safeguards, a small number of community water systems may become polluted, resulting in serious health risks. Cleanup of these sites can be especially challenging if the water comes from an underground source where the pollution is hidden from direct view. Unraveling the cause of these problems requires a thorough understanding of the geology of the groundwater system. Given the potential tragic consequences of groundwater contamination, it is perhaps no surprise that bottled water has become so popular.

A Case of Groundwater Contamination: Woburn, Massachusetts

Groundwater pollution was at the heart of a lawsuit filed on behalf of a group of families from Woburn, Massachusetts. Children of the families were diagnosed with leukemia after their mothers drank water from two polluted wells (wells H and G, Figure 12.1) while they were pregnant.

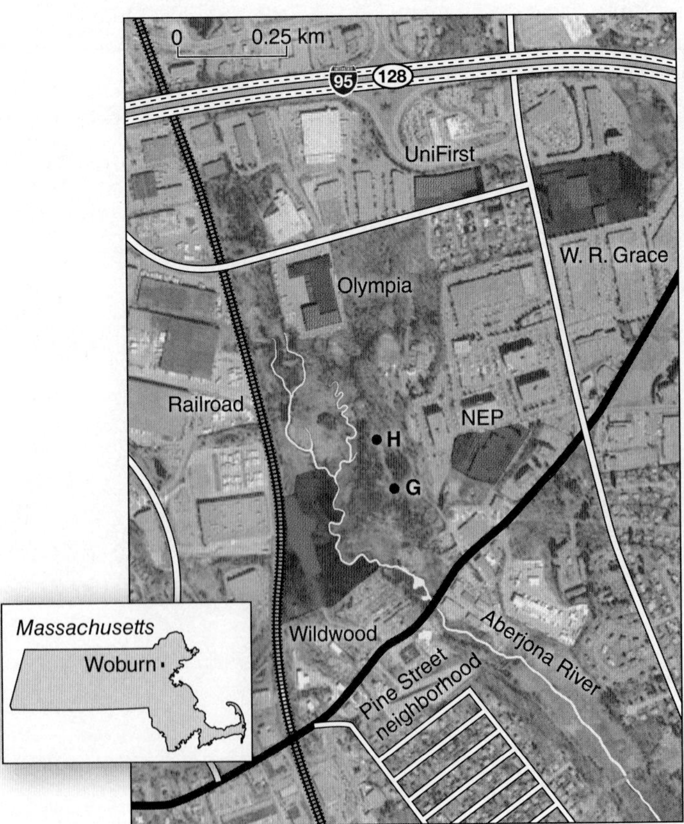

Figure 12.1 Potential sources of pollution at Woburn, Massachusetts. Leukemia cases were clustered in the Pine Street neighborhood south of wells H and G. Properties near the wells included W. R. Grace (chemicals), NEP (plastics), Olympia (trucking), UniFirst (dry cleaning), and Wildwood (a tannery).

In the Woburn case, the families' attorney tried to show that the wells were contaminated by industrial chemicals that had been dumped illegally by nearby companies. The case revolved around how quickly the different chemicals could travel through the groundwater system to reach the wells that supplied drinking water. Lawyers representing the plaintiffs (the families) and the defendants (the accused industries) called geologists as expert witnesses. What do you think these experts were asked to determine? The geologists tried to figure out how a variety of factors had influenced the flow of chemicals from different sources to the polluted wells. Scientists debated how factors such as the local geology, the distance of the industrial facilities from the wells, and the properties of the chemical products themselves could have influenced the rate at which the chemicals traveled through the groundwater to the wells. It was essential to understand the interactions between the rock and water systems to determine the source of contamination at Woburn and to make successful use of groundwater resources in general.

After more than 2 months of often complex scientific testimony, the Woburn jury found that chemicals from one company, W. R. Grace, could have contaminated the wells. Soon thereafter, Grace settled with the families for $8 million. But the other companies were not off the hook. Several of the surrounding landowners paid nearly $70 million to clean up the 133-hectare (330-acre) site (Figure 12.2). (For more on the Woburn case, consider checking out the movie *A Civil Action*.)

Though most municipalities test their water to ensure it is safe, the contamination in Woburn is not an isolated case; there have probably been similar sets of circumstances in your state (check www.epa.gov/superfund/sites/npl/). In 1980, Congress established the Superfund program to clean up the sites most contaminated by hazardous chemicals and where the original polluter was unknown or had gone out of business. Most of the companies responsible for pollution in Woburn could be readily identified, but some could not. More than 1,500 such sites are listed on Superfund's National Priorities List (NPL). In addition, there are thousands of other locations where the pollution is significant yet not extreme enough to make the NPL. One-quarter of the US population lives within 6 kilometers (4 miles) of an NPL site. Cleanup programs that seek to prevent dangerous levels of contamination from reaching drinking water wells must take into account the surrounding geology.

Groundwater is just one element in the hydrologic cycle (see Figure 11.3), which ties together the processes that cause water to change form as it moves between different components of the earth system. Abundant supplies of groundwater in regions that are today characterized by deserts are indicative of an ancient wetter climate. Population growth and global climate change will impact future groundwater supplies. In this chapter, we will describe creative ways to increase water for domestic supplies such as recycling toilet water. In addition, we will revisit some of the topics covered in previous chapters that discussed rocks and minerals, streams, and weathering processes. We will seek answers to such questions as: Is there groundwater below your feet? What does a naked aquifer look like? How did an effort to provide safe drinking water turn into the biggest mass poisoning in history? When is a swamp a good thing?

Self-Reflection Survey: Section 12.1

Answer the following questions as a means of uncovering what you already know about groundwater and wetlands.

1. Where are the nearest caves, springs, and wetlands to you?
2. How often do you drink bottled water? When at home, are you more likely to drink bottled water or water from the tap? Why do you make that choice?
3. List all the things you use water for in and around your home. In addition, what other ways is water used in your community?

a.

b.

Figure 12.2 Evidence of contamination. **a.** Abandoned 55-gallon (208-liter) drum and **b.** excavation of contaminated soil, wells G and H site, Woburn, Massachusetts.

12.2 Holes in Earth Materials

Learning Objectives

- Define the terms *porosity* and *permeability*, and describe how they might vary as sediment is converted to sedimentary rock.
- Explain how the properties of different rocks control the flow of groundwater.

Although it is impossible to get blood from a stone, we can get water from a rock. In fact, there is approximately 70 times more water below ground than in the lakes and streams on Earth's surface. In dramatic cases, water has flowed beneath the surface for so long that it has dissolved vast volumes of rock to form caves. Although caves can be formed by other geologic processes, the dissolving of limestone by weakly acidic groundwater is the most common method of cave formation (Figure 9.10; see Section 9.3, Chemical Weathering, in Chapter 9 for more on how limestone is dissolved). The world's largest known cavern system, Mammoth Cave in Kentucky, is an example of such a cave (Figure 12.3).

The flow of water through caves has also generated one of the most commonly held misconceptions about groundwater. Although parts of some cave systems contain streams (Figure 12.3b), most groundwater is *not* made up of a network of underground rivers and lakes. Rather, *most* **groundwater is present in billions of tiny spaces between mineral grains or in narrow cracks.** Caves are relatively rare, but groundwater resources are present nearly everywhere across the United States. In almost all cases, if you were to drill down below your present location, you would find groundwater, although sometimes not enough for a steady supply of drinking water. The volume and composition of available groundwater at any given location are strongly linked to the physical characteristics of the materials below Earth's surface, particularly, their *porosity* and *permeability.* In most places, these earth materials will include bedrock and overlying unconsolidated materials that may be variously described as regolith (formed by weathering of underlying bedrock), sediment (weathered material that is transported and deposited), or till (glacial deposits).

Porosity

Where does the milk go when it is poured over a bowl of breakfast cereal? It fills in the gaps between the cornflakes or Lucky Charms cereal or whatever is in your bowl that morning. The same thing happens when water enters rock or sediment. Nearly all earth materials contain fractures or small holes (pores) between grains. Those spaces can hold water. **Porosity is a measure of the proportion of a material that is made up of spaces** and depends on the size and arrangement of the grains and the density of fractures (Figure 12.4). Porosity tends to be highest where materials have been well sorted so that their grains are close to the same size and shape. Sorting occurs during transport and deposition of clastic sediments that are later lithified (see Chapter 7). In rare cases, as much as one-half of the total volume of a rock may consist of pores, making the porosity 50 percent.

Do you think porosity is higher in unconsolidated material or in its rock equivalent? Porosity is typically higher in unconsolidated earth materials, such as sand and gravel, than in the equivalent rocks (sandstone and conglomerate). As sediment goes through lithification to form sedimentary rock, its initial porosity

a.

b.

Figure 12.3 Features of Mammoth Cave, Kentucky. **a.** This wide passageway formed when the limestone bedrock dissolved away. **b.** A small stream, formed by the accumulation of groundwater in cave passages, exits a cave in Mammoth Cave National Park.

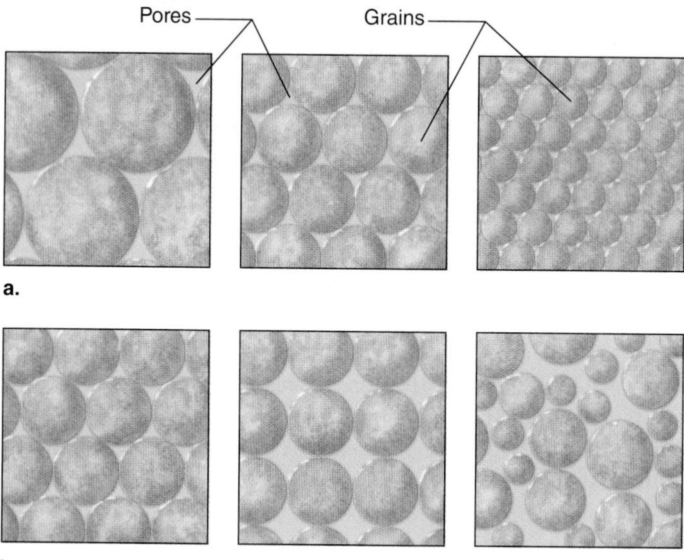

Figure 12.4 Porosity. Different grain sizes and packing arrangements result in different porosity values. **a.** Individual pore spaces decrease in size with decreasing grain size. **b.** Porosity varies with the packing arrangement of the constituent grains.

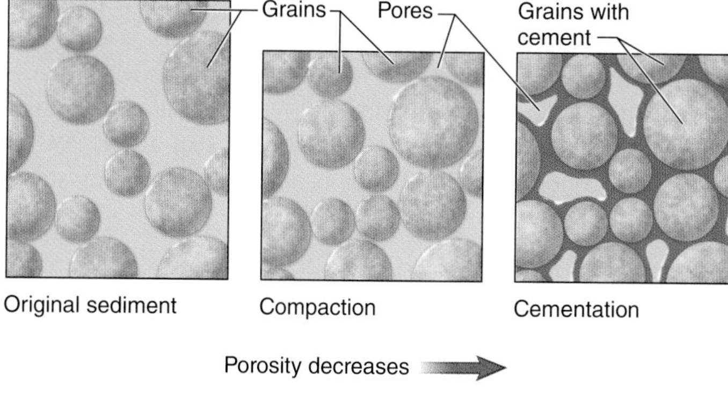

Figure 12.5 Porosity reduction during lithification. The process of converting sediment to rock reduces pore space through compaction, which squeezes the grains closer together, and cementation, which fills in some of the pore spaces.

is reduced (Figure 12.5). First, compaction squeezes grains closer together, collapsing air-filled spaces. Later, some of the remaining spaces are filled with cementing agents that hold the grains of the rock together (cementation). In some cases, porosity may later be enhanced by dissolution in limestone or by fracturing of the rock during faulting and mountain formation.

If you have ever picked a pebble out of a stream, you will have noticed a film of water on the surface of the pebble. Some water adheres (sticks) to the pebble surface. The same is true for grains and groundwater. Some of the groundwater sticks to the grain surfaces. The remainder of the water resides freely in the spaces between grains or in fractures. The volume of **water on the surface of the grains cannot flow through the material and is termed the *specific retention* of the material.** Even if water is allowed to drain from the rock, the water on the surface of the grains will remain. The volume of **groundwater that can drain from rock or sediment is known as the *specific yield*** (or drainable porosity) and is determined by subtracting the specific retention of a material from its porosity:

$$\text{Specific yield} = \text{porosity} - \text{specific retention}$$

Do-It-Yourself Porosity Demonstration. To visualize porosity, fill a 1,000-milliliter beaker with dry sand and pour 500 milliliters of water into another beaker (Figure 12.6a). Slowly pour the water into the sand-filled beaker. Allow the water to move through the sand. Watch the sand change color as it becomes wet. Continue to pour the water until all the sand is completely saturated and water rises to the surface of the sand. The water level in the second beaker will have declined by an amount equivalent to

Checkpoint 12.1

basic | intermediate | advanced | superior

Imagine that you have a box that is 15 centimeters (6 inches) square. You also have a never-ending supply of rectangular sticks measuring 15 × 3 × 3 centimeters (6 × 1.2 × 1.2 inches). How could you arrange the sticks in the box to model a system with very low porosity (~0 percent) and high porosity (~40 percent)?

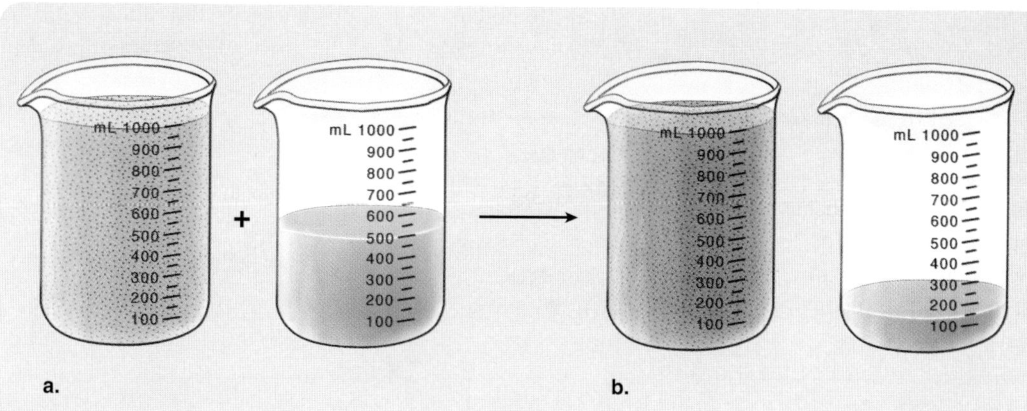

Figure 12.6 Do-it-yourself porosity demonstration. **a.** Beakers are filled with 1,000 milliliters of sand (left) and 500 milliliters of water (right). Then the water is poured into the sand. **b.** The sand-filled beaker now contains an additional 350 milliliters of water. Water filled the original pore spaces, and 150 milliliters of water is left in the other beaker.

the porosity of the sand. In our experiment, the water level declined from 500 milliliters to about 150 milliliters (Figure 12.6b). This indicates that 350 milliliters of water filled the pore spaces between sand grains. Therefore, 350 milliliters of the original 1,000-milliliter sand volume was made up of connected air-filled pore spaces (porosity = 35 percent). Can you suggest another similar type of bench-top experiment that would provide some additional data on porosity?

How could you determine the specific yield of the sand? If specific retention is very low, as is the case in a coarse-grained material such as gravel, specific yield is nearly equal to porosity because the surface film takes up a small proportion of the pore space. In contrast, yield is diminished when specific retention is high in fine-grained materials such as silt or clay because the surface films represent most of the water in the small pores between many grains. Consequently, **the specific yield of fine-grained materials is low even though their porosity may be high.** It turns out that the specific retention of even fine sand is similar to that of gravel so that the specific yield of sand is just a few points less than its total porosity. In contrast, clay can have a porosity approaching 50 percent, but almost all of that is accounted for by its specific retention, leaving just a few percent for specific yield.

Igneous and metamorphic rocks typically have low porosity because their grains grow together during rock formation, leaving no gaps. However, **the presence of fractures in some rocks produces properties suitable for the storage and flow of groundwater.** For example, thousands of meters of ancient lava flows are preserved as the Columbia plateau basalts in the northwestern United States (see the discussion of lava plateaus in Section 6.6 of Chapter 6). The tops of the lava flows were heavily fractured as the lava cooled during formation, and the bases of the adjacent flows often contain abundant gas bubbles. The fractures, and cavities left by the gas bubbles, create zones of high porosity that can yield more than 100 gallons (378.5 liters) of water per second. (For comparison, water flows from your kitchen faucet at 1 to 2 gallons [3.8 to 7.6 liters] per minute.)

Permeability

Think again about the porosity experiment we just described. Water was poured into the sand and passed easily through pore spaces between grains. Suppose you mixed cement with the sand and let the mixture dry. Would the water flow easily through the material? **Permeability is the capacity of water to flow through earth materials.** Water can flow readily through rocks that have connections (pathways) between the pore spaces or have many fractures. In contrast, sediments or rocks with fewer connected spaces have little or no permeability (impermeable). High permeability often goes hand in hand with high porosity and large grain size in unconsolidated materials and clastic sedimentary rocks. Permeability is also high in limestone where rock has been dissolved away and in all types of rocks that contain many fractures. High porosity on its own is no guarantee of high permeability. Connections between pore spaces are narrow in fine-grained materials (clay). Consequently, while clay and other fine-grained materials may have high porosity, the pore spaces are often not connected, and therefore water cannot flow through the material.

What can reduce permeability? Permeability is reduced by the same processes that reduce porosity: compaction and cementation, the processes that close the connections between pores. The surface water films (specific retention) in clay and other fine-grained materials may fill the narrow connections between pore spaces, blocking the passage of groundwater. For this reason clay is often used as a liner in ponds to keep water from seeping out into more permeable surrounding soils.

Groundwater Flow Rates. Groundwater flows much more slowly than the water in streams. While the water in streams typically travels 1 meter (3.3 feet) in a few seconds, **rapid groundwater flow in sand and gravel deposits is typically around 10 meters (33 feet) per day, and slow flow is in the range of 1 to 10 centimeters (up to 4 inches) per day.**

Why does groundwater flow more slowly than water above ground? In Chapter 11, we explained that stream friction increases with increasing channel roughness and results in a decrease in average flow velocity. This situation is magnified for groundwater

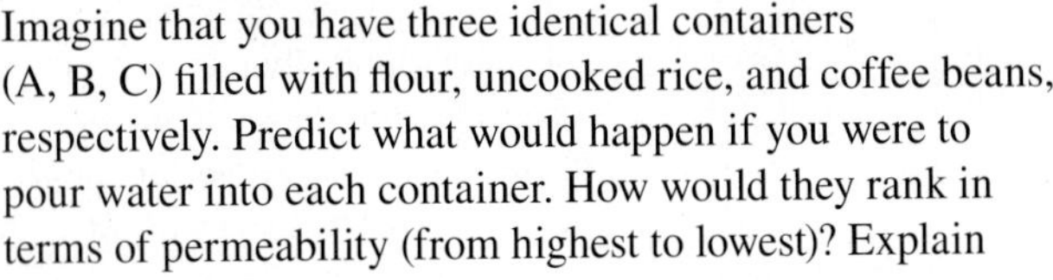

Imagine that you have three identical containers (A, B, C) filled with flour, uncooked rice, and coffee beans, respectively. Predict what would happen if you were to pour water into each container. How would they rank in terms of permeability (from highest to lowest)? Explain your thinking.

A large volume of liquid waste was dumped on the ground at Otis Air Force Base, Massachusetts. Both the base and the nearby city of Falmouth are located over the same deposit of sand and gravel. Examine the following diagram and predict the approximate length of time before the waste would begin to show up in the drinking water wells of the city of Falmouth, assuming a groundwater flow rate of 0.5 meter (1.6 feet) per day.

a) 10 months
b) 8 years
c) 16 years
d) 40 years

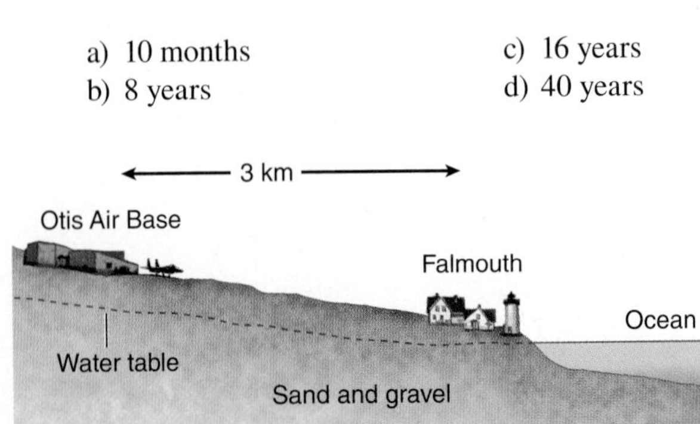

Checkpoint 12.4

basic intermediate advanced superior

Your uncle learns that you are taking an earth science class and shows up at your house with three samples of sediment that he dug up from three different locations on his property. He tells you he is going to drill a well on the property and wants you to help him decide which location would make the best site for the well. How would you analyze the samples to help him pick the best site?

because the water must push through tiny spaces between grains, providing millions of opportunities for friction to slow down flow. This also explains why groundwater flows more rapidly through fractures and more slowly through pore spaces.

The slow velocity of groundwater flow can sometimes be a benefit. For example, if pollutants are discovered, there may be time to take corrective action, such as shutting down drinking water wells, before the contamination reaches underground water supplies.

12.3 Groundwater Systems

Learning Objectives

- Describe the principal characteristics of an aquifer, and identify the materials that make up most US aquifers.
- Explain how streams and groundwater are part of the same system in open aquifers.
- Draw and label a sketch to compare and contrast open and closed aquifers.
- Discuss how water flows into and out of aquifers.
- Describe some of the consequences of the overuse of groundwater.
- Evaluate the best and worst conditions for the development of a groundwater resource.

Water has been found in wells as deep as 9 kilometers (over 5 miles) into Earth's crust, but most usable fresh groundwater is relatively shallow. Unlike oil wells that are often thousands of meters deep, water wells tend to be drilled down less than 100 meters (330 feet) in unconsolidated materials or may be up to several hundred meters deep in bedrock. Deeper waters are more expensive to retrieve and are more likely to contain high concentrations of dissolved minerals.

What types of earth materials make the best sources of groundwater? To identify the best place to drill a well, we need to know the important physical characteristics of groundwater systems (see Chapter Snapshot). Groundwater is stored in bodies of rock or sediment called aquifers. **An aquifer is composed of sufficient saturated permeable material to yield significant quantities of water.** There is nothing particularly mysterious about the rock or sediment that is found in aquifers. The same materials that make up an aquifer can be found in "naked aquifers," sand and gravel deposits and rock outcrops at the surface (Figure 12.7).

The best aquifers hold large amounts of water (high porosity) and allow water to flow freely from one location to another (high permeability). In such aquifers, we can pump out the water we need, and nearby groundwater will flow in to quickly recharge (refill) the aquifer. The most productive aquifers are typically found

a.

b.

c.

Figure 12.7 Naked aquifers. Such aboveground formations are representative of underground aquifers, which can be composed of a variety of materials, including **a.** sand and gravel, **b.** sandstone with good porosity and permeability, and **c.** fractured rocks.

in unconsolidated earth materials. Aquifers composed of sand and gravel (Figure 12.8a) account for approximately 80 percent of all groundwater withdrawals in the United States. Dissolved openings in limestone, fractures in the volcanic rock basalt, and pore spaces in sandstone account for the high permeability of other common aquifer systems (Figure 12.8b, c, d, e). Thick rock units that have lower permeability can be equally productive as a thinner aquifer with high permeability. Many locations have more than one available aquifer. A stacked sequence of multiple aquifers is often interlayered with materials with low permeability.

A low-permeability material such as clay, shale, or an unfractured igneous or metamorphic rock can act as a barrier to groundwater flow and is known as an *aquitard*. The properties of the materials that make up an aquifer or an aquitard may change from place to place so that flow within aquifers may be compartmentalized. As a result, some parts of a producing aquifer may become depleted, while other nearby parts of an aquifer can continue to produce water.

The first half of this section describes the natural flow of water in groundwater systems, and the second half considers how natural systems are altered by use.

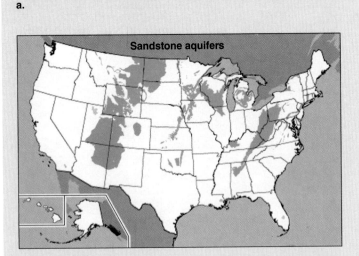

Sand and gravel aquifers

Unconsolidated

Semi-consolidated

a.

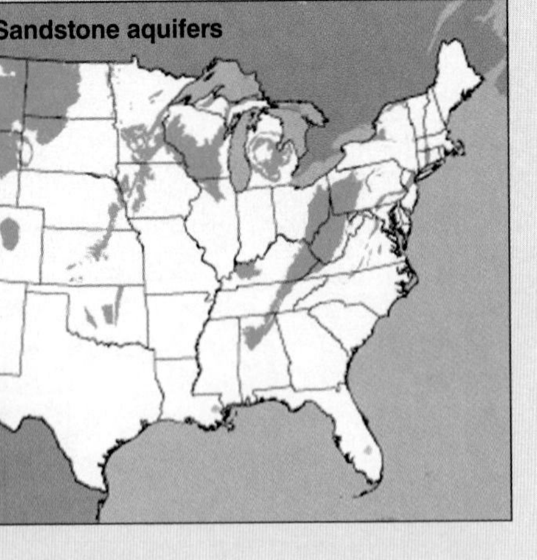

b.

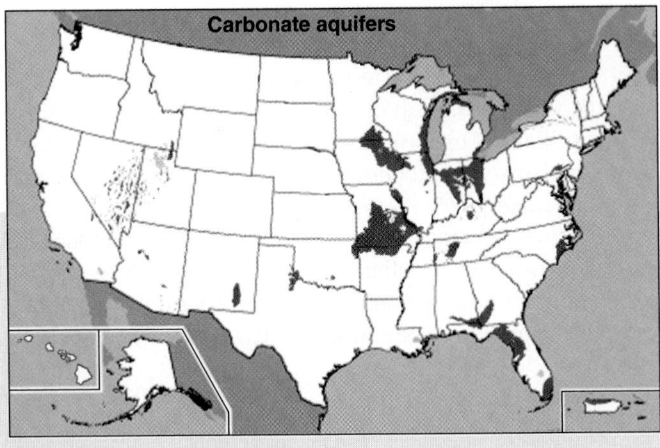

c.

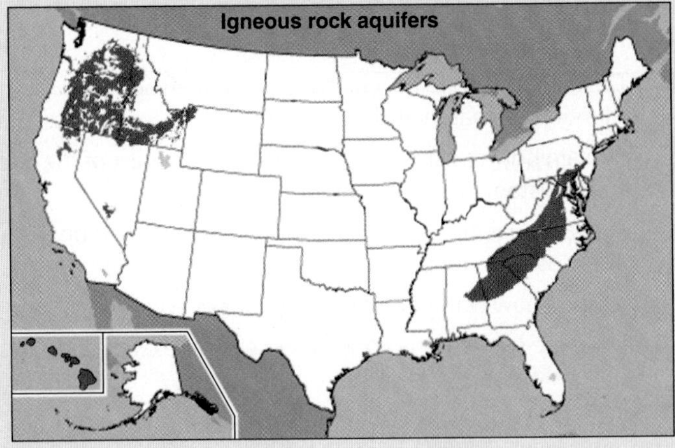

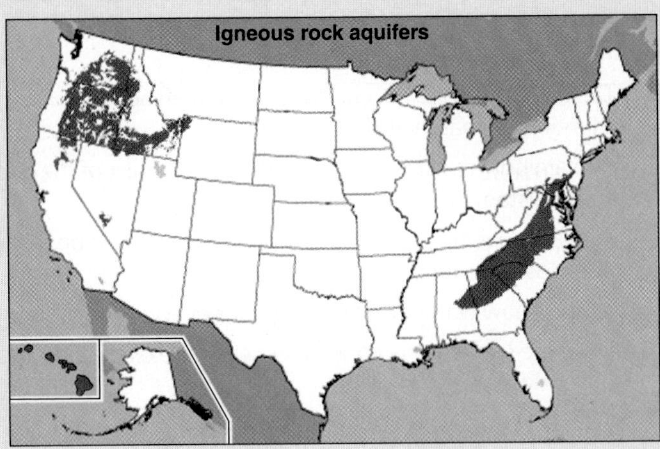

d.

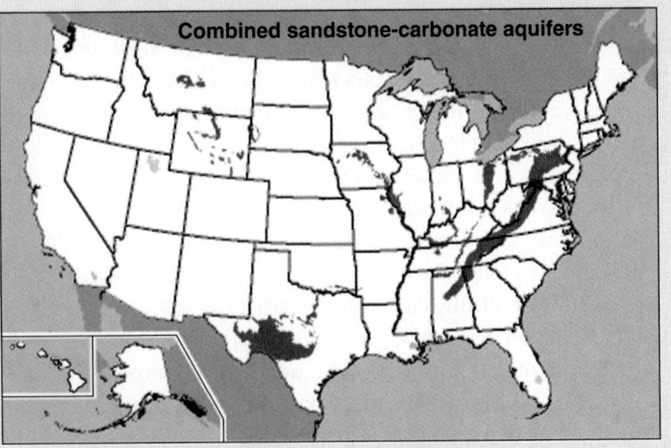

e.

Figure 12.8 Common surface aquifer materials in the United States. The maps show where the major aquifer materials make up the shallowest principal aquifer. Only small areas of some aquifers are illustrated, as they may be covered by other shallower aquifer systems. Other, more productive aquifers may lie below those shown, while some local aquifers are too small to appear on these maps. Most carbonate aquifers are composed of limestone. What types of aquifers are present in your state?

Aquifers

Aquifers may be open to direct recharge from above (open aquifers) or closed to direct recharge (confined or artesian aquifers).

Open (Unconfined) Aquifers. **In an open aquifer, water infiltrates through overlying rock or earth materials directly into the aquifer below.** The water passes downward until it reaches the saturated zone of the aquifer, where all the pore spaces are filled with water. The top of the saturated zone is called the *water table* (Figure 12.9), because in areas of low relief, it is relatively flat over distances of hundreds of meters.

In general, the **water table follows the shape of the land surface**—for example, it is higher under hills and lower in valleys. Groundwater flows down the slope of the water table (called the hydraulic gradient) from areas where the water table is high to areas where it is low (Figure 12.10). The actual flow paths may vary from straight lines to long curves, depending on the geology (for example, rock type, fractures). When the water table intersects the land surface, a stream, lake, or spring will occur. As the depth of the water table fluctuates over time, water levels in these associated surface features will also change.

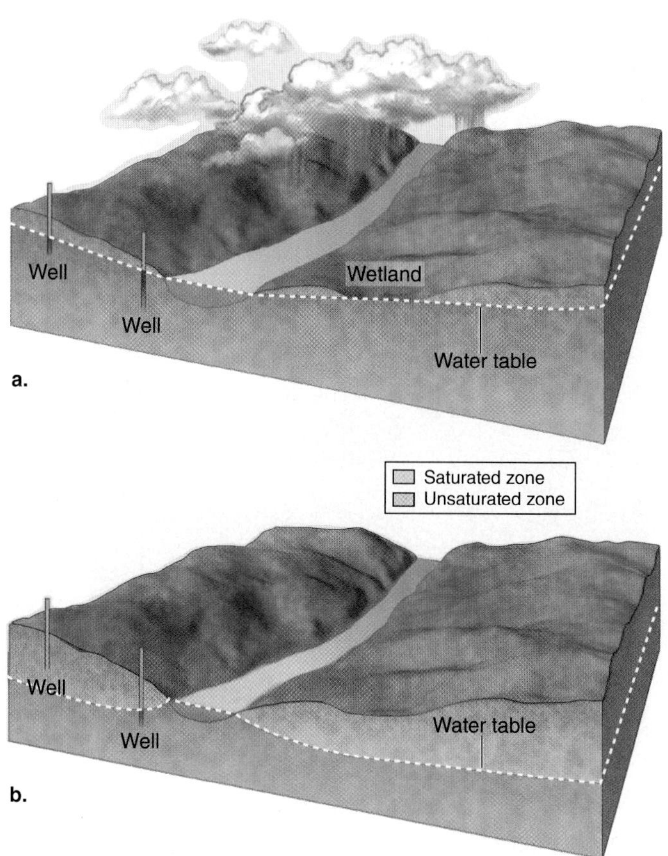

Figure 12.9 An open aquifer. Water levels in an open aquifer vary during periods of wet and dry climate. **a.** Wetlands and springs are present where the water table lies at the ground surface. **b.** Some wells may become dry if they do not penetrate far enough into the saturated zone of the aquifer.

Checkpoint 12.5

basic | intermediate | advanced | superior

The following cross section simplifies the groundwater sources in a county in a midwestern state. Which location would have the potential for the best groundwater production?

a) Location A
b) Location B
c) Location C
d) Location D

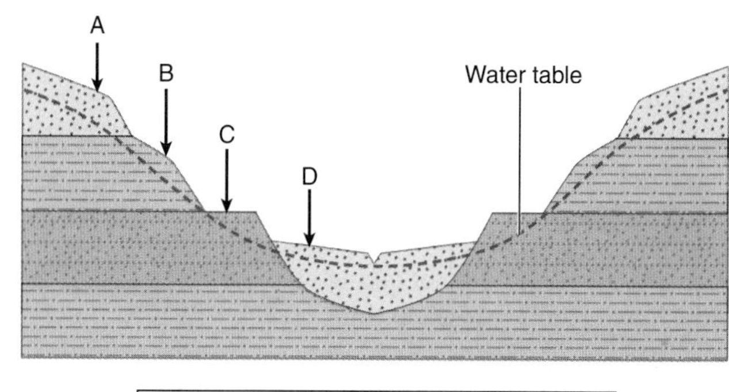

Liquid hazardous waste is disposed of by pumping it down injection wells. On the following diagram, which well location would be most suitable for use as an injection well?

a) Well A
b) Well B
c) Well C

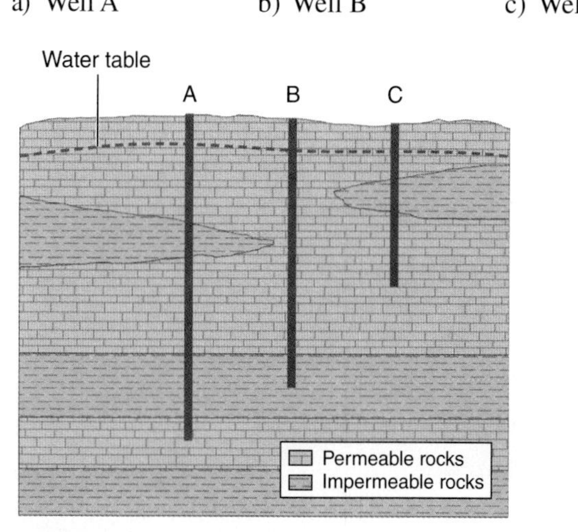

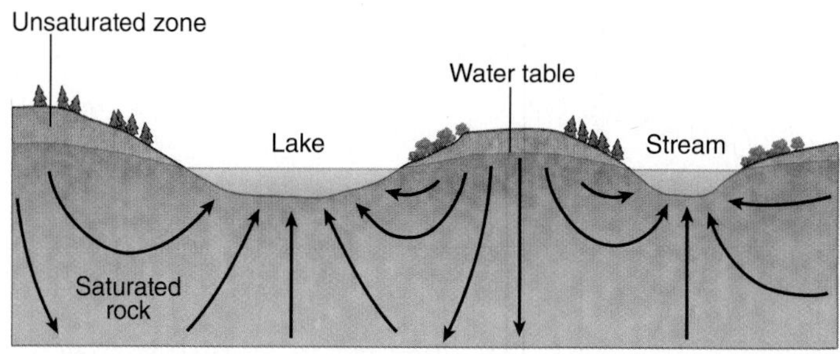

Figure 12.10 Groundwater flow paths. Groundwater flows through rock down the hydraulic gradient (slope) from locations where the water table is at high elevations to places where it has low elevation. The shape of the flow path varies, depending on the local geologic characteristics.

Why does the depth of the water table change? The water table drops during prolonged dry periods and rises again when precipitation is plentiful. Therefore, wells must be drilled far enough into the saturated zone to ensure a year-round supply of water. If a well is not drilled deeply enough, it may stop producing water in periods of drought. On the other hand, money is wasted if the well is drilled too deeply (beyond where the well could conceivably ever go dry).

Confined (Artesian) Aquifers. **A confined aquifer is enclosed above and below by impermeable materials** that prevent water from infiltrating directly down into the top of the aquifer or from flowing out the base of the aquifer. Because the surrounding rocks are aquitards, water can enter the aquifer only if the rock layer is exposed at higher elevations, called the *recharge zone* (Figure 12.11). The water flows down into the aquifer layer, partially filling it up.

Checkpoint 12.6

basic intermediate advanced superior

A ruling in a recent court case between Kansas and Nebraska required farmers in southern Nebraska to pump less water from wells in order to maintain water levels in the Republican River as it flowed south to Kansas (see map). The two states have long had an agreement that prevents Nebraska from drawing too much water from the river because Kansas farmers need to use the river water for irrigation. Nebraska authorities claimed that they were not taking water from the river but from underground aquifers and that this practice is not prohibited by the agreement. Explain why the court ruled that groundwater should be considered part of the surface water system. Draw a sketch to show how the consumption of groundwater would have changed water levels in the river.

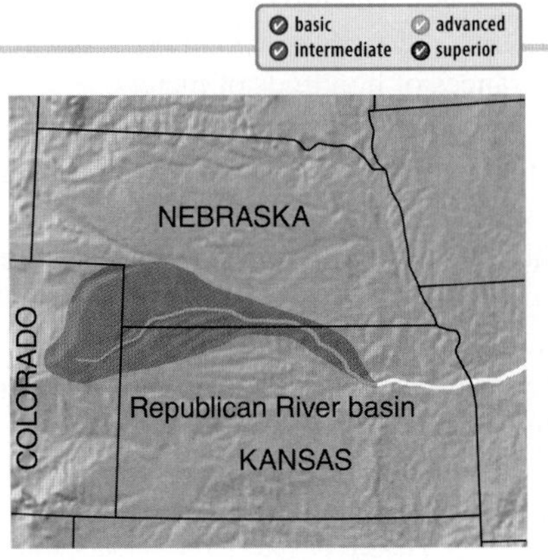

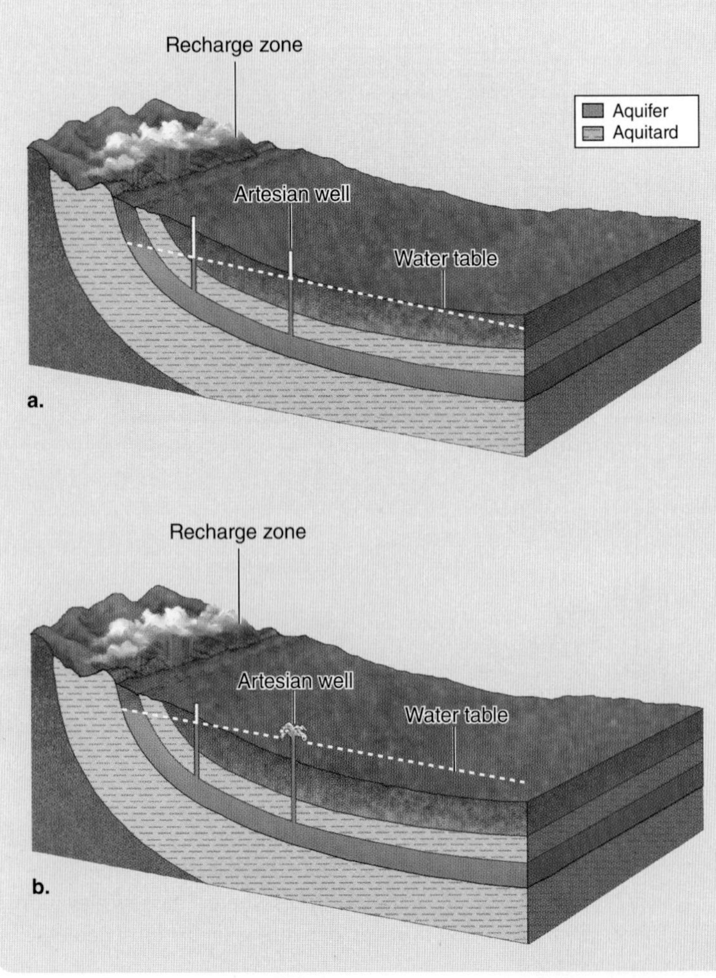

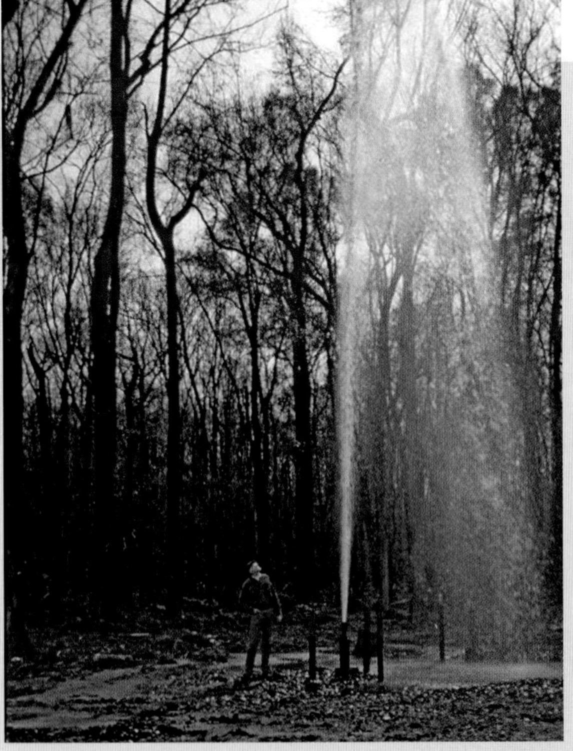

Figure 12.11 A confined aquifer. Changing water levels in a confined aquifer can affect wells drilled into the aquifer. Pressure from the water higher in the aquifer can cause the water table to rise **a.** above the level of the aquifer in a specific well or **b.** all the way to the surface without pumping. **c.** An artesian well shoots water far above the land surface in South Dakota in the early 1900s.

Have you ever let air out of a tire (or had a flat)? Why does the air escape? It escapes through a hole because the air inside the tire is at a higher pressure than the outside air. If geologic conditions are right, water is actually forced out of wells drilled in confined aquifers because that water is under pressure. How is this possible beneath the ground? The weight of the overlying water column in the confined aquifer increases pressure deeper in the aquifer (just as happens to your eardrum when you dive into a pool). That pressure on the aquifer is sufficient to cause the level of water in a well to rise above the depth of the aquifer. Water in these *artesian* wells can rise high enough above the aquifer to reach the surface under pressure and flow without pumping (Figure 12.11).

In the early 1900s, settlers on the plains of South Dakota observed water gushing to 30 meters (100 feet) above the ground from artesian wells drilled into a confined aquifer composed of the Dakota Sandstone (Figure 12.11c). The aquifer was recharged by rainfall in mountains to the west. More than 10,000 wells had been drilled into the aquifer by 1915. Inevitably, as water was withdrawn from the aquifer, the pressure dropped, requiring many of the wells to add pumps to bring water to the surface. ("Artesian" bottled water comes from a confined aquifer.)

Recharge zones do not have to be located in mountainous regions. The Floridian aquifer represents the largest limestone aquifer in the United States and yields 4 billion gallons (15 billion liters) of water each day to supply agriculture and cities throughout Florida. The aquifer is made up of a sequence of limestone and other rocks that are over 900 meters (2,950 feet) thick in places. The uppermost layers in the aquifer tilt gently southward and can be traced from outcrops in northern Florida to depths of more than 300 meters (980 feet) below the southern tip of the state. Recharge occurs where the aquifer rocks come to the surface in Florida and neighboring Georgia and in places where fractures or sinkholes allow water to pass downward through the overlying confining layer.

Natural Groundwater Budget: Inflow Versus Outflow

What would happen to a city or town whose main aquifer stopped producing water? Could that even happen? The answer is yes, and such questions are currently being asked about some of the largest cities in the southwestern United States. To understand these issues, you must understand the water budget of groundwater systems—the balance between inflow (recharge) and outflow (discharge).

Aquifers are usually recharged when precipitation infiltrates into the ground or when water from streams moves into the groundwater system. But there is one other source for groundwater; it may already be present in aquifers, left over from a wetter time in the past. Aquifers in presently dry regions may have stored large volumes of water during previous periods of especially wet climates.

Not only do humans extract water from aquifers, but water eventually leaves the groundwater system at discharge points such as streams, springs or wetlands, and the ocean.

Recharge Through Infiltration. Rainfall infiltrates through permeable earth materials to replenish open aquifers or through recharge zones for closed aquifers. The volume of water that enters an aquifer depends on how and where precipitation occurs and the properties of the aquifer. Steady rainfall replenishes an aquifer more than torrential storms because it takes time for water to infiltrate. If the ground becomes saturated during a storm, water runs off to streams instead of flowing below ground. Likewise, rain falling on farm fields or natural lands is more likely to infiltrate below ground than rain falling in urban areas (why?). Seasonally, relatively small changes take place in the elevation of the water table. The water table can drop significantly during prolonged droughts or rise during periods of long-term (months) precipitation.

Recharge Through Streams. In areas with dry climates, the water table is often far below the ground surface. Streams in dry regions may lose water to groundwater through high-permeability materials that make up the stream banks and streambed. Such streams are termed *losing streams* (Figure 12.12a, Chapter Snapshot). For example, water is lost from unlined canals and surface reservoirs built to supply irrigation waters from the Colorado River in the desert Southwest. Water may flow directly from the stream into the aquifer's saturated zone or may infiltrate down through the unsaturated zone in a disconnected losing stream that holds water temporarily due to infrequent precipitation (Figure 12.12b).

Checkpoint 12.7

basic / intermediate / advanced / superior

Draw a labeled diagram that summarizes how water flows into and out of an open aquifer.

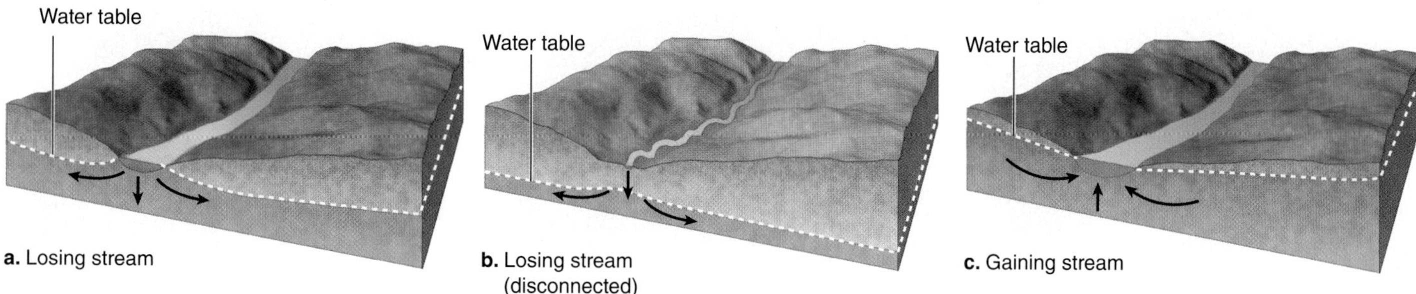

Figure 12.12 Losing and gaining streams. **a.** Losing streams lose water to the saturated zone of an underlying aquifer or **b.** to the unsaturated zone in dry regions. **c.** Water enters gaining streams from the saturated zone of an adjoining aquifer.

Stored Groundwater. The climate of North America is warmer and drier today than it was prior to the end of an interval known as the *last glacial maximum* (ended 12,000 years ago), when much of the northern half of the continent was covered in a thick sheet of ice. A great deal of the water in North American aquifers today began as precipitation during the wetter climate of the glacial maximum. In much of the western United States, there is relatively little present-day inflow of water from precipitation or streams. Since those ice sheets are now gone, the water in many western aquifers represents a finite resource that cannot be replaced once used.

Discharge Through Streams. Groundwater may flow into streams in areas with relatively high water tables. These streams are termed *gaining streams* (Figure 12.12c). Depending on the region, groundwater can account for much of the normal daily flow of a stream. For example, the Sturgeon River, Michigan, flows over permeable sands and gravels, and receives approximately 90 percent of its minimum discharge from groundwater (Figure 12.13). One benefit of this arrangement is that the river rarely runs dry, even during prolonged periods of low rainfall. In contrast, streams with no input from groundwater may experience periods of low flow conditions (Figure 12.13b).

Discharge Through Wetlands and Springs. Water may flow out at the ground surface from **a spring or wetland located where the water table intersects the ground surface.** Springs may form in a single location where fracture systems or cave systems (Figure 12.14a, b) bring water to the ground surface to form

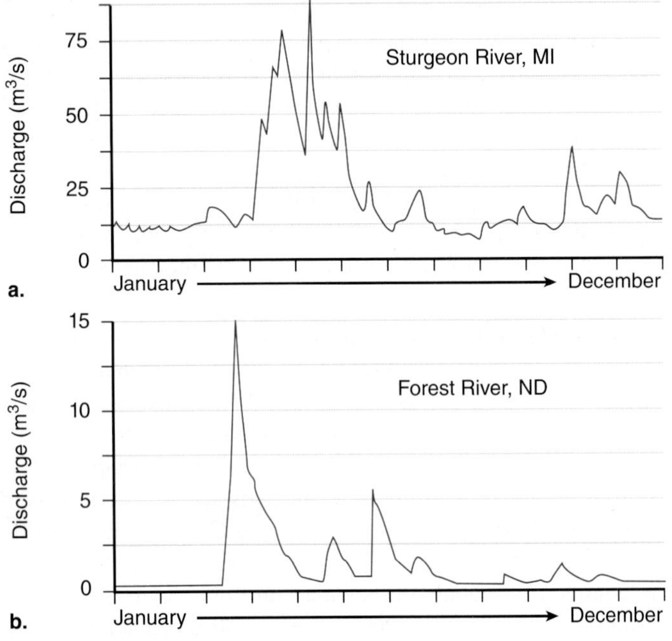

Figure 12.13 Influence of groundwater on stream flow. **a.** Base flow in Sturgeon River does not approach zero, as the river receives much of its discharge from groundwater, making it a gaining stream. **b.** The Forest River flows over less permeable silt and clays, and it receives little of its flow from groundwater sources. The stream discharge is reduced during periods of low precipitation. In what season do both streams have their maximum discharge?

Checkpoint 12.8

basic · intermediate · advanced · superior

Student Answer Evaluation Exercise

In response to the question "Is there groundwater below lakes?" a student wrote the following response. The instructor gave the answer a score of 7 out of 10 and wrote the following comment on the assignment: "You have clearly understood some of the concepts related to this topic, but you have not yet pulled all the key ideas together. Think a little more about the difference between surface water and groundwater systems."

Review the student's answer, and write a more complete response.

Is there groundwater below lakes?
No, there should not be any groundwater below lakes. Groundwater is present in rocks with good porosity and permeability. If rocks under a lake had good porosity and permeability, then the water from the lake would drain out into the groundwater and the lake would be gone. But if the rock under the lake had bad permeability and porosity, then water could not move into it and the lake would not lose water.

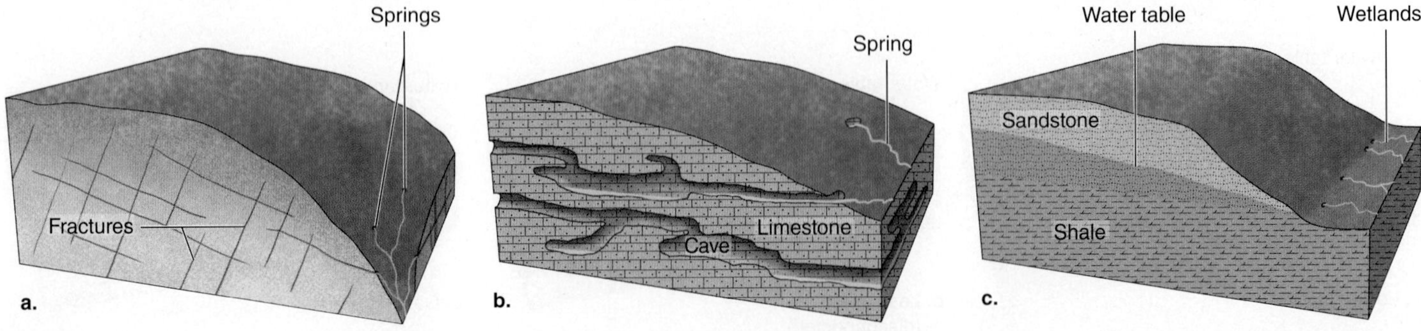

Figure 12.14 Groundwater reaches the land surface at springs and wetlands. Springs form where **a.** fractures or **b.** cave systems intersect the land surface. **c.** Wetlands may form where several small springs distribute water over a region underlain by a low-permeability material such as clay or shale.

a stream. Alternatively, many small springs may form in multiple locations where a permeable layer overlies an aquitard, so that surface flow is dispersed over a wide area to form a wetland (Figure 12.14c).

Discharge to the Ocean. Saltwater along the coast will infiltrate the subsurface just as freshwater does (12.15). Consequently, a layer of fresh groundwater (called a *lens* because of its shape) is typically found floating above denser saltwater in coastal regions. Groundwater flows into the ocean where the lens of freshwater meets the coast. Coastal cities can extract water from the upper part of the freshwater lens. However if the freshwater is pumped faster than it is naturally replenished, saltwater may enter water wells and pollute the water supply, creating a situation known as *saltwater intrusion* (Figure 12.15b). Saltwater intrusion has been a problem for communities in Long Island, New York, and along the southeast coast of Florida. Consequently, many coastal urban communities rely on surface sources of freshwater (reservoirs, streams, and lakes).

Consequences of Human Actions

Groundwater systems were operating for millions of years before humans got involved. Water levels rise during wetter periods and fall during droughts. The rapid growth in human populations and the development of technology to extract large volumes of groundwater place greater stress on underground water sources. In some locations, the groundwater cannot be replenished fast enough to replace water consumed for human activities. This produces a situation known as **groundwater overdraft, where groundwater extraction occurs more rapidly than recharge.** In such circumstances, water is essentially a nonrenewable resource; in fact, some suggest that water has the potential to become the "new oil"! However, some ingenious solutions are also on offer to replenish groundwater supplies. One of these involves the reuse of water that is flushed down toilets in Orange County, California, before being reprocessed and coming back as drinking water. More than 2 million residents will be supplied by this water from the world's largest water purification project. Despite the apparent "yuck factor," the resulting drinking water is of better quality than water flowing from taps in many cities. The domestic wastewater goes through a series of cleaning treatment steps in which it is purified and filtered before being pumped into lakes. The water then infiltrates into groundwater through the permeable sediment and rock at the bottom of the lakes. Finally, years later the groundwater is pumped back to the surface and into homes throughout Orange County.

Two negative consequences of the overuse of groundwater are local changes in the water table and a drop in the land surface as grains in the aquifer collapse when water is removed.

Changes in the Water Table. Although aquifers can store vast volumes of water, the physical flow of water to a well can limit the volume of water available. Extracting groundwater from an aquifer is a little like trying to suck up spilled soda from a table using a straw. Ever tried it? You get the soda that was close to the straw, but most of the drink remains on the table surface (Figure 12.16). As

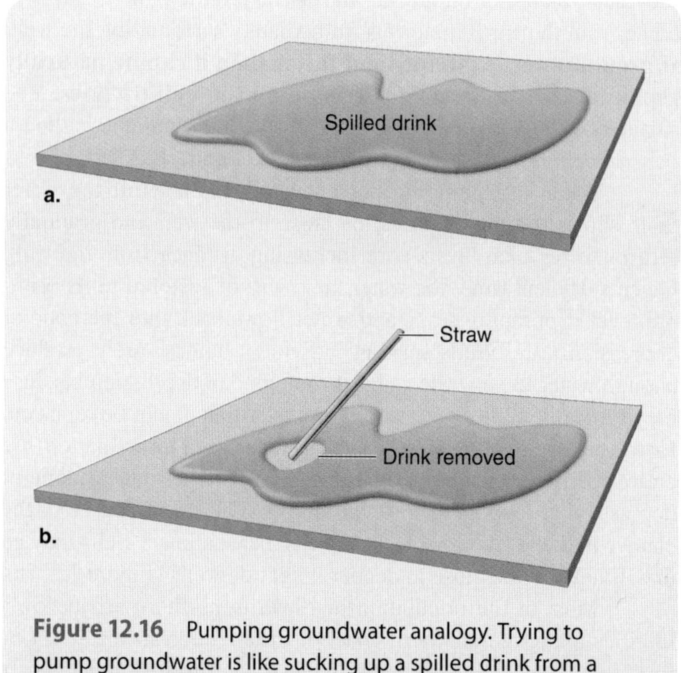

Figure 12.16 Pumping groundwater analogy. Trying to pump groundwater is like sucking up a spilled drink from a table. No matter how big a straw you use, most of the drink stays on the tabletop.

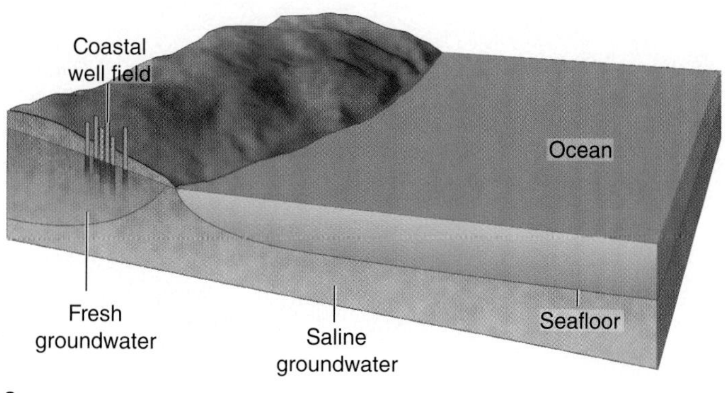

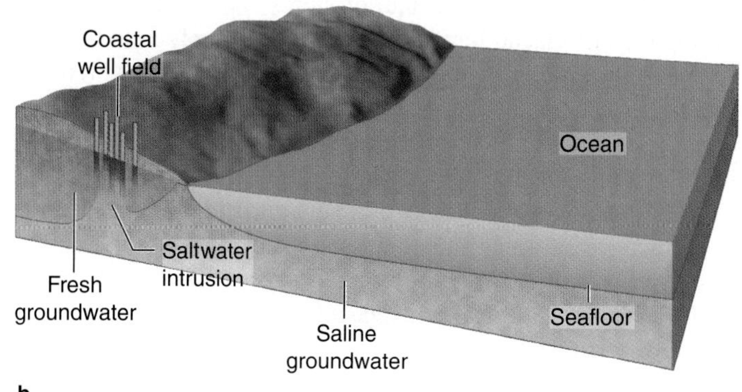

Figure 12.15 Saltwater intrusion. **a.** A layer of fresh groundwater above the layer of saltwater can be pumped and used for domestic water supply. **b.** However, if the freshwater is pumped out faster than it can be replaced, saltwater may intrude and pollute the water supply.

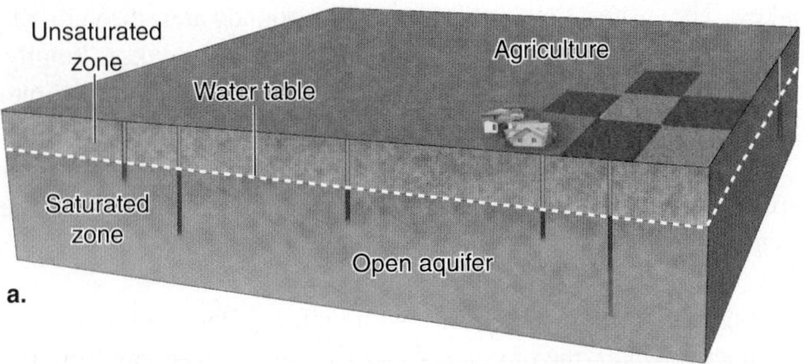

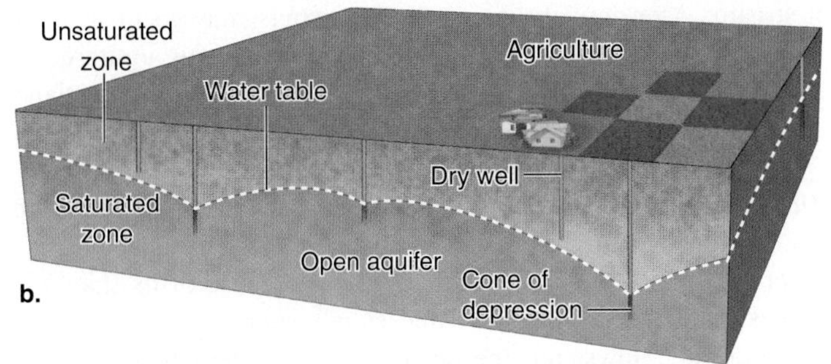

Figure 12.17 Cone of depression. When the water table **a.** is drawn down by overpumping of a well, a cone of depression forms **b.** around the heavily pumped well. Smaller wells located near the cone of depression may lose their water supply. Note that a considerable volume of groundwater remains beyond the reach of the most productive well.

previously stated, groundwater moves very slowly, and it can only come from the earth materials immediately surrounding the well. **If groundwater is pumped out faster than it can be naturally replaced, the water table surrounding the well declines.** The surface of the depleted water table forms a cone shape, called a *cone of depression,* surrounding the well (Figure 12.17).

When overpumping has occurred, the level of the water table shows the greatest decline close to the well and gradually returns to regional levels with increasing distance from the well. Given sufficient time, the water table will be restored to its original level if pumping stops and water flows back into the cone of depression area. Wells serving individual homes rarely produce enough water to generate a significant cone of depression because they typically do not take water out faster than it can be replaced. However, large-volume wells, such as those used for irrigation, are entirely different. These wells may develop a permanent cone of depression as long as they remain in use. One consequence of the change in the level of the water table is that shallower wells may go dry if they are too close to deeper irrigation wells (Figure 12.17b).

Most of the population of Libya in northern Africa lives in cities along the southern coast of the Mediterranean Sea. Unfortunately, excessive groundwater use has resulted in saltwater intrusion into drinking water wells over large areas of the coast. Without the discovery of a giant aquifer up to 4,000 meters (13,000 feet) thick under the desert, Libya's citizens would have been forced to migrate. The aquifer, called the Nubian Sandstone aquifer, extends into the neighboring nations of Egypt, Chad, and Sudan (Figure 12.18). The aquifer stores water that fell as rain during the last glacial maximum. The water supply in the Libyan sector of the aquifer is equivalent to several times the volume of the North American Great Lakes. However, much of this water cannot be used because two widespread cones of depression have developed around the source wells. These cones of depression formed when Libya drilled two great well fields into the aquifer to extract groundwater and pump it 1,600 kilometers (990 miles) to the north to supply its major population centers. The two cones of depression are gradually increasing in size as great volumes of water are extracted and almost none is replaced. Given the huge volume of water available in the aquifer, it will be several decades before water supplies begin to diminish, but much of the water will remain untapped unless Libya establishes more wells in other locations in the desert.

Checkpoint 12.9

Two wells (A and B) are drilled in rocks that have the same porosity, but the rocks around well A have a higher permeability than those around well B. Suppose both wells are pumped at the same rate. Which statement is true?

a) Well A will have a larger cone of depression.
b) Well B will have a larger cone of depression.
c) The cone of depression will be the same for both wells.

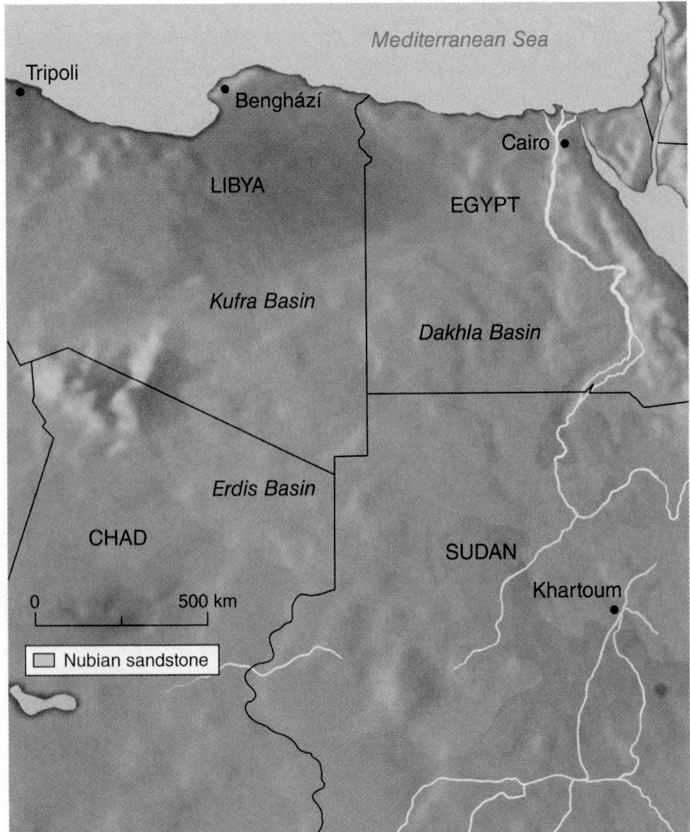

Figure 12.18 The Nubian Sandstone aquifer. The aquifer covers a region that includes parts of Egypt, Libya, Chad, and Sudan.

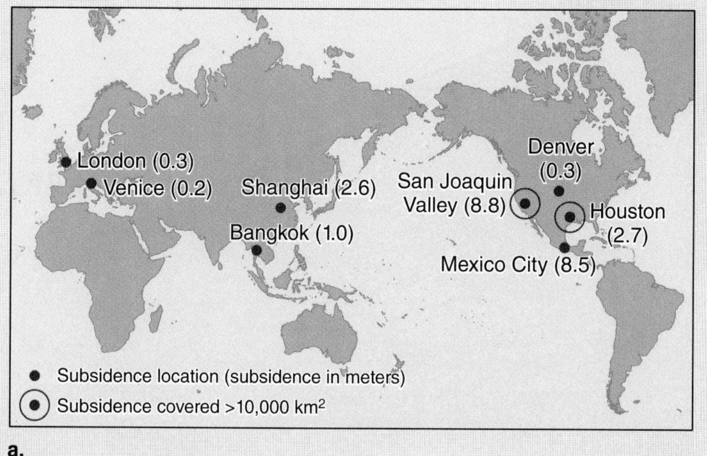

Figure 12.19 Subsidence due to groundwater withdrawal. **a.** Some of the world's largest cities have experienced subsidence. For example, over 12,000 square kilometers (4,630 square miles) were affected in Houston, making some subdivisions vulnerable to subsequent flooding. **b.** Groundwater withdrawal caused more than 10 meters (over 30 feet) of subsidence in some parts of the San Joaquin Valley since the 1920s.

Ground Subsidence. What happens to a sponge left to dry on your kitchen sink? It shrinks, making the pores smaller. A similar thing happens underground when groundwater is rapidly removed from an aquifer made of loose earth materials such as sand or gravel. **When groundwater is removed from unconsolidated material, the grains that make up the sediment may compact as the pore spaces that once held water close.** But, unlike the sponge, once sediment pore spaces collapse, they do not increase in size again when more water is present. Normally, the weight of the overlying material (including structures such as roads and buildings) is supported by both the mineral grains and the water in the pore spaces in the aquifer. Pressure from water on the surrounding grains keeps the pore spaces open. When the water is extracted, mineral grains may collapse inward on the empty pore spaces. This causes a decrease in the volume of the underlying sediment and can result in *subsidence* (a drop) of the ground surface.

Ground subsidence due to groundwater depletion is an international problem (Figure 12.19a). Subsidence induced by groundwater withdrawal from thick sand and gravel deposits has occurred at rates of about 10 centimeters per year (4 inches per year) over an area of approximately 13,000 square kilometers (5,000 square miles) in the San Joaquin Valley, California (Figure 12.19b). Much of that water was used for irrigation and subsequently lost to evaporation or runoff.

Cities built over weak, unconsolidated, fine-grained sediments associated with environments such as floodplains, deltas, or lake beds often show evidence of subsidence. Roads, pipelines, and large buildings may fracture or collapse as a growing city's population extracts more and more groundwater.

Checkpoint 12.10

basic / intermediate / advanced / superior

Who Owns the Groundwater?

Bob, a rancher in Arizona, drilled a 50-meter-deep (164-foot) well that readily supplied all the water his family needed for 10 years. Then his new neighbor, Glen, drilled a network of new and deeper (150-meter-deep; 492-foot-deep) wells to irrigate a cotton crop in response to a jump in cotton prices. Within 6 months, Bob's well had gone dry.

Bob took Glen to court, arguing that rocks stay but water moves—essentially, that a landowner cannot own the groundwater any more than he or she can own the air above the land. Glen's lawyer disagreed, claiming that Glen owned everything under the ground surface of his ranch and could pump as much groundwater as his crops needed as long as the well was on his property.

Draw a diagram to illustrate why Bob's well went dry. How could Bob ensure a continuous supply of groundwater?

Checkpoint 12.11

basic / intermediate / advanced / superior

In the case described in Checkpoint 12.10, the court ruled in Glen's favor. What are the implications of this ruling for dry western states that are currently experiencing rapid population growth that puts a strain on water used for both agricultural and domestic supplies?

GROUNDWATER

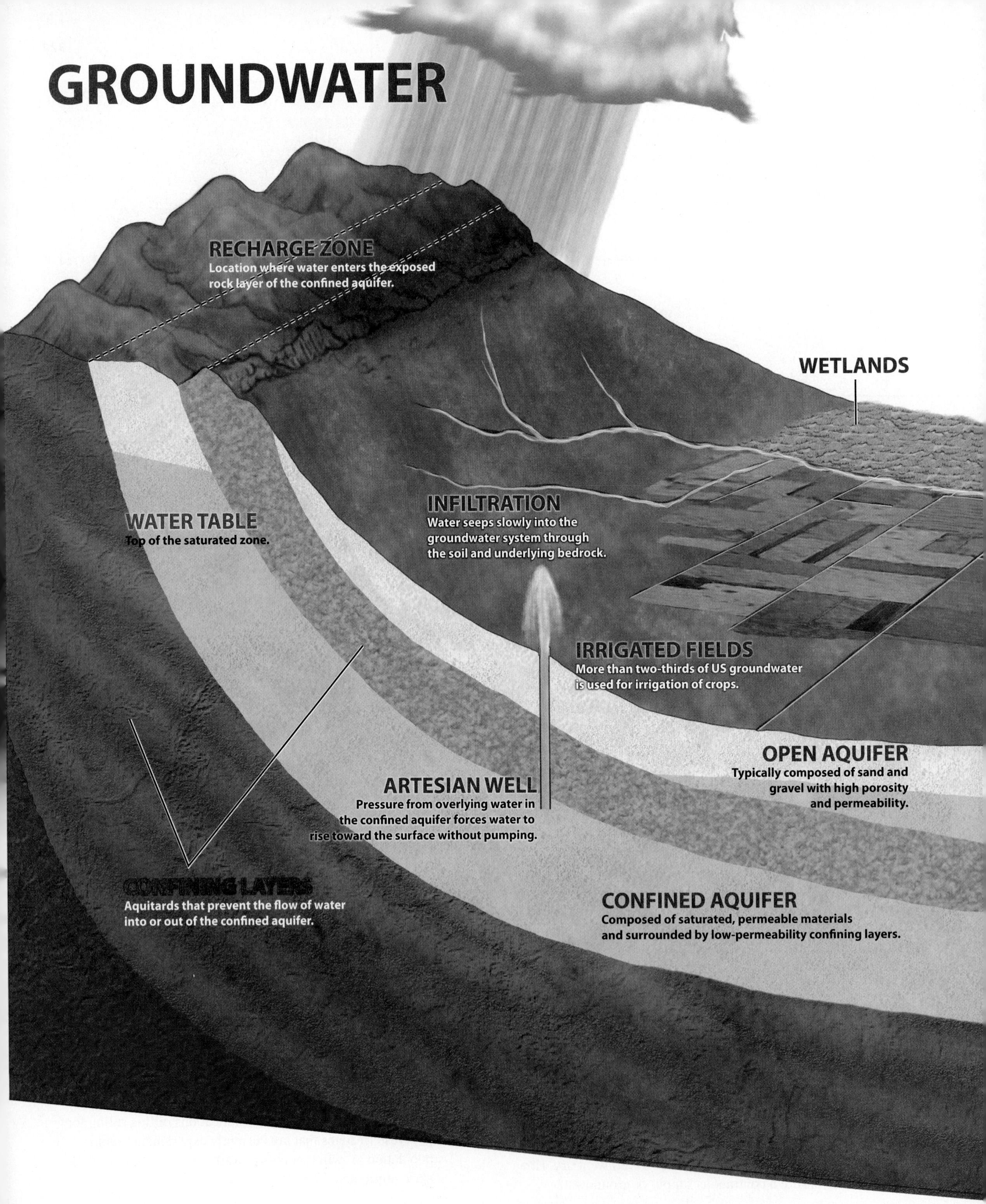

RECHARGE ZONE
Location where water enters the exposed rock layer of the confined aquifer.

WETLANDS

WATER TABLE
Top of the saturated zone.

INFILTRATION
Water seeps slowly into the groundwater system through the soil and underlying bedrock.

IRRIGATED FIELDS
More than two-thirds of US groundwater is used for irrigation of crops.

OPEN AQUIFER
Typically composed of sand and gravel with high porosity and permeability.

ARTESIAN WELL
Pressure from overlying water in the confined aquifer forces water to rise toward the surface without pumping.

CONFINING LAYERS
Aquitards that prevent the flow of water into or out of the confined aquifer.

CONFINED AQUIFER
Composed of saturated, permeable materials and surrounded by low-permeability confining layers.

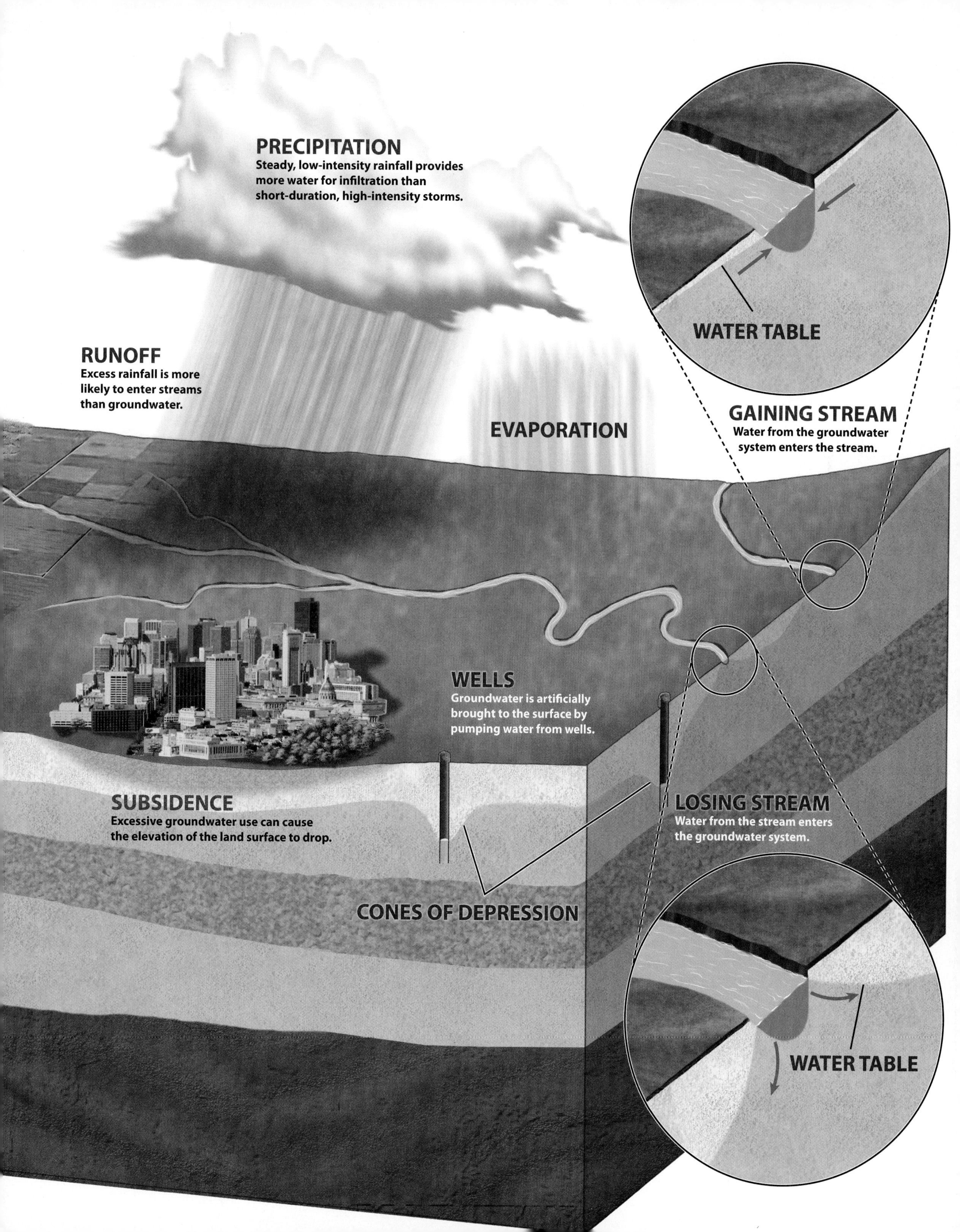

PRECIPITATION
Steady, low-intensity rainfall provides more water for infiltration than short-duration, high-intensity storms.

WATER TABLE

RUNOFF
Excess rainfall is more likely to enter streams than groundwater.

GAINING STREAM
Water from the groundwater system enters the stream.

EVAPORATION

WELLS
Groundwater is artificially brought to the surface by pumping water from wells.

SUBSIDENCE
Excessive groundwater use can cause the elevation of the land surface to drop.

LOSING STREAM
Water from the stream enters the groundwater system.

CONES OF DEPRESSION

WATER TABLE

Checkpoint 12.12

basic · intermediate · advanced · superior

Groundwater Evaluation Rubric

You are asked to help locate a new aquifer that will supply your town with water. In examining the potential sites, you recognize that several different factors will influence groundwater availability and at no single site are all of the factors optimal. You decide to create a scoring scheme to evaluate the most important factors that will influence the availability of groundwater. The location that scores the highest according to the rubric will be selected for the well field. One factor is included as an example in the table below; identify five more.

Factors	Poor (1 point)	Moderate (2 points)	Good (3 points)
Depth to water table	Deep	Intermediate	Shallow

12.4 A Case Study: The High Plains Aquifer

Learning Objectives

- Describe the location and properties of the High Plains aquifer.
- Sketch a graph showing changes in groundwater levels in the aquifer from 1900 to 2000.
- Explain why the High Plains aquifer water table is dropping in some places but is not changing in other locations.

Approximately two-thirds of all the fresh groundwater pumped from aquifers in the United States is used for irrigation (Figure 12.20). Much of this irrigation occurs in agricultural lands west of the Mississippi River in the region occupied by the Great Plains states of Texas, Oklahoma, Kansas, and Nebraska and their neighbors. Early explorers dubbed this region the "Great American

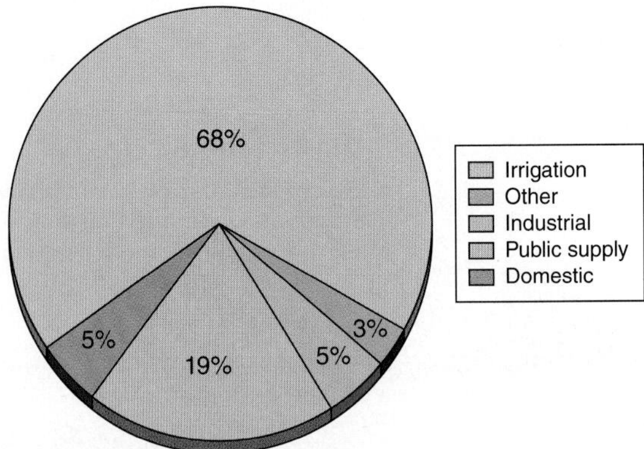

Figure 12.20 Uses of fresh groundwater in the United States, 2005. More than two-thirds of the 80 billion gallons (303 billion liters) of groundwater used per day was used for irrigation; most of this was used in western states with relatively dry climates. Public supplies may be used for community drinking water or other uses. The "other" category represents use of groundwater by mining, utilities, livestock, and aquaculture.

Desert." Leading an expedition through the area in 1819, Major Stephen Long considered it as "almost wholly unfit for cultivation and . . . uninhabited by people depending upon agriculture for their subsistence." Nevertheless, with additional water supplied by irrigation, the Great Plains states have successfully developed an agricultural economy. For example, Nebraska produces about one-half of the nation's corn, but few locations in the state can rely on precipitation alone to produce maximum crop yields. Corn needs approximately 63 centimeters (25 inches) of water to thrive during its 5-month (May–September) growing season in Nebraska. However, the eastern half of the state receives about 45 centimeters (18 inches) of precipitation over this time, while the dryer western part of the state averages just one-half of the necessary rainfall.

Irrigation waters for many of the Great Plains states are taken from a single large aquifer, the High Plains (also known as the Ogallala) aquifer (Figure 12.21). The High Plains aquifer is mainly composed of sand and gravel with some underlying sandstone. Because it is an open aquifer, it is partially recharged by water infiltrating from above, typically from rainfall and snowmelt. The aquifer's saturated zone is up to 425 meters (1,400 feet) thick, reaching its greatest thickness below Nebraska, but aquifer dimensions vary along its length (Figure 12.21a). The water table is always relatively shallow, typically less than 100 meters (330 feet) below the surface.

Groundwater from the aquifer was first widely used in the 1930s in Texas. Today, the aquifer produces more water than any other groundwater source in the nation. It supplies almost one-quarter of the water taken from all US aquifers and 30 percent of groundwater used for crop irrigation. More than 170,000 wells draw water from the aquifer throughout an area of 450,000 square kilometers (174,000 square miles), making it the largest area of irrigation-sustained cropland in the world.

Just as with the Nubian Sandstone aquifer in Libya, there is no contemporary source for water to recharge the whole aquifer. The aquifer is partially replenished through the limited recharge that occurs from precipitation and losing stream inflow. But most of the water in the aquifer entered the groundwater system thousands

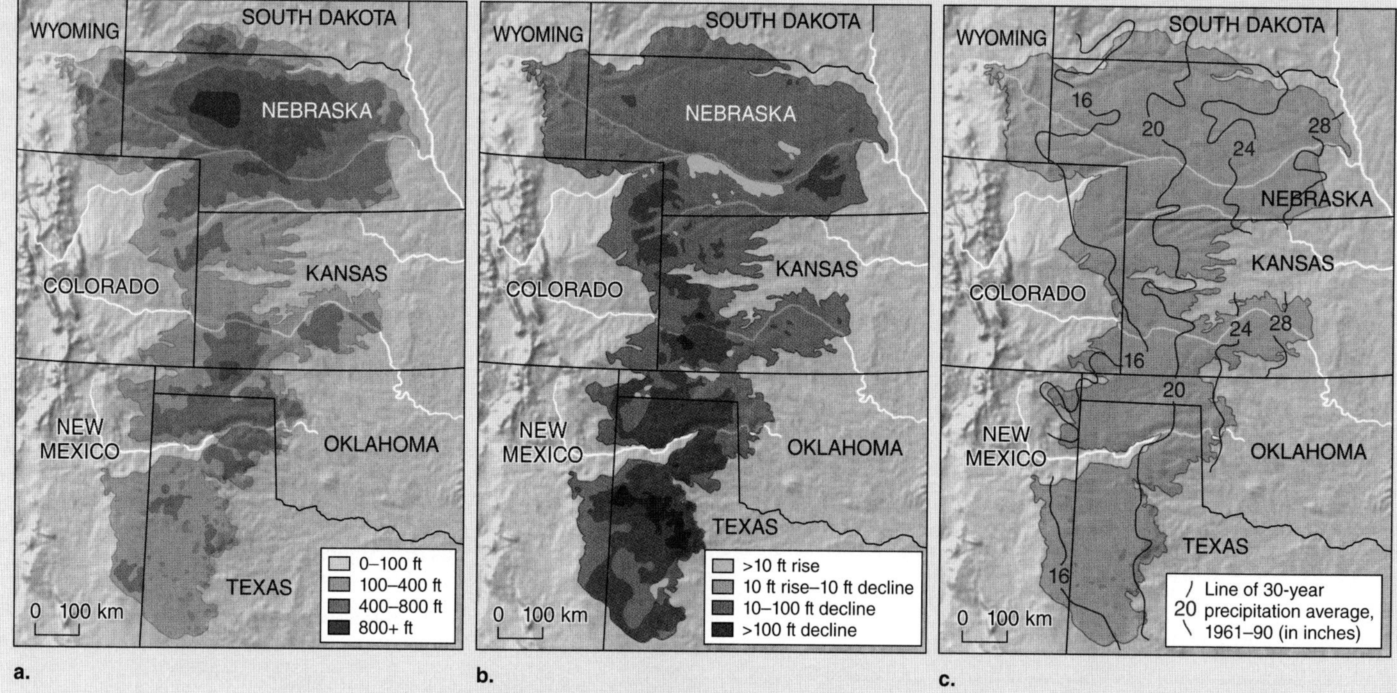

Figure 12.21 Characteristics of the High Plains aquifer. **a.** Thickness of the saturated zone in the aquifer. **b.** Changes in the depth of the water table prior to 1980. **c.** Variations in annual precipitation within the regions of the High Plains aquifer. (Compare with Figure 12.8a.) Note: Values presented in feet and inches on the basis of the original US Geological Survey maps.

of years ago during a wetter climate interval associated with the end of the last glacial maximum. This "fossil" water is being used up faster than it is replenished, and the water table is dropping over large regions of the aquifer (Figure 12.21b). Approximately 11 percent of the total groundwater supply has been extracted.

Over the life of the aquifer, the groundwater overdraft has caused the water table to drop by up to 70 meters (more than 200 feet) in parts of the Texas panhandle (Figure 12.21b); areas in western Kansas have consumed nearly 40 percent of their groundwater. Both of these locations are in areas with some of the lowest annual rainfall totals for the region (Figure 12.21c).

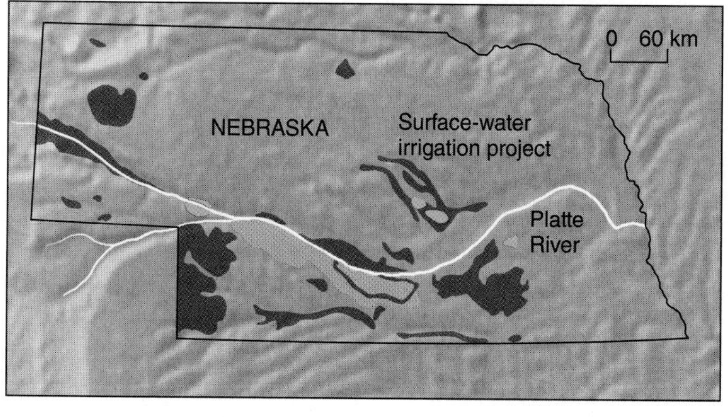

Figure 12.22 Areas of recharge in the High Plains aquifer below Nebraska. The water table in the High Plains aquifer has risen along stretches of the Platte River in central Nebraska, in contrast to a regional decrease in water table elevation.

In contrast, Nebraska has more than 60 percent of the aquifer underlying the state, recharge from the Platte River, less intensive groundwater consumption, and some of the region's highest precipitation (Figures 12.21 and 12.22). Researchers are currently studying the High Plains aquifer to determine how long it can continue to produce groundwater at current consumption rates.

Checkpoint 12.13

basic | intermediate | advanced | superior

Which is the *best* explanation for the changes in water level in the High Plains aquifer? Explain why you chose your answer.

a) The distribution of present-day rainfall on land above the aquifer
b) Differences in the thickness of the rocks that make up the aquifer
c) The number of water wells that have been drilled into the aquifer
d) The type of crops supplied by irrigation from the aquifer

Checkpoint 12.14

basic | intermediate | advanced | superior

Sketch and label a graph illustrating changes in water levels in part of the High Plains aquifer from 1900 to 2000. Assume that wells were not drilled into this part of the aquifer until 1940 and that the area is currently experiencing groundwater overdraft.

Checkpoint 12.15

basic · intermediate · advanced · superior

Is there a relationship between the locations where water levels in the High Plains aquifer have increased or decreased and the distribution of precipitation and the locations of regional river systems? Are these two factors sufficient to explain changes in water levels? Explain your reasoning.

Checkpoint 12.16

basic · intermediate · advanced · superior

Much of the agriculture in the middle United States relies on water from the High Plains aquifer. What are the long-term implications if we continue to use large volumes of groundwater for irrigation faster than it can be replenished?

12.5 Groundwater Quality

Learning Objectives

- Identify multiple examples of types of groundwater pollution from both natural sources and human activities.
- Describe the differences between pollution from point sources and from nonpoint sources.
- Analyze a scenario to evaluate how to respond to a case of groundwater contamination.

Sections 12.1 through 12.4 focused mainly on groundwater availability—the factors that control whether groundwater is present and how it flows in the subsurface. This section will examine the chemistry of groundwater and its implications for water supplies. All the groundwater in the world is of little help if, as in the Woburn example, it is contaminated by pollutants that cause illness or even death. Once contaminated, groundwater becomes essentially unusable because it is too expensive to treat the water. It may take decades or centuries before nature can sufficiently reduce the concentration of the contaminants to a safe level. Natural groundwater is far from pure, but in the United States, it typically contains few chemicals in sufficient quantities to cause harm to humans or ecosystems. However, under specific geologic conditions, harmful elements such as arsenic or mercury may be concentrated in groundwater. An example of widespread groundwater contamination by arsenic that may become the greatest mass poisoning in history is currently unfolding in Bangladesh.

Drinking Yourself to Death, Naturally

Bangladesh is one of the most densely populated nations in the world, with more than 160 million people living in an area about the size of Wisconsin (Figure 12.23). (For comparison, the population of Wisconsin is 5.7 million.) Unfortunately, Bangladesh is also one of the poorest nations in the world, with little significant health care available outside of the larger cities. Just a few decades ago, most citizens obtained their domestic water from surface ponds and streams or wells that were often contaminated by agricultural and industrial pollution. Hundreds of thousands of people died each year from diseases such as cholera and diarrhea contracted from drinking polluted water.

Beginning in the late 1970s, the United Nations sought to clean up the drinking water supply by providing materials for 1 million shallow water wells. So popular were the wells that residents drilled millions of more wells into the near-surface open aquifers for additional drinking water and for use in irrigation. Nearly 10 million wells are now operating in Bangladesh, and 97 percent of the population drinks well water. Unfortunately, no one tested the groundwater for some of the most toxic natural pollutants until the wells were already in use. It now turns out that large concentrations of the element arsenic are present in many wells and therefore in people's drinking water. How did it get there?

Bangladesh is located on the world's largest delta complex, where the Ganges and Brahmaputra rivers enter the Indian Ocean (Figures 7.24a, 12.23b). Both rivers have their origins in the foothills of the Himalaya Mountains, where the rocks contain unusually high natural concentrations of arsenic. These rocks are eroded from the mountains, and the resulting sediment is transported south in the rivers and deposited in the delta complex. Chemical weathering of these sediments releases arsenic into the groundwater. The water quality standard for arsenic in drinking water is 50 parts per billion (ppb, or 0.05 milligrams per liter) in Bangladesh, in contrast to the standard of 10 ppb set by the World Health Organization

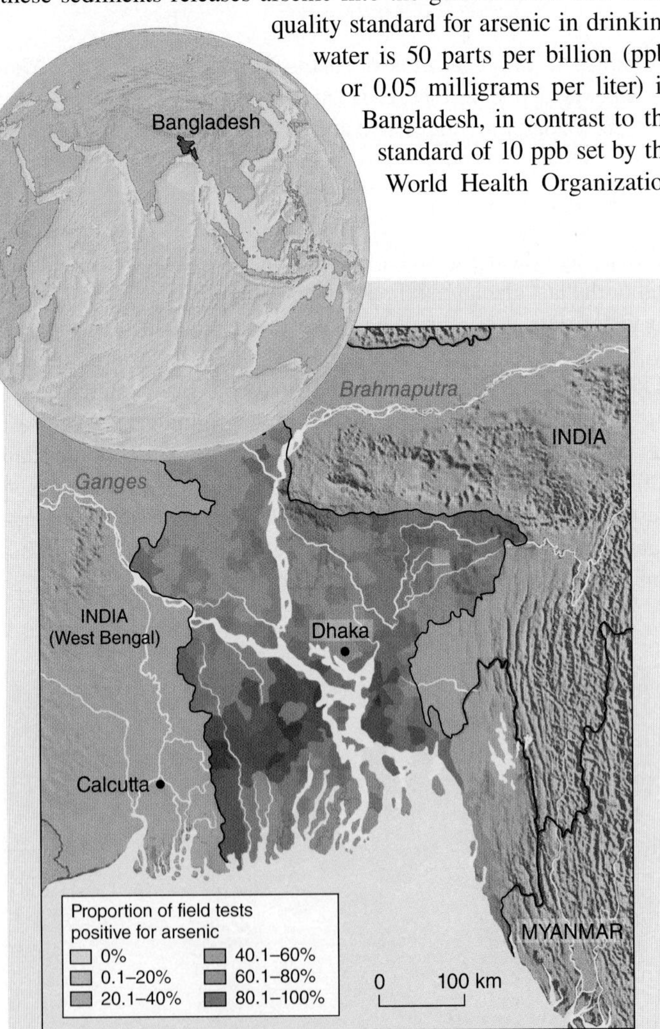

Figure 12.23 Arsenic contamination in wells in Bangladesh. Darker colors indicate locations where a higher proportion of wells are contaminated by arsenic. The Ganges and Brahmaputra rivers drain the Himalaya Mountains to the north. Note that the greatest distribution of arsenic-contaminated wells is in the delta region south of the confluence of the two major rivers.

(WHO). Arsenic levels of 500 to 1,000 ppb are not uncommon in wells in Bangladesh, and levels of over 2,000 ppb have been recorded.

Initially, warnings about the dangers of arsenic poisoning went unheeded because many of the early symptoms were diagnosed as leprosy. The governments of India and Bangladesh were primarily concerned with other diseases that were claiming thousands of lives each year. There were few resources to devote to investigating reports of arsenic poisoning. A survey of several hundred thousand wells found that 40 percent of them are contaminated with arsenic. Researchers estimate that approximately one-half of the population (60 million people) of Bangladesh may be exposed to arsenic levels that are above the WHO standard. In humans, long-term exposure to arsenic shows up as skin lesions, spots, bumps, and warts. Eventually 10 percent of those exposed will be affected by cancer of the lung, kidney, liver, or bladder.

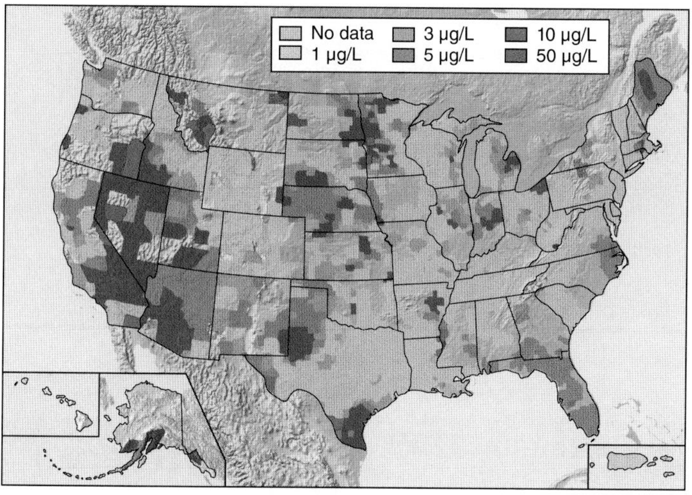

Figure 12.24 US arsenic concentrations. Map showing the range of maximum arsenic concentrations for at least 25 percent of tested samples. Highest arsenic concentrations above the WHO standard are found more frequently in western states. Arsenic in groundwater is a result of chemical weathering processes and is controlled mainly by the local geology and may be locally exaggerated by the effects of mining.

If arsenic poisoning is discovered during the early stages, its effects can be halted by following a nutritious diet and drinking arsenic-free water. A more affluent nation would solve the problem of water contamination by supplying those affected with bottled water or by drilling deeper wells to aquifers that do not contain arsenic-tainted sediment. Sadly, even such straightforward solutions are beyond the means of much of the population in a poor nation such as Bangladesh. One simple but effective effort to address the problem has been to paint contaminated wells red as a warning to residents that the water is unsafe.

In the United States, the standard for arsenic in drinking water was 50 ppb until a few years ago, when it was lowered to 10 ppb. The Environmental Protection Agency (EPA) has estimated that some 13 million US residents may be exposed to arsenic in drinking water at 10 ppb or higher. Just as in Bangladesh, arsenic is derived from chemical reactions between rocks and minerals and groundwater. Higher arsenic levels are more likely to be found in western states (Figure 12.24) that generally have more igneous and metamorphic rocks present on or near the surface. The levels found in the United States are far lower than those in Bangladesh, where economic conditions make remediation difficult.

Do-It-Yourself Groundwater Contamination

Although nature does its part to change the chemistry of the groundwater, most groundwater contamination is more likely to have a human source, especially in heavily populated regions. Human activities have added many potential pollutants to the groundwater supply.

Sources of human and natural contamination are classified as point sources and nonpoint sources. A **point source can be specifically identified and isolated,** such as a leaking gasoline storage tank (Figure 12.25). Once identified, a point source can

Checkpoint 12.17

basic / intermediate / advanced / superior

Did the natural arsenic contamination of groundwater in Bangladesh originate from a point source or a nonpoint source? Explain the reasoning behind your answer.

a) Point source b) Nonpoint source

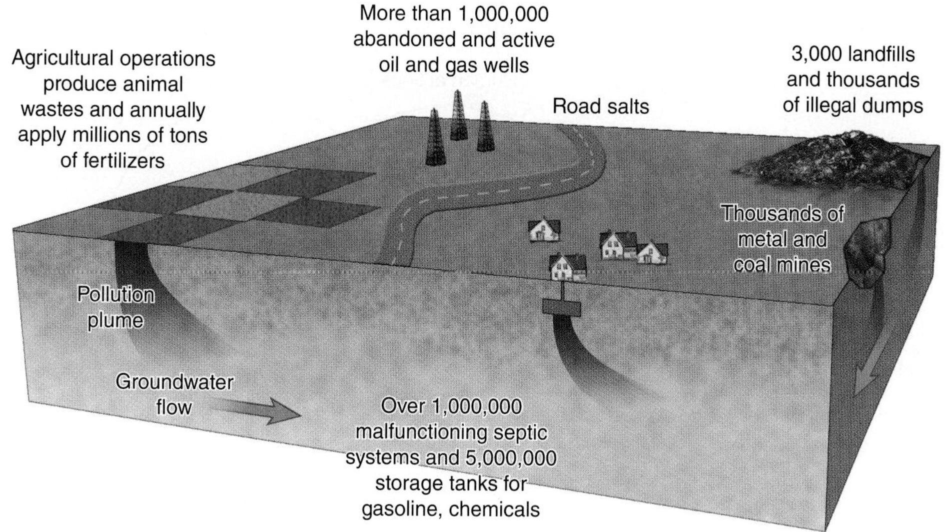

Figure 12.25 Potential pollution sources for groundwater in the United States. Septic systems or landfills represent examples of point sources. Large mines and agricultural operations can be nonpoint sources for pollution.

be readily shut down. By contrast, **nonpoint sources cannot be traced to a single origin but rather occur over a wide area,** such as croplands (Figure 12.25).

Pollutants from both point and nonpoint sources are often chemicals, such as benzene or nitrates (Table 12.1). Many of these come from industrial sources, as in the Woburn case cited in the chapter introduction. Many other potential sources of pollution can be found nearly anywhere in the United States and elsewhere (Figure 12.25). Without careful management, these pollutants can end up in the water supply. Nonpoint sources of pollution in agricultural regions are difficult to detect but can have some of the most far-reaching effects. This is because rural water wells are not monitored in the same fashion as municipal water sources. Testing of wells cannot prevent contamination, but it can identify when and where it has occurred. Pollutants that may be present in rural wells include pesticides and nitrates that are products of fertilizers. Pesticides and fertilizers are applied to crops, and some are washed off into the groundwater and surface water systems.

Not all groundwater contamination is from chemicals. Thousands of people worldwide are killed and millions more get sick every year because groundwater and surface sources are contaminated by harmful microscopic organisms (microbes). Many of those contaminants (for example, fecal coliform, *E. coli, Enterococci*) come from untreated human and animal waste that enters the groundwater or other water source. More than 400,000

Checkpoint 12.18

A farmer drilled a well into an open aquifer composed of sand and gravel. He installed a septic system downslope from the drinking well (see diagram). After a few years, he noticed that the septic system had begun to leak. Water tests showed that the well water was clean and uncontaminated by bacteria present in the septic system. Why did the septic system not contaminate the drinking water supply?

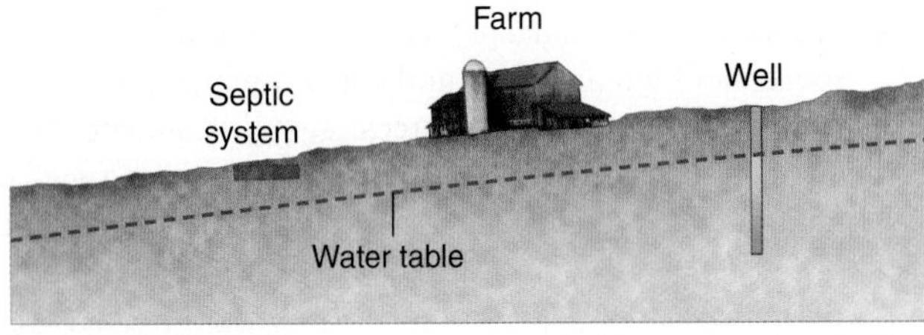

Review the following answers and rank them from 1 to 7, with 1 the most accurate. The most accurate response will be correct and will answer the question fully. Some responses may be wrong; others may include correct statements that are not complete answers.

	Rank
The well is higher than the septic system.	
A leak from the septic system would run downhill, away from the well.	
Water falling as precipitation will reach the well first because it is upslope from the septic system.	
Groundwater flows downslope, carrying any septic system leak away from the well.	
The septic system is above the water table and the well is drawing water from below the water table, so any leak from the septic system can't reach the well.	
There is an aquitard between the septic system and the well.	
It would take the contaminated water too long to travel to the well.	

Table 12.1 Common Pollutants in Drinking Water

Contaminant	Health effect	Selected sources
Benzene	Cancer	Leaking fuel tanks, industrial solvents
Toluene	Kidney disease	Chemical manufacture, industrial solvents
Trichloroethane	Cancer	Dry-cleaning, industrial solvents
Arsenic	Cancer, skin lesions	Rocks, pesticides, industrial wastes
Nitrate	Blue baby syndrome (methemoglobinemia)	Fertilizers, feedlots, sewage
Lead	Nervous system damage	Corrosion of lead pipes
Cryptosporidium	Stomach illness	Human/animal wastes

Checkpoint 12.19

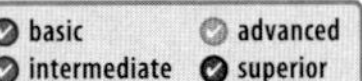

Venn Diagram: Human and Natural Groundwater Contamination Compared

Use the Venn diagram provided here to compare and contrast groundwater pollution from human actions (for example, Woburn) with that arising from natural sources (for example, Bangladesh). Write features unique to either group in the larger areas of the left and right circles; note features they share in the overlap area in the center of the image. Identify at least 10 features.

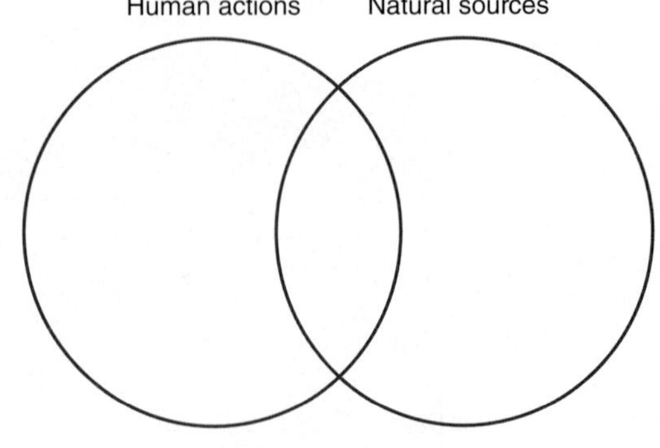

residents of Milwaukee, Wisconsin, became ill and more than 100 died in 1993 during the largest outbreak of waterborne disease in US history when a water plant malfunctioned and the intestinal parasite *Cryptosporidium parvum* contaminated the city water supply. Cholera is a deadly disease caused by a bacterium common in developing countries that forms in brackish, untreated water and can find its way into foods or water supplies.

In the United States, the 1972 Clean Water Act and its amendments banned the most blatant examples of pollution from industrial point sources, but many less obvious pollution sources still exist. The Environmental Protection Agency is responsible for enforcing water quality standards for drinking water. Some of the common pollutants that the EPA recognizes in drinking water are listed in Table 12.1, along with their known damaging health effects and typical sources.

Checkpoint 12.20

basic | advanced | intermediate | superior

Groundwater Case Study

The following map shows the characteristics of a hypothetical rural area about a 30-minute drive from a nearby city. The area has two homes that are over 50 years old (near B and C) and a housing development with expensive houses that were built in the southwest corner of the area 3 years ago. A stream flows through the area and is used for recreation (for example, fishing, boating, swimming). Most of the land is used for grazing or is wooded.

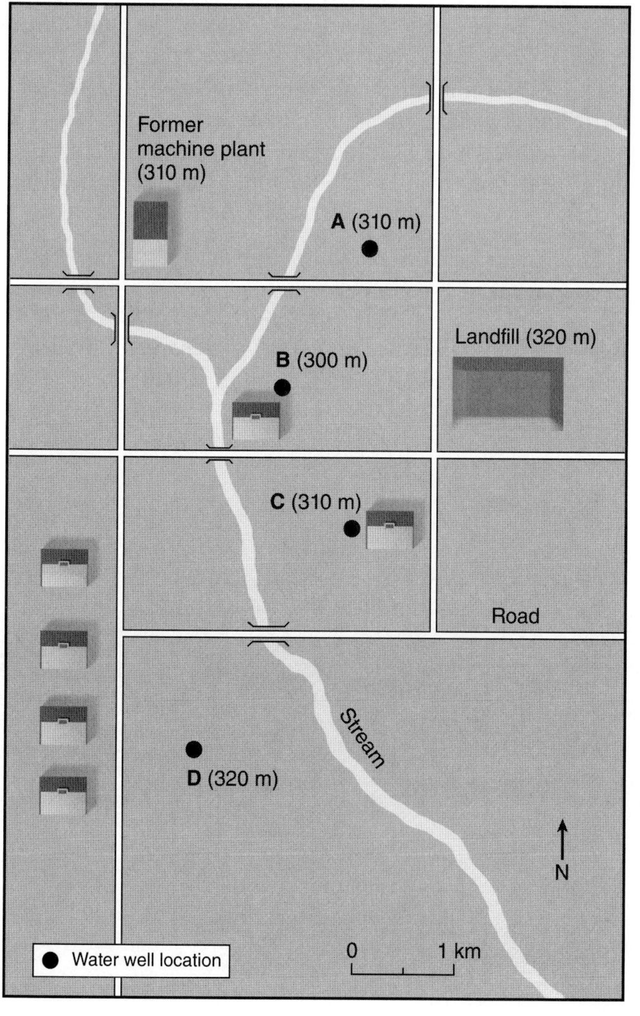

All homes are supplied by water from wells (B, C, or D) that are drilled into a sand and gravel aquifer that underlies the whole area. In addition, the county maintains a monitoring well (A) to track aquifer levels and water quality, but nobody uses well A as a source for drinking water. Multiple wells at D serve all the homes in the housing development. The elevations of the land around each well and at the landfill are indicated on the map.

Residents have just learned that the most recent analysis of water from monitor well A revealed unusually high levels of trichloroethylene (TCE). TCE is widely used as a solvent to remove oil and grease from metal surfaces prior to processes such as painting or machining. Drinking water with TCE may cause nausea, liver and kidney damage, and impaired fetal development. The local water board is suspicious that the TCE was added to the well by a disgruntled employee who was fired a couple of months earlier. However, a review of the industrial history of the area revealed that 40 years ago, a machine plant occupied now-overgrown ground about 1,800 meters (5,900 feet) west of well A. No records from the plant were preserved, so no information on whether it used TCE is available. Further, the landfill is licensed to accept industrial wastes from a variety of companies, some of which may have used TCE.

The county paid for analyses of wells B, C, and D that revealed no contamination at this time. The county water board has recommended that residents continue to use water from the wells for the immediate future. They have hired a geologic consultant to evaluate the risk to future drinking water supplies in wells B, C, and D. The county has made funds available to drill five additional test wells to sample the groundwater. Residents are preparing for a meeting with the geologic consultant to discuss where the wells should be drilled to better constrain the source of TCE.

From what you know about groundwater and the hydrologic cycle, answer the questions that follow:

1. What are the main scientific issues in this scenario?
2. What are the main social issues in this scenario?
3. Where should the additional wells be drilled? Justify your answer.

12.6 Introduction to Wetlands

Learning Objectives

- Describe the conditions that are necessary for an area to be classified as a wetland.
- Explain why wetlands are an essential part of the earth system.
- Discuss how changes to natural systems in the Florida Everglades resulted in ecological problems, and explain how those problems are being corrected.

Characteristics of Wetlands

Wetlands represent an important link among streams, lakes, and groundwater. The same US government that once encouraged filling in of wetlands today joins with other nations in signing the Ramsar Convention, a treaty intended to preserve and protect large areas of wetlands around the world. There are more than thirty Ramsar sites in the United States, in areas ranging from the northern rain forests of the Alaskan coast, to a Nevada desert oasis, to Florida's Everglades National Park. (Is there one in your state?) Our nation's changing view of the role of wetlands illustrates how our thinking has evolved regarding these natural environments.

Wetlands can carry an array of labels (for example, marsh, bog, swamp) and may be covered by water all year or for just a few weeks (Figure 12.26). The Clean Water Act defines *wetlands* as "those areas that are inundated or saturated by surface or groundwater at a frequency and duration sufficient to support, and under normal circumstances do support, a prevalence of vegetation typically adapted for life in saturated soil conditions." **To be classified as a wetland, an area must be saturated with water and have poorly drained soils and specific types of plants.** Wetlands must meet the following conditions:

1. **Hydrologic (water) conditions.** Water must be present on the land surface, or soils in the root zone must be saturated during the growing season or for longer periods.
2. **Hydrophytic (water-tolerant) vegetation.** Specific plants that grow under wet conditions (for example, cattails, wild rice, willows, and sawgrass) must be present. An estimated 9 percent of all plant species in the United States occur exclusively in wetlands.
3. **Hydric soils.** Wetlands are characterized by poorly drained (water-logged) soils that exhibit anaerobic (oxygen-deficient) conditions during the growing season to the degree that microorganisms and vegetation requiring oxygen will not live or thrive in the soils.

The federal government divides wetlands into two general types: coastal and freshwater. Coastal wetlands account for about 5 percent of US wetlands, including mangrove swamps and salt marshes found in places such as the Louisiana Gulf coast. Freshwater wetlands make up the rest; they include forested swamps, inland marshes, bogs, and water-logged areas adjacent to rivers and lakes.

Distribution of Wetlands. Wetlands cover approximately 445,000 square kilometers (172,400 square miles; an area the size of California) in the lower 48 states and an even greater area in Alaska. In the lower 48, Florida, Minnesota, and Texas have the largest areas of wetlands. **Outside of Alaska, wetland areas have declined by approximately 55 percent since the 1600s in the United States.** California, Ohio, and Iowa have experienced the greatest proportional losses of wetland area, with approximately 10 percent or less of the original wetlands remaining in each of these states.

Most wetland losses in recent years have resulted from the draining of wetlands to support agriculture, with the rest attributed

Checkpoint 12.21

basic · intermediate · advanced · superior

Review the 12 soil orders in Table 9.2 in chapter 9. Which soil order is most likely to be present in wetlands?

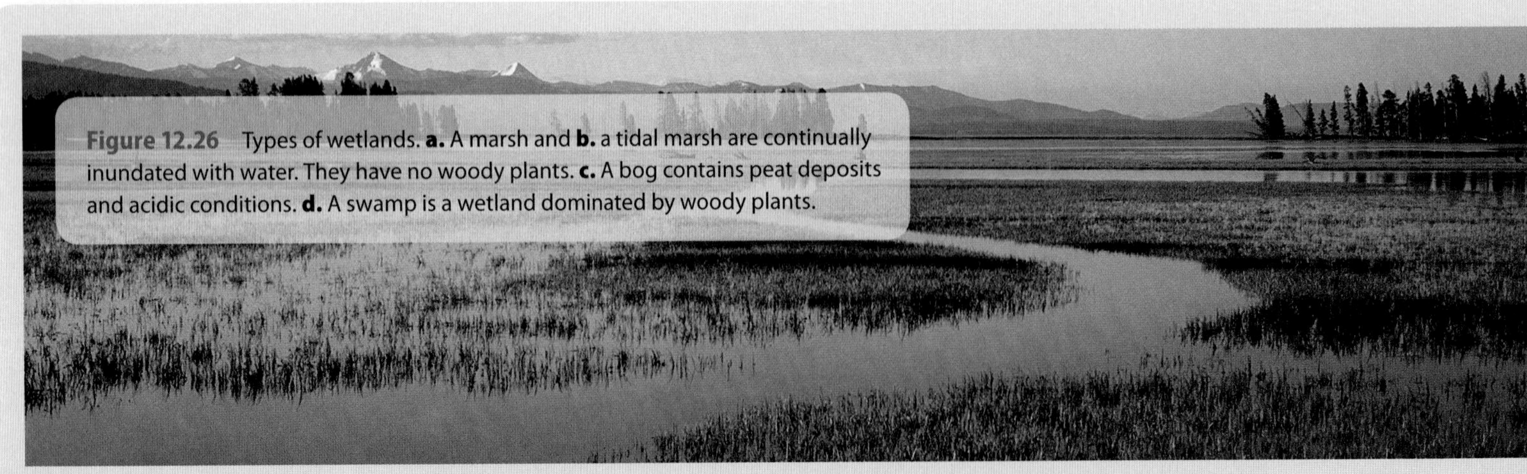

Figure 12.26 Types of wetlands. **a.** A marsh and **b.** a tidal marsh are continually inundated with water. They have no woody plants. **c.** A bog contains peat deposits and acidic conditions. **d.** A swamp is a wetland dominated by woody plants.

a.

to the draining and in-filling of wetlands due to urbanization and development. The rate of wetland loss has declined from nearly 203,000 hectares (500,000 acres) per year in the 1950s to approximately 5,000 hectares (12,355 acres) per year more recently.

Benefits of Wetlands. Studies have revealed that **wetlands perform many positive functions in regulating the natural environment.** Some have labeled wetlands the "kidneys of the earth." Wetlands improve water quality in rivers by filtering out sediments and other contaminants, thus providing better breeding grounds for fish and shellfish and supporting commercial fishing operations. Wetlands also provide ecological habitats for migrating birds and many other species. Wetlands moderate the effects of flooding by slowing runoff, especially downstream from urban centers. Finally, wetlands provide recreation for humans in the form of canoeing, hunting, fishing, and bird watching.

Case Study: The Florida Everglades

The Kissimmee River, Lake Okeechobee, and the Everglades are the key components of the largest drainage basin in southern Florida (Figure 12.27). The Kissimmee River feeds Lake Okeechobee, which supplies water that flows south through the Everglades, one of the world's largest freshwater wetlands. Historically, heavy precipitation and overflow from Lake Okeechobee saturated thick peat soils that filled a shallow basin that occupies much of south-central Florida. The land elevation drops less than 10 meters (33 feet) over the 150 kilometers (96 miles) from the lake to the southern coast.

Checkpoint 12.22

basic | intermediate | advanced | superior

After a series of summer thunderstorms, Cathy's lawn is covered with a shallow pond of water up to 15 centimeters (6 inches) deep in places. The water remains for nearly 10 days. Does the water in Cathy's backyard make it a wetland? Explain your answer.

Conservationist Marjory Stoneman Douglas termed the slow, unchanneled flow of water down the gentle slope through the sawgrass prairie of the Everglades a "river of grass."

Beginning in the early 1900s, development of southern Florida resulted in the construction of thousands of miles of canals

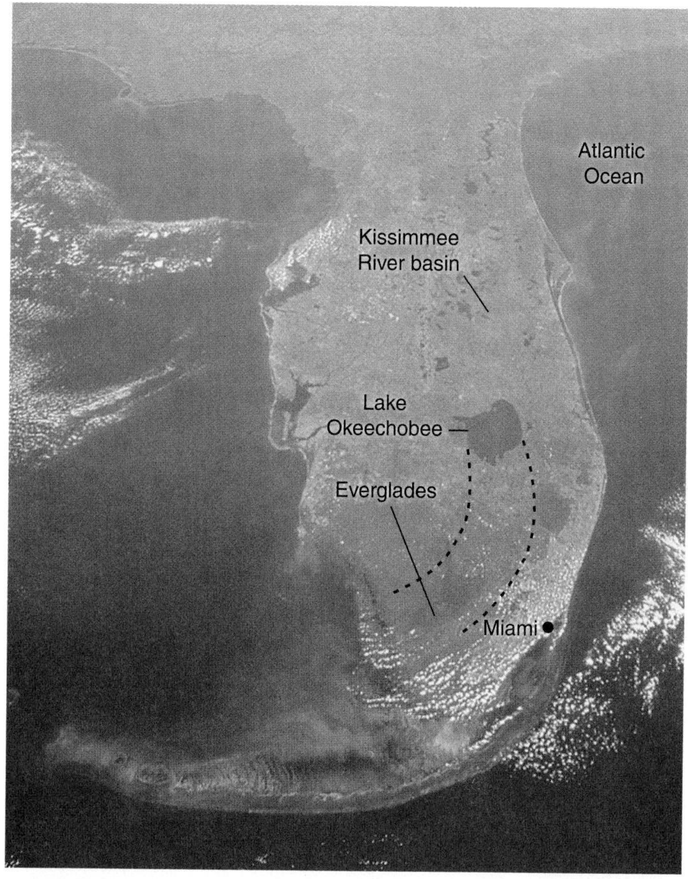

Figure 12.27 Southern Florida drainage basin. Water flows down a gentle slope (3 to 6 centimeters per kilometer; 1.5 to 3 inches per mile) from Lake Okeechobee toward the Everglades.

b.

c.

d.

that diverted water away from wetland environments to the north and south of Lake Okeechobee. With regard to the water, development proceeded with four goals, summarized as the "4 Ds": dike it, dam it, divert it, drain it. This hydraulic engineering feat resulted in greater flood control, expanded agriculture, and improved water supply (Figure 12.28). Although these changes benefited landowners, they often had negative consequences for the local ecosystems. The loss of 50 percent of the original wetlands destroyed fish and wildlife habitats, including feeding and nesting areas for wading birds along the transition from dry land to the wetlands.

Scientists belatedly realized that the Everglades wetland system was very sensitive to the duration that surface water stayed in the wetlands, its depth, and its flow patterns. Changes to the natural system increased the amount of time it took for water to flow through the Everglades, decreased water depth, and changed flow patterns. Lower water levels in the wetlands and underlying open aquifer allowed saltwater to intrude (migrate landward) along the east coast. The lower water levels also lowered groundwater recharge rates, which reduced the flow of water to Florida Bay along the southern coast. The water entering the Everglades became more polluted with fertilizers carried by runoff from agricultural operations such as sugarcane production that had replaced much of the original wetlands (Figure 12.28).

The loss of water to the wetlands caused soil formation to slow, and lower water levels exposed peat (compressed vegetation matter that forms thick layers in wetlands) that then dried. Ironically, those dry materials later created fire hazards during drought conditions in what were formerly saturated wetlands.

To start reversing these negative trends, the South Florida Water Management District has purchased thousands of hectares of land to be used to enhance groundwater recharge in an effort to halt the decline of groundwater levels. The federal government has also purchased large tracts of land in and around the Everglades in an effort to restore it to more natural flow conditions.

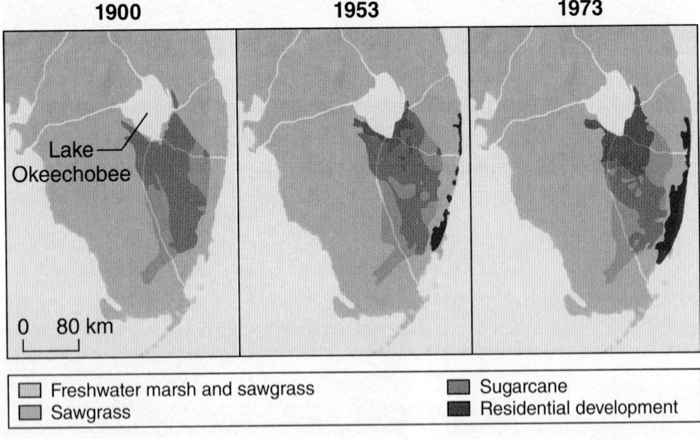

Figure 12.28 Land use patterns in southern Florida from 1900 to the 1970s. Original wetlands represented by marsh and sawgrass were replaced by agriculture dominated by sugarcane. Additional environmental stress was placed on the ecosystem by the increase in population represented by the expansion of coastal cities.

Checkpoint 12.23

basic · intermediate · advanced · superior

Venn Diagram: Wetlands and Groundwater Compared

Use the Venn diagram provided here to compare and contrast wetlands and groundwater systems. Write features unique to either group in the larger areas of the left and right circles; note features that they share in the overlap area in the center of the image. Identify at least 12 features.

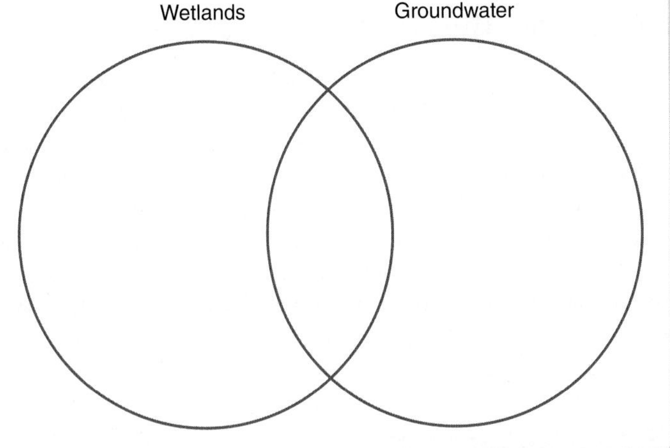

Checkpoint 12.24

basic · intermediate · advanced · superior

Wetland Management Exercise

After graduation, you begin work for your local county government. A few years later, you are put in charge of the local park system. One park includes a wetland area in the meander bend of a river. A boardwalk has been built out into the wetland so that residents can admire the water lilies, watch for fish, and sit at dusk to look for beaver and ducks. Part of your job is to monitor the health of the wetland. There is some concern that it is starting to be overrun by a type of water grass that will push out other vegetation and change the character of the area. What kinds of factors could you survey annually to measure the general health of the wetland? How could you monitor the status of the water grass?

the big picture

One central theme for this chapter has been the social value of science. The cases of contamination we described for aquifers in Massachusetts and Bangladesh illustrate the effect of the earth system on people. Depletion of the water supply from the Nubian Sandstone and High Plains aquifers, as well as the environmental changes resulting from development of the Everglades, show how people can affect the earth system. These cases remind us that we have to be careful in how we interact with Earth's most basic resources. The water in Chapter 11 was surface water—pretty familiar to everyone and easy to observe. Groundwater, as discussed in this chapter, is more obscure and, unless they rely on a well for drinking water, is easily overlooked by most people. This out-of-sight, out-of-mind situation makes it more difficult to identify when groundwater systems are contaminated or are undergoing significant changes.

What priorities should we consider when determining the appropriate use of water resources? Do crops get higher priority than cities? Should development be put on hold if needed wells would reduce water levels in natural springs that are home to endangered species? Who will answer the questions and make the decisions? The issues are complex and involve many components in the earth system. Scientists' findings add to the public debate, but many of the considerations for these decisions lie outside the realm of science and depend on informed citizens balancing economic, cultural, environmental, and scientific factors.

Americans drink more than 9 billion gallons (34 billion liters) of bottled water each year. Whether it is from a bottle or the tap, our drinking water comes from the same place and typically undergoes similar processing and monitoring to make sure it is safe. At the beginning of the chapter we asked: Which is better, bottled water or tap water? In almost all cases, there is no real difference in their quality. The Environmental Protection Agency regulates community water systems, while the Food and Drug Administration regulates bottled water. Municipal water systems are typically required to test for more potential contaminants and to do so more frequently than water bottling plants. One area of clear difference is cost. For the cost of 1 gallon (3.8 liters) of that fancy Swedish mineral water mentioned at the start of the chapter, you could buy thousands of gallons of tap water from most municipal water systems. But beyond the purchasing cost, there are additional costs for Americans to import water from Sweden or any other foreign nation. Consider the fossil fuels consumed to transport crates of water on trucks and ships from the Swedish bottling plant to a grocery store in your city. Or think about the 1.5 million tons (1,361,000 metric tons) of plastic used annually to manufacture water bottles—most of which are not recycled. Of course, tap water is not as portable as bottled water . . . unless you pour it into an empty water bottle.

Groundwater and Wetlands: Concept Map

Complete the following concept map to evaluate your understanding of the interactions between the earth system and groundwater and wetlands. Label as many interactions as you can using the information from this chapter.

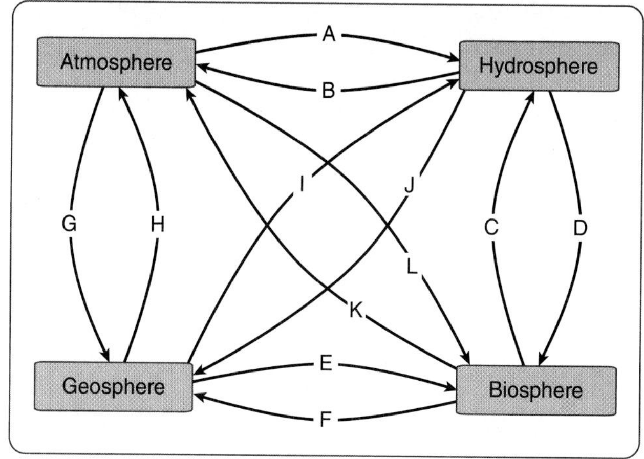

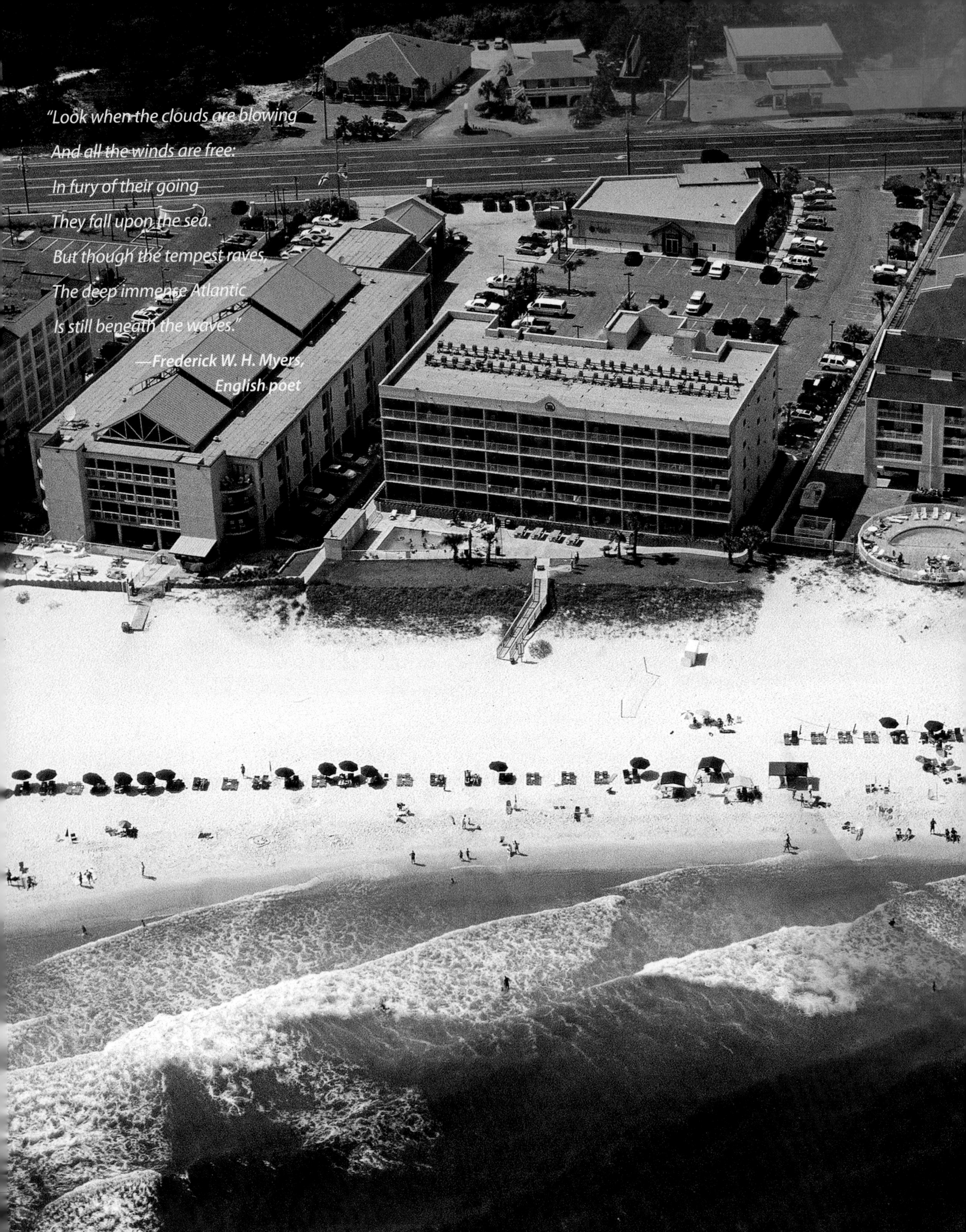

"Look when the clouds are blowing
And all the winds are free:
In fury of their going
They fall upon the sea.
But though the tempest raves,
The deep immense Atlantic
Is still beneath the waves."

—Frederick W. H. Myers,
English poet

Chapter 13

Oceans and Coastlines

Chapter Outline

the big picture

Would you wait out a hurricane in one of these hotels?

See The Big Picture box at the end of this chapter for the full story on this image.

13.1 Our Changing Oceans

Chapter Learning Outcomes

- Students will recognize that Earth is the water planet.
- Students will describe physical characteristics of the seafloor and how they affect oceanic circulation.
- Students will describe the influence of oceanic currents, circulation patterns, and tides on the earth system.
- Students will predict the consequences of human development along shorelines.

Over two-thirds (about 71 percent) of Earth is covered in seawater (Figure 13.1), but we know relatively little about the oceans and seas in comparison to our knowledge of the continents. More than 1,500 people have stood on the summit of Mount Everest, the highest point on Earth (8,850 meters; 29,035 feet), but only two people have dived to a depth of approximately 11,000 meters (33,000 feet) in the deepest region of the ocean floor along the Mariana Trench in the western Pacific Ocean. The oceans represent the final frontier for research on Earth.

Few fields of earth science exhibit as much diversity in research methods as the study of oceans. For example, right now, somewhere in the northern Pacific Ocean, there is an elephant seal with a few electronic data sensors glued to its back. Although elephant seals are big, lumbering creatures that often weigh more than a ton (Figure 13.2), they can do a better job of collecting certain kinds of detailed information about the oceans than sophisticated satellites and research vessels. Satellites yield information about the surface conditions of the oceans, and widely spaced stationary buoys collect data about water conditions at different depths. However, neither of these technologies can provide detailed information about changes in the properties of shallow ocean waters. In contrast, as elephant seals migrate from California to Alaska and back, they dive as deep as 600 meters (1,980 feet) up to 50 times

Figure 13.2 Researchers at rest. A female elephant seal with an electronic data sensor (background) relaxes with her seal pup on a California beach. There is no evidence that sensors harm the animals or interfere with their behavior.

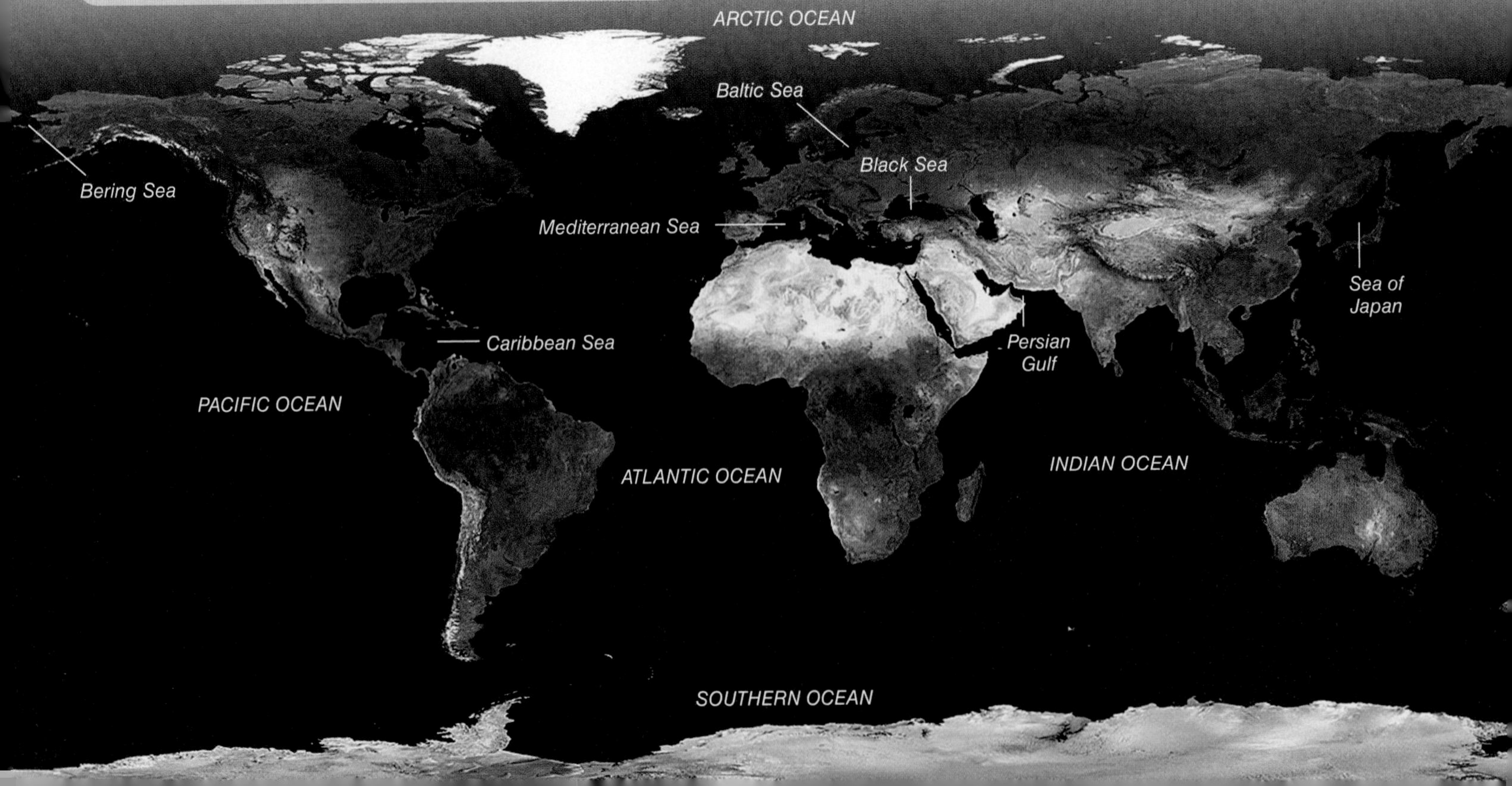

Figure 13.1 The world's oceans and selected seas. Nearly 90 percent of the Southern Hemisphere is covered by oceans.

each day and can provide temperature and salinity (saltiness) profiles of the ocean at relatively closely spaced locations. This adds up to thousands of data sets during a single season. The animals then return to the same California beach, where they molt, shedding their old layer of skin and leaving the data sensors to be collected and downloaded. Alternatively, data may be transmitted directly to satellites when the animals come to the surface.

Thousands of marine animals from many different species collect data for the Tagging of Pacific Pelagics (TOPP) program. (The term *pelagic* refers to the open ocean.) The TOPP data can be used to better understand the physical character of the oceans and animal behavior. The TOPP program and similar research provide scientists with data to help them better understand oceanic circulation systems that help regulate global climate patterns.

The Dynamic Nature of Oceans and Coastlines

The oceans are not just big bathtubs of water that separate the continents. They are dynamic environments in which water is constantly in motion. Water in the oceans moves both horizontally and vertically because of the rotation of Earth, the influence of the moon's orbit, the push of the winds on the surface, and variations in the properties of seawater. Small changes in the positions of the continents can change these circulation patterns to bring warm temperatures to one location or allow massive ice sheets to form in another. **Oceanic and atmospheric circulation patterns redistribute Earth's heat and play a crucial role in controlling our planet's climate.**

Along the margins of the oceans, coastlines advance or retreat, depending on the balance between the supply of sediment from continental interiors and the removal of material by wave erosion. Daily tides and seasonal variations in stream flow and storm activity cause short-term changes in the position of the coastline. Climate cycles measured in decades, centuries, or millennia may result in increasing or decreasing sea levels, shrinking or growing the size of the landmasses. Finally, tectonic cycles measured in thousands or millions of years may revitalize coastlines by periodic episodes of uplift, as exemplified by the contrast between the sandy beaches of the Atlantic shore (passive margin) and the rocky headlands of the Pacific coast (active margin).

Besides these natural processes, human activities have an impact on oceans and coastlines—and vice versa. Today, many people enjoy living along the coastlines of major oceans; more than one-quarter of the US population lives in counties along the Atlantic and Gulf coasts. Many citizens and organizations take a keen interest in preserving the natural coastlines. For example, to document coastline changes caused by either natural processes or human development, Ken and Gabrielle Adelman set out to create a photographic record of the entire California coast from Oregon to Mexico. Gabrielle flew a helicopter a few hundred meters offshore as Ken snapped thousands of photographs, which he then loaded onto their website (www.californiacoastline.org) for anyone to see (Figure 13.3).

The Adelmans had hoped their site would be used by environmental organizations, state agencies, and anyone with an interest in the coastline. But how could they make sure people even knew it existed? That problem was soon solved. Without knowing it, one of the images the Adelmans had snapped showed the back side of a palatial home in Malibu owned by celebrity Barbra Streisand. This displeased the famously reserved singer, who promptly filed a $50 million lawsuit for invasion of privacy. Well, this was big news; suddenly, the website with all those views of the California coast was itself famous and receiving thousands of visitors each day. The suit was eventually thrown out of court, Streisand was required to pay the Adelmans' legal expenses, and the photograph is still on the website.

Legal conflicts are perhaps inevitable when private property borders public coastlines. It is a similar story in international disputes over who owns the seafloor where access to potential mineral and energy resources is up for grabs. In this chapter, we will take a walk across the floor of the Atlantic Ocean to consider various features found there. Later, we will learn how the chemistry of the oceans is slowly changing and how bath toys can be used as a research tool in the Pacific Ocean. Finally, we will focus on how natural processes and human activities affect the ocean margins.

Self-Reflection Survey: Section 13.1

Answer the following questions as a means of uncovering what you already know about oceans and coastlines.

1. How have you interacted with the world's oceans, either directly or indirectly?
2. What are the advantages and disadvantages of living along the coast?
3. Does the ocean have an effect on you where you live today? If so, how? If not, why?

Figure 13.3 Coastal development, Malibu, California. Coastal California communities are home to some of the most expensive real estate in the United States. The rock layers in the cliff provide greater resistance to coastal erosion than sand dunes present along the Atlantic Coast.

13.2 Ocean Basins

Learning Objectives

- Identify the locations of seafloor features associated with active and passive continental margins.
- Interpret the topography of the ocean floor within a major ocean basin.
- Describe the characteristics and processes associated with the four major depth zones of the ocean floor.

The physical characteristics of the ocean basins result from plate tectonic processes. As you learned in Chapter 4, the configurations of most of the wide oceans and smaller seas (for example, Atlantic Ocean and Red Sea) evolved as slabs of continental lithosphere moved apart, while others were created by the closure of previously larger basins (for example, Mediterranean Sea). Even now, oceans and seas continue to slowly increase or decrease in size as plates diverge or converge.

From what you learned about plate tectonics in Chapter 4, would you expect the depths to be the same throughout the world's oceans? **The depth from the ocean surface to the ocean floor varies from 0 meters along shorelines to a maximum of nearly 11 kilometers (7 miles) along the Mariana trench in the western Pacific Ocean.** The highest landform on the continents, Mount Everest, would sit comfortably in the trench with more than 2,000 meters (6,600 feet) to spare. The average elevation of the land surface is less than 1 kilometer (0.6 mile), but the average depth of the oceans is approximately 3.8 kilometers (2.3 miles). The volume of water in the oceans is nearly 10 times the volume of the dry land that lies above sea level. Consequently, if erosion were to lower the level of the continents to sea level, the eroded material would fit easily in the ocean basins with room to spare.

Sea Level

The **elevation at the surface of the ocean is known as sea level** and is assumed to be 0 meters. Sea level has changed by hundreds of meters in the geologic past. Why do you think sea level fluctuates that much? Sea level can rise or fall because of changes in the shapes of the ocean basins or as a result of long-term climate changes that trap water in ice caps or cause ice to melt.

Recent studies have revealed that the sea surface is not actually "level." Rather, it has bumps and low spots, fluctuating by as much as 100 meters (330 feet). National Oceanographic and Atmospheric Administration (NOAA) scientists determined this by precisely measuring the distance from an orbiting satellite to the ocean surface (Figure 13.4a). This technology has also been used to determine the depth and topography of the ocean floor.

Bathymetry of the Ocean Floor

The elevation of the surface of the ocean varies because the elevation of the ocean floor varies (Figure 13.4). The **measurement of the depth to the ocean floor, and the mapping of its features, is termed *bathymetry*.** It took more than a century for ship-based research tools such as sonar (sound wave imaging) to accurately reveal the detailed bathymetry of some parts of the ocean floor. However, satellite technology has made it possible for researchers to determine the character of large sections of the ocean floor much more rapidly. Satellites are needed to unmask the features on the ocean floor because it is impossible to simply guess how deep a segment of ocean floor will be, and it would take many decades to map the ocean floor using shipborne instruments alone. Data collected by ships and submarines are combined with satellite data in locations such as the remote Arctic Ocean, which is shrouded in ice. These data reveal that the ocean floor has changes in elevation similar to the mountains, valleys, and plains found on land.

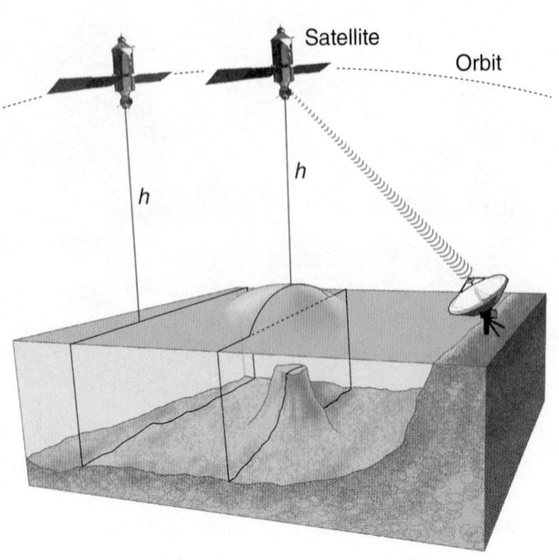

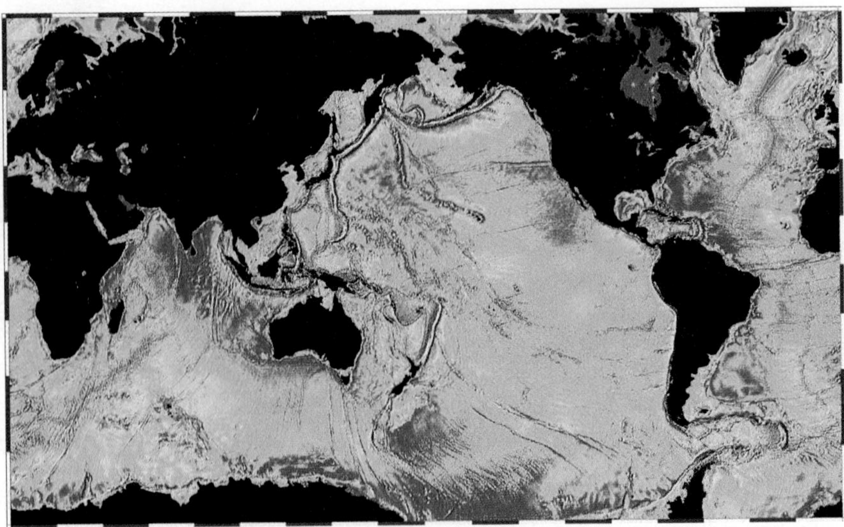

Figure 13.4 Measuring the elevation of the ocean surface. **a.** The gravitational attraction of a volcano on the ocean floor causes water to pile up above the volcano. A satellite measures the difference between the elevation of the ocean surface and the neighboring ocean. **b.** On the basis of the altimeter data, the gravitational field of the ocean basins can be calculated. Those gravity data correlate closely with ocean floor topography.

As we will learn in Section 13.5, the gravitational attraction of the moon causes the sea level to rise and fall, producing tides. A similar gravitational attraction for water can be attributed to features on the seafloor. For example, masses of rock that comprise volcanoes or ridges on the seafloor exert an additional gravitational pull on the water in the ocean, causing it to pile up and form a mound on the ocean surface (Figure 13.4a). Similarly, trenches on the ocean floor are much deeper than surrounding areas. Since the trench is filled with water rather than rock, the gravitational attraction there is lower. This slightly lower gravitational field causes a depression in the overlying ocean surface. These changes in the elevation of the ocean surface can be accurately measured by radars on satellites and used to calculate variations in Earth's gravity field. The gravity data reveal differences in the bathymetry of the ocean floor (Figure 13.4b).

A Walk Across the Ocean Floor: The Four Major Depth Zones

If you set out to walk across much of North America, you would be walking over relatively flat ground in the center of the continent, while regions in the east and west would force you to climb up and down mountains. What would you encounter if you set out to walk from west to east along the floor of the Atlantic Ocean from New Jersey to Portugal along latitude 40°N? (A lot of people yelling and trying to drag you out of the water.) OK, it's a hypothetical walk. What does the ocean floor look like as we cross from one continent to another? Beginning at the edge of the continents, we can recognize four major depth zones in the oceans: continental shelf, abyssal plain, oceanic ridges, and oceanic trenches (Figure 13.5 and Chapter Snapshot).

Continental Shelf. Entering the water along the shore of Normandy Beach, New Jersey, the first depth level you would encounter would be the continental shelf, the shallow ocean floor immediately adjacent to continental landmasses. **The continental shelf is actually submerged continental crust that slopes away from the coast with maximum water depths of a few hundred meters.** The width of the shelf varies, depending on whether it is adjacent to a passive continental margin, where no plate boundary is present (for example, Atlantic coast), or to an active margin adjacent to a trench (for example, south coast of Alaska). The continental shelf is a relatively

Checkpoint 13.1

basic | intermediate | advanced | superior

On the following map, label three active continental margins with an A and three passive continental margins with a P. (The X-Y line is used in Checkpoint 13.3.)

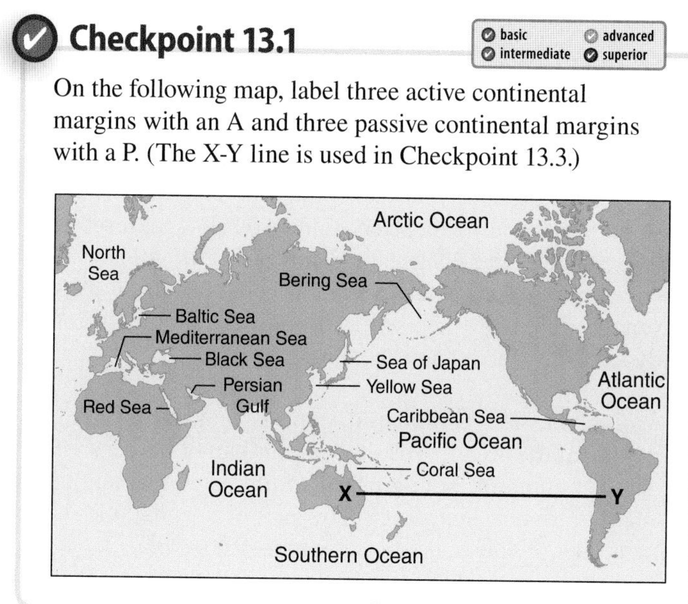

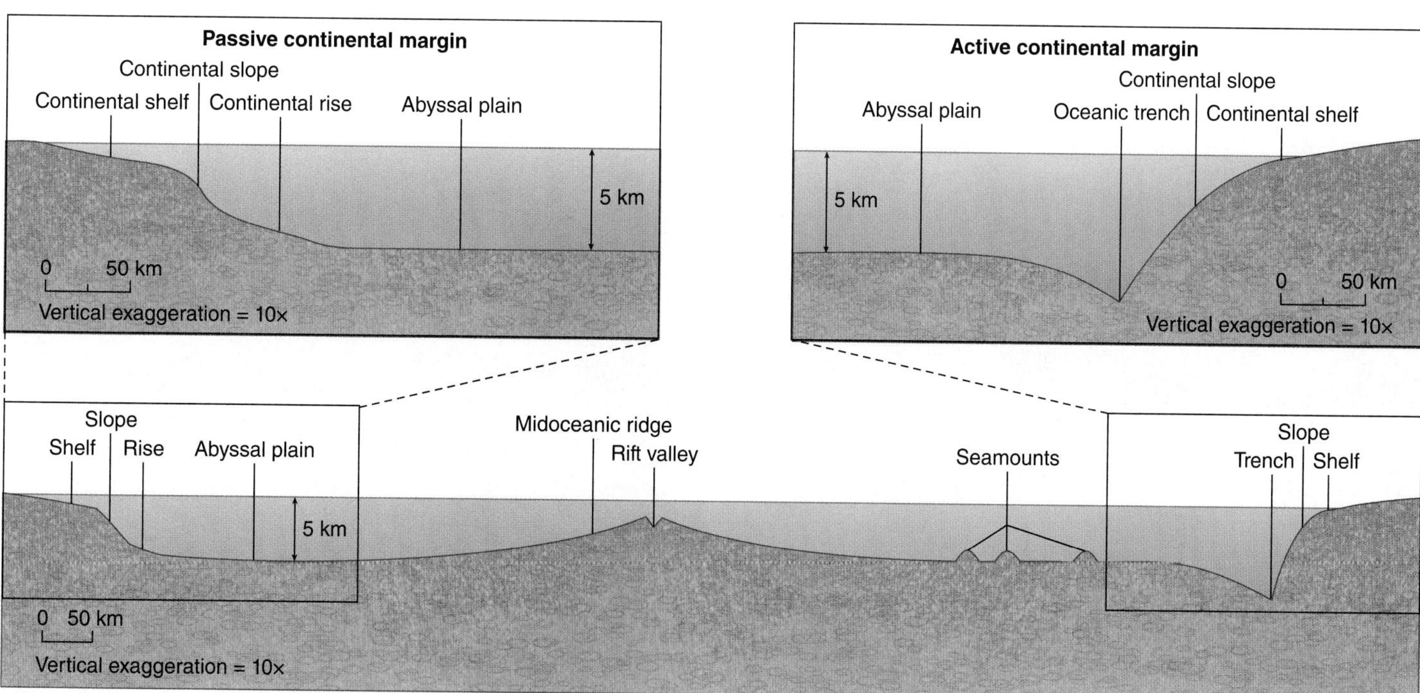

Figure 13.5 Ocean depth zones. The principal topographic elements of the ocean floor include an elevated oceanic ridge; deep, narrow trenches; and a gradual rise of the ocean floor to the continents along passive margins. Note the narrow shelf along the active margin and the broad shelf along the passive margin.

wide zone (up to hundreds of kilometers across) bordering passive margins, whereas it may be little more than a narrow strip a few tens of kilometers wide at active continental margins.

The width of the shelf at any time is related to the sea level. Remember, the shelf is defined by the gently sloping surface *and* an overlying shallow body of seawater. Therefore, the width of the shelf decreases as the sea level falls and increases as the sea level rises. During the time that a portion of the shelf is above sea level, it is exposed to the same kinds of erosional processes we discussed in previous chapters. Rivers that once emptied into the ocean will cut across the exposed continental shelf until they reach the new, lower sea level. Those rivers cut canyons in the continental shelf; the canyons are later submerged when the sea level rises, and we call them *submarine canyons.* For example, a river cut a deep, narrow canyon into the continental shelf off the coast of Monterey, California, during the last interval of glaciation in the Northern Hemisphere (Figure 13.6).

As you continue your seafloor journey across the Atlantic, you would still be just 200 meters (660 feet) below sea level when you were about 150 kilometers (93 miles) east of Atlantic City, New Jersey. Up to this point, except for those occasional river-cut canyons, it would have seemed as if you were walking across a nearly level surface. But now the fun begins.

Abyssal Plain. **The continental slope is marked by a rapid deepening of the ocean.** The shape of this part of the ocean floor is like the bottom of an in-ground swimming pool where there is a sharp transition from the shallow half of the pool (the shelf) to the deeper, diving end (abyssal plain). The *continental slope* is the transition zone. The depth of the seafloor increases by a couple of thousand meters before the slope becomes gentler at a zone termed the *continental rise* as you approach the deep ocean floor (see Figure 13.5). The continental rise marks the base of the continental slope where sediment swept off the shelf accumulates, resulting in a decrease in slope angle. The continental slope may also be interrupted by submarine canyons that are continuations of those on the shelf. The sea level never fell far enough for the rivers to extend out onto the continental slope. These deeper sections of the submarine canyons were carved by dense, sediment-clogged flows of water that cut into the soft sediment of the slope much like an underwater landslide. Such flows are relatively rare and are triggered by earthquakes and other disturbances in the seafloor.

The continental slope and continental rise mark the transition to the second major depth level of the ocean floor, the abyssal plain, or deep ocean floor (see Figure 13.5). The **abyssal plains are generally over 4 kilometers (about 2.5 miles) below the ocean surface and are some of the flattest portions of Earth's surface.** Abyssal plains are simply oceanic crust that has moved away from the midocean ridge and has been covered by very fine sediments deposited over millions of years. Those sediments filled the cracks and crevasses originally present in the oceanic crust. The abyssal plains are sometimes dotted with underwater volcanoes called seamounts. After slogging through the mud of the abyssal plain for hundreds of kilometers, you would notice the ocean floor gradually beginning to rise.

Oceanic Ridge. The seafloor rises from the abyssal plain to a third level representing the **oceanic ridge system, a submarine mountain chain that can be traced around the world** (see Chapter 4).

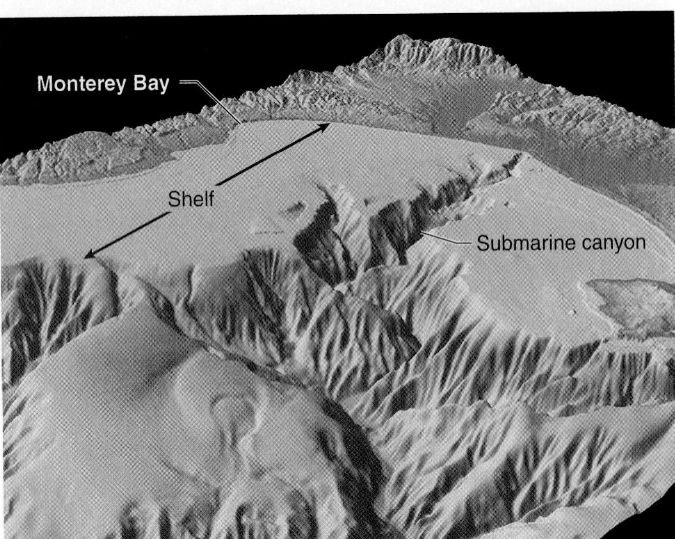

Figure 13.6 Submarine canyon and continental shelf, Monterey Bay, California. The Monterey submarine canyon was cut into the shelf by a river flowing during an interval of lower sea level associated with the last time much of North America was covered with ice.

Checkpoint 13.2

basic | intermediate | advanced | superior

Use the following list to identify and label some of the components of the ocean floor in the image of South America.

Passive margin	Continental slope
Active margin	Abyssal plain
Continental shelf	Oceanic trench

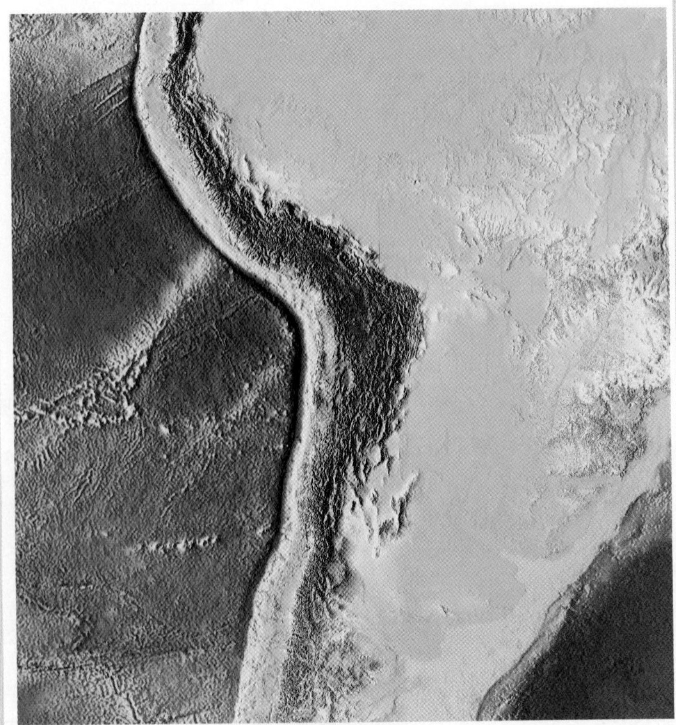

Oceanic ridges are home to more than 80 percent of the world's volcanic activity but do little to change the temperature of the ocean. The heat from the volcanism is quickly dissipated in the vast volume of very cold water found at the depth of the ridges. Researcher Peter Michael of the University of Tulsa compared it to "holding up a small match in a big swimming pool."

The ocean depth is less than 3 kilometers (2 miles) along the crest of the ridge system. The ridge system dominates the floor of the Atlantic Ocean, occupying over one-half its width (see Figure 4.5, Chapter Snapshot). It takes up proportionally less space in the Pacific Ocean. What would happen to the level of the oceans if many midocean ridge systems formed? During the Early Cretaceous period (140 to 100 million years ago), rapid seafloor spreading caused the oceanic ridge system to expand relative to the area of the abyssal plain, decreasing the average depth of the ocean floor, displacing water, and raising the sea level by approximately 170 meters (560 feet). The length of oceanic ridge has steadily declined since the Cretaceous, resulting in a gradual drop in sea level at the agonizingly slow pace of about 1 centimeter every 5,000 years (1 inch per 12,500 years). In the modern ocean, this decline is not apparent as there has been a greater rise in sea level due to the thermal expansion of oceans as a result of global warming.

After you reached the crest of the oceanic ridge system, you would descend into a central rift valley where magma upwells at the spreading center (see Chapter 4). This is a region of submarine hot springs known as *hot smokers* and is home to some of the most peculiar organisms on the planet, living in a unique ocean ecosystem that scientists have only recently begun exploring (Figure 13.7). Once you moved out of the valley, you would begin the second half of your trans-Atlantic walk by descending the other side of the ridge, crossing another section of abyssal plain, and then climbing up the continental slope to reach the shelf and eventually dry land. If you walked due east along latitude 40°N, you would walk out of the ocean along the coast of Portugal onto the lovely sandy beaches of Osso da Baleia.

Oceanic Trenches. The transition from continental shelf to abyssal plain to oceanic ridge (that is, the three depth zones described so far) is continuous along many passive continental margins where the continental and oceanic crusts are part of the same plate. This pattern is not true of active margins where two plates converge, forming an oceanic trench near subduction zones (see Chapter 4). There we find the fourth and final depth level, the **narrow, deep oceanic trenches** (see Figures 4.6 and 13.5). The trenches mark the locations where oceanic lithosphere descends into the mantle. Trenches are consistently the deepest areas on the ocean floor, at depths of 7 to 11 kilometers (4 to 7 miles).

Checkpoint 13.3

basic · intermediate · advanced · superior

Note the line X–Y on the world map used for Checkpoint 13.1. On the basis of your understanding of plate tectonics, which of the profile views most accurately models the bathymetry of the ocean floor along that line?

a) Profile a b) Profile b c) Profile c

a.

b.

c.

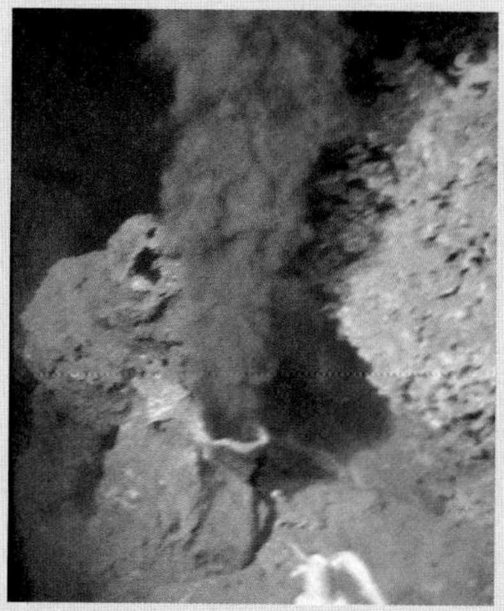

a.

b.

Figure 13.7 Features of the oceanic ridge environment. **a.** A hot smoker vent where mineral-rich heated waters are expelled into the cold waters of the deep Pacific Ocean. **b.** A white crab and tubeworm colony.

Checkpoint 13.4

basic · intermediate · advanced · superior

Sketch a profile of ocean floor bathymetry along line X–Y on the following map. (Hint: See Figures 4.5 and 4.6.)

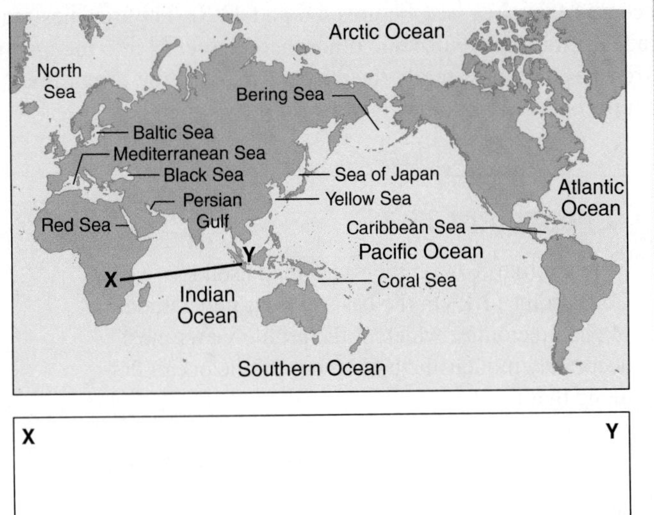

X Y

13.3 Ocean Waters

Learning Objectives

- Identify and explain processes that produce global surface salinity patterns.
- Describe the pattern of surface temperatures in the world's oceans.
- Explain why the salinity and temperature of oceans are variable at the surface but relatively uniform at depth.
- Discuss how and why ocean water salinity, temperature and density change from the ocean surface to the ocean floor.

Where did all that water come from? You will recall from Chapter 2 that Earth's surface was a hostile, hot mass of nearly molten rock for millions of years after its formation. Widespread, violent volcanic activity released gases, including lots of water vapor, into the early atmosphere. This "outgassing" is still observed today, though at a much reduced rate, and is the likely source for most of the water on Earth.

The early atmosphere was much warmer than today and capable of retaining much more water vapor. As the planet slowly cooled, this moisture condensed to form thick clouds, which eventually released their water as precipitation. The rain immediately evaporated from the warm surface, causing a continuous loop of evaporation, condensation, and precipitation. These dismal, hot, overcast, and rainy conditions continued until the planet's surface was sufficiently cool for water to collect in hollows—the beginnings of oceans. The original oceans grew steadily as the planet cooled so that scientists now think that they were mostly in place by 4 billion years ago. As we mentioned in Chapter 3, considerable modern research is aimed at finding out whether comets could also have contributed water to the young Earth. Data from space missions targeting comets were inconclusive because at least some of the small population of studied comets contained water with different properties from that found in Earth's oceans. However, recent examination of a comet originating from the Kuiper belt (see Chapter 3) has revealed water with a similar chemical signature to that of seawater, strenthening the hypothesis that a significant proportion of our oceans were delivered from an extraterrestrial source.

Water Chemistry

Have you ever gone to the ocean and gotten a mouth full of seawater? Do you remember what it tasted like? Seawater contains dissolved salts and minerals (Table 13.1). Where did all those dissolved materials come from and which are most common in seawater? Rainwater falling on the continents reacted with rocks, dissolving minerals that were carried to the oceans in rivers and groundwater. Rivers and streams in the United States alone are estimated to deliver several hundred million tons of sediment and dissolved solids to the ocean annually. It is likely some salts and minerals were also added to the oceans from volcanic activity and atmosphere-ocean reactions. Over time, the dissolved solids increased the salinity of the oceans. For the purposes of this section, think of seawater as a combination of freshwater plus salt (note that common salt, NaCl, accounts for about 85 percent of the dissolved solids in seawater). **Salinity is the measure of the concentration of salt in seawater.** All water contains some salt. Even water in streams on land is likely to have a salinity of up to 0.5 percent. The average salinity of much of the world's major ocean basins is 3.5 percent (35 parts per thousand).

Recently, scientists have noticed a troubling trend in ocean chemistry, *ocean acidification*. The average pH value of the world's oceans has declined from 8.16 to 8.05 since the Industrial Revolution. The pH is slowly decreasing and is expected to decline to around 7.8 by the end of this century (see Chapter 9 for a fuller discussion of pH). The oceans are absorbing some of the excess carbon dioxide from the atmosphere. This forms a weak carbonic acid in the oceans. The more carbon dioxide absorbed, the greater the acidification. Note that the oceans are not technically acidic. A pH value of 7 is neutral; lower values signal acidic solutions, and higher values indicate alkaline solutions. However, pH values are now sufficiently low that the shells of some tiny marine organisms are dissolving and other organisms are growing more slowly; the lower pH levels also can be toxic to eggs and larvae of some species.

Table 13.1 Major Elements and Compounds Found in Seawater

Element or mineral	Percent of dissolved material by weight in seawater
Chloride (Cl)	55.0
Sodium (Na)	30.6
Sulfate (SO_4)	7.7
Calcium (Ca)	1.2
Magnesium (Mg)	3.7
Potassium (K)	1.1
Mn, Pb, Au, Fe, I	Trace

Salinity Patterns. The salinity of the warm, well-mixed surface waters over much of the world's major ocean basins ranges between 3.3 and 3.7 percent (33 and 37 parts per thousand; Figure 13.8). **Increasing the amount of salt in seawater increases the density of the seawater.** What patterns of salinity concentration do you observe in Figure 13.8, and what variables might influence those patterns? Salinity in the world's oceans varies because of temperature and the mixing action of ocean currents, plus freshwater input from rain, streams, and melting ice. The big picture is that salinity is generally highest around 25°N and 25°S latitude (Figure 13.8), where temperatures are relatively high but skies are clear as a result of global air circulation patterns (see Chapter 14). The higher temperatures and lack of rainfall result in more evaporation, which removes water but leaves the salt it contains behind.

Why are the salinity values not highest right at the equator, where annual temperatures are highest? Salinity values are a little lower at the equator because more precipitation occurs over equatorial regions, delivering a regular supply of freshwater. Salinity values are a few points lower at higher latitudes (for example, 60°N and 60°S; Figure 13.8), where cooler temperatures mean less evaporation and where melting ice adds freshwater. Note that the difference in salinity between the coast of Antarctica and the ocean around the Hawaiian Islands in the north-central Pacific Ocean is less than 0.2 percent (2 parts per thousand; ppt). As we will learn in Section 13.4, the ocean currents provide an effective mixing mechanism to even out some of the differences in salinity in the open ocean.

If we want to find more-extreme salinity values, we need to look in smaller, restricted ocean basins and seas. For example, look at the Mediterranean Sea in Figure 13.8. Salinity values there are some of the highest on Earth (over 4.0 percent; 40 ppt). The Mediterranean Sea has only a small opening to the Atlantic Ocean (at the

Checkpoint 13.5

Examine the following map of mean salinity for the Indian Ocean. Provide an explanation of why salinity values are lower for the tropical Bay of Bengal (east of India) than for the cold waters of the Southern Ocean just north of Antarctica.

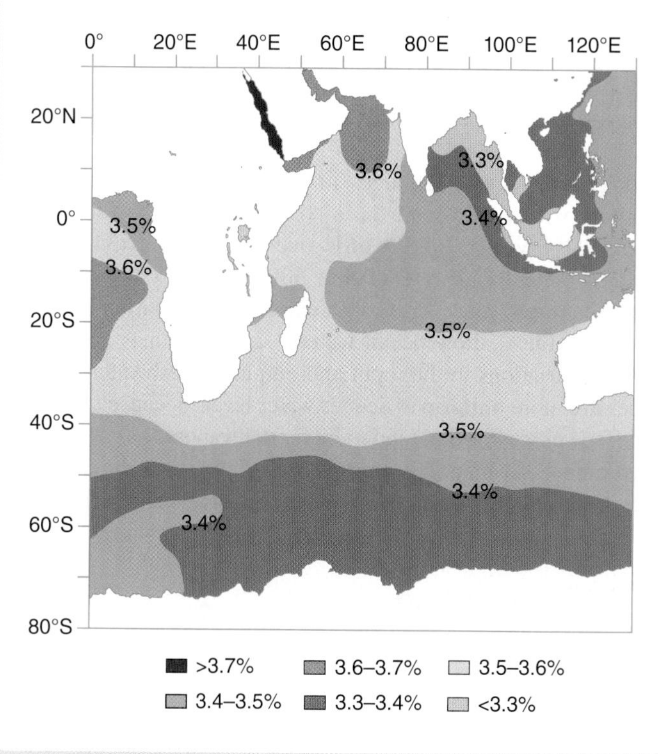

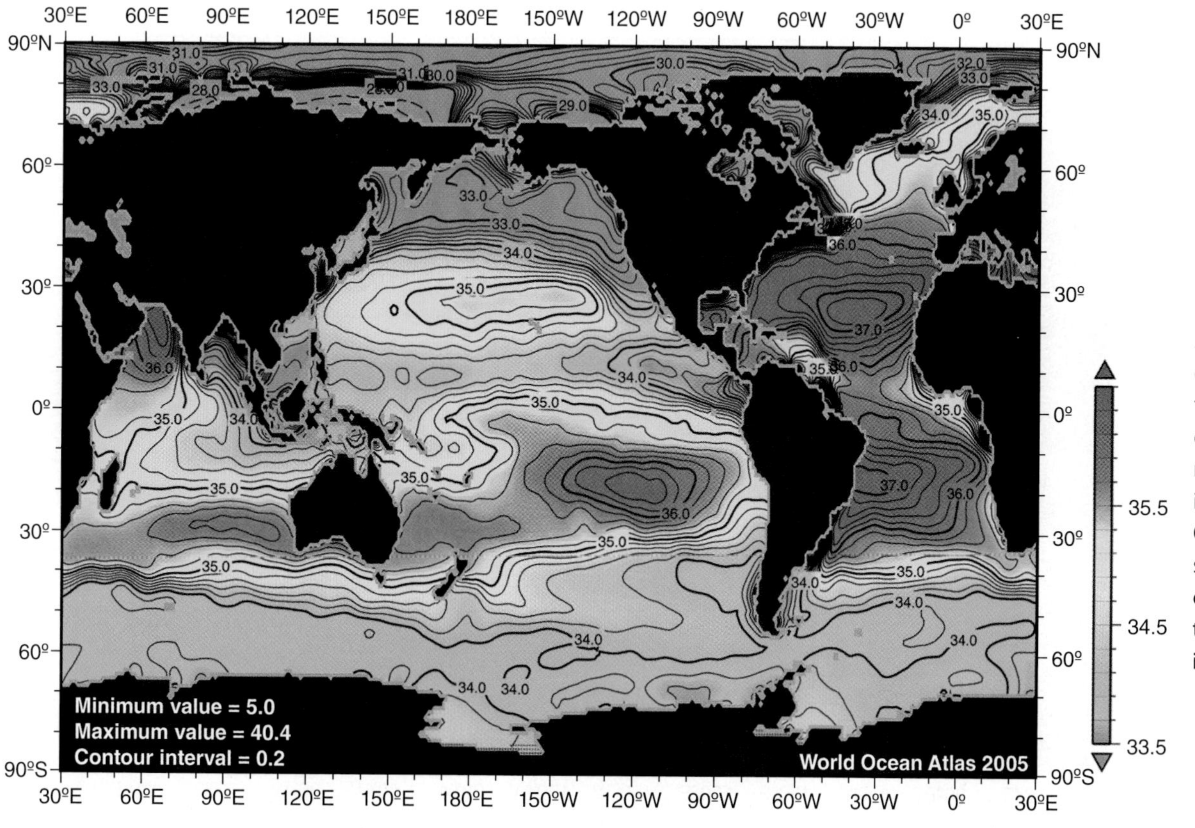

Figure 13.8
Salinity and latitude. Numbers represent annual mean salinity values in parts per thousand (10 parts per thousand equals 1 percent) using a scale known as the Practical Salinity Scale (PSS). Salinity in the open ocean is greatest in tropical regions and decreases in polar regions, being lowest in the isolated Arctic Ocean. Note that salinity values are consistently higher in the Atlantic Ocean than in the Pacific Ocean.

Strait of Gibraltar) and is located just above 30°N. High temperatures coupled with sparse precipitation and very restricted mixing with Atlantic waters all contribute to the high salinity values in the Mediterranean Sea and in other sheltered seas at similar latitudes.

Lower salinity values of 2 to 3 percent are recorded for the high-latitude Arctic Ocean, which connects to the major oceans only through narrow passages in the North Atlantic Ocean and at the Bering Strait between Alaska and Russia (Figures 13.1 and 13.8). Water in the isolated Baltic Sea also undergoes little mixing with the North Atlantic Ocean and consequently has salinity values approaching those of freshwater (0.5 to 1 percent).

Salinity and Depth. Changes in salinity at the ocean surface are only part of the story. Salinity is much more uniform with depth. Salinity values of a little over 3.4 percent (34 ppt) are found consistently in waters below 500 meters (1,640 feet). What processes might cause salinity to exhibit less variation in deeper waters? In the shallow ocean waters, salinity values change because of variations in the input and output of freshwater. Salinity values are more uniform at deeper water levels because the effects of surface processes that tend to increase (evaporation) or decrease (stream flow) salinity are absent. **A well-defined rapid change in salinity with depth is known as a *halocline*.** For most regions of the world's oceans, the halocline—the depth zone where the salinity changes—starts just below the surface and ends at a depth of around 500 meters (1,640 feet).

Checkpoint 13.6

Predict how salinity varies with depth in the tropical Pacific Ocean and in the Arctic Ocean. Sketch a graph of depth versus salinity for both locations.

Water Temperature

As we discussed in Chapter 2, some parts of Earth receive more solar radiation than others. Solar radiation strikes Earth more directly at the equator and tropics than it does in polar regions, transferring 2.5 times more solar energy to the oceans near the equator. As a result, **more heat is available to be absorbed in the tropics than near the poles.** Much of this solar energy raises the temperature of the ocean. The highest average annual ocean temperatures (about 27°C; 81°F) are present along the equator; temperatures decrease symmetrically to the north and south, approaching 0°C (32°F) at high latitudes, the area of the globe characterized by annual ice caps (Figure 13.9). Ocean currents moderate the temperature gradient from the equator to the poles, but this action is not sufficient to overcome the contrast in incoming heat energy.

Water exhibits a key attribute that makes the oceans a great storage reservoir for thermal energy. **Water can absorb a lot of**

Checkpoint 13.7

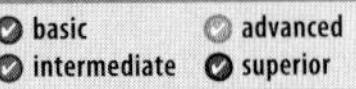

The specific heat of the water in the oceans is about four times that of rock and soil on the continents. In addition, water in the oceans moves, while rock and soil are effectively stationary. What are the implications of these observations for differences in maximum and minimum temperatures for the oceans and continents?

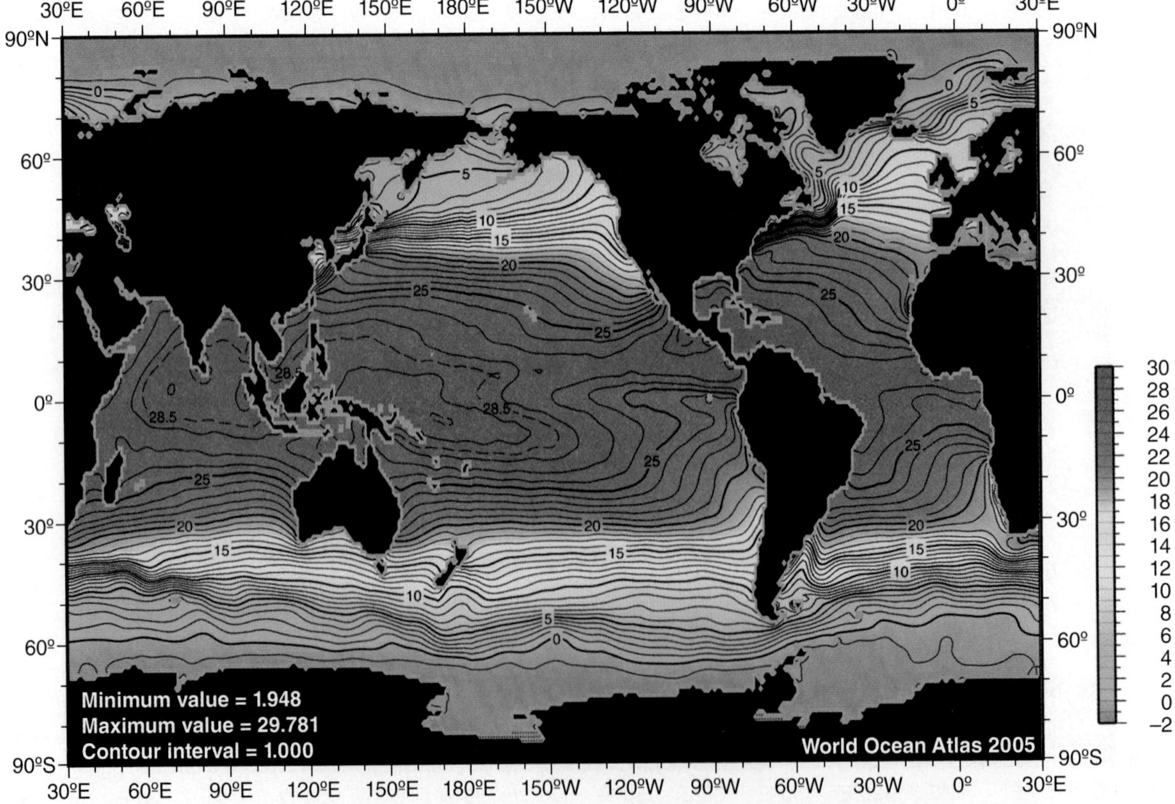

Figure 13.9 Ocean temperatures and latitude. The average annual mean temperature of the world's ocean surface varies with latitude. Numbers represent temperature in degrees Celsius.

thermal energy in comparison to equivalent masses of other materials. The amount of thermal energy required to raise the temperature of 1 gram of a material by 1°C is termed the *specific heat.* Heat is the transfer of thermal energy from one mass to another in a system. The temperatures of materials with low specific heat will rise quickly as thermal energy is applied. For example, the bottom of a pan on a stove will be hot soon after the heating element is switched on. In contrast, materials such as water that have high specific heat can absorb a lot of energy without displaying much of a change in temperature (the measure of thermal energy for a material with a given mass and specific heat).

Figure 13.10 Water density differences. Icebergs float on a bay in Greenland because ice is less dense than water.

For example, suppose you place 1 liter of water (which weighs 1,000 grams) in a 1,000-gram aluminum pan on a stove and heat that water. It takes 200 calories of heat to raise the temperature of the metal pan by 1°C. In contrast, it takes 1,000 calories of heat to raise the temperature of the water by the same amount.

Why are we concerned about water's high specific heat? The ability of water in the oceans to absorb, store, and release heat energy plays a crucial role in regulating global climate patterns. As we will learn in Section 13.4, the high specific heat of water (that is, three to five times greater than that of solids) allows ocean currents to transfer huge amounts of heat from the tropics to higher latitudes. This raises temperatures in cold places such as Canada and northern Europe and has a cooling effect on temperatures in the tropics.

Water's Density, Temperature, and Depth

Have you ever gone swimming in a lake in the summer? If you have, you may have noticed that the water was much colder near the bottom of the lake than at the top. This is a function of another property of water—the density of water decreases with increasing temperature (above 4°C [39°F]). In general, **warm water is less dense than cold water,** and so it "floats" on top of the cold water. Water density increases as water temperature decreases down to a temperature of 4°C (39°F). Below 4°C, the trend reverses, and the density of really cold water decreases, especially when water changes from a liquid to a solid to form ice. Think about ice cubes floating in a drink or icebergs floating in the ocean. Consequently, **dense cold water can sink below less-dense warm water,** but ice will float on the ocean's surface (Figure 13.10).

Because of the effects of temperature on water, the major oceans can be divided vertically into shallow layers of relatively warm water and layers of cold water at greater depths (Figure 13.11).

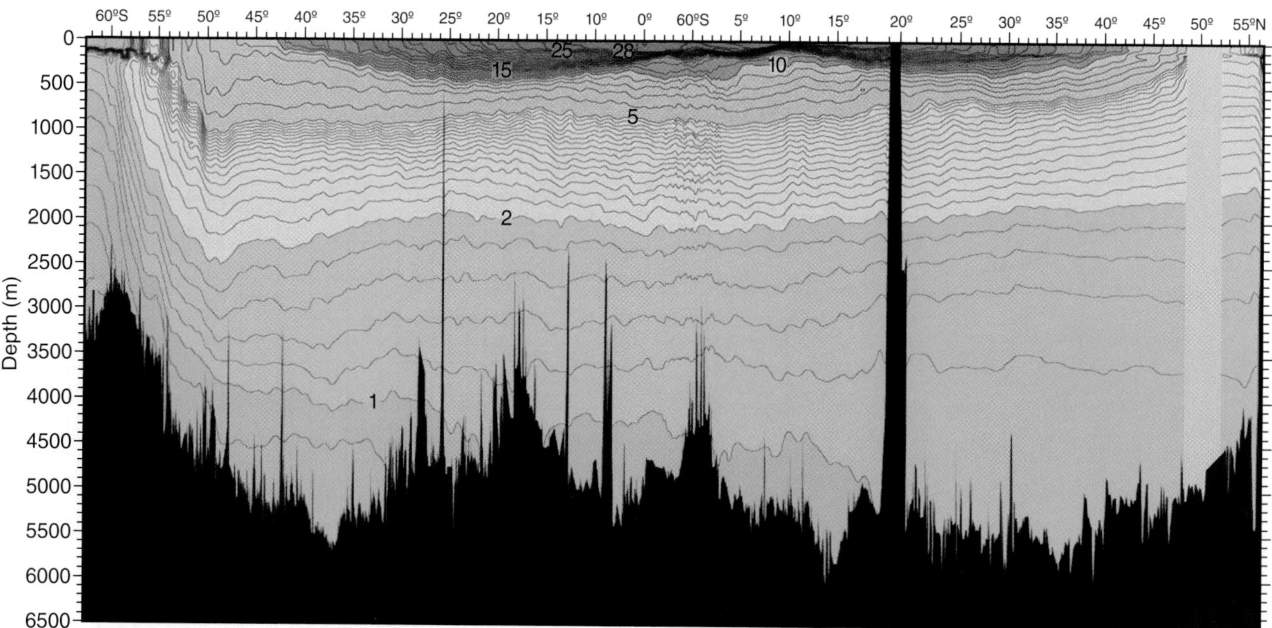

Figure 13.11 Ocean temperature versus depth and latitude. North-south profile through the Pacific Ocean along the 150°W meridian, illustrating the range of temperature with depth and latitude. Temperature decreases with depth. These temperatures are corrected for temperature changes caused by increasing pressure with depth (thus called potential temperature). The most rapid temperature changes (the thermocline) occur between 100 and 500 meters (328 and 1,640 feet). The black pattern represents the seafloor bathymetry along this profile. Temperatures are expressed in degrees Celsius.

Surface waters are warmed by solar radiation (though less and less as one moves north or south from the equator). Near the surface, currents mix the shallow waters, resulting in relatively uniform temperature distributions by latitude. Because sunlight doesn't penetrate far below the ocean surface, solar radiation does not help heat the waters below a few hundred meters, and the temperature stabilizes below these depths.

Temperatures exceed 20°C (68°F) over much of the tropical oceans' surfaces but decline to a uniformly chilly 2°C (36°F) at a depth of about 2,000 meters (6,600 feet). **The depth zone where temperature decreases most rapidly is known as a *thermocline*** (just as the rapid change in salinity was called a *halocline*). The bottom of the thermocline may be as deep as 1,000 meters (3,300 feet).

In addition to salinity and temperature, a third factor affects density. Can you predict what it is? Think about diving into a pool or lake and swimming to the bottom. What is the effect on your ears? Do you remember feeling an increase in pressure? As you go deeper and deeper, pressure increases uniformly with depth. Increasing the pressure by increasing the depth of the overlying column of water slightly increases the density of the water below.

Now think back to our discussion of ocean salinity and temperature. How does each of those factors affect seawater density? Increasing salinity increases seawater density, and decreasing temperature increases seawater density. Take another look at the salinity and temperature data (Figures 13.8, 13.9, and 13.11). Predict what happens to density as the depth of an ocean increases.

Temperature, salinity, and pressure combine to affect the density of seawater at depth in the oceans. Temperature is, by far, the most important factor. One can deduce that by analyzing the idealized density versus depth graph in the oceans (Figure 13.12). Salinity varies depending on latitude and depth. The more salt in the seawater, the denser the seawater. Pressure increases with depth, and therefore deeper water is denser. However, the graph shows there is a rapid increase in density from a depth of about 200 to 1,000 meters (660 to 3,300 feet) (called a pycnocline). Below that, density is relatively constant. That is most similar to the pattern observed in the temperature data (Figure 13.11). Temperatures do not change much below 1,000 meters; therefore density does not change much below that depth. The density differences that do exist separate the ocean into three main vertical density layers: surface (about 2 percent of ocean waters), middle (about 18 percent), and bottom (about 80 percent). As we will see in Section 13.4, these vertically separated density layers play an important role in controlling how heat is transferred between different parts of the world's oceans.

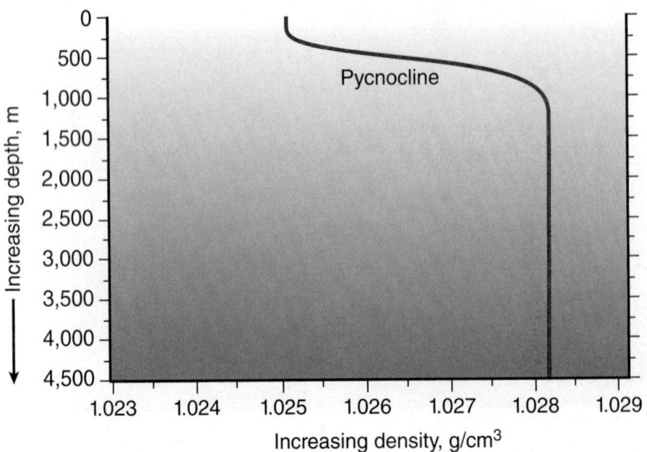

Figure 13.12 Idealized seawater density versus depth graph. Density is lowest at the surface and increases to a depth of approximately 1,000 meters (3,300 feet). Density is generally uniform below the pycnocline.

Checkpoint 13.8

basic / intermediate / advanced / superior

The **thermocline** marks a *zone of relatively rapid temperature change* between the warm surface water and the deeper cold waters. In this exercise, you will attempt to identify the location of the thermocline.

1. Plot the data points from the table on the graph provided here, and sketch a best-fit curve for each data set.
2. These data come from subpolar and tropical oceans. Label your plots as subpolar and tropical.
3. Identify the approximate range of depths for the thermocline on each curve, and label those parts of the curves accordingly.
4. Circle the data points on the graph representing where you would expect to find the highest and lowest salinity values. Explain why.

Depth, meters	Temperature, °C	
	Data set 1	Data set 2
0	12	30
100	12	24
300	11	21
500	9	17
1,000	8	13
1,500	7	10
2,000	6	9
3,000	5	8

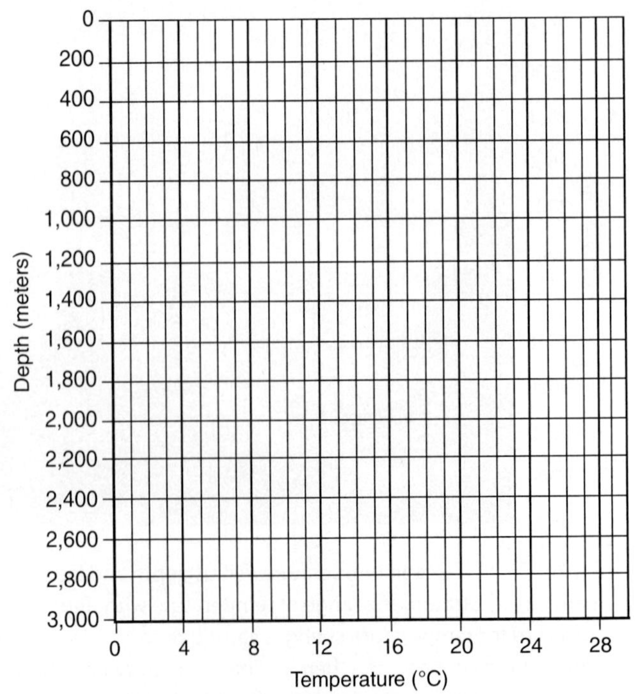

13.4 Oceanic Circulation

Learning Objectives

- Predict the path of currents in the world's oceans.
- Describe the primary factors that control ocean circulation patterns.
- Explain the processes and consequences that result in thermohaline circulation.

Water in the oceans is in constant motion. We will discuss the movements of tides in Section 13.5 and waves in Section 13.6, but this section is concerned with the currents that cross oceans, the same currents that carried explorers from Europe to North America and that allowed the settlement of isolated Pacific islands.

Ocean Currents

Just as we can use elephant seals to learn more about ocean conditions, we can use rubber ducks to learn more about ocean currents. Each year, cargo ships lose an estimated 10,000 containers that fall overboard or are swept away in storms. In 1992, thousands of bright yellow ducks, along with some colorful turtles, beavers, and frogs set off on a heroic ocean journey after some containers were lost overboard in the northern Pacific Ocean. Hundreds of the toys have subsequently been recovered on beaches in Alaska and Canada. (Some may have even been eaten by elephant seals!) The locations where the toys washed ashore have been used by researchers to track the paths of the regional ocean currents. Rest assured, however, that research on ocean currents does not rest exclusively on the analysis of lost bath toys. Scientists also rely on satellites that monitor extensive arrays of floating buoys rigged with instruments at a range of depths.

As we consider how and why water moves around in the oceans, we recall that processes in the oceans are influenced by other components of the earth system, especially the atmosphere, geosphere, and exosphere. Each component plays an important role in determining how and why water moves in the oceans. First, consider what happens near the surface of the ocean where it interacts with the exosphere (space) as represented by the external energy input from the sun. At the simplest level, warm ocean surface water is heated at low latitudes and flows outward from equatorial regions toward the poles. The moving water is affected by a second earth system component, winds, which are themselves influenced by the rotation of Earth. There is friction between the wind and the water surface. As winds blow across water, they generate a force on the water in the same direction that the wind is blowing. This force does not get transmitted completely through the water column but only affects about the upper 100 meters (330 feet) of the oceans. Thus, only about 10 percent of the world's ocean water is actually moving in surface currents at any given time. In general, **ocean currents follow prevailing wind directions except where the current encounters an obstacle, such as a landmass.** The continents represent barriers to ocean currents, deflecting them from their initial course.

On a global scale, **circulation patterns in the atmosphere generate near-circular patterns of ocean currents known as *gyres*** (see Chapter 14 to learn more about why these air circulation patterns occur). The gyres are centered on 30 degrees latitude in each of the major ocean basins (Figure 13.13). Circulation of the

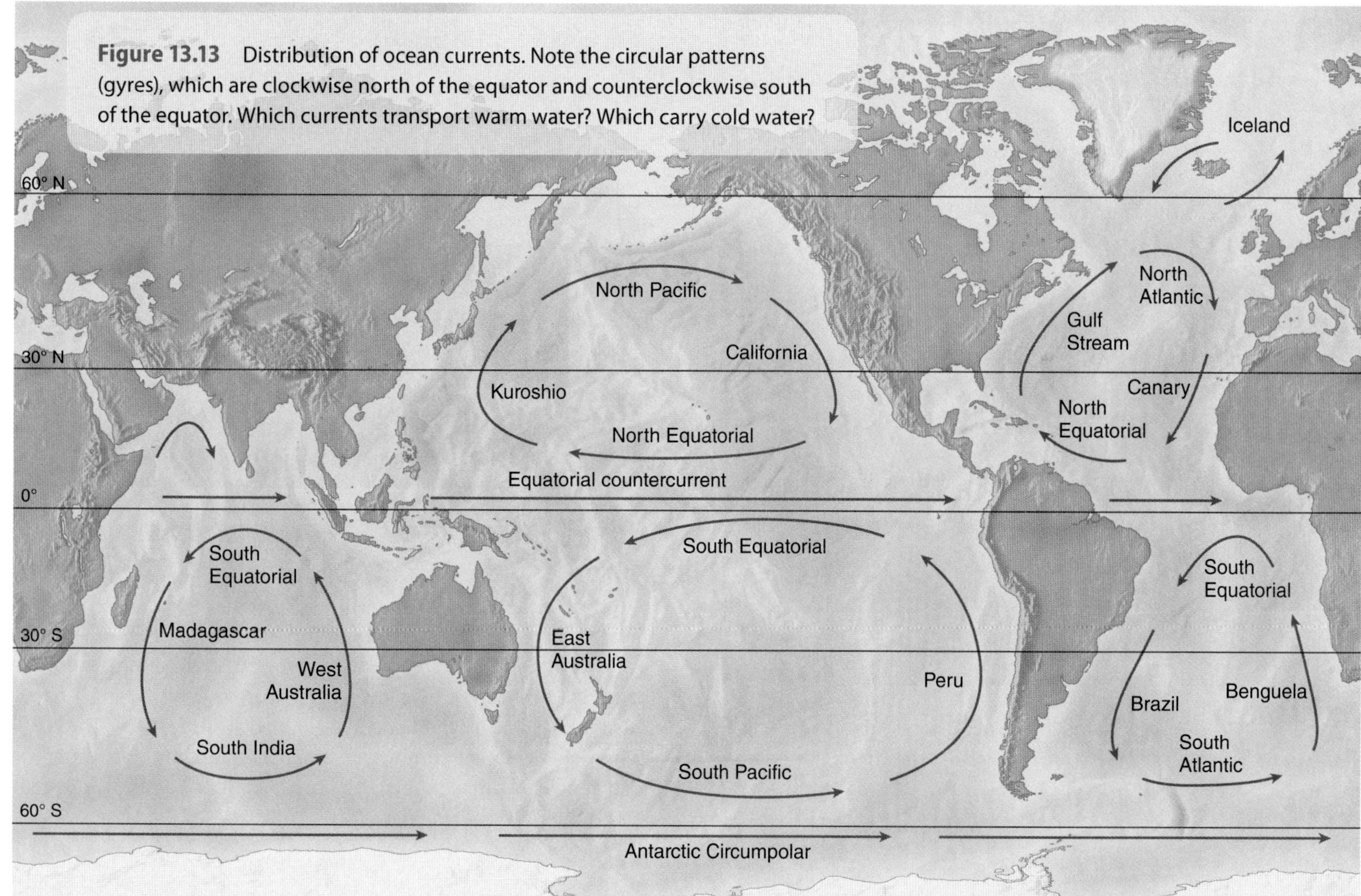

Figure 13.13 Distribution of ocean currents. Note the circular patterns (gyres), which are clockwise north of the equator and counterclockwise south of the equator. Which currents transport warm water? Which carry cold water?

gyres is clockwise in the Northern Hemisphere and counterclockwise in the Southern Hemisphere (more about that later). Surface water might take from several months to a few years to complete the circuit of a given gyre.

Circulation in the western portion of the gyres results in **fast-flowing boundary currents that redistribute warm tropical waters toward the poles.** Boundary currents such as the Gulf Stream and the Kuroshio current (Figure 13.13) can be thought of as marine rivers because they are relatively narrow water masses (less than 100 kilometers [62 miles] across), and they flow at speeds of 100 to 200 kilometers per day (62 to 124 miles per day) for thousands of kilometers. The Gulf Stream (Figure 13.14) can transport over 50 million cubic meters (1.76 billion cubic feet) of warm, tropical water per second, hundreds of times more water than the Amazon, the world's largest river. Some of this water will not make it all the way around the gyre. Some will take a detour toward the Arctic Ocean in the North Atlantic, while part of the rest sinks toward the ocean floor near Iceland. The eastern boundary currents that complete the eastern leg of each gyre (for example, Canary, Benguela; Figure 13.15) are wider and shallower, carry less water, and move more slowly. The Canary current, nearly 1,000 kilometers (620 miles) wide, carries just one-third of the volume of water in the Gulf Stream and travels at tens of kilometers per day. **Circulation in the eastern parts of the gyres produces currents that transport colder water from high latitudes toward the equator.**

Coriolis Effect

Look carefully at Figure 13.13. Why do the gyres in the Northern Hemisphere circulate clockwise whereas the southern gyres circulate counterclockwise? Why don't they both show the same circulation

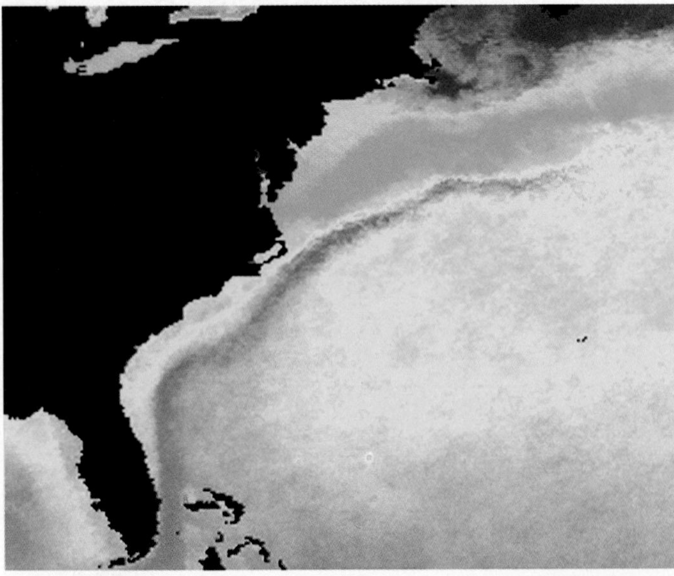

Figure 13.14 Thermal infrared image of the western Atlantic Ocean. Temperatures in orange are about 25°C (77°F), while temperatures in dark blue are about 2°C (36°F). Can you identify the path of the Gulf Stream?

Checkpoint 13.9

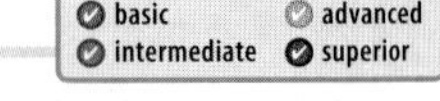

A shipment of rubber elephants falls overboard in the northern Pacific at location A on the following map. What path do the elephants subsequently follow?

a) A–G–B–F–E–A
b) A–E–C–G–A
c) A–G–C–E–A
d) A–E–F–B–G–A

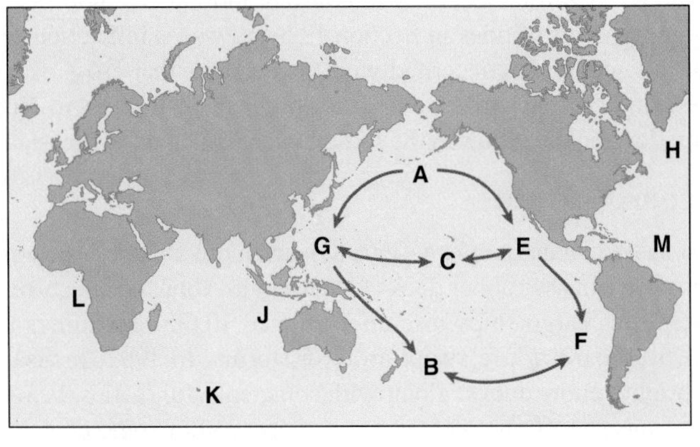

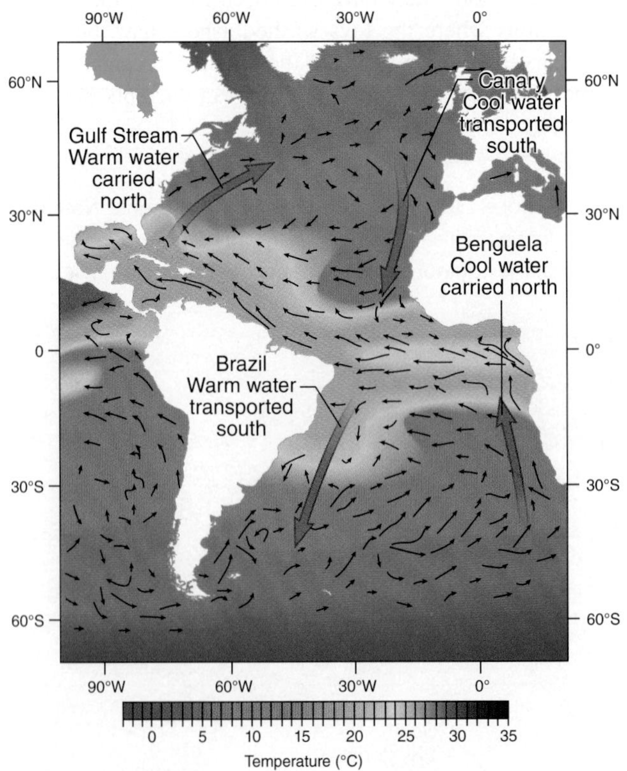

Figure 13.15 Atlantic Ocean currents and sea surface temperature averaged for April–June. The Gulf Stream and Brazil currents represent bands of warm currents (yellow and light-green colors) along the Atlantic coastline of North and South America, respectively. The Canary and Benguela currents are broader regions of cooler water (dark green) that move south and north along the coasts of northern and southern Africa, respectively.

pattern? **Atmospheric and oceanic circulation patterns are deflected to the right of their course in the Northern Hemisphere and to the left of their course in the Southern Hemisphere; this pattern is termed the *Coriolis effect.*** The Coriolis effect is a consequence of the fact that Earth rotates from west to east and that objects traveling at different latitudes on Earth's surface move at different speeds as a result of the rotation of the planet. The winds or currents do not actually change direction, but the planet beneath them changes position. The effect is most easily observed in large air masses or moving bodies of water. The net effect of these deflections is the circular path of shallow ocean currents.

Visualizing the Coriolis Effect. The Coriolis effect can be a difficult concept to grasp because, although we recognize that Earth rotates, it is difficult to accept that most of the US population is hurtling along at over 1,000 kilometers per hour (620 miles per hour). It is much like traveling in a car. Although we are sitting still in a car that is moving at 100 kilometers per hour (62 miles per hour), we recognize that we too are traveling at the same speed. If we were foolish enough to try to exit the vehicle, we would have the same velocity as the car (until we hit something that wasn't moving).

Checkpoint 13.10

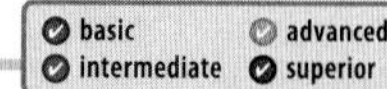

How would the deflection of ocean currents be altered in the Northern Hemisphere if Earth rotated from east to west (instead of from west to east)?

a) Currents stay the same, deflect right of their courses.
b) Currents stay the same, deflect left of their courses.
c) Currents switch directions, deflect right of their courses.
d) Currents switch directions, deflect left of their courses.

To understand how the Coriolis effect works, we must first recall that Earth makes one complete rotation on its axis every 24 hours. An object on the equator travels the circumference of the globe (approximately 40,000 kilometers; 25,000 miles) each day. An object at a higher latitude travels a shorter distance in that same time period. The velocity of a location on Earth's surface decreases with increasing latitude. For example, the city of Columbus, Ohio, located at 40°N (Figure 13.16), is moving at 1,284 kilometers per hour (about 800 miles per hour) compared to US cities located along the 30th parallel (for example, New Orleans), which travels at 1,452 kilometers per hour (about 900 miles per hour). Meanwhile, Seward, Alaska (60°N), pokes along at only 838 kilometers per hour (about 520 miles per hour). Objects at the poles simply rotate around a vertical axis.

So what has all this got to do with the Coriolis effect? This contrast in velocity with latitude causes the travel paths of winds and ocean currents to be deflected to the right of their course in the Northern Hemisphere. Here's why. Imagine yourself in Panama City, Florida, at latitude 30°N and 1,100 kilometers (680 miles) south of Columbus, Ohio (Figure 13.16). At noon, you fire a rocket directly north. The rocket travels north at 1,100 kilometers per hour (680 miles per hour). However, it also has the eastward velocity (1,452 kilometers per hour [902 miles per hour]) of the launch site (Earth is rotating to the east). The rocket would arrive in Columbus at 1 P.M., if Columbus were traveling at the same velocity as Panama City. But Columbus moves eastward only 1,284 kilometers (798 miles) between noon and 1 P.M. Since the rocket moves eastward an additional 168 kilometers, it will land east of Columbus. Your dastardly plans are thwarted.

To the citizens of Columbus, it would appear that the rocket was deflected to the right of its course. If they sought to retaliate by launching a similar missile toward Panama City, they too would miss their target by 168 kilometers (104 miles). Because Panama City moves east more rapidly than Columbus, the rocket would land west of the city, again an apparent deflection to the right of its (southerly) course.

Continents and Oceanic Circulation

We learned in Chapters 4 and 6 that plate tectonics can influence climate through volcanism and the formation of mountains. Now we will consider how plate tectonics can influence climate patterns by altering the distribution of continents and opening and closing avenues for oceanic circulation.

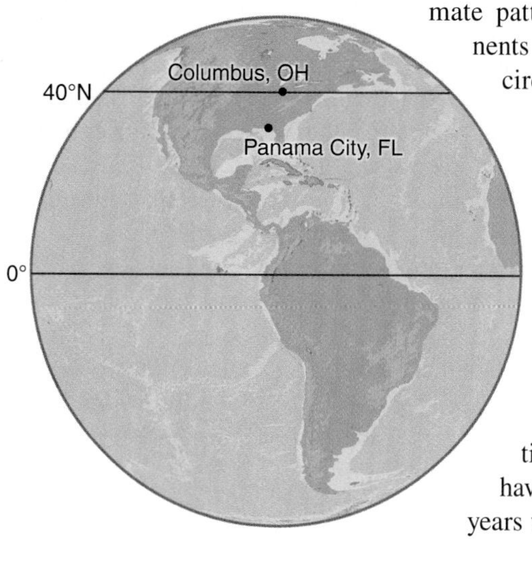

Examination of any world map will reveal that the continents form barriers to circulation in the oceans. The dominant north- and south-directed currents, such as the Gulf Stream or the Canary current, occur along the continental margins. What did oceanic circulation look like at times in the geologic past when the continents were located elsewhere? We don't have to look too far to answer this question. The locations of continents in the Northern Hemisphere have not changed much in the last few million years with the exception of the connection between

40°N
Columbus, OH
Panama City, FL
0°

Figure 13.16 Latitude and rate of rotation. Locations at a latitude of 40°N travel a much shorter distance during one rotation of Earth than places located on the equator. Thus, their rate of rotation is about 400 kilometers per hour (249 miles per hour) slower.

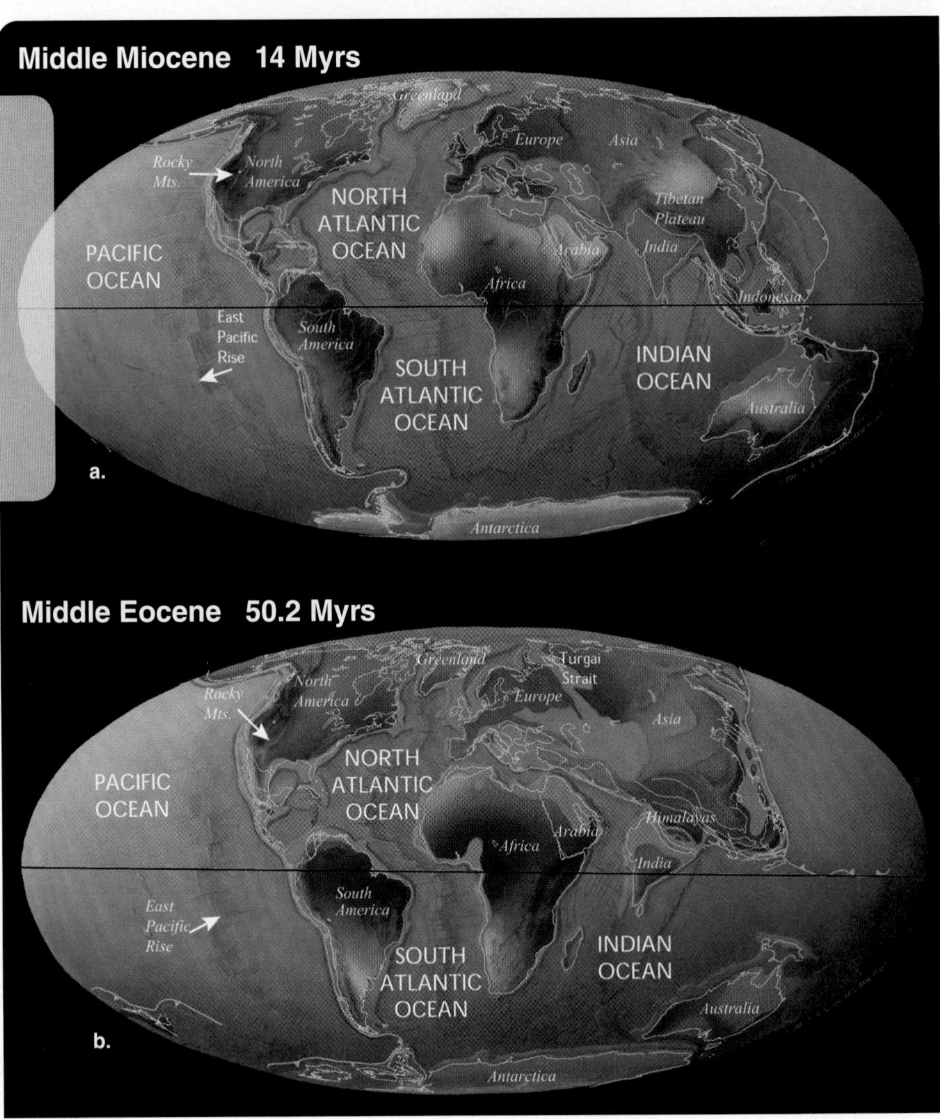

Figure 13.17 Distribution of continents and oceans 14 and 50 million years ago. **a.** The Isthmus of Panama was open 14 million years ago, allowing water to circulate between the oceans. When the isthmus closed, Atlantic Ocean currents shifted north. **b.** The proximity of southern continents to Antarctica kept it ice-free as it was warmed by currents from the Pacific, Atlantic, and Indian oceans. Plate tectonics resulted in the northward migration of Australia, Africa, and South America, and a circumpolar circulation pattern developed that surrounded Antarctica with cold ocean currents.

North and South America (Figure 13.17a). A gap existed between the North and South American continents until around 3 million years ago. Prior to that time, ocean currents traveling through the gap carried water from the Pacific Ocean into the Atlantic Ocean. A chain of volcanic islands gradually filled in the gap along the convergent plate boundary between the Caribbean and Cocos plates (see Chapter 4 Snapshot). Uplift and sedimentation created the narrow strip of land that we now know as the Isthmus of Panama. Once this land bridge was completed, it opened the way for animals from South America to migrate north. Thanks to the Isthmus of Panama, Texans can see the near-prehistoric-looking armadillo roaming through their state.

The closing of the Isthmus of Panama had a significant impact on circulation patterns in the Atlantic Ocean. Currents in the western Atlantic were forced northward, strengthening the Gulf Stream and bringing warmer waters farther into the North Atlantic. The warmer water raised temperatures in Europe, increasing winter temperatures as much as 10 C° (18 F°). One consequence of this change was that winters are much milder in northern Europe than they are in parts of North America located much farther south. Duluth, Minnesota, can go for weeks in winter without getting above freezing, but it is rare for Paris, France, located at a similar latitude, to experience extremely cold temperatures. Spring tulips bloom at the same time in Scotland (60°N) as they do in equivalent settings in Vermont (43°N). Without the influx of less-salty water from the Pacific Ocean (see Figure 13.8), the salinity of Atlantic waters increased. This in turn would help drive the vertical circulation patterns that control global climates (see the discussion of thermohaline circulation in this section).

An opposite change in oceanic circulation in the Southern Hemisphere is responsible for the development of continental ice sheets on Antarctica. This time, plate tectonic processes separated two continents. Antarctica has long been located at the South Pole but did not have any significant ice cover for much of that time. The trigger for large-scale glaciation appears to have been the separation of the South American and Australian continents from the Antarctica landmass about 34 million years ago. Before that time, the warm tropical waters in the Pacific and Atlantic oceans would have been carried south to warm up Antarctica (Figure 13.17b). The separation of Antarctica from South America opened up the strong currents that encircle Antarctica in the Southern Ocean, essentially isolating the southern continent from moderating ocean currents and maintaining deep-freeze conditions year-round that are necessary for sustaining a permanent ice cap.

Thermohaline Circulation

The variations in the temperature and salinity of the oceans play a critical role in oceanic circulation, which in turn influences global climate patterns. **Deep currents that transport water far below the ocean surface are driven by differences in water density that depend on contrasts in both temperature and salinity.** The Gulf Stream carries high-salinity, warm waters from the central Atlantic to higher latitudes (see Figure 13.15). The water slowly cools as it travels northward and is dispersed among the landmasses of the north Atlantic Ocean. The salty Gulf Stream water is cooled by cold winds, and water density increases further as more salt is expelled into the Atlantic as seawater changes state to form ice. This cold, salty water is denser than the surrounding surface waters, so it sinks to the bottom of the ocean near Greenland and Iceland. **The sinking water forms the start of a circulation cell that carries these waters southward along the bottom of the Atlantic Ocean** (Chapter Snapshot). This descending current is known as the North Atlantic Deep Water (NADW) mass and is found at depths of 2 to 4 kilometers (1 to 2.5 miles). When the NADW reaches Antarctica, it is diverted eastward to the Indian ocean.

The deep-water current eventually comes to the surface (upwelling) in the northern Indian and Pacific oceans. Normally, we would not expect these deep waters to rise back to the surface because they should be denser than the warmer surface waters. However, upwelling of dense, deep ocean waters occurs at locations where the deep-water currents meet a landmass and where prevailing winds produce surface currents that carry shallow waters away from the continents. These upwelling locations are extremely important because these currents bring nutrients to the surface. These nutrients are essential for the survival of many marine organisms. Eventually, the water returns to the Atlantic Ocean by a series of surface currents. A complete loop of the cell from the North Atlantic to the Pacific and back takes more than 1,000 years. As a consequence, the deep ocean is today capturing heat from our present climate and releasing heat it absorbed more than a millennium ago. **The pattern of deep currents is termed *thermohaline circulation*** because both temperature (*thermo*) and salinity (*haline*) play a role. The sinking of this cold, dense water in the North Atlantic is a key step in the global conveyor belt. This system moves thermal energy from the tropics to the poles and helps moderate Earth's climate by redistributing heat around the various oceans.

Checkpoint 13.11

basic | intermediate | advanced | superior

A fish tank is filled with water at room temperature. Cold water is added on one side of the tank, and warm water is added to the other side. The water at each temperature is dyed a different color to show its movement through the tank.

1. Predict what will happen when warm and cold water are added to the tank simultaneously. Briefly describe your prediction and sketch it in the drawing of the tank.
2. Label the diagram with features that serve as analogs for the low latitudes, the high latitudes, and the thermocline.

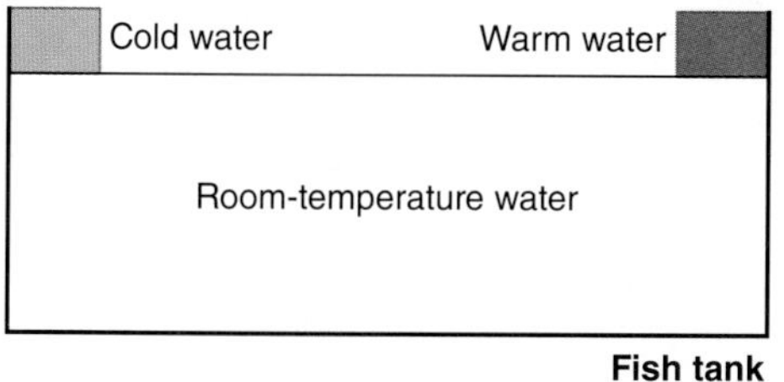

Checkpoint 13.12

basic | intermediate | advanced | superior

Predict what would happen to thermohaline circulation if there were catastrophic melting of the Greenland ice sheet. What would be the consequences for northern Europe?

The El Nino/Southern Oscillation (ENSO): An Example of Earth as a System

El Niño is an unusual oceanic current that forms in the Pacific Ocean and occurs on average every 4 years, each time lasting about 12 months. After the seasons, El Niño and its associated La Niña event are the most significant cause of year-to-year climate variability on Earth. During a normal (non–El Niño) year, ocean waters located near Australia and Indonesia are heated by incoming solar radiation. This leads to an atmospheric circulation pattern that generates trade winds that blow from east to west, causing water to "pile up" in the western Pacific Ocean. Meanwhile, upwelling cold, deep waters carry nutrients to the surface off the coast of Peru, supplying food for one of the greatest fisheries in the world. Those conditions result in warm temperatures and wet conditions in the western Pacific and cooler, dry weather off the coast of South America.

How does this pattern change during an El Niño year? The trade winds disappear, and hot water remains in the eastern Pacific, raising the elevation of the ocean surface (Figure 13.18a). Heavy rain falls on the west coast of South America, resulting in flooding and mudslides in coastal regions. The precipitation contributes to decreasing surface water salinity near the coast and prevents upwelling currents. Drought occurs in western Pacific nations, and warm air moves farther north over parts of North America. El Niño results in altered climate patterns around the world. Some of the results are negative (droughts, floods), whereas others may be positive (milder winters).

Eventually, conditions may either return to normal or swing in the opposite direction, forming La Niña conditions (Figure 13.18b). In the latter case, cold conditions are even more pronounced than usual, causing droughts in South America and the western United States and more severe weather than normal in the western Pacific. These changes are also reflected in atmospheric air pressures with rising pressures in the east accompanied by falling

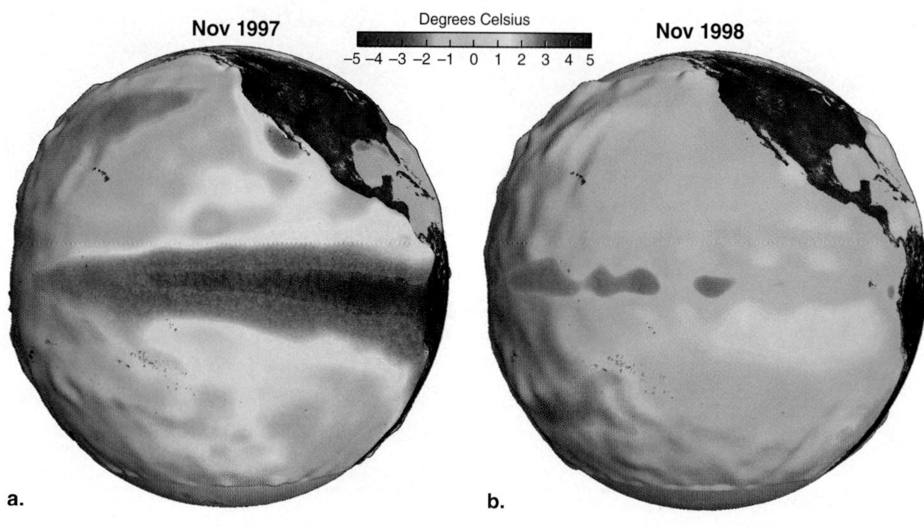

Figure 13.18 Sea surface temperature anomalies and elevation for El Niño and La Niña events. Red colors indicate water temperatures that were up to 5 C° (9 F°) warmer than normal. Blue colors show water temperatures that were up to 5 C° (9 F°) colder than normal. **a.** The 1997 El Niño featured warmer waters (red) and higher seas in the eastern Pacific Ocean. **b.** The subsequent La Niña event had cooler waters (blue) and lower sea surface elevations. How do sea temperatures along the western US coast vary between these events?

GLOBAL CIRCULATION AND TOPOGRAPHY

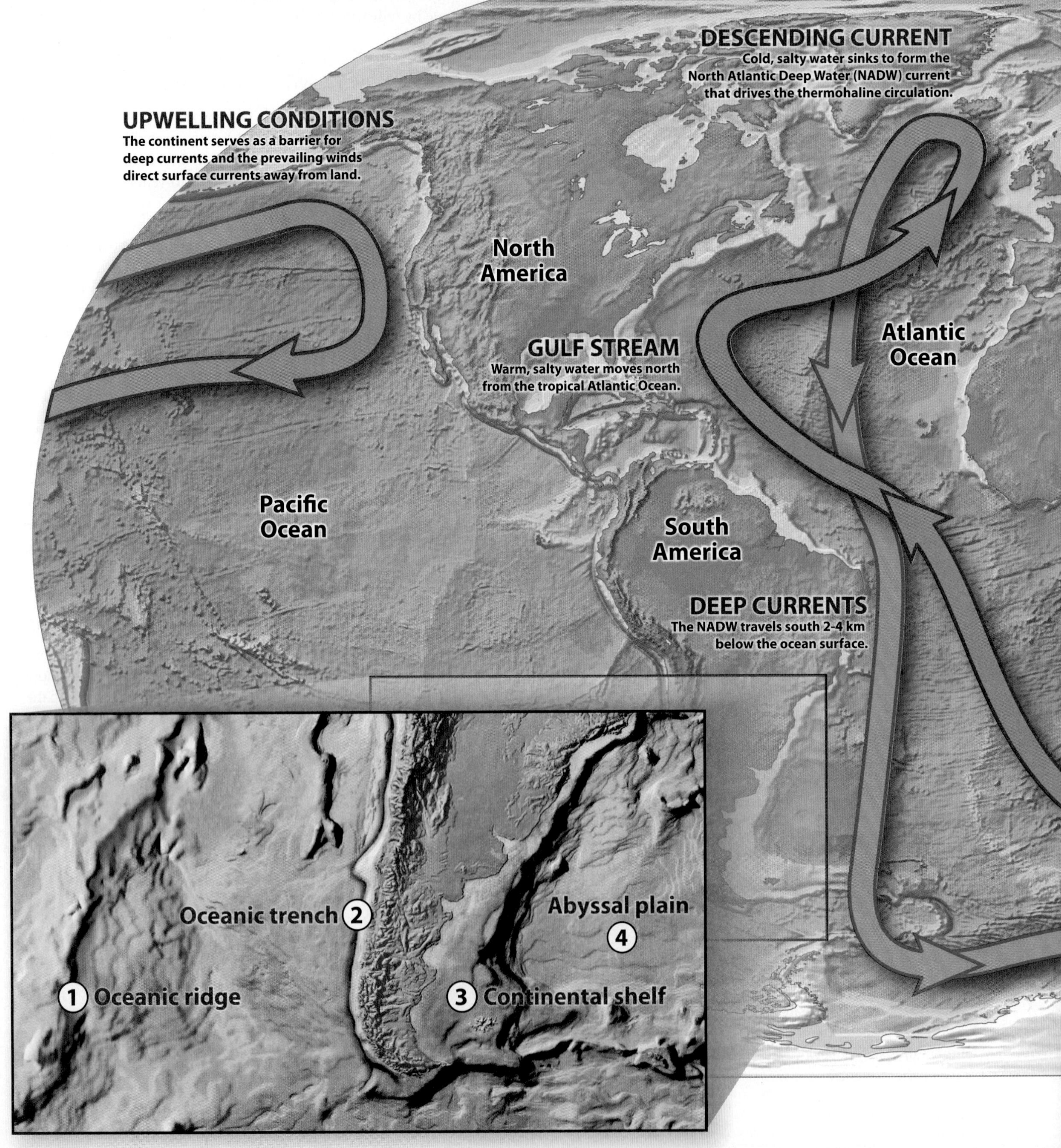
DESCENDING CURRENT
Cold, salty water sinks to form the North Atlantic Deep Water (NADW) current that drives the thermohaline circulation.
UPWELLING CONDITIONS
The continent serves as a barrier for deep currents and the prevailing winds direct surface currents away from land.
North America
Atlantic Ocean
GULF STREAM
Warm, salty water moves north from the tropical Atlantic Ocean.
Pacific Ocean
South America
DEEP CURRENTS
The NADW travels south 2-4 km below the ocean surface.
Oceanic trench 2
Abyssal plain
4
1 Oceanic ridge
3 Continental shelf

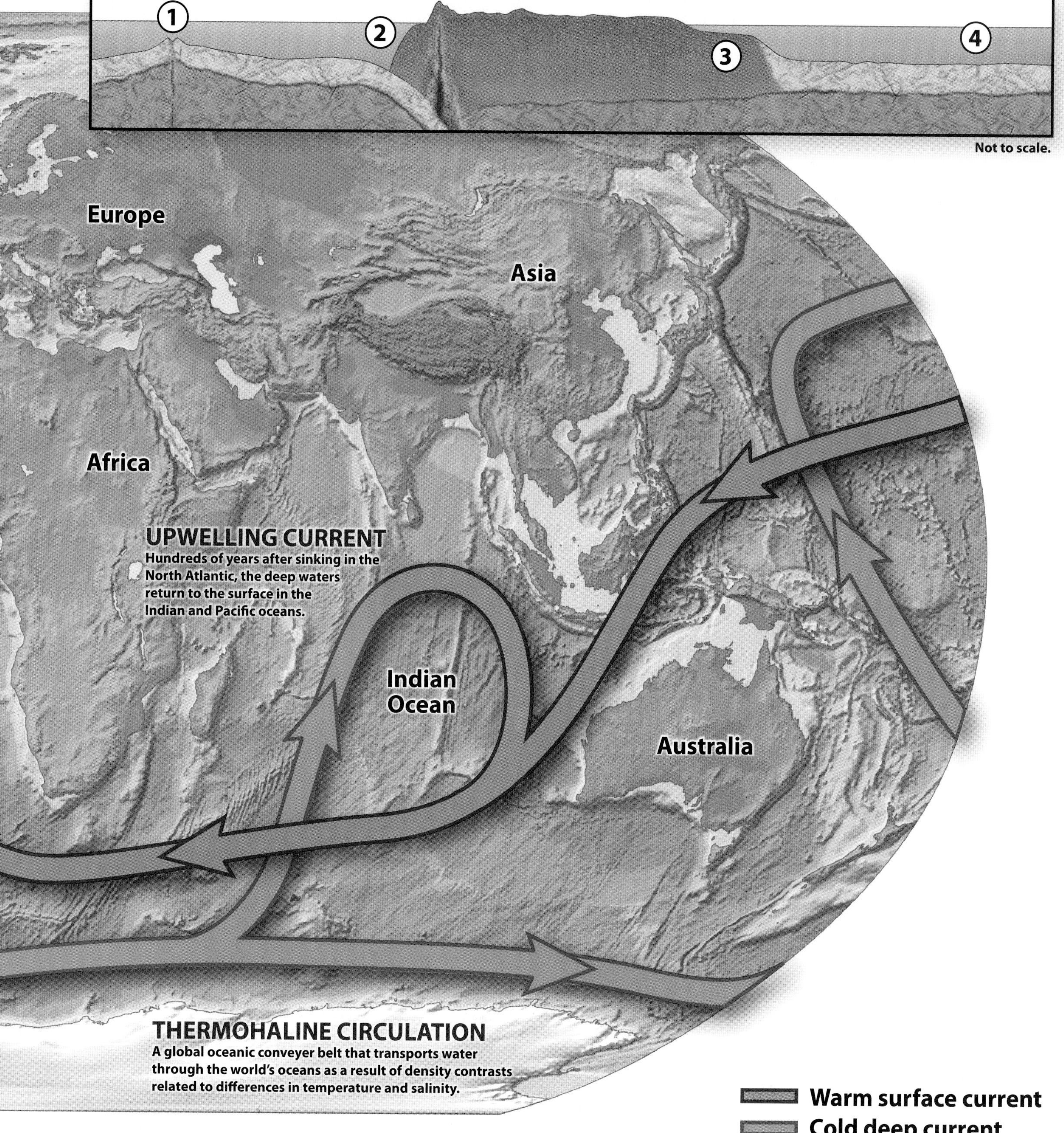
OCEANIC RIDGE
The youngest part of the oceanic floor; high heat flow due to rising magma contributes to higher elevations along the oceanic ridge system.
OCEANIC TRENCH
The deepest section of the ocean floor marks a plate boundary where the oceanic lithosphere of the Nazca plate descends a subduction zone adjacent to South America.
CONTINENTAL SHELF
Shallow, submerged section of continental crust present along *both* passive and active margins but wider at passive margins.
ABYSSAL PLAIN
Some of the oldest and flattest sections of the ocean floor, located landward of the oceanic ridge.
1
2
3
4
Not to scale.
Europe
Asia
Africa
UPWELLING CURRENT
Hundreds of years after sinking in the North Atlantic, the deep waters return to the surface in the Indian and Pacific oceans.
Indian Ocean
Australia
THERMOHALINE CIRCULATION
A global oceanic conveyer belt that transports water through the world's oceans as a result of density contrasts related to differences in temperature and salinity.
Warm surface current
Cold deep current

pressures in the west, and vice versa. The flip-flop in pressure is measured between sites (Darwin, Australia, and Tahiti) in the Southern Hemisphere, thus representing the "SO," or southern oscillation, aspect of ENSO. The cause of the El Niño/La Niña cycle is still under investigation, but it clearly results in a series of interactions among all components of the earth system. Solar radiation from the sun (exosphere) heats the ocean (hydrosphere), which interacts with the atmosphere (winds). The ocean and weather conditions that result affect the biosphere and, through weather-induced erosion and landslides, can also impact the geosphere.

13.5 Tides

Learning Objectives

- Discuss the cause of tides including the difference between spring and neap tides.
- Describe the type of tidal patterns associated with different coastal environments.
- Predict how tidal conditions on Earth would change if we had two moons.

We now turn our attention from the open ocean to the interactions between the oceans and the continents. Interactions among all components of the earth system (hydrosphere, geosphere, atmosphere, biosphere, and exosphere) play important roles where oceans encounter continents. Next we will examine how wave action can modify the coastline (see Section 13.6). But first, we will discuss another phenomenon that contributes to the impact of wave action—the tides.

Have you ever noticed that the beach appears wider—that is, more land is exposed—at some times of day than at others? Such daily changes are the result of tides. **Tides are changes in the sea surface height caused by the gravitational attraction of the moon** and the sun. These changes are visible along the shoreline, where the beach is alternately wider when the tide is out and narrower as part of it is flooded when the tide is in.

Why Tides Occur

Recall from Chapter 2 that the moon orbits Earth and Earth orbits the sun. Figure 13.19 illustrates the relative positions of the moon and Earth during the moon's phases. Tides primarily occur because of the force of gravity exerted by the moon and the sun on the oceans. Since the oceans are a fluid, they are free to respond to gravity. In the simplest case, the side of Earth closest to the moon at any given time during a day will experience the greatest gravitational attraction. Gravity "pulls" water toward the moon, forming a tidal bulge on the side of Earth closest to the moon. (Although the sun has 27 million times more mass than the moon, it is much farther away from Earth, so its gravitational pull is about one-half the pull of the moon.) The tidal bulge on the opposite side of Earth (Figure 13.20) cannot be explained using just the force of gravity from the moon. That equal, but opposite, tidal bulge can be compared to what would happen if you were to take a water balloon tied with a string and swing it around in a circle (trouble just waiting to happen). The water would deform the balloon on the side opposite to where the string is attached. The balloon keeps the water from flying off in one direction (which it would do if the balloon broke). In the case of tides, the tidal bulge occurs through a complex balancing of forces. The gravitational attraction from the moon, coupled with rotation of the Earth-moon system about a common center of mass called the *barycenter*, and rotation of the Earth on its axis combine to cause the bulge. See http://tidesandcurrents.noaa.gov/restles1.html for a more detailed explanation.

Tides are greatest when the moon and the sun align, which occurs at the new moon and the full moon. At these points, gravity forms the largest tidal bulges, which are known as *spring tides*, regardless of the time of the year (Figure 13.20a). In contrast, during the first and last quarters of its orbit, the moon is oriented at right angles to the sun, generating the smallest differences between tides, which are called *neap tides* (Figure 13.20b).

Because Earth rotates much faster than the moon orbits, the location of the tidal bulge changes; the moon is not always over the same spot on Earth. We can visualize the effect of Earth's and the moon's rotation by considering the moon to be stationary while Earth rotates on its axis. Imagine the tidal bulges as stationary, with Earth rotating below the bulges. A coastal site would rotate below two tidal bulges on opposite sides of Earth each day. It would also pass through two minima (where the lowest tides would be recorded). In the simplest case, every coastal location on Earth would experience two clearly defined maximum and minimum tides of similar heights occurring 12 hours apart each day. Of course, the moon is not stationary but moves slowly around Earth.

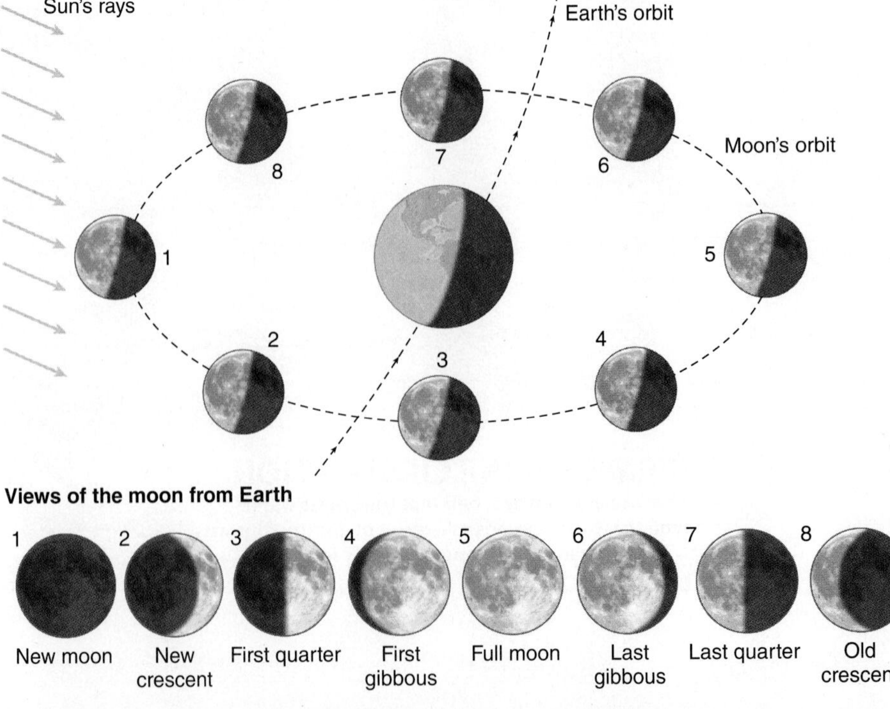

Figure 13.19 Phases of the moon. It takes 29.5 days to complete a cycle of the phases of the moon. The new moon occurs when the moon is between Earth and the sun, and the full moon represents times when Earth is between the sun and the moon.

Checkpoint 13.13

basic · intermediate · advanced · superior

What would happen to spring tides if the moon were farther away from Earth?

a) Tides would be higher.
b) Tides would be lower.
c) There would be no change to spring tides.

A complete cycle of the moon's orbit takes 29.5 days from one new moon to the next. This means that successive high or low tides are separated not by 12 hours but by 12 hours and 25 minutes.

Tidal Patterns

Consider the changes to sea surface heights that occur at Charleston, South Carolina, or Galveston, Texas (Figure 13.21a, b). What do you observe in these data? You should notice that the maximum

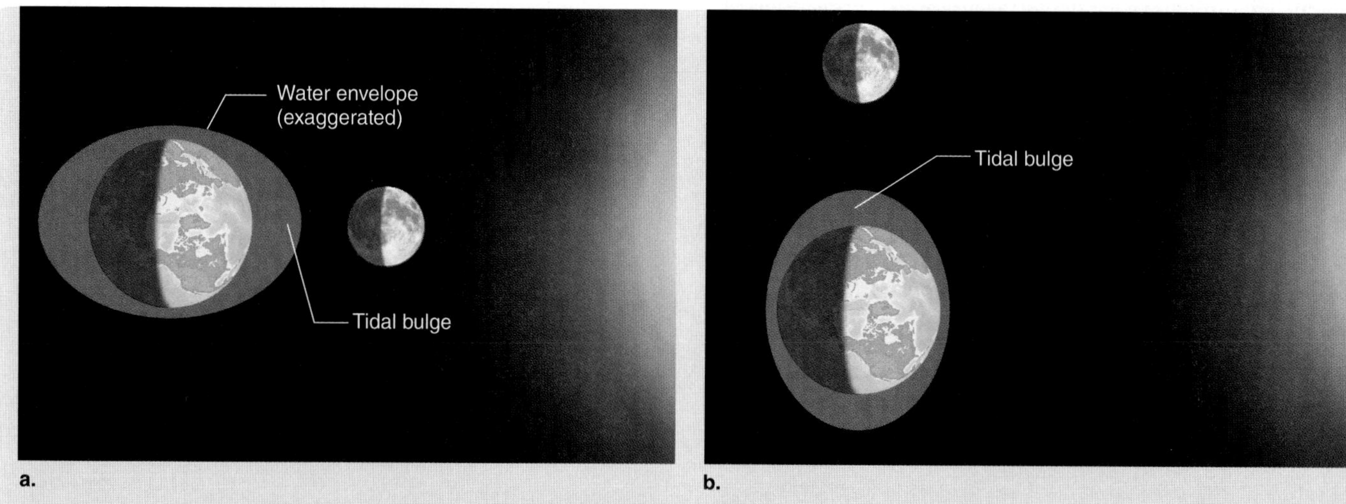

Figure 13.20 Spring and neap tides. **a.** The gravitational pull on the oceans is greatest during spring tides, when the moon and sun are aligned (during the new or full moon). **b.** The gravitational pull is least during neap tides, when the moon and sun pull the oceans in different directions. Not to scale.

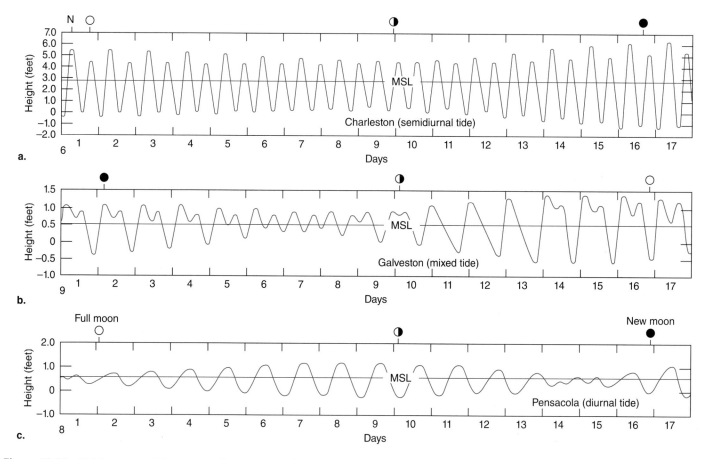

Figure 13.21 Tidal patterns. Tides are classified as **a.** semidiurnal, **b.** mixed, and **c.** diurnal.

heights are different at the two locations and that the relative timing of the tides differs. Why do these changes occur, and why are the data different at different locations?

At Charleston on most days, high tides (and low tides) are just over 12 hours apart. This is known as a *semidiurnal tide (occuring approximately every half-day or 12 hours).* At Galveston, high and low tides may be separated by short or long time intervals and show considerable variations in height; this pattern is known as a *mixed tide.* A third pattern, *diurnal tides,* has only one maximum and minimum each day, as occurs at Pensacola, Florida (Figure 13.21c).

What factors affecting tides could change the "regular" semidiurnal pattern? The position of the moon changes relative to Earth, systematically altering the position and height of the tidal bulge (Figure 13.22). In addition, coastlines have varying shapes, and the bays and inlets along those coasts have different geometries. Tidal patterns can be substantially altered in semi-enclosed seas such as the Gulf of Mexico (Galveston, Pensacola), where the net result is considerable variation in the height of high and low tides.

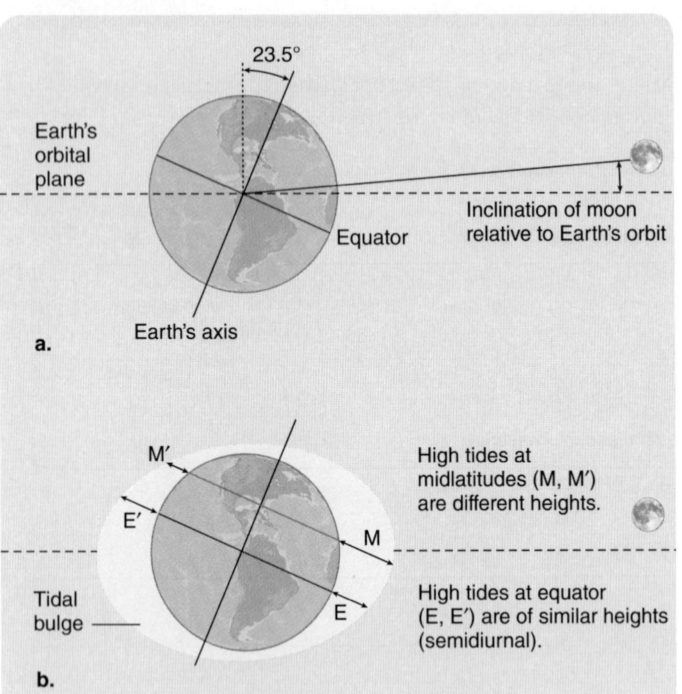

Figure 13.22 Influence of the moon's position **a.** on the location of the tidal bulge on Earth. Depending on the position of the moon relative to Earth and the latitude of a coastal site, the two daily tides may be very similar (semidiurnal) or may exhibit significant variations. **b.** For example, an equatorial location would have a semidiurnal pattern, whereas a location around the midlatitudes would have one high tide considerably higher than the other. How will this pattern change when the moon is on the other side (left) of Earth? Not to scale.

Checkpoint 13.14

basic · intermediate · advanced · superior

Which tidal pattern is represented by the tide data for San Diego, California, shown here?

a) Semidiurnal b) Diurnal c) Mixed

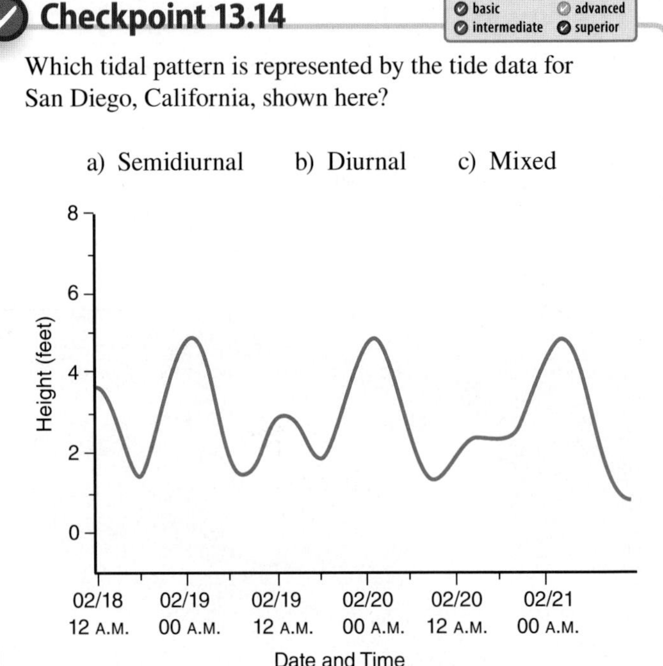

Checkpoint 13.15

basic · intermediate · advanced · superior

Many planets have multiple moons. Discuss how the tides would be affected if Earth had two moons (A and B), each one-half the size of the current moon, in the following two scenarios.

a) Assume the two moons followed the current orbit of the moon and were located on opposite sides of Earth (one-half orbit apart; for example, in the positions of the new moon and full moon).
b) Assume the two moons followed the current orbit of the moon and were located one-quarter orbit apart (for example, in the positions of the new moon and the first quarter moon).

Draw diagrams showing the locations of the moons relative to Earth and the sun and how each scenario would change a typical semidiurnal tidal pattern recorded on a tidal gauge measuring near-shore depth of the ocean.

Checkpoint 13.16

basic · intermediate · advanced · superior

Construct a concept map that shows how the following ideas are related to one another.

sun	coastline	gravity	beach
moon	moon phase	semidiurnal	mixed
neap tide	tides	wave	diurnal
spring tide			

13.6 Wave Action and Coastal Processes

Learning Objectives

- Describe the characteristics of ocean waves and their relationship with atmospheric circulation.
- Explain what happens with a wave approaches the shore.
- Discuss how wave action affects people.

Have you ever attended a sporting event or concert where the crowd performed "the wave"? How did the wave move around the space? The wave passes around a stadium as people in each section of seats stand up and sit down in turn. The individual parts of the wave (the people) remain in the same place, but the shape of the wave (the waveform) can be seen to travel around the stadium. Waveforms in the open ocean move in much the same way.

Wave Motion in the Open Ocean

In the open ocean, water does not travel with the waves but simply bobs up and down, tracing a circular path as a wave passes (Figure 13.23). It is the waveform that moves across the ocean surface, not the water itself. Wave size, speed, and direction are controlled by winds. The waves we observe in the ocean represent wind energy transferred to the surface waters.

Wave size can be measured in terms of wave height and wave length (Figure 13.24). Wave height is the distance between the lowest point of the wave (the *trough*) and the highest point of the wave (the *crest*). Typically, wave height can range up to 30 meters (100 feet). For wind-generated waves, the distance between adjacent wave crests, known as the *wavelength*, generally ranges from a few centimeters up to 600 meters (1,970 feet).

Since wave motion affects only surface waters, a limiting factor in wave movement is water depth. **Wave action decreases downward to a depth equal to about one-half wavelength, called the *wave base*** (Figure 13.24). Below the wave base, water particles no longer undergo any appreciable movement due to the passing wave. Even for the largest ocean waves, water below 300 meters (980 feet)—or most of the water in ocean basins—is rarely affected by surface waves. The elephant seals described in Section 13.1 spend most of their time in still waters, no matter how choppy the surface of the ocean.

The deeper the wave base, the greater the volume of water involved in the wave. Much of the power of tsunami (see Chapter 5) derives from the fact that they travel at the speed of a jet and have very long wavelengths, measured in hundreds of kilometers. Consequently, their wave base is effectively the seafloor. A rapidly approaching tsunami can grow to heights measured in tens of meters as it moves across the continental shelf. The tallest wave ever recorded was generated by a tsunami off the coast of Japan; it measured more than 80 meters (260 feet) in height.

Effect of the Wind on Ocean Waves

What factors control the actual size of a wave? Anyone who has ever blown on a hot bowl of soup will understand that wind-generated waves increase in size with wind speed.

The wind speed and the distance over the ocean that wind blows (called the *fetch*) determine how much frictional force is generated and ultimately how high the waves are. Larger waves are generated by fast winds that blow steadily across a wide region of ocean with no obstructions. Wave height varies around the

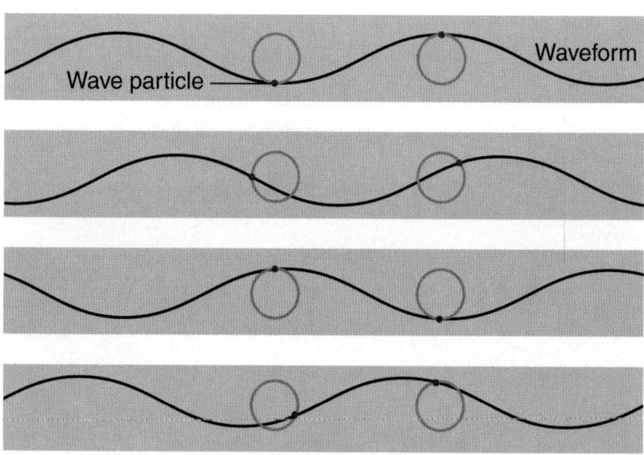

Figure 13.23 Wave motion in open water. The wave shape (waveform) moves, while the water particles follow a circular path and remain in place.

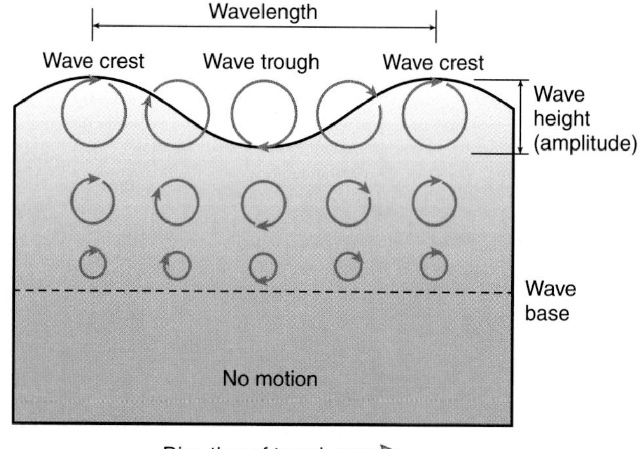

Figure 13.24 Wave height. Wave height is the distance between the lowest point (trough) and the highest point (crest) of the wave. Wavelength is the distance between adjacent crests.

worlds' oceans, but some locations have consistently taller waves (Figure 13.25). It will probably not surprise you that much of the larger Pacific Ocean has consistently taller waves (2 to 3 meters; 6 to 8 feet) than the Atlantic Ocean (about 1 meter; 3 feet), but the largest waves are found in the Southern Ocean (5 to 10 meters; 16 to 33 feet), where there are no continents to interrupt the fetch of the waves. World championship surfing competitions are often held on the beaches of such places as Australia and South Africa to take advantage of these big waves.

Wave Motion Close to Shore

The motion of waves changes as they approach the coastline and the depth of the water decreases. Those wave motions can affect coastal landforms—that is, the erosion, transport, and deposition of sand and other sediment along the shore are tied to the interactions between waves and the coastline.

What happens as a deep ocean wave approaches shallow regions near shore? Because **water in contact with the seafloor is slowed by friction, the wavelength decreases and the wave becomes taller and steeper** (Figure 13.26). When this happens, the water in the wave actually does move forward (unlike the circular motion of deep-water waves). Eventually, the wave collapses because of oversteepening; it is known as a *breaking wave*. A breaking wave forms surf that washes up the shore before flowing back to sea (at which point it is called *backwash*).

When the wave collapses, the water in the wave becomes disrupted (or turbulent). Depending on conditions, turbulent flow

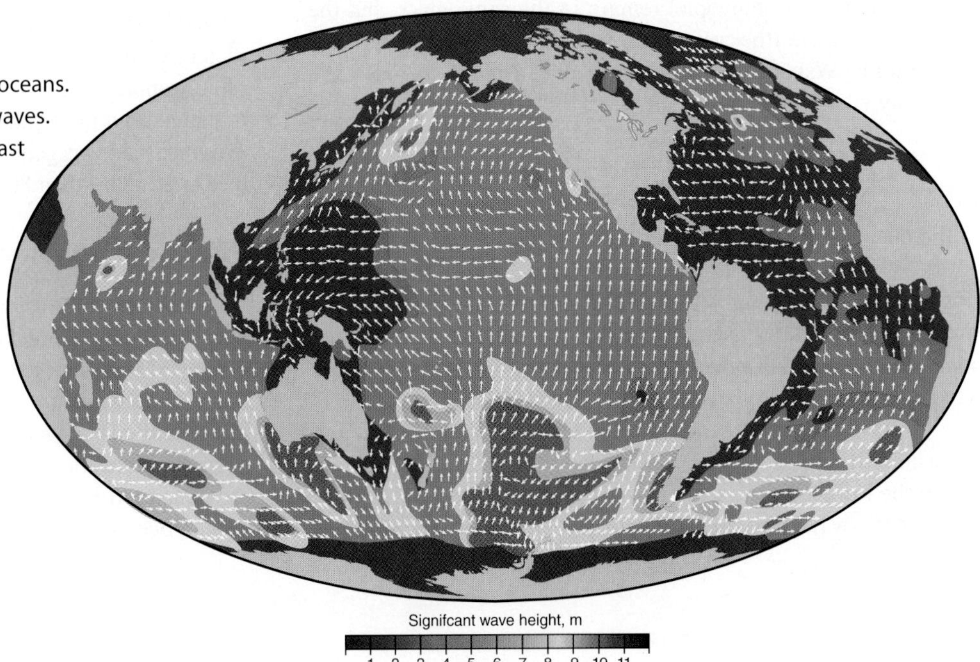

Figure 13.25 Wave heights in the world's oceans. Yellow, orange, and red indicate the tallest waves. How does wave height compare along the east and west coasts of North America? Which ocean has the tallest waves? Significant wave height is the average height of the tallest one-third of the waves over a 12-hour period. The largest individual waves would be approximately twice the significant wave height.

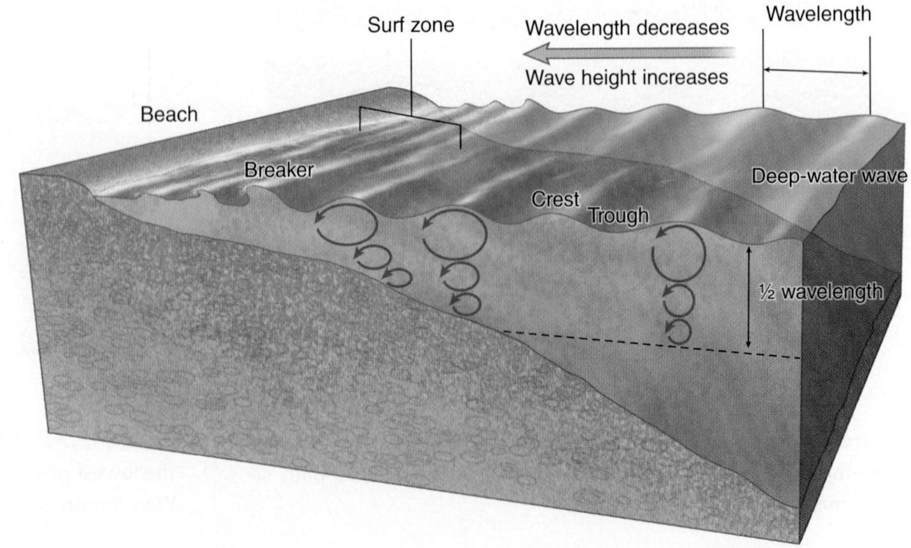

Figure 13.26 Wave motion close to the shore. As waves approach the shore, they steepen and their wavelengths decrease. Steepened waves eventually collapse (break), forming surf that surges up the slope of the shore.

in the *surf zone,* which is **between the line of breaking waves and the shore, can cause erosion or deposition of sand on the beach**. In addition, some of this material may be transported along the shore by currents in the surf zone. Wave erosion associated with large storms can remove large sections of beach in a single storm. In contrast, the redeposition of the beach typically occurs much more slowly over several following months of calmer weather.

Rip Currents. Irregularities such as sandbars and channels along the coastline cause differences in the height of the waves moving parallel to the beach in the surf zone. These variations in the

Checkpoint 13.17

basic · intermediate · advanced · superior

At which location on the following diagram would the waves begin to break farthest from the beach?

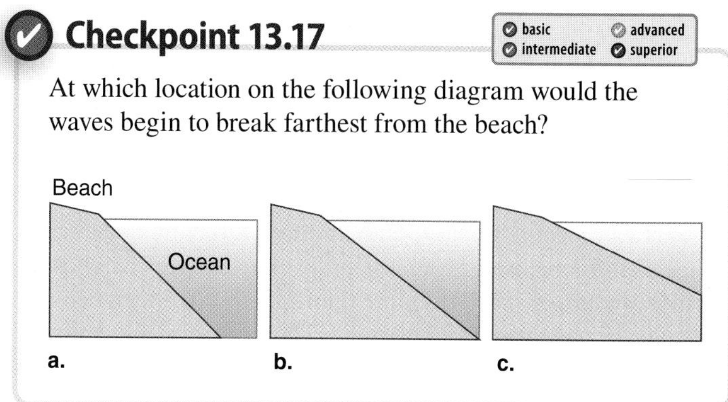

in Further Depth

Waves During Hurricane Katrina

NOAA (National Oceanic and Atmospheric Administration) operates a network of buoys moored around the US coastline and in key sections of the world's oceans. The buoys record a range of data, including wind speed, direction, and wave height. Hourly data from the buoys are transmitted via satellite to a processing site before being uploaded to the Internet (www.ndbc.noaa.gov).

The influence of wind on wave heights is illustrated by analyzing data recorded at buoys located in the Gulf of Mexico during Hurricane Katrina. Station 42040 is a buoy located 120 kilometers (74 miles) off the Alabama coast in the Gulf of Mexico (Figure 1a). Wind speeds of 13 meters per second (29 miles per hour) and large waves with an average height of 5 meters (16 feet) were recorded at station 42040 on August 28, 2005 (Figure 1b). Twenty-four hours later, the eye of Hurricane Katrina was 135 kilometers (84 miles) west of the buoy. Wind speeds had increased to 28 meters per second (62 miles per hour), and the *average* height of the largest waves was nearly 17 meters (55 feet; Figure 1c). The largest individual waves are estimated to have been more than 30 meters (100 feet) tall. Keep in mind that this recording station was some distance from the center of the hurricane. Buoy 42003 was closer to the path of Hurricane Katrina but capsized from the force of the storm—the first time that had happened in the 30 years of the program in the Gulf of Mexico. It should come as no surprise that some of the massive oil platforms that are common along the Gulf were washed ashore by the strength of the storm.

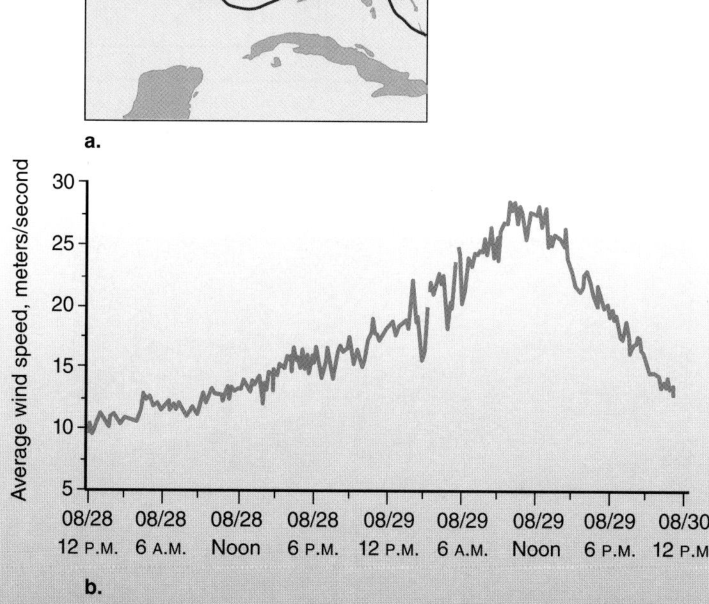

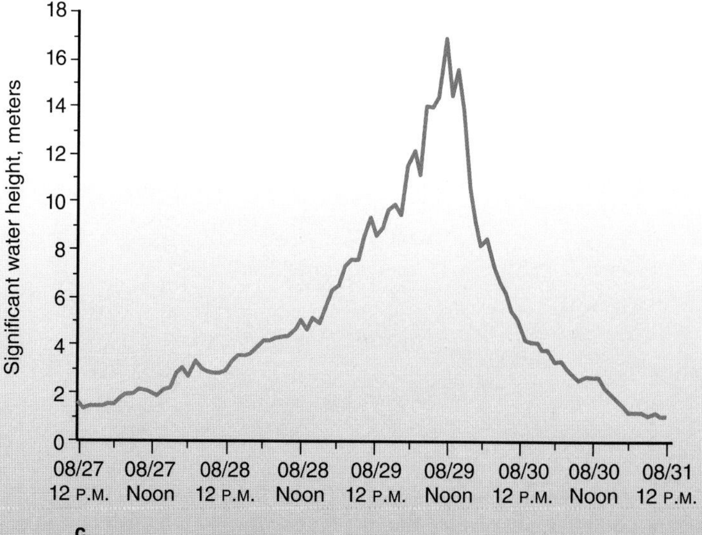

Figure 1 Wind speed and wave height at station 42040 associated with Hurricane Katrina. **a.** Track of Hurricane Katrina and location of station 42040 directly south of Dauphin Island, Alabama. **b.** Average wind speed for 10-minute intervals from August 28–30, 2005, and **c.** significant wave height from August 27–31. How does wave height compare to wiind speed?

surf zone can drive the backwash of water to generate dangerous currents known as rip currents. **Rip currents are narrow currents of water flowing through gaps in sandbars lying just offshore** (Figure 13.27a). Look at Figure 13.27b. Do you notice a location that might prove dangerous if you were swimming there? Do you think you could see the danger if you were on the beach? Similar currents can form adjacent to human-made structures that act as obstacles to funnel water back to the sea. Rip currents can flow at Olympic swimmer speeds of more than 2 meters (6.6 feet) per second and can carry people far offshore. These currents are typically faster where wave size is greater. It is estimated that rip currents annually cause the deaths of about 100 people in the United States, double the average number killed by tornadoes. If you get caught in a rip current, the best escape strategy is to allow the current to take you past the sandbar or structure responsible for its formation. Once past the sandbar, swim parallel to the beach away from the rip current and then back toward shore. People often drown when they get worn out from trying to swim directly against the current.

Checkpoint 13.18

basic · intermediate · advanced · superior

Standing on a beach, you observe a red ball floating about 15 meters (50 feet) offshore. You notice that some well-defined waves are approaching the shore and are just about to reach the ball. The waves are not breaking until they are 5 meters (16 feet) from the beach. Where will the ball be when the last wave reaches the beach?

a) Closer to shore
b) At about the same position
c) Farther offshore

Checkpoint 13.19

basic · intermediate · advanced · superior

What are the potential advantages and disadvantages of the adoption of wave energy technology along the North American coastline?

Checkpoint 13.20

One summer, you get a job as a lifeguard on a beach in a southern state. The previous year, four people died because of rip currents, so the state has mandated that a lifeguard from each section of beach attend a half-day training session on rip currents. During the training session, the participants are divided into teams and asked to create a scoring scheme to estimate the daily risk from rip currents along a stretch of beach. The teams are given one factor (included as an example in the following table) and asked to identify four more. Complete the scoring rubric by adding other factors and identifying the characteristics that make them high-, moderate-, or low-risk.

Factors	Low risk (1 point)	Moderate risk (2 points)	High risk (3 points)
Wave height	Low (<1 meter)	Medium (1–2 meters)	High (> 2 meters)

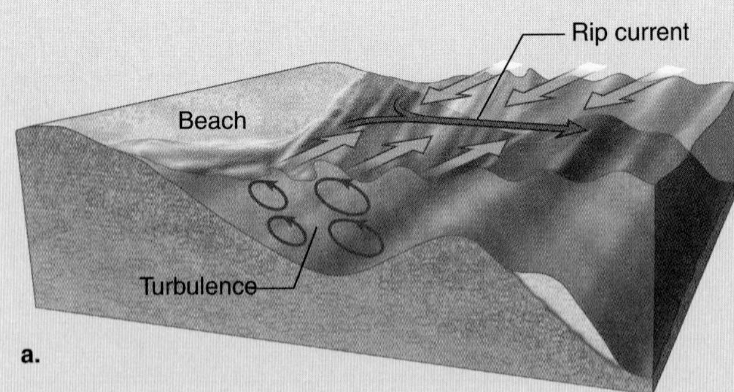

a.

b.

Figure 13.27 Rip currents. **a.** Rip currents form because of irregularities in waves in the surf zone that are often related to changes in the morphology of the seafloor. **b.** Rip current on a Florida beach after Hurricane Jeanne, 2004. The channel in the center of the beach represents the location of a rip current. Water funnels into the head of the channel along the upper part of the beach and flows back to the ocean. The plume of water on the oceanward side of the channel is often an indicator of the presence of a rip current.

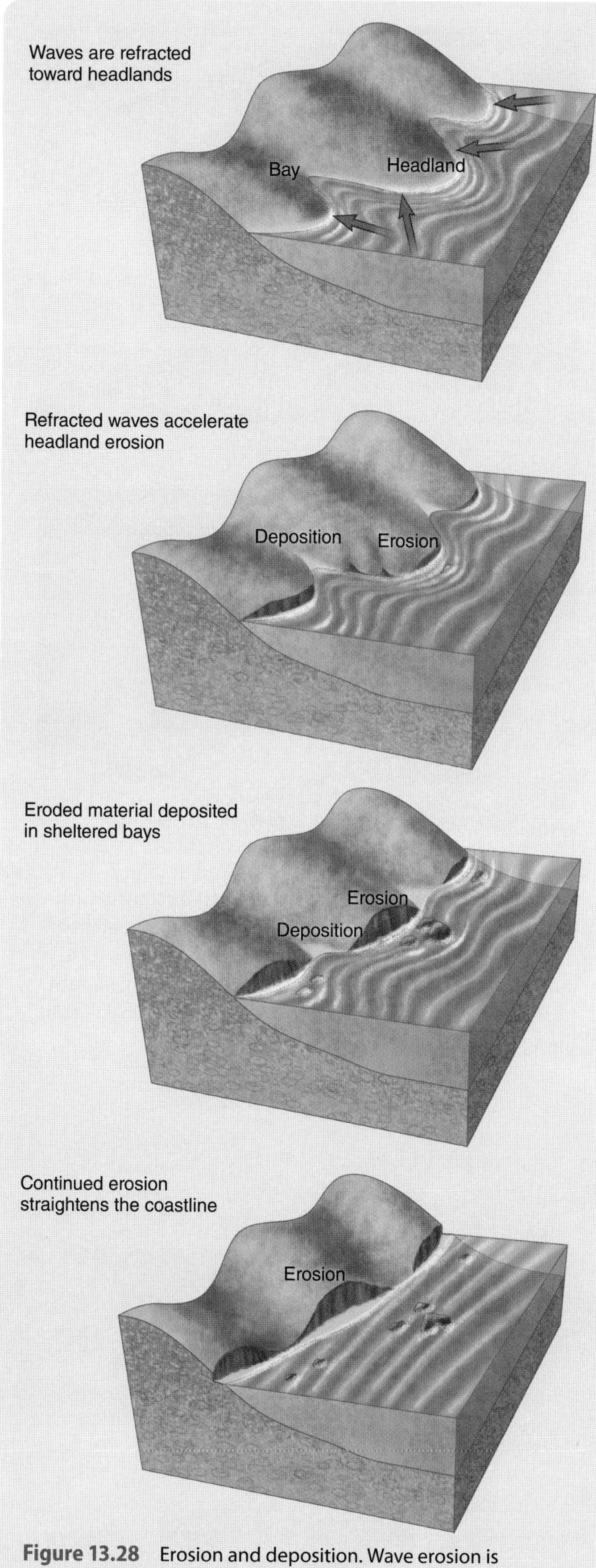

Figure 13.28 Erosion and deposition. Wave erosion is concentrated on headlands, and bays become areas of deposition. Eventually, the coastline becomes straightened as the headlands are eroded back and the bays are filled with sediment.

Headlands and Bays. If there are irregularities in the shoreline or changes in the seafloor, waves approaching a coastline may change direction. The seafloor shallows more rapidly toward headlands that protrude from the coast than toward beaches in adjoining sheltered bays. The wavelength and velocity of the leading edge of the wave decrease because of friction. That part of the wave in deeper water continues to move forward at its original speed. **This causes a "bending" of the wave toward the shore that is called *refraction*** (Figure 13.28). Waves approaching a rugged coastline are refracted toward the resistant headlands, causing erosion to be focused on the headlands and ultimately resulting in the straightening of the coastline.

Wave Energy

New energy companies have proposed harnessing the perpetual motion of waves, tides, and ocean currents to provide a clean, renewable energy source for coastal locations around the world. The principle behind these installations is much the same as that employed by wind energy sites on land. But instead of using moving air to rotate blades, these devices convert the movement of water into electricity (Figure 13.29). A variety of technologies have been proposed that would be anchored to the seafloor or tethered to floating buoys at or near the water's surface in near-shore or offshore locations. Given the characteristics of ocean circulation patterns, it is expected that commercial projects along the East Coast would rely on the powerful flow of the Gulf Stream located just 25 kilometers (15 miles) from the Florida coast. In contrast, West Coast installations would be more likely to utilize the more intense waves of the Pacific Ocean rather than rely on the weaker California current. Projections suggest that a wave farm could generate 50 megawatts of electricity (an average coal-fired power plant generates 600 megawatts).

Figure 13.29 The rotating blades of current turbines look like underwater wind turbines and produce power in much the same way. A generator converts the rotational energy of the blades into electricity that is incorporated into an electrical grid.

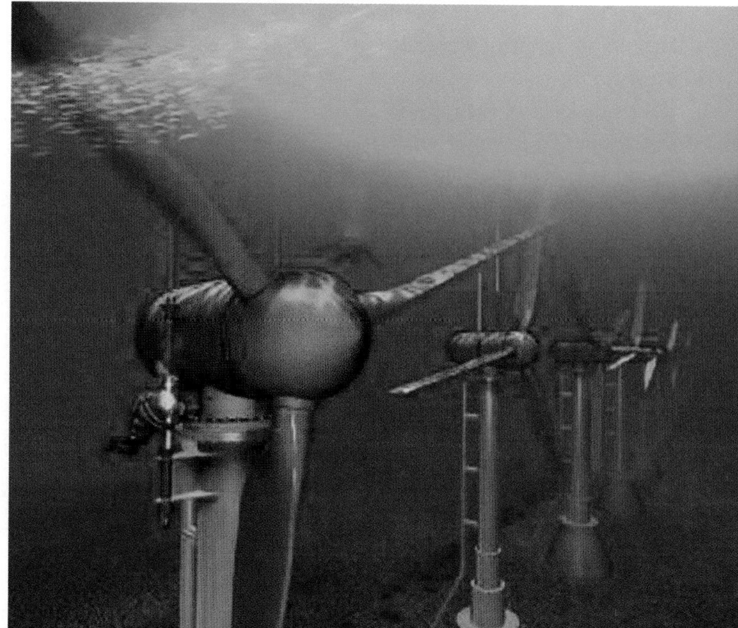

13.7 Shoreline Features

Learning Objectives

- Identify the principal types of coastlines.
- Discuss how the sediment budget is related to erosion or deposition along coastlines.
- Assess the impact of human activity and climate on sediment budgets.

Depending on the geologic characteristics of the continental margin, coastlines may be characterized by rocky cliffs or sandy beaches. Rocky shorelines form where resistant rocks meet the shore, and cliffs are often present where the land is rising relative to sea level. This may be the result of several factors, including an active plate boundary (for example, much of the west coast of North America) or gradual uplift following the removal of a large ice sheet (for example, the eastern coast of Canada). A cliff may be like a wall, or alternating headlands and bays may form from more- and less-resistant rocks (Figures 13.28 and 13.30a, b). Beach shorelines are typically backed by sand dunes, which can range from a few meters to as many as 25 meters (82 feet) in elevation (Figure 13.30c). Much of the Atlantic and Gulf coasts are fronted by long, narrow sandbars known as barrier islands (Figure 13.30d).

a.

b.

Figure 13.31 Moving the Cape Hatteras lighthouse, North Carolina. **a.** The lighthouse was just 40 meters (130 feet) from the shore on June 17, 1999 before **b.** it was moved 800 meters (half a mile) to a final location that was 500 meters (1,640 feet) from the shore (move completed on July 9, 1999). Do you think the lighthouse will need to be moved again?

The Changing of Coastal Landforms

Wave action causes the erosion, transport, and deposition of sand and other materials along coastlines. These processes not only give rise to certain characteristic landforms, but also lead to physical changes in coastlines over time.

Erosion. Waves approaching a rugged coastline are refracted toward the resistant headlands. Sediment eroded from those headlands is typically deposited in the relatively quiet waters of the nearby bays to form beaches. Over time, the coastline is straightened as erosion wears away the headland and the bays are filled with sediment (see Figure 13.28).

Figure 13.30 Types of coastlines. Depending on the geological history of the coast, coastlines may be rocky (a, b) or sandy (c, d). **a.** Giant's Causeway, Northern Ireland, is a rocky coastline dominated by cliffs and low-lying wave-cut platforms. **b.** White Park Bay, Northern Ireland, features rocky headlands separated by beaches. **c.** Strathy, Scotland, is a low-lying coastline dominated by long beaches backed by dunes. **d.** Assateague Island, Maryland, developed a narrow, elongated barrier island that separates the ocean from the mainland.

a.

b.

In 1999, the National Park Service had to pick up the nation's tallest lighthouse at Cape Hatteras, North Carolina, and move it to a site farther inland (Figure 13.31). Why move such a large structure? The lighthouse was originally built 500 meters (1,640 feet) from shore, but erosion had slowly worn away the distance between the waves and the building. One big storm could have destroyed it.

Structures in other coastal regions of the United States are frequently in danger because of eroding shorelines. Twenty-six of the 30 states bordering an ocean or one of the Great Lakes are experiencing a net loss of usable land along their shorelines (Figure 13.32). The Federal Emergency Management Agency estimates that 10,000 structures are located on coastal lands that will erode within a decade. The most rapid erosion rates are along the Gulf coast (Louisiana) and the Atlantic shore (South Carolina, Maryland, New Jersey), where erosion eats away an average of at least 1 meter (3.3 feet) of coastline each year.

Erosion is most effective on loose, unconsolidated sediment such as sandy beaches or on adjacent sand dunes; rocky shorelines are less susceptible. Erosion is accelerated by the actions of surges

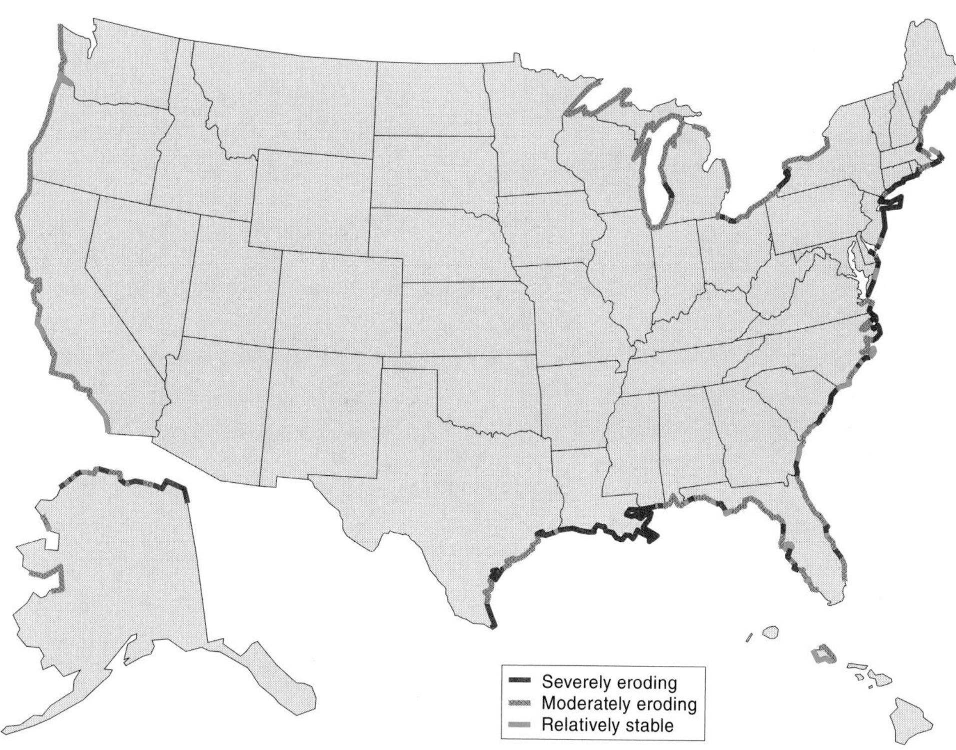

Figure 13.32 US coastal erosion rates. Erosion rates are highest (red) for sections of the Gulf and East coasts.

c.

d.

and violent waves associated with large storms. Hurricanes and other storms generate waves that can destroy beaches and dunes in a matter of hours. Erosion is particularly severe in association with winter storms (Figure 13.33), especially those occurring during times when the sea level is higher than normal. Typical erosion rates for sections of shoreline in parts of California and Massachusetts are 0.2 meter (8 inches) per year, but winter storms can cause these same shorelines to recede 5 to 10 meters (16 to 33 feet) in a single season, up to 50 times faster than normal.

Figure 13.33 Retreating shorelines. Twelve homes were condemned at Pacifica, California, when the cliff retreated 10 meters (33 feet) during a series of winter storms (1997–1998).

Transport and Deposition. While much of a coastline is undergoing erosion, other sections, sometimes along the same stretch of beach, may be growing wider because of deposition of eroded sand.

The relative orientation of the waves to the shoreline controls the distribution of erosion and deposition. Waves that strike the beach head-on move sand up and down the beach. During severe storms waves may transport sand offshore and deposit it in sandbars. If waves strike the beach at an angle, they transport the sediment along the beach in a zigzag pattern. These oblique-angle waves generate **a longshore current, which transports sediment parallel to the beach in the surf zone** (Figure 13.34). As a result, sediment on the beach is transported parallel to the shoreline in the direction of wave motion. Both mechanisms—the physical transport of sand particles along the beach and the movement of sand in suspension in the longshore current—ensure that sediment is transported along the length of the beach and may be added to widen and extend some sections of beaches (Figure 13.35).

The beach is not the final resting place for coastal sediment; it is an intermediate stop on a longer journey. Sediment is transported to the coast by streams and redistributed along the shoreline by longshore currents and wave action. Some of that sand may form offshore sandbars, while some of the remainder may be lost when it flows through submarine canyons off the edge of the

Checkpoint 13.21

Make a labeled sketch or diagram illustrating the four principal types of coastlines.

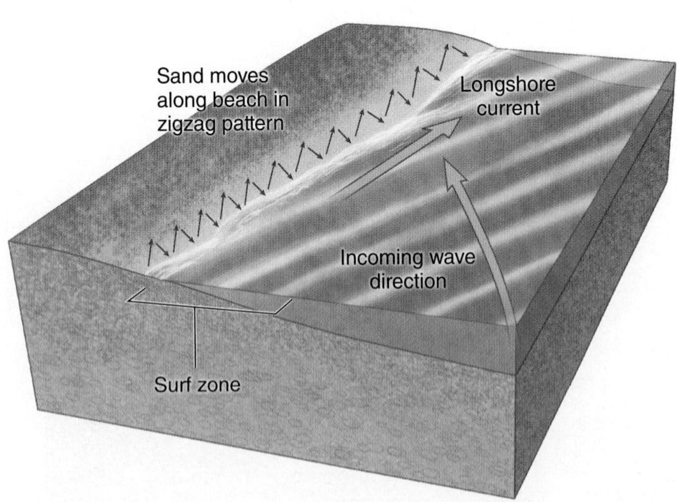

Figure 13.34 Longshore currents. Waves strike a beach obliquely, generating a longshore current that transfers sediment along the shore (left to right, in this case). Sand particles are moved along the beach in a zigzag pattern.

Figure 13.35 Sand deposition. **a.** The bay at Puget Sound, Washington, has been almost cut off from the ocean by a narrow spit that may become a baymouth bar. The spit formed from sediment transported along the coastline by longshore currents from the lower part of the image. **b.** Tomales Bay, north of Point Reyes, California, prior to storms. **c.** After storms, sand was transported along the shoreline from the left side of the image.

a.

Checkpoint 13.22

basic · intermediate · advanced · superior

Examine the section of coastline in the following image taken at Santa Barbara, California. Storms erode sand from the cliff and carry it along the coast. In what direction does the sand travel along the beach?

a) Left (north)   b) Right (south)

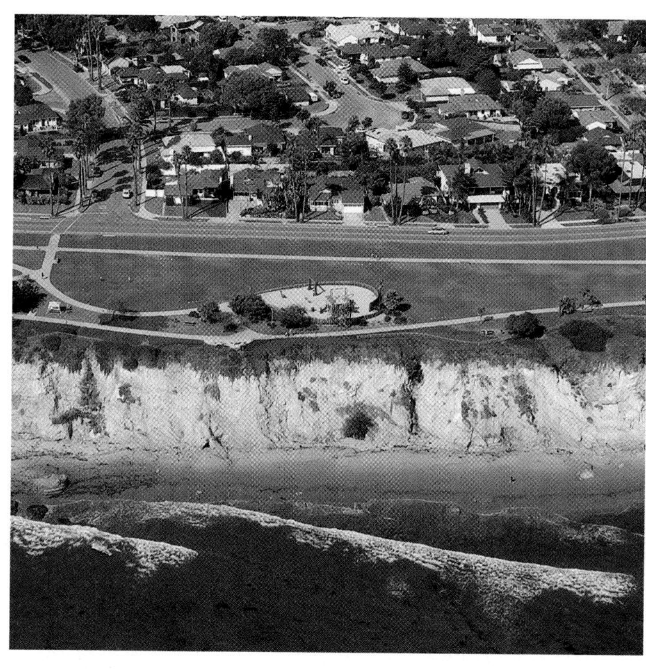

continental shelf. Sediment transported along the beach can give rise to certain characteristic landforms when it is eventually deposited in the calm waters of adjoining bays. Sandbars that partially block the mouth of a stream channel form a *spit* (Figure 13.35a), or they may completely block the channel to form a *baymouth bar*.

The Sediment Budget

Shoreline processes are influenced by the **sediment budget, the balance between the material that is deposited on the shore and the material that is eroded from it.** Sediment added to the shore comes from headland erosion or is delivered to the coast by stream flow. Much like a financial balance sheet, the sediment budget remains in a state of equilibrium as long as the volume of sediment coming in is equal to the material being lost.

The natural cycle of shoreline erosion, transport, and deposition along the coast—and hence, the sediment budget—is often disrupted by human activity, as exemplified by changes to the Louisiana coastline.

Natural and Human-Induced Coastline Changes: The Mississippi River Delta. The Louisiana coastline serves as an example of the combined effects of natural processes and human activity on a coastline. Erosion rates for some sections of the Louisiana coast are as high as 20 meters (66 feet) per year. Much of the eroded sediment is deposited in a delta at the mouth of the Mississippi River (see Figure 11.22). The delta is submerged below sea level as the sediment becomes compacted. The subsidence rate, the rate at which the land surface is sinking, is approximately 1 meter (3.3 feet) per century. In the past, this subsidence was compensated

b.

c.

by additional sediment supplied during floods. However, the construction of flood control levees along the river's channel has robbed the delta of its primary source of sediment, which is now deposited in the Gulf of Mexico. The delta subsides as sediment is no longer being added, and the coastline moves landward, flooding what was once usable land.

Loss of sediment from the continental interior to the coast accelerates coastal erosion and removes a buffer that previously slowed down hurricanes, reducing their power. The construction of dams on major rivers reduces the volume of sediment reaching the coast, resulting in sediment starvation because sediment that would once have been deposited along the shoreline is trapped in upstream reservoirs. For example, dams within the Mississippi River drainage basin (see Chapter 11) have reduced sediment supply to the Mississippi delta by approximately one-half. Natural drought conditions may also reduce stream flow and thus diminish the amount of sediment transported to the shore by streams. Coastal development may further modify the system by resulting in the construction of structures designed to reduce erosion or to prevent the deposition of sediment in navigation channels.

Checkpoint 13.23

Draw a diagram to illustrate how sand is transferred among the components of a shoreline.

Checkpoint 13.24

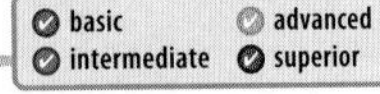

Venn Diagram: Streams and Coastal Systems

Use the Venn diagram provided here to compare and contrast erosion, transport, and deposition in stream systems and in coastal systems. Write features unique to either group in the larger areas of the left and right circles; note features they share in the overlap area in the center of the figure. Identify at least eight features.

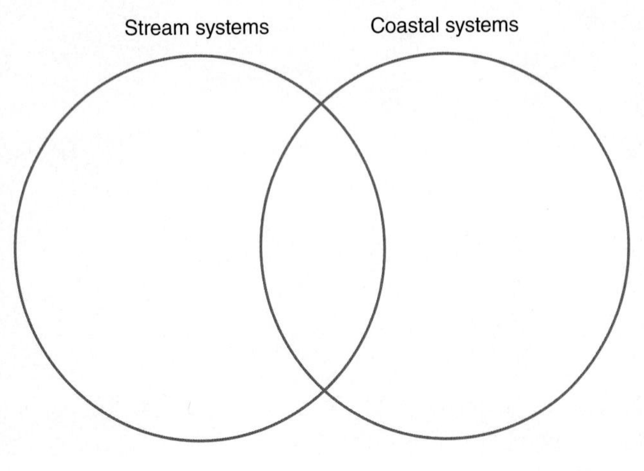

13.8 Shoreline Protection

Learning Objectives

- Compare and contrast the impact of seawalls, breakwaters, and other shoreline protection measures on beach development.
- Evaluate the best prevention and adjustment strategies that could be used to diminish erosion along coastlines.
- Discuss the case for and against funding of beach nourishment programs in states with severe coastal erosion.

Suppose you lived on a shoreline that was eroding at such a rate that it threatened to destroy your house within the next few years. Short of selling your home, what could you do? Three basic strategies are available to the desperate homeowner: (1) stop the cause of the erosion by keeping the waves from reaching the shoreline; (2) change the characteristics of the location to encourage deposition rather than erosion; or (3) add more sand to replace that being lost. While these measures may temporarily provide some protection, you should keep in mind that recent research suggests that many of these measures do more harm than good. In some cases, local governments are taking steps to prevent landowners from make any changes to the shoreline.

Erosion Prevention Strategies

A survey of conditions along a section of Florida shoreline revealed certain natural shoreline features that are most likely to protect coastal residents from erosion:

- Tall, continuous dunes behind beaches protect against large storms.
- Wide, stable beaches (30 meters [100 feet] or more across) absorb and deplete wave energy.
- Exposed offshore sandbars absorb the force of breaking waves before they arrive on the beach.

These features were not universal. Although more than 80 percent of the beaches in the study were wide enough, only about half had dunes, and just 7 percent had an exposed sandbar.

Artificial Barriers. When natural shoreline features do not exist to deter erosion, engineers have sometimes designed artificial barriers to mimic the desired natural effects. For example, a little more than one-tenth of the California coastline is armored with structures intended to prevent erosion. Three types of structures may be built on or near coasts—seawalls, groins, and breakwaters—and a process called *artificial beach nourishment* has also been employed to offset the consequences of erosion.

Seawalls built of rock or concrete are used to protect property from receding shorelines (Figure 13.36). Such a wall serves as a barrier between waves and the shoreline, and it prevents wave erosion from undercutting cliffs below buildings. Waves that strike the seawall are reflected from the wall onto the adjoining beach. Unfortunately, although seawalls may protect the shoreline behind the wall, they typically hasten beach erosion adjacent to the wall because of wave refraction. Erosion is often exaggerated where the seawall ends, causing the shoreline to recede more rapidly on either side of the structure.

Groins are wall-like structures built perpendicular to the shoreline along beaches to act as barriers to longshore currents (Figure 13.37). These structures are intended as obstacles to the transport of sediment along the beach. As a longshore current meets the groin, it loses speed, causing the current to deposit part of its sediment load. That material is deposited on the upcurrent side of the groin, thus building up the adjacent beach. However, as the current passes the groin, it picks up additional sediment on the downcurrent side of the structure, causing local erosion there.

Breakwaters are barriers built offshore to protect part of the shoreline (Figure 13.38). They prevent erosion by slowing the waves and allowing the beach to grow behind the structure. However, the beach behind the breakwater often grows at the expense of the adjacent unprotected shoreline, which is more quickly eroded.

Checkpoint 13.25

basic | intermediate | advanced | superior

Compare and contrast seawalls and breakwaters.

Checkpoint 13.26

basic | intermediate | advanced | superior

Analyze Figure 13.31 and explain why the shoreline erosion/deposition processes at the site of Cape Hatteras required that the lighthouse be moved.

Figure 13.37 A groin. This structure was built adjacent to Los Angeles International Airport, California. What is the direction of sediment transport along the shoreline?

Figure 13.36 A seawall. This wall constructed of rock was placed at the base of an eroding cliff north of Monterey, California. Note how erosion is exaggerated where the seawall ends.

Figure 13.38 Breakwaters. These structures are intended to protect the south shore of Lake Erie at Sims Park, Ohio.

Ultimately, the best buffer for protecting valuable real estate along the shore is a big, wide beach. Waves lose energy and erosive power as they cross a wide beach. **Artificial beach nourishment involves dredging sand offshore and then pumping it onto the beach** (Figure 13.39). The beach will grow as long as material is added faster than natural processes remove it. Artificial beach nourishment is often a temporary fix because the sand continues to erode and must be replaced again a few years later. For example, material added to many East Coast beaches remained for fewer than 2 years before the beaches returned to their prenourishment state. However, a successful beach nourishment effort occurred at Miami Beach, Florida, in the 1970s. The city spent $64 million to stabilize and expand nearly 16 kilometers (10 miles) of beaches to meet the needs of the booming tourism industry. Much of that beach remains today.

Figure 13.39 Beach nourishment. A pipeline pumps sand collected offshore by a dredge onto the Ocean City Beach in Maryland. Note the wider beach near the bottom of the image.

Erosion Adjustment Strategies

As we have just pointed out, nearly all of the artificial methods for preventing erosion have shortcomings. Although they may work in the short term, their effects are local and often lead to other problems elsewhere. Therefore, during the past decade, many coastal states have focused on adjustment strategies instead. For example, state governments have passed legislation aimed at controlling construction adjacent to eroding coasts rather than trying to prevent the erosion. Florida has introduced strict regulations that require buildings constructed near the shoreline to meet rigorous standards so that they are strong enough to resist storm surges and high winds. In 1995, when Hurricane Opal struck the coast, none of the buildings that had been constructed to meet these standards failed, while 56 percent of all other habitable buildings in the storm's path were heavily damaged.

Checkpoint 13.27

In Chapter 11, we discussed the difference between prevention and adjustment for flooding. Describe the difference between prevention and adjustment strategies for shoreline protection.

Checkpoint 13.28

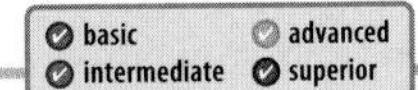

The US Army Corps of Engineers is responsible for spending money to protect the nation's beaches that are most susceptible to erosion. The Corps is funded by the federal government. Should your tax money be used to maintain wide beaches in places such as South Carolina, Florida, or Texas? What are the consequences of not funding these programs? Identify arguments for and against continued federal funding of artificial beach nourishment programs in states with severe coastal erosion.

the big picture

To ancient people, the ocean was a frightening place—a massive, endless body of water possibly inhabited by large, scary monsters and too daunting to even try to investigate. Scientists do not think this way. Science assumes that Earth *is* knowable. People can discover patterns in all of nature by using their intellect, their senses, and instruments that extend their senses. At the same time, scientists realize that the quest for knowledge is eternal, especially in parts of the planet that are hostile or almost unreachable to humans. Exploration and discovery will continue to expand our knowledge of the deep oceans as long as people ask questions and answers remain to be found.

In the United States, over 123 million people live in coastal counties, and coastal property values continue to rise to levels far above those in most of the rest of the nation. Some folks are glad to spend millions to live in elegant homes with a view of the ocean. However, the weather patterns that bring balmy breezes on warm summer days can also generate devastating storms and even hurricanes.

In 2004, four hurricanes (Frances, Ivan, Jeanne, and Charley) barreled across Florida, causing extensive damage and substantial coastal erosion. Damage was greatest where the timing of the hurricane coincided with the local high-tide conditions. Hurricane Ivan entered the Gulf of Mexico as a strong hurricane that made landfall in Alabama. That storm moved inland across the southeast United States, went back out to sea, and then moved south as a tropical storm that also affected southern Florida before entering the Gulf of Mexico again. The associated storm surge eroded beaches and dunes, causing the hotel pictured in the chapter opening photo to collapse when the underlying sand was eroded (Figure 13.40).

However, if you want to take your chances in the beautiful but dynamic coastal environment, home lots are still available. You might have to plunk down a few million dollars just for a lot to build on, but if you leave the lot undeveloped, it may save you the inconvenience of rebuilding after hurricane season.

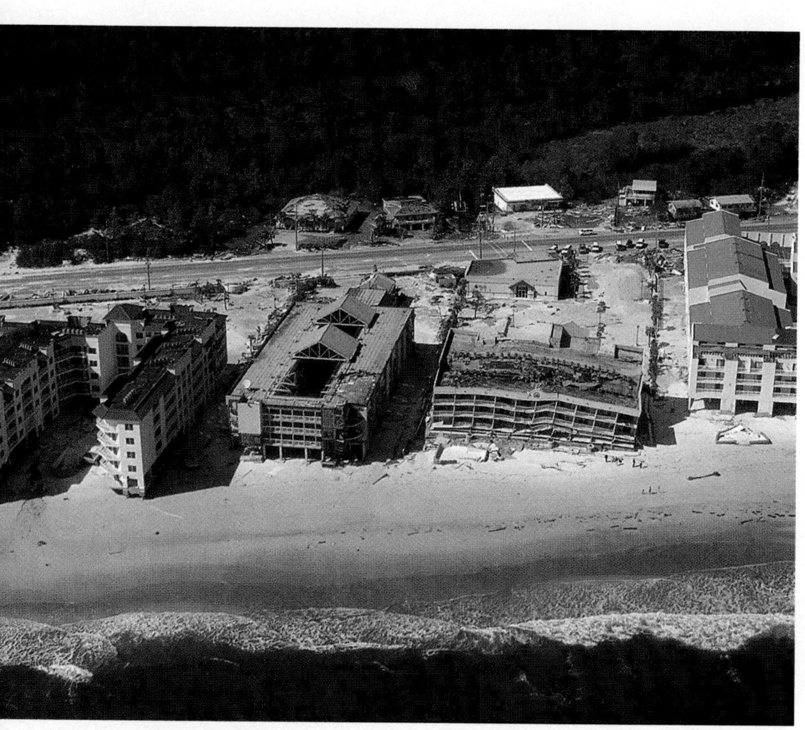

Figure 13.40 Damage from Hurricane Ivan. The five-story building collapsed when the dune below it was eroded by waves generated by the hurricane.

Oceans and Coastlines: Concept Map

Complete the following concept map to evaluate your understanding of the interactions between the earth system and oceans and coastlines. Label as many interactions as you can with information from this chapter.

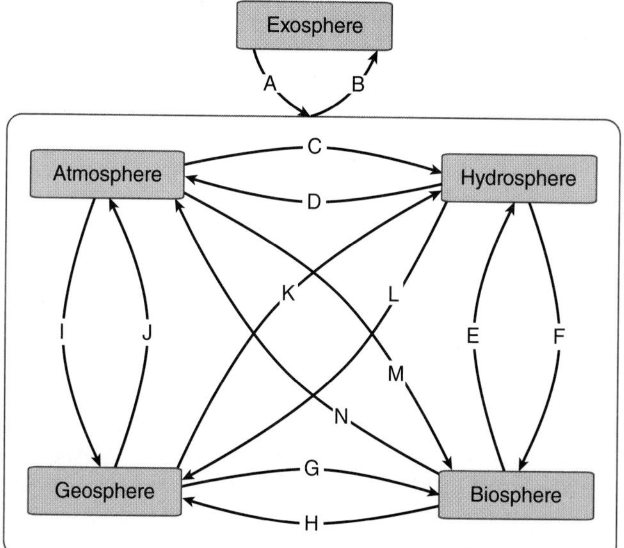

Wrigley Field, Chicago

"For the first time in my life, I saw the horizon as a curved line. It was accentuated by a thin seam of dark blue light—our atmosphere. Obviously, this was not the 'ocean' of air I had been told it was so many times in my life. I was terrified by its fragile appearance."

—Ulf Merbold, astronaut

Coors Field, Denver

Chapter 14

The Atmosphere

Chapter Outline

the big picture

Why is it easier to hit home runs at Coors Field than Wrigley Field?

See The Big Picture box at the end of this chapter for the full story on these images.

14.1 Science and Skydiving

Chapter Learning Outcomes

- Students will apply physical principles to predict atmospheric temperatures.
- Students will apply principles governing the exchange of water in the atmosphere.
- Students will describe and apply the processes and consequences of atmospheric circulation.

Early in the morning of August 16, 1960, while many people were settling in for breakfast, Captain Joe Kittinger was plummeting to Earth at nearly 1,000 kilometers per hour (620 miles per hour). Captain Kittinger was completing the world's highest skydive as part of a special US Air Force (USAF) program designed to test potential escape methods for pilots of high-altitude jets. The USAF was investigating whether pilots could bail out of a troubled aircraft flying at extreme altitudes in the farthest reaches of the atmosphere. Captain Kittinger had already made jumps from ultrahigh altitudes, but on August 16, his balloon capsule ascended above the New Mexico desert to an unprecedented height of 32 kilometers (20 miles). (For reference, Mount Everest is less than 9 kilometers [6 miles] high, and commercial jets fly at about 11 kilometers [7 miles] of altitude.)

As he drifted slowly upward in the capsule, Captain Kittinger watched the balloon expand and the surrounding sky turn black. He wore a pressurized suit to protect him from the subfreezing temperatures that dominated most of his ascent. The thin air also required that his equipment supply him with oxygen. From the target altitude, he could observe the curvature of Earth on the horizon. After a few minutes, he shuffled to the door of the capsule, said a short prayer, and stepped out (Figure 14.1), beginning his long fall. He dropped rapidly, reaching speeds of 980 kilometers per hour (609 miles per hour) with only a small stabilizing parachute to prevent him from going into an uncontrollable spin that would have resulted in certain death. For several minutes, he rushed toward the ground, gradually slowing as he fell. A second, larger parachute opened at about 5 kilometers (16,400 feet) and carried him safely to his landing site, less than 14 minutes after he stepped out of the capsule far above Earth's surface.

Kittinger's skydive record stood for 52 years until Felix Baumgartner jumped from 39 kilometers (24 miles) on October 14, 2012. Baumgartner was supported by a crack team of experts (including Kittinger), the best of modern technology, and commercial sponsors. He fell so quickly during the early part of his descent that he broke the sound barrier, reaching speeds of more than 1,350 kilometers per hour (840 miles per hour). Unlike the original Kittinger jump which is memorialized by a few photographs, Baumgartner's jump was filmed by an array of cameras and watched live on YouTube by millions of viewers. Watching the video clips online will give you a feel for what Earth looks like when viewed from the stratosphere. Why did Captain Kittinger's balloon get bigger as he rose higher, even though the atmospheric temperature had decreased? Why is the sky blue at low altitudes and black at high altitudes? Why did Baumgartner fall at the highest speed early in his dive and then *slow down* as he came closer to the surface of Earth? How close was he to being in space and where exactly *is* space? Of course, Joe Kittinger and Felix Baumgartner would have known the answers to these questions before they stepped into the balloon capsule. You will resolve these and other questions by the end of the chapter.

The atmosphere consists of the air around us and serves as a protective layer against the many potential hazards from space. For example, Earth's atmosphere blocks harmful radiation from the sun and protects us from incoming space projectiles that burn up before reaching the planet's surface (see Chapter 3). **The lower boundary of the atmosphere touches Earth's surface; its upper boundary is a gradational transition into space.** Ironically, to get a clear picture of the thin veil of atmosphere that surrounds Earth, we need to leave the planet. When viewed from beyond Earth, the atmosphere appears as just a thin cover that stretches upward a few hundred kilometers before giving way to the airless vacuum of space (Figure 14.2).

a.

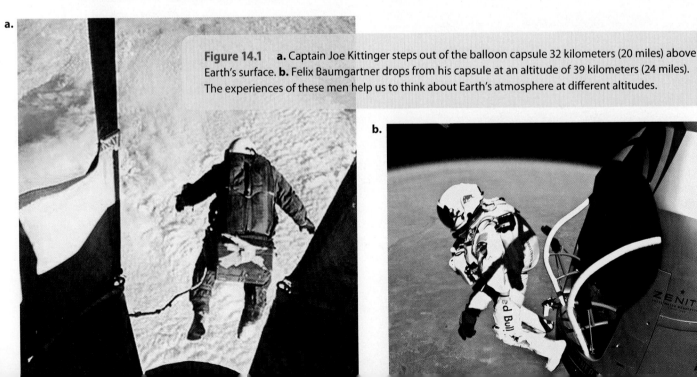

Figure 14.1 **a.** Captain Joe Kittinger steps out of the balloon capsule 32 kilometers (20 miles) above Earth's surface. **b.** Felix Baumgartner drops from his capsule at an altitude of 39 kilometers (24 miles). The experiences of these men help us to think about Earth's atmosphere at different altitudes.

b.

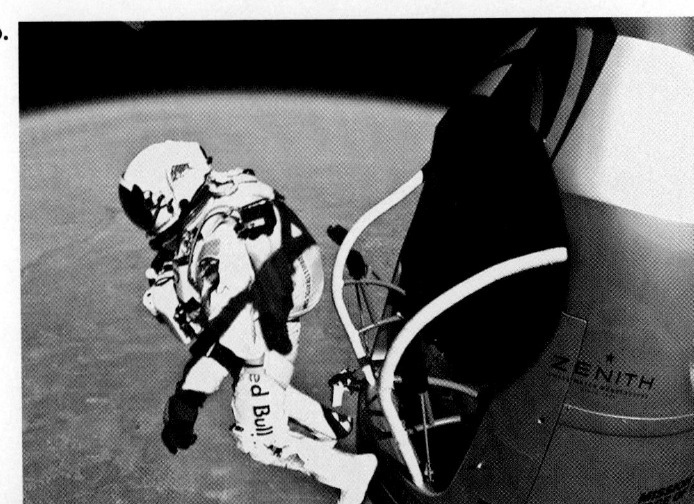

The interaction of the thin layer of air with solar radiation determines the distribution of global temperatures. **Solar radiation supplies the energy necessary for cloud formation, precipitation, and wind generation.** There are extensive plans to harness wind power to provide a larger share of US energy in the future. Most of the nation's current energy needs are supplied by fossil fuels (oil, gas, coal) that carry with them concerns about energy security and global warming. Renewable energy in various forms produces 7 percent of US energy. Renewable energy sources such as wind power have the potential for minimal environmental harm and freedom from reliance on foreign suppliers. Wind power currently supplies less than 0.5 percent of our energy needs, but there is great opportunity for increasing that share in the years ahead.

Self-Reflection Survey: Section 14.1

Answer the following questions as a means of uncovering what you already know about the atmosphere.

1. List the ways in which human beings interact with the atmosphere.
2. Describe the variations you have observed in the atmosphere where you live or from your travels to other locations.
3. What characteristics of the atmosphere are typically described in weather forecasts?

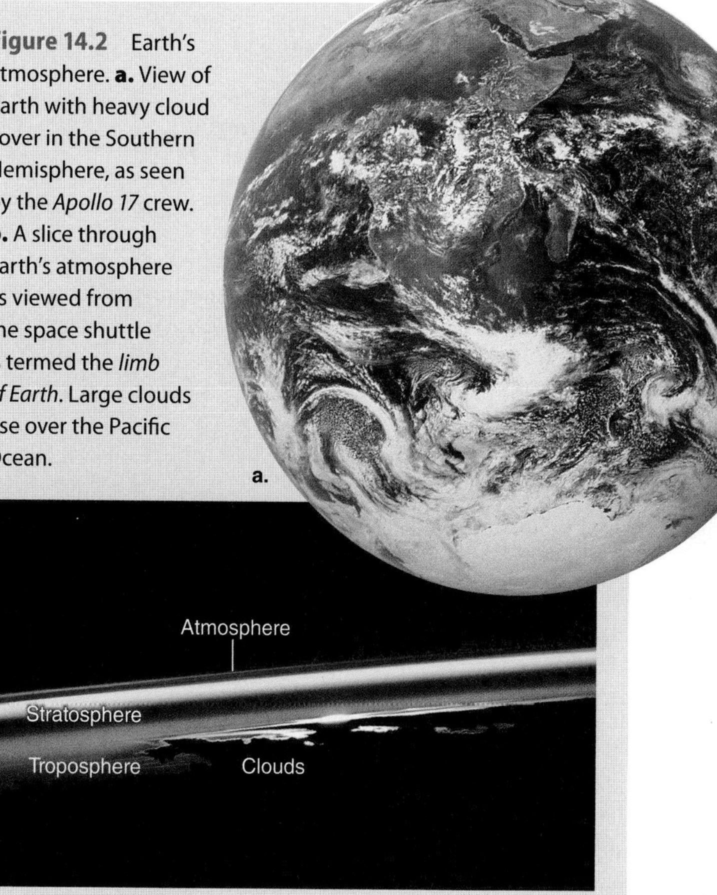

Figure 14.2 Earth's atmosphere. **a.** View of Earth with heavy cloud cover in the Southern Hemisphere, as seen by the *Apollo 17* crew. **b.** A slice through Earth's atmosphere as viewed from the space shuttle is termed the *limb of Earth*. Large clouds rise over the Pacific Ocean.

14.2 Air Evolves

Learning Objectives

- Describe the composition and relative proportions of gases in the atmosphere.
- Explain how the composition of the atmosphere changed during Earth's history.
- Discuss how the proportion of oxygen in the atmosphere is tied to the evolution of Earth's biosphere.

The specific mix of gases in Earth's atmosphere is termed *air*. **Two common gases, nitrogen and oxygen, make up about 99 percent of dry air** (Figure 14.3). Air also contains trace amounts of other gases (argon, carbon dioxide, methane, nitrous oxide, and ozone), as well as microscopic particles and chemicals (collectively called aerosols). The amount of water vapor in the air can range from next to nothing over dry deserts to up to 7 percent (by weight) in humid climates. We observe changes in the atmosphere above our heads in the form of daily weather patterns and seasonal transformations. These relatively short-term changes assume a uniform atmospheric composition—that's what makes them predictable. In contrast, as we will discuss in Chapters 16 and 17, most scientists agree that the composition of the atmosphere is gradually changing because of human activity and that such alterations will inevitably lead to modifications to other components of the earth system.

An Atmosphere Evolves

Atmospheres are common around other planets and even some moons in our solar system. The presence of an atmosphere is an important consideration in the search for life beyond Earth. As described in Chapter 2, our guardian atmosphere has been largely responsible for the development and protection of life on our planet. The early atmospheres of both Venus and Earth began with an initial hot, hazy mixture of gases that included

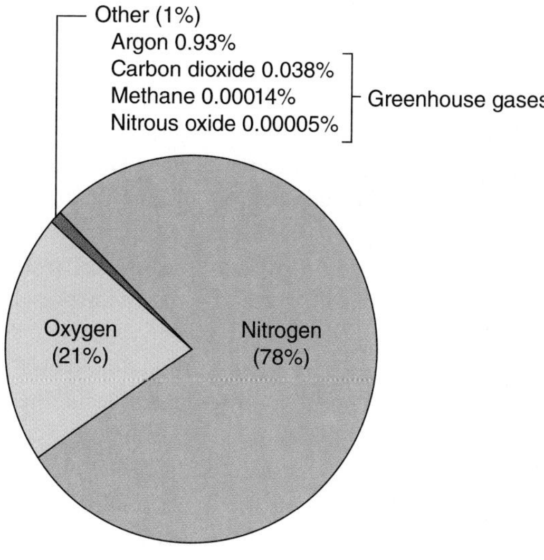

Figure 14.3 Gases in Earth's atmosphere. Concentrations of gases in dry air. Greenhouse gases trap heat near Earth's surface.

Checkpoint 14.1

basic intermediate advanced superior

What proportion of Earth's atmosphere is made up of carbon dioxide?

a) 18 percent
b) 7.3 percent
c) 0.55 percent
d) 0.038 percent

Checkpoint 14.2

basic intermediate advanced superior

When would oxygen have started to accumulate in the atmosphere if the early Earth had less land area?

a) Before 2.5 billion years ago
b) After 2.5 billion years ago
c) 2.5 billion years ago (no change)

abundant carbon dioxide, hydrogen, and water vapor. These **early atmospheres originated from gases expelled from extensive volcanism and from comets and meteorites** that collided with Earth.

The closer proximity of Venus to the sun resulted in higher temperatures on Venus than on Earth. Thus, water on Venus existed almost exclusively in the form of water vapor, a gas. These gas molecules were split into their constituent oxygen and hydrogen atoms by intense solar radiation, and the lighter hydrogen was lost to space. The spare oxygen combined with carbon to form additional carbon dioxide, the dominant atmospheric gas on Venus. The thick, dense atmosphere on Venus is 95 percent carbon dioxide and acts as an insulating blanket over the planet, raising surface temperatures to an ovenlike 464°C (867°F) and making it the hottest planet in the solar system. Luckily for us, Earth underwent a much different scenario!

Around 4 billion years ago, oxygen was absent from Earth's atmosphere, and carbon dioxide made up about one-quarter of atmospheric gases. As early Earth cooled, water vapor condensed to form a nearly continuous rain that eventually collected in oceans (see Chapter 13). The rain stripped carbon dioxide from the atmosphere, forming carbonic acid that weathered the rocks of the primitive continental crust (see Chapter 9). The weathering reactions combined carbon dioxide with minerals, which were subsequently deposited in sedimentary rocks, effectively locking the carbon dioxide into Earth's crust. These ancient rocks can be observed today in old sections of continental crust found at the surface in places such as South Africa and Australia. Later on, primitive organisms (for example, algae) used photosynthesis to consume carbon dioxide and water in the presence of sunlight to produce organic material and oxygen.

While the atmospheric carbon dioxide was gradually being removed by weathering and photosynthesis, there was no corresponding increase in oxygen. None of the oxygen produced by the earliest organisms collected in the atmosphere. Oxygen was steadily accumulating in the ocean, but oxygen is an extremely reactive element and any atmospheric oxygen would instead have combined with other elements (for example, carbon, sulfur, iron) to form minerals found in ancient rocks. Only after these reactions had absorbed all the oxygen they needed could the concentration

Checkpoint 14.3

basic intermediate advanced superior

Use the graph in Figure 14.4b to identify which times characterized by low oxygen levels also correspond to major extinction events (see Figure 8.17).

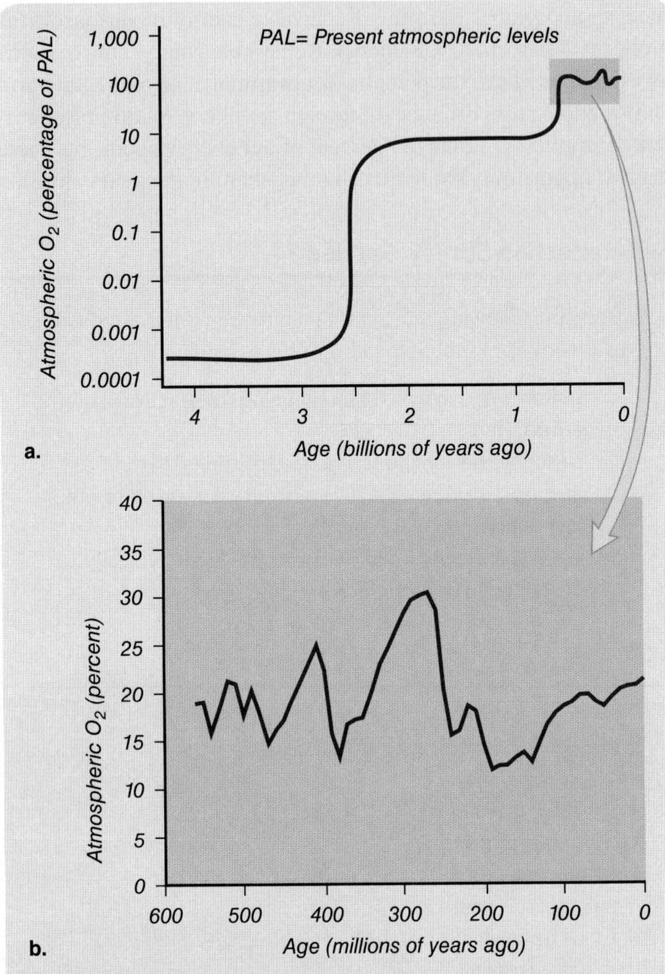

Figure 14.4 Atmospheric oxygen concentrations during Earth's history. **a.** Oxygen levels grew in two steps 2.5 and 0.5 billion years ago. **b.** Oxygen concentrations fluctuated between 12 and 30 percent during the Phanerozoic eon (the last 542 million years).

of oxygen in the atmosphere begin to increase. Scientists estimate that **it took more than 2 billion years before free oxygen began to accumulate in the atmosphere.** Plant life gradually built up oxygen to its current level (21 percent) over millions of years (Figure 14.4). Earth's biosphere therefore moderated the composition of the atmosphere to make it suitable for life. This moderation is an excellent example of earth system interrelationships.

Oxygen levels fluctuated considerably during the Phanerozoic eon (last 542 million years). The earliest animals lived in the oceans until atmospheric oxygen levels became sufficient to support life on land (see Chapter 8). Creatures began colonizing land in earnest about 425 million years ago (Mya) when oxygen concentrations reached levels similar to today. About 40 million years later, during the Devonian period (see Chapter 8), a decline

basic / intermediate / advanced / superior

How would you expect the oxygen content of the atmosphere to vary during snowball Earth conditions (much of Earth's land surface covered by glaciers) and hothouse Earth conditions (no ice, higher average global temperatures)?

in atmospheric oxygen concentration to around 13 percent was matched with declines in the number of arthropod species (for example, insects, crustaceans). Oxygen levels may have plunged as oxygen was taken out of the atmosphere to combine with other elements by weathering or if dry conditions reduced the amount of vegetation available to produce the gas.

Oxygen levels rose to an all time high of around 30 percent during the Permian period, 270 Mya, with the evolution of extensive oxygen-producing forests. This was accompanied by a second wave of species colonizing the land, led by the reptiles. These new oxygen-rich environments supported giant insects such as dragonflies with 75-centimeter (over 2-foot) wingspans or millipedelike creatures that were well over 1 meter (3.3 feet) long. Scientists have found large amounts of charcoal preserved in rocks of this time, indicating that there were extensive wildfires, fed by the high oxygen content of the atmosphere.

During a second episode of low oxygen concentrations (12 percent) in the Triassic period, 230 Mya, a group of organisms appeared with a new and more efficient respiratory system. Today, we know that reptiles such as lizards are rarely found at high altitudes characterized by low oxygen concentrations. It is probable that reptiles were replaced by these new organisms, the dinosaurs, that dominated life during the rest of the Mesozoic era, partially as a function of their ability to survive in a low-oxygen environment. Similar adaptations are seen in birds today, allowing them to fly at high altitudes where oxygen is limited.

14.3 Structure and Processes of the Atmosphere

Learning Objectives

- Predict relationships among altitude, temperature, and air pressure.
- Explain the difference between heat and temperature in Earth's atmosphere.
- Describe the characteristics of the four major layers of the atmosphere.

How much would you pay to fly to the edge of space, and how high would you have to go to get there? Because of the effects of gravity, almost all atmospheric gases (99 percent) lie within 32 kilometers (20 miles) of Earth's surface. **The density of the air rapidly decreases as you travel upward from Earth's surface.** A few stray gas molecules exist at 100 kilometers (62 miles) above the planet's surface, the internationally accepted boundary with space. However, spacecraft exiting Earth's atmosphere still feel the effects of a slight frictional drag until they reach altitudes above 500 kilometers (310 miles); thus some atmospheric gases must extend to that altitude. This does not stop some enterprising entrepreneurs from offering "edge of space flight experiences" that top out at around 25 kilometers (16 miles). These flights, in former Russian fighter jets, do not get as high as Joe Kittinger's balloon ride and cost more than $20,000. (Now you know what to ask for on your next birthday!)

As you traveled upward on your flight through the atmosphere, the temperatures would vary greatly. To understand why this happens, we must first clarify some important concepts: heat, temperature, and heat capacity.

Heat Versus Temperature

Think about heating two pans of water on the stove. Pan 1 has twice as much water as pan 2 (Figure 14.5). Imagine turning on the burner beneath each pan to the same heat level for the same amount of time. If you turn off the burner and measure the temperatures, which pan of water will be warmer? Which pan will contain more heat? To determine what happens here, you need to know that atoms in a substance such as water or air are constantly in motion. The **energy of their motion is known as kinetic energy.** Kinetic energy increases as the speed of atomic motion increases.

- **Heat is the *total* kinetic energy of all the atoms in a substance.** Assuming that all atoms are moving at the same speed, the more atoms present, the greater the heat.
- For a given quantity of a substance, **temperature represents the *average* kinetic energy of all the atoms in a substance.** A few atoms with rapid motion will have a higher temperature than many atoms with slow motion.

Now, back to the pans of water. Since the pans were heated for the same amount of time and at the same heat level, they contain the same amount of heat (all other factors being the same). That heat is now spread among the water molecules in each pan. However, the water in half-full pan 2 would have a higher temperature because the heat would have produced more rapid motion among fewer water molecules.

We can measure how efficiently a substance gains or loses heat by measuring temperature changes. Water, for example, can absorb or release large amounts of heat without a correspondingly large change in temperature. Thus, water is said to have a high **heat capacity, the amount of heat it must absorb (or release) to produce a corresponding temperature increase (or decrease).** The heat capacity of air is one-quarter of that of water. Consequently, if the same amount of heat were added to similar masses of air and water, the air would experience a greater temperature increase.

Figure 14.5 Demonstration of heat versus temperature. Two pans of water are placed on a stove. Pan 1 has twice the volume of water of pan 2, and each pan is heated at the same rate for the same amount of time. Which pan's contents will contain more heat? Which pan will be at a higher temperature?

Checkpoint 14.5

basic / intermediate / advanced / superior

What is the difference between heat and temperature?

a) Heat deals with total kinetic energy, temperature with average kinetic energy.
b) Heat deals with average kinetic energy, temperature with total kinetic energy.
c) There is no difference, since they both deal with kinetic energy.

Checkpoint 14.6

basic / intermediate / advanced / superior

At approximately what temperature did Captain Joe Kittinger begin his descent upon exiting the balloon capsule?

a) 0°C b) –25°C c) –40°C d) –75°C

Next in this section, we will use contrasts in heat and temperature to identify vertical changes in the structure of the atmosphere. In Section 14.5, we will discuss how heat can be transported within the atmosphere and how that affects weather.

The Four Layers of the Atmosphere

How could you measure the temperature of the atmosphere? For many decades, such measurements were taken by using instruments on balloons. Today, as a result of advances in technology, data are also collected using rockets and satellites. All of the data reveal alternating layers of increasing and decreasing temperatures within the atmosphere. These layers are the troposphere, stratosphere, mesosphere, and thermosphere. The boundaries between the layers are actually transition zones labeled with a *-pause* extension (for example, tropopause; Figure 14.6).

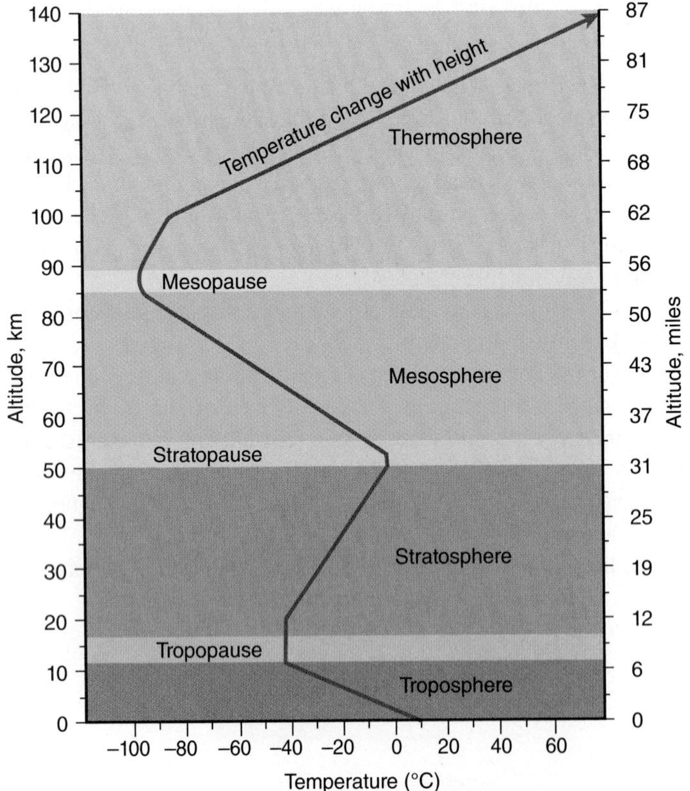

Figure 14.6 The four layers of the atmosphere. There are three areas of higher temperatures (the ground surface, upper stratosphere, and thermosphere) and two of lower temperatures (upper troposphere and upper mesosphere). The thermosphere continues upward to approximately 500 kilometers (310 miles) from Earth's surface.

Troposphere. The lowest atmospheric layer, termed the *troposphere,* displays a **decrease in temperature with increasing altitude.** The temperature of the air near the surface is relatively high because energy radiated from the Earth's surface warms that air. Moving up in elevation, air temperature decreases at a constant rate as we will describe in Section 14.6. Temperatures reach a minimum of approximately −42°C (−44°F) at the top of the troposphere (Figure 14.6). The troposphere contains our weather systems, air pollution, and the bulk of air and aerosols. This layer ends at the tropopause.

Do you think the troposphere is the same thickness everywhere on Earth? On average, where are the coldest surface temperatures—and the warmest? Since the troposphere is defined by its thermal character, we would expect it to be thicker above warm regions and thinner over cold areas. Hence, it should come as little surprise that the thickness of the troposphere increases from 8 kilometers (5 miles) over the poles to as much as 16 kilometers (10 miles) at the equator. Air density, the mass of gas molecules per volume of air, is greatest in the troposphere, which contains three-quarters of all the air in the atmosphere. Rising above the New Mexico plains, Joe Kittinger would have spent nearly one-half of his ascent in the troposphere.

Stratosphere. The stratosphere lies immediately above the troposphere and is **characterized by an increase in temperature with altitude.** The stratosphere is over 40 kilometers (25 miles) thick and contains about 20 percent of the atmosphere's air. Felix Baumgartner stepped out of his balloon capsule about halfway through the stratosphere. Maximum temperatures approach 0°C (32°F) at the top of the stratosphere (Figure 14.6). The stratosphere contains ozone, a compound that blocks ultraviolet (UV) radiation. Ozone is an oxygen molecule that is composed of three oxygen atoms (O_3; see Chapter 17), compared to the two oxygen atoms of the common oxygen molecule (O_2). Ozone keeps harmful UV rays from reaching Earth's surface, but it did not form until oxygen became an abundant gas in the atmosphere. One of the reasons life initially flourished in the oceans is that the water protected organisms from UV radiation. It is probable that life on land would have remained much more primitive without the development of the protective ozone layer.

Temperature increases upward in the stratosphere as ozone molecules absorb UV radiation. Since those molecules have fewer neighbors to bump into (and slow them down), they have higher kinetic energy (and consequently, higher temperatures) than molecules found lower in the stratosphere. The warm cap of the stratosphere overlying the relatively cool troposphere results in a stable atmospheric configuration because the cool air cannot rise into the warm layer. The stratosphere ends at the stratopause.

Checkpoint 14.7

basic · intermediate · advanced · superior

At extremely low temperatures, the thin polyethylene fabric (0.002 inch thick) that made up the balloon carrying Joe Kittinger's capsule would have become nearly brittle. Any small flaws in the fabric could have caused the balloon to spring a leak and deflate. At what location during the ascent would the risk of this potential danger have been most acute?

a) lower troposphere
b) upper troposphere
c) middle stratosphere

Mesosphere. The mesosphere, the third thermal layer in the atmosphere, is **characterized by decreasing air temperatures** that reach a minimum at the boundary with the overlying thermosphere (Figure 14.6). Temperatures decrease with altitude in this layer because there are fewer and fewer ozone molecules to be warmed by incoming UV radiation. While oxygen and nitrogen molecules are rarer at these altitudes, sufficient gas is still present to heat up incoming space debris to form meteors. Most near-Earth objects burn up in this layer (see Chapter 3). The upper mesosphere boundary, the mesopause, is the second temperature minimum in the atmosphere.

Thermosphere. The outermost layer of the atmosphere, **the thermosphere, blocks much of the harmful cosmic radiation,** including X-rays, gamma rays, and some UV radiation, from reaching Earth's surface. All of this incoming solar radiation raises temperatures in the upper thermosphere to more than 1,000°C (1,830°F), but the number of gas molecules is so small at this altitude that heat energy is very low. Such isolated gas

Figure 14.7 Aurora borealis (the Northern aurora) viewed near Anchorage, Alaska. A similar aurora known as the aurora Australis is visible in the Southern Hemisphere.

Checkpoint 14.8

Soon after Joe Kittinger began his jump, he was traveling at over 982 kilometers per hour (610 miles per hour). What characteristic of the atmosphere caused him to slow down as he descended (before opening his main chute)? Why did he not hear the sound of air whistling by him as he began to fall?

molecules in the thermosphere are broken into their constituent ions as solar radiation strips electrons from oxygen and nitrogen molecules. These ionized gases generate spectacular visual effects called *auroras.* Auroras occur when electrons and protons from the sun interact near Earth's magnetic poles (Figure 14.7). The *International Space Station* is actually located in the thermosphere, about 350 kilometers (217 miles) above Earth's surface. There are too few gas molecules at an altitude of 500 kilometers (310 miles) to denote a clear boundary (there is no thermopause) between the thermosphere and space.

14.4 Solar Radiation and the Atmosphere

Learning Objectives

- Describe connections between the sun's output of energy and the earth system.
- Apply the radiation budget to explain temperature fluctuations on Earth.
- Predict how Earth's albedo at various locations will affect atmospheric temperatures.
- Predict how changes in the characteristics of the atmosphere would affect global temperatures.

Life as we know it needs a constant source of energy. For Earth, that source of energy is the radiation we receive from the sun. Solar radiation warms Earth's surface and the atmosphere, providing the habitable planet we live on, and it is the ultimate source that contributes to global warming. In addition, solar radiation supplies the energy necessary for cloud formation, precipitation, and local weather conditions. The relatively pleasant average global temperature of 15°C (59°F) is a direct result of two factors. First, visible light is converted to heat when solar radiation strikes Earth's surface. Second, the heat is absorbed close to the planet's surface by greenhouse gases (carbon dioxide, methane, nitrous oxide, and water vapor) in the atmosphere.

Solar Radiation and the Electromagnetic Spectrum

Our sun emits electromagnetic radiation (EMR). The characteristics of EMR are vital to life on Earth. **Electromagnetic radiation is described by its wavelength and frequency.** Just like an ocean wave (see Chapter 13), an electromagnetic wave has a wavelength, which is the distance between two adjacent wave crests. Frequency represents the number of wave crests that pass a point in 1 second. The sun's EMR spectrum spans wavelengths from short-wavelength (0.000001 micrometer; 0.00000000004 inch) gamma rays to long-wavelength (hundreds of meters) radio waves (Figure 14.8a).

The supply of energy used daily by natural earth systems is derived almost completely from solar radiation. Visible light (wavelength 0.4 to 0.7 micrometer) makes up a little less than one-half (47 percent) of the solar radiation reaching Earth's atmosphere (Figure 14.8b). Infrared radiation is associated with heat and makes up 45 percent of the sun's radiation emissions that make it to Earth. About 8 percent of solar radiation is short-wavelength UV radiation. All short-wavelength radiation (including X-rays, gamma rays, and UV rays) is invisible to the naked eye. Humans cannot affect the amount or type of radiation emitted from the sun. However, any human action or natural event that affects the amount of solar radiation reaching the surface (or retained by the atmosphere) may have an effect on life on Earth.

Earth's Energy Budget

What do you think can happen to EMR that reaches Earth? There are three possible fates for incoming solar radiation as it passes through Earth's atmosphere on its way to the surface of the planet (Figure 14.9):

1. ***EMR can be scattered.*** When EMR strikes small particles and gas molecules in the atmosphere, it may change direction, a process known as *scattering*. Have you ever wondered, "Why is the sky blue?" It is blue because the wavelength of blue light is about the same as the size of the oxygen atoms in the atmosphere. Consequently, blue light is scattered more easily than other colors with longer wavelengths. The blue light that is scattered toward Earth's surface reaches our eyes, making the sky appear blue. In higher sections of the atmosphere with fewer gas molecules to scatter incoming solar radiation, the sky would appear black, much as it appeared to Felix Baumgartner when he reached an altitude of 39 kilometers (24 miles).
2. ***EMR can be reflected.*** Think of your reflection in the mirror. What about your reflection on a shiny table or a brick wall? Different surfaces reflect different amounts of light, making it possible to see our reflection in some surfaces but not in others. In contrast to scattering, **reflection in the atmosphere causes incoming radiation to be returned to space.** Solar radiation is reflected from gas molecules and other particles in the atmosphere as well as from features on Earth's surface. Some natural surfaces on Earth are more reflective than others. For

Checkpoint 14.9

basic / intermediate / advanced / superior

The ozone hole over Antarctica actually represents a thinning of the ozone layer. What are the consequences of the loss of ozone?

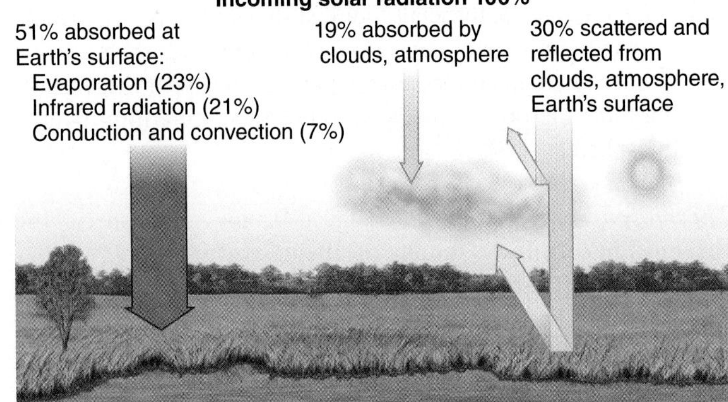

Figure 14.9 Where solar energy goes. Approximately one-half of the incoming solar radiation heats Earth's surface.

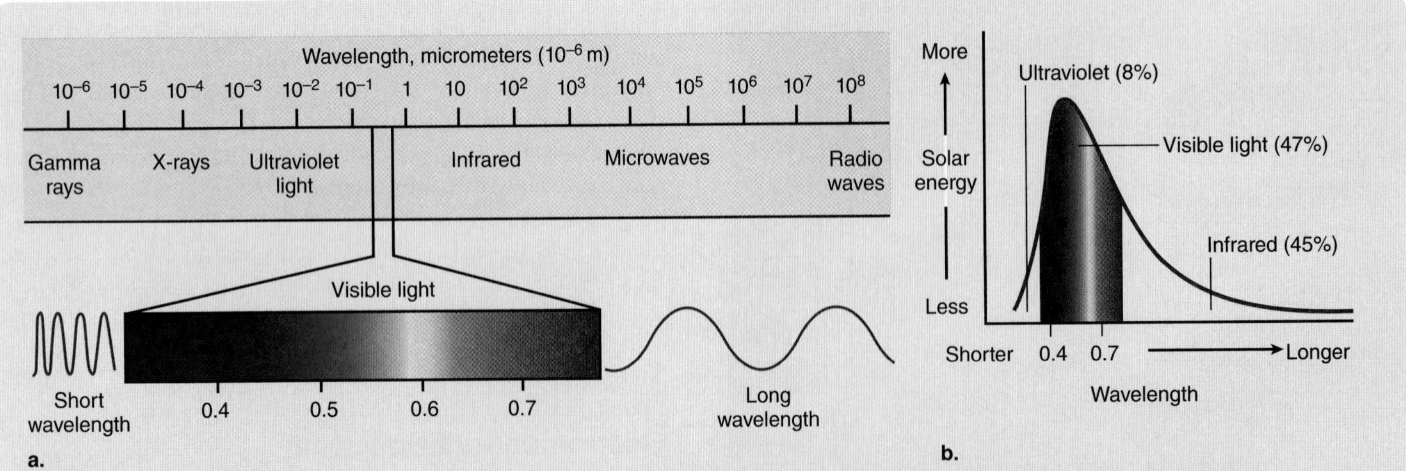

Figure 14.8 The electromagnetic spectrum. **a.** Wavelength of radio waves can be measured in hundreds of meters. In contrast, wavelengths for visible light are less than 1/1,000 millimeter across but 1 million times larger than the wavelengths of gamma rays (1 million micrometers = 1 meter). **b.** The relative proportions of solar radiation reaching Earth. Infrared and visible light make up more than 90 percent of solar radiation at Earth's surface.

instance, more solar radiation is reflected from light-colored surfaces than from dark-colored surfaces. **The measure of the reflectivity of a surface is termed *albedo*** (see Chapter Snapshot). The average albedo of Earth is about 31 percent, though it varies with the region. Light-colored, reflective surfaces and thick cloud cover have a high albedo (80 to 90 percent) because these features reflect more sunlight. In contrast, dark surfaces (for example, forests, water) in the absence of cloud cover have low albedo values (less than 25 percent). Pollution from light-colored aerosols, which are carried downwind from populous regions, produces a haze that blocks and reflects incoming EMR, producing a regional cooling effect. Some have suggested taking advantage of this phenomenon to inject aerosols high into the atmosphere in an effort to reduce incoming radiation and slow rates of global warming.

3. ***Some forms of EMR are absorbed*** in the atmosphere or at Earth's surface. Absorption is the process whereby radiation interacts with a material and is converted to some other form of energy (for example, heat). For absorption, the wavelength of the incoming solar radiation is important because different atmospheric gases absorb specific wavelengths. Forms of radiation with the shortest wavelengths (X-rays, gamma rays) are absorbed by gases in the thermosphere, while ozone in the stratosphere absorbs UV radiation (Figure 14.10). This characteristic of ozone effectively blocks the majority of incoming UV rays that can cause skin cancer. Forest fires and industrial pollution may add black soot particles that will absorb EMR and heat the surrounding atmosphere. Recent calculations have determined that these dark particles are the second most significant contributor to global warming (after carbon dioxide). Dark soot particles that are washed out of the atmosphere by rain can reduce the albedo and increase heat absorbed on the ground. These particles are blamed for some of the increase in melting rates of Himalaya glaciers. Visible light and longer-wavelength infrared radiation reach the troposphere. Water vapor and carbon dioxide in the troposphere absorb infrared radiation in the range of approximately 1 to 2 micrometers. As we will learn in a moment, it is this property that causes the atmosphere to warm up.

Visible light, some ultraviolet rays, and infrared radiation (what we feel as heat from the sun) may reach Earth's surface (see Figure 14.9). **Sunlight is absorbed by land and oceans, warming the planet's surface.** Some of this energy is used to evaporate water or to heat objects on the surface. The energy that is absorbed (regardless of its wavelength) is converted to thermal energy (heat). To illustrate this process, suppose you go the beach in the morning of a sunny summer day. You park your car on a black asphalt parking lot and walk in your bare feet down to the beach. No problem. Later that day, you decide to return to your car—in your bare feet. What will you experience as you walk (or maybe run) to your car? Big problem! The asphalt parking lot is hot, much hotter than it

Checkpoint 14.10

basic | intermediate | advanced | superior

All objects emit heat. The hotter an object, the faster its molecules move and the shorter the wavelength of radiation it emits. This relationship between the heat of a body and wavelength is expressed in Wien's law, which states that the maximum wavelength (in micrometers) of electromagnetic radiation can be determined by dividing 2,897 by the absolute temperature (T). Absolute temperatue T is expressed in kelvins (K). The Kelvin temperature scale begins at absolute 0 (where all particle motion stops), with the freezing point of water (0°C) equivalent to 273 K on this scale (30°C = 303 K).

$$\text{Maximum wavelength} = (2{,}897/T)$$

For example, for the sun, $T = 6{,}000$ K. Substituting the temperature of the sun into the above equation illustrates that incoming solar radiation should have a wavelength of approximately 0.5 micrometer, which lies within the range of visible light.

Use Wien's law to determine the dominant wavelengths emitted from (a) Earth and (b) Venus, and identify the corresponding type of radiation.

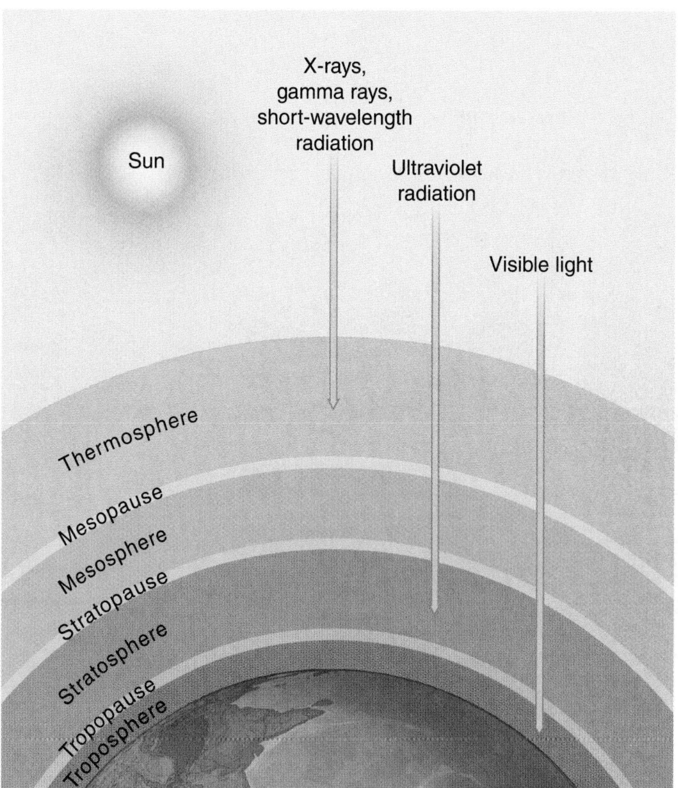

Figure 14.10 Absorption of sun's rays. Most short-wavelength solar radiation, including X-rays, gamma rays, and ultraviolet radiation, is intercepted before it reaches the troposphere. Visible light and radiation associated with heat are the most common wavelengths to reach Earth's surface.

THE EARTH'S ALBEDO

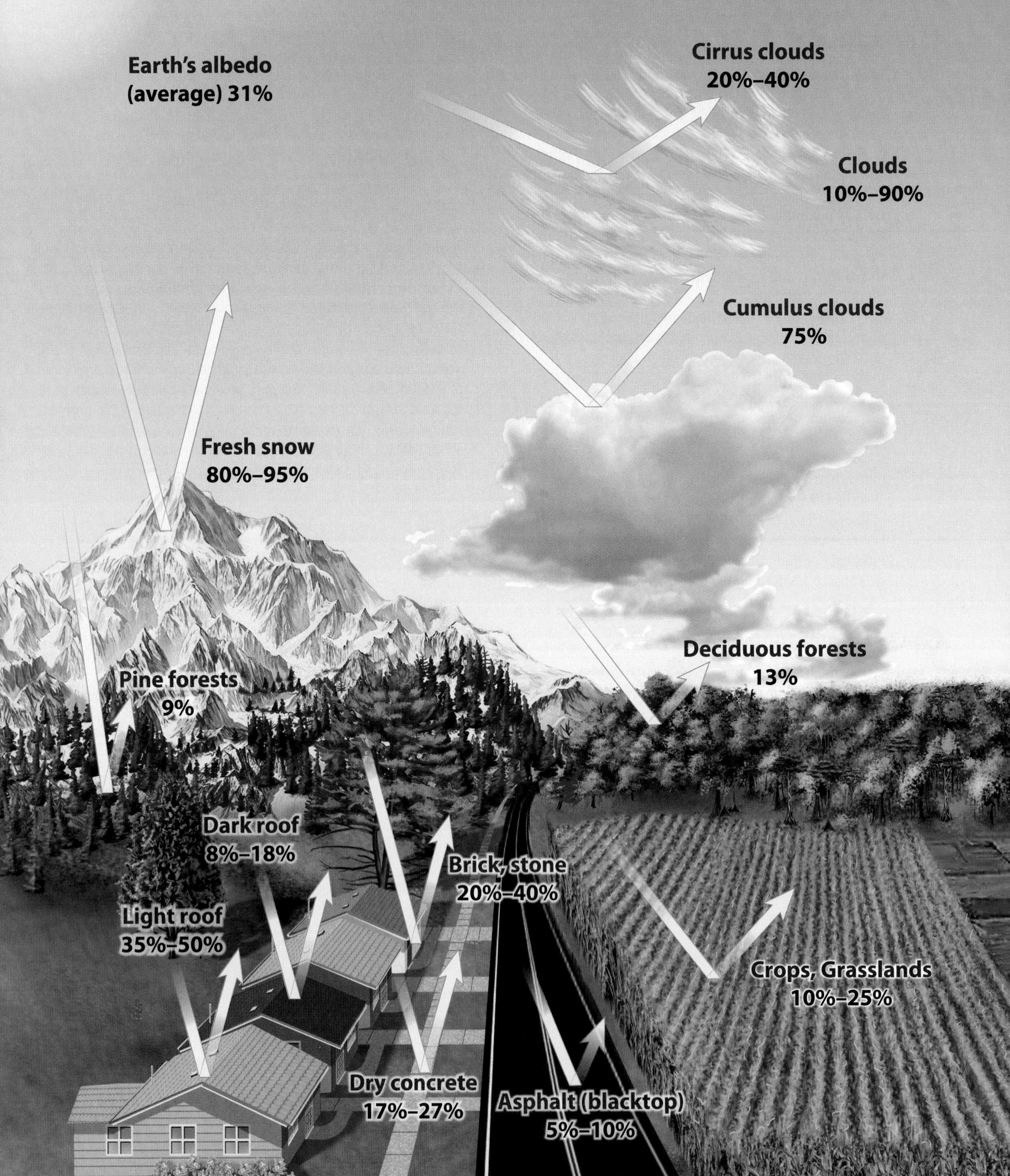

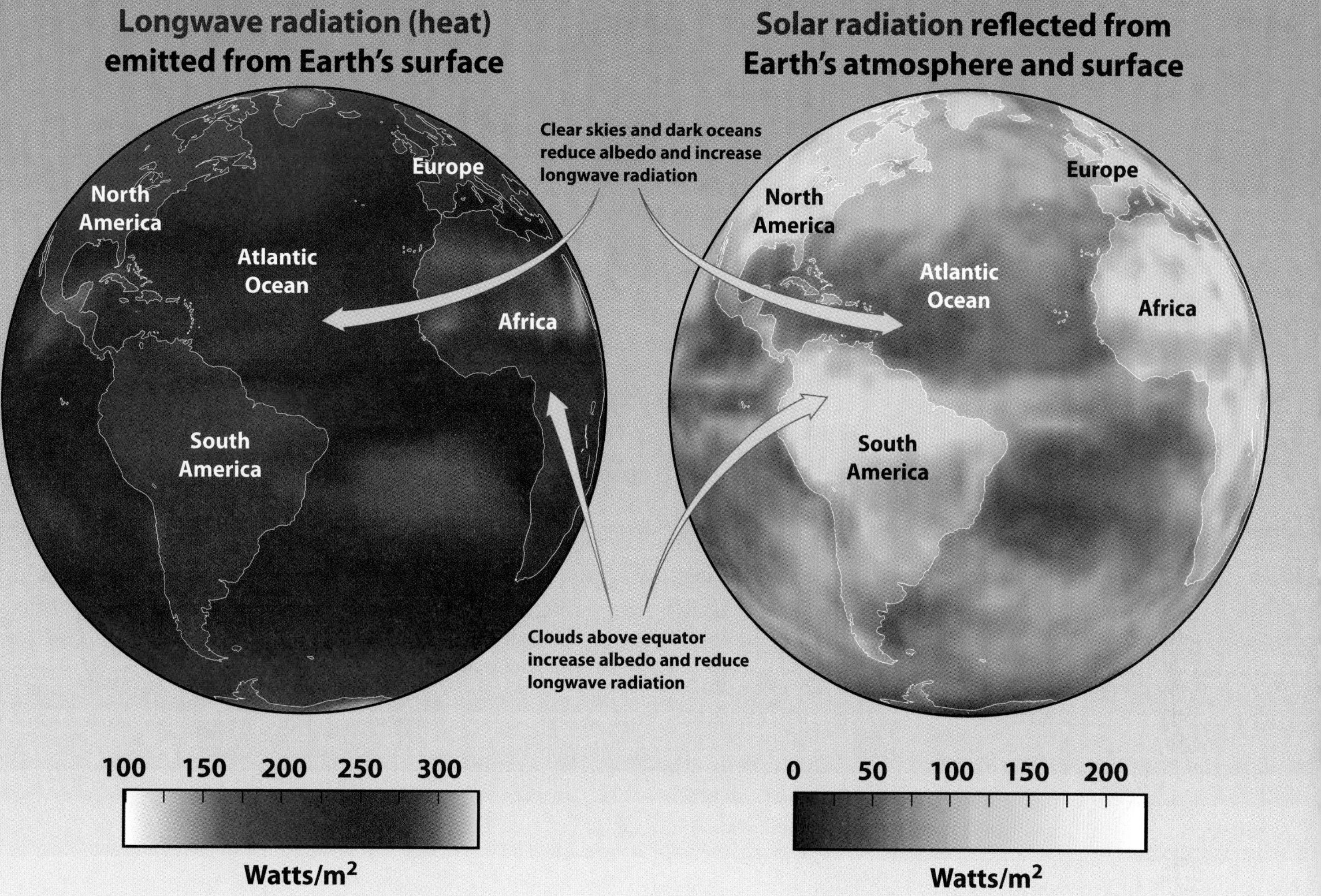

Longwave radiation (heat) emitted from Earth's surface

Solar radiation reflected from Earth's atmosphere and surface

Clear skies and dark oceans reduce albedo and increase longwave radiation

North America

Europe

Atlantic Ocean

Africa

South America

Clouds above equator increase albedo and reduce longwave radiation

100 150 200 250 300

Watts/m^2

0 50 100 150 200

Watts/m^2

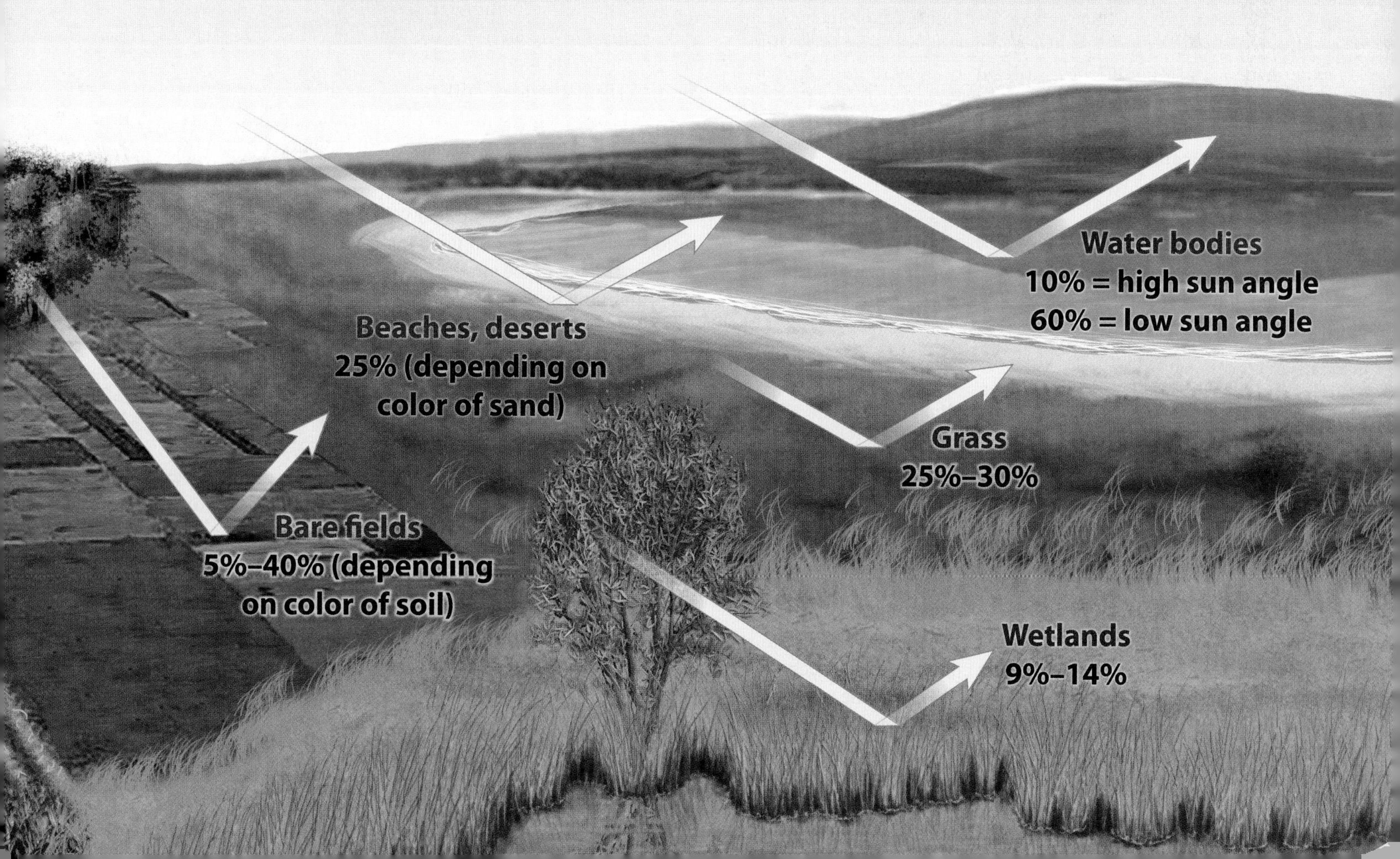

Water bodies
10% = high sun angle
60% = low sun angle

Beaches, deserts
25% (depending on color of sand)

Grass
25%–30%

Bare fields
5%–40% (depending on color of soil)

Wetlands
9%–14%

was earlier, because it absorbed both the heat and much of the incident light from the sun. That heat is then reradiated as longer-wavelength infrared radiation.

The Greenhouse Effect. The air temperature in the troposphere decreases steadily as distance increases from the warm Earth's surface, partly because of Earth's heating and partly because the air is thinner (see Section 14.6). Anyone who has ever been at high elevations knows that temperature decreases with altitude (that's why we go to the mountains to ski). **Air temperature declines at a rate of 6.5 C° per kilometer (12 F° per mile, known as the *normal lapse rate*) through the troposphere,** beginning at an average of 15°C (59°F) at sea level.

Earth's surface warms as it absorbs radiation that has not been scattered, reflected, or absorbed elsewhere in the atmosphere. Surfaces with low albedo would warm up and emit energy as heat in the form of long-wavelength infrared radiation (that asphalt parking lot is still warm long after the sun sets). Surfaces with low albedo emit more heat energy than surfaces with high albedo (see Chapter Snapshot). This radiation is absorbed by water vapor, carbon dioxide, and other trace gases (methane, nitrous oxide) in the troposphere, creating a situation that has come to be known as the *greenhouse effect.* **The greenhouse effect increases temperatures at Earth's surface by 33 C° (59 F°),** ensuring that we have a livable planet. Earth would have an average surface temperature of −18°C (0°F) rather than the present average of +15°C (59°F) without the heat-absorbing property of water vapor and other gases in our atmosphere. In contrast, Venus, with a carbon dioxide–rich atmosphere, has an out-of-balance (runaway) greenhouse effect, resulting in surface temperatures of 464°C (867°F; see Chapter 2).

Changing the surface albedo can change how much heat is absorbed and therefore can affect temperatures. A region could experience a warming effect as a result of a decrease in albedo because the surface retains (and then emits) more energy. For example, at high latitudes, warming could occur when white sea ice melts, leaving behind a dark sea surface.

Checkpoint 14.11

Explain if the interaction of EMR with Earth would cause global temperatures to increase or decrease under the following conditions.

1. The atmosphere is thicker.
2. The atmosphere contains more carbon dioxide.
3. The atmosphere contains more aerosols.
4. The atmosphere contains more black soot.
5. Tree leaves are white.
6. There is no ice on Earth.

Checkpoint 14.12

Construct a concept map to show the connections between solar radiation and the atmosphere, using at least the following terms: incoming radiation, absorption, reflection, scattering, albedo, and the greenhouse effect.

14.5 The Role of Water in the Atmosphere

Learning Objectives

- Apply concepts of latent heat to explain energy transfer during water phase changes.
- Predict how relative and absolute humidity are affected by temperature changes.
- Apply the concept of latent heat to explain processes behind everyday phenomena.

Can you think of any naturally occurring substance that can exist at Earth's surface as a solid, liquid, and gas? Because of its unique atomic structure, **water is the only natural compound that can exist in all three states on Earth's surface: a liquid (water), a gas (water vapor), and a solid (ice).** We experience these forms of water as rainfall, humidity, or hail, respectively. In Chapter 11, we discussed the various places where water is found in the earth system and how those places are linked together by the hydrologic cycle (Figure 11.3).

The atmosphere contains just a small part of the water present on Earth. Imagine a 1-gallon (3.8-liter) jug of milk sitting on a kitchen table. Now, pour out slightly less than 0.5 cup (118 milliliters) of milk and set it down on the table beside the gallon jug. Next, take just 1 drop of milk and place it on a teaspoon.

- The original 1-gallon jug represents all the water on Earth.
- The milk remaining in the gallon jug represents the 97 percent of Earth's water in the oceans (see Figure 11.4a).
- The half-cup of milk represents all the water in streams, lakes, glaciers, and groundwater on the continents.
- That single drop of milk is equivalent to the tiny fraction of the planet's water that is in the atmosphere.

If we were to add up the annual volume of water falling as precipitation, we would find that it is more than 30 times greater than the moisture stored in the atmosphere at any given moment. To sustain such high precipitation volumes, water must be constantly cycled through the atmosphere. Evaporation and condensation of water are continuous in the atmosphere, and the conversion of water from one state to another transfers energy throughout the troposphere.

Three States of Water

The liquid, solid, and gaseous forms of water differ because of water's atomic structure. One molecule of water is composed of two hydrogen atoms and one oxygen atom. The arrangement

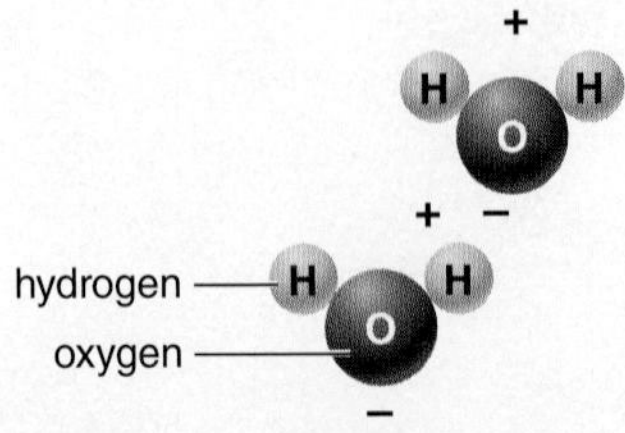

Figure 14.11 Dipolarity of water. Water molecules attract because of the net partial negative charge on the oxygen atom and the net partial positive charge on one of the hydrogen atoms in the two different molecules.

of these atoms results in opposite charges on each end of the molecule; hence, water is termed *dipolar.* The dipolar nature of water molecules results in the formation of a dipole-dipole attraction between their negative and positive ends so that adjacent molecules join together (Figure 14.11). The three states of water are determined by the distance between the individual water molecules and their degree of motion: In solid form, the molecules are joined together in a closely spaced, well-ordered hexagonal structure (Figure 14.12a). Water molecules vibrate in the solid state but do not move around. Ice is the most ordered configuration of water molecules. In liquid form, small groups of molecules remain attached to one another but move rapidly enough to generate semiordered structures (Figure 14.12b). In gaseous form, most individual water molecules move too rapidly to allow the polar attraction to join them together (Figure 14.12c). Water vapor is the least-ordered configuration of water molecules.

Changing States of Water

How does water change from one state to another? Think of a glass of ice water on a hot day. Eventually, the ice will melt and the water will get warmer; you know from your own experience that this happens because the water absorbs heat energy. Conversely, if you put a glass of warm water in the freezer, it will get colder and eventually become solid ice; this happens because the water releases heat energy. As water absorbs or releases heat, it changes state. However, not all heat absorbed is used to raise temperature; some may be required to change the heated material from one state to another.

For example, if a pan containing a mixture of water and ice (initial temperature 0°C [32°F]) were heated on a stove, a thermometer would not record a change in temperature until all the ice had melted. The water-ice mixture would first absorb the heat to convert the ice to water, but there would be no corresponding change in temperature until all the ice had changed state. **The amount of heat absorbed or released as water changes state is termed *latent heat.***

The absorption of latent heat occurs during melting, evaporation, or sublimation (the change from a solid directly to a gas); the release of latent heat results from freezing, condensation, or deposition (Figure 14.13). The amount of heat gained or lost per gram of water is expressed in calories of latent heat. Thus, the conversion of ice to liquid water absorbs 80 calories of heat for each gram of water converted from a solid to a liquid state. In the reverse reaction, 80 calories of latent heat are released into the atmosphere per

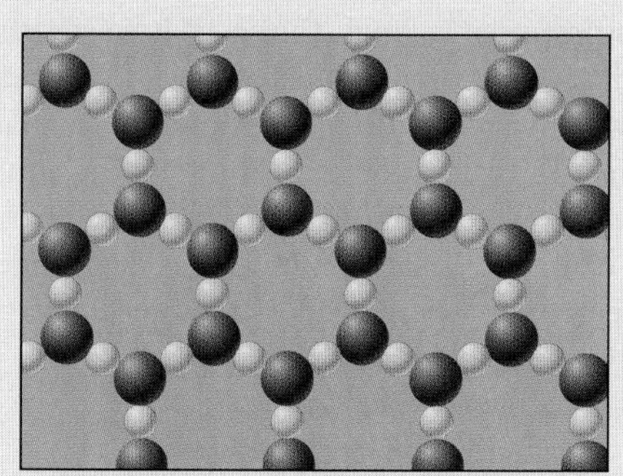
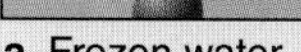

a. Frozen water

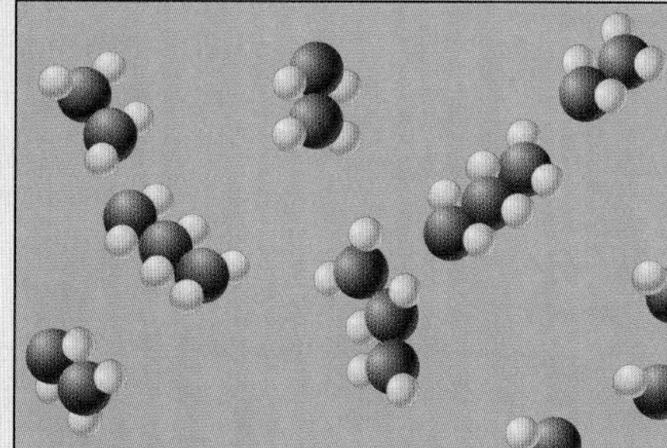

b. Liquid water

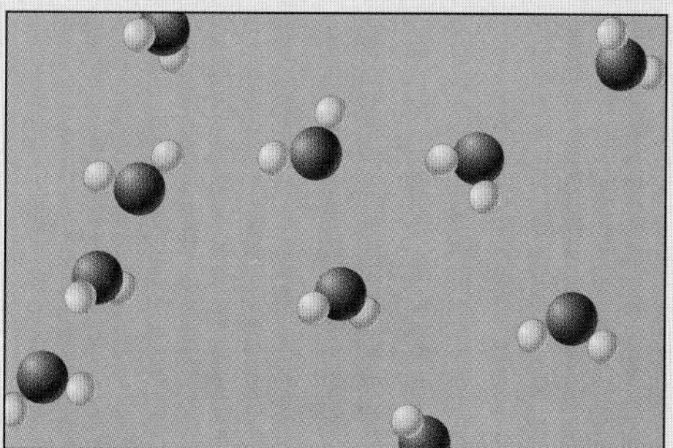

c. Water vapor

Figure 14.12 Molecular structures of the three states of water. Individual water molecules are composed of one atom of oxygen (red) and two atoms of hydrogen (tan). Molecules are more ordered (that is, more closely spaced and move less) in the **a.** solid state and become less ordered progressing to the **b.** liquid and **c.** gaseous states.

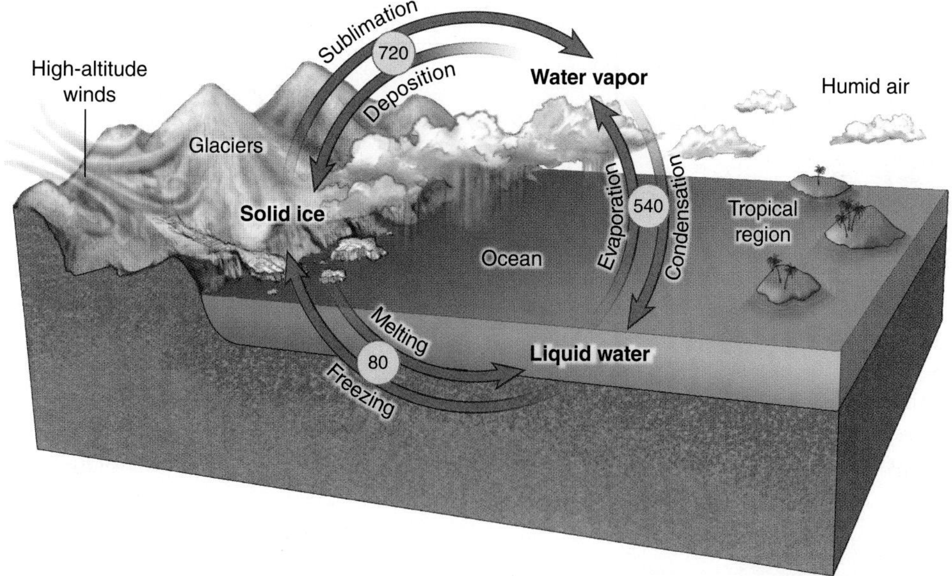

Figure 14.13 Release or absorption of latent heat. Six potential changes in the state of water are possible. The calories per gram of water (heat) that are lost or gained during each change are least for freezing/melting (80) and are greatest for changes between solid and vapor forms (720). The calories released or absorbed during condensation or evaporation are estimated for water at 0°C (32°F). When water changes directly from a solid to a vapor during sublimation, 720 calories per gram of latent heat are required. The latent heat requirements are: 80 calories per gram (solid-liquid) + 100 calories per gram (for liquid-phase temperature rise to 100°C) + 540 cal/g (liquid-vapor).

gram of water that is converted from liquid to solid (Figure 14.13). Much more latent heat is released or absorbed during changes between liquid and gaseous states (evaporation, condensation) than during changes between solid and liquid states. This is a function of the number of bonds that must be broken or modified between water molecules.

Of the six potential changes in the state of water just mentioned, **evaporation and condensation are especially significant.** These two processes occur over large areas of Earth's surface and contribute to weather phenomena and redistribution of heat in the atmosphere.

During evaporation, liquid water is converted to water vapor. You may have noticed this change of state occurring on your skin after you step out of a hot shower into a cool room or when you perspire during exercise. Your body supplies the heat needed for evaporation, and as heat is released, you feel cooler. In the environment, water evaporates from bodies of water and becomes water vapor in the air.

Condensation occurs when a gas is converted to a liquid. During this process, heat is released into the surrounding environment. For example, condensation often causes a film of water to form on the outside of a glass containing an iced drink (Figure 14.14). The air around the glass becomes slightly warmer as a result of the condensation (but not enough for you to notice). In the environment, water vapor in the air releases heat as it condenses and becomes liquid water (rain).

Your body feels cooler when you step out of a warm shower because

a) water evaporates on your skin.
b) water condenses on your skin.
c) water evaporates from the surrounding air.
d) water condenses in the surrounding air.

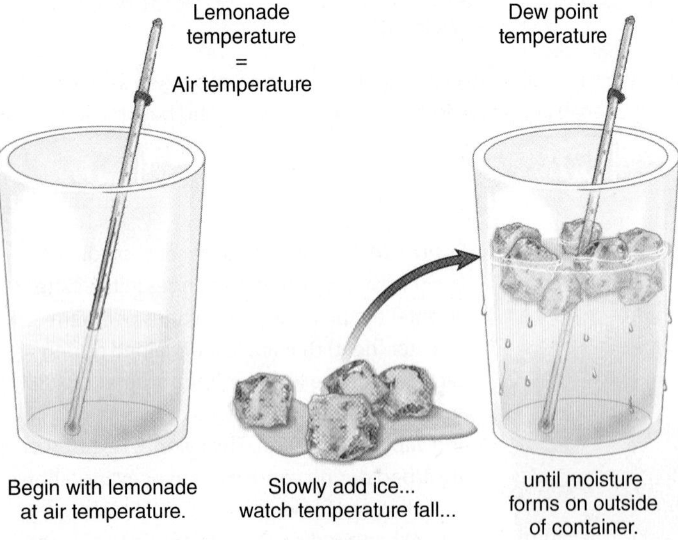

Figure 14.14 Condensation. The lemonade in the glass is initially at room temperature. Ice cubes are added to the glass, causing the temperature of the lemonade and the air surrounding the glass itself to decrease. Moisture forms by condensation on the outside of the glass when the temperature of the air around the glass reaches the dew point.

Humidity

Evaporation and condensation determine **the amount of moisture in the air, termed *humidity*.** Have you ever experienced a hot, humid day? How about a cold, humid day? We rarely experience humid conditions in cold winter weather because cold air is less able to hold moisture than warm air (Figure 14.15). **Air is considered saturated with water if it can hold no more water vapor at a given temperature.** For example, based on Figure 14.15, air at 30°C (86°F) can hold a maximum of 34 grams per cubic meter of water vapor (on the saturation curve), but air with a temperature of 20°C (68°F) can hold only 18 grams per cubic meter.

There are two ways to measure humidity: absolute humidity and relative humidity. **Absolute humidity measures the mass of water (grams) in a volume of air (cubic meters),** essentially the density of water vapor (mass per volume). Absolute humidity is expressed as grams per cubic meter (g/m^3). However, when your local weather forecasters report humidity, they are referring to relative humidity. **Relative humidity measures the amount of water vapor in air in comparison to the maximum mass of water vapor the air could possibly hold (that is, if it were saturated).** Relative humidity is expressed as a percentage. If we were to describe how much water was in a glass, we might say the glass was half full. Relative humidity does the same thing for the atmosphere. It tells us how much water vapor the atmosphere contains and how much more it can hold. Relative humidity is determined by using the following equation, where condensed water in the air and deposited water are not included:

$$\text{Relative humidity} = \frac{\text{actual water vapor density}}{\text{saturated water vapor density}} \times 100\%$$

For instance, according to Figure 14.15, air at 30°C (86°F) can hold a maximum of 34 grams per cubic meter of water vapor (on the

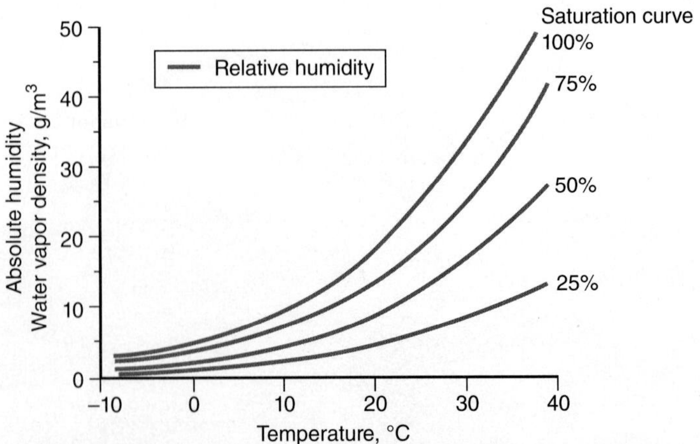

Figure 14.15 Humidity and temperature. The amount of water vapor that air can hold per volume of air increases with increasing temperature.

saturation curve). If the air actually has 17 grams per cubic meter, its relative humidity would be 50 percent. In contrast, if the temperature were 19°C (66°F), air containing 17 grams per cubic meter would have a relative humidity of 100 percent (Figure 14.15).

Changes in Humidity. At any given time, evaporation and condensation are both occurring in air, but one process typically dominates the other. Depending on which process comes out on top, humidity may increase or decrease. **Both absolute and relative humidity can increase when additional water vapor is added to the atmosphere by evaporation if the temperature of the air remains constant.** Absolute and relative humidity may decrease when condensation occurs, converting water vapor to precipitation, which may then leave the atmosphere. Different processes may dominate at different heights in the atmosphere. Therefore, it is possible for condensation to cause precipitation at high altitudes and for the raindrops to evaporate as they fall through a warmer, drier layer of air closer to Earth's surface.

In many instances, air is far enough from sources of water that the actual amount of water vapor in the air does not change as the air moves in the atmosphere. In Section 14.6, we will find that the temperature of air changes when that air rises or falls. What happens to the relative humidity of air if that air gets cooler but no additional water is added to the air? Consider Figure 14.15 again. Imagine a parcel of air with a temperature of 34°C (93°F) and an absolute humidity of 20 grams per cubic meter. If the absolute humidity remains constant and temperature decreases, the relative humidity will increase from 50 percent (move straight left on the chart). The parcel of air would become saturated when the temperature fell to 22°C (72°F). (Note that we are now on the 100 percent saturation curve.) **Relative humidity increases if air temperature decreases** because the saturation level decreases but the amount of water vapor in the air remains unchanged. The reverse is true if the air warms (saturation level increases, relative humidity decreases).

One key difference among the three states of water is the velocity of the water molecules. When cold air moves over a warm water body, evaporation from the water body adds moisture to the air. This is the water vapor (called *steam fog*) that can commonly be seen rising from a warm pond or lake in the early morning during summer, when air temperatures are cooler than water temperatures. The bonds between the water molecules break as the velocity of the molecules increases and some of the liquid is converted to a gas phase. Reducing air temperature decreases the saturated vapor density because colder air "holds" less moisture than warm air. The *dew point* is the temperature the air must attain for that air to become saturated. When a warm air mass moves over cold water, the cold water cools the air. When the air temperature drops, some moisture condenses and may form fog. **Condensation occurs when the relative humidity of the air increases so much that the air becomes saturated with moisture** (relative humidity = 100 percent). As these fog examples illustrate, relative humidity can increase with the addition of more water vapor or as a result of a decrease in temperature.

Checkpoint 14.14

basic intermediate advanced superior

Measurements reveal that 1 cubic meter (35 cubic feet) of air at 12°C (54°F) actually holds 6 grams (0.2 ounce) of water. What happens if the temperature of the air increases? Explain your answer.

a) Absolute and relative humidity both increase.
b) Absolute humidity increases and relative humidity remains constant.
c) Absolute and relative humidity both decrease.
d) Absolute humidity remains constant and relative humidity decreases.

Checkpoint 14.15

Two identical pans sit on a stove. Pan A contains a mixture of water and ice. Pan B contains an equal volume of water that is chilled to the same temperature. How will the temperatures of the contents change as the pans are heated?

Each pan contains a thermometer that records the temperature of the water and of the water-ice mixture. The stove is turned to high, and both pans are heated until boiling occurs. Heat continues to be applied for several more minutes.

Plot separate curves on the following graph to illustrate how temperature changes with time for both pans throughout the experiment. Label the curves.

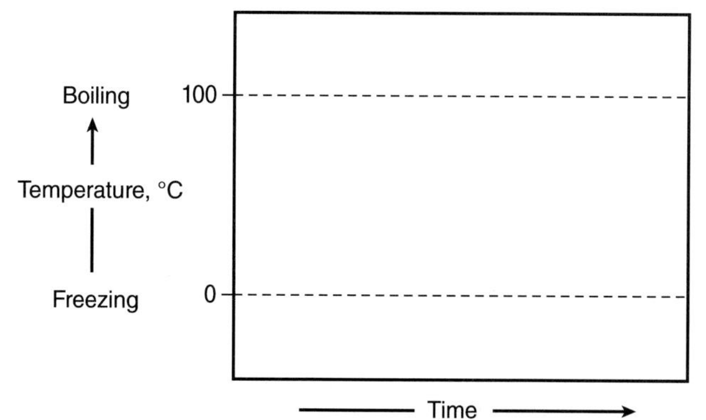

Checkpoint 14.16

basic intermediate advanced superior

1. Explain why people can see their breath on a cold winter's day in terms of water changing state and latent heat.
2. Explain why a hair dryer actually dries your hair rather than leaving you with hot, wet hair. Explain the process taking place.

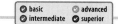

14.6 Air Pressure, Condensation, and Precipitation

Learning Objectives

- Describe the relationship between air pressure and air density.
- Predict how changes to air pressure will affect air temperature.
- Determine the temperature and humidity of air as it rises or falls in the atmosphere.
- Explain connections among lapse rates, cloud formation, and precipitation.

Air Pressure and Air Density

The pressure exerted by the weight of an overlying column of air is atmospheric pressure. For example, the average air pressure on Earth's surface (1,013 millibars, or 14.7 pounds per square inch) is determined by the weight of a 500-kilometer-tall (310-mile) air column. Air pressure declines with increasing altitude (Figure 14.16) because, as altitude increases, there is less and less overlying air.

Air pressure at any point is also influenced by air density. **Air density is determined by measuring the mass of atoms and molecules of nitrogen, oxygen, and other gases per volume of air.** Think of air as a box of gas molecules. The more molecules that are in the box, the greater the density of the air (Figure 14.17a). Air pressure would decrease uniformly with altitude if gases were evenly distributed throughout the atmosphere. However, gravity pulls most gas molecules close to Earth's surface, increasing the air density closer to the surface in the troposphere and decreasing the density in the outer layers of the atmosphere. Consequently, 50 percent of all air lies below 5.5 kilometers (3.4 miles) of altitude; therefore, air pressure at this altitude is one-half the air pressure at sea level (Figure 14.16). Almost all the air in the atmosphere lies below the height where Felix Baumgartner stepped out of his capsule for his big jump.

Checkpoint 14.17

basic, intermediate, advanced, superior

Which of the following images best approximates the distribution of the two principal gases in Earth's atmosphere?

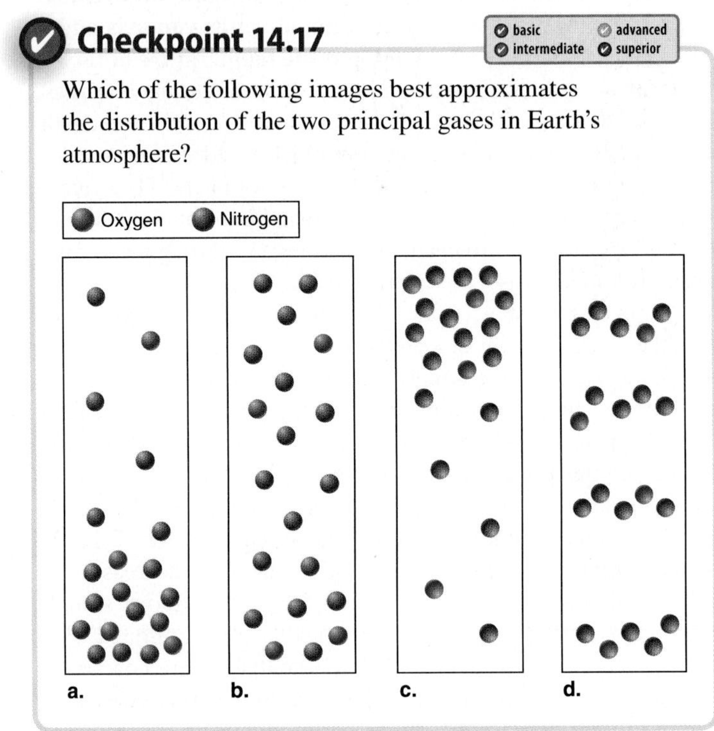

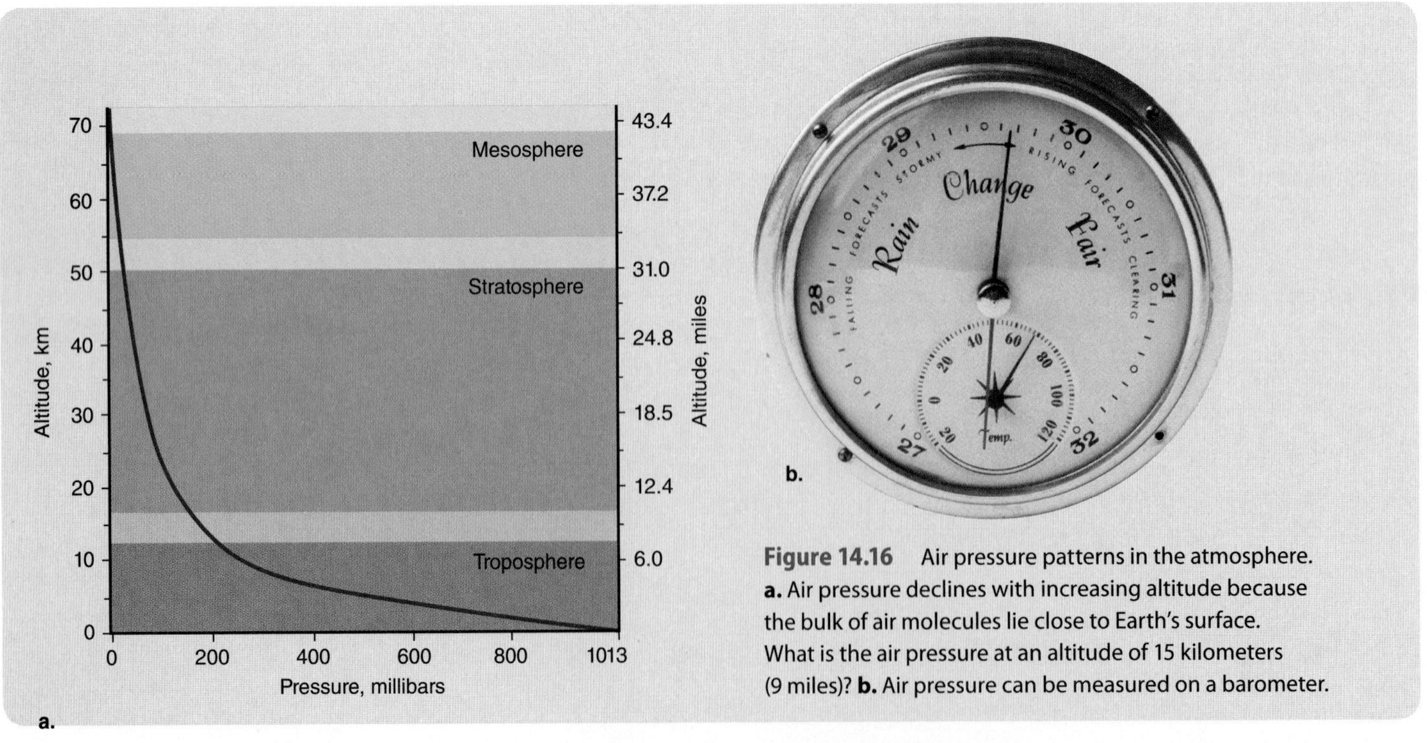

Figure 14.16 Air pressure patterns in the atmosphere. **a.** Air pressure declines with increasing altitude because the bulk of air molecules lie close to Earth's surface. What is the air pressure at an altitude of 15 kilometers (9 miles)? **b.** Air pressure can be measured on a barometer.

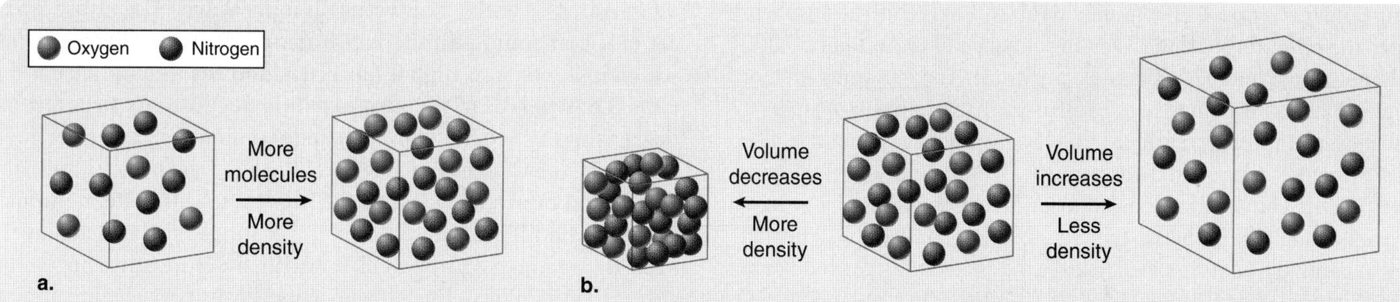

Figure 14.17 Air density. The number of molecules of gas in a given volume of air determines the air density. **a.** In two equivalent volumes of air, the air mass with the greater number of molecules would have greater density. Which box would represent air closer to Earth's surface? **b.** If the number of molecules remains constant, density increases with a decrease in air volume. Assume the bottom of each diagram represents relative distributions of gases at Earth's surface and elevation increases toward the top of the page.

Air density changes as air is warmed or cooled. **Air contracts when cooled,** resulting in gas molecules occupying a smaller volume, consequently increasing density and **generating higher air pressures.** In contrast, **air expands when it is warmed,** resulting in the same gas molecules occupying a larger volume and **generating a decrease in air pressure.** Think of our box of gas molecules: Cooling causes the box to get smaller, but the number of molecules remains the same, so the density increases (Figure 14.17b). In contrast, warming causes the box to expand, so the molecules are farther apart and density is less.

On the basis of temperature alone, where do you predict high- (or low-) pressure air would form on Earth's surface? It should come as no surprise that the highest atmospheric pressures are recorded in cold regions and the lowest pressures are recorded in tropical warm environments (associated with hurricanes and typhoons; Figure 14.18). If we move upward through the atmosphere, **air pressure decreases rapidly at lower altitudes where air density is greatest.** Air pressure decreases more slowly at higher altitudes above 10 kilometers (6 miles) because the air density there is lower (see Figure 14.16). Now back to Joe Kittinger's balloon—why do you think it got larger as he ascended? In his case, the balloon was expandable and filled with helium. As the balloon ascended, it expanded because the air pressure outside was decreasing.

Checkpoint 14.18

basic / intermediate / advanced / superior

In a few sentences, summarize the relationship among density, temperature, and pressure in a mass of air.

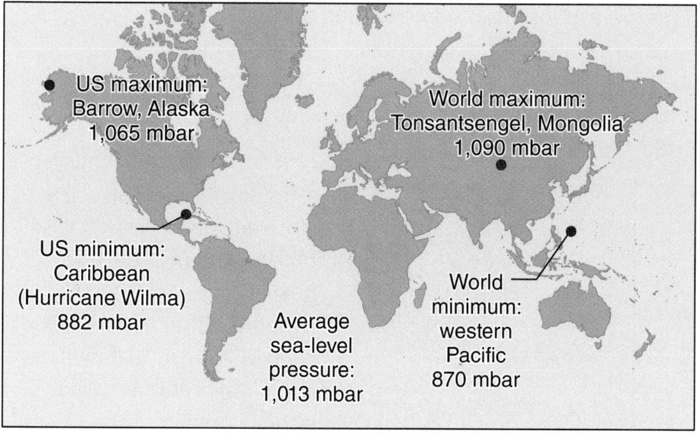

Figure 14.18 Maximum and minimum global pressure measurements. The highest pressures are found in colder regions, while the lowest values are associated with hurricanes (called typhoons in the Pacific Ocean) in tropical oceans. mbar = millibar.

Air pressure varies not only with elevation but also with weather systems. Weather maps typically display air pressure in millibars (mbar). These pressures are measured using a device called a *barometer* (Figure 14.16). Meteorologists must correct for changes in altitude to understand the pressure characteristics of air masses in order to forecast future weather patterns. Pressure data on weather maps are therefore recalculated to reflect pressure measurements for hypothetical sea level. If no effort were made to recalculate air pressures, we would always have low-pressure conditions over mountains due to less-dense air higher in the troposphere and higher pressures along coastlines due to more-dense air at low elevations. This would make accurate weather forecasting nearly impossible because it would not reflect the true difference in the pressures of air masses as they moved to common altitudes.

Effects of Air Pressure on Temperature

How does the pressure of your car tire compare to that of the surrounding atmosphere? Typical tire pressures are about twice that of the outside air. Have you ever noticed that air escaping from a tire is colder than the surrounding air? This temperature change is the result of the drop in pressure as the air expands as it escapes from the high pressure in the tire to the relatively lower pressure in the surrounding atmosphere. **Compressed air becomes warmer; expanding air becomes cooler.** Similar increases or decreases in temperature result when pressure changes in nature. These changes in temperature are called ***adiabatic changes*** **because they occur without any loss or gain of energy to or from the surrounding air** (as if the air were thermally insulated). We can observe a vivid example of this phenomenon when meteors burn up in the upper atmosphere (Figure 14.19). A meteor arrives at Earth's atmosphere after traveling through the vacuum of space at more than 50,000 kilometers per hour (31,000 miles per hour). The incoming meteor slows down as it compresses the gases of the upper atmosphere, causing the air temperature around the meteor to rise rapidly. The heat from the compressed gases raises the temperature of the meteor. (*Note:* The heating does not occur because of friction—there are not enough atoms in the upper atmosphere for friction to be that significant!) The extreme heat causes the meteor to glow red hot at temperatures of more than 1,500°C (2,730°F). Faster-moving objects compress more air and generate greater heat; slower-moving objects push against less air and produce less heat.

Some of the mechanical energy needed to force air molecules into a small space in front of the speeding meteor is converted to heat as air is compressed. These temperature changes represent the conversion of mechanical energy into heat energy and are not associated with the addition or loss of heat from outside sources (for example, the sun). In the same way, the air in a car tire warms up as tire pressure is increased, or the barrel of a hand pump becomes warm as a bicycle tire is pumped up. When air expands adiabatically, it converts heat to mechanical energy as air molecules move apart, so the air escaping from a punctured tire feels cooler than the surrounding air. Why are these changes in temperature associated with changes in pressure important? As we will learn in this chapter and Chapter 15, these adiabatic temperature changes help fuel the processes that cause air to rise and promote the development of extreme weather phenomena such as thunderstorms.

Adiabatic Lapse Rates

Perhaps you have been to the mountains and noticed that the air is cooler at higher altitudes than it is at lower elevations. To understand why, we must consider the air in distinct parcels—volumes that can behave adiabatically and independently of the surrounding air. The temperature of the atmosphere is largely determined by the lapse rate, which is the rate at which air cools as it rises or warms as it falls. **The actual temperature of a rising parcel of air at some altitude is a function of the temperature at which it starts, how high it has risen, and the adiabatic cooling rate. That parcel of air also expands because air pressure is decreasing.** In Section 14.4, we noted that air temperature declines at a rate of 6.5 C° per kilometer through the troposphere (known as the normal or average lapse rate). This is the lapse rate we would expect to find if we measured many locations on Earth and averaged the readings. If a parcel of air in some location begins to rise, an additional cooling effect occurs because of the temperature changes resulting from the expansion of air. The total amount of energy present in the parcel of air will not change, but it can be used either to maintain a constant temperature or to work to expand the size of the parcel of air. As the air expands, the heat energy is converted to mechanical energy and distributed through a larger volume, producing a cooling effect. This results in a **dry adiabatic lapse rate of 10 C° per 1,000 meters (10 C° per kilometer; 29 F° per mile) for unsaturated air** (Figure 14.20).

Figure 14.19 A Perseid meteor from the 2002 shower over the southwestern United States.

Checkpoint 14.19

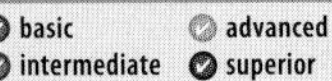

An instructor asked her class to summarize some information about atmospheric processes. Students submitted the following four statements as part of their answers. The instructor returned the statements and told the students that they could correct them for full credit.

Identify what is wrong with each statement, and describe how you would fix these answers to earn full credit.

1. The temperature of a rising parcel of air decreases by the normal lapse rate.
2. The percentage of oxygen in the atmosphere decreases with altitude.
3. When it rains, you have to use the wet adiabatic lapse rate to figure out temperatures at higher elevations.
4. The dry adiabatic lapse rate is higher than the wet adiabatic lapse rate, so air temperatures should be higher in dry air (before condensation occurs) than in wet air (after condensation occurs).

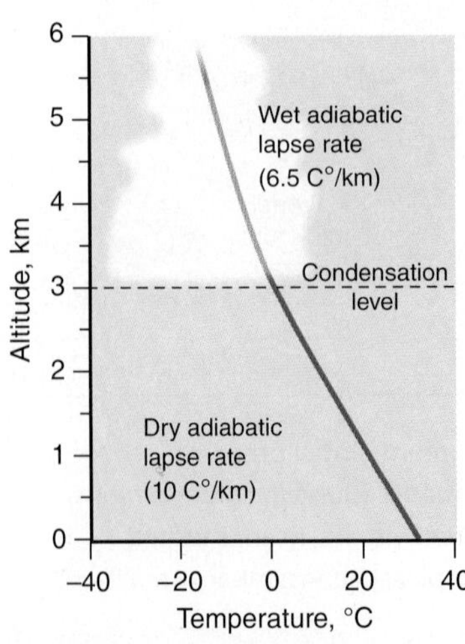

Figure 14.20 Adiabatic lapse rate and altitude. The adiabatic lapse rate decreases because of the addition of the latent heat of condensation for a hypothetical air mass following condensation at 3 kilometers (1.9 miles) of altitude. What is the temperature of a parcel of air rising to 2 kilometers (1.2 miles) of altitude?

As the rising parcel of air cools, its relative humidity increases and the air becomes saturated with water vapor at some elevation. At that point, condensation occurs, and the lapse rate becomes a "wet" adiabatic lapse rate. Latent heat is released in the conversion of water from a vapor to a liquid state, and the addition of the latent heat partially counterbalances the cooling effect of the adiabatic temperature changes. This results in a reduction in the cooling rate. **An average value for the wet adiabatic lapse rate is 6 C° per kilometer (18 F° per mile)** (Figure 14.20). The parcel of air will continue to rise until its temperature (and related density) is the same as that of the surrounding air.

Condensation and Cloud Formation

Rising air gradually cools, slowly increasing the relative humidity of the air mass to the point where saturation occurs, triggering condensation and cloud formation. However, it is not enough just to cool the air for condensation to occur. **To form a cloud, water vapor must also have surfaces upon which to condense.** In the atmosphere, the surfaces come from microscopic aerosol particles (dust, smoke, salt, pollutants) that range in size from 0.001 to 10 micrometers (wider than a billionth, narrower than a thousandth of an inch). For reference, the diameter of human hair is about 50 micrometers. On land, it could be any surface, including grass or your car (dew or frost). In the air, billions of water droplets approximately 20 micrometers in diameter can form on aerosol particles in a single cubic meter of air. The droplets are readily kept airborne by air turbulence until they collide and coalesce (clump) to form larger drops. **Clouds are composed of billions of tiny water droplets that can combine to form rain or hail.**

Precipitation

The larger, heavier **cloud droplets fall, colliding with other droplets and combining with them, a process called *collision coalescence*, to form raindrops** (containing approximately 1 million cloud droplets each). Updrafts in the air can force the droplets upward and recycle them through the cloud, increasing droplet size. Updrafts are relatively common in large clouds as condensation warms the air, causing it to rise. During times of drought, efforts are sometimes made to artificially introduce condensation nuclei into the atmosphere by cloud seeding. Unfortunately, there is little hard evidence that such methods work.

While the collision coalescence process effectively explains rainfall from warm clouds, it is less effective in accounting for the formation of rain from clouds below the freezing point of water. **Air temperatures fall below freezing within a few kilometers of Earth's surface in the lower portion of the troposphere** at normal lapse rates. Consequently, given the typical range of cloud elevations, much of the water in clouds is in the form of supercooled droplets with temperatures ranging from 0°C to −20°C (32°F to −4°F). You might expect that any water at these altitudes would freeze, but it does not. Supercooled water can freeze only with sufficient agitation (strong vibrations) or when there is a surface upon which to freeze (such as a dust particle or ice crystal). Pure water droplets in high clouds can remain in liquid form down to temperatures as low as −39°C (−38°F).

So how do raindrops form in these supercooled clouds? It turns out that most clouds contain ice crystals and supercooled water droplets. When a water droplet comes close to an ice crystal, supercooled water will transfer onto the ice crystal because air on that surface is unsaturated (compared to air at the surface of the water droplet). When that happens, the air immediately next to the water droplet appears subsaturated, and evaporation occurs. That moisture is also added to the ice crystal, thus growing the crystal. The crystal will fall when it becomes too heavy to be supported by air currents, melting along the way to form cold rain. This **mechanism whereby raindrops are formed is called the *Bergeron process*** after the scientist who first described the process.

So, how does snow form? Recall that absolute humidity measures the mass of water in a volume of air. The absolute humidity of air required for saturation decreases with temperature (see Figure 14.15). Cold air with temperatures between 0°C (32°F) and −20°C (−4°F) would be saturated with 5 grams per cubic meter to 1 gram per cubic meter of water vapor, respectively. Below temperatures of about −5°C (23°F), air needs a little less water vapor to be saturated for ice than for water. Under these conditions, condensation preferentially produces ice crystals rather than water droplets. **Miniature ice crystals act as condensation surfaces and gradually increase in size to form snowflakes.** Eventually, precipitation in the form of ice or snow occurs, unless the frozen water is warmed as it falls through the atmosphere, resulting in rain.

Checkpoint 14.20

Describe what would happen to a parcel of air that begins to rise. Your answer should discuss the normal lapse rate, dry and wet adiabatic lapse rates, and humidity. Include a sketch showing the parcel of air at different altitudes.

14.7 Clouds and Frontal Systems

Learning Objectives

- Classify and explain the processes forming clouds of different shape, altitude, and propensity to form precipitation.
- Compare and contrast the four main mechanisms and/or processes that lift air.

More than 30,000 flights cross the skies of the United States on an average day. As jets pass overhead, their exhausts expel water vapor and a mixture of other gases into the chilly air of the upper troposphere. The water vapor immediately condenses to form a contrail (condensation trail), a skinny line of cloud following the path of the flight.

Climatologist David Travis used weather data from the 3 days following September 11, 2001, when all commercial flights in the United States were canceled, to try to unravel the impact of contrails on surface temperatures. Travis compared average temperature data from that 3-day period with data from the previous three decades. He found that night temperatures were generally cooler and daytime temperatures were warmer than during days with normal flight operations. Travis interpreted the data to show that contrails absorbed heat during the night, causing a warming effect, and had a cooling influence during the day as they reflected incoming sunlight. When he examined areas with the heaviest air traffic, the temperatures differed by as much as 3 C° (5 F°). The fact that these thin, transitory, upper-level clouds have an effect on surface temperatures illustrates that the type and distribution of all clouds likely affect solar radiation and temperatures across Earth's surface.

Much of the incoming solar radiation is either absorbed by clouds or reflected into space from cloud surfaces (see Figure 14.9). Increasing the thickness or water vapor content of low-level clouds causes more sunlight to be reflected from the cloud surfaces (see Chapter Snapshot) and thus reduces the heat energy that contributes to the greenhouse effect. In contrast, water vapor in clouds is a key greenhouse gas that absorbs heat energy close to Earth's surface; therefore, increasing cloud cover might act to increase the amount of heat absorbed in the lower atmosphere. Current atmospheric research is attempting to unravel the role of clouds to determine whether they provide a net negative or net positive feedback to global warming processes or whether their warming and cooling effects essentially cancel one another out. Unfortunately, the varied lateral and vertical distribution of clouds and their rapidly changing shape make them difficult to incorporate into climate computer models.

Cloud Classification

Clouds are classified on the basis of their altitude, appearance, and associated precipitation (Figure 14.21). High-level clouds are composed of ice; low-level clouds are usually composed of water droplets. Most cloud types occur at a specific range of altitudes, but some, such as cumulonimbus clouds associated with thunderstorms, may span several levels. The prefix *cirr-* indicates a high-level cloud with a wispy shape; *alto-* designates midlevel clouds; and *-cumulus* represents a heap shape. The prefix *nimbo-* or suffix *-nimbus* denotes rain clouds.

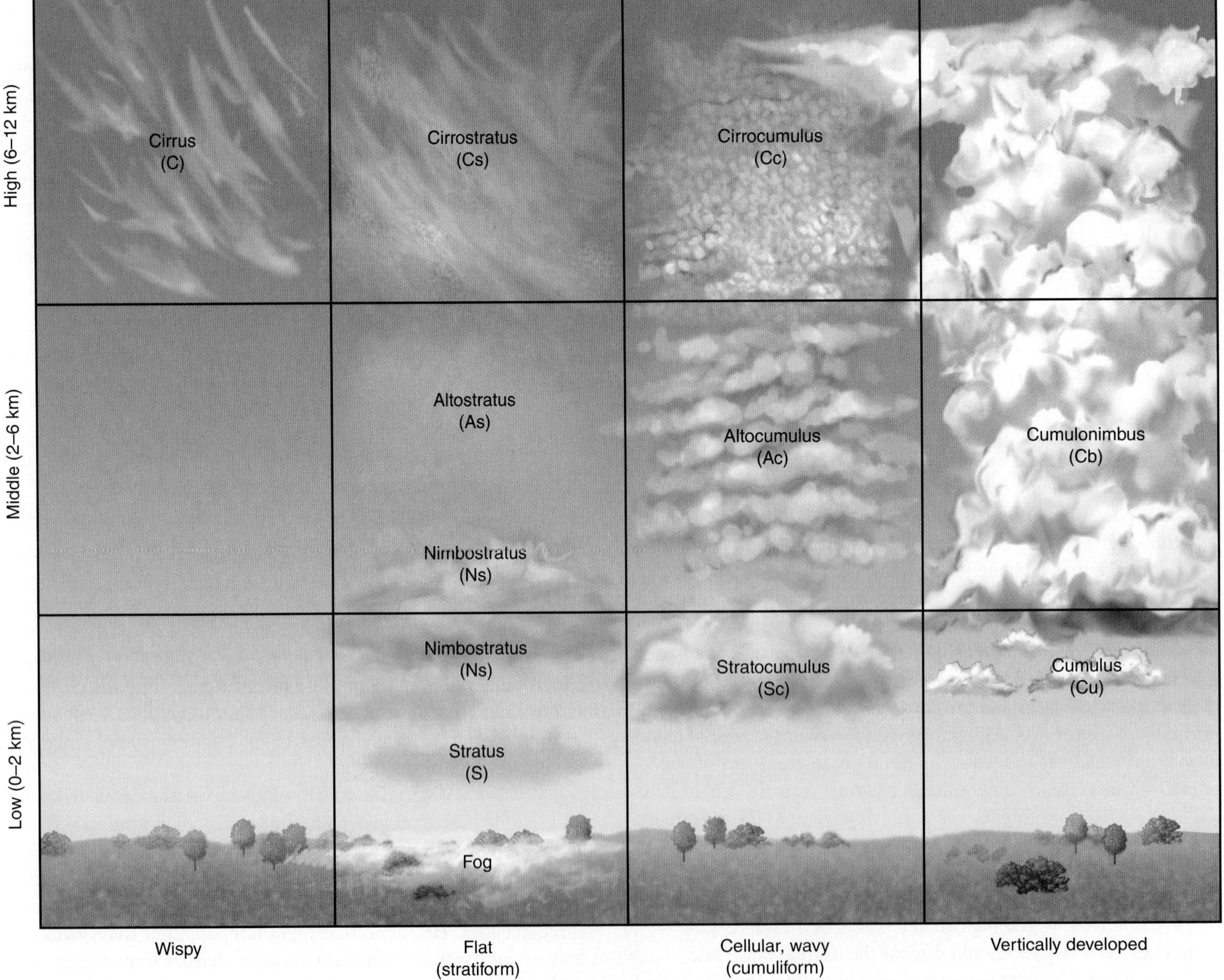

Figure 14.21 Types of clouds. Clouds are classified on the basis of their altitude and appearance. Most clouds occur at a specific range of altitudes, but some, such as cumulonimbus clouds associated with thunderstorms, may span several levels.

Cirrus clouds are thin, wispy clouds that form in the upper troposphere; they would be equivalent to contrails. Although cirrus clouds appear slight and delicate compared to other cloud types, they are effective heat-absorbing devices.

Stratus clouds form sheets that cover the whole sky and may form at any level. Especially thick stratus clouds are termed *nimbostratus*. Looking upward, an air traveler might spy altostratus clouds below high-level cirrostratus clouds.

Puffy, cauliflower-shaped cumulus clouds are the most commonly recognized clouds. If a sufficient number of these clouds form, they may create a low, irregular layer of stratocumulus clouds. Cumulus clouds at higher altitudes are termed *altocumulus* or *cirrocumulus* clouds (Figure 14.21).

Cloud Formation Mechanisms

Varying moisture levels in the air, combined with temperature and related pressure differences, cause imbalances in the Earth system that generate different types of clouds. As described in Section 14.6, clouds form when condensation occurs as rising air cools adiabatically. The propensity of air to rise (or fall) in the atmosphere is closely tied to its density. Just like a hot-air balloon, a parcel of air may rise naturally if it is lighter than the surrounding air masses; this process is called *density* (or *convection*) *lifting*. Otherwise, the parcel of air may be forced upward by a variety of other processes such as frontal lifting, orographic lifting, or convergence lifting.

Checkpoint 14.21

basic · intermediate · advanced · superior

Classify the clouds in each of the following images.

a.

b.

c.

d.

Density Lifting. What happens to warm air in your house, and why? As you have probably experienced, warm air rises. It does so because it is less dense (and therefore lighter) than the colder air. In the atmosphere, warm air is usually located near Earth's surface (why?). If air has a tendency to resist movement, it is considered *stable;* if it has a tendency to rise or fall, it is *unstable*. The temperature of air decreases with altitude at the normal lapse rate. So what should happen to a parcel of warm air near the surface? **Warm air will rise as long as the density of the parcel of warm air remains lower that of the surrounding air.** If the air is unsaturated, warm air will cool according to the dry adiabatic lapse rate (Table 14.1). The air continues to rise until it cools to a temperature equal to that of the surrounding air (Figure 14.22). At that point, the density of the rising air is the same as the surrounding air, and density lifting ends.

Table 14.1 Typical Temperature Changes in Unstable and Stable Air Masses

Altitude, km (mi)	Warm air mass (dry adiabatic lapse rate)	Stable air mass (normal lapse rate)
0	20	11.25
0.5 (0.3)	15	8
1.0 (0.6)	10	4.75
1.5 (0.9)	5	1.5
2.0 (1.2)	0	−1.75
2.5 (1.5)	−5	−5

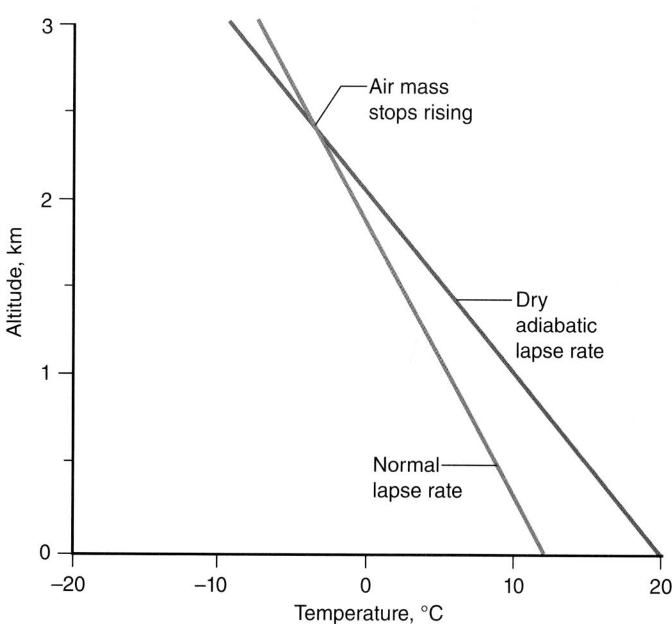

Figure 14.22 Density lifting. Assume that a warm air mass begins to rise at a temperature of 20°C (68°F) and that the surrounding air has a temperature of 11.25°C (52°F). The temperature of the rising air mass and the surrounding air will be equalized at −5°C (23°F) at an altitude of 2.5 kilometers (1.6 miles). At that altitude, all the air has the same temperature; therefore, there is no longer a separate parcel of rising air.

Frontal Lifting. Frontal lifting occurs when two large air masses of contrasting density (temperature, moisture content) meet (Figure 14.23). The boundary between the air masses is termed a *front* and may be 10 to 150 kilometers (6 to 93 miles) across and hundreds of kilometers in length.

A warm front forms when a warm air mass advances to meet a cold air mass. Which air mass should rise? The lighter warm air rises above the heavier colder air. Warm air is also forced upward when cold air advances on a warm air mass along a cold front. Cold fronts form when cold air advances over land previously occupied by warm air. Cold fronts are steeper than warm fronts and cause cloud formation and precipitation to occur across a narrower area. Severe weather (heavy rain and snow) usually occurs along cold fronts in the central portion of the continental United States. (See Chapter 15 for more on these weather patterns.)

Orographic Lifting. **Orographic lifting occurs when air is forced to rise over an obstruction in the landscape, typically a mountain range.** The side of a mountain sloping toward the oncoming wind is known as the *windward slope*. The other side of the range slopes in the direction that the wind is blowing and is known as the *leeward slope*. Condensation occurs as the air cools with increasing altitude as it rises up the windward flank of the range, and precipitation may occur at higher elevations. The air warms up and can absorb moisture as it descends the leeward side of the mountains, creating a **rain shadow effect where precipitation is relatively rare**. Where do you predict a rain shadow effect might occur in North America? Because air masses typically move from west to east across the western half of the continent, the rain shadow effect is often found on the eastern slopes of mountain ranges.

Checkpoint 14.22

On July 2, 1982, truck driver Larry Walters decided to attach 45 helium-filled weather balloons to a lawn chair and go for a ride. Lawn Chair Larry rose to an altitude of nearly 5 kilometers (16,400 feet). The high elevations and lack of oxygen made him dizzy, so he decided it was time to deploy his principal altitude control device, a pellet gun. He shot out several balloons and descended back to Earth. Which process was most significant in Lawn Chair Larry's balloon ride? Explain your reasoning.

a) Density lifting
b) Orographic lifting
c) Convergence lifting
d) Frontal lifting

In the following landscape, how would the amount of rainfall change at location X if the mountain eroded down to the dashed line?

a) Rainfall would increase.
b) Rainfall would decrease.
c) Rainfall would stay the same.

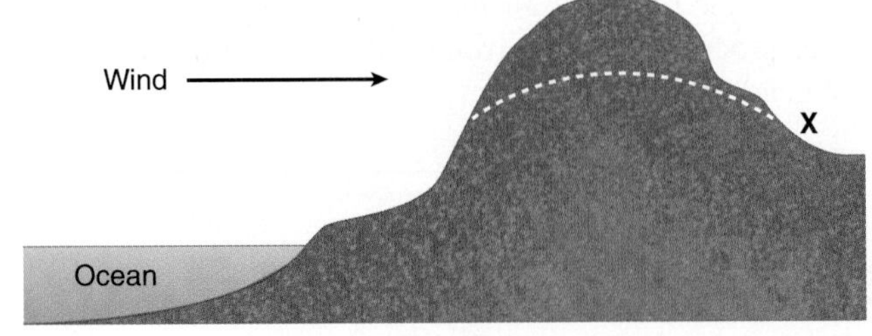

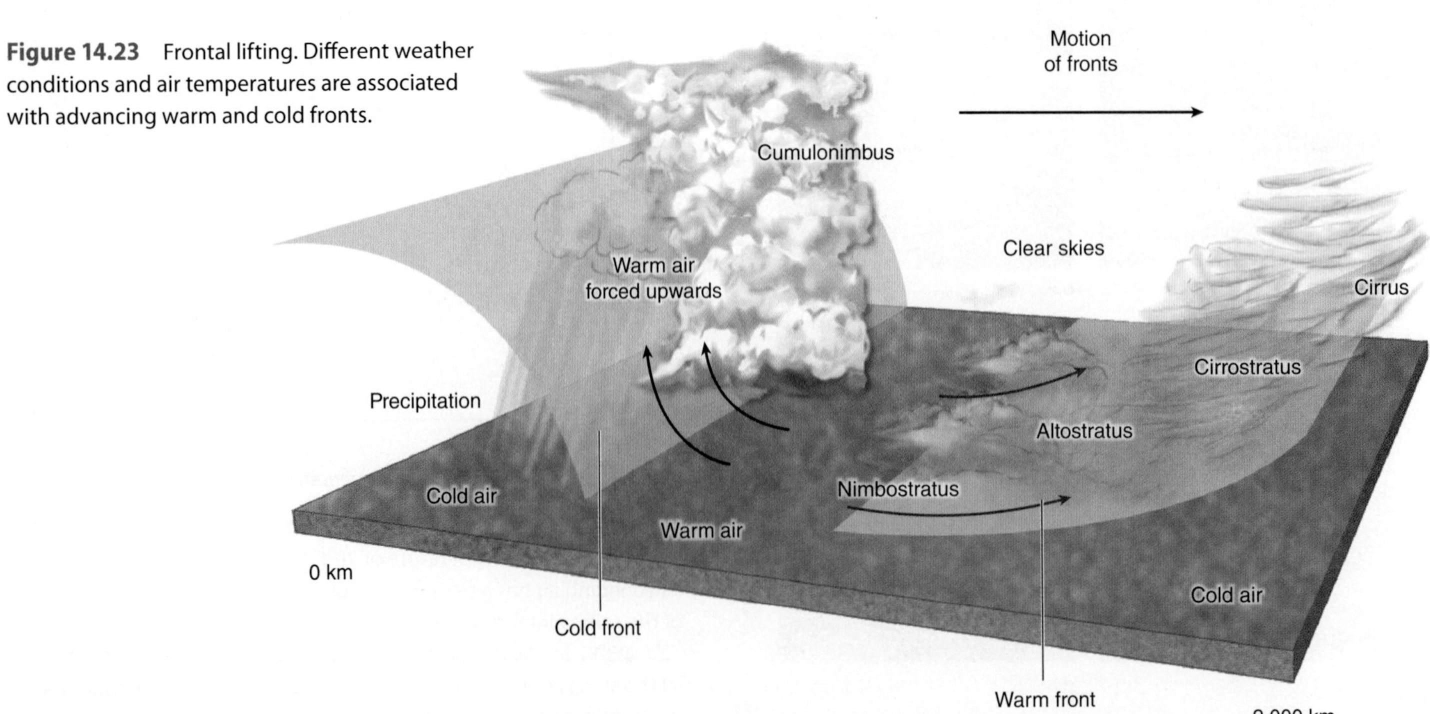

Figure 14.23 Frontal lifting. Different weather conditions and air temperatures are associated with advancing warm and cold fronts.

Convergence Lifting. Convergence lifting occurs when two air masses of similar temperatures collide, forcing some air upward because both air masses cannot occupy the same space (Figure 14.24). This happens regularly over Florida, where air moves westward from the Atlantic Ocean and eastward from the Gulf of Mexico to collide over the Florida peninsula.

Checkpoint 14.23

basic · advanced · intermediate · superior

Use the information from the chapter to explain:

1. Which air-lifting mechanism dominates where you live?
2. Which states might provide examples of the four air-lifting mechanisms?

Checkpoint 14.24

A parcel of warm air with an average temperature of 29°C (84°F) and 50 percent relative humidity is at sea level. The surrounding air has a temperature of 26.5°C (80°F). The parcel of air begins to rise due to density lifting. The relative humidity of the air mass is 100 percent when it reaches 1-kilometer (0.6-mile) elevation. Analyze how the temperature of the air mass would change as it rises, and determine at what elevation it would become stable.

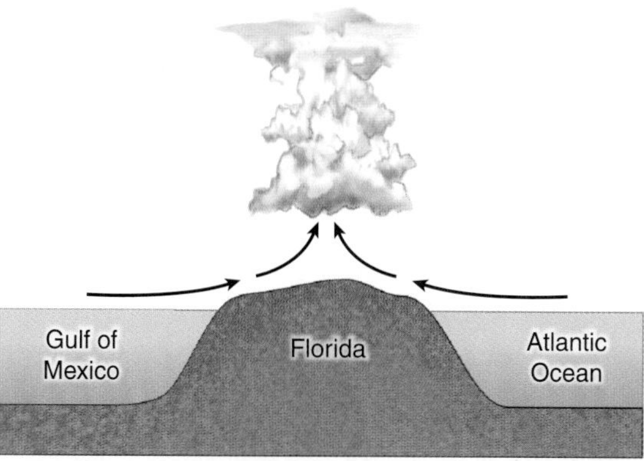

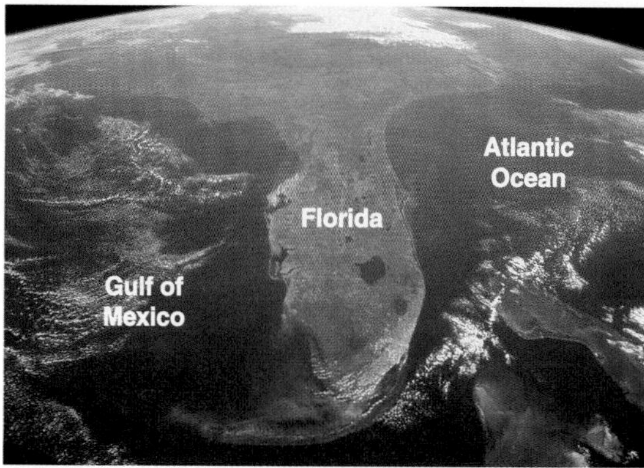

Figure 14.24 Convergence lifting. Condensation above the Florida peninsula may be assisted by density lifting because the air overlying the land mass will be warmer than the air above the adjacent ocean.

14.8 Winds

Learning Objectives

- Predict relative winds speed and direction, using atmospheric pressure data.
- Compare and contrast processes that form high- and low-pressure air masses.
- Explain how the pressure gradient, Coriolis effect, and friction combine to influence wind speed and direction.
- Illustrate how the atmosphere interacts within the earth system.

A cabin on top of Mount Washington in New Hampshire is secured to the ground by chains to prevent it from blowing away (Figure 14.25). At 1,917 meters (6,290 feet), Mount Washington is the highest peak in the northeastern United States. The highest wind speed ever recorded, 371 kilometers per hour (231 miles per hour), was measured at a weather observatory on the mountain's summit on April 12, 1934. Scientists working at the station had to chip the ice off the anemometer (a device for measuring wind speed) to take the reading. On at least one occasion on more than 100 days each year, the observatory experiences winds with speeds equivalent to those of a typical hurricane. What is it about this location that makes it the home for such ferocious winds? To answer that question, we first have to understand a little more about how air pressure varies on Earth's surface.

The Relationship Between Air Pressure and Wind

We have spent much of the second half of this chapter describing the vertical variations in air pressure—basically, how air pressure decreases with increasing altitude. However, examination of

Figure 14.25 Secured building at the Mount Washington Observatory. Note the three sets of chains anchored to the ground to hold the building in place during high winds.

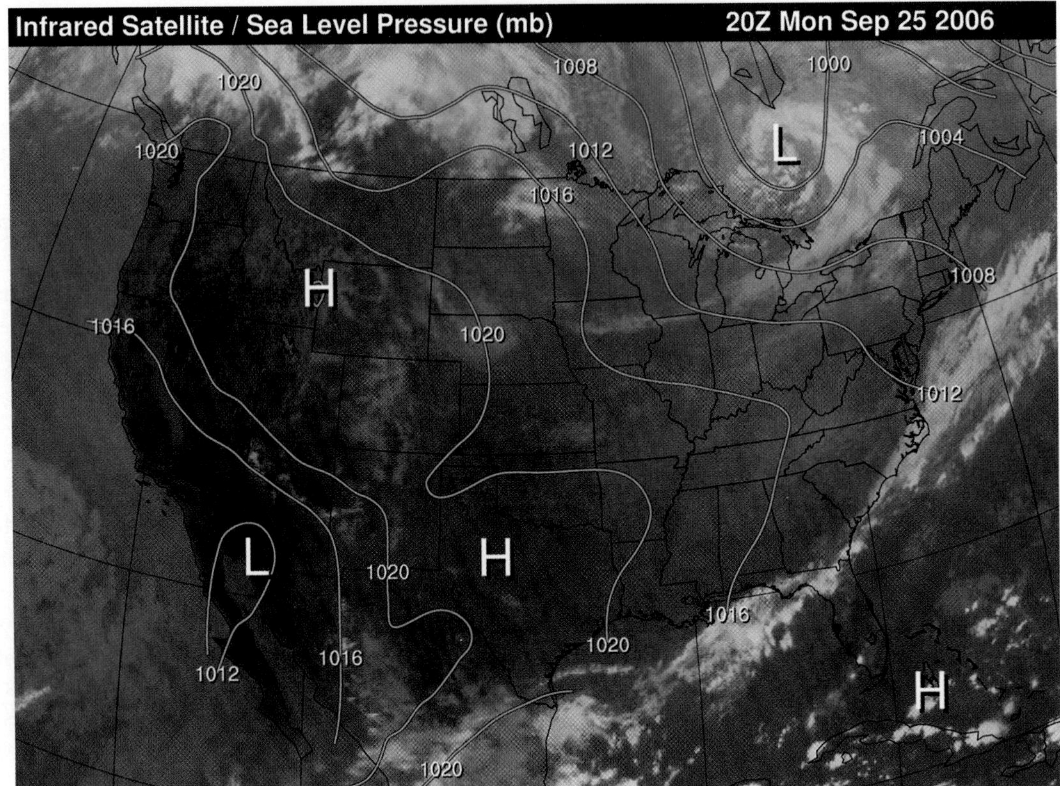

Figure 14.26 Pressure map of North America. Atmospheric pressure for September 25, 2006, is high over the West and low over the Northeast. The lines of equal pressure, called isobars, are spaced 4 millibars apart. What is the range of pressures indicated by the map?

weather maps reveals that air pressure is not constant across the surface of North America. On a typical day, air pressure over the United States, Canada, and Mexico ranges from lows of around 1,000 millibars (mbar) to highs of more than 1,020 millibars (Figure 14.26).

Why are pressures not the same in different parts of the continent at the same latitude? As we will learn in Chapter 15, **air above Earth's surface inherits the thermal characteristics of the underlying surface.** If the surface is cold compared to the air, the air above the surface is cooled; if the surface is warm, the air is warmed. These temperature differences lead to regional differences in air density and, consequently, differences in atmospheric pressure.

Wind is the horizontal movement of air that arises from differences in air pressure. Winds represent moving air as it **flows from areas of high pressure to areas of low pressure.** Mount Washington's high elevation and location—between the cold air sweeping south from Canada and the air moving north from the Gulf of Mexico or east from the Atlantic Ocean—makes it an ideal place to experience large differences in pressure in the transition zone between pressure systems. Large pressure differences produce fast-moving winds.

Winds are characterized by their speed and direction. **Wind direction refers to the direction from which the wind originates rather than where it is going.** For example, a westerly wind blows from west to east. Three factors influence wind speed and direction: the pressure gradient, the Coriolis effect, and friction with the surface.

 **Checkpoint 14.25**

Find your state on the map in Figure 14.26. On the basis of the isobars, in what direction is the wind blowing in your state?

Regional Pressure Gradient

The pressure gradient is the magnitude of the change in pressure between two points, divided by the distance between those points. **The greater the contrast in pressure, the steeper the gradient will be and the faster the wind will blow.** Isobars join points of equal pressure plotted on a map (Figure 14.26). If no other factors influenced wind direction, winds would blow down the pressure gradient, perpendicular to the isobars.

Since isobars indicate differences in pressure, **the closer the isobars are together, the greater the pressure drop with distance and the faster the wind speed.** For example, in Figure 14.26, wind speed would be greatest over southeastern Canada and least in the central and western United States. Wind speed associated with a pressure gradient between high- and low-pressure systems would increase if the systems moved closer together or if atmospheric pressures became more extreme (Figure 14.27). Initial wind directions will be modified by the influence of the Coriolis effect and friction close to the planet's surface.

Coriolis Effect

Recall from Chapter 13 that atmospheric and oceanic circulation patterns are deflected to the right of their course in the Northern Hemisphere and to the left of their course in the Southern Hemisphere. This pattern, termed the *Coriolis effect,* occurs because objects at different latitudes travel on Earth's surface at different speeds as a result of the rotation of the planet.

Checkpoint 14.26

basic / intermediate / advanced / superior

Draw arrows indicating the direction of winds in Figure 14.26 on the basis of the pressure gradient alone. Where would the fastest winds be found?

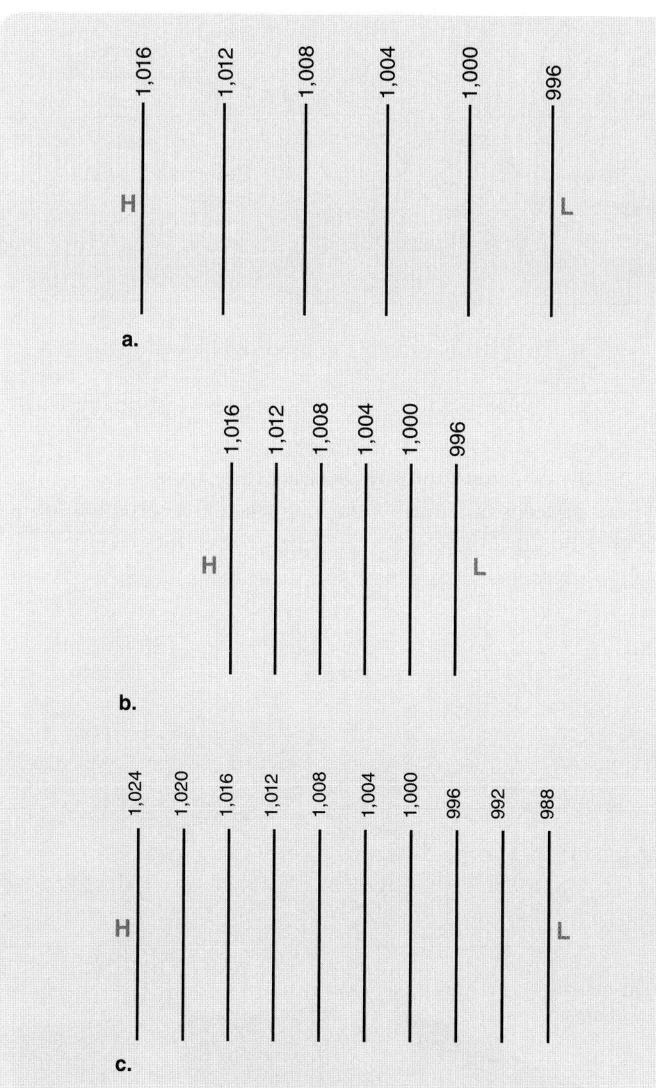

Figure 14.27 Two mechanisms for increasing wind speed between high- and low-pressure systems. Pressure expressed in millibars. **a.** Neighboring high- and low-pressure systems initially have uniform spacing between isobars. **b.** Higher wind speeds are indicated by closer spacing of the isobars as pressure systems converge. **c.** Higher wind speeds are indicated by closer spacing of isobars as pressure systems strengthen.

Winds are deflected to the right of their course in the Northern Hemisphere. These winds continue to be deflected until the pressure gradient effect balances the Coriolis effect and the wind moves parallel to the isobars (Figure 14.28). **Winds blowing parallel to isobars are termed *geostrophic winds.*** Geostrophic winds are disrupted by the influence of friction immediately above Earth's surface but are fully developed a few kilometers above the surface where friction is no longer significant.

Friction

Wind blowing near Earth's surface is slowed by frictional drag from the surface. For instance, friction is most dramatic above rugged land surfaces such as mountain ranges and is least over flat land or open water. An increase in friction reduces the Coriolis effect because the deflection is proportional to wind speed. The diminished Coriolis effect cannot balance the influence of the pressure gradient and typically results in nongeostrophic winds close to Earth's surface (Figure 14.29).

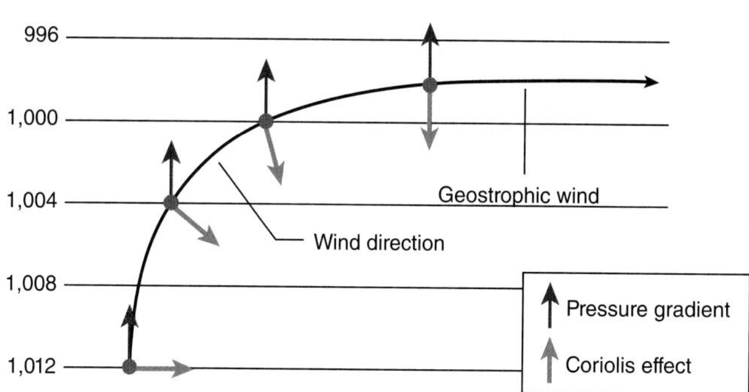

Figure 14.28 Interaction of pressure gradient and Coriolis effect. The pressure gradient causes winds to flow from high pressure toward low pressure. Eventually, the pressure gradient is exactly offset by the Coriolis effect, forming geostrophic winds in the middle to upper troposphere that blow parallel to isobars.

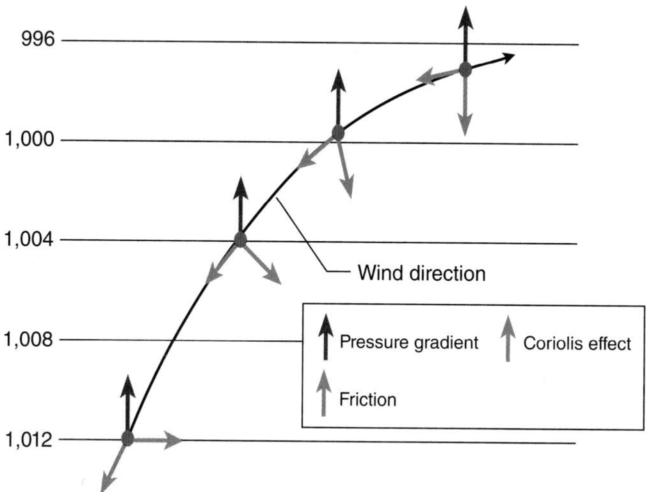

Figure 14.29 Influence of friction. Friction close to Earth's surface reduces pressure gradient and Coriolis effects, resulting in winds that are oriented at an angle to isobars.

Therefore, at the surface, winds are oblique (at an angle, rather than perpendicular) to isobars. The angle between wind direction and the pressure gradient depends on the character of the underlying surface. Generally, wind directions are oriented 40 to 50 degrees from the surface isobars. The more varied the land surface, the greater the friction produced and the larger the angle between wind direction and isobars. Friction is less significant above flat terrain or the ocean surface, but still results in winds that make an angle of 10 to 20 degrees with the regional isobars.

Cyclones and Anticyclones

The combined effect of influences on wind direction is that airflow forms a spiral pattern as it rotates clockwise or counterclockwise around high- or low-pressure systems, respectively. Winds converge toward a low-pressure system from surrounding regions of higher pressure. These converging winds will be deflected to the right of their course, setting up a **counterclockwise airflow, known as a *cyclone*, around the low-pressure center** (Figure 14.30a). In contrast, winds blowing outward, away from a high-pressure center, will be deflected to the right of their course to produce a **clockwise flow of air known as an *anticyclone*** (Figure 14.30b).

Low-pressure systems are rapidly dissipated by converging air unless the inrushing air is balanced by a rising air column. **Rising air in low-pressure systems becomes cooler and may reach saturation, resulting in clouds and rain.** At the top of the rising column of air, there must be divergent airflow to balance the convergent flow at the surface (Figure 14.31). Anticyclones must have convergent flow aloft so that air is continually fed into the high-pressure systems to ensure their survival (Figure 14.31).

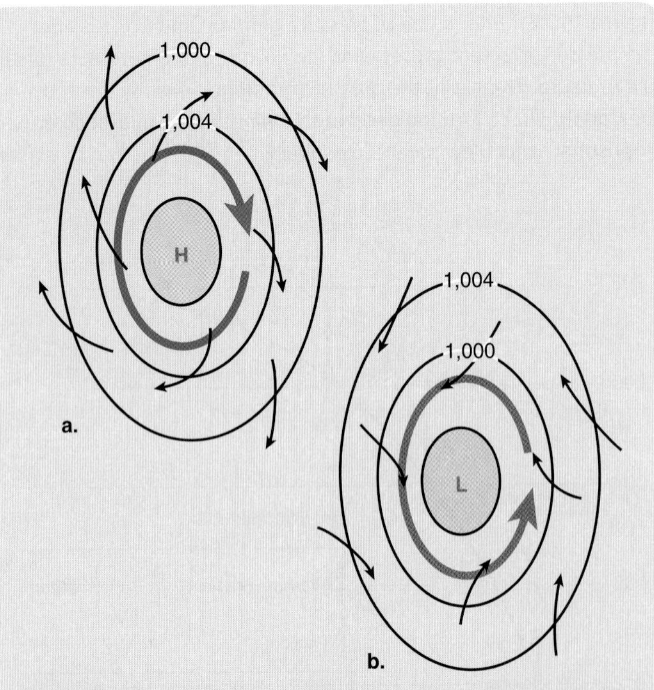

Figure 14.30 Cyclone and anticyclone in low- and high-pressure systems. **a.** Winds diverge from high-pressure systems, generating a clockwise airflow. **b.** Winds converge in low-pressure systems, creating a counterclockwise airflow at the surface.

Checkpoint 14.27

1. Draw an arrow to show the location and direction of the highest wind velocity on the following air pressure map, on the basis of the isobars alone.

2. Which point on the map is at the center of a low-pressure system?
   a) A b) B c) C d) D e) E

3. Rising air is present at which point?
   a) A b) B c) C d) E e) G

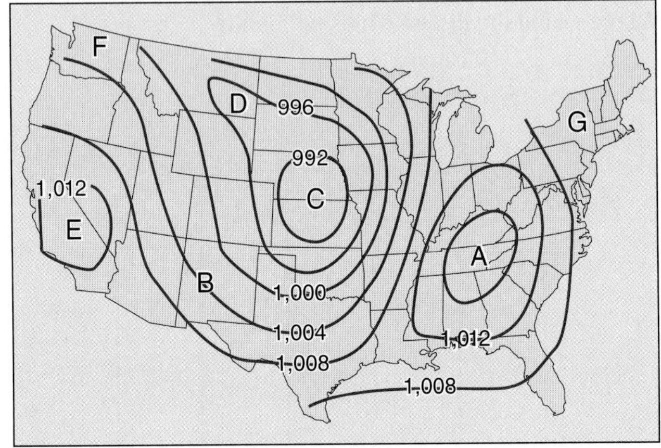

4. If Earth did not rotate, wind would blow directly from ______.
   a) A to C b) C to A c) C to B d) F to E

5. Draw arrows to fully illustrate the circulation patterns of winds on the map when it is corrected for the Coriolis effect.

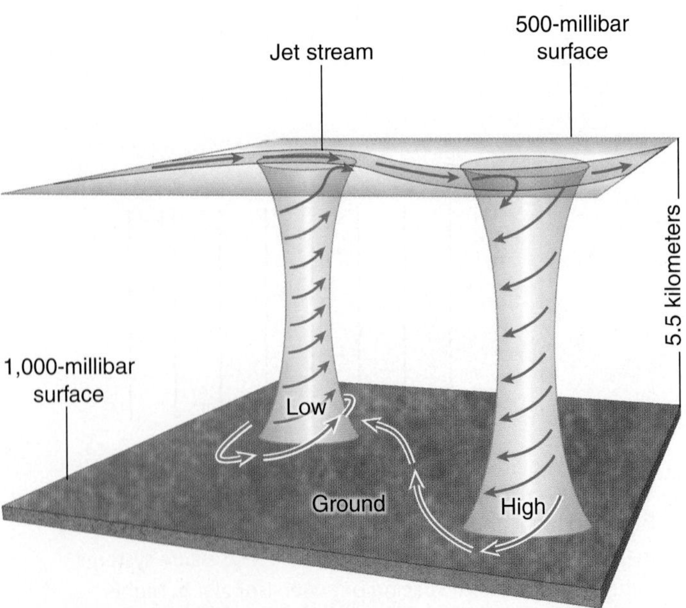

Figure 14.31 Converging and diverging winds. Rising air produces clouds over low-pressure systems, and descending air results in clear skies above high-pressure systems.

Air descends in high-pressure zones, warming as it approaches Earth's surface. As the air becomes warmer, its relative humidity decreases. Hence, high-pressure anticyclones result in clear skies and dry weather (see Figure 14.26), whereas low-pressure cyclones are often associated with rainfall and cloud formation. We will revisit low- and high-pressure systems in Chapter 15 because they play a significant role in controlling regional weather patterns.

Wind Energy

Wind power accounts for more than 3 percent of US electricity (Figure 14.32a), and its share of national electricity generation is growing faster than any energy source. Technological advances are making wind power increasingly competitive, with costs ranging from 1 to 10 times those of fossil fuels. Suitable wind velocities (over 20 kilometers per hour; 12 miles per hour) are consistently

a.

Figure 14.32 Wind energy. **a.** Wind farm of multiple turbines, California. **b.** Map of US wind resources with economic potential. Red and blue colors indicate regions with greatest potential for the development of wind energy resources (more consistent winds, higher wind speeds).

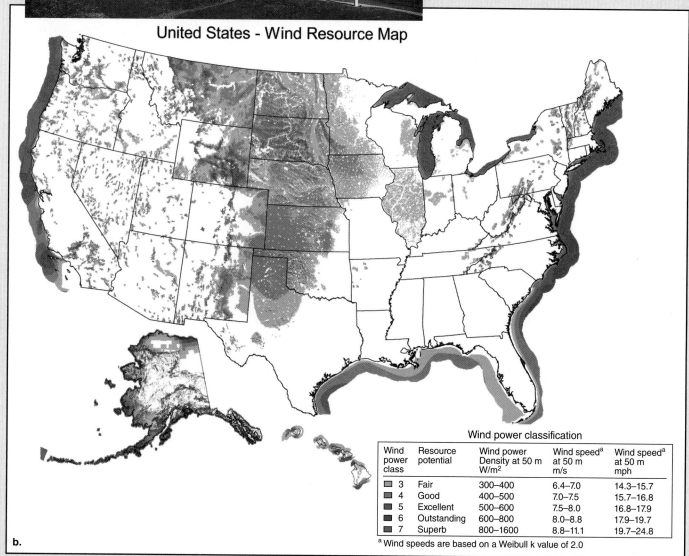

United States - Wind Resource Map

Wind power classification

Wind power class	Resource potential	Wind power Density at 50 m W/m[2]	Wind speed[a] at 50 m m/s	Wind speed[a] at 50 m mph
3	Fair	300–400	6.4–7.0	14.3–15.7
4	Good	400–500	7.0–7.5	15.7–16.8
5	Excellent	500–600	7.5–8.0	16.8–17.9
6	Outstanding	600–800	8.0–8.8	17.9–19.7
7	Superb	800–1600	8.8–11.1	19.7–24.8

[a] Wind speeds are based on a Weibull k value of 2.0

b.

present over about 13 percent of the United States (Figure 14.32b), and some projections estimate that wind power could generate as much as 20 percent of US energy within two decades. One constraint on the use of wind energy is the amount of land required. Power plants using fossil fuels or nuclear power require a relatively small land area. Depending on conditions, a wind energy installation would require about 20 times more land to generate the same power as a single nuclear power plant.

Many early US wind turbines were located in mountain gaps and other particularly windy parts of California. The land areas with the greatest potential for wind power are determined by the patterns of prevailing winds. These winds consistently exhibit sufficient velocity and reliability only over the the region of the continental US covered by the Great Plains states. Texas has the wind farm installations to generate more wind energy than any other state, and South Dakota and Iowa already produce more than 20 percent of their electricty from wind power. An alternative is for coastal states to build offshore wind farms to take advantage of steady coastal winds (Figure 14.32b). However, such plans are often contested on the basis of the potential for lost tourist revenues. It remains unclear if the mix of commercial and government support will be sufficient to continue the rapid growth of wind energy so that it accounts for a substantial proportion of national electricity generation in years to come. One benefit of the growth in this industry is that it has the potential to reduce the volume of greenhouse gases currently pumped out by power plants driven by fossil fuels.

Checkpoint 14.28

Venn Diagram: Atmospheric Versus Oceanic Circulation

Use the Venn diagram provided here to compare and contrast the features of atmospheric circulation discussed in this chapter and the characteristics of oceanic circulation described in Chapter 13. Write features unique to either group in the larger areas of the left and right circles; note features that they share in the overlap area. Identify at least eight features.

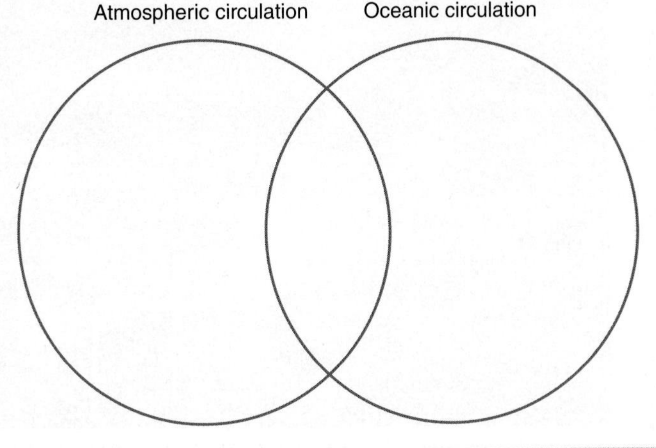

the big picture

While at first you may have thought that this chapter would describe a bunch of features in the atmosphere, you should have realized by now that this is really a chapter about energy, especially in the form of heat. The composition of the atmosphere means that enough heat derived from solar radiation is absorbed near the surface to make our planet livable. The amount of heat that land or water can absorb depends on their heat capacity. Water is a terrific place to store heat. It soaks up heat like a sponge, with only a relatively modest increase in temperature. Heat absorbed by water will either slowly raise the water's temperature or fuel changes in the water's state. These changes in state from solid to liquid to gas and back again do not create or destroy energy; they just transfer it from one place to another. The simple processes of evaporation and condensation result in the vertical transfer of heat energy from the lower troposphere to the upper troposphere. The horizontal transfer of warm and cold air is driven by winds produced by pressure differences between bodies of air with different temperatures. So the vertical and horizontal mixing of the atmosphere acts as a heat-transfer system.

Pity the poor pitchers for the Colorado Rockies baseball team. They consistently give up more of their home runs while playing at their home stadium, Coors Field in Denver. It can be hypothesized that this occurs because of the decrease in air density at high altitudes such as Denver (1,610 meters; 5,280 feet). Moving upward through the atmosphere, air pressure decreases rapidly at lower altitudes where air density is greatest. Gas molecules are more widely spaced at Coors Field than at lower-altitude sites such as Wrigley Field in Chicago (about 200 meters [656 feet] of elevation). Consequently, a home-run baseball will push against fewer gas molecules as it flies through the air at Coors Field and will therefore travel about 10 meters (33 feet) farther than at Wrigley Field. So, to answer the question posed with the chapter opening image: It is easier to hit home runs at Coors Field, because it has lower air pressure than at Wrigley Field. Efforts to offset this effect by keeping the baseballs in a humidor before they were used have been shown to have negligible effect. Increasing the humidity of the baseball does nothing to counteract the high altitude. While the ball becomes a little heavier, the moisture actually appears to improve the way it flies through the air.

As a final note, keep in mind that temperature and heat are not the same thing. Higher temperatures do not necessarily correspond to more heat. The few stray gas molecules in the thermosphere are moving much faster than the molecules in the troposphere and result in much higher temperature readings at the top of the atmosphere. But we should all be very grateful that there is a vast reservoir of heat in the slower motions of the billions and billions of nitrogen and oxygen molecules bouncing around the troposphere.

The Atmosphere: Concept Map

Complete the following concept map to evaluate your understanding of the interactions between the earth system and the atmosphere. Label as many interactions as you can, using information from this chapter.

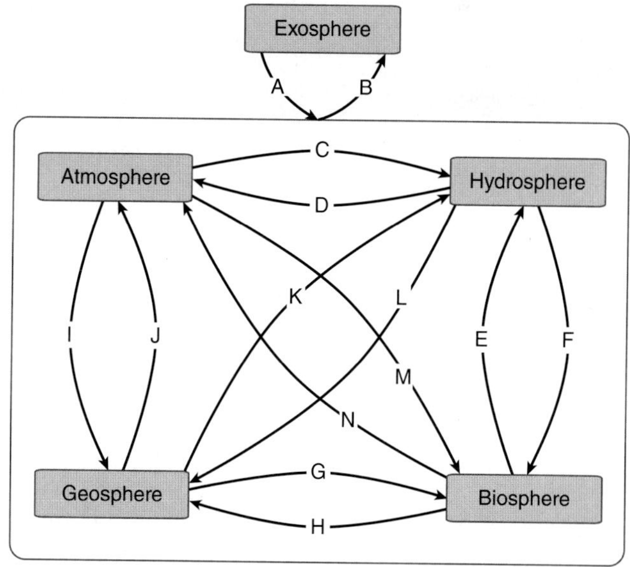

"A great thunderstorm; an extensive flood; a desolating hurricane; a sudden and intense frost; an overwhelming snowstorm; a sultry day—each of these different scenes exhibits singular beauties even in spite of the damage they cause. Often whilst the heart laments the loss to the citizen, the enlightened mind, seeking for the natural causes, and astonished at the effects, awakes itself to surprise and wonder."

—J. Hector St. John de Crèvecoeur, eighteenth-century author

Chapter 15

Weather Systems

Chapter Outline

the big picture

Why would someone chase a tornado?

See The Big Picture box at the end of this chapter for the full story on this image.

15.1 The Weather Around Us

Chapter Learning Outcomes

- Students will predict the impact weather has on humans.
- Students will be able to read and interpret a weather map.
- Students will evaluate the risk to humans from weather-related hazards.

At its best, the weather lifts our spirits. At its worst, it kills people. Our society is partially defined by **weather, the state of the atmosphere at any given time and place.** Texas and Oklahoma are two states in a section of the country nicknamed "tornado alley" because of severe weather that frequently occurs there. Thunderstorms and rain can affect almost any region of the country. Most of the time, weather simply influences what we do, but occasionally, **extreme weather threatens lives, disrupts transportation, and causes billions of dollars of damage in its destruction** (Figure 15.1). If we analyzed weather records from 1980 to 2005, we would find that an average of between two and three weather disasters per year caused damages valued in excess of a billion dollars. However, in just the 2011–2012 period the United States was battered by 25 of these costly weather disasters, culminating in the destruction of Hurricane Sandy at an estimated cost of $50 billion. While this recent increase in costs is unusual, there has been a steady increase in the frequency of these billion-dollar disasters over the last few decades.

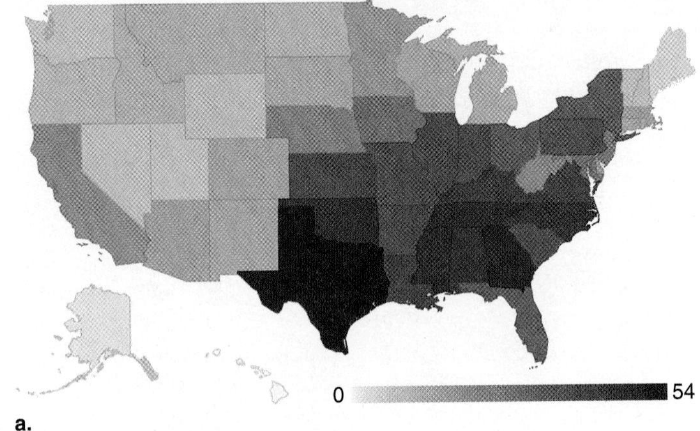

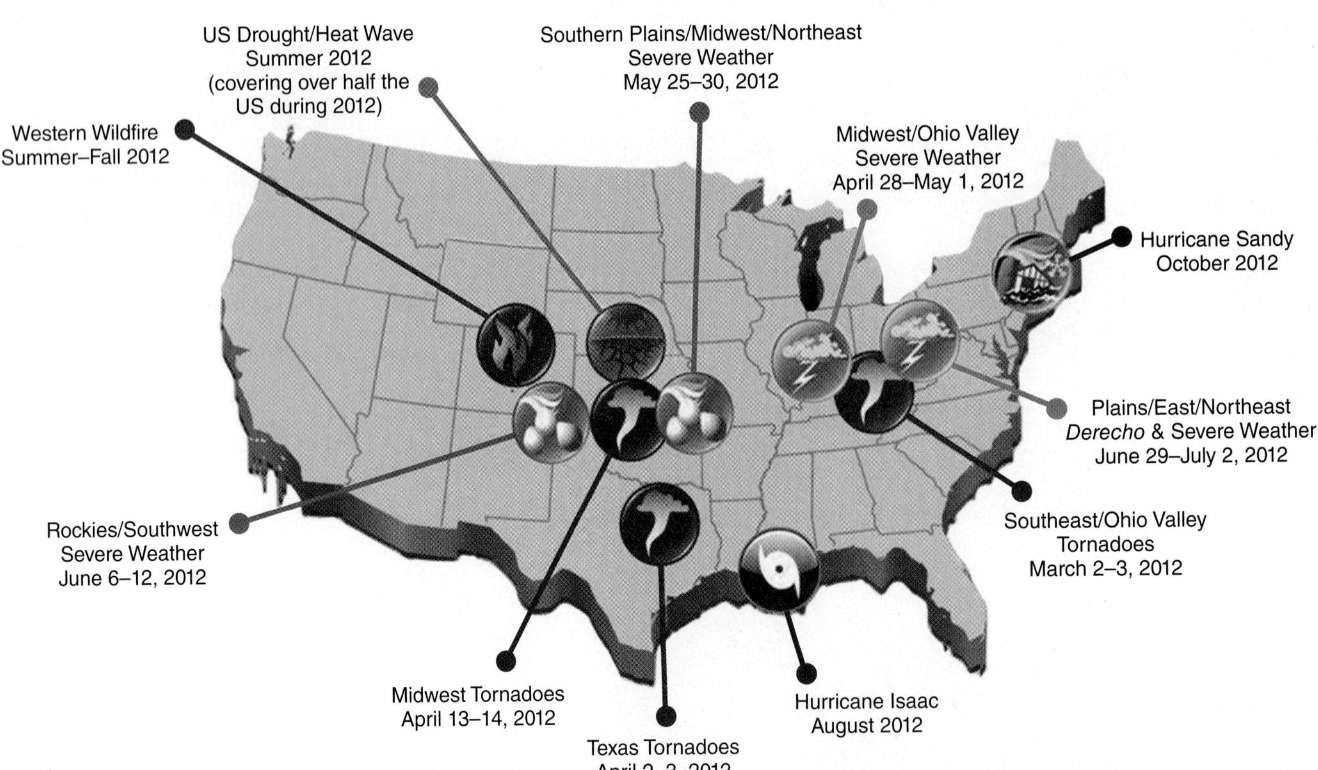

Figure 15.1 US billion-dollar weather/climate disasters. **a.** Disasters from 1980 to 2012 by state (costs of disasters adjusted for 2012 Consumer Price Index). Billion-dollar disasters are more frequent in southern states. **b.** Eleven US billion-dollar disasters for 2012.

Much of this increase is attributed to population growth and development, which are increasing the potential for more costly weather disasters. In 1900, nearly two-thirds of the US population lived in the northeastern and midwestern states. Today, more than 100 million residents make the South the most heavily populated region of the nation. Florida, Georgia, North Carolina, and Texas have seen some of the largest population increases in the last decade. Yet, a cursory glance at the states most likely to experience weather-related disasters shows that these migrants are moving into some of the more disaster-prone regions in America (Figure 15.1a). About one-half of these disasters fall into two types, tropical cyclones (including hurricanes) and severe local storms (Figure 15.1b). Winter storms, wildfires, and droughts together account for another one-third of the total.

Facts About Severe Weather

Which do you think kills more people (on average) every year: lightning, tornadoes, temperature extremes, floods, hurricanes, or snowstorms and ice storms? Data on deaths from natural hazards in the United States for the 25-year span from 1988 to 2012 reveal that the most common cause of death was temperature extremes (summer heat/drought, winter cold), which accounted for more than one-quarter of all deaths (Figure 15.2). On average, more than 500 people a year die from natural hazards, the vast majority of which are related to weather phenomena. Over time, individual high-magnitude events such as Hurricane Sandy are less responsible for deaths than smaller, more frequent events such as storms and lightning. There is some geographical variety to death from natural hazards, as is hinted at by Figure 15.1. Fatalities from natural hazards in southern states are more typically due to severe storms and tornadoes, while extreme heat accounts for more deaths in the northern Great Plains, and winter storms and floods are the greatest hazards in the Mountain West. Midwestern states such as Illinois, Indiana, and Ohio have the lowest mortality rates from all types of natural hazards.

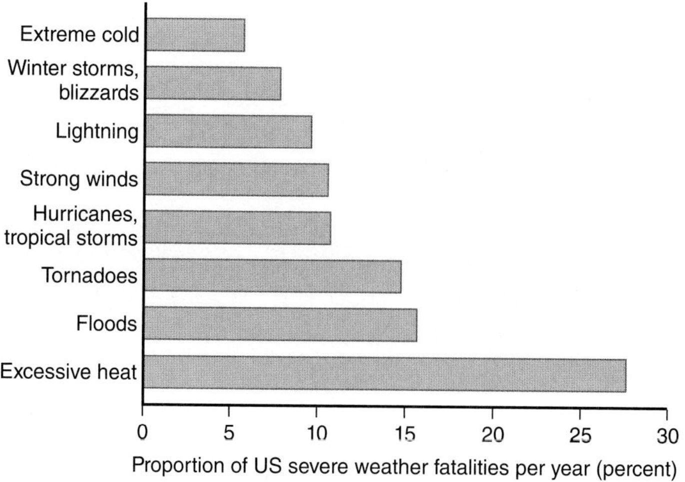

Figure 15.2 US severe weather fatalities, average per year, 1988–2012. The US averages 519 fatalities per year in the 25-year period from 1988 to 2012. Over one-quarter of those fatalities occurred as a result of heat waves. Phenomena associated with severe types of storms (e.g., high winds, lightning, tornadoes) accounted for more than one-half of all deaths.

The National Oceanographic and Atmospheric Administration (NOAA) considers **extreme heat "one of the most underrated and least understood of the deadly weather phenomena"** because, unlike in the situation with a tornado, the victims are not easily recognized and the hazard is less apparent. When a mass of hot air moves over a region, our bodies respond by losing heat through our skin and through perspiration. However, perspiration works to lower body temperatures only when the humidity is low enough to allow sweat to evaporate (the latent heat of evaporation takes heat away from our bodies) and when we replace lost fluids by drinking water. If the body cannot get rid of the excess heat, the core body temperature will rise above the normal 37°C (98.6°F). Heatstroke occurs when body temperatures rise to around 40.5°C (105°F). More than 26,000 people died in Europe in 2003 when the continent experienced its hottest summer on record, with temperatures routinely exceeding 40°C (104°F).

Elderly citizens and young children are most at risk from heatstroke as their bodies are less able to make the needed physiological adjustments to extreme heat conditions. It is not uncommon to read of children who were endangered when they were left in a locked car for more than 15 minutes during a hot day. In July 1993, a 3-day period of record high temperatures in Illinois and Wisconsin resulted in more than 500 heat-related deaths; more than two-thirds of those who died were older than 65. Many were found in hot city apartments that lacked air conditioning or had windows that had been nailed shut to protect against burglaries.

Extreme heat, tornadoes, hurricanes, and other dangerous weather phenomena cannot be stopped, but by relying on detailed observations, meteorologists can provide timely warnings to prepare and protect people from these hazards. The most expensive natural disaster in US history occurred in 2005 when Hurricane Katrina wreaked havoc across the Gulf Coast states, causing over $100 billion in damage. Continued coastal development makes a similar future disaster inevitable. Consequently, scientists primarily focus on collecting and analyzing data, and educating those who use that information. Ultimately, it is up to government agencies to make wise decisions that will protect their citizens from harm. It is also important for citizens, like you, to understand and heed the warnings. As we will see with the example of Hurricane Katrina, sometimes the warning signs are not appreciated until it is too late.

The first half of this chapter describes how common weather systems develop across much of the nation. We will discuss how and why air masses range from dry and cold to warm and humid, and what happens when different air masses interact. We will also explore how meteorologists can successfully use these characteristics to forecast weather for 2 to 5 days in the future. The second half of the chapter focuses on dangerous weather phenomena such as thunderstorms, tornadoes, and hurricanes.

Self-Reflection Survey: Section 15.1

Answer the following questions as a means of uncovering what you already know about weather systems.

1. Explain three examples of how the weather influences your daily life.
2. Describe your ideal annual weather conditions.
3. Describe the most extreme weather event you have experienced.

15.2 The Science of Weather: From Folklore to Forecasting

Learning Objectives

- Explain how advanced technology, such as weather satellites and Doppler radar, allow the accurate prediction of tornadoes, hurricanes, and other dangerous weather systems.
- Describe how the principles of science are evident in the historical development of meteorology.
- Devise a method for assessing the accuracy of weather forecasts.

Modern **meteorology, the study of the atmosphere and its weather,** is humankind's latest attempt to make sense of natural changes in the world around us. The ancient Egyptians believed that the hottest days of summer were related to the appearance of the dog star, Sirius, in the night sky. They reasoned that the higher temperatures of July and early August were due to the extra heat supplied by that star. This belief gave us the term *dog days of summer,* which we still associate with hot summer days. However, it has nothing to do with Sirius, which is too far away to contribute any heat to Earth.

The First Meteorologists

The Greek philosopher Aristotle was the first to publish a text, *Meteorologica,* containing theories about the origins of weather. Although he correctly identified some basic characteristics of weather (for example, that temperature decreases with altitude), most of his explanations for weather phenomena were later found to be incorrect. However, just as the Greek idea about the geocentric Earth held sway for nearly 2,000 years, so too did Aristotle's explanations for the weather.

The first true science text on weather was published in 1637 by the French philosopher and all-around genius René Descartes. Descartes's *Les Météores* was an improvement over Aristotle's interpretations and provided explanations for clouds (made of water droplets) and other weather phenomena. Descartes promoted the use of scientific methods to unravel the characteristics of weather, but he was ahead of his time, since accurate instruments for measuring weather conditions had yet to be invented. Consequently, most of Descartes's science consisted of descriptions rather than numerical data.

A paradigm shift in our understanding of weather occurred during the next century as the development of accurate instruments enabled scientists to make empirical observations of weather phenomena and test hypotheses. In 1643, Evangelista Torricelli developed the first barometer as a way of measuring air pressure, but it was another 200 years before the portable modern form of the device was produced. Two types of mercury thermometers were invented by Gabriel Fahrenheit (1714) and Anders Celsius (1742), and a hygrometer, an instrument to measure humidity, was created a few decades later. So by the 1700s, the technology had caught up with the science, and improved communications systems would soon make it possible to recognize regional weather patterns.

Checkpoint 15.1

basic · intermediate · advanced · superior

Examine Figure 15.1b. Examine the map of weather disasters for 2012. What type of weather disaster was nearest to where you live? On the basis of your own experience, is this the most representative type of severe weather for your location?

Communications Developments

US and European institutions made early attempts to form weather networks toward the end of the eighteenth century. But until the telegraph became widely available in the mid-1800s, scientists could not exchange weather information over large enough distances to make any useful attempts at weather forecasting. **Once information became available from multiple locations simultaneously, patterns in the weather data began to emerge.** Scientists realized that weather systems migrated from west to east, that fair skies were associated with high pressure, and that rain often accompanied low-pressure systems.

The US National Weather Service (NWS) was established in 1870 as part of the Signal Service Corps in the Department of War (now the Department of Defense) and was given the charge "to provide for taking meteorological observations at the military stations in the interior of the continent and at other points in the States and Territories . . . and for giving notice . . . of the approach and force of storms." The fledgling service made its first simultaneous observations at just 24 sites on November 1, 1870. Within 2 years, it was creating national weather maps (Figure 15.3), and by 1878, daily observations were being collected at hundreds of sites and relayed cross-country by telegraph.

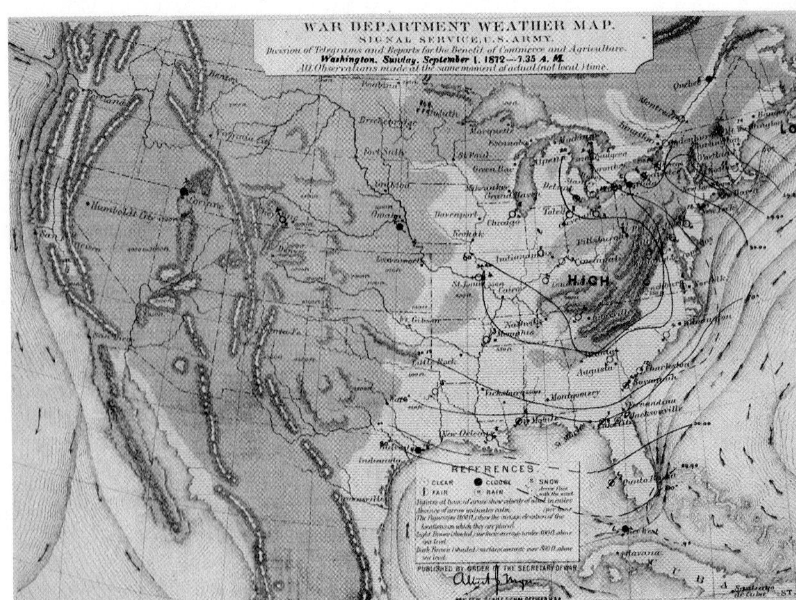

Figure 15.3 Early national weather map. According to this map, there was a high-pressure system over Kentucky and West Virginia on September 1, 1872.

For the next few decades, weather forecasting in North America and Europe typically consisted of identifying what the weather was like to the west and predicting that the same weather would arrive in your area the next day. It was not until the first decades of the 1900s that a talented group of Norwegian scientists identified the concepts of air masses (large volumes of air with similar temperature and moisture content) and frontal systems (locations where air masses interact), which allowed meteorologists to use raw temperature and pressure data to accurately predict weather over large regions. These scientists, who came to be known as the Bergen School, used inductive reasoning. They used individual cases to draw general conclusions on the basis of data telephoned in from teams of weather observers. These observations were analyzed to arrive at some general principles about how weather works. It soon became apparent that the same types of air masses that were recognized over Norway were also present elsewhere in the world. **The recognition of these basic types of air masses and their motions made it possible to develop some rules for the evolution of weather systems.** Scientists anywhere could readily apply these rules to predict future weather patterns, an example of deductive reasoning (that is, applying general rules to specific cases). A few decades later, these deductive rules were transferred into mathematical equations that could be programmed into early computers to produce daily weather forecasts.

Weather Technology Today

Such is our collective fascination with the weather that millions of people tune in to the Weather Channel daily. But regardless of where we get the weather forecast, all of the information and data come from the same place, the National Weather Service. The NWS processes over 1 million surface, air, and satellite weather observations per day. These basic observations may be reprocessed by commercial weather companies (for example, Accuweather) to generate maps and graphics for public distribution to a variety of media. The current NWS mission is to provide "weather, hydrologic, and climate forecasts and warnings for the United States, its territories, adjacent waters and ocean areas, for the protection of life and property and the enhancement of the national economy."

Today, the NWS uses sophisticated satellite technology to keep tabs on developing weather systems worldwide. For example, the Geostationary Operational Environmental Satellite (GOES) program began in 1968 and today has two satellites in synchronous orbit above Earth, providing weather coverage for 60 percent of the planet's surface (Figure 15.4). The NWS also has over 150 Doppler radar sites nationwide that are used to track rapid changes in regional storms. Doppler radar can be used to identify storms with swirling winds that tend to be extremely dangerous. The nationwide expansion of Doppler radar installations has resulted in earlier warnings for sudden weather phenomena such as tornadoes and flash floods. The use of new technology has increased the average lead time for tornado warnings from just a few minutes 20 years ago to 10 or more minutes today. That is plenty of time for folks to grab the kids and head for the basement.

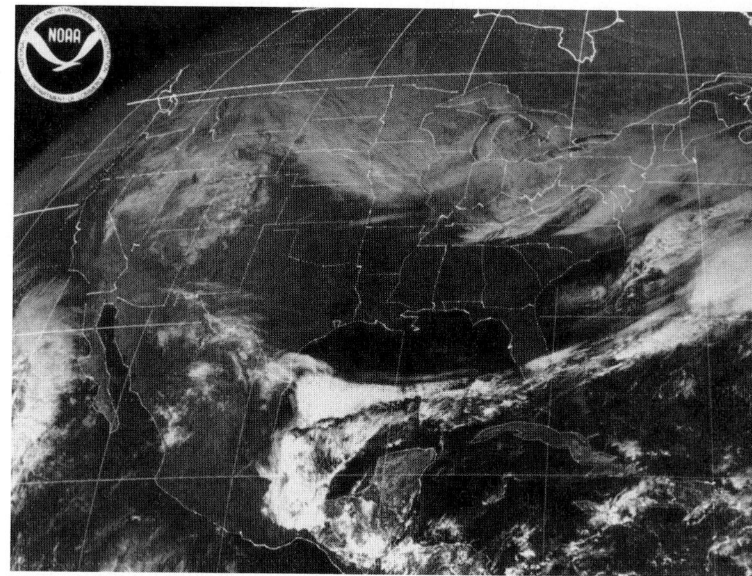

Figure 15.4 A satellite weather image of North America. Geostationary satellites generate thousands of images per day, which are used to monitor the weather over most of Earth's surface.

Checkpoint 15.2

basic · intermediate · advanced · superior

Describe other earth science phenomena where it is necessary to assimilate data on a regional scale to accurately determine patterns.

Checkpoint 15.3

How are the following four key principles of science evident in the brief history of meteorology in this section?

1. Phenomena can be explained by natural causes.
2. Explanations are provisional (tentative).
3. Science is based on empirical observations.
4. Explanations should be testable.

Checkpoint 15.4

basic · intermediate · advanced · superior

Weather Forecast Evaluation

Go to the Weather Channel website (www.weather.com) and enter your city or Zip code. Follow directions at the site to obtain the 10-day forecast for your location.

1. How could you measure the accuracy of the forecast? Create a scoring scheme that anyone could use to determine the accuracy of the forecast. Describe your scheme.
2. Track the weather over the next 10 days and evaluate the accuracy of the forecast.

15.3 Air Masses

Learning Objectives

- Explain why the location where an air mass forms affects the temperature, humidity, and pressure of an air mass.
- Predict the physical characteristics of air masses forming at various locations on Earth.
- Predict how the physical properties of air masses would be modified as they traveled from one location to another.

What kind of weather can delay trains, cause traffic lights to malfunction, and result in hundreds of calls to tow-truck companies? All of these conditions occurred on a January morning when the northeastern United States and neighboring parts of Canada were covered by a blanket of bone-chilling cold air. Temperatures in the Canadian cities of Toronto and Québec rivaled those of Siberia as a single large air mass swept south from northern Canada. The cold conditions extended into states as far south as the Carolinas. **An air mass is a large region of the lower troposphere that has relatively uniform temperature and moisture content.** The characteristics of individual air masses depend on where they form (called the *source area*) and any changes that occur as they move into regions having different conditions. Weather is influenced by changes in the air masses over time and interactions at the boundaries between air masses (called *fronts*), where rising air may lead to condensation and precipitation.

Source Areas

Think about the interaction between the atmosphere and the tropical ocean on a sunny day. The ocean exchanges heat with the overlying air, helping to evaporate water. A warm, moist air mass has developed. How would that air mass differ from one over a desert? An air mass develops when the atmosphere above a relatively uniform land or water surface remains stationary for several days and the lower atmosphere takes on the temperature and moisture properties of the underlying surface. **Air masses are identified by the temperature and the moisture characteristics of the underlying surface.** Cold, polar air masses form at high latitudes (more than 50°N, S), and warm, tropical air masses form closer to the equator (Figure 15.5). Maritime air masses develop above oceans and contain much more moisture than dry air masses formed over the continents. However, because temperatures vary with seasons, there are no firm boundaries that define where air masses form. For example, polar air masses creep farther south during winter and retreat northward during summer in the Northern Hemisphere. As we will see later, heavy rains and severe thunderstorms (or snowstorms) can result from the interaction between a continental polar air mass and a maritime tropical air mass that collide over the eastern United States during summer (or winter).

Types of Air Masses

Meteorologists use a form of scientific shorthand to label air masses. Five common types are shown in Figure 15.5 and can be described as follows:

- ***cA.*** Continental Arctic/Antarctic air forms at high latitudes around the poles above permanently snow-covered ground (or sea ice; Figures 15.5 and 15.6). These air masses are characterized by extremely cold, dry air that may sweep south across Canada and produce days of below-freezing temperatures over much of the central and eastern United States during winter.
- ***cP.*** Continental polar air forms over the northernmost portions of North America, Europe, and Asia (Figures 15.5 and 15.6). This type of air mass shares its basic characteristics (cold, dry) with cA air but without the exceptionally cold temperatures.
- ***cT.*** Continental tropical hot, dry air forms over continental interiors such as the dry lands of northern Mexico and the southwestern United States (Figures 15.5 and 15.6). The moisture content and temperature of the air mass are modified as it moves to the east or north.
- ***mP.*** Maritime polar air masses form in the northern Atlantic and Pacific oceans and over the Southern Ocean (Figures 15.5 and 15.6). These air masses are

Checkpoint 15.5

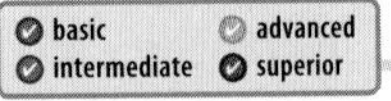

Of the five most common types of air masses, which ones most directly affect the area where you live?

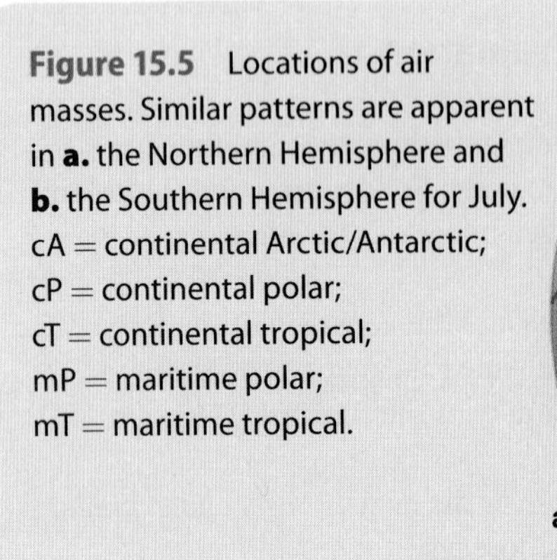

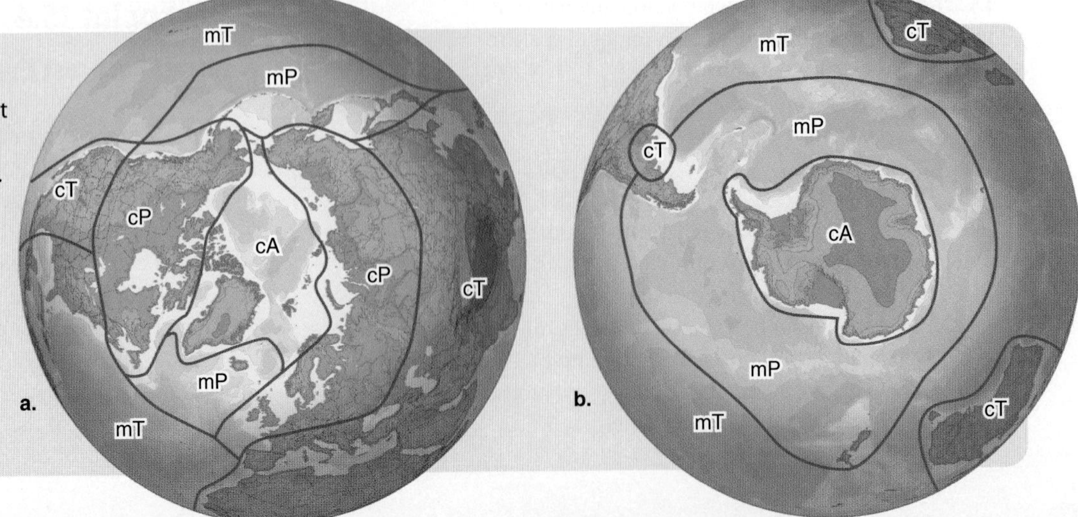

Figure 15.5 Locations of air masses. Similar patterns are apparent in **a.** the Northern Hemisphere and **b.** the Southern Hemisphere for July. cA = continental Arctic/Antarctic; cP = continental polar; cT = continental tropical; mP = maritime polar; mT = maritime tropical.

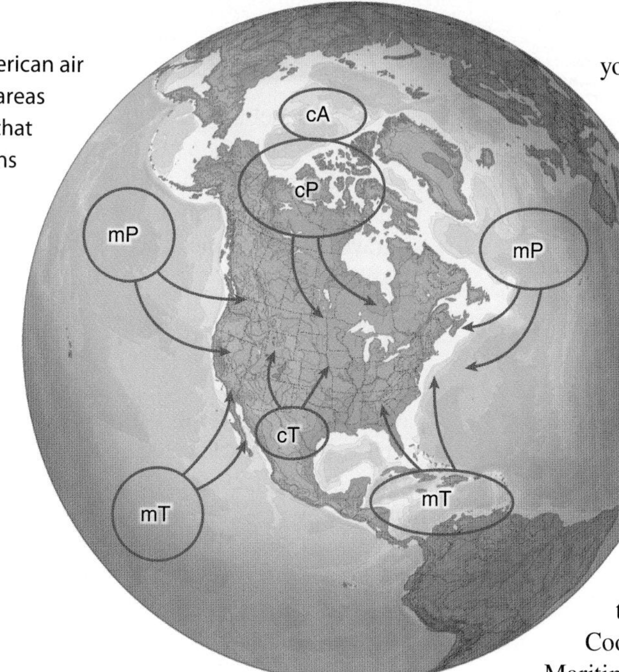

Figure 15.6 North American air masses. Principal source areas and paths for air masses that influence weather patterns across North America. Arrows indicate where the air mass moves to after it forms. For example, cP air masses originate in northern Canada and move to the south and southeast across the United States.

characterized by cool, moist air that affects northeastern and northwestern North America. Temperatures at the ocean surface are less extreme (less cold and less hot) than on land, so mP air is warmer than cP air. Such air masses bring rains to the coasts of Washington and Oregon or snows to the mountains inland.

- ***mT.*** Maritime tropical air masses move inland from the tropical Pacific Ocean, Gulf of Mexico, and tropical Atlantic Ocean (Figures 15.5 and 15.6). They are distinguished by high temperatures and high humidity. The mT air brings hot, humid summers to southeastern states and can form at any time during the year.

Modification of Air Masses

Think about driving to the grocery store on a hot, sunny summer day. On the way there, you run the air conditioning in your car, and the temperature in your vehicle is pleasantly cool. You have created your own personal car-sized air mass. You park and go inside for 20 minutes to stock up on pickles and frozen corn. What will the interior temperature of your car be like when you return? Of course, the temperature of the air inside the car will be warmer because the car has been sitting in the sun. In the same way, air masses can gain or lose heat and moisture as they move from one location to another. Can you think of an example of this that we have discussed in a previous chapter?

The principal factors that can modify an air mass are the temperature and topography of the underlying surface. If the air is heated by a warmer land surface (as occurs with cP air moving south), the air will begin to rise. Such air is unstable—that is, it will continue to rise until its temperature matches that of the surrounding atmosphere (see Chapter 14). Cooling (mT air moving north) has the opposite effect.

Maritime air, whether polar or tropical, can be forced upward over mountain ranges, leading to cooling and condensation. Condensation removes moisture from the air through precipitation and changes the formerly humid air to a much dryer air mass, producing a rain shadow effect. Washington and Oregon provide a good illustration of these trends. Both states can be divided into wet western highlands and dry eastern deserts (Figure 15.7). Each

Checkpoint 15.6

basic | intermediate | advanced | superior

Which air masses would be present during July at each of the locations indicated on the following globe?

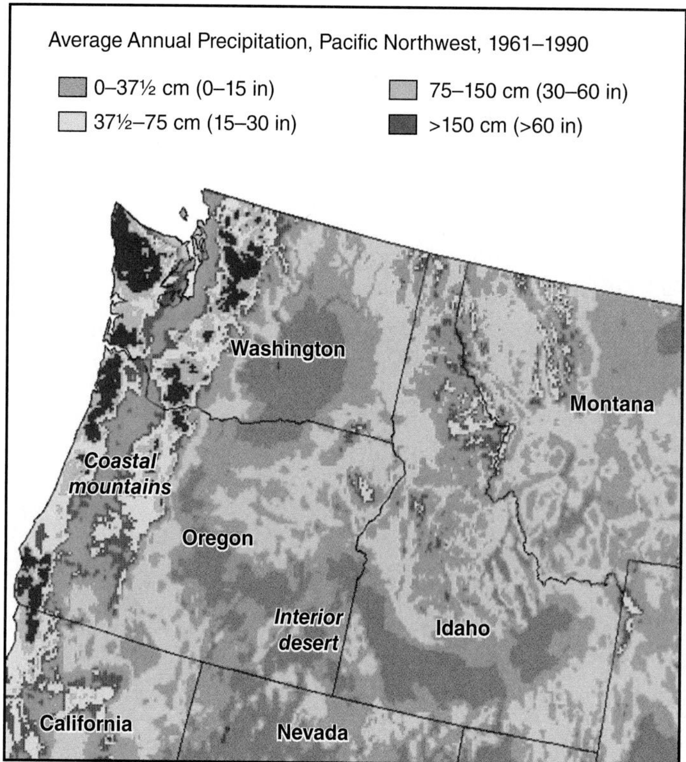

Figure 15.7 Annual precipitation in the Pacific Northwest. Washington and Oregon receive much more rainfall over mountains in the western one-third of the states, where the maritime polar air mass comes onshore. Much of this moisture is lost during orographic lifting, creating a rain shadow effect that contributes to the distribution of desert conditions in the eastern two-thirds of both states.

Checkpoint 15.7

basic · intermediate · advanced · superior

Thunderstorms and tornadoes are most frequently associated with weather patterns that form where maritime tropical air masses interact with continental polar air masses over the United States. Predict where this would be most likely to occur, using Figure 15.6.

Checkpoint 15.8

basic · intermediate · advanced · superior

Draw and label idealized graphs illustrating (a) how the relative humidity of a mP air mass would change as it traveled from the Pacific Ocean to Idaho and (b) how the temperature and humidity of a cP air mass would change as it traveled from Canada across the Great Lakes to produce lake effect snows.

state experiences plentiful annual rainfall near the coast (more than 150 centimeters; 60 inches) due to maritime polar air masses rising over coastal mountain ranges. However, deserts in the dryer rain shadow just to the east of the mountains receive less than 25 centimeters (10 inches) of rainfall.

In contrast, dry continental air masses can pick up moisture as they move over water bodies or saturated soils. Heavy winter snowfalls, labeled "lake effect" snows, are commonplace in states such as Michigan, Ohio, and New York that are south and east (downwind) of the Great Lakes. Dry cP air masses pick up moisture as they cross the warmer waters of the lakes (Figure 15.8). When an air mass arrives at the southern lakeshore, the air cools as it encounters colder land. Cooling results in saturation (see Chapter 14), which produces condensation and precipitation, resulting in heavy snowfalls. It is no surprise that the cities of Buffalo and Syracuse, New York, located downwind from Lake Erie and Lake Ontario, experience some of the heaviest snowfalls in the nation. How and why these air masses move are linked to circulation patterns. We will cover that and how these air masses impact climate in Chapter 16.

Figure 15.8 The lake effect. Michigan and parts of Wisconsin are almost completely obscured below bands of snow on the satellite image of the Great Lakes region. Snow is formed as cold, dry continental air blows over the warmer water of Lake Superior and Lake Michigan and then back over land on the southern shores of the lakes.

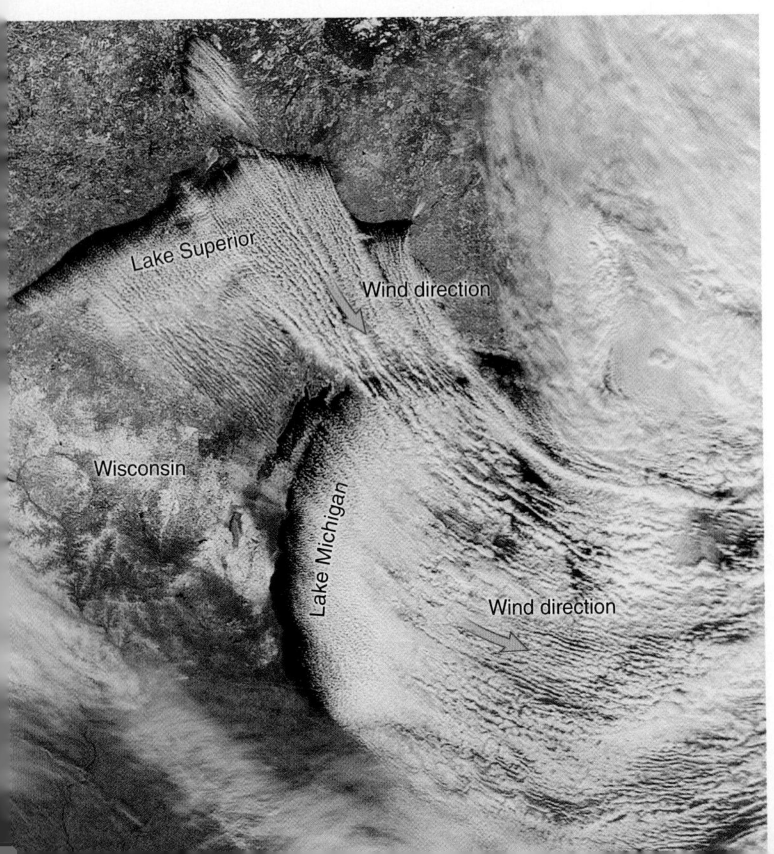

15.4 Midlatitude Cyclones and Frontal Systems

Learning Objectives

- Discuss the role that midlatitude cyclones play in driving weather patterns.
- Compare and contrast cold and warm fronts.
- Predict how weather conditions differ as a result of the passage of frontal systems.

Much of the weather experienced over large sections of North America is a result of the west-to-east migration of **regional-scale low-pressure systems known as *midlatitude cyclones.*** As its name suggests, a midlatitude cyclone forms between 30° and 60°N or S, in the middle latitudes (Figure 15.9). In North America, this covers everything from southern Alaska to Florida. Most midlatitude cyclones begin in this collision zone between air masses, which migrates south during winter and north during summer. Midlatitude cyclones may be 1,000 to 2,000 kilometers (620 to 1,240 miles) across and can affect much of the continental landmass for periods ranging from 3 days to as long as a week.

The boundary between one air mass and another is termed a *front.* Fronts represent meteorological battle lines. As air masses move across Earth's surface, they interact with one another along pairs of relatively narrow, long, slightly curved, regions known as a *frontal system.* For example, a wedge of warm mT air pushes northward into a cold cP air mass that is moving south. This interaction creates transition zones between the warm and cold air masses at the eastern and western margins of the wedge, forming *warm* and *cold fronts,* respectively (Figure 15.10). Where these fronts merge, they form an *occluded front.* Years of careful observations have revealed that weather conditions change in a predictable sequence as warm and cold fronts pass over an area. Advancing frontal systems bring clouds and precipitation, and are accompanied by changes in moisture, temperature, pressure, and wind direction (Table 15.1).

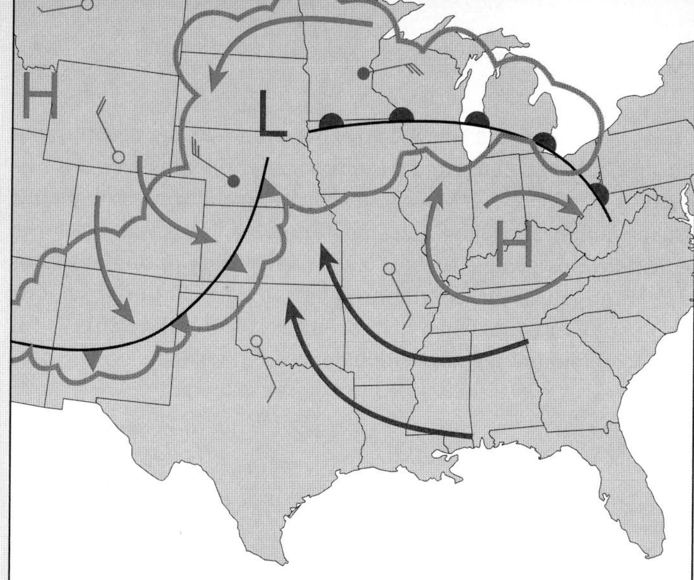

a.

b.

Figure 15.9 A midlatitude cyclone. **a.** At the point where a low-pressure system (cyclone) interacts with neighboring high-pressure systems, a midlatitude cyclone forms. The midlatitude cyclone is centered over the low-pressure system and is characterized by warm (red) and cold (blue) fronts. Cloud cover is concentrated over the fronts and the low-pressure center. **b.** A classic comma-shaped cloud pattern is associated with a midlatitude cyclone in the central United States, Christmas Eve, 1997. The frontal system is composed of a combination of an occluded front and a cold front.

It is useful to discuss the evolution of midlatitude systems because they tend to dominate weather patterns in the United States. The boundary between air masses is initially a stationary (nonmoving) front, with airflow in opposite directions on either side. Midlatitude cyclones develop where surface irregularities, such as mountains or land and water boundaries, cause local shearing (a sideways-acting force) that distorts the front. The front is transformed as warm air pushes northward and cold air moves south, generating the counterclockwise rotation typical of cyclones and forming the warm and cold front pairings.

Cold Fronts

The clash between cP (continental polar) and mT (maritime tropical) air masses over the Great Plains and Midwest is one of the most common sources of frontal systems in the United States. As a frontal system advances from west to east, the relatively warm mT air mass will be replaced by a colder, dryer cP air mass

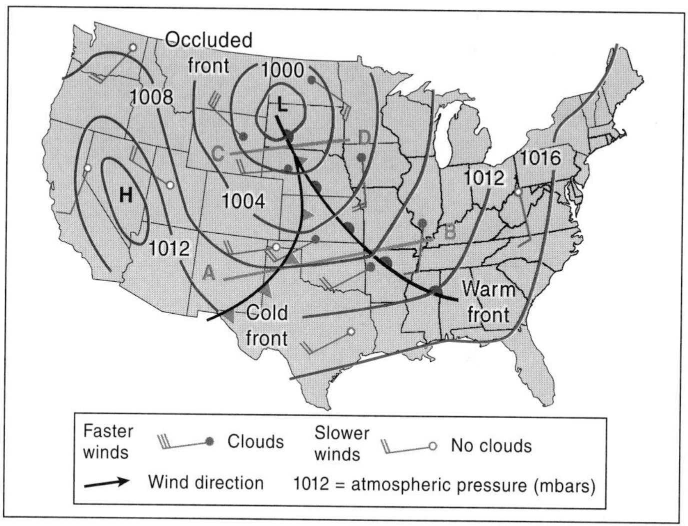

Figure 15.10 A frontal system. These are the weather patterns typically encountered with cold and warm fronts associated with a midlatitude cyclone over the central United States. Warm maritime tropical air from the Gulf of Mexico lies between the two fronts. Note that cloud cover occurs in advance of the cold front, adjacent to the warm front, and around the occluded front. Lines AB and CD represent sections through the frontal system (see Figures 15.11 and 15.13). On weather maps, a cold front is represented by a line with triangles (blue) and a warm front has semicircles (red). Both types of symbols "point" in the direction of air movement.

Table 15.1 Weather Conditions Associated with Passing Frontal System Shown in Figure 15.9

Conditions	Before warm front	Between warm and cold fronts	After cold front
Pressure	Decreasing	Small decrease, then small increase	Increasing
Winds	South, southeast	Southwest	West, northwest
Temperature	Cool	Warm	Cold
Clouds	Cirrus, cirrostratus, altostratus, nimbostratus	Cumulus, cumulonimbus	Cumulus, altostratus
Precipitation	Light to moderate, increasing	None, then heavy rains prior to cold front	Moderate to light, decreasing

(Figure 15.11). People living downwind from the front experience a decrease in temperature and humidity and an increase in atmospheric pressure when the cold front passes (Table 15.1). The slope of the cold front is pictured as relatively steep in Figure 15.11, but its actual inclination is about 1 degree toward the warm air. Another way of expressing this is to consider that the cold front rises 1.75 kilometers (1.1 miles) over a horizontal distance of 100 kilometers (62 miles).

Warm air is displaced up and over the advancing cold front because it is less dense and therefore lighter than the cold air. The rising air undergoes relatively rapid cooling and condensation in a narrow region above the cold front. The condensing water vapor in rising air provides the fuel for cloud formation. **Cooling and condensation cause the development of tall cumulonimbus clouds that usually produce heavy but relatively short-lived precipitation** (commonly known as a thunderstorm).

Checkpoint 15.9

basic · intermediate · advanced · superior

Frontal Systems Exercise 1

Use this map to answer the questions that follow.

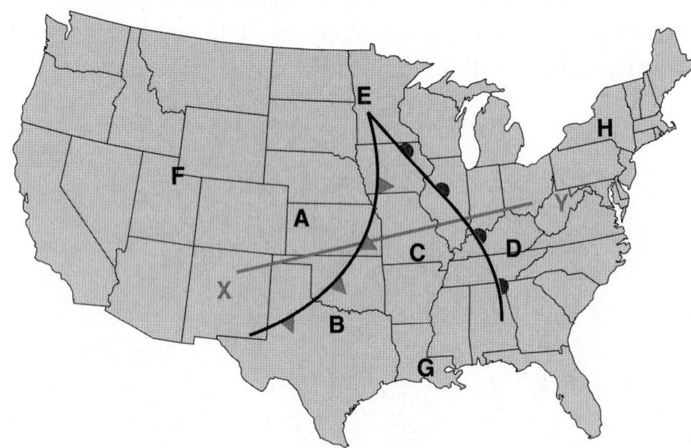

1. The map illustrates the relative positions of a warm front and a cold front. Where is the warm front located?
   a) Between A and B
   b) Between C and D
   c) At E

2. Where is it most likely to be raining?
   a) A and B c) C and D
   b) B and C d) B and D

3. Which location is in a maritime tropical air mass?
   a) A b) G c) E d) H

4. Which location will become warmer in the next 12 hours?
   a) A b) B c) C d) D

5. Which of the following images best represents conditions along the line XY on the map? Explain what is happening along XY for your choice.

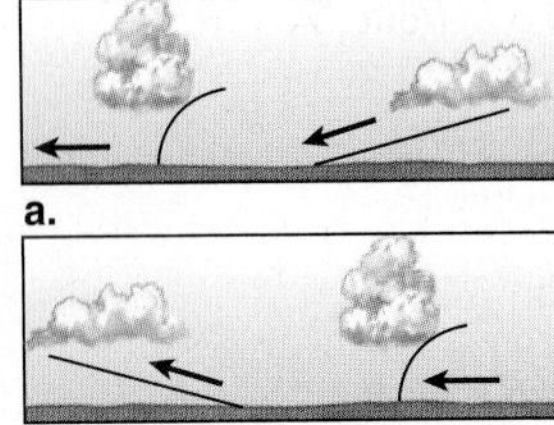

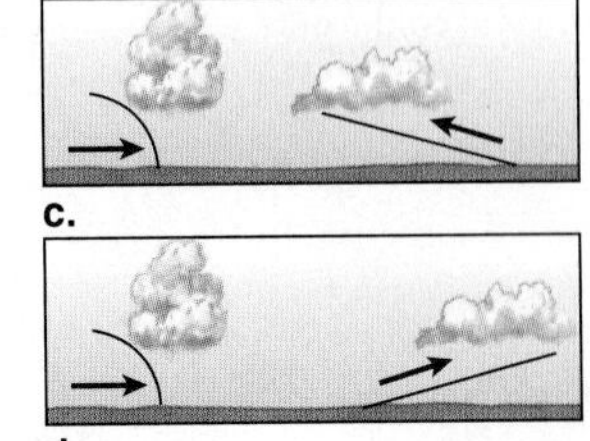

Checkpoint 15.10

Frontal Systems Exercise 2

The data in the accompanying table represent changes in rainfall and temperature at Heathrow Airport, London, over parts of 2 days in October 2000, during the passage of a frontal system. Use the data to answer the questions that follow.

Precipitation and Temperature at Heathrow Airport Weather Station

Date	Time	Precipitation, mm	Temperature, °C
10-29	10.00 A.M.	0	10.0
	12 Noon	0	11.6
	2.00 P.M.	0	11.8
	4.00 P.M.	0	10.5
	6.00 P.M.	1.6	11.4
	8.00 P.M.	2.4	11.7
	10.00 P.M.	1.6	13.2
10-30	12 Midnight	5.4	13.1
	2.00 A.M.	2.4	13.2
	4.00 A.M.	1.6	13.0
	6.00 A.M.	1.4	12.8
	8.00 A.M.	5.0	10.2
	10.00 A.M.	0	8.4
	12 Noon	0.2	7.6
	2.00 P.M.	0.2	10.8
	4.00 P.M.	0	11.0

1. When did the warm front pass the weather station?
2. When did the cold front pass the weather station?
3. Does rainfall or temperature represent a better indicator of the passage of a cold or warm front? Justify your answer.
4. Which exerted the greater influence on temperature? Justify your answer.
   a) Time of day
   b) Passage of the frontal system

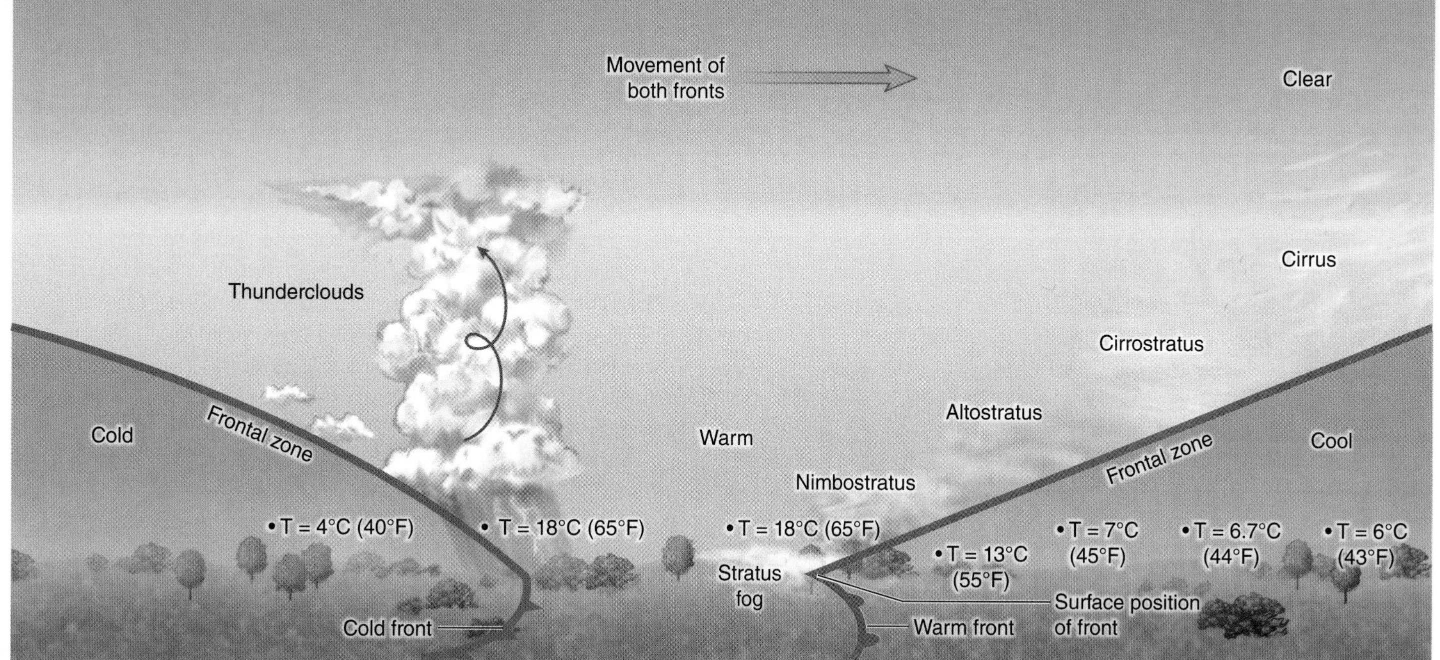

Figure 15.11 Weather conditions associated with cross section AB on Figure 15.10. Warm air (mT) lies between the cold front and the warm front. (Left) The cold front advances more rapidly than the warm front, forcing warm air to rise and forming thunderclouds and heavy rains. (Right) Warm air is forced to rise above the more gently sloping warm front, resulting in the formation of a series of low to high clouds.

Warm Fronts

Weather changes accompanying the passage of a warm front are usually less severe than those at cold fronts, because the warm air does not rise as rapidly. Warm fronts slope gently (about ½ degree inclination) over a distance of up to 1,000 kilometers (620 miles) toward the warm air mass (see Figure 15.11). Clouds formed above the sloping front would be more than 8 kilometers (5 miles) high and 1,000 kilometers (620 miles) ahead of where the warm front intersected the land surface. The first signal of an approaching warm front is the appearance of light, upper-level clouds (cirrus, cirrostratus). Up to 12 hours later, the high clouds are replaced by lower nimbostratus clouds with associated light to moderate precipitation. Rain associated with a warm front may last longer than precipitation accompanying a cold front because the warm front typically moves more slowly and extends over a larger area. Look at Figure 15.10, location B: As the front advances, temperatures and humidity rise, and the warmer, moister air replaces the colder, drier air. Winds also change direction (from south to southwest) in this case because the lines of pressure are deflected at the front (see Figure 15.10).

Occluded Fronts

Warm air continues to advance along the warm front at rates of 15 to 20 kilometers per hour (9 to 13 miles per hour), moving over ground previously covered by colder air. In the scenario shown in Figure 15.12, the cold front lies to the west and moves about twice as fast as the warm front. This convergence forces the warm air trapped between the fronts to move upward, generating additional

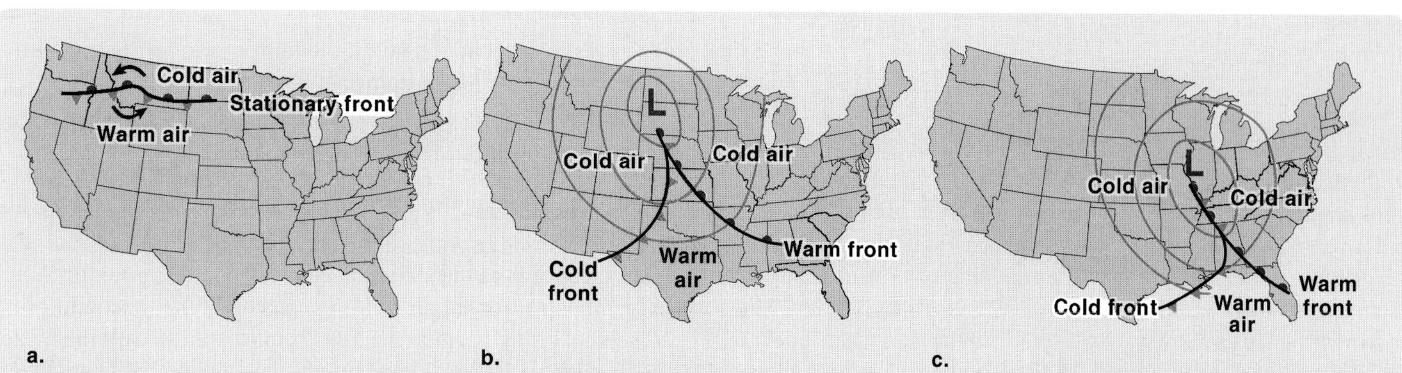

Figure 15.12 Three stages in the development of a midlatitude cyclone over the United States. **a.** A cyclone forms. **b.** The cyclone reaches mature stage, with well-developed warm and cold fronts. **c.** The cyclone begins to weaken as warm and cold fronts merge to form an occluded front. Note the west-to-east track of the cyclone and the merging of warm and cold fronts to form an occluded front in the final image shortly before the cyclone decays.

Checkpoint 15.11

basic · intermediate · advanced · superior

Frontal Systems Changes

Weather conditions change with the passage of warm or cold fronts. These changes are related to changes in air pressure, air temperature, and the state of water. Examine Figure 15.11 and answer the questions below which are related to key concepts from Chapter 14.

1. Describe the change in the state of water associated with the passage of a warm front.
2. Will this change in state cause latent heat to be released or absorbed? Justify your answer.
3. Describe the adiabatic temperature changes associated with the passage of a warm front.

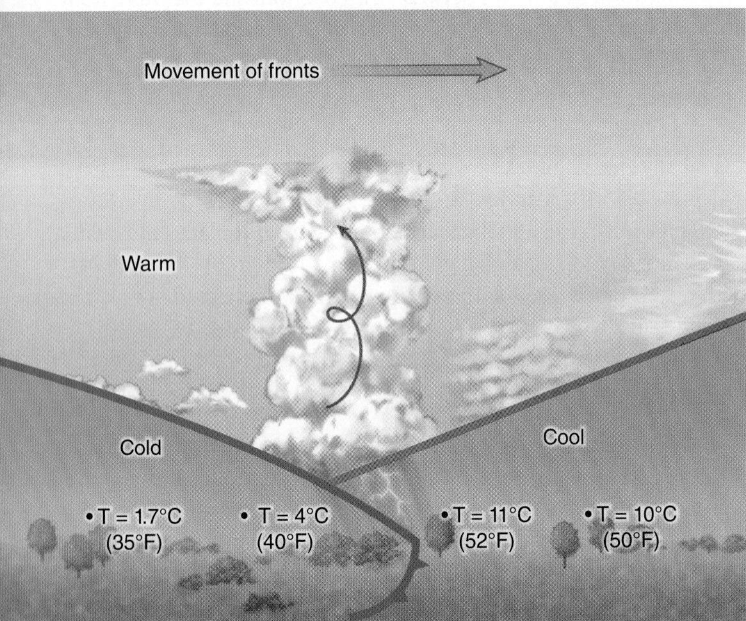

Figure 15.13 An occluded front. When a cold air mass overtakes a warmer air mass, an occluded front forms. Nimbostratus clouds generate precipitation along an occluded front (see section CD in Figure 15.10).

precipitation (Figure 15.13). The cold front will eventually catch up to the warm front. The cold front and warm front merge, mixing cool air that was in advance of the warm front with the cold air mass behind the cold front. Such a weather front is termed an *occluded front* and is represented by a combination of warm and cold front symbols on weather maps (see Figures 15.10, 15.12c). This mixing produces a more stable, layered atmosphere where cool air lies below warmer air (Figure 15.13). The occluded front places two bodies of colder air next to each other, with the warmer of the two masses forced up and over the other. Occluded fronts may be marked by nimbostratus clouds. This more stable configuration signals the start of the decay of the cyclone. The system decays further as it moves northeast toward progressively colder conditions or over the north Atlantic Ocean.

Checkpoint 15.12

basic · intermediate · advanced · superior

Venn Diagram: Cold and Warm Fronts

Use the Venn diagram provided here to compare and contrast the characteristics of warm fronts and cold fronts. Place the numbers corresponding to the listed features in the appropriate places on the diagram. Five items have been provided, identify at least five more.

1. Air becomes warmer after its passing.
2. Air becomes cooler after its passing.
3. Heavy rainfall upon passage.
4. Increasing humidity after passing.
5. Changing wind directions.

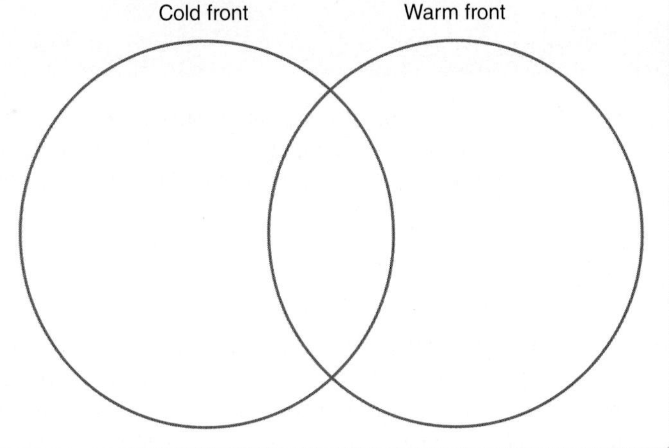

15.5 Severe Weather: Thunderstorms and Tornadoes

Learning Objectives

- Predict the processes associated with thunderstorms, and explain why they are more frequent in some locations than in others.
- Describe the processes of formation and decay of tornadoes.
- Predict locations favorable for tornado formation, and estimate related damage.
- Explain why tornado fatalities have fallen as the number of recorded tornadoes has increased.

The National Weather Service defines *severe weather* as having one or more of the following elements: a tornado, damaging wind speeds (more than 93 kilometers per hour; 58 miles per hour), or penny-sized or larger hail. These conditions occur in about 10 percent of the 100,000 thunderstorms that form over the United States each year. Tornadoes, funnel-shaped spirals of rapidly rotating air, form in association with thunderstorms. Both thunderstorms and tornadoes are short-term events that last just minutes or hours in comparison with other longer-lived weather phenomena such as air masses, midlatitude cyclones, and hurricanes.

Despite their short life span, thunderstorms and tornadoes can cause extensive damage and loss of life (Figure 15.2). Of course, this does not stop people from chasing these storms as if it were the latest extreme sport. In this section, we examine why and where thunderstorms occur and discuss the characteristics of tornadoes.

Thunderstorms

A thunderstorm can be a scary event, especially when accompanied by pounding rain, cracks of thunder and lightning, or the rattle of hailstones. One of my dogs, Molly, has astraphobia (fear of thunderstorms); well, technically, she has brontophobia (fear of thunder). If you do too, you may want to skip the rest of the paragraph. Think about your own experience with thunderstorms. When and where have you been in a thunderstorm? During which season? At what time of the day? Approximately how many thunderstorms do you think occur each year where you live?

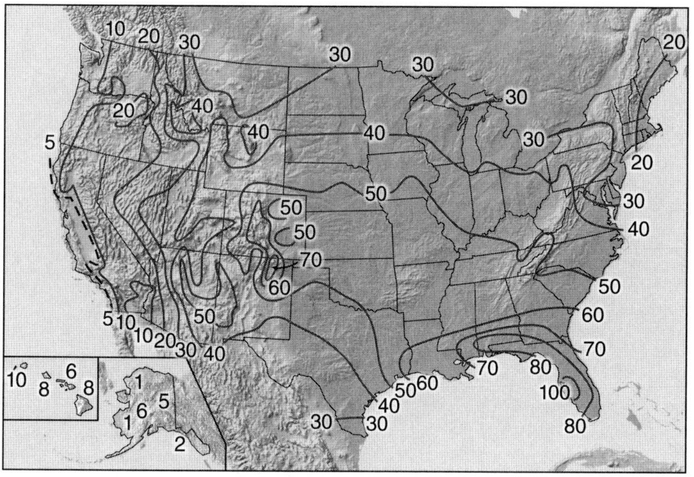

Figure 15.14 Annual US thunderstorm days. Which regions of the nation experience the most days with a thunderstorm? What are the physical characteristics of these regions? How do they relate to the movement of air masses?

Now look at Figure 15.14, a map of thunderstorm days in the United States. How does the map compare with your experience? Where do the most thunderstorms occur? Where should an astraphobe live to avoid having to deal with thunderstorms?

Over 18,000 thunderstorms occur on Earth at any moment. **The basic ingredients for thunderstorms are moisture in the atmosphere, warm air, and a lifting mechanism.** Thunderstorms typically form where warm, humid air is forced upward into the upper half of the troposphere. Condensation occurs as the rising air cools, releasing latent heat and ensuring that the air remains warmer than the surrounding air as it continues to rise. Most thunderstorms are around 25 kilometers (16 miles) in diameter and can form and dissipate within a couple of hours. Depending on the conditions that cause the air to rise, thunderstorms may occur as relatively isolated, short-lived events or as components of longer-lived severe storms.

Thunderstorm Formation. There are three stages during a typical thunderstorm: the cumulus stage, the mature stage, and the dissipating stage.

The cumulus stage is marked by early cloud development and occurs when a cumulus cloud expands horizontally and vertically as warm rising air cools adiabatically (because of decreasing air pressure). Cloud formation is rapid, often requiring just 15 minutes to reach heights of 10 kilometers (6 miles; Figure 15.15a). **Updrafts, rapid upward movement of a column of air within the cloud,** carry humid air to higher, colder levels, where condensation occurs. No precipitation occurs during the cumulus stage.

Checkpoint 15.13

basic | intermediate | advanced | superior

Which airport is most likely to experience flight delays due to thunderstorm activity? Explain your reasoning.

a) Atlanta's Hartsfield airport, Georgia
b) Salt Lake City International airport, Utah
c) Seattle-Tacoma airport, Washington

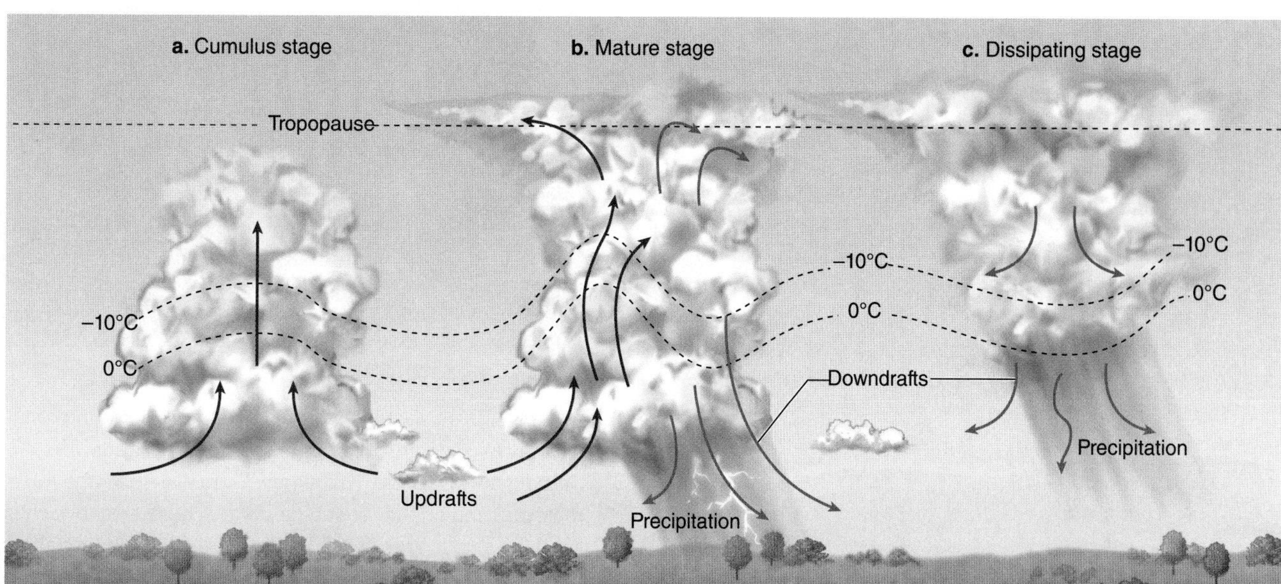

Figure 15.15 The life cycle of a thunderstorm. **a.** The cumulus stage is characterized by rapid vertical growth of the cloud and condensation as updrafts carry warm air aloft. **b.** Downdrafts and precipitation characterize the mature stage. **c.** Loss of moisture leads to the dissipation of the storm.

At the mature stage, the top of a cloud may reach altitudes of up to 15 kilometers (9 miles), the top of the troposphere (Figure 15.15b). The mature stage is more likely to occur when initial updrafts are fortified by additional lifting mechanisms, such as orographic or frontal lifting. Condensation in the uppermost reaches of the towering clouds forms ice crystals that coalesce to become too large to be supported by updrafts and thus fall to the ground. This falling precipitation generates friction within the rising air, creating a **column of downward-moving air called a *downdraft*** (see Figure 15.15b). Descending air warms up, often converting smaller ice particles to rain while larger particles fall as hail. This stage lasts for 15 to 30 minutes. The presence of both updrafts and downdrafts is characteristic of a mature thunderstorm. We may experience these up- or downdrafts as gusts of wind or as unsettling sudden changes in altitude while flying in airplanes. These up- and downdrafts also cause polarization of positive and negative charges between ice particles and droplets within the clouds. When the charges build sufficiently, lightning is discharged (Figure 15.16). Lightning can also occur in ash clouds from volcanic eruptions.

During the dissipating stage, cloud formation ends as the moisture is expended and descending air cools the cloud mass. Some of the descending precipitation during the mature stage undergoes evaporation that absorbs latent heat and cools the cloud. The cooler air provides stability, prevents updrafts needed to sustain the cloud, and signals the start of the dissipation process. This final stage is characterized by diminishing precipitation (light rain) as the cloud dissolves (see Figure 15.15c).

Where Thunderstorms Form. Where and when do you think most thunderstorms occur? Severe storms that bring strong winds, heavy rain, and lightning can occur almost anywhere in the United States, but their causes vary by region.

If you live in the central or eastern portion of the continental United States, you are probably familiar with isolated afternoon thunderstorms. These storms are commonplace in warm summer months when humid air rises. Isolated summer thunderstorms are typically brief storms associated with a single large cloud, termed a *cell.* The sun heats the land surface, raising the temperature to a maximum during the midafternoon. The air needs moisture to produce the conditions necessary to form a cloud. Air may become more humid as it moves over saturated soils or a lake but is most likely to gain moisture where maritime tropical air moves over land along the Gulf Coast states. The warm land surface heats the overlying air, causing it to become unstable enough to rise by density lifting (see Chapter 14), leading to condensation that builds a tall cumulonimbus cloud and generates a thunderstorm.

Thunderstorms associated with midlatitude cyclones are most common over the central and eastern United States. These severe storms are made up of a series of cumulonimbus clouds, or supercells. **These thunderstorms are associated with frontal lifting along the cold front between cP and mT air masses.** They are most common during spring and early summer, when the contrast in temperatures between air masses is greatest and generates the strongest updrafts needed to build cumulonimbus clouds.

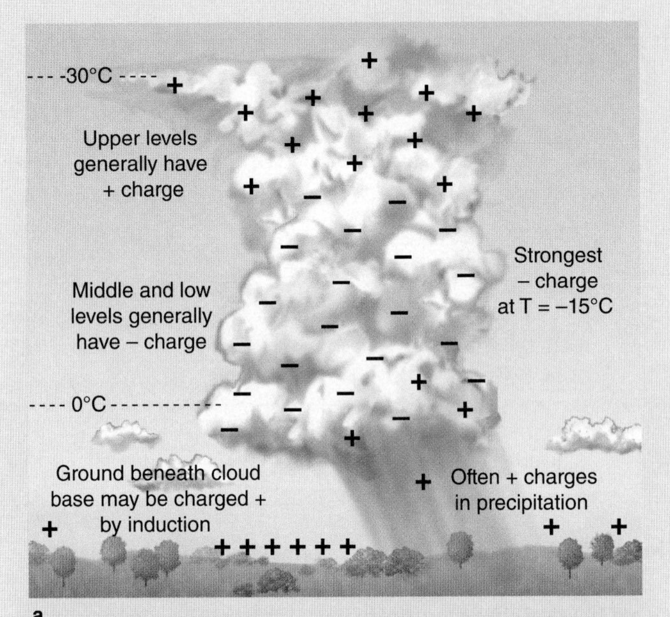

Figure 15.16 The mechanism of lightning. Lightning can join any two centers of opposite charge associated with a thunderstorm. **a.** Positive charges accumulate near the top of the cloud and on the ground below the cloud. Negative charges are present in the lower half of the cloud. **b.** Cloud-to-ground lightning occurs when negative charges in a cloud are connected to positive charges on the ground. Cloud-to-cloud lightning occurs when opposite charges in one or more clouds are connected.

Checkpoint 15.14

basic intermediate advanced superior

Updrafts responsible for the formation of thunderstorm clouds are most likely to occur with which combination of conditions?

a) Low-level warm, moist air; upper-level warm, moist air
b) Low-level cool, dry air; upper-level warm, moist air
c) Low-level warm, moist air; upper-level cool, dry air
d) Low-level cool, dry air; upper-level cool, dry air.

Checkpoint 15.15

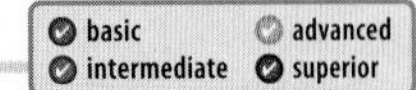

basic intermediate advanced superior

Why do parts of Florida have over 100 thunderstorms a year while parts of Texas at the same latitude have fewer than one-third as many storms?

Thunderstorms are also frequent over the Rocky Mountains, where they result from atmospheric **instability due to orographic lifting** processes that push air upward over the 4,000-meter (13,120-foot) peaks of the Colorado Rockies. Residents of Florida experience more thunderstorm days per year than people living elsewhere in the nation. The warm maritime air rises as a result of a combination of lifting mechanisms, including density lifting, frontal lifting, and the **convergence of winds** from both the Atlantic Ocean and the Gulf of Mexico. Thunderstorms are rare (fewer than 20 per year) in Pacific Coast states (see Figure 15.14), because the weather there is influenced by the cool ocean, which contributes to more-stable air.

Checkpoint 15.16

basic | intermediate | advanced | superior

Rank the three thunderstorm components (air temperature, moisture, a lifting mechanism) in order of their significance in causing thunderstorms. Justify your ranking.

Tornadoes

April 3, 1974, was a typical spring morning across the nation—but that was about to change. At 1 P.M., a tornado touched down in Morris, Illinois. Over the next several hours, tornadoes would be sighted from the Canadian border all the way to the Gulf Coast states, from Illinois in the west to Virginia in the east. A total of 148 tornadoes touched down in 13 states, killing 330 people and injuring over 5,000. This was the worst US tornado outbreak ever recorded. The heaviest destruction occurred when a massive single tornado tore through Xenia, Ohio. Two tractor trailers were tossed onto the roof of a bowling alley as the storm sped through town at 80 kilometers per hour (50 miles per hour). Thirty-three people were killed and 1,300 buildings destroyed, including five of the city's 10 schools. A local forecaster compared the scene to bombed-out buildings he witnessed during World War II.

Tornadoes are narrow, funnel-shaped spirals of rapidly converging and rotating air that form in association with thunderstorms (Figure 15.17). Like hurricanes and midlatitude cyclones, tornadoes are near-circular low-pressure systems that rotate in a

a.

b.

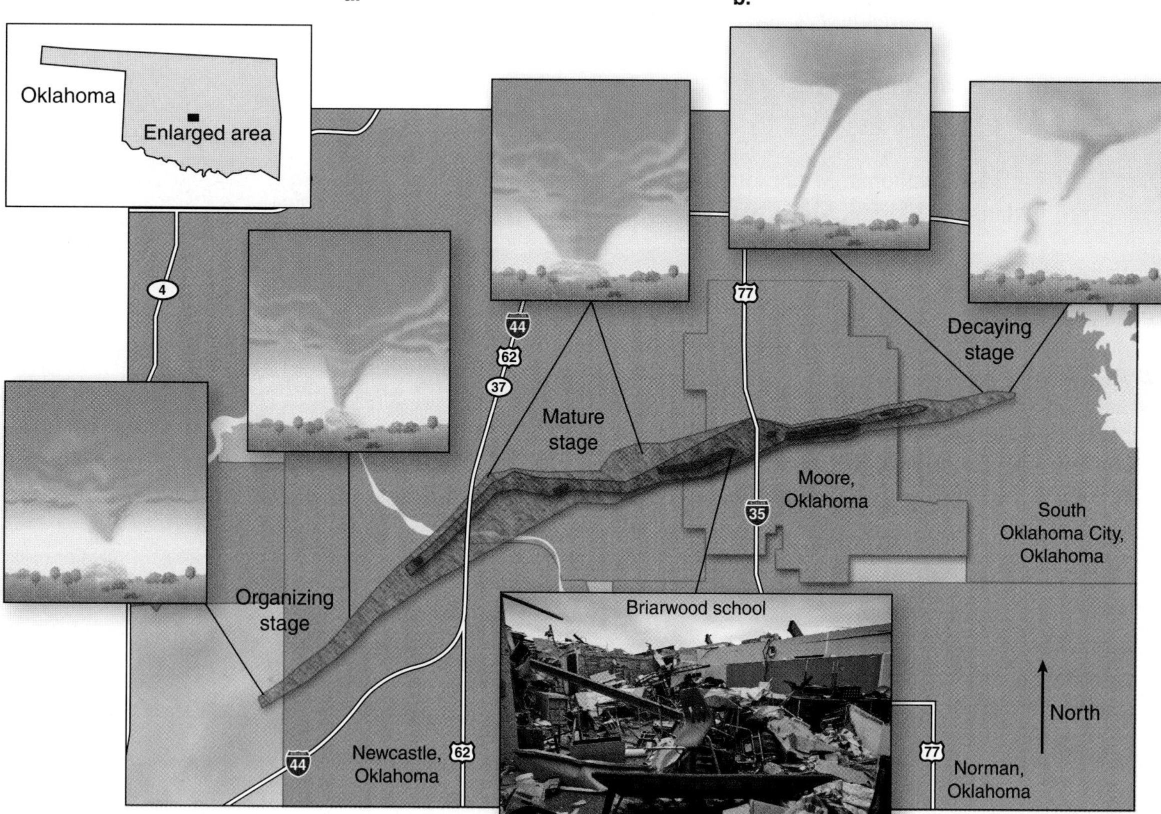

c.

Figure 15.17 Moore tornado, May 20, 2013. **a.** Early organizing phase. **b.** Later mature stage as the 2.1 kilometer wide (1.3 mile diameter) tornado moved through the southern part of Oklahoma City. **c.** Map of the path of the tornado illustrating changes in the strength of the tornado (dark colors indicate greater damage). Also shown are the likely locations of various stages of tornado development. The tornado lasted for 39 minutes and traveled 27 kilometer (17 mile) path, injuring hundreds and killing 23 people including 7 children at Briarwood Elementary school (inset photo).

counterclockwise direction in the Northern Hemisphere. However, the pressure gradient (change in pressure over distance) is much more intense for tornadoes. Recall from Chapter 14 that wind velocity is directly related to the pressure gradient. Pressure differences across midlatitude cyclones are in the range of 20 to 30 millibars (mbar) over hundreds of kilometers (pressure gradient, 0.04 millibar per kilometer). Hurricanes may experience pressure gradients of more than 100 millibars over similar distances (0.2 millibar per kilometer), but the same large pressure differences in tornadoes occur over distances measured in only hundreds of meters. Extreme pressure gradients of up to 100 millibars per kilometer are possible for tornadoes, generating the strongest natural winds on Earth.

Despite their destructive power, tornadoes have a mystique that causes otherwise sensible people to spend their vacations chasing storms down dirt roads from Texas to Iowa in the hopes of getting up close and personal with a twister. Storm chasing is the pursuit of a severe storm in hopes of observing one or more tornadoes. Storm chasing began about 50 years ago as an effort by individual meteorologists to learn more about the development of thunderstorms and tornadoes. Today, the announcement that a tornado is on the way is as likely to send people searching for their video camera as it is to get them heading for the basement. Companies such as Cyclone Tours or Twister Sisters will place grateful tourists in areas with the potential to spawn tornadoes. Unfortunately, sometimes so many vehicles are searching for tornadoes that they can occasionally slow down emergency response personnel from reaching storm victims. Professional meteorologists will occasionally report cheering tour groups piling out of minivans as a tornado rips open a farmhouse. Not surprisingly, the home owner's response is less enthusiastic.

Checkpoint 15.17

basic / intermediate / advanced / superior

Given what you have already learned in this chapter, predict which listing below shows the correct order for states with the greatest number of tornadoes in March, June, and August.

a) Florida, Oklahoma, Minnesota
b) Oklahoma, Florida, Minnesota
c) Minnesota, Oklahoma, Florida

The Enhanced Fujita Intensity Scale. Scientists studying tornadoes had a problem. How could they measure the wind speed of one of these storms? Obviously, the high winds capable of blasting apart whole buildings would make short work of their measuring instruments. Also, how could anyone get close enough to deploy the instruments? In 1971, Dr. Theodore Fujita came up with a way to measure the strength of tornadoes called the Fujita or "F" scale. The F scale (Table 15.2) was initially developed to distinguish between weak and strong tornadoes by using wind speeds estimated from damage caused by the tornadoes. The F scale went from F0 (winds causing light damage at 40–72 miles per hour) to F5 (winds causing incredible damage at 261–318 miles per hour). The wind estimates were based on analysis of aerial photographs, engineering considerations, and educated guesses by on-the-spot observers surveying damage caused by the tornado. As can happen in science, researchers found that the wind estimates by on-ground observers were not

Checkpoint 15.18

basic / intermediate / advanced / superior

Examine the image of tornado damage. Rate the damage, using the Fujita intensity scale. Explain your reasoning.

a) F1 b) F2 c) F3 d) F4

Table 15.2 Fujita (F) and Enhanced Fujita (EF) Intensity Scales for Tornado Damage

F (or EF) Number	F scale wind speed, km/h (mph)	EF 3-Second gust, km/h (mph)	Damage
F0	<116 (<72)	105–137 (65–85)	Light: breaks tree branches, damages signs
F1	116–180 (73–112)	138–177 (86–110)	Moderate: overturns mobile homes, peels tiles off roofs, pushes moving cars off road
F2	181–253 (113–157)	178–217 (111–135)	Considerable: tears roofs off some homes, uproots large trees, demolishes mobile homes; light objects become missiles
F3	254–332 (158–207)	218–265 (136–165)	Severe: tears roofs and walls off houses, overturns trains
F4	333–419 (208–260)	266–322 (166–200)	Devastating: demolishes houses, throws cars through the air; large objects become missiles
F5	>420 (>260)	>322 (>200)	Incredible: picks up houses, throws cars and other large missiles more than 100 meters (0.6 mile); steel-reinforced buildings badly damaged

very precise. Even so, the F scale was used for 33 years; though scientists recognized the system did not include enough damage indicators, it did not account for different types of construction, and there was only a weak correlation between damage and wind speed. So in 1997, the scale was refined to the Enhanced Fujita or "EF" scale.

The Enhanced Fujita intensity scale also places tornadoes in one of six categories (EF0 to EF5; Table 15.2) based on estimated wind speeds. However, the method of determining the wind speed was expanded to include 28 different damage indicators that could be assessed on various different types of buildings, structures, and trees along the path of the tornado. Furthermore, the observers were trained in using the damage indicators to ensure more-uniform reporting. The type and amount of destruction are still used as a proxy (substitute) for wind speed. The scale divides tornadoes into three subgroups based on the most likely maximum 3-second duration wind gust. As a rule of thumb, EF0–1 tornadoes can tear shingles off the roof of a house, EF2–3 tornadoes can tear off the whole roof, and EF4–5 tornadoes can tear up the whole house. More than two-thirds of all tornadoes are classified as EF0 or 1, and only 2 percent rank as EF4–5. However, two-thirds of tornado-related deaths occur as a result of these stronger tornadoes. On the basis of damage, EF5 tornadoes are estimated to generate wind speeds of up to 500 kilometers per hour (over 310 miles per hour). However, on the basis of the intensity of damage, it is difficult to rank winds that touch down in sparsely populated areas where few structures exist to sustain damage.

Venn Diagram: Fujita and Mercalli Scales

Use the Venn diagram provided here to compare and contrast the characteristics of the Fujita and Modified Mercalli (see Chapter 5) scales as measures of tornado and earthquake damage. Place the numbers corresponding to the listed features in the appropriate places on the diagram. Five items have been provided; identify at least five more.

1. Related to damage
2. Wind speed is a factor
3. Relies on observations
4. Has five levels
5. One level describes nearly complete destruction

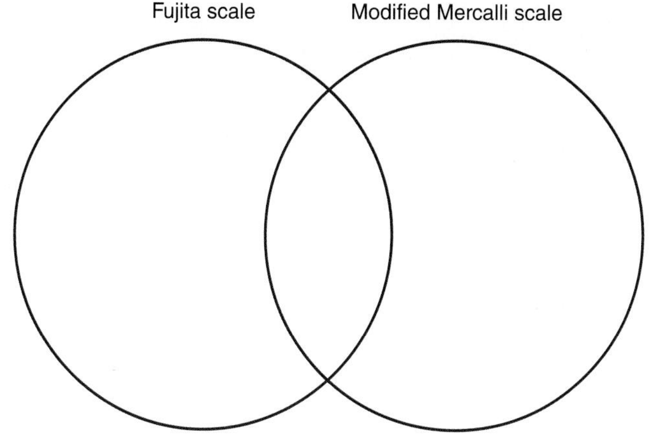

While the winds inside the funnel of a tornado move at very high speeds, the funnel itself moves much more slowly. Funnels are typically less than 600 meters (2,000 feet) wide, and average funnel velocities are 50 kilometers per hour (31 miles per hour), although velocities as high as 200 kilometers per hour (124 miles per hour) have been recorded. Most tornadoes carve out a path of destruction that ranges from 5 to 25 kilometers (3 to 16 miles) in length (see Figure 15.17c), but some more destructive tornadoes may remain on the ground for more than an hour and travel over 100 kilometers (62 miles). Tornado paths follow the direction of movement of their parent thunderstorms, which in turn are often associated with midlatitude cyclones that move toward the east or northeast. Consequently, most tornadoes travel toward the east or northeast. This relatively consistent travel direction enables weather forecasters to issue accurate tornado warnings for affected residents. Careful tornado chasers can use this information to make sure they stay upwind of a funnel cloud. However, local changes in the characteristics of a storm or the underlying topography can cause tornadoes to unexpectedly diverge from these "normal" paths (see Figure 15.17c).

Tornado Development. One of the first organized teams of researchers to investigate tornadoes was working on the Tornado Intercept Project in 1973 when they were able to get video footage of the Union City tornado in Oklahoma. This was the first time that a combination of on-site video, Doppler radar records, and observational data was used to track the complete life cycle of a tornado.

Although meteorologists are able to observe tornadoes once they form, they are unable to go inside thunderclouds to identify the precursor stages. However, they have been able to identify the conditions under which tornadoes occur, and this has led to a hypothesis that breaks down tornado formation into three stages (Figure 15.18):

- ***Early stage.*** Friction slows winds at the ground surface, resulting in higher wind velocity moving upward from the surface. These contrasting vertical wind speeds generate winds that tend to rotate counterclockwise about a central, near-horizontal axis, producing a rolling spiral of wind.

Figure 15.18 Stages in the development of a tornado.

- ***Updraft stage.*** Updrafts below a thunderstorm draw the spiraling horizontal winds upward to form a small cyclone, known as a mesocyclone, within the larger storm cloud. Mesocyclones are essentially rotating thunderstorms that can be observed on radar and may be up to 10 kilometers (6 miles) in diameter.
- ***Tornado stage.*** Rotation within the mesocyclone forms smaller, more intense spiraling winds within a newly formed tornado, which then extend downward from a cloud base toward the ground surface (Figure 15.18). This is the stage seen by storm chasers. Current hypotheses suggest that a tornado forms in response to temperature differences across the downdraft of air, though this has not been confirmed experimentally.

Sometimes smaller funnels appear to skip across the surface, destroying one home while leaving neighboring properties undisturbed. In reality, scientists hypothesize that tornadoes do not skip—that is, form, dissipate, and then re-form. Instead, they have used film of actual tornadoes to identify small, intense minitornadoes embedded in the main funnel. It is these minitornadoes that hit or miss homes.

Improvements in forecasting methods have reduced the number of fatalities associated with tornadoes (Figure 15.19). Nearly one-half of US fatalities occur as a result of tornadoes destroying nonpermanent mobile homes, which are especially susceptible to higher winds as they have little structural strength.

Because cP and mT air masses frequently interact along frontal systems across the central portion of North America, the United States is home to the majority of the world's tornadoes, averaging about 1,200 a year. **Tornadoes occur when thunderstorm activity is at an optimum across much of the nation.** The timing of tornado activity is tied to seasonal movement of the polar front, the boundary between warm, humid air from the Gulf of Mexico and cold, dry air from Canada. Tornadoes are more common in Gulf Coast and southeastern states in early spring. Then tornado activity moves into the Great Plains as the polar front retreats northward in late spring. Tornado activity shifts to the north in summer when these air masses interact in the northern Great Plains and upper Midwest. The highest frequency of tornadoes per area occurs over the Great Plains states (Texas, Oklahoma, Kansas, Nebraska) and parts of the upper Midwest (Iowa, Indiana, Illinois), a region that has come to be known as "tornado alley" (Figure 15.20).

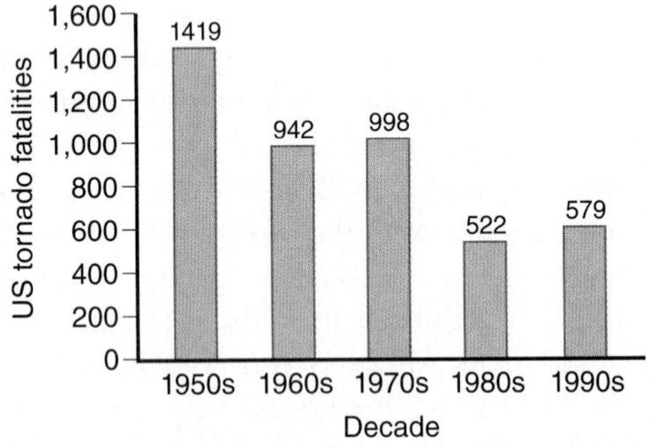

Figure 15.19 US tornado fatalities have declined as forecasting technology has improved to provide better tornado warnings.

Checkpoint 15.20

Explain why the number of tornadoes counted each year has increased, while the number of days with at least one tornado sighting has remained essentially unchanged for several decades.

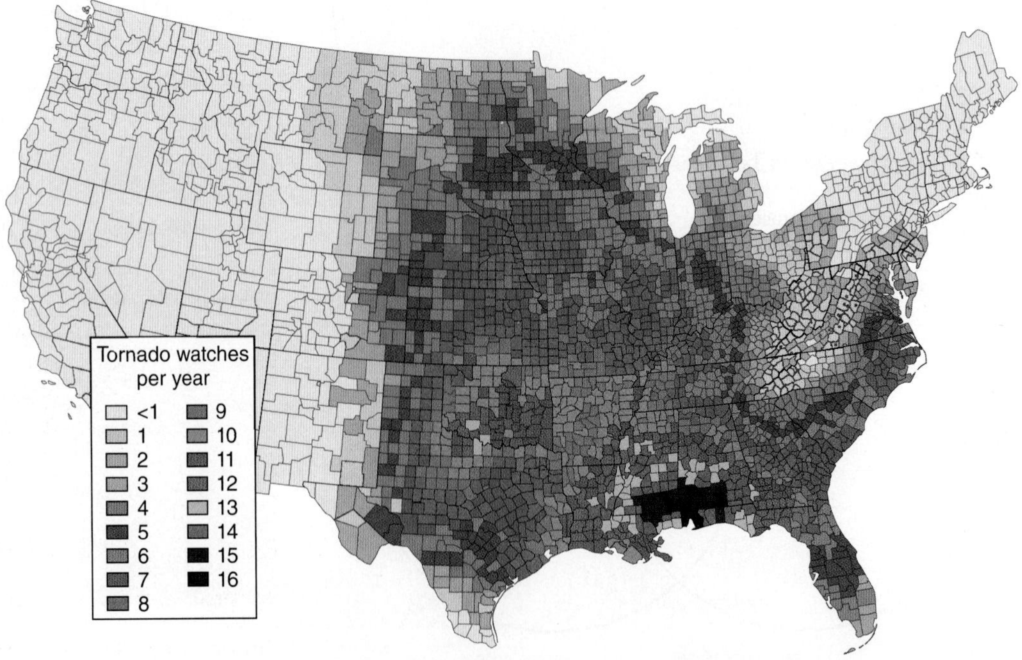

Figure 15.20 Tornado alley. The locations of states with an annual average of more than five strong to violent (F2 to F5) tornadoes, 1950–1995. While Texas experiences more tornadoes, several other states experience more tornadoes per 10,000 square kilometers (for example, Oklahoma, Kansas, Iowa, Louisiana, and Mississippi).

15.6 Severe Weather: Hurricanes

Learning Objectives

- Describe the life cycle of a hurricane.
- Relate different earth system processes to the formation and decay of hurricanes.
- Compare and contrast hurricanes with other types of low-pressure systems.
- Describe factors used to rank the intensity of a hurricane.
- Evaluate the risk to humans for hurricane scenarios.

Philosopher George Santayana famously remarked, "Those who cannot learn from history are doomed to repeat it." Many scientists are concerned that we will see a repetition of the destruction that the Gulf Coast experienced from Hurricane Katrina, but perhaps not where we most expect. Records of hurricanes striking the Atlantic and Gulf coasts of North America stretch back over 150 years. This time span is small in geologic terms, but it is long enough to reveal some important patterns. Hurricanes have affected every location along the coasts from Maine to the southern tip of Texas.

Analyses of past hurricanes indicate that recent population growth will result in increased loss of life and property when similar storms inevitably occur again in the same locations. Lightning *does* strike the same place twice, and so do hurricanes! This was brought home sharply when Hurricane Katrina barreled into the Gulf Coast on August 29, 2005, submerging most of New Orleans, Louisiana, and blowing away large portions of Biloxi, Mississippi, and other coastal cities (Figure 15.21). More than 1,300 people

Figure 15.21 Before and after Hurricane Katrina. **a.** NASA's Terra satellite snapped this view of New Orleans, sandwiched between Lake Pontchartrain and the Mississippi River, 2 days prior to the arrival of Hurricane Katrina. **b.** A day after the hurricane, widespread flooding (blue) was apparent. **c.** Biloxi, Mississippi, before Hurricane Katrina. Note the location of the large mansion, pier, and pier house. **d.** After Katrina. The hurricane destroyed these structures.

a.

b.

c.

d.

died, and the storm caused an estimated $100 billion in damage in communities in Louisiana, Mississippi, and Alabama. **Hurricane Katrina was the worst natural disaster to strike the modern United States.** However, it should not have come as a surprise. Much of the damage from Hurricane Katrina had long been predicted. A dozen major hurricanes had made landfall along the Louisiana coast in the last 100 years; four of those storms resulted in major flooding in New Orleans. It was just a matter of time before the region suffered another hit. Well before Hurricane Katrina struck, government reports, articles in science magazines, and a week-long series of stories in the New Orleans *Times Picayune* newspaper had all predicted in eerie detail what would happen if a major hurricane made landfall near modern New Orleans.

It is one thing to know what might happen; it is another to be able and willing to do something about it. **Measures to protect populations and structures must account for cultural and social factors and are the responsibility of local, regional, and federal governments.** This brings up an important point regarding science and society. Who is best equipped to address the complexity of such hazards? Although scientists and engineers could predict the probability of a disastrous hurricane, they did not have access to funds and resources to do much about it. On the other hand, government officials, who had the power to allocate funds for levee improvements or evacuation plans, may not have known of or understood the seriousness of these predictions. Alternatively, they may have recognized the risks but were unwilling to cut other programs to take measures needed to mitigate these hazards. These contrasting roles for scientists and government officials set up an interesting dialogue. One group seeks to be clear about the potential dangers without being labeled alarmist, while the other tries to protect residents without incurring economic losses. As we discuss the general features of all hurricanes, we will describe the characteristics of Hurricane Katrina and other storms. Ultimately we may conclude, as George Bernard Shaw observed, "We learn from history that we learn nothing from history."

Building a Hurricane

Before we go any further, take a few minutes to examine the image of Hurricane Katrina in Figure 15.22. What do you observe about this hurricane? Keep looking . . . make at least three separate observations.

OK. You probably noticed the large size of the storm; it was large enough to cover several states at once. You may have observed that the hurricane is composed of a dense mass of clouds surrounding a central clear "eye." Also, you probably noted the swirling, spiral pattern in the clouds. Finally, you may have identified that the hurricane was centered over the ocean—in this case, the Gulf of Mexico. We will explain the development of these features next.

Checkpoint 15.21

The costliest hurricanes are not necessarily the most intense. Explain why two category 1 hurricanes (Agnes, 1972; Diane, 1955) are among are the top 10 most costly US hurricanes when adjusted for inflation.

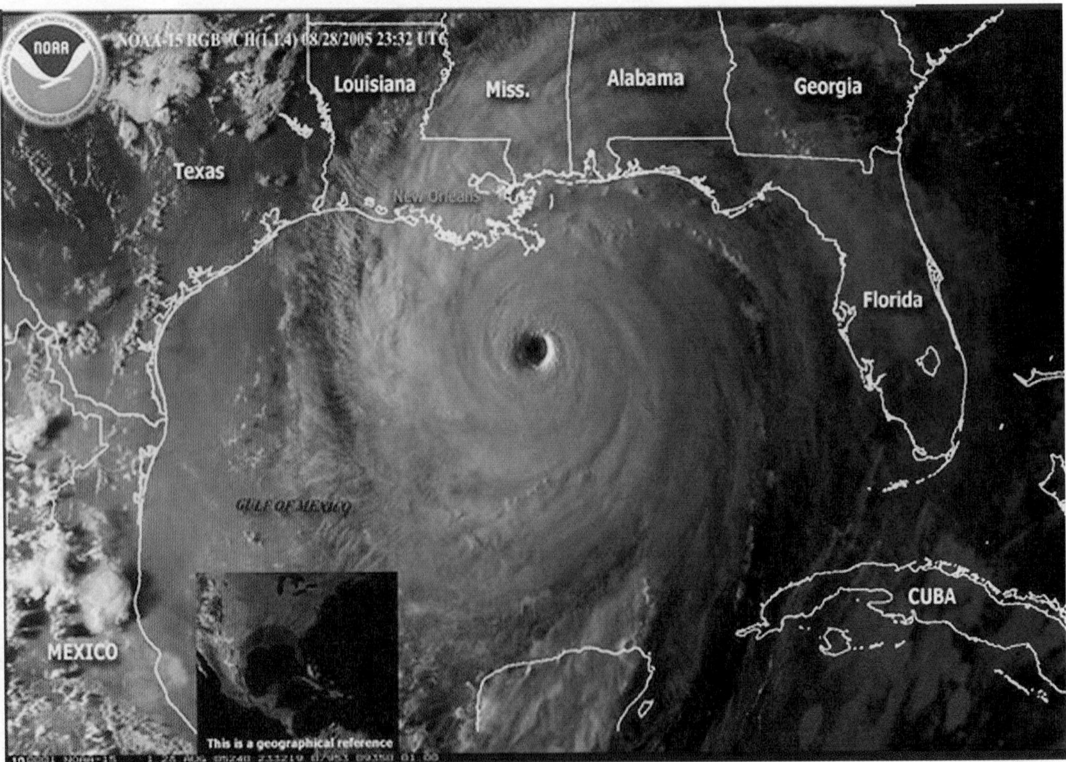

Figure 15.22 Hurricane Katrina approaching the Louisiana coast, August 28, 2005.

Hurricanes are characterized by high winds, heavy rainfall, and storm surges (elevated water levels) along coastlines. Hurricanes can form under appropriate conditions all over the world but are named differently in different oceans; for example, they are called typhoons in the Pacific Ocean and cyclones in the Indian Ocean. Some hurricanes live their short, dramatic lives exclusively over the oceans, noticed only by meteorologists and a few anxious sailors. It is some of the others, those that make landfall, that grab the attention of millions of coastal residents in the United States during the North Atlantic hurricane season from June 1 to November 30. Let us look at how these storms develop, with special reference to Hurricane Katrina.

Two to Three Weeks Before Landfall. Hurricane development requires warm surface waters and a specific set of conditions in the troposphere. **Surface water temperatures must be at least 27°C (81°F) down to a depth of about 60 meters (200 feet).** Only over warm ocean waters can sufficient evaporation and condensation occur to provide the huge volumes of moisture necessary to form the large masses of clouds present in a hurricane. A deep layer of warm surface water prevents cold water from being drawn to the surface, which would slow the evaporation needed to fuel the developing storm. As we learned in Chapter 13, warm surface waters straddle the equator but move farther north during summer in the Northern Hemisphere, creating optimal conditions for hurricanes with the potential to strike North America (Figure 15.23). Most US hurricanes make landfall from mid-August through September. In the 10 years from 1995 through 2004, the North Atlantic region experienced 20 hurricanes during August months. These storms originated all across the North Atlantic, from just off the east coast of Africa to the west coast of Mexico (Chapter Snapshot). Some of these storms made landfall, while others spiraled harmlessly into the North Atlantic Ocean. The amount of time between formation and landfall varies, depending on the birthplace of the storm. Most storms originating in the eastern Atlantic Ocean require about 3 weeks to reach the coast of North America. Hurricane Katrina developed in the western Atlantic Ocean, near the Bahamas, in a similar location to that of Hurricane Dennis in 1999.

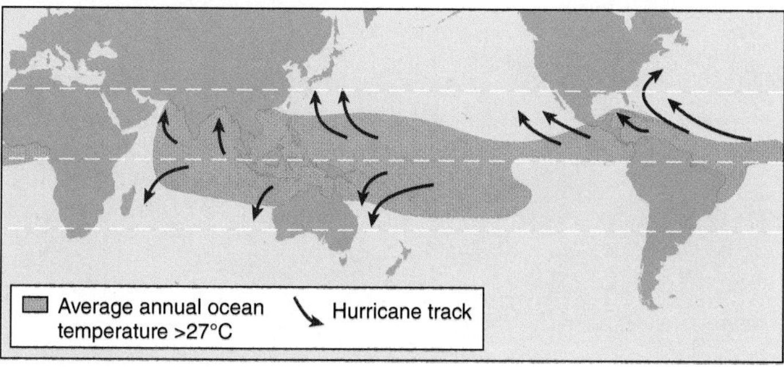

Figure 15.23 Hurricanes originate in areas of the world's oceans where water temperatures are greater than 27°C (81°C). Hurricanes in the Northern Hemisphere are most common during summer and early fall. Southern Hemisphere hurricanes are frequent during our winter (their summer). How do you explain the curved track of these storms?

A hurricane begins to grow when warm, humid air is forced aloft. The rising air cools and condenses to form cumulus clouds that will develop into cumulonimbus cells if the air is sufficiently warm and humid. The inflow of air into the developing low-pressure system must be matched with an outflow of air in the upper troposphere to maintain the pressure gradient in the developing hurricane. If not, the pressure contrast decreases and wind speed declines. A developing storm can be dispersed by fast-blowing winds at high altitudes or by a particularly dry layer of air in the upper atmosphere that impedes condensation. The number of storms decreases during El Niño years (see Chapter 13) and when dust storms blow west out of the Sahara desert in North Africa. El Niño alters the path of the jet stream to produce strong wind shears that destroy growing storms. In contrast, the dust storms block sunlight, decreasing ocean temperatures and reducing the heat supply necessary for storm formation.

Earth's rotation, as exhibited by the Coriolis effect (sideways deflection; see Chapters 13, 14), imparts a clockwise rotation to the growing storm in the Southern Hemisphere and a counterclockwise rotation in the Northern Hemisphere. The magnitude of the Coriolis effect is zero at the equator and increases with increasing latitude. Because there is not sufficient energy in rotation at the equator to produce hurricanes, **the majority of hurricanes originate between 10° and 20°N or S of the equator** (see Figure 15.23).

One to Two Weeks Before Landfall. If the appropriate conditions are met, a tropical depression (wind speed 37–62 kilometers per hour; 23–39 miles per hour) develops from some previously unorganized thunderstorms. If conditions remain favorable, after about 5 days, the tropical depression grows into a tropical storm (wind speed 63–118 kilometers per hour, 39–74 miles per hour) before developing into a hurricane, with wind speeds of at least 119 kilometers per hour (75 miles per hour), a couple of days later. **Air pressure is lowest inside the central eye of the hurricane, where warm air is rising.** Wind speed is related to pressure changes; the lower the pressure inside the hurricane, the higher the wind velocity.

Bands of clouds spiral outward from the eye (Chapter Snapshot), which is surrounded by thick clouds known as the *eye wall.* Beyond the eye of the storm, rising air in the deepening low-pressure system cools and condenses, releasing latent heat. This release of heat helps generate a growing spiral of dense cumulonimbus clouds rotating about the eye. The hurricane continues to grow in size and intensity as long as the underlying water temperature remains above 27°C (81°F). Precipitation is concentrated within a radius of approximately 100 to 200 kilometers (62 to 124 miles) on either side of the eye, releasing up to 20 billion tons of water per day.

Three to Seven Days Before Landfall. Atlantic hurricanes, driven westward by prevailing winds at rates of 10 to 25 kilometers per hour (6 to 16 miles per hour), may turn parallel to the US East Coast or pass south of Florida to strike along the Gulf Coast or the Caribbean islands (Figure 15.23, Chapter Snapshot). Florida and Texas experience more hurricane landfalls than any

HURRICANE ANATOMY

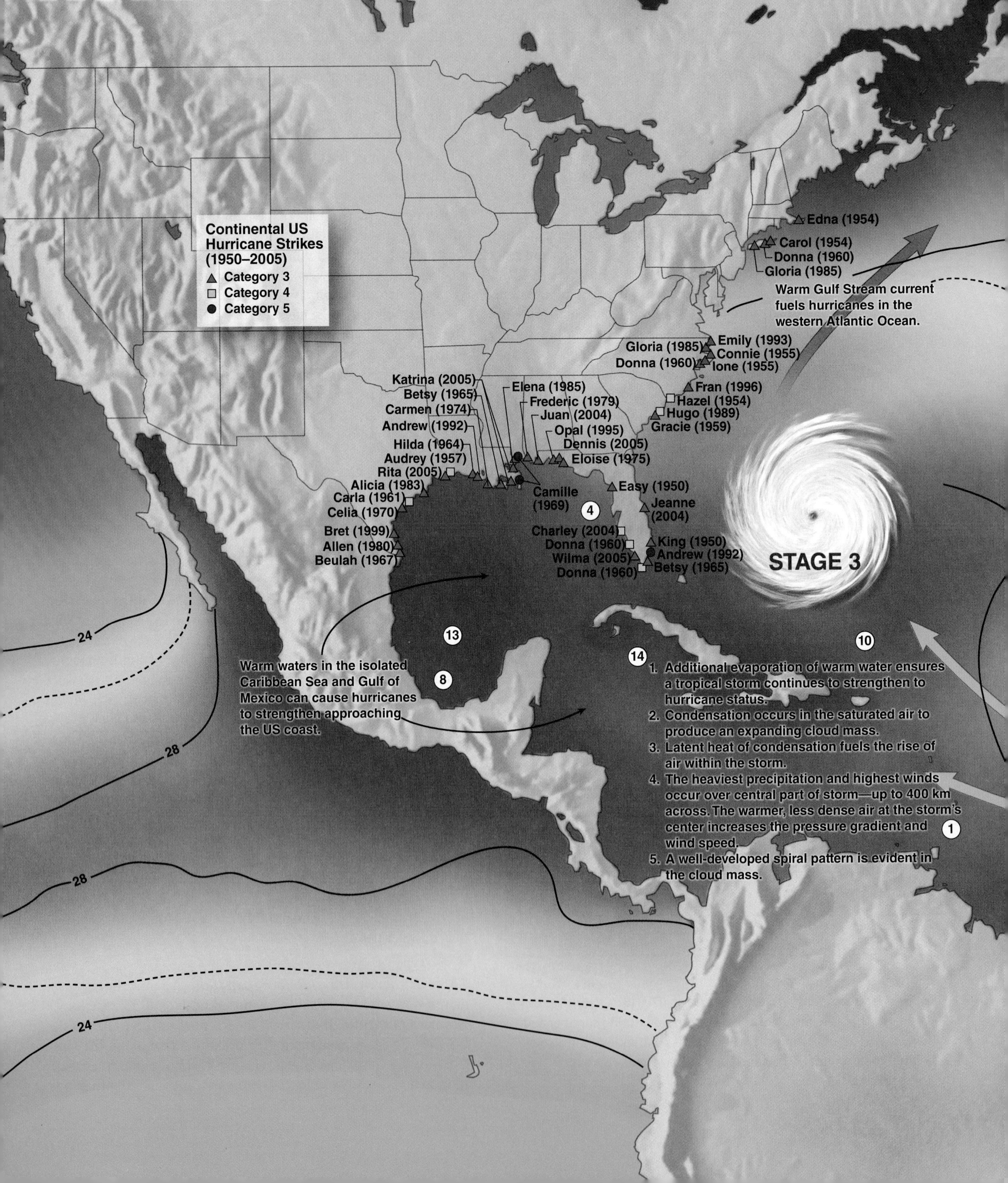

Continental US Hurricane Strikes (1950–2005)
▲ Category 3
□ Category 4
● Category 5

Edna (1954)
Carol (1954)
Donna (1960)
Gloria (1985)

Warm Gulf Stream current fuels hurricanes in the western Atlantic Ocean.

Emily (1993)
Gloria (1985)
Connie (1955)
Donna (1960)
Ione (1955)
Fran (1996)
Hazel (1954)
Hugo (1989)
Gracie (1959)

Katrina (2005)
Betsy (1965)
Carmen (1974)
Andrew (1992)
Hilda (1964)
Audrey (1957)
Rita (2005)
Alicia (1983)
Carla (1961)
Celia (1970)
Bret (1999)
Allen (1980)
Beulah (1967)

Elena (1985)
Frederic (1979)
Juan (2004)
Opal (1995)
Dennis (2005)
Eloise (1975)
Camille (1969)
Easy (1950)
Jeanne (2004)
Charley (2004)
Donna (1960)
Wilma (2005)
Donna (1960)
King (1950)
Andrew (1992)
Betsy (1965)

4

STAGE 3

13

8

14

10

1

24

28

28

24

Warm waters in the isolated Caribbean Sea and Gulf of Mexico can cause hurricanes to strengthen approaching the US coast.

1. Additional evaporation of warm water ensures a tropical storm continues to strengthen to hurricane status.
2. Condensation occurs in the saturated air to produce an expanding cloud mass.
3. Latent heat of condensation fuels the rise of air within the storm.
4. The heaviest precipitation and highest winds occur over central part of storm—up to 400 km across. The warmer, less dense air at the storm's center increases the pressure gradient and wind speed.
5. A well-developed spiral pattern is evident in the cloud mass.

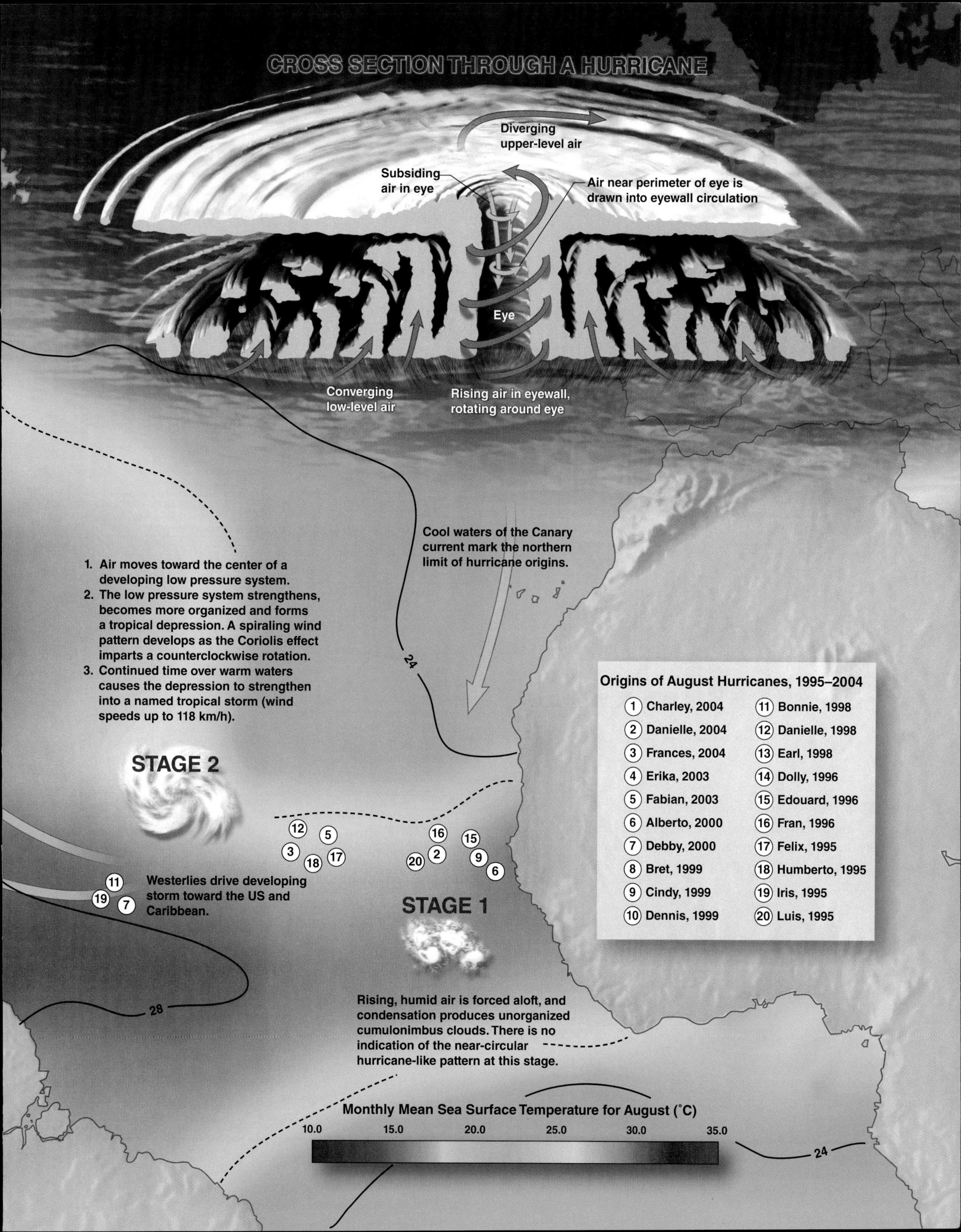

CROSS SECTION THROUGH A HURRICANE

Diverging upper-level air

Subsiding air in eye

Air near perimeter of eye is drawn into eyewall circulation

Eye

Converging low-level air

Rising air in eyewall, rotating around eye

Cool waters of the Canary current mark the northern limit of hurricane origins.

1. Air moves toward the center of a developing low pressure system.
2. The low pressure system strengthens, becomes more organized and forms a tropical depression. A spiraling wind pattern develops as the Coriolis effect imparts a counterclockwise rotation.
3. Continued time over warm waters causes the depression to strengthen into a named tropical storm (wind speeds up to 118 km/h).

24

STAGE 2

12 5 3 18 17

16 15 20 2 9 6

11 19 7

Westerlies drive developing storm toward the US and Caribbean.

STAGE 1

Rising, humid air is forced aloft, and condensation produces unorganized cumulonimbus clouds. There is no indication of the near-circular hurricane-like pattern at this stage.

28

Origins of August Hurricanes, 1995–2004

1. Charley, 2004
2. Danielle, 2004
3. Frances, 2004
4. Erika, 2003
5. Fabian, 2003
6. Alberto, 2000
7. Debby, 2000
8. Bret, 1999
9. Cindy, 1999
10. Dennis, 1999
11. Bonnie, 1998
12. Danielle, 1998
13. Earl, 1998
14. Dolly, 1996
15. Edouard, 1996
16. Fran, 1996
17. Felix, 1995
18. Humberto, 1995
19. Iris, 1995
20. Luis, 1995

Monthly Mean Sea Surface Temperature for August (°C)

10.0 15.0 20.0 25.0 30.0 35.0

24

other states (Figure 15.24). A week before the storm might make landfall, storm forecasters use mathematical models to provide advance warning to states that may be potential landfall sites. The hurricane will have grown in strength over time, allowing meteorologists to use pressure differences to predict the wind speeds and potential damage at landfall, using the Saffir-Simpson scale (Table 15.3).

Two to Three Days Before Landfall. The likely landfall site is usually identified 2 or 3 days before the storm reaches the coast, while the center of the storm is still 500 to 1,000 kilometers (310 to 620 miles) away. Forecasters predicted that Hurricane Katrina would make landfall in southeast Louisiana more than 2 days before the center of the storm came on land (Figure 15.25). To have any hope of clearing out major cities, evacuations should be in full gear by this time. An estimated 1.2 million people living along the Gulf Coast were under evacuation orders prior to the arrival of Hurricane Katrina. With such large numbers of people involved, it can take days to evacuate large cities. Even with several days' advance notice prior to Hurricane Rita in September 2005, highways north of Houston, Texas, became so clogged with traffic that people ran out of gas while their vehicles moved just a few miles per hour.

One Day Before Landfall. The size and relatively slow motion of these giant storms mean that their impact is drawn out over several days. Hurricanes are so large and slow-moving that the first effects of the storm reach the coastline many hours before the strongest winds around the central eye make landfall. The outer edge of a 500-kilometer-wide (310-mile) storm will bring rain, high winds, and large waves to the coast up to a full day ahead of the center of the storm. By that time, forecasters will have collected enough data from buoys in the ocean and from reconnaissance flights to know

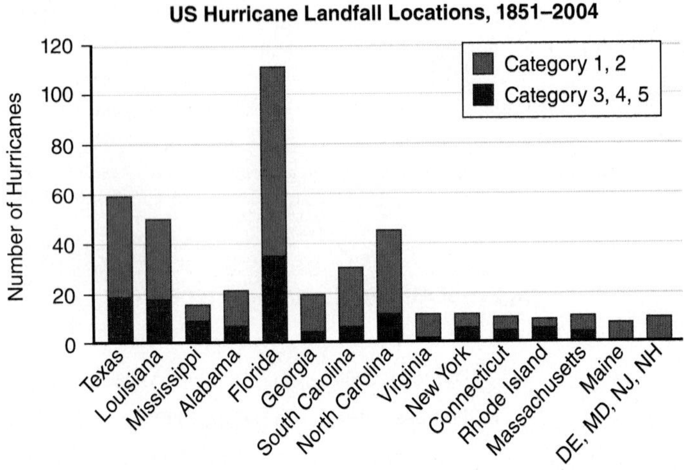

Figure 15.24 US hurricane landfall locations, 1851–2004. Category 1 and 2 hurricanes have made landfall more frequently than more-violent storms. Hurricanes are more likely to make landfall in states with long coastlines along the Gulf of Mexico and Atlantic coast.

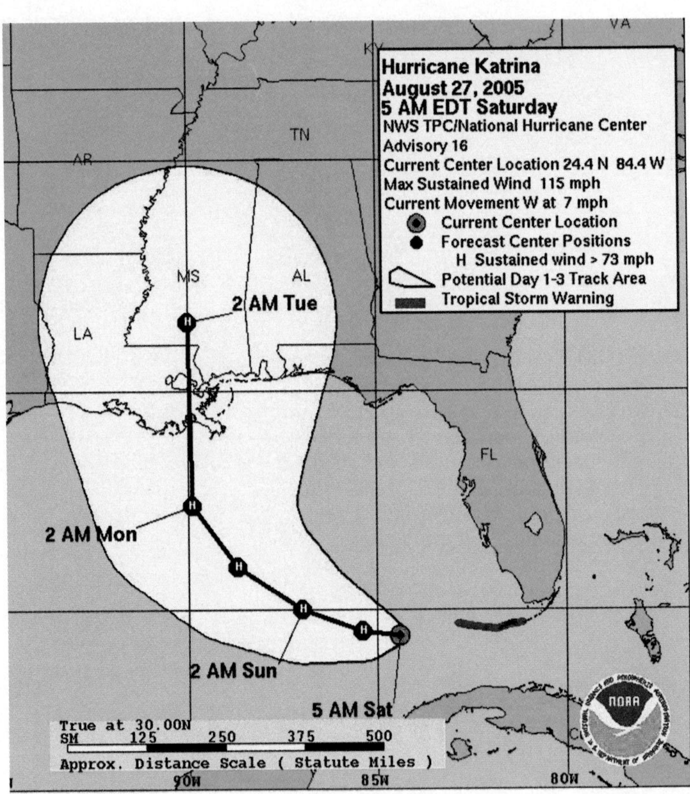

Figure 15.25 Predicted landfall site for Hurricane Katrina. This prediction was made at 5 A.M. on Saturday, 48 hours before the hurricane made landfall just east of the predicted location.

Table 15.3 Saffir-Simpson Hurricane Intensity Scale

Category	Wind speed, km/h (mph)	Pressure, millibars	Storm surge, meters (feet)	Description	Examples
1	119–154 (74–95)	>980	1.2–1.5 (4–5)	Minimal	Stan (2005)
2	155–178 (96–110)	965–980	1.6–2.4 (6–8)	Moderate	Sandy (2012)
3	179–210 (111–130)	945–964	2.5–3.6 (9–12)	Extensive	Karl (2010)
4	211–250 (131–155)	920–944	3.7–5.4 (13–18)	Extreme	Opal (1995)
5	>250 (>155)	<920	>5.4 (>18)	Catastrophic	Mitch (1998)

the strength of the storm and the height of storm-generated waves. Waves up to 30 meters high (100 feet) were present in the Gulf of Mexico during Hurricane Katrina. However, some storms can undergo dramatic strengthening or weakening immediately prior to landfall. People on the west coast of Florida went to sleep before the arrival of Hurricane Opal in 1995, expecting to wake up to a weak hurricane they could ride out. By the next morning, Opal had strengthened to a major storm, and it was too late for most people to evacuate.

The impact of a north-moving storm on Gulf Coast locations varies, depending on whether a location is in the northeast or northwest quadrant of the storm. **As hurricanes approach landfall, winds in the northeast quadrant of the storm are blowing onshore, piling up water in a storm surge.** The rising waters associated with storm surges account for approximately 90 percent of the damage in coastal areas and most deaths. Depending on the strength of the storm, a storm surge can increase water height by more than 5 meters (16 feet). Add another 1 or 2 meters for high tide and an additional 5 to 10 meters for waves, and a storm surge from a major hurricane such as Katrina could easily reach low-lying areas 10 to 20 kilometers (6 to 12 miles) inland along coastlines in Texas and Louisiana. These storm surges cause flooding and severe beach erosion (sometimes completely removing beaches), often dumping sand many kilometers inland. The height of the storm surge from Hurricane Katrina was difficult to determine in many locations because of the widespread failure of tide gauges and the destruction of buildings that could otherwise have revealed high-water marks. However, data indicate that water reached heights of 9 meters (30 feet) above sea level in the northeast quadrant of the storm.

Winds in the northwest quadrant of the storm were blowing offshore, opposite to the direction of the storm's movement. Maximum wind speeds may differ by as much as 50 kilometers per hour (31 miles per hour) on different sides of the storm (Figure 15.26). Offshore winds reduce the local impact of a storm surge. In the case of Katrina, New Orleans was located in the northwest segment of the storm. But New Orleans is surrounded by water on three sides, and the offshore winds were ideally suited to push the water from Lake Pontchartrain (see Figure 15.21a, b) over levees into the city to cause widespread flooding.

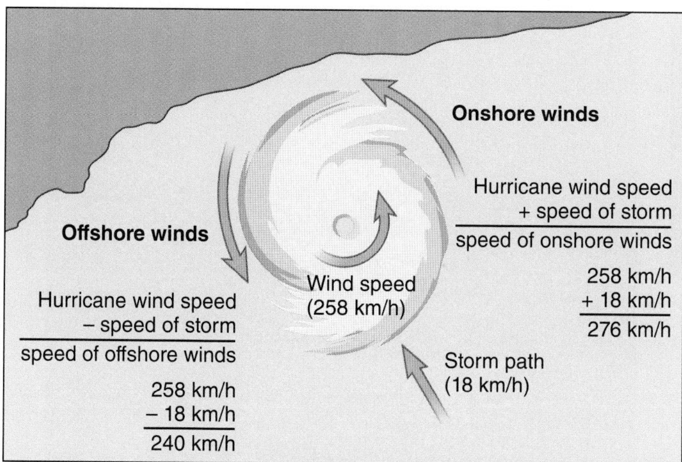

Figure 15.26 Offshore and onshore wind effects. Wind speeds may differ by as much as 50 kilometers per hour (31 miles per hour) in different quadrants of the storm.

Landfall. The more land that lies between people and the storm, the lower the impact of the hurricane's high winds and storm surge. A hurricane begins to weaken when it passes over land because it experiences greater frictional drag and a dramatic decrease in the water supply that is essential for its sustenance.

Checkpoint 15.22

Draw a diagram that illustrates how the four components of the earth system (atmosphere, biosphere, hydrosphere, and geosphere) interact during a hurricane.

Checkpoint 15.23

Venn Diagram: Low-Pressure Systems

Use the Venn diagram provided here to compare and contrast midlatitude cyclones, tornadoes, and hurricanes. Place the numbers corresponding to the listed features in the appropriate places on the diagram. Eight items have been provided; identify at least five more.

1. Low-pressure systems
2. Winds can be in excess of 480 kilometers per hour (>300 miles per hour)
3. Comma-shaped cloud pattern
4. Six intensity levels
5. Involve counterclockwise rotations
6. Two different frontal systems involved
7. Steep pressure gradients
8. Most common in late summer and fall

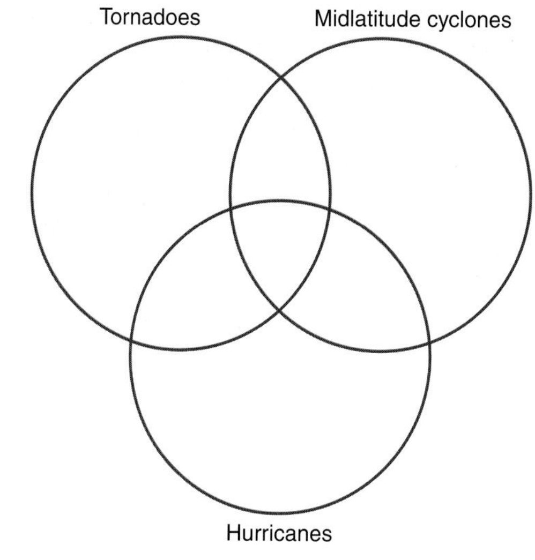

Damaging winds near the core of a hurricane have speeds similar to those of F1 to F3 tornadoes (Tables 15.2 and 15.3) and cause similar destruction—debris blown into the windows of high-rise buildings, roofs torn off homes, large trees uprooted, mobile homes demolished, vehicles overturned, and power lines toppled. Wind speeds are reduced to the level of a tropical storm or depression after landfall, but the storm itself is still capable of dumping large volumes of rain far inland.

Exceptional precipitation can unload 60 centimeters (24 inches) of rain from a single storm system over inland regions in just a few days. For example, heavy rains from Hurricane Mitch in 1998, the most devastating storm to strike North America and Central America in the last two centuries, claimed over 11,000 lives from flooding (both inland and coastal flooding) as well as landslides. Honduras, one of the hemisphere's poorest nations, suffered the brunt of the storm. Entire villages were demolished, nearly 20 percent of the people evacuated their homes, one-quarter of the schools were wrecked, water supplies were cut off, and almost all major roads were damaged. The nation's economy was devastated, and the loss of banana and sugar crops raised prices of those commodities worldwide.

Water was also the major culprit in New Orleans following Hurricane Katrina. The city is located about 50 kilometers (31 miles) inland from the southern margin of the Mississippi delta (see Figure 15.22). Unfortunately, previous hurricanes had eroded land from barrier islands that could have protected the coast from Hurricane Katrina (Figure 15.27). Other protective land has been lost to subsidence in the delta. Normally, the eroded material would be replaced by deposition from the Mississippi River, but modern controls on the river's flow have pushed the sediment out into the Gulf of Mexico, where it is deposited in deeper water. The same processes contributed to the subsidence of the city of New Orleans. Much of the city is below sea level and is consequently below the level of water in nearby Lake Pontchartrain and the Mississippi River. When some of the protective levees burst during the storm, it took just a few hours for 80 percent of the city to disappear under floodwaters.

As devastating as Katrina was on the Gulf Coast, its effects fall far short of the estimated 300,000 deaths (mostly from drowning) that occurred in Bangladesh in 1973 when a cyclone pushed onshore from the Bay of Bengal. That storm generated a 7-meter (22-foot) storm surge and produced widespread flooding of the low-lying coastal plain. The government of Bangladesh subsequently built nearly a thousand concrete shelters in coastal communities, providing sufficient space for a million residents, and improved communications links to try to reduce dangers from future cyclones.

a.

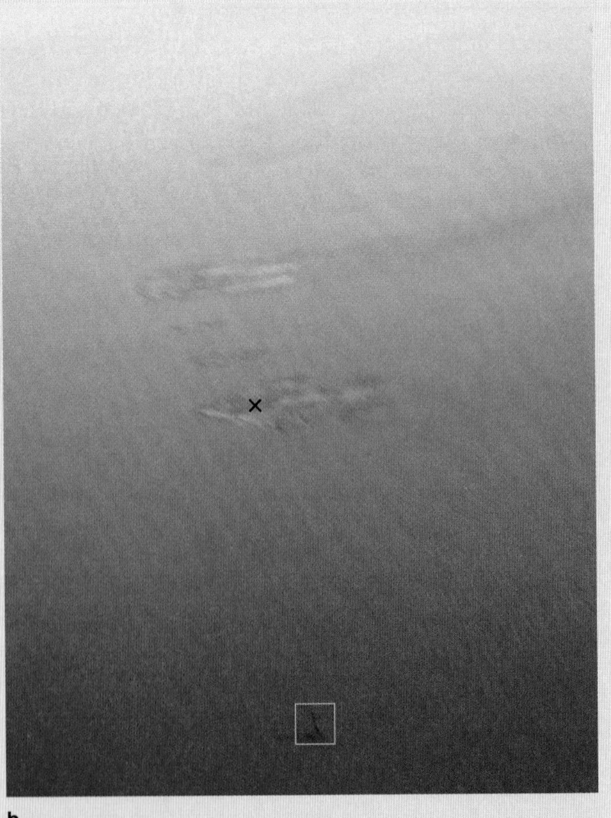

b.

Figure 15.27 Erosion caused by Hurricane Ivan. In September 2004, 1 year prior to Katrina, Hurricane Ivan eroded most of the northern segment of the Chandeleur Islands, Louisiana (view to south). A lighthouse is present in the square in each image. The lighthouse used to be surrounded by an island and had several buildings at its base. Hurricane Katrina demolished the lighthouse. The "x" marks the same location on each image.

Checkpoint 15.24

basic · intermediate · advanced · superior

Hurricane Evaluation Rubric

You work with a team of disaster specialists for the Weather Channel. During discussions about coverage of the upcoming hurricane season, your boss states that she doesn't believe the Saffir-Simpson scale sufficiently reflects the risks associated with hurricanes because it places so much emphasis on the physical characteristics of the storm. The channel wants to create its own scoring system that better evaluates the potential damage from incoming hurricanes.

1. You and your team are assigned to create an evaluation rubric to assess factors influencing the risk of damage from a future hurricane. On the table presented here, identify at least five additional factors; one (wind speed) has been included as an example. When developing your rubric, consider both physical and cultural factors.
2. After completing the rubric, you realize that some factors are more significant than others. Your team decides to double the score of the most important factor. Which factor do they choose? Why?
3. Read the following descriptions of Hurricanes Dennis and Mitch that are abbreviated versions of accounts published by the National Climatic Data Center (www.ncdc.noaa.gov). Do these descriptions cause you to change any of the categories in your scoring rubric? Rank these storms, using your modified rubric.

Hurricane Dennis, August 1999. The coastal areas of North Carolina experienced their fourth tropical storm scare in as many years in late August. Hurricane Dennis developed over the eastern Bahamas on August 26 and drifted northward parallel to the southeast US coast. Dennis became an immediate threat to southeastern North Carolina on August 29. The storm center came to within 97 kilometers (60 miles) of the coast early on August 30 as a strong category 2 hurricane with highest sustained winds of 166 kilometers per hour (103 miles per hour). Rainfall amounts approached 25 centimeters (10 inches) in coastal southeastern North Carolina.

This area is no stranger to hurricane activity. Category 2 Hurricane Bertha and category 3 Hurricane Fran hit Brunswick County in 1996, and Hurricane Bonnie (category 2) followed nearly the same path in 1998. Prior to 1996, the area had been spared from the direct impact of a hurricane since Charlie (category 1) hit Carteret County in 1986.

Because Hurricane Dennis never made landfall, damage was only moderate. However, the storm lingered off the coast for several days, so beach erosion and damage to coastal highways were significant. Residents of Hatteras and Ocracoke Islands were stranded for several days because of severe damage to Highway 12.

Factors	Low risk (1 point)	Moderate risk (2 points)	High risk (3 points)
Wind speed	Low (Category 1, 2)	Intermediate (Category 3)	High (Category 4, 5)

Hurricane Mitch, October/November 1998. Hurricane Mitch will be remembered as the most deadly hurricane to strike the Western Hemisphere in the last two centuries. The death toll was reported as 11,000, with thousands of others missing. More than 3 million people were either left homeless or otherwise severely affected by the storm. In this extremely poor developing region of the globe, estimates of the total damage exceeded $5 billion.

Within 4 days of its origin as a tropical depression on October 22, Mitch had grown into a category 5 storm. On October 26, the monster storm had deepened to a pressure of 905 millibars, with sustained winds of 155 knots (180 miles per hour) and gusts well over 200 miles per hour.

Mitch moved westward, and on October 27, it was about 97 kilometers (60 miles) north of Honduras. Preliminary wave height estimates north of Honduras during this time were as high as 13.5 meters (44 feet), according to one model.

Although the ferocious winds began to abate slowly, it took Mitch 2 days to drift southward and make landfall. Mitch then began a slow westward drift through the mountainous interior of Honduras, finally reaching the border with Guatemala on October 31.

Although the ferocity of the winds decreased during the westward drift, the storm produced enormous amounts of precipitation, caused in part by the mountains of Central America. As moist air from both the Caribbean and the Pacific Ocean to its south fed into Mitch, the stage was set for a disaster of epic proportions. Taking into account the orographic effects of the volcanic peaks of Central America and Mitch's slow movement, rain fell at a rate of 30 to 60 centimeters (12 to 24 inches) per day in many of the mountainous regions. Total rainfall of as much as nearly 2 meters (79 inches) was reported for the entire storm.

Looking to the Future

Although Hurricane Katrina was the most expensive natural disaster in US history, it could have been worse. Katrina was a high-end category 3 hurricane, and its strongest winds missed New Orleans. (Hurricane Katrina did reach category 5 status while over the Gulf of Mexico.) A direct hit on a major city by a category 5 hurricane could result in greater damage, although the lesson of Katrina should be sufficient to encourage evacuations and prevent similar levels of fatalities. Category 5 hurricanes are rare. Only three have struck the United States in the last century. The relative lack of data on these catastrophic storms makes it difficult to measure their threat or to predict how often they will make landfall. However, scientists have discovered the record of these severe storms preserved in sediments of coastal marshlands, far from where we typically think of hurricanes making landfall. Storm surges from hurricanes that occurred hundreds of years ago deposited sand in wetlands all along the Gulf and Atlantic coasts (Chapter Snapshot, Figure 15.28). By drilling through tens of meters of sediment deposits, researchers have discovered layers of sand mixed in with thicker mud. Analysis of these findings suggests that states such as Rhode Island, New Jersey, and New York are likely to experience a major hurricane every few hundred years. We only have to look at the damage from hurricane Sandy, in October 2012, to wonder if New York City might be the next New Orleans.

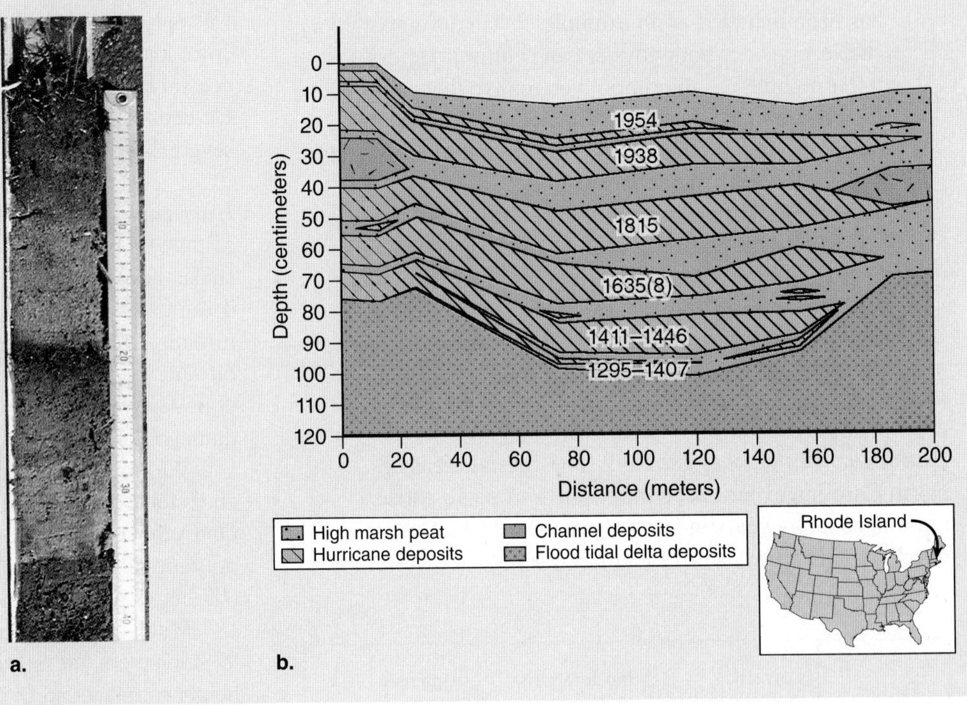

Figure 15.28 Sediment history of past hurricanes. **a.** Sediment cores from marshes reveal storm-generated sand horizons. **b.** The sand forms distinct layers in a coastal marsh in Rhode Island. The dates indicate when the sands were deposited and can be interpreted to represent when major hurricanes struck Rhode Island.

the big picture

Interpreting the weather forecast is as close as most people get to doing science on a daily basis. We look at the weather radar maps online or on television, with their moving bands of precipitation, and try to predict whether we can play the ball game before the rain arrives or whether our flight will be able to take off between snowstorms. In this chapter, we have glanced behind the colorful maps to understand the processes and life cycles that drive the development of our weather. We learned that midlatitude cyclones originate when air masses interact. The relatively short life cycle of these features constrains how far in advance we can accurately predict local weather conditions. Forecasts are tentative because the fronts that will affect our lives next week have probably not even formed yet.

Interpreting satellite data, particularly how a tornado looks on a radar display, enables meteorologists to issue tornado warnings in time to save lives. Few people in the Midwest and Great Plains have not spent time in a basement or interior room waiting out a tornado alert. Our chapter opening image showed a tornado with two vehicles nearby. On the basis of our understanding of how tornadoes work, we can assume that the people in those vehicles would be relatively safe if they were a reasonable distance to the west or southwest of the funnel. Although chasing a tornado is never a "safe" practice, we can be sure that it is better to be behind a tornado than in front of one!

All of our understanding of the science of weather phenomena does no good unless it is applied to help people. When given information about typical weather patterns or extreme events, individuals can make their own decisions about what to do next. However, in the case of hurricanes, the actions necessary to save lives and protect property cannot be made by individuals but must be decided by city councils, governors, members of Congress, and the people who work for them. Some of these decisions, such as when and how to organize evacuations, may be made a matter of days before the arrival of a hurricane but need to be planned months ahead. But other choices need to be made decades before, based on careful analysis of the potential risk. Scientists can help define the risks, but it takes a caring government with the support of its citizens to act responsibly to avert future weather-related tragedies.

Weather Systems: Concept Map

Complete the following concept map to evaluate your understanding of the interactions between the earth system and weather systems. Label as many interactions as you can, using the information from this chapter.

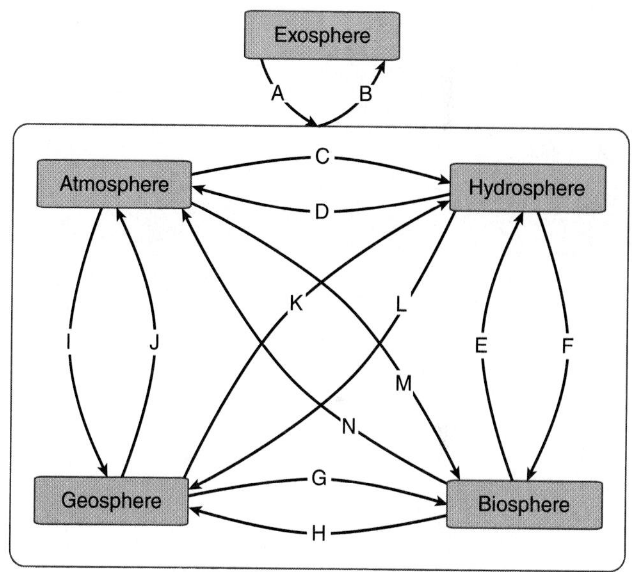

GREAT WHITE BEAR
PRO PHOTO TOURS
675-2781

"What has happened, can happen; climatic history can repeat itself."

—Reid Bryson, professor, University of Wisconsin

"The paleoclimate record shouts out to us that, far from being self-stabilizing, the Earth's climatic system is an ornery beast which overreacts even to small nudges."

—Wallace Broecker, Newberry professor, Columbia University

Chapter 16

Earth's Climate System

Chapter Outline

the big picture

Polar bear and tourist.
Which is in greater danger?

See The Big Picture box at the end of this chapter for the full story on this image.

16.1 Want Ice with That?

Chapter Learning Outcomes

- Students will explain processes related to natural climate variability.
- Students will use climate proxies to decipher past climates.
- Students will describe processes and consequences of extreme climates.
- Students will predict climate changes due to orbital changes.

One of the first things I do every morning is to take my dogs for a walk. It is an easy chore in summer; I just throw on a T-shirt and shorts and head out. By the time September rolls around, I might need a sweatshirt, and by January, I add a heavy coat, lots of layers, and a sturdy pair of boots. Of course, none of this troubles the dogs. They are never happier than when several inches of snow are on the ground. The clothing choices we make each day are determined by the weather conditions, principally the temperature and whether it is raining or not. As we learned in Chapter 15, weather represents the state of the atmosphere at a given place for short periods of time (hours, days).

An assessment of the clothes in your closet would give a pretty good idea of how the weather changes through the year and could define the climate for your region of the world. **Climate is a description of the weather conditions for a region averaged over several decades.** We could get a reasonable picture of climate regions in North America by asking people how many sweatshirts they own. Then we could create a sweatshirt index (SI). SI values would be high for Canada and states such as Minnesota and Maine but would probably hover in the low figures for southern parts of Arizona, Texas, and much of Mexico. Of course, in real life, climate scientists (climatologists) are a lot more careful and typically review weather data for 30-year intervals to get a realistic portrait of an average regional climate.

Weather and climate are the result of a complex series of interactions among all the elements of the earth system (hydrosphere, atmosphere, biosphere, and geosphere), beginning with the interaction between Earth and the sun. **The climate of every region on Earth has changed often.** Frequently, climate changes are relatively subtle (for example, rising temperatures) and occur over centuries. These changes are so gradual that few people would even notice them unless they had access to an extensive database of climate information. An early identification of such trends provides plenty of opportunity to address changes. However, as we will learn in the pages ahead, some climate changes are extremely rapid and take place over time spans so short that it is nearly impossible to take preventive steps. Whether gradual or rapid, climate change has the potential to contribute to significant cultural and environmental changes. Today, examples of these changes are visible in Alaska and other Arctic locations.

Climate Change and the Polar Bear Diet

There is a polar bear jail in Churchill, Manitoba, on the western shore of Canada's Hudson Bay (Figure 16.1). If things keep warming up in the Arctic, there may be a need for a polar bear prison system. Each fall, polar bears wander through Churchill on their way to Hudson Bay to wait for new sea ice to form so that they can hunt seals, their principal food source. The bears must consume enough seals to build up fat reserves to sustain them during summer when the ice melts. An adequate diet of seals is especially important for female bears, since their fertility is tied to their fat storage.

The past few decades have seen a reduction in the Hudson Bay polar bear population and a similar reduction in the average weight of adult bears. These changes have occurred largely because rising temperatures in the Arctic are causing sea ice to melt earlier in the spring and form later in the fall. If fall sea ice forms late, polar bears make a nuisance of themselves around town. In Churchill, bears receive a short jail term that lasts as long as it takes for the bay to freeze over. Other jurisdictions are not always as considerate and just shoot the bears deemed to be causing a nuisance.

The term *Arctic* comes from the ancient Greek word *arktikos,* which means "near the bear," a reference to the visibility of the Ursa Major (Great Bear) constellation in northern latitudes. The *Arctic region* is traditionally defined as the area surrounding the North Pole where temperatures in July average less than 10°C (50°F). This includes much, but not necessarily all, of the planet north of the Arctic Circle (latitude 66.5°N) and parts of Canada, the United States (Alaska), Russia (Siberia), Norway, and Greenland (Figure 16.2). The decline in the polar bear population is just one sign of the changing climate in the Arctic, where average temperatures have risen by about 2 to 3 C° (3 to 5 F°) over the last half-century—more than double the average temperature increases observed throughout the rest of the world. While global warming is still considered hypothetical in some locations, people in places such as Alaska see it as a stark reality in the form of dramatic changes in the physical environment around them. For

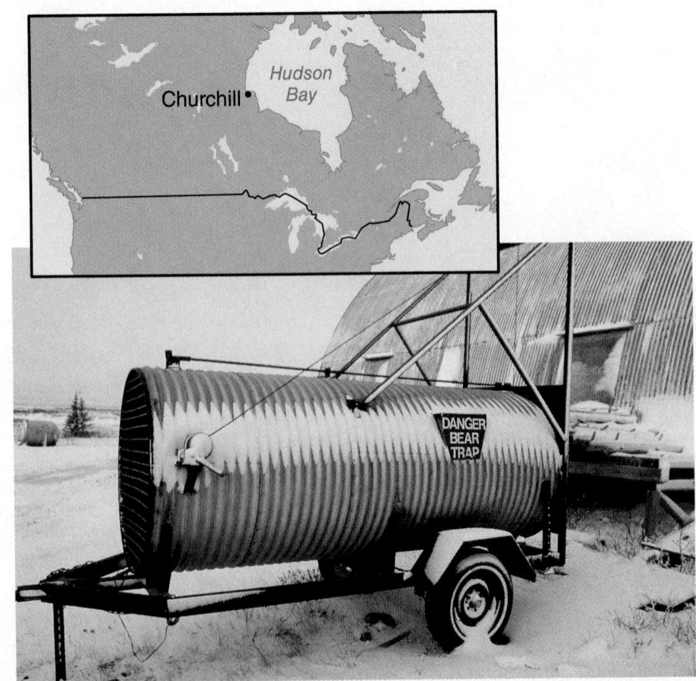

Figure 16.1 Polar bear jail, Churchill, Canada. Polar bears that roam into Churchill (shown on map) are captured in the trap and placed in the "jail" building until Hudson Bay freezes over.

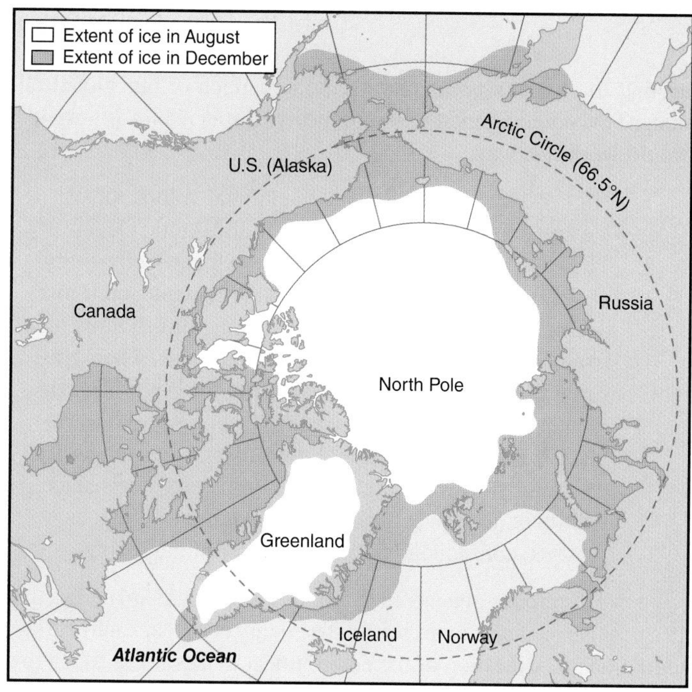

example, US Geological Survey scientist Bruce Molnia collected photographs of Alaskan glaciers; some of these images were more than 100 years old. Molnia visited and rephotographed the same sites (Figure 16.3). His images tell a story of shrinking ice masses.

The sheer remoteness of the Arctic region requires some unconventional research strategies. For example, the European Space Agency was seeking to calibrate readings from its new satellite that would be measuring ice thicknesses across the Arctic region. It needed to know how the depth of the overlying snow would affect the satellite measurements. So it hired two guys to walk the 2,000 kilometers (1,240 miles) across the ice from Siberia to Greenland, stopping every couple of days to measure the thickness

Figure 16.2 The Arctic region. The Arctic represents the region around the North Pole, mostly north of the Arctic Circle at latitude 66.5°N. The Arctic is composed of the Arctic Ocean and surrounding nations, including parts of Canada, the United States, Russia, Norway, and Greenland.

Figure 16.3 Evidence of shrinking glaciers. **a.** The 600-meter-thick (1,970-foot) Riggs glacier, Glacier Bay National Park, Alaska, occupied the bay in 1941. **b.** By 2004, the glacier had undergone significant melting and retreat.

a.

b.

of the snow layer. It took seasoned explorers Alain Hubert and Dixie Dansercoer 106 days to make the journey. Along the way, they saw signs of a changing Arctic. Increasing temperatures have accelerated the melting of land-based glaciers throughout much of the region. The melting ice increased the stream discharge and raised the sea level, which, in turn, picked up the pace of coastal erosion. However, as you will see next, this warming has the potential to lead to much more troublesome global consequences.

The Consequences of Arctic Warming

A widespread and steady decline in sea ice has occurred over the Arctic Ocean during the last few decades (Figure 16.4). In 2012 there was only about one-half of the sea ice during summer months in the Arctic that was typical between 1979 and 2000. This progressive melting could disrupt the oceanic conveyor belt (see Chapter 13 Snapshot), the circulation system that redistributes heat around the world's oceans and regulates much of our global climate. This potential interaction of the hydrosphere and atmosphere would work this way:

1. Higher temperatures reduce the annual volume of sea ice and expose more open water in the Arctic Ocean.
2. The water is darker and has a lower albedo (reflectivity of the surface; see Chapter 14) than the white ice. This decrease in albedo causes the Arctic to absorb more heat from solar radiation. (Even at high latitudes, where the sun is always low in the sky, water in the Arctic Ocean reflects less sunlight than ice.)
3. Sea ice forms later in the fall because it takes longer for the water to become cold enough to freeze. The sea ice is thinner and covers less area than it did in previous seasons; consequently, it melts earlier in the spring.

This process repeats each year as the steps are linked together in a positive feedback loop (part of the energy is fed back into the system). Each modification of the environment makes it less likely that the warming trend will reverse itself. If the trend continues, much of the sea ice will be gone by the end of the twenty-first century.

OK, so there will be less sea ice; beyond the impact on polar bears and seals, why should we care about this? Here is the rest of the story:

4. Salt is concentrated in the water of the Arctic Ocean and North Atlantic as seawater changes state to form ice. A small amount of salt is trapped in the growing ice, but most salt is left behind in the ocean, increasing the salinity and density of the water. Any reduction in the volume of ice would result in a reduction in the salt content and density of the water in the northern oceans.
5. To add insult to injury, freshwater input from melting ice in places such as Greenland further dilutes the salt content in the North Atlantic waters carried by the Gulf Stream. Sinking dense water in the North Atlantic adjacent to Iceland and Greenland drives the oceanic conveyor belt. The combination of less ice formation and increased melting has the potential to reduce water density sufficiently to reduce the volume of sinking water. This, in turn, may slow the conveyor belt.
6. Some scientists argue that sinking waters in the North Atlantic actually "pull" the Gulf Stream. If less warm water is drawn north by the conveyor belt, temperatures in the North Atlantic region would decrease. This could result in a cooling trend that would affect Europe and perhaps much of the world. Recent models suggest that the cooling effect would be too small to offset the overall warming trend. More extreme cooling would require the more unlikely scenario of a catastrophic collapse of the Greenland ice sheet, rapidly releasing vast volumes of freshwater into the North Atlantic.

Of course, there is an alternative scenario that would work this way:

1. Warmer temperatures turn out to be a short-term variation that changes over the next few years as average temperatures return to more traditional levels. Lower temperatures increase the annual volume of sea ice and expose less open water in the Arctic Ocean.

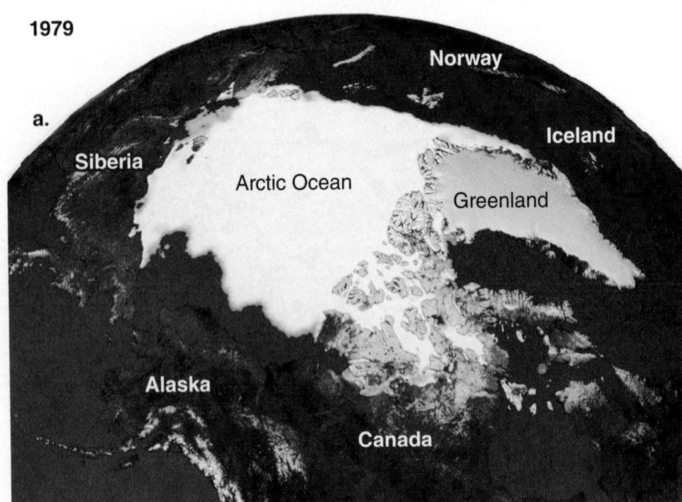

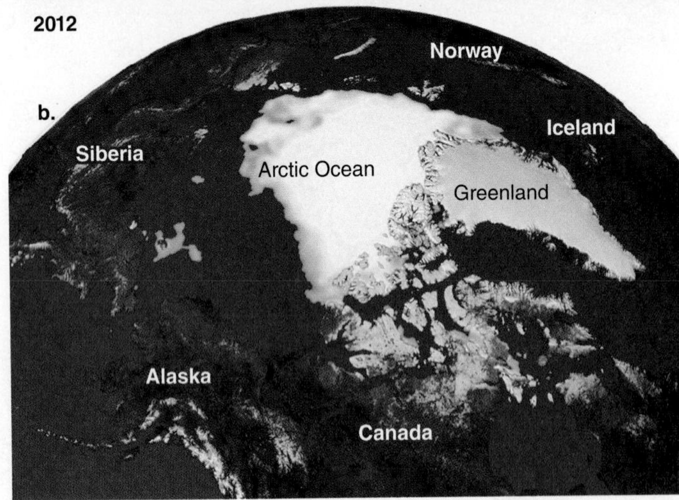

Figure 16.4 Arctic sea ice coverage. **a.** Precise satellite measurement of the sea ice in the Arctic Ocean began in September 1979. **b.** The summer minimum ice coverage in 2012 was about one-half of the average value recorded from 1979 to 2000. Spring melting begins a few weeks earlier than it did in 1979. It is now possible to sail around the Arctic Ocean in summer, a feat that was impossible in 1979.

2. The increasing volume of sea ice raises the albedo, reflecting more solar radiation back to space and causing less heat to be absorbed throughout the Arctic region.
3. Sea ice again forms earlier in the fall as the water becomes cold enough to freeze earlier in the season. The sea ice thickens and covers more area than it did in previous seasons; consequently, it melts later each spring.
4. As the volume of ice increases, more salt is added to seawater, and the salinity (and density) of the Arctic Ocean steadily increases.
5. In addition, increased cooling would result in less melting of the ice sheet on Greenland and elsewhere. The combination of more sea ice formation and a reduction in the melting of continental ice sheets would increase the density of the surface waters and strengthen the circulation of the conveyor belt.
6. If these trends continue, temperatures in the North Atlantic region would gradually decrease and return to levels similar to those of the 1980s.

Unfortunately, as each year passes and the warming trend continues, it becomes less likely that conditions will return to their 1980 levels. The climate changes we observe in the Arctic today may be a temporary aberration typical of normal climate fluctuations, or they may be an early sign of a permanent change in worldwide climates. Which is it? We can review climate records that indicate that 12 of the hottest years on record (for more than 100 years) have occurred since 2000. Is this part of the normal climate cycle or a sign of climate change? **The big question is: Are humans affecting climate change?** The vast majority of climate scientists would tell us that the planet is warming and that human actions are principally responsible. So, what are we going to do about it?

This chapter and Chapter 17 will discuss how Earth's climate system works and present data pertinent to the human-induced climate change question. But before we attempt to analyze recent climate changes, we will learn how the global climate system works and which environments are associated with extreme climates. We can look for indicators of these environments in the geologic record and use this information to unravel our global climate history. Finally, we will examine which factors have had the greatest influence in changing climates.

Self-Reflection Survey: Section 16.1

Answer the following questions as a means of uncovering what you already know about Earth's climate system.

1. Imagine that you were blindfolded and air-dropped onto a continent somewhere on Earth. What features could you use to identify the climate of the region?
2. Any effective climate model must account for regional variations in climate at any moment in time as well as for variations in climate for any location over the course of a year. Make a list of what you know about annual variations in climate in the United States and around the globe.
3. What are some climate-related concepts featured in earlier chapters of this text?

16.2 Global Air Circulation

Learning Objectives

- Explain geographic patterns of temperature, cloud cover, precipitation, and air pressure.
- Relate solar radiation, climate patterns, and atmospheric circulation systems.

In this section, we will revisit several ideas discussed in earlier parts of the book and knit them together to build the components of the global air circulation system. Any system should be able to explain key features of global climate patterns, including the distributions of temperature, precipitation, air pressure, and climate-related environments, such as deserts. What general patterns do you notice about the relationships among these factors in the Chapter Snapshot on the next two pages?

Go on, take a look . . . we will wait; maybe we will do some humming (hmmm . . .).

Several significant patterns can be extracted from a careful analysis of the four maps. First, notice that **temperatures decrease with increasing latitude as one moves north and south from the equator**. Why is this? Recall that solar radiation (sunlight) strikes Earth more directly at the equator and the tropics than at the poles. Locations closer to the equator receive about 2.5 times more solar energy than locations at or near the poles. More energy is transmitted to Earth in the tropics, resulting in higher temperatures at low latitudes. The global atmospheric and oceanic circulation patterns move heat toward the poles and cool air and water toward the tropics.

Second, notice that **the temperature range (highest to lowest) is much greater for the continents than for the oceans** (approximately 60 C° [108 F°] versus 30 C° [54 F°]; Chapter Snapshot). In Chapter 13, you learned that water can store a lot of heat without a correspondingly large rise in temperature because water has a high heat capacity. Also, ocean currents mix waters and moderate temperature over large regions of the ocean surface. In contrast, the soil and rocks of the land surface have a smaller

Checkpoint 16.1

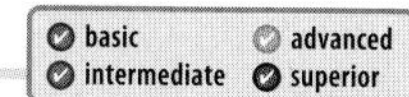

The following map shows two locations in South America. Predict which location has the

a) world's highest average annual precipitation (1,331 centimeters; 524 inches).
b) world's lowest average annual precipitation (0.08 centimeter).

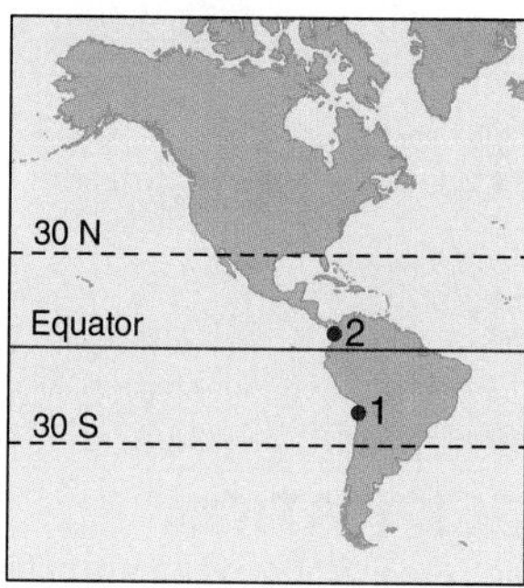

CLIMATE DATA

TEMPERATURE AND CLOUD COVER

Sea and land surface temperatures (°C) and cloud cover. Is the temperature range greater for land or the oceans? Bands of clouds are found along the equator at latitudes 60° N and S and are separated by zones of generally clear skies.

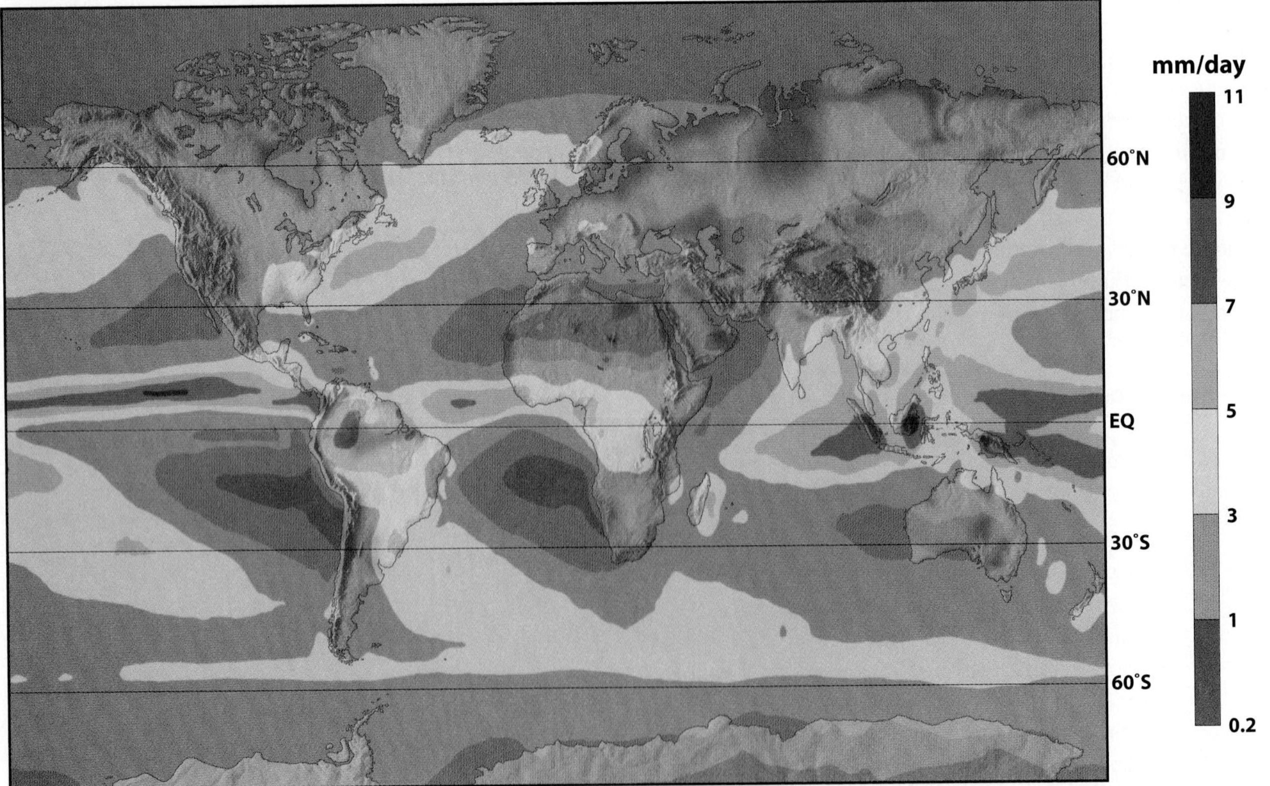

PRECIPITATION

Average precipitation values in millimeters per day averaged from 1979–2004. At what latitudes do you find the highest and lowest precipitation rates?

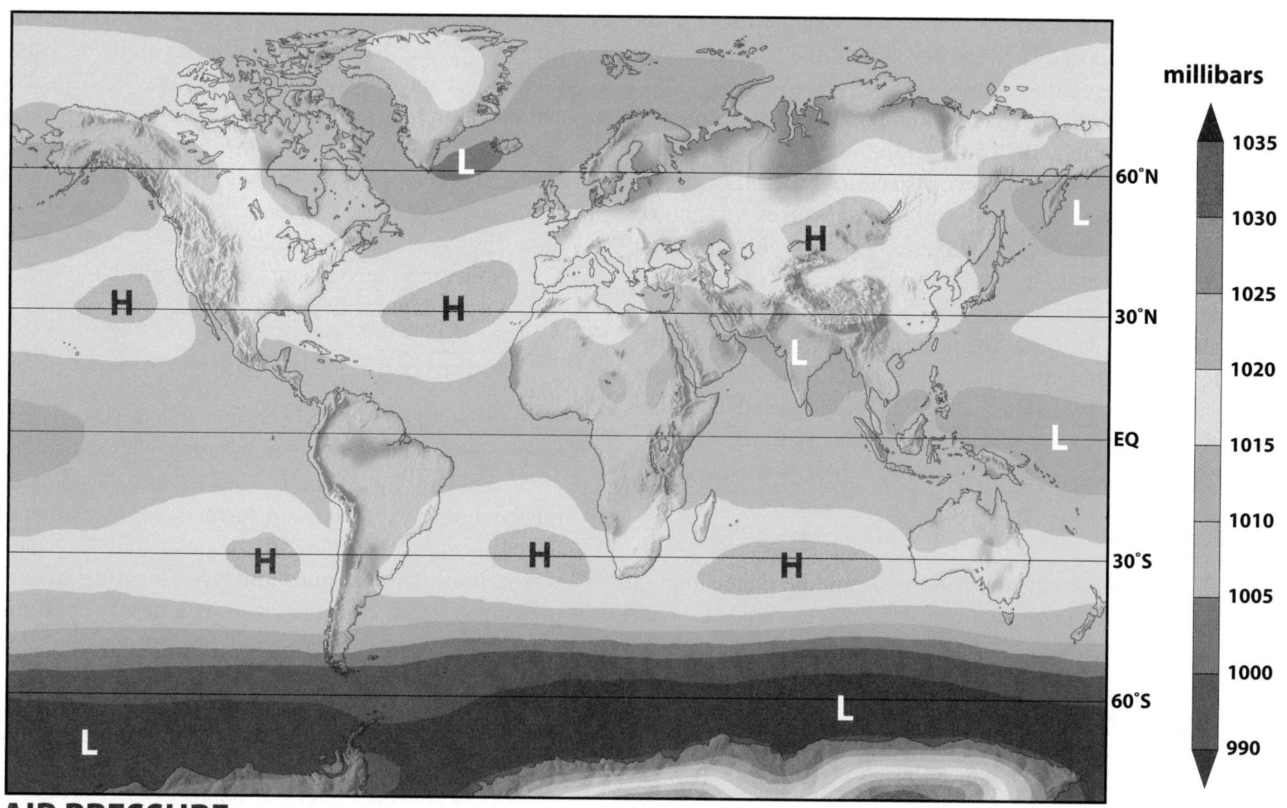

AIR PRESSURE

Annual mean sea level atmospheric pressure (millibars) averaged from 1979–1995.
L = low pressure center, H = high pressure center. At what latitudes do you find the high and low pressure centers?

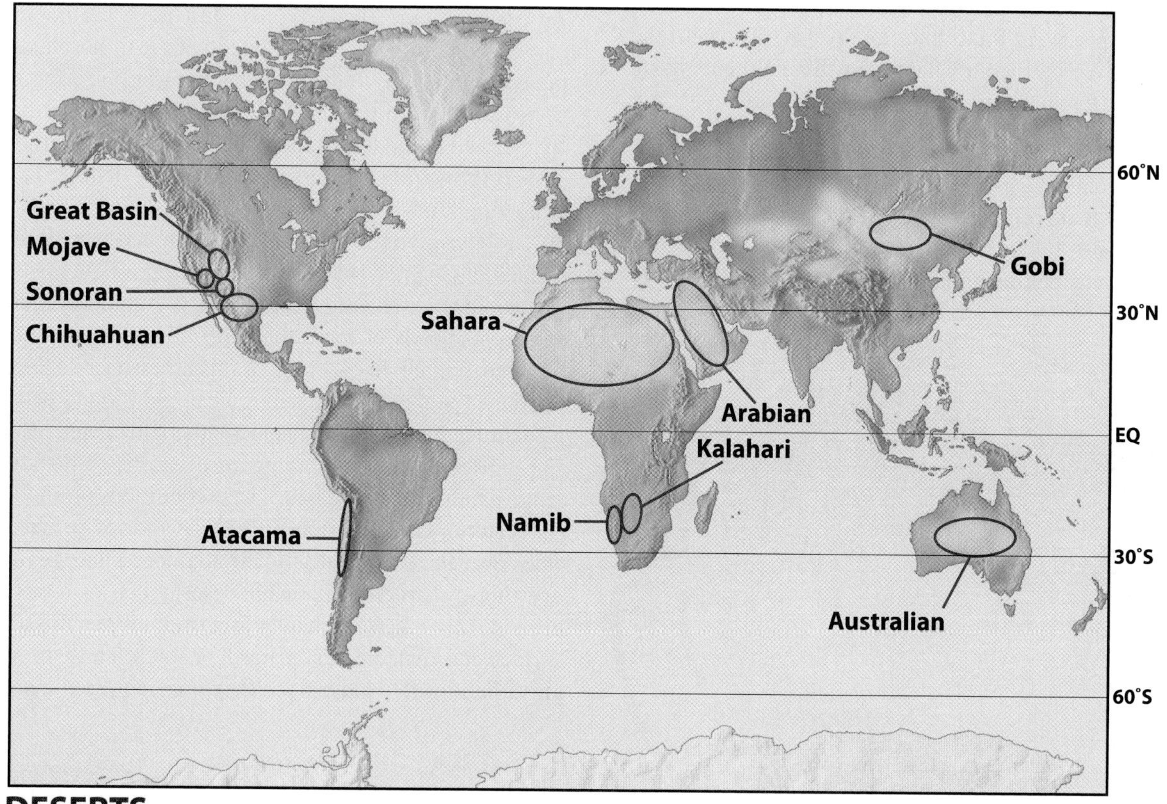

DESERTS

Compare the location of these major deserts with the global temperature and precipitation rate maps.

heat capacity, so they warm to higher temperatures when exposed to a similar amount of solar radiation. Generally, temperatures are more extreme on continents and vary over a smaller range in the oceans. As we will learn later, the capacity of the oceans to store heat is critical in regulating climates.

Third, you may have noted that **clouds are concentrated in irregular bands parallel to the equator and latitudes 60°N and S and that these latitudes are also characterized by higher precipitation rates and low atmospheric pressures**. In contrast, regions around latitude 30°N and S have fair skies with little cloud cover, higher air pressures, and low rainfall rates; these regions also correspond with the locations of almost all the major deserts on Earth (Chapter Snapshot). Temperature shows a steady decrease as one moves away from the equator, but rainfall rates and sea level pressure are distributed in alternating bands of high and low values parallel to latitude (east-west; Chapter Snapshot). In the next sections, we will explain the contrast between these patterns.

The Nonrotating Earth Model

Global atmospheric circulation patterns would be quite simple if Earth did not rotate. The much warmer air at the equator would rise. As the air ascended, it would cool. When it reached the top of the troposphere, it would bump into the base of the warmer stratosphere. Because it could not continue to rise through the stratosphere, the air would have to turn north or south and flow toward the polar regions. Far above Earth's surface, this air would arrive at the pole as a mass of cold, dense air. This cold air would sink (why?) and move outward toward lower latitudes. In the simplest case, **the rising air at the equator and the sinking air at the poles would form opposing limbs (loops) of a large-scale circulation pattern called a *convection cell.*** The convection cell would carry cold air toward the tropics and warm air toward the poles (Figure 16.5). Recall from Chapter 14 that cold and warm air masses have different densities, which would cause pressure differences between the equator and the poles. Rising warm air would generate low pressures at the equator, while sinking cold air would produce high pressures at the poles. Those pressure differences would produce winds that flow parallel to longitude (north-south lines), moving toward the equator. The rising air would cool and condensation would occur, leading to precipitation along the equator. In contrast, sinking dry air at the poles would produce almost no precipitation. So, that is the simple nonrotating model: a difference in solar radiation causes differential heating, which produces big convection cells that drive atmospheric circulation. This model is a place to start, but it is much too simple because it does not account for Earth's rotation.

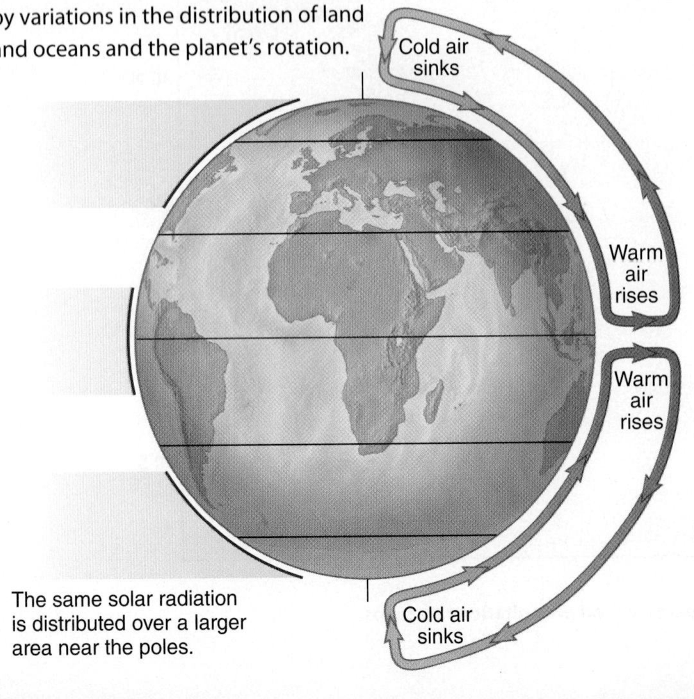

Figure 16.5 Nonrotating Earth model. These patterns are complicated by variations in the distribution of land and oceans and the planet's rotation.

Cold air sinks
Warm air rises
Warm air rises
Cold air sinks
The same solar radiation is distributed over a larger area near the poles.

The Rotating Earth Reality

We have already revisited concepts that deal with air pressure, the properties of water, and solar radiation. Now we will bring in some other ideas we have seen before: latent heat, the Coriolis effect, relative humidity, and ocean currents. More solar energy is transferred to the planet along the equator than anywhere else. This **warm, humid air expands and rises, forming an equatorial low-pressure system characterized by a nearly continuous band of clouds** (see Chapter Snapshot). Cloud formation occurs as a result of the condensation of water vapor in humid air. Through the latent heat of condensation, this process also releases approximately 75 percent of all the energy stored in the atmosphere. Although the location of the equatorial low migrates north and south with the seasons, it is always characterized by increased precipitation and lower-than-average air pressure (see Chapter Snapshot). Because of the heavy rainfall and high temperatures found there, many equatorial regions are characterized by tropical rain forests.

Jet Streams and Trade Winds. Just as in the nonrotating Earth model, the rising air reaches the top of the troposphere before flowing north or south. Remember that the Coriolis effect causes winds to be deflected to the right of their course in the Northern Hemisphere (and to the left of their course in the Southern Hemisphere). The north-moving air mass is continuously deflected to the east and is flowing almost directly eastward by the time it reaches latitude 30°N. (These winds are similar to the geostrophic winds we described in Chapter 14.) **The east-moving air forms a continuous, subtropical jet stream that circles the globe** in the upper troposphere at speeds of more than 100 kilometers per hour (61 miles per hour; Figure 16.6). Some of this air also sinks, forming a high-pressure zone, called the subtropical high, found between 20° and 35° latitude in the Northern and Southern hemispheres (Figure 16.7).

Recall that descending air is warmed adiabatically, and its relative humidity decreases. These conditions promote evaporation (rather than condensation) of water vapor, which results in clear skies over the tropics around 30°N and S. These regions of warm, dry air are characterized by hot deserts such as those in Saudi Arabia and Australia (see Chapter Snapshot). When this air reaches the surface, it is deflected to the north or south. Those prevailing winds are called *trade winds,* since they once powered merchant sailing

Checkpoint 16.2

Florida lies at the same latitude as the Sahara Desert. Why do you think Florida is not a hot, dry desert?

ships across the oceans. In the Northern Hemisphere, the south-flowing air moves toward the equator, producing winds that are deflected to the west—the *northeast trade winds* (Figure 16.7a). It is the northeast trade winds that carried Columbus to the New World in 1492. In the Southern Hemisphere, the *southeast trade winds* flow north and are deflected to the west.

Convection Cells. The columns of rising air and descending air are connected by north- and south-moving winds at high and low altitudes, respectively, to **form a continuous convection cell named a Hadley cell,** after George Hadley (1685–1768), who was the first person to explain the route of the northeast trade winds.

Far to the north or south, the **polar cells are anchored by columns of descending air** (Figure 16.7). This cold, dry air moves away from the poles (cold, high-pressure air toward warmer, low-pressure air). These relatively cold winds collide around latitudes 60°N and S with warm air flowing poleward from lower latitudes. The average location of the convergence zone is equivalent to southern Alaska in the Northern Hemisphere and to the Southern Ocean in the Southern Hemisphere. The converging air masses result in a lifting mechanism that forms a zone of rising air and low pressure known as the *polar front* (Figure 16.7). The polar front is distinguished by cloud formation and elevated rainfall rates but not to the degree seen at the equatorial low.

An intermediate circulation system occurs in the middle latitudes (30° to 60°N and S), separating the thermal circulation cells to the north and south. This **midlatitude convection system is known as a Ferrel cell,** after meteorologist William Ferrel (Figure 16.7). Flow patterns in this system affect much of North America and are substantially modified by the formation of mid-latitude cyclones (described in Chapter 15). As a consequence, this circulation system is not as well defined as the Hadley and polar cells. In the Northern Hemisphere, for reasons that are not yet fully understood, upper-level winds flow northward, eventually forming another jet stream, called the *polar jet stream,* above the polar front. Some of the dry air that descends at the subtropical high-pressure system flows toward the poles in the Ferrel cell to form a belt of winds known as *westerlies*. These winds are deflected to the right due to the Coriolis effect.

When all of these cells are connected, we find that atmospheric circulation can be divided into three convection cells in each hemisphere. However, it will be clear from analysis of the maps in the Chapter Snapshot that these bands of winds are often disrupted by variations in topography across continents, especially in the Northern Hemisphere.

Checkpoint 16.3

basic / intermediate / advanced / superior

Recall that we discussed in Chapter 2 how the seasons were the result of Earth's tilt. Analyze Figure 16.7. Describe how precipitation and temperature would vary along a line of longitude from the equator to a pole over the course of a year.

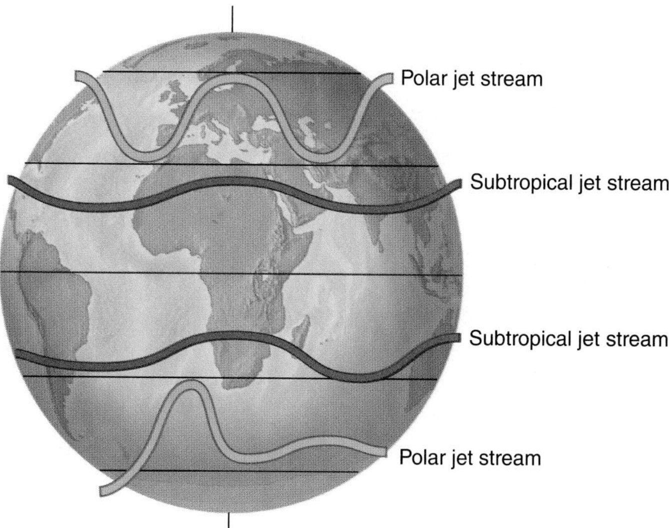

Figure 16.6 Upper-atmosphere jet stream winds. Two jet streams are present in each hemisphere near the top of the troposphere. The subtropical jet stream is around 30°N and S, while the polar jet stream is at 60°N and S. The fastest winds in the core of the jet streams can reach speeds of more than 170 knots (315 kilometers per hour, 195 miles per hour).

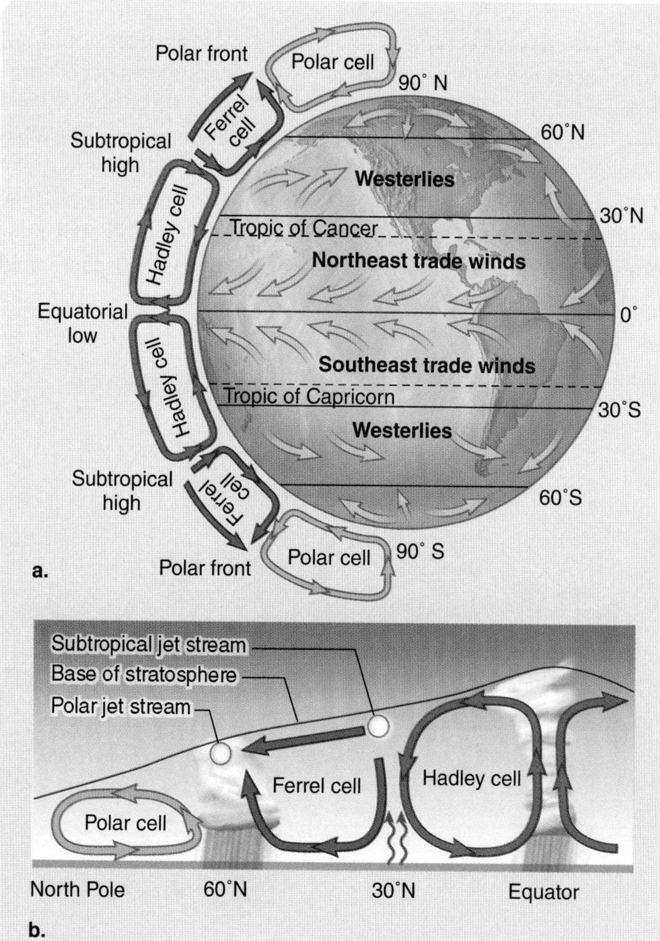

Figure 16.7 A model of atmospheric circulation. **a.** Low- and high-pressure systems and the wind patterns that result from their interaction. The sun is assumed to be overhead at the equator. Earth has been rotated to better illustrate the distribution of circulation cells. **b.** Each convection cell has an ascending and a descending limb connected by upper-level and lower-level winds.

Checkpoint 16.4

basic / intermediate / advanced / superior

Make a concept map of the global circulation system, using at least 10 of the following terms and others of your own choosing, with no more than 15 terms total.

convection cell	Hadley cell	rising air
Coriolis effect	hot deserts	solar radiation
descending air	jet stream	subtropical high
equatorial low	polar front	trade winds
Ferrel cell	precipitation	westerlies

16.3 Global Climate Regions

Learning Objectives

- Predict the type of climate at a given location based on precipitation and temperature data.
- Analyze how climate influences the biosphere across the planet.
- Discuss what type of environmental indicators could be used to identify whether climate change was occurring.

Think about where you live. What is the climate like? What are its dominant characteristics? How far would you have to travel to experience another climate? From your own observations, how would you design a climate classification system? Global temperature and precipitation patterns are related to atmospheric and oceanic circulation patterns that help create different climate regions across the globe. Areas with similar climates are grouped together in climate regions. **Climate regions are differentiated on the basis of temperatures and precipitation and their resulting vegetation.** Since temperatures decrease with altitude (see Chapter 14), elevation also plays a role in defining climate regions.

Köppen-Geiger Classification System

It is useful to develop a classification system that can be applied globally when we discuss climate. Two climatologists, Wladimir Köppen and Rudolph Geiger, developed a system for identifying similar climates, using only temperature and precipitation data, since these two factors most directly affect ecosystems. The **Köppen-Geiger classification system divides Earth into climate regions**, using average monthly temperatures, average monthly precipitation, and total annual precipitation values (Figure 16.8).

The Köppen-Geiger classification system is convenient because the data can be easily collected without expensive equipment, and the system requires only simple calculations. Furthermore, knowing how temperature and precipitation are related to specific habitats makes it possible to reconstruct ancient climates from the record of plant and animal remains (more on that in Section 16.4). However, this climate system has some limitations because the data are averaged. Averaging causes small local climates (called microclimates) to be missed. On its broadest scale, the classification includes six climate regions, each denoted by a letter (Table 16.1).

In addition, the Köppen-Geiger system identifies many subregions on the basis of variations in climate factors. These

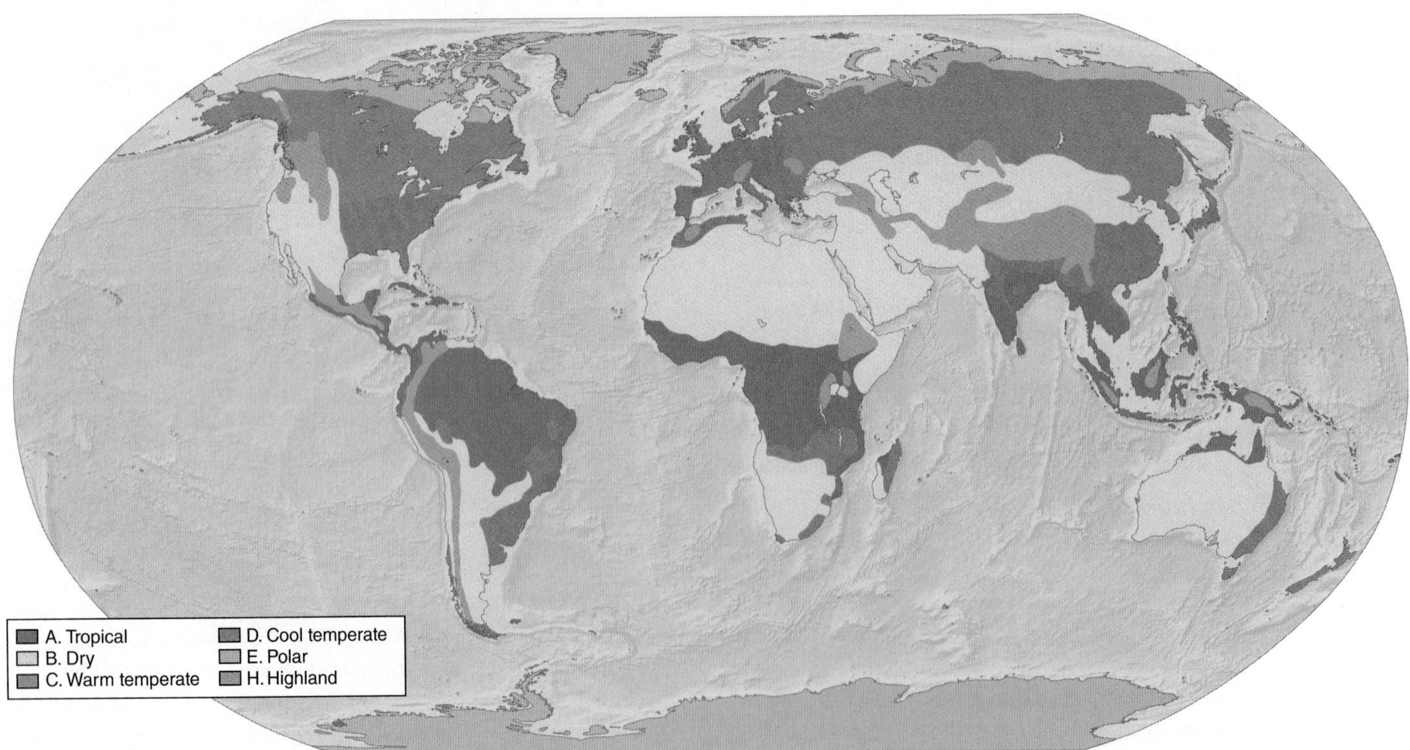

Figure 16.8 Köppen-Geiger climate regions. The Köppen-Geiger classification system divides Earth into climate regions, using average monthly temperatures, average monthly precipitation, and total annual precipitation. The six major climate regions are: A, tropical; B, dry; C, warm temperate; D, cool temperate; E, polar; and H, highland.

smaller divisions are typically named by adding lowercase letters. For example, Af designates a tropical rain forest; Am is a tropical monsoon climate, characterized by a marked dry season lasting one or more months; and Aw is a tropical savanna, which has a winter dry season. However, for the purposes of this discussion, we will use only the six broad classes. All six climate regions are found in North America with the dry (B), warm temperate (C), and cool temperate (D) classes covering the largest areas (Figure 16.8).

Climate and the Biosphere

Where we choose to live is largely determined by the characteristics of our physical environment (for example, climate, landscape) and the availability of basic resources, such as food. Take a look at the map of global population density in Figure 16.9. How does the distribution of people on Earth compare to the temperature, rainfall, and atmospheric circulation patterns discussed in Section 16.2?

Table 16.1 Köppen-Geiger Climate Regions

Letter	Name	Characteristics
A	Tropical	Wet, hot equatorial regions that cover about one-third of Earth's surface. Monthly average temperature above 18°C (64°F). All or most months may have average precipitation of 6 centimeters (2.4 inches). US example: Key West, Florida.
B	Dry	Arid and semiarid deserts and steppes (grassland); evaporation exceeds precipitation. True deserts have little precipitation; semideserts average 25–75 centimeters (10–30 inches) per year. US example: Albuquerque, New Mexico (annual precipitation, 22 centimeters [9 inches]).
C	Warm temperate	Humid subtropical, may have dry summers. Average temperature of warmest month above 10°C (50°F); coldest month below 18°C (64°F) but above 0°C (32°F). US example: New Orleans, Louisiana.
D	Cool temperate	Humid climate with long winters, mild summers. Warmest month above 10°C (50°F); coldest month below 0°C (32°F). US example: Flint, Michigan.
E	Polar	No true summer, warmest month average temperatures below 10°C (50°F); always cold. US example: Barrow, northern Alaska.
H	Highland	Lower temperatures and more precipitation, substantial variation. US example: Blue Canyon, Sierra Nevada mountains, California (annual precipitation, 170 centimeters [67 inches]).

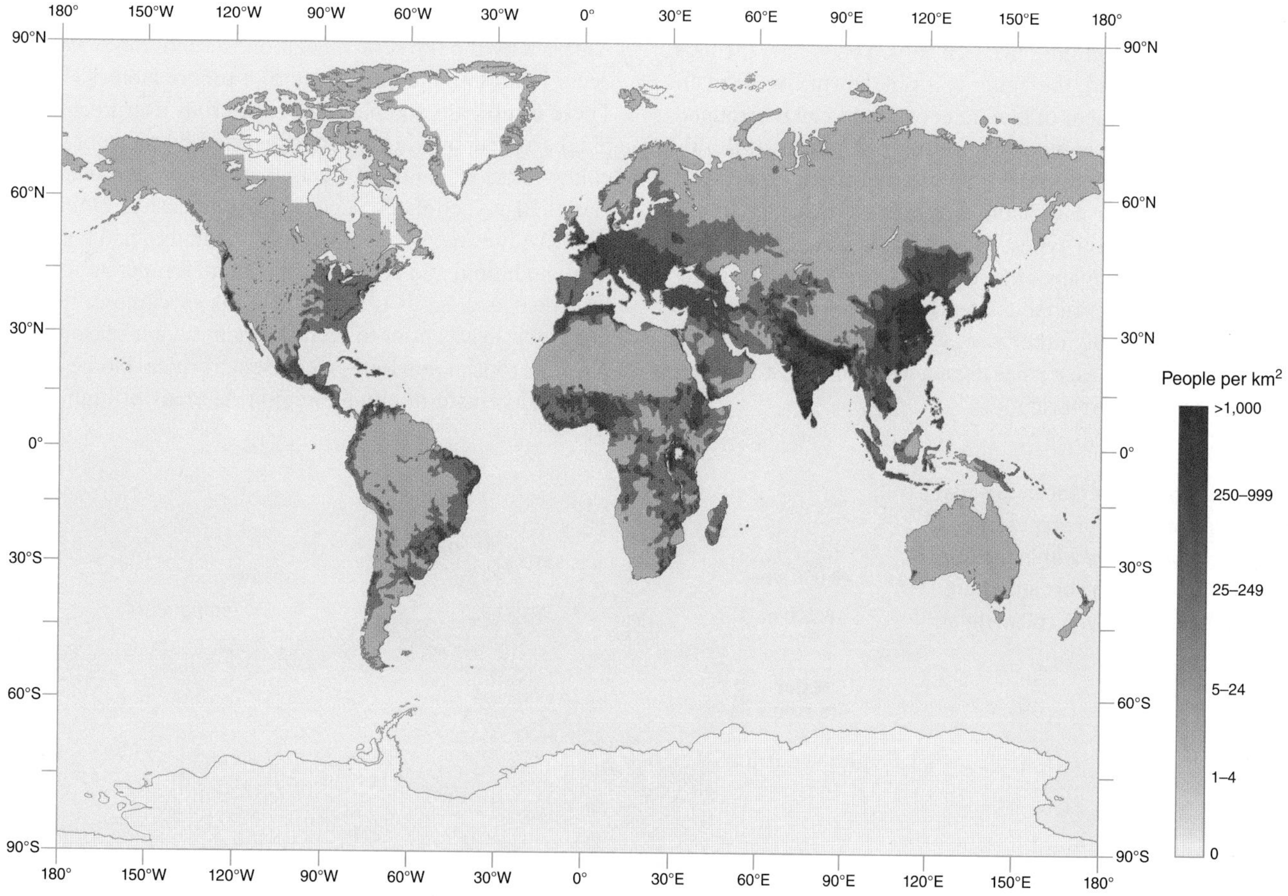

Figure 16.9 Global population density. Darker colors signify regions with the highest population density. Lighter colors show regions with lower population density.

Checkpoint 16.5

basic / intermediate / advanced / superior

The following graph illustrates mean monthly high and low temperatures and the average monthly rainfall for Sydney, Australia. Estimate the average monthly temperatures as halfway between the mean high and low temperatures. On the basis of these data, in what climate region does Sydney belong? Explain your reasoning.

a) Tropical
b) Dry
c) Warm temperate
d) Cool temperate

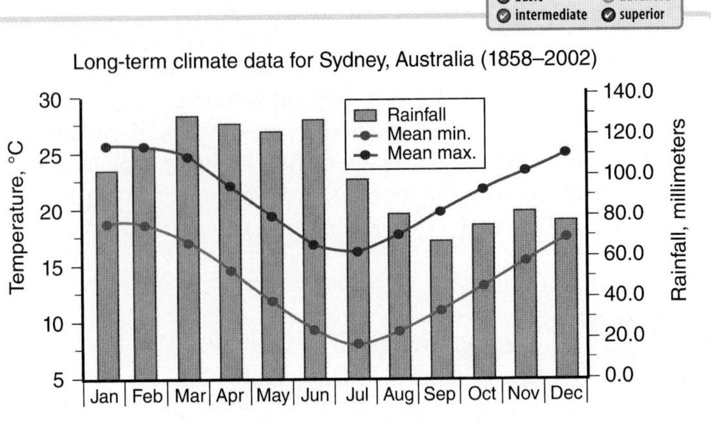

The biosphere represents all life on Earth and is linked to climate. Just like humans, other organisms live in areas where the physical attributes of the environment suit their survival needs. Life can flourish almost anywhere on the planet's surface, from the deepest portion of the ocean floor to the Himalayas' highest mountains. However, in terms of altitude, most organisms live within a narrow vertical band of land and shallow oceans where the climate is warm and moist (Figure 16.10). **Ecosystems are communities of organisms that inhabit specific physical environments,** which are defined primarily by their climate and landforms. **The number of species in an ecosystem is termed *biodiversity.*** Biodiversity varies depending on the characteristics of the individual ecosystems, but is typically greatest in environments with warm temperatures, plentiful rainfall, and rich soils. The greater the biodiversity of an ecosystem, the better it is able to survive through climate extremes. A number of similar ecosystems can be grouped together in a **biome, a regional community of plants and animals named after the dominant type of vegetation.** Biomes are characterized by similar associations of species, comparable climates, and consistent soil types. One of the characteristics that makes us different from many other organisms on this planet is the wide range of biomes we inhabit. Clearly, humans are influenced by more than just climate; otherwise, why have people moved to Las Vegas, a city in the desert, making it one of the fastest-growing urban centers in North America?

Generally, we recognize three major climate-related biome groups: **grasslands, forests,** and **deserts,** each of which can be divided into individual biomes (Figure 16.11). As you might expect, examples of these biomes often can be found in more than one climate region (compare Figures 16.8 and 16.11).

African *savannas,* grasslands characterized by scattered trees, can be found in parts of region A, which has long dry periods, as well as in wetter parts of region B (desert). These warm locations contrast with the treeless, frigid grasslands of the tundra in region E, straddling the Arctic Circle in northern Asia and North America. These polar climate regions are also characterized by low rainfall and can be considered cold deserts.

Two principal forest biomes can be distinguished in North America by the types of vegetation in each. *Boreal forests* (taiga) occur in northern latitudes, south of tundra biomes (Figure 16.11). These forests are composed of coniferous evergreens (spruce, fir, pine) growing in climates characterized by the cool summers and cold winters of region D. *Temperate forests* dominate the regions classified as C climates, such as the southeastern United States and northern Europe, where conditions are humid and hot during summer and humid and cool during winter. Temperate forests contain broadleaf deciduous tree species (oak, sycamore, maple), which drop their leaves prior to their dormant winter season.

Tropical rain forests represent a critical biome. Tropical rain forests dominate in climate region A, areas of high temperature,

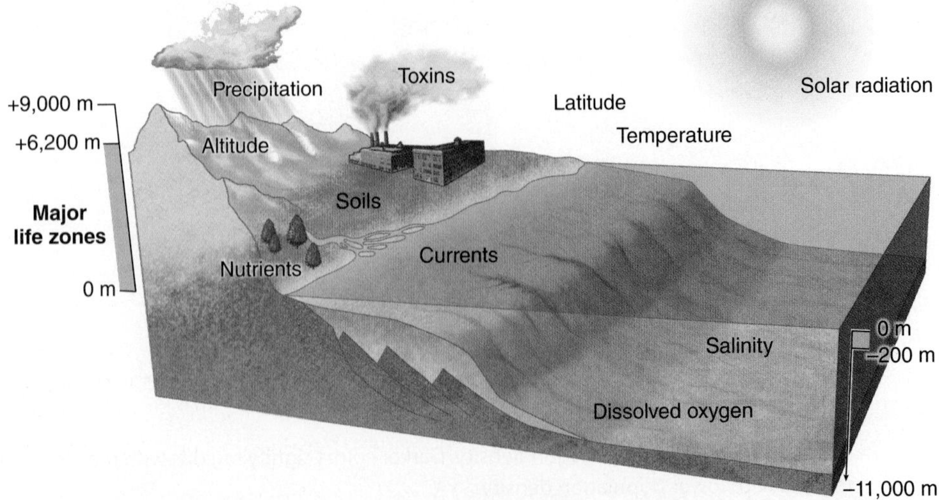

Figure 16.10 Environmental factors that influence the distribution of ecosystems. The majority of species live within the upper 200 meters (660 feet) of the ocean, where light penetrates, and below 6,200 meters (20,500 feet) on land.

high rainfall, plentiful sunlight, and high humidity at low latitudes adjacent to the equator. These forests are home to the most diverse ecosystems on the planet. They contain evergreen, broadleaf, and hardwood trees such as mahogany, teak, and ebony. Tropical rain forests contain approximately 50,000 plant species, in contrast to the 5,000 species typical for temperate forests.

Dry climates are characterized by hot and cold *deserts* and semidesert environments such as the tropical thorn scrub and woodland that surround the Mediterranean Sea and the cold tundra habitat of Arctic regions (Figure 16.11). Hot deserts have high temperatures throughout the year (for example, Sahara, North Africa; Chihuahuan, west Texas, and northern Mexico). There, a few plants cling to life on a substrate of sand or rock. Cacti are relatively common in temperate deserts, which have hot summers but cool winters (for example, the Mojave Desert in southeast California).

Changes in the character of biomes, especially along their margins, may signal a shift in climate. For example, in the warming Arctic region described earlier, the area of tundra has decreased by about 20 percent during the last 20 years, while the edge of the boreal forest has been creeping steadily northward. Similarly, warmer temperatures in highland regions have caused some plant species to migrate to higher elevations.

Checkpoint 16.6

basic / intermediate / advanced / superior

How does the distribution of people on Earth (Figure 16.9) compare to the distribution of climate zones (Figure 16.8)? Which aspects of climate have the greatest influence on the distribution of population?

Energy and the Biosphere

Plants are producers that manufacture the food they need from inorganic compounds in the physical environment. Through photosynthesis, most plants use light energy (from sunlight) to convert atmospheric gases (for example, carbon dioxide), water, and soil nutrients into leaves, branches, and roots.

Animals are consumers that derive their energy from consuming plants or other organisms because they cannot produce their own food from inorganic materials such as air or sunlight. Primary consumers (herbivores; for example, cattle, deer) devour plants, and secondary consumers (for example, wolves, most

Checkpoint 16.7

basic / intermediate / advanced / superior

Mount Kilimanjaro is a 6,000-meter-tall (19,685-foot) mountain in Tanzania, Africa, just 320 kilometers (200 miles) south of the equator. Moving from bottom to top, a hiker passed through savanna grasslands, tropical rain forest, treeless moorland, and alpine desert, ending on a glaciated peak. Explain why the climate conditions and biological environments changed as the hiker climbed up the mountain.

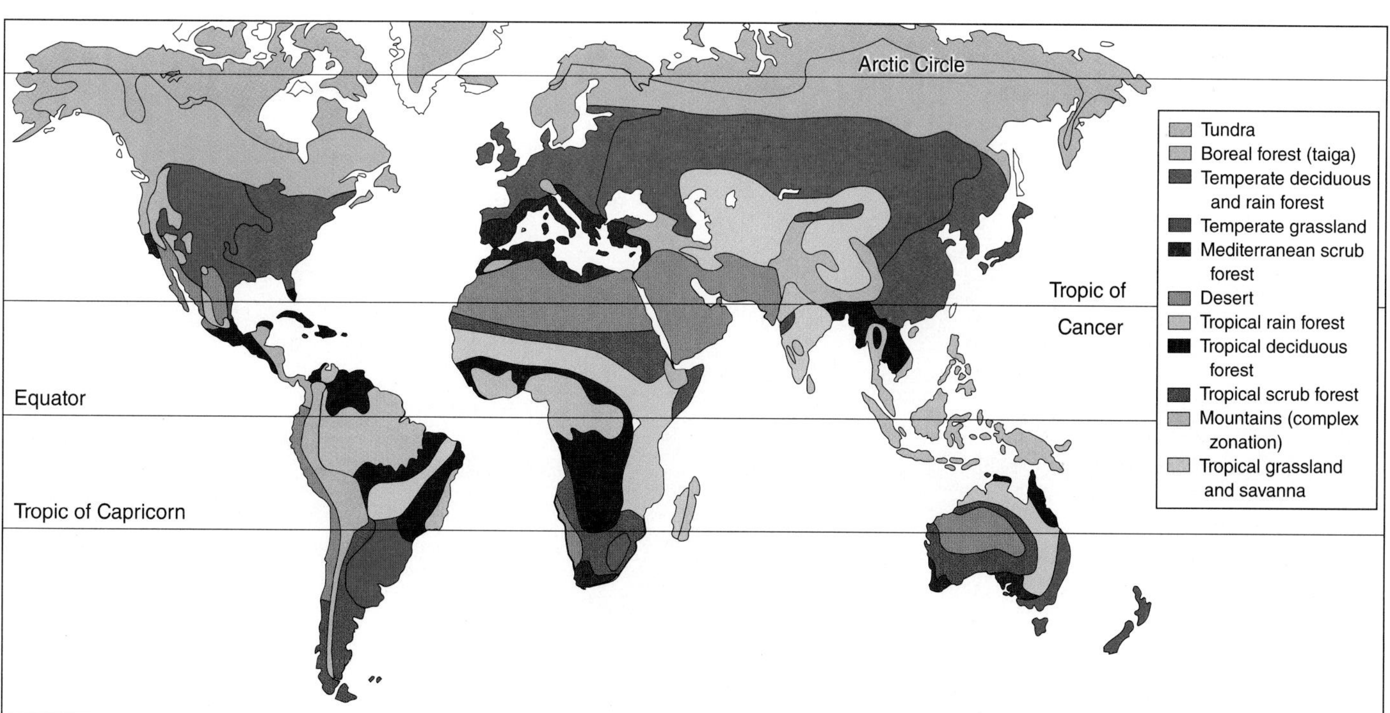

Figure 16.11 Global-scale biomes. Ecologists don't agree on an exact number of ecosystems and biomes because the number varies, depending on how they are defined, for example, which species are used and what climate characteristics are considered. Much of the eastern United States is considered a temperate forest biome (under natural conditions, prior to domestication), but it is composed of several separate ecoregions that include the eastern broadleaf forests of West Virginia, the southeastern mixed forest of the western Carolinas, and the coastal plain mixed forest of northern Florida.

humans) eat primary consumers. Most of the energy represented by the plant material is lost by consumers as heat or is excreted as waste. The consumers and producers are part of a food chain that transfers energy among organisms within an ecosystem. Approximately 5 to 20 percent of energy is transferred with each step up the food chain. The number of organisms at higher levels of the food chain decreases as energy is lost with each step. For example, there are many deer and relatively few wolves.

Certain key elements serve as essential nutrients for life in plants and animals. Six elements (nitrogen, carbon, hydrogen, oxygen, phosphorus, and sulfur) make up approximately 95 percent of the materials in plants and animals (yes, even people) and combine to form compounds that are essential nutrients for life on Earth. These and other **elements pass through the earth system in biogeochemical cycles** that link together processes in the biosphere, atmosphere, hydrosphere, and geosphere (Figure 16.12).

The amount of organic material in an ecosystem is known as the *biomass* and is dominated by plant material (99 percent of all biomass). Tropical rain forests represent one-third of Earth's biomass but cover only 7 percent of the land surface. All of the planet's forests account for approximately three-quarters of all the biomass on one-fifth of the land. In contrast, deserts cover about the same land area but account for less than 2 percent of the biomass.

Biomass Energy Sources. Some of Earth's biomass has been converted over millions of years into concentrated fossil fuel energy sources such as coal, oil, and natural gas. Wood has been burned as a source of energy for thousands of years. Several types of biomass represent renewable energy sources. Wood products and other biomass-related energy sources (for example, biofuels, landfill gases) account for about 3 percent of US energy consumption. Unlike other renewable energy sources, biomass can be converted to transportation fuels. Ethanol and biodiesel are examples of these biofuels that are derived from commonly grown crops such as corn and soybeans, respectively. These liquid bio- fuels have the potential to reduce some of the demand for imported oil. However, given the large areas of land necessary for such crops, it is predicted that they will account for less than 10 percent of US gasoline consumption in the next few decades. In comparison to other forms of renewable energy, biomass demands the greatest land area to generate a given amount of energy. Most land is already dedicated to other uses (agriculture, buildings) or does not have the climate required to develop the fast-growing plant varieties necessary for biomass energy generation. In addition, as all biomass energy ultimately comes from biological sources, burning biomass, whether as a gas, liquid, or solid, produces a comparable volume of greenhouse gases as fossil fuels.

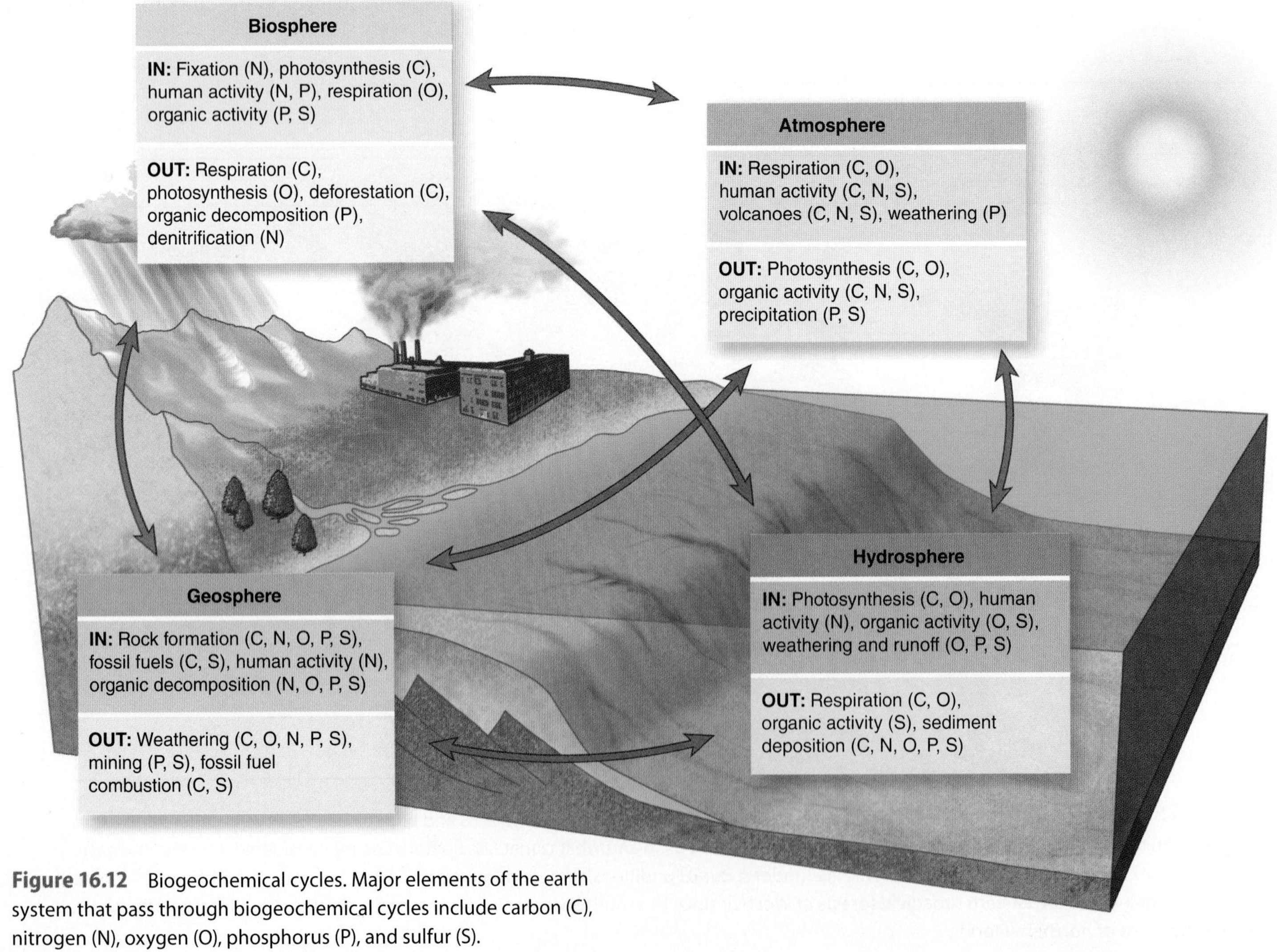

Figure 16.12 Biogeochemical cycles. Major elements of the earth system that pass through biogeochemical cycles include carbon (C), nitrogen (N), oxygen (O), phosphorus (P), and sulfur (S).

Checkpoint 16.8

basic · intermediate · advanced · superior

With all the media reports about climate change, people in your community turn to you to help them figure out if it is really happening. (You are taking an earth science class, after all.) They ask you to generate a commonsense index of climate change that could be used by long-time residents (longer than 20 years) of your community. The index should not be too complicated, so you must identify just three things that an interested resident could observe. For example, they might write down the date that they first observe a robin. Describe three other observations that citizens could make to identify climate change.

It is likely that biomass, like solar, wind, wave, and geothermal renewable energy sources, will expand to meet some of the future demand for energy in the United States and around the world. However, like these other sources, biomass will represent just a modest fraction of overall energy generation. It is unlikely that we will be able to make any significant short-term reductions in the concentrations of atmospheric greenhouse gases simply by using renewable energy sources. Fossil fuels will continue to dominate global energy consumption for decades to come. Consequently, global efforts to battle climate change are more likely to be successful if they address modifications or reductions in the way the world uses fossil fuels rather than simply relying on increased use of renewable energy sources. Biomass may have a more important future role in absorbing excess carbon dioxide from the atmsophere than as a fuel source (more on this in Chapter 17).

16.4 Extreme Climate Environments

Learning Objectives

- Identify and interpret features associated with modern glaciers and ancient glaciated landforms.
- Compare and contrast the characteristics of continental and alpine glaciers.
- Explain the chronological and physical changes of snow as it cycles through a glacier.
- Create plans to track future changes in the size of glaciers and deserts.
- Explain how winds erode, transport, and deposit sediment in desert regions.
- Describe the principal depositional features found in deserts.
- Compare and contrast wind action and glacial processes.

Extreme climate environments are characterized by exceptionally high or low temperatures, a lack of precipitation, or both. (Where are such places found on Figure 16.8?) Typically, such environments exhibit reduced biodiversity and low population densities, and fall within the dry and polar climate regions. In this section, we will examine the geologic features and processes characteristic of these environments and consider how they would be preserved in the geologic record. In particular, we are interested in answering two questions: First, what does the analysis of current extreme environments reveal about the potential for changes to our present climate? Second, how can we use the geologic record of these extreme environments to better understand the long-term distribution and character of ancient climates? The answers to both of these questions can help us understand natural fluctuations in the climate system.

Cold Climates

Visitors to Greenland or Antarctica observe rocky peaks jutting out of vast expanses of snow and ice (Figure 16.13). If you drilled down through the ice, you would find that it is up to thousands of meters thick and rests on bedrock. These two locations represent the world's largest accumulations of ice and snow. **On land, a long-lived mass of slow-moving snow and ice is termed a *glacier.*** Glaciers form where there is a steady supply of snow and cold temperatures year-round. These permanently cold conditions are found near the poles or at high elevations and can produce two major types of glaciers. **Continental glaciers,** or **ice sheets,** form on landmasses at polar latitudes; Antarctica and Greenland are examples. These glaciers may last for millions of years. In contrast, smaller glaciers, called ***alpine glaciers,* can be found at high altitudes in mountainous regions** on every continent except Australia (recall from Chapter 7 that Australia's mountains are lower than those of any other continent). Alpine glaciers may be tens to hundreds of meters thick and typically flow down from high ground to fill lower valleys (see Figure 16.3). Alpine glaciers

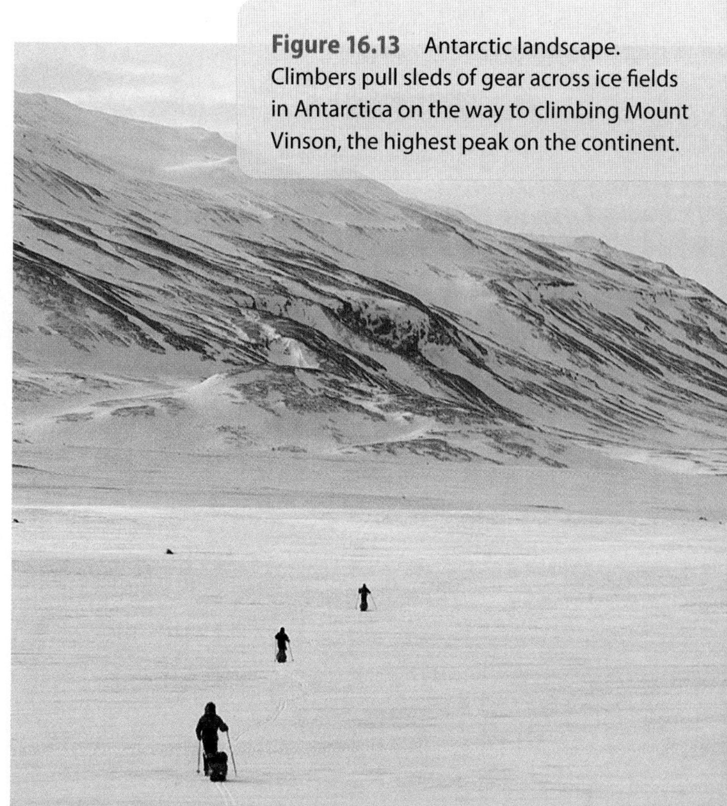

Figure 16.13 Antarctic landscape. Climbers pull sleds of gear across ice fields in Antarctica on the way to climbing Mount Vinson, the highest peak on the continent.

are more susceptible to climate fluctuations than continental glaciers. Individual alpine glaciers may last for decades to thousands of years, depending on their location and size.

Glacier Movement. Glacial ice is thick enough that the weight of the ice alone is sufficient to cause it to move. Glaciers usually move downslope, but massive continental glaciers can actually climb over topography. How is that possible? Imagine pouring thick honey onto a plate. Now, add more honey to the pile. As more is added, the width of the pile will increase as the edges of the pile move outward. Glaciers move in much the same way; as more ice accumulates, the weight of the pile increases, and the influence of gravity causes the ice to flow slowly outward. If there is a low obstruction in the way of the honey, such as a pat of butter, the honey can move up and over the obstruction. Likewise, continental glaciers can move up and over elevated topography. Alpine glaciers move at speeds of 5 to 50 meters (16 to 164 feet) per year, and ice sheets can move as much as hundreds of meters per year. As a rule of thumb, the more ice that accumulates in the glacier, the more rapidly it will flow.

But glaciers are not as fluid as honey; ice is a relatively brittle substance. Pressure from the material above causes the ice in the lower part of the glacier to flow. The uppermost ice is under less pressure and responds to movement as a brittle material would, by fracturing and cracking as the glacier moves forward (Figure 16.14a, b). **Crevasses—steep, narrow cracks in the glacier's surface**—form as the shape of the glacier changes to conform to the shape of the underlying valley floor.

Glacier Composition. Crevasses give us a view into the interior of the glacier. A close look at the side of a crevasse reveals layers in the ice (Figure 16.14c). How did these layers get there? Soon after snow falls on a glacier, the delicate snowflakes are compacted and transformed to small ice particles (Figure 16.14a). These small particles are packed down as they are buried below later snowfalls. Small air bubbles are trapped between compacted ice particles and preserved in the ice crystals. **These tiny bubbles preserve a record of the composition of the atmosphere at the time of their formation.** Ice near the surface of the glacier may undergo daily cycles of partial melting and freezing during warmer summer months. Repeated cycles of partial melting and refreezing plus the pressure of burial force these small pellets of ice together to form larger, dense ice crystals. An original 10-meter-thick (33-foot) layer of snow may end up as a 1-meter-thick (3-foot) layer of ice that will be further compacted to just centimeters in thickness as it is buried deeper below later accumulations. Thus, the **layers in a glacier represent annual accumulations** of snow that were subsequently converted to ice. Scientists studying glaciers can estimate the age of ice in a glacier by counting these annual layers. In Antarctica, the temperatures are cold enough all year long so that little or no melting occurs and all the snow is preserved as annual layers. In contrast, alpine glaciers typically have an accumulation season and a melting season.

A glacier can be divided into zones based on elevation (Figure 16.15). The higher *accumulation zone* is usually the thickest part of the glacier, where the addition of snow exceeds loss by melting. The lower, thinner area is the *ablation zone* (*ablation*

Checkpoint 16.9

Identify three things that are similar and three things that are different between alpine and continental glaciers.

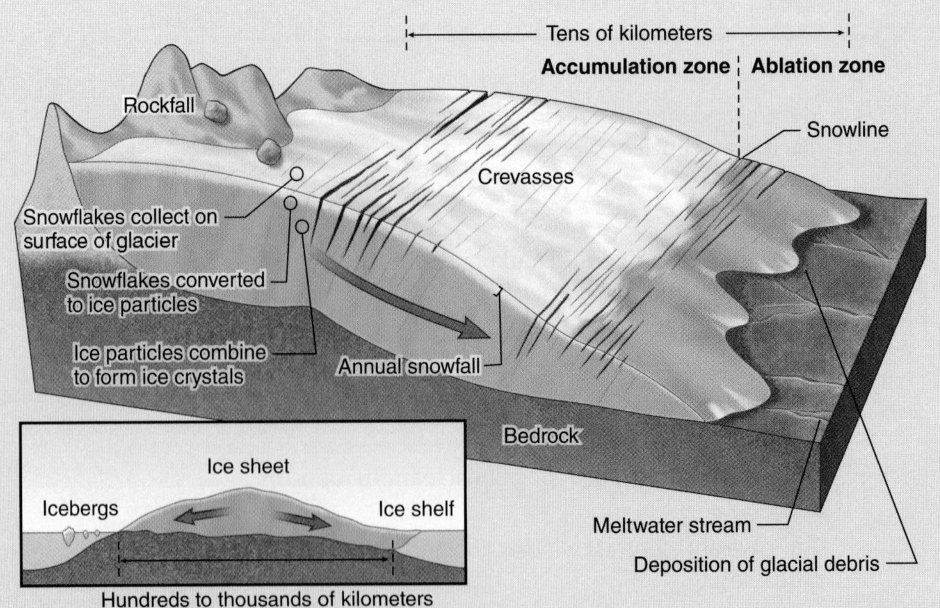

Figure 16.14 Movement of a glacier. **a.** The glacier flows downslope, transporting ice from the high ground in the accumulation zone to lower elevations characterized by the loss of glacial ice (ablation zone). The glacier transports pieces of bedrock that fall off the surrounding cliffs or are torn or worn away from the bedrock floor or the sides of the valley. Inset illustrates a simple view of a continental glacier. **b.** Crevasses form in the upper portion of a glacier as it adjusts to changes in the underlying valley floor. **c.** Snow falling in the accumulation zone of the glacier is compacted to eventually form ice crystals that contain trapped air bubbles. Annual layers of snow and ice can be observed in the walls of a crevasse.

means "removal"), where the seasonal snow melts, causing the glacial ice below to thaw. The boundary between the accumulation and ablation zones of the glacier is indicated by the snow line (Figure 16.15). Ice is lost by melting at the front of the glacier and may be replaced by ice that moves downslope from the accumulation zone.

The mass balance of the glacier is the difference between the amount of snow and ice that accumulates and melts each year. For example, there are more than 700 glaciers in the high peaks of the Cascade Mountains of Washington. About 5 to 15 meters (16 to 49 feet) of snow will accumulate on these glaciers during a 9-month period (mid-October to mid-July) as a result of snowfall, avalanches, and drifting snow driven by winter winds. Melting occurs during the warmer summer months. In the Northern Cascade glaciers, the accumulation zone needs to cover 60 to 70 percent of the surface of a glacier to ensure that it does not lose mass. If accumulation exceeds ablation, the glacier grows in size and moves downslope. If ablation exceeds accumulation, the glacier melts faster than new ice can be added, and the front of the glacier retreats upslope. Scientists who have been analyzing the mass balance of North Cascade glaciers have recorded a steady decline in volume over the last two decades. The glaciers are about one-third smaller today than they were in the mid-1980s.

Erosion and Deposition by Glaciers. Glaciers are not just ice. They also contain rock fragments from large boulders down to sand- and clay-sized particles. Glaciers produce this debris by scraping underlying material and undercutting the adjacent mountainsides, resulting in mass wasting of the landscape onto the glacier. Chunks of rock fall from slopes above the glacier and are plucked out of the underlying bedrock as the glaciers move downslope. Some of these bits of rock are frozen into the sides and base of the glacier where they act like the sand grains in sandpaper, allowing the glacier to

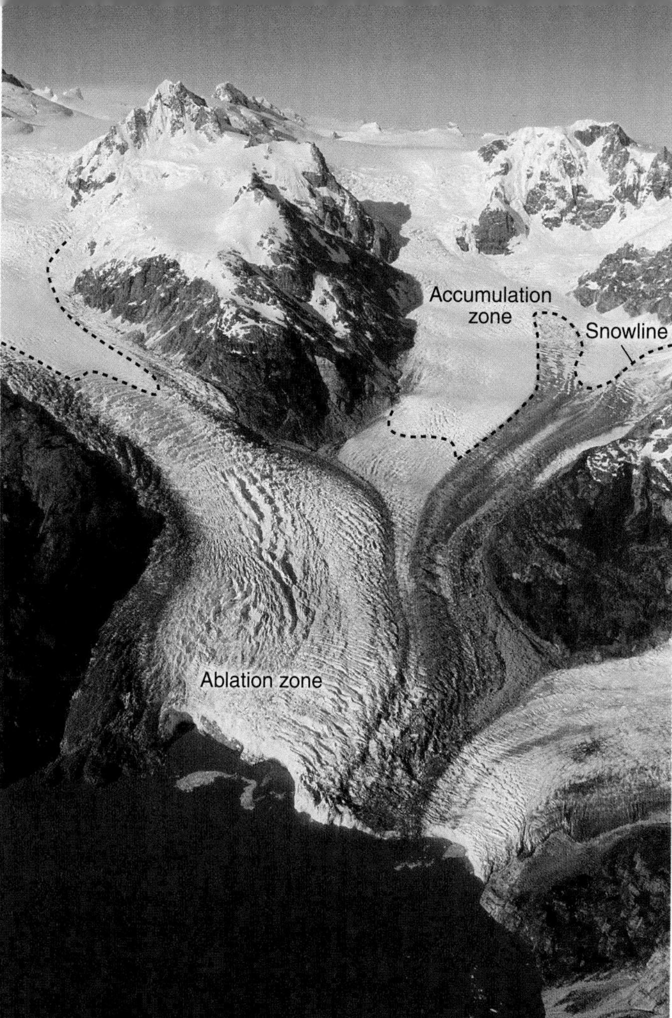

Figure 16.15 Snow line for an alpine glacier. The edge of the snow above indicates the break between the accumulation zone and the ablation zone for a glacier in Alaska.

b.

c.

grind and scratch its way downslope, eroding additional sediment along the way. Unlike streams, glaciers do not sort this debris because the sediment is trapped in the solid ice until the ice melts. On land, unsorted debris collects at the *terminus* (end) of the glacier. Some of this material is transported away by meltwater streams, but much of the remainder forms an unsorted deposit known as *till*, which is unique to glacial regions and characterized by its mixture of sediment sizes. The till may eventually be lithified (converted to rock) to form *tillite* (Figure 16.16).

Melting and accumulation are in balance when ice advances to the terminus of the glacier at the same rate at which it melts. Under these conditions, the terminus of the glacier will

a.

b.

Figure 16.16 Till and tillite. **a.** An unsorted, modern till deposit. **b.** The Gowganda tillite, a 2 billion-year-old glacial deposit near Sudbury, Canada. In each case, note the variety of sizes of blocks.

Figure 16.17 Glacier terminus and moraine. The ridge of moraine represents deposits of till formed when the terminus of the glacier remained in one location.

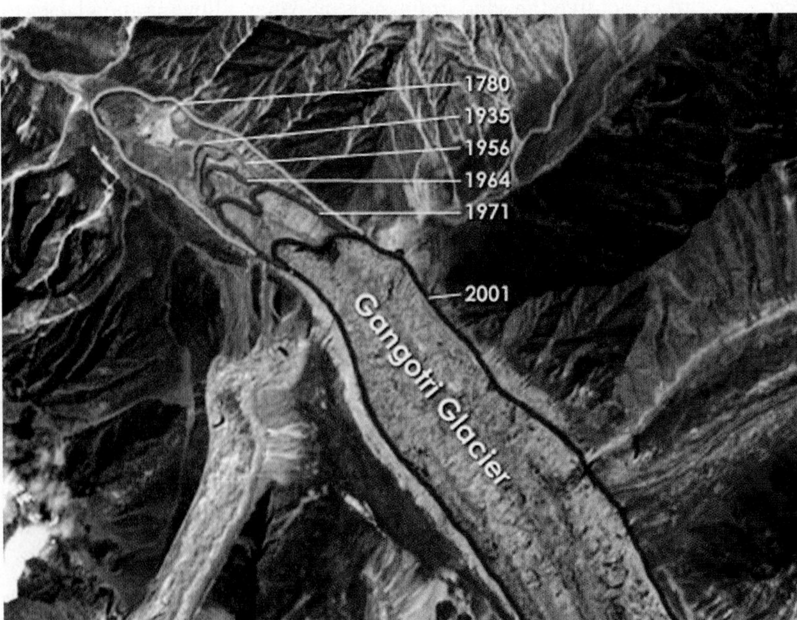

Figure 16.18 Recession of the Gangotri Glacier, India. The blue lines represent the positions of the terminus of the glacier since 1780. The 30-kilometer-long (19-mile) glacier has been receding for the past few centuries but has accelerated its melting in the last few decades. The main portion of the glacier is approximately 1 kilometer (0.6 mile) across.

remain in the same place for a few years, and enough till may be deposited to form a ridge that surrounds the end of the glacier (Figure 16.17). These ridges are known as *moraines.* **The farthest limit of a glacier is marked by a terminal moraine.** Why might that be important? Geologists can determine the rate at which glaciers are melting by mapping the location of the terminal moraines associated with the terminus of the glacier. For example, the Gangotri Glacier, in the foothills of the Himalayan Mountains in India, has receded (melted back) more than 800 meters (0.5 mile) over the last few centuries (Figure 16.18). That observation is based on comparisons of past terminal moraine deposits and the present location of the glacier. Similarly, the presence of moraines in regions without present-day glaciers is evidence for more extensive glaciers in the geologic past. Terminal moraines have been identified throughout the Midwest and Great Plains states, showing that an ice sheet must have covered most of Canada and much of the United States in the relatively recent geologic past (more on this in Section 16.5).

Not all glaciers terminate on land. Many of the ice sheets covering Antarctica and Greenland end in the ocean. Glaciers that reach the ocean may break into icebergs or form a shelf of ice that floats in the protected waters of a shallow bay (Figures 16.14 and 16.19). From time to time, an ice shelf up to the size of Rhode Island may break off and float away from a glacier. Icebergs and ice shelves carry clay, sand, and larger boulders that were eroded

Checkpoint 16.10

basic · intermediate · advanced · superior

Examine the two pictures of glaciers provided.

a) In the photo, label as many features as you can.
b) In the photo, what do you infer is the origin of the dark stripes?

a.

b.

a.

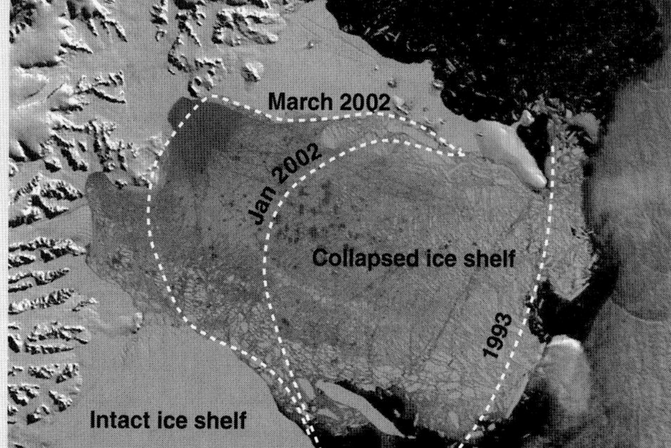

b.

Figure 16.19 The terminus of a glacier adjacent to the ocean. **a.** Icebergs breaking off from a glacier in Greenland. **b.** The Larsen ice shelf, Antarctica. An ice shelf represents the terminus of a glacier floating in shallow water adjacent to a landmass. The size of the ice shelf decreased by a few thousand square kilometers in March 2002, when most of the shelf collapsed into the ocean.

by the glacier. **As the ice melts, these sediments drop into the ocean and descend to the ocean floor to form a *dropstone deposit*** (Figure 16.20). These deposits stand out because most fine-grained sediment deposited on the ocean floor contains little to no sand-sized particles and few examples of larger grains. Iceberg formation is accelerated when the rate of glacial motion increases. Those conditions can occur when glaciers advance more rapidly as thick ice accumulates during cold periods or when ice shelves melt during warming periods. Ocean drilling programs have discovered dropstone layers all across the North Atlantic Ocean, marking cycles of climate changes during the last ice age.

Finally, melting glaciers are a key source of water supplies for farmers and cities around the world. This is especially true in regions with pronounced dry seasons. During the dry season, meltwater streams from the glacier may be the only available source of water. For example, the Gangotri Glacier in northern India (Figure 16.18) supplies about 70 percent of water in the Ganges River during dry periods. Warming climates may result in a short-term increase in stream discharge from glaciers, but as glaciers shrink and recede, water supplies will inevitably decline. This reduction in water supplies will have profound consequences for societies around the world and is arguably the most significant consequence of climate change.

Hot Deserts

What does a desert look like? Examine the images in Figure 16.21; which of the images best represents your idea of a desert?

In most hot deserts, temperatures are high, annual rainfall is less than 25 centimeters (10 inches), and evaporation exceeds precipitation. Sand is the dominant feature in one-half of all desert surfaces, while others have a desert pavement composed of larger pebbles and rocks. In reality, most desert surfaces are a combination of sand, desert pavement, and regions dominated by rock outcrops (Figure 16.21). As we learned previously, many deserts are located below subtropical high-pressure systems around 30°N and S. However, deserts also exist in the centers of some continents simply as a result of their greater distance from the ocean. For example, the Gobi Desert (see Chapter Snapshot) is located far inland from the Indian Ocean where the path of moisture-laden winds is blocked by the high peaks of the Himalayan Mountains.

***Erosion and Deposition in Deserts.* Occasional flash floods and wind action are responsible for eroding, transporting, and depositing material in deserts.** Flash floods, though uncommon, rapidly move large volumes of material and play a major role in shaping desert topography. Slower-acting processes associated with winds act over much longer periods and also affect

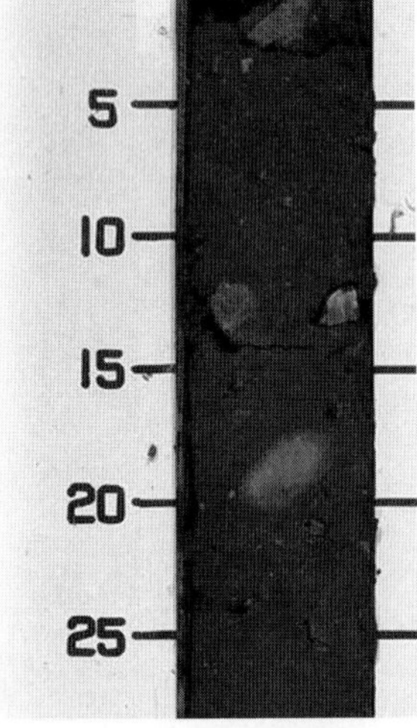

Figure 16.20 Dropstones in fine-grained ocean sediment. The larger grains represent material eroded by glaciers that dropped to the seafloor off the east coast of Greenland when icebergs melted. This core was recovered by scientists of the Ocean Drilling Program (ODP). ODP recovers sediments from the ocean floor around the world as part of a varied research program that analyzes all aspects of the earth system. The numbers represent a centimeter scale.

Figure 16.21 Desert environments. **a.** Sandy desert. **b.** Desert pavement. **c.** Rocky desert. What differences do you notice?

a.

Checkpoint 16.11

basic | advanced | intermediate | superior

Draw a diagram that illustrates the fate of a snowflake that falls in the accumulation zone of a growing alpine glacier. Begin and end the path of the snowflake in the atmosphere.

Checkpoint 16.12

basic | advanced | intermediate | superior

In the introduction to this chapter, we mentioned that glaciers in Alaska are shrinking. Discuss how you would plan a study of glaciers in other places to determine if this was a local, short-term phenomenon or part of a larger, long-term global trend. Include a description of how you would measure changes in the size of a glacier.

deserts. Most deserts lie in the trade winds belt and experience a prevailing northeast or southeast wind. Winds pick up dry sand grains and sandblast objects in their path. Although the sand is removed, larger pebbles are left behind to form a desert pavement (Figure 16.21b).

The effects of wind vary, depending on the size of the sediment particles. Winds carry the finest particles beyond the boundaries of the desert. These light particles can be picked up and transported in suspension, much as the fine particles of pollution are suspended in the air above some cities. For example, dust particles originating in the Sahara are carried across the Atlantic Ocean by vigorous storms (see Figure 7.24b). Increasingly, China, North Korea, South Korea, and Japan must issue health warnings to warn citizens of approaching dust storms originating from the Gobi Desert. Larger sand grains that are too big to be transported in suspension can be skipped along the desert floor by localized wind eddies, a process known as *saltation* (Figure 16.22). Grains that are too heavy to be picked up by winds creep slowly along the surface because of impacts from grains undergoing saltation.

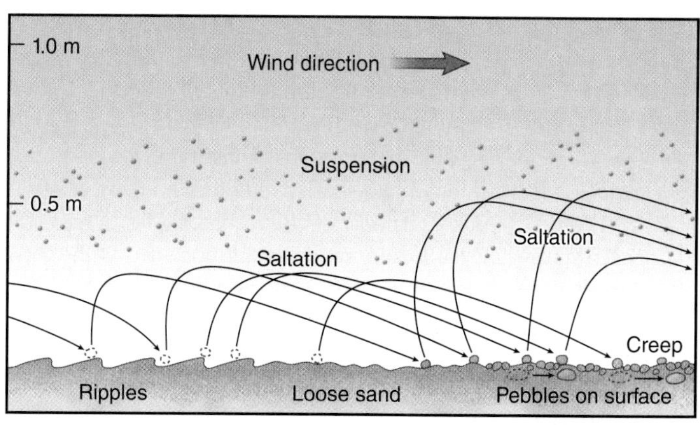

Figure 16.22 Wind action and desert sediment. Winds move sediment by a combination of suspension (fine particles), saltation (sand grains), and creep (larger particles). The wind action results in the formation of ripples in sand and may produce a desert pavement where sand is removed but pebbles are left behind.

The area around the South Pole receives just a few centimeters of snowfall each year. Is the South Pole a desert? Give reasons to support your answer.

How does the transport of sediment by winds (Figure 16.22) compare to transport in streams?

b.

c.

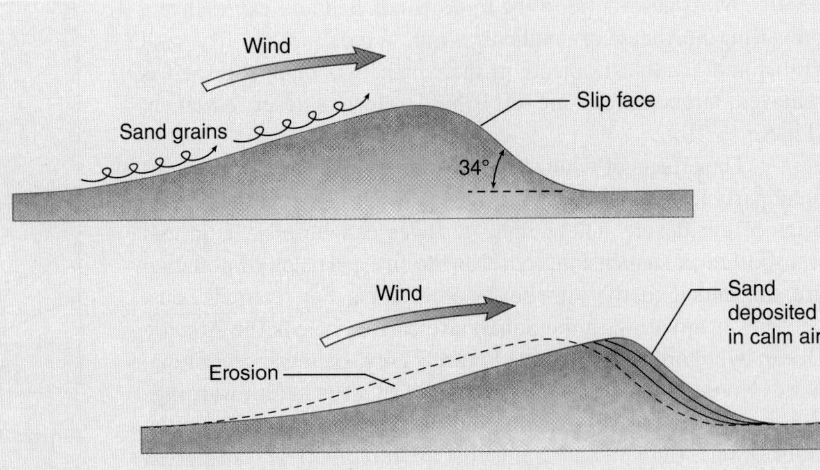

Figure 16.23 Characteristics of sand dunes. **a.** The most common sand dunes have a gentle windward face and a steep lee (slip) face. Sand is removed from the gentle slope and carried to the top of the dune. The dune migrates slowly in the direction of the wind as a result of this process. The sand is deposited on the lee face of the dune. Loose, dry sand cannot support a slope with an angle of more than 34 degrees. The sand is deposited in a series of sloping surfaces known as cross beds on the lee face. **b.** A series of crescent-shaped sand dunes migrating from top right to bottom left, parallel to the wind direction. **c.** Cross bedding in a sand dune. **d.** Cross bedding preserved in the Navajo Sandstone, Zion National Park.

Checkpoint 16.15

Venn Diagram: Wind Action and Glacial Processes

Complete the Venn diagram provided here by placing the six listed descriptions in the appropriate locations on the diagram. Add at least six more characteristics.

1. Transport sediment in direction of movement
2. Can transport large boulders
3. Form dunes
4. Occur most frequently at high latitudes
5. Few associated plants and animals
6. Occur on at least five continents
7.
8.
9.
10.
11.
12.
13.
14.

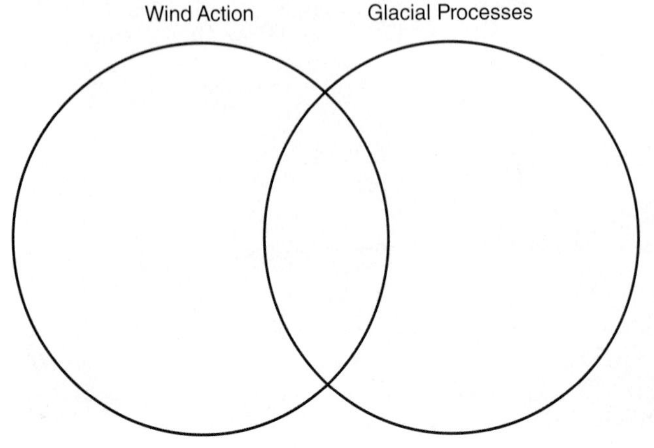

Checkpoint 16.16

Discuss how you would plan a study of desert environments to determine if they are increasing or decreasing in size. Include a description of how you would measure changes in the size of a desert.

Sand grains may be deposited together to form sand dunes. This is especially true in locations where winds converge on their way through a gap in a mountain range. Dunes may be piled up over time to form extensive sheets of sand. The most common forms of dunes are crescent-shaped and asymmetric—they have a short, steep *lee* face and a longer, gently inclined *windward* side (Figure 16.23a, b). Wind transports grains up the windward side of the dune, and the grains tumble down the steep lee slope. These features have characteristic patterns of layering that are preserved when they are converted to rock (Figure 16.23c, d). **Dunes have a pattern of sloping layers, known as cross beds,** that slope in the same direction as the wind blows (Figure 16.23c, d). Cross beds are one of the features that reveal that rocks were formed in an ancient desert environment and can be interpreted to determine the direction of the prevailing winds when the dunes formed.

16.5 Records of Climate Change

Learning Objectives

- Compare, contrast, and apply a proxy record of climate change while interpreting past climate changes.
- Describe major periods of past climate change.

Imagine that you have boarded an airplane with no windows and have flown somewhere far away. The pilot does not tell you where you have landed. What evidence would you use to determine the climate of the region?

b.

c.

d.

You should be able to come up with several indicators, such as the type of plants and animals you see or people's clothing. We term any such indicator as a *proxy* for climate. **A proxy is something that stands in for something else.** For example, the change from native woodlands in Missouri to grasslands in Kansas is a proxy for the westward decrease in precipitation. Climatologists must use the same type of reasoning when investigating past climates because they have complete records for only the last 150 years. Climate fluctuations that occurred on Earth before instrument-based records must be determined by analyzing a variety of cultural, biological, chemical, and geological proxy data sources, including tree rings, oxygen isotopes, and microfossils. Some of these data allow scientists to reconstruct climates stretching back millions of years, while others provide precise annual records that cover the advance of human civilizations.

Weather Records from Instruments

As we learned in Chapter 15, simple weather instruments did not become available until the 1700s, and there were few successful efforts to collect weather data over large regions until more than a century later. Consequently, the **detailed, accurate data on temperature and precipitation that are necessary to characterize climate have only been collected for about 150 years.** These measurements represent the air temperature at the base of the troposphere for stations around the world and show a gradual temperature increase of 0.8 C° (1.4 F°) during the last 130 years (Figure 16.24a). In addition, satellite data from several kilometers up in the troposphere have become available in the past few decades and exhibit a similar global warming trend (Figure 16.24b).

Interpreting the satellite data is much more complex than analyzing the ground-based data. Changes to satellites and adjustments of their orbits require scientists to make numerous calculations in order to accurately calibrate the satellite data. Early interpretations were marred by an error introduced by the drift of the satellites from their projected orbits. This resulted in

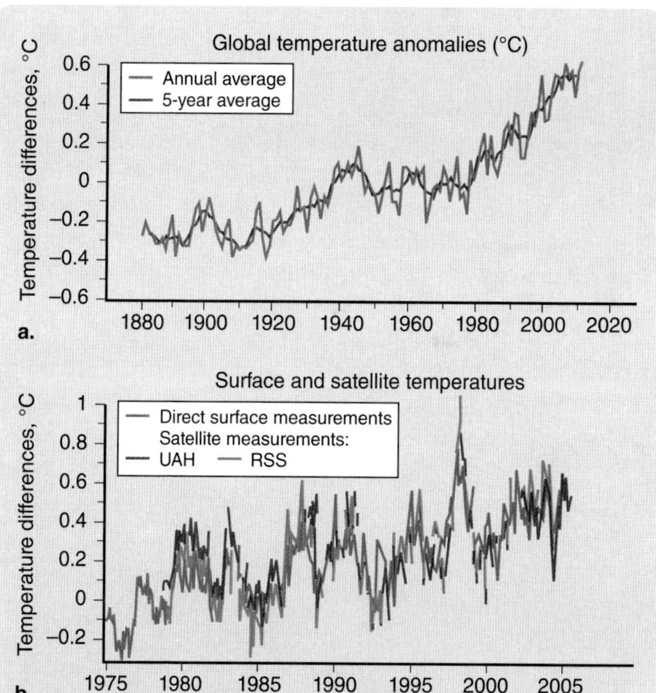

Figure 16.24 Instrumental global temperature record. **a.** Average global temperatures since the 1880s. Temperatures are compared to a standard represented by a 30-year (1951–1980) mean temperature. The difference with this standard is the temperature anomaly. Recent average annual temperatures are about 0.5 C° above the standard, and the temperatures at the beginning of the record are about 0.3 C° cooler. The thin line represents annual data, while the thick line is a 5-year running mean. **b.** Average temperatures from the troposphere taken by satellites since 1979. The plot shows interpretation from two groups of scientists—UAH (University of Alabama, Huntsville) and RSS (Remote Sensing Systems)—that are both working with the same data.

lower-than-anticipated temperature increases and set off a fierce debate about whether warming was actually taking place. However, recent analyses corrected these early problems, and the two data sets now show similar trends—Earth is getting warmer.

Cultural Records

Historical and archeological records indicate that changes in **climate patterns influenced where past civilizations were able to flourish** (or not). One example of this phenomenon is the Viking colonization of the North Atlantic islands. The Vikings migrated from Scandinavia approximately 1,000 years ago, when the North Atlantic pack ice melted during a period of warmer temperatures. Several lines of archeological evidence (ruined buildings, evidence of agriculture, written accounts of voyages) indicate that Viking settlements moved progressively westward from the Scottish islands to Iceland and finally to Greenland between A.D. 850 and 985 (Figure 16.25). Archeologists hypothesize that higher temperatures at that time increased the length of the Greenland growing season sufficiently to support grasses needed for grazing animals.

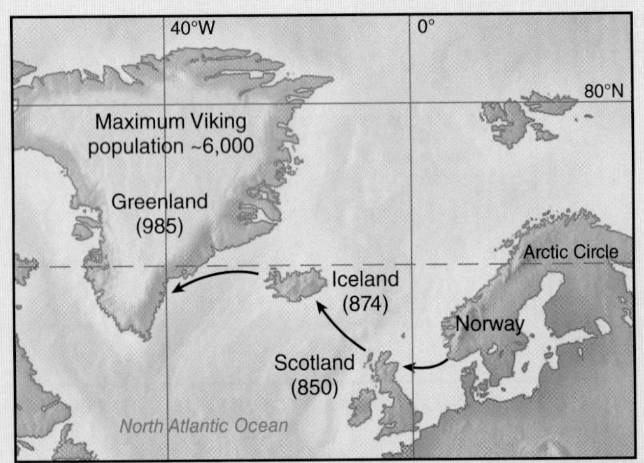

a.

b.

Figure 16.25 Viking colonization of Greenland. **a.** Viking settlers were able to migrate from Norway to Greenland when a warming trend melted pack ice in the North Atlantic. **b.** A few centuries later, when temperatures declined again, so did the Vikings' society, as represented by these ruins on Greenland.

However, historical documents also contain signs that the climate began to reverse itself a few centuries later, beginning a trend from relative warmth to colder temperatures around 1400. As temperatures declined, so did the Vikings' agricultural base. Although they came in contact with native Inuit peoples from the north, the Vikings did not adopt Inuit means of food production (hunting with harpoon), boat building (using skins rather than rare wood), or clothing (wearing skins instead of wool). In addition, increasing pack ice cut them off from supplies via the sea. Consequently, the Vikings became increasingly maladapted to the changing climate, some members left, and the rest of the colony died out in the 1400s. Climate change did not lead to the extinction of the Viking colony on Greenland, but it did make the landscape more hostile, requiring greater social flexibility than was possible in the relatively rigid Viking culture.

Archeological data, written historical records, and even works of art all provide qualitative data that allow us to compare past climates with more recent patterns. For example, an examination of outdoor scenes in thousands of paintings created in the 600 years since 1400 reveals that those painted from 1550 to 1849 are most likely to exhibit cloudy skies and darker scenes. An example is Pieter Bruegel's *Winter Landscape with Bird Trap,* painted in 1565 (Figure 16.26). This painting reflects a time when many European rivers consistently froze solid in winter, a phenomenon that has rarely occurred during the last century.

Such cultural records for the Northern Hemisphere indicate that **the last 1,000 years or so can be divided into three distinct informal climate periods:**

1. European agricultural records and histories of the settlement of Greenland both illustrate that temperatures were relatively warm from A.D. 1000 to around 1450. This period of apparent warmth is known as the "Medieval Warm Period."
2. The next 400 years were marked by some excessively cold periods interspersed with more "normal" temperatures. This period has been labeled the "Little Ice Age," although it was not really an ice age, just a time of relatively cold climate.
3. Finally, toward the end of the nineteenth century, the climate appears to have moderated, leading to our present relatively warm temperatures, which exceed any experienced during the previous 1,000 years.

These qualitative records document that climate changes have affected societies in the recent past. However, such data do not shed much light on how often these changes occur, if they are predictable, or if humans are affecting modern climate. Next, you will learn that the magnitude of the associated changes in temperature or precipitation can be determined by using short-term climate proxies, which can, in turn, be linked to modern records from weather stations.

Figure 16.26 Climate in art. *Winter Landscape with Bird Trap* by Pieter Bruegel the Elder, painted in 1565. It was common for rivers to freeze in Europe at this time, but it is a rare event today.

Short-Term Climate Trends: Annual Cycles

Do you keep a diary or record your thoughts in a journal? If you were to review what you wrote, how many times would weather phenomena appear? How would you have described them? Historical written records tell an incomplete story about climate because people often had more pressing concerns than the weather. Consequently, weather phenomena appear in records only when conditions were unusual, such as during an especially hot or dry episode or a particularly severe winter ("the coldest winter that anyone can remember"). These records lack the detail necessary to compare with modern climate patterns. Just how hot or cold, or wet or dry were these times? To compare past climate patterns with recent temperature and precipitation data, we need a detailed record that overlaps with modern instruments. We can find such records in tree rings, lake sediments, and ice layers because they yield information about annual changes.

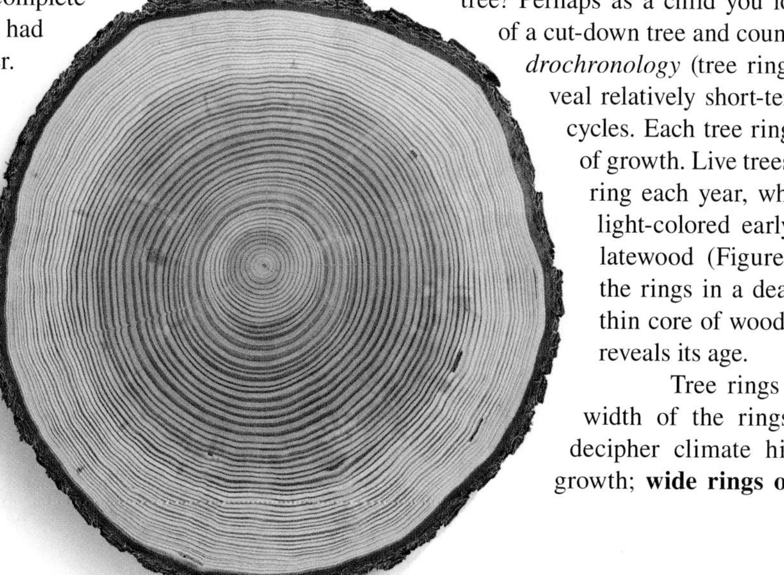

Figure 16.27 Tree rings in a Douglas fir. Note the variation in the spacing between these tree rings. Would this be a good tree species to use for tree ring dating studies?

Tree Rings. Have you ever figured out the age of a tree? Perhaps as a child you looked at the trunk of a cut-down tree and counted the rings. *Dendrochronology* (tree ring research) can reveal relatively short-term climate change cycles. Each tree ring represents 1 year of growth. Live trees add a new growth ring each year, which is made up of light-colored earlywood and darker latewood (Figure 16.27). Counting the rings in a dead tree or taking a thin core of wood from a living tree reveals its age.

Tree rings vary in size. The width of the rings can be used to decipher climate history during tree growth; **wide rings occur during wet,**

warm years, narrow rings during cool or dry years. A single Huron pine tree from Tasmania, Australia, that was found to be more than 2,200 years old revealed the climate fluctuations that occurred from 270 B.C. to A.D. 1973. Few trees are that old. Most dendrochonology records depend on matching partial records from multiple trees. For example, tree rings from 95 trees were used to graph precipitation fluctuations from 1260 to 1998 for northern Wyoming (Figure 16.28). Since these records were anchored in a century of modern precipitation data, scientists were able to correlate ring width to rainfall for tree rings formed in the last century and then use this relationship to predict rainfall values for older rings. This record reveals that droughts were typically longer

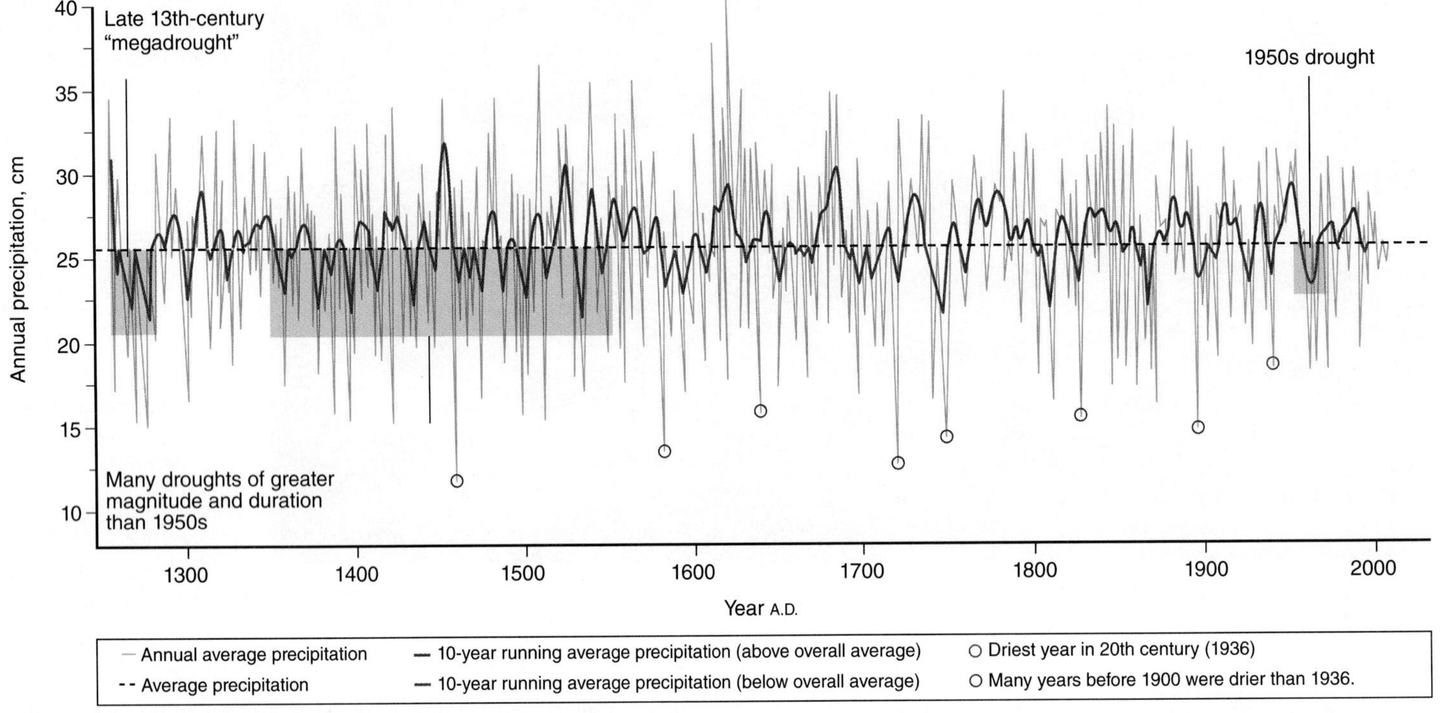

Figure 16.28 Northern Wyoming precipitation record determined from tree ring analysis. Precipitation records from 1895 onward were matched with tree ring widths to estimate precipitation values for pre-1895 tree rings. The dashed horizontal line represents the average precipitation determined from a century-long instrument record database. What patterns can you identify in the graph?

Checkpoint 16.17

Compare the tree ring record of precipitation in New Mexico graphed here with the same interval of the record from Wyoming illustrated in Figure 16.28. How do the records compare, and what does this imply about drought conditions in the West during this time period?

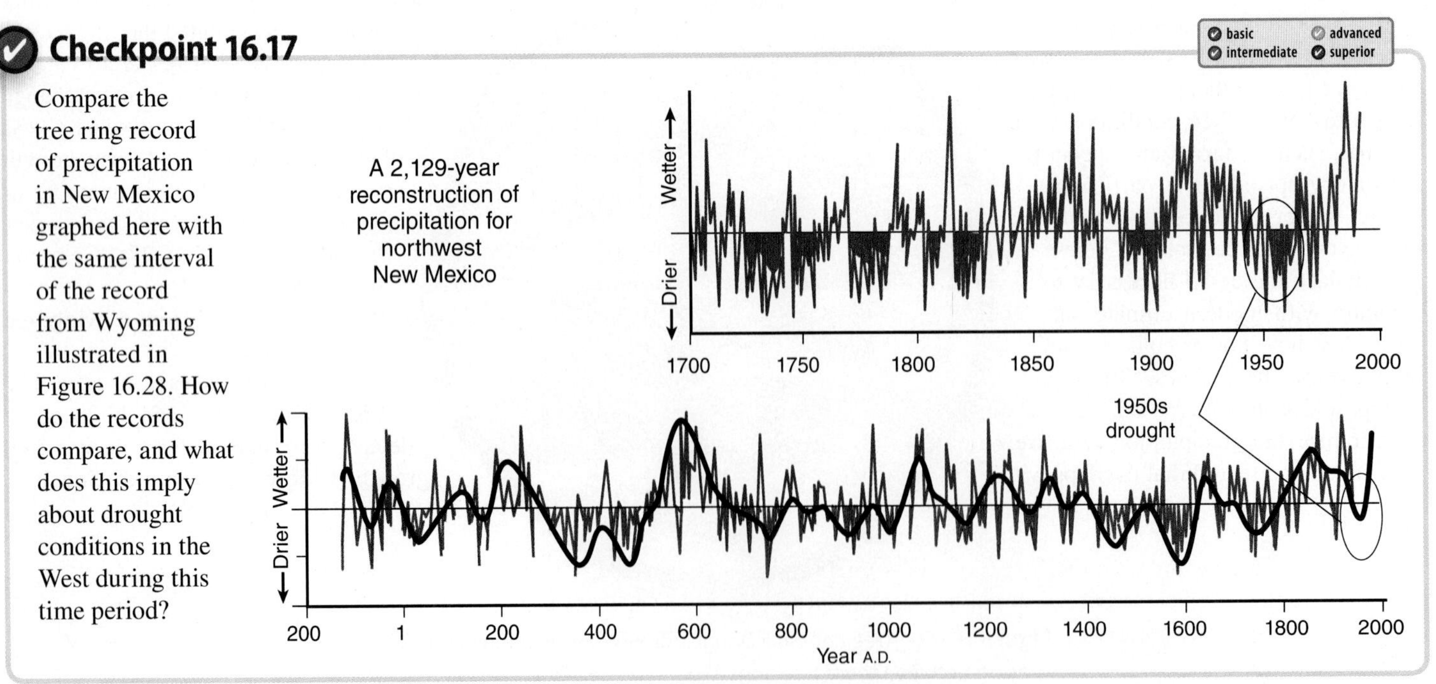

and more severe prior to modern times, with decade-long droughts relatively common. Scientists have been able to use tree rings to establish an 8,000-year climate record for some regions.

Lake Sediments. Much like tree rings, lake sediments are often deposited in pairs of light and dark annual layers, which can be interpreted to reveal the climate history of their region. These **paired layers, known as *varves*, reflect seasonal changes in deposition in the lake.** Algae grow on the lake surface as temperatures warm in late spring and remain until cooling takes hold in the fall. The algae then die and settle to the bottom of the lake, and their organic materials produce a dark layer of sediment. Sediments toward the margins of the lake are deposited by streams and increase in thickness during periods of increased precipitation. During the colder part of the year, much of the sediment deposited in the center of the lake, away from stream sources, comes from windblown silt and clay-sized particles, producing a lighter-colored layer (Figure 16.29). The thickness of varves may indicate the length of the warm season and the presence of wetter or drier (windier) times, respectively.

Unlike tree rings, varves contain additional proxy climate indicators. For example, analysis of a 10,000-year-long varve record from Elk Lake, Minnesota, revealed thicker annual layers produced during a drier interval from 4,000 to 8,000 years ago. These layers contained higher proportions of pollen from dryland plants (sagebrush, grasses), more quartz sediment derived from nearby sand dunes, and a higher sodium content as a result of less chemical weathering. Lake clays may also contain organic remains or ash layers that can be dated by using radioactive isotopes (see Chapter 8) to verify ages.

In the 1930s, scientists discovered something unexpected in pollen samples from lake deposits in Sweden. These samples contained pollen from temperate vegetation (birch and conifer trees) overlain by layers containing the pollen of *Dryas octopetala,* a flower characteristic of the Arctic tundra (Figure 16.30). These sediments were dated as approximately 12,000 years old. This evidence was interpreted to show that a dramatic shift from temperate to polar climates had occurred. Most researchers assumed the climate changed over hundreds or thousands of years. But, as you will soon learn, recent data reveal a much more remarkable story.

Ice Layers. Each year, snow derived from evaporated ocean water falls on Greenland. Because of the low annual temperatures there, very little of that snow melts. As we discussed in Section 16.4, each annual snowfall becomes buried deeper and deeper until the pressure turns it to ice (see Figure 16.14c). Like varves, ice layers can be counted to determine their age. The thickness of the layers is indirectly related to temperature and directly tied to precipitation. In addition, the ice layers contain dust and pollen that can be used to infer climate changes.

Drawing Conclusions from Short-Term Proxy Records. Tree rings, varves, and ice layers can provide considerable information about Earth's climate extending back to more than 10,000 years.

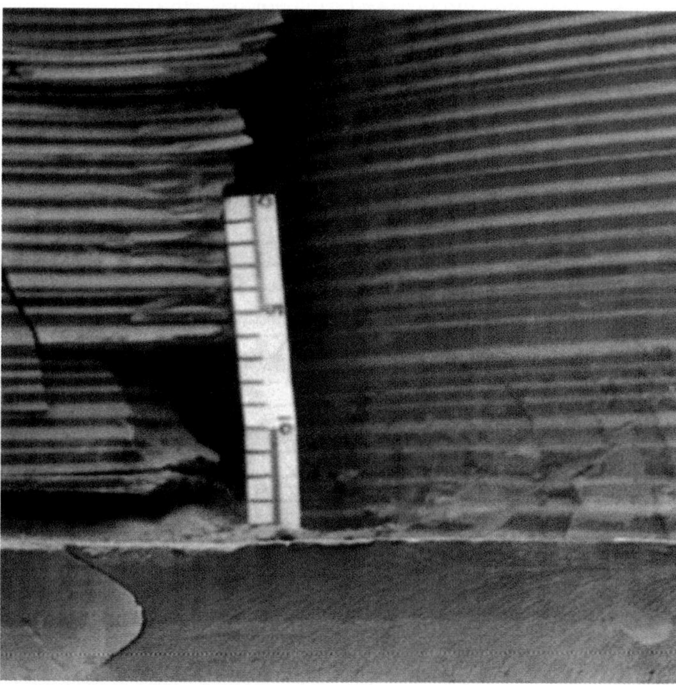

Figure 16.29 Lake sediment record. Sediment deposited in a glacial lake is preserved in paired annual layers. Dark layers represent summer, and lighter layers represent winter. Ruler markings are in centimeters.

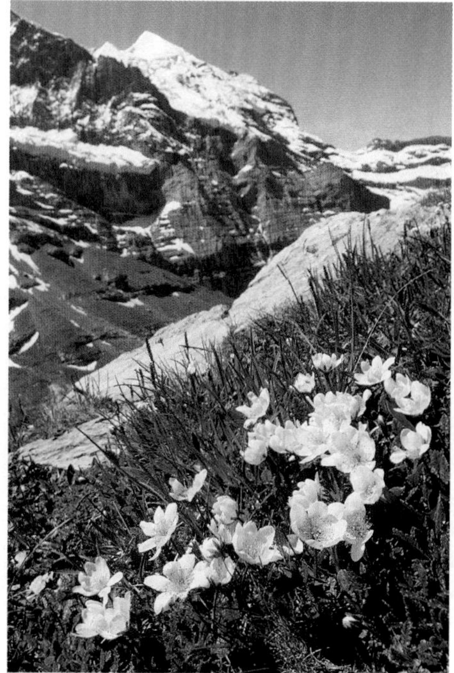

Figure 16.30 *Dryas octopetala* flower. Pollen from this small Arctic flower is a proxy climate indicator for polar climates.

By linking these reconstructions with records of modern instrument readings, scientists are able to recognize intervals such as the Medieval Warm Period and the Little Ice Age in the climate record of the last millennium (Figure 16.31a). Both of these informally recognized intervals are observed in Northern Hemisphere climate records, particularly those in Europe and North America, but their signatures are either absent or less distinct elsewhere. Global climate records show much more extensive and dramatic temperature changes occurred in the last 15,000 years (Figure 16.31b). These proxies provide broad evidence of the character of climate systems around the globe and allow scientists to separate regional from global climate changes and to compare the relative impacts of different events in different locations.

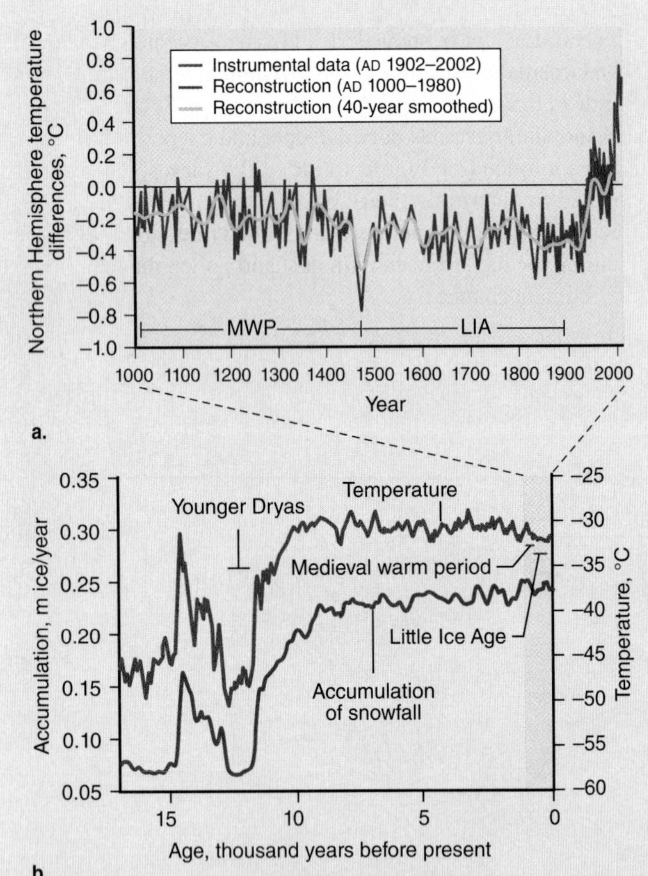

Figure 16.31 Reconstructed temperature records for the Northern Hemisphere. **a.** Temperatures were higher during the Medieval Warm Period (MWP) up to around 1450 and lower during the Little Ice Age (LIA; 1450–1900). The red line on the graph represents the actual instrument record; the blue line is the reconstruction based on climate proxy data. The yellow line is an average of the reconstructed data. Temperatures are presented relative to values for the 30-year period from 1961 to 1990. Note that recent temperatures exceed anything from the last 1,000 years. **b.** Temperature and snowfall record of central Greenland over the last 17,000 years. The global ice age ended around 15,000 years ago, but the Younger Dryas represents a late cooling event that immediately preceded warmer temperatures that persist until today.

Long-Term Climate Trends: Abrupt Change and Millennial Cycles

Only a few physical records stretch further back in time than 10,000 years. Thus, if we are to understand long-term variations in global climates on timescales measured in 100,000-year or longer intervals, we must change our focus to geochemical proxies. This section examines not only the long-term climate record spanning hundreds of thousands of years but also the rate at which these changes occur.

Evidence from Oxygen Isotopes. Oxygen isotopes (oxygen atoms with different numbers of neutrons) serve as a proxy climate indicator for long-term timescales. Water molecules present in the oceans, atmosphere, and ice contain both light isotopes (^{16}O) and heavy isotopes (^{18}O; Figure 16.32). The atoms have 8 and 10 neutrons, respectively. The ^{16}O isotope is dominant, making up 998 of every 1,000 oxygen isotopes. The proportion of each isotope present in the water is expressed as an oxygen isotope ratio of $^{18}O/^{16}O$. Under present conditions, this ratio is very small (approximately 0.002). Any process that causes the number of ^{18}O isotopes to increase or the number of ^{16}O to decrease results in an increase in the ratio. Likewise, more ^{16}O or less ^{18}O produces a decrease in the ratio.

Evaporation of ocean water occurs during both warm and cold periods. Water molecules with ^{16}O preferentially evaporate because ^{16}O is lighter than ^{18}O. During normal (equilibrium) conditions when glaciers are neither expanding nor contracting, ^{16}O is lost by evaporation but is returned to the oceans as precipitation and stream flow, resulting in no net change in the $^{18}O/^{16}O$ ratio (Figure 16.33a). During cold periods, evaporated water that is enriched in ^{16}O is stored in growing ice sheets. This leaves the oceans smaller (by volume) and relatively enriched in ^{18}O compared to when the ice sheets are smaller (Figure 16.33b). Therefore, $^{18}O/^{16}O$ ratios are larger in the ocean during cold periods. During especially warm periods, water in the ice sheets melts, returning ^{16}O isotopes to the ocean. Therefore, ocean $^{18}O/^{16}O$ ratios are smaller in the ocean during warm periods.

These oxygen isotopes are also incorporated into the skeletons of microscopic organisms that dwell in the oceans, such

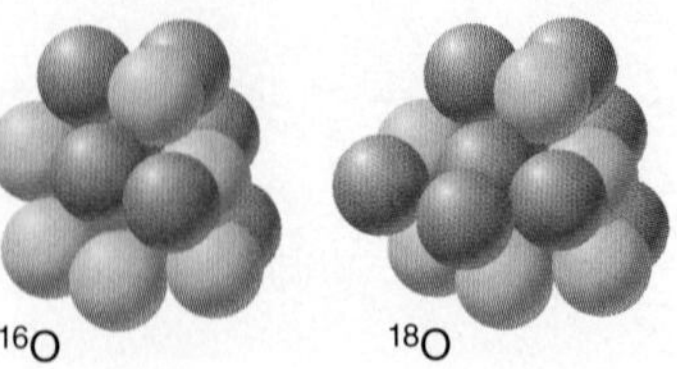

Figure 16.32 Two isotopes of oxygen. ^{18}O has two more neutrons than ^{16}O and is therefore 12.5 percent heavier. ^{16}O is much more abundant in the ocean than ^{18}O. Changes in the ratio of these isotopes indicate different climate states in the ocean.

as *foraminifera* (Figure 16.34). Foraminifera, or forams to their friends, are tiny protozoans that have shells. Often, these shells are made of calcium carbonate, which they secrete in equilibrium with the water in which they are living. If the water has a certain amount of ^{16}O and ^{18}O in it, so will the foram shell. This is also true for other biological proxy indicators, such as microcrustaceans called ostracodes and weird algaelike creatures called acritarchs. Oxygen isotopes as measured in foraminifera carry at least two signals: ice volume and temperature. Ice volume is reflected in the $^{18}O/^{16}O$ ratio, while surface temperatures have the greatest effect on population density (colder temperatures are associated with smaller populations).

The oxygen isotope ratio acts as a paleothermometer for ancient climates. That is, we can compare the oxygen isotope ratio and foram population proxies with modern values. The difference between past and present can be used to estimate the temperature of the water in which the microscopic organisms grew.

In addition to using forams and other organisms as proxy temperature indicators, we can measure the relative abundance of different species to get an idea of how the biosphere was responding to climate changes. Large drops in biodiversity typically correspond with sudden significant climate changes (see "Mass Extinctions" in Section 8.3).

Checkpoint 16.18

basic · intermediate · advanced · superior

Examine the graph and answer two questions.

1. Which letter in the graph corresponds to the highest temperatures on the basis of the $^{18}O/^{16}O$ data?
2. Which letter in the graph represents the time when the oceans were at their lowest levels?

$^{18}O/^{16}O$ ratio: 0.0035, 0.0030, 0.0025, 0.0020, 0.0015, 0.0010
A, B, C, D
1.0, 2.0, 3.0, 4.0
Time before present (millions of years ago)

Figure 16.34 Shells of the microorganism foraminifera. Individual shells are less than 1 millimeter (0.04 inch) across.

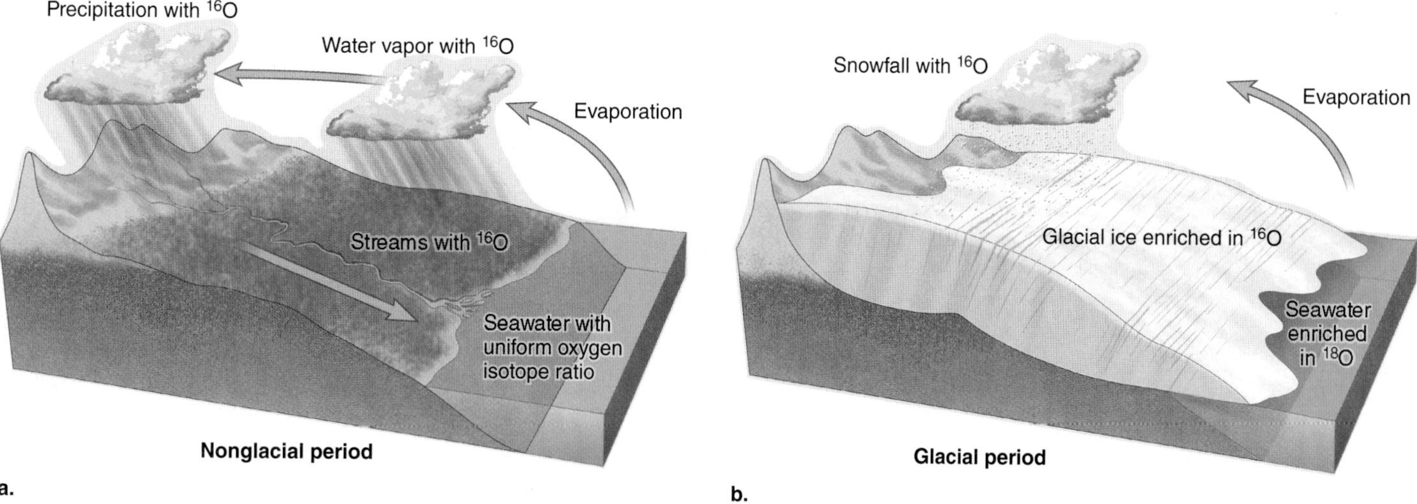

Figure 16.33 Oxygen isotope ratios and the ocean/ice system. **a.** The lighter ^{16}O isotopes evaporate with seawater and are returned to the ocean as a result of precipitation and stream flow. **b.** The ^{16}O isotopes are incorporated into continental ice sheets during glacial intervals. This causes the oceans to become enriched with the heavier ^{18}O isotope.

Interpreting the Climate Record

Widespread evidence such as moraines and other glacial features indicates that **the Northern Hemisphere experienced glaciation during the last 2 million years** (Figure 16.35). This period of extremely cold climate is termed an *ice age*. The climate of North America was dominated by the presence of a 3-kilometer-thick (1.9-mile) continental glacier centered over Canada, which had the following effects.

- Sea level was lower because much of the water was trapped as ice; consequently, the amount of exposed continent was larger.
- Lower sea levels resulted in a land bridge between Siberia and Alaska.

Checkpoint 16.19

basic · intermediate · advanced · superior

The following graph illustrates the relative length of time that each proxy climate indicator can be applied. Match the five proxies listed here with the bars on the graph.

Tree rings
Microfossils
Ice cores
Instrumental records
Archeological data

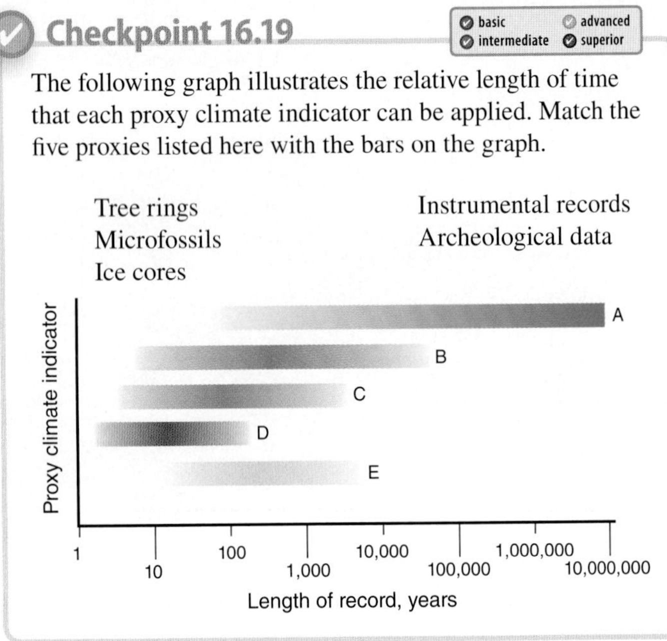

Figure 16.35 Comparison of Northern Hemisphere during and after the last ice age. Large continental ice sheets covered northern regions of North America and Europe 18,000 years ago. Today, an ice sheet is present only on Greenland in the Northern Hemisphere.

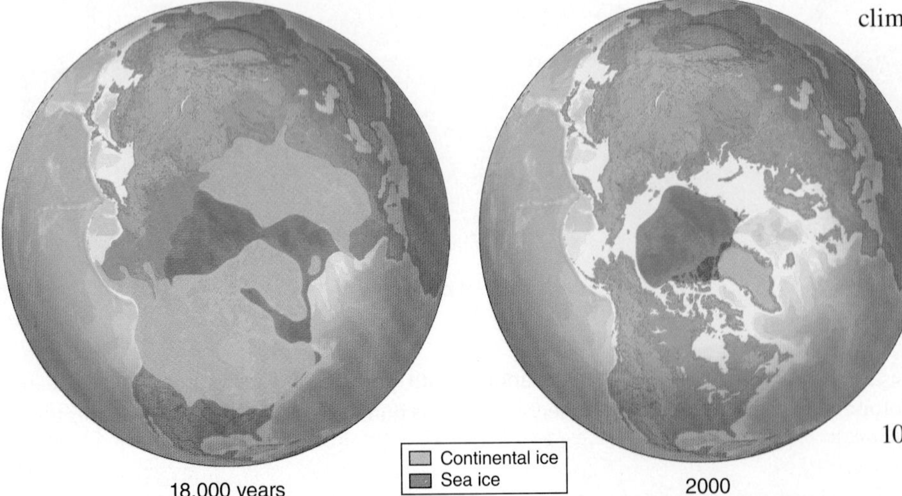

- Eastern North America was forested. Forests extended into present-day grassland regions such as the Great Plains, which experienced higher rainfall (see Section 12.4, A Case Study: The High Plains Aquifer).
- Deserts in the western United States were cooler, and thus less evaporation occurred there during glacial times. This may have coupled with the increased moisture or the same amount of moisture that the region receives today. In either case, the net effect was greater stream discharge and large regional lakes.

The Northern Hemisphere would have looked considerably different than it does today.

Intervals and Rates of Climate Change

Analysis of oxygen isotope ratios and other proxies has been used to divide the ice age into 120,000-year-long cycles featuring **long cold intervals known as *glacials* (about 100,000 years) and short warm intervals termed *interglacials* (about 20,000 years;** Figure 16.36). The temperature range during the glacials and interglacials was an order of magnitude greater than the difference between the Little Ice Age and Medieval Warm Period (see Figure 16.31b). Temperatures show some modest variations but are relatively stable during the interglacials and are typically a few degrees warmer than present-day temperatures. In contrast, **temperatures during the glacials were at least 5 to 10 C° (9 to 18 F°) colder than today** and exhibited considerable fluctuations. On the basis of the typical length of interglacials, we are about 10,000 years away from entering another cold glacial interval.

The ice record shows that temperatures on Greenland increased dramatically approximately 15,000 years ago (see Figure 16.31b) but took a jump back to cold conditions during a relatively brief period, christened the Younger Dryas, from 12,800 to 11,600 years ago. This cold interval was marked by the appearance of the pollen of the polar wildflower *Dryas octopetala* in varves and ice layers (see Figure 16.30). As the Younger Dryas ended, the world entered a warm episode called the Holocene that represents the latest interglacial. Until a few decades ago, it was widely accepted that climate changed gradually, taking hundreds or thousands of years to undergo multidegree temperature changes. While this may be true some of the time, we now know that climate can change much more swiftly.

Analyses of temperature proxies in ice cores drilled from the Greenland ice sheet in the 1990s have allowed scientists to determine exactly how long it took for the climate shift at the end of the Younger Dryas. Greenland went from the cold, polar conditions of the Younger Dryas to a warmer Holocene climate over just a handful of layers in the ice core. **The average annual temperature of much of the Northern Hemisphere increased by 8 C° (12 F°) in less than a decade!** It is as if Chicago, Illinois, were to migrate south to Atlanta, Georgia, over a 10-year period. The transition into the Younger Dryas

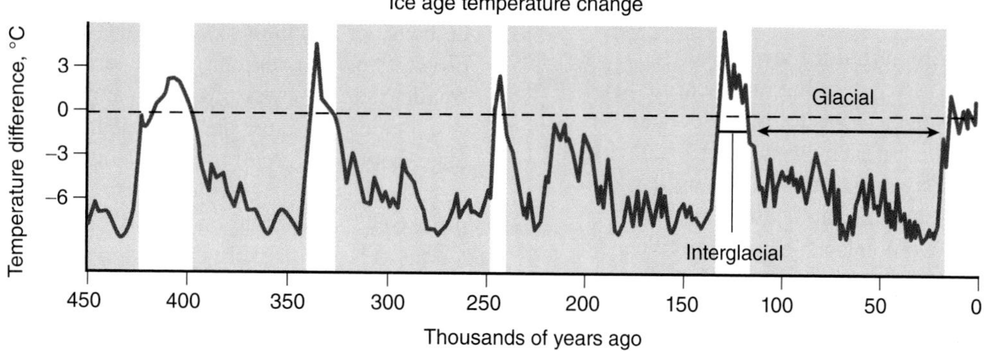

Figure 16.36 Reconstructed long-term temperature records for modern ice sheets. Temperature history is determined from analysis of ice cores from Antarctica. Warmer interglacials occurred at approximate 100,000-year intervals and were separated by colder glacial intervals characterized by significant temperature fluctuations. Glacials were about 5 to 10 C° (9 to 18 F°) colder than today, and interglacials could have been a few degrees warmer.

was also rapid, occurring in a few decade-long steps. The concept of an abrupt switch in climate conditions represented a paradigm shift for scientists studying climate change.

Similar abrupt climate changes occurred throughout the glacials. It is hypothesized that rapid warming at the end of an especially cold interval resulted in the catastrophic failure of the ice sheet that covered much of the Northern Hemisphere. This collapse was marked by fleets of icebergs surging out into the North Atlantic from the Hudson Bay region of Canada or from an ice sheet in northern Scandinavia. In addition, research suggests that a giant freshwater glacial lake, Lake Agassiz, broke through an ice barrier at this time. These events sent huge volumes of freshwater into the North Atlantic Ocean. Imagine adding a layer of freshwater 10 to 20 meters (33 to 66 feet) thick to the surface of the Atlantic Ocean around Greenland and Iceland. Such large volumes of freshwater slowed, or stopped, the thermohaline circulation that drives the oceanic conveyer belt. Once Earth's oceanic heat engine was disrupted, temperatures plummeted across Europe, Asia, and Africa.

Even the longest ice core retrieved from Antarctica extends back less than 1 million years. If we wish to model earlier climate history, we must rely on interpretations of the chemistry of fossils and the characteristics of the sedimentary rock record. These proxies allow scientists to reconstruct a reasonably accurate climate record back to the last major extinction at the beginning of the Cenozoic Era, 65 million years ago (Figure 16.37). What is astonishing is that these data contain evidence that **Earth was 10 to 15 C° (18 to 27 F°) warmer 52 million years ago.** Furthermore, scientists have used these data to infer that the temperature contrast between the equator and the poles was much less than at present. There were no polar ice caps, and rocks on islands in northern Canada contain fossils of alligator-like reptiles from this time. Scientists are still working on a combination of factors that could explain these "hothouse" climate conditions. We will discuss potential factors affecting climate change in Section 16.6.

The climate record of Earth becomes less well constrained as we move back in time. Some scientists hypothesize that the entire Earth was frozen (except in the deepest parts of the ocean) from 580 to 750 million years ago. The ending of this "snowball Earth" period corresponds closely with the explosion of life, marking the

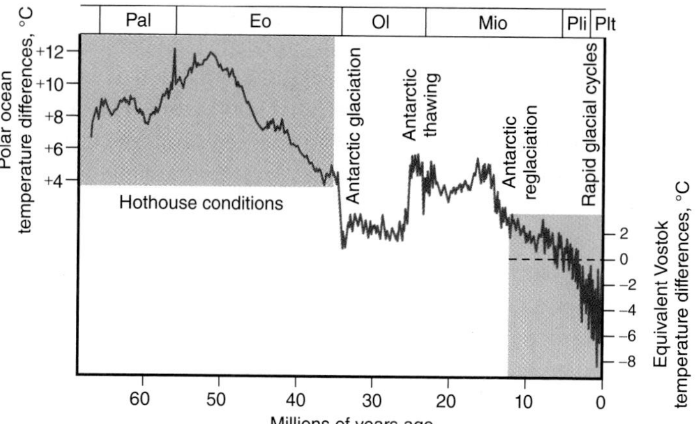

Figure 16.37 Sixty-five million years of climate change. Note that temperature is correlated only with the earliest and latest portions of this curve. Generally, temperatures were 4 to 10 C° (7 to 18 F°) warmer than today. This was the case until the major ice sheet on Antarctica formed around 34 million years ago.

beginning of the Cambrian period (see Chapter 8). Although the details of these climate changes are difficult to discern because they occurred so long ago, there is ample evidence that climate has changed dramatically throughout Earth's history.

Implications for the Future. Dramatic climate changes occurred more than 10,000 years ago because there was a huge accumulation of ice that melted and flooded the northern Atlantic Ocean with freshwater. Could the same thing happen today? Today's potential freshwater sources are represented by smaller continental glaciers, such as the one that occupies Greenland. At current melting rates, the Greenland ice sheet simply cannot produce enough freshwater to generate the dramatic alteration in salinity necessary to cause an abrupt change in thermohaline circulation. Even if the warming were to cause the Greenland ice sheet to melt 100 times faster, it would still take many decades to produce a volume of freshwater equivalent to the ice sheet collapses of the last 100,000 years.

Checkpoint 16.20

basic advanced intermediate superior

Imagine that a 12-year record of ice layers and tree rings corresponded with the rapid warming representing the transition from the Younger Dryas to the Holocene. Draw two diagrams, one for each proxy record, showing the changes in each proxy during the transition. Describe at least one other proxy that might be associated with these records and that would also signal a dramatic warming trend.

Although an abrupt, catastrophic change in climate will not occur (whew!), it remains likely that a steady freshening of the North Atlantic ocean will continue as the Greenland ice sheet shrinks as a result of higher temperatures. Remember, these temperature changes tend to be higher around the Arctic Circle than elsewhere. So, even if the conveyor belt is not shut down, it may well be slowed down.

But we are OK for the time being, aren't we? After all, it will take a long time to switch from our current temperate climate to a polar phase . . . won't it? The rise of civilization as we know it occurred exclusively during the last 11,000 years. For much of this time, Earth was adjusting to warmer temperatures that were causing the ice sheets to melt, the sea level to gradually rise, rivers to build deltas, tundra to be replaced by conifer forests, and conifer forests to be replaced by deciduous woodlands. Our ancestors spread out to occupy the newly temperate lands of Europe, North America, and Asia. The planetary geography and biology that we are so familiar with today represents just the latest stage in this evolving picture. As we discuss the causes of climate change in Section 16.6 and Chapter 17, you should consider how future changes might impact social and cultural conditions on Earth.

16.6 Natural Causes of Climate Change

Learning Objectives

- Identify factors that would cause warming or cooling of Earth.
- Draw a sketch to illustrate how variations in Earth's orbit result in changes in Earth's climate.
- Explain how different natural causes may be related to short- and long-term climate changes.

We have learned that when oceanic and atmospheric circulation patterns change, the result can be dramatic global changes in climate. Data also indicate that Earth has experienced long periods (millions of years) of cold (ice age) conditions when there were permanent ice caps. The ice caps expanded during the coldest parts of the ice age (glacials) and contracted during warmer interglacial periods but did not completely melt. These same data show that Earth has also experienced long warm periods (called hothouse or greenhouse conditions) when there were no ice caps. (Today, we are in an interglacial period, meaning that ice caps are smaller than they were 18,000 years ago.)

Ice ages have occurred at least six times in Earth's history, from the Precambrian (2.7 billion years ago) to the recent past. The causes of such long-term global climate events must be processes that operate on a global scale over very long time intervals. The most likely causes are associated with the changing locations of continents and oceans (plate tectonics), changes in Earth's orbit around the sun, and variations in the composition of the atmosphere (for example, increasing concentrations of greenhouse gases). We will examine the latter characteristic more fully in Chapter 17.

Distribution of the Continents

Remember that glaciers form on land. To build up a long-lived continental ice sheet, there must be a landmass near the poles. **The last three major ice ages occurred when large landmasses were located near one or both poles.** During the most recent ice age, Antarctica was located over the South Pole, and large areas of northern Europe, Asia, and North America were all within the Arctic Circle. We learned in Chapter 4 that parts of all the Southern Hemisphere continents and India bear the imprint of glaciers from when they were located around the South Pole around 300 million years ago during the latter half of the Paleozoic Era (Figure 16.38). However, just having continents near the poles is not enough to create climate change. The organization of the northern continents

Figure 16.38 Distribution of continents and oceans at time of Pangaea. A continental glacier was located on the southern continents that were grouped around the South Pole.

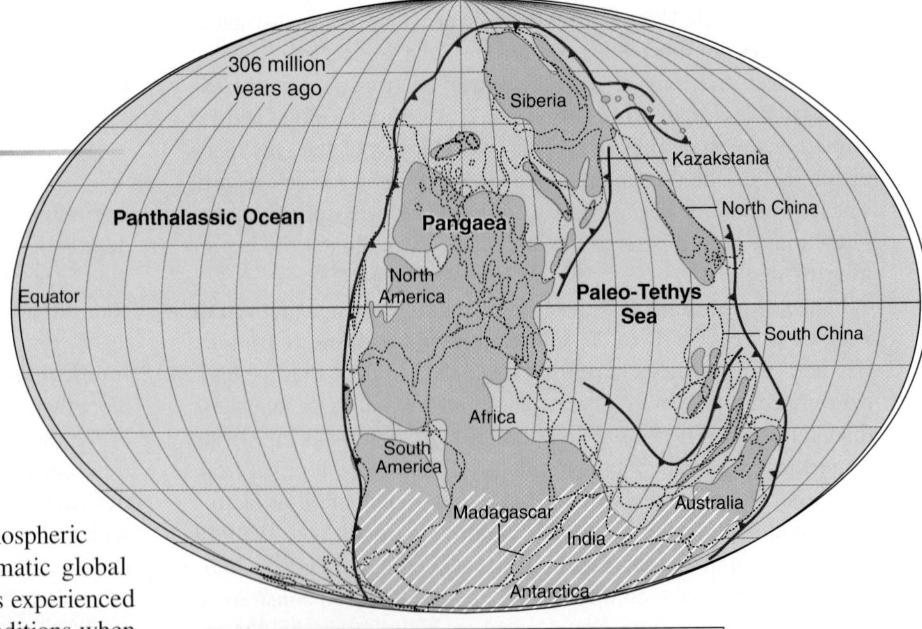

did not change much between 4 million years ago (no ice age) and 2 million years ago (ice age conditions). What other factors might have been important?

Oceanic Circulation Patterns

As discussed in Chapters 4 and 13, a significant change in oceanic circulation patterns took place when the connection between North and South America closed around 3 million years ago. The strengthening of the Gulf Stream brought warmer waters into the North Atlantic region. The warmer water resulted in greater evaporation and more precipitation in the form of snowfall. More snow caused greater reflection of incoming solar radiation and less melting (ablation) of northern glaciers. More accumulation and less ablation led to the long-term formation of ice caps.

Similarly, the trigger for the large-scale glaciation of Antarctica 34 million years ago was the separation of the South American and Australian continents from the Antarctica landmass. This opened up the circular circulation patterns in the Southern Ocean, essentially isolating the southern continent from moderating climate influences from lower latitudes (see Figure 13.17). The workings of Earth's climate involve the interaction of different parts of the earth system—the geosphere (continents), the hydrosphere (oceans, glaciers), the atmosphere (circulation patterns, winds), and the biosphere (photosynthesis, respiration of organisms). We will explore these links again in Chapter 17. But we also have to consider an external agent, the energy received from the sun.

Variations in Earth's Orbit

We have already discussed the major influence of the sun on the seasonality of Earth's annual climate (see Chapter 2). However, overprinted on that pattern are climate fluctuations, which occur in cycles lasting tens of thousands of years and are related to **small changes in Earth's orbit.** These changes, termed ***Milankovitch cycles*** after the astronomer who identified them, are thought to be caused by the interaction of Earth with the gravitational fields of other planets. There are three principal variations:

1. ***Eccentricity of Earth's orbit.*** The exact path of Earth's orbit around the sun changes with time and may range from nearly circular (less eccentric) to more elliptical (more eccentric; Figure 16.39a). These changes occur on a 100,000-year cycle and determine how much solar radiation Earth receives. During the most eccentric orbit, there would be a 20 to 30 percent seasonal difference in the amount of solar radiation reaching Earth.
2. ***Changes in the tilt of Earth's axis.*** Earth's axis is currently tilted at 23.5 degrees. Over a 41,000-year cycle, that axial tilt changes from approximately 22 to 25 degrees (Figure 16.39b). Changing the tilt of the axis changes the angle at which solar radiation strikes Earth and therefore determines which regions receive more or less solar heating. Decreasing the tilt reduces the contrast of insolation associated with the seasons; increasing the tilt exaggerates seasonal differences. Lesser tilt promotes the buildup of ice at the poles; greater tilt allows for more insolation during polar summers, causing more snowmelt.

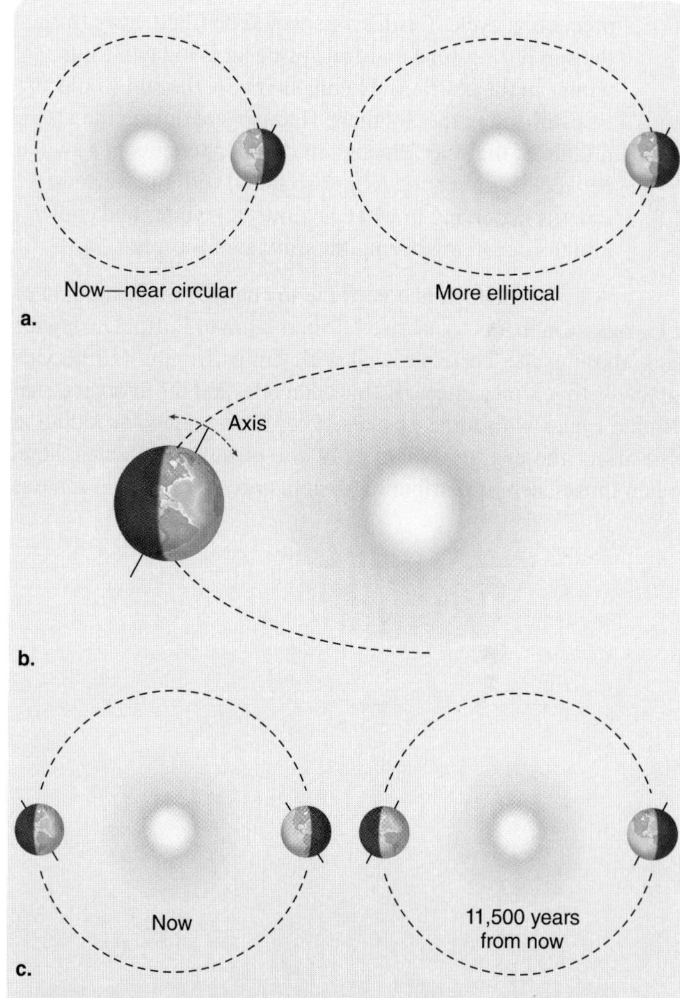

Figure 16.39 Milankovitch cycles. Variations in Earth's orbit account for climate cycles of the last ice age. **a.** Less eccentric orbits are nearly circular, while more eccentric orbits are increasingly elliptical. **b.** An increase in the tilt of Earth's axis increases melting at the poles, while decreasing tilt reduces melting. **c.** The wobble or precession of the orbit is related to whether the axis is tilted away from or toward the sun.

Checkpoint 16.21

Before reading any farther, make a list of factors that affect global climate patterns and that are presented in previous chapters. Where possible, indicate if these factors have a tendency to increase or decrease global temperatures.

Checkpoint 16.22

What combination of changes in the Milankovitch cycle would cause the highest and lowest summer temperatures in North America?

3. ***Precession.*** Have you ever spun a top or dreidel and watched it move? Typically, it rotates or wobbles around its spin axis, instead of staying in one position. That movement of the spin axis is known as *precession.* Earth also rotates (or "wobbles") about its axis. That wobble (precession) changes the direction but not the angle of axial tilt (Figure 16.39c). Precession occurs on a 23,000-year cycle (the length of time required for the axis to make a complete round trip back to where it started). Today, Earth's axis is tilted toward the sun during summer months. After 11,500 years, halfway through the precession cycle, Earth's axis would be tilted away from the sun during June and July, just as it is today during winter in the Northern Hemisphere. So, the sun would be overhead in the Southern Hemisphere during the middle of the year. Precession therefore results in a switch between summer and winter seasons, with the warmest months occurring in what we now call winter and cooler months occurring during the middle of the year.

All of these factors contribute to changes in the amount of solar radiation that is transferred from the sun to Earth. During ice ages, these cycles correlate well with jumps from cold intervals (glacials) to warmer intervals (interglacials) and the short internal climate cycles within the glacials. The changes in solar radiation also affect the energy dynamics of the oceans and atmosphere. When those energy distribution systems are disrupted, climate is affected. Climate fluctuations that occur on even shorter timescales (years to decades) may be linked to short-lived phenomena such as variations in the 11-year sunspot cycle (see Chapter 2), catastrophic volcanic eruptions (see Chapter 6), or variations in ocean circulation patterns such as El Niño (Chapter 13).

Would the amount of incoming solar radiation increase or decrease at the Arctic Circle during July in the Northern Hemisphere if

1. Earth's axis were vertical rather than tilted?
2. Earth's orbit brought it closer to the sun?
3. the tilt of Earth's axis were opposite to its present orientation (away from the sun)?

Make a list of as many factors as you can think of that could cause the temperature of a region of Earth's surface to decrease. Briefly explain each factor.

the big picture

At the beginning of this chapter, we showed you a picture of a polar bear and a tourist and asked which was in greater danger. Obviously, an unarmed person would be in great danger if confronted in the open by an 800-pound (360-kilogram) bear. However, taking a long-term view, polar bears are much more at risk from climate change than humans are. Climate change in the Arctic is destroying the habitat of polar bears and other species. Polar bears do not have the luxury of going to the grocery store to find alternative food sources if the seal population disappears because of a reduction in sea ice. Many scientists doubt that polar bears will survive in the wild beyond this century.

Are the environmental changes we observe in the Arctic today a temporary aberration, or are they the first warning sign of a permanent change in worldwide climates? We cannot make snap judgments about climate change on the basis of data from a few decades. You have learned that climate is defined over relatively long time intervals of at least 30 years. But it is clear from satellite and other temperature records that global temperatures have been warmer over the last few decades than at any time in the past century. Furthermore, proxy data indicate that this century has had hotter temperatures than any in the past 1,000 years. Therefore, evidence suggests that the temperatures we see in the Arctic are not the typical climate fluctuations expected every few decades. On the other hand, if we take a longer view that extends back more than 100,000 years, or even tens of millions of years, Earth has certainly experienced warmer climates than today. Of course, modern civilization did not exist in those times, and many of our major cities were built for a cooler time with lower sea levels.

It is also clear that we are in the middle of a relatively rare period of climate stability, which typically lasts for about 20,000 years between longer, colder episodes of more extreme climate fluctuations such as those that characterized the recent ice age. These warmer interglacial periods are typically followed by colder glacial stages. Past ice age cycles of glacials and interglacials lasted for much longer than the 2 million years of our "current" event.

Before we make any big decisions about how to address the potential for climate changes, we should take a closer look at the potential effects of our apparent warming trend. Chapter 17 builds on the information presented here to examine how we might expect global climate to change during our lifetimes.

Earth's Climate System: Concept Map

Complete the following concept map to evaluate your understanding of the interactions between the earth system and the climate system. Label as many interactions as you can, using the information from this chapter.

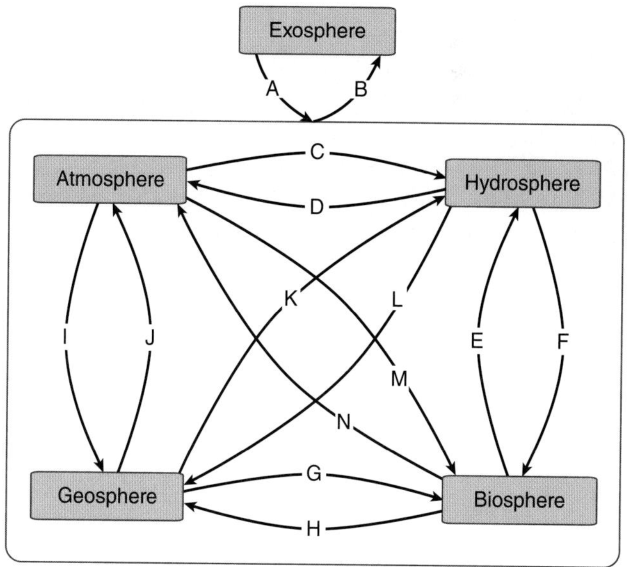

ROM THE DIRECTOR OF INDEPE

"We all grumble about the weather—but—nothing is done about it."

—Mark Twain, author

"Nowadays, everybody is doing something about the weather but nobody is talking about it."

—Stephen Schneider professor, Stanford University

A ROLAND EMMERICH FILM

HE DAY AFTER TOMORROW

WHERE WILL YOU BE?

ENTIETH CENTURY FOX PRESENTS
ENTROPOLIS ENTERTAINMENT / LIONS GATE / MARK GORDON COMPANY PRODUCTION
OLAND EMMERICH FILM "THE DAY AFTER TOMORROW" DENNIS QUAID JAKE GYLLENHAAL IAN HOLM
MY ROSSUM SELA WARD HARALD KLOSER THOMAS M. HAMMEL DAVID BRENNER A.C.E.
BARRY CHUSID UELI STEIGER, A.S.C. STEPHANIE GERMAIN UTE EMMERICH KELLY VAN HORN
MARK GORDON ROLAND EMMERICH ROLAND EMMERICH
ROLAND EMMERICH & JEFFREY NACHMANOFF ROLAND EMMERICH www.thedayaftertomorrow.com

MAY 28

Tomorrow (2004)
oland Emmerich

Chapter 17

Global Change

Chapter Outline

the big picture

A world inundated by water or gripped by ice. How likely is either scenario depicted in these movies?

See The Big Picture box at the end of this chapter for the full story on these images.

17.1 Alternative Climates, Alternative Choices

Chapter Learning Outcomes

- Students will explain the characteristics and significance of a successful global effort to reduce ozone depletion.
- Students will be able to compare and contrast the greenhouse effect and global warming.
- Students will evaluate options for reducing global warming emissions.
- Students will describe how and why carbon cycling is important to the earth system.

Welcome to the great climate experiment. Do you feel lucky? Over the past two centuries, humankind has unintentionally embarked on a global experiment in climate modification that may adversely affect natural systems all over the planet. The scale of the experiment is so large and the time spans involved are so long that some of the basic parameters of the research are still being determined.

Although there had long been speculation that human activity could affect global climate, it wasn't until the 1980s that the general public and national governments became fully aware of just how closely people are linked to atmospheric chemistry. In 1985, the British Antarctic Survey reported a substantial drop in stratospheric ozone concentrations over the South Pole during certain times of the year (Figure 17.1). The same year, a conference of international scientists noted that industrial emissions were changing the atmosphere such that future global warming was inevitable and would require international action. Subsequently, scientists learned that temperatures in recent years were higher than at any time over the last 1,000 years (Figure 17.2).

Curiously, only one of these large-scale environmental issues (ozone) received immediate attention. It is a common misconception that a reduction in ozone over Antarctica is somehow the cause of global warming. This is not true. Both topics came to our attention at about the same time, but they are caused by very different processes (as we will discuss in this chapter). Although a reduction in ozone concentrations and global warming are separate environmental issues, they do share some common elements. Both issues are global in scale. Both are related to specific gases in the atmosphere—ozone in the stratosphere and carbon dioxide (and some other gases) in the troposphere. Both of the principal gases involved make up just a tiny fraction of the atmosphere.

Despite a host of unknowns, the international community forged an agreement (the Montreal Protocol) to reduce the use of ozone-depleting chemicals and took concrete steps to address the threat of ozone loss. Throughout the world, the use of compounds that were known to break down ozone was either greatly reduced or banned. In contrast, policies to address global warming are still subject to debate by governments, industry, and environmental groups. However, we may be nearing consensus on this issue as recognition by citizens, scientists, and politicians is growing that global warming represents a major environmental threat to our modern world. A critical challenge is to find a way to balance future economic growth with strategies that reduce the output of gases that exaggerate the warming effect.

Because changes to global temperatures have potentially adverse effects on the biosphere, scientists are working hard to find definitive answers to important questions such as these: Exactly how much will global temperatures increase? Which regions will see the greatest changes; which will see the least change? What will the impact be on global and national economies? Can nature correct this problem, or must we get off the couch and participate in the fix? Can we sue somebody?

Well, actually, yes . . . you can sue somebody, preferably a large corporation with deep pockets. Public nuisance laws that were originally intended to deal with problems such as dust or noise pollution have been adopted by a variety of groups to try to influence government policies on climate change. For example, here is an abbreviated version of a nuisance law from the California civil code:

> *Anything which is injurious to health . . . or offensive to the senses . . . so as to interfere with the comfortable enjoyment of life or property . . . is a nuisance. A public nuisance is one which affects at the same time an entire community or neighborhood, or any considerable number of persons, although the extent of the annoyance or damage inflicted upon individuals may be unequal.*

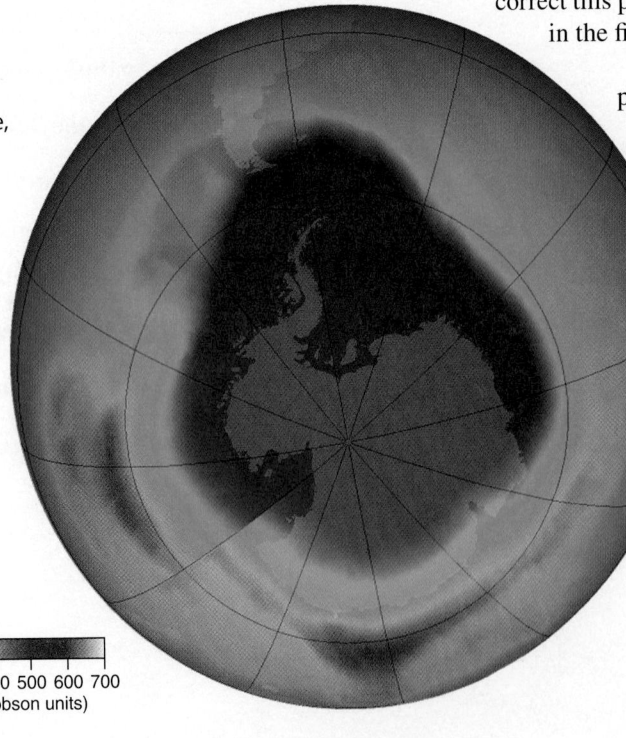

Figure 17.1 Ozone hole, Antarctica. The dark blue and purple area over Antarctica shows the extent of the ozone hole in September 2012. Technically, the "hole" represents a thinning of the ozone in the stratosphere. Concentration is measured in Dobson units. A typical concentration for this region is around 300 Dobson units.

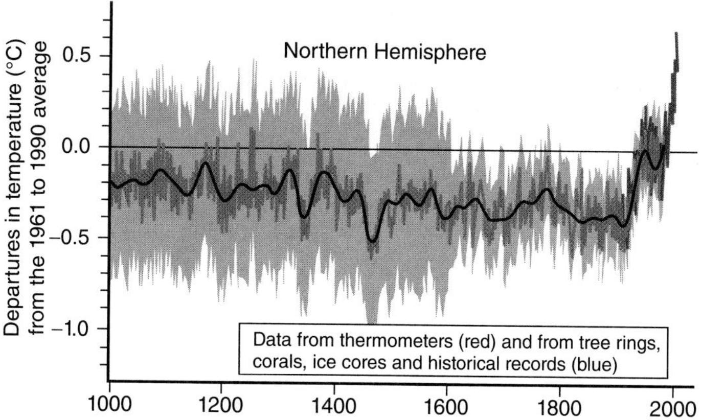

Figure 17.2 Instrumental and proxy global temperature record. Temperatures are compared to a standard represented by a 30-year mean temperature between 1961 and 1990. Recent temperatures have been higher than at any point in the previous 1,000 years.

Can global warming be considered a public nuisance for Florida homeowners visited by three hurricanes in 2 years? How about a ski resort community in the Colorado Rockies where less snow falls than in the past and then melts a month earlier than expected? What about the residents of a coastal city in California threatened by rising seas? In recent years, several states and cities have sued a group of US power companies, claiming that their carbon dioxide emissions represent a public nuisance. These companies produce 10 percent of US emissions, equivalent to the total carbon dioxide production of nations such as Canada or the United Kingdom.

Are power companies liable for the production of greenhouse gases in the same way as cigarette companies have been held responsible for lung cancer? If a company continues to produce greenhouse gases, despite evidence that these gases cause changes to global climate, can it be held legally responsible for the consequences? On the other hand, is everyone who drives a car equally guilty . . . or just SUV owners? The US Supreme Court decision placed responsibility for regulating greenhouse gases in the hands of the Environmental Protection Agency, raising the potential for future rules about how much carbon dioxide and other gases can be released from vehicles, power plants, and industries.

In Chapter 1, we described research on the changes in global climates as an example of "big science" that involved the work of thousands of researchers around the world. There are seven broad research categories—atmospheric composition, the carbon cycle, ecosystems, population and global change, the global water cycle, ancient climates, and Earth's climate system—that are essential to our understanding of global change. Keep that list in mind as you read this chapter, which describes how scientists have worked to understand how all the components of the earth system interact to influence life on this good Earth. We finish the chapter, and the book, by glimpsing at a future, warmer world and examining how Earth might look if the predicted temperature changes occur over the next century. Reader, you will have to decide if these changes are for better or worse.

Self-Reflection Survey: Section 17.1

Answer the following questions as a means of uncovering what you already know about global change.

1. Respond to the following questions taken from recent CNN and Gallup polls, and compare your answers to those of other respondents. (See footnote to compare responses.*)

   i) Which of the following statements comes closest to your view of global warming?

      a. Global warming is a proven fact and is mostly caused by emissions from cars and industrial facilities such as power plants and factories.
      b. Global warming is a proven fact and is mostly caused by natural changes that have nothing to do with emissions from cars and industrial facilities.
      c. Global warming is a theory that has not yet been proved.
      d. Unsure.

   ii) In thinking about the issue of global warming, sometimes called the *greenhouse effect*, how well do you feel you understand this issue?

      a. Very well.
      b. Fairly well.
      c. Not very well.
      d. Not at all.

   iii) Which of the following statements reflects your view of when the effects of global warming will begin to happen?

      a. They have already begun to happen.
      b. They will start happening within a few years.
      c. They will start happening within my lifetime.
      d. They will not happen within my lifetime, but they will affect the future.
      e. They will never happen.

2. Make a list of at least three advantages and three disadvantages of an increase in temperatures in the region where you live. Then do the same for Earth as a whole.

3. Make a list of three reasons that could explain why a larger proportion of the public now believe that global warming is occurring.

*Poll results are: i) a. 54%; b. 22%; c. 23%; d. 1%. ii) a. 21%; b. 59%; c. 18%; d. 2%. iii) a. 61%; b. 4%; c. 10%; d. 13%; e. 11%. (Totals may not add to 100% due to rounding.)

17.2 Ozone and the Stratosphere

Learning Objectives

- Explain how ozone depletion affects humans.
- Describe conditions affecting natural ozone formation and destruction.
- Create a figure to illustrate how ozone formation and destruction is related to the major components of the earth system.

Have you ever looked at the label on sunblock or suntan lotion? What do the letters SPF, UVA, and UVB mean, and why are they important? SPF stands for *sun protection factor.* The higher the SPF number, the more protection the lotion provides. How do you think sunblock lotion manufacturers originally determined the SPF? It turns out, they exposed humans to controlled levels of sunlight, varied the type of lotion applied, and measured how long it took before the participants got a sunburn (ouch!).

There are three types of ultraviolet radiation: UVA, UVB, and UVC. They have slightly different wavelengths. UVC has the shortest wavelength and is the most dangerous form of UV for humans, but it is filtered out effectively by the atmosphere. UVA and UVB are known to cause skin cancer and wrinkles after repeated long-term exposure. UV rays can actually penetrate your clothes. For example, nylon stockings provide SPF 2, hats SPF 3 to 6, and summer-weight clothing SPF 6.5 protection. SPF 15 lotion blocks about 92 percent of the UV that reaches Earth's surface, thus allowing people to remain in sunlight 15 times longer than they could otherwise (SPF 30 blocks an additional 4 percent of UV). **The ozone layer is Earth's own sunblock system that stops 97 to 99 percent of the harmful incoming ultraviolet rays from reaching the planet's surface.** The sunblock you apply just deals with the small percentage of rays that get past the ozone layer.

The Nature of Ozone

The important functions of ozone are related to the structure of Earth's atmosphere and the chemical properties of the ozone molecule. Recall that the atmosphere is divided into four layers based on their thermal characteristics (see Figure 14.6). The ground-level troposphere is overlain by the stratosphere, which extends from approximately 10 to 50 kilometers (6 to 31 miles) in altitude. **The stratosphere is enriched in ozone, a molecule made up of three oxygen atoms (O_3).** In comparison, the more common atmospheric oxygen molecule (O_2) is made up of two oxygen atoms.

Checkpoint 17.1

basic · intermediate · advanced · superior

Which of the following statements is the best analogy?

a) An umbrella is to rain as the stratosphere is to solar radiation.
b) An umbrella is to rain as the stratosphere is to ultraviolet radiation.
c) An umbrella is to rain as the stratospheric ozone is to solar radiation.
d) An umbrella is to rain as the stratospheric ozone is to ultraviolet radiation.

Ozone is formed in a two-step process (Figure 17.3). First, ultraviolet radiation causes an oxygen molecule (O_2) to break apart to form two individual oxygen atoms (O and O):

$$\text{(i) } O_2 + UV \rightarrow O + O$$

Then one of these individual atoms combines with a whole oxygen molecule to form an ozone molecule:

$$\text{(ii) } O_2 + O \rightarrow O_3$$

If these reactions continued, ozone would steadily build up in the atmosphere. However, a third reaction destroys ozone molecules. When UVB strikes the ozone molecule, it breaks down again, forming an oxygen molecule and an oxygen atom:

$$\text{(iii) } O_3 + UV \rightarrow O_2 + O$$

The oxygen atom is then free to recombine with another oxygen molecule to form a new ozone molecule (see Equation ii), and the cycle repeats. No long-term change in the concentration of ozone occurs, as a new ozone molecule is formed to replace each molecule that is destroyed. Since the UV radiation is converted from light to heat by the destruction of oxygen and ozone molecules, temperatures rise in the stratosphere. (To review the four layers of the atmosphere, see Chapter 14.)

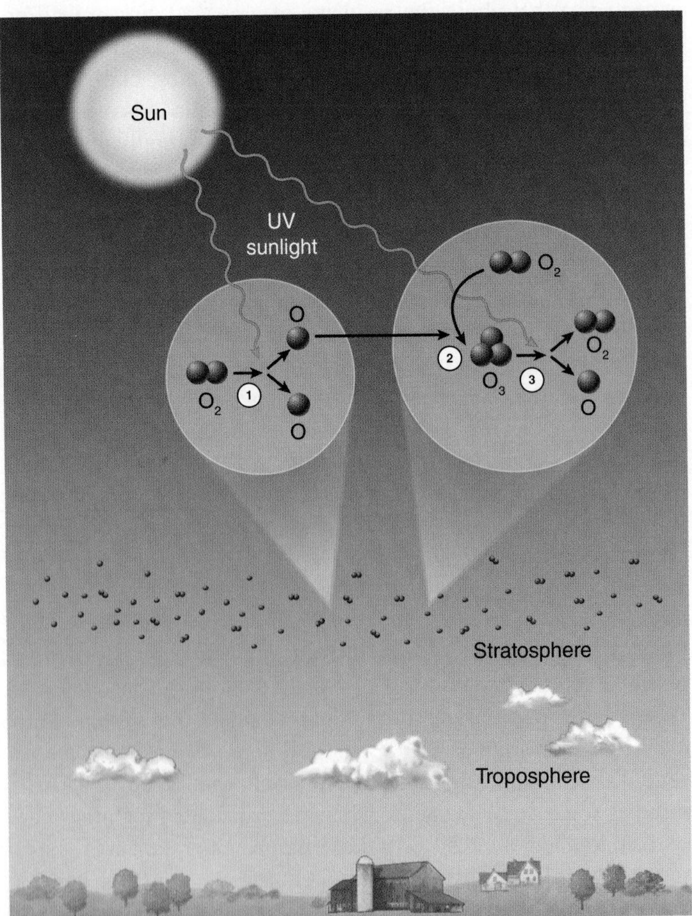

Figure 17.3 Ozone formation and destruction. **1.** Ultraviolet radiation causes oxygen molecules to break into two atoms. **2.** An individual atom combines with an oxygen molecule to form an ozone molecule. **3.** UV radiation breaks the ozone molecule down to an oxygen molecule and an oxygen atom.

Natural Variations in Ozone Concentrations

The greater the concentration of ozone, the more protection we get from incoming ultraviolet radiation. Most ozone is found in the lower third of the stratosphere at altitudes of 10 to 20 kilometers (6 to 12 miles), decreasing with increasing altitude. This part of the stratosphere with higher concentrations of ozone is referred to as the *ozone layer*. The concentration of ozone in the stratosphere is measured in Dobson units (DU), named after G. M. B. Dobson, one of the first scientists to study ozone early in the twentieth century. One Dobson unit (1 DU) is equivalent to a 0.01-millimeter (0.0004-inch) layer of ozone at 0°C (32°F) and air pressures typical of Earth's surface. Therefore, the equivalent average thickness of the ozone in the stratosphere over the United States of about 300 Dobson units would be a layer just 3 millimeters (0.1 inch) thick, about the thickness of two pennies stacked together, if it were at surface atmospheric pressures. Gases are often measured in parts per million for a given volume (ppmv). Even at its most concentrated, ozone makes up just 8 ppmv in the stratosphere. (What percentage of the stratosphere is composed of ozone?)

The ozone layer varies naturally both annually and seasonally. These natural variations are linked to atmospheric circulation patterns, temperature, and variations in the amount of solar radiation reaching Earth.

Since solar radiation is also involved in the creation of ozone, we might expect the ozone concentrations to be greatest in the stratosphere above the equator and lower toward the poles. In fact, it turns out that the opposite is true. Complex stratospheric circulation patterns remove ozone from the tropics and deposit it in the lower stratosphere at higher (polar) latitudes. **Stratospheric ozone has nearly constant concentrations (about 260 Dobson units) above the tropics but becomes more concentrated and subject to larger seasonal variations at higher latitudes** (Figure 17.4). The longest record of ozone measurements is from Arosa, Switzerland, where measurements began in the 1920s. Seasonal variations in the ozone concentration at Arosa range from lows of 280 Dobson units recorded in fall to a maximum around 375 Dobson units during spring. Annual averages at this site have fluctuated between 300 and 360 Dobson units but have declined since the 1970s. That negative trend (about 3 percent per year) results from a reduction in ozone concentrations over Europe.

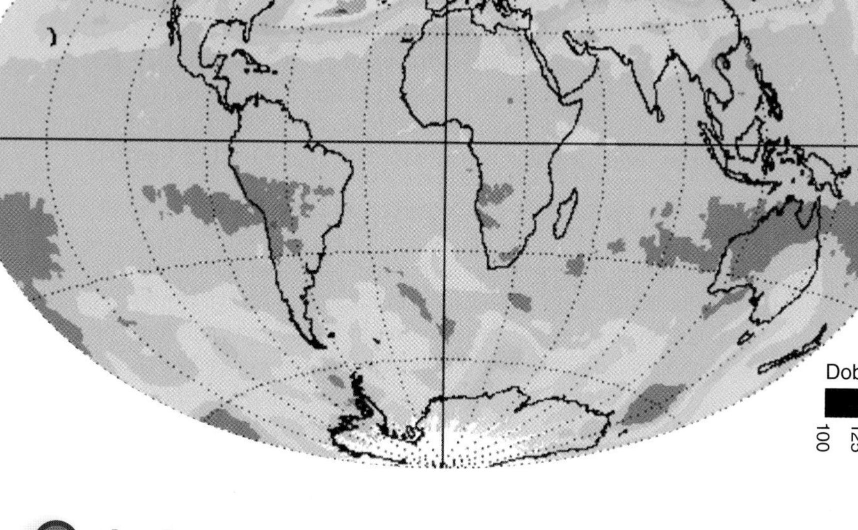

Figure 17.4 Global variations in ozone concentrations, January 2001. Ozone concentrations increase with latitude. Concentrations are greater in the Northern Hemisphere during the Northern Hemisphere spring and greatest in the Southern Hemisphere during the Northern Hemisphere fall.

Dobson units

100 125 150 175 200 225 250 275 300 325 350 375 400 425 450 475 500

Checkpoint 17.2

basic · intermediate · advanced · superior

Which of the following concept maps best represents the sequence of transformations associated with the formation and destruction of ozone in the stratosphere? Explain the reasoning behind your choice.

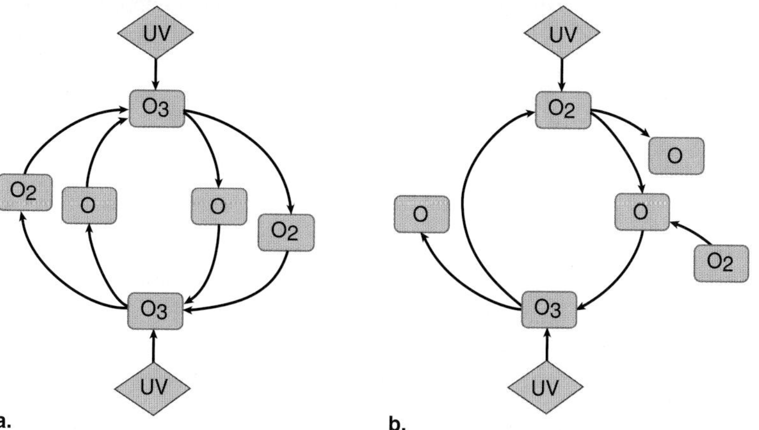

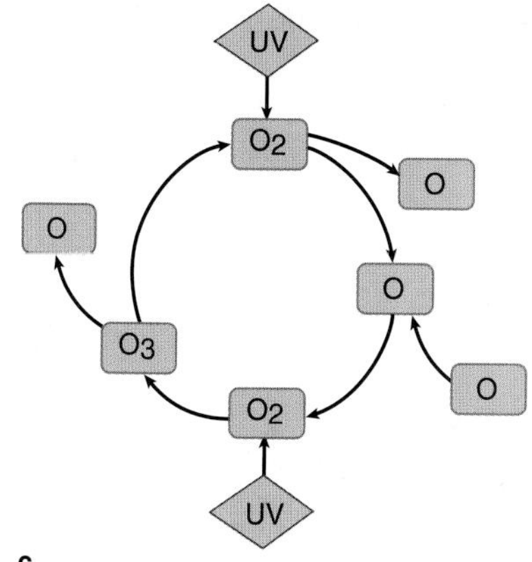

Checkpoint 17.3

basic · intermediate · advanced · superior

The following graph represents the variation in ozone concentrations over four cities: Darwin, Australia, 12°S; Melbourne, Australia, 37°S; San Francisco, USA, 37°N; London, UK, 55°N. Explain which city goes with each of the four plots, and then sketch an estimated plot for the city where you live.

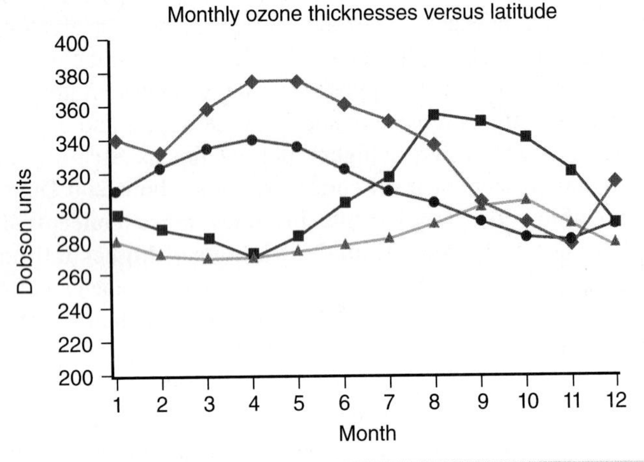

Checkpoint 17.4

basic · intermediate · advanced · superior

Reductions in ozone concentrations involve all aspects of the earth system. UV radiation that interacts with ozone comes from the exosphere (the sun). There is a potential effect on the biosphere, and the hydrosphere, atmosphere, and geosphere all have roles to play. Develop a concept map that illustrates the characteristics of ozone and includes references to different components of the earth system.

17.3 CFCs and Ozone Depletion

Learning Objectives

- Connect natural processes and industrial activity to the concentration of stratospheric ozone.
- Explain the significance of time lags in the earth system response to reductions in CFC production.

In Section 17.2, we described the natural fluctuations in ozone concentrations. Now we will explore the effects of human activities on these phenomena. In 1974, scientists F. Sherwood Rowland and Mario Jose Molina published an article suggesting that the use of a manufactured group of gases known as chlorofluorocarbons (CFCs) was resulting in the depletion of ozone. Two decades later (1995), Rowland and Molina (with Paul Crutzen) received the Nobel Prize in chemistry for their work. **CFCs are volatile organic compounds** that were first patented by Frigidaire in 1928 and were subsequently widely adopted as aerosol propellants and refrigerants. From the 1950s through the 1970s, they were used as propellants in many deodorants and hair sprays. The United States banned aerosol sprays containing CFCs in 1978 and phased out the production of CFCs (replacing them with less ozone-destructive substitutes) by January 1, 1996.

The Nature of CFCs

There are several different types of CFCs. All are inert; that is, they do not react with other gases in the atmosphere. Consequently, they **remain in the atmosphere for periods of 20 to 200 years.** Although they are inert, CFCs can be broken down into their constituent parts by UV radiation (a process known as *photolysis*). Photolysis frees chlorine atoms, which then react with ozone, forming chlorine monoxide (ClO) and oxygen:

$$\text{(i) } Cl + O_3 \rightarrow ClO + O_2$$

The chlorine monoxide (ClO) readily reacts with free oxygen atoms. The result is the formation of another oxygen molecule, leaving the chlorine atom free again:

$$\text{(ii) } ClO + O \rightarrow O_2 + Cl$$

This reaction is significant because an ozone molecule (which is capable of absorbing harmful UV) is destroyed, and a free chlorine atom remains to repeat the process. This sequence is duplicated thousands of times for each chlorine atom. Approximately 80 percent of the chlorine in the stratosphere comes from manufactured compounds (for example, CFCs), and 20 percent comes from natural sources, such as volcanic eruptions and wildfires. In the five decades leading up to the 1990s, the total volume of stratospheric chlorine increased by approximately fourfold.

Reductions in Ozone Concentrations

Remember, the appearance of sunlight starts the reactions that reduce ozone concentrations above Antarctica every spring. However, as we noted in Section 17.1, a reduction in ozone concentrations beyond the expected natural variation was first reported above Antarctica in 1985. Over time, the reductions in ozone at the South Pole gradually became more pronounced. These annual reductions in ozone concentrations over Antarctica have been called the *ozone hole* by the media (see Figure 17.1), even though there is no actual hole—that is, no location in the lower stratosphere where ozone is completely absent. Examination of atmospheric records suggests that ozone levels began to drop in the late 1970s. Recent research has shown that during October (spring in the Southern Hemisphere), ozone concentrations decline to an average of approximately 100 Dobson units above Antarctica (Figure 17.1), with a corresponding increase in the size of the "hole" (Figure 17.5). The levels are down from approximately 320 Dobson units during the late 1950s.

Why Does Ozone Become Depleted over the South Pole?

If the bulk of CFCs are produced and used in the Northern Hemisphere, why are the ozone concentrations so depleted over the unpopulated, ice-covered South Pole? The reasons for the location of the hole are related to the unique weather patterns associated with Antarctica. The ozone hole occupies the polar vortex, a near-circular high-pressure system that isolates the atmosphere over Antarctica during winter in the Southern Hemisphere (June through August). Temperatures in the vortex are below −80°C (−112°F) and lead to the formation of polar stratospheric clouds composed of nitric acid and water. These clouds contain ice particles that provide surfaces upon which chemical reactions can occur. Chlorine compounds in the stratosphere react with cloud particles to release the chlorine needed to destroy ozone molecules. As spring in Antarctica arrives (September through November), temperatures gradually rise in the stratosphere because of increased solar radiation. Ozone formation (and destruction) is a temperature-dependent phenomenon. As long as the temperatures remain cold enough for stratospheric clouds to exist, reactions that destroy ozone will occur. As temperatures warm, processes that form ozone outpace the destructive cycle, and ozone begins to increase in concentration. Consequently, there is more ozone in the stratosphere in summer than in winter in Antarctica. The absence of polar stratospheric clouds elsewhere results in much less ozone loss over the rest of the planet.

These factors result in a concentration of ozone-depleting chemicals and conditions at the South Pole. The global presence of similar chemicals in the atmosphere raises the possibility for ozone depletion to occur elsewhere. Declines in ozone concentrations are also recognized above North America and Europe. However, the reductions observed at these locations are substantially less than those observed over Antarctica (less than 10 percent compared to 70 percent).

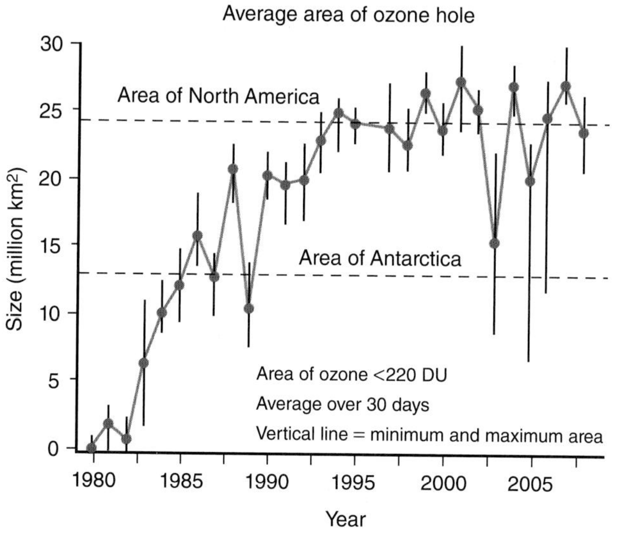

Figure 17.5 Development of the ozone "hole," Antarctica, 1979–2008. Graph of average hole size. Note that these values will be less than the absolute maximum area measured on a given day. The size of the thinned area of the ozone layer is approximately equal to the area of North America.

Our Ozone Future

Production of the more common types of CFCs decreased significantly over a decade ago. This decline has been mirrored by a decrease in the concentrations of atmospheric chlorine (Figure 17.6)

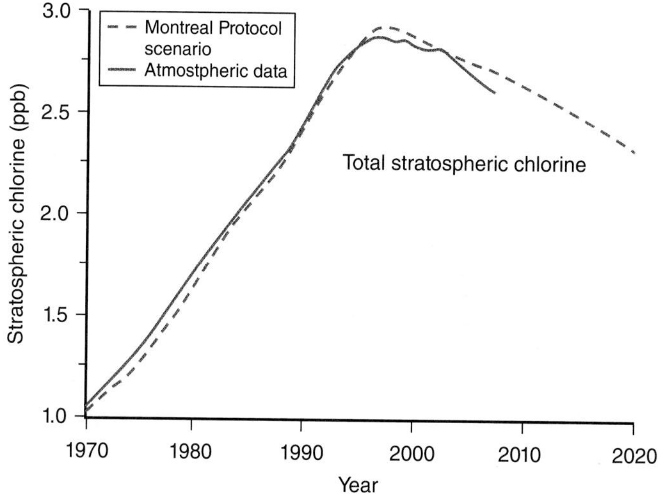

Figure 17.6 Chlorine concentrations in Earth's atmosphere. The concentrations of chlorine and related elements have steadily declined since the early 1990s due to the gradual phaseout of CFCs.

Checkpoint 17.5

basic | intermediate | advanced | superior

Examine Figure 17.5. Describe trends you notice in those data. What future trends would we want to see based on what you have learned so far about ozone?

Checkpoint 17.6

basic | intermediate | advanced | superior

Suppose a volcanic eruption included large volumes of chlorine. What would likely happen to global stratospheric ozone levels and why?

Checkpoint 17.7

basic | intermediate | advanced | superior

Students in a college earth science class were asked on an exam to concisely describe the relationship between CFCs and ozone destruction. Analyze the following four students' responses and rank them from best to worst. Justify your answer choices.

a) CFCs are manufactured gases that destroy the ozone and produce oxygen.
b) Chlorine forms from the disintegration of manufactured gases. The chlorine reacts with ozone molecules to form two gases that cannot block incoming UV radiation.
c) Oxygen molecules are broken down by UV radiation to free oxygen atoms that combine with chlorine monoxide to form new oxygen molecules and free chlorine. Chlorine helps break down CFCs with UV radiation.
d) UV radiation breaks apart complex manufactured chlorofluorocarbons into its constituent atoms, including chlorine. Chlorine destroys ozone molecules as it cycles through a pair of reactions that produce free oxygen molecules.

Finally, write your own answer to the question.

and various other CFCs. Check some labels to see if CFCs are present in any of the household appliances or pressurized sprays (hair spray, deodorant, cleaners) that you use. Did you find any? Some older refrigerators and freezers still contain CFC (for example, a product named Freon), as do a few household products. Several new coolants (hydrochlorofluorocarbons, HCFCs; hydrofluorocarbons, HFCs) have largely replaced CFCs. HFCs have no impact on the ozone layer, and HCFCs have only limited ability to deplete ozone because they do not remain in the atmosphere as long as CFCs.

CFC production in the developed world was effectively halted by the Montreal Protocol. Although not all countries adhere to the agreement, most industrialized nations ratified the treaty. Global CFC production has dropped by more than 98 percent from its peak of over 1 million tons (907,000 tonnes) in 1988. However, this huge drop in production has resulted in a much more modest decrease in atmospheric CFC concentrations (Figure 17.7). The long residence time of CFCs in the atmosphere means that it may take a few decades before ozone levels over Antarctica begin to return to normal levels.

Clearly, it is possible to forge international agreements to tackle global environmental problems. Next we will see how greenhouse gases compare with CFCs.

Checkpoint 17.8

basic · intermediate · advanced · superior

Given the natural variations in climate discussed in Chapter 16, some people argue that governments acted too hastily to ban the use of CFCs just 7 years after the ozone hole had been identified. Summarize three arguments for and three arguments against the ban on CFCs.

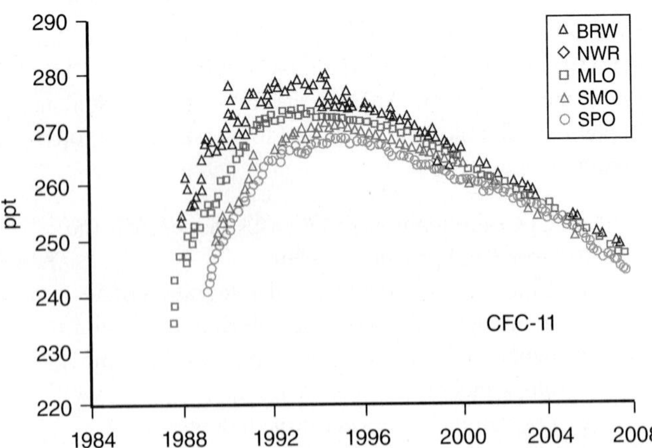

Figure 17.7 Changes in CFC-11 concentrations in parts per trillion (ppt). CFC-11 is the most common CFC in the atmosphere. The increase in CFCs began to slow down in the late 1980s as the chemicals were phased out of production and a decline in CFC concentrations began in the mid-1990s. What is the percentage change in CFC-11 concentration since the early 1990s? Symbols correspond to monitoring stations at a range of latitudes. BRW = Point Barrow, Alaska; NWR = Niwot Ridge, Colorado; MLO = Mauna Loa, Hawaii; SMO = American Samoa, south Pacific Ocean; SPO = South Pole. Compare with Figure 17.6.

17.4 Greenhouse Gases and Global Change

Learning Objectives

- Describe the link between greenhouse gases and temperatures on Earth.
- Explain connections between carbon sources and sinks.
- Identify and explain the flow of carbon within the earth system.
- Discuss associations between humans and atmospheric CO_2 levels.

Infrared radiation emitted from Earth's surface is absorbed by water vapor, carbon dioxide, and other trace gases (methane, nitrous oxide) **in the troposphere, creating a situation that has come to be known as the *greenhouse effect*.** The heat-absorbing effect of greenhouse gases raises the average surface temperature from a frigid −18°C (−0.4°F) to a pleasant +15°C (59°F). But you can have too much of a good thing. As mentioned before, Venus has a greenhouse effect that results in surface temperatures of 464°C (867°F)—pretty hot! (For more on the relationship between solar radiation and the greenhouse effect, see Chapters 1 and 14.)

Thousands of scientists with the Intergovernmental Panel on Climate Change (IPCC) have reviewed vast amounts of climate data and concluded that human activity has had an unequivocal influence on the global climate. The IPCC was established by the World Meteorological Organization and the United Nations Environment Programme to assess scientific, technical, and socioeconomic information relevant to the understanding of human-induced climate change. (For more on past climates, see Chapter 16.) Although there is some debate about the rate and severity of global warming, the basic science behind the warming hypothesis is solid and is linked to one undeniable trend. **The concentration of greenhouse gases in Earth's atmosphere has steadily and measurably increased over the past two centuries.** The signs of a changing climate are becoming apparent at a range of scales affecting multiple components of the earth system:

- The twentieth century was the warmest in the last millennium, and the 10 warmest years in the last 1,000 have occurred since 1998 (Table 17.1; Figure 17.2). Average global temperatures have increased by more than 0.7 C° (1.3 F°) over the last 100 years (Figure 17.8a).
- The average temperature of the upper 3,000 meters (9,840 feet) of the world's oceans has increased over the last 50 years, resulting in an expansion of seawater. The global sea level has increased by 0.17 meter (6.7 inches) over the last century (Figure 17.8b).

Checkpoint 17.9

What is the difference between the greenhouse effect and global warming?

Table 17.1 Hottest Years on Record (1880–2008) for World and United States (Lower 48)

World rank	Year	US rank	Year
1	2010	1	2012
2	2005	2	1998
3	1998	3	2006
4	2003	4	1934
5	2002	5	1999
6=	2006	6	1921
6=	2007	7	2003
6=	2009	8	2007
9	2004	9	2005
10	2012	10	1931

Source: National Oceanic and Atmospheric Administration.

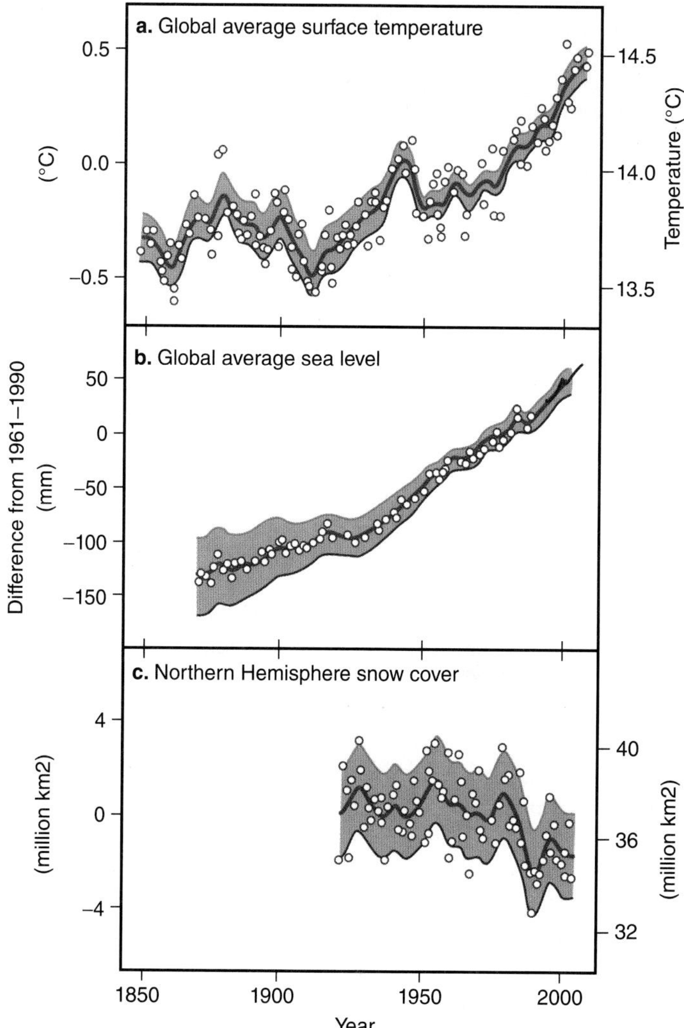

Figure 17.8 Observed changes in **a.** global average temperature; **b.** global average sea level; **c.** Northern Hemisphere spring snow cover. All numbers compared to averages for the period 1961–1990. Smoothed (dark blue) curves represent averaged values; shaded areas indicate range of uncertainty for estimates.

- Temperatures in the troposphere have increased by a similar amount to those for Earth's surface.
- The amount of water vapor in the atmosphere has increased.
- Less land is covered by alpine glaciers and snow. Some glaciers in Greenland and Antarctica are moving faster than in the past and contributing to a loss of ice mass.
- Arctic sea ice has declined steadily for several decades and now covers less area after summer melting than at any time in the recent past (see Chapter 16).
- Arctic permafrost temperatures have warmed by up to 3 C° (5 F°), and the area of seasonally frozen ground has declined by more than 7 percent since 1900. Spring snow cover in the Northern Hemisphere is at some of its lowest levels since records began (Figure 17.8c).
- Over the last century, precipitation has increased in eastern North and South America, northern Europe, and northern and central Asia. Drying has been observed in northern and southern Africa, the Mediterranean region, and parts of southern Asia.
- The salinity of mid- and high-latitude oceans has decreased, while increased salinity has been observed in tropical oceans. These trends are associated with changes in evaporation and precipitation.
- Droughts are longer and more intense, and more-extreme precipitation events have occurred more frequently.
- Cold days and nights are less frequent, and heat waves have occurred more often over the last 50 years.
- Many regions have recorded earlier spring "greening" of vegetation as a result of longer growing seasons.
- Marine scientists have noted shifts in ranges and changes in plankton and fish abundance in high-latitude oceans and earlier fish migrations in rivers.

Observations consistently indicate a warming climate and its consequences. If we are going to try to combat these trends, we need to understand how these greenhouse gases are produced and absorbed by natural systems. In the discussions that follow, we will focus on carbon dioxide because humans have some control over its production and it has a great potential impact on future warming.

The Global Carbon Cycle

Carbon, one of Earth's most basic components, is present in a variety of forms in all parts of the earth system (Chapter Snapshot). Although this discussion considers each component of the earth system separately, it will become clear that carbon is in constant flux among all the components. Each component acts as a temporary holding place (called a *sink*), storing carbon for short or long time intervals. Each component also acts as a source of carbon that can be released into the atmosphere (or to other earth system components) to increase carbon dioxide concentrations. Carbon dioxide is a product of both natural processes and human activities. While the carbon dioxide generated by human actions is just a small fraction of the volume of gas produced by natural processes, it may just be enough to push our climate past a critical tipping point toward a warmer world.

Atmosphere. The atmosphere is a sink and a source of carbon compounds and carbon-based gases such as carbon dioxide (CO_2; Chapter Snapshot). Carbon enters the atmosphere via the biosphere through respiration (for example, animals exhaling CO_2), the burning of forests, and the decay of dead organisms (plants and animals). Carbon dioxide also enters the atmosphere from the burning of fossil fuels, natural volcanic activity, and the release of dissolved gas from the ocean. The atmosphere is a source of carbon dioxide for plant growth, rock formation (limestone), and other biologic activity. Carbon dioxide from the air can also be absorbed by the ocean. Overall, **more carbon enters than exits the atmosphere** (by more than 3 billion tons [2.7 billion tonnes] per year). Atmospheric carbon dioxide levels also fluctuate annually (notice the sawtooth pattern on Figure 17.9). CO_2 levels are lower during the growing season, when plants absorb gas, and higher in winter, when many plants are dormant. The degree of fluctuation in CO_2 levels also varies with latitude. Where do you predict you would find the most pronounced difference in seasonal CO_2 levels due to plant activity?

Hydrosphere. CO_2 is continually exchanged between the atmosphere and the ocean. **Oceans act as a carbon sink by absorbing more gas than they release** to all other components of the earth system (Chapter Snapshot). Oceans retain an additional 2 billion tons (1.8 billion tonnes) of carbon per year. The volume of carbon dioxide that the oceans can absorb increases with decreasing temperature and increasing wind speed. Cooler oceans hold more carbon dioxide. Wave action helps ocean water absorb gas as it produces millions of bubbles that transfer gas from the air to the water. Since carbon dioxide is dissolved in water, it also enters the oceans through precipitation and as runoff from rivers.

Carbon is also carried to the deep ocean in cold, sinking currents such as those in the North Atlantic Ocean that drive the global ocean conveyor system (Chapter 13 Snapshot). These deep waters remain isolated from the atmosphere for more than 1,000 years, representing an effective, long-term carbon reservoir. Carbon dioxide is brought back to the surface and released by upwelling currents such as those in the northeastern Pacific Ocean. How does it get back into the atmosphere? Think about what happens when you open a warm can of soda. Soda contains dissolved CO_2 to provide bubbles. When you open a warm can, the gas rapidly comes out of solution and sprays liquid all over you. A similar, but less violent, process occurs in the oceans. If water temperatures were warm enough, oceans could release more carbon dioxide to the atmosphere than they absorb.

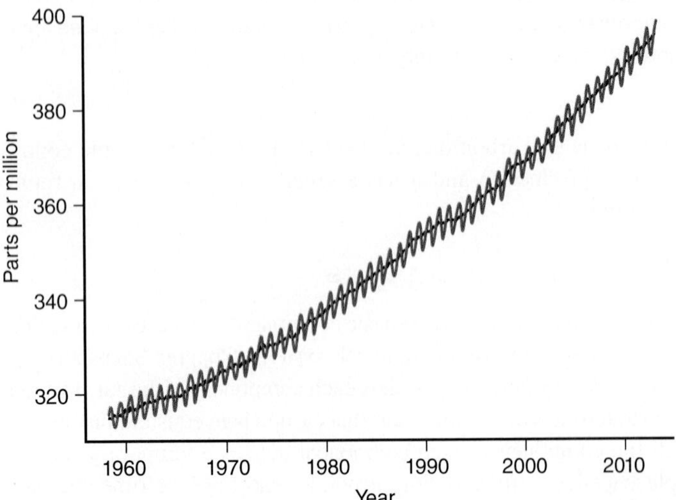

Figure 17.9 Variations in atmospheric carbon dioxide concentrations at Mauna Loa Observatory, Hawaii. The red line represents seasonal carbon dioxide fluctuations and the black line shows annual trends. These data represent the longest record of direct measurements of carbon dioxide in the atmosphere. They were started by C. David Keeling in March 1958.

Checkpoint 17.10

Carbon is stored in carbon reservoirs, or sinks, just as water is stored in a reservoir behind a dam. The carbon is released from these reservoirs and ultimately ends up in another reservoir. Complete the following table to identify how carbon is stored and released from each component of the earth system.

Sink/source	Carbon is stored when . . .	Carbon is released when . . .
Atmosphere		
Biosphere		
Hydrosphere		
Geosphere		

Checkpoint 17.11

basic · advanced · intermediate · superior

Use the concept map provided here to illustrate the natural global carbon cycle by showing how carbon is transferred among the components of the earth system.

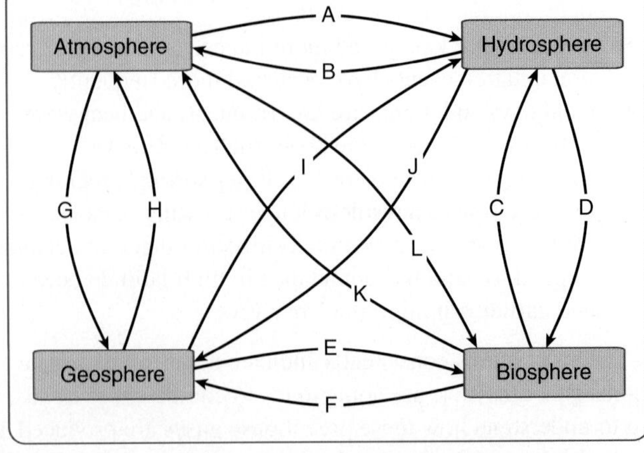

Biosphere. The biosphere interacts with all other parts of the earth system through the carbon cycle. **Plants extract carbon dioxide from the atmosphere by photosynthesis.** Animals eat the plants and consume the carbon. **Carbon is returned to the atmosphere when the organisms die** or may be withheld from the carbon cycle for millions of years if the organic remains are converted to fossil fuel deposits such as coal, oil, or natural gas. Some carbon is retained in soils. Cultivation can release this carbon, but techniques such as conservation tillage (see Chapter 9) can preserve carbon in the soils. Carbon is also retained by marine organisms such as coccolithophores (see Chapter 7) and foraminifera (see Chapter 16) that extract dissolved carbon dioxide from ocean waters to form their shells. Dead organisms sink to the bottom of the ocean where their remains accumulate to form sedimentary rocks that store the carbon within rock layers.

Geosphere. The **largest sinks for carbon on the planet are rocks and minerals of the solid earth** (Chapter Snapshot). Most of that carbon is present as an element of calcium carbonate (limestone). Limestone releases carbon dioxide to the atmosphere when it undergoes chemical weathering at Earth's surface. These processes are most common in mountainous regions undergoing rapid erosion or where hot, wet climates produce conditions suitable for rapid chemical weathering. Some carbon is stored in fossil fuel deposits formed from decayed organic material and is released when these fuels are consumed to produce energy.

Carbon Produced by Human Activity

The volume of carbon in sinks fluctuates in time and space, but overall, CO_2 in the atmosphere remained constant at 280 parts per million (ppm) for several centuries prior to the Industrial Revolution. However, anthropogenic (human-produced) **emissions of greenhouse gases such as carbon dioxide have increased by one-third since the Industrial Revolution** (Figure 17.10). Most anthropogenic greenhouse gases are produced by the consumption of fossil fuels, such as oil for transportation and coal for electrical power, and through the destruction of forests (deforestation). About one-half of the human-produced carbon dioxide is absorbed in carbon sinks in the biosphere, land, and oceans with the remainder left in the atmosphere (see Chapter Snapshot). Since we know that greenhouse gases absorb energy, it seems apparent that continued emissions will result in additional warming.

How might the recent increase in carbon dioxide concentrations compare to population trends? Over the same time period that carbon dioxide and other greenhouse gases have increased, global population has risen dramatically. There were 1.6 billion people living on Earth in 1900; today, there are more than 7 billion. Not only will there be more people in the future, but also economic growth may increase the volume of greenhouse gases that each of these individuals produces.

We next examine where these gases are produced and consider which ones have the greatest potential to raise global temperatures.

Greenhouse Gas Emissions

The major greenhouse gases are water vapor, carbon dioxide (CO_2), methane, halocarbons (HFCs, including CFCs), and nitrous oxide. Water vapor accounts for most of the natural greenhouse effect, but human production of water vapor is not a major factor in global warming. It is the other greenhouse gases produced by human activities that may have tipped the balance toward global warming. Carbon dioxide accounts for more than three-quarters of anthropogenic global greenhouse gas emissions (Figure 17.11). The

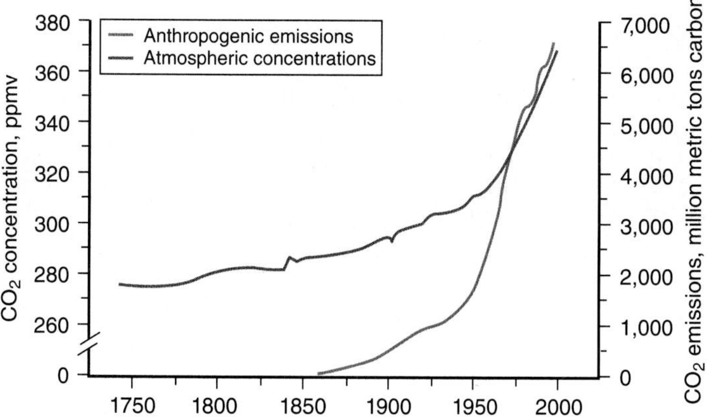

Figure 17.10 Anthropogenic (human-produced) and atmospheric carbon dioxide compared. Human production of carbon dioxide in relatively large concentrations began in the 1800s at the same time as the concentration of carbon dioxide in the atmosphere began to increase. The volume of anthropogenic carbon dioxide can be readily calculated on the basis of the volume of fossil fuels consumed.

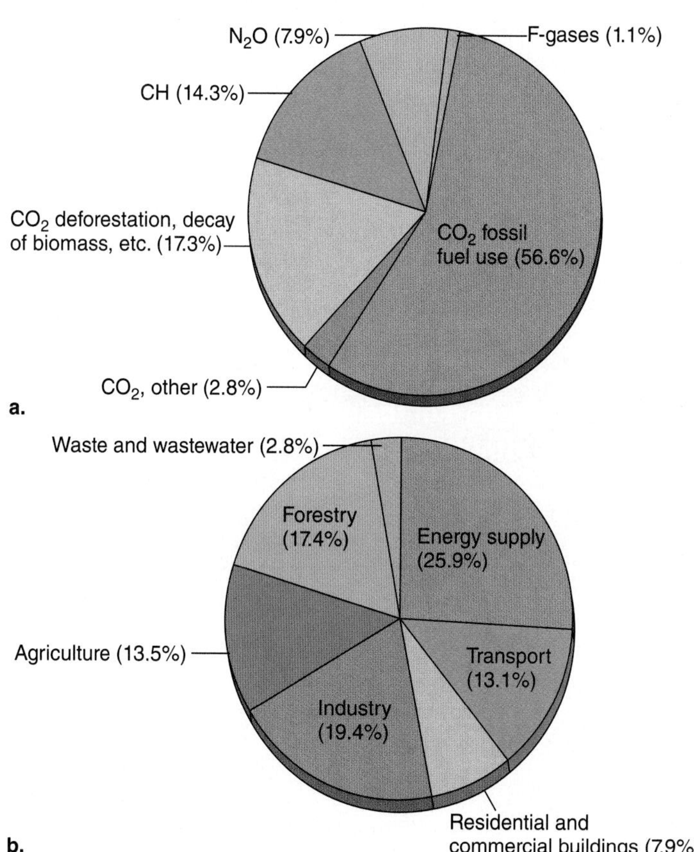

Figure 17.11 Global emissions of anthropogenic greenhouse gases. CH = methane; N_2O = nitrous oxide; F gases = fluorinated gases. **a.** Share of emissions by greenhouse gas; **b.** share of greenhouse gases by sector.

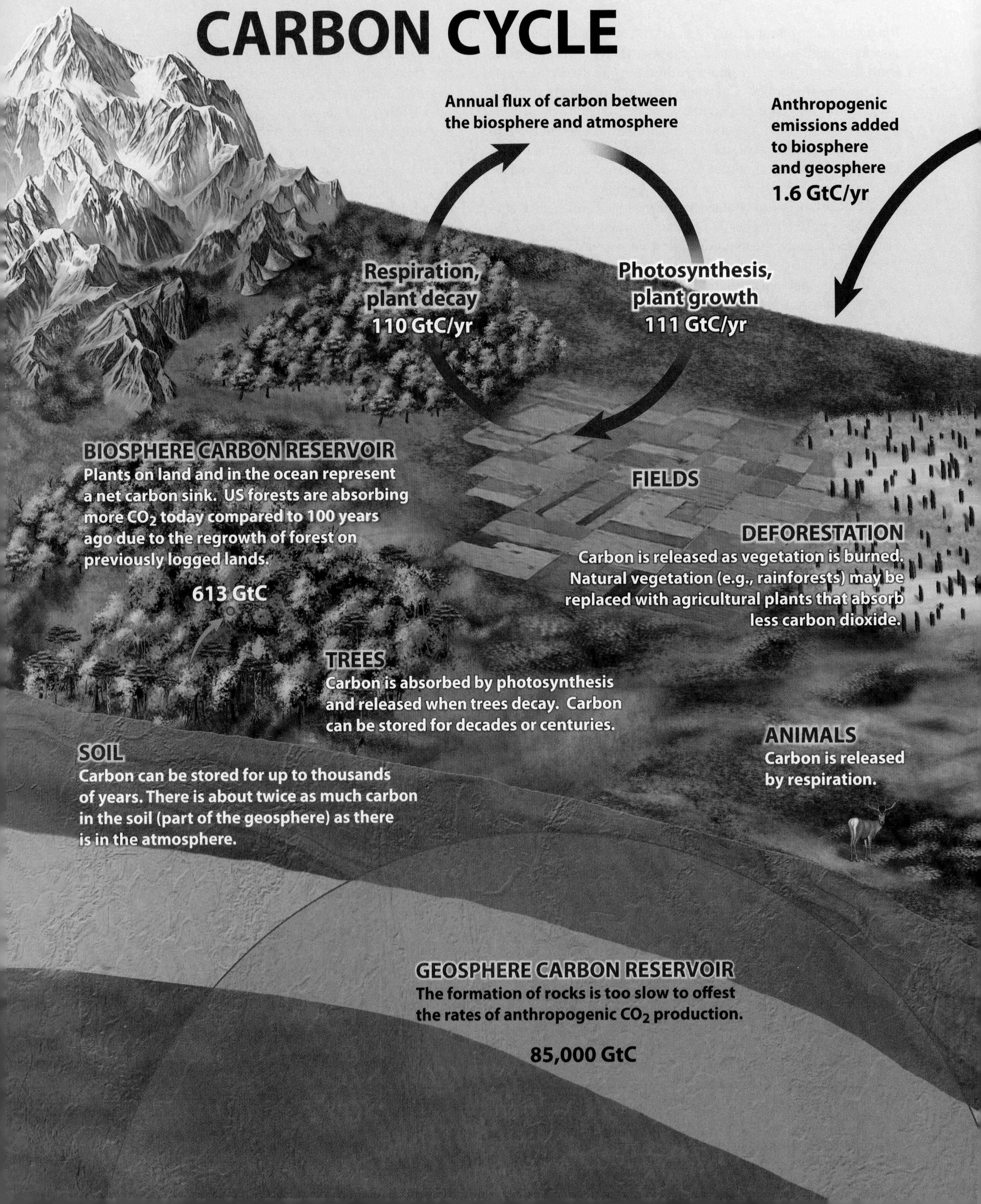

CARBON CYCLE

Annual flux of carbon between the biosphere and atmosphere

Anthropogenic emissions added to biosphere and geosphere
1.6 GtC/yr

Respiration, plant decay
110 GtC/yr

Photosynthesis, plant growth
111 GtC/yr

BIOSPHERE CARBON RESERVOIR
Plants on land and in the ocean represent a net carbon sink. US forests are absorbing more CO_2 today compared to 100 years ago due to the regrowth of forest on previously logged lands.

613 GtC

FIELDS

DEFORESTATION
Carbon is released as vegetation is burned. Natural vegetation (e.g., rainforests) may be replaced with agricultural plants that absorb less carbon dioxide.

TREES
Carbon is absorbed by photosynthesis and released when trees decay. Carbon can be stored for decades or centuries.

ANIMALS
Carbon is released by respiration.

SOIL
Carbon can be stored for up to thousands of years. There is about twice as much carbon in the soil (part of the geosphere) as there is in the atmosphere.

GEOSPHERE CARBON RESERVOIR
The formation of rocks is too slow to offest the rates of anthropogenic CO_2 production.

85,000 GtC

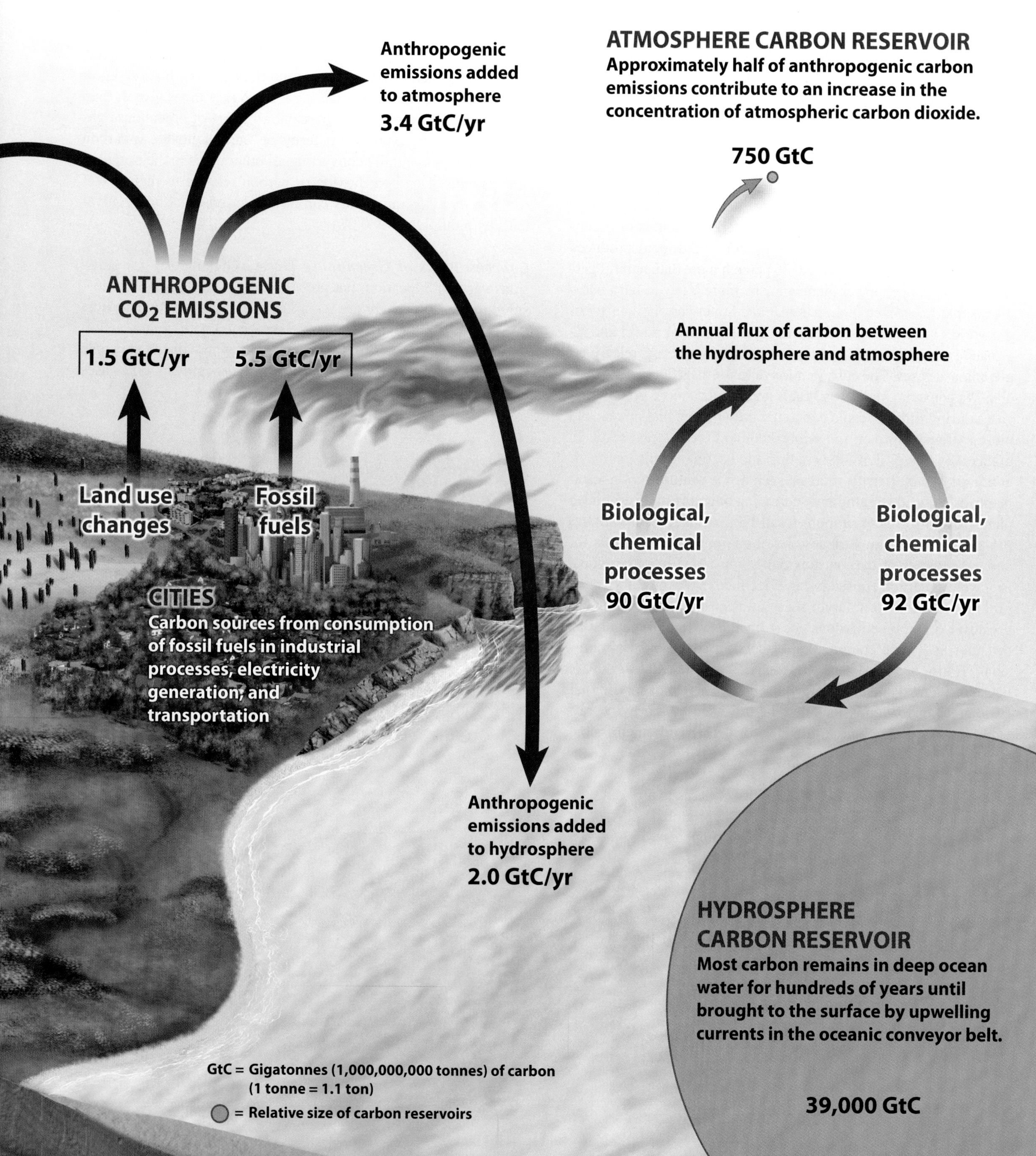

Anthropogenic
emissions added
to atmosphere
3.4 GtC/yr
ATMOSPHERE CARBON RESERVOIR
Approximately half of anthropogenic carbon emissions contribute to an increase in the concentration of atmospheric carbon dioxide.
750 GtC
ANTHROPOGENIC
CO_2 EMISSIONS
1.5 GtC/yr
5.5 GtC/yr
Annual flux of carbon between the hydrosphere and atmosphere
Land use changes
Fossil fuels
Biological, chemical processes 90 GtC/yr
Biological, chemical processes 92 GtC/yr
CITIES
Carbon sources from consumption of fossil fuels in industrial processes, electricity generation, and transportation
Anthropogenic
emissions added
to hydrosphere
2.0 GtC/yr
HYDROSPHERE
CARBON RESERVOIR
Most carbon remains in deep ocean water for hundreds of years until brought to the surface by upwelling currents in the oceanic conveyor belt.
39,000 GtC
GtC = Gigatonnes (1,000,000,000 tonnes) of carbon
(1 tonne = 1.1 ton)
= Relative size of carbon reservoirs

bulk of greenhouse gas emissions results from human use of fossil fuels such as oil, natural gas, and coal. Methane and nitrous oxide come from industrial, agricultural, and waste sources, and they account for much of the remaining gases (Figure 17.11). Fossil fuels are burned to produce electricity for residential heat and light, used as gasoline for transportation, or used to power industry. Electricity generates about one-third of US greenhouse gas emissions, and transportation accounts for a little more than one-quarter.

Different fossil fuels produce different amounts of greenhouse gases. For each unit of energy consumed, coal produces 21 percent more CO_2 than oil and 76 percent more than natural gas. Consequently, new power plants may be more likely to burn natural gas than coal, even though the United States has one-quarter of the world's known coal reserves. In contrast, nuclear power and alternative power sources, such as solar and wind energy, release no greenhouse gases. The consumption of fossil fuels fluctuates annually. We consume more fossil fuels and produce more greenhouse gases during times of rapid economic growth, low fuel prices, and more extreme summer and winter temperatures, which result in higher demand for domestic cooling and heating. Limits on major nonfossil fuel electricity sources, such as the shutdown of nuclear plants or reduced stream flow decreasing output from hydropower plants, can also result in more fossil fuel use. The United States produces a more than 4 tonnes of carbon per person each year (or about 17 tonnes of carbon dioxide), about four times the global average. US carbon emissions have declined in recent years as natural gas accounts for a larger share of electricity generation. The greatest US carbon emissions occur in cities with high population densities and some combination of concentrated industrial emissions (Houston, Texas), heavy traffic (Los Angeles, California), or dense residential and commercial development (Chicago, Illinois). Even areas of low emissions may be traversed by a heavily traveled highway to generate a thin strip of denser carbon emissions (Figure 17.12).

Global Warming Potential. The ability of gases to absorb heat in the atmosphere is known as their global warming potential (GWP). The GWP compares the amount of heat absorbed by a given mass of a greenhouse gas to the heat absorbed by an equivalent mass of carbon dioxide. Individual molecules of methane, nitrous oxide, and HFCs absorb more heat than a molecule of carbon dioxide. **The amount of heat a gas can absorb depends on its composition and the length of time the gas molecules remain in the atmosphere.** The atmospheric lifetimes for key greenhouse gases vary from 1 to 300 years, with carbon dioxide having one of the longer lifetimes (Table 17.2). For example, methane remains in the atmosphere for much less time than carbon dioxide, but it can absorb much more heat, giving it a GWP 21 times greater than that of CO_2. For ease of comparison, all types of greenhouse gas emissions may be expressed in terms of "carbon dioxide or carbon equivalents," essentially converting all other gases to their equivalent amount of carbon dioxide or carbon relative to their GWP. (Carbon dioxide equivalents can be converted to carbon equivalents by multiplying by 0.27.)

Concentrations of Greenhouse Gases. Greenhouse gas emissions vary with location, but greenhouse gas concentrations measured in the atmosphere are remarkably uniform worldwide. How do you account for this? Gases are associated with human activities and are therefore initially concentrated where the most people live, but atmospheric circulation patterns thoroughly mix the gases

Table 17.2 Global Warming Potentials for Greenhouse Gases

Greenhouse gas	Atmospheric lifetime, years	Global warming potential (GWP for 100 years)
Carbon dioxide	50–200	1
Methane	12	21
Nitrous oxide	120	310
HFCs	1–300	140–11,700

Note: Water (H_2O) is a greenhouse gas that accounts for about 35 to 75 percent of the overall greenhouse effect. Water is not listed here because humans do not produce much water vapor and individual water molecules stay in the atmosphere for only a few years.

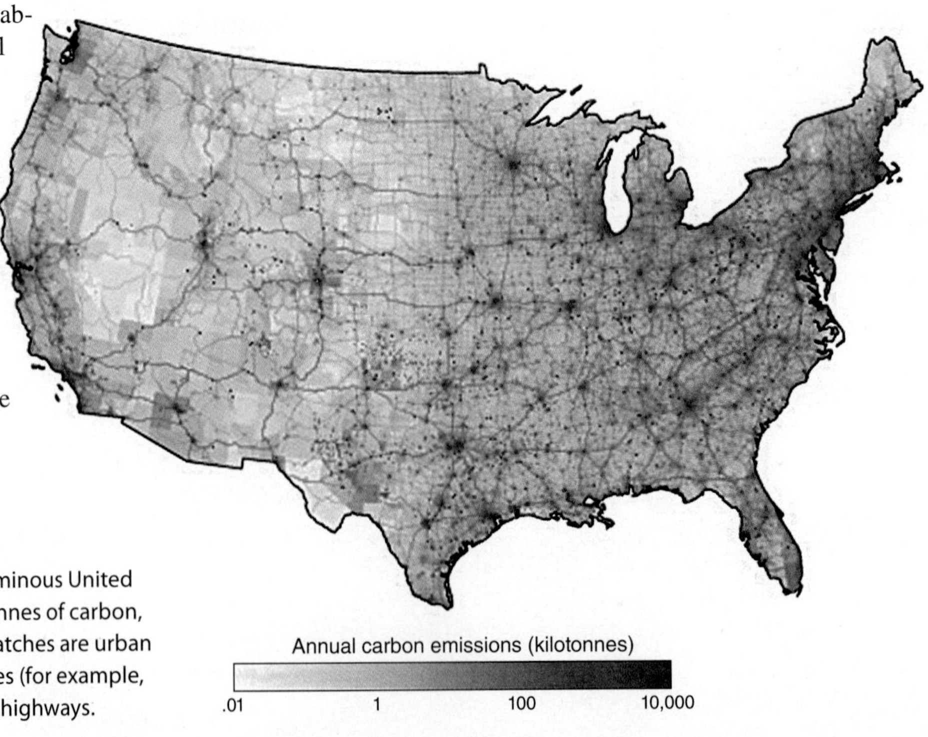

Figure 17.12 Annual carbon emissions for the conterminous United States. The map shows annual emissions in 2002 (kilotonnes of carbon, 1 kilotonne = 1,000 tonnes or metric tons). Larger red patches are urban centers; small red dots represent individual point sources (for example, power stations, factories); and linear patterns are major highways.

and distribute them throughout the troposphere. Global carbon emissions amount to approximately 9 billion tonnes (gigatonnes) of carbon equivalent each year. The principal global sources of greenhouse gases are nations with the greatest economic growth, large populations, or both. Which countries do you predict have the largest carbon footprint (the amount of carbon dioxide equivalent produced annually)? For example, just four nations produce more than one-half of the world's carbon dioxide emissions (data from 2009)—China, 25.2 percent, United States, 17.8 percent; India, 5.3 percent; and Russia, 5.2 percent (Figure 17.13).

Checkpoint 17.12

basic · intermediate · advanced · superior

Review the following three scenarios and discuss the implications for global greenhouse gas emissions. Which scenario will produce the most or least warming? Complete the graph provided here to predict how temperatures would change over the next century for each scenario. We are more interested in the trends than in the absolute values you identify for each scenario.

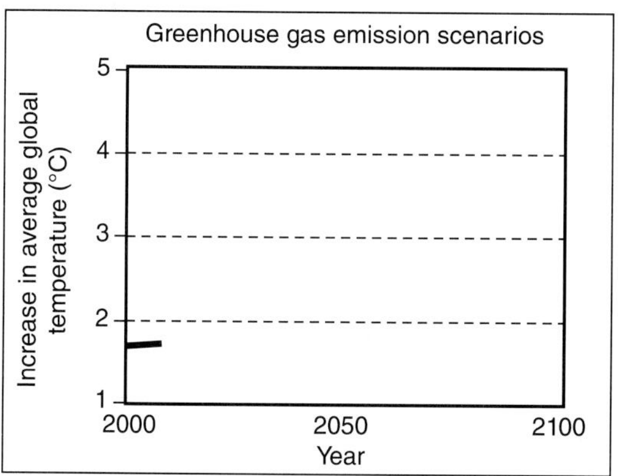

Scenario A: Global population peaks around 9 billion in 2050 and declines thereafter. New technologies are rapidly introduced, and economic disparities between regions are substantially reduced. Fossil fuels continue to supply most of the energy used by humans.

Scenario B: Global population continues to increase throughout the twenty-first century (11 billion by 2050). Disparities in the economic growth of regions persist, and technological change occurs slowly. The energy supply gradually changes from fossil fuels toward alternative energy sources.

Scenario C: Global population peaks at around 9 billion in 2050 and declines thereafter. Economies rapidly become service- and information-oriented, income disparities decrease, and energy technologies that do not rely on fossil fuels are introduced.

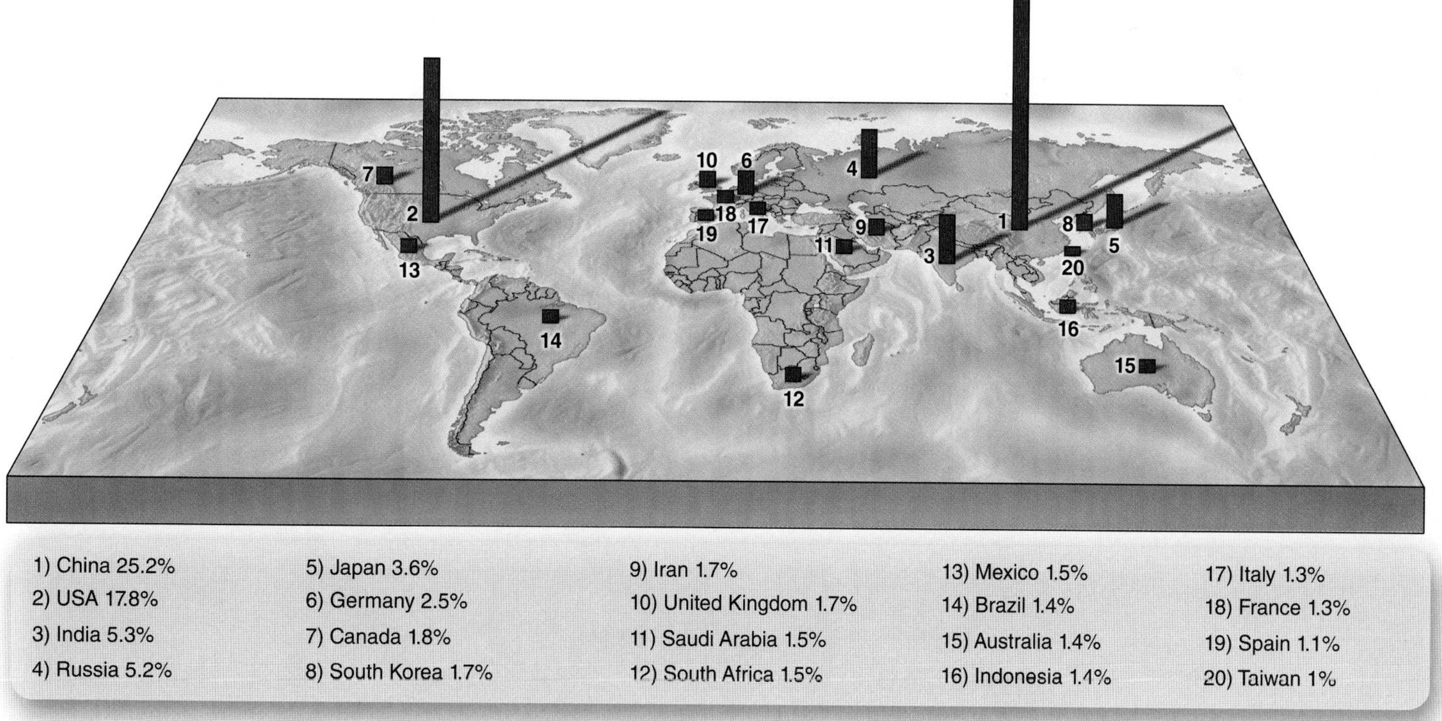

Figure 17.13 Distribution of global carbon emissions. Data for the 20 nations with greatest emissions in 2009 are listed. Emissions of carbon dioxide vary hugely between places due to differences in lifestyle and ways of producing energy. For example, compare emissions from the different continents. These data will change annually; check online to see recent statistics and discover the most significant changes. How do you think this map will look 50 years from now?

17.5 Modeling Global Climate Change

Learning Objectives

- Relate climate forcing and feedback mechanisms in the earth system.
- Illustrate how exceeding tipping points may affect the future climate.
- Explain the significance of computer models to understanding the future climate.

The atmospheric concentrations of carbon dioxide and other greenhouse gases will increase, at least for the next few decades. However, only some of the increase in global temperature is directly attributable to rising greenhouse gas concentrations. A larger proportion is related to feedback systems that alter Earth's energy balance and exaggerate the warming trend. In this section, we will discover positive and negative feedbacks that magnify or diminish average global temperature increases. We will also investigate how scientists use their understanding of the earth system to build sophisticated climate models that are used as tools to predict changes in global climate patterns over the next century.

Forcings and Feedbacks

Have you ever experienced audio feedback? Audio feedback can occur when someone speaks into a microphone which transmits that sound to a speaker. The microphone then picks up the sound from the speaker and amplifies part of that sound back to the speaker. This becomes a loop that can result in a high-pitched squeal. If you have heard that happen, you have experienced a forcing and

Figure 17.14 Radiation balance for Earth. Approximately one-third of incoming solar radiation (342 watts per square meter) is reflected back to space (107 watts per square meter). The rest is absorbed by the atmosphere (67 watts per square meter) and the land and water of Earth's surface (168 watts per square meter). Infrared radiation from the heated surface is trapped by greenhouse gases in the atmosphere, adding more warmth to the atmosphere. An increase in greenhouse gases adds 2 to 3 watts per square meter more warming to the atmosphere. Compare with Figure 14.9.

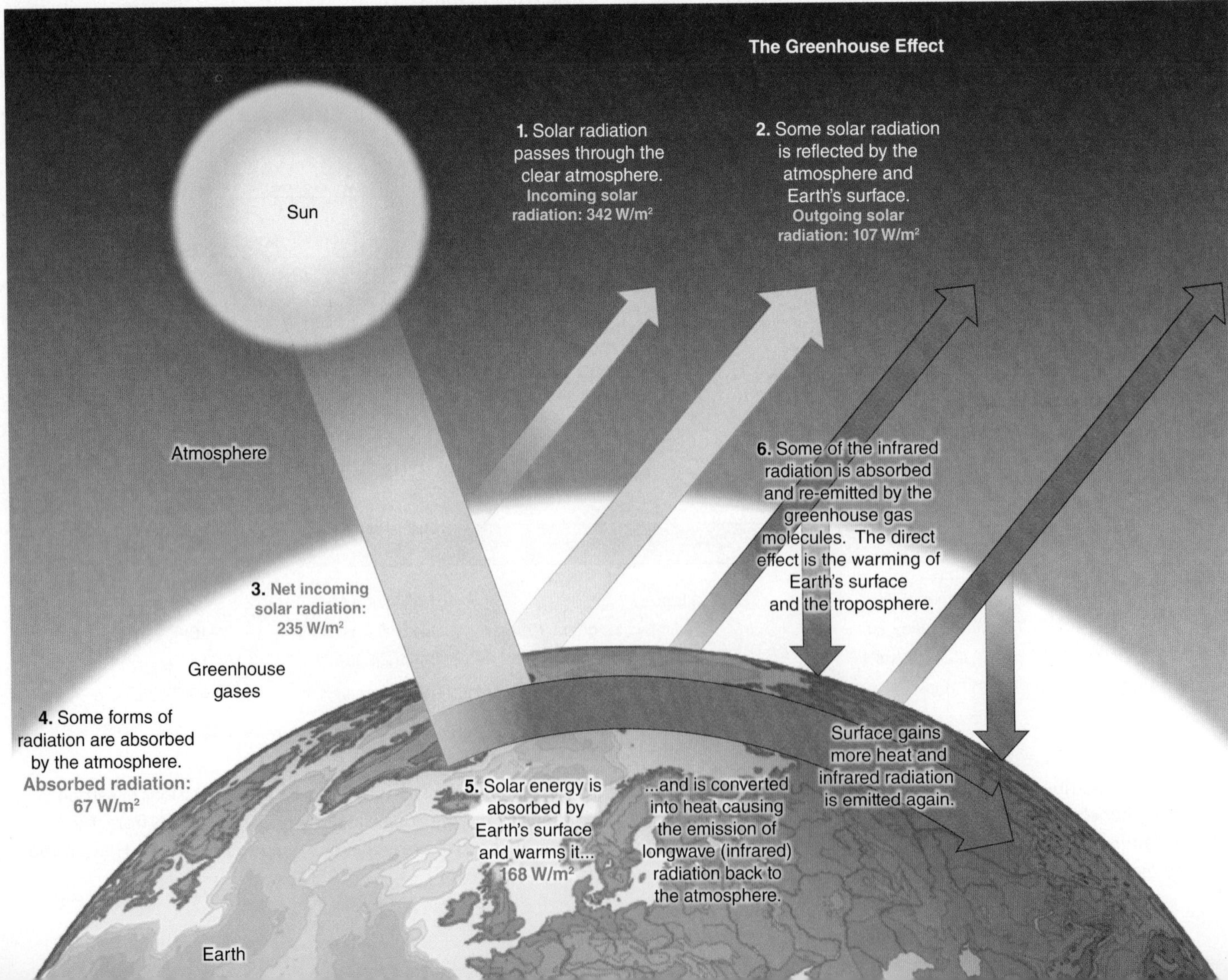

feedback. Feedback results when the output of a system affects the input, in this case the sound. Forcings and feedbacks are very important to our discussion of climate change.

Positive and Negative Climate Forcings. Radiation from the sun is scattered, reflected, or absorbed by Earth (see Chapter 14). The balance between reflected energy and absorbed energy determines the temperature of Earth (Figure 17.14). For instance, the planet cools if more solar radiation is reflected and warms if more radiation is absorbed (see Chapter 14 Snapshot). **Any phenomenon that causes a change in the global solar radiation balance is known as a *climate forcing.*** Forcings are considered positive if they result in global warming and negative if they cause global cooling. We do not mean "good" and "bad" when we use the terms *positive* and *negative* in this context. A positive factor increases an existing trend; a negative factor causes a trend to slow down or reverse. For example, a positive forcing occurs if more solar energy reaches Earth's surface or if more energy is absorbed by the addition of greenhouse gases to the atmosphere. In contrast, a negative forcing occurs if volcanic ash reflects light back into space. The magnitude of a solar forcing mechanism is measured as a change in solar energy in units of watts per square meter (W/m^2) passing through the top of the troposphere. Greenhouse gases represent the most significant agent, with an estimated net forcing of 2.6 watts per square meter (Figure 17.15). Most of this forcing is accounted for by carbon dioxide. Can you think of another forcing that is not related to human activity?

Aerosols: An Example of a Negative Forcing. Aerosols are tiny particles approximately 0.1 micrometer across (about one-thousandth the width of a human hair) produced by the combustion of fossil fuels, burning vegetation, or dust from dry lands. **Most aerosols are particles that reflect sunlight back into space.** Consequently, aerosols cause a negative forcing and diminish the warming effect for some industrialized nations (Figure 17.15).

Checkpoint 17.13

Where would you expect to see the most significant cooling effects due to aerosols? Explain your answer choice.

a) Near the poles
b) Around the equator
c) Between 30° and 60°N latitudes
d) Between 30° and 60°S latitudes

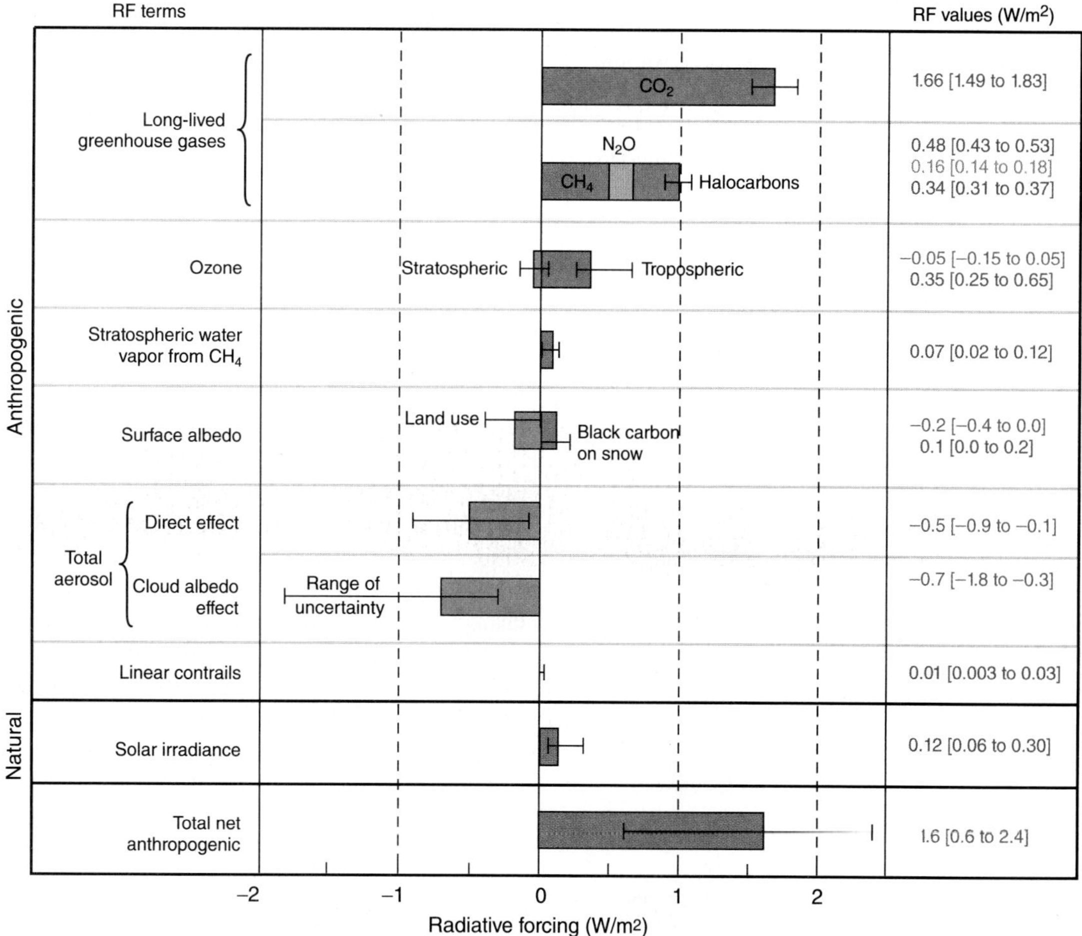

Figure 17.15 Climate forcings. Relative forcings (RFs) are expressed in watts per square meter (W/m^2). The warming effect of greenhouse gases was partially offset by forcings due to human activities, such as industrial pollutants in the form of reflective aerosols. Some forcing agents (for example, clouds) have considerable uncertainty as to their degree of forcing. The net effect of forcing is to warm the planet by 1.6 watts per square meter.

They can even produce a cooling effect in areas of heaviest air pollution (for example, China). Some black particles absorb solar radiation, warming the atmosphere. Ironically, it is only the pollution that produces aerosols that stands between many regions and a more severe warming effect. A surge in pollution associated with rapid industrialization from 1945 to 1975 is thought to have reflected sufficient solar radiation to cause a temporary respite in the steady climb in average global temperatures. Environmental regulations introduced in the early 1970s caused a steep drop in industrial pollution, and the warming trend resumed. Now back to the question at the end of the previous paragraph: What natural geologic event develops aerosols that would provide negative feedback? (See Chapter 6 for a hint.)

Other Feedback Mechanisms. Forcings are only one-half of the story. There are reactions to forcings. The **warming or cooling caused by forcings leads to climatic feedbacks that can exaggerate or reduce temperature changes.** For example, warmer climates add more water vapor to the atmosphere through evaporation. Water vapor is a greenhouse gas that amplifies the warming effect. However, the water vapor may form clouds that reflect more solar radiation. Those same clouds absorb the heat that is radiated from Earth. Consequently, the addition of water vapor may contribute both positive and negative feedback to the warming (see discussion of contrails in Chapter 14). Because of these competing effects, the influences of clouds on global climate are particularly difficult to decipher (Figure 17.15). Additionally, changes in land use result in changes in the heat absorbed at Earth's surface. For instance, warming at high latitudes replaces pale-colored tundra, which reflects much incoming sunlight, with dark forest that absorbs much of the solar radiation. Likewise, melting of sea ice in the Arctic Ocean (see Chapter 16) converts reflective white surfaces to dark ocean water that absorbs more sunlight. Scientists hypothesize that forcing accounts for about 40 percent of global temperature changes and that feedback mechanisms explain the rest. Predicting how climate will change is challenging because these multiple forcings and feedbacks are operating simultaneously in a dynamic climate system. To address these complex problems and understand their implications for future climates, scientists have developed powerful computer models.

basic advanced intermediate superior

Explain the Milankovitch cycles described in Chapter 16 in terms of a climate forcing.

Checkpoint 17.15

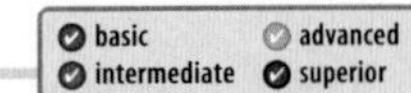

Add labels and arrows to the following figure to illustrate forcings and feedback mechanisms for Earth's climate system. Add features to the figure as needed.

Climate Models

Climatologists cannot conduct a simple experiment to test the effect of increasing carbon dioxide concentrations on temperature. Even if it were possible to vary one climate factor while keeping all the others constant, it would be of limited value. Instead, climate models must deal with multiple forcings and feedbacks that operate simultaneously. One way to explore how climate-forcing variables interact is by building computer climate simulation models. Climatologists have done just that.

The models scientists use today evolved from the first primitive weather models of the 1950s. Early climate models were very limited, because accurate climate-related global data were not available. Such data include detailed information about winds, temperatures, and moisture for all levels of the atmosphere and for all areas of the globe. This information became more readily accessible and available with the introduction of satellite technology in the 1970s. Today, climate scientists use what they understand about forcings and feedbacks in sophisticated climate models. **These computer models use millions of calculations to try to simulate climate factors over the entire earth system.** Modern models seek to represent all the key factors—incoming and outgoing radiation, wind speed and direction, cloud types, precipitation types and amounts, changes in the dimensions of ice sheets, vegetation types, atmospheric gas concentrations, temperature stratification of oceans, and the topography of continents.

How Computer Models Work. Computer models, like weather models, divide Earth's surface into a grid and use data on surface characteristics to set the initial conditions for each square in the grid. In addition, the column of atmosphere above each square is divided into multiple levels, and real-time data on weather conditions are plugged into the model (Figure 17.16a). All of these conditions are represented by a series of mathematical equations for each square of the grid. Since the models are intended to simulate change, these equations take into account the physical laws that control how air and water move and how energy is transferred over time within the earth system. A computer runs a program to predict how conditions within and between each grid square at each level will change over a chosen time interval.

The complexity of the model depends on the size of the grid squares, the number of levels in the modeled atmosphere, and the time steps chosen for the model. Even the fastest supercomputer would get bogged down if the model grid were too small, if the atmosphere had too many levels, or if the model moved forward in time steps that were too short. A typical global climate model divides Earth's surface into a grid with squares about the size of South Carolina or Indiana (Figure 17.16a). The atmosphere has 15 to 30 levels of differing heights, and most are in the troposphere. Finally, the time steps cannot be longer than

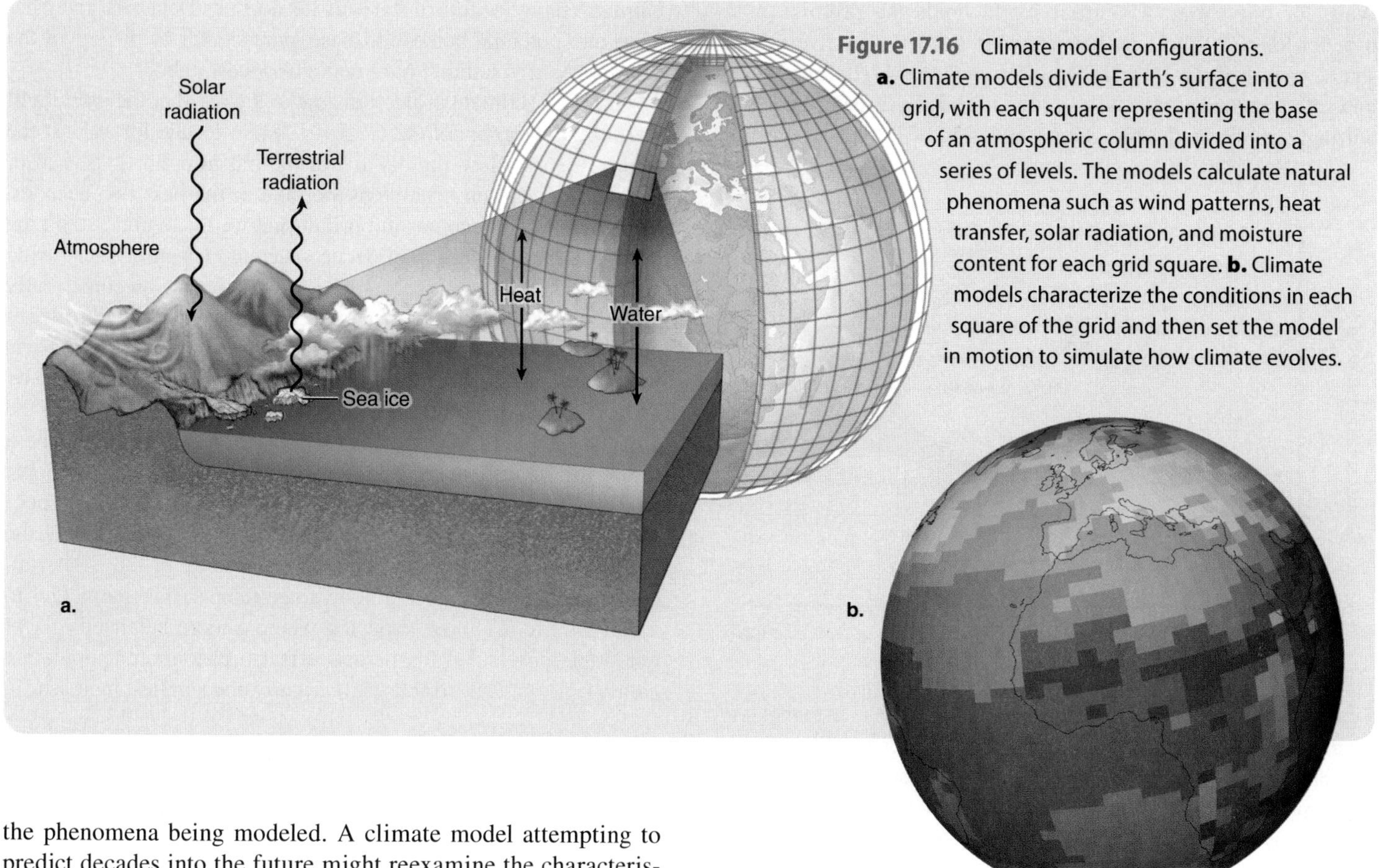

Figure 17.16 Climate model configurations. **a.** Climate models divide Earth's surface into a grid, with each square representing the base of an atmospheric column divided into a series of levels. The models calculate natural phenomena such as wind patterns, heat transfer, solar radiation, and moisture content for each grid square. **b.** Climate models characterize the conditions in each square of the grid and then set the model in motion to simulate how climate evolves.

the phenomena being modeled. A climate model attempting to predict decades into the future might reexamine the characteristics of each level of each grid square for time intervals as short as 30 minutes (Figure 17.16b). So, for a single day, the model performs thousands of calculations for each grid square. This process is repeated for each square for thousands of days into the future. No wonder a staggeringly high number of calculations are involved in these models!

Accuracy of Climate Change Prediction Models. Just as there are different computer games to simulate the same activity, there are several computer models designed to simulate climate. The various climate models have many similar features and typically produce consistent predictions. But just how accurate are these models? One way to evaluate their accuracy is to compare their predictions to the actual record of recent climates (1850 to the present). Such comparisons show that most climate models generate the same major trends that are observed in the climate data. For instance, the models correctly predicted that the Arctic would warm faster than the rest of the planet and that the eruption of Mount Pinatubo in 1991 would generate a global cooling event.

Several independent research groups have generated model results that predicted temperature rises if CO_2 increased. Current best estimates are that forcings and feedbacks associated with doubled CO_2 concentrations would produce an average 2 to 4.5 C° (3.6 to 8.1 F°) temperature increase. The range is due to the uncertainties associated with the impact of aerosol forcing as well as social and economic changes that might occur in the future. Today, these models accurately predict the cooling effect that aerosols appear to have had on the climate over the past century. In Section 17.6, we will examine more model predictions regarding the climate in our future.

Checkpoint 17.16

basic / intermediate / advanced / superior

Explain why computer climate models are really "earth system" models.

17.6 A Warmer World

Learning Objectives

- Describe actions anyone could take to reduce her or his greenhouse gas emissions.
- Use results of climate models to compare and contrast the global warming predictions for different continents.
- Show how various components of the earth system will be impacted by global warming.

To return to our opening question: Do you feel lucky? Will our impact on global climate have positive or negative consequences? The people of the Arctic featured in Chapter 16 might tell us that global warming is already here and has brought some serious problems. But what about the rest of the world? How will the remainder of us be affected by additional temperature changes?

Well, for one thing, a warmer world holds the promise of a 6-meter-long (20-foot) python on your lawn if you live in the southern half of the United States. Pet owners who dumped their unwanted snakes into the wilds of the Florida Everglades have unwittingly established a breeding population of Burmese pythons.

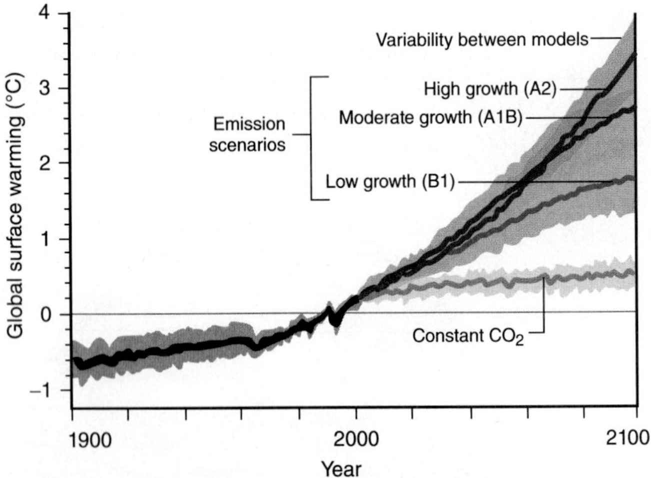

Figure 17.17 Results of global mean temperature change model for three scenarios, 1900–2100. Each plot corresponds to a different greenhouse gas emissions scenario. Scenario A2 assumes slow economic growth, rapid population growth, and slow technological changes favoring the preservation of fossil fuels as the dominant energy source. Scenario A1B presumes a rapid economic growth, moderate population growth, and a balance between old and new energy sources. Scenario B1 involves moderate economic and population growth with emphasis on new technologies available to all nations.

Climate change will turn states as far north as Delaware and Missouri into potential homes for these giant snakes by the end of the century. As if you didn't have enough to worry about.

IPCC climate projections make it clear that the world will warm by an average of 0.2 C° (0.4 F°) per decade for at least the next 15 years. How rapidly it warms and how hot it gets after that will depend on what steps we take in the next two decades. Consequently, it becomes important that we try to understand the potential changes that may occur. Current climate models using various population and economic growth scenarios consistently predict that combinations of **rapid economic growth, significant population increases, and continued reliance of fossil fuels will result in the greatest average temperature increases by 2100** (Table 17.3 and Figure 17.17). The scenarios that produce the greatest temperature increases (A1F1, A2; Table 17.3) assume a future still dependent on fossil fuels with either many more people (15.1 billion in A2) or rapid economic growth that results in peoples of the developing world (for example, China, India, Indonesia) having a standard of living that approaches that of the industrialized nations (A1F1). Moderate average temperature gains (2.4 to 2.8 C°; 4.3 to 5 F°) are expected where a more balanced mix of renewable power and fossil fuels offsets economic or population gains (A1B, A1T, B2; Table 17.3). Finally, the smallest increases in

Checkpoint 17.17

Identify three actions a private citizen could take to reduce the amount of greenhouse gases produced on his or her behalf.

Table 17.3 **Six IPCC Scenarios for Future Global Development Pathways Involving Demographic, Economic, and Technological Factors.** Combinations of rapid economic growth, large populations, and continued reliance on fossil fuel technologies result in the greatest average temperature increases by 2100.

Scenario	Scenario Factors			Consequences
	Economic growth	Population growth rates (population in 2100)[a]	Fossil fuel intensive technology[c]	Temperature change, average (range)
A1F1	High	Low[b] (7.1 billion)	High	High, 4.0 C° (2.4–6.4 C°)
A2	Low	High (15.1 billion)	High	High, 3.4 C° (2.0–5.4 C°)
A1B	High	Low[b] (7.1 billion)	Moderate	Moderate, 2.8 C° (1.7–4.4 C°)
A1T	High	Low[b] (7.1 billion)	Low	Moderate, 2.4 C° (1.4–3.8 C°)
B2	Moderate	Moderate (10.4 billion)	Moderate	Moderate, 2.4 C° (1.4–3.8 C°)
B1	Moderate	Low[b] (7.0 billion)	Moderate	Low, 1.8 C° (1.1–2.9 C°)

[a] Global population in 2013 was 7.1 billion.
[b] In the low population growth cases, population is projected to increase to 8.7 billion by 2050 and then decline to 7.1 billion or less by 2100.
[c] High = more than two-thirds of technology relies on fossil fuels; moderate = approximately one-half to two-thirds of technology relies on fossil fuels; low = approximately one-third or less of technology relies on use of fossil fuels.
Source: Data from IPCC.

global temperatures occur with the B1 scenario, which describes a sustainable world with "clean" technologies, less demand for goods and resources, and a service-based economy.

It is important to remember that temperature changes predicted by these scenario models are *averages*. These average values mask the fact that global temperatures could increase by as much as 8 C° (14 F°) in some places and see little change in other locations. The severity of climate change will depend on which scenario the world follows. Some consequences are described in the next section.

Effects of Warmer Temperatures

The consequences of rising temperatures will vary considerably around the world. What would you do if the summer temperatures where you live increased by an average of 6 C° (10 F°)? Chances are you would simply use more energy to run your air conditioner as people in the desert Southwest are doing. In fact, wealthy nations probably can adapt to small changes in climate that occur over long periods (on human timescales).

Mitigating or adapting to changes will be more difficult in developing nations that do not have the resources of more-affluent countries. Research indicates that changes will be most extreme in the developing world and more moderate in the developed nations that are more responsible for the production of greenhouse gases. Where do you expect that climate change will have the greatest social impact? Projected patterns are similar to those observed over the last few decades. Here are a few predictions from climate models:

- Models predict much greater temperature increases in northernmost latitudes (8 C°; 14 F°) and little change in parts of the Southern Hemisphere and North Atlantic Ocean (Figure 17.18).

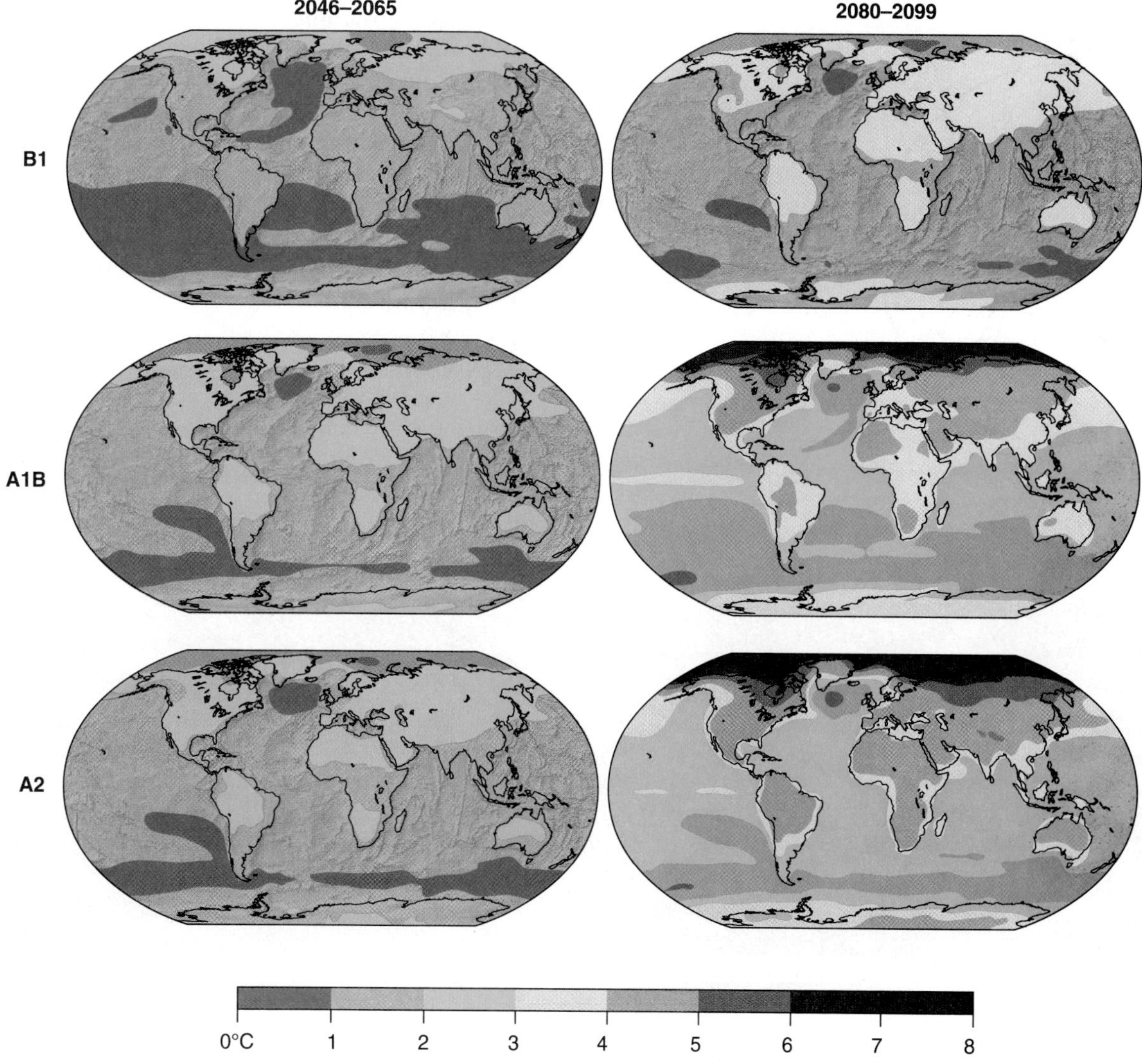

Figure 17.18 Projected annual mean surface warming for the scenarios B1 (top), A1B (middle), and A2 (bottom) for time periods 2046–2065 (left) and 2080–2099 (right). Temperature changes are relative to the average of the period 1980–1999. A warming effect is least for sections of the Southern Ocean and the northern Atlantic Ocean, and greatest for northernmost latitudes. Predicted temperature increases for the United States are between 2 and 6 C° (4 and 11 F°) except for Alaska, where temperatures may increase by 10 C° (18 F°) or more.

- Rising temperatures will increase evaporation and precipitation, but the **extra rainfall will not be distributed evenly around the globe** (Figure 17.19). Precipitation increases will generally be greatest where most rain falls today, around the equator and at high latitudes. A drying trend is evident for latitudes around 20° to 30°N and S that are the sites of modern deserts.

Checkpoint 17.18

basic / intermediate / advanced / superior

Venn Diagram: Climate Change in the Americas

Use Figures 17.18, 17.19, and the Venn diagram provided here to compare and contrast the predicted climate changes in North America with those in South America by 2100.

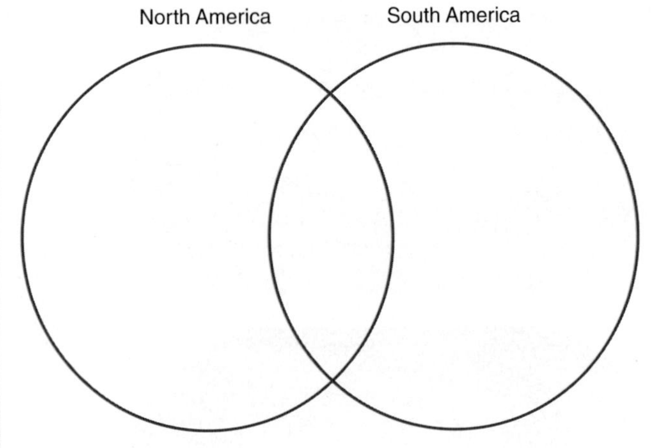

Checkpoint 17.19

basic / intermediate / advanced / superior

During a homework assignment, some students provided the following descriptions of global warming. Review the definitions and rank them in order of accuracy (1 = most accurate; 5 = least accurate).

____ Global warming is the gradual warming of Earth due to the greenhouse effect.

____ Global warming is a gradual, long-term increase in the temperature of Earth due to an increase in the concentration of trace atmospheric gases such as carbon dioxide that absorb heat from Earth's surface.

____ Global warming occurs when the temperature of Earth slowly increases because a layer of chemicals in the atmosphere absorbs more heat than normal.

____ Global warming is the result of the ozone layer in the atmosphere becoming thinner because of certain chemicals we use. This causes more solar radiation to reach Earth's surface, leading to higher average temperatures.

____ Global warming is the warming of climate over many years as a result of natural phenomena, such as volcanic eruptions, that trap heat close to Earth's surface.

Describe the criteria you used to rank the definitions. (For example, why are definitions you rank 1 and 2 better than those you rank 3 and 4?)

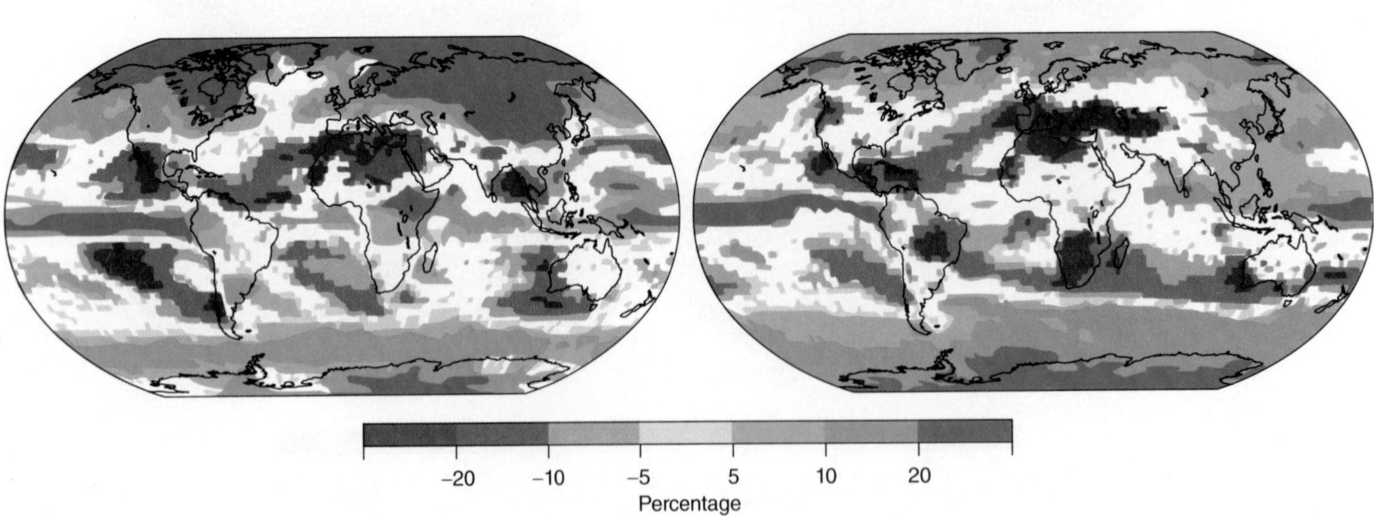

Figure 17.19 Modeled changes in global average precipitation for the period 2090–2099 relative to 1980–1999. Left, changes for the interval December–February; right, changes for July–August. Changes for scenario A1B only. White areas indicate regions where different models do not agree on predictions. Precipitation increases are greatest at equatorial latitudes and for high latitudes. A drying trend is evident for midlatitudes. How will precipitation change over North America?

- Regions with no changes or small precipitation increases may actually become drier because the warmer temperatures will result in higher evaporation rates.
- Changing climate will **increase the severity of storms,** leading to more structural damage and heat-related deaths. We should expect that heavy precipitation events will become more frequent and hurricanes more intense (but perhaps not more frequent).
- Snow cover on land will continue to shrink, and sea ice is projected to disappear entirely from the Arctic in late summer.
- Most scenarios predict that **global sea levels will rise by 20 to 50 centimeters** (8 to 20 inches) over the next century. Thermal expansion of ocean waters and alpine glacial melting will continue for centuries, resulting in a slow increase in sea level, even if global temperatures can be maintained at a constant level after 2100. After 2100, models show a switch to Greenland and Antarctica melting as dominant factors in sea level rise.
- Changes in the dynamics of the atmosphere are expected to shift Earth's great pressure belts, causing a poleward migration of subtropical highs in summer and expansion of polar highs in winter. Differences in heating of land and water will affect monsoons.
- Earth's oceans will become more acidic as CO_2 concentrations increase in the atmosphere. This excess carbon dioxide will change the chemistry of the oceans, facilitating the release of hydrogen ions. This directly affects calcium carbonate minerals in shells of corals, shellfish, and other organisms.

The changing climate has significant consequences for the earth system. A hotter, drier climate could **alter the balance of water resources.** Lower stream discharges would lead to less barge traffic and decrease the potential for generating hydroelectric power. Less water available for irrigation and municipal users could also result in regulations limiting the types of crops grown and/or water rationing programs. People relying on snowmelt for their summer water supply may find that warmer climates replace some of the snow with winter rains and cause the snow to melt earlier in the spring. Billions of people are expected to be affected by reductions in water supplies (Table 17.4).

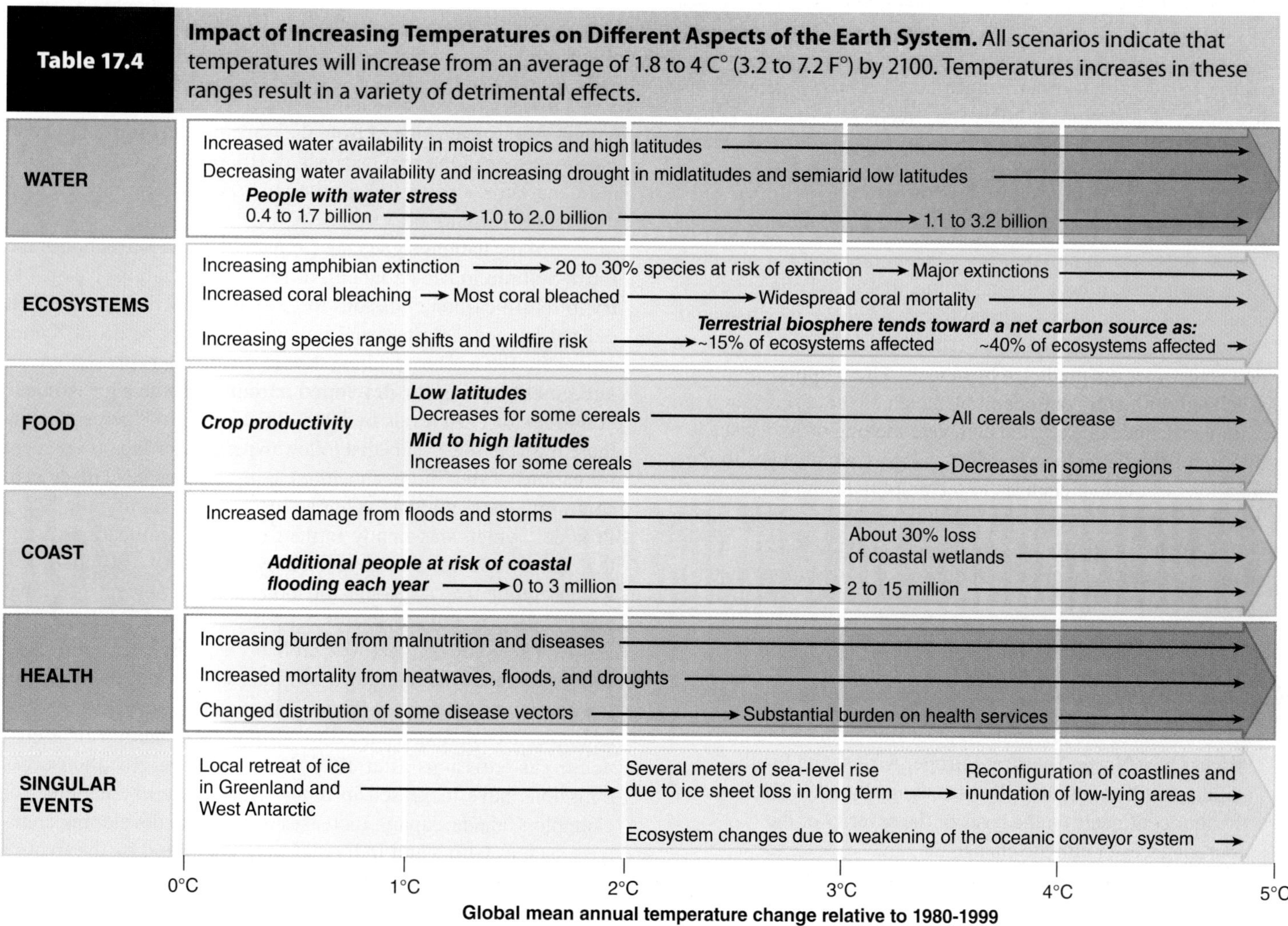

WATER
Increased water availability in moist tropics and high latitudes →
Decreasing water availability and increasing drought in midlatitudes and semiarid low latitudes →
People with water stress
0.4 to 1.7 billion → 1.0 to 2.0 billion → 1.1 to 3.2 billion →
ECOSYSTEMS
Increasing amphibian extinction → 20 to 30% species at risk of extinction → Major extinctions →
Increased coral bleaching → Most coral bleached → Widespread coral mortality →
Increasing species range shifts and wildfire risk → Terrestrial biosphere tends toward a net carbon source as: ~15% of ecosystems affected ~40% of ecosystems affected →
FOOD
Crop productivity
Low latitudes
Decreases for some cereals → All cereals decrease →
Mid to high latitudes
Increases for some cereals → Decreases in some regions →
COAST
Increased damage from floods and storms →
About 30% loss of coastal wetlands →
Additional people at risk of coastal flooding each year → 0 to 3 million → 2 to 15 million →
HEALTH
Increasing burden from malnutrition and diseases →
Increased mortality from heatwaves, floods, and droughts →
Changed distribution of some disease vectors → Substantial burden on health services →
SINGULAR EVENTS
Local retreat of ice in Greenland and West Antarctic → Several meters of sea-level rise due to ice sheet loss in long term → Reconfiguration of coastlines and inundation of low-lying areas →
Ecosystem changes due to weakening of the oceanic conveyor system →
0°C
1°C
2°C
3°C
4°C
5°C
Global mean annual temperature change relative to 1980-1999

Table 17.4 **Impact of Increasing Temperatures on Different Aspects of the Earth System.** All scenarios indicate that temperatures will increase from an average of 1.8 to 4 C° (3.2 to 7.2 F°) by 2100. Temperatures increases in these ranges result in a variety of detrimental effects.

Agricultural production will increase in some regions and decrease in others (Table 17.4). The impact will be minimal in spatially large countries where losses in productivity in one region are balanced by increases elsewhere. The problem is more threatening for smaller countries. Small, developing nations with large populations face famine and the exodus of refugees because of drought conditions. In contrast, northern nations such as Canada and Russia will benefit from longer growing seasons and increased precipitation (see Figures 17.18 and 17.19).

Changing climate will result in shifting ecosystems. Nonnative plants will migrate northward as climate warms there, displacing existing vegetation. Animals that rely on native plants for sustenance may lose their primary source of food. Trees such as oaks typically migrate at a rate of a few hundred meters per year, about one-tenth of the rate of temperature change over land. Most animals also take generations to adjust to changes in climate. Thus, if ecosystems have insufficient time to adapt to the migration of climate belts, more species could become extinct (Table 17.4). Environmental stresses from climate change are expected to reduce the amount of carbon dioxide extracted by plants, increasing future rates of warming. Nearly one-third of species will be at increased risk of extinction with the smallest projected temperature increases.

Higher sea levels will inevitably place additional millions of people at risk of property damages and loss of life from coastal flooding (Table 17.4), especially in association with hurricanes. People living on small islands and along delta regions will be at greatest risk. While warmer temperatures are expected to reduce the number of deaths due to cold exposure, these will be more than offset by deaths due to malnutrition and infectious diseases related to decreasing agricultural output and expanding range of disease-spreading insects (Table 17.4).

As if these consequences from global warming were not enough, there are some scientists who suggest that various parts of the earth system have tipping points that could change everything. A *tipping point* is a critical threshold above which a small change can cause large changes to the future states of a system in ways sometimes not predicted by models. These tipping points are associated with sub-continental elements of the climate system. Examples include unexpected complete melting of Arctic sea ice, collapse of the Greenland ice sheet, large-scale changes to thermohaline circulation, changes to the amplitude and frequency of El Niño–Southern Oscillation (ENSO), instability of the Indian summer monsoon, permanent tundra loss at high latitudes in Russia, or large-scale changes to rain forests or boreal forest dieback.

Checkpoint 17.20

On the basis of the information presented in this section, discuss how North America, Europe, Africa, and South America will experience climate change by 2100. Rank the four continents on the basis of the severity of the consequences of climate change they may experience ($1 =$ most severe negative consequences; $4 =$ least severe consequences). Give reasons to support your rankings.

Any of these changes could have a near-term, regional impact. The greater concern is that any one of these could change other climate-related processes across the entire earth system; hence the term *tipping point*.

17.7 What Can Be Done?

Learning Objectives

- Calculate your carbon footprint.
- Evaluate the effectiveness of recent carbon emission reduction treaties and agreements.
- Create a plan for reducing carbon emissions.

International Agreements to Improve the Environment

The signing of the Montreal Protocol in 1987 demonstrated that governments can agree on common principles to improve the environment. As a result, the use of compounds that were known to break down ozone was greatly reduced or banned throughout the world. However, the ozone issue was relatively simple and directly threatened people with increased incidences of cancer and cataracts if something was not done. Global warming, by contrast, is complex. People do not appear to be as directly threatened by the predicted consequences, After all, some argue, how important are a couple of degrees average increase? Some countries may even benefit from the warming, and Earth has been much warmer in the past (just ask the dinosaurs!). Consequently, reducing emissions of CO_2 in the atmosphere is proving to be a much more difficult effort.

Many nations signed the Climate Change Convention at the Earth Summit in Rio de Janeiro in 1992. These nations pledged to try to return carbon emissions to 1990 levels by the year 2000. It did not happen. International policymakers met again in Kyoto, Japan, in 1997. The resulting **Kyoto Protocol represented an agreement among the developed nations to reduce greenhouse emissions to 1990 levels by 2008–2012,** and 6 to 8 percent below those levels in the 5 years that follow (6 percent for Japan, 8 percent for nations of the European Union). Some nations have made firm commitments to mitigate potential warming by signing the Kyoto Protocol, which was finally ratified in 2005. Ratification means that the agreement was supported by more than 120 countries, which cumulatively account for more than 55 percent of global greenhouse gas emissions.

The Kyoto targets will probably be reached, mainly as a consequence of the shutdown of underperforming industries in the nations of the former Soviet Union and Eastern Europe. For example, Russia experienced a substantial reduction in greenhouse gas emissions after 1990. Unfortunately, greenhouse gas emissions have increased in other Kyoto signatory nations (for example, Canada, Japan). Increases occurred in developing countries such as China and India that were not bound by the treaty's limits. The net result is that global concentrations of greenhouse gases have increased and will continue to rise in the future unless more dramatic steps are taken to mitigate these emissions.

On a more positive note, the Kyoto treaty illustrates that the international community can come together to battle the threat of climate change.

Reducing Greenhouse Gas Emissions

Improvements in technology in the decades since the original Earth Summit now present a variety of options for reducing greenhouse gas emissions that were not available in the 1990s. Take a moment to take another look at Figure 17.17. Notice that the difference between the "constant CO_2," and the A1B or A2 plots (the most likely scenarios) forms a triangle. We present an alternative version of this graph in Figure 17.20. This time, we have replaced the temperatures along the left axis with greenhouse gas emissions, we have shortened the timescale from 1955–2055, and we use a single plot of future emissions that projects current trends to double global greenhouse gas emissions by the middle of the century (Figure 17.20). The section of the graph labeled the stabilization triangle represents all the greenhouse gases we want to avoid producing in the next few decades. The more of this triangle we can get rid of, the less the temperature will increase. We can choose to pursue **mitigation strategies that slow the pace of climate change or adaptation strategies that adjust to its impact**.

Find an online carbon footprint calculator and determine how much carbon you produce annually. What activities produce the most carbon?

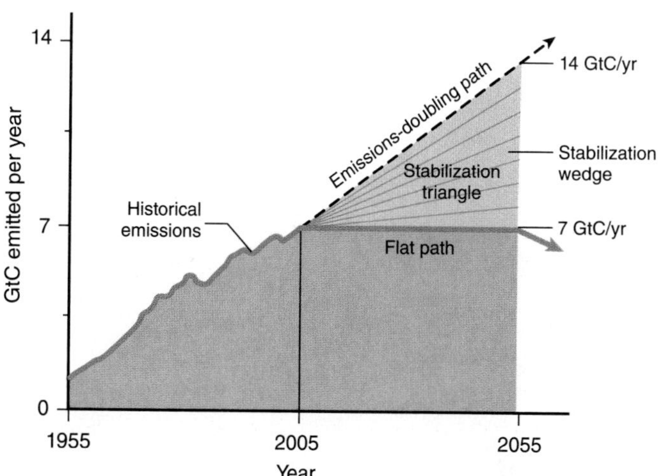

Figure 17.20 Possible carbon emissions through the midcentury. If emissions do not increase, they would follow the "flat path." However, at current trends, it is more likely that we will follow the "emissions-doubling path." The difference in carbon emissions for these two scenarios is represented by the stabilization triangle. Scientists Robert Socolow and Stephen Pacala have suggested that it is useful to divide the triangle into a series of seven stabilization wedges and to pursue a variety of carbon mitigation strategies that could each account for all or part of a wedge, representing 25 gigatonnes of carbon over the time until 2055.

If effective mitigation strategies are not introduced in the next few decades, then adaptation strategies will play a more significant role by midcentury. We discuss a range of mitigation options and outline adjustment choices in Table 17.5.

No single strategy will "fix" global warming. We will have to employ a variety of approaches to cut back greenhouse gas emissions to reduce future temperature increases. If the stabilization triangle is divided into a series of stabilization wedges, each wedge represents a savings of 25 gigatonnes of carbon (GtC) over the 50 years, from 2005 to 2055 (Figure 17.20). Physicist Robert Socolow and ecologist Stephen Pacala of Princeton University developed the concept of stabilization wedges to encourage consideration of a mix of global warming solutions. They have proposed a number of strategies for saving 25 gigatonnes of carbon over the 50-year time interval that fall into four categories: 1) energy efficiency and conservation; 2) changes in electricity energy sources; 3) changes in transportation energy sources; and 4) increasing natural carbon sinks. Any seven stabilization wedges would achieve the aim of minimizing emission increases, but it is more likely that more than seven will be employed with varying degrees of success in the years ahead. If greenhouse gas emissions increased more rapidly, we would need to use more wedges. Which of the options that follow do you think are most likely to be applied in the future?

Efficiency and Conservation Options.

Improved fuel economy: One stabilization wedge would be saved if the number of miles traveled per gallon (mpg) of fuel were to double from an average of 30 to 60 miles per gallon. This assumes the number of passenger vehicles will increase from 600 million to 2 billion by 2055. Hybrid engines and lighter materials will help increase fuel economy.

Reduced reliance on cars: One wedge would be saved if the annual distance traveled by car was were in half from

Table 17.5 Adjustment Strategies for Climate Change

Sector	Adaptation Strategy
Water	Improve irrigation efficiency; reuse domestic water (Chapter 12); build desalination plants; increase water storage capacity.
Agriculture	Change planting dates; introduce alternative plant varieties; improve land management (e.g., conservation tillage methods); change livestock and/or crops.
Infrastructure	Relocate at-risk communities; build/strengthen storm surge barriers and flood levees; expand area of wetland buffers; preserve natural lands; expand climate/weather monitoring systems.
Health	Develop heat-health plans; increase climate-related disease programs; expand health programs in most affected regions.
Transport	Provide incentives for public transportation, telecommuting; relocate at-risk road/rail systems.
Energy	Strengthen electricity transmission lines; provide incentives for renewable resources and energy-efficient technology; expand energy choices.

16,000 kilometers (10,000 miles) to 8,000 kilometers (5,000 miles). Greater use of mass transit and telecommuting can help achieve this goal.

More energy-efficient buildings: A wedge would be saved by employing more efficient technologies for heating, cooling, and lighting to cut emissions from homes and offices by 25 percent. Savings can come from insulation, more-efficient appliances, and passive solar heating of new buildings.

Improved power plant efficiency: Coal power plants operate at about 35 percent efficiency (proportion of coal burned converted to electricity). A wedge would be saved by increasing the efficiency of coal power plants to 60 percent by midcentury. The potential of this strategy is limited as a little over one-third of global electricity is generated from nuclear power and hydropower sources.

Changes in Electricity Energy Sources.

Replacing coal with natural gas: Natural gas power plants produce about one-half as much carbon per unit of electricity as coal power plants. Coal power plants produce about one-quarter of the world's carbon emissions. One wedge would be saved if 1,400 large natural gas plants replaced a similar number of coal power plants.

Carbon capture and storage: Carbon capture and storage (CCS, also termed *carbon sequestration*) technology can trap 90 percent of carbon emissions. The captured carbon dioxide would have to be stored, probably in underground geologic reservoirs, for hundreds of years (Figure 17.21). Installing CCS technology at 800 large coal power plants (or 1,600 gas plants) would save a wedge, but we would need to vastly expand carbon storage by more than 1,000 times its current capacity.

Nuclear energy: Nuclear power is carbon-free and provides about 17 percent of global electricity (Figure 17.22a). Tripling the number of nuclear power plants while phasing out coal plants with similar capacity would save a wedge of emissions. Nuclear power held great promise as a cheap source of electricity in the 1960s but fell out of favor following a series of plant accidents in the 1970s and 1980s. Additional expansion of nuclear power plants could account for a second wedge. However, problems remain with storage of highly radioactive by-products, with safeguarding materials that could be used in nuclear weapons, and with public concerns over recent nuclear plant failures.

Checkpoint 17.22

basic · advanced · intermediate · superior

Review the mitigation strategies described in Section 17.7. Which seven would you choose to add to represent the stabilization wedges in Figure 17.20? Why are these seven the best choices to prevent the most severe future warming?

Terrestrial sequestration
Power station with CO_2 capture
Geologic disposal
Unmineable coal beds
Depleted oil or gas reserve
Enhanced recovery
Deep saline formation

Figure 17.21 Carbon sequestration. An active area of research for sequestering carbon dioxide involves injecting it into oil and gas reservoirs, deep salt formations, or abandoned mines.

a.

b.

Figure 17.22 The expansion of carbon-free energy sources such as **a.** nuclear power and **b.** wind energy could account for two or more stabilization wedges of future carbon emissions.

Wind energy: As we discussed in Chapter 14, carbon-free wind energy is providing an increasing proportion of global electricity (Figure 17.22b). Electricity from wind is growing at 30 percent per year. To account for a wedge of emissions, the future wind power capacity would have to be 30 times greater than current capacity. That could come in the form of 2 million new 1-megawatt wind turbines that would occupy an area about the size of New Mexico.

Solar energy: As we discussed in Chapter 2, carbon-free, renewable solar energy provides a potential source of future electricity. Photovoltaic (PV) cells covering an area about the size of New Jersey would be sufficient to replace a wedge of carbon emissions generated from coal power plants. PV cells are more expensive than other energy sources, and their deployment would require a dramatic increase in efficiency in order to improve their capacity by several hundred times from current levels.

Changes in Transportation Energy Sources.

Hydrogen fuel and carbon capture: Hydrogen has the potential to be a future energy source for passenger vehicles. Hydrogen can be produced from fossil fuels, but the carbon produced during this process must be captured and stored (see the carbon capture and storage option). A 10-fold increase in hydrogen production would represent one stabilization wedge. One drawback of this option is that there is no distribution system for hydrogen fuels.

Synfuels and carbon storage: Synfuels are liquid fuels produced from coal (which the United States has in abundance) as a replacement for oil (which we import). Synfuel production generates twice the carbon emissions as an equivalent volume of fuel refined from petroleum. The largest single source of carbon dioxide emissions in the world is a South African synfuels plant. A wedge would be saved if carbon can be captured from 180 similar coal-to-synfuel plants.

Biofuels: As discussed in Chapter 16, biofuels in the form of ethanol or biodiesel can be made from crops. If produced sustainably, the carbon dioxide produced by burning these fuels is offset by plant growth. Increasing ethanol production by 30 times its current level would be sufficient to replace one wedge of carbon emissions generated from consuming petroleum fuels. It is unlikely that a sufficient area of global cropland can be dedicated to producing the raw materials needed to produce large volumes of these biofuels.

Increasing Natural Carbon Sinks.

Forest management: At least one wedge is available if current deforestation is halted and forests are managed to serve as carbon sinks. Despite deforestation in some tropical rain forests, there is a net trend for land plants to absorb more carbon than they produce. Stopping deforestation will generate one wedge. Increasing the size of current forests will increase carbon storage; but without an end to deforestation, we would need to cover an unreasonably large area of land in trees to generate a wedge of emissions savings.

Conservation tillage: About one-half of the carbon stored in soils is lost when natural lands are converted to cropland. Up to one wedge would be available from widespread deployment of conservation tillage practices (Chapter 9) that preserve crop waste in the soil and increase the soil carbon.

Socioeconomic Options. We can't forget that we live in a society that is dependent on carbon-based energy. Fossil fuels power the global economy, and that is not likely to change in the foreseeable future. Recognizing that, there are socioeconomic options that could be used to alter consumption patterns. For instance, some argue that fossil fuel usage patterns could be changed if governments were to tax carbon consumption. The tax could be imposed for the production, distribution, or use of fossil fuels with allowances for the amount of CO_2 that fuel emits. The imposing of such a tax is based on an economic principle called *negative externalities*. That principle assumes that the cost of the product to society from future warming, acid rain, or pollution is not currently reflected in its price. The revenues could be used to mitigate those effects or seed alternative energy research. Another idea proposed to address the CO_2 emission problem is called cap-and-trade. This approach sets limits on carbon emissions. Those caps are lowered over time. High CO_2 producers, such as coal-fired power plants, can "trade" carbon credits with low- or non-CO_2 producers, such as gas or nuclear power plants. Over time, new technologies are expected to allow the high producers to transition to different fuels. A variation of this approach is carbon offsetting whereby companies seek to become carbon-neutral by adopting energy efficiency projects, using renewable energy sources, or finding other ways to compensate for their emissions. As is the case with any measures, there will be those who support such measures and those who oppose them.

What Else Can Be Done?

Several technological fixes have been suggested that target different climate-forcing agents. Some are relatively untested and will require considerable time until full deployment. One extreme, relatively unproven option is to alter Earth's energy balance by reflecting incoming solar radiation by using fleets of silver balloons or giant mirrors in the atmosphere. This plan assumes that we could reflect more incoming sunlight, thus reducing the solar energy reaching the surface and the impact of solar forcing. Just how many balloons you would need, where they would be located, and how they would get up there still need to be determined.

Some scientists have suggested adding tiny iron particles to the oceans to spur growth of plankton, organisms that extract carbon dioxide from the ocean. Sounds like a good idea, right? Not so fast. The idea of altering one part of the earth system in an attempt to fix human-induced modifications in another part has received relatively little support, and new research suggests that the story is more complicated than it first appears. Also keep in mind that warmer ocean temperatures might improve biologic activity, but warmer water holds less greenhouse gas. Finally, enormous volumes of iron would be needed to have even a minimal effect.

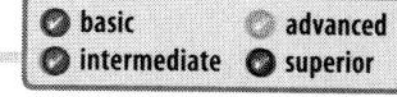

Explain how the comment "Earth can take care of itself" in the context of global warming relates to the concept of "deep time" discussed in Chapter 8.

Making It Personal. If you want to get started on reducing your individual carbon footprint right now, here are some things you can do, in approximate order of increasing impact:

- Replace your lights with compact fluorescent bulbs.
- Wash with cold water and hang clothes to dry.
- Shut off computers and other appliances when not in use.
- Boost car mileage per gallon by tuning engine, replacing air filter, and inflating tires.
- Turn off the air conditioner and open a window.
- Turn down the thermostat in winter and turn it up in summer.
- Buy locally grown produce to reduce food transportation costs.
- Ride the bus, cycle, or walk instead of driving your car.

Checkpoint 17.24

basic | intermediate | advanced | superior

What are some actions that could be taken to diminish the impact of global warming? Different groups of people have different perspectives on this issue. For some, solutions that would avert global warming may not be in their economic interest.

1. List actions each of the following interest groups could take to reduce global warming.
   i. A major oil company
   ii. A utility that burns coal to generate electricity
   iii. A large car manufacturing company
   iv. A company that manufactures wind turbines and solar energy panels
   v. An international insurance company
   vi. The Maldives, a small island nation with an average elevation of 1 meter (3.3 feet)
   vii. A heavily populated developing nation (for example, China, India)
   viii. A heavily populated developed nation with a high standard of living (for example, the United States)
   ix. A family of four in the United States with two vehicles.

Some interest groups are essentially powerless to do anything themselves and must rely on the actions of others. Consider questions 2 and 3 below and evaluate the answer for each group.

2. What would be the impact of global warming on each group?
3. How does the activity of each group impact future global warming?

Choose any three of the groups, and answer these two questions for each group you choose.

4. If applicable, what steps could the group take to diminish its impact on global warming?
5. What incentives would encourage the group to change its habits to reduce its contribution to future warming?

the big picture

Over the last few decades, scientists have learned more about past climates and the factors that influence climate change. It is important to recognize that humans are not the first life-form to affect the planet as a whole. Early organisms, using photosynthesis as a mechanism for sustaining life, gave off oxygen for 2 billion years. Ultimately, that changed the planet. We now have a good idea of the potential range of global temperature increases that human activities could contribute over the next century. Our actions will determine the extent of that increase. In some ways, we are victims of our own success. The faster we grow the global economy, the more greenhouse gases we produce. The more gases we produce, the more extreme future climate changes will be, unless technological advances help us reduce emissions.

In the movie *Waterworld,* the ice caps melted and the planet was inundated with water, flooding almost all the land. We know that won't happen. There just isn't enough ice on the planet to produce the water needed to submerge most of the continents. Increasing rates of melting on Greenland and Antarctica may slowly add a few centimeters a year to global oceans, and in the next few centuries, southern Florida and low-lying parts of coastal nations may slowly submerge. But there is no chance we will be converted to a "waterworld."

In contrast, *The Day After Tomorrow* envisaged much of the Northern Hemisphere locked in the grip of ice sheets. The movie postulated that changes to thermohaline circulation could lead to a shutdown in the global conveyor belt in the North Atlantic that would trigger catastrophic climatic changes in a matter of weeks. Even though current research suggests there may be a slowdown in the conveyor, that process will take many years and the resulting cooling of Europe is likely to be offset by future warming. Like many stories, these movies were each founded on a kernel of truth but quickly lost sight of reality as they spun out their stirring tales. For instance, if you have seen *The Day After Tomorrow*, you know it shows air cooling as it falls from higher altitudes. We learned in Chapter 14 that the opposite actually happens.

Almost all potential concentrations of greenhouse gases are "safe" if you have enough resources to insulate yourself from any negative climate consequences. The greenhouse gases themselves are not a visible threat and pose little direct danger. However, the threat of rising sea levels, decade-long droughts, more severe storms, melting ice caps, bigger floods, and altered ecosystems will make the reality of global change evident in many regions. These changes will be most significant in high latitudes, coastal regions, and areas that already have the potential for severe weather conditions. Computer models of global change are sufficiently sophisticated to give us time to prepare for changes. The clock is ticking. As stewards of planet Earth, it is up to us to propose potential solutions that can diminish the negative effects of global warming. The cure must not be worse than the disease. Nature can provide a solution for this problem, but it may not be to our liking and it probably will not occur on a human timescale.

The signals are clear. Change is inevitable. The question now is just how much we want to control our destiny by playing a role in the solution. Much of what lies ahead for the good Earth is up to us. **Know. Care. Act.**

Global Change: Concept Map

Complete the following concept map to evaluate your understanding of the interactions among the earth system and global change. Label as many interactions as you can. using the information from this chapter.

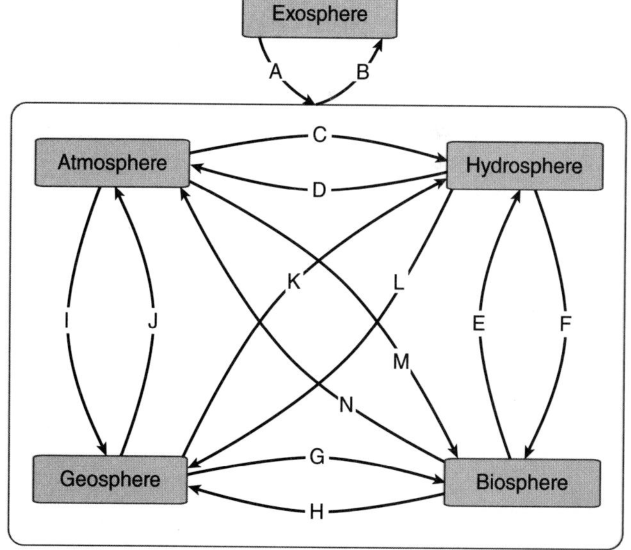

Appendix A
Conversion Factors

To convert	Multiply by	To find
LENGTH		
Inches	2.54	Centimeters
Centimeters	0.39	Inches
Feet	0.30	Meters
Meters	1.09	Yards
Miles	1.61	Kilometers
Kilometers	0.62	Miles
AREA		
Square inches	6.45	Square centimeters
Square centimeters	0.15	Square inches
Square feet	0.09	Square meters
Square meters	10.76	Square feet
Square miles	2.59	Square kilometers
Square kilometers	0.39	Square miles
VOLUME		
Cubic inches	16.38	Cubic centimeters
Cubic centimeters	0.06	Cubic inches
Cubic feet	0.028	Cubic meters
Cubic meters	35.3	Cubic feet
Cubic miles	4.17	Cubic kilometers
Cubic kilometers	0.27	Cubic miles
Liters	1.06	Quarts
Liters	0.26	Gallons
Gallons	3.78	Liters
MASSES AND WEIGHTS		
Ounces	28.35	Grams
Grams	0.035	Ounces
Pounds	0.45	Kilograms
Kilograms	2.205	Pounds
Metric tons (tonnes)	1.102	Tons
Tons	0.907	Metric tons (tonnes)
TEMPERATURE		
To convert degrees Fahrenheit (°F) to degrees Celsius (°C), subtract 32 degrees and divide by 1.8.		
To convert degrees Celsius (°C) to degrees Fahrenheit (°F), multiply by 1.8 and add 32 degrees.		

Appendix B

The Periodic Table of Elements

1 1A	2 2A	3 3B	4 4B	5 5B	6 6B	7 7B	8 8B	9 8B	10 8B	11 1B	12 2B	13 3A	14 4A	15 5A	16 6A	17 7A	18 8A
1 **H** Hydrogen 1.008																	2 **He** Helium 4.003
3 **Li** Lithium 6.941	4 **Be** Beryllium 9.012											5 **B** Boron 10.81	6 **C** Carbon 12.01	7 **N** Nitrogen 14.01	8 **O** Oxygen 16.00	9 **F** Fluorine 19.00	10 **Ne** Neon 20.18
11 **Na** Sodium 22.99	12 **Mg** Magnesium 24.31											13 **Al** Aluminum 26.98	14 **Si** Silicon 28.09	15 **P** Phosphorus 30.97	16 **S** Sulfur 32.07	17 **Cl** Chlorine 35.45	18 **Ar** Argon 39.95
19 **K** Potassium 39.10	20 **Ca** Calcium 40.08	21 **Sc** Scandium 44.96	22 **Ti** Titanium 47.88	23 **V** Vanadium 50.94	24 **Cr** Chromium 52.00	25 **Mn** Manganese 54.94	26 **Fe** Iron 55.85	27 **Co** Cobalt 58.93	28 **Ni** Nickel 58.69	29 **Cu** Copper 63.55	30 **Zn** Zinc 65.39	31 **Ga** Gallium 69.72	32 **Ge** Germanium 72.59	33 **As** Arsenic 74.92	34 **Se** Selenium 78.96	35 **Br** Bromine 79.90	36 **Kr** Krypton 83.80
37 **Rb** Rubidium 85.47	38 **Sr** Strontium 87.62	39 **Y** Yttrium 88.91	40 **Zr** Zirconium 91.22	41 **Nb** Niobium 92.91	42 **Mo** Molybdenum 95.94	43 **Tc** Technetium (98)	44 **Ru** Ruthenium 101.1	45 **Rh** Rhodium 102.9	46 **Pd** Palladium 106.4	47 **Ag** Silver 107.9	48 **Cd** Cadmium 112.4	49 **In** Indium 114.8	50 **Sn** Tin 118.7	51 **Sb** Antimony 121.8	52 **Te** Tellurium 127.6	53 **I** Iodine 126.9	54 **Xe** Xenon 131.3
55 **Cs** Cesium 132.9	56 **Ba** Barium 137.3	57 **La** Lanthanum 138.9	72 **Hf** Hafnium 178.5	73 **Ta** Tantalum 180.9	74 **W** Tungsten 183.9	75 **Re** Rhenium 186.2	76 **Os** Osmium 190.2	77 **Ir** Iridium 192.2	78 **Pt** Platinum 195.1	79 **Au** Gold 197.0	80 **Hg** Mercury 200.6	81 **Tl** Thallium 204.4	82 **Pb** Lead 207.2	83 **Bi** Bismuth 209.0	84 **Po** Polonium (210)	85 **At** Astatine (210)	86 **Rn** Radon (222)
87 **Fr** Francium (223)	88 **Ra** Radium (226)	89 **Ac** Actinium (227)	104 **Rf** Rutherfordium (257)	105 **Db** Dubnium (260)	106 **Sg** Seaborgium (263)	107 **Bh** Bohrium (262)	108 **Hs** Hassium (265)	109 **Mt** Meitnerium (266)	110 **Ds** Darmstadtium (269)	111 **Rg** Roentgenium (272)	112	(113)	114	(115)	116	(117)	(118)

Key: 9 **F** Fluorine 19.00 — Atomic number (9), Atomic mass (19.00)

58 **Ce** Cerium 140.1	59 **Pr** Praseodymium 140.9	60 **Nd** Neodymium 144.2	61 **Pm** Promethium (147)	62 **Sm** Samarium 150.4	63 **Eu** Europium 152.0	64 **Gd** Gadolinium 157.3	65 **Tb** Terbium 158.9	66 **Dy** Dysprosium 162.5	67 **Ho** Holmium 164.9	68 **Er** Erbium 167.3	69 **Tm** Thulium 168.9	70 **Yb** Ytterbium 173.0	71 **Lu** Lutetium 175.0
90 **Th** Thorium 232.0	91 **Pa** Protactinium (231)	92 **U** Uranium 238.0	93 **Np** Neptunium (237)	94 **Pu** Plutonium (242)	95 **Am** Americium (243)	96 **Cm** Curium (247)	97 **Bk** Berkelium (247)	98 **Cf** Californium (249)	99 **Es** Einsteinium (254)	100 **Fm** Fermium (253)	101 **Md** Mendelevium (256)	102 **No** Nobelium (254)	103 **Lr** Lawrencium (257)

Metals

Metalloids

Nonmetals

The 1–18 group designation has been recommended by the International Union of Pure and Applied Chemistry (IUPAC) but is not yet in wide use. In this text we use the standard U.S. notation for group numbers (1A–8A and 1B–8B). No names have been assigned for elements 112, 114, and 116. Elements 113, 115, 117, and 118 have not yet been synthesized.

Appendix C

Answers to Selected Checkpoint Questions

Chapter One

1.5 c; **1.6** 1, 2, and 6; **1.10** b.

Chapter Two

2.1 a; **2.2** b; **2.6** b; **2.9** a; **2.10** a, c; **2.13** b; **2.15** Terrestrial characteristics—smaller, rocky surfaces, closer to the sun, shorter orbits, more closely spaced; Jovian—gas giants, strong gravitational fields, larger, have rings, multiple moons; Both—orbit the sun, have atmospheres, spherical, formed at same time; **2.17** a; **2.18** c; **2.21** d; **2.23** a, b; **2.25** Correct answers from top to bottom: cooler, smaller, different, less, more, less, stronger, thinner, colder.

Chapter Three

3.1 d; **3.3** 1c, 2b, 3a, 4b; **3.4** Rank order, best to worst: B, D, A, C; **3.5** b; **3.9** a; **3.10** b; **3.13** a; **3.15** This concept map would probably rate as a "2" on the grading rubric. Some key items are missing (no reference to two types of craters, no reference to asteroids as a cause for impacts) and the organization could be better.

Chapter Four

4.1 a; **4.2** a, b, c, and e; **4.3** #1 fits with pieces 2a, 3a, 5c, and 6b; #2 fits with pieces 1b and 5d; #3 fits with pieces 1a and 5a; #4 fits with pieces 5b and 6a; #5 fits with pieces 1b, 2b, 3b, 4a, and 6a; #6 fits with pieces 1c, 4b, and 5c; **4.5** b; **4.6** d; **4.8** The key contrasts should be that the contracting Earth model should not exhibit any difference in the age of the ocean floor, the ocean floor topography would not include an oceanic ridge system, heat flow would not be concentrated toward the middle of the oceans, volcanic activity should be along the edges of all oceans or none, and there is no logical explanation for the Wadati-Benioff zone distribution of earthquakes; **4.9** a; **4.10** b; **4.11** c; **4.17** c; **4.19** 1 is divergent, 2 and 3 are convergent; **4.20** part 1, 3.

Chapter Five

5.1 Here are three: Lima, Istanbul, Los Angeles, find the rest; **5.2** d; **5.4** Since general earthquake and engineering principles (for example, frequency of earthquakes, building type and age) are used to derive specific rates for homeowners, the reasoning is general to specific, thus deductive; **5.5** b; **5.7** c; **5.9** 1. a, 2. d; **5.10** 1. b, 2. 5 seconds; **5.12** Hint: 2, 6 for P waves only, 4 for S and P waves only, and 1, 5, 11 for surface waves only; **5.14** a; **5.17** 60–80 kilometers for Northridge, 30–40 kilometers for Whittier.

Chapter Six

6.1 A: milk, B: maple syrup, C: peanut butter, D: frozen yogurt; **6.2** b; **6.3** d; **6.6** Here are five more, find the rest: intermediate silica rock (andesitic), example: Hawaiian Islands (basaltic), contains silica (all three), lowest melting temperature (rhyolitic), most common at convergent plate boundaries (andesitic); **6.7** a) Rainier/Juan de Fuca, Colima/Cocos, Galeras/Nazca, Merapi/Indian-Australian, Sakurajima/Pacific; b) No; c) Mauna Loa or Nyiragongo or Teide; d) Teide; **6.9** a; **6.11** Scientists first became concerned based on observations. Technology was employed to collect more observations used to develop tentative predictions; **6.14** Mount St. Helens: 1, 4, 15, 18, Nyiragongo: 2, 3, 5, 11, 13, 16, Both: 6, 7, 8, 9, 10, 12, 14, 17; **6.17** b; **6.18** 1. b, 2. Rainier, Colima, Santa Maria, Galeras, Taal, Etna, 3. Mauna Loa, Teide; **6.21** b; **6.22** Continental lithosphere of Australia is not near a plate boundary unlike most of the highest mountain ranges on other continents; **6.25** b; **6.26** 1. Alaska is located adjacent to an active convergent boundary, 2. Rocky Mountains are about 200 million years younger than the Appalachians, 3. Former mountains in Canada are more than a billion years old and have been eroded down to their roots.

Chapter Seven

7.1 a; **7.2** b; **7.3** c; **7.5** d; **7.7** Three cleavage planes; **7.9** a; **7.10** From left to right: granite, basalt, rhyolite, and diorite. Individual grains are visible in granite and diorite but not in basalt and rhyolite; granite and rhyolite are light colored, basalt is dark, diorite clearly has both light and dark minerals; **7.11** Four that are true for both volcanic and plutonic igneous rocks: may form from basaltic magma, have texture, form as a result of melting, contains minerals; **7.14** There are distinct layers with different size materials. The size of the grains would relate to the velocity of the water that deposited them and suggest a depositional history characterized by "normal" flow and flood conditions; **7.15** Three common attributes of chemical and biochemical sedimentary rocks: form under shallow marine conditions, form by precipitation from a solution, examples of sedimentary rocks; **7.17** a; **7.18** Three that are true for both regional and contact metamorphism: formed at temperatures above 200°C, may originally have been an igneous rock, marble is a possible example; **7.21** b, a.

Chapter Eight

8.1 Best order from oldest to youngest: B, C, A, E, D; **8.2** 1b, 2a, 3b; **8.3** Rock units from oldest to youngest: H, F, D, I, B, G, A, C, E, J; **8.6** Fossil 2 is the best index fossil because it is always found in limestone B. Fossil 3 is the least appropriate because it is found in multiple units; **8.7** The image shows corals, crinoids, and cephalopods that were only described in the text from the Redwall Limestone; **8.9** 1b, 2b; **8.10** b, a; **8.11** Answer for part A only: Were insect species affected by the asteroid impact thought to have caused extinction of the dinosaurs? **8.14** a, a, c; **8.15** c; **8.16** a) T, A, C, P, G, N, S, K, L, F; **8.19** Rare, high-magnitude events include asteroid impacts and supervolcano eruptions, and frequent, low-magnitude events include processes such as erosion and plate movements.

Chapter Nine

9.1 a; **9.2** a; **9.7** The rocks that are weathered in each area must have different compositions. Pyroxene and olivine have simpler silicate structures than quartz and would therefore weather more rapidly. Transport distances would be shorter in Hawaii, and therefore, these minerals may not have time to break down; **9.11** Macroprocesses generally result in rock breakage whereas microprocesses operate at the mineral level to change composition; **9.13** Chemical weathering likely dominates with high temperatures and high precipitation at location 1. Location 4 is likely too cold to have any weathering. Assume location 2 is the Sahara desert and location 3 is the Rocky Mountains; **9.14** Changes in the type of material, age, and climate conditions could result in different weathering rates from one site to another; **9.15** Kilwa Kisiwani, has the greatest weathering potential, what would come next? **9.17** b; **9.18** Hint: Regolith could not form until there was rock exposed to weathering and soils need biologic materials (land plants); **9.21** a, b.

Chapter Ten

10.1 b; **10.2** Excessive rainfall, steep slopes, weak soils, plants with shallow root systems; **10.3** Where would you place the following: hundreds affected, town at foot of slope, involved mud or soil, thousands affected, town at mouth of canyon, heavy rain, steep slopes, previous events; **10.5** b, d, d; **10.6** a, c.

Chapter Eleven

11.1 Groundwater and infiltration, stream and runoff, rainfall and precipitation, gas and water vapor, ice and melt water, plants and transpiration; **11.2** c; **11.5** b; **11.8** This region displays dendritic stream systems on the Piedmont and Coastal Plain and trellis patterns in the Valley and Ridge province; **11.10** a; **11.11** a; **11.13** d; **11.14** Deposition will occur in the reservoir behind the dam, reducing the downstream stream load and resulting in more erosion, deepening and widening the channel; **11.17** Increased volume, rate, and/or duration of precipitation, ice dams on north flowing rivers, ground saturation leading to increased runoff, landscape modifications that increase impervious cover, dam collapse, flood control structures; **11.18** d; **11.21** Direct losses of life or property, work-related losses, interruption of commerce, losses to crops, and subsequent land changes; **11.22** Prevention measures are construction of levee, dredging, and dam construction; **11.23** Hint: Flood protection information = 66 points, Open space preservation = 900 points.

Chapter Twelve

12.1 ~0% porosity if sticks set side by side in rows to fill the box, ~40% porosity if sticks arranged in overlapping rows or a latticework pattern; **12.2** Highest permeability = coffee beans, lowest permeability = flour; **12.3** c; **12.5** d, a; **12.9** b; **12.11** If Bob had won, groundwater could be considered a shared resource that must be managed to the benefit of all. Since Glen won, groundwater would be a commodity to be used as available. As agriculture demands much more water than do people, the implication is that people in cities could lose out to the needs of agriculture; **12.12** Hint: Two other factors could be the porosity of the rocks and the properties of the surface materials; **12.13** c; **12.15** In general, there is no correlation between the location of the maximum and minimum saturation thickness and the locations of streams; **12.17** b; **12.18** Best answers are d and b, then a and c, and the worst answers are e, f, and g. Answers may be ranked in different orders but should be arranged in these groupings; **12.21** Histosols; **12.22** No, wetlands require hydric conditions and hydrophytic plants, not just occasional wet soils; **12.23** Some common features: part of the hydrologic cycle, involve the slow flow of water, may contain freshwater or saltwater, lower levels during drought conditions, store water.

Chapter Thirteen

13.1 Active margin examples: west coast of South America or Central America, south coast of Alaska, western margin of Pacific Ocean; Passive margin examples: east coasts of North and South America, west coasts of Europe and Africa, coast of Antarctica and Australia; **13.3** a; **13.5** Salinity is lower in the Bay of Bengal because of runoff from the Ganges and Brahmaputra rivers (see Chapters 7, 8, and 11); **13.8** Partial answer: Data set 1 is subpolar based on the lower surface temperature. The thermocline for data set 2 at around 100- to 1,500-meter depth and around 300 to 500 meters for data set 1; **13.9** b; **13.10** d; **13.13** b; **13.14** c; **13.17** c; **13.18** b; **13.20** Hint: two factors—wave height, location of human-made water obstacles; **13.22** b; **13.24** Hint: five features—sediment comes from the continental interior, sand is deposited in bars, wide range of grain sizes and sorting possible, occur at a range of elevations, erosion by wave action; **13.27** Prevention strategies include all the human-engineered armoring techniques, adjustment strategies focus on land management practices (for example, limiting coastal development).

Chapter Fourteen

14.1 d; **14.5** a; **14.6** b; **14.7** b; **14.9** The loss of ozone will allow more UV radiation to penetrate the atmosphere; **14.10** For Earth there is a dominant emission wavelength of ~10 micrometers; **14.11** 1. Decrease, 2. Increase, 3. Decrease; **14.13** a; **14.14** d; **14.17** a; **14.19** Responses for parts 1 and 2: 1. An actual parcel of rising air will decrease its temperature by the dry adiabatic lapse rate if it is unsaturated and the saturated lapse rate if saturated; 2. The percentage of oxygen in the atmosphere does not change with altitude. The available number of molecules of all gases will decrease with altitude; **14.21** a) Cirrus, b) Altocumulus, c) Cumulonimbus, d) Cirrostratus; **14.22** a, a; **14.26** Fastest winds will be north of Lake Superior in Canada (near the "L" symbol); **14.27** 2c, 3c, 4a.

Chapter Fifteen

15.2 For example, hazards associated with earthquakes and floods; **15.6** Three answers: A: mP, B: cT, C: mT; **15.7** Midwestern and eastern United States, where cP and mT air masses collide; **15.9** 1b, 2d, 3b, 4d, 5d; **15.10** 1) 10 P.M., October 29, 2) 10 A.M., October 30, 3) Temperature, 4) b; **15.12** Cold front: air becomes cooler after its passing, heavy rainfall with passage, Warm front: air becomes warmer after its passing, increasing humidity after passing, Both: changing wind directions, warm air rises, moves from west to east; **15.13** a; **15.14** c; **15.17** a; **15.18** b; **15.19** Fujita scale: wind speed is a factor, has five levels, Both: related to damage, relies on observations, one level describes nearly complete destruction; **15.23** Tornadoes: winds can be in excess of 480 kilometers per hour, six intensity levels, Midlatitude cyclones: comma-shaped cloud pattern, two different frontal systems involved, Hurricanes: most common in late summer and fall, All three: low-pressure systems, involve counterclockwise rotations.

Chapter Sixteen

16.1 a) 2, b) 1; **16.5** c; **16.6** Regions with the highest population density often occur in association with climate region C, which is in turn characterized by grasslands and deciduous forests; **16.8** James Hansen at NASA established a Common Sense Climate Index that takes into account factors such as seasonal average temperatures, the frequency of extreme daily temperatures, and record high and low temperatures; **16.10** Find moraines, zone of accumulation, zone of ablation and the snowline. The lines in the lower figure are materials that are scoured off of the sides of the valleys; **16.13** Polar regions are deserts because of low precipitation; **16.15** Wind action: forms dunes, Glacial processes: can transport large boulders, occur most frequently at high latitudes, Both: occur on at least five continents, transport sediment in direction of movement, few associated plants and animals; **16.19** A microfossils, B ice cores, C tree rings, D instrumental records, E archeological records; **16.21** Tilt of Earth's axis, distance from the sun, pollution in atmosphere, cloud cover, albedo of features on Earth's surface, surface and deep ocean currents (thermohaline circulation), locations of continents, composition of the atmosphere, presence/absence of ozone in stratosphere, volcanic eruptions; **16.23** 1. Decrease, 2. Small increase, 3. Decrease.

Chapter Seventeen

17.1 d; **17.2** b; **17.3** Melbourne = green squares, San Francisco = red circles, London = blue diamonds, Darwin = brown triangles; **17.7** From best to worst: d, b, c, a; **17.8** One argument for the ban: there was an immediate health threat, One argument against the ban: natural changes were not sufficiently documented; **17.10** Two examples of each: Carbon is stored in the atmosphere when CO_2 is from burning fossil fuels and in the biosphere when CO_2 is used by plants to form biomass. Carbon is released from the geosphere during weathering of limestone and from the biosphere when forests are burned; **17.13** c; **17.17** Possible responses include (but are not limited to): reduce driving (in any way), purchase alternative-fuel vehicle, use less electricity, recycle, reduce home heating or cooling, add solar panels, plant trees; **17.18** South America: greatest temperature increase in warmest locations, drying in northern part of continent, greater range in precipitation changes. North America: sea level increases greatest along northern coasts. Both: greatest temperature increases in northern parts of continents, greater precipitation over east and west coasts; **17.19** Rank of statements from top to bottom: 4, 1, 2, 5, 3.

Glossary

a

ablation zone Area of a glacier that is losing net snow or ice due to melting.

abrasion The erosion of a stream channel as sand or gravel is transported downstream.

absolute humidity A measure of the actual amount of water vapor in the air at a given time—for example, in grams per cubic meter.

abyssal plain Large, flat regions of the seafloor at depths below 3,000 meters.

accuracy The degree to which a set of measured values matches an accepted or known value.

acid mine drainage Runoff resulting when water reacts with metallic ores.

acid rain Rain with a pH value less than the pH of unpolluted precipitation.

active continental margins A margin located adjacent to a plate boundary.

adiabatic lapse rate Rate of temperature change in the atmosphere due to changing atmospheric pressure.

adiabatic temperature change Cooling or warming of air caused when air is allowed to expand or is compressed, not because heat is added or subtracted.

aerosols Tiny solid and liquid particles suspended in the atmosphere.

aftershock Sequence of earthquakes that follow a larger event.

A horizon Top layer of soil defined by the presence of organic matter mixed in with mineral and rock fragments.

air A mixture of many discrete gases, of which nitrogen and oxygen are most abundant.

air density The mass of gases (for example, nitrogen, oxygen) per volume of air.

air mass A large body of air that is characterized by a near-uniform temperature and humidity.

air pollution The presence of airborne particles and gases in concentrations that endanger the health and well-being of organisms or disrupt the orderly functioning of the environment.

albedo The reflectivity of a surface, usually expressed as a fraction of the incident radiation reflected.

alpine glacier A mountain glacier.

altocumulus clouds A principal cloud type that occurs at middle of the troposphere.

altostratus clouds A high cloud type that appears thin and spread out.

amphibole A member of a mineral group in which all members are double-chain silicates.

andesite Fine-grained igneous rock of intermediate composition. Up to one-half of the rock is plagioclase feldspar, with the rest being ferromagnesian minerals (pyroxene, amphibole).

andesitic magma A magma with intermediate silica content.

angle of repose The steepest angle at which loose material remains stationary without sliding downslope.

anticyclones A high-pressure center characterized by a clockwise flow of air in the Northern Hemisphere.

aquifer A body of rock or regolith with sufficient porosity and permeability to provide water in useful quantities to wells or springs.

aquitard Impermeable beds that hinder or prevent groundwater movement.

Archean A term, meaning ancient, which has been applied to the oldest rocks of the Precambrian.

artesian aquifer An aquifer completely filled with pressurized water and separated from the land surface by a relatively impermeable confining bed, such as shale. Also called confined aquifer.

artificial beach nourishment Addition of sand to a beach by mechanical means.

ash Very fine-grained volcanic material erupted from a volcano that commonly forms a cloud.

asteroid A small, generally rocky, solid body orbiting the sun and ranging in diameter from a few meters to hundreds of kilometers.

asteroid belt Region between Mars and Saturn where majority of solar-orbiting asteroids are found.

asthenosphere A weak layer of Earth's mantle below the lithosphere upon which tectonic plates move.

astronomical unit The average distance from the sun to Earth.

atmosphere The gaseous envelope surrounding a planet or celestial body. One of the four components of the earth system.

atmospheric pressure Pressure exerted by weight of the overlying column of air.

atom Smallest possible particle of an element that retains the properties of that element.

atomic mass number The total number of neutrons and protons in an atom.

atomic number The total number of protons in an atom.

aurora A bright display of ever-changing light caused by charged particles interacting with the upper atmosphere in the polar region.

b

barrier island A long, narrow sandy ridge of land that forms parallel to, but separate from, the mainland coast.

barycenter A common center of mass around which two objects revolve.

basalt A fine-grained, mafic, igneous rock composed predominantly of dark ferromagnesian (olivine, pyroxene) minerals and with lesser amounts of plagioclase feldspar.

basaltic magma Mobile rock material with a composition similar to that of basalt.

base level The lowest level to which a stream can erode its bed. The ultimate base level is usually sea level.

batholith A large body of intrusive igneous rocks.

bathymetry The measurement of ocean depths and the charting of the shape or topography of the ocean floor.

bauxite The principal ore of aluminum.

baymouth bar A sandbar that completely crosses a bay, sealing it off from the open ocean.

bed The smallest division of a series of layered rocks, marked by a well-defined divisional plane from its neighbors above and below.

bed load Sand- and gravel-size particles transported downstream by moving along the bottom of a stream channel.

Bergeron process A theory that relates the formation of precipitation to supercooled clouds, freezing nuclei, and the different saturation levels of ice and liquid water.

B horizon The soil horizon notable for colors and textures indicating accumulation of minerals that are not present in the parent material. Typically found below the A horizon and above the C horizon.

Big Bang The theory that the universe began more than 10 to 20 billion years ago with a massive and rapid inflation.

biochemical sedimentary rock A sedimentary rock made by or from organic matter.

biodiversity The number of species in an ecosystem.

biogeochemical cycles The cycling of chemical constituents that pass through geologic and biologic systems.

biological weathering The processes by which the actions of organisms break down rocks and minerals.

biomass The total mass of an organism or group of organisms in a particular area or ecosystem.

biome The community of plants and animals that characterize a region.

biosphere One of the four components of the earth system consisting of plants and animals.

breakwater An offshore structure built to absorb the force of breaking waves and provide quiet water near shore.

body waves Seismic waves that travel through Earth's interior.

boreal forest Forest that grows just south of the tundra.

Bowen's reaction series The sequence by which minerals crystallize from a cooling basaltic magma.

braided stream A stream consisting of numerous intertwining channels.

c

caldera A large depression typically caused by collapse or ejection of the summit area of a volcano.

Cambrian period The oldest period that contains abundant fossils of animals and the first (earliest) geologic period of the **Paleozoic era** from 542 million to 488 million years ago.

Canary current The prevailing southward flow of water along the northwestern coast of Africa.

carbon sequestration The trapping of carbon in natural or artificial storage reservoirs.

catastrophism The concept that some Earth features are shaped by catastrophic events of a short-term nature.

cementation The filling of pore spaces with precipitated minerals that bind individual grains together; a step in lithification of sediment to form sedimentary rock.

Cenozoic era The most recent of the eras; also known as the Age of Mammals.

chalk A very fine-grained biochemical limestone.

channel roughness A measure of the resistance to flow offered by the material in the stream channel.

chemical sedimentary rock A sedimentary rock consisting of material that was precipitated from water by either inorganic or organic processes.

chemical weathering The processes by which the internal structure of a mineral is altered by the removal or addition of elements.

chlorofluorocarbons Volatile organic compounds that remain in the atmosphere for periods of 20 to 200 years.

C horizon The lowest soil horizon, which is the least weathered horizon and commonly represents the parent material for soil formation.

cinder cone volcano A rather small volcano built primarily of pyroclastics ejected from a single vent.
cirrocumulus A high-altitude cloud often with a rippled appearance.
cirrostratus clouds High-altitude, thin clouds made of ice particles that are barely visible.
cirrus clouds One of three basic cloud forms; also one of the three high cloud types. They are thin, delicate ice-crystal clouds often appearing as veil-like patches or thin, wispy fibers.
clastic sedimentary rock A sedimentary rock made of fragments of preexisting rock.
cleavage The tendency of a mineral to break along planes of weak bonding.
climate A description of aggregate weather conditions that helps characterize a place or region; the average of all statistical weather information over a 30-year period.
climate forcing Any phenomenon that causes a change in the global solar radiation balance.
climatic change The long-term fluctuation in rainfall, temperature, and other aspects of Earth's climate.
climatology The scientific study of the climate.
cloud A form of condensation best described as a dense concentration of suspended water droplets or tiny ice crystals.
coastline The coast's seaward edge. The landward limit of the effect of the highest storm waves on the shore.
cohesion A phenomenon in which the surface tension created by a small amount of water holds sand grains together.
cold front A front along which a cold air mass thrusts beneath a warmer air mass.
collision-coalescence process A theory of raindrop formation in warm clouds (above 0°C) in which large cloud droplets collide and join together with smaller droplets to form a raindrop.
comet Small object in space, a few kilometers in diameter, composed of frozen methane, frozen ammonia, and water-ice, with small solid particles and dust embedded in the ices.
Community Internet Intensity Map (CIIM) A map that shows the distribution of intensities associated with an earthquake. A computer program is used to generate an average intensity for each ZIP code affected on a regional map.
compaction A loss in overall volume and pore space of a rock as the particles are packed closer together by the weight of overlying material.
condensation The change of state from a gas to a liquid.
conduction The transfer of heat through matter by molecular activity. Energy is transferred through collisions from one molecule to another.
cone of depression A cone-shaped lowering of the water table that surrounds a well being pumped faster than it recharges.
confined aquifer Source of groundwater capped by an impermeable material.
conservation tillage Soil conservation strategies designed to reduce water and wind erosion.
contact metamorphism Changes in rock caused by the heat from a nearby magma body.
continental Arctic air mass Very cold, dry air masses that form over the Arctic.
continental crust The rocks that make up most continental land masses.
continental drift theory A concept suggesting that continents move over Earth's surface.
continental polar air mass Cold, dry air masses that form over northern Canada and Alaska.
continental rise The sloping surface at the base of the continental slope.
continental shelf The gently sloping submerged portion of the continental margin, extending from the shoreline to the continental slope.
continental slope The steep gradient that leads to the deep-ocean floor and marks the seaward edge of the continental shelf.
continental tropical air mass Warm, dry air masses that form over continents at low latitudes.
contrail Vapor trail in the atmosphere caused by jet exhaust.
convection The transfer of heat by the movement of a mass or substance.
convection cell Atmospheric circulation due to temperature-density differences.
convergence The movement of two objects toward each other.
convergence lifting Occurs when two air masses collide, forcing some air upward, as both air masses cannot occupy the same space.
convergent plate boundary A boundary in which two plates move together.
coquina A limestone consisting of coarse shells.
core Located beneath the mantle, it is the innermost layer of Earth. The core is divided into an outer core and an inner core.
Coriolis Effect The deflection caused by Earth's rotation on all free-moving objects, including the atmosphere and oceans. Deflection is to the right in the Northern Hemisphere and to the left in the Southern Hemisphere.
covalent bond Union formed when two or more atoms mutually share electrons.
creep A very slow, continuous downslope movement of soil or debris.
creep meter An instrument used to record minute changes in distance between two points on Earth's surface.
crest The high point of a wave.
Cretaceous period The last period of the Mesozoic era, between 145 million and 65 million years ago.
Cretaceous-Tertiary extinction (K-T event) Time about 65 million years ago when 70 percent of species became extinct, including dinosaurs.
crevasse Fracture in a glacier caused by tension.
crosscutting relationships, principle of A principle or law stating that a disrupted pattern is older than the cause of the disruption.
crust The thin outermost layer of Earth (also see **continental crust** and **oceanic crust**).
crystal form The arrangement of various faces on a crystal in a definite geometric relationship to one another.
crystal settling During the crystallization of magma, the earlier-formed minerals are denser than the liquid portion and settle to the bottom of the magma chamber.
cumulonimbus clouds A principal cloud type, exceptionally dense and vertically developed, occurring either as isolated clouds or as a line or wall of clouds.
cumulus clouds One of the three basic cloud forms; also the name given to clouds of vertical development. Cumulus clouds are billowy individual cloud masses that often have flat bases.
cut bank The concave bank of a winding stream where erosion takes place.
cyclone A low-pressure center characterized by a counterclockwise flow of air in the Northern Hemisphere.

d

data Information, often involving measurements, used to describe something.
daughter isotope An elemental isotope produced by decay of a radioactive **parent isotope.**
debris flow A relatively rapid type of mass wasting that involves a flow of rock, soil, and regolith containing a large amount of water.
decompression melting Partial melting of hot rock when it moves upward and the pressure is reduced to the extent that the melting point drops to the melting temperature of some minerals in the rock.
deductive reasoning Occurs when scientists draw specific conclusions based on general principles.
delta An accumulation of sediment formed where a stream enters a lake or ocean.
dendritic drainage Irregular stream branching, with tributaries joining the main stream at angles of less than 90°.
dendrochronologist A scientist who measures time intervals and dates events and environmental changes by reading growth layers of trees as demarcated by annual rings.
dendrochronology The science of measuring time intervals and dating environmental changes by reading growth layers of trees as demarcated by the annual rings.
density lifting Rising air due to density differences.
deposition The settling or coming to rest of transported material.
desert Dry environments characterized by evaporation exceeding precipitation.
Devonian period The fourth period of the Paleozoic era, between 416 million and 360 million years ago.
dew point The temperature at which atmospheric water vapor begins to condense into dew as the air cools.
dike A tabular-shaped intrusive igneous feature that cuts through the surrounding rock.
diorite Large-grained igneous rock solidified from intermediate-composition magma.
dip-slip fault An inclined fault in which the movement is up or down the slope of the fault.
discharge The volume of water in a stream that passes a given point in a period of time.
dissolution Chemical reactions where bonds break between atoms or molecules that then disperse in water.
dissolved load Dissolved materials carried by a stream.
diurnal tide A tidal condition in which only one high tide and one low tide occur each day.
divergent plate boundary A boundary separating two plates moving away from each other.
diversion channel See **floodway.**
Doppler effect An apparent changing frequency of sound or light due to the relative motion of a source and an observer.
Doppler radar In addition to the tasks performed by conventional radar, this new generation of weather radar can directly detect motion and hence greatly improve tornado and severe storm warnings.
downdraft Zone of downward-moving air.
drainage basin The land area that contributes water to a stream.
drainage divide An imaginary line separating the area drained by one stream from the areas drained by adjacent streams.
dropstone deposit Places on the seafloor where rocks carried by melting icebergs are deposited.

earthquake A nearly instantaneous release of stored energy resulting from the breaking and sudden movement of stressed rock.
earthquake intensity A measure of the effects of earthquakes on people and buildings.
earthquake magnitude A single, standard measure used to compare the shaking and energy released from different earthquakes.
earth science The investigation of interactions among the four components of the earth system—the atmosphere, hydrosphere, biosphere, and geosphere.

Earth's magnetic field The magnetic field hypothesized to be generated by convection currents in the outer core.
ecology The study of the interaction of organisms with one another and their environment.
ecosystem A community of organisms that inhabit a specific physical environment, defined primarily by its climate and landforms.
E horizon The soil horizon that is the zone of leaching, characterized by the downward movement of water and removal of fine-grained soil components.
electromagnetic radiation The transfer of energy through space by electromagnetic waves.
electromagnetic spectrum The distribution of electromagnetic radiation by wavelength.
electron A negatively charged subatomic particle that has a negligible mass and is found outside an atom's nucleus.
elements Chemical substances that cannot be split into simpler substances.
empirical observations Observations that can be measured and confirmed by others.
eon The largest unit of geological time.
epicenter The location on Earth's surface that lies directly above the focus of an earthquake.
equinox The time when the vertical rays of the sun are striking the equator. The lengths of daylight and darkness are equal at all latitudes during an equinox.
erosion The physical removal of rock by an agent such as running water, glacial ice, or wind.
evaporation The process of converting a liquid to a gas.
evaporite A sedimentary rock formed of material deposited from solution because of evaporation of the water.
exfoliation dome Large, dome-shaped structure formed by physical weathering, usually composed of granite.
exosphere The sun and other components of space that interact with the earth system.

f

fault A break in a rock mass along which movement has occurred.
fault scarp A cliff created by movement along a fault. It represents the exposed surface of the fault prior to modification by weathering and erosion.
fault system A group of faults that move in a similar fashion.
felsic Describes igneous rocks dominated by quartz, sodium-rich plagioclase, potassium feldspars, and the magmas that these rocks crystallize from; derived from the words *feldspar* and *silica*.
Ferrell cell A secondary circulation cell located between Hadley and polar cells.
fetch The distance over which wind blows to produce waves.
flash flood A flood that occurs with little or no warning.
flood A temporary overflow of a river onto adjacent lands not normally covered by water.
flood control Measures taken to reduce the impact of flooding (for example, building levees and dams).
floodplain The flat, low-lying portion of a stream valley subject to periodic inundation.
floodwall Vertical walls constructed to keep water in its channel.
floodway An artificially constructed diversion channel that will transport floodwaters away from inhabited areas.
flow A type of movement common to mass-wasting processes in which water-saturated material moves downslope as a viscous fluid.
focus The location within Earth where rock displacement produces an earthquake.
foliation Planes of minerals formed in response to stress; a feature of many metamorphic rocks.
foraminifera Tiny shelled animals that live or once lived in marine environments. Oxygen isotope data from shells are used as proxy climate indicators.
fossil The preserved trace, imprint, or remains of a plant or animal.
fossil fuel General term for any hydrocarbon that may be used as a fuel, including coal, oil, and natural gas.
fractures Cracks in rocks and materials.
front The boundary between two adjoining air masses having contrasting characteristics.
frontal lifting Forcing aloft of a low-density air mass over a high-density air mass where the two masses meet.
frost wedging A type of physical weathering in which the expansion of freezing water breaks apart a rock.
Fujita intensity scale Scale used to categorize tornadoes by the level of destruction they cause.
fumarole Vents where volcanic gases escape.

g

gabbro Plutonic rock equivalent to basalt.
gaining stream A stream that receives water from the zone of saturation.
geocentric The concept of an Earth-centered universe.
geochemical cycle During geologic changes, the sequence of stages in the migration of elements between the components of the earth system.
geologic time General reference to long time span that predates human records.
geologic timescale The division of Earth history into blocks of time—eons, eras, and periods. The timescale was created by using relative dating principles.
geology The science of the origin, composition, structure, and history of Earth.
geosphere One of the four components of the earth system, consisting of land and rocks.
geostrophic wind A wind, usually above a height of 500 meters, that blows parallel to the isobars.
geothermal gradient The gradual increase in temperature with depth in the crust.
geyser A fountain of hot water ejected periodically.
glacials Cold intervals during the last ice age when temperatures were 5 to 10 C° colder than today.
glaciation Alteration of any part of Earth's surface by passage of a glacier, chiefly by glacial erosion or deposition.
glacier A large, long-lasting mass of ice, formed on land by the compaction and recrystallization of snow, which moves because of its own weight.
granite A felsic, coarse-grained, intrusive igneous rock containing quartz and composed mostly of potassium- and sodium-rich feldspars.
grassland A biome dominated by grasses.
greenhouse effect The transmission of short-wave solar radiation by the atmosphere, coupled with the selective absorption of longer-wavelength terrestrial radiation, especially by water vapor and carbon dioxide.
groin A short wall built at a right angle to the shore to trap moving sand.
ground subsidence Lowering of land surface that can be caused by removal of groundwater.
groundwater Water in the zone of saturation.
groundwater overdraft The withdrawing of groundwater at rates exceeding recharge.
gyre The large circular surface current pattern found in each ocean.

h

Hadley cell An atmospheric convection cell bounded by rising, humid air at the equator and descending, dry air at latitudes 30°N and S.
half-life The time required for one-half of the atoms of a radioactive substance to decay.
halocline A layer of water in which there is a high rate of change in salinity in the vertical dimension.
heat The kinetic energy of random molecular motion.
heat capacity Change in energy per amount of material divided by the corresponding change in temperature.
heliocentric Describing the view that the sun is at the center of the solar system.
heliosphere Extent to which solar wind from the sun reaches into interstellar space.
highland climate Complex pattern of climate conditions associated with mountains. Highland climates are characterized by large differences that occur over short distances.
Holocene An epoch of the Quaternary period from the end of the Pleistocene, around 12,000 years ago, to the present.
horizon A layer in the soil profile.
hot spot A concentration of heat in the mantle capable of producing magma, which in turn extrudes onto Earth's surface. The intraplate volcanism that produced the Hawaiian Islands is one example.
hot spring Spring with a water temperature warmer than human body temperature.
humidity Term that refers to water vapor in the air but not to liquid droplets of fog, cloud, or rain.
hurricane A tropical cyclonic storm having winds in excess of 119 kilometers (74 miles) per hour.
hydric soils Poorly drained (water-logged) soils that exhibit anaerobic (oxygen-deficient) conditions during the growing season.
hydrograph A graphical representation of stage, flow, velocity, or other characteristics of water at a given point as a function of time.
hydrologic cycle The complete cycle through which water passes, from the oceans, through the atmosphere, to the land, and back to the ocean. Also referred to as **water cycle.**
hydrologist A scientist who studies the quantity, distribution, and flow of water on a planet.
hydrology The study of water's properties, circulation, and distribution.
hydrolysis Process that occurs when minerals react with water to form products such as clay.
hydrosphere One of the four components of the earth system, consisting of water and ice.
hydrothermal mineral deposits Mineral deposits formed when heated water circulates through rock pores or other openings.
hypothesis An idea shaped from scientific observations.

i

ice age A major interval of geologic time during which extensive ice sheets (continental glaciers) formed over many parts of the world.
igneous rock A rock formed by the crystallization of magma or lava.
inclusion A fragment or rock that is distinct from the body of igneous rock in which it is enclosed.
inclusion, principle of Fragments included in a host rock are older than the host rock.
index fossil A fossil that is associated with a particular span of geologic time.
inductive reasoning Process whereby scientists draw general conclusions from specific observations.

infiltration The movement of surface water into rock or soil through cracks and pore spaces.
infrared radiation Radiation with a wavelength from 0.7 to 200 micrometers.
insolation Amount of solar radiation per area (watts/square meter).
interglacial Descriptive of warm intervals, lasting approximately 20,000 years, during the last ice age when temperatures were up to 3 C° warmer than today.
ion An atom or molecule that possesses an electrical charge.
ionic bond Molecular bond formed when two oppositely charged atoms (ions) attract.
ionosphere A complex zone of ionized gases that coincides with the lower portion of the thermosphere.
isobar A line drawn on a map connecting points of equal atmospheric pressure, usually corrected to sea level.
isostasy The principle that the upper parts of Earth's outer layers are in a state of flotational equilibrium.
isotopes Varieties of the same element that have different mass numbers; their nuclei contain the same number of protons but different numbers of neutrons.

j

jet stream Regional currents of fast-moving air in the upper troposphere.
Jovian planets The Jupiter-like planets: Jupiter, Saturn, Uranus, and Neptune. These planets have relatively low densities.
Jurassic period The second period of the **Mesozoic era** of geologic time, between 200 million and 145 million years ago.

k

kinetic energy The energy resulting from the constant motion of atoms in a substance.
Köppen-Geiger classification system Climate classification system based on average temperatures, average precipitation, and vegetation.
Kyoto Protocol An agreement among the developed nations to reduce greenhouse emissions to 1990 levels by 2008–2012.

l

lahar Indonesian term describing the rapid flow of water and loose debris down steep volcanic slopes, a volcanic mudflow.
landslide Perceptible downslope movement of earth materials.
latent heat The energy absorbed or released during a change in state of water.
lateral blast A volcanic eruption where force is directed sideways rather than vertically.
lateral migration A process in which a stream migrates across the valley floor in the direction of the erosion.
laterite soil A type of clay-rich soil that has a tendency to harden into a bricklike solid.
lava Magma that reaches the ground surface.
lava dome A steep-sided lava flow commonly almost as high as it is wide and caused by the extrusion of extremely viscous lava.
lava flow Molten material extruded at, and flowing away from, a volcano or fissure.
lava plateau An elevated tableland or flat-topped highland that is several hundreds to several thousands of square kilometers in area and underlain by a thick succession of lava flows.
leachate A liquid that has percolated through soil and dissolved some soil materials in the process.
leaching The depletion of soluble materials from the upper soil horizons by downward-percolating water.
levee An embankment bordering one or both sides of a channel.
limestone A sedimentary rock composed mostly of calcite.
liquefaction The transformation of loosely packed, water-saturated sediment into a fluid mass. Commonly it occurs when grains settle during earthquake ground shaking, which displaces the water in the intervening pore spaces upward such that the water pressure moves the grains apart and the whole sediment-water mixture loses strength.
lithification The process by which unconsolidated sediments become consolidated rocks.
lithosphere The relatively rigid outer layer of Earth that is distinguished from a plastic zone below called the **asthenosphere.**
Little Ice Age The period extending from A.D. 1450 to 1900 during which global temperatures dropped and glaciers advanced.
longshore current A component of motion in the water directed parallel to the shoreline.
losing streams Streams that lose water to the zone of saturation.
Love waves A type of surface seismic wave that moves sideways but with no vertical movement.
luminosity The brightness of a star. The amount of energy radiated by a star.
luster Appearance taken on by a mineral in reflected light.

m

magma Molten rock that may contain dissolved gases as well as some solid minerals.
magmatic mineral deposits Mineral deposits formed when minerals become concentrated as magma cools and solidifies.
magnetic polarity Direction of Earth's magnetic field.
magnetic polarity reversal Direction of Earth's magnetic field when it is the opposite of the present-day field.
mantle A thick shell of rock that separates Earth's crust above from the core below.
marble A coarse-grained metamorphic rock composed of interlocking calcite crystals.
maritime air mass An air mass that originates over the ocean. These air masses are relatively humid.
maritime polar air mass Polar air initially possessing similar properties to those of continental polar air but that, in passing over water, attains a higher moisture content.
maritime tropical air mass The principal type of tropical air, produced over the tropical and subtropical seas; it is warm and humid and is frequently carried poleward on the western flanks of the subtropical highs.
mass number The number of neutrons and protons in the nucleus of an atom.
mass wasting Movement, caused by gravity, in which earth materials move downslope in bulk.
meander A pronounced sinuous curve along a stream's course.
megathrust earthquake An informal term for the largest of earthquakes occurring on long faults associated with subduction zones.
melting The change of state from solid to liquid.
meltwater Melting snow or ice.
mesocyclone A small cyclone formed when updrafts below a thunderstorm draw the spiraling horizontal winds upward.
mesopause The top of the mesosphere.
mesosphere The layer of the atmosphere immediately above the **stratosphere** and characterized by decreasing temperatures with altitude.
Mesozoic era The era that followed the **Paleozoic era** and preceded the **Cenozoic era.**
metamorphic rock A rock produced by metamorphism.
metamorphism The effects of increasing temperature, pressure, or both within Earth that produce solid-state mineralogical, chemical, and textural changes that alter the features of preexisting rocks.
meteor An asteroid that passes through Earth's atmosphere, sometimes incorrectly called a "shooting" or "falling" star.
meteorite An asteroid that strikes Earth's surface.
meteorology The study of the atmosphere and atmospheric phenomena; the study of weather and climate.
midlatitude cyclone Large center of low pressure with an associated cold front and often a warm front. Frequently it is accompanied by abundant precipitation.
Milankovitch cycles Periodic variations in Earth's rotation axis and in its orbit around the sun that affect the amount of solar energy that reaches Earth.
mineral A crystalline substance that is naturally occurring and is chemically and physically distinctive.
mineral resource All discovered and undiscovered deposits of a useful mineral that can be extracted now or at some time in the future.
Mississippian period The time between 360 million and 318 million years ago during the **Paleozoic era.**
mixed tide A tidal condition in which two high and two low tides of different ranges occur each day.
Modified Mercalli scale Scale expressing intensities of earthquakes (judged on amount of damage done) in Roman numerals ranging from I to XII.
Mohs hardness scale Scale on which 10 minerals are designated as standards of hardness.
mud volcano A conical accumulation formed by precipitation of minerals from hot fluids.
mudflow A flowing mixture of fine-grained debris and water, usually moving down a channel.

n

neap tide Low tidal range occurring when the moon and sun are 90 degrees apart.
near-Earth objects Asteroids and comets that pass close to Earth's orbit.
Neogene A period of the **Cenozoic era**, between 23 million and 2.5 million years ago.
Neptunism The obsolete theory that all rocks of Earth's crust were deposited from, or crystallized out of, water.
neutron An uncharged subatomic particle found in the nucleus of atoms. Its weight is slightly more than that of a proton.
nimbostratus clouds Precipitation-bearing clouds that form a thick, continuous layer.
nonrenewable resources Resources that form or accumulate over such long time spans that they must be considered as fixed in total quantity.
normal fault A fault in which the rocks above the fault plane have moved down relative to the rock below.
normal lapse rate The temperature of stable air (air that is neither rising nor falling) that decreases with altitude at a rate of 6 C° per 1,000 meters (1.8 C° per 1,000 feet).
numerical time Amount of time elapsed as determined by analyses of the decay of naturally occurring radioactive elements.

o

oasis An isolated fertile area, usually limited in extent and surrounded by desert and marked by vegetation and a water supply.
occluded front A front formed when a cold front overtakes a warm front. It marks the beginning of the end of a middle-latitude cyclone.

ocean A major primary subdivision of the intercommunicating body of saltwater occupying the depressions of Earth's surface, bounded by continents and imaginary lines.

ocean floor The seafloor surface of an ocean basin.

oceanic crust The rocks that make up the uppermost part of the lithosphere on oceanic plates.

oceanic ridge A continuous elevated zone on the floor of the entire major ocean basins, varying in width from 500 to 5,000 kilometers. The rifts at the crests of ridges represent divergent plate boundaries.

oceanic trench A narrow, deep trough parallel to the edge of a continent or an island arc.

O horizon Dark-colored soil layer that is rich in organic material and forms just below surface vegetation.

Oort cloud A massive spherical cloud of cosmic debris where long-period comets originate.

open aquifer A groundwater source that is recharged as water infiltrates through permeable material directly above.

open system One in which both matter and energy flow into and out of the system. Most natural systems are of this type.

Ordovician period The second period of the **Paleozoic era,** after the **Cambrian** and before the **Silurian,** from approximately 488 million to 444 million years ago.

ore Naturally occurring material that can be profitably mined.

original horizontality, principle of Most water-laid sediment was deposited in horizontal or near-horizontal layers that are essentially parallel to Earth's surface.

orographic lifting The lifting of an air current caused by its passage up and over surface elevations.

oxidation Chemical reaction between a substance and oxygen that produces new substances.

ozone A molecule containing three oxygen atoms.

ozone layer A region in the stratosphere with a greater concentration of ozone that stops 97 to 99 percent of ultraviolet rays from reaching Earth's surface.

p

Pacific Tsunami Warning System (PTWS) A network of stations designed to identify potentially damaging tsunami caused by earthquakes in or around the Pacific Ocean. The PTWS issues warnings or watches that predict tsunami arrival times for coastal areas.

paleoclimate The climate of a given interval of geologic time.

Paleogene The first period of the **Cenozoic era**, between 66 million and 23 million years ago.

paleomagnetism A study of ancient magnetic fields.

paleontology The systematic study of fossils and the history of life on Earth.

Paleozoic era The era that followed the **Precambrian** and began with the appearance of complex life, as indicated by abundant fossils.

Pangaea A supercontinent that broke apart 200 million years ago to form the present continents.

paradigm A theory that is held with a very high degree of confidence and is comprehensive in scope.

paradigm shift A change in a widely held view in science.

parent isotope Radioactive isotope that decays through time to a **daughter isotope.**

partial melting Melting of the components of a rock with the lowest melting temperatures.

passive continental margin A margin that includes a continental shelf, continental slope, and continental rise whose morphology is not affected greatly by active tectonic processes.

peat A brown, lightweight, unconsolidated or semiconsolidated deposit of plant remains.

pelagic community The plants and animals that live in the open ocean, not on the bottom or along shorelines.

Pennsylvanian period The time between 318 million and 299 million years ago, toward the end of the **Paleozoic era**.

percolation Flow of groundwater through the pore spaces in rock or soil.

peridotite Ultramafic rock composed of pyroxene and olivine present in the upper mantle.

period Each era of the standard geologic timescale is subdivided into periods (for example, the **Cretaceous period**).

permeability The capacity of a rock to transmit a fluid such as water or petroleum.

Permian period The time period between 251 million and 299 million years ago; last period of the **Paleozoic era.**

Permian-Triassic extinction (P-T event) Time about 251 million years ago that marked the largest mass extinction event known on Earth.

pH A measure of the acidity or alkalinity of a solution. Explicitly, pH is the logarithm of the hydrogen-ion concentration. A solution with a pH of 7 is neutral, pH values less than 7 are acidic, and pH values greater than 7 are alkaline.

Phanerozoic The part of geologic time represented by rocks containing abundant fossil evidence. The eon extending from the end of the **Proterozoic eon** to the present.

photolysis A chemical reaction caused by the interaction of a material with sunlight.

photosynthesis The process by which plants and algae produce carbohydrates from carbon dioxide and water in the presence of chlorophyll, using light energy and releasing oxygen.

physical weathering Disaggregation of rocks by mechanical processes.

plate tectonics A theory that Earth's surface is divided into a few large, mobile plates that are slowly moving and changing in size. Intense geologic activity occurs at the plate boundaries.

pluton An igneous body that crystallized deep underground.

plutonic igneous rocks Igneous rock formed at great depth.

Plutonism The obsolete theory that the rocks of Earth's crust formed by solidification of a molten material or were heated.

point bar A place where sediment accumulates on the inside of a bend in a stream, where the water flow is slow.

point source pollution Specific, identifiable sources of pollution, such as leaking storage tanks.

polar cell Circulating air mass ascending at high latitudes (~60°N and S) and descending at the poles.

polar front The stormy frontal zone separating air masses of polar origin from air masses of tropical origin.

polar vortex The large-scale cyclonic circulation in the middle and upper troposphere centered generally in the polar regions.

polarity The orientation of Earth's magnetic field, which is described as either normal or reverse polarity.

pore space The space taken up by openings within a rock.

porosity The percentage of a rock's volume that is taken up by openings.

porphyry An igneous rock composed of large crystals embedded in fine-grained minerals.

Precambrian The vast amount of time that preceded the **Paleozoic era**.

precession The change in the direction of tilt of Earth's axis.

precipitation Water that falls from the atmosphere in the form of ice, snow, hail, or sleet or in the liquid state.

precision The degree to which measurements in a series match one another.

pressure gradient The amount of pressure change occurring over a given distance.

pressure release A significant type of physical weathering that causes rocks to crack when overburden is removed.

Proterozoic eon The eon following the **Archean** and preceding the **Phanerozoic.**

proton A subatomic particle that contributes mass and a single positive electrical charge to an atom.

proxy A substitute.

P wave A compressional wave (seismic wave) in which rock vibrates parallel to the direction of wave propagation.

pycnocline A layer of water in which there is a rapid change of density with depth.

pyroclastic flow A turbulent mixture of tephra and gases flowing down the flank of a volcano.

pyroxene A group of minerals that are single-chain silicates.

q

quartz A common mineral composed of silicon and oxygen, SiO_2.

quartzite The metamorphic equivalent of quartz sandstones.

Quaternary A division of geologic time from 2.5 million years ago to the present day that represents the last period in the **Cenozoic era**.

r

radial drainage A drainage pattern in which streams diverge outward like spokes of a wheel.

radioactive decay The process by which unstable (radioactive) isotopes transform to new elements by a change in the number of protons (and neutrons) in the nucleus.

radiometric dating The procedure of calculating the absolute ages of rocks and minerals that contain certain radioactive isotopes.

rain shadow A region on the downwind side of mountains that has little or no rain because of the loss of moisture on the upwind side of the mountains.

Rayleigh wave A seismic wave that travels at or close to the surface and involves vertical motions.

recharge zone An area in which water is absorbed that eventually reaches the zone of saturation in one or more aquifers.

rectangular drainage A drainage pattern in which tributaries of a river change direction and join one another at right angles.

recurrence interval The time interval between the occurrences of a type of event. Usually it is used in relation to the probability that a floodplain floods each year; a recurrence interval of 100 years means that there is a 1 in 100 (0.01) probability of a flood with a specified discharge each year.

red shift A shift in the wavelength of electromagnetic radiation to a longer wavelength. For visible light, this implies a shift toward the red end of the spectrum.

regional metamorphism The name applied to those metamorphic alterations that affect rocks over large areas and are indicative of widespread environmental changes rather than localized deformation or magmatism.

regolith A general term applied to soil and other unconsolidated material that lie above bedrock.

relative humidity The ratio of the amount of water vapor in the air to the amount that could be in the air if the air were saturated at the same temperature.

relative time The sequence in which events took place (not measured in time units).

renewable resource A resource that is virtually inexhaustible or that can be replenished over relatively short time spans.

reserve Already identified deposits from which minerals can be extracted profitably.
residence time The length of time that a given volume of water remains in part of the hydrologic cycle.
residual mineral deposits Mineral deposit formed from the leaching of soil by water.
resources The total amount of geologic material in all its deposits, discovered and undiscovered.
reverse fault A fault in which the material above the fault plane moves up in relation to the material below.
rhyolitic magma Molten rock with a silica content greater than 68 percent.
Richter scale A scale of earthquake magnitude based on the highest amplitude wave arriving at a seismograph.
rift valley A tensional valley bounded by normal faults. Rift valleys are found at diverging plate boundaries on continents and along the crest of the midoceanic ridge.
rip current Current that moves water directly offshore and away from the coast.
rock cycle A sequence of processes and products that relates each rock type to the others and that describes rocks as continuously forming from preexisting rocks.
rockfall Rock falling freely or bouncing down a cliff.
rock salt An evaporate composed of halite.
rockslide The rapid slide of a mass of rock downslope along planes of weakness.
runoff The portion of precipitation that flows off the ground surface into streams.

S

Saffir-Simpson intensity scale Scale dividing hurricanes into five categories on the basis of wind speed and storm surge.
salinity The proportion of dissolved salts to pure water, usually expressed in parts per thousand or as a percentage.
saltwater intrusion Displacement of fresh groundwater by saltwater due to its greater density.
sand Sediment composed of particles with a diameter between 0.0625 and 2 millimeters.
sand dune A depositional feature consisting of sand.
sandstone A medium-grained sedimentary rock formed by the cementation of sand grains.
saturated zone A subsurface zone in which all rock openings are filled with water.
saturation The maximum quantity of water vapor that the air can hold at any given temperature and pressure.
savanna A tropical biome dominated by grasses but having small trees that do not form a canopy.
schist A metamorphic rock characterized by coarse-grained minerals oriented approximately parallel.
scientific method Process used to systematically and objectively examine and explain a problem or observed phenomenon.
seafloor spreading theory The concept that the ocean floor is moving away from the midoceanic ridge and across the deep-ocean basin, where it is subducted beneath continents and island arcs.
sea level The mean levels of the sea after tidal and wind effects are removed.
seamounts A mountain that rises from the sea floor but does not break the surface of the ocean. Most seamounts are extinct volcanoes.
seawall A wall constructed along the base of retreating cliffs to prevent wave erosion.
sediment Loose, solid particles that can originate by (1) weathering and erosion of preexisting rocks, (2) chemical precipitation from solution, usually in water, and (3) secretion by organisms.
sedimentary rock Rock that has formed from (1) lithification of any type of sediment, (2) precipitation from solution, or (3) consolidation of the remains of plants or animals.
sediment budget The balance between erosion and deposition (can be positive or negative) along a coast.
seismic gap A segment of a fault that has not experienced earthquakes for a long time; such gaps may be the site of large future quakes.
seismic wave The release of energy produced by an earthquake.
seismogram The record made by a seismograph.
seismograph A recording device that produces a permanent record of Earth motion.
seismometer An instrument designed to detect seismic waves or Earth motion.
shale A fine-grained sedimentary rock (grains finer than 0.0039 millimeter in diameter) formed by the cementation of clay (mud). Shale has thin layers (laminations) and an ability to split into small chips.
shield volcano Broad, gently sloping cone constructed of solidified lava flows.
silica A term used for oxygen plus silicon.
silicate A substance that contains silica as part of its chemical formula.
silica tetrahedron The basic building block of the silicate mineral crystal structure, consisting of four oxygen atoms surrounding and bonded to a silicon atom.
sill A tabular intrusive structure concordant with local rock.
Silurian period A period of geologic time of the **Paleozoic era,** covering a time span of between 444 million and 416 million years ago.
sinkhole A closed depression found on land surfaces underlain by limestone.
slope failure The downward and outward movement of earth materials on a natural slope.
slump In mass wasting, movement along a curved surface in which the upper part moves vertically downward while the lower part moves outward.
snowline Hypothetical line marking the transition from the ablation to the accumulation zone on a glacier.
soil A layer of weathered, unconsolidated material on top of bedrock; often also defined as containing organic matter and being capable of supporting plant growth.
soil conservation Practices intended to reduce the loss of soil by erosion.
soil erosion The detachment and movement of topsoil by the action of wind and flowing water.
soil profile A vertical section through a soil showing its succession of horizons and the underlying parent material.
solar radiation Electromagnetic radiation supplied by the sun.
solar system The sun, planets, their moons, and other bodies that orbit the sun.
solstice The time when the vertical rays of the sun are striking either the Tropic of Cancer or the Tropic of Capricorn. Solstice represents the longest or shortest day (length of daylight) of the year.
source area The locality that eroded to provide sediment to form a sedimentary rock or the origin of an air mass.
specific heat Amount of energy required to raise the temperature of a material by 1 C°.
specific retention The volume of water that a given body of rock or soil retains after being drained, divided by its porosity.
specific yield The amount of water that an aquifer will produce as a percentage of the total porosity.
spit A sand bar extending from the shore into an adjoining bay.
spring A place where water flows naturally out of rock onto the land surface.
spring tide Unusually high and low tides that result when the sun and moon lie in line with Earth and the attraction of the two bodies is cumulative.
stage The measured elevation of the water surface in a stream. The stage is defined relative to an arbitrarily selected elevation that is typically close to the elevation of the deepest part of the stream channel.
stalactite Iciclelike structure that hangs from the ceiling of a cave and is formed by precipitation of limestone.
stalagmite Columnlike form that grows upward from the floor of a cavern and is formed by precipitation of limestone.
storm surge Rise in the elevation of the sea surface caused by winds associated with hurricanes or other low-pressure storms that form over oceans.
strainmeters Instruments that measure deformation of Earth's crust.
strata Parallel layers of sedimentary rock.
stratification Rock layering.
stratocumulus A low-altitude cloud with a puffy appearance.
stratopause The transition zone between the **stratosphere** and the **mesosphere.**
stratosphere The layer of the atmosphere immediately above the **troposphere,** characterized by increasing temperatures with altitude, owing to the concentration of ozone.
stratovolcano A volcano formed by combined eruptions of cinder and lavas. Eruptions of cinders build the cone high near the vent. Eruptions of lava may extend the base of the volcano.
stratus cloud One of three basic cloud forms; sheets or layers that cover much or all of the sky.
streak Color of a pulverized substance; a useful property for mineral identification.
stream Flowing water that moves through a channel and simultaneously transports dissolved and particulate products of rock weathering.
stream discharge The volume of water flowing through a stream.
stream gauge Device that measures water level at a location in a stream where the channel cross section has been carefully surveyed. Stream gauge data are used to calculate discharge.
stream gradient Downhill slope of a stream's bed or the water surface.
stream load Suspended and dissolved materials that are removed (eroded), carried (transported), and dropped (deposited) by streams.
stream profile The longitudinal profile of a stream.
stream velocity The speed at which water in a stream travels.
strike-slip fault A fault in which movement is horizontal and parallel to the fault surface.
subduction zone Elongated region where one plate descends below another.
sublimation The conversion of a solid directly to a gas without passing through the liquid state.
submarine canyons Valleys that run across the continental shelf and down the continental slope.
subsidence Sinking or downwarping of a part of Earth's surface.
subtropical high A series of high-pressure centers characterized by descending air and located roughly between latitudes 25 and 35 degrees north and south of the equator.

superposition, principle of A principle stating that within a sequence of undisturbed sedimentary rocks, the oldest layers are on the bottom, the youngest on the top.
surface wave A seismic wave that travels on Earth's surface.
suspended load Fine particles carried in the water column of a stream.
S wave A seismic wave propagated by a shearing motion, which causes rock to vibrate perpendicular to the direction of wave propagation.

t

talus Rock fragments that accumulate as a heap or sheet at the base of a steep rock surface.
tectonic plate Any of the internally rigid crustal blocks of the lithosphere that move horizontally across Earth's surface relative to one another.
temperate forest A midlatitude biome with trees.
temperature A measure of the average kinetic energy of individual atoms or molecules in a substance.
temperature inversion The atmospheric condition in which a layer of warmer air lies above cooler air.
tephra Rock fragments produced by volcanic explosion.
terminal moraine Depositional feature found at the downslope end of a glacier.
terrestrial planets Any of the Earth-like planets, including Mercury, Venus, Mars, and Earth.
Tertiary period A period of geologic time, from the end of the **Mesozoic era** (65 million years ago) to about 2 million years ago. Authorities have recently suggested it be replaced by the Paleogene and Neogene periods.
texture The general physical appearance of rocks. The size, shape, and arrangement of component minerals in sedimentary rocks determine the texture. In igneous and metamorphic rocks, the crystallinity and fabric of the minerals are major determinants of texture.
theory A well-supported explanation of a natural phenomenon.
thermocline A layer of water in which there is a rapid change in temperature in the vertical dimension.
thermohaline circulation Movements of ocean water caused by density differences brought about by variations in temperature and salinity.
thermosphere The upper layer of the atmosphere located above the **mesosphere,** starting at about an 85-kilometer altitude. Temperature increases with altitude in this layer.
thrust fault A compressional feature causing a younger unit to move over an older unit.
thunderstorm A storm produced by a cumulonimbus cloud and accompanied by lightning and thunder. It is of relatively short duration and usually accompanied by strong wind gusts, heavy rain, and sometimes hail.
tide Daily rising of sea level caused by gravitational attraction of the moon and sun on the oceans.
till Sediment from glaciers that is unsorted.
tillite Sedimentary rock formed from till.
topography The natural elevations of land surfaces in regions, treated collectively as to form.
Torino scale Asteroid and comet impact hazard scale.
tornado A small-scale cyclone with very strong winds. At ground level, most tornados are less than 500 meters (1,640 feet) in diameter.
trace fossil Trail, track, or burrow resulting from animal movement, preserved in sedimentary rock.
trade winds Two belts of winds that blow almost constantly from easterly directions and are located on the equatorial sides of the subtropical highs.
transform plate boundary Boundary between two plates that are sliding past each other.
transpiration Process whereby water in plant leaves and stems is released as water vapor.
transportation Movement of materials from one location to another.
trellis drainage A drainage pattern characterized by parallel main streams and secondary tributaries intersected at right angles by tributaries.
Triassic period The first period of the **Mesozoic era,** occurring after the **Permian** and before the **Jurassic,** 200 million to 251 million years ago.
tributary Any stream that contributes water to another stream.
tropical depression By international agreement, a tropical cyclone with maximum winds that do not exceed 61 kilometers (38 miles) per hour.
tropopause The boundary between the **troposphere** and **stratosphere.**
troposphere The portion of the atmosphere that is in contact with Earth's surface. The top of the troposphere occurs where the temperature drop with altitude levels out and then begins to rise.
trough The low point of a wave.
tsunami A long-period sea wave produced by a submarine earthquake or volcanic eruptions. It may travel for thousands of miles.
typhoon Low-pressure tropical storm.

u

ultraviolet radiation Electromagnetic radiation with a wavelength from 0.2 to 0.4 micrometer.
unconformity A surface that represents a break in the geologic record, with the rock unit immediately above it being considerably younger than the rock beneath.
unconsolidated A state in which sediment grains are loose, separate, or unattached to one another.
uniformitarianism Principle that geologic processes operating at present are the same processes that operated in the past. The principle is stated more succinctly as "The present is the key to the past."
unloading Process of overlying rocks being eroded, pressure being removed, and buried rocks expanding upward.
unsaturated zone A subsurface zone in which rock openings are generally filled partly with water; above the saturated zone.
updraft Rapid upward movement of air associated with severe storms.

v

varve A pair of annual layers deposited in a sequence of sediments.
viscosity Measure of the resistance of a fluid to flow.
visible light Electromagnetic radiation with a wavelength from 0.4 to 0.7 micrometer.
volcanic explosivity index A system describing the relative size of volcanic eruptions.
volcanic gases Volatile matter composed principally of about 90 percent water vapor and carbon dioxide, sulfur dioxide, hydrogen, carbon monoxide, and nitrogen, released during an eruption of a volcano.
volcano A mountain formed of lava, tephra, or both.

w

Wadati-Benioff zone The inclined zone of earthquakes' foci characteristic of subduction zones at convergent plate boundaries.
warm front A front along which a warm air mass overrides a retreating mass of cooler air.
water cycle See **hydrologic cycle.**
water table The upper surface of the zone of saturation.
wave base The depth at which sediments are not stirred by wave action, usually about 33 feet (10 meters).
wave height Vertical distance between the crest and trough of a wave.
wavelength The horizontal distance between two wave crests (or two troughs).
wave refraction Change in direction of waves due to slowing as they interact with the shoreline.
weather The state of the atmosphere at any given time.
weathering The group of processes that change rock at or near Earth's surface.
wedging See **frost wedging.**
westerlies Winds between 30° and 60° latitude that move west to east and greatly influence weather in North America.
wet adiabatic lapse rate The rate of adiabatic temperature change in saturated air. The rate of temperature change is variable, but it is always less than the dry adiabatic rate.
wetland An environment characterized by hydric soils that support development of hydrophytic (water-forming) plants.
wetted perimeter The portion of the perimeter of a stream channel cross section that is in contact with the water.
wind Movement of air.
wind erosion Detachment, transportation, and deposition of loose topsoil or sand by the action of wind.

y

Younger Dryas A brief interval of cold temperatures at the close of the last major ice age (13,000–11,600 years ago). Temperatures decreased and increased abruptly at start and end of the interval.

Credits

Text and Line Art Credits

Chapter 1

Page 4: Heasley, Katherine "Personal Account of Tsunami and Aftermath in Japan," *www.Oregonlive.com*, April 13, 2011; Figure 1.1a: Adapted from http://serc.carleton.edu/details/images/27439.html; 1.1b: Adapted from http://www.ngdc.noaa.gov/hazard/icons/2004_1226.jpg; Page 5: Von Feldt, Rick "Surviving the Tsunami-Part 2-The First Hour," *www.phukettsunami.blogspot.com*, December 27, 2004; Checkpoint 1.6: Adapted from Anton E. Lawson, *Journal of College Science Teaching*, May 1999, pp 401-11; Checkpoint 1.7: Adapted from www.kgs.ukans.edu/HydroHutch/GeneralGeology/fig 1.5; 1.14a: Adapted from http://www.nytimes.com/interactive/2010/05/27/us/20100527-oil-landfall.html

Chapter 2

Figure 2.3: Adapted from http://www.physics.carleton.ca/ and http://galileo.rice.edu (Figure 2.5b); 2.9: Adapted from http://meinkechem08.wikispaces.com/Stars and http://mail.colonial.net/~hkaiter/life_cycle_of_a_star.html; 2.12: Adapted from http://solarscience.msfc.nasa.gov/images/ssn_predict_l.gif; 2.13: Adapted from John Fix, *Astronomy: Journey to the Cosmic Frontier* 4e, © 2006, Reprinted with permission of McGraw-Hill Education; 2.20: Adapted from Eric Danielson, *Meteorology* 2e © 2003, reproduced with permission of McGraw-Hill Education; 2.25: Adapted from http://www1.eere.energy.gov/geothermal/pdfs/egs_chapter_1.pdf

Chapter 3

CO3 text: Doolittle, Hilda "The Dancer," *Life & Letters Today*, December 1935; Checkpoint 3.3 text: Sources *New York Times*, March 4, 2003, by Henry Fountain; *Toronto Star*, March 9, 2003; Jet Propulsion Lab, Cal Tech; Figure 3.14: Adapted from NASA; 3.16: Adapted from http://www.dailygalaxy.com/photos/uncategorized/2007/09/06asteroid_impact_map_of_earth.gif; 3.18: Adapted from http://www.1pl.arizona.edu/impacteffects/

Chapter 4

Figure 4.2: Adapted from Burton Guttman, *Biology* 1e © 1999, reproduced with permission of McGraw-Hill Education; 4.3: Adapted from USGS; 4.5: Adapted from Charles (Carlos) Plummer, *Physical Geology* 11e © 2007, reproduced with permission of McGraw-Hill Education; 4.7: NOAA; 4.8a: Adapted from www.geo.lsa.umich.edu; 4.8b: USGS; 4.8c, 4.10-4.12, 4.14: Adapted from Charles (Carlos) Plummer, *Physical Geology* 11e © 2007, reproduced with permission of McGraw-Hill Education; 4.15: Adapted from after http://geomaps.wr.usgs.gov/parks/noca/fig 11.gif; 4.16: Adapted from Charles (Carlos) Plummer, *Physical Geology* 11e © 2007, reproduced with permission of McGraw-Hill Education; 4.17: Adapted from NOAA; 4.19: Adapted from Konrad Krauskopf, *The Physical Universe* 11e © 2006, reproduced with permission of McGraw-Hill Education; 4.20: Adapted from Charles (Carlos) Plummer, *Physical Geology* 11e © 2007, reproduced with permission of McGraw-Hill Education; 4.21: Adapted from *Oceanus Magazine*, Woods Hole Oceanographic Institute; 4.23, 4.24: Adapted from Charles (Carlos) Plummer, *Physical Geology* 11e © 2007, reproduced with permission of McGraw-Hill Education; 4.27: Adapted from Woods Hole Oceanographic Institution; 4.28: Adapted from Charles (Carlos) Plummer, *Physical Geology* 11e © 2007, reproduced with permission of McGraw-Hill Education

Chapter 5

Figure 5.3: Adapted from USGS; 5.6-5.9: Adapted from Charles (Carlos) Plummer, *Physical Geology* 11e © 2007, reproduced with permission of McGraw-Hill Education; 5.10: Adapted from Edgar Spencer, *Earth Science: Understanding Environmental Systems* 1e © 2003, reproduced with permission of McGraw-Hill Education; 5.11: Adapted from http://earthquake.usgs.gov/research/hazmaps/products_data/2008/maps/us/pga.usa.jpg; 5.11: USGS; 5.13: Adapted from http://earthquake.usgs.gov/earthquakes/eqarchives/year/2012/2012_data.php; 5.14-5.16, 5.17a,b: Adapted from Charles (Carlos) Plummer, *Physical Geology* 11e © 2007, reproduced with permission of McGraw-Hill Education; 5.18: Adapted from Edgar Spencer, *Earth Science: Understanding Environmental Systems* 1e © 2003, reproduced with permission of McGraw-Hill Education; Checkpoint 5.16, 5.19: USGS; 5.25: Adapted from USGS; 5.26: Adapted from USGS Circular 1187: http://pubs.usgs.gov/circ/c1187/images/fig17.jpg

Chapter 6

Figure 6.5: Adapted from http://www.volcano.si.edu/world/find_regions.cfm; 6.6: Adapted from Reynolds et al., *Exploring Geology*, © 2008 The McGraw-Hill Companies, Inc., Figure 5.06.al, p. 118; 6.7: Adapted from Charles (Carlos) Plummer, *Physical Geology* 11e © 2007, reproduced with permission of McGraw-Hill Education; 6.9, 6.12: Adapted from USGS; 6.13c: Adapted from Reynolds et al., *Exploring Geology,* © 2008 The McGraw-Hill Companies, Inc., Figure 06.11, mtbl1, p 159 and from U.S. Geological Survey Open-File Report 95-59; 6.13d: Adapted from Reynolds et al., Exploring Geology, © 2008 The McGraw-Hill Companies, Inc., Figure 06.11.c2, p 152 and from U.S. Geological Survey Fact Sheet 2005-3024; 6.16, 6.19-6.23, 6.26, 6.27, 6.31: Adapted from Charles (Carlos) Plummer, *Physical Geology* 11e © 2007, reproduced with permission of McGraw-Hill Education; 6.32: Adapted from William Cunningham, *Environmental Science: A Global Concern* 7e © 2003, reproduced with permission of McGraw-Hill Education; 6.33: USGS

Chapter 7

CO7 text: Excerpt from "The Wall," from "The Awful Rowing Toward Goodbye," Anne Sexton. Copyright © 1975 by Loring Conant, Jr., Executor of the Estate, renewed 2003 by Linda G. Sexton. Reprinted by permission of Houghton Mifflin Harcourt Publishing Company; Figure 7.6, 7.8a,b: Adapted from Charles (Carlos) Plummer, *Physical Geology* 11e © 2007, reproduced with permission of McGraw-Hill Education; 7.8c, 7.9: Adapted from Francis Carey, *Organic Chemistry* 5e © 2003, reproduced with permission of McGraw-Hill Education; 7.10-7.16: Adapted from Charles (Carlos) Plummer, *Physical Geology* 11e © 2007, reproduced with permission of McGraw-Hill Education; 7.18: Adapted from James Zumberge, *Laboratory Manual for Physical Geology* 12e © 2005, reproduced with permission of McGraw-Hill Education; In Further Depth, figure 1: Adapted from Charles (Carlos) Plummer, *Physical Geology* 11e © 2007, reproduced with permission of McGraw-Hill Education

Chapter 8

Checkpoint 8.9 text: LeBlond, Lawrence "Scientists Find Fossil Below K-T Boundary," RedOrbit.com, July 13, 2011. Reprinted by permission. http://www.redorbit.com/news/science/2078377/scientists_find_fossil_below_kt_boundary/; Figure 8.16: Adapted from www.pbs.org; Checkpoint 8.11: Erickson, Jim, *Rocky Mountain News (Denver)*, February 22, 2002, p 7a. Used with permission of The Denver Public Library, Western History Collection WH-2129; 8.20: Adapted from Konrad Krauskopf, *The Physical Universe* 11e © 2006, reproduced with permission of McGraw-Hill Education

Chapter 9

Figure 9.3, 9.8: Adapted from Charles (Carlos) Plummer, *Physical Geology* 11e © 2007, reproduced with permission of McGraw-Hill Education; 9.10: Adapted from USGS; 9.12: Adapted from Charles (Carlos) Plummer, *Physical Geology* 11e © 2007, reproduced with permission of McGraw-Hill Education; 9.22a: Adapted from http://soils.usda.gov/use/worldsoils/mapindex/order.html; 9.26: Adapted from http://www.nrcs.usda.gov/Internet/FSE_MEDIA/stelprdb1041882.png; 9.30: Adapted from Natural Resources Conservation Service: http://www.nrcs.usda.gov/technical/land/

Chapter 10

Figure 10.7, 10.17: Adapted from Charles (Carlos) Plummer, *Physical Geology* 11e © 2007, reproduced with permission of McGraw-Hill Education

Chapter 11

Figure 11.3, 11.6: Adapted from Charles (Carlos) Plummer, *Physical Geology* 11e © 2007, reproduced with permission of McGraw-Hill Education; 11.8: Adapted from Carina Hoorn, "The Birth of the Mighty Amazon," *Scientific American*, April 23, 2006, p. 57; 11.9, 11.15: Adapted from Charles (Carlos) Plummer, *Physical Geology* 11e © 2007, reproduced with permission of McGraw-Hill Education; 11.17: Adapted from USGS Professional Paper 1573: http://vulcan.wr.usgs.gov/projects/sediment_trans/PP1573/HTMLReport/slide35.gif; 11.20, 11.21: Adapted from

Charles (Carlos) Plummer, *Physical Geology* 11e © 2007, reproduced with permission of McGraw-Hill Education; 11.24, 11.25: Adapted from USGS; 11.35: US Army Corps of Engineers Digital Library

Chapter 12

Figure 12.8a: Adapted from http://capp.water.usgs.gov/aquiferBasics/uncon.html#anchorA; 12.8b: Adapted from http://capp.water.usgs.gov/aquiferBasics/sandstone.html#anchorA; 12.8c: Adapted from http://capp.water.usgs.gov/aquiferBasics/carbrock.html; 12.8d: Adapted from http://capp.water.usgs.gov/aquiferBasics/volcan.html#anchorA; 12.8e: Adapted from http://capp.water.usgs.gov/aquiferBasics/sandcarb.html#anchorA; 12.10, 12.12: Adapted from Charles (Carlos) Plummer, *Physical Geology* 11e © 2007, reproduced with permission of McGraw-Hill Education; 12.13: Adapted from USGS; 12.14: Adapted from Charles (Carlos) Plummer, *Physical Geology* 11e © 2007, reproduced with permission of McGraw-Hill Education; 12.18: Adapted from www.grida.no/aeo/fig2e5.htm; 12.20: Adapted from USGS; 12.21a: Adapted from http://water.usgs.gov/GIS/brose/ofr00-300_sattk9697.gif; 12.21b: Adapted from http://ne.water.usgs.gov/html/highplains/hp98_web_report/figures/fig2.htm; 12.21c: Adapted from http://ne.water.usgs.gov/html/highplains/highfig1.htm; 12.22: Adapted from USGS; 12.23: Adapted from http://international.usgs.gov/images/projects/big_pics/bg-map-lrg.png; 12.24: Adapted from http://water.usgs.gov/nawqa/trace/pubs/geo_v46n11/fig2.jpeg; 12.28: Adapted from http://www.cotf.edu/ete/modules/everglades/egremote2.html

Chapter 13

Figure 13.5: Adapted from Charles (Carlos) Plummer, *Physical Geology* 11e © 2007, reproduced with permission of McGraw-Hill Education and http://bulletin.mercator-ocean.fr/images/produits/psy3v1/psy3v1_20080924_21451/ocean/ind/psy3v1_20080924_21451_ind_s_n0_t0.gif; 13.11: Adapted from NOAA; 13.12: Adapted from UCAR; 13.7a: Adapted from http://www.scotese.com/miocene.htm; 13.7b: Adapted from http://www.scotese.com/newpage9.htm; 13.19: Adapted from Konrad Krauskopf, *The Physical Universe* 11e © 2006, reproduced with permission of McGraw-Hill Education; 13.20: Adapted from Carla Montgomery, *Environmental Geology* 7e © 2006, reproduced with permission of McGraw-Hill Education; 13.21: Adapted from Edgar Spencer, *Earth Science: Understanding Environmental Systems* 1e © 2003, reproduced with permission of McGraw-Hill Education; 13.22: Adapted from NOAA; 13.25: Adapted from https://www.fnmoc.navy.mil/ww3_cgi/index.html; 13.26: Adapted from Charles (Carlos) Plummer, *Physical Geology* 11e © 2007, reproduced with permission of McGraw-Hill Education; In Further Depth (all): Adapted from NOAA; 13.27: Adapted from Woods Hole Oceanographic Institution; 13.28: Adapted from Charles (Carlos) Plummer, *Physical Geology* 11e © 2007, reproduced with permission of McGraw-Hill Education; 13.32: Adapted from Coasts In Crisis, USGS; 13.34: Adapted from USGS

Chapter 14

Figure 14.4a: Adapted from L.J. Kump, *Nature*, 2008 (451): 278; 14.4b: Adapted from R. A. Berner, *Geochimica et Cosmochimica Acta*, 2006 (70): 5663; 14.12: Adapted from *Physical Geography Net*; 14.14: Adapted from Eric Danielson, *Meteorology* 2e © 2003, reproduced with permission of McGraw-Hill Education; 14.15: Adapted from Konrad Krauskopf, *The Physical Universe* 11e © 2006, reproduced with permission of McGraw-Hill Education; 14.16: Adapted from American Chemical Society, *Chemistry in Context: Applying Chemistry to Society* 5e © 2006, reproduced with permission of McGraw-Hill Education; 14.21: Adapted from Eric Danielson, *Meteorology* 2e © 2003, reproduced with permission of McGraw-Hill Education

Chapter 15

Figure 15.1a,b: http://www.ncdc.noaa.gov/billions/summary-stats; 15.2: Adapted from http://www.crh.noaa.gov/mkx/?n=taw-part10-usa_fatality_stats; 15.5, 15.6: Adapted from NOAA; 15.11, 15.13, 15.15, 15.16: Adapted from Eric Danielson, *Meteorology* 2e © 2003, reproduced with permission of McGraw-Hill Education; 15.20: NOAA/NWS Storm Prediction Center, Norman, Okla.; 15.25: Adapted from NOAA; 15.28: Adapted from http://web.mit.edu/mit-whoi/www/research/mgg/figs/donnelly_3.jpg

Chapter 16

CO16 text: Broecker, Wallace, "Thermohaline Circulation, The Achilles Heel of Our Climate System: Will Man-Made CO2 Upset the Current Balance?" *Science*, Volume 278 no. 5343, pp. 1582-1588, November 28, 1997; Figure 16.1: Adapted from www.greatcanadianoutdoors.com; 16.2: Edgar Spencer, *Earth Science: Understanding Environmental Systems* 1e © 2003, reproduced with permission of McGraw-Hill Education; 16aa: Adapted from http://www.ssec.wisc.edu/data/comp/latest_cmoll.gif; 16ab: http://www.cpc.ncep.noaa.gov/products/precip/Cwlink/wayne/ann.precip.gif; 16ba: http://www.cpc.ncep.noaa.gov/products/precip/realtime/clim/annual/slp/mean_pentad_slp.79_95.total.gif; 16.6, 16.7: Adapted from NOAA; 16.8: Adapted from Encyclopedia Britannica; 16.9: Adapted from USDA; Checkpoint 16.5: Adapted from http://www.metoffice.com/education/data/climate/temperate/Sydney.gif; 16.11: Adapted from Burton Guttman, *Biology* 1e © 1999, reproduced with permission of McGraw-Hill Education; 16.22, 16.23: Adapted from James Zumberge, *Laboratory Manual for Physical Geology* 12e © 2005, reproduced with permission of McGraw-Hill Education; 16.24: Adapted from http://www.giss.nasa.gov/research/news/20110113/ and http://www.giss.nasa.gov/research/news/20110113/509796main_GISS_annual_temperature_anomalies_running.gif; 16.28: Adapted from http://www.wrds.uwyo.edu/sco/climateatlas/drought.html; Checkpoint 16.17: Adapted from http://www.ngdc.noaa.gov/paleo/ctl/images/grissno.jpg; 16.31a: Adapted from http://www.grida.no/climate/ipcc_tar/wg1/pdf/wg1_tar-front.pdf; 16.31b: Adapted from http://www.ncdc.noaa.gov/paleo/pubs/alley2000/alley2000.gif; 16.32: Adapted from NASA; 16.33: Adapted from Victor di Venere; 16.35: Adapted from NOAA; 16.36: Adapted from http://ipcc-wg1.ucar.edu/wg1/report/ar4wg1_print_ch06.pdf, figure 6.3, page 444; 16.38: Adapted from http://www.scotese.com/late.htm; 16.39: Adapted from Eric Danielson, *Meteorology* 2e © 2003, reproduced with permission of McGraw-Hill Education

Chapter 17

Figure 17.2: Adapted from http://www.grida.no/climate/ipcc_tar/wg1/pdf/wg1_tar-front.pdf, figure 1b, page 3; 17.3: Adapted from Eric Danielson, *Meteorology* 2e © 2003, reproduced with permission of McGraw-Hill Education; 17.5: Adapted from http://toms.gsfc.nasa.gov/eptoms/dataqual/oz_hole_avg_area_v8.jpg; 17.6: Adapted from http://www.environment.gov.au/soe/2006/publications/commentaries/atmosphere/stratospheric-ozone.html; 17.7: Adapted from http://www.esrl.noaa.gov/gmd/hats/graphs/graphs.html; 17.8: Adapted from http://www.ipcc.ch/graphics/graphics/syr/fig 1-1.jpg; 17.9: Adapted from http://www.esrl.noaa.gov/gmd/ccgg/trends/; 17.11: Adapted from http://www.ipcc.ch/graphics/graphics/syr/fig2-1.jpg; 17.12: Adapted from USEPA; 17.13: Adapted from http://www.worldmapper.org/images/largepng/295.png, © Copyright 2006 SASI Group (University of Sheffield) and Mark Newman (University of Michigan); 17.14: Adapted from UNEP/GRID-Arendal Maps and Graphics Library, http://maps.grida.no/go/graphic/greenhouse-effect; 17.15: Adapted from Data from IPCC; 17.16: Adapted from http://www.research.noaa.gov/climate/images/modeling_schematic.gif; 17.17: Adapted from Data from IPCC; 17.18: Adapted from http://www.ipcc.ch/graphics/graphics/ar4-wg1/jpg/fig-10-8.jpg; 17.19: Adapted from http://www.ipcc.ch/graphics/graphics/syr/fig3-3.jpg; 17.20: Adapted from http://www.princeton.edu/~cmi/resources/cmi_resources_new_files/cmi_wedge_game_jan_2007.pdf

Photo Credits

Front Matter

Page ix: © StockTrek/Getty Images; p. xii: © PhotoAlto/Punchstock; p. xix (top): Courtesy of David McConnell; p. xix (bottom): Courtesy of David Steer; p. xx (left): Courtesy of Catharine Knight; p. xx (right): Courtesy of Katharine Owens.

Chapter 1

Opener: © Imago Stock&people/Newscom; 1.2 (before), 1.2 (after): Digital Globe; 1.3 (top left): USGS; 1.3 (top right): NOAA; 1.3 (bottom left): USDA; 1.3 (bottom right): © John Wang/Getty Images; 1.5: NOAA; 1.6: Press Association via AP Images; 1.8: © G. Larso; 1.9: © Sanford/Agliolo/Corbis; 1.11: NOAA; 1.12: USGS; 1.13: NASA; 1.14b: © Zhang Jun/ZUMA Press/Newscom; 1.15: © STR/AFP/Getty Images.

Chapter 2

Opener: NASA; 2.1 (left): © Carlos Davila/Getty Images; 2.1 (right): © Stan Russell; 2.2: © Science Source; 2.3a: © Zlatko Kovacevic/VT-2004 Programme/ESO; 2.6: AURA/STSci; 2.7, 2.8: NASA; 2.9: © Jon Lomberg/Science Source; 2.10, 2.11a, 2.11b, 2.12, 2.14, 2.15, 2.16: NASA; 2.17: NASA/JPL; 2.26: Department of Defense; 2.27: Courtesy of Malin Space Science Systems/NASA; 2.28: NASA.

Chapter 3

Opener: © Denis Scott/Corbis; 3.1: USGS; 3.5: NEAR Project, JHU APL, NASA; 3.6: NASA/JPL-Caltech/UMD; 3.7: Associated Press; 3.8, 3.10: NASA; Checkpoint 3.6: © Corbis; 3.12a, 3.12b, 3.13a, 3.13b: NASA; 3.13c: Courtesy of

Landsat/EOS; 3.15: Institut für Geologie, Mineralogie und Geophysik; Checkpoint 3.11: NASA; 3.17: National Park Service.

Chapter 4

Opener (all): © Chris Scotese; 4.1: © PhotoLink/Getty Images; 4.4, 4.7: NOAA; 4.24b: NASA; 4.27: USGS.

Chapter 5

Opener: USGS; 5.1a, 5.1b: NOAA; 5.2a, 5.2b: USGS; 5.3 (top): © Amanda Clement/Getty Images; 5.3c (middle): © Robert Glusic/Getty Images; 5.3 (bottom): © D. Falconer/PhotoLink/Getty Images; 5.4a, 5.4b, 5.5a, 5.5b: USGS; 5.7a: © Lloyd Cluff/Corbis; 5.7b: G.K. Gilbert/USGS; 5.7c, Checkpoint 5.6a: USGS; Checkpoint 5.6b: © David McConnell; 5.9: © C. C. Plummer; In Further Depth fig. 1, p. 116: Serkan B. Bozkurt/USGS; 5.13, Checkpoint 5.16, 5.19, 5.20: USGS; 5.21 (freeway collapse): © Lloyd Cluff/Corbis; 5.21 (map): USGS; 5.22: AP Photo/Nick Ut; 5.23a: © Martitia Tuttle; 5.23b: National Geophysical Data Center; 5.23c: © Imago Stock&people/Newscom; 5.24 (all), 5.25, 5.27: USGS.

Chapter 6

Opener: © Jupiterimages/Getty Images; 6.1 (top): T. Pierson/USGS; 6.1 (bottom, left): R.J. Janda/USGS; 6.1 (bottom right): N. Banks/USGS; 6.2a: NASA; 6.2b: Two-Worlds; 6.2c: AP Photo/Sayyid Azim; 6.3a: USGS; 6.3b: M.L. Tuttle, USGS Photo Library; 6.8a: Luis Fuste/USGS; 6.8b: Donald A. Swanson/USGS; 6.8c: Peter Lipman/USGS; 6.10, 6.11a, 6.11b: USGS; 6.11 (inset): Harry Glicken/USGS; 6.13a: © C. C. Plummer; 6.13b: AP Photo/Omar Oskarsson; 6.14, 6.15 (all), 6.16 (page 157): USGS; 6.16 (page 158): Elliot Endo/USGS; 6.17a: Chris Newhall/USGS; 6.17b: Underwood & Underwood/Library of Congress; 6.18, Checkpoint 6.16, p. 160: USGS; 6.19a: U.S. Geological Survey, HVO; 6.20a: © Hubert Stadler/Corbis; 6.21: USGS; 6.22: C. Dan Miller/USGS; 6.23a, 6.25 (all), 6.33: USGS.

Chapter 7

Opener: NASA; 7.1a, 7.1b, 7.2 (all): © David McConnell; 7.3: © PhotoLink/Getty Images; 7.4: © Nancy R. Cohen/Getty Images; 7.12a: © The McGraw-Hill Companies, Inc./Doug Sherman, photographer; 7.12b: © The McGraw-Hill Companies, Inc./Bob Coyle, photographer; 7.12c, 7.12d, 7.12e: © The McGraw-Hill Companies, Inc./Doug Sherman, photographer; 7.12f: © The McGraw-Hill Companies, Inc./Bob Coyle, photographer; 7.13a: © Dr. John D. Cunningham/Visuals Unlimited; 7.13b: © Albert J. Copley/Visuals Unlimited; Checkpoint 7.7: © The McGraw-Hill Companies, Inc./Bob Coyle, photographer; 7.15a: © Wally Eberhart/Visuals Unlimited; 7.15b: © Ed Degginger/Color-Pic; 7.16, 7.16 (inset): © C. C. Plummer; 7.17: D. A. Swanson/USGS; 7.19: © David McConnell; 7.20 (granite), 7.20 (diorite): © C. C. Plummer; 7.20 (gabbro): © Larry Davis; 7.20 (bassalt): © C. C. Plummer; 7.20 (andesite): © Albert Copley/Visuals Unlimited; 7.20 (rhyolite): © C. C. Plummer; 7.21: © Bill Church/University of Western Ontario; Checkpoint 7.10 (left): © Randall Fung/Corbis; Checkpoint 7.10 (middle left): © Joyce Photographics/Science Source; Checkpoint 7.10 (middle right): © Wally Eberhart/Visuals Unlimited; Checkpoint 7.10 (right): © Albert J. Copley/Visuals Unlimited; 7.22: © The McGraw-Hill Companies, Inc./Doug Sherman, photographer; 7.23 (all) USGS; 7.24a: NASA; 7.24b: USGS; 7.25: U.S. Park Service; 7.26a: © Scientifica/Visuals Unlimited; 7.26b: © Dr. John D. Cunningham/Visuals Unlimited; Checkpoint 7.14: © Gerald & Buff Corsi/Visuals Unlimited; 7.27: © Dennis Flaherty/Science Source; 7.28: © Purestock/SuperStock; 7.29a: NASA; 7.29b: © Bob Krist/Corbis; 7.29c: NASA; 7.30a: © Gerald & Buff Corsi/Visuals Unlimited; 7.30b: © David McGeary; 7.33a: © C. C. Plummer; 7.33a (inset): © Albert Copley/Visuals Unlimited; 7.33b: : © C. C. Plummer; 7.33b (inset): © Albert Copley/Visuals Unlimited; 7.35a: P.D. Rowley, USGS Photo Library, Denver. CO; 7.35b (top): © Joyce Photographics/Science Source; 7.35b (bottom): © Albert J. Copley/Visuals Unlimited; 7.35c (top): © Joyce Photographics/Science Source; 7.35c (bottom): © Andrew J. Martinez/Science Source; 7.37b: USGS; 7.38b: © Stephen Reynolds.

Chapter 8

Opener: © Charles Bowman/Photolibrary/Getty Images; 8.1a: © Jenny Matthews/Alamy; 8.1d: © Hisham F. Ibrahim/Getty Images; 8.1c, 8.2: © Corbis; 8.3a, 8.3b: National Park Service; 8.5: © C.E. Ford 2004; 8.8: © David McConnell; 8.10a: © The McGraw-Hill Companies, Inc./Doug Sherman, photographer; 8.10b: © Ron Blakey, Department of Geology, Northern Arizona University; 8.11: USGS; 8.12: © Ron Blakey, Department of Geology, Northern Arizona University; p. 218: © John Wang/Getty Images; Checkpoint 8.7: © Tom McHugh/Science Source; 8.13a: © Kevin and Betty Collins/Visuals Unlimited; 8.13b, 8.13c, 8.13d, 8.13e: © Ken Lucas/Visuals Unlimited; 8.13f: © Layne Kennedy/Corbis; 8.13g: © Ken Lucas/Visuals Unlimited; 8.13h: © Layne Kennedy/Corbis; Checkpoint 8.9: © Stocktrek Images, Inc./Alamy; 8.15a: © Ted Daeschler/NSF; 8.15b: Zina Deretsky, National Science Foundation; 8.24a, 8.24b: USGS; 8.25a: © Dr. John D. Cunningham/Visuals Unlimited; 8.25b: © Gerald & Buff Corsi/Visuals Unlimited.

Chapter 9

Opener: USGS; 9.1a: © Atlantide Phototravel/Corbis; 9.1b: © Franz-Marc Frei/Corbis; 9.1c: © Corbis; 9.1d: © Adalberto Rios Szalay/Sexto Sol/Getty Images; 9.1e: © Werner Forman/Corbis; 9.2: © David McConnell; 9.3c: N.K. Huber, USGS; 9.4 (both): © David McConnell; Checkpoint 9.1 (both): USGS; 9.5: © Paul Cowan/iStockphoto; 9.6b: © Martin G. Miller/Visuals Unlimited; 9.7: © David Eppstein; 9.9a: © C. C. Plummer; 9.9b: © David McGeary; 9.10a: National Park Service; 9.11: USGS; 9.13: © Diane Carlson, California State University at Sacramento; Checkpoint 9.7a: © Gary Faber/Photodisc Green/Getty Images; Checkpoint 9.7b: USGS; 9.14: Pennsylvania Department of Environmental Protection; 9.15a: © Diane Carlson, California State University at Sacramento; 9.15b: © Diane R. Nelson; 9.15c: © Robert Marien/Corbis; 9.15d: Courtesy of Lisa Park; 9.16, 9.17a: USGS; 9.17b: © David McConnell; 9.18: © Keren Su/Corbis; 9.19a: © Luca I. Tettoni/Corbis; 9.19b: © Gareth Brown/Corbis; 9.20 (all), 9.21, 9.24, 9.26: USDA; 9.25a: USDA Soil Conservation Service; 9.25b, 9.25c, 9.25d, 9.27, 9.29a, 9.29b, 9.29c, 9.29d, 9.29e: USDA.

Chapter 10

Opener: © Juan Herrero/EPA/Newscom; 10.1a: © Doc Searls; 10.2: © Raymond Gehman/Corbis; 10.4b, 10.5: USGS; 10.7a: © Photogen/Alamy; 10.7b: © C. C. Plummer; 10.8, 10.9, 10.10a, 10.10b: USGS; 10.11, 10.12a, 10.12b, 10.12c, 10.13a, 10.13b, 10.14: © David McConnell; 10.15a: USGS; 10.15b: National Park Service; 10.16a: © Atlantide Phototravel/Corbis; p. 278: © David McConnell; Checkpoint 10.5: USGS; 10.17a: D. A. Rahm; 10.18a: USGS; 10.18b: AP Photo/Kevork Djansezian; 10.19: © Geoff Mackley; Checkpoint 10.8: © Georges Reef; 10.20: © Doc Searls; 10.21: USGS.

Chapter 11

Opener: USGS; 11.2: Jacques Descloitres, MODIS Land Science Team/NASA; 11.5a: © Pat O'Hara/Corbis; 11.5b: © Panoramic Images/Getty Images; 11.11: © Rob Howard/Corbis; 11.14a, 11.14b: USGS; 11.16 (inset), 11.16 (middle): NASA; 11.16 (bottom): © Rob van Glabbeek; 11.18: © Julia Waterlow; Eye Ubiquitous/Corbis; 11.19: © David Mott; 11.21b: © David McConnell; p. 301: NASA; Checkpoint 11.15: NASA JPL; 11.22: NASA; 11.23: FEMA; 11.26: © Donald P. Schwert, North Dakota State University; 11.27 (all): © David McConnell; 11.31a: FEMA; 11.31b: © Erik De Castro/Reuters/Corbis; 11.32a, 11.32b: NASA; 11.33b: US Army Corps of Engineers; 11.34: Reproduced with the permission of Natural Resources Canada 2013, courtesy of the Canada Centre for Remote Sensing; 11.35b: US Army Corps of Engineers.

Chapter 12

Opener (left): © David Aubrey/Corbis, Opener (right): © Corbis; 12.2a, 12.2b: EPA; 12.3a, 12.3b: Phil Stoffer, USGS; 12.7a: © W. K. Fletcher/Science Source; 12.7b: Kevin Dennehy, USGS; 12.7c: © Patrick Ward/Corbis; 12.11c, 12.19b: USGS; 12.26a: © David Stubbs/Aurora/Getty Images; 12.26b: © Raymond Gehman/Corbis; 12.26c: © Andrew Brown; Ecoscene/Corbis; 12.26d: © Wilfried Krecichwost/zefa/Corbis; 12.27: NASA.

Chapter 13

Opener: USGS; 13.1: NASA; 13.2: © Daniel P. Costa/University of California, Santa Cruz; 13.3: © 2002-2009 Kenneth & Gabrielle Adelman, California Coastal Records Project, www.Californiacoastline.org; 13.4b: NOAA; 13.6: William Haxby, of the National Science Foundation-supported Marine Geoscience Data System; Checkpoint 13.2: NOAA; 13.7a: © WHOI/D. Foster; 13.7b, 13.8, 13.9: NOAA; 13.10: © Digital Vision; 13.11: Talley, L. D., Hydrographic Atlas of the World Ocean Circulation Experiment (WOCE). Volume 2: Pacific Ocean (eds. M. Sparrow, P. Chapman and J. Gould), International WOCE Project Office, Southampton, U.K.; 13.14: NASA; 13.17a, 13.17b: © Chris Scotese; 13.18a, 13.18b: NASA; 13.27b: NOAA; 13.29: OCS Alternative Energy and Alternate Use Programmatic EIS Information Center; 13.30a, 13.30b: © David McConnell; 13.30c: © Atlantide Phototravel/Corbis; 13.30d: © Kevin Fleming/Corbis; 13.31a, 13.31b: NCDOT Photogrammetry Unit; 13.33: © Corbis; 13.35a: © D.A. Rahm/Easterbrook Photo/Image Center; 13.35b, 13.35c: USGS; Checkpoint 13.22: © 2002-2009 Kenneth & Gabrielle Adelman, California Coastal Records Project, www.Californiacoastline.org; 13.36: USGS; 13.37: © 2002-2009 Kenneth & Gabrielle Adelman, California Coastal Records Project, www.Californiacoastline.org; 13.38, 13.39: US Army Corps of Engineers; 13.40: USGS.

Chapter 14

Opener (top): © Sandy Felsenthal/Corbis, Opener (bottom): © Dave G. Houser/Post-Houserstock/Corbis; 14.1a: US Centennial of Flight Commission; 14.1b: © Red Bull Stratos/ZUMA Press/Newscom; 14.2a, 14.2b: NASA; 14.7: © Hinrich Bîsemann/dpa/Corbis; 14.16: © Steve Cole/Getty Images; 14.19: © Joyce Photographics/Science Source; Checkpoint 14.21 (all): NOAA; 14.24: NASA; 14.25: Mt. Washington Observatory; 14.26: Department of Atmospheric Sciences, University of Illinois; 14.32a: © Thinkstock/Masterfile; 14.32b: © Department of Energy.

Chapter 15

Opener: © Thomas Augustine; 15.1a, 15.1b, 15.3, 15.4: NOAA; 15.7: US Global Change Research Program; 15.8, 15.9b: NASA; 15.16b: © Stockbyte/PunchStock; 15.17a: © Nick Rutledge; 15.17b: AP Photo/The Oklahoman, Paul Hellstern; 15.17c: © Julie Dermansky/Corbis; Checkpoint 15.18: NOAA National Severe Storms Laboratory; 15.21a, 15.21b: NASA; 15.21c, 15.21d: USGS; 15.22: NOAA; 15.25: National Weather Service; 15.27a, 15.27b: USGS National Wetlands Research Center; 15.28a, 15.28b: Courtesy of Jeffrey P. Donnelly.

Chapter 16

Opener: © Paul Nicklen/National Geographic/Getty Images; 16.1: © Kevin Schafer/Corbis; 16.3a, 16.3b: USGS; 16.4a, 16.4b: NASA; 16.13: © Corbis; 16.14b: © B.S.P.I./Corbis; 16.14c: © Mauri S. Pelto, Nichols College; 16.15: © Larry Dale Gordon/zefa/Corbis; 16.16a: © NSIL/G.R. "Dick" Roberts/Visuals Unlimited; 16.16b: Plate 26 from: Pettijohn, F.J., and Potter, P.E., 1964, Atlas and Glossary of Primary Sedimentary Structures, Springer Verlag, New York, p. 370. With kind permission of Springer Science and Business Media; 16.17: © Leslie Garland Picture Library/Alamy; 16.18: NASA; Checkpoint 16.10 (both): USGS; 16.19a: © Joel W. Rogers/Corbis; 16.19b: NSIDC; 16.20: Integrated Ocean Drilling Program; 16.21a: © Digital Vision; 16.21b: © C. Allan Morgan/Peter Arnold, Inc/Getty Images; 16.21c: © Digital Vision/PunchStock; 16.23b: © Andreas Stirnberg/Getty Images; 16.23c: USGS; 16.23d: © The McGraw-Hill Companies, Inc; 16.25b: © Cindy Hopkins/Alamy; 16.26: Scala/Art Resource, NY; 16.27: © Don Mason/Corbis; 16.29: © Nick Eyles, University of Toronto, Scarborough; 16.30: © Tony Wharton; Frank Lane Picture Agency/Corbis; 16.34: © Dr. Eric Condliffe, Leeds University.

Chapter 17

Opener (both): Photofest; 17.1, 17.4, 17.12: NASA; 17.16b: © climateprediction.net; 17.22a: © Steve Allen/Getty Images; 17.22b: © The McGraw-Hill Companies, Inc./John Flournoy, photographer.

Index

a

d

e

h

i

j

k

l

m

n

O

P

t

x

y

z

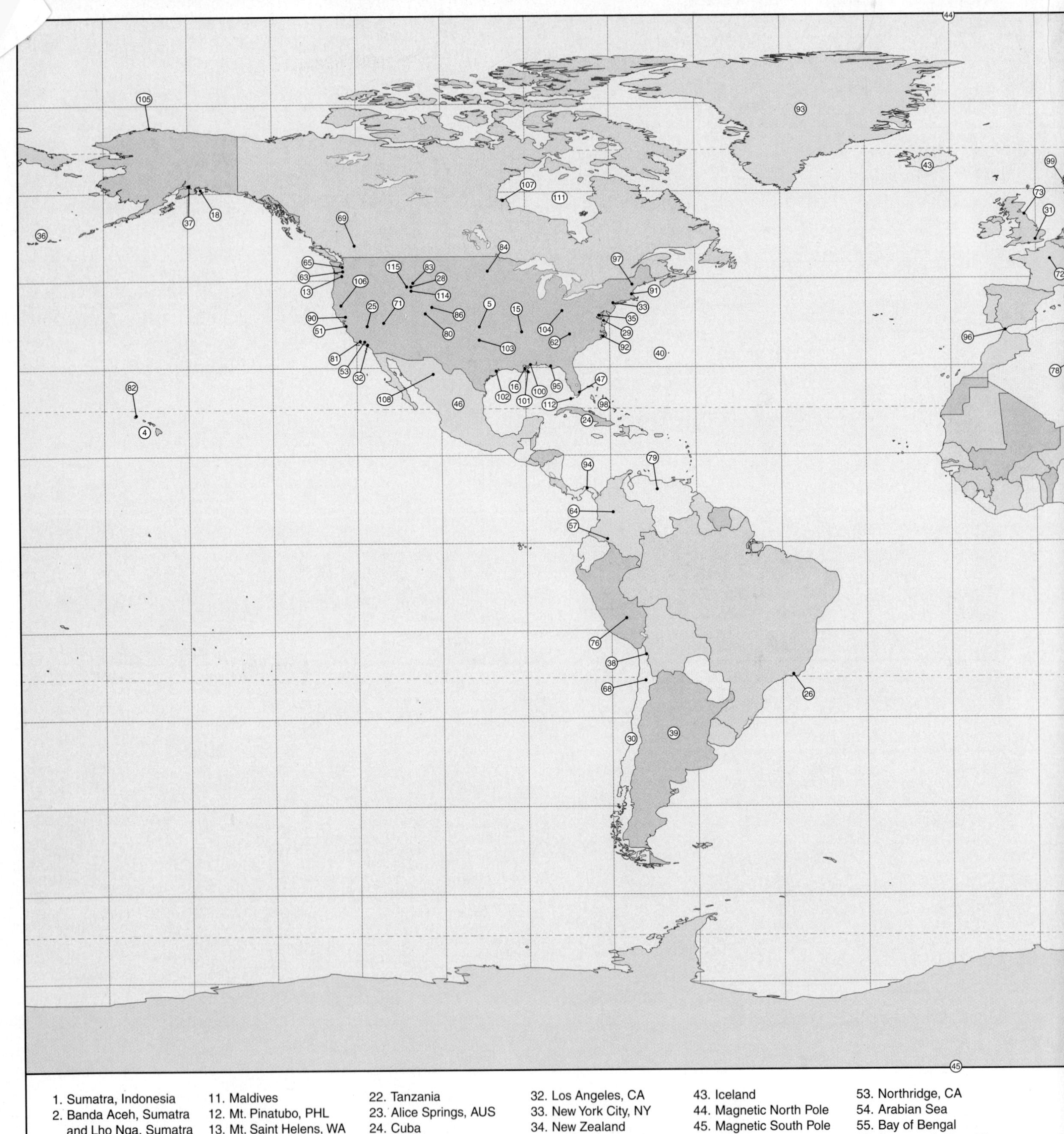

1. Sumatra, Indonesia
2. Banda Aceh, Sumatra and Lho Nga, Sumatra
3. China
4. Hawaii
5. Hutchinson, KS
6. India
7. Japan
8. Kenya
9. Madagascar
10. Malaysia
11. Maldives
12. Mt. Pinatubo, PHL
13. Mt. Saint Helens, WA
14. Myanmar
15. New Madrid, MO
16. New Orleans, LA
17. Phuket Island, Thailand
18. Prince William Sound
19. Seychelles
20. Somalia
21. Sri Lanka
22. Tanzania
23. Alice Springs, AUS
24. Cuba
25. Death Valley, CA
26. Rio de Janiero, Brazil
27. South Africa
28. Yellowstone National Park, WY
29. Chesapeake Bay
30. Chile
31. London, England
32. Los Angeles, CA
33. New York City, NY
34. New Zealand
35. Washington, D.C.
36. Aleutian Islands, AK
37. Anchorage, AK
38. Andes Mountains
39. Argentina
40. Bermuda
41. Gabon
42. Himalayas
43. Iceland
44. Magnetic North Pole
45. Magnetic South Pole
46. Mexico
47. Miami, FL
48. Red Sea
49. Bam, Iran
50. Izmit, Turkey
51. Loma Prieta, CA (epicenter)
52. Kobe, Japan
53. Northridge, CA
54. Arabian Sea
55. Bay of Bengal
56. Democratic Republic of Congo
57. Galeras, Columbia
58. Goma, Congo
59. Lake Nyos, Cameroon
60. Mt. Everest, Nepal
61. Mt. Kilimanjaro, Tanzania
62. Mt. Mitchell, NC